Applied Mathematical Sciences
Volume 166

For further volumes:
http://www.springer.com/series/34

Pavel B. Bochev · Max D. Gunzburger

Least-Squares Finite Element Methods

Springer

Pavel B. Bochev
Sandia National Laboratories
Applied Mathematics and Applications
MS 1320, P.O. Box 5800
Albuquerque NM 87185-1320
USA
pbboche@sandia.gov

Max D. Gunzburger
Florida State University
Department of Scientific Computing
400 Dirac Science Library
Tallahassee FL 32306-4120
USA
gunzburg@fsu.edu

ISBN 978-0-387-30888-3 e-ISBN 978-0-387-68922-7
DOI 10.1007/978-0-387-68922-7

Library of Congress Control Number: 2008943966

Mathematics Subject Classification (2000): 65N30, 65N35, 65N12, 65N21, 65M60, 65M70, 65M12

Printed on acid-free paper

springer.com

To my mother and Biliana

To Janet

Preface

Since their emergence in the early 1950s, finite element methods have become one of the most versatile and powerful methodologies for the approximate numerical solution of partial differential equations. At the time of their inception, finite element methods were viewed primarily as a tool for solving problems in structural analysis. However, it did not take long to discover that finite element methods could be applied with equal success to problems in other engineering and scientific fields. Today, finite element methods are also in common use, and indeed are often the method of choice, for incompressible fluid flow, heat transfer, electromagnetics, and advection-diffusion-reaction problems, just to name a few. Given the early connection between finite element methods and problems engendered by energy minimization principles, it is not surprising that the first mathematical analyses of finite element methods were given in the environment of the classical Rayleigh–Ritz setting. Yet again, using the fertile soil provided by functional analysis in Hilbert spaces, it did not take long for the rigorous analysis of finite element methods to be extended to many other settings. Today, finite element methods are unsurpassed with respect to their level of theoretical maturity.

A finite element method is first and foremost a *quasi-projection scheme.*[1] This truly fundamental property establishes a link with sophisticated mathematical structures and has a tremendous impact on the algebraic problems generated by the method. The paradigm that describes and defines the key properties of what we refer to as a quasi-projection is the marriage of two ingredients: a *variational principle* and a *closed subspace.* Approximate solutions are then characterized as *quasi-projections of the exact (weak) solutions onto the closed subspace.* Finite element approximations are an example of this rule. Indeed, a finite element method and its properties are completely determined by specifying the variational principle and

[1] A projection in a Hilbert space is, of course, defined with respect to an inner product and a closed subspace. Finite element methods do not always involve a true projection because they are not always based on inner products; however, as we show, finite element approximations do, in general, possess many of the important properties of projections.

the approximation subspace.[2] The cornerstone of the great success of finite element methods is the remarkable fact that the combination of these two ingredients prove to be exceptionally well suited for the numerical solution of partial differential equations.

From a mathematical viewpoint, this success is rooted in the existence of strong intrinsic links between differential equations and variational principles; in fact, it is often the case that the latter serve as the primary mathematical model. On the other hand, the practical appeal of finite element methods and their wide acceptance in the engineering community results from the choice of approximating subspaces. These spaces are spanned by locally supported, piecewise polynomial functions defined over a subdivision of a domain into simple geometrical subdomains referred to as finite elements. Such spaces are not only simple to use, but allow for the almost automatic generation of high-order methods[3] with respect to arbitrary unstructured subdivisions, a trait that, if not outright impossible, is not easy to accomplish with other schemes. The popularity of finite element methods is also largely due to the small support of standard finite element basis functions. This property implies that the resulting algebraic problems involve banded and sparse matrices.

Although the choice of approximation space has, especially from a practical point of view, a tremendous influence on the attributes of the resulting discretized systems, all fundamental properties of finite element methods are ultimately governed by the variational principles from which they are defined. It is, therefore, a fortuitous coincidence that the variational foundations of the first finite element methods were provided by unconstrained, quadratic energy minimization principles, a setting that turned out to be, by far, the most attractive one for finite element methods. Such principles search for a point (in a suitable function space) that is the unconstrained minimizer of a convex quadratic functional and give rise to true inner-product projections in Hilbert spaces. This property, exemplified by the classical Rayleigh–Ritz principle, results in distinct, unique, and very desirable computational and analytic advantages for the resulting finite element methods. Most notably, true inner product projections allow for a wide and nonrestrictive choice for the approximating spaces and lead to symmetric and positive definite algebraic systems.

The connection between finite element methods and Rayleigh–Ritz principles was not immediately recognized as the fundamental cause for the remarkable success of the first finite element methods. As a result, some of the early attempts directed at extending finite element methods beyond the class of problems whose solutions can be characterized as global minimizers of convex quadratic functionals

---

[2] There exists a fundamental philosophical difference between finite difference and finite element methods. The former substitutes difference operators for differential operators and leads to approximations defined with respect to a discrete lattice, i.e., the fundamental discretization step is to *approximate operators*. Finite element methods, on the other hand, do not deal directly with the differential operators; instead, discretization is effected using alternative weak formulations of these equations posed over finite-dimensional function spaces. This leads to *functional* approximation as opposed to operator approximation. Finite volume methods occupy a middle ground as they exhibit features shared by both finite element and finite difference methods.

[3] This is due to the possibility of using internal degrees of freedom that enrich the finite element space inside each element.

led to disappointing results and a reluctance (by some) to apply the methods outside the field of structural analysis. However, with the rapid development of the mathematical theory of finite element methods and the rapid accumulation of a wealth of practical experience in applying the methods, there came the understanding for what were the reasons behind these early setbacks. It was realized that differential equations not associated with unconstrained minimization principles lead to two other classes of variational principles with strikingly different properties. In one of them, the quasi-projection is induced by a search for a stationary point of an indefinite functional; constrained minimization problems provide an example of this class. The other type offers even less mathematical structure and defines the quasi-projection by a formal *residual orthogonalization* process. In both cases, the variational principle does not lead to a true inner-product projection; this causes the associated finite element methods to operate in a much less favorable setting as compared to those based on Rayleigh–Ritz principles. The computation of stationary points that is the paradigm of *mixed-Galerkin methods* demands strict compatibility conditions on the discrete spaces, if stable and accurate approximations are desired. Likewise, the formal residual orthogonalization process that provides the template for *Galerkin methods* may require onerous stability conditions on the approximation spaces. In both cases, one is confronted by the problem of solving indefinite and/or non-symmetric algebraic equations.

Not surprisingly, there have been many efforts aimed at developing finite element methods that, for problems not connected to unconstrained energy principles, share some, if not all, of the attractive mathematical and algorithmic properties of the Rayleigh–Ritz setting. Broadly speaking, there are two different ways to approach the construction of better projections or quasi-projections than those afforded by the original problem. The first one improves the quasi-projection by penalization or regularization of the original variational principle. In this approach, the principal role of the naturally occurring variational principle[4] is retained, but the principle itself is modified so as to behave more like a true inner-product projection.[5] Galerkin least-squares, stabilized Galerkin, penalized Lagrangian methods, artificial diffusion, and upwind Petrov–Galerkin methods are all examples of finite element methods resulting from this approach.

A second approach that leads to better quasi-projections, indeed to true projections, is to abandon completely the naturally occurring variational principle and to devise a Rayleigh–Ritz-like environment by *formulating an artificial, externally defined energy-type principle.* This energy principle most often takes the form of *residual minimization* in a suitable Hilbert space; thus, there arises the commonly used adjective *least-squares* to denote the resulting finite element methods. Residual

[4] "Naturally occurring" variational principles include Galerkin principles that are based on residual orthogonalization and may or may not have any physical interpretation.

[5] Often the same effect can be achieved by posing the original variational principle on a *modified* finite element space. One example is the SUPG method which can be viewed both as resulting from the use of a modified test space or a modified variational principle. Another example is finite element spaces "enriched" by bubble functions. In many cases, enriched spaces lead to exactly the same formulations as modified Galerkin principles posed on conventional finite element spaces.

minimization is as universal as residual orthogonalization so that, in principle, least-squares finite element methods have the same wide range of applicability as do Galerkin finite element methods, i.e., they are both applicable to virtually any partial differential equation problem. However, residual minimization differs fundamentally from other variational settings, including modified and formal Galerkin principles. Unlike the case for other settings, *least-squares finite element methods are consistently capable of recovering almost all of the advantages of the Rayleigh–Ritz setting* over a wide range of problems and, with some additional effort, they can often create a completely analogous variational environment for finite element methods. This is what makes least-squares finite element methods stand apart from the rest of the methods in the finite element universe.

Mostly over the last ten to fifteen years, efforts focused on least-squares finite element methods have achieved tremendous success in making the methods viable alternatives to existing schemes. There is a wealth of theoretical and practical experience with the methods and they are steadily gaining a solid reputation and popularity among researchers and practitioners for providing robust, efficient, and practical computational tools. Nevertheless, compared with established finite element techniques, such as mixed-Galerkin methods, the theory for least-squares finite element methods remains much less unified and is not always well understood. Because such methods are based on inner-product projections, they tend to be exceptionally robust and stable. As a result, one is often tempted to forego analyses and proceed with the seemingly most natural and simple least-squares formulations. This sometimes leads to unsatisfactory results and methods that cannot fully exploit the advantages of the least-squares approach.

This book is motivated by the premise that there exists a real need to put least-squares finite element methods on a common, mathematically sound foundation and also to discuss important implementation issues that are critical to their success in practice. It is intended to give both the researcher and the practitioner a guide to the theory and practice of least-squares finite element methods, their strengths and weaknesses, caveats to be followed, and established successes.

The factors that set least-squares finite element methods apart from other finite element methods also call for a different approach in the algorithmic development of these methods, if truly reliable, efficient, and accurate schemes are desired. Most notably, reliance of a least-squares method on externally defined variational principles makes the choice of this principle the single most important step in the design of a method. Unlike traditional finite element methods for which the variational principle is almost always dictated by the given problem, least-squares finite element methods dictate the variational principle and then fit the problem into the principle. Thus, the flexibility afforded by the freedom to choose the variational principle places the principal responsibility for the success of the method on the "fitting" process. There are two, often opposing, forces that affect this process. One is the desire to keep the method as simple as possible so as to develop a scheme that is easy to implement. The other is the need to adhere to the basic premises of a Rayleigh–Ritz-like framework, namely, to work with projections defined by inner products that are equivalent to the natural inner products in suitable Hilbert spaces. The interaction between these two competing forces is, in our opinion, the key to understanding least-squares finite element methods; we make it the central theme of this book. Indeed, we show over and over again that the choices made in reconciling ease of implementation with adherence to the Rayleigh–Ritz framework affect all aspects of least-squares finite element methods, from condition numbers of the algebraic problems and their efficient preconditioning to the existence of quasioptimal asymptotic error estimates.

An overview of the book

Throughout the book, careful attention is paid to not only the rigorous analysis of least-squares finite element methods, but also to practical issues that arise in their implementation.

For those who wish to gain a full understanding of the mathematical analyses connected with least-squares finite element methods, space limitations necessitate the requirement of certain prerequisites. In particular, a basic familiarity with functional analysis in Hilbert spaces and with the theory and implementation of standard finite element methods is assumed at the level of, e.g., [250] and [76, 123, 188], respectively. More advanced background material on functional analysis, partial differential equations, and finite element methods is contained within the book, especially in the appendices. Those who merely wish to learn about least-squares finite element algorithms and their implementation can still use the book by focusing on those sections that consider these topics. In this case, space limitations necessitate familiarity with the algorithmic and implementation aspects of standard finite element methods.

The book is organized into parts, each containing several chapters. In Part I of the book, we provide a necessarily brief review of the finite element universe. It is not meant to give a comprehensive treatment, but rather to provide a context for understanding where least-squares finite element methods fit into that universe. In Chapter 1, "classical" finite element methods, i.e., those based on Rayleigh–Ritz, mixed-Galerkin, and Galerkin variational principles, are discussed. Then, in Chapter 2, a discussion is provided about several of the attempts that have been made to try to recover some of the advantageous features of finite element methods in the Rayleigh–Ritz setting through the definition of modifications to the "classical" principles. Along the way, the strengths and weaknesses of the specific finite element methods encountered are pointed out. At that point, one is ready for the discussion provided in the latter part of Chapter 2 about why and how least-squares finite element methods are meant to improve on the three "classical" approaches and their modifications.

In Part II, we provide the theoretical core of the book that is repeatedly referred to in the rest of the book. Chapter 3 is devoted to an abstract theory of least-squares finite element methods for systems of elliptic partial differential equations. The abstract framework can itself be divided into three classes of least-squares finite element methods that are characterized by discrete least-squares principles having different derivations and properties. The first is defined by simply restricting the minimization of a continuous least-squares functional for the partial differential equations to a finite element space. The other two classes additionally involve a discretization of the least-squares functional itself and differ from each other by the type of equivalence relations that exist (or do not exist) between least-squares functionals and the norms used to measure the size of the solution and the data for the problem. All the possible scenarios emanating from this classification of least-squares finite element methods are analyzed. For a specific problem, one can usually define several different realizations of least-squares finite element methods. Most of these can be viewed as particular applications of one or another of the different scenarios covered by the abstract framework. In addition, several additional basic ingredients needed to define and analyze least-squares finite element methods but which transcend the specific context to which those methods are applied are also discussed. In Chapter 4, the Agmon–Douglis–Nirenberg theory for elliptic partial differential equations is exploited to provide "automated" mechanisms for using solution and data spaces and norms for the definition of least-squares variational principles.

The abstract frameworks developed and analyzed in Part II can be used to define and analyze many least-squares finite element methods for a number of specific settings. In these cases, one need only show how the setting fits into the abstract frameworks and that the hypotheses invoked in those frameworks hold. Thus, there is no need to repeat over and over again proofs specialized to each setting. Using this approach, Part III of the book is devoted to the application of the abstract framework developed in Part II to concrete elliptic partial differential equations.[6] In Chapters 5 and 6, least-squares finite element methods for scalar and vector elliptic partial differential equations are considered, respectively, and, in Chapter 7, least-squares finite element methods for the Stokes equations are considered.

In Part IV, least-squares finite element methods in contexts that do not fit into the abstract theory of Part II, but nevertheless rely on some aspects of that theory, are examined. In Chapter 8, the nonlinear stationary Navier–Stokes equations are considered. In addition to the intrinsic interest engendered by the Navier–Stokes equations, this setting provides an example of how least-squares finite element methods can be used for nonlinear problems. In Chapters 9 and 10, parabolic and hyperbolic (time-dependent) partial differential equations are considered, respectively. The application of least-squares finite element methods to optimization and control problems for systems governed by elliptic partial differential equations is treated in Chapter 11.

Then, in Chapter 12, other settings to which least-squares finite element methods have been applied, but that are not discussed in Chapters 5 through 11, are briefly considered.[7] In that chapter, some variations on the least-squares finite element

[6] There are a few settings considered in Part III that do not fit into the abstract frameworks of Part II; these are treated directly on a case-by-case basis.

[7] Providing a full treatment of all the topics discussed in Chapter 12 would have easily doubled the length of the book. However, we believe these topics should be included in the book, even in a somewhat cursory manner, not only because they are important, but because they serve to further

methods are also briefly discussed; these are methods that have a least-squares character or that use least-squares principles in different ways or for different purposes than those discussed in the rest of the book. Included in Chapter 12 are discussions of boundary condition treatments, LL* least-squares methods, mimetic reformulations of least-squares methods, least-squares collocation methods, restricted least-squares methods, optimization-based least-squares methods, advection–diffusion–reaction problems, higher-order problems, div–grad–curl systems, domain decomposition least-squares methods, multi-physics problems, problems with singular solution, Treffetz least-squares methods, a posteriori error estimation and mesh refinement, least-squares wavelet methods, and meshless least-squares methods.

In the four Appendices, results from functional analysis, partial differential equations, and finite element theory that are used in various places in the rest of the book are provided. In particular, we define a consistent and unified notational system that does not vary from setting to setting, even though much of the source material used for the book is not consistent in this regard. This not only greatly reduces the notational definitions introduced, but, more important, facilitates making connections between the different settings treated in the book. We also include, in addition to the expected index, a list of often used acronyms and a glossary of often used notations.

Acknowledgements

Achi Dosanjh, our editor at Springer, has shown great patience and support throughout the process of getting this book to print. It has been a pleasure to work with her.

Much of this book and much of our own work on least-squares finite element methods would not have come to pass without the help of our collaborators who have worked with us on this subject, especially students and many colleagues with whom we had pleasant and productive interactions. Thus, a great deal of thanks is owed to Dana Bedivan, John Burkardt, Zhiqiang Cai, Yanzhao Cao, Ching-Lung Chang, Jungmin Choi, Christopher Cox, Jennifer Deang, Raytcho Lazarov, Hyung-Chun Lee, Hugh MacMillan, Tom Manteuffel, Steve McCormick, Roy Nicolaides, and Panayot Vassilevski. Both authors especially acknowledge George Fix who was instrumental to their progress in the early stages of their careers and who, directly in the case of Max Gunzburger and indirectly in the case of Pavel Bochev, introduced them to the world of least-squares finite element methods.

This book is dedicated, by Pavel Bochev, to his wife, Biliana, for her unyielding support, kindness and exceptional patience, and for nurturing his resolve in those moments when the sheer magnitude of this endeavor made its success seem distant and unattainable. A special debt of gratitude is owed by Pavel to his mother Dora for believing in his dreams and making the many sacrifices that helped turn these dreams into reality.

This book is also dedicated, by Max Gunzburger, to his wife, Janet Peterson, not so much for the real support she provided throughout the long and arduous path he

illustrate the very significant progress that has recently been made in algorithmic, theoretical, and application aspects of least-squares finite element methods.

took to the completion of the book, but for the constant and enduring patience she has shown, support and help she has provided, sacrifices she has made, and love she has given that have been instrumental to his success, not only in his professional life, but, more important, in his personal life.

Pavel Bochev and Max Gunzburger
Albuquerque and Tallahassee
November, 2008

Contents

Part III Least-Squares Finite Element Methods for Elliptic Problems

Part I
Survey of Variational Principles and Associated Finite Element Methods

Chapter 1
Classical Variational Methods

In order to provide a context for least-squares finite element methods (LSFEMs) and a means for judging their effectiveness, we briefly review, in this chapter and the next, other finite element approaches. In this chapter, we focus on what we refer to as "classical" finite element methods, a designation that is admittedly somewhat arbitrary, but that is nonetheless generally accepted for the methods we include in the so-named class.

Finite element methods belong to the class of *variational methods*. As such, they are *quasi*-projections[1] whose properties are intrinsically connected to an underlying variational formulation. Thus, it is instructive to establish, in an abstract setting, taxonomies of variational formulations that provide the basis for defining variational methods, including finite element methods. We begin the chapter with a review of variational formulations for abstract linear operator equations. We distinguish between weakly and strongly coercive problems and also subdivide the formulations into those that are related to an optimization problem and those that are not. The taxonomy of variational formulations induces a like taxonomy of the associated variational methods, i.e., the quasi-projections that serve to define approximations of solutions of the variational equations. We explain how the properties of the variational formulations affect those of the associated variational methods. We apply our taxonomy to examine finite element methods for several canonical examples of partial differential equation (PDE) problems. We use these examples to show the influence that the underlying variational setting has on each discretization method and to explain the reasons that led to interest in the use of alternative variational formulations, including least-squares principles.

Because the aim of this chapter is to explain and fix important notions that surface regularly throughout the book, we keep the discussion at a fairly informal level.

---

[1] We explain what we mean by a quasi-projection in Remark 1.1.

P.B. Bochev, M.D. Gunzburger, *Least-Squares Finite Element Methods*,
Applied Mathematical Sciences 166, DOI: 10.1007/b13382_1,

1.1 Variational Methods for Operator Equations

Given Banach spaces X and Y, a bounded linear operator[2] $\mathcal{Q}$ from X into Y, and an element $f \in Y$, consider the abstract, linear operator equation

$$\textit{find } u \in X \quad \textit{such that} \quad \mathcal{Q}u = f \quad \textit{in } Y. \tag{1.1}$$

Typically, computational solutions of (1.1) rely on a process, called *discretization*, that converts this problem into a parameterized family[3]

$$\mathrm{Q}^h \vec{u}^h = \vec{f}^h \tag{1.2}$$

of systems of linear *algebraic* equations for the unknown vector $\vec{u}^h$.[4] The main goal in the design of a discretization process is to obtain parametrized families (1.2) of discrete problems whose solution sequence $\{\vec{u}^h\}_{h>0}$ converges, in some sense as $h \to 0$, to an exact solution $u \in X$ of (1.1).

For variational methods, the conversion of (1.1) into a discrete problem (1.2) begins with the reformulation of (1.1) into a suitable *variational equation*. An abstract variational equation or *variational formulation* is defined in terms of two Hilbert spaces U and W, a continuous bilinear form $Q(\cdot,\cdot) : U \times W \mapsto \mathbb{R}$, and a bounded linear functional $F(\cdot) : W \mapsto \mathbb{R}$, and is given by

$$\textit{find } u \in U \quad \textit{such that} \quad Q(u,w) = F(w) \qquad \forall\, w \in W. \tag{1.3}$$

The space U in which we seek the solution u is referred to as the *trial* space and the space W is referred to as the *test* space. Variational equations such as (1.3) are often called *weak formulations* or *generalized formulations*[5] of operator equations such as (1.1).

An operator equation such as (1.1) may be reformulated into several different variational equations of the type (1.3); the specifics of which particular variational equation one uses, i.e., the relationships between the spaces X,Y and U,W and between the operator $\mathcal{Q}$ and the bilinear form $Q(\cdot,\cdot)$, usually depend on the nature of the operator equation and practicality issues. In Section 1.4, we provide examples of the reformulation process.

---

[2] For the class of problems of interest to us, i.e., PDEs, $\mathcal{Q}$ is a partial differential operator.

[3] Usually, h is related to the number of degrees of freedom used in the definition of an approximate solution. In the finite element context, it is some measure of the size of the grid used to define the finite element spaces.

[4] The connection between the components of the vector $\vec{u}^h$ and the solution u of (1.1) depends on the specific discretization process used. For finite difference methods, the components of $\vec{u}^h$ are associated with approximations of u on a set of grid points. For finite element methods, those components are the coefficients of an expansion of an approximate solution u^h in terms of a basis for a finite-dimensional approximating subspace. For finite volume methods, these components are usually some averaged quantities related to u and associated with the edges, faces, or cells in a grid.

[5] The origin of these terminologies is that, in the PDE setting, variational equations often have solutions that are not smooth enough to be pointwise solutions of the corresponding PDE.

On the other hand, the continuous bilinear form $Q(\cdot;\cdot)$ implies the existence of an operator[6,7] $\widetilde{Q} : U \mapsto W^*$ through the relation

$$Q(u,w) = \langle \widetilde{Q}u, w\rangle_{W^*,W} \qquad \forall u \in U, w \in W.$$

Similarly, the bounded linear functional $F(\cdot)$ implies the existence of an element $\widetilde{f} \in W^*$ through the relation

$$F(w) = \langle \widetilde{f}, w\rangle_{W^*,W} \qquad \forall w \in W.$$

Then, (1.3) is equivalent to the problem

$$\textit{find } u \in U \quad \textit{such that} \quad \widetilde{Q}u = \widetilde{f} \quad \textit{in } W^*. \tag{1.4}$$

Because the pairs of spaces X,Y and U,W^* are not necessarily the same, it is clear that, in general, $\widetilde{Q} \neq Q$ and $\widetilde{f} \neq f$.[8]

One of the key reasons for the popularity of variational methods in general and finite element methods in particular is the ease with which variational equations such as (1.3) can be used to obtain parameterized discrete systems such as (1.2). The discretization process is simple to describe. A discrete variational equation consists of the same bilinear form and functional[9] as in (1.3), but restricted to two families of finite-dimensional subspaces[10] $U^h \subset U$ and $W^h \subset W$. Thus, we obtain the family of *discrete weak formulations*

$$\textit{find } u^h \in U^h \quad \textit{such that} \quad Q(u^h, w^h) = F(w^h) \qquad \forall w^h \in W^h. \tag{1.5}$$

To connect (1.5) with the parameterized linear algebraic system (1.2), we choose bases $\{\phi_j^h\}_{j=1}^N$ and $\{\psi_i^h\}_{i=1}^M$ for U^h and W^h, respectively. We then write $u^h = \sum_{j=1}^N \vec{u}_j^h \phi_j^h$ and set $\mathrm{Q}_{ij}^h = Q(\phi_j^h, \psi_i^h)$ and $\vec{f}_i^h = F(\psi_i^h)$. Then, $\vec{u}^h \in \mathbb{R}^N$, $\mathrm{Q}^h \in \mathbb{R}^{M\times N}$, $\vec{f}^h \in \mathbb{R}^M$, and, for every $h > 0$, (1.5) is equivalent to the linear system (1.2).

---

[6] Throughout, W^* denotes the dual space of a Hilbert space W, i.e., W^* is the space of all bounded linear functionals on W. Then, W^* itself is a Hilbert space. Also, $\langle\cdot,\cdot\rangle_{W^*,W}$ denotes the duality pairing between elements of W and its dual space W^*. The inner product on W is denoted by $(\cdot,\cdot)_W$.

[7] We adopt the notation that, e.g., $\widetilde{Q}$ denotes the operator induced by the bilinear form appearing in a weak formulation of a PDE that is defined in terms of the operator Q.

[8] For example, (1.1) could represent a PDE for $u \in X$ that holds pointwise in a domain so that Q involves classical derivatives. On the other hand, $\widetilde{Q}$ could involve generalized derivatives and then $u \in U$ is viewed as a generalized solution of (1.1), i.e., one that does not satisfy (1.1) in a pointwise manner. Thus, although formally (1.1) and (1.4) may look alike, the operators $\widetilde{Q}$ and Q have different domains and ranges.

[9] It is possible to define discretizations based on bilinear forms and functionals that are different from those in (1.3), e.g., mesh-dependent bilinear forms arising from the use of quadrature rules to approximate integrals. However, for brevity, we do not discuss such discretizations here.

[10] Whenever the inclusion $U^h \subset U$ holds, the resulting discrete approximations are referred to as being *conforming*. Nonconforming approximations for which $U^h \not\subset U$ are also used in practice. Unless explicitly noted, in this book, we confine our presentation to conforming approximations.

Remark 1.1 Because $W^h \subset W$, (1.3) implies that

$$Q(u,w^h) = F(w^h) \qquad \forall w^h \in W^h \subset W\,.$$

Together with (1.5), this relation implies that

$$Q(u^h,w^h) = Q(u,w^h) \qquad \forall w^h \in W^h\,. \tag{1.6}$$

Thus, if $u \in U$ and $u^h \in U^h$ are solutions of (1.3) and (1.5), respectively, then (1.6) has the *appearance* of defining a projection $u^h \in U^h \subset U$ of $u \in U$. This is what we mean when we say that finite element methods define *quasi*-projections. In fact, in general, (1.6) does not define a true projection; indeed that is the case only if the bilinear form $Q(\cdot,\cdot)$ defines an inner product, a situation for which a necessary but not sufficient condition is $W = U$. Note that (1.6) can be rewritten as

$$Q(u-u^h,w^h) = 0 \qquad \forall w^h \in W^h\,. \tag{1.7}$$

This relation is often referred to as a *Galerkin orthogonality relation* for the error $u-u^h$; of course, it defines a true orthogonality relation only if the bilinear form $Q(\cdot,\cdot)$ defines an inner product. □

Finite element methods are distinguished from other variational methods by the manner in which one defines the approximation spaces U^h and W^h. Construction of a finite element space begins with partitioning the problem domain into *finite elements*.[11] A *locally supported* basis is then obtained by mapping polynomial functions defined on a standard *reference* element to every element in the domain partition. For further discussion of finite element spaces, see Appendix B.

An important special class of variational problems is provided by *mixed variational problems*. In this setting, we have two Hilbert spaces V and S, two continuous bilinear forms

$$a(\cdot,\cdot) : V \times V \mapsto \mathbb{R} \qquad \text{and} \qquad b(\cdot,\cdot) : V \times S \mapsto \mathbb{R}\,, \tag{1.8}$$

two bounded linear functionals $D(\cdot) : V \mapsto \mathbb{R}$ and $G(\cdot) : S \mapsto \mathbb{R}$, and the mixed variational problem

$$\begin{array}{c}\textit{find } \{v,p\} \in V \times S \\ \textit{such that}\end{array} \qquad \begin{cases} a(v,r) + b(r,p) = D(r) & \forall r \in V \\ b(v,s) \qquad\qquad\;\; = G(s) & \forall s \in S\,. \end{cases} \tag{1.9}$$

---

[11] In two dimensions, triangles or quadrilaterals are used; in three dimensions, tetrahedrons, rectangular parallelepipeds, or hexahedrons are among the finite elements commonly employed. Although the explicit use of finite elements as building blocks for the approximation spaces is still the prevailing way of developing and using finite element methods, finite element methods have a much broader scope and encompass any quasi-projection scheme that uses approximation spaces with locally supported bases. This notion includes, e.g., meshless methods for which a traditional subdivision of a domain into finite elements is not explicitly used in the definition of the approximation space.

Using the duality pairings between the spaces V and S and their dual spaces V^* and S^*, respectively, the continuous bilinear forms in (1.8) give rise to the bounded linear operators $\widetilde{\mathcal{A}}: V \mapsto V^*$, $\widetilde{\mathcal{B}}: V \mapsto S^*$, and $\widetilde{\mathcal{B}}^*: S \mapsto V^*$ defined by

$$a(v,r) = \langle \widetilde{\mathcal{A}}v, r\rangle_{V^*,V} \qquad \forall v,r \in V$$

and

$$b(v,s) = \langle \widetilde{\mathcal{B}}v, s\rangle_{S^*,S} = \langle \widetilde{\mathcal{B}}^* s, v\rangle_{V^*,V} \qquad \forall v \in V,\, s \in S,$$

respectively. The bounded linear functionals $D(\cdot)$ and $G(\cdot)$ respectively give rise to the functions $\widetilde{d} \in V^*$ and $\widetilde{g} \in S^*$ defined by

$$D(r) = \langle \widetilde{d}, r\rangle_{V^*,V} \quad \forall v \in V \qquad \text{and} \qquad G(s) = \langle \widetilde{g}, s\rangle_{S^*,S} \quad \forall s \in S.$$

Then, (1.9) is equivalent to

$$\begin{cases} \widetilde{\mathcal{A}}v + \widetilde{\mathcal{B}}^* p = \widetilde{d} & \text{in } V^* \\ \widetilde{\mathcal{B}}v \qquad\quad = \widetilde{g} & \text{in } S^*. \end{cases} \tag{1.10}$$

Remark 1.2 Analogous to the relation between (1.1) and (1.4), (1.10) or, equivalently, (1.9), can be viewed as a weak or generalized formulation of a system of PDEs of the form

$$\begin{cases} \mathcal{A}v + \mathcal{B}^* p = d \\ \mathcal{B}v \qquad\quad = g, \end{cases} \tag{1.11}$$

where these equations are thought of as holding pointwise. □

By setting $U = W = V \times S$, $u = \{v,p\}$, $w = \{r,s\}$, $F(\{r,s\}) = D(r) + G(s)$, and

$$Q(\{v,p\},\{r,s\}) = a(v,r) + b(r,p) + b(v,s), \tag{1.12}$$

(1.9) assumes the form (1.3) and, by setting

$$\widetilde{\mathcal{Q}} = \begin{pmatrix} \widetilde{\mathcal{A}} & \widetilde{\mathcal{B}}^* \\ \widetilde{\mathcal{B}} & 0 \end{pmatrix} \qquad \text{and} \qquad \widetilde{f} = \begin{pmatrix} \widetilde{d} \\ \widetilde{g} \end{pmatrix}, \tag{1.13}$$

(1.10) takes the form (1.4).

The parameterized linear system (1.2) also has a special structure in the case of mixed variational problems. To define this system, we choose conforming subspaces $V^h \subset V$ and $S^h \subset S$ and restrict (1.9) to these subspaces, i.e., we solve the problem:

$$\begin{array}{l} \textit{find } \{v^h,p^h\} \in \\ V^h \times S^h \textit{ such that} \end{array} \quad \begin{cases} a(v^h,r^h) + b(r^h,p^h) = D(r^h) & \forall r^h \in V^h \\ b(v^h,s^h) \qquad\qquad\quad = G(s^h) & \forall s^h \in S^h. \end{cases} \tag{1.14}$$

If $\{\xi_j^h\}_{j=1}^N$ and $\{\eta_j^h\}_{j=1}^M$ denote bases for V^h and S^h, respectively, then (1.14) is equivalent to the linear system

$$\begin{pmatrix} \mathsf{A}^h & \mathsf{B}^{h\,\mathsf{T}} \\ \mathsf{B}^h & 0 \end{pmatrix} \begin{pmatrix} \vec{v}^h \\ \vec{p}^h \end{pmatrix} = \begin{pmatrix} \vec{d}^h \\ \vec{g}^h \end{pmatrix}, \tag{1.15}$$

where $v^h = \sum_{j=1}^{N} \vec{v}^h_j \xi^h_j$ and $p^h = \sum_{j=1}^{M} \vec{p}^h_j \eta^h_j$ and $(\cdot)^\mathsf{T}$ denotes the transpose matrix. In (1.15), we have that

$$\mathsf{A}^h_{ij} = a(\xi^h_j, \xi^h_i) \quad \text{for } i,j = 1,\dots,N,$$
$$\mathsf{B}^h_{ij} = b(\xi^h_j, \eta^h_i) \quad \text{for } i = 1,\dots,M,\ j = 1,\dots,N,$$
$$\vec{d}^h_i = D(\xi^h_i) \quad \text{for } i = 1,\dots,N, \qquad \text{and} \qquad \vec{g}^h_i = G(\eta^h_i) \quad \text{for } i = 1,\dots,M.$$

Thus, the coefficient matrix Q^h in the parameterized linear system (1.2) is given by

$$\mathsf{Q}^h = \begin{pmatrix} \mathsf{A}^h & \mathsf{B}^{h\,\mathsf{T}} \\ \mathsf{B}^h & 0 \end{pmatrix}.$$

1.2 A Taxonomy of Classical Variational Formulations

We now discuss the basic classes of variational formulations upon which classical finite element methods are based and the conditions that ensure the well-posedness of the weak problems (1.3) and (1.9). We then explore the connections between the variational formulations of Section 1.1 and optimization problems.

1.2.1 Weakly Coercive Problems

A bilinear form $Q(\cdot,\cdot) : U \times W \mapsto \mathbb{R}$ is called *continuous* if

$$|Q(u,w)| \le \alpha \|u\|_U \|w\|_W \qquad \forall u \in U,\, w \in W, \tag{1.16}$$

where $\alpha < \infty$, and is called *weakly coercive* if[12]

$$\inf_{u \in U} \sup_{w \in W} \frac{Q(u,w)}{\|u\|_U \|w\|_W} \ge \beta \tag{1.17}$$

and

[12] We should more precisely write (1.17) and similar expressions as

$$\inf_{u \in U,\, u \neq 0}\ \sup_{w \in W,\, w \neq 0} \frac{Q(u,w)}{\|u\|_U \|w\|_W} \ge \beta.$$

For the sake of economy of notation, we omit the obvious necessity to have $u \neq 0$, and so on.

$$\inf_{w \in W} \sup_{u \in U} \frac{Q(u,w)}{\|u\|_U \|w\|_W} > 0, \qquad (1.18)$$

where $\beta > 0$.[13] The following theorem, often referred to as the Nečas theorem (see [280,281]), shows that (1.16)–(1.18) are sufficient[14] to guarantee that the variational problem (1.3) is well posed.

Theorem 1.3 *Let $F(\cdot)$ denote a bounded linear functional on W and let $Q(\cdot,\cdot)$ denote a continuous and weakly coercive bilinear form on $U \times W$, i.e., (1.16)–(1.18) hold. Then, the abstract variational problem (1.3) has a unique solution $u \in U$. Moreover, that solution satisfies*

$$\|u\|_U \le \frac{1}{\beta} \|F\|_{W^*}$$

so that it depends continuously on the data.[15] □

1.2.2 Strongly Coercive Problems

If $W \neq U$, the most one can expect from a bilinear form is weak coercivity. However, if $W = U$, i.e., if the test space W and the trial space U are the same, a stronger notion of coercivity can be defined. Note that if $W = U$, then (1.16) becomes

$$|Q(u;v)| \le \alpha \|u\|_U \|v\|_U \qquad \forall u, v \in U. \qquad (1.19)$$

---

[13] It is clear why (1.17) and (1.18) are referred to as *inf–sup conditions*. An equivalent way to express, e.g., (1.17) is

$$\sup_{w \in W} \frac{Q(u,w)}{\|w\|_W} \ge \beta \|u\|_U \qquad \forall u \in U.$$

Because of its equivalence to (1.17), we also refer to conditions of this type as inf–sup conditions.

[14] The necessity of these conditions can also be demonstrated.

[15] It is instructive to consider an example in finite-dimensional spaces for which $U = \mathbb{R}^N$, $W = W^* = \mathbb{R}^M$, $\widetilde{Q} \in \mathbb{R}^{M \times N}$, $\vec{f} \in \mathbb{R}^M$, $Q(\vec{u};\vec{w}) = \vec{w}^T \widetilde{Q} \vec{u}$, and $F(\vec{w}) = \vec{w}^T \vec{f}$. With these identifications, one can then verify that the abstract problem (1.3) is equivalent to the linear system $\widetilde{Q}\vec{u} = \vec{f}$. In this case, the weak coercivity conditions (1.17) and (1.18) imply that $\widetilde{Q}$ has full row and column ranks, i.e., it is a square invertible matrix.

Note the distinction between this example, where the problem-defining operator is finite dimensional, and a parameterized linear system such as (1.2). In the former case, the given problem is a linear system with a coefficient matrix having fixed dimensions. In the case of (1.2), the dimensions of the coefficient matrix grow as the parameter h tends to zero; we have emphasized this point by attaching the superscript h to the matrix Q^h in (1.2). This distinction between the coefficient matrices is important because, in the case of (1.2), it is not enough to know that the coefficient matrix is invertible for any fixed value of h; one also needs to know about the uniformity of the invertibility, e.g., the uniformity of a bound on the inverse matrix as $h \to 0$. Of course, this last point does not apply to the example discussed in the previous paragraph.

A bilinear form $Q(\cdot,\cdot): U \times U \mapsto \mathbb{R}$ is called *strongly coercive* or *U-elliptic* if, for some $\beta > 0$,

$$Q(u;u) \geq \beta \|u\|_U^2 \qquad \forall u \in U\,. \tag{1.20}$$

It is easy to verify that (1.20) implies the weak coercivity conditions (1.17) and (1.18), i.e., strong coercivity implies weak coercivity; the converse is not true. The classical Lax–Milgram lemma shows that strongly coercive problems are well posed.[16]

Corollary 1.4 *Let $F(\cdot)$ denote a bounded linear functional on U and $Q(\cdot,\cdot)$ denote a continuous, strongly coercive bilinear form on $U \times U$, i.e.,* (1.19) *and* (1.20) *hold. Then, the abstract variational problem* (1.3) *has a unique solution. Moreover, that solution satisfies*

$$\|u\|_U \leq \frac{1}{\beta}\|F\|_{U^*}$$

so that it depends continuously on the data.[17] □

1.2.3 Mixed Variational Problems

Mixed variational problems provide an example of weakly coercive problems and therefore they can be analyzed using Theorem 1.3.[18] However, due to their special structure, it is advantageous to state the hypotheses needed to guarantee the well-posedness of problems such as (1.9) in terms of the constituent bilinear forms and linear functionals. To this end, we need to use the subspace

$$Z = \{v \in V \mid b(v,s) = 0\ \forall s \in S\} \subset V \tag{1.21}$$

consisting of those elements $v \in V$ that belong to the null space of the operator $\widetilde{\mathcal{B}}$.

The following generalization of the Brezzi theorem (see [83,87,183]) establishes hypotheses for which the mixed variational problem (1.9) is well posed.

Theorem 1.5 *Given the Hilbert spaces V and S, the bounded linear functionals $D(\cdot)$ and $G(\cdot)$ on V and S, respectively, and the bilinear forms $a(\cdot,\cdot)$ and $b(\cdot,\cdot)$ on*

---

16 Obviously, because strong coercivity implies weak coercivity, Corollary 1.4 is simply a special case of Theorem 1.3. Here, we state it as a separate result because of its historical importance.

17 We again visit the finite-dimensional setting. Let $U = W = W^* = \mathbb{R}^N$, $\widetilde{\mathbb{Q}} \in \mathbb{R}^{N\times N}$, $\vec{f} \in \mathbb{R}^N$, $Q(\vec{u};\vec{w}) = \vec{w}^\mathsf{T}\widetilde{\mathbb{Q}}\vec{u}$, and $F(\vec{w}) = \vec{w}^\mathsf{T}\vec{f}$. A simple argument shows that the strong coercivity of $Q(\cdot,\cdot)$ is equivalent to $\widetilde{\mathbb{Q}}$ being a real matrix with a positive definite symmetric part. A real, positive definite matrix is always invertible. This is the algebraic version of the statement that strong coercivity implies weak coercivity. On the other hand, because a matrix does not have to be positive definite to be invertible, it is clear that strong coercivity is a sufficient but not necessary condition for guaranteeing the results of Theorem 1.3.

18 Indeed, if $a(\cdot,\cdot)$ and $b(\cdot,\cdot)$ satisfy the hypotheses of Theorem 1.5 below, then it can be shown that the compound form $Q(\cdot,\cdot)$ defined in (1.12) satisfies (1.16)–(1.18).

$V \times V$ and $V \times S$, respectively. Assume that the bilinear forms $a(\cdot,\cdot)$ and $b(\cdot,\cdot)$ are continuous, i.e., for some constants $\alpha_a < \infty$ and $\alpha_b < \infty$,

$$a(v,r) \le \alpha_a \|v\|_V \|r\|_V \qquad \forall\, v,r \in V \tag{1.22}$$

and

$$b(v,s) \le \alpha_b \|v\|_V \|s\|_S \qquad \forall\, v \in V,\ s \in S\,. \tag{1.23}$$

Assume further that $a(\cdot,\cdot)$ is weakly coercive on the null space of $\widetilde{\mathcal{B}}$, i.e.,

$$\begin{aligned} \inf_{v\in Z}\sup_{r\in Z} \frac{a(v,r)}{\|v\|_V \|r\|_V} &\ge \beta_a \\ \inf_{r\in Z}\sup_{v\in Z} \frac{a(v,r)}{\|v\|_V \|r\|_V} &> 0\,, \end{aligned} \tag{1.24}$$

where $\beta_a > 0$, and that $b(\cdot,\cdot)$ satisfies the inf–sup condition

$$\inf_{s\in S}\sup_{v\in V} \frac{b(v,s)}{\|v\|_V \|s\|_S} \ge \beta_b\,, \tag{1.25}$$

where $\beta_b > 0$. Then, the mixed variational problem (1.9) has a unique solution $\{v,p\} \in V \times S$ and that solution satisfies

$$\|v\|_V + \|p\|_S \le C(\|D\|_{V^*} + \|G\|_{S^*})$$

so that it depends continuously on the data.[19] □

Null space methods

The problem (1.9) can, in principle, be "simplified" by restricting it to an affine space on which the second equation is satisfied. For the sake of simplicity, we assume that $G(\cdot) = 0$ in (1.9)[20] so that now we have that $b(v,s) = 0$ for all $s \in S$, i.e., $v \in Z$. We then use, in the first equation in (1.9), test functions $r \in Z \subset V$ so that $b(r,p) = 0$. Then, (1.9) is reduced to the problem

---

[19] Considering again the finite-dimensional setting, let $V = V^* = \mathbb{R}^N$, $S = S^* = \mathbb{R}^M$, $\widetilde{\mathcal{A}} \in \mathbb{R}^{N\times N}$, $\widetilde{\mathcal{B}} \in \mathbb{R}^{M\times N}$, $\vec{v} \in \mathbb{R}^N$, $\vec{p} \in \mathbb{R}^M$, $a(\vec{v},\vec{r}) = \vec{r}^{\,T}\widetilde{\mathcal{A}}\vec{v}$, $b(\vec{v},\vec{s}) = \vec{s}^{\,T}\widetilde{\mathcal{B}}\vec{v}$, $D(\vec{r}) = \vec{r}^{\,T}\vec{d}$, and $G(\vec{s}) = \vec{s}^{\,T}\vec{g}$. It is not difficult to show that (1.22)–(1.25) imply that the matrix $\widetilde{\mathcal{B}}$ has full row rank M and that the matrix $\widetilde{\mathcal{A}}$ is invertible on the null space of $\widetilde{\mathcal{B}}$, i.e., that the problem $\mathcal{Z}^T\widetilde{\mathcal{A}}\mathcal{Z}\vec{q} = \mathcal{Z}^T\vec{d}$ has a unique solution $\vec{q} \in \mathbb{R}^{N-M}$ for any $\vec{d} \in \mathbb{R}^N$, where $\mathcal{Z} \in \mathbb{R}^{N\times(N-M)}$ is a matrix whose columns form a basis for $N(\widetilde{\mathcal{B}})$, where $N(\cdot)$ denotes the null space. These conditions on the matrices $\widetilde{\mathcal{A}}$ and $\widetilde{\mathcal{B}}$ in turn guarantee that the matrix $\widetilde{\mathcal{Q}} \in \mathbb{R}^{(M+N)\times(M+N)}$ defined in (1.13) is invertible. Because clearly that matrix is not positive definite, we see that in this case $\widetilde{\mathcal{Q}}$ satisfies the weak coercivity conditions (1.17) and (1.18).

[20] The more general case of $G(\cdot) \ne 0$ can, in principle, be reduced to this case by subtracting from v any particular solution $\widehat{v} \in V$ satisfying the equation $b(\widehat{v},s) = G(s)$ for all $s \in S$.

$$\text{find } v \in Z \subset V \quad \text{such that} \quad a(v,r) = D(r) \quad \forall r \in Z. \tag{1.26}$$

Note that (1.26) is of the form (1.3) with $U = W = Z$, $Q(\cdot,\cdot) = a(\cdot,\cdot)$, and $F(\cdot) = D(\cdot)$.

Under the same hypotheses as those of Theorem 1.5, the problem (1.26) is a weakly coercive variational problem; indeed, (1.22) implies that the bilinear form $a(\cdot,\cdot)$ is continuous and (1.24) is exactly the definition of weak coercivity whenever that form is restricted to act on $Z \times Z$. Then, by Theorem 1.3, we have that (1.26) has a unique solution $v \in Z$, and, because $Z \subset V$ and $V^* \subset Z^*$, that solution satisfies $\|v\|_V \leq \frac{1}{\beta_a}\|D\|_{V^*}$.

If one wishes to also determine the variable p in (1.9), one need only solve, after determining $v \in Z$ from (1.26), the problem:

$$\text{find } p \in S \quad \text{such that} \quad b(r,p) = D(r) - a(v,r) \quad \forall r \in Z^{\perp} \subset V. \tag{1.27}$$

The continuity condition (1.23) and the inf–sup condition (1.25) not only guarantee that $Z^{\perp}$ is well defined, but also that the problem (1.27) has a unique solution $p \in S$ and that solution satisfies $\|p\|_S \leq C\|D\|_{V^*}$; see, e.g., [183], for details.[21]

1.2.4 *Relations Between Variational Problems and Optimization Problems*

If we make further assumptions on the bilinear form $Q(\cdot,\cdot)$ of the problem (1.3) and the bilinear form $a(\cdot,\cdot)$ of the problem (1.9), we can relate those two problems to optimization problems.

Unconstrained optimization problems (the Rayleigh–Ritz setting)

We first consider the classical *Rayleigh–Ritz setting* in which strongly coercive variational problems can be associated with the minimization of energy functionals.

Proposition 1.6 *Assume that the hypotheses of Corollary* 1.4 *hold. Assume further that the bilinear form $Q(\cdot,\cdot)$ is symmetric, i.e., that $Q(u,w) = Q(w,u)$ for all $u,w \in U$. Then, the problem* (1.3) *is equivalent*[22] *to the* unconstrained *optimization problem*

[21] With $G(\cdot) = 0$, in the finite-dimensional setting we have been following, we have that (1.26) has the form $\mathcal{Z}^{\mathsf{T}}\widetilde{\mathcal{A}}\mathcal{Z}\vec{q} = \mathcal{Z}^{\mathsf{T}}\vec{d}$ for $\vec{q} \in \mathbb{R}^{N-M}$. The hypotheses of Theorem 1.5 imply that the matrix $\mathcal{Z}^{\mathsf{T}}\widetilde{\mathcal{A}}\mathcal{Z} \in \mathbb{R}^{(N-M)\times(N-M)}$ is invertible. Then, after solving for $\vec{q}$, $\vec{v} \in \mathbb{R}^N$ satisfying (1.9) is found by setting $\vec{v} = \mathcal{Z}\vec{q}$. If the matrix $\mathcal{Z}_{\perp} \in \mathbb{R}^{N\times M}$ has columns that form a basis for $Z^{\perp}$, then the problem (1.27) for $\vec{p} \in \mathbb{R}^M$ takes the form $\mathcal{Z}_{\perp}^{\mathsf{T}}\widetilde{\mathcal{B}}^{\mathsf{T}}\vec{p} = \mathcal{Z}_{\perp}^{\mathsf{T}}\vec{d} - \mathcal{Z}_{\perp}^{\mathsf{T}}\widetilde{\mathcal{A}}\vec{v}$; the hypotheses of Theorem 1.5 imply that the matrix $\mathcal{Z}_{\perp}^{\mathsf{T}}\widetilde{\mathcal{B}}^{\mathsf{T}} \in \mathbb{R}^{M\times M}$ is invertible.

[22] Equivalence of two problems means that solutions of one problem solve the other and conversely.

$$\textit{find } u \in U \quad \textit{that minimizes} \quad J(u;F) = \frac{1}{2}Q(u;u) - F(u)\,. \qquad \square \tag{1.28}$$

In fact, (1.3) is the first-order necessary condition that solutions $u \in U$ of the unconstrained minimization problem (1.28) are required to satisfy. In the Rayleigh-Ritz setting, the operator $\widetilde{Q}$ associated with the bilinear form $Q(\cdot,\cdot)$ is symmetric and positive definite. Note also that, in this setting, the discrete problem (1.5) is equivalent to the discrete optimization problem

$$\textit{find } u^h \in U^h \quad \textit{that minimizes} \quad J(u^h;F) = \frac{1}{2}Q(u^h;u^h) - F(u^h)\,.$$

It is important to note that in the Rayleigh-Ritz setting, the bilinear form $Q(\cdot,\cdot)$ defines an equivalent inner product[23] on the Hilbert space U and *the functional* $J(\cdot;0)$ *defines an equivalent norm*[24] *on* U. These observations follow from the symmetry of $Q(\cdot,\cdot)$ and the relation

$$\beta\|u\|_U^2 \le 2J(u;0) = Q(u;u) \le \alpha\|u\|_U^2$$

which in turn follows easily from (1.19) and (1.20).

Constrained optimization problems

We now consider the relation between mixed variational principles and constrained optimization problems.

Proposition 1.7 *Assume that the hypotheses of Theorem* 1.5 *hold. Assume further that the bilinear form* $a(\cdot,\cdot)$ *is symmetric, nonnegative, and strongly coercive on the null space of* $\widetilde{\mathcal{B}}$*, i.e., that*

$$a(v,r) = a(r,v) \qquad \textit{and} \qquad a(v,v) \ge 0 \qquad \textit{for all } v,r \in V \tag{1.29}$$

and

$$a(v,v) \ge \beta_a\|v\|_V^2 \qquad \textit{for all } v \in Z\,, \tag{1.30}$$

where $\beta_a > 0$. *Then, the problem* (1.9) *is equivalent to the* constrained *optimization problem*

$$\begin{aligned} \textit{find } v \in V \textit{ that minimizes} \quad & J(v;D) = \frac{1}{2}a(v,v) - D(v) \\ \textit{subject to} \quad & b(v,s) = G(s) \quad \forall s \in S\,. \qquad \square \end{aligned} \tag{1.31}$$

---

[23] Thus, in the Rayleigh-Ritz setting, (1.6) implies that the approximate solution $u^h \in U^h$ is a true projection of $u \in U$ with respect to the inner product $Q(\cdot,\cdot)$. Similarly, in that case, (1.7) implies that the error $u - u^h$ is truly orthogonal to all elements of the approximating space $W^h = U^h$, again with respect to the inner product $Q(\cdot,\cdot)$. In the more general setting of Section 1.1, (1.6) does not define a true projection and (1.7) does not define a true orthogonality relation.

[24] The norm induced by the inner product $Q(\cdot,\cdot)$ is often referred to as the *energy norm*.

In fact, if we enforce the constraint in (1.31) by introducing the Lagrange multiplier $p \in S$ and the Lagrangian functional

$$L(v,p;D,G) = J(v;D) - b(v,p) + G(p),$$

then the equations in (1.9) are the first-order necessary conditions that saddle points $\{v,p\} \in V \times S$ of $L(\cdot,\cdot;\cdot,\cdot)$ are required to satisfy. In the setting of Proposition 1.7, the operator $\widetilde{\mathcal{A}}$ associated with the bilinear form $a(\cdot,\cdot)$ is symmetric and positive semi-definite; it is also positive definite with respect to the null space of the operator $\widetilde{\mathcal{B}}$ associated with the bilinear form $b(\cdot,\cdot)$. However, even in this case, the composite operator $\widetilde{\mathcal{Q}}$ defined in (1.13), although symmetric, is indefinite so that the associated bilinear form $Q(\cdot,\cdot)$ remains weakly coercive. This, of course, is the nature of saddle point problems. Note also that in the setting of constrained optimization problems, the discrete problem (1.14) is equivalent to the discrete constrained optimization problem

$$\begin{aligned} &\textit{find } v^h \in V^h \textit{ that minimizes} \quad J(v^h;D) = \frac{1}{2}a(v^h,v^h) - D(v^h) \\ &\qquad\qquad\qquad\qquad \textit{subject to} \quad b(v^h,s^h) = G(s^h) \quad \forall s^h \in S^h. \end{aligned}$$

A second approach for handling the constraint in the problem (1.31) is to restrict the minimization problem to an affine space on which the constraint is satisfied, i.e, to use the null space method described in Section 1.2.3. With $G(\cdot) = 0$, the hypotheses of Theorem 1.7 imply that (1.26) are the first-order necessary conditions for $v \in Z$ to solve the reduced, unconstrained optimization problem

$$\textit{find } v \in Z \quad \textit{that minimizes} \quad J(v;D) = \frac{1}{2}a(v,v) - D(v) \tag{1.32}$$

that is equivalent to the constrained optimization problem (1.31). Under the same hypotheses as those of Theorem 1.7, the problem (1.26) is now a strongly coercive variational problem; this should be contrasted to the situation for Lagrange multiplier methods for which the same hypotheses lead to the weakly coercive variational problem involving the bilinear form (1.12). The unconstrained optimization problem (1.32) is of the Rayleigh-Ritz form (1.28) and therefore the former may be discretized in the same manner as the latter. One chooses a conforming subspace $\widetilde{Z}^h \subset Z$ and then restricts the problem (1.32) to that subspace, i.e., one solves the discrete unconstrained optimization problem

$$\left.\begin{aligned} &\textit{find } v^h \in \widetilde{Z}^h \subset Z \subset V \\ &\qquad \textit{that minimizes} \end{aligned}\right\} \quad J(v^h;D) = \frac{1}{2}a(v^h,v^h) - D(v^h).$$

The first-order necessary conditions corresponding to this problem are given by (1.45).

1.3 Approximation of Solutions of Variational Problems

Problems such as (1.3) are generally referred to as *Galerkin*[25] *formulations* of the operator equation (1.1); finite element methods based on (1.5) or, equivalently, (1.2), are referred to as *Galerkin finite element methods*. As seen in Sections 1.2.1 and 1.2.2, Galerkin formulations can involve either a strongly or weakly coercive bilinear form $Q(\cdot,\cdot)$. The *Rayleigh–Ritz* setting refers to the special case for which this bilinear form is not only strongly coercive, but is also symmetric. *Mixed-Galerkin formulations* of the operator equation (1.11) refer to the system (1.9); finite element methods based on (1.14) or, equivalently, (1.15), are therefore known as *mixed-Galerkin finite element methods* or, more simply, as *mixed finite element methods*. There is less need to distinguish between problems of the form (1.9) that are related to a constrained optimization problem and those that are not because in both cases one is led to weakly coercive problems.

Recall that discretized variational problems are restrictions of variational problems to finite-dimensional subspaces so that their well-posedness is subject to similar conditions as were imposed for the parent problems. In this section, we apply this observation to state sufficient stability conditions, first for discrete problems spawned from weakly and strongly coercive variational problems, and then for those spawned from mixed variational problems. It is important to note that, in some cases, a condition imposed on the discretized problem is automatically inherited from the corresponding condition imposed on the parent problem and, in other cases, it is not.

1.3.1 Weakly and Strongly Coercive Variational Problems

For the discretization of weakly coercive problems, we have the following theorem that is sometimes referred to as Babuška's theorem [19, 20].

Theorem 1.8 *Let the spaces U and W, the bilinear form $Q(\cdot,\cdot)$, and the functional $F(\cdot)$ satisfy the hypotheses of Theorem 1.3. Let $U^h \subset U$ and $W^h \subset W$ be closed subspaces.*[26] *Then, $Q(\cdot,\cdot)$ is continuous on $U^h \times W^h$. Assume also that the bilinear form $Q(\cdot,\cdot)$ satisfies the discrete inf–sup conditions*

$$\inf_{u^h\in U^h}\sup_{w^h\in W^h}\frac{Q(u^h,w^h)}{\|u^h\|_U\|w^h\|_W}\geq\beta^h \tag{1.33}$$

and

$$\inf_{w^h\in W^h}\sup_{u^h\in U^h}\frac{Q(u^h,w^h)}{\|u^h\|_U\|w^h\|_W}>0, \tag{1.34}$$

---

[25] Strictly speaking, the term "Galerkin formulation" refers to variational problems for which the test and the trial spaces coincide. In the general case when $U \neq W$, one speaks of "Petrov–Galerkin variational formulations." For brevity, we use the former terminology in all cases.

[26] Note that it is not required for the subspaces to be finite dimensional.

where $\beta^h \geq \widehat{\beta} > 0$. *Then, for every* $h > 0$, *the discretized problem* (1.5) *has a unique solution* $u^h \in U^h$. *Moreover, that solution satisfies the stability estimate*

$$\|u^h\|_U \leq \frac{1}{\widehat{\beta}}\|F\|_{W*}$$

and, if $u \in U$ *denotes the unique solution of* (1.3), *the optimal error estimate*[27,28]

$$\|u - u^h\|_U \leq \left(1 + \frac{\alpha}{\widehat{\beta}}\right) \inf_{v^h \in U^h} \|u - v^h\|_U . \qquad \square \tag{1.35}$$

In Section 1.1 it is seen that the discretized problem (1.5) is equivalent to a parameterized family of linear systems (1.2). From this theorem, it follows that for every $h > 0$, the matrix Q^h in (1.2) is square and nonsingular. This is basically all that can be inferred about the discretized equations that result from weakly coercive variational formulations.

For the discretization of strongly coercive problems, we have the following result that is sometimes referred to as Céa's lemma [82, 108, 123].

Theorem 1.9 *Let the space* U, *the bilinear form* $Q(\cdot,\cdot)$, *and the linear functional* $F(\cdot)$ *satisfy the hypotheses of Corollary* 1.4. *Let* $U^h \subset U$ *be a closed subspace. Then,* $Q(\cdot,\cdot)$ *is continuous on* $U^h \times U^h$ *and satisfies*

$$Q(u^h, u^h) \geq \beta \|u^h\|_U^2 \qquad \forall u^h \in U^h \tag{1.36}$$

and, for every $h > 0$, *the discretized problem* (1.5) *has a unique solution* $u^h \in U^h$. *Moreover, that solution satisfies the stability estimate*

$$\|u^h\|_U \leq \frac{1}{\beta}\|F\|_{U*}$$

and, if $u \in U$ *denotes the unique solution of* (1.3), *the optimal error estimate*

$$\|u - u^h\|_U \leq \frac{\alpha}{\beta} \inf_{w^h \in U^h} \|u - w^h\|_U . \qquad \square \tag{1.37}$$

Remark 1.10 Remarkably, if, in addition, the bilinear form $Q(\cdot,\cdot)$ is symmetric, e.g., in the Rayleigh-Ritz setting of unconstrained optimization problems, the error estimate (1.37) can be improved to

---

27 An *optimal error estimate* is one for which the error in a finite-dimensional approximation is bounded from above by a constant multiple of the error of the best approximation out of the same finite-dimensional subspace. An approximate solution that satisfies an optimal error estimate is referred to as being *optimally accurate.*

28 It is important to keep in mind that optimality of an error estimate of a numerical method is not sufficient to guarantee *convergence* of the numerical solution u^h to the exact solution u. The latter requires an additional *approximability property* of the discrete spaces. See Appendix B for further details.

$$\|u-u^h\|_U \le \sqrt{\frac{\alpha}{\beta}} \inf_{v^h \in U^h} \|u-v^h\|_U\,; \tag{1.38}$$

see, e.g., [123]. That this is indeed an improvement over (1.37) follows from the fact that necessarily $\alpha \ge \beta$. □

This theorem also asserts that, for every $h > 0$, the matrix Q^h in (1.2) is square and nonsingular. However, the discrete coercivity (1.36) implies that, in addition to these properties, strongly coercive variational formulations give rise to *positive definite* linear systems. Furthermore, in the Rayleigh–Ritz setting, these systems are also *symmetric*.

A cursory inspection of Theorems 1.8 and 1.9 points out an important difference between the conditions that guarantee the stability of discretizations of weakly and strongly coercive variational problems. In both cases, the continuity property of the bilinear form $Q(\cdot,\cdot)$ with respect to conforming discrete subspaces is inherited from the continuity property that holds on the parent spaces. Likewise, the strong coercivity property (1.36) with respect to $U^h \subset U$ automatically follows from the strong coercivity property (1.20) with respect to U. Thus,

> *for strongly coercive variational problems, the conformity of the approximating subspace $U^h \subset U$ is all that is needed to guarantee the stability of the discrete variational problem and the optimal accuracy of the approximate solution, i.e., conformity of the spaces is necessary and sufficient for their compatibility.*

However, this is not true for weakly coercive variational problems;

> *the inclusions $U^h \subset U$ and $W^h \subset W$ are not by themselves sufficient to guarantee that the discrete weak coercivity conditions* (1.33) *and* (1.34) *hold, even if the conditions* (1.17) *and* (1.18) *are known to hold, i.e., conformity of the spaces is necessary but not sufficient for their compatibility.*

Thus, in Theorem 1.8, the discrete weak coercivity conditions must be stated as additional hypotheses on the bilinear form $Q(\cdot,\cdot)$ and the discrete subspaces U^h and W^h.

Let us examine these issues in more detail. Assume first that $Q(\cdot,\cdot)$ is continuous on $U \times W$. Let u^h and w^h denote arbitrary elements belonging to the conforming discrete subspaces $U^h \subset U$ and $W^h \subset W$, respectively. Obviously, $u^h \in U$ and $w^h \in W$ as well and (1.16) holds for u^h and w^h considered as elements of the parent spaces. Because u^h and w^h are arbitrary, we conclude that (1.16) holds on $U^h \times W^h$. The same argument can be used to show that if $Q(\cdot,\cdot)$ is strongly coercive on $U \times U$, then (1.36) holds on any conforming subspace $U^h \subset U$.

To understand why the weak coercivity properties with respect to the discrete spaces are not inherited from the analogous property on the parent spaces, note that an equivalent way to express (1.17) is that, given $u \in U$, there exists $w_u \in W$ such that

$$Q(u,w_u) \ge \beta \|u\|_U \|w_u\|_W\,. \tag{1.39}$$

Thus, given $u^h \in U^h \subset U$, we obviously have that $u^h \in U$ and so (1.39) implies that there exists a $w_{u^h} \in W$ such that

$$Q(u^h, w_{u^h}) \geq \beta \|u^h\|_U \|w_{u^h}\|_W .$$

However, the function w_{u^h} is not guaranteed to belong to the discrete subspace W^h. Unfortunately, it is this last property that is required for (1.33) to hold; see Figure 1.1.

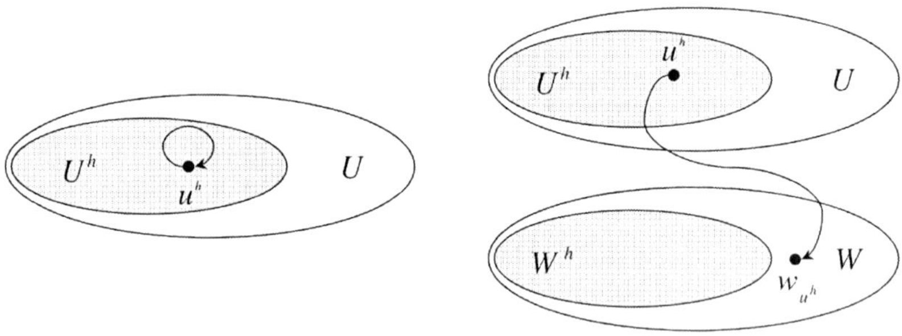

Fig. 1.1 The strong coercivity property is inherited on the subspace U^h (left). The weak coercivity property is not inherited by the subspaces U^h and W^h (right) because the function w_{u^h} whose existence follows from (1.17) is not guaranteed to belong to W^h.

1.3.2 Mixed Variational Problems

In the case of mixed variational problems, the well-posedness of (1.14) is subject to the conditions (1.24) and (1.25) but restricted to the subspaces $V^h \subset V$ and $S^h \subset S$. As is the case in Section 1.3.1 for weakly coercive variational problems, the conformity of the subspaces V^h and S^h is not sufficient for the discrete versions of (1.24) and (1.25) to follow directly from the corresponding conditions on the parent spaces. For this reason, the discrete versions of these conditions have to be stated as assumptions in the following generalization of a result found in, e.g., [83, 87, 183].

Theorem 1.11 *Let the spaces V and S, the linear functionals $D(\cdot)$ and $G(\cdot)$, and the bilinear forms $a(\cdot,\cdot)$ and $b(\cdot,\cdot)$ satisfy the hypotheses of Theorem* 1.5. *Let $V^h \subset V$ and $S^h \subset S$ denote conforming subspaces. Then, $a(\cdot,\cdot)$ and $b(\cdot,\cdot)$ are continuous on $V^h \times V^h$ and $V^h \times S^h$, respectively. Define $Z^h \subset V^h$ by*

$$Z^h = \{v^h \in V^h \mid b(v^h, s^h) = 0 \quad \forall s^h \in S^h\} \tag{1.40}$$

and assume that $a(\cdot,\cdot)$ is weakly coercive on Z^h, i.e.,

$$
\begin{aligned}
&\inf_{v^h\in Z^h}\sup_{r^h\in Z^h}\frac{a(v^h,r^h)}{\|v^h\|_V\|r^h\|_V}\geq\beta_a^h\\
&\inf_{r^h\in Z^h}\sup_{v^h\in Z^h}\frac{a(v^h,r^h)}{\|v^h\|_V\|r^h\|_V}>0,
\end{aligned}
\tag{1.41}
$$

where $\beta_a^h\geq\widehat{\beta}_a>0$. Assume that $b(\cdot,\cdot)$ verifies the LBB[29] *or discrete inf–sup condition*

$$
\inf_{s^h\in S^h}\sup_{v^h\in V^h}\frac{b(v^h,s^h)}{\|v^h\|_V\|s^h\|_S}\geq\beta_b^h,
\tag{1.42}
$$

where $\beta_b^h\geq\widehat{\beta}_b>0$. Then, for every $h>0$, the discrete problem (1.14) *has a unique solution $\{v^h,p^h\}\in V^h\times S^h$. Moreover, for some $C>0$ whose value is independent of h, that solution satisfies the stability estimate*

$$
\|v^h\|_V+\|p^h\|_S\leq C(\|D\|_{V^*}+\|G\|_{S^*})
$$

and, if $\{v,p\}\in V\times S$ denotes the solution of (1.9), *the optimal error estimate*

$$
\|v-v^h\|_V+\|p-p^h\|_S\leq C\Big(\inf_{r^h\in V^h}\|v-r^h\|_V+\inf_{s^h\in S^h}\|p-s^h\|_S\Big).\qquad\Box
\tag{1.43}
$$

The mixed discrete problem (1.14) is equivalent to the parameterized family of linear systems (1.15). Similarly to Theorems 1.8 and 1.9, Theorem 1.11 also asserts that, for every $h>0$, the 2×2 block matrix in this system is square and invertible. If, in addition, the parent mixed variational formulation is related to a constrained minimization problem, this matrix is also symmetric. However, in contrast with the strongly coercive setting, the matrix in (1.15) is always *indefinite*.

It is instructive to examine (1.42) more closely. Just as is the case for (1.17)–(1.18) and (1.33)–(1.34), the conformity of the subspaces $V^h\subset V$ and $S^h\subset S$ is not sufficient to guarantee that the inf–sup condition (1.25) automatically implies that the discrete inf–sup condition (1.42) holds; thus, the latter must be explicitly assumed in Theorem 1.11; see Figure 1.2.

Clearly, the weak coercivity conditions in (1.24) for the bilinear form $a(\cdot,\cdot)$ with respect to the space Z do not immediately imply the like conditions in (1.41) with respect to the space Z^h. But there is more to this particular issue. Suppose we replace the weak coercivity assumptions in (1.24) with the strong coercivity assumption (1.30) and then replace, in Theorem 1.11, the discrete weak coercivity assumptions in (1.41) by the discrete strong coercivity assumption

$$
a(v^h,v^h)\geq\beta_a^h\|v^h\|_V^2\qquad\forall v^h\in Z^h,
\tag{1.44}
$$

[29] The terminology "LBB" originates from the facts that this condition was first explicitly discussed in the finite element setting for saddle point problems by Brezzi [83] and that it is a special case of the general weak-coercivity condition (1.17) first discussed for finite element methods by Babuška [19] and that, in the continuous setting of the Stokes equation, this condition was first proved to hold by Ladyzhenskaya; see [253].

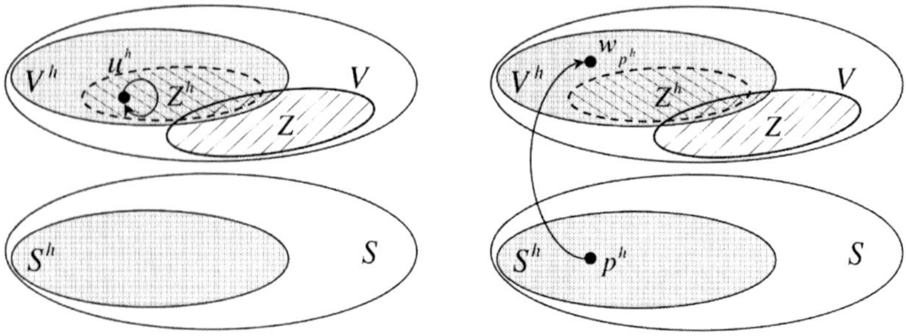

Fig. 1.2 Conditions for the well-posedness of discrete saddle-point problems. Coercivity on the null space (left) and the discrete inf–sup condition (right) must be imposed independently on the discrete spaces because 1) in general $Z^h \not\subset Z$ and 2) the function w_{p^h} whose existence follows from the discrete inf–sup condition is not guaranteed to belong to V^h.

where $\beta_a^h \geq \widehat{\beta}_a > 0$. Strong coercivity is inherited on subspaces; thus, the inclusion of (1.44) as a hypothesis for Theorem 1.11 might seem redundant in light of (1.30). However, although the latter condition does imply that the bilinear form $a(\cdot,\cdot)$ is coercive on any subspace of Z, it is generally the case that $Z^h \not\subset Z$. As a result, even for the case for which the bilinear form $a(\cdot,\cdot)$ is strongly coercive on the null space Z, e.g., for constrained optimization problems, (1.44) must also be imposed as a hypothesis on the discrete space Z^h in Theorem 1.11; see Figure 1.2.

Null space methods

We are now in a position to discuss two approaches for using null space methods to "simplify" the discretization of (1.9). One can first reduce to the null space, i.e., derive the reduced problem (1.26), then choose a subspace $\widetilde{Z}^h \subset Z$, and then restrict (1.26) to that subspace to obtain the discrete problem

$$\textit{find } v^h \in \widetilde{Z}^h \subset Z \subset V \qquad \textit{such that} \qquad a(v^h, r^h) = D(r^h) \quad \forall\, r^h \in \widetilde{Z}^h. \qquad (1.45)$$

Alternately, one can discretize (1.9) using the subspaces $V^h \subset V$ and $S^h \subset S$ to obtain (1.14) and then reduce the latter (discrete) problem using the discrete null space $Z^h \subset V^h$ defined in (1.40). The discrete problems resulting from the two approaches are not the same. For one thing, we cannot choose $\widetilde{Z}^h = Z^h$ because we want $\widetilde{Z}^h \subset Z$ but, in general, $Z^h \not\subset Z$. For another, in (1.45), one directly chooses a subspace $\widetilde{Z}^h$ of the null space Z without any need to introduce the intermediate spaces V^h and S^h.

The distinction between the reduce-then-discretize and discretize-then-reduce approaches becomes even more evident when we consider the corresponding parameterized linear systems. If $\{\widetilde{\xi}_j^h\}_{j=1}^{\widetilde{N}_z}$ denotes a basis for $\widetilde{Z}^h \subset Z$, then the solution $v^h \in \widetilde{Z}^h$ of (1.45) can be expressed as $v^h = \sum_{j=1}^{\widetilde{N}_z} \widetilde{z}_j^h \widetilde{\xi}_j^h$ and then (1.45) is equivalent to the linear system

$$\widetilde{\mathsf{A}}^h \vec{z}^h = \widetilde{d}^h, \tag{1.46}$$

where $\vec{z}^h \in \mathbb{R}^{\widetilde{N}_z}$ and where $\widetilde{\mathsf{A}}^h \in \mathbb{R}^{\widetilde{N}_z \times \widetilde{N}_z}$ and $\widetilde{d}^h \in \mathbb{R}^{\widetilde{N}_z}$ are respectively given by

$$\widetilde{\mathsf{A}}^h_{ij} = a(\widetilde{\xi}^h_j;\widetilde{\xi}^h_i) \qquad \text{and} \qquad \widetilde{d}^h_i = D(\widetilde{\xi}^h_i) \qquad \text{for } i,j = 1,\dots,\widetilde{N}_z .$$

Now, let Z^h denote a matrix whose columns form a basis for $N(\mathsf{B}^h)$. For the sake of simplicity, we assume that (1.42) is satisfied so that B^h has full row rank, $\dim(Z^h) = N - M$, and $\mathsf{Z}^h \in \mathbb{R}^{N\times(N-M)}$. Then, from (1.15), we easily obtain that

$$\mathsf{Z}^{h^\mathsf{T}} \mathsf{A}^h \vec{v}^h = \mathsf{Z}^{h^\mathsf{T}} \vec{d}^h . \tag{1.47}$$

Furthermore, the vector of coefficients $\vec{v}^h \in \mathbb{R}^N$ (with respect to a chosen basis for V^h) corresponding to any function $v^h \in Z^h$ can be expressed in the form $\vec{v}^h = \mathsf{Z}^h \vec{q}^h$ for some $\vec{q}^h \in \mathbb{R}^{N-M}$. Then, it follows from (1.47) that

$$\widehat{\mathsf{A}}^h \vec{q}^h = \widehat{d}^h, \tag{1.48}$$

where

$$\widehat{\mathsf{A}}^h = \mathsf{Z}^{h^\mathsf{T}} \mathsf{A}^h \mathsf{Z}^h \qquad \text{and} \qquad \widehat{d}^h = \mathsf{Z}^{h^\mathsf{T}} \vec{d}^h .$$

Clearly, the systems (1.46) and (1.48) are not the same.

It would seem that, in the case for which the bilinear form $a(\cdot,\cdot)$ is strongly coercive on Z, finding approximate solutions of constrained optimization problems by solving the reduced problem (1.45) is preferable to solving (1.14). The application of the Lagrange multiplier rule results in the discrete mixed-Galerkin formulation (1.14) that is necessarily weakly coercive.[30] However, because, in general, the discrete null space $Z^h \not\subset Z$, it is clear that it cannot be used in (1.45). Thus, $\widetilde{Z}^h$ must be constructed separately, even if a stable pair $\{V^h, S^h\}$ for (1.14) is already available. On the other hand, if (1.30) holds and if one has available a subspace $\widetilde{Z}^h \subset Z$, we have that the bilinear form $a(\cdot,\cdot)$ is strongly coercive with respect to $\widetilde{Z}^h$ so that the problem (1.45) has a unique solution, that solution depends continuously on the data, and satisfies the error estimate

$$\|v - v^h\|_V \le C \inf_{r^h \in \widetilde{Z}^h} \|v - r^h\|_V ,$$

where v denotes the solution of (1.26). Thus, the reduced formulation has the advantage of offering a strongly coercive setting for discrete variational methods for the constrained optimization problem (1.31). Unfortunately, in practice, it is often not an easy matter to construct subspaces of the space Z defined by (1.21).

[30] An additional advantage of (1.45) over (1.14) is that the former does not involve p^h, i.e., it involves fewer degrees of freedom.

1.4 Examples

We now provide examples of variational problems that form the basis for classical finite element methods. We begin with the classification of these problems according to the taxonomy given in Section 1.2, and then infer the properties of the associated finite element methods from the discussions given in Section 1.3.

We first consider some PDEs related to optimization problems, starting from the optimization problems themselves and working our way to the equivalent PDE formulations. The optimization problems are turned into weak formulations of the related PDEs by the application of standard techniques from the calculus of variations.

For PDEs such as conservation laws and advection-diffusion-reaction problems that are not related to optimization problems, we cannot use the calculus of variations to obtain variational formulations. Instead, variational methods can be based on *Galerkin principles* that are derived directly from the PDEs. The paradigm of a Galerkin principle is formal *residual* "*orthogonalization*." This process can, in principle, be applied to any PDE, even if there is no underlying optimization problem. Because of their universality, Galerkin principles have provided a natural and popular choice for extending finite element methods beyond PDE problems associated with optimization problems. On the other hand, if such an optimization problem exists, the associated optimality system can be recovered through the use of the Galerkin paradigm.

In Sections 1.4.1 and 1.4.2 we respectively consider the Poisson equation and the equations of linear elasticity; both of these problems can be associated with unconstrained optimization problems. In Section 1.4.3, we consider the Stokes equations that can be associated with a constrained optimization problem. In Section 1.4.4 we consider the Helmholtz equation and in Section 1.4.5 a linear advection-diffusion-reaction equation; neither of these problems can be related to an optimization problem. Finally, in Section 1.4.6, we consider the nonlinear example of the Navier–Stokes equations which also cannot be associated with an optimization problem.

1.4.1 The Poisson Equation

We consider the Dirichlet and Kelvin principles which give rise to two different variational formulations for the Poisson equation. The Dirichlet and Kelvin principles provide examples of unconstrained and constrained, respectively, optimization problems; the former leads to a symmetric, strongly coercive, Rayleigh–Ritz setting for the Poisson problem and the latter leads to a symmetric, weakly coercive, mixed-Galerkin formulation for the same problem.

We assume that the domain $\Omega \subset \mathbb{R}^d$ has boundary $\partial\Omega$ which is made up of two disjoint, nonempty parts Γ and Γ^*.

The Dirichlet principle

Given[31] $f \in (H^1_\Gamma(\Omega))^*$, consider the functional

$$J_D(\phi;f) = \frac{1}{2}\int_\Omega |\nabla\phi|^2\,d\Omega - \int_\Omega f\phi\,d\Omega \tag{1.49}$$

and the *Dirichlet principle*

$$\min_{\phi\in H^1_\Gamma(\Omega)} \; J_D(\phi;f)\,. \tag{1.50}$$

Clearly, (1.50) defines an unconstrained optimization setting.

The Euler–Lagrange condition corresponding to this principle is given by

$$\int_\Omega \nabla\phi\cdot\nabla\psi\,d\Omega = \int_\Omega f\psi\,d\Omega \qquad \forall\,\psi\in H^1_\Gamma(\Omega)\,. \tag{1.51}$$

Letting $U = W = H^1_\Gamma(\Omega)$ and

$$Q(\phi,\psi) = \int_\Omega \nabla\phi\cdot\nabla\psi\,d\Omega \qquad \text{and} \qquad F(\psi) = \int_\Omega f\psi\,d\Omega \qquad \forall\,\phi,\psi\in H^1_\Gamma(\Omega)\,,$$

we see that (1.51) is of the form (1.3). Furthermore, it is well known [82, 123] that, in this case, the bilinear form $Q(\cdot,\cdot)$ is symmetric and strongly coercive on $H^1_\Gamma(\Omega)\times H^1_\Gamma(\Omega)$ so that we find ourselves in the Rayleigh–Ritz unconstrained optimization setting discussed in Section 1.2.4. Furthermore, the error estimate (1.38) applies to solutions of the finite element discrete problem (1.5) corresponding to (1.51) for *any conforming finite element approximation space* $U^h \subset H^1_\Gamma(\Omega)$.

Equivalent PDE. Formally, by integration by parts, (1.51) is equivalent to the Poisson problem

$$\begin{cases} \qquad\qquad -\Delta\phi = f \quad \text{in } \Omega\,, \\ \phi = 0 \quad \text{on } \Gamma\,, \qquad \text{and} \qquad \dfrac{\partial\phi}{\partial n} = 0 \quad \text{on } \Gamma^*\,. \end{cases} \tag{1.52}$$

Thus, (1.51) provides a symmetric, strongly coercive weak formulation for the Poisson problem (1.52).

The Kelvin principle

Consider the functional

$$J_K(\mathbf{v}) = \frac{1}{2}\int_\Omega |\mathbf{v}|^2\,d\Omega$$

[31] $H^1_\Gamma(\Omega)$ is the space $\{\psi\in H^1(\Omega)\mid \psi = 0 \text{ on } \Gamma\}$. When $\Gamma = \partial\Omega$ this space is denoted by $H^1_0(\Omega)$. In later chapters, we adopt the unified notation for function spaces defined in Appendix A so that $H^1_\Gamma(\Omega)$ and $H^1_0(\Omega)$ are instead denoted respectively by $G_\Gamma(\Omega)$ and $G_0(\Omega)$.

and the *Kelvin principle*[32,33,34]

$$\min_{\mathbf{v}\in H_{\Gamma^*}(\text{div};\Omega)} J_K(\mathbf{v}) \qquad \text{subject to} \qquad \nabla\cdot\mathbf{v} = g, \tag{1.53}$$

where $g \in L^2(\Omega)$ is a given function. Clearly, (1.53) is a constrained optimization problem.

Lagrange multiplier method. With the help of a Lagrange multiplier[35] $\phi \in L^2(\Omega)$ and the Lagrangian functional

$$L_K(\mathbf{v},\phi;g) = \frac{1}{2}\int_\Omega |\mathbf{v}|^2\,d\Omega - \int_\Omega \phi\,(\nabla\cdot\mathbf{v} - g)\,d\Omega$$

to enforce the constraint, the Kelvin principle (1.53) can be transformed into the unconstrained problem of determining saddle points $\{\mathbf{v},\phi\} \in H_{\Gamma^*}(\text{div};\Omega)\times L^2(\Omega)$ of $L_K(\mathbf{v},\phi;g)$. The optimality system, obtained by setting the first variations of $L_K(\mathbf{v},\phi;g)$ to zero, is given by: *seek* $\{\mathbf{v},\phi\} \in H_{\Gamma^*}(\text{div};\Omega)\times L^2(\Omega)$ *such that*

$$\begin{cases} \displaystyle\int_\Omega \mathbf{v}\cdot\mathbf{r}\,d\Omega - \int_\Omega \phi\nabla\cdot\mathbf{r}\,d\Omega = 0 & \forall\,\mathbf{r}\in H_{\Gamma^*}(\text{div};\Omega) \\ \displaystyle -\int_\Omega \psi\nabla\cdot\mathbf{v}\,d\Omega \qquad\qquad = -\int_\Omega g\psi\,d\Omega & \forall\,\psi\in L^2(\Omega)\,. \end{cases} \tag{1.54}$$

Letting $V = H_{\Gamma^*}(\text{div};\Omega)$, $S = L^2(\Omega)$, $D(\cdot) = 0$,

$$G(\psi) = -\int_\Omega g\psi\,d\Omega \qquad \forall\psi\in L^2(\Omega),$$

$$a(\mathbf{v},\mathbf{r}) = \int_\Omega \mathbf{v}\cdot\mathbf{r}\,d\Omega \qquad \forall\,\mathbf{v},\mathbf{r}\in H_{\Gamma^*}(\text{div};\Omega),$$

and

$$b(\mathbf{v},\psi) = -\int_\Omega \psi\nabla\cdot\mathbf{v}\,d\Omega \qquad \forall\,\mathbf{v}\in H_{\Gamma^*}(\text{div};\Omega),\ \psi\in L^2(\Omega),$$

one easily sees that (1.54) is a problem of the type (1.9), i.e., it falls into the class of mixed variational problems.

---

32 Setting $g = 0$ and allowing for an inhomogeneous boundary condition for $\mathbf{u}\cdot\mathbf{n}$, the Kelvin principle for inviscid flows states that among all incompressible velocity fields, the one that minimizes the kinetic energy is irrotational. In structural mechanics (where $\mathbf{v}$ is a tensor), (1.53) is known as the *complementary energy principle*.

33 In Chapter 5, we have much more to say about the Kelvin and Dirichlet principles and their relation to each other.

34 $H_{\Gamma^*}(\text{div};\Omega)$ is the space $\{\mathbf{v}\in[L^2(\Omega)]^d \mid \nabla\cdot\mathbf{v} \in L^2(\Omega)$ and $\mathbf{v}\cdot\mathbf{n} = 0$ on $\Gamma^*\}$. In later chapters, using the notation defined in Appendix A, we instead denote $H_{\Gamma^*}(\text{div};\Omega)$ by $\mathbf{D}_{\Gamma^*}(\Omega)$.

35 If $\Gamma = \emptyset$ so that we have $\partial\Omega = \Gamma^*$, we use $L^2_0(\Omega) = \{\phi\in L^2(\Omega) \mid \int_\Omega \phi\,d\Omega = 0\}$ instead of $L^2(\Omega)$ for the Lagrange multiplier space. In later chapters, using the notation defined in Appendix A, we instead denote $L^2_0(\Omega)$ by $S_0(\Omega)$.

For (1.54), we have that the null space Z defined by (1.21) is given by

$$Z = \left\{ \mathbf{v} \in H_{\Gamma^*}(\text{div};\Omega) \;\middle|\; \int_\Omega \psi \nabla \cdot \mathbf{v}\, d\Omega = 0 \quad \forall \psi \in L^2(\Omega) \right\},$$

i.e., Z consists of (weakly) solenoidal functions. It is well known [87] that, for the Kelvin principle, the hypotheses of Theorem 1.5 and Proposition 1.7 are satisfied so that we find ourselves in the constrained optimization setting discussed in Section 1.2.4. Furthermore, the error estimate (1.43) applies to solutions of the finite element discrete problem (1.14) corresponding to (1.54), *provided that the conforming finite element spaces $V^h \subset H_{\Gamma^*}(div;\Omega)$ and $S^h \subset L^2(\Omega)$ are chosen so that the discrete stability conditions* (1.42) *and* (1.44) *are satisfied.* For a discussion of finite element spaces that satisfy these conditions, see, e.g., [87]. Here, we merely note that defining finite element subspaces that satisfy these conditions is no easy matter.[36]

Equivalent PDEs. Formally, by integration by parts, (1.54) is equivalent to the first-order PDE system

$$\begin{cases} \nabla \cdot \mathbf{v} = f \\ \mathbf{v} + \nabla \phi = 0 \end{cases} \text{in } \Omega \qquad \text{and} \qquad \begin{array}{rl} \phi = 0 & \text{on } \Gamma \\ \mathbf{v} \cdot \mathbf{n} = 0 & \text{on } \Gamma^*. \end{array} \tag{1.55}$$

We can easily eliminate $\mathbf{v}$ from this system to yield the Poisson problem (1.52) for ϕ. Thus, (1.54) provides a symmetric, weakly coercive variational formulation that gives rise to a mixed-Galerkin method for the Poisson problem.

1.4.2 The Equations of Linear Elasticity

A classical example that provided the variational foundation for the first finite element methods is the *minimum energy principle* of linear elasticity. Let $\mathbf{u}$ denote the displacement field of an elastic body Ω that is subjected to a body force[37] $\mathbf{f} \in [H^{-1}(\Omega)]^d$. For simplicity, we assume that the displacement field vanishes along the boundary of Ω. Then, the energy is given by[38]

$$J_E(\mathbf{u};\mathbf{f}) = \frac{1}{2} \int_\Omega \left(2\mu\, \varepsilon(\mathbf{u}) : \varepsilon(\mathbf{u}) + \lambda (\nabla \cdot \mathbf{u})^2 \right) d\Omega - \int_\Omega \mathbf{f} \cdot \mathbf{u}\, d\Omega\,,$$

---

[36] Recall that for the null space method of Section 1.2.3, we assume that the linear functional $G(\cdot) = 0$, i.e., $g = 0$ in (1.54). Because in this case (1.54) is a problem with purely homogeneous data, we do not here discuss the null space method in the context of the Kelvin principle. Such a discussion is given in Chapter 5 where generalizations of the Kelvin principle are considered.

[37] We use the standard notation $H^{-1}(\Omega)$ for the dual of $H_0^1(\Omega)$, i.e., $H^{-1}(\Omega) = (H_0^1(\Omega))^*$.

[38] We have that $J_E(\cdot;\cdot)$ is the sum of the strain energy and the potential energy due to the external force $\mathbf{f}$.

where $\varepsilon(\mathbf{u}) = \frac{1}{2}(\nabla\mathbf{u} + (\nabla\mathbf{u})^T)$ is the strain tensor and λ and μ are the Lamé coefficients. The minimum energy principle postulates that the displacement field $\mathbf{u}$ is the solution of the unconstrained optimization problem

$$\min_{\mathbf{u}\in[H_0^1(\Omega)]^d} J_E(\mathbf{u};\mathbf{f}).$$

The Euler–Lagrange equation corresponding to this optimization problem is given by

$$\int_\Omega \Big(2\mu\,\varepsilon(\mathbf{u}) : \varepsilon(\mathbf{v}) + \lambda(\nabla\cdot\mathbf{u})(\nabla\cdot\mathbf{v})\Big)\, d\Omega = \int_\Omega \mathbf{f}\cdot\mathbf{v}\, d\Omega \quad \forall\mathbf{v}\in[H_0^1(\Omega)]^d. \tag{1.56}$$

Letting $U = W = [H_0^1(\Omega)]^d$,

$$Q(\mathbf{u},\mathbf{v}) = \int_\Omega \Big(2\mu\,\varepsilon(\mathbf{u}) : \varepsilon(\mathbf{v}) + \lambda(\nabla\cdot\mathbf{u})(\nabla\cdot\mathbf{v})\Big)\, d\Omega \qquad \forall\mathbf{u},\mathbf{v}\in[H_0^1(\Omega)]^d,$$

and

$$F(\mathbf{v}) = \int_\Omega \mathbf{f}\cdot\mathbf{v}\, d\Omega \qquad \forall\mathbf{v}\in[H_0^1(\Omega)]^d,$$

we see that (1.56) is of the form (1.3). Furthermore, it is well known (see, e.g., [123]) that, in this case, the bilinear form $Q(\cdot,\cdot)$ is symmetric and strongly coercive on $[H_0^1(\Omega)]^d \times [H_0^1(\Omega)]^d$ so that we find ourselves in the Rayleigh–Ritz unconstrained optimization setting discussed in Section 1.2.4. Furthermore, the error estimate (1.38) applies to solutions of the finite element discrete problem (1.5) corresponding to (1.56) for *any conforming finite element approximation space* $U^h \subset [H_0^1(\Omega)]^d$.

Equivalent PDEs. Formally, by integration by parts, (1.56) is equivalent to the problem

$$\begin{cases} -\mu\Delta\mathbf{u} - (\lambda+\mu)\nabla(\nabla\cdot\mathbf{u}) = \mathbf{f} & \text{in } \Omega \\ \mathbf{u} = 0 & \text{on } \partial\Omega\,. \end{cases} \tag{1.57}$$

Thus, (1.56) provides a symmetric, strongly coercive weak formulation for (1.57) which are known as the equations of linear elasticity.[39]

1.4.3 The Stokes Equations

Given $\mathbf{f}\in[H^{-1}(\Omega)]^d$, consider the quadratic functional

$$J_S(\mathbf{v};\mathbf{f}) = \frac{1}{2}\int_\Omega \nabla\mathbf{v} : \nabla\mathbf{v}\, d\Omega - \int_\Omega \mathbf{f}\cdot\mathbf{v}\, d\Omega$$

[39] The complementary energy principle mentioned in Section 1.4.1 in connection with the Kelvin principle leads to a symmetric, weakly coercive mixed-Galerkin formulation for the equations of linear elasticity.

and the minimization problem

$$\min_{\mathbf{v}\in[H_0^1(\Omega)]^d} J_S(\mathbf{v};\mathbf{f}) \qquad \text{subject to} \qquad \nabla\cdot\mathbf{v} = 0 \text{ in } \Omega. \tag{1.58}$$

Clearly, (1.58) is a constrained optimization problem.

Lagrange multiplier method. We enforce the constraint in (1.58) by introducing a Lagrange multiplier $p \in L_0^2(\Omega)$ and the Lagrangian functional

$$L_S(\mathbf{v},p;\mathbf{f}) = J_S(\mathbf{v};\mathbf{f}) - \int_\Omega p\nabla\cdot\mathbf{v}\,d\Omega\,.$$

Then, the problem (1.58) is transformed into the unconstrained optimization problem of finding the saddle points $\{\mathbf{v},p\} \in [H_0^1(\Omega)]^d \times L_0^2(\Omega)$ of $L_S(\mathbf{v},p;\mathbf{f})$. The corresponding first-order necessary conditions are given by: *seek* $\{\mathbf{v},p\} \in [H_0^1(\Omega)]^d \times L_0^2(\Omega)$ *such that*

$$\begin{cases} \displaystyle\int_\Omega \nabla\mathbf{v} : \nabla\mathbf{r}\,d\Omega - \int_\Omega p\nabla\cdot\mathbf{r}\,d\Omega = \int_\Omega \mathbf{f}\cdot\mathbf{r}\,d\Omega & \forall\,\mathbf{r}\in[H_0^1(\Omega)]^d \\ \displaystyle\int_\Omega s\nabla\cdot\mathbf{v}\,d\Omega = 0 & \forall\,s\in L_0^2(\Omega)\,. \end{cases} \tag{1.59}$$

We refer to a solution $\{\mathbf{v},p\}$ of (1.59) as a *velocity–pressure* pair. With the identifications $V = [H_0^1(\Omega)]^d$, $S = L_0^2(\Omega)$, $G(\cdot) = 0$,

$$D(\mathbf{v}) = \int_\Omega \mathbf{f}\cdot\mathbf{v}\,d\Omega\,, \qquad a(\mathbf{v},\mathbf{r}) = \int_\Omega \nabla\mathbf{v} : \nabla\mathbf{r}\,d\Omega \qquad \forall\,\mathbf{v},\mathbf{r}\in[H_0^1(\Omega)]^d\,,$$

and

$$b(\mathbf{v},s) = -\int_\Omega s\nabla\cdot\mathbf{v}\,d\Omega \qquad \forall\,\mathbf{v}\in[H_0^1(\Omega)]^d,\ s\in L_0^2(\Omega)\,,$$

one easily sees that (1.58) is a problem of the type (1.31), i.e., it falls into the class of mixed variational problems.

For (1.58), we have that the null space Z defined by (1.21) is given by

$$Z = \left\{\mathbf{v}\in[H_0^1(\Omega)]^d \;\Big|\; \int_\Omega s\nabla\cdot\mathbf{v}\,d\Omega = 0 \quad \forall\,s\in L_0^2(\Omega)\right\}, \tag{1.60}$$

i.e., Z consists of (weakly) solenoidal functions. It is well known [83, 87, 183, 191] that, for the Stokes problem, the hypotheses of Theorem 1.5 and Proposition 1.7 are satisfied so that we find ourselves in the constrained optimization setting discussed in Section 1.2.4. Furthermore, the error estimate (1.43) applies to solutions of the finite element discrete problem (1.14) corresponding to (1.54), *provided that the conforming finite element spaces* $V^h \subset [H_0^1(\Omega)]^d$ *and* $S^h \subset L_0^2(\Omega)$ *are chosen so that the discrete inf–sup stability condition* (1.42) *is satisfied.*[40] For a discussion of

[40] For the Stokes problem, we have that $a(\cdot,\cdot)$ is strongly coercive on all of $[H_0^1(\Omega)]^d$ and not just on $Z \subset [H_0^1(\Omega)]^d$. Thus, in this case, because Z^h defined by (1.40) is a subspace of $V =$

finite element spaces that satisfy this condition, see, e.g., [83, 87, 183, 191]. Here, we again note that defining finite element subspaces that satisfy this condition is no easy matter.

Null space method. An alternate means for handling the constraint in (1.58) is given by the null space method introduced in Section 1.2.3. The reduced optimization problem (1.32) is now given by

$$\min_{\mathbf{v}\in Z} J_S(\mathbf{v};\mathbf{f})$$

and the corresponding first-order necessary condition (1.26) is now given by: *seek* $\mathbf{v} \in Z$ *such that*

$$\int_\Omega \nabla\mathbf{v} : \nabla\mathbf{r}\, d\Omega = \int_\Omega \mathbf{f}\cdot\mathbf{r}\, d\Omega \qquad \forall \mathbf{r} \in Z. \tag{1.61}$$

The bilinear form $a(\cdot,\cdot)$ is coercive on all of $[H_0^1(\Omega)]^d \times [H_0^1(\Omega)]^d$; therefore, it is certainly coercive with respect to any finite element subspace $\widetilde{Z}^h \subset Z \subset V = [H_0^1(\Omega)]^d$ so that (1.61) provides a (strongly coercive) Rayleigh–Ritz setting for (1.58). However, the conformity requirement $\widetilde{Z}^h \subset Z$ implies that a function belonging to the finite element space $\widetilde{Z}^h$ must be (at least weakly) divergence free; in practice, such spaces are difficult to construct; see [191].

Equivalent PDEs. Formally, by integration by parts, (1.59) is equivalent to the classical *Stokes system* in primitive variable form:

$$\begin{cases} -\Delta\mathbf{v} + \nabla p = \mathbf{f} & \text{in } \Omega \\ \nabla\cdot\mathbf{v} = 0 & \text{in } \Omega \\ \mathbf{v} = 0 & \text{on } \partial\Omega\,. \end{cases} \tag{1.62}$$

Thus, (1.59) provides a symmetric, weakly coercive, mixed-Galerkin formulation for the Stokes problem (1.62). Similarly, (1.61) is formally equivalent to the problem $-\Delta\Pi_Z\mathbf{v} = \Pi_Z\mathbf{f}$, where $\Pi_Z\mathbf{v}$ denotes the projection of $\mathbf{v}$ onto the space of (weakly) solenoidal functions. Thus, (1.61) provides a strongly coercive, although impractical, Rayleigh–Ritz type formulation for the Stokes problem (1.62).

1.4.4 The Helmholtz Equation

The first example of a PDE that cannot be related to an optimization problem is the Helmholtz equation

---

$[H_0^1(\Omega)]^d$, the discrete strong coercivity condition (1.44) automatically holds for any conforming finite element subspace $V^h \subset [H_0^1(\Omega)]^d$. This should be contrasted with the form $a(\cdot,\cdot)$, arising in the Kelvin principle (see Section 1.4.1) that is not coercive on all of $H_{\Gamma^*}(\text{div};\Omega)$.

$$
\begin{cases} -\Delta\phi - k^2\phi = f & \text{in } \Omega \\ \phi = 0 & \text{on } \partial\Omega \,. \end{cases} \tag{1.63}
$$

Because we cannot apply calculus of variation techniques to an associated optimization problem, weak formulations of (1.63) are usually derived using the Galerkin paradigm. To this end, we multiply the differential equation in (1.63) by a test function $\psi \in H_0^1(\Omega)$, then integrate the result over the domain Ω, and then apply the divergence theorem to equilibrate the order of the highest derivatives applied to the trial function $\phi \in H_0^1(\Omega)$ and the test function ψ. The boundary integral that appears as a result of the application of the divergence theorem vanishes by virtue of the boundary condition in (1.63), i.e., because $\phi \in H_0^1(\Omega)$.

The result of applying the formal Galerkin procedure to (1.63) is the weak formulation

$$
\int_\Omega (\nabla\phi \cdot \nabla\psi - k^2\phi\psi)\, d\Omega = \int_\Omega f\psi\, d\Omega \qquad \forall\, \psi \in H_0^1(\Omega)\,. \tag{1.64}
$$

Clearly, with the associations $U = W = H_0^1(\Omega)$, $F(\psi) = \int_\Omega f\psi\, d\Omega$ for all $\psi \in H_0^1(\Omega)$, and

$$
Q(\phi,\psi) = \int_\Omega (\nabla\phi \cdot \nabla\psi - k^2\phi\psi)\, d\Omega \qquad \forall \phi,\, \psi \in H_0^1(\Omega)\,,
$$

(1.64) is a problem of the form (1.3). It is easy to show that the bilinear form $Q(\cdot,\cdot)$ is symmetric but, if k^2 is larger than the smallest eigenvalue of the operator[41] $-\Delta$ with zero boundary conditions, it is not strongly coercive, i.e., it does not define an inner product on $H_0^1(\Omega) \times H_0^1(\Omega)$. As a result, proving the existence and uniqueness of weak solutions is not as simple a matter as it is for the Poisson equation case.[42]

Because, for k^2 larger than the smallest eigenvalue of the operator $-\Delta$ with zero boundary conditions, the bilinear form $Q(\phi,\psi)$ is at best weakly coercive, according to Theorem 1.8, any conforming finite element space $U^h \subset H_0^1(\Omega)$ used in defining discretized problems of the type (1.5) is required to satisfy the discrete inf–sup conditions (1.33) and (1.34). Fortunately, if $-k^2$ is bounded away from an eigenvalue of the operator $-\Delta$ with zero boundary conditions, these requirements turn out to be satisfied by any conforming subspace $U^h \subset H_0^1(\Omega)$; in this case, finite element approximations of the solution of (1.63) or, equivalently, (1.64), satisfy the error estimate (1.35).

---

41 Δ denotes the Laplace operator.

42 In fact, solutions of (1.63) or, equivalently, (1.64) are not unique whenever $-k^2$ equals an eigenvalue of the operator $-\Delta$ with zero boundary conditions. In the case for which the solution of (1.63) is unique, the weak coercivity of the bilinear form $Q(\cdot,\cdot)$ can be proved with the help of the Gårding inequality. To obtain, in (1.17), a stability constant β that is uniformly bounded away from zero, $-k^2$ has to be bounded away from any eigenvalue of the operator $-\Delta$ with zero boundary conditions.

1.4.5 A Scalar Linear Advection-Diffusion-Reaction Equation

Our next example of a PDE that cannot be obtained as necessary conditions for an optimization problem is the linear scalar advection-diffusion-reaction problem

$$\begin{cases} -\varepsilon\Delta\phi + \mathbf{b}\cdot\nabla\phi + c\phi = f & \text{in } \Omega \\ \phi = 0 & \text{on } \partial\Omega\,, \end{cases} \tag{1.65}$$

where $\varepsilon > 0$ is usually small, $\mathbf{b}$ is a given velocity field, and c is a given scalar-valued function.

Following a standard Galerkin procedure for (1.65) results in the Galerkin weak formulation

$$\int_\Omega \Big(\varepsilon\nabla\phi\cdot\nabla\psi + \psi\mathbf{b}\cdot\nabla\phi + c\phi\psi\Big)\,d\Omega = \int_\Omega f\psi\,d\Omega \qquad \forall\,\psi\in H_0^1(\Omega)\,. \tag{1.66}$$

Letting $U = W = H_0^1(\Omega)$, $F(\psi) = \int_\Omega f\psi\,d\Omega$ for all $\psi\in H_0^1(\Omega)$, and

$$Q(\phi,\psi) = \int_\Omega \Big(\varepsilon\nabla\phi\cdot\nabla\psi + \psi\mathbf{b}\cdot\nabla\phi + c\phi\psi\Big)\,d\Omega \qquad \forall\,\phi,\,\psi\in H_0^1(\Omega)\,,$$

we see that (1.66) is of the form (1.3). In general, the bilinear form $Q(\phi,\psi)$ is weakly coercive and thus, Theorems (1.3) and (1.8) apply.[43]

1.4.6 The Navier–Stokes Equations

A *nonlinear* example of a problem that cannot be related to an optimization problem but for which a weak formulation may be defined through the Galerkin process is the Navier–Stokes system for incompressible viscous flows:

$$\begin{cases} -\nu\Delta\mathbf{u} + \mathbf{u}\cdot\nabla\mathbf{u} + \nabla p = \mathbf{f} & \text{in } \Omega \\ \nabla\cdot\mathbf{u} = 0 & \text{in } \Omega \\ \mathbf{u} = 0 & \text{on } \partial\Omega\,, \end{cases} \tag{1.67}$$

[43] We have that

$$\int_\Omega \psi\mathbf{b}\cdot\nabla\phi\,d\Omega = -\int_\Omega (\phi\mathbf{b}\cdot\nabla\psi + \psi\phi\nabla\cdot\mathbf{b})\,d\Omega + \int_{\partial\Omega}\psi\phi\mathbf{b}\cdot\mathbf{n}\,d\Gamma\,.$$

If $\mathbf{b}$ is solenoidal, we then have that $\int_\Omega \phi\mathbf{b}\cdot\nabla\phi\,d\Omega = 0$ so that

$$Q(\phi,\phi) = \int_\Omega \Big(\varepsilon|\nabla\phi|^2 + c\phi^2\Big)\,d\Omega \qquad \forall\,\phi\in H_0^1(\Omega)\,.$$

Thus, if $c\geq 0$ in addition to $\nabla\cdot\mathbf{b} = 0$, the bilinear form $Q(\cdot,\cdot)$ is strongly coercive on $H_0^1(\Omega)\times H_0^1(\Omega)$. However, because it is not a symmetric bilinear form, even in this special case the problem (1.65) cannot be associated with an optimization problem.

where $\mathbf{u}$ and p respectively denote the velocity and pressure fields and the constant ν denotes the kinematic viscosity. A standard weak formulation analogous to (1.59) but containing an additional nonlinear term is given by: *seek* $\{\mathbf{u}, p\} \in [H_0^1(\Omega)]^d \times L_0^2(\Omega)$ *such that*

$$\begin{cases} \nu \int_\Omega \nabla \mathbf{u} : \nabla \mathbf{v}\, d\Omega - \int_\Omega p \nabla \cdot \mathbf{v}\, d\Omega \\ \qquad + \int_\Omega \mathbf{u} \cdot \nabla \mathbf{u} \cdot \mathbf{v}\, d\Omega = \int_\Omega \mathbf{f} \cdot \mathbf{v}\, d\Omega & \forall \mathbf{v} \in [H_0^1(\Omega)]^d \\ \int_\Omega q \nabla \cdot \mathbf{u}\, d\Omega = 0 & \forall q \in L_0^2(\Omega). \end{cases} \tag{1.68}$$

Despite the close resemblance between (1.59) and (1.68), these two problems are strikingly different in their variational origins. Specifically, the second problem does not represent an optimality system, i.e., there is no optimization problem associated with these weak equations. As a result, (1.68) cannot be derived in any other way than through the Galerkin process.

1.5 A Comparative Summary of Classical Finite Element Methods

The theories reviewed in Section 1.3 and the examples given in Section 1.4 serve to show that the theoretical and practical difficulties encountered in defining a finite element method and solving the corresponding algebraic systems increase as their underlying variational foundation becomes more and more estranged from that of a true inner-product projection, i.e., from the Rayleigh–Ritz setting. This is the price one pays for the increasing generality of the applicability of the methods we considered as we moved from Section 1.4.1 through Section 1.4.4. Unconstrained quadratic optimization problems lead to the "ideal" situation of symmetric, strongly coercive variational problems which give rise to symmetric and positive definite algebraic systems. Linearly constrained quadratic optimization problems lead to saddle-point problems and restrictive stability conditions on the finite element spaces. The resulting algebraic equations are symmetric but indefinite.

If a problem possesses no associated optimization principle so that one is led to the formal Galerkin method, then things can get even worse[44] and not much can be said about the associated algebraic problems, except that they are nonsingular provided the discrete inf–sup conditions in Theorem 1.8 are satisfied.

A comparative summary of the features of classical finite element methods is given in Table 1.1, where the methods are classified according to their relation, or lack thereof, to optimization problems. Clearly, in every respect, the Rayleigh–Ritz

---

[44] We emphasize that the difficulties encountered by Galerkin finite element methods in general settings should not be viewed as a defect of those methods; these problems are difficult to solve and any standard discretization scheme encounters equal or even greater complications.

setting of unconstrained optimization problems has the most desirable properties. This has led to many attempts directed at developing nonstandard finite element methods that can recover at least some of the desirable features of the Rayleigh–Ritz setting for problems where Galerkin methods fail to do so; we review some of these attempts in the next chapter. Least-squares finite element methods represent another such attempt; in fact, the main motivation for developing least-square finite element methods and their most interesting feature is that they can succeed at recovering a Rayleigh–Ritz type setting for wide classes of problems for which Galerkin and other standard and nonstandard methods fail to do so.

	Related to optimization	
	Unconstrained Rayleigh-Ritz	Constrained mixed-Galerkin
Properties of bilinear forms	strongly coercive and symmetric	weakly coercive and symmetric
Stability requirements for existence and uniqueness	none	strong coercivity on the null space and inf–sup conditions
Requirements on the discrete spaces	conformity	conformity, strong coercivity on the discrete null space, and discrete inf–sup condition
Properties of discrete problems	symmetric and positive definite	symmetric but indefinite

	Not related to optimization		
	Strongly coercive Galerkin	Weakly coercive Galerkin	Mixed Galerkin
Properties of bilinear forms	strongly coercive but not symmetric	weakly coercive and not symmetric	weakly coercive and not symmetric
Stability requirements for existence and uniqueness	none	inf–sup conditions	weak coercivity on the null space and inf–sup conditions
Stability requirements on the discrete spaces	conformity	conformity and discrete inf–sup conditions	conformity, weak coercivity on the discrete null space, and discrete inf–sup condition
Properties of discrete problems	not symmetric but positive definite	not symmetric and indefinite	not symmetric and indefinite

Table 1.1 Comparison of different settings for classical finite element methods in their most general sphere of applicability.

Chapter 2
Alternative Variational Formulations

In general, the finite element methods based on Galerkin and mixed-Galerkin variational principles described in Section 1.3 do not possess all of the theoretical and practical advantages held by finite element methods in the Rayleigh–Ritz setting. As a result, many alternative variational formulations have been proposed with the goal of recovering at least some of these advantages in more general settings. Generally speaking, there are two classes of alternative variational formulations that have been introduced for this purpose. First, for a given partial differential equation (PDE) problem, one may *modify* the "naturally" occurring variational principle with the goal of defining better quasi-projections. This approach usually allows for the recovery of some, but not all, of the advantages possessed by finite element methods in the Rayleigh–Ritz setting. Methods based on stabilized, penalty, and augmented Lagrangian variational formulations are members of this class. Although they are not within the focus of this book, in Section 2.1, we provide a concise summary of the corresponding finite element methods. There are several reasons for doing so. First, such formulations lead to important classes of methods that are often used in practice. Second, these methods sometimes use least-squares type terms to improve an existing quasi-projection scheme and thus provide additional examples of applications of least-squares notions. Third, our catalogue of modified variational formulations also assists in drawing comparisons between least-squares finite element methods (LSFEMs) and other methods.

The second approach is to *replace* the naturally occurring variational formulation by an externally defined one based on minimizing the residuals of the PDE problem. This idea ultimately leads to *bona fide*[1] *least-squares* variational principles and finite element methods that can, in principle and in fact, recover all the advantages of the Rayleigh–Ritz setting.[2] In Section 2.2, we offer a brief and informal first look

[1] We use the adjective "bona fide" to distinguish the methods that are the main subject of this book from some modified variational principles that are also commonly referred to using the adjective "least-squares;" see Section 2.1.1. In the sequel, we drop the adjective bona fide when referring to the former class of methods.

[2] From a general perspective, the two approaches are examples of *regularization* techniques. Modification of a given variational formulation can be viewed as *regularization by penalty* and/or

P.B. Bochev, M.D. Gunzburger, *Least-Squares Finite Element Methods*, Applied Mathematical Sciences 166, DOI: 10.1007/b13382_2,

at least-squares variational principles and the associated finite element methods. This discussion is intended to serve as an overture to our detailed examination of LSFEMs and to familiarize the reader with the basic notions and ideas that enter into the construction of modern LSFEMs, including potential pitfalls and how they can be avoided or circumvented.

2.1 Modified Variational Principles

2.1.1 *Enhanced and Stabilized Methods for Weakly Coercive Problems*

The main source of difficulties in Galerkin methods based on weakly coercive principles is that *weak coercivity is not automatically inherited on proper subspaces*; see Figure 1.1. As a result, the conformity of the discrete spaces is not sufficient to guarantee the stability of the resulting discrete problem or the optimal accuracy of approximations; those spaces are also required to satify the discrete inf–sup conditions (1.33) and (1.34); see Theorem 1.8. In addition, Galerkin finite element methods based on weakly coercive variational formulations generally result in linear systems that are indefinite and nonsymmetric and thus pose additional difficulties.

With regard to weakly coercive problems, we use the term *stabilized methods* for modifications that render the associated (modified) bilinear form $Q_M(\cdot,\cdot)$ strongly coercive, or at least allow it to satisfy[3] the discrete inf–sup conditions (1.33) and (1.34) for *arbitrary* discrete conforming subspaces $U^h \subset U$ and $W^h \subset W$. Modifications of $Q(\cdot,\cdot)$ that do not remove the requirement of satisfying (1.33) and (1.34) but make the resulting algebraic systems easier to solve are referred to as *enhanced methods*.

Many stabilized and enhanced methods in use today debuted in the context of mixed-Galerkin variational formulations so that we use (1.9) as an abstract setting for a brief overview of the many modifications that have been proposed. For simplicity, we assume that $G = 0$ in (1.9).

Enhanced methods

Enhanced methods for the mixed-Galerkin problem (1.9) seek to make the discrete systems "easier" to solve. In come cases, they also result in a reduction of the

---

Tikhonov regularization for which the original problem is *perturbed* into a problem that is well-posed and/or easier to solve. Least-squares principles, on the other hand, provide examples of *regularization by selection* (see, e.g., [337]), a process that involves the *construction of an externally defined problem*.

[3] Throughout this section, we use the notations introduced in Chapter 1.

number of degrees of freedom. The two most well-known enhanced methods are penalty methods and augmented Lagrangian methods, which we now review.

Penalty methods. Penalty methods were among the first examples of finite element methods based on modified variational formulations; see, e.g., [31, 158, 273, 290]. A penalty method can be obtained by modifying (1.9) to

$$\begin{cases} a(v,r)+b(r,p) = D(r) & \forall r \in V \\ b(v,s) \qquad\qquad\;\; = \varepsilon c(p,s) & \forall s \in S, \end{cases} \tag{2.1}$$

where $\varepsilon > 0$ is a specified penalty parameter and $c(\cdot,\cdot)$ is a symmetric, continuous, *strongly coercive* bilinear form on $S \times S$, i.e., we have that, for all $p,s \in S$,

$$c(p,s) = c(s,p), \quad |c(p,s)| \le \alpha_c \|p\|_S \|s\|_S, \quad \text{and} \quad c(p,p) \ge \beta_c \|p\|_S^2 \tag{2.2}$$

for some constants $\alpha_c > 0$ and $\beta_c > 0$. The operator form of (2.1) is

$$\begin{cases} \widetilde{\mathcal{A}}v + \widetilde{\mathcal{B}}^* p = \widetilde{d} & \text{in } V^* \\ \widetilde{\mathcal{B}}v - \varepsilon\widetilde{\mathcal{C}}p \;= 0 & \text{in } S^*, \end{cases} \tag{2.3}$$

where $\widetilde{\mathcal{C}} : S \mapsto S^*$ is the operator corresponding to the bilinear form $c(\cdot,\cdot)$, i.e., we have that

$$c(p,s) = \langle \widetilde{\mathcal{C}}p, s \rangle_{S^*,S} \qquad \forall p,s \in S.$$

Due to (2.2), $\widetilde{\mathcal{C}}$ is a self-adjoint and positive definite operator so that we can eliminate p from (2.3) to obtain

$$\Big(\widetilde{\mathcal{A}} + \frac{1}{\varepsilon}\widetilde{\mathcal{B}}^*\widetilde{\mathcal{C}}^{-1}\widetilde{\mathcal{B}}\Big)v = \widetilde{d} \quad \text{in } V^*. \tag{2.4}$$

This can be viewed as the operator form of the weak formulation

$$a(v,r) + \frac{1}{\varepsilon}\big\langle \widetilde{\mathcal{B}}v, \widetilde{\mathcal{C}}^{-1}\widetilde{\mathcal{B}}r \big\rangle_{S^*,S} = D(r) \quad \forall r \in V. \tag{2.5}$$

In case (1.9) corresponds to the first-order necessary conditions for a constrained optimization problem so that (1.29) and (1.30) hold, (2.1) and (2.5) are also the first-order necessary conditions corresponding to modifications of the Lagrangian functional $L(v,p;D)$ and the functional $J(v;D)$, respectively, defined in Section 1.2.4.

Remark 2.1 First, consider the modified problem of finding the saddle-point $\{v,p\} \in V \times S$ of the penalized Lagrangian

$$L_\varepsilon(v,p;D) = L(v,p;D) + \frac{\varepsilon}{2}c(p,p). \tag{2.6}$$

One recognizes (2.1) as the first-order necessary conditions corresponding to the saddle points of $L_\varepsilon(\cdot,\cdot;D)$.

Next, consider the modified problem of finding the unconstrained minimizer $v \in V$ of the penalized energy functional

$$J_\varepsilon(v;D) = J(v;D) + \frac{1}{2\varepsilon}\|\widetilde{\mathcal{B}}v\|_{S^*}^2 . \tag{2.7}$$

As $\varepsilon \to 0$, the last term in (2.7) enforces the constraint without using a Lagrange multiplier. The first-order necessary condition corresponding to the minimization of $J_\varepsilon(\cdot;D)$ is given by (2.5) with $\widetilde{\mathcal{C}}$ the identity operator.

An important case that often arises in practice is $S = S^*$ and $c(\cdot,\cdot) = \|\cdot\|_S^2$. Then, the optimization problems for (2.6) and (2.7) are equivalent as are the corresponding weak formulations (2.1) and (2.5), respectively, in the sense that, for a given value of ε, all four have the same solution v. □

Discretization of the modified system (2.1) by a pair of conforming subspaces $V^h \subset V$ and $S^h \subset S$ yields the parameterized modified mixed-Galerkin system

$$\begin{pmatrix} \mathsf{A}^h & \mathsf{B}^{h^T} \\ \mathsf{B}^h & -\varepsilon\mathsf{C}^h \end{pmatrix} \begin{pmatrix} \vec{v}^h \\ \vec{p}^h \end{pmatrix} = \begin{pmatrix} \vec{d}^h \\ 0 \end{pmatrix}, \tag{2.8}$$

where $\mathsf{C}^h \in \mathbb{R}^{M\times M}$ is the symmetric, positive definite matrix obtained from the bilinear form $c(\cdot,\cdot)$ in the usual manner. Note that $\vec{p}^h$ can be eliminated from (2.8) to obtain the reduced parameterized system

$$\left(\mathsf{A}^h + \frac{1}{\varepsilon}\mathsf{B}^{h^T}\mathsf{C}^{h^{-1}}\mathsf{B}^h\right)\vec{v}^h = \vec{d}^h . \tag{2.9}$$

Discretization of (2.5) requires only a space V^h for v^h. In this case, the parameterized linear system is given by

$$\left(\mathsf{A}^h + \frac{1}{\varepsilon}\mathsf{G}^h\right)\vec{v}^h = \vec{d}^h , \tag{2.10}$$

where G^h is the symmetric, semi-definite matrix obtained in the usual manner from the second term in (2.5). In general, $\mathsf{G}^h \neq \mathsf{B}^{h^T}\mathsf{C}^{h^{-1}}\mathsf{B}^h$ so that (2.9) and (2.10) are not equivalent, even though their continuous parent equations (2.1) and (2.5), respectively, yield identical solutions.[4] However, the coefficient matrices in both (2.9)

---

[4] The lack of equivalence between (2.9) and (2.10) can be explained by comparing the order of the discretization and elimination steps in the two problems. The system (2.9) is obtained by discretization of the saddle-point problem (2.1) followed by elimination of the *discrete* Lagrange multiplier p^h. In contrast, (2.10) is a discretization of (2.7) that can be obtained from (2.1) by elimination of the Lagrange multiplier p, i.e., in this problem the elimination step precedes the discretization step. Changing the order of the discretization and elimination steps leads to different discrete problems having different solutions. A related example that underscores the lack of commutativity between the elimination of the Lagrange multiplier and the discretization of the variational formulation is the null space method discussed in Section 1.3.2.

and (2.10) are positive definite so that this advantageous feature of the Rayleigh–Ritz setting is recovered by using a penalty method.[5]

The two forms (2.1) and (2.5) of the classical penalty method do not allow for the circumvention of the discrete inf–sup stability conditions (1.33) and (1.34). Therefore, according to our nomenclature, penalty methods are not *stabilized* formulations. Their proper interpretation is as improved solution methods for the discrete mixed problem (1.15) that allow for the elimination of $\vec{p}^h$ and for the replacement of the indefinite saddle-point matrix by a positive definite matrix.

Remark 2.2 Let us briefly discuss the well-posedeness of (2.8)–(2.10). For small values of the penalty parameter ε, the system (2.8) approaches the unmodified problem (1.15). As a result, despite the presence of the symmetric positive definite matrix C^h, the stability of (2.8), or equivalently (2.9), remains subject to (1.33) and (1.34).

On the other hand, because (2.10) is defined by a single discrete space, it seems reasonable to assume that that problem is well-posed as long as $V^h \subset V$. Unfortunately, this is not true. If G^h is positive definite, we can assume the existence of a $\widetilde{\mathsf{B}}^h$ such that $\widetilde{\mathsf{B}}^{h^\mathsf{T}}\widetilde{\mathsf{B}}^h = \mathsf{G}^h$ and then, letting $\varepsilon\vec{p}^h = \widetilde{\mathsf{B}}^h\vec{v}^h$, (2.10) can be converted to

$$\begin{pmatrix} \mathsf{A}^h & \widetilde{\mathsf{B}}^{h^\mathsf{T}} \\ \widetilde{\mathsf{B}}^h & -\varepsilon\mathsf{I}^h \end{pmatrix} \begin{pmatrix} \vec{v}^h \\ \vec{p}^h \end{pmatrix} = \begin{pmatrix} \vec{d}^h \\ 0 \end{pmatrix}. \tag{2.11}$$

Therefore, (2.10) remains associated with a saddle-point formulation in which the space S^h is implicitly defined from V^h by $S^h = \widetilde{\mathsf{B}}^h V^h$; see [273]. Consequently, the stability of (2.11) and, by extension, of (2.10) is contingent upon the compatibility of V^h and the implicitly defined space S^h. In some extreme cases the nullspace Z^h implied by the implicit space S^h can be empty, causing the finite element method to *lock*. A classical example of locking occurs when (2.5) is used to solve the Stokes equations (1.62) by piecewise linear elements. Then, for small values of the penalty parameter ε, the piecewise linear solution of (2.5) is identically zero. Locking can be avoided if one changes the definition of the implicitly defined space S^h by evaluating the penalty term in (2.5) using, e.g., a reduced-order integration rule; see [158, 220, 290]. □

Solutions of penalty formulations contain an $O(\varepsilon)$ penalty error, e.g., the differences between solutions of (1.9) and (2.1) are of $O(\varepsilon)$. Unfortunately, in practice, one cannot choose small values for the penalty parameter ε because the condition numbers of the matrices in (2.9) and (2.10) increase with decreasing ε. At additional cost, penalty methods can be implemented in an iterative manner to reduce the penalty error and allow for the use of larger values of ε; see, e.g., [172, 192].

Augmented Lagrangian methods. The contradictory demands placed on ε by accuracy and efficiency requirements are a serious weakness of penalty methods. Augmented Lagrangian methods [173] are an alternative modification technique in

[5] In the specialized setting of constrained optimization problems, the matrix A is symmetric so that, in that case, the coefficient matrices in (2.9) and (2.10) are symmetric as well as positive definite.

which the first equation in (1.9) is modified instead of the second; the modification term is chosen to be of the form of the second term in (2.5). Thus, instead of (1.9), one solves the modified problem

$$\begin{cases} a(v,r) + \dfrac{1}{\varepsilon}\langle \widetilde{\mathcal{B}}v, \widetilde{\mathcal{C}}^{-1}\widetilde{\mathcal{B}}r\rangle_{S^*,S} + b(r,p) = D(r) & \forall r \in V \\ b(v,s) \qquad\qquad\qquad\qquad\qquad\qquad\quad = 0 & \forall s \in S, \end{cases} \tag{2.12}$$

where $\varepsilon > 0$ is given constant. This approach is called an *augmented Lagrangian method* because, in case (1.9) corresponds to a constrained optimization problem for which $S^* = S$, (2.12) is the first-order necessary condition corresponding to saddle points of the augmented Lagrangian

$$L_\varepsilon(v,p;D) = L(v,p;D) + \frac{1}{2\varepsilon}\|\widetilde{\mathcal{B}}v\|^2_{S^*}\,. \tag{2.13}$$

Choosing a pair of conforming subspaces $V^h \subset V$ and $S^h \subset S$, the discretization of (2.12) results in the parameterized linear system

$$\begin{pmatrix} \mathsf{A}^h + \dfrac{1}{\varepsilon}\mathsf{G}^h & \mathsf{B}^{h\,T} \\ \mathsf{B}^h & 0 \end{pmatrix} \begin{pmatrix} \vec{v}^h \\ \vec{p}^h \end{pmatrix} = \begin{pmatrix} \vec{d}^h \\ 0 \end{pmatrix} \tag{2.14}$$

that can be viewed as another modification of the mixed-Galerkin system (1.15). In contrast to (2.8), the modification now affects the first equation in (1.15), where the $(1,1)$ block is replaced by the same matrix as in (2.10), and leaves the second equation unchanged. As a result, solutions of (2.14) satisfy the discrete constraint equation and are not subject to a penalty error. Consequently, small values of ε are unnecessary for accuracy and the value of this parameter can be optimized for solver efficiency. The stability of (2.14) remains subject to the inf–sup conditions (1.33) and (1.34) so that, again, (2.14) does not define a *stabilized* method.

Compared with penalty formulations, augmented Lagrangian methods improve the efficiency of iterative solvers by decreasing the condition number of the $(1,1)$ block of the discrete system. This is of particular interest for problems where $a(\cdot,\cdot)$ is coercive strictly on the null space Z, defined in (1.21). As is explained in Section 1.3.2, the discrete null-space Z^h is generally not a subspace of Z so that the discrete strong coercivity assumption (1.44) has to be imposed on the spaces in addition to the discrete inf–sup condition (1.42). Augmented Lagrangian methods make it possible to use discrete spaces that satisfy only the latter and produce algebraic systems for which A^h may not be invertible. The drawback of augmented Lagrangian methods is that (2.14) is an indefinite linear system and a reduction in the number of unknowns cannot be effected.

Both penalty and augmented Lagrangian methods extend, in a natural way, to nonlinear problems such as the Navier–Stokes equations (1.68).

Stabilized methods

In contrast to enhanced methods, stabilized methods for (1.31) seek to modify the weakly coercive mixed bilinear form[6] $Q(\cdot,\cdot)$ of (1.12) in such a way that the discrete inf–sup conditions (1.33) and (1.34) follow automatically for any conforming choice of the discrete spaces $V^h \subset V$ and $S^h \subset S$. Nearly all stabilized mixed formulations proposed thus far stem from two basic approaches which we now review.

Residual stabilization. In this method class, the mixed weak formulation (1.9) or, equivalently, (1.3) with $Q(\cdot,\cdot)$ given by (1.12), is modified by adding terms involving the residuals $\widetilde{\mathcal{A}}v + \widetilde{\mathcal{B}}^* p - \widetilde{d}$ and[7] $\widetilde{\mathcal{B}}v$ corresponding to (1.10). Specifically, (1.3) is replaced by

$$\begin{aligned} Q(\{v,p\},\{r,s\}) - \big(\widetilde{\mathcal{A}}v + \widetilde{\mathcal{B}}^* p - \widetilde{d}, \tau_1 \mathcal{W}_1(\{r,s\})\big)_{V^*} & \\ + \big(\widetilde{\mathcal{B}}v, \tau_2 \mathcal{W}_2(\{r,s\})\big)_{S^*} = D(r) \qquad \forall \{r,s\} \in V \times S\,, & \end{aligned} \tag{2.15}$$

where $\mathcal{W}_1(\{r,s\}) : V \times S \mapsto V^*$ and $\mathcal{W}_2(\{r,s\}) : V \times S \mapsto S^*$ are weighting operators and τ_1 and τ_2 are *stabilization parameters*.[8]

The use of equation residuals in the modification results in (2.15) being *consistent*, i.e., if $\{v,p\} \in V \times S$ is a solution of (1.10) so that it also satisfies (1.9), then $\{v,p\}$ is also a solution of (2.15).[9] This consistency feature is a main appeal of residual stabilization methods because the definition of high-order stabilized methods becomes straightforward.

Of course, we also want the modified weak formulation (2.15) to be stable after a discretization based on any conforming pair $V^h \subset V$ and $S^h \subset S$ of approximating subspaces. Thus, letting

$$\begin{aligned} Q_M(\{v,p\},\{r,s\}) &= Q(\{v,p\},\{r,s\}) \\ &\quad - \big(\widetilde{\mathcal{A}}v + \widetilde{\mathcal{B}}^* p, \tau_1 \mathcal{W}_1(\{r,s\})\big)_{V^*} + \big(\widetilde{\mathcal{B}}v, \tau_2 \mathcal{W}_2(\{r,s\})\big)_{S^*} \end{aligned}$$

and

$$F_M(\{r,s\}) = D(r) - \big(\widetilde{d}, \tau_1 \mathcal{W}_1(\{r,s\})\big)_{V^*}$$

for all $\{v,p\}$ and $\{r,s\}$ in $V \times S$ so that (2.15) is equivalent to

$$Q_M(\{v,p\},\{r,s\}) = F_M(\{r,s\}) \qquad \forall \{r,s\} \in V \times S\,, \tag{2.16}$$

---

[6] Recall that the inclusions $V^h \subset V$ and $S^h \subset S$ are not sufficient for (1.33) and (1.34) to hold for $Q(\cdot,\cdot)$ defined in (1.12).

[7] Recall that we are assuming that $G(\cdot) = 0$ so that $g = 0$ as well.

[8] A stabilization parameter τ_i is a function that is either identically zero or assumes strictly positive values. Generally, τ_1 and τ_2 can be scalar or tensor functions. For simplicity, we restrict attention to the former case. Usually, in finite element methods these functions are defined on an element-by-element basis. In some cases, e.g., shape regular mesh partitions, stabilization parameters can be constant on all of Ω.

[9] For this reason, stabilized methods based on residual stabilization are often referred to as *consistently stabilized methods*.

we want to choose the weighting operators $\mathcal{W}_1(\cdot)$ and $\mathcal{W}_2(\cdot)$ and the stabilization parameters τ_1 and τ_2 in such a way that $Q_M(\cdot,\cdot)$ satisfies the discrete inf–sup conditions (1.33) and (1.34) for any $V^h \subset V$ and $S^h \subset S$. The bilinear form $Q_M(\cdot,\cdot)$ is referred to as being *absolutely stable* if (1.33) and (1.34) hold for τ_1 and τ_2 whose range is unbounded, and *conditionally stable* otherwise.

The key to effecting stabilization is to choose $\mathcal{W}_1(\cdot)$ so that the term

$$-\left(\widetilde{\mathcal{B}}^* p, \tau_1 \widetilde{\mathcal{B}}^* s\right)_{V^*} \tag{2.17}$$

is present in (2.16). Known choices for the weighting operators, all of which include the additional term (2.17), take the form

$$\mathcal{W}_1(\{r,s\}) = \gamma_1 \widetilde{\mathcal{A}} r + \widetilde{\mathcal{B}}^* s \qquad \text{and} \qquad \mathcal{W}_2(\{r,s\}) = \gamma_2 \widetilde{\mathcal{B}} r, \tag{2.18}$$

where γ_2 is usually chosen to be zero[10] and γ_1 is chosen from among $\{-1,0,1\}$.[11] Substituting the choices in (2.18) into (2.16) change (1.14) to the modified mixed variational formulation: *find* $\{v^h, p^h\} \in V^h \times S^h$ *such that*

$$\begin{cases} a(v^h,r^h) + b(r^h,p^h) - \gamma_1 \left(\widetilde{\mathcal{A}} v^h + \widetilde{\mathcal{B}}^* p^h, \tau_1 \widetilde{\mathcal{A}} r^h\right)_{V^*} & \\ \qquad + \gamma_2 \left(\widetilde{\mathcal{B}} v^h, \tau_2 \widetilde{\mathcal{B}} r^h\right)_{S^*} = D(r^h) - \gamma_1 \left(\widetilde{d}, \tau_1 \widetilde{\mathcal{A}} r^h\right)_{V^*} & \forall r^h \in V^h \\ b(v^h,s^h) - \left(\widetilde{\mathcal{A}} v^h + \widetilde{\mathcal{B}}^* p^h, \tau_1 \widetilde{\mathcal{B}}^* s^h\right)_{V^*} = -\left(\widetilde{d}, \tau_1 \widetilde{\mathcal{B}}^* s^h\right)_{V^*} & \forall s^h \in S^h. \end{cases} \tag{2.19}$$

Equivalently, we have the parameterized linear system

---

[10] The reason for this is that the operator $\mathcal{W}_2(\cdot)$ plays no role in stabilization.

[11] In the case of constrained optimization problems for which (1.29) and (1.30) hold, some, but not all, stabilized methods defined through (2.18) are also optimality systems for modifications of $L(v,p,D)$. For example, choosing $\gamma_1 = \gamma_2 = 1$ makes (2.15) identical with the optimality system of the modified Lagrangian

$$L_{FLS}(v,p;D) = L(v,p;D) - \frac{1}{2}\|\sqrt{\tau_1}(\widetilde{\mathcal{A}} v + \widetilde{\mathcal{B}}^* p - \widetilde{d})\|_{V^*}^2 + \frac{1}{2}\|\sqrt{\tau_2}\widetilde{\mathcal{B}} v\|_{S^*}^2$$

and choosing $\gamma_1 = 1$ and $\gamma_2 = 0$ corresponds to finding saddle-points of another modified Lagrangian

$$L_{PLS}(v,p) = L(v,p;D) - \frac{1}{2}\|\sqrt{\tau_1}(\widetilde{\mathcal{A}} v + \widetilde{\mathcal{B}}^* p - \widetilde{d})\|_{V^*}^2.$$

However, there is no modified Lagrangian whose optimality system is given by (2.15) with $\gamma_1 = -1$ and $\gamma_2 = 0$. As a result, the symmetry of the original saddle-point variational problem is lost for this case.

We call L_{FLS} and L_{PLS} *full* and *partial* least-squares modifications of the Lagrangian. This terminology should not be confused with the later use, e.g., in Section 2.2, of "least-squares" which then refers to the *replacement* of the Lagrangian by a least-squares functional. Note that the augmented Lagrangian functional (2.13) uses the second component of the residual $\widetilde{\mathcal{B}} v$ and so it itself is a type of partial least-squares modification for which $\tau_1 = 0$. However, this modification lacks the critical term $(\widetilde{\mathcal{B}}^* p, \widetilde{\mathcal{B}}^* s)_{V^*}$ needed for (1.33) and (1.34) to hold for $Q_M(\cdot,\cdot)$ on arbitrary conforming subspaces.

$$\left\{ \begin{pmatrix} \mathsf{A}^h & \mathsf{B}^{h\mathsf{T}} \\ \mathsf{B}^h & 0 \end{pmatrix} + \begin{pmatrix} \gamma_1 \mathsf{A}_1^h(\tau_1) & \gamma_1 \mathsf{B}_1^h(\tau_1)^\mathsf{T} \\ -\mathsf{B}_1^h(\tau_1) & -\mathsf{D}^h(\tau_1) \end{pmatrix} \right.$$
$$\left. + \gamma_2 \begin{pmatrix} \mathsf{G}^h(\tau_2) & 0 \\ 0 & 0 \end{pmatrix} \right\} \begin{pmatrix} \vec{v}^h \\ \vec{p}^h \end{pmatrix} = \begin{pmatrix} \vec{d}^h \\ \vec{0} \end{pmatrix} - \begin{pmatrix} \gamma_1 \vec{d}_1^h(\tau_1) \\ \vec{g}_1^h(\tau_1) \end{pmatrix}. \tag{2.20}$$

The first block matrix on the left-hand side is generated by the unmodified bilinear form in (1.9); see (1.15). The terms added using the residuals contribute to the second and third matrices and to the right-hand side. The matrices $\mathsf{B}_1^h(\tau_1)$, $\mathsf{D}^h(\tau_1)$, and $\mathsf{A}_1^h(\tau_1)$ correspond to the terms $\left(\widetilde{\mathcal{B}}^* p^h, \tau_1 \widetilde{\mathcal{A}} r^h\right)_{V^*}$, $\left(\widetilde{\mathcal{B}}^* p^h, \tau_1 \widetilde{\mathcal{B}}^* s^h\right)_{V^*}$, and $\left(\widetilde{\mathcal{A}} v^h, \tau_1 \widetilde{\mathcal{A}} r^h\right)_{V^*}$, respectively; the vectors $\vec{d}_1^h(\tau_1)$ and $\vec{g}_1^h(\tau_1)$ correspond to the terms $\left(\widetilde{d}, \tau_1 \widetilde{\mathcal{A}} r^h\right)_{V^*}$ and $\left(\widetilde{d}, \tau_1 \widetilde{\mathcal{B}}^* s^h\right)_{V^*}$, respectively.

Stabilization is effected by the $(2,2)$ block of the second matrix in (2.20), i.e., by the matrix $-\mathsf{D}^h(\tau_1)$ that corresponds to the term (2.17). The other terms in this matrix as well as the additional terms in the right-hand side are needed to ensure the consistency of the method and may have a destabilizing effect. Thus, the role of the parameter τ_1 is to prevent destabilization by properly balancing the stabilization and consistency terms.[12]

Remark 2.3 Conforming implementation of the additional stabilization terms in (2.20) can be impractical for some problems. One example is the Stokes equations (1.62) for which V^* is the negative-order Sobolev space $[H^{-1}(\Omega)]^d$ and $\widetilde{\mathcal{A}}$ is the second-order differential operator $-\Delta$ with zero boundary conditions. According to Theorem A.1, $\|u\|_{-1} = \|(-\Delta)^{-1/2} u\|_0$ which means that conforming implementation of (2.18) requires computation of impractical terms such as

$$\left(\widetilde{\mathcal{A}} v + \widetilde{\mathcal{B}}^* p, \tau_1 \mathcal{W}_1(\{r,s\})\right)_{V^*} = \left(-\Delta \mathbf{u} + \nabla p, \tau_1 (-\Delta)^{-1}(-\gamma_1 \Delta \mathbf{v} + \nabla p)\right)_0.$$

In most cases, the minus one norm is replaced by a *quasi-equivalent* approximation (B.90) so that the stabilization term assumes the form

---

[12] Most of the stabilized methods that have the abstract form (2.19) were originally developed in the context of the mixed Stokes formulation (1.59). The choice $\gamma_1 = 1$ and $\gamma_2 = 0$ in (2.18) gives the *Galerkin least-squares* method [217] for (1.62), whereas choosing $\gamma_1 = 0$ and $\gamma_2 = 0$ gives the *pressure–Poisson* stabilized Galerkin method [218]. For these methods, (1.33) and (1.34) provably hold only if the stabilization parameters are bounded from above by a constant that depends on the shape of Ω, the Poincaré inequality constant (see Section B.3.2), and an inverse inequality constant (see Section B.3.4), i.e., these methods are conditionally stable [176]. There is some computational evidence [23] that this result is sharp for the Galerkin least-squares method, but that the pressure-Poisson method is actually unconditionally stable. Although no proof exists of the latter result, an inconsistent but optimally convergent modification of the pressure-Poisson method has been proven to be unconditionally stable; see [57]. The choice $\gamma_1 = -1$ and $\gamma_2 = 0$ in (2.18) results in the absolutely stable *Douglas–Wang* stabilized Galerkin method [147] for (1.62). For a further discussion and taxonomy of these methods and their extensions to other mixed problems, see [23, 176, 177].

$$\sum_{\kappa\in\mathcal{T}_h}\int_\kappa \tau_1 h_\kappa^2\Big(-\Delta\mathbf{u}^h+\nabla p^h\Big)\Big(-\gamma_1\Delta\mathbf{v}+\nabla p\Big)\,d\Omega\,, \tag{2.21}$$

where h_κ is the diameter of the finite element κ defined in (B.9). Note that even on shape-regular finite element partitions we cannot use the simpler quasi-equivalent norm (B.91) because standard finite element functions are not continuously differentiable and thus $-\Delta$ cannot be applied globally in Ω. Other alternatives are to use the *norm-equivalent* discrete minus one norm $\|\cdot\|_{-h}$ defined in (B.99) (see [94]), or to replace $-\Delta$ by the discrete operator $-\Delta_G^h$ defined in (B.59) (see [57].) Strictly speaking, in the latter case, the resulting method is slightly inconsistent in the sense that the modified residual term $-\Delta_G^h\mathbf{u}+\nabla p-\mathbf{f}$ does not vanish for the exact solution. However, the inconsistency is within the approximation order and so convergence rates do not suffer.

Residual stabilization of finite element methods based on the Kelvin principle (1.54) does not experience the same type of problems as those just described for the Stokes equations. The relevant residual $\mathbf{v}+\nabla p$ in (1.55) contains only first-order derivatives. As a result, the stabilizing term can be implemented using a standard $L^2(\Omega)$ inner product. A stabilized method that uses the Galerkin least-squares form of the weight function (2.18), i.e., that uses $\gamma_1=1$ and $\gamma_2=0$, was developed in [275]. An interesting feature of this method is that one may set $\tau_1=1/2$, i.e., τ_1 may be chosen to be independent of the mesh size. □

Non-residual stabilization. As the name "non-residual" suggests, this class of methods stabilizes the bilinear form $Q(\cdot,\cdot)$ given by (1.12) without resorting to residual terms.[13] Most of the known non-residual stabilizing methods share a common abstract structure. However, usually the construction of such methods depends on the specific choice made for the finite element subspaces and thus can vary greatly from case to case. For this reason, after briefly reviewing the abstract approach, we move on to illustrate the main ideas using concrete examples.

A typical non-residual stabilization method modifies $Q(\cdot,\cdot)$ to[14]

$$Q_M(\{v,p\},\{r,s\})=Q(\{v,p\},\{r,s\})-c^h(p,s)\,, \tag{2.22}$$

where $c^h(\cdot,\cdot)$ is a discrete stabilizing bilinear form. Let $\widetilde{S}^h$ be a subspace of S that forms a stable pair with V^h, i.e., such that the discrete inf–sup conditions (1.33) and (1.34) hold for the bilinear form $Q(\cdot,\cdot)$ and the subspaces V^h and $\widetilde{S}^h$. A general rule of a thumb for rendering (2.22) stable, i.e., at least weakly coercive for at least

[13] As a result, this class of methods is *inconsistent* in the sense that exact solutions do not in general exactly satisfy the discrete equations.

[14] Equivalently, in the constrained optimization case for which (1.29) and (1.30) hold, one can modify the Lagrangian $L(r,s;D)$ to

$$L_{NR}(r,s)=L(r,s;D)-\frac{1}{2}c^h(s,s)\,.$$

some conforming subspaces $V^h \subset V$ and $S^h \subset S$, is to choose $c^h(\cdot,\cdot)$ to be positive definite on $S \setminus \widetilde{S}^h$ and vanish on $\widetilde{S}^h$, i.e., the stabilizing term in (2.22) acts as a filter that is "activated" only for the unstable components of p^h. This makes (2.22) fundamentally different from a penalty method where $c(\cdot,\cdot)$ is positive definite.

Compared with (2.20), the linear system associated with (2.22) has the much simpler form

$$\left\{ \begin{pmatrix} \mathsf{A}^h & \mathsf{B}^{h^\mathsf{T}} \\ \mathsf{B}^h & 0 \end{pmatrix} + \begin{pmatrix} 0 & 0 \\ 0 & -\widetilde{\mathsf{C}}^h \end{pmatrix} \right\} \begin{pmatrix} \vec{v}^h \\ \vec{p}^h \end{pmatrix} = \begin{pmatrix} \vec{d}^h \\ 0 \end{pmatrix}, \tag{2.23}$$

where $\widetilde{\mathsf{C}}^h$ is obtained from $c^h\left(p^h, s^h\right)$ in the usual manner. Note that if A^h is symmetric, then symmetry is maintained in (2.23).[15]

We now proceed to examine some typical non-residual stabilized methods for the Stokes equations (1.62). The *pressure gradient projection method* [33] relaxes the incompressibility constraint by the difference between the pressure gradient and its $L^2(\Omega)$ projection onto the velocity subspace V^h. For this method,

$$c^h\left(p^h, s^h\right) = \alpha\left((\pi_V - \mathcal{I})\nabla p^h, (\pi_V - \mathcal{I})\nabla s^h\right)_0,$$

where π_V is the $L^2(\Omega)$ projection $[L^2(\Omega)]^d \mapsto V^h$ and $\alpha > 0$ is a constant. The pressure gradient projection method is motivated by fractional-step solution techniques [124] for transient incompressible flow problems. Because ∇p^h is projected onto a C^0 velocity space, computation of the stabilization term requires the solution of a global problem. In practice, the method is implemented in a mixed form by introducing the projected gradient as an additional dependent variable.

The *polynomial pressure projection method* [49, 142] stabilizes the mixed Stokes formulation by using the difference between the pressure and its local $L^2(\Omega)$ projection onto a suitable finite element space. The stabilizing term is

$$c^h\left(p^h, s^h\right) = \alpha\left((\pi_S - \mathcal{I})p^h, (\pi_S - \mathcal{I})s^h\right)_0,$$

where π_S is a locally defined projection operator acting on S^h. One version of this method is designed to stabilize equal-order velocity–pressure subspaces V^h and S^h. In this version, π_S maps the pressure space S^h onto a discontinuous piecewise polynomial space $\cup_{\kappa \in \mathcal{T}_h} P(\kappa)$ which contains polynomials of one degree less than those used to construct S^h. A second version of the polynomial pressure projection method is designed specifically for low-order pairs such as the linear (or bilinear, or trilinear) velocity and constant pressure pairs. In this case, π_S maps the discontinuous

[15] The matrix in (2.23) has the exact same structure as the matrix (2.8) for the penalty method. One key difference is that $\widetilde{\mathsf{C}}^h$ is symmetric but is only semi-definite, a fact that prevents the elimination of $\vec{p}^h$ from (2.23); thus, as already mentioned, (2.22) does not define a penalty method. On the other hand, a second, more important difference is that the form (2.22) satisfies (1.33) and (1.34) for at least some conforming subspaces $V^h \subset V$ and $S^h \subset S$ that are not stable for the penalty method.

pressure space into a C^0 piecewise polynomial space of one degree higher than S^h. In both cases, the stabilization term can be computed locally on each element.

Two other non-residual stabilized formulations designed specifically for low-order finite elements with discontinuous pressure spaces S^h are the *global pressure jump* and the *local pressure jump* stabilization methods [154,315,316]. Let Γ_h denote the set of all internal interfaces σ between the elements in a finite element partition $\mathcal{T}_h$ of Ω and let $[p^h]$ denote the jump of a function $p^h \in S^h$ across an interface σ. In the first method, the stabilizing term is given by

$$c^h(p^h, s^h) = \tau h \sum_{\sigma \in \Gamma_h} \int_\sigma [p^h][s^h] dS,$$

where τ is a stabilization parameter. To define the second method, suppose that the elements in $\mathcal{T}_h$ can be arranged into non-overlapping patches $\mathcal{T}_h^i$, all containing approximately the same number of elements. The stabilizing term for the local pressure jump stabilization method is given by

$$c^h(p^h, s^h) = \tau h \sum_{\mathcal{T}_h^i} \sum_{\sigma \in \Gamma_h^i} \int_\sigma [p^h][s^h] dS,$$

where Γ_h^i is the set of all internal interfaces in the patch $\mathcal{T}_h^i$. In this method, jumps are integrated only along those $\sigma \in \Gamma_h$ that are internal to a patch $\mathcal{T}_h^i$. As a result, this stabilized formulation enforces local conservation on each patch.

2.1.2 Stabilized Methods for Strongly Coercive Problems

At first glance, the modification of strongly coercive variational formulations appears to be a wasteful endeavor. Unlike weak coercivity, strong coercivity is inherited on proper subspaces and thus any conforming discrete space is automatically compatible with the variational formulation. Furthermore, Lemma 1.9 asserts that finite element solutions of strongly coercive problems satisfy the optimal error estimate (1.37). However, "improving" this arguably good variational setting for the finite element method is sometimes useful because of the asymptotic nature of (1.37). If, for a given PDE problem, the naturally occurring variational formulation is such that the ratio α/β in (1.37) is large, then solving this problem accurately may require an impractically small value of the mesh size h.

A canonical example of such a setting is provided by the scalar advection-diffusion-reaction equation (1.65) whenever the diffusion coefficient ε is small compared to the magnitude of the advection vector $\mathbf{b}$ and/or the reaction coefficient c. In this case, α/β behaves as does $O(\varepsilon^{-1})$ which, in practice, requires that h must be of $O(\varepsilon)$ to obtain a meaningful solution.[16]

[16] We can reach the same conclusion without the aid of (1.37) by observing that for small ε, (1.65) is a singularly perturbed problem that can develop internal and/or boundary layers whose

Thus, although in theory we can always choose h sufficiently small so that (1.37) holds, in practice, computations are always subject to time and/or memory limitations that make such mesh sizes impractical. As a result, modifications of strongly coercive problems are in widespread use; see e.g., [4, 106, 107, 215, 221, 244].

It is important to keep in mind that the goals pursued by modification of strongly coercive formulations are quite different from those pursued in Section 2.1.1, where, at least for stabilized methods, the objective was to extend weak coercivity to general conforming finite element pairs. In the present case, conforming spaces are already compatible with the strongly coercive problem and the goal is to enable accurate approximations in the pre-asymptotic regime, i.e., without resorting to excessively small values of h.

Despite these differences, the structure of the stabilizing terms used for strongly coercive problems bears a strong similarity with the terms used for mixed variational formulations. The general form is given by

$$Q(u;v) + c^h(u,v) = F(v) \qquad \forall v \in V\,, \tag{2.24}$$

where $c^h(\cdot,\cdot)$ is a stabilizing bilinear form that usually depends on the mesh parameter h. Because the majority of the modifications in use today originated with the advection-reaction-diffusion equation (1.65), we use this problem to describe a few of the commonly used stabilized methods.

The main purpose of $c^h(\cdot,\cdot)$ is to enhance the ellipticity of the bilinear form in (1.66) for grids in the pre-asymptotic regime. The easiest way to accomplish this is to set $c^h(\phi,\psi) = (\mathcal{D}\phi, \tau\mathcal{D}\psi)$, where $\mathcal{D}$ is a first-order differential operator and τ is a stabilization parameter that the user has to choose. Typical examples are the *artificial diffusion method* for which

$$\mathcal{D} = \nabla \quad \text{and} \quad \tau = h$$

and the *streamline diffusion method* for which[17]

$$\mathcal{D} = \mathbf{b}\cdot\nabla \quad \text{and} \quad \tau = h - \varepsilon\,.$$

The streamline diffusion method tends to be more accurate because it adds diffusion only in the streamwise direction. However, in both cases, the modified problem is

---

thickness is proportional to $\sqrt{\varepsilon}$ or ε [243, p. 174]. Even though the associated bilinear form in (1.66) is strongly coercive for solenoidal $\mathbf{b}$ and nonnegative c (see Section 1.4.5), to obtain accurate finite element solutions the mesh has to be fine enough to resolve these layers. It is well known [215, 243, 244] that a finite element method based on the Galerkin weak formulation (1.66) only works well for $h \le \varepsilon$ and may result in highly oscillatory and inaccurate solutions if $\varepsilon \ll h$.

Another related explanation [189] of the troubles experienced by the formally strongly coercive formulation (1.66) in the case for which $\varepsilon \ll h$ is that restriction of (1.66) to a finite element space is effectively equivalent to a discretization of the advection–reaction problem $\mathbf{b}\cdot\nabla\phi + c\phi = f$ with Dirichlet data imposed on all of $\partial\Omega$. Because, in the absence of diffusion, Dirichlet data can only be specified on the inflow part of $\partial\Omega$, this amounts to an ill-posed problem.

[17] In the pre-asympototic regime $\varepsilon \ll h$ so that $\tau > 0$ as required of a stabilization parameter.

an $O(h)$ perturbation of the original equations and so both are inconsistent modifications of the original variational formulation.

We can, however, again use residual terms to add the stabilizing term, i.e., $(\mathbf{b}\cdot\nabla\phi, \mathbf{b}\cdot\nabla\psi)$, in the streamline diffusion stabilization case, to (1.66) without accruing consistency errors. A general form of such a *consistently* stabilized method is given by

$$\mathcal{Q}(\phi,\psi)+(\mathcal{R}(\phi),\tau\mathcal{W}(\psi))_h = F(\psi) \qquad \forall\,\psi\in H^1_0(\Omega),$$

where $\mathcal{R}(\phi) = -\varepsilon\Delta\phi + \mathbf{b}\cdot\nabla\phi + c\phi - f$ is the residual of (1.65) and $\mathcal{W}(\cdot)$ is a suitable weighting operator. Similarly to (2.18), there are three basic choices for this operator:

$$\mathcal{W}(\psi) = -\gamma\varepsilon\Delta\psi + \mathbf{b}\cdot\nabla\psi + c\psi$$

with γ chosen from among $\{-1,0,1\}$. In the literature, the choice $\gamma = 1$ is again known as the *Galerkin–least-squares* method [219], $\gamma = -1$ as the *unusual, multi-scale*, or *adjoint* stabilized Galerkin method [174, 175, 213, 216], and $\gamma = 0$ as the *streamline upwind Petrov–Galerkin method*[18] method [215].

We note that, in many cases, exactly the same stabilized methods can be derived by different approaches. For example, when $\varepsilon = 0$, the three consistently stabilized methods for (1.65) can be interpreted as Petrov–Galerkin[19] methods with a weighting function given by $\psi + \tau\mathcal{W}(\psi)$. The corresponding stabilized weak form can be written as

$$(\mathcal{R}(\phi),\psi+\tau\mathcal{W}(\psi)) = 0 \qquad \forall\,\psi\in H^1(\Omega) \text{ such that } \psi = 0 \text{ on } \Gamma_-,$$

where $\Gamma_- = \{\mathbf{x}\in\partial\Omega \,|\, \mathbf{n}\cdot\mathbf{b} < 0\}$ denotes the inflow[20] part of the boundary $\partial\Omega$.

Remark 2.4 Each of the modified variational formulations considered in this section have their adherents and are used in practice. Perhaps the most serious drawback of these methods is that their stability and accuracy depend strongly on mesh-dependent stabilization parameters that must be re-calibrated from application to application; see, e.g., [88, 212, 214, 313, 335]. Improper selection of these parameters in, e.g., consistently stabilized methods, may lead to the loss of stability properties. In enhanced methods, improper selection of parameters may affect the accuracy and/or the condition numbers of the algebraic systems. Another detriment is that the analysis of many of these methods remains an open problem for important nonlinear equations such as the Navier-Stokes equations. A final observation about modified variational principles is that they do not enable full and in some cases even partial recovery of the advantageous features of the Raleigh–Ritz setting. □

---

[18] This method is usually referred to by simply using its acronym SUPG.

[19] This is one of the few cases in this book where this term is helpful to draw the distinction between a formulation where test and trial functions are in the same space and another one where they belong to different approximation spaces.

[20] When $\varepsilon = 0$, equation (1.65) is a purely hyperbolic PDE meaning that boundary values of ϕ can only be prescribed on Γ_-.

2.2 Least-Squares Principles

By now we know that the Rayleigh–Ritz setting corresponding to unconstrained minimization problems for convex quadratic functionals provides the best variational setting for finite element methods because, in this case, finite element quasi-projections are true projections. Of course, this is a very restricted class of problems so that one naturally asks:

> *can Rayleigh–Ritz-type finite element methods be developed for any linear or nonlinear PDE problem?*

Clearly, this goal cannot be realized by using the Galerkin methods or the modified variational formulations described in Section 1.3 and Section 2.1, respectively. Galerkin methods either recover a naturally existing variational principle, if one exists, or lead to a formal weak problem. Modified methods can ameliorate some of the deficiencies that occur naturally in variational formulations but cannot convert those formulations into a Rayleigh–Ritz-like principle.

Knowing that Rayleigh–Ritz-type finite element methods correspond to linear PDEs related to a class of unconstrained minimization problems, the above question can be split into two connected parts. First, we would like to know if[21]

> *for any given* linear *PDE problem, can one construct a convex, quadratic functional whose unconstrained minimizers are also solutions of the PDE problem?*

If this is possible, then we also would like to know if

> *for any given* linear *or* nonlinear *PDE problem, can one construct an unconstrained minimization problem whose minimizers coincide with the solutions of the PDE and can those solutions be approximated by solving a sequence of unconstrained minimization problems with convex quadratic functionals?*

Affirmative answers to the last two questions are provided by the idea of *residual minimization.* Just as with *residual orthogonalization* that is the basis of the very general Galerkin method, the residual minimization idea is applicable to virtually any linear or nonlinear PDE problem. In particular, for linear PDEs, residual minimization can lead to unconstrained optimization problems for convex quadratic functionals even if the original equations were not at all associated with optimization. If the PDE problem is nonlinear, then properly executed residual minimization leads to unconstrained minimization problems whose linearization[22] gives rise to unconstrained minimization problems with convex quadratic functionals. In either

[21] We do not care if the functional in question has physical relevance; we are happy if it is merely a mathematical construction.

[22] The same result can be achieved by exchanging the order of the residual minimization and linearization steps.

case, residual minimization has the potential to define a true inner-product projection or a sequence of such projections that recover the solutions of any[23] given PDE.

Because the inner-product property of the linearized formulation is not affected by the order in which one takes the residual minimization and linearization steps, it is clear that recovery of a Rayleigh–Ritz-type setting for any given PDE rests on our ability to do so for *linear* problems. Thus, unless stated otherwise, this setting is implicitly assumed for the remainder of this chapter.

To use the residual minimization paradigm to define a finite element method, we start by interpreting some measure of the PDE problem residual as an "energy," i.e., an externally defined convex quadratic functional, that must be minimized, with the exact solution of the PDE problem having zero "energy." This naturally leads to the notion of measuring residuals in $L^2(\Omega)$-based norms;[24] hence, the descriptor *least-squares functional* for the artificially defined "energy" becomes appropriate. The next step is to formulate a *least-squares variational principle* by choosing the space over which to minimize the residual energy. A finite element method can now be defined as in the Rayleigh–Ritz case by restricting the least-squares principle, i.e., the minimization of the least-squares functional, to a finite-dimensional subspace.

The single most important prerequisite for the success of this plan for recovering a Rayleigh–Ritz-like setting is to ensure that the least-squares variational principle corresponds to a true inner-product projection. From Section 1.2.4, we know that this is the same as saying that the (externally defined) least-squares "energy" functional is *norm equivalent*. If this turns out to be the case, then we have completely succeeded in associating the PDE problem with a well-posed unconstrained minimization problem for a convex quadratic functional.

It should be clear that this plan can be executed in many possible ways. This, of course, is due to the fact that least-squares principles are external to the PDE problem, which in turn is caused by the fact that least-squares principles minimize residuals rather than physical energies. Because residual "energy" can be measured in many different ways, there is virtually no limit to the number of different least-squares principles that can be associated with a given PDE problem. This "ambiguity" permits the adaptation of least-squares principles to specific objectives and/or practical constraints and so it should be considered as an asset rather than a liability of the approach.

Using the notation established in Chapter 1, we next describe the most straightforward means for carrying out the plan.

---

[23] This should be again contrasted with residual orthogonalization which leads to true projections if and only if the original PDE problem is itself related to an unconstrained optimization problem for a convex quadratic functional.

[24] In some situations, it may be more appropriate to consider residual minimization with respect to other norms, e.g., $L^1(\Omega)$-based norms. In such cases, residual minimization does not lead to true inner-product projections. However, $L^1(\Omega)$ minimization problems can be approximated by a sequence of standard $L^2(\Omega)$ minimization problems, resulting in more practical finite element methods. Examples of such "least-squares" principles are discussed in Chapter 10.

2.2.1 A Straightforward Least-Squares Finite Element Method

We begin with the abstract operator equation (1.1), where for simplicity X and Y are taken to be Hilbert spaces. We assume it to be well posed so that there exist positive constants α and β such that

$$\beta\|u\|_X \le \|\mathcal{Q}u\|_Y \le \alpha\|u\|_X \qquad \forall u \in X\,. \tag{2.25}$$

Then, consider the least-squares functional[25]

$$J(u;f) = \|\mathcal{Q}u - f\|_Y^2 \tag{2.26}$$

and the unconstrained minimization problem

$$\min_{u\in X} J(u;f)\,. \tag{2.27}$$

Note that the functional (2.26) measures the residual of (1.1) using the data space norm $\|\cdot\|_Y$ and the minimization problem (2.27) seeks a solution in the solution space X for which (2.25) is satisfied. It is clear that the problems (1.1) and (2.27) are equivalent in the sense that $u \in X$ is a solution of (2.27) if and only if it is also a solution, perhaps in a generalized sense, of (1.1).

Combining (2.25) and (2.26), one sees that the functional $J(\cdot;\cdot)$ is *norm equivalent* in the sense that

$$\beta^2\|u\|_X^2 \le J(u;0) \le \alpha^2\|u\|_X^2 \qquad \forall u \in X\,. \tag{2.28}$$

Choosing least-squares functionals so that they are, in some sense, norm equivalent is of crucial importance in the design of least-squares finite element methods (LSFEMs) and is a recurring theme throughout the book.

A LSFEM can be defined by choosing a family of finite element subspaces $X^h \subset X$ parameterized by h tending to zero and then restricting the minimization problem (2.27) to the subspaces. Thus, the LSFEM approximation $u^h \in X^h$ to the solution $u \in X$ of (1.1) or (2.27) is the solution of the problem

$$\min_{u^h\in X^h} J(u^h;f)\,. \tag{2.29}$$

The Euler–Lagrange equations corresponding to the minimization problems (2.27) and (2.29) are respectively given by (1.3) and (1.5), where $U = W = X$, $U^h = W^h = X^h$, and

$$Q(u,v) = \big(\mathcal{Q}u, \mathcal{Q}v\big)_Y \quad \text{and} \quad F(v) = \big(\mathcal{Q}v, f\big)_Y \qquad \forall u,v \in X \tag{2.30}$$

---

[25] The factor of $1/2$ that appears in physical energy principles is not needed in least-squares functionals because the latter include the data f in the definition of the artificial "energy". Formally, the corresponding optimality system then has a factor of 2 appearing on its left- and right-hand sides; clearly, this factor can be omitted without changing that system.

with $(\cdot,\cdot)_Y$ denoting the inner product[26] on Y.

One easily finds that the operator form of (1.3) is given by (1.4) with $\widetilde{\mathcal{Q}} = \mathcal{Q}^*\mathcal{Q}$ and $\widetilde{f} = \mathcal{Q}^* f$. Thus, in the least-squares setting, (1.3) is equivalent to the normal equations

$$\mathcal{Q}^*\mathcal{Q}u = \mathcal{Q}^* f \qquad \text{in } X \tag{2.31}$$

for (1.1), where the operator $\mathcal{Q}^*$ from Y to X is the Hilbert space adjoint of $\mathcal{Q}$.[27]

If, as in Section 1.1, we choose a basis $\{\phi_j^h\}_{j=1}^N$ for $X^h = U^h = W^h$, then the elements of the matrix $\mathsf{Q}^h \in \mathbb{R}^{N\times N}$ and the vectors $\vec{f} \in \mathbb{R}^N$ appearing in the linear system (1.2) are now given, for $i,j = 1,\ldots,N$, by

$$\mathsf{Q}_{ij}^h = \left(\mathcal{Q}\phi_j^h, \mathcal{Q}\phi_i^h\right)_Y \qquad \text{and} \qquad \vec{f}_i^h = \left(\mathcal{Q}\phi_i, f\right)_Y.$$

The results of the following theorem follow directly from (2.25).

Theorem 2.5. *Assume that* (2.25)*, or equivalently,* (2.28)*, holds and that* $X^h \subset X$*. Then,*

– the bilinear form $Q(\cdot,\cdot)$ *defined in* (2.30) *is continuous, symmetric, and strongly coercive*

– the linear functional $F(\cdot)$ *defined in* (2.30) *is continuous*

– the problem (1.3) *has a unique solution* $u \in X$ *that is also the unique solution of the minimization problem* (2.27)

– the problem (1.5) *has a unique solution* $u^h \in X^h$ *that is also the unique solution of the minimization problem* (2.29)

– for some constant $C > 0$*,* u *and* u^h *satisfy the error estimate*[28]

$$\|u - u^h\|_X \le C \inf_{v^h \in X^h} \|u - v^h\|_X \tag{2.32}$$

– the matrix Q^h *of* (1.2) *is symmetric and positive definite.* □

It is important to note that

> *the only hypotheses made in Theorem* (2.5) *are that the operator equation* (1.1) *is well posed and that the finite element space* X^h *is conforming.*

In particular, it is not assumed that $\mathcal{Q}$ in (1.1) is self-adjoint or positive definite as it would have to be in the Rayleigh–Ritz setting. Furthermore, the finite element

---

26 Because we work in a Hilbert space setting, instead of the Banach space setting used in Chapter 1, bilinear forms are now defined in terms of inner products instead of duality pairings. Of course, because the dual of a Hilbert space can be identified with the space itself, one can move back and forth between the two settings as is convenient.

27 It is also easily seen that (1.3), which now represents the variational form of (1.1) resulting from the least-squares principle (2.27), can also be viewed as a Galerkin formulation of the normal equation (2.31). It is important to note that because $\mathcal{Q}$ is a differential operator, (1.3) involves, compared to (1.1), the higher-order differential operator $\mathcal{Q}^*\mathcal{Q}$. This observation has a profound effect on how practical LSFEMs are defined.

28 Note that (2.32) shows that LSFEM approximations are optimally accurate with respect to solution norm $\|\cdot\|_X$ for which (1.1) is well posed.

space X^h is not required to satisfy compatibility conditions of the type (1.33) and (1.34) as would, in general, be needed in a Galerkin finite element setting. Despite the generality allowed for in (1.1),

> *the LSFEM based on* (2.29) *recovers* all *the desirable features possessed by finite element methods in the Rayleigh–Ritz setting.*

This is what makes LSFEMs intriguing and attractive.

2.2.2 Practical Least-Squares Finite Element Methods

In general, it is not a difficult task[29] to define a norm-equivalent least-squares functional for a given PDE problem, i.e., to choose spaces X and Y and a functional $J(\cdot;\cdot)$ so that (2.28) holds. For example, several norm-equivalent least-square functionals can be defined for each of the problems discussed in Section 1.4. As a result, it is tempting to conclude that we are just about done, i.e., for just about any problem, we know how to construct, using the recipe provided in Section 2.2.1, a LSFEM that recovers all the desirable features of the Rayleigh–Ritz setting. Although this is true, it does not tell the whole story becuase *LSFEMs produced by the straightforward recipe are not necessarily practical.*[30]

Indeed, to succeed in formulating a viable and competitive LSFEM, one needs more than formal mathematical well posedness. After all, the ultimate goal is to devise *a robust, efficient, and accurate computational methodology*; thus, a (mathematically well-posed) discretization scheme must also be *practical*. Admittedly, practicality is a rather subjective characterization, but if we want LSFEMs to be competitive, existing methods for the same PDE problem offer a natural "practicality gauge." Thus, to deem a LSFEM as being practical, one should at least be able to obtain the discrete system without difficulty, certainly with no more difficulty than that encountered for a typical Galerkin or a mixed-Galerkin method for the same problem. To meet this requirement,

> *the matrices and right-hand sides of the LSFEM discrete problem should be "easily" computable,*

e.g., inner products for fractional or negative order Sobolev spaces should not occur in their definition because they contradict this requirement. Second,

---

[29] This is thanks to the vast PDE literature at our disposal.

[30] This, of course, is due to the obvious fact that a Rayleigh–Ritz setting does not have to be practical; just because it has attractive theoretical properties does not mean that these properties can always be taken advantage of in a practical manner. The null-space method for the Stokes equations, encountered in Section 1.4.3, is one good example. Conforming finite element spaces for (1.61) are more difficult to construct than for the usual mixed variational formulation (1.59). As a result, the theoretically superior Rayleigh–Ritz setting of (1.61) is rarely used for the Stokes equations, whereas the theoretically inferior saddle-point setting of (1.59) is the basis for most of the existing standard finite element methods for this problem.

discretization of a LSFEM should not call for finite element spaces that are more difficult to work with than those normally encountered in typical Galerkin and mixed-Galerkin methods for the same PDE problem.

For example, Galerkin and mixed-Galerkin methods for the PDE problems in Section 1.4 use finite element subspaces of $H^1(\Omega)$, $H(\text{div};\Omega)$, and $L^2(\Omega)$; a practical LSFEM for any of these problems should therefore not use, e.g., finite element spaces of continuously differentiable functions, which are more difficult to define than any of the aforementioned spaces. One must, of course, also solve the discrete problem and thus a third requirement for practicality, especially when iterative solution methods are used, is that

the discrete LSFEM problem should have a "manageable" condition number that is comparable to the condition numbers of typical Galerkin and mixed-Galerkin methods for the same PDE problem.[31]

We use the Poisson problem[32]

$$\begin{cases} -\Delta\phi = f & \text{in } \Omega \\ \phi = 0 & \text{on } \partial\Omega \end{cases} \tag{2.33}$$

as a setting for discussing how well the straightforward LSFEM of Section 2.2.1 satisfies these three practicality criteria.

Impractical Straightforward LSFEMs for the Poisson Equation

Consider the problem (2.33), where we assume that Ω is either a convex Lipschitz domain or has a smooth boundary. Of course, this is a problem that fits into the Rayleigh–Ritz framework so that there is no apparent need[33] to use any other type of finite element method. However, let us proceed and use the straightforward LSFEM method anyway, and see what happens. Here, we have that (2.25) holds with[34,35] $X = H^2(\Omega)\cap H_0^1(\Omega)$, $Y = L^2(\Omega)$, and $\mathcal{Q} = -\Delta$ so that we can define the norm-equivalent least-squares functional

$$J(\phi;f) = \|\Delta\phi + f\|_0^2 \qquad \forall\phi \in H^2(\Omega)\cap H_0^1(\Omega) \tag{2.34}$$

[31] Practicality is not necessarily limited to the three requirements listed. It may also involve considerations such as utilization of legacy codes, existing data structures, algebraic solvers, and so on. These criteria, however, are specific to individual implementations, whereas the three requirements listed are more universal in nature.

[32] The practicality of LSFEMs for other problems, including those discussed in Section 1.4, are a subject of interest throughout the book.

[33] Inhomogeneous Dirichlet boundary conditions provide a situation in which one might want to use LSFEMs even for the Poisson problem (2.33) in as much as LSFEM allow for the weak imposition of essential boundary conditions.

[34] This is thanks to the assumptions on the domain Ω; see [183, Theorem 1.8, p.12].

[35] $H^2(\Omega)$ is the Sobolev space of functions having two square integrable derivatives.

and the associated Euler-Lagrange equation (1.3) with

$$Q(\phi,\psi) = \int_\Omega \Delta\phi\,\Delta\psi\,d\Omega \qquad \text{and} \qquad F(\psi) = \int_\Omega f\Delta\psi\,d\Omega$$

for all $\phi,\psi \in H^2(\Omega)\cap H_0^1(\Omega)$. The straightforward LSFEM of Section 2.2.1 is then defined by choosing a subspace $X^h \subset X = H^2(\Omega)\cap H_0^1(\Omega)$ and then posing the problem (1.5).

To gauge the practicality of this LSFEM, we compare it to the standard Galerkin method for (2.33). The latter requires a finite element subspace of $H_0^1(\Omega)$ and leads to linear systems that are easy to assemble and whose condition numbers are $O(h^{-2})$ [156, Theorem 9.11, p. 388]. In contrast, our straightforward LSFEM requires finite element subspaces of $H^2(\Omega)$ which, as is well known, are much more difficult to construct than subspaces of $H^1(\Omega)$: the latter only need be continuous (see Theorem B.3) whereas, in practice, the former have to consist of continuously differentiable functions [123, Theorem 2.1.2, p. 40]. This fact greatly complicates[36] the construction of bases and the assembly of the matrix problem (1.2). Furthermore, it is well known that the condition number of the matrix problem is $O(h^{-4})$ [156, Theorem 9.11]. Thus, in this setting, the straightforward LSFEM fails all three practicality tests. The failure of this method can be explained by noting that the minimization of the least-squares functional has turned the second-order Poisson problem into a fourth-order problem. Indeed, it is easy to see that the problem

$$\begin{aligned} &\textit{find} \quad \phi \in H^2(\Omega)\cap H_0^1(\Omega) \quad \textit{such that} \\ &Q(\phi,\psi) = F(\psi) \qquad \forall\psi \in H^2(\Omega)\cap H_0^1(\Omega) \end{aligned}$$

is a weak formulation of the biharmonic problem

$$\begin{cases} \Delta^2\phi = -\Delta f & \text{in } \Omega \\ \phi = 0 \quad \text{and} \quad \Delta\phi = -f & \text{on } \partial\Omega\,. \end{cases}$$

[36] The $C^1(\Omega)$ regularity requirement complicates finite element spaces in several ways. First, it cannot be satisfied unless the reference polynomial space is of a sufficiently high degree. For triangles this degree is 5 and the lowest-order $C^1(\Omega)$ elements are the quintic *Argyris* [82, p. 74] and *Bell* [123, p. 74] triangles with 21 and 18 degrees of freedom, respectively. One also has the cubic *Hsieh–Clough–Tocher macro element* [123, p. 340] for which linear systems must be solved to define the basis functions. For rectangles, the lowest-order $C^1(\Omega)$ element is the bicubic *Bogner–Fox–Schmit* element [123, p. 76] with 16 degrees of freedom. Second, unisolvency sets of $C^1(\Omega)$ elements include both values of a function and its derivatives. For instance, the 21 degrees of freedom for the Argyris triangle are the values of a function, its gradient, and all second derivatives at the vertices of the triangle, plus the values of the normal derivative at edge midpoints. Finally, owing to their use of, e.g., normal derivative values, $C^1(\Omega)$ elements are not necessarily *affine equivalent* [123, p. 85] because affine mappings do not necessarily preserve the normal direction. See Section B.2 for the definition of normal-preserving mappings.

This system constitutes the normal equations (2.31) for the second-order Poisson problem (2.33); as indicated earlier, normal equations are of higher order compared to the given PDEs.[37]

A practical straightforward LSFEM for the Poisson problem

Consider the problem

$$\begin{cases} \nabla \cdot \mathbf{v} = f & \text{in } \Omega \\ \mathbf{v} + \nabla \phi = 0 & \text{in } \Omega \\ \phi = 0 & \text{on } \partial\Omega\,. \end{cases} \tag{2.35}$$

One easily sees that, formally, this problem is equivalent to the Poisson problem (2.33).

With the spaces $X = H_0^1(\Omega) \times H(\mathrm{div};\Omega)$ and $Y = L^2(\Omega) \times [L^2(\Omega)]^d$, we have that (2.25) is satisfied for the system (2.35) so that we can define a straightforward LSFEM as in Section 2.2.1 based on the least-squares functional

$$J(\phi, \mathbf{v}; f) = \|\nabla \cdot \mathbf{v} - f\|_0^2 + \|\mathbf{v} + \nabla \phi\|_0^2. \tag{2.36}$$

The Euler-Lagrange equation corresponding to the minimization of this functional is given by (1.3) with

$$Q(\{\phi, \mathbf{v}\}, \{\psi, \mathbf{w}\}) = \int_\Omega (\nabla \cdot \mathbf{v})(\nabla \cdot \mathbf{w})\, d\Omega + \int_\Omega (\nabla \phi + \mathbf{v}) \cdot (\nabla \psi + \mathbf{w})\, d\Omega$$

and

$$F(\{\psi, \mathbf{w}\}) = \int_\Omega f \nabla \cdot \mathbf{w}\, d\Omega\,.$$

Now, for any conforming subspace $X^h \subset X = H_0^1(\Omega) \times H(\mathrm{div};\Omega)$, Theorem 2.5 holds. Moreover, all three practicality criteria are met. Any finite element spaces that are subspaces of $H_0^1(\Omega)$ and $H(\mathrm{div};\Omega)$ are allowable;[38] thus, it follows that implementation of our LSFEM does not call for any finite element spaces beyond those that are already required in a Galerkin or a mixed-Galerkin method for the Poisson equation (2.33). The assembly of the matrix system (1.2) is uncomplicated and its condition number is $O(h^{-2})$.

---

[37] In fact, the Dirichlet problem for the biharmonic operator provides another well-known example for which the Rayleigh–Ritz setting is impractical for exactly the same reasons given here for the impracticality of the LSFEM resulting from the least-squares functional (2.34). In fact, this is the reason the Dirichlet problem for the biharmonic operator provided one of the major impetuses for the development of mixed finite element methods.

[38] This also allows us to define X^h using the same standard $C^0(\Omega)$ finite element space for all solution components; this feature of LSFEMs has long been considered to be one of the key advantages of LSFEMs. Implementations of (2.36) using standard equal-order nodal elements are examined in detail in Chapter 5 where it is seen that, in some cases, such spaces may not be the best choice to discretize this functional.

The keys to the practicality of LSFEMs

The above examples provide some guidance about what makes a LSFEM practical. Perhaps the most important observation that can be made is that the practical LSFEM for the Poisson problem example involved a first-order system of PDEs. In fact, useful LSFEMs are based on first-order formulations of PDE systems. Of course, many if not most PDEs of practical interest are not usually posed as a first-order system. Thus,

> *the first key to defining a* practical *LSFEM is to recast the given PDE system into first-order system form.*[39]

Unfortunately, there is no unique way to do this. For example, the three problems

$$\begin{cases} \nabla\cdot\mathbf{v} = f & \text{in } \Omega \\ \mathbf{v}+\nabla\phi = 0 & \text{in } \Omega \\ \phi = 0 & \text{on } \partial\Omega\,, \end{cases} \qquad \begin{cases} \nabla\cdot\mathbf{v} = f & \text{in } \Omega \\ \mathbf{v}+\nabla\phi = 0 & \text{in } \Omega \\ \nabla\times\mathbf{v} = 0 & \text{in } \Omega \\ \phi = 0 & \text{on } \partial\Omega\,, \end{cases} \qquad \begin{cases} \nabla\cdot\mathbf{v} = f & \text{in } \Omega \\ \nabla\times\mathbf{v} = 0 & \text{in } \Omega \\ \mathbf{n}\times\mathbf{v} = 0 & \text{on } \partial\Omega \end{cases}$$

are all first-order systems that are equivalent[40] to the Poisson problem (2.33). Each happens to be norm equivalent, but with respect to different norms that may also change depending on the type of the domain Ω.[41]

Besides the non-uniqueness of the transformation process, what is even more important is that, by itself, having a first-order system is not enough to define a practical LSFEM in all possible circumstances. For example, if solutions of (2.33) are not sufficiently regular, then the least-squares functional

$$J(\phi,\mathbf{v};f) = \|\nabla\cdot\mathbf{v}-f\|_{-1}^2 + \|\mathbf{v}+\nabla\phi\|_0^2 \tag{2.37}$$

that is norm equivalent with respect to $X = H_0^1(\Omega)\times[L^2(\Omega)]^d$, may be a better choice than (2.39) for defining a least-squares method. However, despite the fact that (2.37) is based on a first-order system, it is no more practical than (2.39) because it

[39] In addition to having a crucial role for defining practical LSFEMs, transformation to first-order systems has the added advantage of allowing direct approximations of physically important quantities that appear in the reformulated equations.

[40] The second equation in the rightmost system implies that $\mathbf{v} = -\nabla\phi$.

[41] For example, in Chapter 5, we see that if Ω has smooth boundary or is a convex polyhedron, the solution space X in (2.25) for the above three problems is respectively given by $H_0^1(\Omega)\times H(\mathrm{div};\Omega)$, $H_0^1(\Omega)\times[H^1(\Omega)]^d$, and $\mathbf{H}_\tau^1(\Omega)$, where $H(\mathrm{div};\Omega) = \{\mathbf{v}\in[L^2(\Omega)]^d \mid \nabla\cdot\mathbf{v}\in L^2(\Omega)\}$ and $\mathbf{H}_\tau^1(\Omega) = \{\mathbf{v}\in[H^1(\Omega)]^d \mid \mathbf{n}\times\mathbf{v} = 0 \text{ on } \partial\Omega\}$. If, instead, Ω is non-convex, the first space is unchanged but the last two spaces weaken to $H_0^1(\Omega)\times[H(\mathrm{div};\Omega)\cap H_0(\mathrm{curl};\Omega)]$ and $H(\mathrm{div};\Omega)\cap H_0(\mathrm{curl};\Omega)$, respectively, where, $H_0(\mathrm{curl};\Omega) = \{\mathbf{v}\in[L^2(\Omega)]^d \mid \nabla\times\mathbf{v}\in[L^2(\Omega)]^d \text{ and } \mathbf{n}\times\mathbf{v} = 0 \text{ on } \partial\Omega\}$. In Chapter 5 it is shown that this has a profound effect on the properties of LSFEMs based on these formally equivalent first-order systems.

involves the *negative order* Sobolev norm $\|\cdot\|_{-1}$ which leads to difficulties[42] in the assembly of the matrix system.

To expose another situation that can wreck the practicality of first-order systems in the LSFEM setting, consider the slight generalization

$$\begin{cases} \nabla\cdot\mathbf{v} = f & \text{in } \Omega \\ \mathbf{v}+\nabla\phi = 0 & \text{in } \Omega \\ \phi = g & \text{on } \partial\Omega \end{cases} \tag{2.38}$$

of the problem (2.35).[43] In Chapter 5 it is shown that a norm-equivalent functional for (2.38) that includes the boundary condition residual is given by

$$J(\phi,\mathbf{v};f,g) = \|\nabla\cdot\mathbf{v}-f\|_0^2 + \|\mathbf{v}+\nabla\phi\|_0^2 + \|\phi-g\|_{1/2,\partial\Omega}^2\,. \tag{2.39}$$

Unfortunately, this functional is not practical because the fractional trace norm used to measure the boundary equation residual also leads to difficulties[44] in the assembly of the matrix system.[45]

The main conclusion that can be gleaned from these examples is that, if one wants to be able to assemble the matrix system using standard finite element techniques, one should avoid least-squares functionals involving negative or fractional Sobolev norms and, in fact, norms other than $L^2(\Omega)$ and $L^2(\partial\Omega)$ norms so that

> *the second key to defining a* practical *LSFEM is to use $L^2(\Omega)$ and $L^2(\partial\Omega)$ norms of equation residuals in the definition of the least-squares functional.*

2.2.3 Norm-Equivalence Versus Practicality

In Section 2.2.1, we learned how to define a *norm-equivalent* least-squares functional that results in a LSFEM that recovers all the desirable properties of the Rayleigh–Ritz setting and, in Section 2.2.2, we identified two key steps towards

---

[42] The source of these difficulties is explained in Section 2.1.1.

[43] The inhomogeneous boundary condition in (2.38) allows us to introduce another advantage of LSFEMs over, say, Galerkin and mixed-Galerkin finite element methods. Satisfying inhomogeneous boundary conditions by restricting the finite element space is a difficult and error-prone endeavor, not only for Galerkin finite element methods, but for LSFEMs as well. However, in the LSFEM context, we can treat $\phi - g$ as another equation residual and include it in the least-squares functional as is done in (2.39). Having done this, one does not have to restrict the finite element space to satisfy any boundary conditions. This general means for treating boundary conditions, even essential ones, is an important advantage of LSFEMs over both Galerkin and mixed-Galerkin methods. See Section 12.1.

[44] This is because definition (A.10) of the trace norm is not constructive. As a result, it cannot be computed by standard finite element means.

[45] In Section 12.1, we provide a brief discussion of how practical LSFEMs may be defined that account for boundary conditions in the least-squares functional.

making LSFEMs *practical*. Unfortunately, it is not always the case that LSFEMs based on norm-equivalent functionals are practical nor is it always the case that practical LSFEMs involve norm-equivalent functionals. As a result, the straightforward LSFEM of Section 2.2.1 has a limited range of applicability if one insists on the method being practical.

Clearly, if one wants to extend LSFEMs beyond those that are both practical and norm equivalent, some compromise has to be achieved. Ultimately, because we want LSFEMs to be competitive and even superior to other finite element methods, we do not want to give up on practicality. Instead, we look to generalize the notion of norm equivalence of least-squares functionals so that resulting LSFEMs are practical and at the same time the advantageous features of the Rayleigh–Ritz setting are still preserved. How such generalizations are accomplished is a central subject of this book. Here, we provide a taste of what is to come by exploring the conflict between norm equivalence and practicality in a little more detail.

Let us return to the example of (2.39), where impracticality is caused by the fractional order norm $\|\cdot\|_{1/2,\partial\Omega}$. In order to obtain a practical LSFEM, we may redefine this functional by using only appropriate $L^2(\Omega)$ norms of the residuals of all the equations in (2.38). Thus, we are led to the following *practical* version of (2.39):

$$J(\phi,\mathbf{v};f,g) = \|\nabla\cdot\mathbf{v}-f\|_0^2 + \|\mathbf{v}+\nabla\phi\|_0^2 + \|\phi-g\|_{0,\partial\Omega}^2 \tag{2.40}$$

along with the solution space $X = H^1(\Omega)\times H(\text{div};\Omega)$. Unfortunately, although this functional does indeed satisfy both of our keys to practicality, it is not norm equivalent. In other words, achieving practicality by the simple means of swapping impractical norms by practical ones comes with the cost of having to give up the norm equivalence of the least-squares functional, i.e., the property that was crucial to the full recovery of the Rayleigh–Ritz setting.

The easiest remedy for this conflict between norm equivalence and practicality is simply to omit, as in (2.36), the boundary equation residual in the functional and instead enforce the boundary condition on the trial finite element space; this is not always possible or convenient, especially for inhomogeneous boundary conditions.

Another remedy is to replace the fractional norm in (2.39) by a *mesh-dependent weighted* $L^2(\partial\Omega)$-norm so that we are led to the functional

$$J(\phi,\mathbf{v};f,g) = \|\nabla\cdot\mathbf{v}-f\|_0^2 + \|\mathbf{v}+\nabla\phi\|_0^2 + h^{-1}\|\phi-g\|_{0,\partial\Omega}^2 \tag{2.41}$$

that involves only practical $L^2(\partial\Omega)$ norms.[46] Of course, the functional (2.41) is not norm equivalent. However, it turns out that LSFEMs based on the functional (2.41) are not only practical, but also behave in every respect as does the straightforward LSFEM of Section 2.2.1.

[46] The choice of the weight in (2.41) is motivated by the proper scaling factor between $\|\cdot\|_{0,\partial\Omega}$ and $\|\cdot\|_{1/2,\partial\Omega}$ considered as norms over finite element spaces. As a result, the weighted functional (2.41) and its continuous prototype (2.39) are comparable only when restricted to finite element spaces.

In the case of (2.37), impracticality is caused by the minus one norm $\|\cdot\|_{-1}$. This situation parallels the one already encountered in Section 2.1.1 for which this norm rendered impractical a stabilized method for the Stokes equations. Of course, the same idea of replacing an impractical norm by a discrete equivalent is applicable in the present context as well. For example, we can use the *quasi-equivalent* approximation (B.91) of the negative norm to obtain the mesh-weighted functional[47]

$$J(\phi,\mathbf{v};f) = h^2\|\nabla\cdot\mathbf{v}-f\|_0^2 + \|\mathbf{v}+\nabla\phi\|_0^2. \tag{2.42}$$

Also, as mentioned in Section 2.1.1, instead of weighted $L^2(\Omega)$ norms, we could use any other available discrete replacement $\|\cdot\|_{-h}$ for the negative norm, including the norm-equivalent approximation from (B.99). Thus, in actuality, there is a whole class of discrete functionals

$$J(\phi,\mathbf{v};f) = \|\nabla\cdot\mathbf{v}-f\|_{-h}^2 + \|\mathbf{v}+\nabla\phi\|_0^2 \tag{2.43}$$

that can be associated with the same prototype functional (2.37). The many possibilities that exist in the choice of $\|\cdot\|_{-h}$ are yet another example of a factor that contributes to the "ambiguity" or "flexibility" of LSFEMs.

2.2.4 *Some Questions and Answers*

The ideas touched upon in Sections 2.2.1–2.2.3 form the nucleus of modern LFEMSs. We saw that the basic ingredients for the design of such methods consist of:

- a (quadratic, convex) least-squares functional that measures the size of the equation residuals in appropriate norms;
- a minimization principle for the least-squares functional;
- a discretization step in which one minimizes the functional over a finite element trial space.[48]

We also saw that the least-squares approach can be applied to any linear PDE problem so that, in theory, it is always possible to devise a finite element method based on an attractive Rayleigh–Ritz-type setting. This raises the first question.

When is the use of LSFEMs justified?

---

[47] The weight in this functional is motivated by the proper scaling factor between $\|\cdot\|_0$ and $\|\cdot\|_{-1}$ considered as norms over finite element spaces. As for (2.39) and (2.41), the weighted functional (2.42) and its prototype (2.37) are neither defined nor equivalent on the same Hilbert space. The prototype functional is meaningful for any $\mathbf{u}\in[L^2(\Omega)]^d$ but the discrete functional is not meaningful for functions $\mathbf{u}$ whose divergence does not belong to $L^2(\Omega)$.

[48] It is possible to reverse the order of these last two steps, i.e., to choose the trial space of candidate solutions first, then set up an *overdetermined* algebraic system by testing the candidate solution against a large number of test functions, and finally, define a least-squares functional for the resulting algebraic system. Such methods are traditionally referred to as *point-matching* or *collocation* least-squares methods and are in use in the engineering community; see [153] and Section 12.4.

The answer is quite obvious: the attractiveness of LSFEMs depends on the type of quasi-projection that can be associated with the Galerkin method. In particular, the appeal of a LSFEM increases with the deviation of the naturally occurring variational principle from the Rayleigh–Ritz setting. Thus, for example, if one needs to solve (1.52) only for ϕ, the standard finite element method already operates in the desirable Rayleigh–Ritz setting and there is no discernible advantage in the use of LSFEMs. However, the moment one wants to directly approximate not only ϕ, but also its gradient, or the moment that one wants to satisfy inhomogeneous essential boundary conditions, the situation changes completely. In these cases, LSFEMs become an attractive alternative to both Galerkin methods and mixed-Galerkin methods.

We also saw that there are many forks on the road to applying the LSFEM methodology to a given PDE problem, leading to many different possible settings for such methods. This gives rise to the next question.

Which least-squares setting is the "best" and how does one quantify this mathematically?

The answer to this question is also not difficult and has already been stated. Let us repeat it one more time: becuase we wish to create a Rayleigh–Ritz-like setting, the variational equation must correspond to a true inner-product projection. This is the same as saying that the least-squares functional must be norm equivalent in some function space. This criterion is now widely accepted (even though not always practiced) as the guiding principle in choosing a least-squares functional.

Having reached an agreement on the "best" choice of a least-squares principle and when to use it, we are immediately led to a considerably more daunting question.

Are the "best" LSFEMs, as dictated by analyses, also the ones that are most convenient to use in practice?

Our examples show that often the answer to this question is no because certain attributes of norm-equivalent functionals such as high-order derivatives and fractional and negative norms put them on a collision course with the practicality of the associated LSFEM. On the other hand, seemingly practical, naively defined functionals are not always norm equivalent.

The conflict between norm equivalence and practicality captures what represents the crux of the matter in the quest for viable and competitive LSFEMs.

How does one reconcile the "best" and the "most convenient" principles?

This question has been the focus of least-squares practitioners and analysts since the inception of modern LSFEMs. Although the two groups routinely differ in their interpretation of this question, their combined efforts have led to a number of tools that can help answer it and which are nowadays standard staples of modern LSFEMs. One example is the use of equivalent first-order systems[49] illustrated above

[49] Some authors use the terminology FOSLS (for *first-order system least squares*) to refer to such methods.

for the Poisson problem (2.33). Since its emergence in the late 1970s (see [162, 163, 227]), this idea has become a powerful and by now standard tool in LSFEM methodologies; see [95–100, 111–116, 119, 162–164, 167, 227, 228, 234, 237], among others. Weighted [18, 53, 54, 259] and discrete negative norms [36, 39, 68, 77–79], as well as LL*-LSFEMs [100, 258, 274] are other examples of tools and approaches intended to bridge the gap between "best" and "most convenient" principles.

These tools may help to define a "most convenient" functional, improve its norm-equivalence, or deal with less regular solutions; however, it is not immediately obvious that they do so in an efficient manner. This gives rise to the last question.

How competitive are LSFEMs compared to other discretization approaches?

In other words, if reconciling "best" with "most convenient" necessitates the use of sophisticated and complex tools, are we instead better off using an alternative, mathematically inferior but practically more simple functional, or perhaps even a Galerkin method? This question is neither trivial nor is its answer completely settled. A good example is the use of weighted versus discrete negative norms. The former are simpler but mathematically inferior in the sense that the weighted functionals are only quasi-norm-equivalent; see Section B.4.1. The latter are more complicated but mathematically superior because they lead to discretely norm-equivalent least-squares functionals; see Section B.4.2.

2.3 Putting Things in Perspective and What to Expect from the Book

By now, the reader is probably perplexed by the seemingly endless profusion of difficulties and remedies that seem to pile up in the development of LSFEMs. Even worse, she or he is probably under the impression that coming up with a good LSFEM is a hopelessly messy and chaotic process. To a casual observer, this impression can only be further sharpened by perusing even a small sample of the references cited so far and therein finding many and varied approaches with often conflicting lists of pros and cons.

So, before we venture any further, let us try to dispel this pessimistic notion and show that, despite their seemingly varied nature, almost all LSFEMs do actually fit into a relatively compact classification scheme. Once this is accomplished, the reader should have no problems in identifying the classes to which various methods belong, as well as their weaknesses and strengths.

Continuous and discrete least-squares principles

We can imagine the realm of least-squares methods as a tree whose roots represent the mathematical prerequisites for a least-squares principle to be a true inner-product projection; the trunk represents the class of all such least-squares principles and the branches are the LSFEMs related to these principles. In this analogy, the farther a branch is from the root, the weaker is its connection to the mathematical standards located at the roots of the tree.

The true inner-product projection property depends on the least-squares functional being norm equivalent; thus, norm equivalence is the root of our least-squares tree. In Section 3.2.2, we elaborate on the material of Section 2.2.1 to show that norm-equivalent functionals are engendered in a natural way by a priori estimates for the PDE problem; the *data spaces* indicated by the estimate provide the appropriate metrics for measuring the residual "energy" and the *solution spaces* provide the candidate minimizers, i.e., define the trial spaces. In this manner, one achieves a mathematically correct "energy balance" between minimizers and data. This identification leads to external "energy" functionals and *continuous least-squares principles* (CLSPs) that are in every way analogous to a Rayleigh–Ritz principle and that form the trunk of our least-squares tree; see Figure 2.1. CLSPs are the mathematically Utopian setting for LSFEMs because they provide true inner-product projections. If we were to restrict ourselves to this setting, the task of growing our least-squares tree would be completed by adding a *single major* branch representing *compliant DLSPs* that result from restricting the minimization of a given CLSP to a *conforming* finite element subspace.

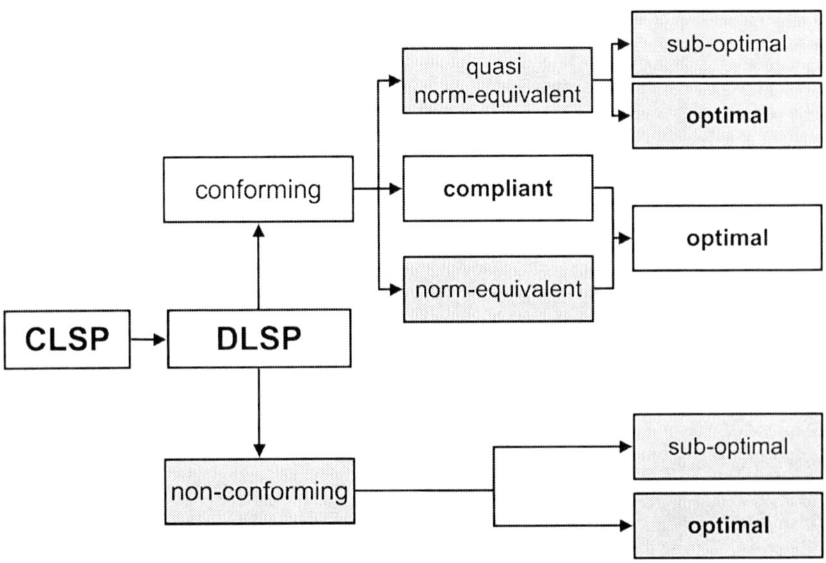

Fig. 2.1 The discrete least-squares principle tree. The path through the white boxes corresponds to LSFEMs that completely recover the Rayleigh-Ritz setting; any other path necessarily entails compromises required to make the LSFEMs practical.

In practice, one finds soon enough that many CLSPs are unusable because their straightforward restriction to conforming finite element subspaces results in LSFEMs that are in conflict with one or more of the practicality requirements. To make LSFEMs competitive, we cannot give up on practicality and so the remedy is either to enlarge[50] the CLSP class until it contains a satisfactory (practical) one, or to use non-compliant DLSPs having changed prototype functionals. In both cases, our tree grows some additional branches. Those corresponding to compliant methods obtained after an enlargement of the original CLSP class is added to the already existing major branch. However, the use of non-compliant transformation necessitates some restructuring of our tree.

Such transformations can lead to two possible outcomes. In the first scenario, we obtain a *non-conforming* DLSP, i.e., a problem for which the approximating space is *not a proper subspace* of the function space in its parent CLSP. To accommodate such DLSPs, we have to add a *second major* branch; see Figure 2.1. Non-conforming DLSPs are difficult to classify any further because their norm-equivalence properties cannot be directly compared to those of their parent CLSPs. Thus, the only possible further refinement to the non-conforming branch is to add the layers for *optimal* and *suboptimal* DLSPs.

In the second scenario, we have *conforming* but *non-compliant* DLSPs. To accommodate such DLSPs, we have to split the major conforming branch into further sub-branches that represent various non-compliant DLSPs. The sub-branches that are closest to the compliant layer consist of *norm-equivalent* DLSPs. Such principles use a discrete metric for the residual that reproduces the correct "energy balance" on a finite-dimensional space *without any dependence* on the parameters defining this space and its dimension. They represent the next best choice if a compliant method is unavailable or is not entirely usable, i.e., is impractical.

The final layer of branches consists of *quasi-norm-equivalent* DLSPs. In contrast to norm-equivalent DLSPs, these principles use a discrete metric for the residual that *cannot reproduce* the correct "energy balance" without dependence on the parameters defining this space, i.e., the equivalence bounds change with the dimension of the approximation space. Because of this dependence, standard elliptic arguments lead to suboptimal error estimates for this class of methods, which does not necessarily reflect their true behavior. In fact, it turns out that this branch contains some methods that are *capable of producing optimally convergent solutions* and other methods that are provably suboptimal. The latter class of DLSPs is the farthest from the root of the least-squares tree; see Figure 2.1.

Our classification scheme does not formally separate LSFEMs for which the least-squares step precedes the discretization step from those for which this ordering of these steps is reversed. The main reason is that the latter form of residual minimization can still be implicitly related to an existing CLSP, even though it is commonly derived directly from strong forms of PDE problems. As a result, the

[50] The enlargement of the CLSP class is accomplished by using equivalent reformulated problems. Often one gains additional tangible benefits such as being able to obtain direct approximations of physically relevant variables. Enlargement of the CLSP class may also call for lots of ingenuity and usually must be carried out on a case-by-case basis.

proper placement of the discretize-then-least-squares methods in the least-squares tree can be determined by comparing them with the implicit continuous principle.

At this point, the classification scheme represented by Figure 2.1 is largely an informal and intuitive construction. Not before Section 3.5 do we have enough background to make this scheme mathematically precise and provide formal definitions for each class of DLSPs.

What to expect from this book

Even though hardly complete, the list of issues raised above should give the reader a pretty good idea of the wide range of mathematical ideas, numerical tools, and algorithmic prowess that is involved in the design of a good LSFEM and, indeed, any successful numerical method. The goal of this book is to address as completely as possible the theoretical aspects of LSFEMs and also addressing, albeit in a less specific manner, their algorithmic design. This choice acknowledges the real and existing need to place the development of LSFEMs on a unified, solid mathematical foundation. It also reflects our philosophy that investment in mathematical analyses and the formal development of algorithms is worth the effort. Analyses clearly lead to a better understanding of the methods and the scope of their applicability. But analysis can have algorithmic benefits as well; it tells us how far one can push the limits of various "variational crimes" in the implementation process before suffering a major breakdown of an algorithm. In the subsequent chapters, we show that this philosophy is justified even though LSFEMs tend to be much more robust and less susceptible to variational crimes compared to other discretization schemes. In fact, this does not diminish the role of rigorous analysis but makes it even more important as the breakdowns of LSFEMs tend to be more subtle and less dramatic than with other methodologies.

Part II
Abstract Theory of Least-Squares Finite Element Methods

Chapter 3
Mathematical Foundations of Least-Squares Finite Element Methods

In Section 2.2, we introduced many of the ideas that form the core of modern least-squares finite element methods (LSFEMs). In this chapter, we develop a mathematical theory that makes precise the key ideas and provides a rigorous framework for the application of least-squares principles. At the center of our framework is an abstract least-squares theory for solving operator equations in Hilbert spaces. The specialization of this framework to partial differential equation (PDE) problems provides a template for LSFEMs that is used throughout the book.

Our theory emphasizes the use of least-squares principles as external variational formulations that replace the naturally occurring formulation of a given PDE problem. Consequently, we focus on methods for which the least-squares minimization step precedes the discretization step.[1] The fact that this approach is by far the most popular in practice further justifies our choice of emphasis.

The basic framework for solving operator equations by residual minimization is developed in Section 3.1, using the results collected in Appendix C. We apply the framework to an abstract PDE problem in Section 3.2 and, in Section 3.2.2, obtain the external, synthetic energy minimization principles that constitute the continuous least-squares principle (CLSP) class for the PDE problem.

The remainder of the chapter is devoted to the formulation and analysis of the discrete least-squares principles (DLSPs) that define least-squares finite element approximations of the solution of the PDE problem. We present the theory in two stages. The first stage (see Section 3.3) examines what can be expected from a discrete residual minimization principle if *no connection* to a CLSP class is assumed. This stage not only helps to explain the remarkable robustness of LSFEMs, but also reveals the limitations of this very general setting. In the second stage, the analysis is extended to include DLSPs obtained from a CLSP class associated with the PDE

[1] Methods for which discretization precedes least-squares minimization can also fit into our framework by linking them to a companion continuous least-squares principle. For example, a discrete least-squares principle based on collocating the PDE at Gauss points and then applying an algebraic least-squares principle to the resulting discrete equations can be viewed as resulting from the approximation of integrals in some CLSP by a quadrature rule. This process can be realized in many different ways; examples of such methods are provided in Section 12.4.

P.B. Bochev, M.D. Gunzburger, *Least-Squares Finite Element Methods*, Applied Mathematical Sciences 166, DOI: 10.1007/b13382_3,

problem. In Section 3.4, we examine the transformation of a CLSP into a DLSP and show that this process consists of choosing approximate *norm-generating* and *differential* operators.[2] In Section 3.5, the three basic types of DLSPs, e.g., *compliant*, *norm-equivalent*, and *quasi-norm-equivalent*, are shown to result from specific approximation choices. Using the link between CLSPs and DLSPs, we develop there an approximation theory for least-squares finite element approximations of solutions of PDE problems.

Throughout this chapter, we use the notation established in Appendix C.

3.1 Least-Squares Principles for Linear Operator Equations in Hilbert Spaces

Given two Hilbert spaces X and Y, a Fredholm operator[3] $\mathcal{Q} \in L(X,Y)$, and a function $f \in Y$, consider the operator equation

$$\textit{find} \quad u \in X \quad \textit{such that} \quad \mathcal{Q}u = f\,. \tag{3.1}$$

In this section, we develop and analyze variational methods that "solve" (3.1) by minimizing some norm of the *residual* $\mathcal{R}v = \mathcal{Q}v - f$. The quotation marks indicate that, depending on the deficiency and the nullity of $\mathcal{Q}$, the problem (3.1) may or may not be solvable for arbitrary $f \in Y$ or may have multiple solutions. On the other hand, it turns out that a properly formulated residual minimization principle, i.e., a least-squares based-method, always has a unique minimizer so that the sense in which this minimizer "solves" (3.1) must be made clear. To this end, we show that a properly formulated least-squares principle recovers the *exact* solution of one of the "nearby" *auxiliary* operator equations defined in Appendix C.[4]

The key to constructing a Rayleigh–Ritz setting for (3.1) is to find a solution space S and a data space H such that the energy balance (C.2) holds, possibly after a redefinition of the operator $\mathcal{Q}$. Then, the least-squares functional

$$J(v;f) = \|\mathcal{R}v\|_H^2 = \|\mathcal{Q}v - f\|_H^2 \tag{3.2}$$

is *norm equivalent*, i.e., it satisfies

$$C_1\|v\|_S \le J(v;0) \le C_2\|v\|_S \qquad \forall\, v \in S \tag{3.3}$$

for some positive constants C_1 and C_2 having values independent of v. Thus, we see that the estimates in (C.2) for the operator equation (3.1) constitute the fundamental prerequisites for the norm-equivalence of the functional $J(\cdot;\cdot)$.

---

[2] Concrete examples of discrete operators and norms that are useful in LSFEMs are provided in Sections B.3.1 and B.4.

[3] See Appendix C for the definition of a Fredholm operator and of $L(X,Y)$.

[4] From a regularization perspective, this means that least-squares principles provide *regularization by selection*; see [337].

Positive nullity and positive deficiency have different impacts on least-squares principles. The former affects the norm equivalence of least-squares functionals and can be dealt with by changing the solution and data spaces to something other than the standard choices. The latter requires us to find a nearby problem whose solution coincides with the minimizer of (3.2). However, in both cases, we are led to consider auxiliary operator equations in lieu of (3.1). The energy balances developed in Section C.2 for those equations provide us with all the necessary tools to formulate least-squares principles for (3.1) and to interpret their minimizers as "solutions" of that abstract operator equation. We first consider problems with trivial null spaces and then proceed to the general case of problems with positive nullity.

3.1.1 Problems with Zero Nullity

If $\mathrm{null}(\mathcal{Q}) = 0$, then the spaces X^N, X^C, and $X^\perp$ defined in Section C.1 coincide with the standard space X and (C.9) defaults to the original setting for $\mathcal{Q}$. For simplicity, we write X instead of $\check{X}$. In this case, the energy balance required to formulate a least-squares principle for (3.1) is given by (C.13). This balance gives rise to the external, residual energy functional

$$J(u;f) = \|\mathcal{Q}u - f\|_Y^2 \tag{3.4}$$

and the least-squares principle for (3.1)

$$\min_{u\in X} J(u;f). \tag{3.5}$$

With (3.4), we associate the "energy" inner product

$$((u,v)) = (\mathcal{Q}u, \mathcal{Q}v)_Y \tag{3.6}$$

and "energy" norm

$$|||v||| = J(v;0)^{1/2} = (\mathcal{Q}v, \mathcal{Q}v)_Y^{1/2}. \tag{3.7}$$

The following theorem shows that (3.5) is a well-posed problem.

Theorem 3.1 *Assume that* $\mathrm{null}(\mathcal{Q}) = 0$*. Then, the least-squares minimization problem* (3.5) *has a unique minimizer* $u_{LS} \in X$ *for any* $f \in Y$*.*

Proof. The first-order necessary condition for (3.5) is the variational problem

$$\textit{seek } u \in X \textit{ such that} \quad Q(u;v) = F(v) \qquad \forall v \in X, \tag{3.8}$$

where

$$Q(u;v) = ((u,v)) \qquad \text{and} \qquad F(v) = (\mathcal{Q}v, f)_Y.$$

From (C.13), it follows that the least-squares functional (3.4) is norm equivalent and that (3.6) is an equivalent inner product on X. Therefore, $Q(\cdot;\cdot)$ is a continuous and strongly coercive bilinear form on $X \times X$; it is also not difficult to show that $F(\cdot)$ is a continuous linear functional. As a result, the unique solvability of the problem (3.8) follows from Corollary 1.4 (the Lax–Milgram lemma). □

If, in addition to having a trivial null space, $\mathcal{Q}$ also has zero deficiency, then it is clear that u_{LS} is also a solution of the operator equation (3.1). However, if $\mathcal{Q}$ has positive deficiency, then (3.1) is not solvable unless $f \in R(\mathcal{Q})$ whereas, according to Theorem 3.1, the least-squares principle (3.5) has a unique minimizer u_{LS} for *any* $f \in Y$.

The following theorem shows that u_{LS} solves the auxiliary equation (C.16). This makes it possible to interpret u_{LS} as a "solution" of (3.1), even when f does not belong to the range of $\mathcal{Q}$.

Theorem 3.2 *Assume that* $\text{null}(\mathcal{Q}) = 0$ *and let* $u_{LS} \in X$ *denote the unique minimizer of* (3.5). *Furthermore, for a given* $f \in Y$, *let*[5] $\vec{a} = (f, \vec{v})_V$. *Then, the pair* $\{u_{LS}, \vec{a}\}$ *is the unique solution of the modified problem* (C.16).

Proof. From Lemma C.8, we know that (C.16) has a unique solution $\{u, \vec{a}\}$, where $\vec{a}$ is as in the statement of that lemma. Consider now a least-squares principle for (C.16) derived from the auxiliary operator (C.15). The energy balance for this operator, given by (C.19), gives rise to the minimization problem

$$\min_{\{u,\vec{a}\} \in X \times \mathbb{R}^{K^*}} \|\mathcal{Q}u + \vec{v} \cdot \vec{a} - f\|_Y^2 . \tag{3.9}$$

Thanks to (C.19), the functional in (3.9) is norm equivalent. Therefore, the arguments from the proof of Theorem 3.1 can be invoked to show that (3.9) has a unique minimizer. But the unique solution $\{u, \vec{a}\}$ of the auxiliary problem is also a minimizer of (3.9) so that they must coincide. Now that we have established this fact, the theorem follows if we can also prove that u that solves (3.9) is also solution of the minimization problem (3.5), i.e., that $u = u_{LS}$.

To this end, consider the first-order optimality condition for (3.9): *seek* $\{u, \vec{a}\} \in X \times \mathbb{R}^{K^*}$ *such that*

$$Q(\{u, \vec{a}\}; \{w, \vec{c}\}) = F(\{w, \vec{c}\}) \qquad \forall \{w, \vec{c}\} \in X \times \mathbb{R}^{K^*}, \tag{3.10}$$

where

$$Q(\{u, \vec{a}\}; \{w, \vec{c}\}) = (\mathcal{Q}u + \vec{v} \cdot \vec{a}, \mathcal{Q}w + \vec{v} \cdot \vec{c})_Y \quad \text{and} \quad F(\{w, \vec{c}\}) = (f, \mathcal{Q}w + \vec{v} \cdot \vec{c})_Y .$$

The unique solution of (3.10) is, of course, the unique solution $\{u, \vec{a}\}$ of the auxiliary problem. Because $\{v_i\}_{1=1}^{K^*}$ is a (finite) basis for the co-range, we have that $(\mathcal{Q}w, v_i)_V = 0$ for $i = 1, \ldots, K^*$ and any $w \in X$. This simplifies (3.10) to

$$(\mathcal{Q}u, \mathcal{Q}w)_Y + (\vec{v} \cdot \vec{a}, \vec{v} \cdot \vec{c})_Y = (f, \mathcal{Q}w + \vec{v} \cdot \vec{c})_Y .$$

Moving all co-range terms to the right-hand side yields

[5] $\vec{v} = (v_1, \ldots, v_{K^*})^T$ is a basis for the co-range of $\mathcal{Q}$; see Section C.1.

$$\begin{aligned}(\mathcal{Q}u,\mathcal{Q}w)_Y &= (f,\mathcal{Q}w+\vec{v}\cdot\vec{c})_Y - (\vec{v}\cdot\vec{a},\vec{v}\cdot\vec{c})_Y \\ &= (f,\mathcal{Q}w)_Y + (f-\vec{v}\cdot\vec{a},\vec{v}\cdot\vec{c})_Y = (f,\mathcal{Q}w)_Y \qquad \forall w\in X\,.\end{aligned}$$

The last identity follows from the definition of $\vec{a}$ that implies that $f-\vec{v}\cdot\vec{a}\in R(\mathcal{Q})$. Therefore, (3.10) is reduced to exactly the same variational equation as in (3.8) and we can conclude that $u = u_{LS}$. □

Corollary 3.3 *Assume that* $\text{null}(\mathcal{Q})=0$ *and let* $u_{LS}\in X$ *denote the unique minimizer of* (3.4).

1. *If* $f\in R(\mathcal{Q})$, *then* u_{LS} *is also a solution of the operator equation* (3.1).
2. *If the data belong to the co-range, i.e., if* $f\in span\{v_i\}_{i=1}^{K^*}$, *then* $u_{LS}\equiv 0$.
3. *If the data are neither entirely in the range nor the co-range, i.e.,* $f\notin span\{v_i\}_{i=1}^{K^*}$ *and* $|\vec{a}|\neq 0$, *then the pair* $\{u_{LS},\vec{a}\}$ *solves* (C.16). □

This corollary shows that least-squares principles for (3.1) always select the "best" possible solution for any given right-hand side f in the sense that u_{LS} coincides with the unique solution of the equation

$$\mathcal{Q}u = f^{\perp},$$

where $f^{\perp}$ is the orthogonal projection of f onto $R(\mathcal{Q})$. Thus, in the ideal case for which $f\in R(\mathcal{Q})$, the least-squares principle simply recovers the solution of (3.1). In the extreme case when f is in the co-range, minimization of (3.5) returns zero. In intermediate situations, the least-squares solution u_{LS} is a function that is mapped to the part of f that belongs to $R(\mathcal{Q})$.[6]

The fact that least-squares principles have a unique, well-defined minimizer, even when the data fail to be compatible, is undoubtedly a valuable computational advantage of the approach. Note that least-squares principles offer this advantage "for free" because the computation of u_{LS} does not require either knowledge of the co-range basis $\vec{v}$ or the use of the auxiliary problem (C.16).

3.1.2 Problems with Positive Nullity

When $\mathcal{Q}$ has positive nullity, the setting of (C.9) involves non-standard spaces that are more difficult to approximate. In this case, it is preferable to work in the bijective setting provided by (C.10) which uses only the standard space X. The relevant energy balance for (C.10) is given by (C.14) and leads to the following residual energy functional:

---

[6] The results of Corollary 3.3 also mean that least-squares principles can handle, in a natural way, data errors introduced by discretization. For instance, a perturbation of f in (3.1) caused by, e.g., approximation, may turn a formerly solvable equation into one with incompatible data f^h; see [67] for an example. Some variational methods break down under these circumstances; however, least-squares principles automatically generate a solution $\{u_{LS},\vec{a}\}$ in which u_{LS} matches the part of f^h that is in the range of $\mathcal{Q}$ and $\vec{a}$ is proportional to the approximation error $f-f^h$.

$$J(u;f,\vec{c}) = \|\mathcal{Q}u - f\|_Y^2 + |\vec{\ell}(u) - \vec{c}|^2 \tag{3.11}$$

and the least-squares principle for (3.1)

$$\min_{u \in X} J(u;f,\vec{c})\,. \tag{3.12}$$

The "energy" inner product and norm associated with (3.11) are

$$((u,w)) = (\mathcal{Q}u, \mathcal{Q}w)_Y + \vec{\ell}(u) \cdot \vec{\ell}(w) \tag{3.13}$$

and

$$|||u||| = J(u;0,\vec{0})^{1/2} = ((u,u))^{1/2}\,, \tag{3.14}$$

respectively. It follows from the energy balance (C.14) that the functional (3.11) is norm equivalent; as a result, (3.13) is an equivalent inner product on X. The following theorem extends the results of Theorem 3.1 to the present case.

Theorem 3.4 *The least-squares minimization problem* (3.12) *has a unique minimizer $u_{LS} \in X$ for any $f \in Y$ and $\vec{c} \in \mathbb{R}^K$. Moreover, $\vec{\ell}(u_{LS}) = \vec{c}$.*

Proof. The existence and uniqueness of the least-squares minimizer u_{LS} follows from the Lax–Milgram lemma (Corollary 1.4) along the same lines as in the proof of Theorem 3.1.

To prove the second part of the theorem, note that u_{LS} satisfies the first-order optimality condition

$$(\mathcal{Q}u_{LS}, \mathcal{Q}w)_Y + \vec{\ell}(u_{LS}) \cdot \vec{\ell}(w) = (f, \mathcal{Q}w)_Y + \vec{c} \cdot \vec{\ell}(w) \qquad \forall w \in X\,. \tag{3.15}$$

Testing with the basis $\{u_k\}_{k=1}^K$ of the null space $N(\mathcal{Q})$ reduces (3.15) to the algebraic equation

$$\vec{\ell}(u_{LS}) \cdot \vec{\ell}(u_k) = \vec{c} \cdot \vec{\ell}(u_k) \qquad \text{for } k = 1, \dots, K$$

that, in terms of the matrix L defined in (C.6), takes the form

$$\mathsf{L}^T \vec{\ell}(u_{LS}) = \mathsf{L}^T \vec{c}\,.$$

By assumption, the ℓ_k's are such that L is nonsingular so that $\vec{\ell}(u_{LS}) = \vec{c}$. □

The interpretation of u_{LS} as a solution to (3.1) can be derived from the second auxiliary problem (C.18).

Theorem 3.5 *Let $f \in Y$ and $\vec{c} \in \mathbb{R}^K$ be given and let u_{LS} be the unique minimizer of* (3.12)*. Then, the pair $\{u_{LS}, \vec{a}\}$, where $\vec{a} = (f, \vec{v})_Y$, solves the auxiliary problem* (C.18).

Proof. According to Lemma C.9, the problem (C.18) has a unique solution $\{u, \vec{a}\}$ for any $f \in Y$ and $\vec{c} \in \mathbb{R}^K$. In this solution, $\vec{a}$ is the same as in the statement of the theorem and $\vec{\ell}(u) = \vec{c}$. From Theorem 3.4, we already know that $\vec{\ell}(u_{LS}) = \vec{c}$, i.e., u_{LS} satisfies the second equation in the auxiliary problem. Thus, it remains to show that $\{u_{LS}, \vec{a}\}$ also satisfies the first equation in (C.18). With slight modifications, the proof in Theorem 3.2 can be used to this end. □

3.2 Application to Partial Differential Equations

In this section, we specialize the abstract least-squares theory given in Section 3.1 to operator equations (3.1) that represent PDE problems. In what follows, $\mathcal{L}(\mathbf{x}, D)$ is a linear differential operator that acts on functions u, defined on a bounded open region $\Omega \subset \mathbb{R}^d$, and $\mathcal{B}(\mathbf{x}, D)$ is a linear operator acting on functions u defined on the boundary $\partial\Omega$ of Ω.

For simplicity, we often write $\mathcal{L}u$ and $\mathcal{B}u$ whenever the meaning of these symbols is clear from the context. It is also convenient to separate the data space into the two spaces $Y = Y(\Omega)$ and $B = B(\partial\Omega)$ corresponding to the data for the PDE and boundary condition, respectively. Then, (3.1) specializes to the following boundary value problem: given $f \in Y(\Omega)$ and $g \in B(\partial\Omega)$, find $u \in X = X(\Omega)$ such that

$$\begin{cases} \mathcal{L}u = f & \text{in } Y(\Omega) \\ \mathcal{B}u = g & \text{in } B(\partial\Omega). \end{cases} \tag{3.16}$$

Remark 3.6 The concrete forms of $\mathcal{L}$ and $\mathcal{B}$ depend on the arrangement of the dependent variables in u. One possibility is to divide u into scalar and/or vector fields corresponding to various physical quantities modeled by the variables, e.g., currents, fluxes, concentrations, or potentials. Alternatively, one can view u as a vector comprising of the scalar coordinate functions of the physical fields relative to some coordinate system. Of course, this only changes the appearance of the PDE problem (3.16) but not the problem itself, nor its solution.

For an example, consider the first-order Poisson system (1.55) in $\mathbb{R}^2$ endowed with Cartesian coordinates $\{x, y\}$. The variables in that system can be arranged as a pair $u = \{\phi, \mathbf{v}\}$ of a scalar field and a vector field, or as a triple $u = \{\phi, v_1, v_2\}$ of scalar functions, where v_1 and v_2 denote the components of $\mathbf{v}$. The forms of $\mathcal{L}(\mathbf{x}, D)$ corresponding to these two ways of expressing the variables are given by[7]

$$\mathcal{L} = \begin{pmatrix} 0 & \nabla\cdot \\ \nabla & \mathcal{I} \end{pmatrix} \quad \text{and} \quad \mathcal{L} = \begin{pmatrix} 0 & \partial_x & \partial_y \\ \partial_x & 1 & 0 \\ \partial_y & 0 & 1 \end{pmatrix},$$

respectively. In the first case $\mathcal{L}$ is a matrix of coordinate-independent differential operators, whereas in the second case $\mathcal{L}$ is a matrix of partial derivatives with respect to the assumed Cartesian coordinate system. We have more to say about these two viewpoints in the subsequent chapters. □

[7] ∂_x denotes $\partial/\partial x$.

To formulate least-squares principles for (3.16) and interpret their minimizers as "solutions" of that boundary value problem, we apply the template developed for the abstract equation (3.1). Recall that the energy balances in Section C.2 are the key to defining well-posed residual minimization problems and that the auxiliary problems in Section C.1 are the key to interpreting their minimizers as solutions of (3.1). Therefore, we begin by specializing the results of those sections to (3.16).

To apply the abstract theory of Section 3.1, it is necessary to make the following assumption about $\mathcal{L}$ and $\mathcal{B}$ which we tacitly assume holds throughout this chapter.

Assumption 3.7 *There exist Hilbert spaces* $X = X(\Omega)$, $Y = Y(\Omega)$, *and* $B = B(\partial\Omega)$ *such that the mapping* $\mathcal{Q} : X \mapsto Y \times B$ *defined by* $u \mapsto \{\mathcal{L}u, \mathcal{B}u\}$ *is a Fredholm operator.*[8] □

We retain the same notation for the finite-dimensional basis of the null space so that we write $N\{\mathcal{L},\mathcal{B}\} = \text{span}\{u_1, u_2, \ldots, u_K\}$, where $u_k \in X$. If $\mathcal{Q} = \{\mathcal{L},\mathcal{B}\}$ has a nontrivial co-range, a basis for the co-range consists of K^* linearly independent functions $v_i = \{r_i, b_i\} \in Y \times B$ such that

$$(\{\mathcal{L}u, \mathcal{B}u\}, \{r_i, b_i\})_{Y\times B} = (\mathcal{L}u, r_i)_Y + (\mathcal{B}u, b_i)_B = 0 \qquad \text{for } i = 1, 2, \ldots, K^* .$$

We set $\vec{v} = (\vec{r}, \vec{b})^\top \in \mathbb{R}^{2K^*}$, where $\vec{r} = (r_1, \ldots, r_{K^*})^\top \in \mathbb{R}^{K^*}$ and $\vec{b} = (b_1, \ldots, b_{K^*})^\top \in \mathbb{R}^{K^*}$.

3.2.1 Energy Balances

Whenever the boundary value problem (3.16) has a unique solution, i.e., the null space of $\{\mathcal{L},\mathcal{B}\}$ is trivial, well-posed least-squares principles can be defined without any modifications to the spaces and operators. When $\{\mathcal{L},\mathcal{B}\}$ has positive nullity, we use the bijective setting provided by the augmented operator

$$\{\mathcal{L},\mathcal{B},\vec{\ell}\} : X \mapsto Y \times B \times \mathbb{R}^K; \qquad \{\mathcal{L},\mathcal{B},\vec{\ell}\}u = \begin{pmatrix} \mathcal{L}u \\ \mathcal{B}u \\ \vec{\ell}(u) \end{pmatrix} . \tag{3.17}$$

The following theorem is a direct consequence of Theorem C.7.[9]

Theorem 3.8 *If* $\{\mathcal{L},\mathcal{B}\}$ *has zero nullity, there exist positive constants* C_1 *and* C_2 *such that*

$$C_1\|u\|_X \le \|\mathcal{L}u\|_Y + \|\mathcal{B}u\|_B \le C_2\|u\|_X \qquad \forall\, u \in X . \tag{3.18}$$

If $\{\mathcal{L},\mathcal{B}\}$ *has finite nullity K, then*

$$C_1\|u\|_X \le \|\mathcal{L}u\|_Y + \|\mathcal{B}u\|_B + |\vec{\ell}(u)| \le C_2\|u\|_X \qquad \forall\, u \in X . \qquad \square \tag{3.19}$$

---

[8] In the sequel we simply write X, Y, and B instead of $X(\Omega)$, $Y(\Omega)$, and $B(\partial\Omega)$.

[9] Recall that throughout this chapter, we assume that Assumption 3.7 holds.

Specialization of the auxiliary problems of Section C.1 to (3.16) is straightforward. For operators with trivial null spaces, the auxiliary problem is

$$\begin{cases} \mathcal{L}u + \vec{r}\cdot\vec{a} = f & \text{in } \Omega \\ \mathcal{B}u + \vec{b}\cdot\vec{a} = g & \text{on } \partial\Omega \end{cases} \tag{3.20}$$

and for operators with positive nullity, that problem is

$$\begin{cases} \mathcal{L}u + \vec{r}\cdot\vec{a} = f & \text{in } \Omega \\ \mathcal{B}u + \vec{b}\cdot\vec{a} = g & \text{on } \partial\Omega \\ \vec{\ell}(u) = \vec{c} & \text{in } \mathbb{R}^K. \end{cases} \tag{3.21}$$

The following theorem is a direct consequence of Lemma C.9 and Theorem C.10.

Theorem 3.9 *Let* $K^* = \text{def}\{\mathcal{L},\mathcal{B}\}$ *and* $K = \text{null}\{\mathcal{L},\mathcal{B}\}$. *The problem* (3.20) *has a unique solution* $\{u,\vec{a}\} \in X \times \mathbb{R}^{K^*}$ *for any* $\{f,g\} \in Y \times B$. *The problem* (3.21) *has a unique solution* $\{u,\vec{a}\} \in X \times \mathbb{R}^{K^*}$ *for any* $\{f,g,\vec{c}\} \in Y \times B \times \mathbb{R}^K$. *Moreover, there exist positive constants* C_1 *and* C_2 *such that*

$$C_1\big(\|u\|_X + |\vec{a}|\big) \le \|\mathcal{L}u + \vec{r}\cdot\vec{a}\|_Y + \|\mathcal{B}u + \vec{b}\cdot\vec{a}\|_B \le C_2\big(\|u\|_X + |\vec{a}|\big) \tag{3.22}$$

for (3.20) *and*

$$C_1\big(\|u\|_X + |\vec{a}|\big) \le \|\mathcal{L}u + \vec{r}\cdot\vec{a}\|_Y + \|\mathcal{B}u + \vec{b}\cdot\vec{a}\|_B + |\vec{\ell}(u)| \le C_2\big(\|u\|_X + |\vec{a}|\big) \tag{3.23}$$

for (3.21). □

3.2.2 Continuous Least-Squares Principles

In this section, we use residual minimization to develop external, *least-squares* variational formulations for (3.16). These formulations replace the naturally occurring and/or formal Galerkin variational principles as the basis for developing finite element methods for this boundary value problem. Because the least-squares principles considered in this section operate in infinite dimensional Hilbert spaces, we refer to them as *continuous least-squares principles* (CLSPs), a terminology that is first encountered in Section 2.3.

For boundary value problems with zero nullity, the fundamental energy balance is given by (3.18) which leads to the residual energy functional

$$J(u;f,g) = \|\mathcal{L}u - f\|_Y^2 + \|\mathcal{B}u - g\|_B^2 \tag{3.24}$$

and the corresponding CLSP

$$\min_{u\in X} J(u;f,g). \tag{3.25}$$

If $\{\mathcal{L},\mathcal{B}\}$ has positive nullity, then the relevant energy balance is given by (3.19), the residual energy functional is given by

$$J(u;f,g,\vec{c}) = \|\mathcal{L}u - f\|_Y^2 + \|\mathcal{B}u - g\|_B^2 + |\vec{\ell}(u) - \vec{c}|^2\,, \tag{3.26}$$

and the CLSP is given by

$$\min_{u\in X} J(u;f,g,\vec{c})\,. \tag{3.27}$$

The energy inner products corresponding to (3.25) and (3.27) are given by

$$((u,w)) = (\mathcal{L}u,\mathcal{L}w)_Y + (\mathcal{B}u,\mathcal{B}w)_B \tag{3.28}$$

and

$$((u,w)) = (\mathcal{L}u,\mathcal{L}w)_Y + (\mathcal{B}u,\mathcal{B}w)_B + \vec{\ell}(u)\cdot\vec{\ell}(w)\,, \tag{3.29}$$

respectively. In both cases, the energy norm is

$$|||u||| = J(u;0,\dots,0)^{1/2} = ((u,u))^{1/2}\,,$$

where J is one of the functionals (3.24) or (3.26). The following theorem reveals the relationship between the minimizers of problems (3.25) and (3.27) and the solutions of the auxiliary boundary value problems (3.20) and (3.21), respectively.

Theorem 3.10 *Assume that* $\{\mathcal{L},\mathcal{B}\}$ *has a trivial kernel. Then,* (3.25) *has a unique minimizer* $u_{LS}\in X$ *for any* $\{f,g\}\in Y\times B$. *The pair* $\{u_{LS},\vec{a}\}$, *where* $\vec{a} = (f,\vec{r})_Y + (g,\vec{b})_B$, *solves the auxiliary problem* (3.20). *If* $\{\mathcal{L},\mathcal{B}\}$ *has positive nullity,* (3.27) *has a unique minimizer* u_{LS} *for any* $\{f,g,\vec{c}\}\in Y\times B\times\mathbb{R}^K$. *The pair* $\{u_{LS},\vec{a}\}$ *solves the auxiliary problem* (3.21). *In particular,* $\vec{\ell}(u_{LS}) = \vec{c}$. □

We omit the proof of this theorem as it is basically a repetition of the arguments found in the proofs of the abstract results in Section 3.1. Specialization of the statement of Corollary 3.3 to least-squares principles for the boundary value problem (3.16) is also straightforward. The key point here is that the main message of Section 3.1 carries over unchanged to the present setting: if (3.16) has a unique solution $u\in X$, then it is recovered by the least-squares principle; otherwise, the unique least-squares minimizer solves an auxiliary problem.

Remark 3.11 We have just described a formal procedure for defining external variational formulations for the boundary value problem (3.16) that provide a Rayleigh–Ritz-like variational setting for a finite element method. These formulations are based on least-squares residual minimization and are unrelated and independent of any naturally occurring variational formulations for (3.16). □

Remark 3.12 Because least-squares principles are external to the PDE problem, the associated weak problem

$$\textit{seek } u\in X \textit{ such that} \quad Q(u;v) = F(v) \qquad \forall v\in X \tag{3.30}$$

is not a Galerkin weak formulation of (3.16). For example, assuming that (3.30) is the Euler–Lagrange equation corresponding to (3.24) and that $\mathcal{L}$ is such that the Green's formula[10]

$$(u,\mathcal{L}v)_Y - \langle \mathcal{L}_Y^* u, v\rangle_\Omega = \langle \mathcal{B}_Y^* u, v\rangle_{\partial\Omega} \tag{3.31}$$

holds, where $\langle\cdot,\cdot\rangle$ denotes an appropriate duality pairing, then smooth solutions of (3.30) are not directly solutions of (3.16); instead, they solve the *strong* PDE

$$\mathcal{L}_Y^*\mathcal{L}u = \mathcal{L}^* f \qquad \text{in } \Omega \tag{3.32}$$

along with the essential boundary condition

$$\mathcal{B}u = g \qquad \text{on } \partial\Omega \tag{3.33}$$

and the natural boundary condition

$$\mathcal{B}_Y^*\mathcal{L}u = \mathcal{B}_Y^* f \qquad \text{on } \partial\Omega\,. \tag{3.34}$$

The system (3.32)–(3.34) forms the boundary value problem for which the least-squares functional (3.24) is the naturally occurring convex quadratic, energy functional that provides the Rayleigh–Ritz setting.[11] In other words, the strong problem (3.32)–(3.34) is the PDE problem whose weak Galerkin formulation coincides with the *least-squares* variational problem (3.30). Thus, it is conceivable to develop a least-squares principle for (3.16) by immersion[12] of these equations into the appropriate strong least-squares PDE problem followed by a standard Galerkin procedure. Of course, this is hardly the most efficient or lucid way to proceed. □

Remark 3.13 To provide a final bird's-eye view of the least-squares framework, let us revert again to formal operator notation with the understanding that $\mathcal{Q}$ is one of $\{\mathcal{L},\mathcal{B}\}$ or $\{\mathcal{L},\mathcal{B},\vec{\ell}\}$. By $D(\mathcal{Q})$ and $R(\mathcal{Q})$ we denote the appropriate domain and range, respectively. The starting point in the development of continuous least-squares principles is the *energy balance*

$$C_1\|u\|_{D(\mathcal{Q})} \le \|\mathcal{Q}u\|_{R(\mathcal{Q})} \le C_2\|u\|_{D(\mathcal{Q})} \tag{3.35}$$

that gave rise to the residual energy functional

$$J(u;f) = \|\mathcal{Q}u - f\|^2_{R(\mathcal{Q})}, \tag{3.36}$$

the least-squares minimization principle

$$\min_{u\in D(\mathcal{Q})} J(u;f), \tag{3.37}$$

[10] Note that $\mathcal{L}_Y^*$ coincides with the standard adjoint of $\mathcal{L}$ only if $Y \equiv L^2(\Omega)$; in general, $\mathcal{L}_Y^*$ is the (Hilbert space) adjoint of $\mathcal{L}$ with respect to the inner product on Y.

[11] The system (3.32)–(3.34) can be viewed as the normal equations for the original system (3.16).

[12] In general, the problem (3.32)–(3.34) can be determined from (3.16) thorough differentiation and linear combinations that account for the norm structure of Y.

the energy norm

$$|||u||| = J(u;0)^{1/2}, \tag{3.38}$$

and the energy inner product

$$((u,v)) = (\mathcal{Q}u, \mathcal{Q}v)_{R(\mathcal{Q})}. \tag{3.39}$$

The fact that (3.36) was derived from the energy balance (3.35) made certain that (3.38) and (3.39) are an equivalent norm and inner product, respectively. This was essential in proving that (3.37) has a unique minimizer because the bilinear form $Q(\cdot;\cdot)$ appearing in

$$\textit{seek } u \in D(\mathcal{Q}) \textit{ such that } \quad Q(u;w) = F(w) \qquad \forall\, w \in D(\mathcal{Q}) \tag{3.40}$$

that is the Euler–Lagrange equation corresponding to (3.37) turned out to be identical to the energy inner product (3.39). It is clear that the process described so far establishes

> *a mathematical framework that associates a well-posed unconstrained minimization problem with any linear PDE problem that satisfies Assumption* 3.7.

Minimization problems constructed through this process are completely defined by the *function space X* and the *norm-equivalent functional* $J(\cdot)$ being minimized over that space, i.e., the pair $\{J,X\}$. The set of all such pairs that can be associated with a given PDE problem constitutes the class of its *continuous least-squares principles* (CLSPs).[13] □

Remark 3.14 As a final note, before we move on to DLSPs, let us mention that the abstract CLSPs developed in this section did not assume any specific connection between the PDE and the spaces X, Y, and B, except that they verify Assumption 3.7. In Section 12.2, we briefly describe an approach, referred to as LL* least-squares methods, in which least-squares principles are defined using norms induced by the differential operator $\mathcal{L}$ itself. □

3.3 General Discrete Least-Squares Principles

CLSPs[14] $\{J,X\}$ offer an alternative, external variational formulation of (3.16). By using $\{J,X\}$ instead of the naturally occurring and/or formal Galerkin formula-

---

[13] As we have already mentioned in Section 2.2, for some PDE problems, stability properties are more naturally stated using norms in Banach spaces. Formal application of residual minimization in a Banach space setting gives rise to unconstrained optimization problems with non-differentiable functionals. Examples of such "least-squares" formulations and strategies for their solution are considered in Chapter 10.

[14] $\{J,X\}$ is shorthand for the principle $\min_{u\in X} J(u)$. In fact, because the pair $\{J,X\}$ completely defines the least-square principle, from now on we state least-squares principles by providing this pair and not writing, in every instance, $\min_{u\in X} J(u)$.

tions for (3.16), we can develop finite element methods for this problem using a Rayleigh–Ritz like setting *even if the partial differential equation problem itself is not associated with an unconstrained minimization problem for a convex quadratic functional.*

LSFEMs for which the least-squares step precedes the discretization step have the necessary prerequisite of identifying a CLSP class for the given boundary value problem, i.e., to find the data and solution spaces that verify (3.35). For some problems, this task may be far from trivial and, as we saw in Section 2.2, the CLSP class can include impractical norms and inner products. These issues, critical to the formulation of LSFEMs, are dealt with in Section 3.4; here, we take some time to explore what happens if one decides not to bother with CLSPs and instead defines *discrete least-squares principles* (DLSPs) directly.

A general DLSP for (3.16) is a parameterized family of unconstrained minimization problems

$$\textit{seek } u^h \textit{ in } X^h \textit{ such that} \quad J^h(u^h; f,g) \le J^h(w^h; f,g) \qquad \forall\, w^h \in X^h\,, \tag{3.41}$$

where X^h is a finite dimensional space, parameterized by h, whose dimension is proportional to h^{-d} and $J^h : X^h \mapsto \mathbb{R}$ is a convex functional. Following the established shorthand notation, we denote DLSPs by $\{J^h, X^h\}$.

To guarantee that $\{J^h, X^h\}$ has a unique minimizer that, in some sense, approximates solutions of (3.16), it is necessary to make the following two nonrestrictive assumptions.

Assumption 3.15 There exists a *discrete energy inner product*

$$((\cdot,\cdot))_h \,:\, X^h \times X^h \mapsto \mathbb{R} \tag{3.42}$$

and a *discrete energy norm* $|||\cdot|||_h = ((\cdot,\cdot))_h^{1/2}$ such that

$$J^h(u^h; 0,0) = ((u^h,u^h))_h = |||u^h|||_h^2 \qquad \forall\, u^h \in X^h\,. \quad \square \tag{3.43}$$

Assumption 3.16 The discrete energy norm can be extended to all smooth functions $u \in X$. Furthermore, there exist positive semi-definite bilinear forms $\mathrm{e}(\cdot,\cdot)$ and[15] $\varepsilon(\cdot,\cdot)$ such that, for all $u^h \in X^h$ and for all smooth functions $u \in X$,

$$J^h(u^h; \mathcal{L}u, \mathcal{B}u) = \frac{1}{2}\Big(((u^h,u^h))_h + ((u,u))_h + \varepsilon(u,u)\Big) - ((u,u^h))_h - \mathrm{e}(u,u^h)\,. \;\square \tag{3.44}$$

The first assumption guarantees that $\{J^h, X^h\}$ is a well-posed minimization problem with a unique minimizer out of X^h. The second assumption links the DLSP with the PDE problem that we are trying to solve.

---

[15] The term $\varepsilon(u,u)$ in (3.44) has no effect on the minimization of $J^h(u^h;\cdot,\cdot)$; it merely adjusts the minimum value of the functional. It is introduced so that the definitions of discrete least-squares functionals are of a relatively simple form.

3.3.1 Error Analysis

The following theorem shows that if the value of the *truncation error* $\mathrm{e}(u,u^h)$ is "small," then the minimization of J^h selects a function $u^h \in X^h$ that is "close" to the exact solution u of the PDE problem.

Theorem 3.17 *Assume that Assumptions* 3.15 *and* 3.16 *hold for the DLSP* $\{J^h, X^h\}$ *and let* $u \in X$ *denote a sufficiently smooth solution of* (3.16). *Then, the discrete problem* (3.41) *has a unique minimizer* $u^h \in X^h$. *Moreover, that minimizer satisfies the error estimate*

$$|||u-u^h|||_h \le \inf_{w^h \in X^h} |||u-w^h|||_h + \sup_{w^h \in X^h} \frac{\mathrm{e}(u,w^h)}{|||w^h|||_h}, \tag{3.45}$$

i.e., up to a truncation error term, u^h *is the orthogonal projection of* u *with respect to the discrete "energy" inner product* (3.42).

Proof. Using (3.42) and (3.44), it is easy to see that the first-order necessary condition that minimizers of (3.41) satisfy is given by

$$\textit{seek } u^h \textit{ in } X^h \quad \textit{such that} \quad Q^h(u^h;w^h) = F^h(w^h) \qquad \forall\, w^h \in X^h, \tag{3.46}$$

where

$$Q^h(\cdot;\cdot) = ((\cdot,\cdot))_h \qquad \text{and} \qquad F^h(\cdot) = ((u,\cdot))_h + \mathrm{e}(u,\cdot).$$

To show the unique solvability of (3.46), let $\{\phi_j^h\}_{j=1}^J$ denote a basis for X^h and $\vec{w} = (w_1,\dots,w_J)^T$ denote the coefficient vector of a function $w^h \in X^h$ with respect to this basis so that $w^h = \sum_{j=1}^J \vec{w}_j \phi_j^h$. Define the matrix A and the vector $\vec{f}$ by

$$\mathsf{A}_{ij} = ((\phi_j^h,\phi_i^h))_h \qquad \text{and} \qquad \vec{f}_i = ((u,\phi_i^h))_h + \mathrm{e}(u,\phi_i^h), \tag{3.47}$$

respectively. Then, (3.46) is equivalent to the linear system

$$\mathsf{A}\vec{u} = \vec{f} \tag{3.48}$$

of algebraic equations for the unknown coefficient vector $\vec{u}$ corresponding to u^h. From Assumption 3.15, it follows that A is a Gramm matrix corresponding to the basis $\{\phi_j^h\}_{j=1}^J$ relative to $((\cdot,\cdot))_h$ so that A is symmetric and positive definite, the system (3.48) has a unique solution $\vec{u}$, and (3.41) has a unique minimizer $u^h = \sum_{j=1}^J \vec{u}_j \phi_j^h$.

To prove the error estimate (3.45), let $u_\perp^h$ denote the orthogonal projection of u onto X^h with respect to the discrete energy inner product $((\cdot,\cdot))_h$. Using the triangle inequality, we obtain

$$|||u^h-u|||_h \le |||u_\perp^h - u|||_h + |||u^h-u_\perp^h|||_h \le \inf_{w^h \in X^h} |||w^h-u|||_h + |||u^h-u_\perp^h|||_h.$$

To complete the proof, it is necessary to show that $|||u^h - u^h_\perp|||_h$ is bounded by the truncation error. Because u^h solves (3.46),

$$((u^h - u, w^h))_h = \mathrm{e}(u, w^h) \qquad \forall w^h \in X^h$$

and because $u^h_\perp$ is an orthogonal projection,

$$((u^h_\perp - u, w^h))_h = 0 \qquad \forall w^h \in X^h .$$

Subtracting the last two equations yields

$$((u^h_\perp - u^h, w^h))_h = ((u - u^h, w^h))_h = \mathrm{e}(u, w^h) \qquad \forall w^h \in X^h . \tag{3.49}$$

The error estimate (3.45) then follows from

$$|||u^h_\perp - u^h|||_h = \sup_{w^h \in X^h} \frac{((u^h_\perp - u^h, w^h))_h}{|||w^h|||_h} = \sup_{w^h \in X^h} \frac{\mathrm{e}(u, w^h)}{|||w^h|||_h} . \qquad \square$$

Definition 3.18 The discrete least-squares functional J^h is *order r-consistent*, or simply *r-consistent*, if there exists a positive number r such that for all sufficiently smooth functions $u \in X$,

$$\sup_{w^h \in X^h} \frac{\mathrm{e}(u, w^h)}{|||w^h|||_h} \le C(u) h^r , \tag{3.50}$$

where $C(u)$ is a positive number whose value may depend on u but not on h. A discrete least-squares functional is called *consistent* if the truncation error is zero, i.e., (3.44) holds with $\mathrm{e}(\cdot,\cdot) \equiv 0$.[16] $\square$

We then have the following result.

Corollary 3.19 *Let the hypotheses of Theorem* 3.17 *hold. Let $u^h \in X^h$ denote the unique solution of the DLSP $\{J^h, X^h\}$, i.e., of* (3.41). *If J^h is order r-consistent, then*

$$|||u - u^h|||_h \le \inf_{w^h \in X^h} |||u - w^h|||_h + C(u) h^r . \tag{3.51}$$

If $\{J^h, X^h\}$ is consistent, then its solution u^h minimizes the discrete energy norm error, i.e.,

---

[16] If J^h is a consistent least-squares functional,

$$J^h(u, \mathcal{L}u, \mathcal{B}u) = \frac{1}{2} \varepsilon(u, u)$$

for any sufficiently smooth solution u of (3.16). On the other hand, given that $X^h \subset X$, it is clear that

$$J^h(u, f, g) \le J^h(u^h, f, g) .$$

Therefore, if $\varepsilon(u, u) = 0$, consistent least-squares functionals provide natural error monitors that can be used in mesh refinement algorithms. See Section 12.14.

$$|||u-u^h|||_h = \inf_{w^h \in X^h} |||u-w^h|||_h\,. \tag{3.52}$$

Proof. The proof follows from (3.45) by using (3.50) in the r-consistent case and by noting that $\mathrm{e}(\cdot,\cdot)\equiv 0$ for a consistent principle. □

Corollary 3.19 implies that minimizers of r-consistent functionals are orthogonal projections of the exact solutions up to $O(h^r)$ terms, and that minimizers of consistent functionals are true orthogonal projections of these solutions.

3.3.2 The Need for Continuous Least-Squares Principles

Theorem 3.17 shows that reasonable DLSPs can be defined under a minimal set of assumptions. This explains why LSFEMs tend to be much more robust than their mixed and formal Galerkin brethren and why even naively defined least-squares principles rarely fail in a spectacular way. Indeed, chances are that a naively defined mixed or a formal Galerkin method almost surely violate the delicate variational compatibility conditions in (1.41) and (1.42) or (1.33) and (1.34). On the other hand, one must be extremely unlucky to stumble upon a pair $\{J^h, X^h\}$ that does not satisfy Assumptions 3.15 and 3.16.

However, the fact that least-squares principles can produce reasonable results assuming only a loose connection with the PDE problem leads to the erroneous (and precarious) conclusion that LSFEMs can be safely developed using Assumptions 3.15 and 3.16 as the sole guidelines. The fallacy of this thinking is that neither Theorem 3.17 nor Corollary 3.19 can give us any further information about the asymptotic behavior of the error or the condition number of the least-squares algebraic problem (3.48). Let us conduct a "thought experiment" that shows how things can go wrong if we do not establish a tighter bond between a DLSP $\{J^h, X^h\}$ and a well-posed artificial energy principle $\{J, X\}$.

Let $\{J, X\}$ denote a well-posed CLSP for (3.16) and assume that $\{J^h, X^h\}$ is a DLSP for the same problem that satisfies Assumptions 3.15 and 3.16. Assume that both the continuous energy norm $|||\cdot|||$ and the norm $\|\cdot\|_X$ are meaningful for $u^h \in X^h$ so that their restrictions are well-defined norms on X^h. The discrete energy norm $|||\cdot|||_h$ is another norm on this finite-dimensional space so that it must be equivalent to the restrictions of $|||\cdot|||$ and $\|\cdot\|_X$ to X^h. As a result, for every fixed $h>0$, there exist constants $C_1(h)$ and $C_2(h)$ such that

$$C_1(h)\|u^h\|_X \le |||u^h|||_h \le C_2(h)\|u^h\|_X\,.$$

This explains the remarkable robustness of least-squares principles; *a version of the correct energy balance continues to hold for any fixed h.* Likewise, there are two other constants $\delta_1(h)$ and $\delta_2(h)$ such that

$$\delta_1(h)|\vec{u}|^2 \le \vec{u}^T \mathsf{A}\vec{u} \le \delta_2(h)|\vec{u}|^2$$

for the coefficient vector $\vec{u}$ of u^h. This is the whole story as far as an individual, fixed minimization problem from $\{J^h, X^h\}$ is concerned.

However, for a *parameterized family* of problems, what matters is the asymptotic behavior of these "constants" that depends entirely on the relation between the *discrete* and *continuous* energy norms; neither Assumptions 3.15 or 3.16 can control the growth (or decay) of $C_i(h)$ and $\delta_i(h)$. As a result, although each individual minimization problem may appear perfectly adequate, the family as a whole may experience a decline in convergence rates and/or sharp increase in the condition number of A as h becomes smaller and smaller.

A LSFEM for which these ratios deteriorate asymptotically may continue to perform well and without apparent indication of a failure. One reason is that realistic values of h used in practice are often not "small" enough for the deterioration of $C_2(h)/C_1(h)$ to be noticeable. Deterioration of $\delta_2(h)/\delta_1(h)$, on the other hand, tends to be more noticeable, especially when iterative methods are used to solve the linear systems. Users of such LSFEMs who have tried to switch from direct to iterative solvers have experienced a steep growth of iterations as the problem size increases.

The conclusion from this thought experiment is unambiguous: despite the initial appeal of simple conditions such as Assumptions 3.15 and 3.16, they alone cannot guarantee robust and efficient LSFEMs. To accomplish this, a DLSP $\{J^h, X^h\}$ must "mimic" a well-posed CLSP $\{J, X\}$ for (3.16). The level of association between discrete and continuous principles and how it affects the properties of $\{J^h, X^h\}$ are the subject of the next section.

3.4 Binding Discrete Least-Squares Principles to Partial Differential Equations

In defining a DLSP for a given PDE problem, two steps are taken; one must define a least-squares principle and one must discretize the problem, i.e., transform it from a infinite-dimensional to a finite-dimensional problem. The order in which one takes the discretization and least-squares steps corresponds to two fundamentally different ways of associating PDEs with DLSPs.

In the *discretize and then minimize* approach, the PDE is first discretized, i.e., replaced by a possibly overdetermined system of algebraic equations. This system is then solved by an algebraic least-squares method.[17] The lack of a clear connection with a well-posed residual energy principle for (3.16) is a serious drawback of the discretize and then minimize strategy. In general, LSFEMs obtained by this approach are not easily amenable to stability and error analysis beyond that which is already established in Theorem 3.17 and Corollary 3.19.[18] For this reason, the

[17] A typical example is collocation LSFEMs in which the initial reduction of the PDE to an overdetermined linear system is accomplished by collocation. This and other examples are considered in Section 12.4.

[18] In some special cases, e.g., when collocation points coincide with quadrature nodes, a connection can be developed and exploited in the analysis.

discretize and then minimize approach is not pursued in this book, other than the brief discussion in Section 12.4.

Instead, our main focus continues to be on methods where the least-squares step precedes the discretization step. Such *minimize and then discretize* approaches maintain a close connection between CLSPs and DLSPs and result in a much more satisfying mathematical structure and also in methods that are more easily amenable to rigorous stability and error analyses. The main prerequisite for the success of these methods is the careful execution of the discretization step, i.e., the transition from a CLSP $\{J,X\}$ to a DLSP $\{J^h,X^h\}$.

3.4.1 Transformations from Continuous to Discrete Least-Squares Principles

The transformation of a CSLP $\{J,X\}$ into a DLSP $\{J^h,X^h\}$ should ultimately be guided by practicality considerations. In an ideal situation, that transformation would only entail restriction of the minimization process to a subspace X^h of X, i.e., we would thus obtain the DLSP $\{J,X^h\}$.[19] In less ideal situations, a practical DLSP $\{J^h,X^h\}$ may differ substantially from its continuous prototype. In either case, properties, both attractive and unattractive, of DLSPs depend on their deviation from the energy balance prescribed by $\{J,X\}$. Thus, to evaluate the outcome of a particular transition to a discrete principle, we must be able to asses how well J^h mimics the norm equivalence of J.

To facilitate this task, we rewrite the energy balance of (3.16) using norm-generating operators for the data and the solution spaces. Then, the transition to a discrete principle can be regarded as a process wherein norm-generating and problem-defining operators are replaced by discrete approximations. This viewpoint allows one to easily track the impact that different approximation choices for the original operators have on the resulting DLSP. In particular, it is shown below that depending what choice one makes, the transformation of CLSPs to DLSPs can take three distinct routes that lead to three different categories of DLSPs.

To formalize the discussion, we focus on problems with trivial null spaces. This setting is sufficient because norm equivalence is unaffected by the terms used to remove the null spaces. We consider the problem (3.16) and assume that the energy balance (3.18) holds for some spaces X, Y, and B.

Let $\mathcal{S}_{(*)}$ for $* \in \{X,Y,B\}$ denote norm-generating operators for X, Y, and B, respectively, with $L^2(\Omega)$ acting as a pivot space, i.e., we have that

$$\|u\|_X = \|\mathcal{S}_X u\|_0, \qquad \|w\|_Y = \|\mathcal{S}_Y w\|_0, \qquad \text{and} \qquad \|b\|_B = \|\mathcal{S}_B b\|_{0,\partial\Omega}\,.$$

Using these operators, the energy balance (3.18) takes the form

$$C_1\|\mathcal{S}_X u\|_0 \le \|\mathcal{S}_Y \circ \mathcal{L}u\|_0 + \|\mathcal{S}_B \circ \mathcal{B}u\|_{0,\partial\Omega} \le C_2\|\mathcal{S}_X u\|_0\,. \tag{3.53}$$

[19] This is exactly the "straightforward" LSFEM discussed in Section 2.2.1.

Norm-generating operators[20] depend on the spaces involved in (3.18). Because (3.16) may admit multiple energy balances, we treat $\mathcal{S}_{(*)}$ as a family of operators that represent all admissible space combinations in (3.18). In the setting relevant to LSFEMs, X, Y, and B belong to Hilbert scales induced by $L^2(\Omega)$ or $L^2(\partial\Omega)$ and a generating, self-adjoint, positive definite operator; see, e.g., [249]. As a result, each family $\mathcal{S}_{(*)}$ can be identified with selected powers of the generating operators for these Hilbert scales. Consequently, $\mathcal{S}_{(*)}$ is a family of self-adjoint, positive definite operators.

Armed with the operator form (3.53) of the energy balance (3.18), we are now prepared to talk about transitions to DLSPs. To make the discussion as straightforward as possible, it is convenient to make the following assumption.

Assumption 3.20 A practical subspace[21] X^h of X that satisfies the *approximability assumption* (B.8) exists for the CLSP $\{J,X\}$.

This assumption means that the rest of this chapter targets primarily *conforming* LSFEMs, i.e., we have that $X^h \subset X$.[22]

In terms of norm-generating operators, Assumption 3.20 means that $\mathcal{S}_X$ is such that its domain $D(\mathcal{S}_X)$ contains "practical" discrete subspaces. As a result,

> *the practicality of the CSLP $\{J,X\}$ depends solely on the effort required to compute $\mathcal{S}_Y \circ \mathcal{L}u^h$ and $\mathcal{S}_B \circ \mathcal{B}u^h$.*

If this effort is deemed reasonable, the original energy norm

$$|||u||| = \|\mathcal{S}_Y \circ \mathcal{L}u\|_0 + \|\mathcal{S}_B \circ \mathcal{B}u\|_{0,\partial\Omega}$$

can be retained and the transition process is complete. Otherwise, we proceed to replace the composite operators $\mathcal{S}_Y \circ \mathcal{L}$ and $\mathcal{S}_B \circ \mathcal{B}$ by computable discrete approximations $\mathcal{S}_Y^h \circ \mathcal{L}^h$ and $\mathcal{S}_B^h \circ \mathcal{B}^h$, respectively. Finally, we may need *projection* operators π_Ω^h and $\pi_{\partial\Omega}^h$ that act on the data f and g so as to place them in the domains of $\mathcal{S}_Y^h$ and $\mathcal{S}_B^h$, respectively. In both cases, the conversion process and the key properties of the resulting DSLP can be encoded by the *transition* diagram

[20] Concrete examples of norm-generating operators are given at the end of this section.

[21] Recall that a practical subspace is one for which basis functions are easily constructed, i.e., with no more difficulty than one would encounter for Galerkin and mixed-Galerkin finite element methods for the same problem.

[22] Conformity of LSFEMs is not a crippling restriction because, for a large number of PDEs, Assumption 3.20 can be fulfilled and conforming least-squares methods can be defined through a judicious choice of $\{J,X\}$, including a possible reformulation of (3.16) into an equivalent problem, e.g., a first-order system. One notable exception are PDEs whose solution space is the intersection of the spaces $\mathbf{C}(\Omega)$ and $\mathbf{D}(\Omega)$ (see Appendix B for definition of these spaces). In Section B.2.2, we explain why, in this case, conforming finite element subspaces are unsatisfactory and why non-conforming approximations are preferable. Typical representatives of such PDEs are the div–curl systems that are considered in Section 6.3. There, and also in Section 7.7, we encounter examples of *non-conforming* LSFEMs for which $X^h \not\subset X$. Analysis of these methods relies on specific properties of X^h which makes it less amenable to generalizations. This is why the bulk of the abstract least-squares theory developed in this chapter deals with conforming LSFEMs.

$$
\begin{array}{ccccccc}
J(u;f,g) & = & \|\mathcal{S}_Y\circ(\mathcal{L}u-f)\|_0^2 & + & \|\mathcal{S}_B\circ(\mathcal{B}u-g)\|_0^2 & \rightarrow & |||u||| \\
\downarrow & & \downarrow & & \downarrow & & \downarrow \\
J^h(u^h;f,g) & = & \|\mathcal{S}_Y^h\circ(\mathcal{L}^h u^h-\pi_\Omega^h f)\|_0 & + & \|\mathcal{S}_B^h\circ(\mathcal{B}^h u^h-\pi_{\partial\Omega}^h g)\|_0 & \rightarrow & |||u^h|||_h
\end{array}
\tag{3.54}
$$

and the companion *norm-equivalence* diagram[23,24]

$$
\begin{array}{ccccc}
C_1\|u\|_X & \le & |||u||| & \le & C_2\|u\|_X \\
\downarrow & & \downarrow & & \downarrow \\
C_1(h)\|u^h\|_X & \le & |||u^h|||_h & \le & C_2(h)\|u^h\|_X\,.
\end{array}
\tag{3.55}
$$

Remark 3.21 The scope of (3.54) obviously extends to non-conforming LSFEMs for which $X^h \not\subset X$. However, for such methods nothing much can be said without further information about the discrete spaces and operators involved. In particular, for non-conforming LSFEMs, we do not have the companion diagram (3.55) because $\|\cdot\|_X$ may not be defined for $u^h \in X^h$. The absence of such a diagram makes it more difficult to estimate the level of "norm equivalence" of non-conforming methods because they lack the clear reference point provided by the upper row in (3.55) that is available in the conforming case. □

The truncation error of discrete least-squares functionals obtained by (3.54) is easily quantifiable.

Lemma 3.22 *Assume that* J^h *is transformed from* J *according to* (3.54). *Then, for all sufficiently smooth* $u \in X$,

$$
\begin{aligned}
\mathrm{e}(u,w^h) = &-\big(\mathcal{S}_Y^h\circ(\pi_\Omega^h\mathcal{L}-\mathcal{L}^h)u\,,\;\mathcal{S}_Y^h\circ\mathcal{L}^h w^h\big)_{0,\Omega}\\
&-\big(\mathcal{S}_B^h\circ(\pi_{\partial\Omega}^h\mathcal{B}-\mathcal{B}^h)u\,,\;\mathcal{S}_B^h\circ\mathcal{B}^h w^h\big)_{0,\partial\Omega}
\end{aligned}
\tag{3.56}
$$

and

$$
\varepsilon(u,u) = \frac{1}{2}\|\mathcal{S}_Y^h\circ(\pi_\Omega^h\mathcal{L}-\mathcal{L}^h)u\|_{0,\Omega}^2 + \frac{1}{2}\|\mathcal{S}_B^h\circ(\pi_{\partial\Omega}^h\mathcal{B}-\mathcal{B}^h)u\|_{0,\partial\Omega}^2\,. \quad \square \tag{3.57}
$$

---

[23] Together, (3.54) and (3.55) represent a useful and important tool for understanding LSFEMs. In addition to providing a systematic process for *deriving* DLSPs, they can also be used to establish a reverse *association* between a given DLSP and a continuous prototype. As a tool for the design of LSFEMs, (3.54) and (3.55) highlight (in combination with (3.53)) the different roles played by $\mathcal{L}$, $\mathcal{B}$, and $\mathcal{S}_Y$, $\mathcal{S}_B$, provide guidelines for choosing approximations compatible with these roles, and allow one to assess the impact of these choices. On the other hand, the possibility to always relate DLSPs to well-posed CLSPs is convenient when trying to assess the qualities of DLSPs defined in an ad hoc manner.

[24] In principle, a reverse association with a well-posed CLSP $\{J,X\}$ may be sought for any DLSP, including those obtained by the discretize and then minimize approach, so as to enable their variational analysis. However, that process entails many more ambiguities compared to reverse associations for DLSPs obtained by the minimize and then discretize approach.

The proof of this lemma is straightforward.

$\mathcal{L}$ and $\mathcal{B}$ define the problem that is being solved so that the main objective is to choose $\mathcal{L}^h$ and $\mathcal{B}^h$ that bind $\{J^h, X^h\}$ to this problem, i.e., that make J^h as close as possible to J for the exact solution of (3.16). An appropriate choice is to use operators that lead to truncation errors of order r in (3.44), i.e., $\mathcal{L}^h$ and $\mathcal{B}^h$ are such that (3.50) holds for some positive r.

On the other hand, $\mathcal{S}_Y$ and $\mathcal{S}_B$ define the energy balance of (3.16), i.e., the proper scaling between data and solution. As a result, the main objective in the choice of $\mathcal{S}_Y^h$ and $\mathcal{S}_B^h$ is to ensure that the scaling induced by J^h is as close as possible to (3.18), i.e., to "bind" $\{J^h, X^h\}$ to the energy balance of $\{J, X\}$.

To obtain robust and efficient DLSPs, approximations of the problem-defining operators and the norm-generating operators must work in concert to provide good accuracy and reasonable norm equivalence for $\{J^h, X^h\}$. The scaling between discrete data and solution spaces

$$C_1(h)\|u^h\|_X \leq |||u^h|||_h \leq C_2(h)\|u^h\|_X \tag{3.58}$$

induced by $\{J^h, X^h\}$ depends on the discrete energy norm

$$|||u^h|||_h = \left(\|\mathcal{S}_Y^h \circ \mathcal{L}^h u^h\|_0^2 + \|\mathcal{S}_B^h \circ \mathcal{B}^h u^h\|_{0,\partial\Omega}^2\right)^{1/2}$$

which in turn depends on the choices made for $\mathcal{S}_Y^h$, $\mathcal{S}_B^h$, $\mathcal{L}^h$, and $\mathcal{B}^h$. In general, the upper and lower bounds in (3.58) are mesh-dependent and this may affect performance of DLSPs in several ways; see Section 3.3.2. The severity of the performance drop as measured by the decrease in convergence rates and/or increase in condition numbers can serve as a precise measure for the deviation of $\{J^h, X^h\}$ from the optimal energy balance. Based on how well $|||\cdot|||_h$ represents this balance, DLSPs can be divided into three distinct categories. These categories, their properties, and their formal error analysis are presented in Section 3.5.

Examples of norm-generating operators

The following examples illustrate the notion of norm-generating operators.

Example 3.23 The Laplace operator $-\Delta$ with homogeneous boundary conditions is self-adjoint and positive definite. As a result, all powers of this operator are defined and we have that

$$\|\phi\|_{-1} = \|(-\Delta)^{-1/2}\phi\|_0, \quad \|\phi\|_0 = \|(-\Delta)^0\phi\|_0, \quad \text{and} \quad |\phi|_1 = \|(-\Delta)^{1/2}\phi\|_0 .$$

For proof of the first identity, see Theorem A.1. □

Example 3.24 The first-order Poisson equation (1.55) admits the following energy balance[25] (see Theorem 5.9):

$$C(\|\phi\|_1 + \|\mathbf{v}\|_0) \le \|\nabla\cdot\mathbf{v}\|_{-1} + \|\nabla\phi + \mathbf{v}\|_0 .$$

Using the characterizations from Example 3.23, we can rewrite this balance as

$$C\Big(\big(\|(-\Delta)^0\phi\|_0^2 + \|(-\Delta)^{-1/2}\phi\|_0^2\big)^{1/2} + \|(-\Delta)^0\mathbf{v}\|_0\Big)$$
$$\le \|(-\Delta)^{-1/2}\nabla\cdot\mathbf{v}\|_0 + \|(-\Delta)^0(\nabla\phi + \mathbf{v})\|_0 .$$

Therefore, the operator $\mathcal{S}_X$ which acts on the solution $\{\phi,\mathbf{v}\}$ and the operator $\mathcal{S}_Y$ which acts on the data can be identified with

$$\mathcal{S}_X = \Big((-\Delta)^0 + (-\Delta)^{1/2}\,,\,(-\Delta)^0\Big)^T \qquad \text{and} \qquad \mathcal{S}_Y = \Big((-\Delta)^{-1/2}\,,\,(-\Delta)^0\Big)^T,$$

respectively.

3.5 Taxonomy of Conforming Discrete Least-Squares Principles and their Analysis

Broadly speaking, there are two fundamental types of LSFEMs. LSFEMs for which Assumption 3.20 is satisfied belong to the class of conforming finite element methods. For such methods, restriction of their parent CLSP to the conforming discrete subspace $X^h \subset X$ (which may or may not be practical), provides a natural reference point that allows one to assess the level of deviation of conforming LSFEMs from the ideal Rayleigh–Ritz setting. This leads to the three distinct subclasses of conforming LSFEMs discussed in this section for which it is possible to develop an abstract approximation theory under a reasonable set of assumptions.

The second fundamental type consists of non-conforming LSFEMs, i.e., methods for which $X^h \not\subset X$. In this case, solution norms from X cannot be applied to functions in X^h so that we lack an obvious reference point to estimate the deviation from the ideal Rayleigh–Ritz setting; see Remark 3.21. For this reason, further subdivision of non-conforming LSFEMs, based on their "norm equivalence," and an abstract approximation theory is not attempted; we deal with non-conforming LSFEMs on a case by case basis.

Apart from the ideal *compliant* class for which LSFEMs reproduce the classical Rayleigh–Ritz principle, there are two other kinds of DLSPs that gradually drift away from this setting, primarily by *simplifying the approximations* of the norm-generating operators. The *norm-equivalent* class retains virtually all attractive prop-

[25] Throughout the book we often state only the lower bound in the energy balance because, as a rule, it is much harder to prove than the upper bound. The latter in most cases follows directly from the triangle inequality.

erties of the Rayleigh–Ritz setting, including identical convergence rates and matrix condition numbers. The *quasi-norm-equivalent* class admits the broadest range of DLSPs.[26] However, the generality of this class also makes its analysis much more complex and involved, and, for this reason, we present it last.

In what follows, we assume that the dependent variables are arranged as $u = \{u_1, u_2, \ldots, u_M\}$ and that $\mathcal{L}$ and $\mathcal{B}$ are in a form that corresponds to that arrangement (see Remark 3.6), i.e., $\mathcal{L}$ and $\mathcal{B}$ are given by

$$\mathcal{L} = \begin{pmatrix} \mathcal{L}_{11} & \cdots & \mathcal{L}_{1M} \\ \vdots & \vdots & \vdots \\ \mathcal{L}_{M1} & \cdots & \mathcal{L}_{MM} \end{pmatrix} \quad \text{and} \quad \mathcal{B} = \begin{pmatrix} \mathcal{B}_{11} & \cdots & \mathcal{B}_{1M} \\ \vdots & \vdots & \vdots \\ \mathcal{B}_{L1} & \cdots & \mathcal{B}_{LM} \end{pmatrix},$$

where $\mathcal{L}_{ij}$ and $\mathcal{B}_{ij}$ are differential operators acting on Ω and $\partial\Omega$, respectively. We use X_i to denote the component space for the ith element[27] of u, i.e., $X = X_1 \times \cdots \times X_M$. Likewise, $Y = Y_1 \times \cdots \times Y_M$ and $B = B_1 \times \cdots \times B_L$, where Y_j and B_l are the data spaces for the jth differential equation and the lth boundary condition in (3.16).

Some additional hypotheses about X^h are necessary to discuss the least-squares classes. First, X^h is expected to satisfy the following inverse assumptions.

Assumption 3.25 There exist non-negative weights $\omega_1, \ldots, \omega_M$ such that, for all components $u_i^h \in X_i^h$, $i = 1, \ldots, M$, of $u^h \in X^h = X_1^h \times X_2^h \times \cdots \times X_M^h$, either

$$\|u_i^h\|_0 \le \|u_i^h\|_{X_i} \le C_I h^{-\omega_i} \|u_i^h\|_0 \tag{3.59}$$

or

$$\|u_i^h\|_0 \ge \|u_i^h\|_{X_i} \ge C_I h^{\omega_i} \|u_i^h\|_0 . \quad \Box \tag{3.60}$$

We write (3.59) and (3.60) in the compact forms

$$\|u^h\|_0 \le \|u^h\|_X \le C_I h^{-\vec{\omega}} \|u^h\|_0$$

and

$$\|u^h\|_0 \ge \|u^h\|_X \ge C_I h^{\vec{\omega}} \|u^h\|_0 ,$$

respectively, where $\vec{\omega} = (\omega_1, \ldots, \omega_M)^\top$ and $h^{\pm\vec{\omega}} = (h^{\pm\omega_1}, \ldots, h^{\pm\omega_M})^\top$. We define $\omega_{max} = \max\{\omega_j\}$.

The second assumption deals with the relation between the $L^2(\Omega)$ norm of a finite element function u^h and the Euclidean norm of the corresponding vector of coefficients $\vec{u}$.

Assumption 3.26 There exist positive constants μ_1 and μ_2 such that, for all $u^h \in X^h$,

$$\mu_1 h^d |\vec{u}|^2 \le \|u^h\|_0^2 \le \mu_2 h^d |\vec{u}|^2 , \tag{3.61}$$

[26] In fact, the quasi-norm-equivalent class contains the compliant and norm-equivalent classes as special cases.

[27] Recall that each element u_i can be a scalar or a vector field.

where d denotes the space dimension. □

Both (3.59) and (3.61) hold for a wide range of finite element spaces under some mild restrictions on the partition of the domain Ω into finite elements; see Theorems B.26 and B.27 in Section B.3.4.

In what follows, we use u, u^h, and w^h to denote a solution of (3.16), its least-squares approximation out of X^h, and an arbitrary element of X^h, respectively. For simplicity, we continue to restrict attention to problems with trivial null spaces for which the relevant energy balance is (3.18). Also for simplicity, we consider only homogeneous boundary conditions and assume that X and X^h are constrained by the boundary condition $\mathcal{B}u = 0$.

Our last assumption deals with the possibility that the discrete operator $S_Y^h \circ \mathcal{L}^h$ may not necessarily be defined for all functions in X.

Assumption 3.27 There exist subspaces $\widetilde{X} \subseteq X$ and $\widetilde{Y} \subseteq Y$ such that $S_Y^h \circ \mathcal{L}^h u$ is defined for every $u \in \widetilde{X}$ and the energy balance (3.18) holds with $\widetilde{X}$ and $\widetilde{Y}$. □

3.5.1 Compliant Discrete Least-Squares Principles

We say that[28] $\{J^h, X^h\}$ is a *compliant* DLSP for $\{J, X\}$ if

- X^h is a finite dimensional subspace of X
- $J^h \equiv J$.

LSFEMs corresponding to compliant least-squares principles are exactly the straightforward LSFEMs discussed in Section 2.2.1. A compliant DLSP is obtained by restricting the minimization process in $\{J, X\}$ to a finite dimensional subspace of X, i.e., the transition process is completed by simply choosing $X^h \subset X$. As a result, $\{J^h, X^h\}$ inherits[29] the energy norm and the energy balance of its continuous counterpart, i.e. $|||\cdot|||_h = |||\cdot|||$ and

$$C_1\|w^h\|_X \le |||w^h|||_h \le C_2\|w^h\|_X \qquad \forall w^h \in X^h \tag{3.62}$$

with the same C_1 and C_2 as in (3.18).

Theorem 3.28 *Assume that $\{J, X\}$ is a well-posed least-squares principle for* (3.16). *Every compliant DLSP $\{J^h, X^h\}$ has a unique minimizer u^h. Moreover, that minimizer satisfies the error estimate*

$$\|u - u^h\|_X \le \frac{C_2}{C_1} \inf_{w^h \in X^h} \|u - w^h\|_X . \tag{3.63}$$

---

[28] Recall that the notation $\{J^h, X^h\}$ implies the variational principle $\min_{X^h} J^h$.

[29] This justifies the use of the term *compliant* as a way to indicate that approximation in such DLSPs is confined to the space X^h and that otherwise they are indistinguishable from their parent CLSP.

Proof. Given that $|||\cdot|||_h$ is simply the energy norm $|||\cdot|||$, Assumption 3.15 holds trivially for $\{J^h, X^h\}$. For the same reason, Assumption 3.16 holds with[30] $\mathrm{e}(\cdot,\cdot) \equiv 0$. From (3.52) in Corollary 3.19,

$$|||u-u^h|||_h \le |||u-w^h|||_h \qquad \forall w^h \in X^h .$$

Using again $|||\cdot|||_h = |||\cdot|||$, we see that in actuality, (3.62) holds not only on X^h but on all of X. Therefore,

$$C_1\|u-u^h\|_X \le |||u-w^h|||_h \le C_2\|u-w^h\|_X .$$

The theorem follows by taking the infimum over X^h. □

To estimate the condition numbers of resulting linear systems, we use the inverse inequalities (3.59) and (3.60) as well as (3.61).

Theorem 3.29 *Let $\{J^h, X^h\}$ be a compliant DLSP and A the matrix of the associated linear system of algebraic equations* (3.48). *Let Assumptions* 3.25 *and* 3.26 *hold. Then,*[31]

$$\mathrm{cond}(\mathsf{A}) \le C h^{-2\omega_{max}} . \tag{3.64}$$

Proof. Assume that (3.59) holds; the proof in the case of (3.60) is similar. Using (3.59), (3.61), (3.62), and the definition (3.47) of A,

$$\begin{aligned} h^d|\vec{w}|^2 &\le \frac{1}{\mu_1}\|w^h\|_0^2 \le \frac{1}{\mu_1}\|w^h\|_X^2 \le \frac{1}{\mu_1 C_1}|||w^h|||_h^2 \\ &= \frac{1}{\mu_1 C_1}((w^h,w^h))_h = \frac{1}{\mu_1 C_1}\vec{w}^T\mathsf{A}\vec{w}. \end{aligned}$$

Therefore,

$$\lambda_{\min} \ge h^d \mu_1 C_1 .$$

Using the upper bounds in the same inequalities,

$$\begin{aligned} \vec{w}^T\mathsf{A}\vec{w} &= ((w^h,w^h))_h = |||w^h|||_h^2 \le C_2\|w^h\|_X^2 \\ &\le C_2 C_I^2 h^{-2\omega_{max}}\|w^h\|_0^2 \le \mu_2 C_2 C_I^2 h^{-2\omega_{max}} h^d |\vec{u}|^2 . \end{aligned}$$

Therefore,

$$\lambda_{\max} \le \mu_2 C_2 C_I^2 h^{-2\omega_{max}+d}$$

and thus

$$\mathrm{cond}(\mathsf{A}) = \frac{\lambda_{\max}}{\lambda_{\min}} \le \frac{\mu_2 C_2 C_I^2}{\mu_1 C_1} h^{-2\omega_{max}} . \qquad \square$$

---

[30] As a result, we see that compliant least-squares principles are consistent.

[31] We denote the condition number of the matrix A by $\mathrm{cond}(\mathsf{A})$.

3.5.2 Norm-Equivalent Discrete Least-Squares Principles

We say that $\{J^h, X^h\}$ is a *norm-equivalent* DLSP for $\{J, X\}$ if

- X^h is a finite dimensional subspace of X
- J^h is such that

$$\widehat{C}_1 \|w^h\|_X \leq |||w^h|||_h \leq \widehat{C}_2 \|w^h\|_X \qquad \forall\, w^h \in X^h \tag{3.65}$$

with positive constants $\widehat{C}_1$ and $\widehat{C}_2$ independent of h.

Every compliant DLSP is trivially norm equivalent because (3.62) holds for *all* functions in X, including functions from its proper subspace X^h. The converse is not true because (3.65) is only required to hold for discrete functions. This difference turns out to be far less important for the analysis than the requirement that $\widehat{C}_1$ and $\widehat{C}_2$ are independent of h.

From (3.18) and (3.65), it is easy to show that $|||\cdot|||_h$ and $|||\cdot|||$ are equivalent norms on X^h.

Lemma 3.30 *Assume that* (3.65) *holds for* $\{J^h, X^h\}$*. Then,*

$$\frac{C_1}{\widehat{C}_2} |||w^h|||_h \leq |||w^h||| \leq \frac{C_2}{\widehat{C}_1} |||w^h|||_h \qquad \forall w^h \in X^h. \qquad \square \tag{3.66}$$

This result implies that norm-equivalent functionals preserve, uniformly in h, the correct energy balance between the discrete data and solution spaces; thus the terminology *norm-equivalent* is fully justified. The following theorem shows that this is enough to guarantee that norm-equivalent functionals enjoy virtually the same computational properties, with regard to accuracy and matrix conditioning, as their compliant cousins.

Theorem 3.31 *Let* $\{J^h, X^h\}$ *be a norm-equivalent DLSP associated with a well-posed CLSP* $\{J, X\}$ *for* (3.16)*. Assume that the minimizer* u *of* $\{J, X\}$ *belongs to the space* $\widetilde{X}$ *from Assumption* 3.27*. Then, for every* $h > 0$*,* $\{J^h, X^h\}$ *has a unique minimizer* u^h*. Moreover, that minimizer satisfies the error estimate*

$$\|u - u^h\|_X \leq \frac{1}{C_1} \inf_{w^h \in X^h} |||u - w^h||| \\ + \frac{C_2}{C_1 \widehat{C}_1} \left(2 \inf_{w^h \in X^h} |||u - w^h|||_h + \sup_{w^h \in X^h} \frac{\mathrm{e}(u, w^h)}{|||w^h|||_h} \right). \tag{3.67}$$

Proof. A norm-equivalent functional trivially satisfies Assumption 3.15 so that, as a result, the existence and uniqueness of the discrete least-squares solution follows from Theorem 3.17.

Regarding the error estimate, there are two factors that prevent us from reusing the proof of Theorem 3.28. The first is that the error $u - u^h$ does not belong to X^h and thus (3.65) is not applicable to this function. The second factor is that u^h minimizes the *discrete* rather than the continuous energy norm. This means that the inequality

$$|||u - u^h||| \leq |||u - w^h|||$$

does not hold for all $w^h \in X^h$.

So, instead, we begin by bounding the error from above by using the continuous energy balance (3.18):

$$\|u - u^h\|_X \leq \frac{1}{C_1} |||u - u^h||| . \tag{3.68}$$

A repeated use of the triangle inequality, application of the norm equivalence (3.66), and the error estimate (3.45) from Theorem 3.17 yield the following estimate of the continuous energy norm: for all $w^h \in X^h$,

$$\begin{aligned}
|||u - u^h||| &\leq |||u - w^h||| + |||w^h - u^h||| \\
&\leq |||u - w^h||| + \frac{C_2}{\widehat{C}_1} |||w^h - u^h|||_h \\
&\leq |||u - w^h||| + \frac{C_2}{\widehat{C}_1} \left(|||u - u^h|||_h + |||u - w^h|||_h \right) \\
&\leq |||u - w^h||| + \frac{C_2}{\widehat{C}_1} |||u - w^h|||_h \\
&\quad + \frac{C_2}{\widehat{C}_1} \left(\inf_{w^h \in X^h} |||u - w^h|||_h + \sup_{w^h \in X^h} \frac{\mathrm{e}(u, w^h)}{|||w^h|||_h} \right) .
\end{aligned}$$

Because the last inequality holds for an arbitrary $w^h \in X^h$, it follows that

$$|||u - u^h||| \leq \inf_{w^h \in X^h} |||u - w^h||| + \frac{C_2}{\widehat{C}_1} \left(2 \inf_{w^h \in X^h} |||u - w^h|||_h + \sup_{w^h \in X^h} \frac{\mathrm{e}(u, w^h)}{|||w^h|||_h} \right) .$$

The theorem follows by combining the last bound and (3.68). □

The next result shows that, for a given CLSP $\{J, X\}$, the conditioning of the systems generated by compliant and norm-equivalent discrete principles is essentially the same.

Theorem 3.32 *Let* $\{J^h, X^h\}$ *be a norm-equivalent DLSP and* A *the matrix of the associated linear system of algebraic equations* (3.48). *Let Assumptions* 3.25 *and* 3.26 *hold. Then,*

$$\mathrm{cond}(\mathsf{A}) \leq C h^{-2\omega_{max}} . \tag{3.69}$$

Proof. In contrast to Theorem 3.31, the proof of Theorem 3.29 carries over to the norm-equivalent case without modifications. The reason for this is that estimates of the largest and the smallest eigenvalues of A require only the discrete energy balance (3.65), the inverse inequalities (3.59) and (3.60), and assumption (3.61). □

Before moving on to the next class of discrete principles, let us stress again that norm-equivalent DLSPs *are not compliant principles*. In particular, a norm-equivalent functional J^h is *different* from J, may not be meaningful outside X^h, and its construction can be rather complicated.

3.5.3 Quasi-Norm-Equivalent Discrete Least-Squares Principles

We say that $\{J^h, X^h\}$ is a *quasi-norm-equivalent* DLSP for $\{J, X\}$ if

- X^h is a finite dimensional subspace of X;
- J^h is such that

$$\widehat{C}_1(h)\|w^h\|_X \leq |||w^h|||_h \leq \widehat{C}_2(h)\|w^h\|_X \qquad \forall\, w^h \in X^h, \tag{3.70}$$

  where $\widehat{C}_1(h) > 0$ and $\widehat{C}_2(h) > 0$ for all $h > 0$.

Virtually every conforming DLSP obtained by the *minimize and then discretize* approach, including compliant and norm-equivalent DLSPs, falls within this definition and thus can be deemed quasi-norm-equivalent. However, (3.70) admits DLSPs that are neither compliant nor norm-equivalent. This necessitates some fundamental changes in the way one approaches the analysis of quasi-norm-equivalent principles.

Because the lower and the upper bounds in (3.70) involve mesh-dependent comparability constants, standard elliptic arguments based on continuity and coercivity inevitably lead to suboptimal error estimates. This does not necessarily reflect the true behavior of quasi-norm-equivalent DLSPs; as seen below, such methods *can* produce optimally accurate approximations despite the mesh dependence of the constants in (3.70). However, to reveal such behavior, error estimates must be based on carefully constructed duality arguments. This, of course, requires additional assumptions about the regularity of solutions of (3.16). Before stating the assumptions, recall that $Y = Y_1 \times \cdots \times Y_M$ and $X = X_1 \times \cdots \times X_M$. Let Y_i^* denote the dual of Y_i with $L^2(\Omega)$ acting as a pivot space, and d_i the number of scalar components of $y \in Y_i$.

Assumption 3.33 For all $i = 1, \ldots, M$, $Y_i^* \subseteq [L^2(\Omega)]^{d_i} \subseteq Y_i$. □

Assumption 3.34 There exist spaces $X^{(i)} \subset X$ and $Y^{(i)} \subset Y$, $i = 1, \ldots, M$, such that[32]

$$Y^{(i)} = Y_1^{(i)} \times \cdots \times Y_i^* \times \cdots Y_M^{(i)}$$

and, for all $u \in X^{(i)}$, we have the "stronger" energy balance

$$C_1\|u\|_{X^{(i)}} \leq \|\mathcal{L}u\|_{Y^{(i)}} \leq C_2\|u\|_{X^{(i)}}\,. \qquad \square \tag{3.71}$$

[32] Note that the ith component space of $Y^{(i)}$ is required to coincide with the ith component of the dual space Y^*. However, the remaining components of $Y^{(i)}$ are in no way restricted to be components of Y^* or the original space Y.

Assumption 3.35 For all $i = 1,\dots,M$, the discrete norm-generating operators $\mathcal{S}^h_{Y_i}$ can be extended to symmetric, positive definite operators $Y_i^* \mapsto Y_i^*$. □

Assumption 3.36 For all $i = 1,\dots,M$, the problem (3.16) is $X^{(i)}$-regular in the sense that, for every $f_i \in Y_i^*$, the system

$$\begin{cases} \mathcal{L}v = f^{(i)} & \text{in } \Omega \\ \mathcal{B}v = 0 & \text{on } \partial\Omega\,, \end{cases} \tag{3.72}$$

where $f^{(i)} = (0,\dots,0,f_i,0,\dots,0)^\mathsf{T} \in Y^{(i)}$, has a solution $v^{(i)} \in X^{(i)}$ that satisfies (3.71). □

As a consequence of (3.71) and (3.72), we have that

$$C_1\|v^{(i)}\|_{X^{(i)}} \le \|f^{(i)}\|_{Y_i} = \|f\|_{Y_i^*}\,. \tag{3.73}$$

The key to the analysis of quasi-norm-equivalent principles is the ability to bound the continuous energy norm associated with the CLSP $\{J,X\}$ by the discrete energy norm associated with the quasi-norm-equivalent DLSP. The following lemma establishes a general result of this type that is valid under the additional assumptions made above.

Lemma 3.37 *Let $w \in \widetilde{X}$ and let Assumptions* 3.15, 3.16, *and* 3.33–3.36 *hold. Then,*

$$\begin{aligned} |||w||| \le\ & \|(\mathcal{L}-\mathcal{L}^h)w\|_Y \\ & + \sum_{i=1}^M \sup_{f_i\in Y_i^*} \left\{ \frac{\sum_{k=1}^M \left(\mathcal{L}^h_k w, (\mathcal{S}^h_{Y_k})^2 \circ (\mathcal{L}_k - \mathcal{L}^h_k)v^{(i)}\right)}{\|f_i\|_{Y_i^*}} \right\} \\ & + \sum_{i=1}^M \sup_{f_i\in Y_i^*} \frac{((w,v^{(i)}))_h}{\|f_i\|_{Y_i^*}}\,, \end{aligned} \tag{3.74}$$

where, for $i = 1,\dots,M$, the functions $v^{(i)} \in Y^{(i)}$ solve the problems

$$\begin{cases} \mathcal{L}v^{(i)} = f^{(i)}_{\mathcal{S}} & \textit{in } \Omega \\ \mathcal{B}v^{(i)} = 0 & \textit{on } \partial\Omega \end{cases} \tag{3.75}$$

for

$$f^{(i)}_{\mathcal{S}} = (0,\dots,0,(\mathcal{S}^h_{Y_i})^{-2}f_i,0,\dots,0) \in Y^{(i)}\,.$$

Proof. For the sake of brevity, we use $\mathcal{L}_k u$ to denote $\sum_{j=1}^M \mathcal{L}_{kj}u_j$. Recall that

$$|||w||| = \|\mathcal{L}w\|_Y = \sum_{k=1}^M \|\mathcal{L}_k w\|_{Y_k}$$

and

$$|||w|||_h = \|\mathcal{S}_Y^h \circ \mathcal{L}^h w\|_0 = \sum_{k=1}^{M} \|\mathcal{S}_{Y_k}^h \circ \mathcal{L}_k^h w\|_0 .$$

Let $i = 1,\dots,M$. Using duality,

$$\|\mathcal{L}_i w\|_{Y_i} = \sup_{f_i \in Y_i^*} \frac{(\mathcal{L}_i w, f_i)}{\|f_i\|_{Y_i^*}} = \sup_{f_i \in Y_i^*} \frac{(\mathcal{L}_i^h w, f_i) + ((\mathcal{L}_i - \mathcal{L}_i^h)w, f_i)}{\|f_i\|_{Y_i^*}}$$

$$\leq \sup_{f_i \in Y_i^*} \frac{(\mathcal{L}_i^h w, f_i)}{\|f_i\|_{Y_i^*}} + \|(\mathcal{L}_i - \mathcal{L}_i^h)w\|_{Y_i} .$$

According to (3.75), we have that $f_i = (\mathcal{S}_{Y_i}^h)^2 \circ \mathcal{L}_i v^{(i)}$ and $\mathcal{L}_k v^{(i)} = 0$ for $k \neq i$. Using this and the definition of the discrete inner product $((\cdot,\cdot))_h$ gives

$$\frac{(\mathcal{L}_i^h w, f_i)}{\|f_i\|_{Y_i^*}} = \frac{\left(\mathcal{L}_i^h w, (\mathcal{S}_{Y_i}^h)^2 \circ \mathcal{L}_i v^{(i)}\right)}{\|f_i\|_{Y_i^*}} = \frac{\sum_{k=1}^{M} \left(\mathcal{L}_k^h w, (\mathcal{S}_{Y_k}^h)^2 \circ \mathcal{L}_k v^{(i)}\right)}{\|f_i\|_{Y_i^*}}$$

$$= \frac{\sum_{k=1}^{M} \left(\mathcal{S}_{Y_k}^h \circ \mathcal{L}_k^h w, \mathcal{S}_{Y_k}^h \circ \mathcal{L}_k^h v^{(i)}\right)}{\|f_i\|_{Y_i^*}} + \frac{\sum_{k=1}^{M} \left(\mathcal{L}_k^h w, (\mathcal{S}_{Y_k}^h)^2 \circ (\mathcal{L}_k - \mathcal{L}_k^h) v^{(i)}\right)}{\|f_i\|_{Y_i^*}}$$

$$= \frac{((w, v^{(i)}))_h}{\|f_i\|_{Y_i^*}} + \frac{\sum_{k=1}^{M} \left(\mathcal{L}_k^h w, (\mathcal{S}_{Y_k}^h)^2 \circ (\mathcal{L}_k - \mathcal{L}_k^h) v^{(i)}\right)}{\|f_i\|_{Y_i^*}} .$$

Combining with the previous inequality,

$$\|\mathcal{L}_i w\|_{Y_i} \leq \|(\mathcal{L}_i - \mathcal{L}_i^h)w\|_{Y_i} + \sup_{f_i \in Y_i^*} \frac{((w, v^{(i)}))_h}{\|f_i\|_{Y_i^*}}$$

$$+ \sup_{f_i \in Y_i^*} \left\{ \frac{\sum_{k=1}^{M} \left(\mathcal{L}^h w, (\mathcal{S}_{Y_k}^h)^2 \circ (\mathcal{L}_k - \mathcal{L}_k^h) v^{(i)}\right)}{\|f_i\|_{Y_i^*}} \right\} .$$

The lemma follows by summing the bounds for $i = 1,\dots,M$. □

The first two terms in the upper bound (3.74) represent consistency errors. For consistent least-squares functionals, Lemma 3.37 can be substantially simplified.

Corollary 3.38 *Assume that $\{J^h, X^h\}$ is consistent. For every $w \in \widetilde{X}$*

$$|||w||| \leq \sum_{i=1}^{M} \sup_{f_i \in Y_i^*} \frac{((w, v^{(i)}))_h}{\|f_i\|_{Y_i^*}} , \tag{3.76}$$

where the $v^{(i)}$ are defined by (3.75). □

The key idea in the analysis of quasi-norm-equivalent DSLPs is to use bounds such as (3.74) or (3.76) instead of the mesh-dependent inequalities in (3.70) to estimate the error in the norm of X. For consistent discrete functionals, this approach is illustrated in the next theorem. The general case differs by the presence of truncation error terms and, for brevity, is not considered.

Theorem 3.39 *Assume that $\{J^h, X^h\}$ is a consistent quasi-norm-equivalent DLSP associated with a well-posed CLSP $\{J, X\}$ for* (3.16)*. Let Assumptions* 3.15, 3.16, *and* 3.33–3.36 *hold and assume that the minimizer u of $\{J, X\}$ belongs*[33] *to $\widetilde{X}$. Then, for every $h > 0$, $\{J^h, X^h\}$ has a unique minimizer u^h. Moreover, that minimizer satisfies the error estimate*

$$\|u - u^h\|_X \leq \frac{1}{C_1} \inf_{w^h \in X^h} |||u - w^h|||_h \sum_{i=1}^{M} \sup_{f_i \in Y_i^*} \left\{ \frac{\inf_{w^h \in X^h} |||v^{(i)} - w^h|||_h}{\|f_i\|_{Y_i^*}} \right\}. \tag{3.77}$$

Proof. Because every quasi-norm-equivalent DLSP satisfies Assumptions 3.15 and 3.16, the existence and uniqueness of u^h follow from the general result in Theorem 3.17. Using the continuous energy balance (3.18) and (3.76), we bound the error in the norm of X as follows:

$$\|u - u^h\|_X \leq \frac{1}{C_1} |||u - u_h||| \leq \frac{1}{C_1} \sum_{i=1}^{M} \sup_{f_i \in Y_i^*} \frac{((u - u^h, v^{(i)}))_h}{\|f_i\|_{Y_i^*}}.$$

By assumption, $\{J^h, X^h\}$ is consistent so that

$$((u - u^h, v^{(i)}))_h = ((u - u^h, v^{(i)} - w^h))_h \qquad \forall w^h \in X^h.$$

As a result, using the Cauchy inequality,

$$((u - u^h, v^{(i)}))_h \leq |||u - u^h|||_h \, |||v^{(i)} - w^h|||_h \qquad \forall w^h \in X^h.$$

Because the last inequality holds for all w^h, it is easy to see that

$$\frac{((u - u^h, v^{(i)}))_h}{\|f_i\|_{Y_i^*}} \leq |||u - u^h|||_h \frac{\inf_{w^h \in X^h} |||v^{(i)} - w^h|||_h}{\|f_i\|_{Y_i^*}}.$$

The theorem follows by taking supremum over Y_i^* and using (3.52). □

Remark 3.40 First, we note that the discrete energy balance (3.70) is not used at all in the proof of the error estimates. As a result, the mesh-dependence of its lower and upper bounds is prevented from entering explicitly into the error bounds. This, however, does not automatically imply that (3.77) is optimal with respect to the norm of X. Whether this is true ultimately depends on the structure of the discrete

---

[33] See Assumption 3.27 for the definition of the space $\widetilde{X}$.

energy norm and the order of the approximation error for *a set of sufficiently regular functions* as measured by this norm. In other words, the quality of the quasi-norm-equivalent least-squares approximation is strongly affected by the properties of the term

$$\sum_{i=1}^{M} \sup_{f_i \in Y_i^*} \frac{\inf_{w^h \in X^h} |||v^{(i)} - w^h|||_h}{\|f_i\|_{Y_i^*}}. \tag{3.78}$$

In Chapter 4, we consider examples of quasi-norm-equivalent methods for first-order elliptic systems that illustrate the importance of this term in the error analysis. There, we also discuss in more detail the conditioning of the algebraic systems engendered by quasi-norm-equivalent methods, as this requires more detailed knowledge of the spaces and operators involved in setting up the method.

In particular, in Section 4.5, we present two methods that have identical mesh-dependent energy balances but differ tremendously in their asymptotic order of accuracy. One of them uses trivial approximations of norm-generating operators that causes (3.78) to contribute negative powers of h, thereby reducing the order of accuracy in (3.77). By using a judicious choice of discrete operators $\mathcal{S}_Y^h$, the other method simultaneously keeps this term of $O(1)$ and ensures that $|||u - u^h|||_h$ is of optimal-order of accuracy for all sufficiently smooth functions u. As a result, for this method, (3.77) is optimal. □

3.5.4 Summary Review of Discrete Least-Squares Principles

This chapter provides the necessary tools to imbue the informal least-squares classification scheme given in Figure 2.1 with precise mathematical meaning. In particular, analyses of compliant, norm-equivalent, and quasi-norm-equivalent DLSPs show that recovery of a Rayleigh–Ritz-like setting for the finite element method becomes increasingly difficult as DLSPs deviate more and more from the compliant setting of Section 3.5.1. This observation underscores the key role of norm-equivalence in LSFEMs and validates our decision to base the classification scheme in Figure 2.1 on this key property. A concise summary of the properties of the three basic types of conforming DLSPs is presented in Table 3.1.

DLSP→ Property↓	Conforming			Non-conforming
	compliant	norm-equivalent	quasi-norm-equivalent	
Energy balance	independent of h	independent of h	dependent on h	varies
Error estimate	standard elliptic argument	uses norm-equivalence	uses duality argument	varies
Condition number	$h^{-2\omega_{max}}$	$h^{-2\omega_{max}}$	varies	varies

Table 3.1 Comparison of different classes of discrete least-squares principles.

Chapter 4
The Agmon–Douglis–Nirenberg Setting for Least-Squares Finite Element Methods

The least-squares theory formulated in the previous chapter shows that norm equivalence is at the root of successful least-squares finite element methods (LSFEMs). Therefore, the first task in the formulation of LSFEMs is to obtain a collection of energy balances for the given partial differential equation (PDE) problem.

As far as linear elliptic PDEs are concerned, in this book we use two different settings to accomplish this task. These settings correspond to the two different ways of arranging the dependent variables illustrated in Remark 3.6. In the first, the unknowns in the PDE problem are thought of as belonging to either the domain or the range of the basic *vector calculus* differential operators grad, curl, and div; we refer to this setting as the *vector-operator setting*. This places the PDE problem in the context of the differential De Rham complex (A.52) and gives rise to energy balances in the norms of the spaces $G(\Omega)$, $\mathbf{C}(\Omega)$, $\mathbf{D}(\Omega)$, and $S(\Omega)$ that form that complex, or in the norms of spaces such as $\mathbf{C}(\Omega)\cap\mathbf{D}(\Omega)$. The vector-operator setting is particularly well suited for elliptic PDE problems involving generalized scalar and vector Laplace operators because their structure can be encoded using a primal and a dual De Rham complex [7, 8, 13, 75, 139, 206, 225, 314, 338, 339]. For such problems, this setting allows us to define *compatible* LSFEMs, i.e., LSFEMS that have local conservation properties similar to those of mixed Galerkin methods for the same equations [58, 59, 63]. Examples of compatible LSFEMs are presented in Chapters 5–7.

The principal drawback of the vector-operator setting is its specialized scope. First, we must be able to "fit" the PDE problem[1] into the De Rham complex, and second, the energy balance can be expressed only in terms of the norms of the spaces forming that complex or their intersections such as $\mathbf{C}(\Omega)\cap\mathbf{D}(\Omega)$. Conforming

[1] The Stokes equations (1.62) are an example of a PDE problem that is not easy to handle using the vector-operator setting because the velocity field naturally belongs to the space $[H^1(\Omega)]^d$ which is not part of the De Rham complex. In particular, neither one of the available vector-field spaces in the De Rham complex, i.e., $\mathbf{C}(\Omega)$ and $\mathbf{D}(\Omega)$, can be constrained by the velocity boundary condition in (1.62). This is why, in Chapter 7, we use this setting only with non-standard boundary conditions for the Stokes equations.

P.B. Bochev, M.D. Gunzburger, *Least-Squares Finite Element Methods*, Applied Mathematical Sciences 166, DOI: 10.1007/b13382_4,

approximations of the latter that satisfy Assumption 3.20 are not easy to construct[2] so that the vector-operator setting also produces some *non-conforming* LSFEMs. For these reasons, we do not attempt to provide a general framework for developing LSFEMs in the vector-operator setting. Instead, we focus, in Chapters 5–7, on PDEs for which this setting works best, using the abstract least-squares theory of Chapter 3 to analyze LSFEMs for which it is applicable, and deal directly with LSFEMs for which that theory is not applicable.

The second setting for deriving energy balances is provided by the Agmon–Douglis–Nirenberg (ADN) elliptic regularity theory [2]. The *ADN setting* is substantially more general than the vector-operator setting because the ADN theory is applicable to virtually any elliptic boundary value problem.[3] In addition, the ADN setting allows us to "automate" the process of establishing the energy balances because their proof is reduced to the verification of algebraic conditions involving the principal parts of the differential and boundary operators. Specifically, the theory of [2] asserts that Assumption 3.7 holds for the boundary value problem (3.16), whenever the algebraic ADN conditions are satisfied.[4] Consequently, for any given elliptic PDE problem, correct energy balances, norm-equivalent functionals, and the associated continuous least-squares principles (CLSPs) can be quickly determined by a series of tedious but routine algebraic calculations. Thus, the ADN setting provides a formal framework for developing least-squares principles and remains widely used[5] in modern LSFEMs.

A further advantage of the ADN setting is that, in conjunction with first-order[6] systems, it gives rise to energy balances for which the data spaces can always be approximated by finite element subspaces that satisfy Assumption 3.20, i.e., non-conforming LSFEMs can be altogether avoided. As a result, by using the ADN setting and first-order system formulations, we are always led to one of the three classes of *conforming* LSFEMs described in Section 3.5. This fact makes it worth investing some efforts to specialize the abstract results given in Chapter 3 to the ADN setting. In this chapter, we provide such a specialization and formulate a simple and straightforward framework for developing LSFEMs in the ADN setting using first-order systems.

---

[2] The reasons for this are explained in Section B.2.2.

[3] The ADN setting is indeed quite general with respect to the elliptic PDE and boundary condition. However, there is one exception to the improved generality of the ADN setting: it can only treat problems with one type of boundary condition (general as it may be), i.e., it cannot handle the case of mixed boundary conditions in which one imposes different types of boundary conditions, e.g., Dirichlet and Neumann, on different parts of the boundary.

[4] In Appendix D.2, examples are provided that show how, for concrete PDEs, one can verify the *complementing condition*, a key aspect of the ADN theory.

[5] After the pioneering work in [18], the ADN theory has been put to regular use in LSFEMs; see e.g., [35,41,53,55,115,247]. A somewhat simpler analogue of the ADN theory exists for first-order elliptic systems in the plane; see [342]. This theory relies on another algebraic condition that bears the name of *Lopatinskii* to establish relations such as (3.18) and (3.19). For further specific details about this condition and its application in LSFEMs, see, e.g., [112,116,342].

[6] Recall that, according to Section 2.2.2, recasting a PDE problem into a first-order system form is the first key to practicality for LSFEMs.

Of course, the ADN setting is not without its faults because a price must be paid for the extreme generality of the underlying ADN elliptic theory. For one thing, there are higher smoothness requirements on the boundary of the domain Ω and the coefficients of the PDE problem. For another, the ADN theory views differential and boundary operators as matrices of partial derivatives operating on *independent scalar functions* belonging to the scalar Sobolev spaces[7] $H^k(\Omega)$, i.e., $\mathcal{L}$ and $\mathcal{B}$ have the forms given in (D.1) and (D.2), respectively. As a result, data and solution spaces appearing in energy balances derived in the ADN setting are always given by products of scalar Sobolev spaces. Among other things, this makes it impossible to obtain energy balances in the norms of the spaces $\mathbf{D}(\Omega)$ or $\mathbf{C}(\Omega)$ that treat vector field components collectively. Ultimately, because treating vector components as independent scalars often does not tell the whole story about preserving key conservation properties of a PDE, the ADN setting cannot be used to derive the compatible LSFEMs discussed in Chapters 5–7.

A summary of the ADN theory is given in Section D.1. The verification of the assumptions of that theory is a straightforward but somewhat tedious process; concrete examples of that process are provided in Section D.2. This chapter begins with a brief discussion of a general procedure, due to [2], for transforming PDEs to first-order system form. Then, we show that the ADN theory breaks up first-order elliptic systems into two distinct classes of problems that admit two fundamentally different types of ADN energy balances. For *homogeneous* elliptic problems, all data can be measured using the same norm, wheras for *non-homogeneous* elliptic problems, they have to be measured in different norms. This difference is a consequence of the way ADN theory treats the dependent variables and has a profound impact on the formulation of practical LSFEMs. In particular, it implies that *practical compliant* LSFEMs are not available for the *non-homogeneous elliptic* class of first-order systems. Ultimately, this means that, when using the ADN setting to develop LSFEMs,

> *recasting a given PDE problem into a first-order system is enough to enable the formulation of conforming LSFEMs but is not enough to guarantee that compliant LSFEMs are practical.*

4.1 Transformations to First-Order Systems

Any ADN-elliptic system of order higher than one can be transformed into an equivalent first-order system that remains elliptic in the same sense.[8] This transformation can be carried out using the following algorithm described in [2].

---

[7] The ADN theory also includes results for L^p-based Sobolev spaces and Hölder spaces. Because the theory of Chapter 3 is set in Hilbert spaces, we restrict attention to ADN results for $L^2(\Omega)$-based Sobolev spaces.

[8] This is not true if the usual definition of ellipticity, involving only differentiated terms, is used; see Remark D.3.

Consider the problem (3.16) where $\mathcal{L}$ and $\mathcal{B}$ have the forms given by (D.1) and (D.2), respectively. First, all variables are divided into two sets according to their indices: a set $\{u_{k'}\}$ containing all variables for which $t_j > 1$ and a set $\{u_{k''}\}$ of all variables for which $t_j \leq 1$. Then, new dependent variables are introduced as

$$u_{k',j} = \partial_j u_{k'}$$

and these equations that define these variables are added to the differential operator. The original operator $\mathcal{L}$ itself also undergoes a transformation. All terms in which $u_{k'}$ is not differentiated remain unchanged. A term in which $u_{k'}$ is differentiated is substituted according to the rule

$$D^\alpha(\partial_j u_{k'}) \mapsto D^\alpha(u_{k',j}).$$

Although this process of transforming $\mathcal{L}$ is not unique, it can be shown (see [2]) that the new system is always elliptic in the sense of ADN and that $\max t_j \leq 2$ and $\min s_i \geq -1$.

The original boundary conditions in (3.16) are transformed into equivalent boundary conditions for the first-order system using a similar process. Again, this process is not unique and can be accomplished in several possible ways; the important fact is that if the complementing condition is satisfied by the original boundary operator $\mathcal{B}$, then it is also satisfied by the new boundary operator; see [2]. As a result, we are guaranteed that the new differential operator augmented with the new boundary operator is well posed, so that the a priori bound (D.6) remains valid.

4.2 Energy Balances

Suppose that the PDE problem (3.16) has already been transformed into a first-order system. To streamline the notation, we reuse the symbols $\mathcal{L}$ and $\mathcal{B}$ for the new differential and boundary operators. To avoid some nonessential details, we make the following assumptions:

- $\{\mathcal{L},\mathcal{B}\}$ has zero nullity, i.e., (3.16) has a unique solution for all sufficiently smooth data
- the boundary conditions are such that, after transforming (3.16) into a first-order system, the order of all terms $\mathcal{B}_{lj}$ in the transformed boundary operator $\mathcal{B}$ equals zero, i.e.,

$$\deg \mathcal{B}_{lj}(\mathbf{x},\vec{\xi}) = 0 \qquad \forall\, l = 1,\ldots,m,\ \ j = 1,\ldots,M \tag{4.1}$$

- Assumption D.2 holds for the first-order system.

4.2.1 Homogeneous Elliptic Systems

According to Definition D.3, for a homogeneous elliptic system $s_1 = \cdots = s_M = 0$ and $t_1 = \cdots = t_M$. In the case of a first-order system, this implies that $t_j = 1$ for $j = 1, \ldots, M$, whereas assumption (4.1) on the boundary operator implies that $r_l = -1$ for $l = 1, \ldots, m$. As a result, for first-order homogeneous elliptic systems, $t' = 1$ and $\widehat{r} = 0$, whereas the solution and data spaces in Theorem D.1 specialize to

$$X_q = \prod_{j=1}^{M} H^{q+1}(\Omega) \tag{4.2}$$

and

$$Y_q = \prod_{i=1}^{M} H^q(\Omega) \qquad \text{and} \qquad B_q = \prod_{l=1}^{m} H^{q+1/2}(\partial\Omega)\,, \tag{4.3}$$

respectively. Accordingly, the abstract energy balance (D.6) from that theorem assumes the form

$$\sum_{j=1}^{M} \|u_j\|_{q+1} \le C\Big(\sum_{i=1}^{M} \| \sum_{j=1}^{M} \mathcal{L}_{ij} u_j \|_q + \sum_{l=1}^{m} \| \sum_{j=1}^{M} \mathcal{B}_{lj} u_j \|_{q+1/2} \Big)\,, \tag{4.4}$$

provided the coefficients of $\mathcal{L}$ and $\mathcal{B}$ are of class $C^q(\overline{\Omega})$ and $C^{q+1}(\partial\Omega)$, respectively, and Ω is of class C^{q+1}, where $q \ge 0$ is an integer. Recall that Assumption D.2 requires (4.4) to hold for all smooth functions and all real q.

4.2.2 Non-Homogeneous Elliptic Systems

If (3.16) is not homogeneous elliptic, then there is at least one equation index $s_i = -1$. Because all $\mathcal{L}_{ij}$ are at most of order one, it follows that there must be at least one unknown index $t_j = 2$. Without loss of generality, we can assume that for some integer k, $1 \le k \le M$,

$$s_1 = \cdots = s_k = 0 \qquad \text{and} \qquad s_{k+1} = \cdots = s_M = -1 \tag{4.5}$$

and for some integer l, $1 \le l \le M$,

$$t_1 = \cdots = t_l = 1 \qquad \text{and} \qquad t_{l+1} = \cdots = t_M = 2\,. \tag{4.6}$$

Also, owing to assumption (4.1), there is an integer p, $1 \le p \le m$, such that

$$r_1 = \cdots = r_p = -1 \qquad \text{and} \qquad r_{p+1} = \cdots = r_m = -2\,. \tag{4.7}$$

It follows that for first-order non-homogeneous elliptic systems $t' = 2$ and $\widehat{r} = 0$, whereas the solution and data spaces in Theorem D.1 specialize to

$$X_q = \prod_{j=1}^{l} H^{q+1}(\Omega) \times \prod_{j=l+1}^{M} H^{q+2}(\Omega); \tag{4.8}$$

and

$$\begin{aligned} Y_q &= \prod_{i=1}^{k} H^{q}(\Omega) \times \prod_{i=k+1}^{M} H^{q+1}(\Omega) \\ B_q &= \prod_{l=1}^{p} H^{q+1/2}(\partial\Omega) \times \prod_{l=p+1}^{m} H^{q+3/2}(\partial\Omega), \end{aligned} \tag{4.9}$$

respectively. As a result, for such first-order systems the abstract energy balance (D.6) in Theorem D.1 assumes the form

$$\begin{aligned} &\sum_{j=1}^{l} \|u_j\|_{q+1} + \sum_{j=l+1}^{M} \|u_j\|_{q+2} \\ &\quad \le C\Big(\sum_{i=1}^{k} \| \sum_{j=1}^{M} \mathcal{L}_{ij} u_j \|_q + \sum_{i=k+1}^{M} \| \sum_{j=1}^{M} \mathcal{L}_{ij} u_j \|_{q+1} \\ &\qquad + \sum_{l=1}^{p} \| \sum_{j=1}^{M} \mathcal{B}_{lj} u_j \|_{q+1/2} + \sum_{l=p+1}^{m} \| \sum_{j=1}^{M} \mathcal{B}_{lj} u_j \|_{q+3/2} \Big), \end{aligned} \tag{4.10}$$

provided the coefficients of $\mathcal{L}$ and $\mathcal{B}$ are of class $C^{q+1}(\overline{\Omega})$ and $C^{q+2}(\partial\Omega)$, respectively, and Ω is of class C^{q+2}; $q \ge 0$ an integer. As in the last section, Assumption D.2 requires (4.10) to hold for all smooth functions and all real q.

4.3 Continuous Least-Squares Principles

Once the structure of the energy balance, i.e., the solution and data spaces for the two classes of first-order ADN systems are known, the next step is to specialize the results of Section 3.2.2 to these systems. Recall that the data spaces Y and B provide the correct norms in which to measure the residual energy of the differential equations and boundary conditions, respectively, in (3.16), whereas the solution space X provides the space of candidate minimizers over which to minimize that energy.

4.3.1 Homogeneous Elliptic Systems

For homogeneous first-order elliptic systems, the data spaces are given by Y_q and B_q as defined in (4.3). Therefore, for such systems, (3.24) gives rise to the following family of norm-equivalent least-squares functionals:

$$J_q(u;f,g) = \sum_{i=1}^{M} \Big\| \sum_{j=1}^{M} \mathcal{L}_{ij} u_j - f_i \Big\|_q^2 + \sum_{l=1}^{m} \Big\| \sum_{j=1}^{M} \mathcal{B}_{lj} u_j - g_l \Big\|_{q+1/2}^2 . \tag{4.11}$$

The continuous least-squares principle (CLSP) $\{J_q, X_q\}$ is a family of unconstrained minimization problems

$$\min_{u \in X_q} J_q(u;f,g), \tag{4.12}$$

where the solution space X_q is the space from (4.2). Because the least-squares functional (4.11) is norm-equivalent, the least-squares principle (4.12) satisfies all the assumptions of Theorem 3.10. Therefore, according to that theorem, for every value of the parameter q, the unconstrained minimization problem (4.12) has a unique minimizer $u_q \in X_q$ that depends continuously on the data $f \in Y_q$ and $g \in B_q$.

The families of minimization problems $\{J_q, X_q\}$ just defined account for all possible CLSPs that can be obtained for homogeneous elliptic first-order systems in the ADN setting. Of course, the point at which one has assembled a collection of CLSPs is also the point at which one has to start worrying about the practicality of the available CLSPs. By transforming the given PDE problem into a first-order system, we have already fulfilled the conditions in the first key to practicality of LSFEMs as formulated in Section 2.2.2. We next need to show how to fulfill the conditions from the second key, i.e., how to use only $L^2(\Omega)$ norms of equation residuals in the definition of the least-squares functional. It turns out that for homogeneous elliptic systems, there are CLSPs that readily satisfy this condition so that, for such systems, *practical compliant* discrete least-squares principles (DLSPs) exist.[9]

To simplify matters, we consider $g = 0$ in (3.16) and assume that Ω has a simple enough boundary so that homogeneous boundary conditions can be easily imposed[10] on the solution space X_q. This eliminates the need to deal with the fractional-order norms of the boundary residuals in (4.11).

If the regularity index $q = 0$ in (4.3), then the data space Y_q is a product of $L^2(\Omega)$ spaces. Thus, for homogeneous elliptic systems, we always choose $q = 0$ in (4.11), thereby satisfying the second key to practicality of LSFEMs as formulated in Section 2.2.2. This gives rise to the least-squares principle[11,12]

---

[9] This is not the case for non-homogeneous elliptic systems where we have to consider *norm-equivalent* and *quasi-norm-equivalent* discrete principles; see Section 4.3.2.

[10] Section 12.1 provides additional information about LSFEMs with weakly enforced boundary conditions.

[11] To economize notation, from now on we present least-squares principles by giving the pair $\{J, X\}$, where J is a least-squares functional and X is a solution space. Without stating so, we implicitly have the principle $\min_X J$. Thus, the pair $\{J_0(u;f), X_{0,0}\}$ given in (4.13) implies the least-squares principle

$$\min_{u \in X_{0,0}} J_0(u;f).$$

[12] Recall that we are assuming that X_q is constrained by the boundary condition so that the boundary terms appearing in (4.11) can be deleted.

$$\begin{cases} J_0(u;f) = \sum_{i=1}^{M} \| \sum_{j=1}^{M} \mathcal{L}_{ij} u_j - f_i \|_0^2 \\ X_{0,0} = \left\{ u \mid u \in X_0 = \prod_{j=1}^{M} H^1(\Omega); \quad \mathcal{B}u = 0 \quad \text{on } \partial\Omega \right\}. \end{cases} \tag{4.13}$$

The Euler–Lagrange equation for (4.13) is given by the variational problem

$$\textit{seek } u \in X_{0,0} \quad \textit{such that} \quad Q(u;v) = F(v) \qquad \forall v \in X_{0,0}\,, \tag{4.14}$$

where

$$Q(u;v) = \sum_{i=1}^{M} \Big(\sum_{j=1}^{M} \mathcal{L}_{ij} u_j \,, \sum_{j=1}^{M} \mathcal{L}_{ij} v_j \Big)_0 \quad \text{and} \quad F(v) = \sum_{i=1}^{M} \Big(f_i \,, \sum_{j=1}^{M} \mathcal{L}_{ij} v_j \Big)_0 . \tag{4.15}$$

The functional in (4.13) is norm-equivalent so that $\{J_0, X_{0,0}\}$ satisfies all the hypotheses of Theorem 3.10. Thus, for every $f \in Y_0$, the least-squares principle (4.13) has a unique minimizer $u \in X_{0,0}$ that depends continuously on the data; equivalently, the unique minimizer $u \in X_{0,0}$ is also the unique solution of the variational equation (4.14). Regarding the practicality of $\{J_0, X_{0,0}\}$, note that the functional J_0 uses only $L^2(\Omega)$ norms of equation residuals.

4.3.2 Non-Homogeneous Elliptic Systems

For non-homogeneous first-order elliptic systems, the data spaces are given by Y_q and B_q as defined in (4.9). In this case, (3.24) specializes to the following family of norm-equivalent least-squares functionals:

$$\begin{aligned} J_q(u;f,g) = & \sum_{i=1}^{k} \| \sum_{j=1}^{M} \mathcal{L}_{ij} u_j - f_i \|_q^2 + \sum_{i=k+1}^{M} \| \sum_{j=1}^{M} \mathcal{L}_{ij} u_j - f_i \|_{q+1}^2 \\ & + \sum_{l=1}^{p} \| \sum_{j=1}^{M} \mathcal{B}_{lj} u_j - g_l \|_{q+1/2}^2 + \sum_{l=p+1}^{m} \| \sum_{j=1}^{M} \mathcal{B}_{lj} u_j - g_l \|_{q+3/2}^2 . \end{aligned} \tag{4.16}$$

The corresponding least-squares principle is given by the pair $\{J_q(u;f,g), X_q\}$, where $J_q(u;f,g)$ and X_q are defined in (4.16) and (4.8), respectively. Because the least-squares functionals in (4.16) is norm-equivalent, the least-squares principle defined by (4.16) and (4.8) satisfies all the assumptions of Theorem 3.10. Therefore, according to this theorem, for every value of the parameter q, that principle has a unique minimizer $u_q \in X_q$ that depends continuously on the data $f \in Y_q$ and $g \in B_q$.

The families of minimization problems just defined account for all possible CLSPs that can be obtained for non-homogeneous elliptic first-order systems in the ADN setting. Again, the first key to practicality of LSFEMs is taken care of because we are dealing with first-order systems. However, unlike what is the case for homogeneous elliptic systems, for non-homogeneous elliptic first-order systems, compliant DLSPs do not meet the second key to practicality and we instead have to consider *norm-equivalent* and *quasi-norm-equivalent* discrete principles; see Section 4.3.2. Let us see why this is the case.

We again consider $g = 0$ and assume that homogeneous boundary conditions are imposed on the solution space X_q so that the fractional-order norms of the boundary residuals in (4.16) are omitted. We can proceed as in the last section and set $q = 0$. This gives rise to the CLSP

$$\begin{cases} J_0(u;f) = \sum_{i=1}^{k} \| \sum_{j=1}^{M} \mathcal{L}_{ij} u_j - f_i \|_0^2 + \sum_{i=k+1}^{M} \| \sum_{j=1}^{M} \mathcal{L}_{ij} u_j - f_i \|_1^2 \\ X_{0,0} = \left\{ u \mid u \in X_0 = \prod_{j=1}^{l} H^1(\Omega) \times \prod_{j=l+1}^{M} H^2(\Omega), \quad \mathcal{B}u = 0 \quad \text{on } \partial\Omega \right\}. \end{cases} \tag{4.17}$$

In contrast to the homogeneous elliptic case for which $X_{0,0}$ is a product of $H^1(\Omega)$ spaces, here the last $M - l$ components of $X_{0,0}$ are the Sobolev spaces[13] $H^2(\Omega)$. The appearance of $H^2(\Omega)$ spaces in $X_{0,0}$ is linked to another sign that the second key to practicality has been violated in (4.17), namely the fact that the last $M - k$ residuals in (4.17) are measured in the norm of $H^1(\Omega)$. Because of this, the *effective* order of some terms in (4.17) is two even though all the differential operators appearing in $\mathcal{L}_{ij}$ are of order at most one.

Noting that violations of the second key to practicality in (4.17) are caused by the fact that in non-homogeneous elliptic systems $q - s_i = 1$ for $i = k+1, \ldots, M$ when $q = 0$, we can try to set $q = -1$, thereby making $q - s_i$ equal to zero for the second group of residuals. The resulting CLSP is given by[14]

$$\begin{cases} J_{-1}(u;f) = \sum_{i=1}^{k} \| \sum_{j=1}^{M} \mathcal{L}_{ij} u_j - f_i \|_{-1}^2 + \sum_{i=k+1}^{M} \| \sum_{j=1}^{M} \mathcal{L}_{ij} u_j - f_i \|_0^2 \\ X_{-1,0} = \left\{ u \mid u \in X_{-1} = \prod_{j=1}^{l} L^2(\Omega) \times \prod_{j=l+1}^{M} H^1(\Omega), \quad \mathcal{B}u = 0 \quad \text{on } \partial\Omega \right\}. \end{cases} \tag{4.18}$$

Now, the last $M - k$ residuals are measured by $L^2(\Omega)$ norms, but the first k residuals are measured by the minus one norm (A.9). Consequently, the CLSP $\{J_{-1}, X_{-1,0}\}$ still does not satisfy the second key to practicality.

Remark 4.1 Although for non-homogeneous elliptic systems both $\{J_0, X_{0,0}\}$ and $\{J_{-1}, X_{-1,0}\}$ are impractical, their impracticality is manifested differently. For $\{J_0, X_{0,0}\}$, what prevents it from being practical is the need to approximate, in a

---

[13] In Section 2.2.2 it is explained why conforming finite element subspaces of $H^2(\Omega)$ are more difficult to construct than for $H^1(\Omega)$ and should be deemed impractical in the context of first-order systems.

[14] This, of course, assumes that the boundary condition $\mathcal{B}u = 0$ is well defined for functions belonging to X_{-1}.

conforming manner, functions belonging to the Sobolev space $H^2(\Omega)$ that enters into the definition of X_0. In contrast, the impracticality of $\{J_{-1},X_{-1,0}\}$ is caused by the need to compute minus one inner products in the assembly process; conforming approximation of the space X_{-1} is not a problem. Therefore, $\{J_{-1},X_{-1,0}\}$ satisfies Assumption 3.20.

It is clear that any other choice of q would lead to even more impractical[15] CLSPs $\{J_q,X_q\}$. Therefore, to obtain practical LSFEMs from $\{J_q,X_q\}$, we must explore the other two alternatives to compliant DLSP: norm-equivalent and quasi-norm-equivalent discrete least-squares principles. □

4.4 Least-Squares Finite Element Methods for Homogeneous Elliptic Systems

Regarding the practicality of $\{J_0,X_{0,0}\}$, we see that, in addition to having a functional J_0 that uses only $L^2(\Omega)$ norms, the solution space $X_{0,0}$ of that CLSP is a product of $H^1(\Omega)$ spaces which can be approximated by a product of the *same standard* C^0 *nodal spaces* G^r of degree $r \geq 1$:

$$X^h_{r,0} = \{u^h \mid u^h \in \prod_{j=1}^{M} G^r, \quad \mathcal{B}u^h = 0 \quad \text{on } \partial\Omega\}\,. \tag{4.19}$$

It follows that $\{J_0,X_{0,0}\}$ satisfies both key requirements for practicality, as formulated in Section 2.2.2 and so, the compliant DLSP $\{J_0,X^h_{r,0}\}$ and the corresponding LSFEM

$$\textit{seek } u^h \in X^h_{r,0} \quad \textit{such that} \quad Q(u^h;v^h) = F(v^h) \qquad \forall\, v^h \in X^h_{r,0}\,, \tag{4.20}$$

derived from (4.13),[16] are practical. Of course, (4.20) is simply the first-order optimality condition for the minimizer of $\{J_0,X^h_{r,0}\}$. The next theorem confirms that such LSFEMs are also optimally accurate and lead to easy to precondition matrices with condition numbers comparable to those in Galerkin methods.

Theorem 4.2 *Assume that* (3.16) *is a first-order homogeneous elliptic system, the coefficients of* $\mathcal{L}$ *and* $\mathcal{B}$ *are of class* $C^q(\overline{\Omega})$ *and* $C^{q+1}(\partial\Omega)$*, respectively,* Ω *is of class* C^{q+1}*, where* $q \geq 0$ *is an integer, and* $\mathcal{T}_h$ *is a uniformly regular partition of* Ω *into*

---

[15] The fact that, for non-homogeneous elliptic systems, $\{J_q,X_q\}$ does not contain any practical CLSPs does not mean that such principles do not exist. Examples of practical CLSPs for non-homogeneous elliptic systems are presented in Chapter 5. The catch is that the energy balances leading to such practical CLSPs are stated in the norms of the spaces $\mathbf{D}(\Omega)$ and $\mathbf{C}(\Omega)$ of *vector-valued functions* and can only be identified in the *vector-operator* setting. In other words, the feature that makes the ADN setting so general, i.e., its exclusive reliance on products of scalar Sobolev spaces in the energy balances, is also responsible for preventing the ADN setting from uncovering practical energy balances that cannot be stated in terms of such product spaces.

[16] $Q(\cdot;\cdot)$ and $F(\cdot)$ in (4.20) are given in (4.15).

finite elements. Let $X^h_{r,0}$, $r \geq 1$, be the finite element approximating space defined in (4.19). *Then, we have the following results.*

1. *The compliant DLSP $\{J_0, X^h_{r,0}\}$ has a unique minimizer u^h; moreover,*

$$\|u - u^h\|_1 \leq C \inf_{v^h \in X^h_{r,0}} \|u - v^h\|_1 . \tag{4.21}$$

2. *Assume that the minimizer of the parent CLSP $\{J_0, X_{0,0}\}$ belongs to the space $X_p \cap X_{0,0}$ for some $p \geq 0$ and let $\widetilde{r} = \min\{r, p\}$. Then,*

$$\|u - u^h\|_1 \leq C h^{\widetilde{r}} \|u\|_{\widetilde{r}+1} . \tag{4.22}$$

   If, in addition Ω is such that the first-order system has full elliptic regularity, then we also have the optimal $L^2(\Omega)$ estimate

$$\|u - u^h\|_0 \leq C h^{\widetilde{r}+1} \|u\|_{\widetilde{r}+1} . \tag{4.23}$$

3. *Let $\{\vec{\phi}_i\}$ and $\{\phi_i\}$ denote standard bases for $X^h_{r,0}$ and G^r, respectively, and let* Q *denote the least-squares discretization matrix associated with* (4.20), *having elements $\mathsf{Q}_{ij} = Q(\vec{\phi}_j; \vec{\phi}_i)$. Then,* $\mathrm{cond}(\mathsf{Q}) = O(h^{-2})$ *and* Q *is spectrally equivalent to the $M \times M$ block-diagonal matrix*

$$\mathsf{K} = \mathrm{diag}(\mathsf{D}, \ldots, \mathsf{D}) \tag{4.24}$$

   with diagonal blocks $\mathsf{D}_{ij} = (\phi_j, \phi_i)_1$.

Proof. The first assertion of the theorem follows directly from Theorem 3.28; the a priori error estimate (4.21) is a specialization of the error bound (3.63) for an abstract compliant LSFEM, and the asymptotic error estimate (4.22) follows from (4.21) and the approximation property (B.20) of nodal spaces; see Theorem B.6. Assuming that the PDE problem has full elliptic regularity, the $L^2(\Omega)$ estimate (4.23) follows by a standard duality argument based on the Aubin–Nitsche lemma [123, Theorem 3.2.4, p. 137].

To prove the last part of the theorem, note that

$$(\vec{u}^h)^\top \mathsf{Q} \vec{u}^h = Q(u^h; u^h) \qquad \text{and} \qquad (\vec{u}^h_i)^\top \mathsf{D} \vec{u}^h_i = (u^h_i, u^h_i)_1 .$$

Because $Q(\cdot;\cdot)$ is continuous and coercive on $X_{0,0}$, we have that

$$C^{-1} \sum_{i=1}^{M} (\vec{u}^h_i)^\top \mathsf{D} \vec{u}^h_i \leq (\vec{u}^h)^\top \mathsf{Q} \vec{u}^h \leq C \sum_{i=1}^{M} (\vec{u}^h_i)^\top \mathsf{D} \vec{u}^h_i .$$

Therefore, Q and $\mathrm{diag}(\mathsf{D}, \ldots, \mathsf{D})$ are spectrally equivalent.

To prove the condition number estimate, note that, from (B.85), it follows that $X^h_{r,0}$ satisfies the inverse assumption (3.59) in Section 3.5.1 with $\omega_i = 1$, $i = 1, \ldots, M$. Therefore, we can apply Theorem 3.29 with $\omega_{max} = 1$ and that establishes that $\mathrm{cond}(\mathsf{Q}) = O(h^{-2})$. □

Theorem 4.2 asserts that the CLSP class of a first-order homogeneous elliptic system always contains a principle that gives rise to a practical compliant DLSP. The bilinear form associated with that CLSP is equivalent to a product of $H^1(\Omega)$ norms so that the resulting linear system has the same condition numbers and can be preconditioned by the same preconditioners as in the standard Galerkin method for the Poisson equation (1.52). Indeed, suppose that T is one such preconditioner. We can assume that T is spectrally equivalent to the matrix D. Using the spectral equivalence of the least-squares matrix Q and the block-diagonal matrix K in (4.24), it follows that the block-diagonal matrix

$$\widetilde{\mathsf{K}} = \mathrm{diag}(\mathsf{T},\dots,\mathsf{T}) \tag{4.25}$$

is a spectrally equivalent preconditioner for Q. In other words, for first-order homogeneous elliptic systems, LSFEMs offer the same attractive setting as do Galerkin methods[17] for the Poisson equation.

4.5 Least-Squares Finite Element Methods for Non-Homogeneous Elliptic Systems

In the ADN setting, practical compliant DLSPs are not available for non-homogeneous elliptic systems (see Remark 4.3.2), so that we turn to quasi-norm-equivalent and norm-equivalent DLSPs in our search for practical DLSPs.

4.5.1 *Quasi-Norm-Equivalent Discrete Least-Squares Principles*

We now apply the transformation process, described in Section 3.4.1, to generate quasi-norm-equivalent DLSPs for non-homogeneous elliptic systems. We use $\{J_{-1}, X_{-1,0}\}$ as a parent CLSP because it satisfies Assumption 3.20. In particular, the space $X_{-1,0}$ is approximated by a direct product of nodal spaces whose degrees are specified by a multi-index $(r) = (r_1,\dots,r_M)$, $r_j \ge 1$:

$$X^h_{(r),0} = \{u^h \mid u^h \in \prod_{j=1}^{M} G^{r_j}, \quad \mathcal{B}u^h = 0 \quad \text{on } \partial\Omega\}. \tag{4.26}$$

Remark 4.3 The definition of $X^h_{(r),0}$ requires some further clarification. The different approximation orders in (4.26) are needed to define LSFEMs that are optimally accurate with respect to the solution space components. On the other hand, considering that the first l components of $X_{-1,0}$ are $L^2(\Omega)$ spaces, using C^0 finite elements

[17] Recall that in the Poisson equation setting, Galerkin methods coincide with Rayleigh–Ritz methods.

to approximate them may seem unnecessary. The main reason to use these elements is the same as the reason that prompted us to make Assumption 3.20, namely, by doing so we can restrict the potential impracticalities of CLSPs to the approximation of norms and inner products only. Indeed, in general, conversion of an impractical CLSP to a conforming practical DLSP via the transition diagram (3.54) entails approximation of both the norm-generating and the problem-defining operators so that $\mathcal{S}_Y \circ \mathcal{L}$ is replaced by $\mathcal{S}_Y^h \circ \mathcal{L}^h$; see Section 3.4.1. By using C^0 elements for the $L^2(\Omega)$ components of $X_{-1,0}$, approximation of $\mathcal{L}$ can be avoided and we can focus on approximating the norm-generating operator[18] $\mathcal{S}_Y$ for the minus one norm. □

The most straightforward approximation of $\mathcal{S}_Y$ is by the identity operator $\mathcal{I}$. With this choice, the transition diagram (3.54) specializes to

$$\begin{array}{ccccc} J_{-1}(u;f) &=& \sum_{i=1}^{k} \left\| \sum_{j=1}^{M} \mathcal{L}_{ij}u_j - f_i \right\|_{-1}^2 &+& \sum_{i=k+1}^{M} \left\| \sum_{j=1}^{M} \mathcal{L}_{ij}u_j - f_i \right\|_0^2 \\ \downarrow & & \downarrow & & \downarrow \\ J_{\mathcal{I}}(u;f) &=& \sum_{i=1}^{k} \left\| \sum_{j=1}^{M} \mathcal{L}_{ij}u_j - f_i \right\|_0^2 &+& \sum_{i=k+1}^{M} \left\| \sum_{j=1}^{M} \mathcal{L}_{ij}u_j - f_i \right\|_0^2 . \end{array} \tag{4.27}$$

Effectively, this transition reduces the least-squares functional in (4.18) to the same least-squares functional as in (4.13) so that we can use the same finite element space $X_{r,0}^h$, defined in (4.19), to discretize it. The resulting DLSP $\{J_{\mathcal{I}}, X_{r,0}^h\}$ is identical in appearance to the compliant DLSP $\{J_0, X_{r,0}^h\}$ analyzed in Theorem 4.2. However, the two discrete principles have fundamentally different mathematical properties.

The key assumption in the definition of J_0 is that the first-order system is homogeneous elliptic and this is what made this functional norm-equivalent. In contrast, its identical twin $J_{\mathcal{I}}$ is derived as an *approximation* of the norm-equivalent functional J_{-1} for a non-homogeneous elliptic system. As a result, even though $\{J_0, X_{r,0}^h\}$ and $\{J_{\mathcal{I}}, X_{r,0}^h\}$ appear identical, the former is *compliant* whereas the latter is generally only *quasi-norm-equivalent*.[19]

---

[18] Recall that, according to Theorem A.1, for the minus one norm $\mathcal{S}_Y = (-\Delta)^{-1/2}$.

[19] We always talk about the norm equivalence of a particular functional in the context of an energy balance established by a choice of a parent CLSP. Therefore, by saying that a functional identical to the one in (4.13) (which earlier was declared norm-equivalent) is not norm-equivalent, we mean this with respect to the spaces X_{-1} and Y_{-1} prescribed by the parent principle $\{J_{-1}, X_{-1,0}\}$. However, a non-homogeneous elliptic system can possess another CLSP with respect to which $J_{\mathcal{I}}$ is norm-equivalent. Obviously, such a CLSP necessarily has to prescribe a different combination of data and solution spaces. The main implication, as far as LSFEMs are concerned, is that the choice of the parent principle prescribes the solution norm with respect to which we seek to obtain optimally convergent approximations. Thus, if we seek optimal error estimates with respect to the norm of X_{-1}, then $J_{\mathcal{I}}$ is not the right functional to minimize. However, the same functional could still work if we were to seek optimal error estimates in other norms. We encounter examples of this situation in Chapter 5, where a functional that is deemed non-norm-equivalent by the *ADN setting* turns out to be norm-equivalent in the *vector-operator* setting. We have already explained that the root cause for this ostensible discord is the exclusive reliance of the ADN setting on products of scalar Sobolev spaces in the energy balances.

The most straightforward approach assigns the same relative importance to all discrete residual terms. On the other hand, the presence of different norms in (4.18) means that this norm-equivalent functional assigns different relative importance to different residual terms. To change the relative importance of the terms in $J_{\mathcal{I}}$ in a way that mimics their correct hierarchy in J_{-1}, recall that the norm generating operator for the minus one norm is $(-\Delta)^{-1/2}$; see Theorem A.1 and Example 3.23. Thus, $\mathcal{S}_Y$ can be approximated by the scaled identity operator[20] in (B.89) with $\alpha = -1/2$, i.e., $\mathcal{S}_Y^h = h\mathcal{I}$. Then, the transition diagram (3.54) specializes to

$$\begin{array}{ccc} J_{-1}(u;f) = \sum_{i=1}^{k} \Big\| \sum_{j=1}^{M} \mathcal{L}_{ij}u_j - f_i \Big\|_{-1}^2 & + & \sum_{i=k+1}^{M} \Big\| \sum_{j=1}^{M} \mathcal{L}_{ij}u_j - f_i \Big\|_0^2 \\ \downarrow \qquad\qquad \downarrow & & \downarrow \\ J_{-1}^h(u;f) = h^2 \sum_{i=1}^{k} \Big\| \sum_{j=1}^{M} \mathcal{L}_{ij}u_j - f_i \Big\|_0^2 & + & \sum_{i=k+1}^{M} \Big\| \sum_{j=1}^{M} \mathcal{L}_{ij}u_j - f_i \Big\|_0^2 . \end{array} \tag{4.28}$$

The choice of (B.89) to approximate $\mathcal{S}_Y$ replaces the $H^{-1}(\Omega)$ norm in (4.17) by its quasi-norm-equivalent approximation $\|\cdot\|_{-1,h}$ defined in (B.91). The resulting quasi-norm-equivalent DLSP is given by[21]

$$\{J_{-1}^h(u;f), X_{(r),0}^h\}, \tag{4.29}$$

where $J_{-1}^h(u;f)$ is defined in (4.28) and $X_{(r),0}^h$ is the space defined in (4.26).

Remark 4.4 The same idea can be applied to replace the $H^1(\Omega)$ norms in (4.17) by practical approximations. The norm-generating operator for the $H^1(\Omega)$ norm is $(-\Delta)^{1/2}$ (see Example 3.23) and its approximation by a scaled identity is given also by (B.89), but with $\alpha = 1/2$. This approximation gives rise to the weighted least-squares functional

$$J_0^h(u;f) = \sum_{i=1}^{k} \Big\| \sum_{j=1}^{M} \mathcal{L}_{ij}u_j - f_i \Big\|_0^2 + h^{-2} \sum_{i=k+1}^{M} \Big\| \sum_{j=1}^{M} \mathcal{L}_{ij}u_j - f_i \Big\|_0^2 \tag{4.30}$$

in which $H^1(\Omega)$ norms are replaced by their quasi-norm-equivalent approximation $\|\cdot\|_{1,h}$, defined in (B.91). Note that (4.28) and (4.30) differ by the common (and unimportant for the minimization) factor h^2 so that minimization of (4.30) over a space X^h is equivalent to the minimization of J_{-1}^h over the same space. Again, the

[20] In the context of least-squares methods this idea was first used in [18] to develop optimally accurate LSFEMs for second-order elliptic boundary value problems. Application to first-order formulations of the Stokes equations can be found in [53, 55] and extension to general first-order ADN elliptic systems was provided in [41].

[21] For the last time we note that, in the notation we are using, the pair (4.29) implies the minimization principle

$$\min_{u^h \in X_{(r),0}^h} J_{-1}^h(u^h;f).$$

reason to consider $\{J_{-1}, X_{-1,0}\}$ rather than $\{J_0, X_{0,0}\}$ as a parent CLSP for this minimization problem is that $\{J_0, X_{0,0}\}$ fails to satisfy Assumption 3.20 by formally requiring subspaces of $H^2(\Omega)$. □

Because the choices of $\mathcal{S}_Y^h$ in (4.27) and (4.28) replace negative norms by $L^2(\Omega)$ and weighted $L^2(\Omega)$ norms, respectively, the discrete functionals are not defined for all functions in X_{-1}. However, it is easy to see that Assumption 3.27 is satisfied by setting $\widetilde{X} = X_q$ and $\widetilde{Y} = Y_q$ for any $q \geq 0$. The choice of these spaces imposes additional regularity assumptions on the minimizer of $\{J_{-1}, X_{-1,0}\}$ and so it is convenient to choose the pair with the lowest regularity, i.e., $\widetilde{X} = X_0$ and $\widetilde{Y} = Y_0$, where X_0 and Y_0 are the spaces defined in (4.8) and (4.9).

Let $|||\cdot|||_{\mathcal{I}}$ and $|||\cdot|||_{-1,h}$ denote the discrete energy norms associated with $J_{\mathcal{I}}$ and J_{-1}^h, respectively. The following technical result proves useful later on.

Lemma 4.5 *For every* $w \in \widetilde{X}$,

$$|||w|||_{\mathcal{I}} \leq C \sum_{j=1}^{M} \|w_j\|_1 \tag{4.31}$$

and

$$|||w|||_{-1,h} \leq C \left(\sum_{j=1}^{l} \Big(\|w_j\|_0 + h\|w_j\|_1 \Big) + \sum_{j=l+1}^{M} \|w_j\|_1 \right). \tag{4.32}$$

Proof. The choice $\widetilde{X} = X_0$ guarantees that all component functions w_j of $w \in X_0$ are at least in $H^1(\Omega)$. The first inequality then follows by noting that each $\mathcal{L}_{ij}$ is at most a first-order differential operator and that $|||w|||_{\mathcal{I}}$ is a sum of expressions $\|\mathcal{L}_{ij} w_j\|_0$ that are bounded from above by the $H^1(\Omega)$ norm of w_j.

To prove (4.32), it is necessary to take under consideration the weights assigned to each operator $\mathcal{L}_{ij}$. Using the triangle inequality, we can break $|||\cdot|||_{-1,h}$ into four groups of terms:

$$\begin{aligned} C|||w|||_{-1,h} \leq & \sum_{i=1}^{k} h \Big(\underbrace{\sum_{j=1}^{l} \|\mathcal{L}_{ij} w_j\|_0}_{(I)} + \underbrace{\sum_{j=l+1}^{M} \|\mathcal{L}_{ij} w_j\|_0}_{(II)} \Big) \\ & + \sum_{i=k+1}^{M} \Big(\underbrace{\sum_{j=1}^{l} \|\mathcal{L}_{ij} w_j\|_0}_{(III)} + \underbrace{\sum_{j=l+1}^{M} \|\mathcal{L}_{ij} w_j\|_0}_{(IV)} \Big). \end{aligned}$$

Recall that the weights are chosen so that the order of each operator $\mathcal{L}_{ij}$ is bounded by $s_i + t_j$. If $\mathcal{L}_{ij}$ belongs to group (I), then $s_i = 0$, $t_j = 1$ and its order is at most one. If $\mathcal{L}_{ij}$ belongs to group (II), $s_i = 0$, $t_j = 2$, but the order of $\mathcal{L}_{ij}$ still cannot exceed one because we are dealing with a first-order system. If $\mathcal{L}_{ij}$ is in group (III), then $s_i = -1$, $t_j = 1$ and its order is zero. Finally, for $\mathcal{L}_{ij}$ in group (IV), $s_i = -1$, $t_j = 2$ and its order is bounded by one. Therefore,

$$\|\mathcal{L}_{ij}w_j\|_0 \le C \begin{cases} \|w_j\|_1 & \text{for groups } (I), (II), \text{ and } (IV) \\ \|w_j\|_0 & \text{for group } (III) \end{cases}$$

and (4.32) easily follows. □

Lemma 4.5 can be extended to discrete functions.

Lemma 4.6 *The inequalities* (4.31) *and* (4.32) *hold for all functions* $u^h \in X^h_{(r),0}$. *If the inverse inequality* (B.85) *holds for the components of* $X^h_{(r),0}$, *then*

$$|||u^h|||_{-1,h} \le C\Big(\sum_{j=1}^{l} \|u^h_j\|_0 + \sum_{j=l+1}^{M} \|u^h_j\|_1\Big). \tag{4.33}$$

Proof. The choice of $X^h_{(r),0}$ in (4.26) guarantees that all components u^h_j of $u^h \in X^h_{(r),0}$ belong to $H^1(\Omega)$. Using the same arguments as in the proof of Lemma 4.5, we see that (4.31) and (4.32) hold for all $u^h \in X^h_{(r),0}$. The inequality (4.33) is a direct consequence of (4.32) and the inverse inequality (B.85). □

The following theorem shows that the straightforward and weighted least-squares functionals satisfy the same mesh-dependent energy balance.

Theorem 4.7 *Assume that* $X^h_{(r),0}$ *is as defined in* (4.26). *Then, there exist positive constants* C_1 *and* C_2 *such that* (3.70) *holds for* $J_{\mathcal{I}}$ *and* J^h_{-1} *with*

$$\widehat{C}_1(h) = C_1 \qquad \text{and} \qquad \widehat{C}_2(h) = h^{-1}C_2\,.$$

Proof. Let $u^h \in X^h_{(r),0}$. Consider first the straightforward principle $\{J_{\mathcal{I}}, X^h_{(r),0}\}$. Using the inclusion $X^h_{(r),0} \subset X_{-1}$, the energy balance (4.10) with $q=-1$, the fact that $H^{-1}(\Omega)$ is continuously embedded in $L^2(\Omega)$, and Lemma 4.6 gives

$$\begin{aligned}
\sum_{j=1}^{l}\|u^h_j\|_0 + \sum_{j=l+1}^{M}\|u^h_j\|_1 &\le C\Big(\sum_{i=1}^{k}\Big\|\sum_{j=1}^{M}\mathcal{L}_{ij}u^h_j\Big\|_{-1} + \sum_{i=k+1}^{M}\Big\|\sum_{j=1}^{M}\mathcal{L}_{ij}u^h_j\Big\|_0\Big)\\
&\le C\Big(\sum_{i=1}^{k}\Big\|\sum_{j=1}^{M}\mathcal{L}_{ij}u^h_j\Big\|_0 + \sum_{i=k+1}^{M}\Big\|\sum_{j=1}^{M}\mathcal{L}_{ij}u^h_j\Big\|_0\Big) = C|||u^h|||_{\mathcal{I}}\\
&\le C\sum_{j=1}^{M}\|u^h_j\|_1 \le C\Big(C_I h^{-1}\sum_{j=1}^{l}\|u^h_j\|_0 + \sum_{j=l+1}^{M}\|u^h_j\|_1\Big)\\
&\le h^{-1}C\Big(\sum_{j=1}^{l}\|u^h_j\|_0 + \sum_{j=l+1}^{M}\|u^h_j\|_1\Big).
\end{aligned}$$

This proves the assertion for $\{J_{\mathcal{I}}, X^h_{(r),0}\}$. The proof for $\{J^h_{-1}, X^h_{(r),0}\}$ is similar and uses (4.33) from Lemma 4.6:

$$\sum_{j=1}^{l}\|u_j^h\|_0+\sum_{j=l+1}^{M}\|u_j^h\|_1\le C\Big(\sum_{i=1}^{k}\|\sum_{j=1}^{M}\mathcal{L}_{ij}u_j^h\|_{-1}+\sum_{i=k+1}^{M}\|\sum_{j=1}^{M}\mathcal{L}_{ij}u_j^h\|_0\Big)$$
$$\le C\Big(\sum_{i=1}^{k}\|\sum_{j=1}^{M}\mathcal{L}_{ij}u_j^h\|_0+\sum_{i=k+1}^{M}\|\sum_{j=1}^{M}\mathcal{L}_{ij}u_j^h\|_0\Big)$$
$$\le h^{-1}C\Big(h\sum_{i=1}^{k}\|\sum_{j=1}^{M}\mathcal{L}_{ij}u_j^h\|_0+\sum_{i=k+1}^{M}\|\sum_{j=1}^{M}\mathcal{L}_{ij}u_j^h\|_0\Big)=h^{-1}C\,|||u^h|||_{-1,h}$$
$$\le h^{-1}C\Big(\sum_{j=1}^{l}\|u_j^h\|_0+\sum_{j=l+1}^{M}\|u_j^h\|_1\Big). \qquad \Box$$

The fact that both the straightforward $J_{\mathcal{I}}$ and weighted J_{-1}^h functionals satisfy identical energy balances does not really tell the whole story with regard to their accuracy. As we have seen from the abstract error analysis in Theorem 3.39, the optimality of such methods is contingent upon the structure of the discrete energy norm and the order of the approximation error for a set of sufficiently regular functions as measured by this norm.

Our next result shows that the choice of weights in (4.28) provides the proper balance between the error terms in the abstract estimate (3.77) and, with a proper choice of the finite element space $X_{(r),0}^h$, leads to an optimally accurate LSFEM defined by the first-order optimality condition for (4.29):

$$\textit{seek } u^h\in X_{(r),0}^h \quad \textit{such that} \quad Q_{-1}^h(u^h;v^h)=F_{-1}^h(v^h) \qquad \forall v^h\in X_{(r),0}^h, \tag{4.34}$$

where

$$Q_{-1}^h(u^h;v^h)=h^2\sum_{i=1}^{k}\Big(\sum_{j=1}^{M}\mathcal{L}_{ij}u_j^h,\sum_{j=1}^{M}\mathcal{L}_{ij}v_j^h\Big)_0+\sum_{i=k+1}^{M}\Big(\sum_{j=1}^{M}\mathcal{L}_{ij}u_j^h,\sum_{j=1}^{M}\mathcal{L}_{ij}v_j^h\Big)_0$$

and

$$F_{-1}^h(v^h)=h^2\sum_{i=1}^{k}\Big(f_i,\sum_{j=1}^{M}\mathcal{L}_{ij}v_j^h\Big)_0+\sum_{i=k+1}^{M}\Big(f_i,\sum_{j=1}^{M}\mathcal{L}_{ij}v_j^h\Big)_0 .$$

Theorem 4.8 *Assume that* (3.16) *is a first-order non-homogeneous elliptic system, the coefficients of $\mathcal{L}$ and $\mathcal{B}$ are of class $C^{q+1}(\overline{\Omega})$ and $C^{q+2}(\partial\Omega)$, respectively, Ω is of class C^{q+2}, where $q\ge 0$ is an integer, and $\mathcal{T}_h$ is a uniformly regular partition of Ω into finite elements. Let the indices s_i and t_j be given by* (4.5) *and* (4.6), *respectively and, for $r\ge 1$,*

$$X_{(r),0}^h=X_{r+,0}^h:=\Big\{u^h\mid u^h\in\prod_{j=1}^{l}G^r\times\prod_{j=l+1}^{M}G^{r+1};\quad \mathcal{B}u^h=0\quad on\ \partial\Omega\Big\}. \tag{4.35}$$

Then, the following results hold.

1. *The quasi-norm-equivalent DLSP $\{J_{-1}^h,X_{r+,0}^h\}$ has a unique minimizer u^h.*

2. *Assume that the minimizer of the parent CLSP* $\{J_{-1}, X_{-1,0}\}$ *belongs to the space*

$$X_{p,0} = \{u \mid u \in \prod_{j=1}^{l} H^{p+1}(\Omega) \times \prod_{j=l+1}^{M} H^{p+2}(\Omega); \quad \mathcal{B}u = 0 \ \ on \ \partial\Omega\},$$

for some $p \geq 0$ *and let* $\widetilde{r} = \min\{r, p\}$*. Then,*

$$\sum_{j=1}^{l} \|u_j - u_j^h\|_0 + \sum_{j=l+1}^{M} \|u_j - u_j^h\|_1 \leq Ch^{\widetilde{r}+1}\Big(\sum_{j=1}^{l} \|u_j\|_{\widetilde{r}+1} + \sum_{j=l+1}^{M} \|u_j\|_{\widetilde{r}+2}\Big) \quad (4.36)$$

and

$$\sum_{j=1}^{l} \|u_j - u_j^h\|_1 \leq Ch^{\widetilde{r}}\Big(\sum_{j=1}^{l} \|u_j\|_{\widetilde{r}+1} + \sum_{j=l+1}^{M} \|u_j\|_{\widetilde{r}+2}\Big). \quad (4.37)$$

3. *The condition number of the linear system corresponding to the LSFEM* (4.34) *is bounded by* $O(h^{-4})$.

Proof. Assume that (4.36) has already been established. We now prove that (4.37) holds. Let $j = 1, \ldots, l$. Using the triangle inequality, the inverse inequality (B.85), the approximation property (B.20), and (4.36) itself, we obtain

$$\begin{aligned}
\|u_j - u_j^h\|_1 &\leq \|u_j - v_j^h\|_1 + Ch^{-1}\|u_j^h - v_j^h\|_0 \\
&\leq Ch^{-1}\big(\|u_j - v_j^h\|_0 + h\|u_j - v_j^h\|_1\big) + Ch^{-1}\|u_j - u_j^h\|_0 \\
&\leq Ch^{-1}\left(h^{\widetilde{r}+1}\|u_j\|_{\widetilde{r}+1} + h^{\widetilde{r}+1}\Big(\sum_{j=1}^{l} \|u_j\|_{\widetilde{r}+1} + \sum_{j=l+1}^{M} \|u_j\|_{\widetilde{r}+2}\Big)\right)
\end{aligned}$$

from which (4.37) easily follows.

Let us now prove the rest of the theorem. From (4.34), it is clear that $\{J_{-1}^h, X_{r^+,0}^h\}$ is a consistent principle. Therefore, the abstract approximation result in Theorem 3.39 applies provided we can verify that Assumptions 3.33–3.36 hold for $\{J_{-1}^h, X_{r^+,0}^h\}$.

The parent principle of $\{J_{-1}^h, X_{r^+,0}^h\}$ is $\{J_{-1}, X_{-1,0}\}$. Therefore, the spaces X and Y from the abstract theory are given by $X_{-1,0}$ and Y_{-1}, defined in (4.18) and (4.9), respectively, and

$$Y^* = (Y_{-1})^* = \prod_{i=1}^{k} H^1(\Omega) \times \prod_{i=k+1}^{M} L^2(\Omega).$$

As a result, Assumption 3.33 trivially holds.

To verify Assumption 3.34, we define the spaces $X^{(i)}$ and $Y^{(i)}$ as follows. For $i = 1, \ldots, k$,

$$X^{(i)} = \prod_{i=1}^{l} H^2(\Omega) \times \prod_{i=l+1}^{M} H^3(\Omega) \qquad \text{and} \qquad Y^{(i)} = \prod_{i=1}^{k} H^1(\Omega) \times \prod_{i=k+1}^{M} H^2(\Omega)$$

and, for $i = k+1, \ldots, M$,

$$X^{(i)} = \prod_{i=1}^{l} L^2(\Omega) \times \prod_{i=l+1}^{M} H^1(\Omega) \qquad \text{and} \qquad Y^{(i)} = \prod_{i=1}^{k} H^{-1}(\Omega) \times \prod_{i=k+1}^{M} L^2(\Omega)\,.$$

Succinctly,

$$X^{(i)} = X_{q_i} \qquad \text{and} \qquad Y^{(i)} = Y_{q_i} \qquad \text{with} \qquad q_i = \begin{cases} 1 \text{ for } i = 1, \ldots, k \\ -1 \text{ for } i = k+1, \ldots, M\,, \end{cases}$$

where X_q and Y_q are the spaces defined in (4.8) and (4.9), respectively. The choice of $Y^{(i)}$ meets the first requirement of Assumption 3.34 that the ith component space in $Y^{(i)}$ is the same as the ith component of the dual space Y^*. The choice of $X^{(i)}$, on the other hand, ensures that the "stronger" energy balance (3.71) in Assumption 3.34 follows from the energy balance (4.10) with $q = 1$ or $q = -1$.

Assumption 3.35 is trivially satisfied because $\mathcal{S}_Y^h$ is a scaled identity operator. The final Assumption 3.36 is a regularity hypothesis whose validity depends on the smoothness of $\partial\Omega$ and the equation coefficients. We take it for granted that all necessary prerequisites for this assumption to hold are satisfied for our problem.

This completes the verification of the assumptions needed to apply Theorem 3.39. The first conclusion from this theorem is that $\{J_{-1}^h, X_{r+,0}^h\}$ has a unique minimizer u^h. The abstract error estimate (3.77) also holds for $\{J_{-1}^h, X_{r+,0}^h\}$ with $X^{(i)}$, $Y^{(i)}$, and Y_i^* being the spaces defined above. To obtain (4.36) from the abstract result, we need to estimate the best approximation error as measured by the discrete energy norm and the terms appearing in (3.78).

We begin with the bound for the discrete energy norm. Let $I(v)$ denote finite element approximation of a function $v \in \widetilde{X}$ defined by applying the bounded projection operator Π_G (see (B.15)) to its component functions. Using (4.32) in Lemma 4.5 and the fact that (B.20) holds component-wise for $I(u)$ gives that

$$\begin{aligned}
&\inf_{w^h \in X_{r+,0}^h} |||u - w^h|||_{-1,h} \\
&\quad \le C\left(\sum_{j=1}^{l} \Big(\|u_j - I(u_j)\|_0 + h\|u_j - I(u_j)\|_1 \Big) + \sum_{j=l+1}^{M} \|u_j - I(u_j)\|_1 \right) \qquad (4.38)\\
&\quad \le Ch^{\widetilde{r}+1}\left(\sum_{j=1}^{l} \|u_j\|_{\widetilde{r}+1} + \sum_{j=l+1}^{M} \|u_j\|_{\widetilde{r}+2} \right).
\end{aligned}$$

To estimate the terms appearing in (3.78), it is necessary to consider two cases. If the index i is between 1 and k, then $Y_i^* = H^1(\Omega)$, the discrete norm-generating operator $\mathcal{S}_{Y_i}^h$ is given by the scaled identity $h\mathcal{I}$, and $(\mathcal{S}_{Y_i}^h)^{-2} = h^{-2}\mathcal{I}$. In this case, the problems (3.75) that define the functions $v^{(i)}$ specialize to

$$\begin{cases} \mathcal{L}_i v^{(i)} = h^{-2} f_i & \text{in } \Omega \\ \mathcal{L}_j v^{(i)} = 0 & \text{in } \Omega \text{ for } j \neq i \\ \mathcal{B} v^{(i)} = 0 & \text{on } \partial\Omega \,, \end{cases}$$

where $f_i \in Y_i^* = H^1(\Omega)$. Thanks to Assumption 3.36, each $v^{(i)}$ belongs to $X^{(i)} = X_1$ and, from the energy balance (4.10) with $q = 1$, it follows that

$$\sum_{j=1}^{l} \|v_j^{(i)}\|_2 + \sum_{j=l+1}^{M} \|v_j^{(i)}\|_3 \le C h^{-2} \|f_i\|_1 \,.$$

Because $v^{(i)} \in X_1$ and $r \ge 1$, from (4.32), (B.20), and this inequality, we find that

$$\begin{aligned} &\inf_{v^h \in X^h_{r+,0}} |||v^{(i)} - v^h|||_{-1,h} \\ &\le C\Big(\sum_{j=1}^{l} \Big(\|v_j^{(i)} - I(v_j^{(i)})\|_0 + h\|v_j^{(i)} - I(v_j^{(i)})\|_1 \Big) + \sum_{j=l+1}^{M} \|v_j^{(i)} - I(v_j^{(i)})\|_1 \Big) \\ &\le C h^2 \Big(\sum_{j=1}^{l} \|v_j^{(i)}\|_2 + \sum_{j=l+1}^{M} \|v_j^{(i)}\|_3 \Big) \le C \|f_i\|_1 \,. \end{aligned}$$

As a result, for the first k terms in (3.78),

$$\sup_{f_i \in Y_i^*} \frac{\inf_{v^h \in X^h_{r+,0}} |||v^{(i)} - v^h|||_{-1,h}}{\|f_i\|_{Y_i^*}} \le \sup_{f_i \in H^1(\Omega)} \frac{|||v^{(i)} - I(v^{(i)})|||_{-1,h}}{\|f_i\|_1} \le C \sup_{f_i \in H^1(\Omega)} \frac{\|f_i\|_1}{\|f_i\|_1} = C \,.$$

For the second group of terms, i.e., when $k+1 \le i \le M$, we have that $Y_i^* = L^2(\Omega)$ and $\mathcal{S}^h_{Y_i}$ is the identity operator $\mathcal{I}$. In this case, the functions $v^{(i)}$ belong to $X^{(i)} = X_{-1}$ and solve the problems

$$\begin{cases} \mathcal{L}_i v^{(i)} = f_i & \text{in } \Omega \\ \mathcal{L}_j v^{(i)} = 0 & \text{in } \Omega \text{ for } j \neq i \\ \mathcal{B} v^{(i)} = 0 & \text{on } \partial\Omega \,, \end{cases}$$

where $f_i \in Y_i^* = L^2(\Omega)$. Because $v^{(i)} \in X_{-1}$, the finite element approximation $I(v^{(i)})$ may be undefined for some of the components of $v^{(i)}$. Fortunately, this is not a problem because the inclusion $0 \in X^h_{r+,0}$ implies the upper bound

$$\inf_{v^h \in X^h_{r+,0}} |||v^{(i)} - v^h|||_{-1,h} \le |||v^{(i)}|||_{-1,h} \,,$$

whereas the fact that $\mathcal{L}_j v^{(i)} = 0$ for $j \neq i$ implies the identity

$$|||v^{(i)}|||_{-1,h} = \|\mathcal{L}_i v^{(i)}\|_0 = \|f_i\|_0 .$$

Consequently, for $k+1 \leq i \leq M$,

$$\sup_{f_i \in Y_i^*} \frac{\inf_{v^h \in X^h_{r+,0}} |||v^{(i)} - v^h|||_{-1,h}}{\|f_i\|_{Y_i^*}} \leq \sup_{f_i \in L^2(\Omega)} \frac{|||v^{(i)}|||_{-1,h}}{\|f_i\|_0} \leq C \sup_{f_i \in L^2(\Omega)} \frac{\|f_i\|_0}{\|f_i\|_0} = C .$$

Thus, we have shown that

$$\sup_{f_i \in Y_i^*} \frac{\inf_{v^h \in X^h_{r+,0}} |||v^{(i)} - v^h|||_{-1,h}}{\|f_i\|_{Y_i^*}} \leq C$$

for all $i = 1, \ldots, M$. The error estimate (4.36) follows by using this bound and (4.38) in the abstract error estimate (3.77).

The discrete weak equation (4.34) gives rise to a matrix Q in the usual manner. To estimate the condition number of this matrix, we use the discrete energy balance of $\{J^h_{-1}, X^h_{r+,0}\}$. According to Lemma 4.7,

$$C_1 \|u^h\|^2_{X_{-1}} \leq Q^h_{-1}(u^h;u^h) \leq C_2 h^{-2} \|u^h\|_{X_{-1}} \qquad \forall u^h \in X^h_{r+,0} .$$

Using the continuity of the embedding $H^1(\Omega) \subset L^2(\Omega)$ and the inverse inequality (B.85), we obtain

$$\|u^h\|^2_0 \leq \|u^h\|^2_{X_{-1}} = \sum_{j=1}^{l} \|u^h_j\|^2_0 + \sum_{j=l+1}^{M} \|u^h_j\|^2_1 \leq Ch^{-2} \|u^h\|^2_0 .$$

Therefore,

$$C_1 \|u^h\|^2_0 \leq Q^h_{-1}(u^h;u^h) \leq C_2 h^{-4} \|u^h\|_0$$

which implies that cond(Q) is bounded by $O(h^{-4})$. □

The key to proving the optimal error estimates for $\{J^h_{-1}, X^h_{r+,0}\}$ is the ability to show that the first k terms in (3.78) are bounded by a constant; for the last $M-k$ terms this is trivial. It can be shown, although we do not prove it here, that for the straightforward principle $\{J_{\mathcal{I}}, X^h_{r,0}\}$, the last $M-k$ terms in (3.78) are also bounded by a constant, but that the first k are not. In fact, it is not difficult to see that the best upper bound that one can obtain for these terms is Ch^{-1}. As a result, for the straightforward method, the abstract error estimate from Theorem 3.39 yields at best suboptimal convergence rates.

One may, of course, argue that the upper bounds implied by Theorem 3.39 for the straightforward principle $\{J_{\mathcal{I}}, X^h_{r,0}\}$ are not necessarily sharp and that in practice this method may perform much better. Unfortunately, these hopes prove futile. In

Chapter 7, we present computational examples that clearly demonstrate the suboptimal convergence of the straightforward LSFEM for a non-homogeneous elliptic first-order Stokes system.

4.5.2 Norm-Equivalent Discrete Least-Squares Principles

The transformations of $\{J_{-1}, X_{-1,0}\}$ into the two discrete quasi-norm-equivalent least-squares principles just considered used fairly simple approximations of the norm-generating operator $\mathcal{S}_Y = (-\Delta)^{-1/2}$. Thanks to this simplicity, the resulting LSFEMs are very easy to implement. However, for the same reason, these methods are far from flawless. The most straightforward method $\{J_{\mathcal{I}}, X^h_{r,0}\}$ has attractive condition numbers but is not provably optimal. The weighted method $\{J^h_{-1}, X^h_{r+,0}\}$ is provably optimal but has higher condition numbers.

We now consider a third transformation of $\{J_{-1}, X_{-1,0}\}$ into a DLSP that uses the same type of a finite element space $X^h_{r+,0}$ as $\{J^h_{-1}, X^h_{r+,0}\}$, but employs a much more sophisticated approximation[22] $\mathcal{S}^h_Y$ of the norm-generating operator. The resulting DLSP turns out to be norm-equivalent and avoids the pitfalls of its simpler cousins. However, the associated finite element method is not as easy to implement as are the quasi-norm-equivalent methods just discussed.

The starting point in the development of this *discrete negative-norm* method is Lemma B.1 which asserts that the mesh-dependent norm $\|\cdot\|_{-h}$ defined in (B.99) is, under some assumptions, equivalent to $\|\cdot\|_{-1}$ on $X^h_{r+,0}$. Recall that $\|\cdot\|_{-h}$ corresponds to $\mathcal{S}^h_Y = h\mathcal{I} + (\mathcal{K}^h)^{1/2}$, where $\mathcal{K}^h$ is a spectrally equivalent approximation of the Galerkin solution operator for $-\Delta$ with homogeneous Dirichlet boundary condition. The latter term is critical for the definition of $\mathcal{S}^h_Y$ because, without it, this operator reduces to the *quasi-norm-equivalent* scaled identity approximation $\mathcal{S}^h_Y = h\mathcal{I}$ of the norm-generating operator.

Using $\mathcal{S}^h_Y = h\mathcal{I} + (\mathcal{K}^h)^{1/2}$ in the transition diagram (3.54), with $\{J_{-1}, X_{-1,0}\}$ as a parent CLSP, gives the *discrete negative-norm* least-squares functional:

$$
\begin{array}{ccccc}
J_{-1}(u;f) = & \sum_{i=1}^{k} \Big\| \sum_{j=1}^{M} \mathcal{L}_{ij}u_j - f_i \Big\|_{-1}^2 & + & \sum_{i=k+1}^{M} \Big\| \sum_{j=1}^{M} \mathcal{L}_{ij}u_j - f_i \Big\|_0^2 \\
\downarrow & \downarrow & & \downarrow \\
J_{-h}(u;f) = & \sum_{i=1}^{k} \Big\| \sum_{j=1}^{M} \mathcal{L}_{ij}u_j - f_i \Big\|_{-h}^2 & + & \sum_{i=k+1}^{M} \Big\| \sum_{j=1}^{M} \mathcal{L}_{ij}u_j - f_i \Big\|_0^2
\end{array}
\tag{4.39}
$$

and the discrete negative-norm least-squares principle

$$\{J_{-h}(u;f), X^h_{r+,0}\}, \tag{4.40}$$

[22] This approximation was introduced in [77] and subsequently used by many other authors; see, e.g., [36, 39, 41, 68, 78, 79].

where $X^h_{r^+,0} \subset X_{-1,0}$ is defined in (4.35). The first-order optimality system for (4.40) is given by

$$\textit{seek } u^h \in X^h_{r^+,0} \quad \textit{such that} \quad Q_{-h}(u^h;v^h) = F_{-h}(v^h) \qquad \forall v^h \in X^h_{r^+,0}, \tag{4.41}$$

where

$$Q_{-h}(u^h;v^h) = \sum_{i=1}^{k}\Big(\sum_{j=1}^{M}\mathcal{L}_{ij}u^h_j, \sum_{j=1}^{M}\mathcal{L}_{ij}v^h_j\Big)_{-h} + \sum_{i=k+1}^{M}\Big(\sum_{j=1}^{M}\mathcal{L}_{ij}u^h_j, \sum_{j=1}^{M}\mathcal{L}_{ij}v^h_j\Big)_0$$

and

$$F_{-h}(v^h) = \sum_{i=1}^{k}\Big(f_i, \sum_{j=1}^{M}\mathcal{L}_{ij}v^h_j\Big)_{-h} + \sum_{i=k+1}^{M}\Big(f_i, \sum_{j=1}^{M}\mathcal{L}_{ij}v^h_j\Big)_0.$$

The weak problem (4.41) defines the *discrete negative-norm* LSFEM. Our first result is to show that $\{J_{-h}, X^h_{r^+,0}\}$ satisfies the discrete norm-equivalence condition (3.65) stipulated in Section 3.5.2.

Theorem 4.9 *Let $|||u|||_{-h} = J_{-h}(u;0)^{1/2}$ be the discrete energy norm associated with $J_{-h}(u;f)$. Then,*

$$C_1\|u^h\|_{X_{-1}} \le |||u^h|||_{-h} \le C_2\|u^h\|_{X_{-1}}. \tag{4.42}$$

Proof. Because $u^h_j \in G^r$ or $u^h_j \in G^{r+1}$ and the order of each $\mathcal{L}_{ij}$ is at most one, $\mathcal{L}_{ij}u^h_j \in L^2(\Omega)$ for all $i,j = 1,\dots,M$. Using the lower bound in (B.101), the norm equivalence of (4.18) and the fact that $X^h_{r^+,0}$ is a subspace of X_{-1} yields

$$\begin{aligned}
|||u^h|||^2_{-h} &= \sum_{i=1}^{k}\Big\|\sum_{j=1}^{M}\mathcal{L}_{ij}u^h_j\Big\|^2_{-h} + \sum_{i=k+1}^{M}\Big\|\sum_{j=1}^{M}\mathcal{L}_{ij}u^h_j\Big\|^2_0 \\
&\ge C_1\Big(\sum_{i=1}^{k}\Big\|\sum_{j=1}^{M}\mathcal{L}_{ij}u^h_j\Big\|^2_{-1} + \sum_{i=k+1}^{M}\Big\|\sum_{j=1}^{M}\mathcal{L}_{ij}u^h_j\Big\|^2_0\Big) = C_1J_{-1}(u^h;0) \\
&\ge C_1\Big(\sum_{j=1}^{l}\|u^h_j\|^2_0 + \sum_{j=l+1}^{M}\|u^h_j\|^2_1\Big) = C_1\|u^h\|^2_{X_{-1}}.
\end{aligned}$$

To prove the upper bound, we proceed as in the proof of Lemma 4.5 and break the discrete energy norm into four groups of terms:

$$C|||u^h|||_{-h} \le \sum_{i=1}^{k}\Big(\underbrace{\sum_{j=1}^{l}\|\mathcal{L}_{ij}u^h_j\|_{-h}}_{(I)} + \underbrace{\sum_{j=l+1}^{M}\|\mathcal{L}_{ij}u^h_j\|_{-h}}_{(II)}\Big)$$

$$+\sum_{i=k+1}^{M}\Big(\underbrace{\sum_{j=1}^{l}\|\mathcal{L}_{ij}u_j^h\|_0}_{(III)}+\underbrace{\sum_{j=l+1}^{M}\|\mathcal{L}_{ij}u_j^h\|_0}_{(IV)}\Big).$$

Recall that the order of all operators $\mathcal{L}_{ij}$ from groups (I), (II), and (IV) is at most one and that the order of all operators from group (III) is zero. Using this and the upper bound in (B.101), we see that

$$\|\mathcal{L}_{ij}u_j^h\|_{-h}\le C\Big(h\|\mathcal{L}_{ij}u_j^h\|_0+\|\mathcal{L}_{ij}u_j^h\|_{-1}\Big)\le C\Big(h\|u_j^h\|_1+\|u_j^h\|_0\Big)\le C\|u_j^h\|_0$$

for all terms in groups (I) and (II),

$$\|\mathcal{L}_{ij}u_j^h\|_0\le C\|u_j^h\|_0$$

for all terms in group (III), and

$$\|\mathcal{L}_{ij}u_j^h\|_0\le C\|u_j^h\|_1$$

for all terms in group (IV). It is now easy to see that

$$|||u^h|||_{-h}\le C_2\Big(\sum_{j=1}^{l}\|u_j^h\|_0+\sum_{j=l+1}^{M}\|u_j^h\|_1\Big)=C_2\|u^h\|_{X_{-1}}.\qquad\Box \tag{4.43}$$

The norm-equivalence result in this theorem shows that the abstract approximation theory from Section 3.5.2 is applicable to the discrete negative-norm least-squares principle. It turns out that the norm-equivalence of J_{-h} also allows us to reduce the minimal approximation order of the finite element space that is necessary to prove optimal error estimates. Specifically, the next theorem shows that, instead of the space $X^h_{r+,0}$ for which $r\ge 1$, we can use the space

$$X^h_{\widehat{r},0}=\Big\{u^h \mid u^h\in\prod_{j=1}^{l}G^{\widehat{r}}\times\prod_{j=l+1}^{M}G^{r+1};\quad \mathcal{B}u^h=0\quad\text{on }\partial\Omega\Big\}, \tag{4.44}$$

where $\widehat{r}=\max\{1,r\}$ and $r\ge 0$. Note that with this definition and $r=0$, the first l components of $X^h_{\widehat{r},0}$ default to the lowest-order C^0 space G^1 rather than to the lowest-order $L^2(\Omega)$ finite element subspace S^0. Although the latter would still yield a conforming subspace of X_{-1}, the reasons to use this definition of $X^h_{\widehat{r},0}$ are the same as in Remark 4.3. We have more to say about this topic in Remark 4.11.

We now specialize the abstract result in Theorem 3.31 to $\{J_{-h},X^h_{\widehat{r},0}\}$.

Theorem 4.10. *Assume that* (3.16) *is a first-order non-homogeneous elliptic system, the coefficients of $\mathcal{L}$ and $\mathcal{B}$ are of class $C^{q+1}(\overline{\Omega})$ and $C^{q+2}(\partial\Omega)$, respectively, Ω is of class C^{q+2}, where $q\ge 0$ is an integer, and $\mathcal{T}_h$ is a uniformly regular partition of Ω into finite elements. Let the indices s_i and t_j be given by* (4.5) *and* (4.6), *respectively,* *and* $X^h_{\widetilde{r},0}$, $r \geq 0$ *be the finite element approximating space defined in* (4.44). *Then, we have the following results.*

1. *The discrete negative-norm DLSP* $\{J_{-h}, X^h_{\widetilde{r},0}\}$ *has a unique minimizer* u^h.
2. *Assume that the minimizer of the parent CLSP* $\{J_{-1}, X_{-1,0}\}$ *belongs to the space* $X_{p,0}$, $p \geq 0$, *defined in Theorem* 4.8, *and let* $\widetilde{r} = \min\{r, p\}$. *Then,*

$$\sum_{j=1}^{l} \|u_j - u_j^h\|_0 + \sum_{j=l+1}^{M} \|u_j - u_j^h\|_1 \leq Ch^{\widetilde{r}+1}\Big(\sum_{j=1}^{l} \|u_j\|_{\widetilde{r}+1} + \sum_{j=l+1}^{M} \|u_j\|_{\widetilde{r}+2}\Big) \quad (4.45)$$

and

$$\sum_{j=1}^{l} \|u_j - u_j^h\|_1 \leq Ch^{\widetilde{r}}\Big(\sum_{j=1}^{l} \|u_j\|_{\widetilde{r}+1} + \sum_{j=l+1}^{M} \|u_j\|_{\widetilde{r}+2}\Big). \quad (4.46)$$

3. *Let* Q *be the matrix associated with* (4.41), $\mathsf{G} = (\phi_j, \phi_i)_0$ *be the Gramm matrix for the basis of* $G^{\widehat{r}}$, *and* $\mathsf{D} = (\phi_j, \phi_i)_1$ *be the Dirichlet matrix for the basis of* G^{r+1}. *Then, the condition number of* Q *is bounded by* $O(h^{-2})$ *and this matrix is spectrally equivalent to the* $M \times M$ *block-diagonal matrix*

$$\mathsf{K}_{-h} = (\underbrace{\mathsf{G}, \ldots, \mathsf{G}}_{l}, \underbrace{\mathsf{D}, \ldots, \mathsf{D}}_{M-l}). \quad (4.47)$$

Proof. The proof that (4.46) holds provided (4.45) is true follows exactly the same argument as in the proof of Theorem 4.8. Therefore, we proceed with the rest of the assertions in the present theorem.

Theorem 4.9 established that J_{-h} is norm-equivalent; this is the sole prerequisite for applying the abstract theory of Section 3.5.2. Therefore, all conclusions of Theorem 3.31 hold for $\{J_{-h}, X^h_{\widetilde{r},0}\}$. In particular, this DLSP has a unique minimizer $u^h \in X^h_{\widetilde{r},0}$ that depends continuously on the data and the abstract error estimate (3.67) holds with $X = X_{-1,0}$ and $|||\cdot|||_h = |||\cdot|||_{-h}$.

To prove (4.45), note that (4.41) implies that $\{J_{-h}, X^h_{\widetilde{r},0}\}$ is consistent. Therefore, the last term in (3.67) vanishes and the abstract error estimate (3.67) specializes to

$$\|u - u^h\|_{X_{-1}} \leq C\Big(\inf_{v^h \in X^h_{\widetilde{r},0}} |||u - v^h|||_{-1} + \inf_{v^h \in X^h_{\widetilde{r},0}} |||u - v^h|||_{-h}\Big), \quad (4.48)$$

where $|||\cdot|||_{-1} = J_{-1}(u;0)^{1/2}$. Let $I(u) \in X^h_{\widetilde{r},0}$ be the finite element approximation of u, defined as in Theorem 4.8. Then, (4.45) follows from (4.48), the inequalities

$$\inf_{v^h \in X^h_{\widetilde{r},0}} |||u - v^h|||_{-1} \leq |||u - I(u)|||_{-1}, \qquad \inf_{v^h \in X^h_{\widetilde{r},0}} |||u - v^h|||_{-h} \leq |||u - I(u)|||_{-h},$$

and the approximation properties (B.20).

The spectral equivalence between Q and the matrix K_{-h} follows from the identities $(\vec{u}^h)^T \mathsf{Q} \vec{u}^h = Q_{-h}(u^h; u^h)$, $(\vec{u}^h_j)^T \mathsf{D} \vec{u}^h_j = (u^h_j, u^h_j)_1$, $(\vec{u}^h_j)^T \mathsf{G} \vec{u}^h_j = (u^h_j, u^h_j)_0$, and (4.42). This also implies that $\mathrm{cond}(\mathsf{Q}) = O(h^{-2})$. □

Remark 4.11 For $r \geq 1$, the finite element space defined in (4.44) is identical with the finite element space (4.35) in Theorem 4.8. However, Theorem 4.8 is valid only for $r \geq 1$. This *minimal approximation condition* requires that all variables associated with indices t_j whose value equals 2 must be approximated by at least G^2 spaces such as piecewise quadratic, biquadratic, or triquadratic elements. The reason for this condition is the duality argument used in the proof of Theorem 4.8. In contrast, in Theorem 4.10, the minimal approximation condition for these variables is relaxed to $r \geq 0$. Therefore, by using a discrete negative-norm, the variables whose indices t_j equal 2 can be approximated by G^1 spaces such as piecewise linear, bilinear, or trilinear elements. This means that the first group of variables (those whose indices t_j equal 1) could, in principle, be approximated by subspaces of $L^2(\Omega)$. However, the functional obtained through the transition diagram (4.39) is not appropriate for such spaces because it still uses the original first-order differential operators that are not defined for finite element subspaces of $L^2(\Omega)$. For this reason, the finite element space (4.44) used in Theorem 4.10 is defined in a such a way that it defaults to a product of equal-order G^1 elements if $r = 0$.

If, instead, we use a space such as

$$X^h_{0+,0} = \left\{ u^h \mid u^h \in \prod_{j=1}^{l} S^h \times \prod_{j=l+1}^{M} G^1; \quad \mathcal{B}u^h = 0 \quad \text{on } \partial\Omega \right\}, \tag{4.49}$$

where S^h is a discontinuous subspace of $L^2(\Omega)$ (see Section B.2.1), we would have to modify (4.39) to the transition diagram

$$\begin{array}{ccccc} J_{-1}(u;f) & = & \sum_{i=1}^{k} \left\| \sum_{j=1}^{M} \mathcal{L}_{ij} u_j - f_i \right\|_{-1}^2 & + & \sum_{i=k+1}^{M} \left\| \sum_{j=1}^{M} \mathcal{L}_{ij} u_j - f_i \right\|_0^2 \\ \downarrow & & \downarrow & & \downarrow \\ J_{-h}(u;f) & = & \sum_{i=1}^{k} \left\| \sum_{j=1}^{M} \mathcal{L}^h_{ij} u_j - f_i \right\|_{-h}^2 & + & \sum_{i=k+1}^{M} \left\| \sum_{j=1}^{M} \mathcal{L}_{ij} u_j - f_i \right\|_0^2, \end{array} \tag{4.50}$$

where, for $i = 1, \ldots, k$, $j = 1, \ldots, M$, the discrete operators $\mathcal{L}^h_{ij}$ approximate $\mathcal{L}_{ij}$ on subspaces of $L^2(\Omega)$. Such methods are still conforming in the sense that $X^h_{0+,0}$ is a proper subspace of X_{-1}; however, their analysis is outside the scope of this chapter. □

Remark 4.12 Because in the ADN setting the unknowns are broken down into independent scalar components u_i, every component can be approximated by an independently chosen finite element subspace of $H^1(\Omega)$. This permits, among other things, the use of equal-order C^0 interpolation spaces defined with respect to the same partition of Ω into finite elements, even for non-homogeneous elliptic systems. Of course, for such systems, the equal-order finite element space still has to satisfy the minimal approximation condition stated in Remark 4.11. This means that the quasi-norm-equivalent DLSP (4.29) can be implemented using the equal-order nodal space of *second degree*

$$X_{2,0}^h = \{u^h \mid u^h \in \prod_{j=1}^{M} G^2, \quad \mathcal{B}u^h = 0 \quad \text{on } \partial\Omega\}$$

but not using the equal-order nodal space of *first degree*

$$X_{1,0}^h = \{u^h \mid u^h \in \prod_{j=1}^{M} G^1, \quad \mathcal{B}u^h = 0 \quad \text{on } \partial\Omega\}. \qquad \square$$

Theorem 4.9 shows that a norm-equivalent DLSP is optimally accurate and gives rise to linear systems with condition numbers comparable to those arising in standard Galerkin methods. Block preconditioners for these systems can be designed by using the fact that (4.47) is spectrally equivalent to the discretization matrix Q. To define the preconditioner, the blocks G are replaced by any matrix that is spectrally equivalent to the Gramm matrix and the blocks D are replaced by any standard preconditioner T for the Poisson equation. For example, G can be replaced by a *diagonally lumped* mass matrix $h^d\,\mathsf{I}$. Then, the block diagonal matrix

$$\widetilde{\mathsf{K}}_{-h} = \operatorname{diag}(\underbrace{h^d\,\mathsf{I},\dots,h^d\,\mathsf{I}}_{l},\underbrace{\mathsf{T},\dots,\mathsf{T}}_{M-l}) \tag{4.51}$$

is a preconditioner for (4.41).

The existence of good preconditioners is critical for discrete negative-norm methods because, as a rule, the matrix representation of the operator $\mathcal{K}^h$ that enters the definition of $\mathcal{S}_Y^h$ is dense. This rules out solution methods that require assembly of the actual matrix and makes iterative solvers the only viable option for computing the finite element solutions.

4.6 Concluding Remarks

In this chapter, we presented a comprehensive least-squares framework, based on the ADN setting, to derive the energy balances that are applicable to virtually any elliptic boundary value problem. The process of formulating a LSFEM started with recasting the given problem into an equivalent first-order system so as to satisfy the first key to practicality. A general transformation procedure is described in Section 4.1. The second step is to obtain a reasonable collection of CLSPs for the first-order system. By using the ADN theory, we are able to "automate" this process by reducing it to the verification of algebraic conditions on the principal parts of the differential and boundary operators. The two steps in this process: identification of the energy balances for the first-order systems and formulation of well-posed minimization problems were explained in Sections 4.2 and 4.3, respectively.

The price one pays for the generality of the ADN setting is that it forces us to treat any first-order system as either homogeneous elliptic or non-homogeneous elliptic. In the former case, there always exists a compliant DLSP that satisfies all practicality requirements set forth in Section 2.2.2: the matrices and the right-hand sides are easily computable, discretization can be accomplished with easy to use C^0 finite element spaces, and discrete problems have condition numbers comparable with those of standard Galerkin methods.

If, however, the first-order system have the misfortune of being classified by the ADN theory as non-homogeneous elliptic, then a completely different picture emerges. Most important, for such systems, the ADN theory does not allow us to measure all equation residuals using the same norm and leads to CLSPs *not having* practical compliant DLSPs. As a result, for such systems, the ADN setting forces us to consider quasi-norm-equivalent or norm-equivalent DLSPs. In Sections 4.5, we applied the abstract transition process to a parent CLSP that uses negative norms. We used three different approximations of the negative-norm generating operator that varied in their level of "binding" to that operator. Table 4.1 summarizes the properties of the three LSFEMs. Among other things, this table shows that if we decide to use the ADN setting to develop LSFEMs and our problem is classified as non-homogeneous elliptic, then we are led to LSFEMs that cannot be simultaneously optimally accurate, easy to implement, and conditioned as are standard Galerkin methods.[23]

DLSP	$\mathcal{S}_Y^h$	Norm-equivalence	Coding	Accuracy	Condition
$\{J_{\mathcal{I}}, X_{r,0}^h\}$	$\mathcal{I}$	quasi-norm-equivalent	simple	suboptimal	$O(h^{-2})$
$\{J_{-1}^h, X_{r^+,0}^h\}$	$h\mathcal{I}$	quasi-norm-equivalent	simple	optimal	$O(h^{-4})$
$\{J_{-h}, X_{\hat{r},0}^h\}$	$h\mathcal{I} + (\mathcal{K}^h)^{1/2}$	norm-equivalent	not as simple	optimal	$O(h^{-2})$

Table 4.1 Comparison of three conforming DLSPs obtained from the same parent principle $\{J_{-1}, X_{-1,0}\}$ by using three different approximations $\mathcal{S}_Y^h$ of the $H^{-1}(\Omega)$ norm-generating operator.

[23] We did see, however, that if one uses iterative solution methods for the associated matrix problem, the third choice in Table 4.1 is not so difficult to implement.

Part III
Least-Squares Finite Element Methods for Elliptic Problems

Chapter 5
Scalar Elliptic Equations

In this and the next chapter, we study least-squares finite element methods (LSFEMs) for two classes of second-order elliptic partial differential equations (PDEs) that are prototypes of many mathematical models of practical importance in science and engineering. Chapter 6 focuses on *vector* equations whereas in this chapter we consider *scalar* second-order elliptic boundary value problems. In the latter case, we consider two forms of such equations. The first is cast in terms of a scalar-valued function ϕ that solves the problem

$$\begin{cases} -(\Theta_0^{-1}\nabla\cdot\Theta_1)\nabla\phi+\gamma\phi=f & \text{in } \Omega \\ \phi=0 \quad \text{on } \Gamma \qquad \text{and} \qquad \mathbf{n}\cdot\Theta_1\nabla\phi=0 & \text{on } \Gamma^*. \end{cases} \tag{5.1}$$

In the second setting, we seek a scalar-valued function ψ that satisfies the complementary system

$$\begin{cases} -\nabla\cdot(\Theta_2^{-1}\nabla\Theta_3)\psi+\gamma\psi=f & \text{in } \Omega \\ \Theta_3\psi=0 \quad \text{on } \Gamma \qquad \text{and} \qquad \mathbf{n}\cdot(\Theta_2^{-1}\nabla\Theta_3)\psi=0 & \text{on } \Gamma^*. \end{cases} \tag{5.2}$$

We refer to (5.1) and (5.2) as *complementary* because ϕ and ψ can be thought of as mathematical entities that describe complementary, or dual, properties of the same physical field. In what follows, we refer to ϕ and ψ as the *potential* and the *density*, respectively.[1] We return to this topic in Section 5.3, where we discuss first-order reformulations of (5.1) and (5.2).

In (5.1) and (5.2), $\gamma\in\mathbb{R}$ is a non-dimensional parameter or function and Θ_i are non-degenerate weights, i.e., there exist real constants $\alpha_i>0$, $i=0,\dots,3$, such that

$$\frac{1}{\alpha_i}\le\Theta_i\le\alpha_i,\quad i=0,3 \qquad \text{and} \qquad \frac{1}{\alpha_i}\vec{\xi}^{\,T}\vec{\xi}\le\vec{\xi}^{\,T}\Theta_i\vec{\xi}\le\alpha_i\vec{\xi}^{\,T}\vec{\xi},\quad i=1,2. \tag{5.3}$$

[1] Formally, by letting $\Theta_0=\Theta_3^{-1}$, $\Theta_1=\Theta_2^{-1}$, $\phi=\Theta_3\psi$, and letting γ and f in (5.2) correspond to $\Theta_0\gamma$ and $\Theta_0 f$, respectively, in (5.1), these two problems are the same. The advantages of considering the two complementary descriptions are made apparent below.

P.B. Bochev, M.D. Gunzburger, *Least-Squares Finite Element Methods*, Applied Mathematical Sciences 166, DOI: 10.1007/b13382_5,

Other assumptions on Θ_i, γ, and the data f are made as needed.

The PDEs in (5.1) and (5.2) are often referred to as (linear) scalar *diffusion–reaction* equations because they model physical phenomena that combine diffusion and reaction processes. In this context, the scalar Θ_0 in (5.1) is a positive reaction coefficient and the tensor Θ_1 describes material properties such as the permeability of a porous medium, the thermal conductivity of a heat conductor, or the electron and hole mobility in a semiconductor. Similar identifications can be made for the tensor Θ_2 and scalar Θ_3.

For the purposes of formulating and analyzing LSFEMs, in this chapter it suffices to only consider either $\gamma = 0$ or $\gamma = 1$. The case $\gamma = 1$ arises in two settings: a steady-state diffusion–reaction process and temporal discretizations of parabolic problems. In the latter setting, γ can be identified with the inverse of the time step Δt, suitably normalized to 1, used to effect the temporal discretization; see Chapter 9. The case $\gamma = 0$ corresponds to a generalized Poisson equation.[2] Thus, when $\gamma = 0$ and all weights equal one, both (5.1) and (5.2) reduce to (1.52). On several occasions, we use this simpler problem, instead of the more general ones, to introduce and explain ideas without the distraction of purely technical details.[3]

Viewed as generalizations of the Poisson equation (1.52), the boundary value problems (5.1) and (5.2) are canonical examples of PDEs that naturally correspond to the unconstrained minimization of quadratic functionals. This is why the utility or need for LSFEMs for these problems is often misjudged. After all, Galerkin methods for both (5.1) and (5.2) recover the whole attractive Rayleigh–Ritz setting whereas a straightforward LSFEM, applied to the same equations, fails[4] all three practicality tests from Section 2.2.2.

It turns out that impracticality of straightforward LSFEMs can be easily avoided by switching to a different energy balance, or by using a different transition diagram to obtain the associated discrete least-squares principle (DLSP). This possibility is explored in Section 5.2 for the Poisson equation, where we see that the resulting *practical* DLSPs coincide with the Galerkin method for (1.52). Although interesting, this equivalence between Galerkin and least-squares finite element methods simply underscores the fact that the weak Galerkin problem (1.51) can be arrived at in more than one way, without giving us a truly new method.

A far more important setting for LSFEMs arises when the second-order problems (5.1) and (5.2) are replaced by equivalent first-order systems.[5] This changes the naturally occurring optimization setting from unconstrained to constrained min-

---

[2] The differential operators in (5.1) and (5.2), i.e., $-(\Theta_0^{-1}\nabla \cdot \Theta_1)\nabla$ and $-\nabla \cdot (\Theta_2^{-1}\nabla\Theta_3)$, are generalizations of the usual Laplacian operator $-\Delta = -\nabla \cdot \nabla$ acting on scalar functions.

[3] A third possibility that corresponds to a generalization of the Helmholtz equation (1.63), is $\gamma = -1$. In this case, certain combinations of the scalar weight functions in (5.1) and (5.2) can be identified with a wave number or frequency.

[4] This is due to the need to use approximations that are continuously differentiable and to a too large condition number for the resulting linear algebraic systems.

[5] There are many reasons one may choose to use a first-order form for a second-order elliptic partial differential equation. For example, as is mentioned in Chapter 1, the use of a first-order system formulation enables the direct approximation of solution derivatives and can be used to convert essential boundary conditions into natural ones and vice versa.

imization. For example, switching from the second-order Poisson equation (1.52) to the first-order system (1.55) changes the optimization setting from that of the Dirichlet principle (1.50) to that of the Kelvin principle (1.53). As a result, when applied to (1.55), Galerkin methods recover the weak optimality equations (1.54) which are saddled with onerous stability requirements for the finite element spaces.

For first-order reformulations of (5.1) and (5.2), least-squares principles become an attractive alternative to the naturally occurring optimization problem because, in addition to offering a better variational setting, they already satisfy the first key to practicality in Section 2.2.2. In the sections that follow, we use the abstract least-squares theory developed in the earlier chapters to formulate a collection of LSFEMs for (5.1) and (5.2) that are practical and offer a range of valuable computational properties. Among other things, in Section 5.8, we see that by using balances in suitable Hilbert spaces of vector-valued functions and compatible finite element spaces, it is possible to define LSFEMs that provide approximate solutions that inherit the best properties of both Galerkin and mixed-Galerkin methods. In particular, such compatible least-squares methods can be *locally conservative.*

5.1 Applications of Scalar Poisson Equations

Applications of (5.1) and (5.2) are so numerous that it is out of the question to include examples from all possible science and engineering fields, even if we were to restrict ourselves to a single case per application area. Thus, we content ourselves with three classical examples that involve the generalized Laplace operator from (5.1) and that provide an inkling to the scope of this prototype equation as a modeling tool. Whenever applicable, we briefly explain the physical meaning of the variables ϕ, $\nabla\phi$, and $\Theta_1\nabla\phi$ and the weights Θ_0 and Θ_1.

Heat transfer

The basic model of heat transfer is given by

$$\begin{cases} -\nabla\cdot\kappa\nabla T = q & \text{in } \Omega \\ T = 0 \quad \text{on } \Gamma \quad \text{and} & \mathbf{n}\cdot\kappa\nabla T = 0 \quad \text{on } \Gamma^*, \end{cases} \tag{5.4}$$

where T denotes the *temperature*, κ the *thermal conductivity* of the medium, and q is a given internal *heat source*. Equation (5.4) is a consequence of the Fourier heat-conduction law which states that *heat flux* $\mathbf{Q}$ is proportional to the negative temperature gradient, i.e., $\mathbf{Q} = -\kappa\nabla T$, and satisfies the energy balance $\nabla\cdot\mathbf{Q} = q$. Consequently, $\mathbf{Q}$ is a variable of independent interest. In a homogeneous isotropic media, κ is constant; for non-homogeneous anisotropic materials, $\kappa = \kappa(\mathbf{x})$ is a $d\times d$ symmetric positive definite matrix-valued function. The boundary condition on Γ models situations for which a prescribed temperature is maintained on a portion of the surface of the material sample. The Neumann condition on Γ^* prescribes the heat flux entering or leaving the body.

For time-dependent problems, we replace the PDE in (5.4) by

$$-\nabla \cdot \kappa \nabla T + \gamma c T = q$$

where c denotes the *specific heat capacity*, $\gamma = 1/\Delta t$, and we add an initial condition for T. Note also that, in this case, q also contains an additive term involving γ, c, and possibly κ as well as the values of T and possibly ∇T at one or more previous time levels.

Electrostatics

Physical phenomena driven by stationary electric charges can be modeled by the boundary value problem

$$\begin{cases} -\nabla \cdot \varepsilon_0 \nabla \phi = q & \text{in } \Omega \\ \phi = 0 \quad \text{on } \Gamma \qquad \text{and} \qquad \mathbf{n} \cdot \varepsilon_0 \nabla \phi = 0 \quad \text{on } \Gamma^*, \end{cases} \tag{5.5}$$

where ϕ is the *electrostatic potential* or *voltage*, ε_0 is the *vacuum permittivity*,[6] and q is the *electric charge density*. A variable of practical interest is the *electric field* $\mathbf{E} = -\nabla \phi$. Therefore, electrostatics models phenomena for which the electric field $\mathbf{E}$ can be assumed irrotational, i.e., $\nabla \times \mathbf{E} = 0$. The boundary conditions can be interpreted as the specification of the voltage and electric flux on the Dirichlet and Neumann parts of the boundary, respectively.

Magnetostatics

In contrast to electrostatics for which the charge is assumed stationary, *magnetostatics* describes phenomena associated with *stationary* or *direct* currents. If all currents $\mathbf{J}$ in the system are known, then the basic magnetostatics model is given by

$$\begin{cases} -\nabla \cdot \mu_0 \nabla \phi = \nabla \cdot \mu_0 \mathbf{H}_J & \text{in } \Omega \\ \phi = 0 \quad \text{on } \Gamma \qquad \text{and} \qquad \mathbf{n} \cdot \mu_0 \nabla \phi = 0 \quad \text{on } \Gamma^*, \end{cases} \tag{5.6}$$

where now ϕ denotes the (scalar) *magnetic potential,* μ_0 is the *vacuum permeability,*[7] and the source field $\mathbf{H}_J$ is determined from the given current field by the Biot–Savart law

[6] Also referred to as the electric constant, the dielectric constant of vacuum, or the permittivity of free space.

[7] Also referred to as the magnetic constant or the permeability of free space.

$$\mathbf{H}_J = \frac{1}{4\pi}\int_{\mathbb{R}^3} \frac{\mathbf{J}\times\mathbf{r}}{|\mathbf{r}|^2}dV\,.$$

Variables of practical interest are the *magnetic field intensity* $\mathbf{H} = \nabla\phi + \mathbf{H}_J$ and the *magnetic flux density* $\mathbf{B} = \mu_0\nabla\phi + \mu_0\mathbf{H}_J$. In Section 6.2.1, we consider examples of generalized *vector* Laplacian equations that model the magnetostatics problem directly in terms of these two fields.

Table 5.1 summarizes the above applications and their identification with the prototype equation (5.1). Other classical examples of applications that are modeled by the generalized scalar Poisson equations include, among many others, irrotational flow of an ideal fluid, groundwater flows, single phase flows in porous media, and deformations of elastic membranes. Further examples and details for some of these applications can be found in, e.g., [75, 76, 186, 303, 350].

Application	ϕ	Θ_0	Θ_1	Other variables
Heat transfer	temperature	heat capacity	thermal conductivity	heat flux
Electrostatics	electric potential or voltage		electric permittivity	electric field
Magnetostatics	magnetic potential		magnetic permeability	magnetic field and flux

Table 5.1 Applications of generalized scalar Laplacian equations. The absence of an item indicates that the model can be expressed without that item.

5.2 Least-Squares Finite Element Methods for the Second-Order Poisson Equation

In Section 2.2.2, we use the second-order Poisson equation with homogeneous Dirichlet boundary conditions

$$\begin{cases} -\Delta\phi = f & \text{in } \Omega \\ \phi = 0 & \text{on } \partial\Omega \end{cases} \tag{5.7}$$

to spotlight the inability of straightforward LSFEMs, based on measuring residuals in $L^2(\Omega)$ norms, to deliver practical methods for second-order equations and to motivate their transformation into first-order systems as the first key to obtaining a practical LSFEMs.

In turns out that by using a suitable energy balance, or a suitable transformation diagram to obtain the DLSP, this key can be bypassed. The catch is that in both cases we end up with a least-squares method that coincides with the standard Galerkin method for (5.7). In other words, practical DLSPs for (5.7) default to methods derived from the naturally occurring minimization principle (1.49). Although the absence of genuinely new LSFEMs for the second-order formulation of Poisson-type problems is somewhat disappointing, the fact that we are able to recover the standard Galerkin method from a *least-squares* principle is intriguing enough to warrant a closer examination.

5.2.1 Continuous Least-Squares Principles

For inhomogeneous Dirichlet boundary conditions and domains with sufficiently smooth boundaries, the energy balance for the Poisson equation can be obtained from Theorem D.1. In the case of homogeneous boundary conditions, an energy balance for a larger class of domains can be derived from a general result in [187].

Theorem 5.1 *Assume that Ω is a bounded convex polygon in $\mathbb{R}^2$ or polyhedron in $\mathbb{R}^3$. Then, there exists a positive constant C such that*

$$C\|\phi\|_1 \leq \|\Delta\phi\|_{-1} \qquad \forall\phi \in H_0^1(\Omega) \tag{5.8}$$

and

$$C\|\phi\|_2 \leq \|\Delta\phi\|_0 \qquad \forall\phi \in H^2(\Omega)\cap H_0^1(\Omega). \qquad \square \tag{5.9}$$

This theorem implies that the abstract energy balance (3.18) holds for $\mathcal{L} = -\Delta$ and

$$X_{-1} = H_0^1(\Omega) \qquad \text{and} \qquad Y_{-1} = H^{-1}(\Omega)$$

or

$$X_0 = H^2(\Omega)\cap H_0^1(\Omega) \qquad \text{and} \qquad Y_0 = L^2(\Omega).$$

As in Section 3.2.2, we define continuous least-squares principles (CLSPs) for the Poisson equation using these data spaces to measure the residual energy and these solution spaces to describe the set of candidate minimizers. The first energy balance following from Theorem 5.1 then leads to the well-posed minimization problem

$$\begin{cases} J_{-1}(\phi;f) = \|\Delta\phi + f\|_{-1}^2 \\ \\ X_{-1} = H_0^1(\Omega). \end{cases} \tag{5.10}$$

The second energy balance corresponds to another well-posed minimization problem:

$$\begin{cases} J_0(\phi;f) = \|\Delta\phi + f\|_0^2 \\ \\ X_0 = H^2(\Omega)\cap H_0^1(\Omega). \end{cases} \tag{5.11}$$

We have already encountered, in Section 4.5, negative-norm CLSPs in the context of first-order non-homogeneous elliptic Agmon–Douglis–Nirenberg (ADN) systems. There, we obtained practical DLSPs by either using weighted L^2 norms or discrete negative norms. The following theorem shows that this is unnecessary for the negative norm functional in (5.10) because its optimality condition coincides with (1.51). In other words, the least-squares principle (5.10) and the Dirichlet principle (1.50) for (1.52) have identical minimizers.

Theorem 5.2 *Let* $\phi \in H_0^1(\Omega)$ *denote the minimizer of* (5.10). *Then,* ϕ *solves the weak Galerkin formulation* (1.51), *i.e.,*

$$(\nabla\phi, \nabla\psi)_0 = (f, \psi)_0 \qquad \forall\, \psi \in H_0^1(\Omega).$$

Proof. The first-order optimality condition for (5.10) is given by: seek $\phi \in H_0^1(\Omega)$ such that

$$(\Delta\phi, \Delta\psi)_{-1} = (f, -\Delta\psi)_{-1} \qquad \forall\, \psi \in H_0^1(\Omega). \tag{5.12}$$

From Theorem A.1, it follows that

$$(\Delta\phi, \Delta\psi)_{-1} = ((-\Delta)^{-1}\Delta\phi, \Delta\psi)_0 = (-\phi, \Delta\psi)_0 = (\nabla\phi, \nabla\psi)_0$$

for all $\phi, \psi \in H_0^1(\Omega)$. Applying the same theorem to the right-hand side of (5.12) yields

$$(f, -\Delta\psi)_{-1} = (f, \psi)_0$$

for all $\psi \in H_0^1(\Omega)$. It follows that (5.12) is equivalent to (1.51).[8] □

5.2.2 Discrete Least-Squares Principles

Because the optimality condition (5.12) is equivalent to (1.51), it is clear that conforming discretizations of (5.10) do not require the use of negative norms and results in a practical LSFEM that coincides with the standard Galerkin method for (5.7). On the other hand, obtaining a practical DLSP from (5.11) seems hopeless; after all, in Section 2.2.2, we used[9] that CLSP to argue that the first key to practicality of LSFEMs is transformation of second-order problems into equivalent first-order systems.

---

[8] The fact that the least-squares principle (5.10) and the Dirichlet principle (1.49) are equivalent can also be inferred from the following identity, obtained using Theorem A.14:

$$\begin{aligned} J_{-1}(\phi; f) &= \|(-\Delta)\phi\|_{-1}^2 - 2(f, (-\Delta)\phi)_{-1} + \|f\|_{-1}^2 \\ &= \|\nabla\phi\|_0^2 - 2(f, \phi)_0 + \|f\|_{-1}^2 = 2J_D(\phi; f) + \|f\|_{-1}^2 . \end{aligned}$$

We see that the least-squares functional J_{-1} and the Dirichlet functional J_D differ by the unimportant (for the purpose of minimization) term $\|f\|_{-1}^2$ and factor 2 so that they must have the same minimizer. The identity $\|(-\Delta)\phi\|_{-1} = \|\nabla\phi\|_0$ is also used for the *least-squares optimization methods* considered in Section 12.6.

[9] Recall that the main reason (5.11) is deemed impractical is the order of the derivatives involved.

However, this argument is true only when we look for *conforming* discretizations of (5.11). Recall that conforming DLSPs are obtained via the transition diagram (3.54) in which the problem-defining operators $\mathcal{L}$ and $\mathcal{B}$ are not approximated. If, instead, we use the version of (3.54)

$$\begin{array}{ccccc} J_0(\phi;f) & = & \|(\Delta\phi+f)\|_0^2 & \rightarrow & |||\phi||| \\ \downarrow & & \downarrow & & \downarrow \\ J_0^h(\phi^h;f) & = & \|(\Delta^h\phi^h+f)\|_0^2 & \rightarrow & |||\phi^h|||_h \end{array} \tag{5.13}$$

in which Δ is approximated by some discrete operator Δ^h, then (5.11) can be associated with a practical DLSP. For example, we can use (5.13) with the weak discrete Laplace operator (B.59) to define the following practical DLSP:

$$\begin{cases} J^h(\phi^h;f) = \|\Delta_G^h\phi^h + f\|_0^2 \\ X_{r,0}^h = G^r \cap H_0^1(\Omega)\,. \end{cases} \tag{5.14}$$

Interestingly enough, (5.14) also turns out to coincide with the standard Rayleigh–Ritz and Galerkin methods for (5.7).

Theorem 5.3 *Let* $\phi^h \in X_{r,0}^h$ *denote a minimizer of* (5.14). *Then,* ϕ^h *solves the discrete equation*

$$\left(\nabla\phi^h,\nabla\psi^h\right)_0 = \left(f,\psi^h\right)_0 \qquad \forall\,\psi^h \in X_{r,0}^h\,. \tag{5.15}$$

Proof. The first-order optimality condition for (5.14) is given by: seek $\phi^h \in X_{r,0}^h$ such that

$$\left(\Delta_G^h\phi^h,\Delta_G^h\psi^h\right)_0 = \left(f,\Delta_G^h\psi^h\right)_0 \qquad \forall\,\psi^h \in X_{r,0}^h\,. \tag{5.16}$$

From (B.59), it follows that, considered as a mapping $X_{r,0}^h \mapsto X_{r,0}^h$, the operator Δ_G^h is invertible so that (5.16) is equivalent to

$$\left(\Delta_G^h\phi^h,\psi^h\right)_0 = \left(f,\psi^h\right)_0 \qquad \forall\,\psi^h \in X_{r,0}^h\,.$$

Using again (B.59), we obtain

$$\left(\Delta_G^h\phi^h,\psi^h\right)_0 = \left(\nabla\phi^h,\nabla\psi^h\right)_0 \qquad \forall\,\psi^h \in X_{r,0}^h\,,$$

i.e., (5.16) is equivalent to (5.15). □

5.3 First–Order System Reformulations

In Section 5.2, we saw that practical LSFEMs, defined directly for the second-order Poisson equation (5.7), default to a standard Galerkin method for that problem. In this section, the generalized forms (5.1) and (5.2) of the Poisson equation are transformed into equivalent first-order systems that eventually yield a larger pool of LS-FEMs.

5.3.1 The Div–Grad System

The second-order problem (5.1) can be transformed into an equivalent first-order system using the algorithm given in Section 4.1. According to this algorithm, all first derivatives of ϕ have to be employed as new dependent variables. Therefore, we can choose between[10] $-\Theta_1\nabla\phi$ or $-\nabla\phi$ as new variables. We refer to $-\Theta_1\nabla\phi$ and $-\nabla\phi$ as the *flux* and the *intensity*, respectively, to emphasize the fact that they model *complementary* or *dual* properties of vector-valued physical fields, just as the potential and density modeled dual properties of scalar fields in (5.1) and (5.2).

Remark 5.4 The distinction between $-\Theta_1\nabla\phi$ and $-\nabla\phi$ is not formal because, in practical applications, they have different units. Consider, for example, the electrostatics equations (5.5). In that model, ϕ is the electric potential (with units of volts V), the intensity $-\nabla\phi$ is the electric field $\mathbf{E}$ (with units of volts per meter V/m), and the flux $-\varepsilon_0\nabla\phi$ is the electric displacement[11] field $\mathbf{D}$ (units of coulomb per meter squared C/m^2). The vacuum permittivity ε_0 has units of F/m (farads per meter) and "converts"[12] $\mathbf{E}$ into $\mathbf{D}$. Similarly, in the magnetostatics application (5.6), the intensity is the magnetic field strength $\mathbf{H}$ (with units of amperes per meter A/m) and the flux is magnetic flux density $\mathbf{B}$ (with units of webers per meter squared W/m^2). In a vacuum, conversion of $\mathbf{H}$ into $\mathbf{B}$ is effected by the vacuum permeability μ_0.

Mathematically, the distinction between $-\Theta_1\nabla\phi$ and $-\nabla\phi$ is encoded by regarding the intensity as belonging to the *range* of the gradient in a *primal* De Rham complex (A.52), whereas the flux is viewed as an element in the *domain* of the divergence in a *dual* complex (A.53); see Section A.2.2. In the context of (5.1), the primal complex is $\{G_0(\Omega,\Theta_0),\mathbf{C}_0(\Omega,\Theta_1),\mathbf{D}_0(\Omega,\Theta_2),S_0(\Omega,\Theta_3)\}$, the dual complex is $\{G(\Omega,\Theta_3^{-1}),\mathbf{C}(\Omega,\Theta_2^{-1}),\mathbf{D}(\Omega,\Theta_1^{-1}),S(\Omega,\Theta_0^{-1})\}$, the potential ϕ belongs to $G_0(\Omega,\Theta_0)$, the intensity belongs to $\mathbf{C}_0(\Omega,\Theta_1)$, and the flux belongs to $\mathbf{D}(\Omega,\Theta_1^{-1})$. Intuitively, Θ_1 is the operator $L^2(\Omega,\Theta_1)\mapsto L^2(\Omega,\Theta_1^{-1})$ that converts units of intensity into units of flux; see Remarks A.3 and A.4. □

Remark 5.5 The observations in Remark 5.4 can be further formalized using the language of exterior calculus [101,168,343] that respectively identifies $\{G,\mathbf{C},\mathbf{D},S\}$ with a complex of differential 0, 1, 2, and 3-forms and ∇, $\nabla\times$, and $\nabla\cdot$ with the

[10] A third choice that does not follow from the reduction algorithm in Section 4.1 but is considered in this chapter is to use *both* $-\Theta_1\nabla\phi$ and $-\nabla\phi$ as new variables.

[11] Strictly speaking, $\mathbf{D}=\varepsilon_0\mathbf{E}+\mathbf{P}$, where $\mathbf{P}$ is the polarization density of the material.

[12] The units of the intensity and flux variables can vary from application to application and are not restricted to have the length scalings $1/m$ and $1/m^2$, respectively. For example, in Fick's first law, $J=-\mu\nabla\phi$, the *concentration* ϕ has units of mol/m^3 and so the intensity has units mol/m^4. The flux is the *diffusion flux* J which has units of $mol/(m^2s)$. Conversion between the two fields is effected by the *diffusivity* μ having units of m^2/s.

exterior derivative acting on these spaces. In this setting, the potential is a 0-form, the density is a 3-form, the intensity is a 1-form, the flux is a 2-form, Θ_1 is the Hodge-$\star$ operator acting on 1-forms, and Θ_0 is the Hodge-$\star$ operator acting on 0-forms. However, the use of differential forms requires significant prerequisites in subject matters that are not within the main focus of the book. In lieu of this abstract approach, we have adopted a more familiar setting of weighted Hilbert spaces, traditional vector calculus notation, and informal discussion of the structure of PDEs in terms of these entities. This is sufficient to capture the key ideas relevant to LSFEMs on an intuitive level without wandering into a lengthy detour. For those interested in a rigorous and in-depth discussion of how exterior calculus, differential forms, and algebraic topology can be used to encode PDEs and to solve them numerically, see, e.g., [7,8,13,58,66,69,122,137,139,204,206,223–225,276,314,332,338,339] and the references cited therein. □

We proceed with the reformulation of the model equations into first-order systems. Let us first use the flux $\mathbf{u} = -\Theta_1 \nabla\phi$ as a new dependent variable for (5.1). This transforms the second-order equation into the following *potential–flux* div–grad system:

$$\begin{cases} \nabla\cdot\mathbf{u} + \gamma\Theta_0\phi = \Theta_0 f \\ \nabla\phi + \Theta_1^{-1}\mathbf{u} = 0 \end{cases} \text{in } \Omega \quad \text{and} \quad \begin{matrix} \phi = 0 & \text{on } \Gamma \\ \mathbf{n}\cdot\mathbf{u} = 0 & \text{on } \Gamma^* . \end{matrix} \tag{5.17}$$

When $\gamma = 0$, this system reduces to

$$\begin{cases} \nabla\cdot\mathbf{u} = \Theta_0 f \\ \nabla\phi + \Theta_1^{-1}\mathbf{u} = 0 \end{cases} \text{in } \Omega \quad \text{and} \quad \begin{matrix} \phi = 0 & \text{on } \Gamma \\ \mathbf{n}\cdot\mathbf{u} = 0 & \text{on } \Gamma^* . \end{matrix} \tag{5.18}$$

Remark 5.6 In the sequel, we view the dependent variables appearing in (5.17) and (5.18) and in other first-order systems related to (5.1) and (5.2) in two different ways. In particular, for (5.17) and (5.18), we have the following distinction.

- The scalar-valued variable ϕ and the d components of the vector-valued variable $\mathbf{u}$ are viewed as a collection of $d+1$ scalar variables; in particular, the vector nature of $\mathbf{u}$ or of the vector operators, e.g., the divergence, appearing in the PDE system are not recognized. Each of the $d+1$ variables belongs to a scalar Sobolev space so that the solution spaces for the corresponding least-squares principles are the product of these spaces. This viewpoint is consistent with the application of the ADN theory for the analysis of (5.17) and (5.18) and of the corresponding LSFEMs; see Chapter 4. Naturally, we refer to this setting as the *ADN setting*.
- The vector-valued nature of $\mathbf{u}$ is recognized in that the solution spaces for the corresponding least-squares principles include spaces such as $\mathbf{D}(\Omega)$ and perhaps, for other systems, $\mathbf{C}(\Omega)$ and the vector operators, e.g., the divergence, acting on these Hilbert spaces are taken into account. This viewpoint does not allow for the use of the ADN theory so that the analyses of (5.17) and (5.18) and of LSFEMs must proceed in a direct manner; see Chapter 3. We refer to this setting as the *vector-operator setting*. □

Viewed as problems with $d+1$ scalar unknowns, the systems (5.17) and (5.18) are both square systems of PDEs with three equations and unknowns in $\mathbb{R}^2$ and four equations and unknowns in $\mathbb{R}^3$. Consequently, the ADN theory can be applied without any further modifications of (5.17) and (5.18). Regarding the vector-operator setting, the natural solution spaces for (5.17) and (5.18) are $\phi \in G(\Omega,\Theta_0)$ and $\mathbf{u} \in \mathbf{D}(\Omega,\Theta_1^{-1})$.

A second possibility for transforming (5.1) into a first-order system is to use the intensity $\mathbf{v} = -\nabla\phi$ as a new dependent variable. In this case, the second-order system (5.1) is replaced by a *potential–intensity* version of the first-order div–grad system:

$$\begin{cases} \nabla\cdot\Theta_1\mathbf{v}+\gamma\Theta_0\phi = \Theta_0 f \\ \nabla\phi+\mathbf{v}=0 \end{cases} \quad \text{in } \Omega \qquad \text{and} \qquad \begin{matrix} \phi = 0 & \text{on } \Gamma \\ \mathbf{n}\cdot\Theta_1\mathbf{v}=0 & \text{on } \Gamma^*\,. \end{matrix} \tag{5.19}$$

Natural solution spaces for (5.19) are $\phi \in G(\Omega,\Theta_0)$ and $\mathbf{u} \in \mathbf{C}(\Omega,\Theta_1)$. This system is not used as often as the potential–flux equations (5.17) because, in applications for which the scalar field is less important, the quantity of primary physical interest is the flux,[13] rather than the intensity.

A third possibility, not implied by the reduction algorithm in Section 4.1, is to use *both* $-\nabla\phi$ and $-\Theta_1\nabla\phi$ as new dependent variables. This approach calls for the inclusion of an *algebraic* "constitutive law" which relates the flux and intensity and gives rise to a *potential–flux–intensity* first-order system with three unknown fields:

$$\begin{cases} \nabla\cdot\mathbf{u}+\gamma\Theta_0\phi = \Theta_0 f \\ \nabla\phi+\mathbf{v}=0 \\ \mathbf{v}-\Theta_1^{-1}\mathbf{u}=0 \end{cases} \quad \text{in } \Omega \qquad \text{and} \qquad \begin{matrix} \phi = 0 & \text{on } \Gamma \\ \mathbf{n}\cdot\mathbf{u}=0 & \text{on } \Gamma^*\,. \end{matrix} \tag{5.20}$$

This system uses the two complementary descriptions of the vector field but only one of the two possible descriptions of the scalar field. Recall that this description, i.e., the potential ϕ, is inherited from the parent second-order problem (5.1) and is in the *domain* $G(\Omega,\Theta_0)$ of the gradient in the primal De Rham complex (A.52); see Remark 5.4. By duality, its complementary variable ψ is a density field in the *range* $S(\Omega,\Theta_0^{-1})$ of the divergence in the dual De Rham complex (A.53). The two variables are related by a similar "constitutive law" $\psi = \Theta_0\phi$. Adding the density variable to (5.20) yields a *potential–density–flux–intensity* first–order system having four unknowns:

[13] For example, in oil reservoir simulations, the main goal is to obtain accurate flux approximations. This is the reason why methods for porous media flows are often based on the Kelvin principle (1.53) rather than the Dirichlet principle (1.50); see, e.g., [350] and the references cited therein. In Section 5.3.3, we further discuss the choice of flux versus intensity, using the application examples from Section 5.1.

$$\begin{cases} \nabla\cdot\mathbf{u}+\gamma\psi=\Theta_0 f \\ \nabla\phi+\mathbf{v}=0 \\ \mathbf{v}-\Theta_1^{-1}\mathbf{u}=0 \\ \psi-\Theta_0\phi=0 \end{cases} \quad \text{in } \Omega \qquad \text{and} \qquad \begin{array}{ll} \phi=0 & \text{on } \Gamma \\ \mathbf{n}\cdot\mathbf{u}=0 & \text{on } \Gamma^*. \end{array} \tag{5.21}$$

Natural solution spaces for the complementary pairs of scalar and vector variables are $\{\phi,\psi\}\in G(\Omega,\Theta_0)\times S(\Omega,\Theta_0^{-1})$ and $\{\mathbf{u},\mathbf{v}\}\in \mathbf{D}(\Omega,\Theta_1^{-1})\times\mathbf{C}(\Omega,\Theta_1)$. Their relationships and the structure of (5.21) can be encoded in terms of a *Tonti diagram* [338, 339]:

$$\begin{array}{ccccc} & \text{potential } \phi & \xrightarrow{\nabla} & \mathbf{v} \text{ intensity} & \\ \Theta_0 & \downarrow & & \uparrow & \Theta_1^{-1} \\ & \text{density } \psi & \xleftarrow{\nabla\cdot} & \mathbf{u} \quad \text{flux}. & \end{array} \tag{5.22}$$

Finally, because the intensity $\mathbf{v}$ is in the range of the gradient, (5.21) can be augmented by the redundant equation $\nabla\times\mathbf{v}=0$:

$$\begin{cases} \nabla\cdot\mathbf{u}+\gamma\psi=\Theta_0 f \\ \nabla\phi+\mathbf{v}=0 \\ \nabla\times\mathbf{v}=0 \\ \mathbf{v}-\Theta_1^{-1}\mathbf{u}=0 \\ \psi-\Theta_0\phi=0 \end{cases} \quad \text{in } \Omega \qquad \text{and} \qquad \begin{array}{ll} \phi=0 & \text{on } \Gamma \\ \mathbf{n}\cdot\mathbf{u}=0 & \text{on } \Gamma^*. \end{array} \tag{5.23}$$

Natural solution spaces for (5.23) are the same as for (5.21).

Exactly the same types of first-order systems can be obtained from the complementary second-order problem (5.2) in which the scalar field is represented by the density variable $\psi\in S(\Omega,\Theta_3)$. For example, the counterpart of (5.23) is another potential–density–flux–intensity first-order system

$$\begin{cases} \nabla\cdot\mathbf{u}+\gamma\psi=f \\ \nabla\phi+\mathbf{v}=0 \\ \nabla\times\mathbf{v}=0 \\ \mathbf{v}-\Theta_2\mathbf{u}=0 \\ \psi-\Theta_3^{-1}\phi=0 \end{cases} \quad \text{in } \Omega \qquad \text{and} \qquad \begin{array}{ll} \phi=0 & \text{on } \Gamma \\ \mathbf{n}\cdot\mathbf{u}=0 & \text{on } \Gamma^* \end{array} \tag{5.24}$$

in which $\{\phi,\psi\}\in G(\Omega,\Theta_3^{-1})\times S(\Omega,\Theta_3)$ and $\{\mathbf{u},\mathbf{v}\}\in\mathbf{D}(\Omega,\Theta_2)\times\mathbf{C}(\Omega,\Theta_2^{-1})$. Clearly, problems (5.23) and (5.24) have identical structures and differ only by the placement of their variables in a primal and dual De Rham complex. In (5.24), the potential and intensity are in the dual complex and the density and flux are in the primal complex, whereas in (5.23) the opposite is true, i.e., in (5.24) the roles of the primal and dual complexes are reversed. It follows that any first-order system obtained from (5.1) can be mapped to a like first-order system obtained from (5.2). Consequently, any method and/or theoretical result developed using any of the systems (5.17)–(5.23) trivially extends to both (5.1) and (5.2). To avoid unnecessary duplication, throughout this chapter we work exclusively with first–order systems obtained from (5.1). From all the possible choices of first-order systems listed above, we restrict attention to the potential–flux first-order system (5.17) and the four-field system (5.23). The systems (5.20) and (5.21) were included only to show the stages that led to the final system (5.23), whereas the potential–intensity problem (5.19) is "included" in (5.23).

5.3.2 *The Extended Div–Grad System*

In Section 5.4.1, we verify that the potential–flux div–grad system is not homogeneous elliptic. According to Section 4.5, such systems are not the "best" possible setting[14] for least-squares principles in the ADN setting. Historically, in the least-squares literature, this issue has been dealt with by adding the "redundant" equations[15]

$$\nabla \times \Theta_1^{-1}\mathbf{u} = 0 \quad \text{in } \Omega \qquad \text{and} \qquad \mathbf{n} \times \Theta_1^{-1}\mathbf{u} = 0 \quad \text{on } \Gamma$$

to the potential–flux first-order system (5.17); see, e.g., [95, 96, 103, 105, 110, 112, 115, 238]. The resulting *extended div–grad* system

$$\begin{cases} \nabla \cdot \mathbf{u} + \gamma\Theta_0\phi = \Theta_0 f \\ \nabla\phi + \Theta_1^{-1}\mathbf{u} = 0 \\ \nabla \times \Theta_1^{-1}\mathbf{u} = 0 \end{cases} \quad \text{in } \Omega \qquad \text{and} \qquad \begin{aligned} \phi &= 0 \quad \text{on } \Gamma \\ \mathbf{n} \times \Theta_1^{-1}\mathbf{u} &= 0 \quad \text{on } \Gamma \\ \mathbf{n} \cdot \mathbf{u} &= 0 \quad \text{on } \Gamma^* \end{aligned} \tag{5.25}$$

has four scalar equations and three scalar unknowns in $\mathbb{R}^2$ and seven scalar equations and four scalar unknowns in $\mathbb{R}^3$. Thus, formally, (5.25) is not elliptic in the sense of ADN because it has more equations than unknowns. An equivalent ADN elliptic system can be obtained by adding the curl of a slack variable plus an appropriate boundary condition for that variable. In two dimensions, the slack variable w is a scalar field that is set to zero on Γ (see [112]), resulting in the following ADN version[16] of the extended system (5.25):

---

[14] The reason for this is that practical conforming DLSPs for non-homogeneous elliptic systems require either discrete minus one or weighted norms.

[15] We refer to these equations as being "redundant" because the first one follows from the identity $\nabla \times \nabla\phi = 0$ and the second one is implied by the homogeneous Dirichlet condition on Γ: when $\phi = 0$ on Γ all tangential derivatives of ϕ vanish so that $\mathbf{n} \times \Theta_1^{-1}\mathbf{u} = \mathbf{n} \times \nabla\phi = 0$.

[16] Recall that in two dimensions we have two curl operators: $\nabla\times$ takes vectors into scalars and $\nabla^\perp$ takes scalars into vectors.

$$
\begin{cases} \nabla\cdot\mathbf{u}+\gamma\Theta_0\phi = \Theta_0 f \\ \nabla\phi+\Theta_1^{-1}\mathbf{u}+\nabla^{\perp}w=0 \\ \nabla\times\Theta_1^{-1}\mathbf{u}=0 \end{cases} \quad \text{in } \Omega \quad \text{and} \quad \begin{array}{rl} \phi=0 & \text{on } \Gamma \\ \mathbf{n}\times\Theta_1^{-1}\mathbf{u}=0 & \text{on } \Gamma \\ w=0 & \text{on } \Gamma^* \\ \mathbf{n}\cdot\mathbf{u}=0 & \text{on } \Gamma^* \end{array} \tag{5.26}
$$

that is homogeneous elliptic. It can be shown that w is identically zero so that the slack variable can be ignored and LSFEMs can be formulated using the system (5.25). Similar arguments hold in three dimensions.

The main purpose of the "redundant" curl equation in the extended div–grad system (5.25) is to obtain a homogeneous elliptic system. However, when the assumptions of the ADN theory fail to hold, the addition of the redundant equation turns out to be harmful for LSFEMs based on (5.25). The reason for this is that the natural solution space of (5.25) reverts from[17] $\{\phi,\mathbf{u}\}\in H^1(\Omega)\times[H^1(\Omega)]^d$ to $\phi\in G(\Omega,\Theta_0)$ and $\mathbf{u}\in\mathbf{D}(\Omega,\Theta_1^{-1})\cap\mathbf{C}(\Omega,\Theta_1)$. In Section B.2.2, it is explained why finite element approximations of $\mathbf{D}(\Omega,\Theta_1^{-1})\cap\mathbf{C}(\Omega,\Theta_1)$ are problematic and should be avoided.

Remark 5.7 On an intuitive level, the difficulty with (5.25) stems from the fact that, in this system, a single mathematical entity (the flux variable) is saddled with representing dual properties of the same physical field. In contrast, the four-field system (5.23) avoids this pitfall by using both the flux and the intensity of the physical field. Mathematically, the distinction between (5.25) and (5.23) is reflected in the structure of their natural solution spaces: in (5.25) the physical field is modeled by the *intersection space* $\mathbf{D}(\Omega,\Theta_1^{-1})\cap\mathbf{C}(\Omega,\Theta_1)$ whereas in (5.23) the same field is modeled by the *direct product space* $\mathbf{D}(\Omega,\Theta_1^{-1})\times\mathbf{C}(\Omega,\Theta_1)$. The latter can be approximated without much difficulty using the compatible finite element spaces discussed in Section B.2. □

5.3.3 Application Examples

In this section, we provide some first-order reformulations of the application examples presented in Section 5.1 and briefly discuss the relative importance of intensity and flux variables in these applications.

For the heat–transfer equation (5.4), the potential–flux first-order system (5.17) assumes the form of a *temperature–heat flux* problem

$$
\begin{cases} \nabla\cdot\mathbf{Q}+\gamma cT=q \\ \nabla T+\kappa^{-1}\mathbf{Q}=0 \end{cases} \quad \text{in } \Omega \quad \text{and} \quad \begin{array}{rl} T=0 & \text{on } \Gamma \\ \mathbf{n}\cdot\mathbf{Q}=0 & \text{on } \Gamma^* . \end{array} \tag{5.27}
$$

For the same application, the four fields in (5.23) specialize to temperature, heat density, heat flux, and temperature gradient, respectively, and the first-order system assumes the form

[17] For clarity of notation, boundary conditions are left out from the function space designations.

$$\left\{\begin{aligned} \nabla\cdot\mathbf{Q}+\gamma\rho &= q \\ \nabla T+\mathbf{S} &= 0 \\ \nabla\times\mathbf{S} &= 0 \\ \mathbf{S}-\kappa^{-1}\mathbf{Q} &= 0 \\ \rho - cT &= 0 \end{aligned}\right. \quad \text{in } \Omega \qquad \text{and} \qquad \begin{aligned} T &= 0 \quad \text{on } \Gamma \\ \mathbf{n}\cdot\mathbf{Q} &= 0 \quad \text{on } \Gamma^*. \end{aligned} \tag{5.28}$$

For comparison, the extended div–grad version of the heat–transfer problem corresponding to (5.27) is given by

$$\left\{\begin{aligned} \nabla\cdot\mathbf{Q}+\gamma cT &= q \\ \nabla T+\kappa^{-1}\mathbf{Q} &= 0 \\ \nabla\times\kappa^{-1}\mathbf{Q} &= 0 \end{aligned}\right. \quad \text{in } \Omega \qquad \text{and} \qquad \begin{aligned} T &= 0 \quad \text{on } \Gamma \\ \mathbf{n}\times\kappa^{-1}\mathbf{Q} &= 0 \quad \text{on } \Gamma \\ \mathbf{n}\cdot\mathbf{Q} &= 0 \quad \text{on } \Gamma^*. \end{aligned} \tag{5.29}$$

The relative importance of the heat flux $\mathbf{Q}$ versus the temperature gradient $\mathbf{S}$ varies from application to application. In typical heat–transfer applications such as heating and cooling, the heat flux $\mathbf{Q}$ is of more interest than the temperature gradient. The latter, having units K/m (kelvin per meter), is more relevant in weather and climate models where differences in air temperatures drive large masses of air to equilibrate the temperature. A related concept is the meteorological notion of the *temperature lapse rate* that gives the drop in air temperature with altitude and can be thought of as a negative directional derivative of the temperature along a line perpendicular to the earth's surface.

Electrostatics, e.g., (5.5), is an example of an application where the intensity variable $\mathbf{E}$ is deemed to be more interesting in practice compared to its dual variable $\mathbf{D}$ (see Remark 5.4 for its definition.) For this application, the potential–intensity first-order system (5.19) specializes to a *voltage–electric field* problem

$$\left\{\begin{aligned} \nabla\cdot\varepsilon_0\mathbf{E} &= q \\ \nabla\phi+\mathbf{E} &= 0 \end{aligned}\right. \quad \text{in } \Omega \qquad \text{and} \qquad \begin{aligned} \phi &= 0 \quad \text{on } \Gamma \\ \mathbf{n}\cdot\varepsilon_0\mathbf{E} &= 0 \quad \text{on } \Gamma^*. \end{aligned} \tag{5.30}$$

This system is seldom used in practice and, instead, approximations of $\mathbf{E}$ are derived from a potential computed by a Rayleigh–Ritz method applied to the equivalent second-order system (5.5).

Finally, magnetostatics, e.g. (5.6), is an example of a problem where both the flux and intensity fields $\mathbf{B}$ and $\mathbf{H}$, respectively, are of equal practical interest.

5.4 Energy Balances

We derive two different families of energy balances for the potential–flux and extended div–grad systems. The first family uses the ADN theory and treats $\{\phi,\mathbf{u}\}$ as a collection of $d+1$ scalar-valued functions belonging to the Sobolev spaces

$H^{q+t_j}(\Omega)$. This viewpoint has the advantage that it allows one to "automate" the formulation and analysis of LSFEMs by using the abstract theory of Chapter 4 to derive well-posed CLSPs and then convert them into conforming practical DLSPs. The drawbacks are the smoothness of $\partial\Omega$ required by the ADN theory and the limitations imposed by Theorem D.1 on the boundary conditions, i.e., only one type of boundary condition can be imposed on the whole boundary.

The second family of energy balances regards ϕ and $\mathbf{u}$ in (5.17) or (5.18) as elements of the Hilbert spaces $G(\Omega,\Theta_0)$ and $D(\Omega,\Theta_1^{-1})$, respectively. These energy balances have to be established directly, which takes more effort but allows one to treat mixed boundary conditions and less regular domains. Ultimately, their main advantage turns out to be that they give rise to well-posed CLSPs for (5.17) or (5.18) which use $L^2(\Omega)$ norms to measure all residuals, even though these systems are classified as non-homogeneous elliptic by the ADN theory. As we show, such balances also lead to *compatible* LSFEMs which offer valuable computational properties.

For the potential–density–flux–intensity first-order system (5.23), we only consider energy balances in the vector–operator setting. The reason for this is that energy balances in the ADN setting cannot take advantage of the complementary variables because they cannot distinguish between, e.g., fluxes and intensities. Thus, from the viewpoint of the ADN theory, (5.23) remains non-homogeneous elliptic. As a result, ADN energy balances for (5.23) would lead to LSFEMs that are not much different from those based on (5.17), but with twice the number of variables.

Finally, when stating the spaces entering in the energy balances we adhere to the rule that data space components are listed in the order in which the equations appear in the various PDE formulations and that the solution space components follow the order specified by the list of the variables, e.g., $\{\phi,\mathbf{u}\}$ or $\{\phi,\psi,\mathbf{u},\mathbf{v}\}$.

5.4.1 Energy Balances in the Agmon–Douglis–Nirenberg Setting

Let us start with the potential–flux div–grad systems (5.17) and (5.18). It is easy to see that the "reaction" term in (5.17) does not enter into the principal part so that it does not affect the choice of ADN weights for these systems. Therefore, any results established with $\gamma=0$ automatically extend to the case $\gamma=1$ so that we may as well work with the simpler system (5.18). This setting can be streamlined even further by observing that, as long as (5.3) holds, the ADN weights depend only on the order of the derivatives in the principal part and not on the coefficients[18] multiplying these derivatives. Thus, without loss of generality, we may assume that $\Theta_1 = \mathbb{I}$ and $\Theta_0 = 1$ are unitless.

To verify the assumptions of the ADN theory, we make the identifications $M = d+1$, $L=1$, $u=\{\phi,u_1,\dots,u_d\}$,

[18] In other words, the units are irrelevant to the ADN weights and can be stripped from the variables.

$$\mathcal{L} = \begin{pmatrix} 0 & \partial_x & \partial_y \\ \partial_x & 1 & 0 \\ \partial_y & 0 & 1 \end{pmatrix} \qquad \text{in two dimensions},$$

and

$$\mathcal{L} = \begin{pmatrix} 0 & \partial_x & \partial_y & \partial_z \\ \partial_x & 1 & 0 & 0 \\ \partial_y & 0 & 1 & 0 \\ \partial_z & 0 & 0 & 1 \end{pmatrix} \qquad \text{in three dimensions}.$$

Recall that the complementing condition (Definition D.4) requires one to assume that a single type of boundary condition is imposed on all of $\partial\Omega$, i.e., either $\Gamma = \emptyset$ or $\Gamma^* = \emptyset$. In the former case (Neumann condition), the div–grad operator has positive nullity[19] because the scalar variable ϕ is determined only up to an arbitrary additive constant. Following the approach given in Section C.1, we redefine the div–grad differential operator by adding the relation

$$\ell(\phi) = \int_\Omega \phi \, d\Omega = 0. \tag{5.31}$$

This changes (5.18) to a problem with zero nullity having an operator that has the abstract form (C.10).

Remark 5.8 Another possibility is to let $\ell(\phi) = \langle \delta(\mathbf{x}_0), \phi \rangle = 0$, where $\mathbf{x}_0$ is a point in $\overline{\Omega}$ and $\delta(\mathbf{x})$ is the Dirac delta function. This choice of ℓ corresponds to the common practice of fixing the value of ϕ at a point [24]. Computationally, this approach is not as good as the zero–mean constraint (5.31) because it leads to an increase in the condition number of the linear systems; see [67]. □

The following theorem states the ADN energy balance for the div–grad system.

Theorem 5.9 *Assume that Ω is a bounded domain of class C^{q+2} for $q \geq 0$. Then, there exists a positive constant C_q such that for any collection $u = \{\phi, u_1, \ldots, u_d\}$ of smooth functions,*

$$C_q\big(\|\phi\|_{q+2} + \|\mathbf{u}\|_{q+1}\big) \leq \|\nabla \cdot \mathbf{u}\|_q + \|\nabla\phi + \mathbf{u}\|_{q+1} + \|\phi\|_{q+3/2,\Gamma} \tag{5.32}$$

when $\Gamma = \partial\Omega$, i.e., the Dirichlet boundary condition case, and

$$C_q\big(\|\phi\|_{q+2} + \|\mathbf{u}\|_{q+1}\big) \leq \|\nabla \cdot \mathbf{u}\|_q + \|\nabla\phi + \mathbf{u}\|_{q+1} + \|\mathbf{u}\cdot\mathbf{n}\|_{q+1/2,\Gamma} + |\ell(\phi)| \tag{5.33}$$

when $\Gamma^* = \partial\Omega$, i.e., the Neumann boundary condition case. Furthermore, assumption D.2 *holds for the div–grad system with both types of boundary conditions, i.e.,* (5.32) *and* (5.33) *can be extended to all real values of q.*

Proof. To prove that (5.32) and (5.33) hold for all non-negative real q, we use Theorem D.1. In Section D.2.1, all assumptions of this theorem are verified for the

[19] Recall that we are considering the div–grad system *without* the reaction term. If $\gamma = 1$, then the Neumann problem has a trivial null-space.

div–grad system with the weights from (D.11). Therefore, Theorem D.1 implies the validity of (5.32) and (5.33) for all *integers* $q \geq r'$, where $r' = \max\{0, \max r_l + 1\}$. For the div–grad system, $l = 1$ and r_1 is either -2 (for Dirichlet boundary conditions) or -1 (for Neumann boundary conditions) so that $r' = 0$. This verifies (5.32) and (5.33) for all non-negative integer values of q. Using interpolation [265], we can extend these bounds to all real non-negative values of q.

To prove that (5.32) and (5.33) hold for all real $q < 0$, we first show that these bounds are valid for $q = -2$. Let $u = \{\phi, u_1, \ldots, u_d\}$ denote a collection of smooth functions in Ω. We use a duality argument based on the following characterization of the norms in the left-hand sides of (5.32) and (5.33) when $q = -2$:

$$\|\phi\|_0 + \|\mathbf{u}\|_{-1} = \sup_{f \in L^2(\Omega);\ \mathbf{h} \in [H_0^1(\Omega)]^d} \frac{(\phi, f) + (\mathbf{u}, \mathbf{h})}{\|f\|_0 + \|\mathbf{h}\|_1}.$$

To prove the assertion for Dirichlet boundary conditions, it is necessary to show that if $q = -2$, the supremum is bounded by the norms on the right-hand side of (5.32). Assuming that the boundary of Ω is smooth enough, the ADN theory implies that the solution of the div–grad system

$$\begin{cases} -\nabla \cdot \mathbf{v} = f & \text{in } \Omega \\ -\nabla \psi + \mathbf{v} = \mathbf{h} & \text{in } \Omega \\ \psi = 0 & \text{on } \partial\Omega \end{cases} \tag{5.34}$$

satisfies

$$\|\psi\|_2 + \|\mathbf{v}\|_1 \leq C(\|f\|_0 + \|\mathbf{h}\|_1). \tag{5.35}$$

Using the Green's identities (A.47) and (A.49) and the fact that $\psi = 0$ on $\Gamma = \partial\Omega$, we obtain

$$\begin{aligned}
(\phi, f) + (\mathbf{u}, \mathbf{h}) &= (\phi, -\nabla \cdot \mathbf{v}) + (\mathbf{u}, -\nabla\psi + \mathbf{v}) \\
&= (\nabla \cdot \mathbf{u}, \psi) + (\nabla\phi + \mathbf{u}, \mathbf{v}) - \langle \phi, \mathbf{v} \cdot \mathbf{n} \rangle_{\partial\Omega} \\
&\leq \|\nabla \cdot \mathbf{u}\|_{-2}\|\psi\|_2 + \|\nabla\phi + \mathbf{u}\|_{-1}\|\mathbf{v}\|_1 + \|\phi\|_{-1/2,\Gamma} \sum_{i=1}^{d} \|\mathbf{v}_i \mathbf{n}_i\|_{1/2,\Gamma} \\
&\leq \left(\|\nabla \cdot \mathbf{u}\|_{-2} + \|\nabla\phi + \mathbf{u}\|_{-1} + \|\phi\|_{-1/2,\Gamma}\right)\left(\|\psi\|_2 + \|\mathbf{v}\|_1 + \sum_{i=1}^{d} \|\mathbf{v}_i \mathbf{n}_i\|_{1/2,\Gamma}\right) \\
&\leq C\left(\|\nabla \cdot \mathbf{u}\|_{-2} + \|\nabla\phi + \mathbf{u}\|_{-1} + \|\phi\|_{-1/2,\Gamma}\right)\left(\|\psi\|_2 + \|\mathbf{v}\|_1\right).
\end{aligned}$$

The last bound follows from the trace inequality (A.75) (Theorem A.16) applied to the components[20] of $\mathbf{v}$ with $s = 1$. Combining this bound with (5.35) shows that

$$C(\|\phi\|_0 + \|\mathbf{u}\|_{-1}) \leq \|\nabla \cdot \mathbf{u}\|_{-2} + \|\nabla\phi + \mathbf{u}\|_{-1} + \|\phi\|_{-1/2,\Gamma},$$

[20] Recall that we treat $\mathbf{v}$ as a set of d independent scalar functions.

i.e., (5.32) holds with $q=-2$. The proof that (5.33) holds for $q=-2$ is similar and uses (5.34) with $\mathbf{v}\cdot\mathbf{n}=0$ on $\partial\Omega$.

Using identical duality arguments and interpolation, we can extend (5.32) and (5.33) to all integer values $q\le -2$ and then to all real $q<-2$. Therefore, the div–grad operator is a bounded operator

$$H^{q+2}(\Omega)\times[H^{q+1}(\Omega)]^d \mapsto H^q(\Omega)\times[H^{q+1}(\Omega)]^d\times\begin{cases} H^{q+3/2}(\partial\Omega)\\ H^{q+1/2}(\partial\Omega)\end{cases}$$

for all $q\le -2$ and $0\le q$, where the upper and lower choices correspond to the Dirichlet and Neumann boundary condition cases, respectively. Using interpolation it follows that it is also a bounded operator for $-2<q<0$. □

The next theorem establishes ADN energy balances for the extended div–grad system. We state the results in terms of (5.25) instead of its ADN version (5.26) because, as already mentioned, the slack variable is identically zero.

Theorem 5.10 *Assume that Ω is a bounded domain of class C^{q+1} for $q\ge 0$. Then, there exists a positive constant C_q such that for any collection $u=\{\phi,u_1,\dots,u_d\}$ of smooth functions,*

$$\begin{aligned} &C_q\big(\|\phi\|_{q+1}+\|\mathbf{u}\|_{q+1}\big)\\ &\qquad\le \|\nabla\cdot\mathbf{u}\|_q+\|\nabla\phi+\mathbf{u}\|_q+\|\nabla\times\mathbf{u}\|_q+\|\phi\|_{q+1/2,\Gamma} \end{aligned}\tag{5.36}$$

when $\Gamma=\partial\Omega$, i.e., the Dirichlet boundary condition case, and

$$\begin{aligned} &C_q\big(\|\phi\|_{q+1}+\|\mathbf{u}\|_{q+1}\big)\\ &\qquad\le \|\nabla\cdot\mathbf{u}\|_q+\|\nabla\phi+\mathbf{u}\|_q+\|\nabla\times\mathbf{u}\|_q+\|\mathbf{u}\cdot\mathbf{n}\|_{q+1/2,\Gamma}+|\ell(\phi)| \end{aligned}\tag{5.37}$$

when $\Gamma^*=\partial\Omega$, i.e., the Neumann boundary condition case. The bounds (5.36) *and* (5.37) *can be extended to all real values of q.*

Proof. Proceeding as in Section D.2.1, it is possible to show that all the assumptions of the ADN theory hold for the full extended div–grad system (5.26) with $s_1=\cdots=s_N=-1$, $t_1=\cdots=t_N=1$, and $r_1=-1$, where $N=3d-2$. Therefore, this system is homogeneous elliptic and, according to [308, Theorem 4], it has a complete set of homeomorphisms, i.e., (5.36) and (5.37) hold for all q. □

Remark 5.11 The energy balances (5.33) and (5.37) specialize the abstract energy balance (3.19) for problems with positive nullity when the div–grad and the extended div–grad systems are considered without the reaction term. Because this term does not affect the principal parts of the div–grad and extended div–grad operators, Theorems 5.9 and 5.10 remain valid when $\gamma=1$. In this case, the div–grad and extended div–grad problems have zero nullity and $|\ell(\phi)|$ can be omitted from (5.33) and (5.37). □

Theorems 5.9 and 5.10 assert that the div–grad system is non–homogeneous elliptic, wheras the extended div–grad system is homogeneous elliptic. As a result, for (5.17) and (5.18), the spaces (D.5) specialize to

$$X_q = H^{q+2}(\Omega) \times [H^{q+1}(\Omega)]^d \qquad Y_q = H^q(\Omega) \times [H^{q+1}(\Omega)]^d$$

$$B_q = \begin{cases} H^{q+3/2}(\partial\Omega) & \text{if } \Gamma^* = \emptyset \\ H^{q+1/2}(\partial\Omega) & \text{if } \Gamma = \emptyset \,. \end{cases} \tag{5.38}$$

For the extended div–grad system (5.25), these spaces are given by

$$X_q = [H^{q+1}(\Omega)]^{d+1} \qquad Y_q = [H^q(\Omega)]^{3d-2} \qquad B_q = H^{q+1/2}(\partial\Omega)\,. \tag{5.39}$$

5.4.2 *Energy Balances in the Vector-Operator Setting*

We start this section with energy balances for the potential–flux systems (5.17) and (5.18) in which ϕ and $\mathbf{u}$ are regarded as elements of the weighted Hilbert spaces $G(\Omega,\Theta_0)$ and $\mathbf{D}(\Omega,\Theta_1^{-1})$, respectively. Because the a priori bounds are obtained directly, it is not necessary to assume that either $\Gamma = \emptyset$ or $\Gamma^* = \emptyset$, i.e., we can consider mixed boundary conditions.

Remark 5.12 One of the key differences between energy balances in the ADN and vector-operator settings is the way we think of the gradient and divergence operators and the functions upon which they act. In Section 5.4.1, these operators were treated as a collection of unrelated partial derivatives acting on a collection of independent scalar-valued functions[21] $u = \{\phi, u_1, \dots, u_d\}$. In contrast, in this section, the gradient and divergence are regarded as operators acting on the Hilbert spaces[22] $G(\Omega,\Theta_0)$ and $\mathbf{D}(\Omega,\Theta_1^{-1})$ from a primal and dual De Rham complexes, respectively. This viewpoint requires us to measure residuals of (5.17) using norms from the ranges of these operators in the appropriate differential complex. In particular, the range of ∇ in the primal complex (A.52) is a subspace of $\mathbf{C}(\Omega,\Theta_1)$ so that the residual of the second equation in (5.17) has to be measured in the norm of $\mathbf{L}^2(\Omega,\Theta_1)$. The range of $\nabla\cdot$ in the dual complex (A.53) is $L^2(\Omega,\Theta_0^{-1})$ that supplies the correct norm to measure the residual of the first equation. □

Remark 5.13 In practical applications, the weights in $G(\Omega,\Theta_0)$ and $\mathbf{D}(\Omega,\Theta_1^{-1})$ are endowed with units that are consistent with the physical quantities modeled by these spaces. As a result, once the weights are properly identified, energy balances in the vector-operator setting are guaranteed to be dimensionally consistent. To illustrate this point, consider the first-order electrostatics system (5.30), but written in

[21] The form of the div–grad operator corresponding to this viewpoint is a square $(d+1)\times(d+1)$ matrix of *partial derivative* operators; see Remark 3.6.

[22] The form of the div–grad operator corresponding to this viewpoint is a 2×2 matrix of *vector calculus* operators; see Remark 3.6.

terms of the dual electric displacement field **D** (see Remark 5.4 for its definition):

$$\nabla \cdot \mathbf{D} = 0 \qquad \text{and} \qquad \nabla \phi + \varepsilon_0^{-1}\mathbf{D} = 0 .$$

This system is an example of the potential–flux problem (5.18) with $\Theta_1 = \varepsilon_0$ and $\Theta_0 \sim 1/m^3$ because it converts ϕ (volts) into units consistent with charge density (coulombs per meter cubed). Theorem 5.15 shows that

$$\begin{aligned} &\|\nabla \cdot \mathbf{D}\|^2_{0,\Theta_0^{-1}} + \|\nabla \phi + \varepsilon_0^{-1}\mathbf{D}\|^2_{0,\varepsilon_0} \\ &\quad = \int_\Omega \Theta_0^{-1}(\nabla \cdot \mathbf{D})^2 d\Omega + \int_\Omega \varepsilon_0 |\nabla \phi + \varepsilon_0^{-1}\mathbf{D}|^2 d\Omega \end{aligned}$$

provides an appropriate notion of residual energy in the relevant Hilbert spaces. Recall from Remark 5.4 that ϕ, **D**, and ε_0 have units of V, C/m^2, and F/m, respectively. It is not difficult to see that the terms under the two integrals above have units proportional to $1/m^3$ so that the two residuals are dimensionally consistent.

In contrast, energy balances in the ADN setting do not posses intrinsic dimensional consistency. For example, for $q = 0$, the residual energy in (5.32) is

$$\begin{aligned} &\|\nabla \cdot \mathbf{D}\|_0^2 + \|\nabla \phi + \varepsilon_0^{-1}\mathbf{D}\|_0^2 + |\nabla \phi + \varepsilon_0^{-1}\mathbf{D}|_1^2 \\ &= \int_\Omega (\nabla \cdot \mathbf{D})^2 d\Omega + \int_\Omega |\nabla \phi + \varepsilon_0^{-1}\mathbf{D}|^2 \, d\Omega + \int_\Omega |\nabla(\nabla \phi + \varepsilon_0^{-1}\mathbf{D})|^2 \, d\Omega . \end{aligned} \tag{5.40}$$

In this definition, the terms under the first, second, and third integrals have units proportional to $1/m^6$, $1/m^2$, and $1/m^4$, respectively. Intuitively, this lack of dimensional consistency stems from the fact that ADN theory breaks down all variables into independent scalar components which are stripped of their physical units.[23] □

Besides dimensional consistency, choosing norms as suggested in Remark 5.12 also provides for an attractive "splitting" property of the energy balance for (5.17) when $\gamma = 1$ that greatly simplifies the proof of the energy balance in the following theorem. Section 5.8.1 shows that this "splitting" also imparts least-squares methods in the vector-operator setting with valuable computational properties.

Theorem 5.14 *Let $\gamma = 1$ and assume that Ω is a bounded region in $\mathbb{R}^d$ with a Lipschitz-continuous boundary. Then, there exists a positive constant C such that*

$$C\big(\|\phi\|_G + \|\mathbf{u}\|_D\big) \le \|\nabla \cdot \mathbf{u} + \Theta_0 \phi\|_{0,\Theta_0^{-1}} + \|\nabla \phi + \Theta_1^{-1}\mathbf{u}\|_{0,\Theta_1} \tag{5.41}$$

for all $\{\phi, \mathbf{u}\} \in G_\Gamma(\Omega, \Theta_0) \times \mathbf{D}_{\Gamma^*}(\Omega, \Theta_1^{-1})$.

Proof. Expanding the right-hand side of (5.41) gives

[23] Of course, in practice, the integrals in (5.40) can be multiplied by constant factors to make them dimensionally consistent.

$$
\begin{aligned}
&\|\nabla\cdot\mathbf{u}+\Theta_0\phi\|^2_{0,\Theta_0^{-1}}+\|\nabla\phi+\Theta_1^{-1}\mathbf{u}\|^2_{0,\Theta_1}\\
&\quad=\|\phi\|^2_{0,\Theta_0}+\|\nabla\phi\|^2_{0,\Theta_1}+\|\nabla\cdot\mathbf{u}\|^2_{0,\Theta_0^{-1}}+\|\mathbf{u}\|^2_{0,\Theta_1^{-1}}\\
&\qquad\qquad+2\big((\nabla\cdot\mathbf{u},\Theta_0\phi)_{0,\Theta_0^{-1}}+(\Theta_1^{-1}\mathbf{u},\nabla\phi)_{0,\Theta_1}\big)\\
&\quad=\|\phi\|^2_G+\|\mathbf{u}\|^2_D+2\big((\nabla\cdot\mathbf{u},\Theta_0\phi)_{0,\Theta_0^{-1}}+(\Theta_1^{-1}\mathbf{u},\nabla\phi)_{0,\Theta_1}\big).
\end{aligned}
$$

From the definition (A.18) of the weighted $L^2(\Omega)$ inner product, we see that

$$(\nabla\cdot\mathbf{u},\Theta_0\phi)_{0,\Theta_0^{-1}}=\int_\Omega(\nabla\cdot\mathbf{u})\Theta_0^{-1}(\Theta_0\phi)d\Omega=(\nabla\cdot\mathbf{u},\phi)_0$$

and

$$(\Theta_1^{-1}\mathbf{u},\nabla\phi)_{0,\Theta_1}=\int_\Omega(\Theta_1^{-1}\mathbf{u})^\mathsf{T}\Theta_1(\nabla\phi)d\Omega=(\mathbf{u},\nabla\phi)_0.$$

Because $\phi=0$ on Γ and $\mathbf{n}\cdot\mathbf{u}=0$ on Γ^*,

$$(\nabla\cdot\mathbf{u},\Theta_0\phi)_{0,\Theta_0^{-1}}+(\Theta_1^{-1}\mathbf{u},\nabla\phi)_{0,\Theta_1}=(\nabla\cdot\mathbf{u},\phi)_0+(\mathbf{u},\nabla\phi)_0=\int_{\partial\Omega}\phi\,\mathbf{n}\cdot\mathbf{u}\,dS=0.$$

It follows that residual energy equals[24] the solution energy:

$$\|\nabla\cdot\mathbf{u}+\Theta_0\phi\|^2_{0,\Theta_0^{-1}}+\|\nabla\phi+\Theta_1^{-1}\mathbf{u}\|^2_{0,\Theta_1}=\|\phi\|^2_G+\|\mathbf{u}\|^2_D.$$

Therefore, (5.41) holds with $C=1/\sqrt{2}$. □

Without the reaction term, i.e., when $\gamma=0$, the proof of the energy balance is somewhat more involved.

Theorem 5.15 *Let $\gamma=0$ and assume that Ω is a bounded region in $\mathbb{R}^d$ with a Lipschitz-continuous boundary. Then, there exists a positive constant C such that*

$$C\big(\|\phi\|_G+\|\mathbf{u}\|_D\big)\le\|\nabla\cdot\mathbf{u}\|_{0,\Theta_0^{-1}}+\|\nabla\phi+\Theta_1^{-1}\mathbf{u}\|_{0,\Theta_1}\tag{5.42}$$

for all $\{\phi,\mathbf{u}\}\in G_\Gamma(\Omega,\Theta_0)\times\mathbf{D}_{\Gamma^*}(\Omega,\Theta_1^{-1})$.

Proof. Let β be a positive real parameter whose value is to be determined later. Splitting the first term on the right-hand side of (5.42) in two, adding and subtracting $\beta\Theta_0\phi$ to $1/2\|\nabla\cdot\mathbf{u}\|_{0,\Theta_0^{-1}}$, and then expanding terms gives

[24] For $\Theta_1=\mathsf{I}$ and $\Theta_0=1$, the splitting property was first used for LSFEMs in [102] in the context of continuous methods for transmission problems for (12.81); see Section 12.10.

$$
\begin{aligned}
&\|\nabla\cdot\mathbf{u}\|^2_{0,\Theta_0^{-1}} + \|\nabla\phi + \Theta_1^{-1}\mathbf{u}\|^2_{0,\Theta_1} \\
&= \frac{1}{2}\left[\|\nabla\cdot\mathbf{u} - \beta\Theta_0\phi\|^2_{0,\Theta_0^{-1}} + \beta^2\|\phi\|^2_{0,\Theta_0} + 2\big(\nabla\cdot\mathbf{u} - \beta\Theta_0\phi, \beta\Theta_0\phi\big)_{0,\Theta_0^{-1}}\right] \\
&\quad + \frac{1}{2}\|\nabla\cdot\mathbf{u}\|_{0,\Theta_0^{-1}} + \|\mathbf{u}\|^2_{0,\Theta_1^{-1}} + \|\nabla\phi\|^2_{0,\Theta_1} + 2\big(\nabla\phi, \Theta_1^{-1}\mathbf{u}\big)_{0,\Theta_1} \\
&\ge \frac{1}{2}\|\nabla\cdot\mathbf{u}\|_{0,\Theta_0^{-1}} + \|\mathbf{u}\|^2_{0,\Theta_1^{-1}} + \|\nabla\phi\|^2_{0,\Theta_1} - \frac{\beta^2}{2}\|\phi\|^2_{0,\Theta_0} \\
&\quad + 2\left[\big(\nabla\phi, \Theta_1^{-1}\mathbf{u}\big)_{0,\Theta_1} + \frac{\beta}{2}\big(\nabla\cdot\mathbf{u}, \Theta_0\phi\big)_{0,\Theta_0^{-1}}\right] \equiv (I).
\end{aligned}
$$

Using that $\phi = 0$ on Γ and $\mathbf{n}\cdot\mathbf{u} = 0$ on Γ^*, we obtain

$$
\big(\nabla\phi, \Theta_1^{-1}\mathbf{u}\big)_{0,\Theta_1} + \frac{\beta}{2}\big(\nabla\cdot\mathbf{u}, \Theta_0\phi\big)_{0,\Theta_0^{-1}} = \left(1 - \frac{\beta}{2}\right)\big(\nabla\phi, \Theta_1^{-1}\mathbf{u}\big)_{0,\Theta_1}.
$$

The ε-inequality

$$
2\left(1 - \frac{\beta}{2}\right)\big(\nabla\phi, \Theta_1^{-1}\mathbf{u}\big)_{0,\Theta_1} \le \frac{1}{\varepsilon}\left(1 - \frac{\beta}{2}\right)^2\|\nabla\phi\|^2_{0,\Theta_1} + \varepsilon\|\mathbf{u}\|^2_{0,\Theta_1^{-1}}
$$

together with the Poincaré inequality (A.63) yield the lower bound

$$
(I) \ge \frac{1}{2}\|\nabla\cdot\mathbf{u}\|_{0,\Theta_0^{-1}} + (1-\varepsilon)\|\mathbf{u}\|^2_{0,\Theta_1^{-1}} + \left(1 - \frac{1}{\varepsilon}\left(1 - \frac{\beta}{2}\right)^2 - \frac{\beta^2 C_P^2}{2}\right)\|\nabla\phi\|^2_{0,\Theta_1}.
$$

Setting $\varepsilon = 1/2$ and expanding terms,

$$
(I) \ge \frac{1}{2}\left[\|\nabla\cdot\mathbf{u}\|_{0,\Theta_0^{-1}} + \|\mathbf{u}\|^2_{0,\Theta_1^{-1}} + \left(1 + \beta - \beta^2\left(\frac{1}{4} + C_P^2\right)\right)\|\nabla\phi\|^2_{0,\Theta_1}\right].
$$

The maximum of the function $f(\beta) = 1 + \beta - \beta^2(1/4 + C_P^2)$ is attained at the point $\beta_0 = 1/(1/2 + C_P^2)^{-1}$ and

$$
f(\beta_0) = 1 + \frac{1}{(1 + 4C_P^2)} > 0.
$$

This proves the theorem. □

Remark 5.16 The proof of Theorem 5.15 can be modified to show that

$$
\begin{aligned}
&C\big(\|\phi\|_G + \|\mathbf{u}\|_D\big) \\
&\le \|\nabla\cdot\mathbf{u}\|_{0,\Theta_0^{-1}} + \|\nabla\phi + \Theta_1^{-1}\mathbf{u}\|_{0,\Theta_1} + \|\phi\|_{1/2,\Gamma} + \|\mathbf{n}\cdot\mathbf{u}\|_{-1/2,\Gamma^*}
\end{aligned}
\tag{5.43}
$$

for $\{\phi,\mathbf{u}\} \in G(\Omega) \times \mathbf{D}(\Omega)$; see [95, 318] for details. This energy bound can be used to define LSFEMs for which boundary conditions are enforced in a weak, least-squares sense through their residuals. Section 12.1 provides further information about the enforcement of boundary conditions in LSFEMs. □

From Theorems 5.14 and 5.15 and Remark 5.16, it follows that the abstract energy balance (3.18) holds for the div–grad system with

$$X = G(\Omega,\Theta_0) \times \mathbf{D}(\Omega,\Theta_1^{-1}) \qquad Y = L^2(\Omega,\Theta_0^{-1}) \times \mathbf{L}^2(\Omega,\Theta_1)$$
$$B = H^{1/2}(\Gamma) \times H^{-1/2}(\Gamma^*)\,. \tag{5.44}$$

Let us compare and contrast (5.41) and (5.42) with the energy balances for the ADN setting in Theorem 5.9. Because Theorem 5.9 does not allow mixed boundary conditions, let us set $\Gamma^* = \emptyset$ and $\phi = 0$ on $\partial\Omega$, i.e., the case of homogeneous Dirichlet boundary conditions. For simplicity, assume that $\Theta_1 = \mathbb{I}$, $\Theta_0 = 1$, $\gamma = 0$ (no reaction term), and the boundary condition is imposed on the solution space. Then, the energy balance in the ADN setting (5.32) with $q = -1$ specializes to

$$C\big(\|\phi\|_1 + \|\mathbf{u}\|_0\big) \le \|\nabla\cdot\mathbf{u}\|_{-1} + \|\nabla\phi + \mathbf{u}\|_0$$

whereas (5.42) takes the form

$$C\big(\|\phi\|_G + \|\mathbf{u}\|_D\big) \le \|\nabla\cdot\mathbf{u}\|_0 + \|\nabla\phi + \mathbf{u}\|_0\,.$$

The negative norm in the first balance appears due to the fact that the ADN theory, used to obtain that balance, classifies the div–grad system as non-homogeneous elliptic. This means that $L^2(\Omega)$ norms are *not appropriate* for all residuals and the resulting CLSPs *do not admit* practical compliant DLSPs.

Theorems 5.14 and 5.15 show that this problem can be circumvented by switching to energy balances in the vector-operator setting. With such balances, $L^2(\Omega)$ norms are *appropriate* for all residuals of the div–grad system, which means that the corresponding CLSPs *do admit* practical compliant DLSPs.

We now consider energy balances for the potential–density–intensity–flux first-order system (5.23). For simplicity, we only consider this system with $\gamma = 1$.

Theorem 5.17 *Let $\gamma = 1$ and assume that Ω is bounded a region in $\mathbb{R}^d$ with a Lipschitz-continuous boundary. Then, there exists a positive constant C such that*

$$\begin{aligned} C\big(\|\phi\|_G + \|\psi\|_S + \|\mathbf{u}\|_D + \|\mathbf{v}\|_C\big) \qquad & \\ \le \|\nabla\cdot\mathbf{u} + \psi\|_{0,\Theta_0^{-1}} + \|\nabla\phi + \mathbf{v}\|_{0,\Theta_1} + \|\nabla\times\mathbf{v}\|_{0,\Theta_2} & \\ + \|\mathbf{v} - \Theta_1^{-1}\mathbf{u}\|_{\Theta_1} + \|\psi - \Theta_0\phi\|_{\Theta_0^{-1}} & \end{aligned} \tag{5.45}$$

for all $\{\phi,\psi\} \in G_\Gamma(\Omega,\Theta_0) \times S(\Omega,\Theta_0^{-1})$ and $\{\mathbf{u},\mathbf{v}\} \in \mathbf{D}_{\Gamma^*}(\Omega,\Theta_1^{-1}) \times \mathbf{C}(\Omega,\Theta_1)$.

Proof. We add and subtract terms to the right-hand side in (5.45) and then expand:

$$
\begin{aligned}
(I) &= \|\nabla\cdot\mathbf{u}+\psi\|^2_{0,\Theta_0^{-1}} + \|\nabla\phi+\mathbf{v}\|^2_{0,\Theta_1} + \|\nabla\times\mathbf{v}\|^2_{0,\Theta_2} \\
&\quad + \|\mathbf{v}+\nabla\phi-(\nabla\phi+\Theta_1^{-1}\mathbf{u})\|^2_{0,\Theta_1} + \|\psi+\nabla\cdot\mathbf{u}-(\nabla\cdot\mathbf{u}+\Theta_0\phi)\|^2_{0,\Theta_0^{-1}} \\
&= 2\Big[\|\nabla\cdot\mathbf{u}+\psi\|^2_{0,\Theta_0^{-1}} + \|\nabla\phi+\mathbf{v}\|^2_{0,\Theta_1}\Big] + \|\nabla\times\mathbf{v}\|^2_{0,\Theta_2} \\
&\quad + \|\nabla\cdot\mathbf{u}+\Theta_0\phi\|^2_{0,\Theta_0^{-1}} + \|\nabla\phi+\Theta_1^{-1}\mathbf{u}\|^2_{0,\Theta_1} \\
&\quad -2\Big[\big(\mathbf{v}+\nabla\phi,\nabla\phi+\Theta_1^{-1}\mathbf{u}\big)_{0,\Theta_1} + \big(\psi+\nabla\cdot\mathbf{u},\nabla\cdot\mathbf{u}+\Theta_0\phi\big)_{0,\Theta_0^{-1}}\Big].
\end{aligned}
$$

To bound the last two terms, we use the Cauchy inequality followed by the ε-inequality with $\varepsilon = 2/3$:

$$
\begin{aligned}
2\big(\mathbf{v}+\nabla\phi,\nabla\phi+\Theta_1^{-1}\mathbf{u}\big)_{0,\Theta_1} &\le \frac{3}{2}\|\mathbf{v}+\nabla\phi\|^2_{0,\Theta_1} + \frac{2}{3}\|\nabla\phi+\Theta_1^{-1}\mathbf{u}\|^2_{0,\Theta_1}; \\
2\big(\psi+\nabla\cdot\mathbf{u},\nabla\cdot\mathbf{u}+\Theta_0\phi\big)_{0,\Theta_0^{-1}} &\le \frac{3}{2}\|\psi+\nabla\cdot\mathbf{u}\|^2_{0,\Theta_0^{-1}} + \frac{2}{3}\|\nabla\cdot\mathbf{u}+\Theta_0\phi\|^2_{0,\Theta_0^{-1}}.
\end{aligned}
$$

From these inequalities, it follows that

$$
\begin{aligned}
(I) \ge \frac{1}{2}&\Big[\|\nabla\cdot\mathbf{u}+\psi\|^2_{0,\Theta_0^{-1}} + \|\nabla\phi+\mathbf{v}\|^2_{0,\Theta_1}\Big] + \|\nabla\times\mathbf{v}\|^2_{0,\Theta_2} \\
&+\frac{1}{3}\Big[\|\nabla\cdot\mathbf{u}+\Theta_0\phi\|^2_{0,\Theta_0^{-1}} + \|\nabla\phi+\Theta_1^{-1}\mathbf{u}\|^2_{0,\Theta_1}\Big] = (II).
\end{aligned}
$$

The last two terms are residuals of the potential–flux first-order system (5.17) which can be "split" as in Theorem 5.14. Therefore, after expanding all terms in (II),

$$
\begin{aligned}
(II) = \frac{1}{2}&\Big[\|\psi\|^2_{0,\Theta_0^{-1}} + \|\mathbf{v}\|^2_{0,\Theta_1}\Big] + \|\nabla\times\mathbf{v}\|^2_{0,\Theta_2} + \big(\nabla\cdot\mathbf{u},\psi\big)_{0,\Theta_0^{-1}} + \big(\nabla\phi,\mathbf{v}\big)_{0,\Theta_1} \\
&+\frac{5}{6}\Big[\|\nabla\cdot\mathbf{u}\|^2_{0,\Theta_0^{-1}} + \|\nabla\phi\|^2_{0,\Theta_1}\Big] + \frac{1}{3}\Big[\|\phi\|^2_{0,\Theta_0} + \|\mathbf{u}\|^2_{0,\Theta_1^{-1}}\Big].
\end{aligned}
$$

To bound the last two terms, we use again the ε-inequality with $\varepsilon = 3/4$:

$$
\begin{aligned}
\big(\nabla\cdot\mathbf{u},\psi\big)_{0,\Theta_0^{-1}} &\le \frac{4}{6}\|\nabla\cdot\mathbf{u}\|^2_{0,\Theta_0^{-1}} + \frac{3}{8}\|\psi\|^2_{0,\Theta_0^{-1}}; \\
\big(\nabla\phi,\mathbf{v}\big)_{0,\Theta_1} &\le \frac{4}{6}\|\nabla\phi\|^2_{0,\Theta_1} + \frac{3}{8}\|\mathbf{v}\|^2_{0,\Theta_1}.
\end{aligned}
$$

After gathering all results together, we have

$$
\begin{aligned}
(II) \ge \frac{1}{8}&\big(\|\psi\|^2_{0,\Theta_0^{-1}} + \|\mathbf{v}\|^2_{0,\Theta_1}\big) + \|\nabla\times\mathbf{v}\|^2_{0,\Theta_2} \\
&+\frac{1}{6}\big(\|\nabla\cdot\mathbf{u}\|^2_{0,\Theta_0^{-1}} + \|\nabla\phi\|^2_{0,\Theta_1}\big) + \frac{1}{3}\big(\|\phi\|^2_{0,\Theta_0} + \|\mathbf{u}\|^2_{0,\Theta_1^{-1}}\big).
\end{aligned}
$$

This proves the theorem. □

Theorem 5.17 implies that the abstract energy balance (3.18) holds for the four-field system with

$$
\begin{aligned}
X &= \{G_\Gamma(\Omega,\Theta_0)\times S(\Omega,\Theta_0^{-1})\}\times\{\mathbf{D}_{\Gamma^*}(\Omega,\Theta_1^{-1})\times\mathbf{C}(\Omega,\Theta_1)\}\\
Y &= L^2(\Omega,\Theta_0^{-1})\times\mathbf{L}^2(\Omega,\Theta_1)\times\mathbf{L}^2(\Omega,\Theta_2)\times\mathbf{L}^2(\Omega,\Theta_1)\times L^2(\Omega,\Theta_0^{-1}).
\end{aligned}
\tag{5.46}
$$

The last energy balance we consider in this section is for the extended potential–flux system. Using Hilbert spaces allows one to relax the conditions on Ω in Theorem 5.10 but requires the solution energy to be measured in the norm of the intersection space $\mathbf{C}(\Omega)\cap\mathbf{D}(\Omega)$.[25]

Theorem 5.18 *Assume that $\partial\Omega$ is Lipschitz continuous. Then, there exists a positive constant C such that*

$$
\begin{aligned}
&C\big(\|\phi\|_G+\|\mathbf{u}\|_{DC}\big)\\
&\qquad\le\|\nabla\cdot\mathbf{u}+\gamma\Theta_0\phi\|_{0,\Theta_0^{-1}}+\|\nabla\phi+\Theta_1^{-1}\mathbf{u}\|_{0,\Theta_1}+\|\nabla\times\Theta_1^{-1}\mathbf{u}\|_{0,\Theta_2}
\end{aligned}
\tag{5.47}
$$

for $\{\phi,\mathbf{u}\}\in G_0(\Omega,\Theta_0)\times\mathbf{C}_0(\Omega,\Theta_1)\cap\mathbf{D}(\Omega,\Theta_1^{-1})$ *or* $\{\phi,\mathbf{u}\}\in G(\Omega,\Theta_0)\times\mathbf{C}(\Omega,\Theta_1)$ $\cap\mathbf{D}_0(\Omega,\Theta_1^{-1})$.

Proof. The proof follows directly from (5.41), (5.42), and Definition (A.61). □

Theorem 5.18 implies that the abstract energy balance (3.18) holds for the extended div–grad system with

$$
\begin{aligned}
X &= \begin{cases} G_0(\Omega,\Theta_0)\times\mathbf{C}_0(\Omega,\Theta_1)\cap\mathbf{D}(\Omega,\Theta_1^{-1})\\ G(\Omega,\Theta_0)\times\mathbf{C}(\Omega,\Theta_1)\cap\mathbf{D}_0(\Omega,\Theta_1^{-1})\end{cases}\\
Y &= L^2(\Omega,\Theta_0^{-1})\times\mathbf{L}^2(\Omega,\Theta_1)\times\mathbf{L}^2(\Omega,\Theta_2).
\end{aligned}
\tag{5.48}
$$

Let us compare (5.47) with the a priori bounds from Theorem 5.10 in the ADN setting. There, we assumed that $\partial\Omega$ is of class C^{q+1} for some non-negative q, wheras Theorem 5.18 only required $\partial\Omega$ to be Lipschitz continuous. As a results, using Theorem A.8, it is possible to extend the energy balances obtained in the ADN setting to a larger class of domains.

Corollary 5.19 *Assume that Ω satisfies the hypotheses of Theorem* A.8. *Then, the energy balance* (5.36) *with $q=0$ holds for the extended div–grad system* (5.25) *and* $\Gamma=\partial\Omega$. *The energy balance* (5.37) *with* $q=0$ *holds for* (5.25) *and* $\Gamma^*=\partial\Omega$. *If the boundary conditions are imposed on the solution spaces, i.e.,* $X=H_0^1(\Omega)\times\mathbf{C}_0(\Omega)\cap\mathbf{D}(\Omega)$ *or* $X=H^1(\Omega)\times\mathbf{C}(\Omega)\cap\mathbf{D}_0(\Omega)$, *respectively, then*

$$
C\big(\|\phi\|_1+\|\mathbf{u}\|_1\big)\le\|\nabla\cdot\mathbf{u}+\Theta_0\phi\|_0+\|\nabla\phi+\Theta_1^{-1}\mathbf{u}\|_0+\|\nabla\times\Theta_1^{-1}\mathbf{u}\|_0 \tag{5.49}
$$

[25] We have already mentioned in Section 5.3.2 and Remark 5.7 that this space causes some problems for LSFEMs based on (5.25).

for all $\{\phi, \mathbf{u}\} \in X$.

Proof. Theorem A.8 asserts that $C\|\mathbf{u}\|_1 \le \|\mathbf{u}\|_{DC}$ for all $\mathbf{u} \in \mathbf{D}_0(\Omega) \cap \mathbf{C}(\Omega)$ or for all $\mathbf{u} \in \mathbf{D}(\Omega) \cap \mathbf{C}_0(\Omega)$. This inequality and (5.47) imply (5.49). □

It is important to keep in mind that equivalence of energy balances in the ADN and vector-operator settings cannot be extended to arbitrary domains. For instance, Theorem 5.18 remains valid when Ω is a non-convex polyhedron. In this case, Theorem A.9 asserts that $[H^1(\Omega)]^d$ vector fields are a closed, infinite co-dimensional subspace of the intersection spaces $\mathbf{C}_0(\Omega) \cap \mathbf{D}(\Omega)$ and $\mathbf{C}(\Omega) \cap \mathbf{D}_0(\Omega)$ so that the equivalence result from Corollary 5.19 does not extend to such domains.

5.5 Continuous Least-Squares Principles

In this section, we specialize the abstract least-squares principles from Sections 3.2.2 and 4.3 to div–grad and extended div–grad systems. The relevant energy balances were established in Theorems 5.9–5.18 and Corollary 5.19. To streamline the process, we make the following assumptions:

- $\gamma = 0$ or $\gamma = 1$, except for (5.23) for which $\gamma = 1$ always;
- either[26,27] $\Gamma = \emptyset$ or $\Gamma^* = \emptyset$;
- the boundary conditions are homogeneous;
- the solution space X is constrained by the homogeneous boundary conditions.

Of course, we also continue to assume that the Θ_i are non-degenerate, i.e., they satisfy (5.3).

Let us start with CLSPs for the potential–flux div–grad system (5.17). The first set of CLSPs for this problem follows from the energy balances in the ADN setting in Theorem 5.9. According to Section 4.5, for non-homogeneous elliptic systems, the preferred regularity index is $q = -1$ because it allows for the satisfaction of Assumption 3.20. In the case of homogeneous Dirichlet boundary conditions, i.e., $\Gamma = \partial\Omega$, this choice gives the following CLSP:[28]

$$\begin{cases} J_{-1}(\phi, \mathbf{u}; f) = \|\nabla \cdot \mathbf{u} + \gamma\Theta_0\phi - \Theta_0 f\|_{-1}^2 + \|\nabla\phi + \Theta_1^{-1}\mathbf{u}\|_0^2 \\ X_{-1} = H_0^1(\Omega) \times [L^2(\Omega)]^d . \end{cases} \tag{5.50}$$

[26] The CLSPs formulated under these assumptions can be easily extended to mixed boundary conditions when they are admissible by the energy balance. Also note that, in Section 12.1, further details are provided about LSFEMs in which boundary residuals are included in the functionals.

[27] This assumption facilitates comparison of LSFEMs based on energy balances in the ADN setting for which one can apply only a single type of boundary condition with LSFEMs based on energy balances in the vector-operator setting for which mixed boundary conditions are admissible.

[28] For brevity, in this chapter and several others, we do not write out the first-order necessary conditions corresponding to continuous or discrete least-squares principles, except when doing so is needed for clarity.

For the homogeneous Neumann condition case, i.e, $\Gamma^* = \partial\Omega$, the choice $q = -1$ is not so convenient if the boundary condition is to be imposed on the solution space for $\mathbf{u}$. Instead, we use $q = 0$ in (5.33) to obtain the following CLSP when $\gamma = 1$:

$$\begin{cases} J_0(\phi, \mathbf{u}; f) = \|\nabla \cdot \mathbf{u} + \Theta_0 \phi - \Theta_0 f\|_0^2 + \|\nabla \phi + \Theta_1^{-1} \mathbf{u}\|_1^2 \\ X_0 = H^2(\Omega) \times [H^1(\Omega)]^d \cap \mathbf{D}_0(\Omega). \end{cases} \tag{5.51}$$

When $\gamma = 0$, the potential is determined only up to a constant. In this case, (5.51) is replaced by

$$\begin{cases} J_0(\phi, \mathbf{u}; f) = \|\nabla \cdot \mathbf{u} - \Theta_0 f\|_0^2 + \|\nabla \phi + \Theta_1^{-1} \mathbf{u}\|_1^2 + |\ell(\phi)|^2 \\ X_0 = H^2(\Omega) \times [H^1(\Omega)]^d \cap \mathbf{D}_0(\Omega). \end{cases} \tag{5.52}$$

This CLSP corresponds to the abstract least-squares setting (3.26) for problems with positive nullity.

Remark 5.20 Alternatively, the one-dimensional null-space can be removed by using the algebraic complement of X_0 with respect to $\ell(\phi)$. Because the null-space only concerns the scalar ϕ, the complement of X_0 is obtained by changing its first component to $H^2(\Omega) \cap L_0^2(\Omega)$. The resulting CLSP

$$\begin{cases} J_0(\phi, \mathbf{u}; f) = \|\nabla \cdot \mathbf{u} - \Theta_0 f\|_0^2 + \|\nabla \phi + \Theta_1^{-1} \mathbf{u}\|_1^2 \\ X_0^C = H^2(\Omega) \cap L_0^2(\Omega) \times [H^1(\Omega)]^d \cap \mathbf{D}_0(\Omega) \end{cases} \tag{5.53}$$

corresponds to the abstract least-squares setting (3.24) for problems with zero nullity. □

Note that, for the div–grad system, CLSPs for the ADN setting employ a mixture of $L^2(\Omega)$ norms and *impractical* minus one or $H^1(\Omega)$ norms; this is a consequence of the fact that the ADN theory classifies this system as non-homogeneous elliptic. Because such CLSPs do not admit practical compliant DLSPs and lead to LSFEMs that are more difficult to implement and solve (see Table 4.1), it is desirable to develop a pool of alternative CLSPs in which all residuals are measured using $L^2(\Omega)$ norms.

Basically, there are two ways to approach this task. If we decide to retain the ADN setting for the energy balances, we then have to switch to the extended div–grad system and restrict Ω to the kind of domains admissible in Corollary 5.19. If instead, we prefer to retain the original div–grad system or use the four-field formulation (5.23), we need to switch to energy balances in the vector-operator setting.

Let us examine the second approach first. For the potential–flux div–grad system with $\gamma = 1$ and homogeneous Dirichlet condition, the energy balance from Theorem 5.14 gives the following CLSP:

$$
\begin{cases}
J(\phi,\mathbf{u};f) = \|\nabla\cdot\mathbf{u}+\Theta_0\phi-\Theta_0 f\|^2_{0,\Theta_0^{-1}} + \|\nabla\phi+\Theta_1^{-1}\mathbf{u}\|^2_{0,\Theta_1} \\
X = G_0(\Omega,\Theta_0)\times\mathbf{D}(\Omega,\Theta_1^{-1})\,.
\end{cases}
\tag{5.54}
$$

Shifting to a homogeneous Neumann condition only changes the definition of the minimization space:

$$
\begin{cases}
J(\phi,\mathbf{u};f) = \|\nabla\cdot\mathbf{u}+\Theta_0\phi-\Theta_0 f\|^2_{0,\Theta_0^{-1}} + \|\nabla\phi+\Theta_1^{-1}\mathbf{u}\|^2_{0,\Theta_1} \\
X = G(\Omega,\Theta_0)\times\mathbf{D}_0(\Omega,\Theta_1^{-1})\,.
\end{cases}
\tag{5.55}
$$

Of course, if $\gamma=0$, $|\ell(\phi)|^2$ must be added[29] to the least-squares functional in (5.55).

For the potential–density–intensity–flux system,[30] a shift from Dirichlet to Neumann boundary conditions again only changes the solution space. Therefore, the two CLSPs implied by the energy balance in Theorem 5.17 can be stated together as follows:

$$
\begin{cases}
J(\phi,\psi,\mathbf{u},\mathbf{v};f) = \|\nabla\cdot\mathbf{u}+\psi-\Theta_0 f\|^2_{0,\Theta_0^{-1}} + \|\nabla\phi+\mathbf{v}\|^2_{0,\Theta_1} + \|\nabla\times\mathbf{v}\|_{0,\Theta_2} \\
\qquad\qquad + \|\mathbf{v}-\Theta_1^{-1}\mathbf{u}\|^2_{0,\Theta_1} + \|\psi-\Theta_0\phi\|^2_{0,\Theta_0^{-1}} \\
X = \\
\qquad \begin{cases}
G_0(\Omega,\Theta_0)\times S(\Omega,\Theta_0^{-1})\times\mathbf{D}(\Omega,\Theta_1^{-1})\times\mathbf{C}_0(\Omega,\Theta_1) & \text{if } \Gamma=\partial\Omega \\
G(\Omega,\Theta_0)\times S(\Omega,\Theta_0^{-1})\times\mathbf{D}_0(\Omega,\Theta_1^{-1})\times\mathbf{C}(\Omega,\Theta_1) & \text{if } \Gamma^*=\partial\Omega\,.
\end{cases}
\end{cases}
\tag{5.56}
$$

Remark 5.21 The least-squares principle (5.56) can also be obtained by preceding the least-squares step by a transformation step in which the four-field system is replaced by the equivalent *optimal control* problem

$$
\begin{cases}
\min\limits_{\{\phi,\psi,\mathbf{u},\mathbf{v}\}\in X} \left(\|\mathbf{v}-\Theta_1^{-1}\mathbf{u}\|^2_{0,\Theta_1} + \|\psi-\Theta_0\phi\|^2_{0,\Theta_0^{-1}}\right) \\
\text{subject to} \\
\nabla\cdot\mathbf{u}+\psi=\Theta_0 f\,, \qquad \nabla\phi+\mathbf{v}=0\,, \qquad \text{and} \qquad \nabla\times\mathbf{v}=0
\end{cases}
\tag{5.57}
$$

and then using the LSFEMs for optimal control and optimization problems discussed in Chapter 11. Specifically, using the method (11.49) that is based on the direct penalization of the objective functional by a least-squares formulation of the constraint equations, we obtain the CLSP (5.56). Alternatively, LSFEMs can be

---

[29] In view of Remark 5.20, $|\ell(\phi)|$ can be omitted if minimization is carried over the complement space $X^C = G(\Omega)\cap L^2_0(\Omega)\times\mathbf{D}_0(\Omega)$.

[30] Recall that, in this setting, we only consider the case $\gamma=1$.

defined using the methods[31] discussed in Section 11.2.2, i.e., by using a LSFEM for the optimality system corresponding to (5.57). We do not pursue this approach because it does not lead to LSFEMs that are better in any way than those obtained from (5.56). □

Let us now examine least-squares principles obtained by keeping the ADN setting for the energy balance and using the extended div–grad system (5.25). For brevity, we restrict attention to homogeneous Dirichlet conditions ($\Gamma = \partial\Omega$) because CLSPs for (5.25) with homogeneous Neumann boundary conditions differ only in the definition of the minimization space.

Assuming that $\partial\Omega$ is sufficiently smooth, Theorem 5.10 implies that the extended div–grad system is homogeneous elliptic. Following Section 4.4, we choose $q = 0$ in (5.36) to obtain the following well-posed CLSP:

$$\begin{cases} J_0(\phi,\mathbf{u};f) = \\ \qquad \|\nabla\cdot\mathbf{u}+\gamma\Theta_0\phi-\Theta_0 f\|_0^2+\|\nabla\phi+\Theta_1^{-1}\mathbf{u}\|_0^2+\|\nabla\times\Theta_1^{-1}\mathbf{u}\|_0^2 \\ X_0 = H_0^1(\Omega)\times[H^1(\Omega)]^d\cap\mathbf{C}_0(\Omega)\,. \end{cases} \tag{5.58}$$

Corollary 5.19 shows that (5.58) remains well-posed for a wider range of problem configurations than those that were admitted by the ADN theory in Theorem 5.10.

Remark 5.22 Unfortunately, Theorem A.9 implies that the scope of (5.58) cannot be extended any further. Thus, for problems that do not meet the conditions of Corollary 5.19, such as when Ω is a non-convex polyhedral domain, the options are to go back to the div–grad formulations (5.54) or (5.55), or switch to the more general vector-operator setting energy balance (5.47) in Theorem 5.18. The latter approach yields the following well-posed least-squares principle:

$$\begin{cases} J(\phi,\mathbf{u};f) = \\ \qquad \|\nabla\cdot\mathbf{u}+\gamma\Theta_0\phi-\Theta_0 f\|_{0,\Theta_0^{-1}}^2+\|\nabla\phi+\Theta_1^{-1}\mathbf{u}\|_{0,\Theta_1}^2+\|\nabla\times\Theta_1^{-1}\mathbf{u}\|_{0,\Theta_2}^2 \\ X = G_0(\Omega,\Theta_0)\times\mathbf{C}_0(\Omega,\Theta_1)\cap\mathbf{D}(\Omega,\Theta_1^{-1})\,. \end{cases} \tag{5.59}$$

Conforming discretization of this CLSP requires finite dimensional subspaces of the intersection space $\mathbf{C}_0(\Omega,\Theta_1)\cap\mathbf{D}(\Omega,\Theta_1^{-1})$. In Section 5.6.2, we see that this leads to some undesirable consequences for compliant DLSPs based on (5.59). □

[31] The optimal control problem (5.57) can be solved using the methods from Section 11.4 in which least-squares principles are used to impose the constraints. This converts (5.57) to a bilevel minimization problem with the abstract structure of (11.77). However, this approach does not result in a bona fide CLSP for the four-field system because the bilevel problem is still a constrained optimization formulation.

5.6 Discrete Least-Squares Principles

In this section, we examine transformations of the pool of CLSPs defined in Section 5.5 into LSFEMs for (5.1) or (5.2). Recall that (5.50)–(5.53) and (5.58) were derived from energy balances in the ADN setting and that the origins of (5.54)–(5.56) and (5.59) are energy balances in the vector-operator setting. To avoid repetition of almost identical formulations, DLSPs are stated only for homogeneous Dirichlet boundary conditions.

5.6.1 The Div–Grad System

For the div–grad system with Dirichlet boundary conditions, the options are to define LSFEMs using either the CLSP (5.50) for the ADN setting or the vector-operator setting version (5.54). The least-squares functional in (5.50) uses the impractical minus one norm. To obtain a practical DLSP, we have the choice of the two transition diagrams from Section 4.5. Using (4.28) gives the *weighted* $L^2(\Omega)$ norm DLSP

$$\begin{cases} J^h_{-1}(\phi^h,\mathbf{u}^h;f) = h^2\|\nabla\cdot\mathbf{u}^h+\gamma\Theta_0\phi^h-\Theta_0 f\|_0^2+\|\nabla\phi^h+\Theta_1^{-1}\mathbf{u}^h\|_0^2 \\ X^h_{r+,0} = G^{r+1}(\Omega)\cap H^1_0(\Omega)\times[G^r(\Omega)]^d, \qquad r\geq 1. \end{cases} \tag{5.60}$$

This method belongs to the class of quasi-norm-equivalent DLSPs; see Section 3.5.3. The other choice is to transform (5.50) according to (4.39). The result is a *discrete negative norm* DLSP:

$$\begin{cases} J_{-h}(\phi^h,\mathbf{u}^h;f) = \|\nabla\cdot\mathbf{u}^h+\gamma\Theta_0\phi^h-\Theta_0 f\|_{-h}^2+\|\nabla\phi^h+\Theta_1^{-1}\mathbf{u}^h\|_0^2 \\ X^h_{\widehat{r},0} = G^{r+1}(\Omega)\cap H^1_0(\Omega)\times[G^{\widehat{r}}(\Omega)]^d, \qquad \widehat{r}=\max\{1,r\},\ r\geq 0. \end{cases} \tag{5.61}$$

This method is an example of a norm-equivalent DLSP; see Section 3.5.2.

The solution spaces in (5.60) and (5.61) are constructed using nodal finite element spaces; see Section B.2.1. The one in (5.60) specializes the abstract approximating space (4.35), whereas the solution space for (5.61) is defined as in (4.44) in order to avoid the need to approximate the divergence in the first residual by a discrete weak operator; see Remark 4.11.

The restriction on the polynomial degree r in the weighted method (5.60) comes from the *minimal approximation condition*; see Remark 4.11. For example, this means that, on a simplicial mesh, the potential variable has to be approximated by at least piecewise quadratic finite elements. In the discrete negative norm method (5.61), this condition is relaxed so that linear elements are admissible for ϕ^h.

For both DLSPs, using the same nodal finite element space for all components of the solution is perfectly admissible, as long as the minimal approximation condition is met; see Remark 4.12. As a result, (5.60) and (5.61) can be implemented using the equal interpolation order space $X^h_{r,0} = G^r(\Omega) \cap H^1_0(\Omega) \times [G^r(\Omega)]^d$, where $r \geq 2$ for the first method[32] and $r \geq 1$ for the second one.

The vector-operator setting CLSP (5.54) uses only $L^2(\Omega)$ norms and satisfies *all* practicality criteria from Section 2.2.2. Its conforming discretization is given by

$$\begin{cases} J(\phi^h, \mathbf{u}^h; f) = \|\nabla \cdot \mathbf{u}^h + \gamma\Theta_0\phi^h - \Theta_0 f\|^2_{0,\Theta_0^{-1}} + \|\nabla\phi^h + \Theta_1^{-1}\mathbf{u}^h\|^2_{0,\Theta_1} \\ X^h = G^h_0(\Omega) \times \mathbf{D}^h(\Omega), \end{cases} \tag{5.62}$$

where $G^h_0(\Omega)$ and $\mathbf{D}^h(\Omega)$ are *compatible* finite element subspaces of $G_0(\Omega, \Theta_0)$ and $\mathbf{D}(\Omega, \Theta_1^{-1})$, respectively; see Section B.2 for definitions for these spaces. The LSFEM (5.62) is an example of a *compliant* DLSP; see Section 3.5.

For G^h_0 we can take any nodal finite element space because, for non-degenerate weights Θ_0 and Θ_1 that satisfy (5.3), the space $G_0(\Omega, \Theta_0)$ is equivalent to $H^1_0(\Omega)$; see Section B.3. It follows that (5.60)–(5.62) all use the same type of finite element spaces to approximate the potential ϕ. However, the *compatible* space $\mathbf{D}^h(\Omega)$ employed in (5.62) is quite different from the *direct product* space $[G^r(\Omega)]^d$ used in (5.60) and (5.61). The latter contains piecewise smooth vector fields that are continuous across element faces, whereas vector fields in $\mathbf{D}^h(\Omega)$ are only normally continuous across these faces; see Theorem B.3.

Remark 5.23 Because the vector-operator setting CLSP (5.54) contains only $L^2(\Omega)$ norms of first-order operators, one can argue that it can be discretized using the equal-order nodal finite element space $X^h_{r,0}$ defined above, instead of the compatible space $X^h = G^h_0(\Omega) \times \mathbf{D}^h(\Omega)$. The resulting DLSP

$$\begin{cases} J(\phi^h, \mathbf{u}^h; f) = \|\nabla \cdot \mathbf{u}^h + \gamma\Theta_0\phi^h - \Theta_0 f\|^2_{0,\Theta_0^{-1}} + \|\nabla\phi^h + \Theta_1^{-1}\mathbf{u}^h\|^2_{0,\Theta_1} \\ X^h_{r,0} = G^r(\Omega) \cap H^1_0(\Omega) \times [G^r(\Omega)]^d, \qquad r \geq 1 \end{cases} \tag{5.63}$$

is formally compliant because $[G^r(\Omega)]^d$ is a proper subspace of $\mathbf{D}(\Omega)$. To distinguish between (5.62) and (5.63), we refer to the former as a *compatible* LSFEM and the latter as a *nodal* LSFEM. □

Remark 5.24 There are at least three reasons why the nodal LSFEM (5.63) should not be used despite its formal compliance and the fact that admissibility of equal-order $C^0(\Omega)$ elements is often viewed as one of the principal computational advantages of LSFEMs.

1. The use of the product space $[G^r(\Omega)]^d$ is consistent with energy balances in the ADN setting in which the variables are treated as a set of unrelated scalar functions that can be approximated *independently*; see Remark 4.12. However, this viewpoint is incompatible with energy balances in the vector-operator setting

[32] Note that this precludes the use of piecewise linear elements for all variables in (5.60).

that view the variables as elements from functions spaces forming a differential De Rham complex.

2. Theorem B.13 asserts that $[G^r(\Omega)]^d$ is too "small" to well approximate minimizers of the parent CLSP (5.54) that belong to $\mathbf{D}(\Omega)$ but not to $[H^1(\Omega)]^d$. In Section 5.9, we present examples that show how using $[G^r(\Omega)]^d$ degrades the accuracy and physical fidelity of the nodal LSFEM.
3. The most important argument[33] in favor of (5.62) is that the compatible LSFEM has valuable computational properties, including local conservation and optimal $L^2(\Omega)$ convergence of the flux variable, that are not achievable by the nodal method. Bearing in mind that lack of local conservation is the main criticism leveled at LSFEMs, this by itself should be reason enough to seriously consider the compatible method. From this vantage point, the possibility of using equal order $C^0(\Omega)$ finite element spaces for all variables is arguably the least important advantage of LSFEMs. □

The least-squares functional for the four-field first-order system also uses only $L^2(\Omega)$ norms and satisfies all practicality criteria. Conforming discretization of (5.56) gives rise to the following compliant DLSP:

$$\left\{\begin{array}{l} J(\phi^h,\psi^h,\mathbf{u}^h,\mathbf{v}^h;f) = \\ \qquad \|\nabla\cdot\mathbf{u}^h+\psi^h-\Theta_0 f\|^2_{0,\Theta_0^{-1}} + \|\nabla\phi^h+\mathbf{v}^h\|^2_{0,\Theta_1} + \|\nabla\times\mathbf{v}^h\|_{0,\Theta_2} \\ \qquad\quad +\|\mathbf{v}^h-\Theta_1^{-1}\mathbf{u}^h\|^2_{0,\Theta_1} + \|\psi^h-\Theta_0\phi^h\|^2_{0,\Theta_0^{-1}} \\ X^h = G^h_0(\Omega)\times S^h(\Omega)\times\mathbf{D}^h(\Omega)\times\mathbf{C}^h_0(\Omega)\,. \end{array}\right. \tag{5.64}$$

The solution space of this DLSP uses all four types of compatible spaces[34] described in Section B.2 which justifies calling it *compatible* as well. It is important to note that the solution space components in (5.62) and (5.64) are not subject to any joint stability conditions such as the discrete inf–sup conditions (1.33) and (1.34) that arise in mixed methods so that they can be chosen completely independently of each other.[35] Of course, to optimize the resulting LSFEMs, it makes sense to pick component spaces that have equilibrated, or at least comparable, approximation orders.

---

[33] Theoretical analyses in support of this argument are given in Section 5.8.

[34] As in Remark 5.23, it is possible to discretize (5.56) using the same nodal space for all solution space components. We do not state the nodal version of (5.64) because all arguments against the use of nodal spaces stated in Remark 5.24 remain valid in the context of this DLSP as well.

[35] Formally, the absence of joint stability conditions enables implementation of (5.62) and (5.64) with different grids for each component space in X^h. However, the optimal control example (5.76) and Theorem 5.37 show that in some cases it is profitable to use finite element spaces from the same discrete De Rham complex.

Comparison of compatible DLSPs

We now compare and contrast the potential–flux DLSP (5.62) with the four-field DLSP (5.64). First, it is clear that if $\{\phi^h, \mathbf{u}^h\} \in G_0^h(\Omega) \times \mathbf{D}^h(\Omega)$ is a potential–flux pair determined by (5.62), then the pair $\{\psi^h, \mathbf{v}^h\} \in S^h(\Omega) \times \mathbf{C}_0^h(\Omega)$ defined by[36]

$$\psi^h = \Pi_S \Theta_0 f - \nabla \cdot \mathbf{u}^h \qquad \text{and} \qquad \mathbf{v}^h = -\nabla \phi^h \,, \tag{5.65}$$

approximates the complementary density–intensity variables. Therefore, both (5.62) and (5.64) can be used to approximate all four fields. The difference is that, in (5.64), these fields are computed *directly* whereas in (5.62) approximations of the complementary variables are *derived* from the primal pair. Because the approximation spaces in (5.64) can be selected independently of each other, we have that, in general, the density–intensity pair from this method need not be the same as the one defined in (5.65), even if the two methods use identical spaces for the potential–flux pair.

Yet, a valid question is what would happen if, in addition to the same potential–flux spaces, the density–intensity variables in (5.64) are approximated using the compatible spaces implied by (5.65)? For example, assume that (5.62) is implemented using the lowest-order nodal space $G^1(\Omega)$ and the lowest-order Raviart–Thomas space $\mathbf{D}^{1\text{-}}(\Omega)$. From (B.45), it follows that the pair $\{\psi^h, \mathbf{v}^h\}$ defined in (5.65) belongs to $S^0(\Omega) \times \mathbf{C}^{1\text{-}}$. Suppose now that the four-field method is implemented using these spaces, i.e., we set

$$X^h = G_0^1(\Omega) \times S^0(\Omega) \times \mathbf{D}^{1\text{-}}(\Omega) \times \mathbf{C}_0^{1\text{-}}(\Omega)$$

in (5.64). If the solution of (5.64) coincides with the pairs produced by (5.62), we can use the latter to obtain the approximations much more cheaply. The following theorem shows that this is possible only in some very special cases.

Theorem 5.25 *Assume that the solution spaces in* (5.62) *and* (5.64) *are defined using the same finite element spaces* $G_0^h(\Omega)$ *and* $\mathbf{D}^h(\Omega)$. *Furthermore, assume that* $S^h(\Omega)$ *and* $\mathbf{C}_0^h(\Omega)$ *in* (5.64) *are such that* $\{G_0^h(\Omega), \mathbf{C}_0^h(\Omega)\}$ *and* $\{\mathbf{D}^h(\Omega), S^h(\Omega)\}$ *belong to two (possibly different) finite element De Rham complexes. Furthermore, let* $\{\phi_1^h, \mathbf{u}_1^h\}$ *denote a minimizer of* (5.62), $\{\psi_1^h, \mathbf{v}_1^h\}$ *denote functions determined from* $\{\phi_1^h, \mathbf{u}_1^h\}$ *by* (5.65), *and* $\{\phi_2^h, \psi_2^h, \mathbf{u}_2^h, \mathbf{v}_2^h\}$ *denote a minimizer of* (5.64). *Then, for arbitrary data* f, *the sets* $\{\phi_1^h, \psi_1^h, \mathbf{u}_1^h, \mathbf{v}_1^h\}$ *and* $\{\phi_2^h, \psi_2^h, \mathbf{u}_2^h, \mathbf{v}_2^h\}$ *are not the same.*

Proof. It suffices to prove the assertion for unit weights. Being a minimizer of (5.64), the set $\{\phi_2^h, \psi_2^h, \mathbf{u}_2^h, \mathbf{v}_2^h\}$ satisfies the first-order optimality condition

$$\begin{aligned}(\nabla \cdot \mathbf{u}_2^h + \psi_2^h, \nabla \cdot \widehat{\mathbf{u}}^h + \widehat{\psi}^h) + (\nabla \phi_2^h + \mathbf{v}_2^h, \nabla \widehat{\phi}^h + \widehat{\mathbf{v}}^h) + (\nabla \times \mathbf{v}_2^h, \nabla \times \widehat{\mathbf{v}}^h) \\ + (\phi_2^h - \psi_2^h, \widehat{\phi}^h - \widehat{\psi}^h) + (\mathbf{u}_2^h - \mathbf{v}_2^h, \widehat{\mathbf{u}}^h - \widehat{\mathbf{v}}^h) = (f, \nabla \cdot \widehat{\mathbf{u}}^h + \widehat{\psi}^h)\end{aligned}$$

---

[36] In (5.65), Π_S denotes the projection onto $S^h(\Omega)$.

for all $\{\widehat{\phi}^h, \widehat{\psi}^h, \widehat{\mathbf{u}}^h, \widehat{\mathbf{v}}^h\} \in X^h$. Setting all test functions except $\widehat{\psi}^h$ to zero and then repeating the same with $\widehat{\mathbf{v}}^h$ yields the identities

$$\big(\nabla\cdot\mathbf{u}_2^h + \psi_2^h, \widehat{\psi}^h\big) - \big(\phi_2^h - \psi_2^h, \widehat{\psi}^h\big) = \big(f, \widehat{\psi}^h\big) \tag{5.66}$$

$$\big(\nabla\phi_2^h + \mathbf{v}_2^h, \widehat{\mathbf{v}}^h\big) + \big(\nabla\times\mathbf{v}_2^h, \nabla\times\widehat{\mathbf{v}}^h\big) - \big(\mathbf{u}_2^h - \mathbf{v}_2^h, \widehat{\mathbf{v}}^h\big) = 0 \tag{5.67}$$

for all $\widehat{\psi}^h \in S^h(\Omega)$ and $\widehat{\mathbf{v}}^h \in \mathbf{C}_0^h(\Omega)$.

The optimality condition for the minimizer $\{\phi_1^h, \mathbf{u}_1^h\}$ of (5.62) is given by

$$\big(\nabla\cdot\mathbf{u}_1^h + \phi_1^h, \widehat{\phi}^h\big) + \big(\nabla\phi_1^h + \mathbf{u}_1^h, \nabla\widehat{\phi}^h\big) = \big(f, \widehat{\phi}^h\big) \tag{5.68}$$

$$\big(\nabla\cdot\mathbf{u}_1^h + \phi_1^h, \nabla\cdot\widehat{\mathbf{u}}^h\big) + \big(\nabla\phi_1^h + \mathbf{u}_1^h, \widehat{\mathbf{u}}^h\big) = \big(f, \nabla\cdot\widehat{\mathbf{u}}^h\big) \tag{5.69}$$

for all $\widehat{\phi}^h \in G_0^h(\Omega)$ and $\widehat{\mathbf{u}}^h \in \mathbf{D}^h(\Omega)$.

Suppose that the density ψ_1^h and the intensity $\mathbf{v}_1^h$ are defined from ϕ_1^h and $\mathbf{u}_1^h$ as in (5.65). Then, $\nabla\phi_1^h + \mathbf{v}_1^h = 0$ and $\nabla\times\mathbf{v}_1^h = 0$ so that

$$\big(\nabla\phi_1^h + \mathbf{v}_1^h, \widehat{\mathbf{v}}^h\big) + \big(\nabla\times\mathbf{v}_1^h, \nabla\times\widehat{\mathbf{v}}^h\big) = 0 \qquad \forall\widehat{\mathbf{v}}^h \in \mathbf{C}_0^h(\Omega) \tag{5.70}$$

and likewise

$$\big(\nabla\cdot\mathbf{u}_1^h + \psi_1^h, \widehat{\psi}^h\big) = \big(f, \widehat{\psi}^h\big) \qquad \forall\,\widehat{\psi}^h \in S^h(\Omega)\,. \tag{5.71}$$

Comparing (5.66) with (5.71) and (5.67) with (5.70) shows that

$$\big(\phi_1^h - \psi_1^h, \widehat{\psi}^h\big) = 0 \qquad \forall\,\widehat{\psi}^h \in S^h(\Omega) \tag{5.72}$$

$$\big(\mathbf{u}_1^h - \mathbf{v}_1^h, \widehat{\mathbf{v}}^h\big) = 0 \qquad \forall\,\widehat{\mathbf{v}}^h \in \mathbf{C}_0^h(\Omega) \tag{5.73}$$

are necessary conditions for $\{\phi_1^h, \psi_1^h, \mathbf{u}_1^h, \mathbf{v}_1^h\}$ and $\{\phi_2^h, \psi_2^h, \mathbf{u}_2^h, \mathbf{v}_2^h\}$ to coincide.

Suppose that (5.72) and (5.73) are indeed satisfied. From (5.71), (5.72), and the inclusion $\nabla\cdot\widehat{\mathbf{u}}^h \in S^h(\Omega)$, we have

$$\big(\phi_1^h, \nabla\cdot\widehat{\mathbf{u}}^h\big) = \big(\psi_1^h, \nabla\cdot\widehat{\mathbf{u}}^h\big) = \big(f - \nabla\cdot\mathbf{u}_1^h, \nabla\cdot\widehat{\mathbf{u}}^h\big) \qquad \forall\widehat{\mathbf{u}}^h \in \mathbf{D}^h(\Omega)\,.$$

This identity and (5.69) imply that

$$\big(\nabla\phi_1^h + \mathbf{u}_1^h, \widehat{\mathbf{u}}^h\big) = 0 \qquad \forall\widehat{\mathbf{u}}^h \in \mathbf{D}^h(\Omega)\,. \tag{5.74}$$

Next, from (5.73), the definition of $\mathbf{v}_1^h$, i.e., $\mathbf{v}_1^h = -\nabla\phi_1^h$, and the fact that $\nabla\widehat{\phi}^h \in \mathbf{C}_0^h(\Omega)$ for all $\widehat{\phi}^h \in G_0^h(\Omega)$, we see that

$$\big(\nabla\phi_1^h + \mathbf{u}_1^h, \nabla\widehat{\phi}^h\big) = 0 \qquad \forall\widehat{\phi}^h \in G_0^h(\Omega)\,. \tag{5.75}$$

Adding (5.74) and (5.75) results in

$$\big(\nabla\phi_1^h + \mathbf{u}_1^h, \nabla\widehat{\phi}^h + \widehat{\mathbf{u}}^h\big) = 0 \qquad \forall\{\widehat{\phi}^h, \widehat{\mathbf{u}}^h\} \in G_0^h(\Omega)\times\mathbf{D}^h\,.$$

In particular,
$$\|\nabla\phi_1^h + \mathbf{u}_1^h\|_0 = 0,$$
i.e., $\nabla\phi_1^h = -\mathbf{u}_1^h$ a.e. Then, because $\mathbf{u}^h \in \mathbf{D}^h(\Omega)$ and $\nabla\phi^h \in \mathbf{C}_0^h(\Omega)$, it follows that $\nabla\phi_1^h \in \mathbf{C}_0^h(\Omega) \cap \mathbf{D}^h(\Omega)$. Therefore, $\nabla\phi_1^h$ is a piecewise smooth polynomial field that is both tangentially and normally continuous, i.e., it must be of class $C^0(\Omega)$ and so, ϕ^h must be of class $C^1(\Omega)$. Clearly this cannot hold for arbitrary data f. □

To illustrate the conclusions of Theorem 5.25, consider the case when (5.62) is implemented using the lowest-order nodal and div-conforming spaces on simplices, i.e., $X^h = G_0^1(\Omega) \times \mathbf{D}^{1\text{-}}(\Omega)$. Let $\{\phi_1^h, \mathbf{u}_1^h\} \in X^h$ denote the solution of (5.62). Then, the necessary condition in Theorem 5.25 requires $\nabla\phi_1^h$ to be of class $C^0(\Omega)$ and ϕ_1^h to be of class $C^1(\Omega)$. On the other hand, because $\phi_1^h \in G_0^1(\Omega)$, its gradient is constant on each element, i.e., $\nabla\phi_1^h$ is a piecewise constant field of class $C^0(\Omega)$. It follows that $\mathbf{u}_1^h = -\nabla\phi_1^h$ is constant in Ω and ϕ_1^h is a linear function that satisfies a homogeneous Dirichlet condition on $\partial\Omega$. Obviously, the only such linear function is $\phi_1^h \equiv 0$, which means that all functions in the set $\{\phi_1^h, \psi_1^h, \mathbf{u}_1^h, \mathbf{v}_1^h\}$ obtained from (5.62) must be identically zero. Therefore, for the lowest-order elements, solutions of (5.64) can coincide with $\{\phi_1^h, \psi_1^h, \mathbf{u}_1^h, \mathbf{v}_1^h\}$ only in the trivial case.

Remark 5.26 It is instructive to compare "compatibility" of the potential–flux DLSP (5.62) and the four-field DLSP (5.64) with the "compatibility" of the classical Rayleigh–Ritz and mixed-Galerkin methods for (5.1) and (5.2).

The solution $\{\mathbf{u}^h, \psi^h\}$ of the mixed-Galerkin method belongs to $\mathbf{D}^h(\Omega) \times S^h(\Omega)$ so that this method approximates the *flux–density* pair of variables. On the other hand, given a solution $\phi^h \in G^h(\Omega)$ of the Rayleigh–Ritz method, it is clear that the pair $\{\phi^h, \mathbf{v}^h = -\nabla\phi^h\}$ approximates the complementary *potential–intensity* variables. This serves to further underscore the duality of Rayleigh–Ritz and mixed-Galerkin methods that is inherited from the duality of their parent Dirichlet (1.50) and Kelvin (1.53) principles, respectively. Thus, we can say that the Rayleigh–Ritz method is "compatible" with the *primal* potential–intensity pair, whereas the mixed-Galerkin method is "compatible" with the *dual* flux–density pair of variables.

By the same token, we see that the four-field DLSP (5.64) is "compatible" with *both* pairs of variables. The same can be said about the potential–flux DLSP (5.62) even though, formally, this method approximates directly only one variable from each pair. However, as we saw earlier, the "missing" intensity and density variables can be recovered by (5.65). In particular, the formula for the intensity variable in (5.65) is exactly the same as for the Rayleigh–Ritz method. In Section 5.8, we show that this is not an accident and that the potential–flux DLSP (5.62) is intimately related to *both* the Rayleigh–Ritz and the mixed-Galerkin methods. □

Direct discretization of the optimal control formulation

According to Remark 5.21, the four-field compatible DLSP (5.64) can be reinterpreted as a least-squares formulation of the optimal control problem (5.57). A question not unlike the one posed in Theorem 5.25 is what would happen if (5.57) is discretized directly without passing through a least-squares minimization step? To answer this question, we restrict (5.57) to compatible finite element spaces and obtain the following discrete optimization problem:

$$\begin{cases} \min\limits_{\substack{\{\phi^h,\psi^h\}\in G_0^h\times S^h\\ \{\mathbf{u}^h,\mathbf{v}^h\}\in\mathbf{D}^h\times\mathbf{C}_0^h}} \left(\|\mathbf{v}^h-\Theta_1^{-1}\mathbf{u}^h\|_{0,\Theta_1}^2+\|\psi^h-\Theta_0\phi^h\|_{0,\Theta_0^{-1}}^2\right) \\ \text{subject to} \\ \nabla\cdot\mathbf{u}^h+\psi^h=\Pi_S\Theta_0 f\,, \qquad \nabla\phi^h+\mathbf{v}^h=0\,, \qquad \text{and} \qquad \nabla\times\mathbf{v}^h=0\,. \end{cases} \tag{5.76}$$

Now assume, as in Theorem 5.25, that the pairs $\{\mathbf{D}^h,S^h\}$ and $\{G_0^h,\mathbf{C}_0^h\}$ belong to two, possibly different, finite element De Rham complexes. Then, S^h coincides with the range of divergence on $\mathbf{D}^h$ and $\mathbf{C}_0^h$ contains the range of the gradient on G_0^h. As a result, we can solve the first two constraint equations in (5.76) for the range variables:

$$\psi^h=\Pi_S\Theta_0 f-\nabla\cdot\mathbf{u}^h \qquad \text{and} \qquad \mathbf{v}^h=-\nabla\phi^h\,.$$

This makes the third constraint trivially satisfied and allows one to eliminate $\mathbf{v}^h$ and ψ^h from (5.76). The result is the following *unconstrained* minimization problem:

$$\min_{\{\phi^h,\mathbf{u}^h\}\in G_0^h\times\mathbf{D}^h}\left(\|\nabla\cdot\mathbf{u}^h+\Theta_0\phi^h-\Pi_S\Theta_0 f\|_{0,\Theta_0^{-1}}^2+\|\nabla\phi^h+\Theta_1^{-1}\mathbf{u}^h\|_{0,\Theta_1}^2\right). \tag{5.77}$$

We see that, with the exception of the term involving the projection Π_S, this problem is the same as the *two-field* potential–flux compatible DLSP (5.62) with $\gamma=1$. Two factors are responsible for making the transition from (5.76) to (5.77) possible. First, we must have $\gamma\neq 0$ in order to solve the first constraint for the density variable ψ^h. A second, equally important, precondition is that $\{\mathbf{D}^h,S^h\}$ and $\{G_0^h,\mathbf{C}_0^h\}$ are taken from the same finite element complexes.[37]

5.6.2 The Extended Div–Grad System

Assuming that all the hypotheses of Corollary 5.19 hold, we can use (5.58) to define the following DLSP:

$$\begin{cases} J_0(\phi^h,\mathbf{u}^h;f)= \\ \qquad \|\nabla\cdot\mathbf{u}^h+\gamma\Theta_0\phi^h-\Theta_0 f\|_0^2+\|\nabla\phi^h+\Theta_1^{-1}\mathbf{u}^h\|_0^2+\|\nabla\times\mathbf{u}^h\|_0^2 \\ X_{r,0}^h=G^r\cap H_0^1(\Omega)\times[G^r]^d\cap\mathbf{C}_0(\Omega)\,, \qquad r\geq 1\,. \end{cases} \tag{5.78}$$

[37] The fact that (5.76) can be reduced to almost the same problem as (5.62) does not contradict Theorem 5.25 because there we compared the potential–flux DLSP with the four-field DLSP (5.64) which is *not equivalent* to (5.76).

This compliant DLSP offers all the usual advantages present in LSFEMs for homogeneous elliptic systems; see Table 4.1.

However, as was already explained in Remark 5.22, the scope of the parent CLSP (5.25) that engenders (5.78) cannot be extended to domains that are not admissible in Corollary 5.19. Thus, (5.78) should not be used if Ω is, e.g., a non-convex polyhedral domain. In that case, we can either fall back to a div–grad formulation or use the more general vector-operator setting formulation (5.59) as the parent CLSP. Formally, a conforming discretization of (5.59) is given by:

$$\begin{cases} J(\phi^h,\mathbf{u}^h;f) = \\ \qquad \|\nabla\cdot\mathbf{u}^h+\gamma\Theta_0\phi^h-\Theta_0 f\|^2_{0,\Theta_0^{-1}}+\|\nabla\phi^h+\Theta_1^{-1}\mathbf{u}^h\|^2_{0,\Theta_1}+\|\nabla\times\mathbf{u}^h\|^2_{0,\Theta_2} \\ X^h = G_0^h(\Omega)\times(\mathbf{C}_0\mathbf{D})^h(\Omega), \end{cases} \tag{5.79}$$

where $(\mathbf{C}_0\mathbf{D})^h(\Omega)$ is a conforming finite element subspace of the intersection space $\mathbf{C}_0(\Omega)\cap\mathbf{D}(\Omega)$. Unfortunately, Proposition B.14 implies that $(\mathbf{C}_0\mathbf{D})^h(\Omega)$ is necessarily a subspace of the direct product Sobolev space $[H^1(\Omega)]^d$, i.e., it has to be a nodal $C^0(\Omega)$ space. Thus, despite the fact that the mathematical setting of (5.59) is more general than that of the ADN setting CLSP (5.58), their conforming discretizations lead to essentially the same LSFEM in which the flux is approximated by nodal $C^0(\Omega)$ finite element spaces. It follows that using the vector-operator setting CLSP for the extended system *fails to achieve* our objective of obtaining LSFEMs that work[38] in situations where the hypotheses of Corollary 5.19 are violated.

The two key lessons that can be learned from this setback are as follows. First, the *modification* of the div–grad system in Section 5.3.2 changes the natural solution space for the flux from $\mathbf{D}(\Omega)$ to the intersection space $\mathbf{C}(\Omega)\cap\mathbf{D}(\Omega)$; the latter is difficult to approximate and should be avoided in variational settings for finite element methods. Second, LSFEMs obtained from the extended div–grad system are not safe to use unless the assumptions of Corollary 5.19 are satisfied.

This should be contrasted with LSFEMs based on the four-field system (5.23) that also includes a redundant curl equation. The key difference between (5.23) and the extended div–grad system (5.25) is that the natural solution space for the flux in the former includes the *direct product* space $\mathbf{C}(\Omega)\times\mathbf{D}(\Omega)$ which, in contrast to $\mathbf{C}(\Omega)\cap\mathbf{D}(\Omega)$, can be approximated well by finite element spaces. As a result, LSFEMs derived from (5.23) are safe to use in configurations for which LSFEMs obtained from (5.25) should be avoided.

---

[38] This does not contradict the best approximation property (3.63) which does hold for (5.79). Indeed, Theorem 3.28 implies that a LSFEM determines the best possible approximation out of the specified finite element space. However, to show the *convergence* of the approximate solution, one also needs for the finite element space to posses the *approximability* property (B.8). Theorem B.13 shows that, in general, nodal finite element spaces do not have this property with respect to $\mathbf{C}_0(\Omega)\cap\mathbf{D}(\Omega)$.

5.7 Error Analyses

In this section, we provide theoretical error estimates for LSFEM approximations of solutions of (5.1) or (5.2). We first focus, in Section 5.7.1, on estimates of the error measured in the norms of the solution spaces of the corresponding least-squares principles. Then, in Section 5.7.2, we consider estimates for the error measured in $L^2(\Omega)$ norms for cases where this norm is not a natural norm for the solution space.

5.7.1 Error Estimates in Solution Space Norms

The LSFEMs discussed in the last section originate from parent CLSPs posed in the ADN or vector-operator setting and result in quite different finite element spaces and error analyses. Compatible DLSPs require curl- and/or div-conforming finite element spaces. Theorems B.7 and B.9 disclose that optimal error estimates for such spaces are not guaranteed on general non-affine grids. As a result, for compatible LSFEMs, specialization of abstract error results is restricted to affine simplicial grids. For DLSPs whose origins are in the ADN setting, this precondition is not necessary because their solution spaces are direct products of $C^0(\Omega)$ nodal finite element spaces. Recall that, for such elements, the optimal approximation property (B.20) holds on non-affine grids as long as they are shape-regular; see Theorem B.6. This is also true for the nodal LSFEM (5.63) that has the same parent CLSP as the compatible method (5.62) but uses nodal elements.

Error Estimates in the ADN setting

The *quasi-norm-equivalent*, *norm-equivalent*, and *compliant* DLSPs defined by (5.60), (5.61), and (5.78), respectively, are examples of methods whose parent CLSPs were derived with the help of the ADN elliptic theory. As a result, the error analysis of these methods follows from the general LSFEM theory for ADN systems given in Chapter 4. The first two methods share a common parent CLSP based on the non-homogeneous elliptic div–grad system. Therefore, error estimates for (5.60) and (5.61) follow from Theorems 4.8 and 4.10, respectively.

Theorem 5.27 *Assume that Ω satisfies the hypotheses in Theorem 5.9, $\mathcal{T}_h$ is a uniformly regular partition of Ω into finite elements such that the inverse inequality* (B.85) *holds for $G^s(\Omega)$, $s \geq 1$, and that the solution $\{\phi, \mathbf{u}\}$ of the div–grad system* (5.17) *with Dirichlet boundary condition, i.e., with $\Gamma = \partial\Omega$, belongs to the space $X_q = H^{q+2}(\Omega) \cap H^1_0(\Omega) \times [H^{q+1}(\Omega)]^d$ for some $q \geq 0$.*

Let $\{\phi^h, \mathbf{u}^h\} \in X^h_{\tilde{r},0}$, $r \geq 0$ *denote a solution of the norm-equivalent DLSP* (5.61). *Then, there exists a positive constant C such that, for*[39] $\tilde{r} = \min\{r, q\}$,

$$\|\phi - \phi^h\|_1 + \|\mathbf{u} - \mathbf{u}^h\|_0 \leq Ch^{\tilde{r}+1}\left(\|\phi\|_{\tilde{r}+2} + \|\mathbf{u}\|_{\tilde{r}+1}\right) \tag{5.80}$$

and

$$\|\mathbf{u} - \mathbf{u}^h\|_1 \leq Ch^{\tilde{r}}\left(\|\phi\|_{\tilde{r}+2} + \|\mathbf{u}\|_{\tilde{r}+1}\right). \tag{5.81}$$

The error estimates (5.80) *and* (5.81) *also hold for the solution* $u^h \in X^h_{r^+,0}$, $r \geq 1$ *of the quasi-norm-equivalent DLSP* (5.60). *The assertions of the theorem remain valid for the div–grad system with the Neumann boundary condition, i.e., with* $\Gamma^* = \partial\Omega$. □

The last method in this group, i.e., the compliant DLSP (5.78), is based on the extended div–grad system (5.25). Because this system is homogeneous elliptic, the error estimates for (5.78) follow from Theorem 4.2.

Theorem 5.28 *Assume that* Ω *satisfies the hypotheses in Corollary* 5.19, $\mathcal{T}_h$ *is a uniformly regular partition of* Ω, *and the solution* $\{\phi, \mathbf{u}\}$ *of* (5.25) *with the Dirichlet boundary condition, i.e., with* $\Gamma = \partial\Omega$, *belongs to the space* $X_q = H^{q+1}(\Omega) \cap H^1_0(\Omega) \times [H^{q+1}(\Omega)]^d \cap \mathbf{C}_0(\Omega)$ *for some* $q \geq 0$.

Let $\{\phi^h, \mathbf{u}^h\} \in X^h_{r,0}$, $r \geq 1$, *denote a solution of the compliant DLSP* (5.78). *Then, there exists a positive constant C such that, for* $\tilde{r} = \min\{r, q\}$,

$$\|\phi - \phi^h\|_1 + \|\mathbf{u} - \mathbf{u}^h\|_1 \leq Ch^{\tilde{r}}\left(\|\phi\|_{\tilde{r}+1} + \|\mathbf{u}\|_{\tilde{r}+1}\right). \tag{5.82}$$

If Ω *has smooth*[40] *boundary,*

$$\|\phi - \phi^h\|_0 + \|\mathbf{u} - \mathbf{u}^h\|_0 \leq Ch^{\tilde{r}+1}\left(\|\phi\|_{\tilde{r}+1} + \|\mathbf{u}\|_{\tilde{r}+1}\right). \tag{5.83}$$

The above error estimates remain valid for the extended div–grad system with Neumann boundary conditions, i.e., with $\Gamma^* = \partial\Omega$. □

Remark 5.29 The errors of the potential and flux variables in Theorem 5.27 are measured in different Sobolev space norms. This is typical of LSFEMs obtained from non-homogeneous elliptic systems. In contrast, the error estimates in Theorem 5.28 use the same Sobolev space norm for both variables because the compliant DLSP (5.78) is derived from a homogeneous elliptic first-order system. □

Error estimates in the vector-operator setting

The DLSPs (5.62)–(5.64) and (5.79) are examples of methods whose parent CLSPs are set in the vector-operator setting so that the LSFEM theory from Chapter 4 does

[39] The definition of the minimization space $X^h_{\tilde{r},0}$ in (5.61) is such that, for $r = 0$, its flux components default to $G^1(\Omega)$. This makes the $H^1(\Omega)$ norm error estimate for $\mathbf{u}^h$ meaningful. See Remark 4.11 for more details.

[40] This condition can be relaxed to an assumption that $\partial\Omega$ is of class $C^{2,1}$; see [105].

not apply. Instead, we use the more general theory from Chapter 3, specifically, the abstract approximation results in Section 3.5 for compliant DLSPs.

Let us first examine the convergence of the *compatible* DLSP (5.62) and its *nodal* version (5.63). Both methods share the same parent CLSP (5.54) and are compliant (see Remark 5.23) so that the abstract results in Theorem 3.28 are applicable. In particular, the abstract error estimate (3.63) implies that the "energy" norm of the error in (5.62) or (5.63) is bounded by the best approximation error out of the respective finite element spaces. This statement is formalized in the following theorem.

Theorem 5.30 *Let $\Omega \subset \mathbb{R}^d$, $d = 2,3$, be a bounded domain with Lipschitz continuous boundary, and let $\mathcal{T}_h$ denote a uniformly regular partition of Ω into finite elements. Assume that X^h is either the compatible space $X^h = G_0^h(\Omega) \times \mathbf{D}^h(\Omega)$ for* (5.62) *or the equal-order nodal space $X^h_{r,0}$ for* (5.63). *Let $\{\phi^h, \mathbf{u}^h\} \in X^h$ denote a solution of the compatible LSFEM* (5.62) *or the nodal LSFEM* (5.63). *Then,*

$$\|\phi - \phi^h\|_G + \|\mathbf{u} - \mathbf{u}^h\|_D \le C \inf_{\{\widehat{\phi}^h, \widehat{\mathbf{u}}^h\} \in X^h} \left(\|\phi - \widehat{\phi}^h\|_G + \|\mathbf{u} - \widehat{\mathbf{u}}^h\|_D \right), \tag{5.84}$$

where $\{\phi, \mathbf{u}\} \in G(\Omega) \times \mathbf{D}(\Omega)$ denotes a solution of (5.17).

Proof. The proof follows from (3.63) in Theorem 3.28. □

If the solution of the div–grad system is sufficiently smooth, this result can be further refined.

Corollary 5.31 *Assume that Ω is as in Theorem* 5.30, *$\mathcal{T}_h$ is a uniformly regular partition of Ω into affine simplicial elements, and the solution $\{\phi, \mathbf{u}\}$ of* (5.17) *belongs to $H^{q+1}(\Omega) \times [H^{q+1}(\Omega)]^d$. Also assume that X^h is given by*[41] *$X^h = G_0^r(\Omega) \times \mathbf{D}^{(r)}(\Omega)$ for* (5.62) *or $X^h_{r,0} = G_0^r(\Omega) \times [G^r(\Omega)]^d$ for* (5.63) *for some $r \ge 1$. Let $\{\phi^h, \mathbf{u}^h\}$ denote a solution of the compatible LSFEM* (5.62) *or the nodal LSFEM* (5.63). *There exists a positive constant C such that, for $\tilde{r} = \min\{r, q\}$,*

$$\|\phi - \phi^h\|_G + \|\mathbf{u} - \mathbf{u}^h\|_D \le C h^{\tilde{r}} \left(\|\phi\|_{\tilde{r}+1} + \|\mathbf{u}\|_{\tilde{r}+1} \right). \quad \Box \tag{5.85}$$

The error estimate (5.85) *holds for the nodal LSFEM on uniformly regular but not necessarily affine partitions $\mathcal{T}_h$.*

Proof. The result follows from the interpolation error estimates in Theorems B.6–B.9 and the error bounds in Theorem 5.30. □

Note that neither the approximation orders of the component spaces in X^h have to be the same nor is one restricted to the equal-order space $X^h_{r,0}$ for the nodal LSFEM. However, these choices make it easier to compare nodal and compatible LSFEMs. In particular, when the approximation orders of X^h and $X^h_{r,0}$ are matched, as in the statement of Corollary 5.31, the energy norm of the error in both LSFEMs converges at the same rate.

---

[41] Recall that $\mathbf{D}^{(r)}(\Omega)$ stands for a div-conforming space of the 1st or the 2nd kind, respectively; see Theorem B.9.

Remark 5.32 Recall that the error estimate (5.85) holds for the nodal LSFEM on non-affine grids because the approximation property (B.20) of nodal $C^0(\Omega)$ elements remains valid on such grids. However, if the compatible LSFEM is implemented using the standard div-conforming elements (B.29)–(B.34), the estimate (5.85) cannot be extended to non-affine grids because the accuracy of these elements degrades on such grids; see [9] and Theorem B.9. Numerical examples in Section 5.9.2 show that the accuracy of such implementations does indeed suffer on non-affine grids and that, in the lowest-order case, convergence can be completely lost. Fortunately, this can be easily fixed by using alternative div-conforming spaces such as the ABF_r space [9] or by invoking ideas [71] from mimetic finite difference methods [223, 224, 226]. The resulting *mimetic reformulation* of (5.62) is discussed in Section 12.3. □

The abstract results of Theorem 3.28 are also applicable to the four-field compatible LSFEM (5.64).

Theorem 5.33 *Assume that Ω and $\mathcal{T}_h$ are as in Theorem* 5.30. *Let* $\{\phi^h, \psi^h, \mathbf{u}^h, \mathbf{v}^h\} \in X^h = G_0^h(\Omega) \times S^h(\Omega) \times \mathbf{D}^h(\Omega) \times \mathbf{C}_0^h(\Omega)$ *denote a minimizer of* (5.64) *and let* $\{\phi, \psi, \mathbf{u}, \mathbf{v}\} \in G_0(\Omega) \times S(\Omega) \times \mathbf{D}(\Omega) \times \mathbf{C}_0(\Omega)$ *denote the solution of* (5.23). *Then,*

$$
\begin{aligned}
&\|\phi-\phi^h\|_G + \|\psi-\psi^h\|_S + \|\mathbf{u}-\mathbf{u}^h\|_D + \|\mathbf{v}-\mathbf{v}^h\|_C \\
&\quad\le C\Big(\inf_{\widehat{\phi}^h\in G_0^h}\|\phi-\widehat{\phi}^h\|_G + \inf_{\widehat{\psi}^h\in S^h}\|\psi-\widehat{\psi}^h\|_S \\
&\qquad\qquad + \inf_{\widehat{\mathbf{u}}^h\in \mathbf{D}^h}\|\mathbf{u}-\widehat{\mathbf{u}}^h\|_D + \inf_{\widehat{\mathbf{v}}^h\in \mathbf{C}_0^h}\|\mathbf{v}-\widehat{\mathbf{v}}^h\|_C\Big). \qquad \square
\end{aligned}
\tag{5.86}
$$

When exact solutions of (5.64) are sufficiently smooth and $\mathcal{T}_h$ is affine, we have an analogue of Corollary 5.31.

Corollary 5.34 *Assume that Ω is as in Theorem* 5.30 *and that $\mathcal{T}_h$ is a uniformly regular partition of Ω into affine simplicial elements. Let* $X^h = G_0^p(\Omega) \times S^{r-1}(\Omega) \times \mathbf{D}^{(t)}(\Omega) \times \mathbf{C}_0^{(s)}(\Omega)$, *where p, r, s, and t are integers greater than or equal to one. Let* $q = \max\{p, r, s, t\}$ *and assume that the solution of* (5.64) *belongs to the product space* $H^{q+1}(\Omega) \times H^{q+1}(\Omega) \times [H^{q+1}(\Omega)]^d \times [H^{q+1}(\Omega)]^d$. *Then,*

$$
\begin{aligned}
&\|\phi-\phi^h\|_G + \|\psi-\psi^h\|_S + \|\mathbf{u}-\mathbf{u}^h\|_D + \|\mathbf{v}-\mathbf{v}^h\|_C \\
&\quad\le C\Big(h^p\|\nabla\phi\|_p + h^r\|\psi\|_r + h^t\|\nabla\cdot\mathbf{u}\|_t + h^s\|\nabla\times\mathbf{v}\|_s\Big). \qquad \square
\end{aligned}
\tag{5.87}
$$

The estimates (5.85) and (5.87) show that it pays to equilibrate approximation orders of the component spaces in LSFEMs. Of course, the observations in Remark 5.32 remain in full force for the four-field compatible LSFEM.

The last method that is formally derived from an energy balance in the vector-operator setting is the compliant DLSP (5.79). It was already explained in Section

5.6.2 that this LSFEM defaults[42] to the same type of a nodal LSFEM (5.78) that is derived from energy balances in the ADN setting. This, however, does not allow us to extend the conclusions of Theorem 5.28 to (5.79) because its parent CLSP (5.59) is well-posed in settings that are not admissible in Corollary 5.19 which is one of the prerequisites for that theorem to hold.

5.7.2 $L^2(\Omega)$ Error Estimates

Theorem 5.30 and Corollary 5.31 may leave one with the impression that nodal and compatible LSFEMs for the potential–flux div–grad system are not much different, at least with regard to their approximation properties. However, the error estimates (5.84) and (5.85) do not tell the whole story about the two methods because they provide no information about convergence with respect to $L^2(\Omega)$ norms.

It turns out that in the compatible LSFEM, *both* the potential and the flux approximation converge optimally in $L^2(\Omega)$. This result, established in [59], is stated in the following theorem.

Theorem 5.35 *Assume that $\Omega \subset \mathbb{R}^d$, $d = 2,3$, is a bounded contractible domain having a Lipschitz continuous boundary, $\mathcal{T}_h$ is a uniformly regular partition of Ω into affine simplicial elements, and, in* (5.62), *$X^h = G^r_\Gamma(\Omega) \times \mathbf{D}^{(p)}_{\Gamma^*}(\Omega)$. Furthermore, assume that* (5.1) *has full elliptic regularity, i.e., $\{\phi,\mathbf{u}\} \in G_\Gamma(\Omega) \cap H^{q+1}(\Omega) \times \mathbf{D}_{\Gamma^*}(\Omega) \cap [H^{q+1}(\Omega)]^d$ for some integer $q \geq 1$ and that $r = p = q$, i.e., the approximation orders of $G^r_\Gamma(\Omega)$ and $\mathbf{D}^{(p)}_{\Gamma^*}(\Omega)$ are equilibrated with respect to the solution regularity. Then, in addition to* (5.85), *the solution $\{\phi^h,\mathbf{u}^h\} \in X^h$ of the compatible LSFEM satisfies the error estimates*

$$\|\phi - \phi^h\|_0 \leq Ch^{q+1}\big(\|\phi\|_{q+1} + \|\mathbf{u}\|_{q+1}\big) \tag{5.88}$$

$$\|\mathbf{u} - \mathbf{u}^h\|_0 \leq \begin{cases} Ch^q\,\big(\|\phi\|_q + \|\mathbf{u}\|_q\big) & \text{if } \mathbf{D}^{(q)}_{\Gamma^*} = \mathbf{D}^{q-}_{\Gamma^*} \\[2ex] Ch^{q+1}\big(\|\phi\|_{q+1} + \|\mathbf{u}\|_{q+1}\big) & \text{if } \mathbf{D}^{(q)}_{\Gamma^*} = \mathbf{D}^{q}_{\Gamma^*}\,. \end{cases} \tag{5.89}$$

Of course, the solution space components in the compatible method are not required to be connected in any way, i.e., Theorem 5.35 holds even if $G^h_\Gamma(\Omega)$ and $\mathbf{D}^h_{\Gamma^*}(\Omega)$ are defined on different grids.

Regarding the nodal LSFEM, the following result is proved in [227].

Theorem 5.36 ([227, Theorem 4.1]) *Assume that Ω has a smooth boundary $\partial\Omega$, $\mathcal{T}_h$ is a uniformly regular finite element partition, and the solution $\{\phi,\mathbf{u}\}$ of the div–grad system* (5.17) *with Dirichlet boundary conditions, i.e., with $\Gamma^* = \partial\Omega$, belongs to $X_q = H^{q+1}(\Omega) \cap H^1_0(\Omega) \times [H^{q+1}(\Omega)]^d$ for some $q \geq 0$. Let $\{\phi^h,\mathbf{u}^h\} \in X^h_{r,0}$, $r \geq$*

[42] Recall that this is because the only conforming finite element subspaces of the solution space $\mathbf{C}(\Omega) \cap \mathbf{D}(\Omega)$ are nodal $C^0(\Omega)$ elements.

1, *denote a solution of the nodal DLSP* (5.63) *and define* $\tilde{r} = \min\{r, q\}$. *Then, in addition to* (5.85),

$$\|\phi - \phi^h\|_0 \le Ch^{\tilde{r}+1}\left(\|\phi\|_{\tilde{r}+1} + \|\mathbf{u}\|_{\tilde{r}+1}\right). \tag{5.90}$$

Assuming full elliptic regularity, (5.85) *and* (5.90) *remain valid for the div–grad system with Neumann* ($\Gamma^* = \partial\Omega$) *and mixed* ($\Gamma \cup \Gamma^* = \partial\Omega$) *boundary conditions.*

However, in [227], no optimal $L^2(\Omega)$ convergence result for the flux is provided. The result in Theorem 5.36 was further refined in [163, 164], where it is shown that optimal $L^2(\Omega)$ error estimates for both variables are possible *if and only if* the nodal flux space satisfies the *grid decomposition property* (GDP). The only nodal space that so far has been shown to possess this property is the piecewise linear finite element space defined on a uniform *criss-cross* grid[43] [163]; see Figure 5.1. For any other finite element partition, optimal $L^2(\Omega)$ rates can only be asserted for the potential approximation. Numerical examples in Section 5.9.1 confirm that this result is sharp, i.e., on general grids, the $L^2(\Omega)$ error of the flux approximation in the nodal LSFEM is suboptimal.

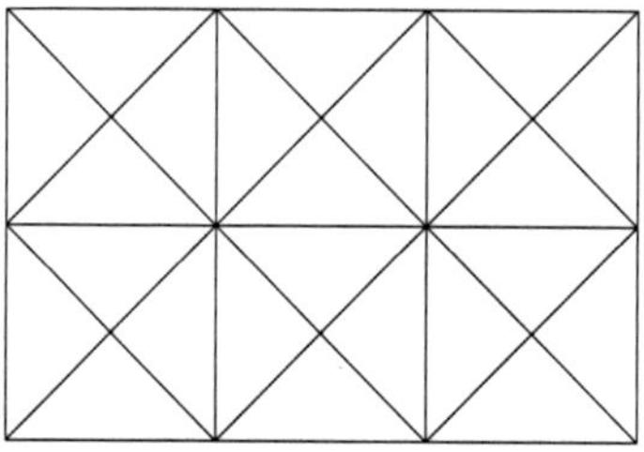

Fig. 5.1 The uniform criss-cross grid (also known as the "Union Jack" grid) is the only known example for which the GDP holds for $C^0(\Omega)$ piecewise linear approximations of the flux [163]. As a result, this grid provides the only known setting for the nodal LSFEM (5.63) in which the $L^2(\Omega)$ convergence rates for both the potential and the flux approximations are guaranteed to be optimal.

5.8 Connections Between Compatible Least-Squares and Standard Finite Element Methods

In Section 5.7.2, it is established that the compatible LSFEM (5.62) has better convergence properties in $L^2(\Omega)$ than its nodal cousin (5.63). In this section, we dis-

[43] Because the GDP is essentially a statement of a discrete Hodge decomposition, it is reasonable to expect that it holds for the div-compatible space used in (5.62). This fact was established in [59]. Because the proof of Theorem 5.35 also relies on a discrete Hodge decomposition property, it can be viewed as an extension of the original results in [163,164] to more general finite element spaces.

close further advantages of the compatible LSFEM that corroborate the arguments made in Remark 5.24 in favor of this method. In so doing, we also fulfill the promise made at the end of Remark 5.26 and demonstrate that the compatible LSFEM is intimately connected to Rayleigh–Ritz and mixed-Galerkin methods for (5.1) and, in fact, *inherits their best computational properties.*

5.8.1 The Compatible Least-Squares Finite Element Method with a Reaction Term

We first examine (5.1) with the reaction term included. The case when $\gamma = 0$ is considered separately in Section 5.8.2 where we show that, with a simple post-procesing step, the above statement extends to formulations without the reaction term.

Assuming that $\gamma = 1$, the Rayleigh–Ritz method for (5.1) is given by: *seek* $\phi^h \in G^h_\Gamma(\Omega)$ *such that*

$$\int_\Omega \nabla\phi^h \Theta_1 \nabla\widehat{\phi}^h \, d\Omega + \int_\Omega \phi^h \Theta_0 \widehat{\phi}^h \, d\Omega = \int_\Omega f \Theta_0 \widehat{\phi}^h \, d\Omega \qquad \forall \widehat{\phi}^h \in G^h_\Gamma(\Omega)$$

which can be written in the compact form

$$\left(\nabla\phi^h, \nabla\widehat{\phi}^h\right)_{\Theta_1} + \left(\phi^h, \widehat{\phi}^h\right)_{\Theta_0} = \left(f, \widehat{\phi}^h\right)_{\Theta_0} \qquad \forall \widehat{\phi}^h \in G^h_\Gamma(\Omega). \tag{5.91}$$

The mixed-Galerkin method approximates the flux–density pair so we set $\psi = \Theta_0\phi$. Then, it is not difficult to see that the mixed method for (5.1) assumes the form: *seek* $\{\mathbf{u}^h, \psi^h\} \in D^h_{\Gamma*}(\Omega) \times S^h(\Omega)$ *such that*

$$\begin{cases} \int_\Omega \nabla\cdot\mathbf{u}^h \Theta_0^{-1} \widehat{\psi}^h \, d\Omega + \int_\Omega \psi^h \Theta_0^{-1} \widehat{\psi}^h \, d\Omega = \int_\Omega f \widehat{\psi}^h \, d\Omega & \forall \widehat{\psi}^h \in S^h(\Omega) \\ \int_\Omega \mathbf{u}^h \Theta_1^{-1} \widehat{\mathbf{u}}^h \, d\Omega - \int_\Omega \psi^h \Theta_0^{-1} \nabla\cdot\widehat{\mathbf{u}}^h \, d\Omega = 0 & \forall \widehat{\mathbf{u}}^h \in D^h_{\Gamma*}(\Omega) \end{cases}$$

which we write compactly as

$$\begin{cases} \left(\nabla\cdot\mathbf{u}^h, \widehat{\psi}^h\right)_{\Theta_0^{-1}} + \left(\psi^h, \widehat{\psi}^h\right)_{\Theta_0^{-1}} = \left(\Theta_0 f, \widehat{\psi}^h\right)_{\Theta_0^{-1}} & \forall \widehat{\psi}^h \in S^h(\Omega) \\ \left(\mathbf{u}^h, \widehat{\mathbf{u}}^h\right)_{\Theta_1^{-1}} - \left(\psi^h, \nabla\cdot\widehat{\mathbf{u}}^h\right)_{\Theta_0^{-1}} = 0 & \forall \widehat{\mathbf{u}}^h \in D^h_{\Gamma*}(\Omega). \end{cases} \tag{5.92}$$

The duality of (5.91) and (5.92) is reflected in their approximation properties. In the Rayleigh–Ritz method, the flux is approximated indirectly by $\Theta_1\nabla\phi^h$, is less accurate than ϕ^h, and is not locally conservative. In the mixed method, the flux is approximated directly, is locally conservative, but the density is less accurate than the potential approximation in (5.91). Of course, in the mixed method, the finite element spaces $D^h_{\Gamma*}(\Omega)$ and $S^h(\Omega)$ cannot be chosen independently and must satisfy the discrete inf–sup stability conditions (1.42) and (1.44).

We can already see some potential advantages of the compatible LSFEM over the two standard methods. The compatible LSFEM uses the same nodal space $G^h_\Gamma(\Omega)$ as in (5.91) to approximate the potential and the same div-compatible space $D^h_{\Gamma*}(\Omega)$ as in (5.92) to approximate the flux, but these spaces are not subject to a joint inf–sup stability condition and we are free to choose them completely independently from each other. Second, from Corollary 5.31 and Theorem 5.35, it follows that the accuracy of the compatible LSFEM matches the *accuracy of the Rayleigh–Ritz method with respect to the potential variable* and the *accuracy of the mixed-Galerkin method with respect to the flux variable*. Of course, we still have the advantage that (5.62) involves only symmetric and positive definite linear systems and can be implemented with weakly imposed boundary conditions.

Finally, it turns out that the "splitting" property of the parent CLSP of (5.62), first encountered in the proof of Theorem 5.14, brings about a remarkable "splitting" property of the corresponding compatible LSFEM:

> *if $\gamma = 1$, the compatible LSFEM determines exactly the same potential approximation as the Rayleigh–Ritz method and exactly the same flux approximation as the mixed-Galerkin method.*

Coupled with the absence of joint compatibility conditions, this means that the compatible LSFEM does indeed combine the best approximation properties of (5.91) and (5.92).

Theorem 5.37 *Let $\gamma = 1$ and assume that $\Omega \subset \mathbb{R}^d$ is a bounded contractible domain, that $\mathcal{T}_h$ is a regular partition of Ω into affine simplicial elements, and that $G^h_\Gamma(\Omega)$, $\mathbf{D}^h_{\Gamma*}(\Omega)$, and $S^h(\Omega)$ are compatible finite element subspaces of $G_\Gamma(\Omega,\Theta_0)$, $D_{\Gamma*}(\Omega,\Theta_1^{-1})$, and $S(\Omega,\Theta_0^{-1})$, respectively. Assume that the pair $\{\mathbf{D}^h_{\Gamma*}(\Omega),S^h(\Omega)\}$ belongs to one of the finite element De Rham complexes in* (B.45). *Let $\{\phi^h_{LS},\mathbf{u}^h_{LS}\} \in G^h_\Gamma(\Omega)\times\mathbf{D}^h_{\Gamma*}(\Omega)$, $\phi^h_G \in G^h_\Gamma(\Omega)$, and $\{\mathbf{u}^h_M,\psi^h_M\} \in \mathbf{D}^h_{\Gamma*}(\Omega)\times S^h(\Omega)$ denote solutions of the compatible LSFEM, the Rayleigh–Ritz, and the mixed-Galerkin methods* (5.62), (5.91), *and* (5.92), *respectively. Then,*

$$\phi^h_{LS} = \phi^h_G \qquad \textit{and} \qquad \mathbf{u}^h_{LS} = \mathbf{u}^h_M\,.$$

Moreover, there exists a derived density approximation $\psi^h \in S^h(\Omega)$ such that $\{\mathbf{u}^h_{LS},\psi^h\}$ solves the mixed problem (5.92).

Proof. From the assumption on $\{\mathbf{D}^h_{\Gamma*}(\Omega),S^h(\Omega)\}$ and (B.1), it follows that the divergence is bounded surjection[44] $\mathbf{D}^h_{\Gamma*}(\Omega)\mapsto S^h(\Omega)$; this is sufficient to guarantee the well-posedness of the mixed method; see [87, Proposition 1.1, p.139].

The first-order optimality condition for the compatible LSFEM (5.62) is given as follows: *seek $\{\phi^h_{LS},\mathbf{u}^h_{LS}\} \in G^h_\Gamma(\Omega)\times\mathbf{D}^h_{\Gamma*}(\Omega)$ such that*

[44] In other words, for every $\psi^h \in S^h(\Omega)$, there exists $\mathbf{v}^h \in \mathbf{D}^h_{\Gamma*}(\Omega)$ such that $\psi^h = \nabla\cdot\mathbf{v}^h$ and $\|\nabla\cdot\mathbf{v}^h\|_0 \le C\|\psi^h\|_0$.

$$Q\big(\{\phi_{LS}^h,\mathbf{u}_{LS}^h\},\{\widehat{\phi}^h,\widehat{\mathbf{u}}^h\}\big) = F\big(\{\widehat{\phi}^h,\widehat{\mathbf{u}}^h\}\big) \tag{5.93}$$

for all $\{\widehat{\phi}^h,\widehat{\mathbf{u}}^h\} \in G_\Gamma^h(\Omega)\times \mathbf{D}_{\Gamma*}^h(\Omega)$, where

$$\begin{aligned} &Q\big(\{\phi_{LS}^h,\mathbf{u}_{LS}^h\},\{\widehat{\phi}^h,\widehat{\mathbf{u}}^h\}\big)\\ &\quad = \big(\nabla\cdot\mathbf{u}_{LS}^h+\Theta_0\phi_{LS}^h, \nabla\cdot\widehat{\mathbf{u}}^h+\Theta_0\widehat{\phi}^h\big)_{\Theta_0^{-1}} + \big(\nabla\phi_{LS}^h+\Theta_1^{-1}\mathbf{u}_{LS}^h, \nabla\widehat{\phi}^h+\Theta_1^{-1}\widehat{\mathbf{u}}^h\big)_{\Theta_1}\end{aligned}$$

and

$$F\big(\{\widehat{\phi}^h,\widehat{\mathbf{u}}^h\}\big) = \big(\Theta_0 f,\nabla\cdot\widehat{\mathbf{u}}^h\big)_{\Theta_0^{-1}} + \big(\Theta_0 f,\Theta_0\widehat{\phi}^h\big)_{\Theta_0^{-1}}.$$

The existence and uniqueness of the least-squares solution follows from Theorem 5.14 which asserts that $Q(\cdot,\cdot)$ is coercive on $G_\Gamma^h(\Omega)\times \mathbf{D}_{\Gamma*}^h(\Omega)$ and the abstract result in Theorem 3.10. Proceeding as in the proof of Theorem 5.14, we see that (5.93) simplifies to

$$\begin{aligned} &\big(\nabla\cdot\mathbf{u}_{LS}^h,\nabla\cdot\widehat{\mathbf{u}}^h\big)_{\Theta_0^{-1}} + \big(\mathbf{u}_{LS}^h,\widehat{\mathbf{u}}^h\big)_{\Theta_1^{-1}} + \big(\nabla\phi_{LS}^h,\nabla\widehat{\phi}^h\big)_{\Theta_1} + \big(\phi_{LS}^h,\widehat{\phi}^h\big)_{\Theta_0}\\ &\qquad = \big(\Theta_0 f,\nabla\cdot\widehat{\mathbf{u}}^h\big)_{\Theta_0^{-1}} + \big(\Theta_0 f,\Theta_0\widehat{\phi}^h\big)_{\Theta_0^{-1}}\end{aligned}$$

for all $\{\widehat{\phi}^h,\widehat{\mathbf{u}}^h\} \in G_\Gamma^h(\Omega)\times \mathbf{D}_{\Gamma*}^h(\Omega)$. Therefore, the least-squares weak problem (5.93) for $\{\phi_{LS}^h,\mathbf{u}_{LS}^h\}$ splits into the independent variational equation

$$\big(\nabla\phi_{LS}^h,\nabla\widehat{\phi}^h\big)_{\Theta_1} + \big(\phi_{LS}^h,\widehat{\phi}^h\big)_{\Theta_0} = \big(\Theta_0 f,\Theta_0\widehat{\phi}^h\big)_{\Theta_0^{-1}} \qquad \forall\widehat{\phi}^h\in G_\Gamma^h(\Omega) \tag{5.94}$$

for the potential approximation ϕ_{LS}^h and another independent equation

$$\big(\nabla\cdot\mathbf{u}_{LS}^h,\nabla\cdot\widehat{\mathbf{u}}^h\big)_{\Theta_0^{-1}} + \big(\mathbf{u}_{LS}^h,\widehat{\mathbf{u}}^h\big)_{\Theta_1^{-1}} = \big(\Theta_0 f,\nabla\cdot\widehat{\mathbf{u}}^h\big)_{\Theta_0^{-1}} \quad \forall\widehat{\mathbf{u}}^h\in \mathbf{D}_{\Gamma*}^h(\Omega) \tag{5.95}$$

for the discrete flux $\mathbf{u}_{LS}^h$. From the identity

$$\big(\Theta_0 f,\Theta_0\widehat{\phi}^h\big)_{\Theta_0^{-1}} = \big(f,\widehat{\phi}^h\big)_{\Theta_0},$$

it follows that (5.94) and (5.91) are the same problem, i.e., the scalar component ϕ_{LS}^h of the least-squares solution coincides with the Rayleigh–Ritz approximation ϕ_G^h.

To show that $\mathbf{u}_{LS}^h=\mathbf{u}_M^h$, we use the assumption that the divergence is a surjection $\mathbf{D}_{\Gamma*}^h(\Omega)\mapsto S^h(\Omega)$ to write the first equation in (5.92) in the form

$$\big(\nabla\cdot\mathbf{u}_M^h,\nabla\cdot\widehat{\mathbf{u}}^h\big)_{\Theta_0^{-1}} + \big(\psi_M^h,\nabla\cdot\widehat{\mathbf{u}}^h\big)_{\Theta_0^{-1}} = \big(\Theta_0 f,\nabla\cdot\widehat{\mathbf{u}}^h\big)_{\Theta_0^{-1}} \quad \forall\widehat{\mathbf{u}}^h\in \mathbf{D}_{\Gamma*}^h(\Omega).$$

From the second equation in (5.92), we have that

$$\big(\psi_M^h,\nabla\cdot\widehat{\mathbf{u}}^h\big)_{\Theta_0^{-1}} = \big(\mathbf{u}_M^h,\widehat{\mathbf{u}}^h\big)_{\Theta_1^{-1}} \qquad \forall\widehat{\mathbf{u}}^h\in \mathbf{D}_{\Gamma*}^h(\Omega).$$

Therefore, ψ_M^h can be eliminated from the mixed method to obtain an equivalent formulation solely in terms of the flux:

$$\left(\nabla\cdot\mathbf{u}_M^h,\nabla\cdot\widehat{\mathbf{u}}^h\right)_{\Theta_0^{-1}}+\left(\mathbf{u}_M^h,\widehat{\mathbf{u}}^h\right)_{\Theta_1^{-1}}=\left(\Theta_0 f,\nabla\cdot\widehat{\mathbf{u}}^h\right)_{\Theta_0^{-1}}\qquad\forall\widehat{\mathbf{u}}^h\in\mathbf{D}^h_{\Gamma*}(\Omega).$$

This equation is identical to the weak problem (5.95) for $\mathbf{u}_{LS}^h$ that resulted from the least-squares formulation (5.93). It follows that $\mathbf{u}_{LS}^h=\mathbf{u}_M^h$. Finally, it is easy to see that if ψ^h is defined by (5.65), the pair $\{\mathbf{u}_{LS}^h,\psi^h\}$ solves the mixed-Galerkin problem. □

Remark 5.38 The flux approximation $\mathbf{u}_M^h$ obtained via the mixed method is locally conservative in the sense that

$$\int_\kappa \Theta_0^{-1}\left(\nabla\cdot\mathbf{u}_M^h+\psi_M^h-\Theta_0 f\right)d\Omega=0\qquad\forall\kappa\in\mathcal{T}_h. \tag{5.96}$$

Theorem 5.37 asserts that the least-squares flux $\mathbf{u}_{LS}^h$ and the derived density approximation ψ^h have the exact same property; this justifies calling the compatible LSFEM *locally conservative.*[45] □

Let us stress again that, in the proof of Theorem 5.37, the spaces $G^h_\Gamma(\Omega)$ and $\mathbf{D}^h_{\Gamma*}(\Omega)$ are not required to be connected in any way. The only prerequisite for the least-squares solution to coincide with ϕ_G^h and $\mathbf{u}_M^h$ is that, individually, $G^h_\Gamma(\Omega)$ and $\mathbf{D}^h_{\Gamma*}(\Omega)$ are the same as in (5.91) and (5.92). Effectively, this means that when the hypotheses of Theorem 5.37 are satisfied, the compatible LSFEM *simultaneously solves* the Rayleigh–Ritz (5.91) and the mixed-Galerkin (5.92) equations.

In contrast, for the nodal LSFEM (5.63), the choice of finite element spaces for ϕ^h is compatible with (5.91) but the choice of finite element spaces for $\mathbf{u}^h$ is not compatible with (5.92). In Section 5.9.1, it is shown that this can compromise the computational properties of the nodal LSFEM.

[45] The conclusions of Theorem 5.37 extend to other affine families of finite element spaces such as those based on parallelograms, rectangles, and parallelepipeds. Formally, the theorem can also be extended to non-affine grids provided that $S^h(\Omega)$ is taken to be the range of divergence restricted to $\mathbf{D}^h_{\Gamma*}(\Omega)$. However, in practice, the mixed-Galerin finite element method is seldom implemented on non-affine grids by using $\nabla\cdot(\mathbf{D}^h_{\Gamma*}(\Omega))$ to approximate the density variable. The reason for this is that the accuracy of this space degrades on such grids up to a complete loss of convergence when $\mathbf{D}^h$ is the lowest-order div-conforming space; see Section B.2.3. Instead, the density is approximated using the polynomial spaces (B.42) introduced in Remark B.12. As explained in Section B.2.3, the resulting flux–density pair satisfies a weak commuting diagram property and is stable, but the divergence is no longer onto $S^h(\Omega)$. The latter property is crucial for the proof of Theorem 5.37.

On the other hand, if the density is first *eliminated* from the mixed variational problem and the resulting equation for the flux is discretized by div-conforming elements, the outcome is a problem that is identical to (5.95), i.e., the least-squares flux coincides with the flux from a "reduced" version of the mixed-Galerkin method. Thus, even when the original statement of Theorem 5.37 does not fully hold, a connection between the compatible LSFEM and the mixed-Galerkin finite element method still remains in force.

5.8.2 The Compatible Least-Squares Finite Element Method Without a Reaction Term

When $\gamma = 0$, (5.62) reduces to

$$\begin{cases} J(\phi^h, \mathbf{u}^h; f) = \|\nabla \cdot \mathbf{u}^h - \Theta_0 f\|^2_{0,\Theta_0^{-1}} + \|\nabla \phi^h + \Theta_1^{-1} \mathbf{u}^h\|^2_{0,\Theta_1} \\ X^h = G^h_{\Gamma}(\Omega) \times \mathbf{D}^h_{\Gamma^*}(\Omega). \end{cases} \tag{5.97}$$

This compliant DLSP inherits all the attractive features of (5.62) except local conservation in the sense defined in Remark 5.38. The reason is that, with the reaction term absent, the optimality system corresponding to (5.97) does not split into independent equations for the potential and flux which is critical for (5.96) to hold. Thus, instead of (5.96), the flux in (5.97) satisfies

$$\int_\kappa \Theta_0^{-1} (\nabla \cdot \mathbf{u}^h - \Theta_0 f) d\Omega \approx 0 \qquad \forall \kappa \in \mathcal{T}_h.$$

Fortunately, local conservation can be recovered by a simple post-processing step. Below we describe a *local* flux-correction algorithm that replaces $\mathbf{u}^h$ by a corrected flux $\widetilde{\mathbf{u}}^h \in \mathbf{D}^h_{\Gamma^*}(\Omega)$ which satisfies this equation exactly.

For simplicity, we describe the procedure for $\Theta_0 = 1$, $f = 0$, and the lowest-order div-compatible space $\mathbf{D}^1\text{-}(\Omega)$ on tetrahedrons; see (B.29). Let κ denote an arbitrary tetrahedral element. The unisolvent set of $\mathbf{D}^1\text{-}(\Omega)$ on κ is given by (B.32). As a result, any $\mathbf{u}^h \in \mathbf{D}^1\text{-}(\Omega)$ can be written on this element in the form

$$\mathbf{u}^h = \sum_{i=1}^{4} \Phi_{C_2^i} \mathbf{u}_i^h,$$

where C_2^i are the (oriented) faces of κ (see Section B.1), $\Phi_{C_2^i}$ is the flux of $\mathbf{u}^h$ across face C_2^i, and $\mathbf{u}_i^h$ is the basis function associated with that face. Using the divergence theorem, it is easy to see that, for $\mathbf{D}^1\text{-}(\Omega)$ elements,

$$\int_\kappa \nabla \cdot \mathbf{u}^h d\Omega = \sigma_1 \Phi_{C_2^1} + \sigma_2 \Phi_{C_2^2} + \sigma_3 \Phi_{C_2^3} + \sigma_4 \Phi_{C_2^4},$$

where $\sigma_i = 1$ if the orientation of C_2^i coincides with the outer normal on $\partial\kappa$ and $\sigma_i = -1$ otherwise. We seek modified flux values $\widetilde{\Phi}_{C_2^i} = \Phi_{C_2^i} - \Delta\Phi_{C_2^i}$ such that

$$\sigma_1 \widetilde{\Phi}_{C_2^1} + \sigma_2 \widetilde{\Phi}_{C_2^2} + \sigma_3 \widetilde{\Phi}_{C_2^3} + \sigma_4 \widetilde{\Phi}_{C_2^4} = 0.$$

Assuming that these fluxes have been found, the desired locally conservative replacement of $\mathbf{u}^h$ can be defined as follows:

$$\widetilde{\mathbf{u}}^h = \sum_{i=1}^{4} \widetilde{\Phi}_{C_2^i} \mathbf{u}_i^h .$$

To define the new flux values for an arbitrary element κ, we proceed as follows.[46] If, for a given face C_2^i, the value of $\Phi_{C_2^i}$ has already been corrected, set $\Delta\Phi_{C_2^i} = 0$. If the flux on C_2^i has not yet been corrected, set

$$\Delta\Phi_{C_2^i} = \frac{\sigma_1\Phi_{C_2^1} + \sigma_2\Phi_{C_2^2} + \sigma_3\Phi_{C_2^3} + \sigma_4\Phi_{C_2^4}}{n(\kappa)},$$

where $n(\kappa) > 0$ is the number of faces in $\mathcal{C}_2(\kappa)$ whose fluxes have not been corrected. Consider now a partition $\mathcal{T}_h$ of Ω into n_h elements. As in (B.4),

$$\mathcal{C}_2(i_1,\ldots,i_k) = \bigcup_{i=i_1,\ldots,i_k} \mathcal{C}_2(\kappa_i),$$

denotes the union of all faces belonging to a set of elements indexed by $i_1,\ldots,i_k$. It is not hard to see that the number of faces in κ_{i_k} that are not shared with any other elements in that set is given by

$$n(i_1,\ldots,i_k) = \dim\Big\{\mathcal{C}_2(i_k)/\{\mathcal{C}_2(i_1,\ldots,i_{k-1})\cap\mathcal{C}_2(i_k)\}\Big\} \geq 0.$$

Given $\mathbf{u}^h \in \mathbf{D}^1\text{-}(\Omega)$, the following algorithm finds $\widetilde{\mathbf{u}}^h \in \mathbf{D}^1\text{-}(\Omega)$ that is divergence free:

1. Define a permutation $\pi = \{i_1, i_2, \ldots, i_{n_h}\}$ of all elements in $\mathcal{T}_h$, such that

$$n(i_1,\ldots,i_k) > 0 \quad \text{for } k = 1,\ldots,n_h$$

2. for $k = 1,\ldots,n_h$, apply the element flux correction to element κ_{i_k}.

5.9 Practicality Issues

The theoretical analyses given in Sections 5.6–5.8 pinpoint a number of potential advantages and disadvantages associated with compatible LSFEMs. Some of our conclusions challenge commonly held views about LSFEMs such as:

- the main reason to use LSFEMs, besides having symmetric and positive definite linear systems, is the admissibility of equal-order, nodal $C^0(\Omega)$ spaces;
- the conservation properties of LSFEMs are inferior to those of mixed-Galerkin finite element methods;

[46] Because local conservation is a topological property that is separate from the accuracy of the finite element space, affinity of $\mathcal{T}_h$ is not a prerequisite for the application of the flux-correction algorithm. Extensions to higher-order div-conforming elements are also possible, and can be facilitated by judicious choice of basis functions such as projection or hierarchical bases [136, 137, 317].

- LSFEMs for the extended div–grad system are superior to compatible methods based on the non-extended first-order problem.

In this section, we present computational evidence that corroborates our theoretical conclusions that refute the above statements and demonstrates previously unexplored computational advantages of compatible LSFEMs. Other examples provided in this section serve as warnings about settings in which compatible LSFEMs may experience difficulties such as

- reduced convergence rates on non-affine grids.

We present examples that illustrate these difficulties and point out strategies that can be used to resolve them.

To save space and avoid repetitions, in this section we focus mainly on practical issues pertaining to the compatible LSFEM and its nodal counterpart. Examination of issues arising in LSFEMs obtained from energy balances in the ADN setting is postponed until Sections 7.6 and 8.4.3 because the bulk of the LSFEMs in those chapters are of that type. There, we investigate topics such as whether deletion of weights or violation of the minimal approximation condition (see Remark 4.11) in quasi-norm-equivalent DLSPs is detrimental to their accuracy and provide some practical advice on the implementation of discrete negative norm LSFEMs.

Sections 5.9.1 and 5.9.2 are devoted to numerical studies of the compatible LSFEM (5.62). To highlight the computational advantages of this method, we pit it against its nodal counterpart (5.63) and the Rayleigh–Ritz and mixed-Galerkin finite element methods. Then, we study the performance of (5.62) on non-affine quadrilateral grids. Section 5.9.3 deals with the compliant method (5.78) for the extended div–grad system and compares it with the nodal method (5.63) for the non-extended div–grad system.

The methods considered in this section require the following finite element spaces:

- $G^h_\Gamma(\Omega) \times \mathbf{D}^h_{\Gamma^*}(\Omega)$ for the compatible LSFEM (5.62);
- $G^h_\Gamma(\Omega) \times [G^h(\Omega)]^2 \cap \mathbf{D}_{\Gamma^*}(\Omega)$ for the nodal LSFEM (5.63);
- $G^h_0(\Omega) \times [G^h(\Omega)]^2 \cap \mathbf{C}_0(\Omega)$ or $G^h(\Omega) \times [G^h(\Omega)]^2 \cap \mathbf{D}_0(\Omega)$ for the compliant LSFEM (5.78);
- $\mathbf{D}^h_{\Gamma^*}(\Omega) \times S^h(\Omega)$ for the mixed-Galerkin finite element method (5.92);
- $G^h_\Gamma(\Omega)$ for the Rayleigh–Ritz finite element method (5.91).

The methods are implemented as follows. We take Ω to be the unit square in $\mathbb{R}^2$, $\mathcal{T}_h$ to be a logically Cartesian (but not necessarily affine) partition of Ω into quadrilaterals, and $G^h(\Omega) = G^1(\Omega) = Q_1$, $\mathbf{D}^h(\Omega) = \mathbf{D}^1\text{-}(\Omega)$, and $S^h(\Omega) = S^0(\Omega) = P_0$, i.e., bilinear elements, the lowest-order Raviart-Thomas elements, and piecewise constant elements, respectively. For definitions of these spaces, see Section B.2.1. In all cases, essential boundary conditions are imposed on the approximating spaces. The linear systems are assembled using 2×2 Gauss quadrature on each element and solved "exactly" by direct solvers from the LAPACK library [6].

With these element choices and assuming uniformly regular affine finite element partitions, the theoretical error estimates of (5.85) and (5.89) for the compatible LSFEM (5.62) specialize to

$$\begin{gathered}\|\phi-\phi^h\|_G+\|\mathbf{u}-\mathbf{u}^h\|_D \le Ch\,,\\ \|\phi-\phi^h\|_0 \le Ch^2\,, \qquad \text{and} \qquad \|\mathbf{u}-\mathbf{u}^h\|_0 \le Ch\,.\end{gathered} \tag{5.98}$$

For the nodal LSFEM (5.63), the affinity assumption can be dropped and the error estimates (5.85) and (5.90) specialize to

$$\|\phi-\phi^h\|_G+\|\mathbf{u}-\mathbf{u}^h\|_D \le Ch \qquad \text{and} \qquad \|\phi-\phi^h\|_0 \le Ch^2\,, \tag{5.99}$$

respectively. For the compliant LSFEM (5.78), the error estimates (5.82) and (5.83) respectively specialize to

$$\begin{gathered}\|\phi-\phi^h\|_1+\|\mathbf{u}-\mathbf{u}^h\|_1 \le Ch\,,\\ \|\phi-\phi^h\|_0 \le Ch^2\,, \qquad \text{and} \qquad \|\mathbf{u}-\mathbf{u}^h\|_0 \le Ch^2\,.\end{gathered} \tag{5.100}$$

5.9.1 Practical Rewards of Compatibility

The goals of this section are twofold. First, by backing up the theoretical arguments made in Remark 5.24 with numerical examples, we hope to convince least-squares adherents that the formal admissibility of equal-order nodal elements in LSFEMs based on first-order systems is a two-edged sword. Specifically, we show that, unless the exact solution of the div–grad system is sufficiently smooth, using such elements can lead to unpredictable (and unreliable) numerical solutions.

Another set of examples illustrates the conclusions of Section 5.8 and addresses opponents of LSFEMs by showing that these methods can achieve the same local conservation property as mixed-Galerkin methods but produce significantly more accurate approximations of the potential variable.

Suboptimal convergence of the flux in the nodal method

The first example shows that the absence of optimal $L^2(\Omega)$ error estimates for the flux in Theorem 5.36 is not accidental. For this purpose, we carry out an order of convergence study for (5.63), using the div–grad system (5.18) with $\Theta_1 = \mathbb{I}$, $\Theta_0 = 1$, Dirichlet, Neumann, and mixed boundary conditions, and the manufactured solution[47]

$$\phi = -e^x \sin y \qquad \text{and} \qquad \mathbf{u} = -\nabla\phi\,. \tag{5.101}$$

[47] For this solution $\nabla\cdot\mathbf{u} = 0$.

Orders of convergence are estimated using data on uniform 33×33 and 65×65 grids with 1,024 and 4,096 square elements, respectively. Results are collected in Table 5.2. The data in this table show that the order of convergence of the nodal method with respect to the energy norm $|||\cdot|||_{GD} = (\|\cdot\|_G^2 + \|\cdot\|_D^2)^{1/2}$ is consistent with the theoretical order in (5.99) for all three choices of boundary conditions. The same is true for the $L^2(\Omega)$ error of the potential approximation which is second-order accurate. However, the table clearly shows that the $L^2(\Omega)$ order of convergence for the flux is suboptimal. In all three cases, the observed rate is half an order less than the best approximation rate for bilinear elements.

Variable	Potential		Flux	
BC ↓ Error →	$L^2(\Omega)$	$H^1(\Omega)$	$L^2(\Omega)$	$\mathbf{D}(\Omega)$
Dirichlet	2.01	1.00	**1.45**	0.95
Mixed	1.92	1.00	**1.46**	0.97
Neumann	1.96	0.99	**1.49**	1.00
BA	2.00	1.00	**2.00**	1.00

Table 5.2 Convergence rates of the nodal LSFEM (5.63) with Dirichlet, Neumann, and mixed boundary conditions versus best approximation rates (BA).

Failure of nodal LSFEMs for rough solutions

It could be argued that suboptimal $L^2(\Omega)$ orders of convergence of flux approximations is not a strong enough reason to dismiss the nodal LSFEM because the rest of the errors, including the energy norm error, converge as predicted by Theorem 5.36. Our next test exposes the fallacy of this argument by showing that suboptimal flux convergence in the nodal LSFEM is symptomatic of a far more serious deficiency which surfaces when the exact solution of (5.18) is such that $\mathbf{u} \in \mathbf{D}(\Omega)$ but $\mathbf{u} \notin [H^1(\Omega)]^d$.

For this test, we use an example problem from [222]. The manufactured solution of the div–grad system (5.18) is defined by dividing Ω into the five identical horizontal strips

$$\Omega_i = \{(x,y) \mid 0.2(i-1) \le y < 0.2i;\ 0 \le x \le 1\}, \quad i = 1,\dots,5,$$

and setting

$$\phi = 1 - x; \quad \mathbf{u}_\mu = (\mu, 0)^\mathsf{T}, \quad \Theta_0 = 1; \quad \text{and} \quad \Theta_1 = \mu \mathbb{I}, \tag{5.102}$$

where μ is a piecewise constant function taking a positive value μ_i on strip Ω_i. The so-defined flux $\mathbf{u}_\mu$ is a piecewise constant vector field whose normal component is continuous across the interface between any two of the strips, but whose tangential component has jumps along these interfaces. From Theorem B.3, it follows that $\mathbf{u}_\mu \in \mathbf{D}(\Omega)$ but $\mathbf{u}_\mu \notin [H^1(\Omega)]^d$. In the test, we solve (5.18) with a Neumann boundary condition, $\mu_1 = 16$, $\mu_2 = 6$, $\mu_3 = 1$, $\mu_4 = 10$, and $\mu_5 = 2$, using a uniform 21×21 grid (400 square elements) aligned with the jumps of μ.

Figure 5.2 shows contour plots of the compatible and the nodal LSFEM solutions. Surface plots of the flux components in these solutions are compared in Figure 5.3. From these plots, it is clear that the compatible LSFEM recovers the exact solution of the test problem. In contrast, the flux computed by the nodal LSFEM is grossly inaccurate. The error data given in Table 5.3 confirm these conclusions.

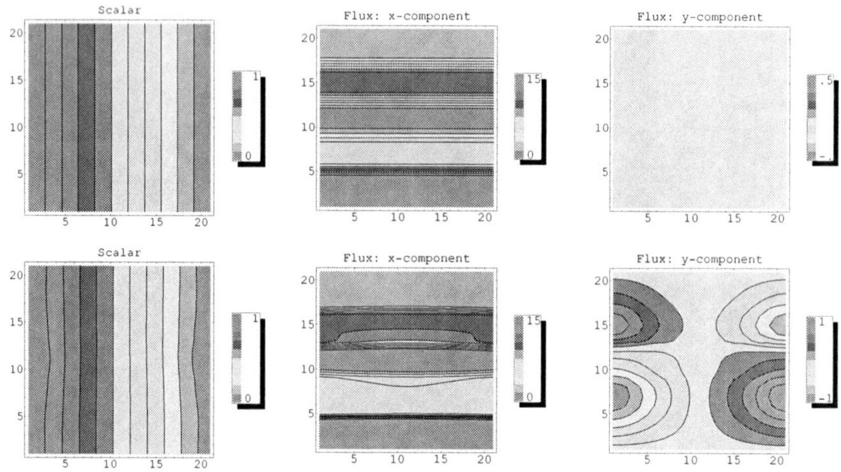

Fig. 5.2 Contour plots of ϕ^h and the components of $\mathbf{u}^h$ computed by the compatible LSFEM (5.62) (top) and the nodal LSFEM (5.63) (bottom).

Variable	Potential		Flux	
LSFEM ↓ Error →	$L^2(\Omega)$	$H^1(\Omega)$	$L^2(\Omega)$	$\|\nabla \cdot \mathbf{u}^h\|_0$
Compatible	0.1622E-14	0.6420E-14	0.1768E-10	0.2041E-12
Nodal	0.1416E-01	0.1422E+00	0.1925E+01	0.4392E+00

Table 5.3 Errors of compatible and nodal LSFEM solutions.

The poor quality of the flux approximation is not strange at all considering that the nodal LSFEM uses $C^0(\Omega)$ elements to approximate a vector field whose x-component is a discontinuous piecewise constant function. However, the failure of (5.63) to recover ϕ is somewhat unexpected because, unlike the flux, this variable *does belong* to the approximating space Q_1 employed in our implementation.

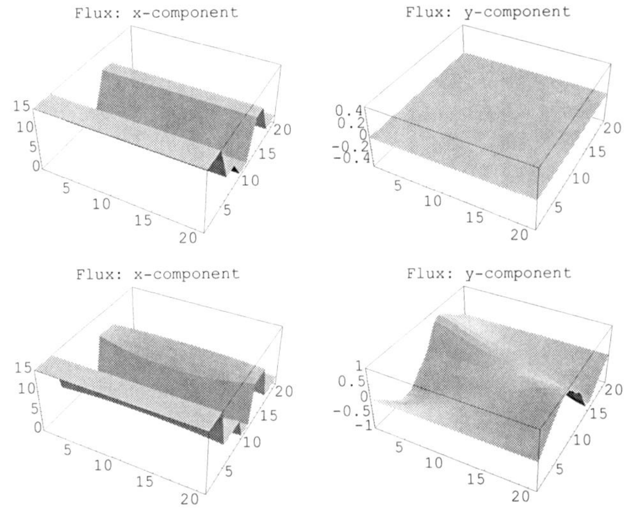

Fig. 5.3 Surface plots of the flux components computed by the compatible LSFEM (5.62) (top) and the nodal LSFEM (5.63) (bottom).

The reason for this failure is very simple. On the vertical parts of $\partial\Omega$, the unit normal is $\mathbf{n} = (\pm 1, 0)^{\mathsf{T}}$ so that there $\mathbf{u}_\mu \cdot \mathbf{n} = \mu$ is piecewise constant. On the other hand, the normal component of a vector field $\mathbf{u}^h \in [G^h(\Omega)]^2$ used in our implementation is a restriction of a Q_1 scalar function to $\partial\Omega$, i.e., it is a $C^0(\partial\Omega)$ piecewise linear function. As a result, the flux in the nodal LSFEM satisfies the Neumann boundary condition *approximately*. In contrast, vector fields from the div-conforming space $\mathbf{D}^1$-(Ω) used in the compatible LSFEM reproduce this boundary condition *exactly*.

Remark 5.39 This problem disappears if the test problem is driven by a pure Dirichlet boundary condition on ϕ. In this case, the nodal LSFEM produces a better flux approximation and recovers ϕ exactly. Coupled with the sensitivity of (5.63) to the implementation of $\mathbf{u}^h \cdot \mathbf{n}$ at the corner points[48] of Ω, this only serves to underscore the lack of robustness of this method and should be contrasted to the compatible LSFEM that gives the same answer regardless of the type of the boundary condition. □

[48] At the corners, $\mathbf{n}$ is discontinuous and has to be approximated. The nodal approximation of $\mathbf{u}_\mu$ changes noticeably with the choice for the "normal" at these points. This problem does not exist in the compatible LSFEM because it uses the unit normals on the *faces* of the elements; these are uniquely defined.

Compatible LSFEMs inherit the best properties of Rayleigh–Ritz and mixed-Galerkin methods

The computational studies discussed so far present irrefutable evidence that, in terms of robustness and accuracy, the compatible LSFEM is superior to the nodal LSFEM. We now use computational experiments to illustrate the results of Theorem 5.37, thereby confirming in practice that (5.62) combines the best computational properties of Rayleigh–Ritz and mixed-Galerkin finite element methods. We use the manufactured solution[49]

$$\phi = -\sin(\pi x^{1.25})^2 \sin(\pi y^{1.5})^2 \qquad \text{and} \qquad \mathbf{u} = -\nabla\phi$$

and set $\gamma = 1$, $\Theta_1 = \mathrm{I}$, and $\Theta_0 = 1$ in (5.17). Because eyeballing plots of solutions could be deceiving, as an alternative we choose to compare solution errors of the compatible LSFEM, Rayleigh–Ritz, and mixed-Galerkin finite element methods (5.62), (5.91), and (5.92), respectively. Table 5.4 shows these errors on 33×33, 65×65, and 128×128 uniform grids with 1,024, 4,098, and 16,384 elements, respectively.

Error	Method	33×33	65×65	129×129
$L^2(\Omega)$	CLS	0.7192623E-01	0.3523105E-01	0.1745720E-01
flux	MG	0.7192623E-01	0.3523105E-01	0.1745720E-01
$\mathbf{D}(\Omega)$	CLS	0.1397179E+01	0.6894290E+00	0.3426716E+00
flux	MG	0.1397179E+01	0.6894290E+00	0.3426716E+00
	RG	0.9378368E-03	0.2274961E-03	0.5621838E-04
$L^2(\Omega)$ potential	CLS	0.9378368E-03	0.2274961E-03	0.5621838E-04
	MG	0.1778803E-01	0.8750616E-02	0.4340574E-02
$H^1(\Omega)$	RG	0.1296329E+00	0.6383042E-01	0.3166902E-01
potential	CLS	0.1296329E+00	0.6383042E-01	0.3166902E-01

Table 5.4 Solution errors of compatible LSFEM (CLS), Rayleigh–Ritz (RG), and mixed-Galerkin (MG) methods.

The data in this table reveal identical flux errors on all three grids for the compatible and the mixed-Galerkin methods. Likewise, we see identical errors for ϕ on all three grids for the compatible and the Rayleigh–Ritz methods. This confirms that (5.62) and (5.92) compute identical flux approximations, whereas the potential approximation is the same for (5.62) and (5.91). A point-by-point examination of

[49] This solution satisfies homogeneous Dirichlet and Neumann boundary conditions. As a result, Rayleigh–Ritz and mixed-Galerkin finite element solutions are not affected by numerical approximation of their respective natural boundary conditions. We emphasize that in the compatible LSFEM, both types of boundary conditions can be treated as being essential.

solution values further validates this conclusion but is not presented here to save space.

Flux correction procedure

Recall that Theorem 5.37 does not extend to div–grad systems without a reaction term. To restore local conservation, we formulated, in Section 5.8.2, a local flux-correction algorithm for (5.97). The tests below show that this local procedure is capable of correcting the least-squares solution without any significant degradation in accuracy.

To make the tests more transparent, we use the manufactured solution in (5.101) for which $\nabla \cdot \mathbf{u} = 0$ and solve for (5.18) with $\Theta_1 = I$ and $\gamma = 0$, i.e., no reaction term. The first two rows in Table 5.5 show the $L^2(\Omega)$ and divergence errors of the flux computed by (5.97) before and after application of the flux-correction algorithm. The last row contains the same data for the benchmark mixed-Galerkin finite element solution. Errors are reported for 31×31 uniform and randomly perturbed grids with 900 quadrilateral elements. From the data in the first two rows of Table 5.5, it is clear that the flux-correction algorithm recovers local conservation without any negative effects on the $L^2(\Omega)$ accuracy of the flux. The divergence of $\mathbf{u}^h$ drops to machine precision and the $L^2(\Omega)$ errors before and after the correction are virtually identical.

Grid	31×31 uniform grid		31×31 random grid	
Error	$\lVert\mathbf{u}-\mathbf{u}^h\rVert_0$	$\lVert\nabla\cdot(\mathbf{u}-\mathbf{u}^h)\rVert_0$	$\lVert\mathbf{u}-\mathbf{u}^h\rVert_0$	$\lVert\nabla\cdot(\mathbf{u}-\mathbf{u}^h)\rVert_0$
CLS	0.2145816E-01	0.1235067E-03	0.2342838E-01	0.1668777E-03
CLS-FC	0.2145813E-01	0.4612541E-14	0.2342836E-01	0.5281635E-14
MG	0.2145805E-01	0.1446273E-13	0.2352581E-01	0.6658244E-14

Table 5.5 Compatible LSFEM fluxes before (CLS) and after (CLS-FC) flux correction compared to the flux in the mixed-Galerkin finite element method (MG).

The last two rows in Table 5.5 show that, with respect to accuracy and local conservation, (5.97) with flux correction is identical in performance to the mixed-Galerkin finite element method. The $L^2(\Omega)$ errors of the flux approximations obtained by the two methods are the same to five digits on the uniform mesh and to three digits on the random mesh. The divergence of the corrected flux approximation is actually closer to machine precision than that for the mixed-Galerkin solution. This conclusion is further upheld by comparing the two methods on a sequence of uniform grids. Error values for selected grid sizes are collected in Table 5.6. The data show not only identical convergence rates for the flux variable approximations in both methods, but also asymptotically identical values of the errors for that

Error	Method	33 × 33	65 × 65	129 × 129	Order
$L^2(\Omega)$	CLS-FC	2.007374E-02	9.877587E-03	4.899909E-03	1.01
flux	MG	2.007368E-02	9.877579E-03	4.899908E-03	1.01
$\mathbf{D}(\Omega)$	CLS-FC	5.181631E-15	1.032372E-14	2.124581E-14	exact
flux	MG	2.738185E-14	1.817327E-14	4.383082E-14	exact
$L^2(\Omega)$	CLS-FC	8.436735E-05	2.042191E-05	5.025008E-06	2.02
potential	MG	1.664313E-02	8.189706E-03	4.062637E-03	1.01

Table 5.6 Compatible LSFEM with flux-corrected solution (CLS-FC) versus mixed-Galerkin method (MG) finite element solution.

variable. As before, the potential approximation in the compatible method is much more accurate (second versus first-order) than in the mixed method.

5.9.2 *Compatible Least-Squares Finite Element Methods on Non-Affine Grids*

In this section, we follow up on the comment made in Remark 5.32 that the optimal error estimate (5.85) for compatible LSFEMs cannot be extended to non-affine grids if the method is implemented using the standard div-conforming spaces (B.29)–(B.34). Actually, in the light of the splitting property established in Theorem 5.37, this comment can be further refined. Indeed, if $\gamma \neq 0$, that theorem tells us that the potential approximation in the compatible LSFEM is the same as in the Rayleigh–Ritz method so that its accuracy should not be affected on non-affine grids. However, if $\gamma = 0$, the potential and flux approximations remain coupled and it is very likely that the reduced accuracy of the latter on non-affine grids spills over to the former.

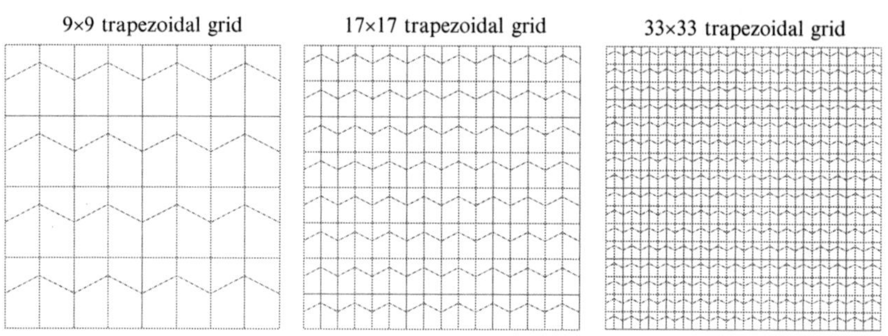

Fig. 5.4 Example trapezoidal grids used in the computational experiments.

To test these conjectures, we carry out an order of convergence study of the compatible LSFEM, with and without the reaction term, using a sequence of "trapezoidal" grids suggested in [9]; see Figure 5.4. Clearly, these grids are uniformly regular but non-affine. We solve (5.17) with $\Gamma = \partial\Omega$, i.e., with Dirichlet boundary conditions. We set

$$\Theta_1 = \begin{bmatrix} \exp((x+y/2)/2) & \sin(2\pi x) \\ \sin(2\pi x) & \exp((x/2+y)/2) \end{bmatrix}, \qquad \Theta_0 = 1,$$

and the right-hand side and boundary data generated from the manufactured solution

$$\phi = -\exp(x)\sin(y) \qquad \text{and} \qquad \mathbf{u} = -\Theta_1 \nabla\phi .$$

The orders of convergence for solutions of (5.62) with $\gamma = 1$ and $\gamma = 0$ are estimated using solution errors on 65×65 and 129×129 grids with 4,096 and 16,384 finite elements, respectively. Error data and the estimated rates of convergence are collected in Table 5.7. From the data in this table, it is clear that, regardless of the value of γ, the accuracy of the flux approximation has significantly deteriorated. The order of the $L^2(\Omega)$ error is reduced by a third, whereas loss of convergence in the divergence approximation is nearly complete. A careful analysis of the root causes for this behavior of div-conforming elements is presented in [9].

Error	div-grad system	65×65	129×129	Order
$\|\phi - \phi^h\|_0$	$\gamma = 1$	0.2162460E-06	0.5412190E-07	1.9984
	$\gamma = 0$	0.1497621E-02	0.1436716E-02	0.0599
$\|\nabla(\phi - \phi^h)\|_0$	$\gamma = 1$	0.7994094E-02	0.4005637E-02	0.9969
	$\gamma = 0$	0.1358298E-01	0.1138904E-01	0.2542
$\|\mathbf{u} - \mathbf{u}^h\|_0$	$\gamma = 1$	0.3620386E-01	0.2254463E-01	0.6834
	$\gamma = 0$	0.3630391E-01	0.2269303E-01	0.6779
$\|\nabla \cdot (\mathbf{u} - \mathbf{u}^h)\|_0$	$\gamma = 1$	0.1089084E+01	0.1073177E+01	0.0212
	$\gamma = 0$	0.1089083E+01	0.1073176E+01	0.0212

Table 5.7 Error data and estimated orders of convergence for the compatible LSFEM with and without the reaction term in (5.17) on trapezoidal grids.

On the positive side, our convergence study upholds the conjecture about the beneficial effect of the reaction term on the potential approximation. Table 5.7 shows that when $\gamma = 1$, the $L^2(\Omega)$-norm and $H^1(\Omega)$-seminorm errors of this variable converge at the best possible rates, i.e., the loss of accuracy has been confined to the flux approximation. However, as anticipated, in the absence of the reaction term, the loss of accuracy spreads to the potential variable. Table 5.7 shows that when $\gamma = 0$, the $L^2(\Omega)$ order of the potential approximation is completely ruined, whereas the $H^1(\Omega)$-seminorm order is reduced by a half.

It is important to realize that these results are not symptomatic of an inherent flaw in the compatible LSFEM but rather spotlight a problem that afflicts any numerical method that uses standard div-conforming elements on non-afffine grids; see [9, 279]. Thus, although disappointing, the loss of accuracy on such grids is not a verdict against the compatible LSFEM, but a statement about the deficiency of the underlying div-conforming space. As a result, one can recover from this situation by simply switching to more accurate *non-standard* div-conforming elements, such as the ABF_r element [9]. In the lowest-order case, an even simpler fix can be obtained by a *reformulation* of the compatible LSFEM prompted by mimetic finite difference methods. We refer to Section 12.3 for a discussion about the reformulated LSFEM.

5.9.3 Advantages and Disadvantages of Extended Systems

The use of extended first-order systems in lieu of the original equations is often rationalized by the fact that homogeneous elliptic systems are a better foundation for LSFEMs than non-homogeneous elliptic problems. This argument rings true for LSFEMs derived from energy balances in the ADN setting, but it does not account for the possibilities afforded by compatible LSFEMs. As a result, the merits of extended systems cannot be judged fairly solely on the basis of comparing the compliant LSFEM (5.78) with the quasi-norm-equivalent and norm-equivalent LSFEMs (5.61) and (5.60), respectively, for the non-extended div–grad system.

Because the results of these comparisons are easy to foresee, in this section, we proceed directly to a comparison of (5.78) and the compatible LSFEM (5.62) and its nodal version (5.63), both of which share the same parent CLSP (5.54) in the vector-operator setting.

Compliant LSFEM for smooth solutions

We limit this study to an order of convergence comparison between (5.78) and (5.63). Because the outcome from this test is also easy to foresee, the real goal is to confirm the theoretical convergence rates in (5.100). To estimate the orders of convergence of the two methods, we solve the extended div–grad (5.25) and div–grad (5.18) systems with $\Theta_1 = \mathbb{I}$, $\Theta_0 = 1$, and a pure Neumann boundary condition,[50] using the manufactured solution defined in (5.101).

Orders of convergence are estimated from error data on uniform 33×33 and 65×65 grids. The results are collected in Table 5.8 and show that the orders of the compliant LSFEM are as predicted by the theory in (5.100). As expected, the $H^1(\Omega)$ norm-equivalence of (5.78), achieved by the addition of the redundant curl equation, translates into optimal $L^2(\Omega)$ and $H^1(\Omega)$ orders of convergence for all variables. Thus, for problems with sufficiently smooth solutions, the compliant LSFEM is

[50] To satisfy assumptions of Corollary 5.19, we do not consider mixed boundary conditions for this test. Results with Dirichlet conditions are identical and are not included.

clearly better than the nodal LSFEM. Of course, both of these methods give flux approximations that are not locally conservative.

System	potential		flux	
div–grad	$L^2(\Omega)$	$H^1(\Omega)$	$L^2(\Omega)$	$\mathbf{D}(\Omega)$
33×33	0.1012E-03	0.1191E-01	0.1007E-02	0.1603E-01
65×65	0.2595E-04	0.5953E-02	0.3580E-03	0.8039E-02
Order	1.96	0.99	**1.49**	1.00
ext-div–grad	$L^2(\Omega)$	$H^1(\Omega)$	$L^2(\Omega)$	$H^1(\Omega)$
33×33	0.1012E-03	0.1191E-01	0.2289E-04	0.2280E-01
65×65	0.2595E-04	0.5953E-02	0.5717E-05	0.1140E-01
Order	1.96	0.99	**2.00**	1.00

Table 5.8 Solution errors and orders of convergence of the nodal LSFEM (5.63) and the compliant LSFEM (5.78).

Compliant LSFEM for rough solutions

As our past experience with the nodal LSFEM shows, orders of convergence do not always tell the whole story about a numerical method. Although good convergence is certainly a desirable trait for any method, the tests in Section 5.9.1 indicate that it may not be sufficient to assert that the method is robust. The latter is what really matters in practical applications for which, in many cases, the regularity assumptions needed in the approximation theory cannot be fulfilled.

To see how the compliant LSFEM performs under duress, we subject it to the same test involving the non-smooth manufactured solution (5.102). The objective is to compare this method to both the nodal LSFEM (which did not do very well in this test) and the compatible LSFEM (which gave excellent results.) For this test, we use the same Θ_1, Θ_0, and a 21×21 uniform grid as before. The difference is that the compliant LSFEM solves the extended div–grad system whereas the other two methods solve the original div–grad system.

Computational results are presented in Figures 5.5 and 5.6. The plots in these figures clearly show that the compliant LSFEM has failed the test. Moreover, the surface plots in Figure 5.6 indicate that inclusion of the redundant curl equation has actually *worsened* the flux approximation relative to the nodal LSFEM. In other words, the same modification that helped the compliant LSFEM perform so well for smooth solutions turned out to be a huge liability for rough solutions. It is not difficult to see why this is the case: the compliant LSFEM (5.78) is designed to compute $H^1(\Omega)$ projections onto finite element subspaces of $H^1(\Omega) \times [H^1(\Omega)]^d \cap \mathbf{D}_0(\Omega)$ and this is what it continues to do, regardless of the available solution regularity.

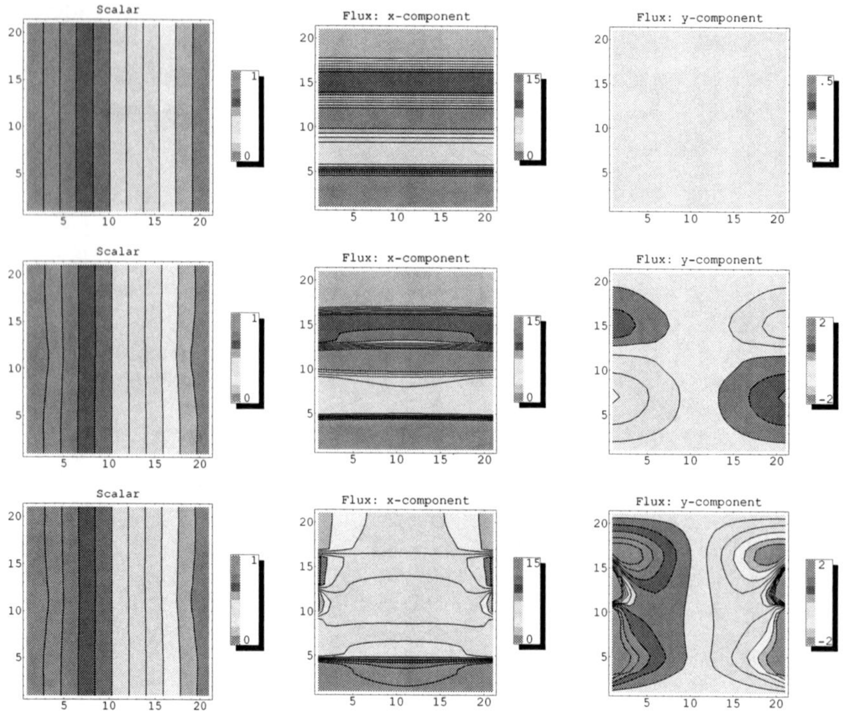

Fig. 5.5 Contour plots of ϕ^h and the components of $\mathbf{u}^h$ determined from the compatible LSFEM (5.62) (top), the nodal LSFEM (5.63) (middle), and the compliant LSFEM (5.78) (bottom).

5.10 A Summary of Conclusions and Recommendations

When solving a div–grad system, our first choice would be the compatible LSFEM. This method is robust for less regular solutions and at the same time offers optimal convergence rates when solutions are sufficiently smooth. Non-affine elements are the only setting where caution needs to be exercised when using this method. For implementations with the lowest-order div-conforming elements on quadrilaterals, we recommend the *mimetic reformulation* of this method presented in Section 12.3. If a higher-order version of the compatible LSFEM on such grids is needed, one can employ the ABF_r elements of [9] in lieu of the standard div-conforming elements given in (B.31) and (B.34). Finally, to avoid the loss of accuracy in the compatible LSFEM on hexahedral grids, one can use the spaces from [279].

For div–grad problems with a reaction term ($\gamma \neq 0$), the compatible LSFEM effectively delivers a *simultaneous* solution of the Rayleigh–Ritz and the mixed-Galerkin methods and a flux approximation that is *locally conservative*. For the div–grad system without a reaction term ($\gamma = 0$), the compatible LSFEM yields flux approximations that are very close to those in the mixed method. An additional flux-

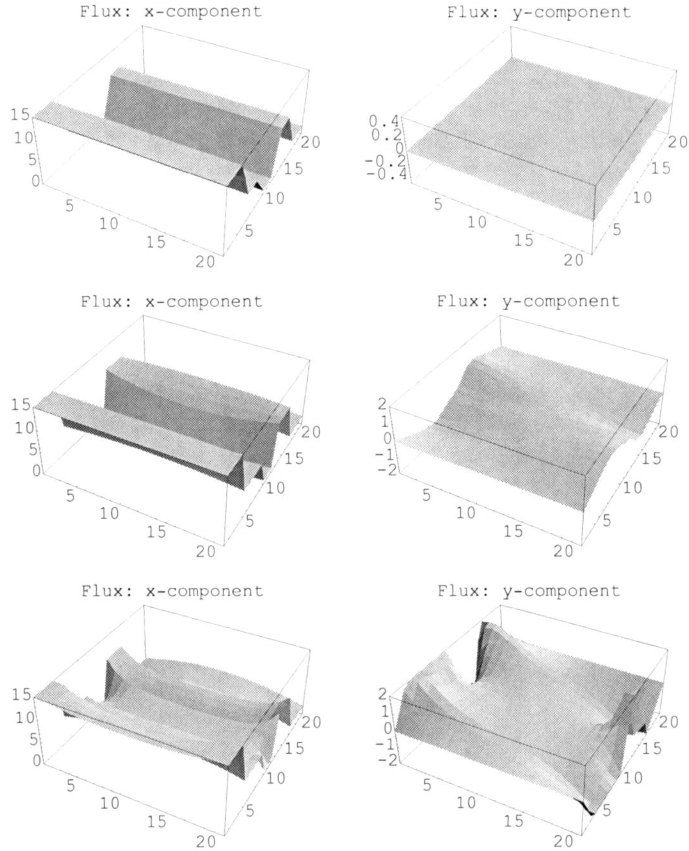

Fig. 5.6 Surface plots of the flux components determined from the compatible LSFEM (5.62) (top), the nodal LSFEM (5.63) (middle), and the compliant LSFEM (5.78) (bottom).

correction step can be applied to make the flux approximation locally conservative without compromising its $L^2(\Omega)$ accuracy.

If accurate approximations of all four fields, i.e., potential, density, flux, and intensity, are required, we recommend the four-field compatible LSFEM (5.64). The advantage of this method over the two-field compatible LSFEM is that it computes approximations of these variables directly using independently chosen approximation spaces; recall that in (5.62), density and intensity approximations are *derived* using the formulas in (5.65).

Clearly, norm-equivalence in $G(\Omega) \times \mathbf{D}(\Omega)$ or $G(\Omega) \times S(\Omega) \times \mathbf{D}(\Omega) \times \mathbf{C}(\Omega)$ makes linear systems in the two compatible LSFEMs somewhat more challenging to solve by iterative methods; it has been argued that this voids the advantages of such LSFEMs. Considering the significant progress in the design of fast and efficient preconditioners in $\mathbf{D}(\Omega)$ and $\mathbf{C}(\Omega)$ (see, e.g., [11–13, 203, 208]), it is clear that this argument cannot be accepted as a serious enough reason[51] to avoid using compatible least-squares formulations.

Of course, one can consider the nodal implementation (5.63) in lieu of (5.62). This choice offers a somewhat simpler implementation and is compatible with respect to the origins of the potential variable. As a result, it leads to optimally accurate approximations for that variable. However, considering that the principal reason to introduce the flux variable is to obtain more accurate flux approximations, it is clear that the nodal LSFEM falls short of meeting this goal. Not only does the nodal flux approximation converge at suboptimal $L^2(\Omega)$ rates, but it also fails to be locally conservative and to approximate well solutions that are only $\mathbf{D}(\Omega)$-regular.

Finally, we also have LSFEMs based on the extended div–grad system. These methods are the easiest to solve by iterative methods because the associated linear systems can be preconditioned by block Poisson preconditioners; see Theorem 4.2. Unfortunately, the scope of these methods is limited to solutions with at least $H^1(\Omega)$ regular flux fields which offsets the advantages of their fast solution. In other words, the strategy of using the extended div–grad system in lieu of (5.18) backfires unless the solution of the original, non-extended div–grad system is such that $\mathbf{u}$ is at least in $[H^1(\Omega)]^d$. Overall, it is safe to say that inclusion of the redundant curl equation is detrimental to the robustness of the LSFEM and should be avoided unless solution regularity is known a priori.

Table 5.9 provides a concise summary of our conclusions and recommendations. Of course, all three LSFEMs compared in this table enjoy the desirable property of engendering symmetric and positive definite linear systems.

Property↓ Method→	Compatible (5.62)	Nodal (5.63)	Compliant (5.78)
Energy balance holds provided	$\partial\Omega$ is Lipschitz	$\partial\Omega$ is Lipschitz	Ω is convex polygon or C^1
optimal energy error	√	√	√
optimal $\|\phi - \phi^h\|_0$	√	√	√
optimal $\|\mathbf{u} - \mathbf{u}^h\|_0$	√	–	√
holds on non-affine $\mathcal{T}_h$	–	√	√
local conservation (5.96)	√	–	–
robust for $\mathbf{u} \notin [H^1(\Omega)]^d$	√	–	–

Table 5.9 Summary properties of selected LSFEMs for scalar, second-order elliptic equations.

[51] In Section 6.5.1, further details about solution and preconditioning of linear systems involving div- and curl-conforming elements are provided.

Chapter 6
Vector Elliptic Equations

In this chapter, we develop least-squares finite element methods (LSFEMs) for two complementary families of second-order *vector* elliptic boundary value problems in $\mathbb{R}^d$, $d = 2,3$. In the first problem, we seek $\mathbf{u}$ that solves

$$\begin{cases} (\Theta_1^{-1}\nabla\times\Theta_2)\nabla\times\mathbf{u} + \nabla(-\Theta_0^{-1}\nabla\cdot\Theta_1)\mathbf{u} + \gamma\mathbf{u} = \mathbf{f} \\ \mathbf{n}\times\mathbf{u} = 0 \quad \text{on } \Gamma \qquad \text{and} \qquad \mathbf{n}\cdot\Theta_1\mathbf{u} = 0 \quad \text{on } \Gamma^*. \end{cases} \tag{6.1}$$

The second problem is to find $\mathbf{u}$ that solves the complementary set of equations

$$\begin{cases} \nabla\times(\Theta_1^{-1}\nabla\times\Theta_2)\mathbf{u} + (-\Theta_2^{-1}\nabla\Theta_3)\nabla\cdot\mathbf{u} + \gamma\mathbf{u} = \mathbf{f} \\ \mathbf{n}\cdot\mathbf{u} = 0 \quad \text{on } \Gamma \qquad \text{and} \qquad \mathbf{n}\times\Theta_2\mathbf{u} = 0 \quad \text{on } \Gamma^*. \end{cases} \tag{6.2}$$

These problems are analogues[1] of the scalar elliptic equations (5.1) and (5.2). As in Chapter 5, we refer to (6.1) and (6.2) as complementary because their solutions are mathematical entities that describe dual properties of the same physical field. Following the nomenclature established in that chapter, we refer to the vector field in (6.1) as the *intensity* and the vector field in (6.2) as the *flux*. The application examples in Section 6.1 show that this choice of terminology is not unfounded.

We continue to assume, as in Chapter 5, that all weights are non-degenerate, i.e., there exist real constants $\alpha_i > 0$, $i = 0,\dots,3$, such that

$$\frac{1}{\alpha_i} \le \Theta_i \le \alpha_i\,,\ i = 0,3\,, \qquad \text{and} \qquad \frac{1}{\alpha_i}\xi^\mathsf{T}\xi \le \xi^\mathsf{T}\Theta_i\xi \le \alpha_i\xi^\mathsf{T}\xi\,,\ i = 1,2\,. \tag{6.3}$$

---

[1] The differential operators in these equations, i.e.,

$$(\Theta_1^{-1}\nabla\times\Theta_2)\nabla\times\ + \nabla(-\Theta_0^{-1}\nabla\cdot\Theta_1) \qquad \text{and} \qquad \nabla\times(\Theta_1^{-1}\nabla\times\Theta_2)\ + (-\Theta_2^{-1}\nabla\Theta_3)\nabla\cdot\,,$$

are generalizations of the operator $\nabla\times\nabla\times - \nabla\nabla\cdot$ that, for sufficiently smooth functions, coincides with the vector Laplace operator $-\vec{\Delta}$; see Remark B.15. The operators in Chapter 5 were generalizations of the scalar Laplace operator $-\Delta$.

P.B. Bochev, M.D. Gunzburger, *Least-Squares Finite Element Methods*, Applied Mathematical Sciences 166, DOI: 10.1007/b13382_6,

In addition, throughout this chapter, we assume that Ω is a bounded contractible region in $\mathbb{R}^d$. In general, $\gamma \geq 0$ is a real-valued function, but, for simplicity, we assume that $\gamma = 0$ or 1.

The problems (6.1) and (6.2) can be augmented with other combinations of boundary conditions, including Neumann and inhomogeneous boundary conditions. For brevity, we only consider LSFEMs for $\Gamma = \partial\Omega$, i.e., (6.1) is considered with the tangential components boundary condition on all of $\partial\Omega$ and (6.2) is considered with the normal component boundary condition on all of $\partial\Omega$. The extension of LSFEMs to more general boundary conditions is fairly straightforward.

LSFEMs for (6.1) and (6.2) are formulated in two settings that correspond to practically important models of physical phenomena. In the first setting, we restrict the range of admissible solutions to divergence-free[2] vector fields $\mathbf{u}$ so that (6.1) with the tangential boundary condition is equivalent to the boundary value problem

$$\begin{cases} (\Theta_1^{-1}\nabla \times \Theta_2)\nabla \times \mathbf{u} + \gamma\mathbf{u} = \mathbf{f} & \text{in } \Omega \\ (\Theta_0^{-1}\nabla \cdot \Theta_1)\mathbf{u} = 0 & \text{in } \Omega \end{cases} \quad \text{and} \quad \mathbf{n} \times \mathbf{u} = 0 \quad \text{on } \Gamma. \tag{6.4}$$

The right-hand side in (6.4) is subject to compatibility condition[3]

$$(\Theta_0^{-1}\nabla \cdot \Theta_1)\mathbf{f} = 0. \tag{6.5}$$

The equivalent form of (6.2) is

$$\begin{cases} \nabla \times (\Theta_1^{-1}\nabla \times \Theta_2)\mathbf{u} + \gamma\mathbf{u} = \mathbf{f} & \text{in } \Omega \\ \nabla \cdot \mathbf{u} = 0 & \text{in } \Omega \end{cases} \quad \text{and} \quad \mathbf{n} \cdot \mathbf{u} = 0 \quad \text{on } \Gamma \tag{6.6}$$

along with the compatibility condition

$$\nabla \cdot \mathbf{f} = 0. \tag{6.7}$$

In the sequel, we refer to (6.4) and (6.6) as the *reduced-intensity* and the *reduced-flux* equations, or simply the reduced equations.[4]

We are also interested in solving the reduced equations under the additional assumption that $\gamma = 0$. In this case, (6.4) and (6.6) undergo some additional simplifications. From the compatibility conditions (6.5) and (6.7) and the Poincaré lemma (see Section A.2.2), it follows that, for some $\mathbf{g}$ and $\mathbf{h}$, $\mathbf{f} = (\Theta_1^{-1}\nabla \times \Theta_2)\mathbf{g}$ for (6.4) and $\mathbf{f} = \nabla \times \mathbf{h}$ for (6.6), respectively. Therefore,

$$(\Theta_1^{-1}\nabla \times \Theta_2)(\nabla \times \mathbf{u} - \mathbf{g}) = 0 \quad \text{and} \quad \nabla \times ((\Theta_1^{-1}\nabla \times \Theta_2)\mathbf{u} - \mathbf{h}) = 0.$$

---

[2] Divergence-free is interpreted in a generalized sense; see the second equation in (6.4).

[3] Recall that we are assuming that γ is constant.

[4] There are other settings, e.g., for the Maxwell's equations, that lead to problems with the structure of (6.4) or (6.6) but for which γ is not a non-negative real number or function. For time-harmonic eddy-current problems [75, p. 219], $\gamma = i\omega$ and for time-harmonic Maxwell's equations $\gamma = -\omega^2$, where ω is the angular frequency; see [75, p. 247]. For brevity, we do not treat these cases here.

Using again Poincaré's lemma, there exist p and q such that

$$\nabla\times\mathbf{u}-\mathbf{g}=(-\Theta_2^{-1}\nabla\Theta_1)p \qquad \text{and} \qquad (\Theta_1^{-1}\nabla\times\Theta_2)\mathbf{u}-\mathbf{h}=-\nabla q\,.$$

From [183, Theorem 3.4, p. 45], we know that the vector potentials $\mathbf{g}$ and $\mathbf{h}$ can be chosen divergence free, which together with the choice of boundary conditions implies that p and q are identically zero. As a result, (6.4) assumes the form

$$\begin{cases} \nabla\times\mathbf{u}=\mathbf{g} & \text{in } \Omega \\ \Theta_0^{-1}\nabla\cdot\Theta_1\mathbf{u}=0 & \text{in } \Omega \end{cases} \qquad \text{and} \qquad \mathbf{n}\times\mathbf{u}=0 \quad \text{on } \Gamma \tag{6.8}$$

whereas (6.6) reduces to

$$\begin{cases} (\Theta_1^{-1}\nabla\times\Theta_2)\mathbf{u}=\mathbf{h} & \text{in } \Omega \\ \nabla\cdot\mathbf{u}=0 & \text{in } \Omega \end{cases} \qquad \text{and} \qquad \mathbf{n}\cdot\mathbf{u}=0 \quad \text{on } \Gamma\,. \tag{6.9}$$

In what follows, we refer to (6.8) and (6.9) as the *intensity-div–curl* equations and *flux-div–curl* equations, respectively, or simply as the *div–curl* equations.

Remark 6.1 Another possibility would be to restrict the range of admissible solutions for (6.1) and (6.2) to irrotational vector fields. Then, (6.1) transforms into the boundary value problem

$$\begin{cases} -\nabla(\Theta_0^{-1}\nabla\cdot\Theta_1)\mathbf{u}+\gamma\mathbf{u}=\mathbf{f} & \text{in } \Omega \\ \nabla\times\mathbf{u}=0 & \text{in } \Omega \end{cases} \qquad \text{and} \qquad \mathbf{n}\times\mathbf{u}=0 \quad \text{on } \Gamma \tag{6.10}$$

with the compatibility condition $\nabla\times\mathbf{f}=0$. Therefore, $\mathbf{u}=\nabla p$ for some[5] $p\in G_0(\Omega)$ and $\mathbf{f}=\nabla g$ for some $g\in G(\Omega)$. Using this ansatz in the first equation of (6.10) gives

$$\nabla\big((\Theta_0^{-1}\nabla\cdot\Theta_1)\mathbf{u}-\gamma p-g\big)=0$$

and, because[6] $N(\nabla)=\{constants\}$, it follows that $(\Theta_0^{-1}\nabla\cdot\Theta_1)\mathbf{u}-\gamma p=g$. As a result, looking for irrotational solutions of (6.10) is the same as looking for solutions of the *div–grad* boundary value problem

$$\begin{cases} (\Theta_0^{-1}\nabla\cdot\Theta_1)\mathbf{u}-\gamma p=g & \text{in } \Omega \\ \mathbf{u}-\nabla p=0 & \text{in } \Omega \end{cases} \qquad \text{and} \qquad p=0 \quad \text{on } \Gamma\,. \tag{6.11}$$

Up to a sign, (6.11) is identical with one of the div–grad systems considered in Chapter 5. Therefore, we see that the vector equations (6.1) in terms of the *intensity* reduce to the *potential–intensity* first-order system (5.19) for the scalar elliptic

[5] Using the ansatz $\mathbf{u}=\nabla p$ in the boundary condition implies that $\mathbf{n}\times\nabla p=0$, i.e., all tangential derivatives of p vanish on Γ, so that $p=$ constant there. It is convenient to set this constant to zero, i.e., choose $p\in G_0(\Omega)$.

[6] Recall that N denotes null space of an operator; see Section A.2.2.

equations considered in that chapter. Repeating this exercise for the vector elliptic equation (6.2) in terms of the *flux* would yield instead the complementary *flux–density* first-order system (5.17). Consequently, the assumption that **u** is irrotational reduces (6.1) and (6.2) to the same first-order systems as in Chapter 5. Thus, it is not necessary to consider this case any further. □

6.1 Applications of Vector Elliptic Equations

In order to capture *dual* properties of magnetic and electric fields, mathematical models of electromagnetism rely on two pairs of complementary variables [75, 240, 330, 332, 339]. For the electric field, this pair consists of the electric field strength **E** (having units of V/m, volts per meter) and the *current density* **J** (having units of A/m^2, amperes per square meter). Magnetic fields are described using *magnetic field strength* **H** (having units of A/m, amperes per meter) and *magnetic flux density* **B** (having units of W/m^2, webers per square meter). The variables in each pair are related by the constitutive laws

$$\mathbf{J} = \sigma \mathbf{E} \qquad \text{and} \qquad \mathbf{B} = \mu \mathbf{H}, \tag{6.12}$$

where σ (having units of S/m, siemens per meter) and μ (having units of H/m, henrys per meter) are the conductivity and the magnetic permeability of the material, respectively. Using (6.12), mathematical models of various electromagnetic phenomena can be cast in a form of reduced or div–curl systems.

In particular, the reduced equations (6.4) and (6.6) are representative of the type of partial differential equations (PDEs) that arise in eddy-current problems; see [75, p. 219]. They are prototypes for two sets of complementary eddy-current equations in terms of either $\{\mathbf{H}, \mathbf{B}\}$ or $\{\mathbf{E}, \mathbf{J}\}$. For example, using the magnetic variables results in a model in terms of **H**:

$$\begin{cases} \partial_t \mathbf{H} + \mu^{-1} \nabla \times \sigma^{-1} \nabla \times \mathbf{H} = \mathbf{f} & \text{in } \Omega \\ \nabla \cdot \mu \mathbf{H} = 0 & \text{in } \Omega \end{cases} \quad \text{and} \quad \mathbf{n} \times \mathbf{H} = 0 \quad \text{on } \Gamma \tag{6.13}$$

whose prototype is the first reduced system (6.4) and a model in terms of **B**:

$$\begin{cases} \partial_t \mathbf{B} + \nabla \times \sigma^{-1} \nabla \times \mu^{-1} \mathbf{B} = \mathbf{f} & \text{in } \Omega \\ \nabla \cdot \mathbf{B} = 0 & \text{in } \Omega \end{cases} \quad \text{and} \quad \mathbf{n} \cdot \mathbf{B} = 0 \quad \text{on } \Gamma \tag{6.14}$$

whose prototype is (6.6). If, instead, one decides to use the electric variables, the first reduced system, i.e., (6.4), is a prototype of an eddy-current model

$$\begin{cases} \partial_t \mathbf{E} + \sigma^{-1} \nabla \times \mu^{-1} \nabla \times \mathbf{E} = \mathbf{f} & \text{in } \Omega \\ \nabla \cdot \sigma \mathbf{E} = 0 & \text{in } \Omega \end{cases} \quad \text{and} \quad \mathbf{n} \times \mathbf{E} = 0 \quad \text{on } \Gamma \tag{6.15}$$

in terms of $\mathbf{E}$, whereas (6.6) is prototype of a model in terms of $\mathbf{J}$:

$$\begin{cases} \partial_t \mathbf{J} + \nabla \times \mu^{-1} \nabla \times \sigma^{-1} \mathbf{J} = \mathbf{f} & \text{in } \Omega \\ \nabla \cdot \mathbf{J} = 0 & \text{in } \Omega \end{cases} \quad \text{and} \quad \mathbf{n} \cdot \mathbf{J} = 0 \quad \text{on } \Gamma. \tag{6.16}$$

After implicit semi-discretization in time, (6.13)–(6.16) reduce to elliptic boundary value problems with the structure of (6.4) or (6.6) with $\gamma = 1/\Delta t$.

Remark 6.2 In practice, the two most-often used eddy-current models are (6.13) and (6.15). As a result, the prototype reduced *intensity* system (6.4) is of greater interest to us than the reduced *flux* system (6.6). □

The div–curl systems (6.8) and (6.9) are also prototypes of PDEs arising in the modeling of the linear *magnetostatics* problem. Specifically, they correspond to two complementary forms of that problem; see [75, p. 23]. The problem (6.8) is a prototype of the magnetostatics equations in terms of $\mathbf{H}$:

$$\begin{cases} \nabla \times \mathbf{H} = \mathbf{J}_0 & \text{in } \Omega \\ \nabla \cdot \mu \mathbf{H} = 0 & \text{in } \Omega \end{cases} \quad \text{and} \quad \mathbf{n} \times \mathbf{H} = 0 \quad \text{on } \Gamma, \tag{6.17}$$

and (6.6) is a prototype of PDEs that describe the same phenomena in terms of the dual magnetic variable $\mathbf{B}$:

$$\begin{cases} \nabla \times \mu^{-1} \mathbf{B} = \mathbf{J}_0 & \text{in } \Omega \\ \nabla \cdot \mathbf{B} = 0 & \text{in } \Omega \end{cases} \quad \text{and} \quad \mathbf{n} \cdot \mathbf{B} = 0 \quad \text{on } \Gamma. \tag{6.18}$$

In (6.17) and (6.18), $\mathbf{J}_0$ is a given function that specifies the imposed current density.

6.2 Reformulation of Vector Elliptic Problems

The reduced equations (6.4) and (6.6) are second-order elliptic boundary value problems. To satisfy the first key to practicality of LSFEMs (see Section 2.2.2), they need to be replaced by equivalent first-order systems before the application of a least-squares principle. The transformation of reduced equations is considered in Section 6.2.2.

The div–curl systems (6.8) and (6.9) are already given in first-order system form so that they satisfy the first key to practicality by default. The problem with these equations is that the resulting continuous least-squares principles (CLSPs) are norm-equivalent with respect to the intersection spaces $\mathbf{C}_0(\Omega) \cap \mathbf{D}(\Omega)$ or $\mathbf{C}(\Omega) \cap \mathbf{D}_0(\Omega)$. We have already encountered the first one of these spaces in Section 5.5 and explained, in Remark 5.7, some of the reasons why it should be avoided for the scalar elliptic equation case. Unfortunately, for the vector elliptic equation case, we cannot so easily avoid using intersection spaces. Recall that $\mathbf{C}_0(\Omega) \cap \mathbf{D}(\Omega)$ appeared in (5.59) because that CLSP is based on the extended div–grad system (5.25). However, in Chapter 5, we could always evade dealing with $\mathbf{C}_0(\Omega)\cap\mathbf{D}(\Omega)$ by simply going back to least-squares principles formulated in terms of the original un-extended div–grad system (5.17). This is not possible for (6.4) and (6.6) because $\mathbf{C}_0(\Omega)\cap\mathbf{D}(\Omega)$ and $\mathbf{C}(\Omega)\cap\mathbf{D}_0(\Omega)$ arise naturally in the energy balance of the div–curl systems so that the attendant complications cannot be avoided so easily.

One way to deal with the difficulties caused by the "double-faced" nature of the spaces $\mathbf{C}_0(\Omega)\cap\mathbf{D}(\Omega)$ or $\mathbf{C}(\Omega)\cap\mathbf{D}_0(\Omega)$ is to give up conformity[7] with respect to one of their "faces" and approximate $\mathbf{u}$ by either div-conforming or curl-conforming elements. Another alternative is to split $\mathbf{u}$ into two independent vector variables that can be approximated separately by curl-conforming and div-conforming elements, respectively. The splitting of $\mathbf{u}$ can be effected by introducing an algebraic "constitutive equation"[8] that defines a second vector field in terms of $\mathbf{u}$. This reformulation is discussed in Section 6.2.1.

To obtain dimensionally and unit consistent least-squares functionals, the weights Θ_i in the reduced and div–curl systems must be properly accounted for when measuring equation residuals. This can be arranged by using the spaces defined in (A.33)–(A.36) to set up the least-squares principles. To make the subsequent choice of norms more transparent, in what follows, we track these weights by including them explicitly in the space designations.

6.2.1 Div–Curl Systems

First consider the div–curl system (6.8). A natural solution space that accounts for the weights in this system is $\mathbf{C}_0(\Omega,\Theta_1)\cap\mathbf{D}(\Omega,\Theta_1^{-1})$. To "split" $\mathbf{u}$, we introduce the new dependent variable $\mathbf{v}=\Theta_1\mathbf{u}$ that transforms (6.8) into the boundary value problem

$$\begin{cases} \nabla\times\mathbf{u}=\mathbf{g} & \text{in } \Omega \\ \mathbf{v}-\Theta_1\mathbf{u}=0 & \text{in } \Omega \\ \nabla\cdot\mathbf{v}=0 & \text{in } \Omega \end{cases} \quad\text{and}\quad \mathbf{n}\times\mathbf{u}=0 \quad \text{on } \Gamma. \tag{6.19}$$

The natural solution space for (6.19) is $\{\mathbf{u},\mathbf{v}\}\in\mathbf{C}_0(\Omega,\Theta_1)\times\mathbf{D}(\Omega,\Theta_1^{-1})$, i.e., we were able to separate the two components of the solution space of (6.8). In the reformulated div–curl system (6.19), one of the equations is an algebraic relation rather than a differential equation. However, this is irrelevant for least-squares methods as long as one can establish an energy balance for the residuals of (6.19).

For the second div–curl system (6.9), the natural solution space is $\mathbf{C}(\Omega,\Theta_2^{-1})\cap\mathbf{D}_0(\Omega,\Theta_2)$. Now, we use $\mathbf{v}=\Theta_2\mathbf{u}$ as a new dependent variable to obtain the system

[7] This is similar to what happens in mixed finite element methods; see Remark B.16. Of course, when used in conjunction with least-squares principles, this idea leads to finite element methods in the attractive Rayleigh–Ritz setting.

[8] A similar idea is used in Chapter 5 to define the potential–density–flux–intensity first-order systems (5.21) and (5.23).

$$\begin{cases} \nabla \times \mathbf{v} = \Theta_1 \mathbf{h} & \text{in } \Omega \\ \mathbf{v} - \Theta_2 \mathbf{u} = 0 & \text{in } \Omega \\ \nabla \cdot \mathbf{u} = 0 & \text{in } \Omega \end{cases} \quad \text{and} \quad \mathbf{n} \cdot \mathbf{u} = 0 \quad \text{on } \Gamma \tag{6.20}$$

for which the natural solution space is $\{\mathbf{v}, \mathbf{u}\} \in \mathbf{C}(\Omega, \Theta_2^{-1}) \times \mathbf{D}_0(\Omega, \Theta_2)$.

The approach used to obtain (6.19) and (6.20) resembles the idea used in Chapter 5 to derive the four-field system (5.23). The difference is that there the constitutive relation is used to include, in the first-order system, flux and density variables derived from the same potential field whereas in (6.19) and (6.20), this relation is used to create a new, complementary variable $\mathbf{v}$ from $\mathbf{u}$.[9]

6.2.2 Curl–Curl Systems

In view of Remark 6.2, we focus solely on the reduced intensity system (6.4); with the equation involving the divergence operator discarded, (6.4) is given by

$$\begin{cases} (\Theta_1^{-1} \nabla \times \Theta_2) \nabla \times \mathbf{u} + \gamma \mathbf{u} = \mathbf{f} & \text{in } \Omega \\ \mathbf{n} \times \mathbf{u} = 0 & \text{on } \Gamma . \end{cases} \tag{6.22}$$

We exclude the divergence equation for two reasons. Obviously, it is implied by the first equation in (6.4) and so it is redundant. A second and more compelling reason is to prevent the emergence of $\mathbf{C}(\Omega) \cap \mathbf{D}(\Omega)$ as a natural solution space for the first-order systems we are about to define. Fortunately, for the reduced system, the appearance of $\mathbf{C}(\Omega) \cap \mathbf{D}(\Omega)$ can be avoided by simply dropping the redundant equation. Later, using the same idea of "constitutive" laws introduced in the last section, we are able to bring this equation back, along with another redundant equation.

As usual, transformation of a second-order problem into an equivalent first-order system requires new dependent variables. For (6.22), we choose $\mathbf{v} = \Theta_2 \nabla \times \mathbf{u}$, resulting in the first-order system

$$\begin{cases} \gamma \Theta_1 \mathbf{u} + \nabla \times \mathbf{v} = \Theta_1 \mathbf{f} & \text{in } \Omega \\ \Theta_2^{-1} \mathbf{v} - \nabla \times \mathbf{u} = 0 & \text{in } \Omega \end{cases} \quad \text{and} \quad \mathbf{n} \times \mathbf{u} = 0 \quad \text{on } \Gamma . \tag{6.23}$$

[9] The algebraic equations in (6.19) and (6.20) are prototypes of "constitutive laws" that relate two different physical variables. For example, the two magnetostatics models (6.17) and (6.18) can be reduced to a single model in terms of both $\mathbf{H}$ and $\mathbf{B}$:

$$\begin{cases} \nabla \times \mathbf{H} = \mathbf{J}_0 & \text{in } \Omega \\ \nabla \cdot \mathbf{B} = 0 & \text{in } \Omega \end{cases} \quad \text{and} \quad \mathbf{B} = \mu \mathbf{H} \quad \text{or} \quad \mathbf{H} = \mu^{-1} \mathbf{B} . \tag{6.21}$$

The first form of the constitutive law corresponds to transformation of the magnetostatic model (6.17) in terms of $\mathbf{H}$ into (6.21). The second form of this law gives the transformation of the complementary model (6.18) into (6.21).

The natural solution space for (6.23) is $\{\mathbf{u},\mathbf{v}\} \in \mathbf{C}_0(\Omega,\Theta_1) \times \mathbf{C}(\Omega,\Theta_2^{-1})$. For obvious reasons, we refer to (6.23) as a *curl–curl* system.[10]

In Section 6.2.1, new dependent variables were introduced through algebraic "constitutive" laws. The same idea can be used here by setting $\mathbf{w} = \Theta_1\mathbf{u}$ and $\mathbf{y} = \Theta_2^{-1}\mathbf{v}$, resulting in the first-order system

$$\begin{cases} \gamma\mathbf{w} + \nabla\times\mathbf{v} = \Theta_1\mathbf{f} & \text{in } \Omega \\ \mathbf{y} - \nabla\times\mathbf{u} = 0 & \text{in } \Omega \\ \mathbf{w} - \Theta_1\mathbf{u} = 0 & \text{in } \Omega \\ \mathbf{y} - \Theta_2^{-1}\mathbf{v} = 0 & \text{in } \Omega \end{cases} \quad \text{and} \quad \mathbf{n}\times\mathbf{u} = 0 \quad \text{on } \Gamma. \tag{6.24}$$

The variables in this four-field system form two complementary pairs given by $\{\mathbf{u},\mathbf{w}\} \in \mathbf{C}_0(\Omega,\Theta_1) \times \mathbf{D}(\Omega,\Theta_1^{-1})$ and $\{\mathbf{v},\mathbf{y}\} \in \mathbf{C}(\Omega,\Theta_2^{-1}) \times \mathbf{D}(\Omega,\Theta_2)$, respectively. In these pairs, $\mathbf{u}$ and $\mathbf{v}$ are mathematical entities describing intensities of two different physical fields, whereas the dual variables $\mathbf{w}$ and $\mathbf{y}$ represent their fluxes.

The makeup of (6.24) closely resembles the four-field system (5.21), except that in the latter one of the complementary pairs is comprised of two scalar fields. The similarities of the two systems are further underscored by the fact that the structure of (6.24) can be represented by the Tonti diagram

$$\begin{array}{ccccc} & \text{intensity} & \mathbf{u} \xrightarrow{\nabla\times} \mathbf{y} & \text{flux} & \\ \Theta_1 & \downarrow & & \uparrow & \Theta_2^{-1} \\ & \text{flux} & \mathbf{w} \xleftarrow{\nabla\times} \mathbf{v} & \text{intensity} & \end{array} \tag{6.25}$$

that is analogous to the diagram (5.22) for (5.21). The parallels between (6.24) and (5.21) can be made even more transparent by using the exterior calculus formalism described in Remark 5.5. However, the prerequisites to do this are beyond the scope of this book; further details can be obtained from references cited in Remark 5.5.

Now that the variables are "split," the formerly discarded divergence equation can be brought back[11] in the form $\nabla\cdot\mathbf{w} = 0$. Also, in view of the second equation in (6.24), one can add another "redundant" equation $\nabla\cdot\mathbf{y} = 0$ to this system; this parallels the inclusion of the "redundant" curl equation in (5.23). With these modifications, (6.24) becomes[12]

---

[10] The second-order system (6.22) could also be justifiably referred to as a curl–curl equation, being that it involves the composition of two curl operators. However, that moniker is more commonly applied to (6.23) so that, to avoid confusion, we refer to (6.22) as a reduced system, based on the observation that it is a "reduced" version of (6.1).

[11] Recall that $\mathbf{f}$ is subject to the compatibility condition (6.5) that is required for the redundant equation to hold.

[12] Although the component equations in (6.26) involve the divergence operator and that system looks like a double div–curl system, we continue to include (6.26) in the curl–curl class of equations and refer to it as one of the members of that class. This is justified by the "redundant" nature of the divergence equations appearing in that system; in "true" div–curl systems such as (6.19) and

$$\begin{cases} \gamma \mathbf{w} + \nabla \times \mathbf{v} = \Theta_1 \mathbf{f} & \text{in } \Omega \\ \mathbf{y} - \nabla \times \mathbf{u} = 0 & \text{in } \Omega \\ \nabla \cdot \mathbf{y} = 0 & \text{in } \Omega \\ \nabla \cdot \mathbf{w} = 0 & \text{in } \Omega \\ \mathbf{w} - \Theta_1 \mathbf{u} = 0 & \text{in } \Omega \\ \mathbf{y} - \Theta_2^{-1} \mathbf{v} = 0 & \text{in } \Omega \end{cases} \quad \text{and} \quad \mathbf{n} \times \mathbf{u} = 0 \quad \text{on } \Gamma . \tag{6.26}$$

The natural solution space for this system is the same as for (6.24). The similarity between this four-field system and (5.23) for div–grad systems is obvious.

Applications of curl–curl systems

The reduced system (6.4) is a prototype for two different versions of the eddy-current problem. Using the identifications

$$\begin{matrix} \mathbf{u} \longrightarrow \mathbf{H} \\ \mathbf{v} \longrightarrow \mathbf{E} \end{matrix} \quad \text{and} \quad \begin{matrix} \Theta_1 \longrightarrow \mu \\ \Theta_2 \longrightarrow \sigma_1^{-1} , \end{matrix} \tag{6.27}$$

we see that (6.23) corresponds to the eddy-current equations in terms of the electric variable **E** and the magnetic variable **H**:

$$\begin{cases} \partial_t \mu \mathbf{H} + \nabla \times \mathbf{E} = \mathbf{f} & \text{in } \Omega \\ \sigma \mathbf{E} - \nabla \times \mathbf{H} = 0 & \text{in } \Omega \end{cases} \quad \text{and} \quad \mathbf{n} \times \mathbf{H} = 0 \quad \text{on } \Gamma . \tag{6.28}$$

Not surprisingly, the systems (6.24) and (6.26) also turn out to be prototypes for another form of the eddy-current problem. Making the identifications

$$\begin{matrix} \mathbf{u} \longrightarrow \mathbf{H} & \mathbf{w} \longrightarrow \mathbf{B} \\ \mathbf{v} \longrightarrow \mathbf{E} & \mathbf{y} \longrightarrow \mathbf{J} \end{matrix} \quad \text{and} \quad \begin{matrix} \Theta_1 \longrightarrow \mu \\ \Theta_2 \longrightarrow \sigma_1^{-1} , \end{matrix} \tag{6.29}$$

we see that (6.26) is a prototype of the eddy-current equations in terms of all four magnetic and electric variables:[13]

(6.20), the divergence equation is not redundant, i.e., it is not implied by the other equations in those systems.

[13] Obviously, (6.24) is a prototype of the same problem without the redundant divergence equations.

$$\begin{cases} \partial_t \mathbf{B} + \nabla \times \mathbf{E} = \mathbf{f} & \text{in } \Omega \\ \mathbf{J} - \nabla \times \mathbf{H} = 0 & \text{in } \Omega \\ \nabla \cdot \mathbf{J} = 0 & \text{in } \Omega \\ \nabla \cdot \mathbf{B} = 0 & \text{in } \Omega \\ \mathbf{B} - \mu \mathbf{H} = 0 & \text{in } \Omega \\ \mathbf{J} - \sigma \mathbf{E} = 0 & \text{in } \Omega \end{cases} \quad \text{and} \quad \mathbf{n} \times \mathbf{H} = 0 \quad \text{on } \Gamma . \tag{6.30}$$

We recognize the first two equations in (6.30) as the Faraday law of induction and the Ampere law that states that currents are the source of magnetic fields. The last two equations are the constitutive laws from (6.12).

6.3 Least-Squares Finite Element Methods for Div–Curl Systems

In this section, our established least-squares routine is applied to div–curl systems. First, we derive energy balances for the original div–curl systems (6.8) and (6.9) and their reformulated versions (6.19) and (6.20). Each energy balance gives rise to a well-posed parent CLSP from which we obtain discrete least-squares principles (DLSPs) by following the transformation rules discussed in Section 3.4.1.

6.3.1 Energy Balances

As usual, we consider two types of energy balances for div–curl systems. The first one uses the Agmon–Douglis–Nirenberg (ADN) theory and treats the vector variable as a collection of d independent scalar functions u_j, $j = 1,\dots,d$, belonging to the Sobolev spaces $H^{q+t_j}(\Omega)$. The advantages and the drawbacks of this approach were explained in Section 5.4. In particular, the scope of such energy balances is restricted to domains that satisfy the assumptions of Theorem A.8. Consequently, such balances play a marginal role in the development of useful LSFEMs for div–curl systems. For this reason, we only provide a limited discussion of this case.

More important to us is the second kind of energy balance for which the variables in the div–curl systems are treated as vector fields in $\mathbf{C}(\Omega)$, $\mathbf{D}(\Omega)$, or $\mathbf{C}(\Omega) \cap \mathbf{D}(\Omega)$. To save space, these balances are discussed only for problems posed in $\mathbb{R}^3$.

A Priori Bounds in the Agmon–Douglis–Nirenberg Setting

The application of the ADN theory to three-dimensional div–curl systems requires the addition of a slack variable; see Section D.2.3. Thus, we first state energy balances for the augmented div–curl system (D.14).[14] For this problem, all assumptions of the ADN theory have been verified in Section D.2.3 with $u = (u_1, u_2, u_3, \phi)$ and t_j, s_i, and r_l given by (D.20). Using the notation of Theorem D.1, $t' = 1$ and $r' = 0$. The following theorem is a direct consequence of Theorem D.1.

Theorem 6.3 *Assume that $\Omega \subset \mathbb{R}^3$ is a bounded domain of class C^{q+1} for $q \geq 0$. Then, there exists a positive constant C_q such that, for any collection $u = \{u_1, u_2, u_3, \phi\}$ of smooth functions,*

$$C_q\big(\|\mathbf{u}\|_{q+1} + \|\phi\|_{q+1}\big) \leq \|\nabla\times\mathbf{u} + \nabla\phi\|_q + \|\nabla\cdot\mathbf{u}\|_q + \|\mathbf{n}\times\mathbf{u}\|_{q+1/2,\Gamma} \tag{6.31}$$

for the problem (6.8) *and*

$$\begin{aligned} &C_q\big(\|\mathbf{u}\|_{q+1} + \|\phi\|_{q+1}\big) \\ &\quad \leq \|\nabla\times\mathbf{u} + \nabla\phi\|_q + \|\nabla\cdot\mathbf{u}\|_q + \|\mathbf{n}\cdot\mathbf{u}\|_{q+1/2,\Gamma} + \|\phi\|_{q+1/2,\Gamma} \end{aligned} \tag{6.32}$$

for the problem (6.9), *where* $\mathbf{u} = (u_1,\, u_2,\, u_3)^T$. □

Because the right-hand sides $\mathbf{g}$ and $\mathbf{h}$ in (6.8) and (6.9) are divergence-free, Proposition D.10 asserts that the slack variable is identically zero. Therefore, we can completely ignore ϕ in the formulation of least-squares principles for these systems.[15]

A Priori Bounds in the vector-operator Setting

Energy balances for the original div–curl systems (6.8) and (6.9) follow from the Poincaré-Friedrichs inequality (A.69) and measure solution "energy" in terms of the norm (A.61). For future reference, we state two versions of this energy balance that account for the weighting differences in (6.8) and (6.9).

Theorem 6.4 *Assume that $\Omega \subset \mathbb{R}^3$ is a contractible bounded domain with a Lipschitz-continuous boundary. Then, there exists a positive constant C such that*

$$C\|\mathbf{u}\|_{DC} \leq \|\nabla\times\mathbf{u}\|_{0,\Theta_2} + \|\nabla\cdot\Theta_1\mathbf{u}\|_{0,\Theta_0^{-1}} \tag{6.34}$$

for all $\mathbf{u} \in \mathbf{C}_0(\Omega,\Theta_1)\cap\mathbf{D}(\Omega,\Theta_1^{-1})$ and

---

[14] Of course, the augmented div–curl system is exactly the div–grad–curl system considered in Section 12.9.

[15] In very special cases, e.g., when Ω is a rectangle [165], one can show that for all smooth $\mathbf{u}$ with $\mathbf{n}\cdot\mathbf{u} = 0$

$$\int_\Omega \nabla\mathbf{u} : \nabla\mathbf{u}\,dx = \int_\omega |\nabla\cdot\mathbf{u}|^2 + |\nabla\times\mathbf{u}|^2\,dx \tag{6.33}$$

which implies (6.32). However, if Ω is not a rectangle, validity of (6.33) cannot be assured; see Remark B.15 for counterexamples. Theorem A.8 allows one to extend (6.31) and (6.32) to a wider range of admissible domains.

$$C\|\mathbf{u}\|_{DC} \leq \|\nabla \times \Theta_2 \mathbf{u}\|_{0,\Theta_1^{-1}} + \|\nabla \cdot \mathbf{u}\|_{0,\Theta_3} \tag{6.35}$$

for all $\mathbf{u} \in \mathbf{C}(\Omega, \Theta_2^{-1}) \cap \mathbf{D}_0(\Omega, \Theta_2)$.

Regarding the reformulated div–curl systems (6.19) and (6.20), we have the following result.

Theorem 6.5 *Assume that* $\Omega \subset \mathbb{R}^3$ *is a contractible bounded domain with a Lipschitz-continuous boundary. Then, there exists a positive constant* C *such that*

$$C\big(\|\mathbf{u}\|_C + \|\mathbf{v}\|_D\big) \leq \|\nabla \times \mathbf{u}\|_{0,\Theta_2} + \|\nabla \cdot \mathbf{v}\|_{0,\Theta_0^{-1}} + \|\mathbf{v} - \Theta_1 \mathbf{u}\|_{0,\Theta_1^{-1}} \tag{6.36}$$

for all $\{\mathbf{u}, \mathbf{v}\} \in \mathbf{C}_0(\Omega, \Theta_1) \times \mathbf{D}(\Omega, \Theta_1^{-1})$ *and*

$$C\big(\|\mathbf{u}\|_D + \|\mathbf{v}\|_C\big) \leq \|\nabla \times \mathbf{v}\|_{0,\Theta_1^{-1}} + \|\nabla \cdot \mathbf{u}\|_{0,\Theta_3} + \|\mathbf{v} - \Theta_2 \mathbf{u}\|_{0,\Theta_2^{-1}} \tag{6.37}$$

for all $\{\mathbf{u}, \mathbf{v}\} \in \mathbf{D}_0(\Omega, \Theta_2) \times \mathbf{C}(\Omega, \Theta_2^{-1})$. □

Proof. The proof follows some ideas of [150]. To prove (6.36), let $\mathbf{u} \in \mathbf{C}_0(\Omega)$ and $\mathbf{v} \in \mathbf{D}(\Omega)$, where for clarity all weights are assumed equal to one. Let $\alpha > 0$ be a real number that is determined later. Then,

$$\|\mathbf{v} - \mathbf{u} - \alpha \mathbf{v}\|_0^2 = \|\mathbf{v} - \mathbf{u}\|_0^2 + \alpha(\alpha - 2)\|\mathbf{v}\|_0^2 + 2\alpha(\mathbf{u}, \mathbf{v})$$

$$\|\mathbf{v} - \mathbf{u} + \alpha \mathbf{u}\|_0^2 = \|\mathbf{v} - \mathbf{u}\|_0^2 + \alpha(\alpha - 2)\|\mathbf{u}\|_0^2 + 2\alpha(\mathbf{u}, \mathbf{v}).$$

Summing these identities and solving for $\|\mathbf{v} - \mathbf{u}\|_0^2$ yields

$$\begin{aligned} \|\mathbf{v} - \mathbf{u}\|_0^2 = \frac{1}{2}\Big[& \|\mathbf{v} - \mathbf{u} - \alpha \mathbf{v}\|_0^2 + \|\mathbf{v} - \mathbf{u} + \alpha \mathbf{u}\|_0^2 \\ & + \alpha\left(1 - \frac{\alpha}{2}\right)\left(\|\mathbf{u}\|_0^2 + \|\mathbf{v}\|_0^2\right) - 2\alpha(\mathbf{u}, \mathbf{v})\Big]. \end{aligned} \tag{6.38}$$

To estimate the inner product of $\mathbf{u}$ and $\mathbf{v}$, we use the Hodge decomposition (A.71) for $\mathbf{C}_0(\Omega)$ functions. Recall that $\mathbf{u} = \nabla p + \mathbf{u}_{N\perp}$, where $p \in G_0(\Omega)$. Using this ansatz and that $p = 0$ on $\partial\Omega$,

$$\begin{aligned} (\mathbf{u}, \mathbf{v}) &= (\nabla p + \mathbf{u}_{N\perp}, \mathbf{v}) = (p, -\nabla \cdot \mathbf{v}) + (\mathbf{u}_{N\perp}, \mathbf{v}) \\ &\leq \|p\|_0 \|\nabla \cdot \mathbf{v}\|_0 + \|\mathbf{u}_{N\perp}\|_0 \|\mathbf{v}\|_0 \\ &\leq C_H\left(\|\mathbf{u}\|_0 \|\nabla \cdot \mathbf{v}\|_0 + \|\nabla \times \mathbf{u}\|_0 \|\mathbf{v}\|_0\right), \end{aligned}$$

where the last inequality follows from (A.72). Using the ε-inequality,

$$2\alpha(\mathbf{u}, \mathbf{v}) \leq \left(\frac{\alpha^2}{\varepsilon}\left(\|\mathbf{u}\|_0^2 + \|\mathbf{v}\|_0^2\right) + \varepsilon C_H^2\left(\|\nabla \cdot \mathbf{v}\|_0^2 + \|\nabla \times \mathbf{u}\|_0^2\right)\right).$$

Together with (6.38), we then obtain the lower bound

$$\|\mathbf{v}-\mathbf{u}\|_0^2 \geq \frac{\alpha}{2}\left(1-\frac{\alpha}{2}-\frac{\alpha}{\varepsilon}\right)\left(\|\mathbf{u}\|_0^2+\|\mathbf{v}\|_0^2\right) - \frac{\varepsilon C_H^2}{2}\left(\|\nabla\cdot\mathbf{v}\|_0^2+\|\nabla\times\mathbf{u}\|_0^2\right).$$

Therefore,

$$\|\nabla\times\mathbf{u}\|_0^2+\|\nabla\cdot\mathbf{v}\|_0^2+\|\mathbf{v}-\mathbf{u}\|_0^2$$
$$\geq \left(1-\frac{\varepsilon C_H^2}{2}\right)\left(\|\nabla\times\mathbf{u}\|_0^2+\|\nabla\cdot\mathbf{v}\|_0^2\right) + \frac{\alpha}{2}\left(1-\frac{\alpha}{2}-\frac{\alpha}{\varepsilon}\right)\left(\|\mathbf{u}\|_0^2+\|\mathbf{v}\|_0^2\right).$$

Setting $\varepsilon = \min\{1, 1/C_H^2\}$ and $\alpha = \varepsilon/2$ guarantees that

$$1-\frac{\varepsilon C_H^2}{2} \geq \frac{1}{2} \quad \text{and} \quad 1-\frac{\alpha}{2}-\frac{\alpha}{\varepsilon} \geq \frac{1}{4}.$$

This proves (6.36). The proof of (6.37) follows along the same lines, but uses the Hodge decomposition of $\mathbf{D}_0(\Omega)$ from Theorem A.15 and inequalities (A.73). □

6.3.2 Continuous Least-Squares Principles

We state the CLSPs generated by the energy balances of Theorems 6.3–6.5 under the usual simplifying assumptions:

- the boundary conditions are homogeneous
- the solution space X is constrained by the homogeneous boundary conditions[16]
- the weights Θ_i are non-degenerate, i.e., they satisfy (6.3), and $\gamma = 1$.

Let us begin with the original div–curl systems (6.8) and (6.9). First, we have CLSPs implied by the energy balances[17] from Theorem 6.3. If $\partial\Omega$ is sufficiently smooth, this theorem asserts that (6.8) and (6.9) are homogeneous elliptic. According to Section 4.4, the preferred regularity index to develop LSFEMs for such systems is $q = 0$. With this choice, the CLSP associated with (6.8) is[18]

$$\begin{cases} J_0(\mathbf{u};\mathbf{g}) = \|\nabla\times\mathbf{u}-\mathbf{g}\|_0^2 + \|\nabla\cdot\Theta_1\mathbf{u}\|_0^2 \\ X_0 = \mathbf{H}_t^1(\Omega) = \{\mathbf{u}\in[H^1(\Omega)]^3 \mid \mathbf{n}\times\mathbf{u} = 0 \ \text{ on } \Gamma\}. \end{cases} \tag{6.39}$$

Again choosing $q = 0$, the CLSP associated with (6.9) is

[16] Recall that the formulation and analysis of LSFEMs in this chapter is restricted to the case $\Gamma = \partial\Omega$.

[17] Recall that Proposition D.10 allows us to ignore the slack variable ϕ from the CLSPs.

[18] For brevity, again we do not write out the first-order necessary conditions corresponding to continuous or discrete least-squares principles, except when doing so is needed for clarity.

$$\begin{cases} J_0(\mathbf{u};\mathbf{h}) = \|\nabla \times \Theta_2\mathbf{u} - \mathbf{h}\|_0^2 + \|\nabla \cdot \mathbf{u}\|_0^2 \\ X_0 = \mathbf{H}^1_n(\Omega) = \{\mathbf{u} \in [H^1(\Omega)]^3 \mid \mathbf{n}\cdot\mathbf{u} = 0 \text{ on } \Gamma\}. \end{cases} \tag{6.40}$$

Next, we have CLSPs implied by the energy balances from Theorem 6.4. The CLSP for the div–curl system (6.8) with tangential boundary condition follows from (6.34):

$$\begin{cases} J(\mathbf{u};\mathbf{g}) = \|\nabla \times \mathbf{u} - \mathbf{g}\|_{0,\Theta_2}^2 + \|\nabla \cdot \Theta_1\mathbf{u}\|_{0,\Theta_0^{-1}}^2 \\ X = \mathbf{C}_0(\Omega,\Theta_1) \cap \mathbf{D}(\Omega,\Theta_1^{-1}). \end{cases} \tag{6.41}$$

The CLSP for (6.9) follows from (6.35):

$$\begin{cases} J(\mathbf{u};\mathbf{h}) = \|\nabla \times \Theta_2\mathbf{u} - \mathbf{h}\|_{0,\Theta_1^{-1}}^2 + \|\nabla \cdot \mathbf{u}\|_{0,\Theta_3}^2 \\ X = \mathbf{C}(\Omega,\Theta_2^{-1}) \cap \mathbf{D}_0(\Omega,\Theta_2). \end{cases} \tag{6.42}$$

Energy balances for the reformulated div–curl systems were stated in Theorem 6.5. The CLSP for (6.19) follows from (6.36):

$$\begin{cases} J(\mathbf{u},\mathbf{v};\mathbf{g}) = \|\nabla \times \mathbf{u} - \mathbf{g}\|_{0,\Theta_2}^2 + \|\nabla \cdot \mathbf{v}\|_{0,\Theta_0^{-1}}^2 + \|\mathbf{v} - \Theta_1\mathbf{u}\|_{0,\Theta_1^{-1}}^2 \\ X = \mathbf{C}_0(\Omega,\Theta_1) \times \mathbf{D}(\Omega,\Theta_1^{-1}). \end{cases} \tag{6.43}$$

On the other hand, (6.37) implies the following CLSP for (6.20):

$$\begin{cases} J(\mathbf{u},\mathbf{v};\mathbf{h}) = \|\nabla \times \mathbf{v} - \mathbf{h}\|_{0,\Theta_1^{-1}}^2 + \|\nabla \cdot \mathbf{u}\|_{0,\Theta_3}^2 + \|\mathbf{v} - \Theta_2\mathbf{u}\|_{0,\Theta_2^{-1}}^2 \\ X = \mathbf{C}(\Omega,\Theta_2^{-1}) \times \mathbf{D}_0(\Omega,\Theta_2). \end{cases} \tag{6.44}$$

A few comments are now in order. First, assuming that the conditions of Theorem A.8 hold for Ω, it follows that, up to a weight factor, the CLSPs (6.39) and (6.40) based on the ADN setting are equivalent to the CLSPs (6.41) and (6.42) based on the vector-operator setting. Of course, the comment made in Remark 5.22 that this equivalence *cannot* be extended to general domains remains in full force.

A second observation is that the reformulation of div–curl systems changes the solution spaces from $\mathbf{C}_0(\Omega)\cap\mathbf{D}(\Omega)$ and $\mathbf{C}(\Omega)\cap\mathbf{D}_0(\Omega)$ for (6.8) and (6.9), respectively, to $\mathbf{C}_0(\Omega)\times\mathbf{D}(\Omega)$ and $\mathbf{C}(\Omega)\times\mathbf{D}_0(\Omega)$ for (6.19) and (6.20), respectively. According to Section B.2.2, approximation of the former by finite elements is problematic[19] and should be avoided as a setting[20] for LSFEMs. In contrast, $\mathbf{C}_0(\Omega)\times\mathbf{D}(\Omega)$ and $\mathbf{C}(\Omega)\times\mathbf{D}_0(\Omega)$ can be easily approximated by tensor products of the compatible

[19] Recall that, from Proposition B.14, conforming finite element approximations of the intersection spaces are always proper subspaces of $[H^1(\Omega)]^d$. On the other hand, Theorem A.9 tells us that, for some domains, $\mathbf{C}_0(\Omega)\cap[H^1(\Omega)]^d$ and $\mathbf{D}_0(\Omega)\cap[H^1(\Omega)]^d$ are closed subspaces of $\mathbf{C}_0(\Omega)\cap\mathbf{D}(\Omega)$ and $\mathbf{C}(\Omega)\cap\mathbf{D}_0(\Omega)$, respectively. As a result, compliant DLSPs for div–curl systems can only be guaranteed to work when Ω satisfies all assumptions of Theorem A.8.

[20] The numerical results presented in Section 5.9.3 lend further credibility to this conclusion.

finite element spaces described in Section B.2.1. Therefore, reformulated div–curl systems provide a better setting for the development of compliant LSFEMs.

Remark 6.6 In Remark 5.21, it is revealed that for first-order systems with a "constitutive law," identical CLSPs can be obtained by first transforming these systems into equivalent PDE-constrained optimization problems and then applying methods from Chapter 11. This observation also extends to the present context: both (6.43) and (6.44) are related to first-order systems with "constitutive laws" and can be derived in this manner. For brevity, we illustrate this idea only for (6.19). The equivalent optimal control formulation of this system is

$$\begin{cases} \min\limits_{\{\mathbf{u},\mathbf{v}\}\in\mathbf{C}_0(\Omega)\times\mathbf{D}(\Omega)} \|\mathbf{v}-\Theta_1\mathbf{u}\|^2_{0,\Theta_1^{-1}} \\ \text{subject to} \qquad \nabla\times\mathbf{u}=\mathbf{g} \qquad \text{and} \qquad \nabla\cdot\mathbf{v}=0\,. \end{cases} \tag{6.45}$$

This problem has the canonical form (11.6) of the abstract control formulation studied in Section 11.1. To solve it, we can use any of the methods in Chapter 11. The most straightforward is the one based on direct penalization of the objective functional by a least-squares formulation of the constraint equations; see Section 11.3. Specialized to (6.45), the method (11.49) gives exactly the same CLSP as in (6.43). This approach has been used in [150] to develop LSFEMs for the magnetostatics equations (6.21). However, as long as one recognizes that least-squares principles allow one to treat the residual of the "constitutive" equation in (6.19) in exactly the same way as the rest of the residuals, it is clear that the optimal control reformulation (6.45) is not necessary to set up a well-posed CLSP.[21] □

6.3.3 Discrete Least-Squares Principles

The final step in the formulation of LSFEMs for div–curl systems is the transformation of CLSPs into practical, well-posed discrete minimization problems. All of the CLSPs in Section 6.3.2 satisfy the two keys to practicality stated in Section 2.2.2 because they involve only $L^2(\Omega)$ residuals of at most first-order differential equations. As a result, the main challenge faced in this section deals with CLSPs having a solution space of the type $\mathbf{C}(\Omega)\cap\mathbf{D}(\Omega)$. The numerical results given in Section 5.9.3 provide clear evidence that DLSPs that are conforming with respect to this space are not necessarily the best discretization choice for such CLSPs.

---

[21] The optimal control reformulation, in conjunction with the Lagrange multiplier method to enforce the constraints, has been studied in [89] and analyzed in [5] for magnetostatics problems. Apparently, the first-ever use of this idea is found in [306], where it was referred to as the "error-based" formulation. The rationale for the error-based approach is that constitutive laws in the Maxwell's equations are obtained from experimental measurements and are less reliable than the rest of the equations. Therefore, they can be enforced in a weaker sense than the differential equations. Note that application of least-squares principles to the reformulated div–curl systems (6.19) and (6.20) achieves the same feat.

Conforming discrete least-squares principles

Of course, when CLSPs for the div–curl systems are based on the ADN setting, conforming DLSPs are perfectly admissible because the constraints placed on Ω in the ADN theory ensure that the intersection space $\mathbf{C}(\Omega)\cap\mathbf{D}(\Omega)$ is equivalent to the Sobolev space $[H^1(\Omega)]^d$. In this case, we have two such CLSPs: (6.39) and (6.40). Conforming discretization of (6.39) results in the compliant DLSP

$$\begin{cases} J_0(\mathbf{u}^h;\mathbf{g}) = \|\nabla\times\mathbf{u}^h-\mathbf{g}\|_0^2+\|\nabla\cdot\Theta_1\mathbf{u}^h\|_0^2 \\ X_0^h = [G^h(\Omega)]^d\cap\mathbf{H}_t^1(\Omega)\,. \end{cases} \tag{6.46}$$

Conforming discretization of (6.40) gives rise to a similar compliant DLSP:

$$\begin{cases} J_0(\mathbf{u}^h;\mathbf{h}) = \|\nabla\times\Theta_2\mathbf{u}^h-\mathbf{h}\|_0^2+\|\nabla\cdot\mathbf{u}^h\|_0^2 \\ X_0^h = [G^h(\Omega)]^d\cap\mathbf{H}_n^1(\Omega)\,. \end{cases} \tag{6.47}$$

LSFEMs of this kind were first considered in [165] for rectangular domains and in [115,116] for more general settings. The analysis in [165] uses the special relation (6.33), whereas the other two papers rely on the ADN theory.

The second setting for which conforming DLSPs are acceptable is provided by CLSPs based on the reformulated div–curl systems (6.19) and (6.20) because they use direct product spaces of the type $\mathbf{C}(\Omega)\times\mathbf{D}(\Omega)$ instead of the intersection space $\mathbf{C}(\Omega)\cap\mathbf{D}(\Omega)$. We have two such CLSPs: (6.43) and (6.44), respectively. Conforming discretization of the first one results in the compliant DLSP

$$\begin{cases} J(\mathbf{u}^h,\mathbf{v}^h;\mathbf{g}) = \\ \qquad \|\nabla\times\mathbf{u}^h-\mathbf{g}\|_{0,\Theta_2}^2+\|\nabla\cdot\mathbf{v}^h\|_{0,\Theta_0^{-1}}^2+\|\mathbf{v}^h-\Theta_1\mathbf{u}^h\|_{0,\Theta_1^{-1}}^2 \\ X^h = \mathbf{C}_0^h(\Omega)\times\mathbf{D}^h(\Omega)\,. \end{cases} \tag{6.48}$$

The compliant DLSP obtained from (6.44) is very similar:[22]

$$\begin{cases} J(\mathbf{u}^h,\mathbf{v}^h;\mathbf{h}) = \\ \qquad \|\nabla\times\mathbf{v}^h-\mathbf{h}\|_{0,\Theta_1^{-1}}^2+\|\nabla\cdot\mathbf{u}^h\|_{0,\Theta_3}^2+\|\mathbf{v}^h-\Theta_2\mathbf{u}^h\|_{0,\Theta_2^{-1}}^2 \\ X^h = \mathbf{C}^h(\Omega)\times\mathbf{D}_0^h(\Omega)\,. \end{cases} \tag{6.49}$$

Remark 6.7 The solution spaces in (6.48) and (6.49) are direct products of two compatible finite element spaces for the dependent variables. The two components can be chosen completely independently of each other. Recall that this is also the case for the compliant DLSPs (5.62) and (5.64). □

[22] A version of (6.49) for the magnetostatics system (6.21), derived from an optimal control form of this system, was studied in [150]; see Remark 6.6.

Non-conforming discrete least-squares principles

The last two members of our CLSP pool for div–curl systems are (6.41) and (6.42) that have solution spaces given by $\mathbf{C}_0(\Omega)\cap\mathbf{D}(\Omega)$ and $\mathbf{C}(\Omega)\cap\mathbf{D}_0(\Omega)$, respectively. Consequently, conforming DLSPs are not a suitable option[23] for their discretization. The approach that we instead take is to abandon[24] conforming DLSPs and *conformally approximate only one of the component spaces* in $\mathbf{C}(\Omega)\cap\mathbf{D}(\Omega)$ so that (6.41) and (6.42) result in non-conforming DLSPs. This calls for transition diagrams (3.54) in which either the curl or divergence is replaced by suitable discrete approximations.

First consider (6.41). For this CLSP, it is more profitable to approximate the solution space $\mathbf{C}_0(\Omega,\Theta_1)\cap\mathbf{D}(\Omega,\Theta_1^{-1})$ by a curl-conforming finite element space $\mathbf{C}_0^h(\Omega)$ and give up on being divergence conforming. This allows us to impose exactly the tangential boundary condition in (6.8) on candidate minimizers. To complete the transition from (6.41) to a non-conforming DLSP, we replace the analytic operator $\nabla\cdot:\mathbf{D}(\Omega,\Theta_1^{-1})\mapsto S(\Omega,\Theta_0^{-1})$ by the discrete weak divergence $\nabla_h^*\cdot:\mathbf{C}_0^h(\Omega,\Theta_1)\mapsto G_0^h(\Omega,\Theta_0)$ defined in (B.51), resulting in the non-conforming (and non-compliant) DLSP

$$\begin{cases} J^h(\mathbf{u}^h;\mathbf{g}) = \|\nabla\times\mathbf{u}^h-\mathbf{g}\|_{0,\Theta_2}^2+\|\nabla_h^*\cdot\mathbf{u}^h\|_{0,\Theta_0}^2 \\ X^h=\mathbf{C}_0^h(\Omega,\Theta_1)\,. \end{cases} \tag{6.50}$$

As far as (6.42) is concerned, it is more appropriate to approximate the solution space $\mathbf{C}(\Omega,\Theta_2^{-1})\cap\mathbf{D}_0(\Omega,\Theta_2)$ by div-conforming elements because they allow us to impose exactly the normal boundary condition from (6.9) on the candidate minimizers. To complete the transition to a non-conforming DLSP, we now replace the analytic curl $\nabla\times:\mathbf{C}(\Omega,\Theta_2^{-1})\mapsto\mathbf{D}(\Omega,\Theta_1^{-1})$ by the discrete weak operator $\nabla_h^*\times:\mathbf{D}_0^h(\Omega,\Theta_2)\mapsto\mathbf{C}_0^h(\Omega,\Theta_1)$ defined in (B.50), resulting in the non-conforming (and non-compliant) DLSP

$$\begin{cases} J^h(\mathbf{u}^h;\mathbf{h}) = \|\nabla_h^*\times\mathbf{u}^h-\mathbf{h}\|_{0,\Theta_1}^2+\|\nabla\cdot\mathbf{u}^h\|_{0,\Theta_3}^2 \\ X^h=\mathbf{D}_0^h(\Omega,\Theta_2)\,. \end{cases} \tag{6.51}$$

Note that, in addition to the primary curl- or div-conforming spaces that define X^h, the DLSPs (6.50) and (6.51) require a choice for an *auxiliary* finite element space for the range of the weak discrete operator acting on the corresponding X^h. Recall that (see Remark B.17) the domains and ranges of weak discrete operators

[23] Recall that the reason for this is that such DLSPs simply default to (6.46) and (6.47) whose parent CLSPs were derived in the ADN setting and whose scope cannot be extended to the vector-operator setting of (6.41) and (6.42).

[24] We again contrast this situation with the one encountered in Section 5.6.2. There, we could always abandon the extended div–grad system because (5.25) is not our primary model. This is not an option for (6.8) and (6.9) because they are the primary equations we want to solve.

must be "matched" in the sense that they should belong to the same finite element De Rham complex.

To illustrate this important point, suppose that we want to implement (6.50) on a simplicial grid, using the Nédélec edge elements of the first kind (see (B.21)), i.e., $X^h = \mathbf{C}_0^{r-}(\Omega)$, $r \geq 1$. To complete the definition of the LSFEM, it is necessary to choose a range $G_0^h(\Omega)$ for $\nabla_h^* \cdot$ which is "matched" to X^h. Upon examination of the four possible discrete complexes in (B.45), we see that $\mathbf{C}_0^{r-}(\Omega)$ is included in two of them, both of which start with $G_0^r(\Omega)$. This nodal space is the correct range of the weak divergence for the given choice of X^h.

For a concrete example, if X^h is the lowest-order Nédélec element of the first kind (see (B.22)), the range of $\nabla_h^* \cdot$ that is matched to this choice is the lowest-order nodal space $G_0^1(\Omega)$, i.e., piecewise linear elements on simplices.[25]

Recall once more that the classification of DLSPs into compliant, norm-equivalent, or quasi-norm-equivalent (see Section 3.5) is not applicable to (6.50) and (6.51) because these methods are non-conforming discretizations of their parent CLSPs. In particular, their stability and error estimates cannot be inferred from the abstract results of Section 3.5. Nevertheless, in Section 6.3.5, we see that they retain all the attractive computational properties ordinarily associated with norm-equivalent methods, including stability and optimal error estimates.

6.3.4 Analysis of Conforming Least-Squares Finite Element Methods

We have defined a total of four conforming DLSPs for the div–curl systems, all of which are compliant. Therefore, error analyses are simply a matter of specializing the appropriate abstract result by taking into consideration whether the origins of their parent CLSPs are in the ADN or vector-operator setting.

First we have the DLSPs (6.46) and (6.47) whose parent CLSPs originate from the ADN setting. Assuming that all hypotheses of Theorem 6.3 are satisfied, the div–curl system is homogeneous elliptic so that error estimates for these LSFEMs follow from Theorem 4.2 in Section 4.4.

Theorem 6.8 *Let $\Omega \subset \mathbb{R}^3$ denote a bounded domain that satisfies all the hypotheses of Theorem* 6.3*, $\mathcal{T}_h$ denote a uniformly regular partition of Ω into finite elements, and $\mathbf{u}^h$ denote a minimizer of* (6.46) *or* (6.47)*. Assume that* (6.46) *and* (6.47) *are defined using the nodal space $G^r(\Omega)$ for some integer $r \geq 1$. Assume that the solution of* (6.8) *satisfies $\mathbf{u} \in \mathbf{C}_0(\Omega) \cap [H^{q+1}(\Omega)]^d$ and the solution of* (6.9) *satisfies $\mathbf{u} \in \mathbf{D}_0(\Omega) \cap [H^{q+1}(\Omega)]^d$ for some integer $q \geq 0$. Then, there exists a positive constant C such that*

[25] Both (6.50) and (6.51) can be easily extended to mixed boundary conditions. In this case, one of the boundary conditions is imposed strongly on all candidate minimizers by constraining the compatible finite element space, whereas the other boundary condition is imposed weakly via the definition of the discrete differential operator; see Section B.3.1.

$$\|\mathbf{u}-\mathbf{u}^h\|_1 \le Ch^{\widetilde{r}}\|\mathbf{u}\|_{\widetilde{r}+1}\,, \tag{6.52}$$

and, if Ω is such that the div–curl system has full elliptic regularity,

$$\|\mathbf{u}-\mathbf{u}^h\|_0 \le Ch^{\widetilde{r}+1}\|\mathbf{u}\|_{\widetilde{r}+1}\,, \tag{6.53}$$

where $\widetilde{r} = \min\{r,q\}$. □

The second group of conforming DLSPs consists of the compliant DLSPs (6.48) and (6.49) for the reformulated div–curl systems (6.19) and (6.20). Their parent CLSPs originate in the vector-operator setting so that error estimates follow from the more general abstract result in Theorem 3.28.

However, a few comments are in order before stating the main results. First, recall that, in Remarks 5.32, B.8, and B.10, we point out that the approximation theory for standard curl- and div-conforming elements given in Theorems B.7 and B.9 does not extend to general non-affine finite element partitions. This fact is also confirmed by the numerical experiments in Section 5.9.2. For this reason, the error analysis of (6.48) and (6.49) is restricted to affine simplicial grids. Such grids also provide the most favorable setting for finite element approximations of the De Rham complex. A second comment is that, according to Remark 6.7, the components of the solution spaces in (6.48) and (6.49) can be chosen independently of each other. However, to optimize the performance of finite element methods, approximation orders of these spaces should be matched.

Theorem 6.9 *Let $\Omega \subset \mathbb{R}^3$ denote a bounded contractible domain with a Lipschitz continuous boundary, $\mathcal{T}_h$ denote a regular partition of Ω into affine simplicial elements, $\{\mathbf{u}_C^h,\mathbf{v}_D^h\} \in \mathbf{C}_0^h(\Omega)\times\mathbf{D}^h(\Omega)$ denote the minimizer of* (6.48), *and $\{\mathbf{u}_D^h,\mathbf{v}_C^h\} \in \mathbf{D}_0^h(\Omega)\times\mathbf{C}^h(\Omega)$ denote the minimizer of* (6.49). *Then,*

$$\|\mathbf{u}_C-\mathbf{u}_C^h\|_C + \|\mathbf{v}_D-\mathbf{v}_D^h\|_D \le \inf_{\widehat{\mathbf{u}}^h\in\mathbf{C}_0^h}\|\mathbf{u}_C-\widehat{\mathbf{u}}^h\|_C + \inf_{\widehat{\mathbf{v}}^h\in\mathbf{D}^h}\|\mathbf{v}_D-\widehat{\mathbf{v}}^h\|_D \tag{6.54}$$

$$\|\mathbf{u}_D-\mathbf{u}_D^h\|_D + \|\mathbf{v}_C-\mathbf{v}_C^h\|_C \le \inf_{\widehat{\mathbf{u}}^h\in\mathbf{D}_0^h}\|\mathbf{u}_D-\widehat{\mathbf{u}}^h\|_D + \inf_{\widehat{\mathbf{v}}^h\in\mathbf{C}^h}\|\mathbf{v}_C-\widehat{\mathbf{v}}^h\|_C\,, \tag{6.55}$$

where $\{\mathbf{u}_C,\mathbf{v}_D\} \in \mathbf{C}_0(\Omega)\times\mathbf{D}(\Omega)$ denotes the solution of (6.19) *and $\{\mathbf{u}_D,\mathbf{v}_C\} \in \mathbf{D}_0(\Omega)\times\mathbf{C}(\Omega)$ is solution of* (6.20).

Proof. The results follow from (3.63) in Theorem 3.28. □

For sufficiently regular solutions of the reformulated div–curl systems, we then have the following result.

Corollary 6.10 *Assume the hypotheses of Theorem* 6.9. *Assume that $\mathcal{T}_h$ is uniformly regular, $X^h = \mathbf{C}_0^{(r)}\times\mathbf{D}^{(p)}$ in* (6.19), *and $X^h = \mathbf{D}_0^{(r)}\times\mathbf{C}^{(p)}$ in* (6.20), *where $r\ge 1$ and $p\ge 1$ are integers.*[26] *Let $s=\max\{r,p\}$. Assume that the solutions of* (6.19) *and* (6.20) *belong to $[H^s(\Omega)]^d\times[H^s(\Omega)]^d$. Then, there exists a constant C such that*

[26] Recall that $\mathbf{C}^{(r)}(\Omega)$ and $\mathbf{D}^{(p)}(\Omega)$ stand for curl- and div-conforming spaces of the 1st or the 2nd kind, respectively; see Theorems B.7 and B.9.

$$\|\mathbf{u}_C - \mathbf{u}_C^h\|_C + \|\mathbf{v}_D - \mathbf{v}_D^h\|_D \leq C\big(h^r\|\nabla\times\mathbf{u}_C\|_r + h^p\|\nabla\cdot\mathbf{v}_D\|_p\big) \tag{6.56}$$

$$\|\mathbf{u}_D - \mathbf{u}_D^h\|_D + \|\mathbf{v}_C - \mathbf{v}_C^h\|_C \leq C\big(h^r\|\nabla\cdot\mathbf{u}_D\|_r + h^p\|\nabla\times\mathbf{v}_C\|_p\big). \tag{6.57}$$

Proof. The results follow from the interpolation error estimates in Theorems B.7 and B.9 and the error bounds in Theorem 6.9. □

Corollary 6.10 implies that, in order to optimize the performance of (6.48) and (6.49), the solution spaces should be defined using curl- and div-conforming spaces with the same index $p = r \geq 1$, i.e., $\mathbf{C}^{(r)}(\Omega)\times\mathbf{D}^{(r)}(\Omega)$. However, because curl- and div-conforming elements of the first kind are designed to yield improved accuracy with respect to the curl and divergence, the combinations $X^h = \mathbf{C}_0^{r}\text{-}(\Omega)\times\mathbf{D}^{r}\text{-}(\Omega)$ and $X^h = \mathbf{D}_0^{r}\text{-}(\Omega)\times\mathbf{C}^{r}\text{-}(\Omega)$ are perhaps the best choices for (6.48) and (6.49), respectively.

Most of the results in this section can be specialized to div–curl systems in two dimensions with minimal modifications. For the sake of brevity, we do not repeat their statements.

6.3.5 Analysis of Non-Conforming Least-Squares Finite Element Methods

Unlike the LSFEMs from the last section, (6.50) and (6.51) are non-conforming. As a result, their well-posedness cannot be inferred directly from the well-posedness of their parent CLSPs, nor can we use Theorem 3.28 to obtain error estimates. The stability of the parent CLSPs for both (6.50) and (6.51) are with respect to the norm (A.61). However, $\|\cdot\|_{DC}$ is not meaningful for finite element methods that are only curl- or div-conforming. Thus, the first step is to find discrete analogues of this norm that are suitable for (6.50) and (6.51). Using the same weak operators as appear in (6.50) and (6.51) gives rise to two natural discrete replacements for (A.61):

$$\|\mathbf{u}^h\|_{D^*C}^2 = \|\mathbf{u}^h\|_{0,\Theta_1}^2 + \|\nabla\times\mathbf{u}^h\|_{0,\Theta_2}^2 + \|\nabla_h^*\cdot\mathbf{u}^h\|_{0,\Theta_0}^2 \tag{6.58}$$

$$\|\mathbf{u}^h\|_{DC^*}^2 = \|\mathbf{u}^h\|_{0,\Theta_2}^2 + \|\nabla_h^*\times\mathbf{u}^h\|_{0,\Theta_1}^2 + \|\nabla\cdot\mathbf{u}^h\|_{0,\Theta_3}^2\,. \tag{6.59}$$

Recall that the key ingredient in the proof of Theorem 6.4 that established the energy balance for the parent CLSPs is the Poincaré-Friedrichs inequality (A.69). Owing to the use of compatible finite element spaces, we have two discrete versions (B.83) and (B.84) of this inequality that play the same roles in the proof of the stability of (6.50) and (6.51), respectively.

Theorem 6.11 *There exists a positive constant C, independent of h, such that*

$$C\|\mathbf{u}^h\|_{D^*C} \leq \|\nabla\times\mathbf{u}^h\|_{0,\Theta_2}^2 + \|\nabla_h^*\cdot\mathbf{u}^h\|_{0,\Theta_0}^2 \qquad \forall\mathbf{u}^h\in\mathbf{C}_0^h(\Omega) \tag{6.60}$$

and

$$C\|\mathbf{u}^h\|^2_{DC^*} \le \|\nabla_h^* \times \mathbf{u}^h\|^2_{0,\Theta_1} + \|\nabla \cdot \mathbf{u}^h\|^2_{0,\Theta_3} \qquad \forall \mathbf{u}^h \in \mathbf{D}_0^h(\Omega)\,. \tag{6.61}$$

Proof. The results follow directly from the discrete Poincaré inequalities (B.83) and (B.84). □

The bounds in (6.60) and (6.61) provide the proper interpretation of the analytic energy balances (6.34) and (6.35), respectively, for curl- and div-conforming finite elements. Therefore, the stability of (6.50) and (6.51) follows from Theorem 6.11 in the same way as the stability of the corresponding parent CLSPs (6.41) and (6.42) follows from Theorem 6.4.

We proceed with convergence analysis of (6.50) and (6.51). As in the last section, in order to avoid complications with standard div- and curl-conforming elements on general grids, error estimates are stated only for affine simplicial grids. The main results can be extended to other affine spaces as well.

Theorem 6.12 *Let $\Omega \subset \mathbb{R}^3$ denote a bounded, contractible domain having a Lipschitz continuous boundary and $\mathcal{T}_h$ denote a regular partition of Ω into affine simplicial elements. Assume that* (6.50) *is defined using a pair $\{G_0^h(\Omega), X^h = \mathbf{C}_0^h(\Omega)\}$ from the same finite element De Rham complex and let $\mathbf{u}^h \in \mathbf{C}_0^h(\Omega)$ denote the minimizer of* (6.50). *Then,*

$$\begin{cases} \|\nabla \times (\mathbf{u}-\mathbf{u}^h)\|_0 \le \inf\limits_{\mathbf{v}^h \in \mathbf{C}_0^h(\Omega)} \|\nabla \times (\mathbf{u}-\mathbf{v}^h)\|_0 \\ \|\nabla \cdot \mathbf{u} - \nabla_h^* \cdot \mathbf{u}^h\|_0 \le \|\nabla \cdot \mathbf{u} - \pi_G(\nabla \cdot \mathbf{u})\|_0\,, \end{cases} \tag{6.62}$$

where $\mathbf{u} \in \mathbf{C}_0(\Omega) \cap \mathbf{D}(\Omega)$ is a solution of (6.8) *and π_G denotes the L^2 projection to the auxiliary nodal space used to define* (6.50), *i.e., see the first equation in* (B.16).

Proof. For convenience, in the proof all weights are set to one. As usual, minimizers of (6.50) are subject to a first-order necessary optimality condition which here takes the form: *seek $\mathbf{u}^h \in \mathbf{C}_0^h(\Omega)$ such that*

$$\left(\nabla \times \mathbf{u}^h, \nabla \times \mathbf{v}^h\right) + \left(\nabla_h^* \cdot \mathbf{u}^h, \nabla_h^* \cdot \mathbf{v}^h\right) = \left(\mathbf{g}, \nabla \times \mathbf{v}^h\right) \qquad \forall \mathbf{v}^h \in \mathbf{C}_0^h(\Omega)\,. \tag{6.63}$$

According to Theorem B.22 (the discrete Hodge decomposition for $\mathbf{C}_0^h(\Omega)$) and Remark B.24, $\mathbf{v}^h = \mathbf{v}_N + \mathbf{v}_{N\perp}$, where

$$\nabla \times \mathbf{v}_N = 0 \qquad \text{and} \qquad \nabla_h^* \cdot \mathbf{v}_{N\perp} = 0\,.$$

This allows us to write (6.63) as

$$\begin{cases} \left(\nabla \times \mathbf{u}^h, \nabla \times \mathbf{v}_{N\perp}\right) = (\mathbf{g}, \nabla \times \mathbf{v}_{N\perp}) \\ \left(\nabla_h^* \cdot \mathbf{u}^h, \nabla_h^* \cdot \mathbf{v}_N\right) = 0 \end{cases} \qquad \forall \mathbf{v}^h \in \mathbf{C}_0^h(\Omega)\,. \tag{6.64}$$

Noting that

$$\nabla \times \mathbf{v}^h = \nabla \times \mathbf{v}_{N\perp} \qquad \text{and} \qquad \nabla_h^* \cdot \mathbf{v}^h = \nabla_h^* \cdot \mathbf{v}_N\,,$$

it follows that (6.64) splits into two independent equations:

$$\begin{cases} \left(\nabla \times \mathbf{u}^h, \nabla \times \mathbf{v}^h\right) = \left(\mathbf{g}, \nabla \times \mathbf{v}^h\right) \\ \left(\nabla_h^* \cdot \mathbf{u}^h, \nabla_h^* \cdot \mathbf{v}^h\right) = 0 \end{cases} \quad \forall \mathbf{v}^h \in \mathbf{C}_0^h(\Omega). \tag{6.65}$$

Because $\nabla \times \mathbf{u} = \mathbf{g}$ and $\nabla \cdot \mathbf{u} = 0$ for a solution of (6.8), (6.65) provides two error orthogonality equations:

$$\begin{cases} \left(\nabla \times \mathbf{u}^h - \nabla \times \mathbf{u}, \nabla \times \mathbf{v}^h\right) = 0 \\ \left(\nabla_h^* \cdot \mathbf{u}^h - \nabla \cdot \mathbf{u}, \nabla_h^* \cdot \mathbf{v}^h\right) = 0 \end{cases} \quad \forall \mathbf{v}^h \in \mathbf{C}_0^h(\Omega). \tag{6.66}$$

Using the first orthogonality equation in (6.66),

$$\begin{aligned} \|\nabla \times \mathbf{u}^h - \nabla \times \mathbf{u}\|_0^2 &= \left(\nabla \times \mathbf{u}^h - \nabla \times \mathbf{u}, \nabla \times \mathbf{u}^h - \nabla \times \mathbf{u}\right) \\ &= \left(\nabla \times \mathbf{u}^h - \nabla \times \mathbf{u}, \nabla \times \mathbf{v}^h - \nabla \times \mathbf{u}\right) \\ &\le \|\nabla \times \mathbf{u}^h - \nabla \times \mathbf{u}\|_0 \|\nabla \times \mathbf{v}^h - \nabla \times \mathbf{u}\|_0 \end{aligned}$$

for all $\mathbf{v}^h \in \mathbf{C}_0^h(\Omega)$. This yields an upper bound for the error in the curl:

$$\|\nabla \times \mathbf{u}^h - \nabla \times \mathbf{u}\|_0 \le \inf_{\mathbf{v}^h \in \mathbf{C}_0^h(\Omega)} \|\nabla \times \mathbf{v}^h - \nabla \times \mathbf{u}\|_0 .$$

The error in the divergence can be estimated with the help of the second equation in (6.66):

$$\begin{aligned} \|\nabla_h^* \cdot \mathbf{u}^h - \nabla \cdot \mathbf{u}\|_0^2 &= \left(\nabla_h^* \cdot \mathbf{u}^h - \nabla \cdot \mathbf{u}, \nabla_h^* \cdot \mathbf{u}^h - \nabla \cdot \mathbf{u}\right) \\ &= \left(\nabla_h^* \cdot \mathbf{u}^h - \nabla \cdot \mathbf{u}, \nabla_h^* \cdot \mathbf{u} - \nabla \cdot \mathbf{u}\right) \\ &\le \|\nabla_h^* \cdot \mathbf{u}^h - \nabla \cdot \mathbf{u}\|_0 \|\nabla_h^* \cdot \mathbf{u} - \nabla \cdot \mathbf{u}\|_0 . \end{aligned}$$

Using the characterization of $\nabla_h^* \cdot \mathbf{u}$ from Theorem B.19 (see the last identity in (B.55)) gives the following upper bound:

$$\|\nabla_h^* \cdot \mathbf{u}^h - \nabla \cdot \mathbf{u}\|_0 \le \|\pi_G(\nabla \cdot \mathbf{u}) - \nabla \cdot \mathbf{u}\|_0 .$$

This proves the theorem.[27] □

[27] The divergence error bound in Theorem 6.12 can be further sharpened using the result in Proposition B.18. According to this proposition, the discrete weak divergence is a surjection $\mathbf{C}_0^h(\Omega) \mapsto G_0^h(\Omega)$ so that the second equation in (6.65) can be written as

$$\left(\nabla_h^* \cdot \mathbf{u}^h, q^h\right) = \left(\nabla \cdot \mathbf{u}, q^h\right) \qquad \forall q^h \in G_0^h(\Omega).$$

This identity and the definition of π_G imply that $\nabla_h^* \cdot \mathbf{u}^h = \pi_G(\nabla \cdot \mathbf{u})$. As a result, the last inequality in the proof of Theorem 6.12 is actually an equality: $\|\nabla_h^* \cdot \mathbf{u}^h - \nabla \cdot \mathbf{u}\|_0 = \|\pi_G(\nabla \cdot \mathbf{u}) - \nabla \cdot \mathbf{u}\|_0$.

To complete analysis of (6.50), we show that the LSFEM approximation converges, at the best possible rate, to all sufficiently smooth solutions of (6.8).

Corollary 6.13 *Assume that the hypotheses of Theorem* 6.12 *hold. Let* $\mathbf{u}$ *denote the solution of* (6.8), $\mathbf{u}^h$ *denote its LSFEM approximation computed using the DLSP* (6.50), *and* $r \geq 1$ *denote an integer. Assume that* $\mathcal{T}_h$ *is uniformly regular, the pair* $\{G_0^h(\Omega), \mathbf{C}_0^h(\Omega)\}$ *used in the definition of* (6.50) *is either* $\{G_0^{r+1}(\Omega), X^h = \mathbf{C}_0^r(\Omega)\}$ *or* $\{G_0^r(\Omega), X^h = \mathbf{C}_0^{r-}(\Omega)\}$, *and* $\mathbf{u} \in \mathbf{C}_0(\Omega) \cap [H^{r+1}(\Omega)]^d$. *Then, in both cases,*

$$\|\nabla \times (\mathbf{u} - \mathbf{u}^h)\|_0 + \|\nabla \cdot \mathbf{u} - \nabla_h^* \cdot \mathbf{u}^h\|_0 \leq Ch^r \big(\|\nabla \times \mathbf{u}\|_r + \|\nabla \cdot \mathbf{u}\|_r \big) . \tag{6.67}$$

If (6.50) *is defined using the first pair and* $\mathbf{u} \in \mathbf{C}_0(\Omega) \cap [H^{r+3}(\Omega)]^d$ *or* (6.50) *is defined using the second pair and* $\mathbf{u} \in \mathbf{C}_0(\Omega) \cap [H^{r+2}(\Omega)]^d$, *the error estimate can be further improved to*

$$\|\nabla \cdot \mathbf{u} - \nabla_h^* \cdot \mathbf{u}^h\|_0 \leq C \begin{cases} h^{r+2} \|\nabla \cdot \mathbf{u}\|_{r+2} & \text{if} \quad G_0^h(\Omega) = G_0^{r+1}(\Omega) \\ h^{r+1} \|\nabla \cdot \mathbf{u}\|_{r+1} & \text{if} \quad G_0^h(\Omega) = G_0^r(\Omega) . \end{cases} \tag{6.68}$$

Proof. According to (B.45), the pair $\{G_0^{r+1}(\Omega), \mathbf{C}_0^r(\Omega)\}$ belongs to the first two of the four possible finite element De Rham complexes and $\{G_0^r(\Omega), \mathbf{C}_0^{r-}(\Omega)\}$ belongs to the last two. Thus, the key assumption of Theorem 6.12 is satisfied and the error bound (6.62) holds for $\mathbf{u}^h$ and $\mathbf{u}$. If $\mathbf{u} \in \mathbf{C}_0(\Omega) \cap [H^{r+1}(\Omega)]^d$, Theorem B.7 implies that

$$\inf_{\mathbf{v}^h \in \mathbf{C}_0^h(\Omega)} \|\nabla \times (\mathbf{u} - \mathbf{v}^h)\|_0 \leq Ch^r \|\nabla \times \mathbf{u}\|_r$$

for both choices of the solution space X^h and (B.20) in Theorem B.6 implies that

$$\|\nabla \cdot \mathbf{u} - \pi_G(\nabla \cdot \mathbf{u})\|_0 \leq Ch^r \|\nabla \cdot \mathbf{u}\|_r$$

for both choices of the auxiliary nodal space. This proves (6.67). When $\mathbf{u}$ has the additional regularity stipulated in the statement of the corollary, the order of the curl approximation does not improve because it is already the best possible order for $\mathbf{C}_0^r(\Omega)$ and $\mathbf{C}_0^{r-}(\Omega)$. However, the order of the divergence approximation does increase according to (B.20). □

The analysis of (6.51) follows along the same lines.

Theorem 6.14 *Let* Ω *and* $\mathcal{T}_h$ *be the same as in Theorem* 6.12 *and let* $\mathbf{u}^h \in \mathbf{D}_0^h$ *denote the minimizer of* (6.51). *Assume that* (6.51) *is defined using a pair* $\{\mathbf{C}_0^h(\Omega), X^h = \mathbf{D}_0^h(\Omega)\}$ *from the same finite element De Rham complex. Then,*

$$\begin{aligned} \|\nabla \cdot (\mathbf{u} - \mathbf{u}^h)\|_0 &\leq \inf_{\mathbf{v}^h \in \mathbf{D}_0^h(\Omega)} \|\nabla \cdot (\mathbf{u} - \mathbf{v}^h)\|_0 \\ \|\nabla \times \mathbf{u} - \nabla_h^* \times \mathbf{u}^h\|_0 &\leq \|\nabla \times \mathbf{u} - \pi_C(\nabla \times \mathbf{u})\|_0 , \end{aligned} \tag{6.69}$$

where $\mathbf{u} \in \mathbf{C}(\Omega) \cap \mathbf{D}_0(\Omega)$ *denotes the solution of* (6.9).

Proof. Using the Hodge decomposition for $\mathbf{D}_0^h(\Omega)$ (Theorem B.23) allows one to write the test function as $\mathbf{v}^h = \mathbf{v}_N + \mathbf{v}_{N\perp}$, where

$$\nabla \cdot \mathbf{v}_N = 0 \qquad \text{and} \qquad \nabla_h^* \times \mathbf{v}_{N\perp} = 0 .$$

The rest of the proof repeats the same steps as in Theorem 6.12. □

Similar to the analysis of (6.50), the requirement that X^h and the range of $\nabla_h^* \times$ belong to the same finite element complex is a key assumption in Theorem 6.14. However, in contrast to the previous DLSP, for (6.51), this condition is satisfied by more combinations of solution and auxiliary spaces. As before, we have two choices of solution spaces: X^h could be either $\mathbf{D}_0^{r^-}(\Omega)$ or $\mathbf{D}_0^r(\Omega)$ (the Nédélec div-conforming spaces of the first and second kind, respectively.) The difference is that, for each one of these spaces, there are two admissible choices for the auxiliary spaces $\mathbf{C}_0^h(\Omega)$, giving a total of four different possible pairs:

$$\mathbf{D}_0^{r^-}(\Omega) \times \begin{cases} \mathbf{C}_0^{r^-}(\Omega) & \text{(I)} \\ \mathbf{C}_0^{r}(\Omega) & \text{(II)} \end{cases} \qquad \mathbf{D}_0^{r}(\Omega) \times \begin{cases} \mathbf{C}_0^{(r+1)^-}(\Omega) & \text{(III)} \\ \mathbf{C}_0^{r+1}(\Omega) & \text{(IV)}. \end{cases} \tag{6.70}$$

For sufficiently smooth solutions of (6.9), we have an analogue of Corollary 6.13.

Corollary 6.15 *Assume that the hypotheses of Theorem* (6.14) *hold. Let* $\mathbf{u}$ *denote a solution of* (6.9), $\mathbf{u}^h$ *denote its least-squares approximation computed by the DLSP* (6.51), *and* $r \geq 1$ *denote an integer. Assume that* $\mathcal{T}_h$ *is uniformly regular, the pair* $\{\mathbf{C}_0^h(\Omega), \mathbf{D}_0^h(\Omega)\}$ *used in the definition of* (6.51) *is one of the four admissible pairs in* (6.70), *and* $\mathbf{u} \in \mathbf{D}_0(\Omega) \cap [H^{r+1}(\Omega)]^d$. *Then, for all four possible realizations of* (6.51),

$$\|\nabla \cdot (\mathbf{u} - \mathbf{u}^h)\|_0 + \|\nabla \times \mathbf{u} - \nabla_h^* \times \mathbf{u}^h\|_0 \leq Ch^r \left(\|\nabla \cdot \mathbf{u}\|_r + \|\nabla \times \mathbf{u}\|_r \right). \tag{6.71}$$

If (6.51) *is defined using* (II) *or* (III) *and* $\mathbf{u} \in \mathbf{D}_0(\Omega) \cap [H^{r+2}(\Omega)]^d$ *or* (6.51) *is defined using* (IV) *and* $\mathbf{u} \in \mathbf{D}_0(\Omega) \cap [H^{r+3}(\Omega)]^d$, *the error estimate for the curl can be improved to*

$$\|\nabla \times \mathbf{u} - \nabla_h^* \times \mathbf{u}^h\|_0 \leq C \begin{cases} h^{r+1} \|\nabla \times \mathbf{u}\|_{r+1} & \text{for (II) and (III)} \\ h^{r+2} \|\nabla \times \mathbf{u}\|_{r+2} & \text{for (IV)}. \end{cases} \tag{6.72}$$

Proof. The key assumption of Theorem 6.14 is satisfied for the four pairs in (6.70) so that the error bound (6.69) holds for the LSFEM solution. Regarding the discrete solution space, Theorem B.9 implies that

$$\inf_{\mathbf{v}^h \in \mathbf{D}_0^h(\Omega)} \|\nabla \cdot (\mathbf{u} - \mathbf{v}^h)\|_0 \leq Ch^r \|\nabla \cdot \mathbf{u}\|_r$$

for all choices of X^h wheras Theorem B.7 implies that

$$\|\nabla\times\mathbf{u}-\pi_C(\nabla\times\mathbf{u})\|_0\le Ch^r\|\nabla\times\mathbf{u}\|_r$$

for all admissible auxiliary curl-conforming spaces. This establishes the first part of the corollary. For the second assertion, note that regardless of increased regularity, the order of divergence approximation cannot improve because it is already the best possible for div-conforming spaces with index $r\ge 1$; this mirrors the situation with the curl approximation in Corollary 6.13. However, if $\mathbf{u}$ has the additional regularity specified in the statement of the corollary, using (B.28) for the vector field $\nabla\times\mathbf{u}$ results in

$$\|\nabla\times\mathbf{u}-\pi_C\nabla\times\mathbf{u}^h\|_0\le C\begin{cases} h^{r+1}\|\nabla\times\mathbf{u}\|_{r+1} & \text{if}\quad \mathbf{C}_0^h=\mathbf{C}_0^r \text{ or } \mathbf{C}_0^{(r+1)-}\\ h^{r+2}\|\nabla\times\mathbf{u}\|_{r+2} & \text{if}\quad \mathbf{C}_0^h=\mathbf{C}_0^{r+1}\,. \end{cases}$$

This establishes the second part of the corollary. □

6.4 Least-Squares Finite Element Methods for Curl–Curl Systems

Removal of the redundant divergence equation from (6.4) effectively blocks the development of energy balances in the ADN setting for the first-order reformulations of (6.22). This is not a terrible loss because such balances did not lead to very satisfactory CLSPs for the div–curl systems and the situation for curl–curl systems is no different. Thus, the main focus of this section is on LSFEMs derived from energy balances for (6.23) and (6.24) in the vector-operator setting. For economy, all methods and results are stated for three space dimensions.

6.4.1 *Energy Balances*

Without loss of generality, throughout this section we set $\gamma=1$. Regarding (6.23), we have the following result that is an analogue of Theorem 5.14 for the potential–flux div–grad system (5.17).

Theorem 6.16 *Assume that $\Omega\subset\mathbb{R}^3$ is a bounded domain with a Lipschitz-continuous boundary. Then, there exists a positive constant C such that*

$$C\big(\|\mathbf{u}\|_C+\|\mathbf{v}\|_C\big)\le\|\Theta_1\mathbf{u}+\nabla\times\mathbf{v}\|_{0,\Theta_1^{-1}}+\|\Theta_2^{-1}\mathbf{v}-\nabla\times\mathbf{u}\|_{0,\Theta_2} \qquad (6.73)$$

for all $\{\mathbf{u},\mathbf{v}\}\in\mathbf{C}_0(\Omega,\Theta_1)\times\mathbf{C}(\Omega,\Theta_2^{-1})$.

Proof. The proof of (6.73) relies on a cancellation property that is analogous to the the one used in the proof of Theorem 5.14. Expanding the right-hand side of (6.73) results in

$$\begin{aligned}
&\|\Theta_1\mathbf{u}+\nabla\times\mathbf{v}\|^2_{0,\Theta_1^{-1}}+\|\Theta_2^{-1}\mathbf{v}-\nabla\times\mathbf{u}\|^2_{0,\Theta_2}\\
&\qquad=\|\mathbf{u}\|^2_{0,\Theta_1}+\|\nabla\times\mathbf{u}\|^2_{0,\Theta_2}+\|\mathbf{v}\|^2_{0,\Theta_2^{-1}}+\|\nabla\times\mathbf{v}\|^2_{0,\Theta_1^{-1}}\\
&\qquad\quad+2\Big(\big(\nabla\times\mathbf{v},\Theta_1\mathbf{u}\big)_{0,\Theta_1^{-1}}-\big(\nabla\times\mathbf{u},\Theta_2^{-1}\mathbf{v}\big)_{0,\Theta_2}\Big).
\end{aligned}$$

Using the definition (A.20) of weighted $L^2(\Omega)$ inner products, we have that

$$\big(\nabla\times\mathbf{v},\Theta_1\mathbf{u}\big)_{0,\Theta_1^{-1}}=\int_\Omega(\nabla\times\mathbf{v})\Theta_1^{-1}(\Theta_1\mathbf{u})dx=\big(\nabla\times\mathbf{v},\mathbf{u}\big)$$

and

$$\big(\nabla\times\mathbf{u},\Theta_2^{-1}\mathbf{v}\big)_{0,\Theta_2}=\int_\Omega(\nabla\times\mathbf{u})\Theta_2(\Theta_2^{-1}\mathbf{v})dx=\big(\nabla\times\mathbf{u},\mathbf{v}\big).$$

Because $\mathbf{n}\times\mathbf{u}=0$ on $\partial\Omega$, it follows that

$$(\nabla\times\mathbf{v},\mathbf{u})-(\nabla\times\mathbf{u},\mathbf{v})=0.$$

Therefore,

$$\|\Theta_1\mathbf{u}+\nabla\times\mathbf{v}\|^2_{0,\Theta_1^{-1}}+\|\Theta_2^{-1}\mathbf{v}-\nabla\times\mathbf{u}\|^2_{0,\Theta_2}=\|\mathbf{u}\|^2_C+\|\mathbf{v}\|^2_C. \tag{6.74}$$

This proves the theorem. □

A key element in the proof of this theorem (and the similar Theorem 5.14) is the proper choice of weights in the function spaces. Without these weights, the cancellation property does not hold and although it may still be possible to prove an energy balance, this task is considerably more complicated.

The energy balance for the four-field first-order system (6.26) that included constitutive laws also relies on proper weight selection for the function spaces.

Theorem 6.17 *Assume the hypotheses of Theorem* 6.16 *hold. Then, there exists a positive constant C such that*

$$\begin{aligned}
&C\big(\|\mathbf{u}\|_C+\|\mathbf{w}\|_D+\|\mathbf{v}\|_C+\|\mathbf{y}\|_D\big)\\
&\quad\le\|\mathbf{w}+\nabla\times\mathbf{v}\|_{0,\Theta_1^{-1}}+\|\mathbf{y}-\nabla\times\mathbf{u}\|_{0,\Theta_2}+\|\nabla\cdot\mathbf{y}\|_{0,\Theta_3}+\|\nabla\cdot\mathbf{w}\|_{0,\Theta_0^{-1}}\\
&\qquad+\|\mathbf{w}-\Theta_1\mathbf{u}\|_{0,\Theta_1^{-1}}+\|\mathbf{y}-\Theta_2^{-1}\mathbf{v}\|_{0,\Theta_2}
\end{aligned} \tag{6.75}$$

for all $\{\mathbf{u},\mathbf{w}\}\in\mathbf{C}_0(\Omega,\Theta_1)\times\mathbf{D}(\Omega,\Theta_1^{-1})$ *and* $\{\mathbf{v},\mathbf{y}\}\in\mathbf{C}(\Omega,\Theta_2^{-1})\times\mathbf{D}(\Omega,\Theta_2)$.

Proof. Adding and subtracting terms and expanding the right-hand side in (6.75) results in

$$
\begin{aligned}
\text{RHS} &= \|\mathbf{w}+\nabla\times\mathbf{v}\|^2_{0,\Theta_1^{-1}} + \|\mathbf{y}-\nabla\times\mathbf{u}\|^2_{0,\Theta_2} + \|\nabla\cdot\mathbf{y}\|^2_{0,\Theta_3} + \|\nabla\cdot\mathbf{w}\|^2_{0,\Theta_0^{-1}} \\
&\quad + \|(\mathbf{w}+\nabla\times\mathbf{v}) - (\nabla\times\mathbf{v}+\Theta_1\mathbf{u})\|^2_{0,\Theta_1^{-1}} \\
&\quad + \|(\mathbf{y}-\nabla\times\mathbf{u}) + (\nabla\times\mathbf{u}-\Theta_2^{-1}\mathbf{v})\|^2_{0,\Theta_2} \\
&= 2\big(\|\mathbf{w}+\nabla\times\mathbf{v}\|^2_{0,\Theta_1^{-1}} + \|\mathbf{y}-\nabla\times\mathbf{u}\|^2_{0,\Theta_2}\big) + \|\nabla\cdot\mathbf{y}\|^2_{0,\Theta_3} + \|\nabla\cdot\mathbf{w}\|^2_{0,\Theta_0^{-1}} \\
&\quad + \|\nabla\times\mathbf{v}+\Theta_1\mathbf{u}\|^2_{0,\Theta_1^{-1}} + \|\nabla\times\mathbf{u}-\Theta_2^{-1}\mathbf{v}\|^2_{0,\Theta_2} \\
&\quad + 2\Big(\big(\mathbf{y}-\nabla\times\mathbf{u}, \nabla\times\mathbf{u}-\Theta_2^{-1}\mathbf{v}\big)_{0,\Theta_2} - \big(\mathbf{w}+\nabla\times\mathbf{v}, \nabla\times\mathbf{v}+\Theta_1\mathbf{u}\big)_{0,\Theta_1^{-1}}\Big).
\end{aligned}
$$

Using the ε-inequality with $\varepsilon = 2/3$ for the last two terms results in

$$
2\big(\mathbf{y}-\nabla\times\mathbf{u}, \nabla\times\mathbf{u}-\Theta_2^{-1}\mathbf{v}\big)_{0,\Theta_2} \le \frac{3}{2}\|\mathbf{y}-\nabla\times\mathbf{u}\|^2_{0,\Theta_2} + \frac{2}{3}\|\nabla\times\mathbf{u}-\Theta_2^{-1}\mathbf{v}\|^2_{0,\Theta_2}
$$

$$
2\big(\mathbf{w}+\nabla\times\mathbf{v}, \nabla\times\mathbf{v}+\Theta_1\mathbf{u}\big)_{0,\Theta_1^{-1}} \le \frac{3}{2}\|\mathbf{w}+\nabla\times\mathbf{v}\|^2_{0,\Theta_1^{-1}} + \frac{2}{3}\|\nabla\times\mathbf{v}+\Theta_1\mathbf{u}\|^2_{0,\Theta_1^{-1}}.
$$

Therefore,

$$
\begin{aligned}
\text{RHS} &\ge \frac{1}{2}\Big(\|\mathbf{w}+\nabla\times\mathbf{v}\|^2_{0,\Theta_1^{-1}} + \|\mathbf{y}-\nabla\times\mathbf{u}\|^2_{0,\Theta_2}\Big) + \|\nabla\cdot\mathbf{y}\|^2_{0,\Theta_3} + \|\nabla\cdot\mathbf{w}\|^2_{0,\Theta_0^{-1}} \\
&\quad + \frac{1}{3}\Big(\|\nabla\times\mathbf{v}+\Theta_1\mathbf{u}\|^2_{0,\Theta_1^{-1}} + \|\nabla\times\mathbf{u}-\Theta_2^{-1}\mathbf{v}\|^2_{0,\Theta_2}\Big).
\end{aligned}
$$

The last two terms in the above expression are the same as in Theorem 6.16. Expanding all terms in this expression and using (6.74) yields

$$
\begin{aligned}
\text{RHS} &\ge \frac{1}{2}\Big(\|\mathbf{w}\|^2_{0,\Theta_1^{-1}} + \|\mathbf{y}\|^2_{0,\Theta_2}\Big) + \|\nabla\cdot\mathbf{y}\|^2_{0,\Theta_3} + \|\nabla\cdot\mathbf{w}\|^2_{0,\Theta_0^{-1}} \\
&\quad + \frac{5}{6}\Big(\|\nabla\times\mathbf{v}\|^2_{0,\Theta_1^{-1}} + \|\nabla\times\mathbf{u}\|^2_{0,\Theta_2}\Big) + \frac{1}{3}\Big(\|\mathbf{u}\|^2_{0,\Theta_1} + \|\mathbf{v}\|^2_{0,\Theta_2^{-1}}\Big) \\
&\quad + \big(\mathbf{w}, \nabla\times\mathbf{v}\big)_{0,\Theta_1^{-1}} - \big(\mathbf{y}, \nabla\times\mathbf{u}\big)_{0,\Theta_2}.
\end{aligned}
$$

Using again the ε inequality with $\varepsilon = 3/4$ shows that

$$
\big(\mathbf{w}, \nabla\times\mathbf{v}\big)_{0,\Theta_1^{-1}} \le \frac{3}{8}\|\mathbf{w}\|^2_{0,\Theta_1^{-1}} + \frac{2}{3}\|\nabla\times\mathbf{v}\|^2_{0,\Theta_1^{-1}}
$$

and

$$
\big(\mathbf{y}, \nabla\times\mathbf{u}\big)_{0,\Theta_2} \le \frac{3}{8}\|\mathbf{y}\|^2_{0,\Theta_2} + \frac{2}{3}\|\nabla\times\mathbf{u}\|^2_{0,\Theta_2}.
$$

After gathering all results together it follows that

$$\text{RHS} \geq \frac{1}{8}\left(\|\mathbf{w}\|^2_{0,\Theta_1^{-1}} + \|\mathbf{y}\|^2_{0,\Theta_2}\right) + \|\nabla\cdot\mathbf{y}\|^2_{0,\Theta_3} + \|\nabla\cdot\mathbf{w}\|^2_{0,\Theta_0^{-1}}$$
$$+\frac{1}{6}\left(\|\nabla\times\mathbf{v}\|^2_{0,\Theta_1^{-1}} + \|\nabla\times\mathbf{u}\|^2_{0,\Theta_2}\right) + \frac{1}{3}\left(\|\mathbf{u}\|^2_{0,\Theta_1} + \|\mathbf{v}\|^2_{0,\Theta_2^{-1}}\right)$$

which proves the theorem. □

Obviously, this theorem is the counterpart of Theorem 5.17 for the four-field potential–density–intensity-flux system (5.23). The similarities between that system and (6.26) were noted in Section 6.2.2.[28]

Note that by using constitutive laws in the reformulation of the curl–curl system, the natural solution space of (6.26) contains the product spaces $\mathbf{C}(\Omega)\times\mathbf{D}(\Omega)$ as opposed to the intersection spaces $\mathbf{C}(\Omega)\cap\mathbf{D}(\Omega)$. The latter would have cropped up if reformulations were based on the original reduced system (6.4) *with* the redundant divergence equation.

6.4.2 Continuous Least-Squares Principles

We formulate CLSPs under the same assumptions that were stated at the beginning of Section 6.3.2. For the curl–curl system (6.23), the energy balance (6.73) leads to the following well-posed minimization problem:

$$\begin{cases} J(\mathbf{u},\mathbf{v};\mathbf{f}) = \|\Theta_2^{-1}\mathbf{v}-\nabla\times\mathbf{u}\|^2_{0,\Theta_2} + \|\Theta_1\mathbf{u}+\nabla\times\mathbf{v}-\Theta_1\mathbf{f}\|^2_{0,\Theta_1^{-1}} \\ X = \mathbf{C}_0(\Omega,\Theta_1)\times\mathbf{C}(\Omega,\Theta_2^{-1}). \end{cases} \tag{6.76}$$

We also have the first-order system (6.26) with "constitutive laws." For this system, the relevant energy balance is (6.75) and the corresponding well-posed minimization problem is

$$\begin{cases} J(\mathbf{u},\mathbf{w},\mathbf{v},\mathbf{y};\mathbf{f}) = \|\mathbf{w}+\nabla\times\mathbf{v}-\Theta_1\mathbf{f}\|^2_{0,\Theta_1^{-1}} + \|\mathbf{y}-\nabla\times\mathbf{u}\|^2_{0,\Theta_2} \\ \qquad +\|\nabla\cdot\mathbf{y}\|^2_{0,\Theta_3} + \|\nabla\cdot\mathbf{w}\|^2_{0,\Theta_0^{-1}} + \|\mathbf{w}-\Theta_1\mathbf{u}\|^2_{0,\Theta_1^{-1}} + \|\mathbf{y}-\Theta_2^{-1}\mathbf{v}\|^2_{0,\Theta_2} \\ X = \mathbf{C}_0(\Omega,\Theta_1)\times\mathbf{D}(\Omega,\Theta_1^{-1})\times\mathbf{C}(\Omega,\Theta_2^{-1})\times\mathbf{D}(\Omega,\Theta_2). \end{cases} \tag{6.77}$$

We already explained the benefits of having the redundant equations included in (6.26). For this reason, CLSPs for the first-order system (6.24) *with* constitutive laws but *without* redundant equations is not considered.

---

[28] Theorem 6.17 makes clear the role of the "redundant" equations in (6.26): without them, the residual energy of the differential equations is only able to control the $L^2(\Omega)$ norm of $\mathbf{y}$ and $\mathbf{w}$. Therefore, using the redundant equations in the first-order system *with* constitutive laws is justified because they improve the strength of the energy balance.

Remark 6.18 The optimization-based approach (see Remark 6.6) can also be applied to the systems with constitutive laws (6.24) and (6.26). Structurally, the equivalent optimal control problem for (6.26) is very similar to the optimal control formulation (5.57) for the potential–density–flux–intensity first-order system (5.23):

$$\begin{cases} \min\limits_{\substack{\{\mathbf{u},\mathbf{w}\}\in\mathbf{C}_0(\Omega)\times\mathbf{D}(\Omega)\\ \{\mathbf{v},\mathbf{y}\}\in\mathbf{C}(\Omega)\times\mathbf{D}(\Omega)}} \left(\|\mathbf{w}-\Theta_1\mathbf{u}\|^2_{0,\Theta_1^{-1}} + \|\mathbf{y}-\Theta_2^{-1}\mathbf{v}\|^2_{0,\Theta_2}\right) \\ \text{subject to} \\ \gamma\mathbf{w}+\nabla\times\mathbf{v}=\Theta_1\mathbf{f}, \quad \mathbf{y}-\nabla\times\mathbf{u}=0, \quad \nabla\cdot\mathbf{y}=0, \quad \text{and} \quad \nabla\cdot\mathbf{w}=0. \end{cases} \tag{6.78}$$

This problem can be solved by any of the least-squares methods for the abstract control problem (11.6). In particular, using the method (11.49) based on direct penalization of the objective functional by a least-squares form of the constraint equations, we obtain precisely the same CLSP as in (6.77). As is the case for the div–curl system (6.43), we see that conversion to an optimal control formulation is not necessary to obtain a well-posed CLSP. □

6.4.3 Discrete Least-Squares Principles

Our pool of CLSPs for the curl–curl class of equations is comprised of (6.76) and (6.77). None of these CLSPs uses the intersection space $\mathbf{C}(\Omega)\cap\mathbf{D}(\Omega)$ and, as a result, their translation into well-posed discrete minimization problems does not face the obstacles we had in Section 6.3.3 with some of the CLSPs for the div–curl systems. In particular, we can restrict our attention to conforming choices for the finite element spaces.

From (6.76), without much difficulty, we obtain the compliant DLSP

$$\begin{cases} J(\mathbf{u}^h,\mathbf{v}^h;\mathbf{f}) = \|\Theta_1\mathbf{u}^h+\nabla\times\mathbf{v}^h-\Theta_1\mathbf{f}\|^2_{0,\Theta_1^{-1}} + \|\Theta_2^{-1}\mathbf{v}^h-\nabla\times\mathbf{u}^h\|^2_{0,\Theta_2} \\ X^h = \mathbf{C}_0^h(\Omega)\times\mathbf{C}^h(\Omega) \end{cases} \tag{6.79}$$

for the curl–curl system (6.23). The discrete solution space in (6.79) is a direct product of two curl-conforming spaces and the comments from Remark 6.7 also apply here, i.e., the component spaces can be chosen independently of each other. However, it is obvious that for reasons of efficiency and ease of implementation, it is best to choose the two components of X^h to be the same curl-conforming space.

From (6.77) we obtain another compliant DLSP:

$$\begin{cases} J(\mathbf{u}^h,\mathbf{w}^h,\mathbf{v}^h,\mathbf{y}^h;\mathbf{f}) = \|\mathbf{w}^h+\nabla\times\mathbf{v}^h-\Theta_1\mathbf{f}\|^2_{0,\Theta_1^{-1}} + \|\mathbf{y}^h-\nabla\times\mathbf{u}^h\|^2_{0,\Theta_2} \\ \quad +\|\nabla\cdot\mathbf{y}^h\|^2_{0,\Theta_3} + \|\nabla\cdot\mathbf{w}^h\|^2_{0,\Theta_0^{-1}} + \|\mathbf{w}^h-\Theta_1\mathbf{u}^h\|^2_{0,\Theta_1^{-1}} + \|\mathbf{y}^h-\Theta_2^{-1}\mathbf{v}^h\|^2_{0,\Theta_2} \\ X^h = \mathbf{C}_0^h(\Omega)\times\mathbf{D}^h(\Omega)\times\mathbf{C}^h(\Omega)\times\mathbf{D}^h(\Omega). \end{cases} \tag{6.80}$$

The discrete solution space in this DLSP uses two curl-conforming and two div-conforming spaces, all of which can be chosen independently of each other. However, it is clear that the most efficient methods result from a choice that matches the approximation properties of all spaces and, on the other hand, uses the same curl-conforming space for $\mathbf{u}^h$ and $\mathbf{v}^h$ and the same div-conforming space for $\mathbf{w}^h$ and $\mathbf{y}^h$. Of course, in both the DLSPs (6.79) and (6.80), we have to use the same weights in the norm definitions as in their respective parent CLSPs. Below, we examine the two DLSPs for the curl–curl class of systems more closely. In particular, we show that the relationship between (6.79) and (6.80) is a lot like the relationship between the potential–flux CLSP (5.63) and the four-field formulation (5.64).

Comparison of DLSPs for curl–curl systems

The first observation is that, after finding a minimizer $\{\mathbf{u}^h, \mathbf{v}^h\} \in \mathbf{C}_0^h(\Omega) \times \mathbf{C}^h(\Omega)$ of (6.79), approximations of the dual variables can be computed according to

$$\mathbf{y}^h = \nabla \times \mathbf{u}^h \qquad \text{and} \qquad \mathbf{w}^h = \pi_D \mathbf{f} - \nabla \times \mathbf{v}^h , \tag{6.81}$$

respectively. Obviously, this formula is an analogue of (5.65) in Section 5.6.1. From the properties of compatible finite element spaces, it follows that $\mathbf{y}^h$ and $\mathbf{w}^h$ belong to a div-conforming space; this is also the case for (6.80) for which these variables are approximated directly. One important distinction though, is that the div-conforming spaces in (6.80) can be chosen *independently* of the curl-conforming spaces for the primal variables, whereas the variables computed in (6.81) belong to the range of the curl restricted to $\mathbf{C}^h(\Omega)$, i.e., they are dependent upon the choice of this space. For example, if $\mathbf{C}^h(\Omega)$ is the Nédélec curl-conforming space of the first kind $\mathbf{C}^{r}\text{-}(\Omega)$, from (B.45) we see that the possibilities for div-conforming spaces in (6.81) are restricted to $\mathbf{D}^{r}\text{-}(\Omega)$ or $\mathbf{D}^{r-1}(\Omega)$.

Nevertheless, assuming that (6.80) is implemented using the same curl-conforming space as (6.79) and a div-conforming space that provides the correct range of curl, it is of practical interest to know whether the two DLSPs produce the same finite element approximations of all four variables. If this turns out to be the case, the more expensive method (6.80) only is needed to assert optimal error estimates for $\mathbf{y}^h$ and $\mathbf{w}^h$, wheras their actual computation can be done much more economically by using (6.81). The following theorem represents an analogue of Theorem 5.25 and shows that, just as in Section 5.6.1, this shortcut is not feasible.

Theorem 6.19 *Assume that the solution spaces in* (6.79) *and* (6.80) *are defined using the same curl-conforming elements for* $\mathbf{C}_0^h(\Omega)$ *and* $\mathbf{C}(\Omega)$. *Furthermore, assume that the div-conforming spaces are such that the pairs* $\{\mathbf{C}^h(\Omega), \mathbf{D}^h(\Omega)\}$ *for* $\{\mathbf{u}^h, \mathbf{w}^h\}$ *and* $\{\mathbf{v}^h, \mathbf{y}^h\}$ *belong to two (possibly different) finite element De Rham complexes. Let* $\{\mathbf{u}_1^h, \mathbf{v}_1^h\}$ *denote the minimizer of* (6.79), $\{\mathbf{y}_1^h, \mathbf{w}_1^h\}$ *denote functions computed by* (6.81), *and* $\{\mathbf{u}_2^h, \mathbf{w}_2^h, \mathbf{v}_2^h, \mathbf{y}_2^h\}$ *the minimizer of* (6.80). *Then, for arbitrary data* $\mathbf{f}$, *the sets* $\{\mathbf{u}_1^h, \mathbf{w}_1^h, \mathbf{v}_1^h, \mathbf{y}_1^h\}$ *and* $\{\mathbf{u}_2^h, \mathbf{w}_2^h, \mathbf{v}_2^h, \mathbf{y}_2^h\}$ *are not the same.*

Proof. It is clear that if the assertion of the theorem can be shown for unit weights, it also hold for general weight functions. Thus, for clarity, in what follows we assume unit weights. The set $\{\mathbf{u}_2^h, \mathbf{w}_2^h, \mathbf{v}_2^h, \mathbf{y}_2^h\}$ satisfies the usual first-order optimality condition which, for (6.80), takes the form:

$$
\begin{aligned}
&\big(\mathbf{w}_2^h + \nabla\times\mathbf{v}_2^h, \widehat{\mathbf{w}}^h + \nabla\times\widehat{\mathbf{v}}^h\big) + \big(\mathbf{y}_2^h - \nabla\times\mathbf{u}_2^h, \widehat{\mathbf{y}}^h - \nabla\times\widehat{\mathbf{u}}^h\big) \\
&\quad + \big(\mathbf{w}_2^h - \mathbf{u}_2^h, \widehat{\mathbf{w}}^h - \widehat{\mathbf{u}}^h\big) + \big(\mathbf{y}_2^h - \mathbf{v}_2^h, \widehat{\mathbf{y}}^h - \widehat{\mathbf{v}}^h\big) \\
&\quad + \big(\nabla\cdot\mathbf{y}_2^h, \nabla\cdot\widehat{\mathbf{y}}^h\big) + \big(\nabla\cdot\mathbf{w}_2^h, \nabla\cdot\widehat{\mathbf{w}}^h\big) = \big(\mathbf{f}, \widehat{\mathbf{w}}^h + \nabla\times\widehat{\mathbf{v}}^h\big)
\end{aligned}
$$

for all $\{\widehat{\mathbf{u}}^h, \widehat{\mathbf{w}}^h, \widehat{\mathbf{v}}^h, \widehat{\mathbf{y}}^h\} \in X^h$. Extracting the equations that correspond to variations with respect to the dual variables results in

$$
\big(\mathbf{w}_2^h + \nabla\times\mathbf{v}_2^h, \widehat{\mathbf{w}}^h\big) + \big(\nabla\cdot\mathbf{w}_2^h, \nabla\cdot\widehat{\mathbf{w}}^h\big) + \big(\mathbf{w}_2^h - \mathbf{u}_2^h, \widehat{\mathbf{w}}^h\big) = \big(\mathbf{f}, \widehat{\mathbf{w}}^h\big) \tag{6.82}
$$

$$
\big(\mathbf{y}_2^h - \nabla\times\mathbf{u}_2^h, \widehat{\mathbf{y}}^h\big) + \big(\nabla\cdot\mathbf{y}_2^h, \nabla\cdot\widehat{\mathbf{y}}^h\big) + \big(\mathbf{y}_2^h - \mathbf{v}_2^h, \widehat{\mathbf{y}}^h\big) = 0 \tag{6.83}
$$

for all $\widehat{\mathbf{w}}^h \in \mathbf{D}^h(\Omega)$ and $\widehat{\mathbf{y}}^h \in \mathbf{D}^h(\Omega)$.

The optimality condition for $\{\mathbf{u}_1^h, \mathbf{v}_1^h\}$ can be written as

$$
\big(\mathbf{u}_1^h + \nabla\times\mathbf{v}_1^h, \widehat{\mathbf{u}}^h\big) - \big(\mathbf{v}_1^h - \nabla\times\mathbf{u}_1^h, \nabla\times\widehat{\mathbf{u}}^h\big) = \big(\mathbf{f}, \widehat{\mathbf{u}}^h\big) \tag{6.84}
$$

$$
\big(\mathbf{u}_1^h + \nabla\times\mathbf{v}_1^h, \nabla\times\widehat{\mathbf{v}}^h\big) + \big(\mathbf{v}_1^h - \nabla\times\mathbf{u}_1^h, \widehat{\mathbf{v}}^h\big) = \big(\mathbf{f}, \nabla\times\widehat{\mathbf{v}}^h\big) \tag{6.85}
$$

for all $\widehat{\mathbf{u}}^h \in \mathbf{C}_0^h(\Omega)$ and $\widehat{\mathbf{v}}^h \in \mathbf{C}^h(\Omega)$. The function $\mathbf{y}_1^h$, defined as in (6.81), satisfies $\mathbf{y}_1^h - \nabla\times\mathbf{u}_1^h = 0$ and $\nabla\cdot\mathbf{y}_1^h = 0$. Therefore,

$$
\big(\mathbf{y}_1^h - \nabla\times\mathbf{u}_1^h, \widehat{\mathbf{y}}^h\big) + \big(\nabla\cdot\mathbf{y}_1^h, \nabla\cdot\widehat{\mathbf{y}}^h\big) = 0 \qquad \forall\widehat{\mathbf{y}}^h \in \mathbf{D}^h(\Omega). \tag{6.86}
$$

Likewise, from the definition of $\mathbf{w}_1^h$, it follows that

$$
\big(\mathbf{w}_1^h + \nabla\times\mathbf{v}_1^h, \widehat{\mathbf{w}}^h\big) + \big(\nabla\cdot\mathbf{w}_1^h, \nabla\cdot\widehat{\mathbf{w}}^h\big) = \big(\mathbf{f}, \widehat{\mathbf{w}}^h\big) \qquad \forall\widehat{\mathbf{w}}^h \in \mathbf{D}^h(\Omega). \tag{6.87}
$$

Comparing (6.82) with (6.87) and (6.83) with (6.86), we see that

$$
\big(\mathbf{w}_1^h - \mathbf{u}_1^h, \widehat{\mathbf{w}}^h\big) = 0 \qquad \forall\,\widehat{\mathbf{w}}^h \in \mathbf{D}^h(\Omega) \tag{6.88}
$$

$$
\big(\mathbf{y}_1^h - \mathbf{v}_1^h, \widehat{\mathbf{y}}^h\big) = 0 \qquad \forall\,\widehat{\mathbf{y}}^h \in \mathbf{D}^h(\Omega) \tag{6.89}
$$

are necessary conditions for $\{\mathbf{u}_1^h, \mathbf{w}_1^h, \mathbf{v}_1^h, \mathbf{y}_1^h\}$ and $\{\mathbf{u}_2^h, \mathbf{w}_2^h, \mathbf{v}_2^h, \mathbf{y}_2^h\}$ to coincide. Let us assume that this is the case, i.e., that (6.88) and (6.89) are satisfied. Using again the definitions of $\mathbf{y}_1^h$ and $\mathbf{w}_1^h$ in conjunction with (6.88) and (6.89) yields

$$
\big(\mathbf{u}_1^h + \nabla\times\mathbf{v}_1^h, \widehat{\mathbf{w}}^h\big) = \big(\mathbf{f}, \widehat{\mathbf{w}}^h\big) \qquad \forall\widehat{\mathbf{w}}^h \in \mathbf{D}^h(\Omega) \tag{6.90}
$$

$$
\big(\mathbf{v}_1^h - \nabla\times\mathbf{u}_1^h, \widehat{\mathbf{y}}^h\big) = 0 \qquad \forall\widehat{\mathbf{y}}^h \in \mathbf{D}^h(\Omega). \tag{6.91}
$$

By assumption, $\mathbf{D}^h(\Omega)$ contains the range of the curl applied to $\mathbf{C}^h(\Omega)$. Therefore, in (6.90) and (6.91), we can set $\widehat{\mathbf{w}}^h = \nabla\times\widehat{\mathbf{v}}^h$ and $\widehat{\mathbf{y}}^h = \nabla\times\widehat{\mathbf{u}}^h$ and thus obtain

$$\left(\mathbf{u}_1^h + \nabla\times\mathbf{v}_1^h, \nabla\times\widehat{\mathbf{v}}^h\right) = \left(\mathbf{f}, \nabla\times\widehat{\mathbf{v}}^h\right) \qquad \forall\,\widehat{\mathbf{v}}^h \in \mathbf{C}^h(\Omega) \tag{6.92}$$

$$\left(\mathbf{v}_1^h - \nabla\times\mathbf{u}_1^h, \nabla\times\widehat{\mathbf{u}}^h\right) = 0 \qquad \forall\widehat{\mathbf{u}}^h \in \mathbf{C}_0^h(\Omega)\,. \tag{6.93}$$

Together, (6.92) and (6.85) imply that

$$\left(\mathbf{v}_1^h - \nabla\times\mathbf{u}_1^h, \widehat{\mathbf{v}}^h\right) = 0 \qquad \forall\widehat{\mathbf{v}}^h \in \mathbf{C}^h(\Omega)\,.$$

After subtracting (6.93) from this equation, we see that

$$\left(\mathbf{v}_1^h - \nabla\times\mathbf{u}_1^h, \widehat{\mathbf{v}}^h - \nabla\times\widehat{\mathbf{u}}^h\right) = 0 \qquad \forall\,\{\widehat{\mathbf{u}}^h, \widehat{\mathbf{v}}^h\} \in \mathbf{C}_0^h(\Omega)\times\mathbf{C}^h(\Omega)\,.$$

Setting $\widehat{\mathbf{u}}^h = \mathbf{u}_1^h$ and $\widehat{\mathbf{v}} = \mathbf{v}_1^h$ in this equation results in

$$\|\mathbf{v}_1^h - \nabla\times\mathbf{u}_1^h\|_0 = 0\,.$$

Therefore, $\mathbf{v}_1^h = \nabla\times\mathbf{u}_1^h$ a.e. in Ω is a necessary conditions for $\{\mathbf{u}_1^h, \mathbf{w}_1^h, \mathbf{v}_1^h, \mathbf{y}_1^h\}$ and $\{\mathbf{u}_2^h, \mathbf{w}_2^h, \mathbf{v}_2^h, \mathbf{y}_2^h\}$ to coincide. On the other hand, because $\mathbf{v}_1^h \in \mathbf{C}^h(\Omega)$ and $\nabla\times\mathbf{u}_1^h \in \mathbf{D}^h(\Omega)$, it follows that $\mathbf{v}_1^h \in \mathbf{C}^h(\Omega)\cap\mathbf{D}^h(\Omega)$. Arguing as in the proof of Theorem 5.25, we see that $\mathbf{v}_1^h$ must then be of class C^0. Clearly, this cannot be true for all $\mathbf{f}$.□

A simple illustrative example is provided by considering the implementation of (6.79) using the lowest-order Nédélec space of the first kind (B.22), i.e., $\mathbf{C}^1\text{-}(\Omega)$. Then, $\nabla\times\mathbf{u}_1^h$ is constant on each element and the necessary condition in Theorem 6.19 implies that $\mathbf{v}_1^h$ must be a piecewise constant $C^0(\Omega)$ field. It follows that $\mathbf{v}_1^h$ is constant on Ω. Unless $\mathbf{f}$ is such that this condition holds, $\{\mathbf{u}_1^h, \mathbf{w}_1^h, \mathbf{v}_1^h, \mathbf{y}_1^h\}$ and $\{\mathbf{u}_2^h, \mathbf{w}_2^h, \mathbf{v}_2^h, \mathbf{y}_2^h\}$ cannot coincide.

Splitting property

A second observation is that (6.79) has the same "splitting" property as shown in Theorem 5.37 for the div–grad system. In other words, the weak equation representing the first-order optimality condition of (6.79) uncouples into two independent equations for $\mathbf{u}^h$ and $\mathbf{v}^h$, respectively. This statement is formalized in the following theorem.

Theorem 6.20 *Let* $\{\mathbf{u}^h, \mathbf{v}^h\} \in \mathbf{C}_0^h(\Omega)\times\mathbf{C}^h(\Omega)$ *be the minimizer of* (6.79). *Then,* $\mathbf{u}^h$ *solves the problem*

$$\left(\nabla\times\mathbf{u}^h, \nabla\times\widehat{\mathbf{u}}^h\right)_{0,\Theta_2} + \left(\mathbf{u}^h, \widehat{\mathbf{u}}^h\right)_{0,\Theta_1} = \left(\mathbf{f}, \widehat{\mathbf{u}}^h\right)_{0,\Theta_1} \qquad \forall\widehat{\mathbf{u}}^h \in \mathbf{C}_0^h(\Omega) \tag{6.94}$$

and $\mathbf{v}^h$ *solves the independent equation*

$$\left(\nabla\times\mathbf{v}^h, \nabla\times\widehat{\mathbf{v}}^h\right)_{0,\Theta_1^{-1}} + \left(\mathbf{v}^h, \widehat{\mathbf{v}}^h\right)_{0,\Theta_2^{-1}} = \left(\Theta_1\mathbf{f}, \nabla\times\widehat{\mathbf{v}}^h\right)_{0,\Theta_1^{-1}} \quad \forall\widehat{\mathbf{v}}^h \in \mathbf{C}^h(\Omega)\,. \tag{6.95}$$

Proof. The proof of this theorem relies on the same cancellation property as in Theorem 6.16. The details are omitted. □

It is instructive to examine the splitting property in the context of the eddy-current system (6.28) in terms of $\mathbf{H}$ and $\mathbf{E}$. Suppose that this problem is advanced in time using an implicit Euler method with time step $\Delta t = 1$ and the initial data are $\mathbf{H}_0$. Using the identifications (6.27) and $\Theta_1 \mathbf{f} = \mu \mathbf{H}_0$, we see that (6.79) specializes to

$$\begin{cases} J(\mathbf{H}^h, \mathbf{E}^h; \mathbf{H}_0) = \\ \qquad \|\mu \mathbf{H}^h + \nabla \times \mathbf{E}^h - \mu \mathbf{H}_0\|^2_{0,\mu^{-1}} + \|\nabla \times \mathbf{H}^h - \sigma \mathbf{E}^h\|^2_{0,\sigma^{-1}} \\ X^h = \mathbf{C}^h_0(\Omega) \times \mathbf{C}^h(\Omega)\,. \end{cases} \tag{6.96}$$

From Theorem 6.20, it follows that minimizers of (6.96) can be computed independently of each other. To determine $\mathbf{H}^h$, we solve

$$\big(\nabla \times \mathbf{H}^h, \nabla \times \widehat{\mathbf{H}}^h\big)_{0,\sigma_1^{-1}} + \big(\mathbf{H}^h, \widehat{\mathbf{H}}^h\big)_{0,\mu} = \big(\mathbf{H}_0, \widehat{\mathbf{H}}^h\big)_{0,\mu} \qquad \forall \widehat{\mathbf{H}}^h \in \mathbf{C}^h_0(\Omega)\,. \tag{6.97}$$

The weak equation for $\mathbf{E}^h$ is very similar:

$$\big(\nabla \times \mathbf{E}^h, \nabla \times \widehat{\mathbf{E}}^h\big)_{0,\mu^{-1}} + \big(\mathbf{E}^h, \widehat{\mathbf{E}}^h\big)_{0,\sigma} = \big(\mu \mathbf{H}_0, \nabla \times \widehat{\mathbf{E}}^h\big)_{0,\mu^{-1}} \quad \forall \widehat{\mathbf{E}}^h \in \mathbf{C}^h(\Omega)\,. \tag{6.98}$$

We recognize (6.97) as a Rayleigh–Ritz method for the second-order eddy-current formulation (6.13) that is discretized in time by an implicit Euler method. Of course, (6.98) is the Rayleigh–Ritz method for the complementary model (6.15) in terms of the electric variable. This situation mirrors the case of the div–grad system in Theorem 5.37 which also splits into the Rayleigh–Ritz formulations (5.94) and (5.95) for the scalar variable and vector-valued variables, respectively.

Direct discretization of the optimal control formulation

The optimal control setting (6.78) for the four-field system (6.24) parallels that of (5.57). Not surprisingly, if we approach the solution of (6.78) directly instead of by using LSFEMs for optimal control problems, the results are very similar. More precisely, the restriction of (6.78) to suitable compatible finite element spaces makes it possible to eliminate the constraints and obtain a DLSP that is almost identical to (6.79). Let us briefly review the details. Direct discretization of (6.78) gives the following finite-dimensional optimization problem:

$$\begin{cases} \min\limits_{\substack{\{\mathbf{u}^h,\mathbf{w}^h\} \in \mathbf{C}^h_0 \times \mathbf{D}^h \\ \{\mathbf{v}^h,\mathbf{y}^h\} \in \mathbf{C}^h \times \mathbf{D}^h}} \big(\|\mathbf{w}^h - \Theta_1 \mathbf{u}^h\|^2_{0,\Theta_1^{-1}} + \|\mathbf{y}^h - \Theta_2^{-1} \mathbf{v}^h\|^2_{0,\Theta_2}\big) \\ \text{subject to} \quad \mathbf{w}^h + \nabla \times \mathbf{v}^h = \Pi_D \Theta_1 \mathbf{f}\,, \qquad \mathbf{y}^h - \nabla \times \mathbf{u}^h = 0\,, \\ \qquad\qquad\quad \nabla \cdot \mathbf{y}^h = 0\,, \qquad \text{and} \qquad \nabla \cdot \mathbf{w}^h = 0\,. \end{cases} \tag{6.99}$$

Suppose that, as in Theorem 6.19, $\mathbf{C}^h$ and $\mathbf{D}^h$ belong to the same finite element De Rham complex. Then, $\mathbf{D}^h$ contains the range of the curl and the first two constraints in (6.99) can be solved for $\mathbf{w}^h$ and $\mathbf{y}^h$, respectively:

$$\mathbf{w}^h = \Pi_D \Theta_1 \mathbf{f} - \nabla \times \mathbf{v}^h \qquad \text{and} \qquad \mathbf{y}^h = \nabla \times \mathbf{u}^h .$$

Clearly, the so-defined $\mathbf{w}^h$ and $\mathbf{y}^h$ are automatically divergence-free[29] so that the last two constraints can be eliminated from (6.99). As a result, this problem transforms into (compare with (5.77))

$$\min_{\{\mathbf{u}^h,\mathbf{v}^h\} \in \mathbf{C}_0^h \times \mathbf{C}^h} \left(\|\Pi_D \Theta_1 \mathbf{f} - \nabla \times \mathbf{v}^h - \Theta_1 \mathbf{u}^h\|^2_{0,\Theta_1^{-1}} + \|\nabla \times \mathbf{u}^h - \Theta_2^{-1} \mathbf{v}^h\|^2_{0,\Theta_2} \right) .$$

With the exception of the projection Π_D, this unconstrained optimization problem is identical with (6.79).

Similar to Section 5.6.1, the two factors that make this transformation possible are that $\gamma \neq 0$ and $R(\nabla \times) \subset \mathbf{D}^h(\Omega)$. If one of these factors is not fulfilled, the dual variables cannot be eliminated from (6.99). Because $\mathbf{D}^h(\Omega)$ can always be matched with the curl-conforming space for the primal variables, the only actual instance for which (6.99) cannot be reduced is when $\gamma = 0$, which is typical of static problems. We see that this is indeed the case for (6.45) which is representative of the type of optimal control formulations one obtains in magnetostatics; see Remark 6.6.

6.4.4 Error Analysis

The two DLSPs that we defined for curl–curl systems are compliant and based on energy balances in the vector-operator setting. As a result, the analysis of their errors is a straightforward application of the abstract results in Theorem 3.28. We specialize that theorem under the same restrictions on the grid as in Sections 6.3.4 and 6.3.5, i.e., $\mathcal{T}_h$ must be an affine family of simplicial finite elements. We begin with error estimates for (6.79).

Theorem 6.21 *Let $\Omega \subset \mathbb{R}^3$ denote a bounded domain with a Lipschitz continuous boundary, $\mathcal{T}_h$ denote a regular partition of Ω into affine simplicial elements, $\{\mathbf{u}^h, \mathbf{v}^h\} \in X^h = \mathbf{C}_0^h(\Omega) \times \mathbf{C}^h(\Omega)$ denote the minimizer of* (6.48), *and $\{\mathbf{u}, \mathbf{v}\} \in \mathbf{C}_0(\Omega) \times \mathbf{C}(\Omega)$ denote a solution of* (6.23). *Then, there is a positive constant C such that*

$$\|\mathbf{u} - \mathbf{u}^h\|_C + \|\mathbf{v} - \mathbf{v}^h\|_C \le C \inf_{\{\widehat{\mathbf{u}}^h, \widehat{\mathbf{v}}^h\} \in X^h} \left(\|\mathbf{u} - \widehat{\mathbf{u}}^h\|_C + \|\mathbf{v} - \widehat{\mathbf{v}}^h\|_C \right) . \quad \Box \tag{6.100}$$

For sufficiently regular solutions of (6.23), the error bound can be further refined.

---

[29] For $\mathbf{w}^h$, this follows from the compatibility condition (6.5) and the commuting diagram (B.1) which implies that $\nabla \cdot (\Pi_D \Theta_1 \mathbf{f}) = 0$.

Corollary 6.22 *Assume that the hypotheses of Theorem 6.21 hold. Also assume that $\mathcal{T}_h$ is uniformly regular, $X^h = \mathbf{C}_0^{(r)} \times \mathbf{C}^{(r)}$, $r \geq 1$, and that the solution of (6.23) belongs to $[H^{r+1}(\Omega)]^d \times [H^{r+1}(\Omega)]^d$. Then,*

$$\|\mathbf{u}-\mathbf{u}^h\|_C + \|\mathbf{v}-\mathbf{v}^h\|_C \leq Ch^r\left(\|\nabla\times\mathbf{u}\|_r + \|\nabla\times\mathbf{v}\|_r\right). \quad \Box \tag{6.101}$$

For the second compliant DLSP we have similar results.

Theorem 6.23 *Assume that Ω and $\mathcal{T}_h$ are as in Theorem 6.21. Let $\{\mathbf{u}^h,\mathbf{w}^h,\mathbf{v}^h,\mathbf{y}^h\} \in X^h = \mathbf{C}_0^h(\Omega)\times\mathbf{D}^h(\Omega)\times\mathbf{C}^h(\Omega)\times\mathbf{D}^h(\Omega)$ denote the minimizer of (6.80), and $\{\mathbf{u},\mathbf{w},\mathbf{v},\mathbf{y}\} \in \mathbf{C}_0(\Omega)\times\mathbf{D}(\Omega)\times\mathbf{C}(\Omega)\times\mathbf{D}(\Omega)$ denote the solution of (6.26). Then,*

$$\begin{aligned}
&\|\mathbf{u}-\mathbf{u}^h\|_C + \|\mathbf{w}-\mathbf{w}^h\|_D + \|\mathbf{v}-\mathbf{v}^h\|_C + \|\mathbf{y}-\mathbf{y}^h\|_D \\
&\quad \leq C\Big(\inf_{\widehat{\mathbf{u}}^h\in\mathbf{C}_0^h}\|\mathbf{u}-\widehat{\mathbf{u}}^h\|_C + \inf_{\widehat{\mathbf{w}}^h\in\mathbf{D}^h}\|\mathbf{w}-\widehat{\mathbf{w}}^h\|_D + \\
&\qquad \inf_{\widehat{\mathbf{v}}^h\in\mathbf{C}^h}\|\mathbf{v}-\widehat{\mathbf{v}}^h\|_C + \inf_{\widehat{\mathbf{y}}^h\in\mathbf{D}^h}\|\mathbf{y}-\widehat{\mathbf{y}}^h\|_D\Big). \quad \Box
\end{aligned} \tag{6.102}$$

We also have an analogue of Corollary 6.22.

Corollary 6.24 *Assume the hypotheses of Theorem 6.23. Let $X^h = \mathbf{C}_0^{(p)}\times\mathbf{D}^{(r)}\times\mathbf{C}^{(s)}\times\mathbf{D}^{(t)}$, where p, r, s, and t are integers greater than or equal to one. Let $q = \max\{p,r,s,t\}$ and assume that $\mathcal{T}_h$ is uniformly regular and that the solution of (6.26) belongs to $[H^{q+1}(\Omega)]^d\times[H^{q+1}(\Omega)]^d\times[H^{q+1}(\Omega)]^d\times[H^{q+1}(\Omega)]^d$. Then,*

$$\begin{aligned}
&\|\mathbf{u}-\mathbf{u}^h\|_C + \|\mathbf{w}-\mathbf{w}^h\|_D + \|\mathbf{v}-\mathbf{v}^h\|_C + \|\mathbf{y}-\mathbf{y}^h\|_D \\
&\quad \leq C\left(h^p\|\nabla\times\mathbf{u}\|_p + h^r\|\nabla\cdot\mathbf{w}\|_r + h^s\|\nabla\times\mathbf{v}\|_s + h^t\|\nabla\cdot\mathbf{y}\|_t\right). \quad \Box
\end{aligned} \tag{6.103}$$

Although the solution space components in (6.80) can be chosen independently, Corollary 6.24 shows that it pays to choose the same curl-conforming space for $\mathbf{u}^h$ and $\mathbf{v}^h$ and the same div-conforming space for $\mathbf{w}^h$ and $\mathbf{y}^h$. To further optimize the performance of (6.80), the approximation orders of $\mathbf{C}^h(\Omega)$ and $\mathbf{D}^h(\Omega)$ should be equilibrated. However, these spaces do not have to belong to the same finite element complex. For instance, we can choose $\mathbf{C}^h(\Omega) = \mathbf{C}^{r}\text{-}(\Omega)$ and $\mathbf{D}^h(\Omega) = \mathbf{D}^r(\Omega)$ despite the fact that this pair does not belong to any of the four complexes in (B.45).

6.5 Practicality Issues

In this section, we briefly touch upon two practical aspects of LSFEMs for div–curl and curl–curl systems. We focus on the solution of the resulting linear systems and provide concrete details and suggestions about the implementation of non-conforming methods. Because the compatible LSFEMs from Chapter 5 also rely on curl- and div-conforming elements, the first part of our discussion is relevant to them too.

6.5.1 Solution of Algebraic Equations

To avoid repeating over and over almost identical matrix definitions, in what follows, all weights are set to one. After selecting bases $\{\mathbf{c}_i^h\}$ and $\{\mathbf{d}_i^h\}$ for $\mathbf{C}^h(\Omega)$ and $\mathbf{D}^h(\Omega)$, respectively, one can derive, in the usual manner, equivalent algebraic representations of the optimality systems involved in finding minimizers of the DLSPs. Here, we examine linear systems associated with compliant methods for div–curl and curl–curl systems; the non-conforming div–curl formulations (6.50) and (6.51) give rise to somewhat different problems and are considered separately.

Linear systems resulting from LSFEMs are guaranteed to be symmetric and positive definite which makes preconditioned conjugate gradient methods a natural solver choice. Consequently, fast and efficient solution of the linear systems relies on the existence of good preconditioners. For LSFEMs, such preconditioners can be devised by examining the energy balance of each DLSP to come up with spectrally equivalent matrices with simpler structures.

Let us begin with the compliant DLSPs (6.48) and (6.49) for div–curl systems with constitutive laws. It is easy to see that their optimality systems give rise to similar systems of linear algebraic equations having 2×2 block structure:

$$\begin{pmatrix} \mathsf{M}_C+\mathsf{K}_C & \mathsf{M}_{CD} \\ \mathsf{M}_{DC} & \mathsf{M}_D+\mathsf{K}_D \end{pmatrix}. \tag{6.104}$$

The blocks in this matrix are defined as follows:

$$\begin{gathered} (\mathsf{M}_C)_{ij} = \int_\Omega \mathbf{c}_i\cdot\mathbf{c}_j\,d\Omega\,, \qquad (\mathsf{K}_C)_{ij} = \int_\Omega (\nabla\times\mathbf{c}_i)\cdot(\nabla\times\mathbf{c}_j)\,d\Omega\,, \\ (\mathsf{M}_D)_{ij} = \int_\Omega \mathbf{d}_i\cdot\mathbf{d}_j\,d\Omega\,, \qquad (\mathsf{K}_D)_{ij} = \int_\Omega (\nabla\cdot\mathbf{d}_i)(\nabla\cdot\mathbf{d}_j)\,d\Omega\,, \\ (\mathsf{M}_{CD})_{ij} = \int_\Omega \mathbf{c}_i\cdot\mathbf{d}_j\,d\Omega\,, \end{gathered} \tag{6.105}$$

and $\mathsf{M}_{DC} = \mathsf{M}_{CD}^T$. The energy balances for (6.48) and (6.49) are given in Theorem 6.5. From (6.36) and (6.37), it follows that the matrix in (6.104) is spectrally equivalent to the 2×2 block-diagonal matrix

$$\begin{pmatrix} \mathsf{M}_C+\mathsf{K}_C & \\ & \mathsf{M}_D+\mathsf{K}_D \end{pmatrix}. \tag{6.106}$$

Linear systems for curl–curl problems and their spectrally equivalent matrices are comprised of similar blocks. Due to the splitting property established in Theorem

6.20, the matrix for (6.79) has the particularly simple structure

$$\begin{pmatrix} \mathsf{M}_C + \mathsf{K}_C & 0 \\ 0 & \mathsf{M}_C + \mathsf{K}_C \end{pmatrix} \tag{6.107}$$

and coincides with the spectrally equivalent matrix implied by the energy balance (6.73) in Theorem 6.20.

The most complicated matrix arises from (6.80). However, the relevant energy balance (6.75) in Theorem 6.17 implies that this matrix is spectrally equivalent to the block-diagonal matrix

$$\begin{pmatrix} \mathsf{M}_C + \mathsf{K}_C & & & \\ & \mathsf{M}_D + \mathsf{K}_D & & \\ & & \mathsf{M}_C + \mathsf{K}_C & \\ & & & \mathsf{M}_D + \mathsf{K}_D \end{pmatrix}. \tag{6.108}$$

From (6.106)–(6.108), it follows that the efficient solution of LSFEM algebraic equations is contingent upon the existence of efficient (and robust) preconditioners for $\mathsf{M}_C + \mathsf{K}_C$ and $\mathsf{M}_D + \mathsf{K}_D$. As a result, solution of LSFEMs that use curl- and div-conforming spaces is not as straightforward as solution of LSFEMs derived from energy balances in the ADN setting for which all that is required is a good preconditioner for the Poisson equation. However, in the past two decades, there has been significant progress in developing both the theory and software for preconditioning the matrices that result from the use of curl-conforming and div-conforming spaces.

Geometric preconditioners rely on a sequence of geometrically nested grids to provide definitions of the appropriate prolongation and restriction operators. Geometric preconditioning for curl- and div-conforming spaces is studied in, e.g., [12,203,205]. The drawback of geometric preconditioners is their reliance on nested grids which are usually not available for large-scale applications and domains with complex geometries whose meshing may take considerable time. As a result, significant attention has been devoted to *algebraic* preconditioners that require minimal additional grid information and work by "aggregating" the matrix into coarser and coarser algebraic problems; see Figure 6.1. Various approaches exist that differ in the manner this aggregation is effected and how prolongation and restriction operators are defined between the fine and coarse algebraic problems. For some early work, see [50, 64, 211]. Recent approaches include preconditioners based on the idea of *auxiliary* spaces [207, 208, 248] and compatible discrete-gauge reformulations [65].

Highly efficient software implementations already exist for many of these algebraic preconditioners that make implementation and efficient solution of LSFEMs for div–curl and curl–curl systems much less challenging than they were just few years ago. Of particular note are the algebraic preconditioners available in the ML

Fig. 6.1 Aggregation for curl-conforming algebraic multigrid on a mesh with 1,994 edges and two cuts [211]. Each sphere represents a nodal grid point and the grey levels of the spheres give an indication of the nodal aggregates used on the finest level.

package of the Trilinos library [200] developed at Sandia National Laboratories, and the *hypre* library [157] from Lawrence Livermore National Laboratory.

6.5.2 Implementation of Non-Conforming Methods

The structure of algebraic problems in non-conforming methods assumes a particularly simple and illustrative form when using the lowest-order compatible spaces. For such spaces, we have simple and elegant factorizations of discrete operators into topological and metric components; see (B.68)–(B.72). Thus, below, we focus exclusively on implementations of (6.50) and (6.51) that use the Whitney complex (B.46).

Let us begin with the curl-conforming formulation (6.50) which uses the discrete weak divergence operator (B.51). It is not hard to see that the first-order optimality condition corresponding to (6.50) gives rise to the matrix

$$\mathsf{K}_{CD^*} = \mathsf{K}_C + \mathsf{K}_{D^*}\,, \tag{6.109}$$

where K_C is the "curl–curl" matrix defined in (6.105) and K_{D^*} is "stiffness" matrix corresponding to the weak divergence operator. Using the factorizations from Section B.3, it follows that

$$\mathsf{K}_{CD^*} = \mathsf{D}_1^T \mathsf{M}_D \mathsf{D}_1 + \mathsf{M}_C \mathsf{D}_0 \mathsf{M}_G^{-1} \mathsf{D}_0^T \mathsf{M}_C\,, \tag{6.110}$$

i.e., $\mathsf{K}_{CD^*} = \mathsf{M}_C \vec{\Delta}_C^h$, where $\vec{\Delta}_C^h$ is the discrete vector Laplacian operator (B.71) acting on curl-conforming finite element functions.

The div-conforming formulation (6.51) uses the discrete weak curl operator (B.51) and leads to the matrix

$$\mathsf{K}_{C^*D} = \mathsf{K}_{C^*} + \mathsf{K}_D\,. \tag{6.111}$$

In (6.111), K_D is the "div-div" matrix defined in (6.105) and K_{C^*} is the "stiffness" matrix corresponding to the weak curl operator. An equivalent form of (6.111) is given by

$$\mathsf{K}_{C^*D} = \mathsf{M}_D \mathsf{D}_1 \mathsf{M}_C^{-1} \mathsf{D}_1^T \mathsf{M}_D + \mathsf{D}_2^T \mathsf{M}_S \mathsf{D}_2\,, \tag{6.112}$$

i.e., $\mathsf{K}_{C^*D} = \mathsf{M}_D \vec{\Delta}_D^h$, where $\vec{\Delta}_D^h$ is the discrete vector Laplacian operator (B.72) acting on div-conforming finite element functions.

The discrete energy balances (6.60) and (6.61) guarantee that both (6.109) and (6.111) are symmetric and positive definite. Therefore, to solve the linear systems by preconditioned conjugate gradient methods, we again need preconditioners for the discrete vector Laplace operators $\vec{\Delta}_C^h$ and $\vec{\Delta}_D^h$, respectively. Algebraic preconditioners for $\vec{\Delta}_C^h$ based on discrete Hodge decomposition for curl-conforming finite elements were recently developed in [65]. Similar ideas have been used in [70] to obtain algebraic preconditioners for the div-conforming Laplacian. Thus, in principle, the efficient solution of linear systems engendered by non-conforming methods is no more complicated than that for their compliant brethren.

There is, however, one formal complication in the assembly of (6.109) and (6.111) that does not arise with conforming methods. Specifically, the weak discrete divergence and curl operators lead to the appearance of an *inverse* mass matrix as one of the factors in K_{CD^*} and K_{C^*D}. Formally, this means that K_{CD^*} and K_{C^*D} are dense which makes their assembly and storage impossible for all but very small problems.

However, in practice, we do not have to use consistent mass matrices. In the lowest-order case, it suffices to replace M_G and M_C by diagonal *lumped-mass* versions $\widetilde{\mathsf{M}}_G$ and $\widetilde{\mathsf{M}}_C$, respectively, as long as the latter are $O(h)$ approximations of the consistent mass matrices. This does not destroy the accuracy of LSFEMs and leads to modified linear systems for which K_{CD^*} and K_{C^*D} are replaced by

$$\widetilde{\mathsf{K}}_{CD^*} = \mathsf{D}_1^T \mathsf{M}_D \mathsf{D}_1 + \mathsf{M}_C \mathsf{D}_0 \widetilde{\mathsf{M}}_G^{-1} \mathsf{D}_0^T \mathsf{M}_C$$

and

$$\widetilde{\mathsf{K}}_{C^*D} = \mathsf{M}_D \mathsf{D}_1 \widetilde{\mathsf{M}}_C^{-1} \mathsf{D}_1^T \mathsf{M}_D + \mathsf{D}_2^T \mathsf{M}_S \mathsf{D}_2\,,$$

respectively. The modified matrices have the usual sparse structure and their assembly and storage do not pose any problems. Moreover, the analyses in [65,70] include inexact mass matrices so that the algebraic preconditioners from these papers can be applied to solve the modified systems.[30]

6.6 A Summary of Conclusions

As has been our custom, a brief summary of our observations about the LSFEMs for div–curl and curl–curl systems is presented in Table 6.1.

	Methods for div–curl systems	
Property↓ method→	compliant (6.48), (6.49)	non-conforming (6.50), (6.51)
Optimal error	√	√
Sparse matrix	√	–
Preconditioner	√	√
Dependent variables	2 vectors	1 vector

	Methods for curl–curl systems	
Property↓ method→	compliant (6.79)	compliant (6.80)
Optimal error	√	√
Sparse matrix	√	√
Preconditioner	√	√
Dependent variables	2 vectors	4 vectors

Table 6.1 Summary of properties of selected least-squares finite element methods for div–curl and curl–curl systems.

[30] An alternative approach that can be used in more general settings for non-conforming elements is to implement iterative solvers in an assembly-free manner. In this case, we only need the action of K_{CD^*} and K_{C^*D} and not the matrices themselves. Computing the action of these matrices requires the inversion of M_G^{-1} and M_C^{-1} that can be done using an internal conjugate gradient loop. Because M_G^{-1} and M_C^{-1} are well-conditioned and we do not need their inverses computed to machine precision, this requires only a few conjugate gradient iterations. Broadly speaking, this implementation approach resembles the way one implements negative norm LSFEMs for which the action of the least-squares matrix requires an approximate inverse of the discrete Poisson operator which is replaced by a preconditioner for this operator; see Section B.4.2.

Chapter 7
The Stokes Equations

The primitive variable formulation of the Stokes system is given by[1,2]

$$\begin{cases} -\nu\Delta \mathbf{u} + \nabla p = \mathbf{f} & \text{in } \Omega \\ \nabla\cdot\mathbf{u} = 0 & \text{in } \Omega \end{cases} \tag{7.1}$$

along with the velocity boundary condition[3]

$$\mathbf{u} = 0 \qquad \text{on } \partial\Omega \tag{7.2}$$

and the zero mean pressure constraint[4]

$$\int_\Omega p\, d\Omega = 0. \tag{7.3}$$

Throughout this chapter, we assume that $\mathbf{u}$ and p are non-dimensional variables, i.e., that (7.1) has been properly non-dimensionalized.[5]

---

[1] In (7.1)–(7.3), $\mathbf{u}$, p, ν, and $\mathbf{f}$ denote the velocity vector, pressure, kinematic viscosity, and body force per unit mass, respectively.

[2] In Chapter 6, we use the notation $\vec{\Delta}$ to denote the Laplace operator acting on vector-valued functions; we do so there so as to diffrentiate between that operator and the Laplace operator acting on scalar-valued functions that is denoted by Δ in Chapter 5. From this point on, we do not need to keep track of this ambiguity so that we simply use Δ for both cases.

[3] For simplicity, we consider homogeneous boundary conditions; inhomogeneous boundary conditions can be treated using, e.g., the approaches outlined in Section 12.1. Also, the Stokes equations and their equivalent first-order system formulations can be augmented by other types of boundary conditions, some of which are introduced in Section 7.1.

[4] A single scalar constraint on the pressure is needed because (7.1) and (7.2) determine the pressure only up to an additive constant, i.e., the Stokes operator with the velocity boundary condition has positive nullity. The zero mean constraint (7.3) is one way to fix that constant; setting the pressure value at a point is another way to remove the nullspace. The impact of the particular method employed to eliminate the nullspace on the numerical properties of least-squares finite element methods is discussed in Section 7.6.4.

[5] In this case, ν is the inverse of the Reynolds number *Re*.

P.B. Bochev, M.D. Gunzburger, *Least-Squares Finite Element Methods*, Applied Mathematical Sciences 166, DOI: 10.1007/b13382_7,

The Stokes system (7.1) is first introduced in Section 1.4.3 as an example of a partial differential equation (PDE) problem that is connected to the constrained optimization of a quadratic functional. Therefore, a Galerkin method for (7.1) recovers the naturally existing variational formulation, i.e., the mixed problem (1.59). According to Section 1.3.2, the setting of (1.59) is less attractive for finite element methods than the Rayleigh–Ritz setting because it requires the discrete inf–sup conditions (1.41) and (1.42) to hold; also, the corresponding discrete systems are not positive definite.

An alternative to the mixed formulation that offers a Rayleigh–Ritz setting for the Stokes equations is the null-space method (1.61). However, implementation of (1.61) requires divergence-free finite element subspaces of $[H^1(\Omega)]^d$; such spaces are not easy to construct. Therefore, the answer to the first question from Section 2.2.4 is affirmative: the use of least-squares principles for the Stokes equations is justified by the absence of a practical natural Rayleigh–Ritz setting for them.

To satisfy the first key to practicality of least-squares finite element methods (LSFEMs), we transform the primitive variable formulation of the Stokes system into equivalent first-order problems. As we noted in Section 2.2.2, this is not a uniquely defined process, so that we derive, in Section 7.1, three equivalent first-order formulations of the Stokes system (7.1)–(7.3). To obtain the corresponding LSFEMs, we follow the steps outlined in Chapter 3, i.e., start with the formulation of classes of continuous least-squares principles (CLSPs) for each one of the first-order Stokes systems and then follow by transforming select CLSPs into practical discrete least-squares principles (DLSPs).

The key juncture in this process is the identification of the energy balances for the first-order systems. The Agmon–Douglis–Nirenberg (ADN) and the vector-operator settings, already encountered in Chapters 5 and 6, are used here as well. For CLSPs defined in the ADN setting, transition to DLSPs relies on the abstract least-squares theory for ADN systems formulated in Chapter 4. Accordingly, error analysis of the resulting LSFEMs is simply a matter of specializing that theory to the first-order Stokes systems. LSFEMs in the ADN setting for the Stokes system are considered in Sections 7.2–7.6. Formulation and analysis of LSFEMs in the vector-operator setting is somewhat more involved because this setting results in non-conforming methods similar to those encountered in Section 6.3.3. However, the vector-operator setting is well worth the effort because it allows us to formulate, for some sets of non-standard boundary conditions (but, unfortunately, not for the velocity boundary condition), *locally conservative* LSFEMs for the Stokes equations, a feat that is very difficult, if not impossible, to accomplish in the ADN setting. LSFEMs in the vector-operator setting for the Stokes system are considered in Section 7.7.

7.1 First-Order System Formulations of the Stokes Equations

There are many possible ways to transform the primitive variable formulation (7.1) of the Stokes system into a first-order system. In what follows, we focus on three approaches that have been used extensively in connection with LSFEMs. As we saw in Section 2.2.2, first-order reformulation of higher-order PDEs necessarily involves the addition of new variables; the three reformulations of the Stokes system we consider involve different sets of additional variables. For the first reformulation we consider (see Section 7.1.1), the new variable is the axial vector of the *skew-symmetric part* $(\nabla\mathbf{u}-\nabla\mathbf{u}^\top)/2$ of the velocity gradient tensor $\nabla\mathbf{u}$; this approach leads to a *vorticity*-based first-order system. The second reformulation (see Section 7.1.2) uses the *symmetric part* $(\nabla\mathbf{u}+\nabla\mathbf{u}^\top)/2$ of the velocity gradient tensor; this approach gives rise to a *stress*-based Stokes system. The third reformulation (see Section 7.1.3) uses the *whole* velocity gradient tensor $\nabla\mathbf{u}$ as a new dependent variable; essentially, this is the reduction algorithm from [2] (see Section 4.1) applied to the Stokes system (7.1)–(7.3). The first two choices correspond to the introduction of physically meaningful additional variables, namely the vorticity and viscous stress, respectively.

7.1.1 The Velocity–Vorticity–Pressure System

The axial vector of the skew-symmetric part of the velocity gradient is referred to as the *vorticity*[6] $\omega=\nabla\times\mathbf{u}$. Using ω as a new dependent variable, the identity

$$\nabla\times\nabla\times\mathbf{u}=-\Delta\mathbf{u}+\nabla(\nabla\cdot\mathbf{u}),$$

and the incompressibility constraint $\nabla\cdot\mathbf{u}=0$, the primitive variable formulation of the Stokes problem (7.1) can be transformed into the equivalent *velocity–vorticity–pressure* (VVP) first-order Stokes system

$$\begin{cases} \nu\nabla\times\omega+\nabla p=\mathbf{f} & \text{in } \Omega \\ \nabla\times\mathbf{u}-\omega=0 & \text{in } \Omega \\ \nabla\cdot\mathbf{u}=0 & \text{in } \Omega \end{cases} \tag{7.4}$$

along with the velocity boundary condition (7.2) and the zero mean constraint (7.3). This system is, by a wide margin, the most popular[7] setting for LSFEMs for the Stokes (and Navier–Stokes) problems.

---

[6] Recall that, in two dimensions, the curl operator $\nabla\times$ can be given two meanings, namely

$$\nabla^\perp\phi=\begin{pmatrix}\phi_y\\-\phi_x\end{pmatrix} \qquad \text{and} \qquad \nabla\times\mathbf{u}=u_{2x}-u_{1y};$$

see (A.2) and (A.3). The context should make clear which operator is relevant. Also note that in this case the vorticity $\omega=u_{2x}-u_{1y}$ can be viewed as a scalar-valued variable.

[7] The VVP system was introduced in [234] and then explored by a number of researchers in [51,52,228,235–237,239,261], among others. Theoretical analyses of LSFEMs based on the VVP system are found in [34,35,39,53,54].

In this chapter, we use the ADN and the vector-operator settings to develop LSFEMs for (7.4). Depending on the setting, the VVP Stokes operator assumes different forms without, of course, changing the problem itself; see Remark 3.6. In the ADN setting, we view (7.4) as a square 4×4 or 7×7 system in $\mathbb{R}^2$ and $\mathbb{R}^3$, respectively. The VVP Stokes operator corresponding to the vector-operator setting is a 3×3 matrix of differential operators in both $\mathbb{R}^2$ and $\mathbb{R}^3$.

Non-standard boundary conditions

Owing to the inclusion of the vorticity as a dependent variable in (7.4), that system provides a convenient setting for the imposition of *non-standard*[8] boundary conditions that specify $(\nabla \times \mathbf{u}) \cdot \mathbf{n}$ or $(\nabla \times \mathbf{u}) \times \mathbf{n}$ on $\partial\Omega$. Such non-standard boundary conditions have been studied in the context of mixed-Galerkin methods in [32, 81, 152, 181] and are of theoretical and, in some cases, practical interest. Least-squares variational settings educe additional reasons to include non-standard boundary conditions in the formulation of LSFEMs. First, they provide further mathematical enlightenment in the ADN setting by exposing the nuanced dependence of the properties of the VVP system on the boundary operator, including the fact that, in two dimensions, the VVP system admits two different principal parts. Second, some non-standard boundary conditions turn out to be particularly well-suited for the vector-operator setting where they enable formulation of locally conservative LSFEMs.

We consider two sets of non-standard boundary conditions along with the VVP Stokes equations (7.4). The first specifies, in two dimensions,

$$\mathbf{u} \cdot \mathbf{n} = 0 \quad \text{and} \quad p = 0 \qquad \text{on } \partial\Omega \tag{7.5}$$

and, in three dimensions,

$$\mathbf{u} \cdot \mathbf{n} = 0, \quad \omega \cdot \mathbf{n} = 0, \quad \text{and} \quad p = 0 \qquad \text{on } \partial\Omega\,. \tag{7.6}$$

For obvious reasons, the boundary conditions (7.5) and (7.6) are referred to as the *normal velocity–pressure* boundary condition. Because (7.5) and (7.6) specify the pressure, the zero mean constraint (7.3) does not have to be imposed.

The second set of non-standard boundary conditions specifies, in two dimensions,

$$\mathbf{u} \cdot \mathbf{n} = 0 \quad \text{and} \quad \omega = 0 \qquad \text{on } \partial\Omega \tag{7.7}$$

and, in three dimensions,

$$\mathbf{u} \cdot \mathbf{n} = 0 \quad \text{and} \quad \omega \times \mathbf{n} = 0 \qquad \text{on } \partial\Omega\,. \tag{7.8}$$

[8] We refer to the velocity boundary condition $\mathbf{u} = 0$ on $\partial\Omega$ or its non-homogeneous version as the *standard* boundary condition for the Stokes equations.

We refer to the set (7.7) and (7.8) as the *normal velocity–tangential vorticity* boundary condition. This set does not specify the pressure so that the zero mean constraint (7.3) has to be included in the problem definition. The normal velocity–tangential vorticity boundary condition turns out to be of particular importance for the formulation of locally conservative LSFEMs for the Stokes equations.

The fact that, in two dimensions, (7.4) admits two different principal parts is formally verified in Section 7.2.1. One of these principal parts corresponds to a homogeneous elliptic operator and satisfies the ADN complementing condition (see Definition D.4) with both (7.5) and (7.7), but not with the standard velocity boundary condition (7.2). The other principal part corresponds to a non-homogeneous elliptic operator and satisfies the complementing condition with all three boundary conditions.

A similar dependence of the properties of the VVP system on the boundary conditions also exists in three dimensions. In Sections 7.2.1 and 7.7.1, we show that the ellipticity of (7.4), when equipped with one of the non-standard boundary conditions (7.6) or (7.8), can be strengthened (from non-homogeneous to homogeneous elliptic) by adding to (7.4) the "redundant" (in view of the definition of ω) equation

$$\nabla \cdot \omega = 0 \qquad \text{in } \Omega . \tag{7.9}$$

The resulting extended first-order system[9]

$$\begin{cases} \nu \nabla \times \omega + \nabla p = \mathbf{f} & \text{in } \Omega \\ \nabla \cdot \omega = 0 & \text{in } \Omega \\ \nabla \times \mathbf{u} - \omega = 0 & \text{in } \Omega \\ \nabla \cdot \mathbf{u} = 0 & \text{in } \Omega \end{cases} \tag{7.10}$$

has seven scalar unknowns and eight equations. This is not an issue when energy balances are sought in the vector-operator setting, but poses a formal difficulty for the ADN setting, where $\mathcal{L}$ is required to be a square $M \times M$ operator; see Section D.1. Fortunately, the system (7.10) can be easily brought back into compliance with this requirement by adding the gradient of a "slack" variable ϕ to the third equation in (7.10) and the homogeneous Dirichlet boundary condition

$$\phi = 0 \quad \text{on } \partial\Omega \tag{7.11}$$

to (7.6) or (7.8). In summary, whenever energy balances for the three-dimensional VVP Stokes equations with the non-standard boundary conditions are sought in the ADN setting, we have to formally work[10] with the extended VVP system

---

[9] Of course, (7.10) can also be considered along with the standard velocity boundary condition (7.2). However, this combination is not used in the formulation of LSFEMs because ellipticity of (7.4) with (7.2) cannot be improved by the addition of (7.9).

[10] Note that the addition of the slack variable term $\nabla\phi$ is innocuous because one easily sees that (7.11) and (7.12) imply that $\phi = 0$ in Ω.

$$\begin{cases} \nu\nabla\times\omega+\nabla p=\mathbf{f} & \text{in } \Omega \\ \nabla\cdot\omega=0 & \text{in } \Omega \\ \nabla\times\mathbf{u}-\omega+\nabla\phi=0 & \text{in } \Omega \\ \nabla\cdot\mathbf{u}=0 & \text{in } \Omega \end{cases} \tag{7.12}$$

and the extended versions[11]

$$\mathbf{u}\cdot\mathbf{n}=0, \quad \omega\cdot\mathbf{n}=0, \quad p=0 \quad \text{and} \quad \phi=0 \qquad \text{on } \partial\Omega \tag{7.13}$$

and

$$\mathbf{u}\cdot\mathbf{n}=0, \quad \omega\times\mathbf{n}=0, \quad \text{and} \quad \phi=0 \qquad \text{on } \partial\Omega \tag{7.14}$$

of (7.6) and (7.8), respectively. Clearly, the zero mean constraint (7.3) is not required for (7.13) but must be added to (7.14).

7.1.2 The Velocity–Stress–Pressure System

To obtain the *velocity–stress–pressure* (VSP) system, we use the scaled viscous stress tensor[12]

$$\underline{\mathbf{T}}=\sqrt{2\nu}\varepsilon(\mathbf{u}),$$

where[13]

$$\varepsilon(\mathbf{u})=\frac{1}{2}\left(\nabla\mathbf{u}+(\nabla\mathbf{u})^{\mathsf{T}}\right),$$

as a new dependent variable. In this case, we have the vector identity

$$\nabla\cdot\underline{\mathbf{T}}=\sqrt{2\nu}\left(\Delta\mathbf{u}+\nabla(\nabla\cdot\mathbf{u})\right),$$

where $\nabla\cdot\underline{\mathbf{T}}$ denotes a vector whose components are the divergences of the corresponding rows of $\underline{\mathbf{T}}$. Using this identity and the continuity equation $\nabla\cdot\mathbf{u}=0$, the primitive variable Stokes equations (7.1) can be transformed into the equivalent first-order system

$$\begin{cases} \sqrt{2\nu}\nabla\cdot\underline{\mathbf{T}}-\nabla p=\mathbf{f} & \text{in } \Omega \\ \nabla\cdot\mathbf{u}=0 & \text{in } \Omega \\ \underline{\mathbf{T}}-\sqrt{2\nu}\,\varepsilon(\mathbf{u})=0 & \text{in } \Omega \end{cases} \tag{7.15}$$

along with the boundary condition (7.2) and the zero mean constraint (7.3). The first documented use of the VSP system in LSFEMs is [55]. Although (7.15) has

---

[11] These boundary conditions are listed under (BC2) and (BC2A) in Table D.4.

[12] The stress tensor is given by $-p\underline{\mathbf{I}}+\nu(\nabla\mathbf{u}+(\nabla\mathbf{u})^{\mathsf{T}})$, where $\underline{\mathbf{I}}$ denotes the identity tensor. The viscous stress tensor is given by $\nu(\nabla\mathbf{u}+(\nabla\mathbf{u})^{\mathsf{T}})$.

[13] $\varepsilon(\mathbf{u})$ is referred to as the *rate of strain* tensor.

found numerous applications in stabilized methods (see, e.g., [26, 27, 176] and the references therein), its utilization in LSFEMs has remained limited.

Remark 7.1 Clearly, the VSP system is not as convenient as the VVP system for imposing boundary conditions that specify $(\nabla \times \mathbf{u}) \cdot \mathbf{n}$ or $(\nabla \times \mathbf{u}) \times \mathbf{n}$ on $\partial\Omega$. Moreover, in Section 7.2.1, we show that (7.15) admits only one principal part in both two and three dimensions, and that this principal part corresponds to a non-homogeneous elliptic operator. It follows that, in contrast to the VVP system, ellipticity of (7.15) cannot be improved by changing the boundary condition.

Thus, for the VSP system, non-standard boundary conditions do not provide any additional mathematical enlightenment, nor do they offer a better setting for LSFEMs. Accordingly, our consideration of LSFEMs for (7.15) is restricted to the standard velocity boundary condition (7.2) and the ADN setting. The assumptions of that setting for (7.15) are verified in Section D.2.4. Because the VSP operator corresponding to the ADN setting is a square 6×6 or 10×10 matrix of partial derivatives in $\mathbb{R}^2$ and $\mathbb{R}^3$, respectively, it does not have to be "padded" by slack variables to make it compliant with the ADN theory. □

7.1.3 The Velocity Gradient–Velocity–Pressure System

Section 4.4 asserts that the best possible case for developing LSFEMs in the ADN setting arises when the first-order system is homogeneous elliptic. Unfortunately, the VSP system is not homogeneous elliptic and the VVP system is homogeneous elliptic only with select boundary operators (see Table D.4 for the ellipticity classification of (7.4) and (7.12) with different sets of boundary conditions.) It turns out that in order to define a first-order Stokes system that is homogeneous elliptic *for any admissible boundary condition*, we have to use as new dependent variables *all* first derivatives of the velocity field, and then augment the differential equations and the boundary conditions by additional *redundant* equations and boundary conditions. Because the first derivatives of $\mathbf{u}$ are given by the non-symmetric velocity gradient tensor $\nabla \mathbf{u}$, we refer to these systems as velocity-gradient-based Stokes equations.

To define the first *velocity gradient–velocity–pressure formulation* (VGVP), we follow the transformation process described in Section 4.1 and use all first derivatives of the velocity as new dependent variables. We set $\underline{\mathbf{U}} = (\nabla \mathbf{u})^{\mathsf{T}}$ so that $\underline{\mathbf{U}}_{ij} = (\partial u_i / \partial x_j)$, $i, j = 1, \ldots, d$ for $d = 2$ or $d = 3$. Then,

$$\nabla \cdot \underline{\mathbf{U}} = \Delta \mathbf{u},$$

where the divergence operator is applied to the rows of $\underline{\mathbf{U}}$. Therefore, the primitive variable formulation of the Stokes equations (7.1) can be transformed into the following first-order system in terms of $\{\mathbf{u}, \underline{\mathbf{U}}, p\}$:

$$\begin{cases} -\nu\nabla\cdot\underline{\mathbf{U}}+\nabla p=\mathbf{f} & \text{in } \Omega \\ \underline{\mathbf{U}}-(\nabla\mathbf{u})^{\top}=0 & \text{in } \Omega \\ \nabla\cdot\mathbf{u}=0 & \text{in } \Omega \end{cases} \tag{7.16}$$

along with the velocity boundary condition (7.2) and the zero mean pressure constraint (7.3).

The system (7.16) was proposed in [98], where the new tensor-valued variable $\underline{\mathbf{U}}$ is referred to as the "velocity flux." Because that terminology is usually reserved for a different physical quantity, we prefer to refer to $\underline{\mathbf{U}}$ as the "velocity gradient." In Section 7.2, we see that (7.16) is non-homogeneous elliptic. The main idea of [98] was to augment (7.16), (7.2) and (7.3) with "redundant" equations and boundary conditions until the boundary value problem becomes homogeneous elliptic so that the advantages of these systems, explained in Chapter 4, can be exploited in the formulation of LSFEMs.

The "redundant" equations are defined by assuming that all fields are sufficiently smooth and then applying basic vector calculus identities. Specifically, in view of the identity[14] $tr\underline{\mathbf{U}}=\nabla\cdot\mathbf{u}=0$, the definition of $\underline{\mathbf{U}}$, and the boundary condition (7.2), one can add to (7.16), (7.2), and (7.3) the equations

$$\nabla(tr\underline{\mathbf{U}})=0 \qquad \text{in } \Omega\,, \tag{7.17}$$

$$\nabla\times\underline{\mathbf{U}}=0 \qquad \text{in } \Omega\,, \tag{7.18}$$

and the boundary condition

$$\underline{\mathbf{U}}\times\mathbf{n}=0 \qquad \text{on } \partial\Omega\,. \tag{7.19}$$

In (7.18), $\nabla\times\underline{\mathbf{U}}$ denotes the vector whose components are the curls of the corresponding rows of $\underline{\mathbf{U}}$ and, in (7.19), $\underline{\mathbf{U}}\times\mathbf{n}$ is the vector whose components are the vector products of the rows of $\underline{\mathbf{U}}$ with the unit outer normal vector $\mathbf{n}$. The *extended* VGVP boundary value system thus consists of the first-order system

$$\begin{cases} -\nu\nabla\cdot\underline{\mathbf{U}}+\nabla p=\mathbf{f} & \text{in } \Omega \\ \underline{\mathbf{U}}-(\nabla\mathbf{u})^{\top}=0 & \text{in } \Omega \\ \nabla\times\underline{\mathbf{U}}=0 & \text{in } \Omega \\ \nabla\cdot\mathbf{u}=0 & \text{in } \Omega \\ \nabla(tr\underline{\mathbf{U}})=0 & \text{in } \Omega \end{cases} \tag{7.20}$$

along with the velocity boundary condition (7.2), the "redundant" boundary condition (7.19), and the zero mean pressure constraint (7.3).

[14] $tr\underline{\mathbf{U}}$ denotes the trace of the tensor $\underline{\mathbf{U}}$, i.e., $tr\underline{\mathbf{U}}=\sum_{i=1}^{d}(\underline{\mathbf{U}})_{ii}$.

The extended VGVP system (7.20), (7.2), (7.19), and (7.3) is consistent but overdetermined[15] so that, formally, it is not ready for the ADN setting. As in the case of the extended VVP system (7.10), the number of equations and unknowns in (7.20) can be equilibrated by adding a sufficient number of "slack" variables, thereby making the problem compliant with the ADN notion of ellipticity stated in Definition D.1. However, owing to the large number of equations and unknowns, verification of the algebraic conditions in the ADN theory becomes extremely tedious without adding much insight beyond what could be gleaned from the examples already presented in Section D.2.4.

Thus, to derive the energy balance of (7.20), we forego the formal application of the ADN theory and simply quote the relevant results from the literature. Because the available energy balances have all the hallmarks of the ADN theory, i.e., the variables are treated as sets of unrelated scalar fields and the a priori bounds are stated in terms of products of Sobolev spaces, we continue to refer to this setting for (7.20) as the ADN setting[16] and formulate the LSFEMs using the least-squares process for ADN homogeneous elliptic systems developed in Section 4.4.

Remark 7.2 Because (7.20) is already homogeneous elliptic for any well-posed boundary conditions, the non-standard boundary conditions introduced in Section 7.1.1 do not improve the setting for LSFEMs for the extended VGVP system. Likewise, there are no reasons to seek energy balances for that system in the vector-operator setting because the main purpose of (7.20) is to break the dependencies in the solution components so that they can be treated as independent scalar fields in the a priori estimates. This is exactly the opposite of what the vector-operator setting is supposed to achieve. □

Remark 7.3 The first example of using the velocity gradient to reformulate the Stokes equations is given in [111]. The key difference between that approach and the approach of [98] is that the new variables are the entries of the velocity gradient explicitly *constrained* by the equation $\nabla\cdot\mathbf{u}=0$. In two dimensions the new variables are

$$\underline{\mathbf{G}} = \begin{pmatrix} v_1 & v_2 \\ v_3 & -v_1 \end{pmatrix}, \tag{7.21}$$

where

$$v_1 = \frac{\partial u_1}{\partial x_1} = -\frac{\partial u_2}{\partial x_2}, \qquad v_2 = \frac{\partial u_1}{\partial x_2}, \qquad \text{and} \qquad v_3 = \frac{\partial u_2}{\partial x_1}.$$

Using (7.21) and the equality of mixed second derivatives, the Stokes problem (7.1) in two dimensions can be transformed into the first-order system

[15] In two dimensions, the number of scalar unknowns equals 7 and the number of equations equals 11. In three dimensions, we have 13 scalar unknowns and 25 equations.

[16] This is also justified by the fact that the ADN setting would have resulted in precisely the same energy balances, except for the inclusion of additional slack variables.

$$\begin{cases} -\nu\nabla\cdot\underline{\mathbf{G}}+\nabla p=\mathbf{f} & \text{in } \Omega \\ \nabla\times\underline{\mathbf{G}}=0 & \text{in } \Omega \\ \underline{\mathbf{G}}\times\mathbf{n}=0 & \text{on } \partial\Omega\,. \end{cases} \tag{7.22}$$

In [111], the new variables (7.21) are called "accelerations" and the system (7.22) the "acceleration–velocity" formulation of the Stokes equations. However, the new variables are not components of the acceleration vector so that, instead, we call the system (7.22) the *constrained velocity gradient–velocity–pressure* formulation of the Stokes problem.

The planar system (7.22) has four equations and four scalar unknowns; one can show that it is homogeneous elliptic in the sense of ADN. The velocity has been eliminated from (7.22); it can be recovered by solving the div–curl system

$$\begin{cases} \nabla\times\mathbf{u}=v_3-v_2 & \text{in } \Omega \\ \nabla\cdot\mathbf{u}=0 & \text{in } \Omega \\ \mathbf{u}\cdot\mathbf{n}=0 & \text{on } \partial\Omega\,. \end{cases} \tag{7.23}$$

Although it is not obvious that the solution of (7.23) satisfies the boundary condition (7.2), it can be shown that this is indeed the case.

Elimination of **u** in (7.22) can be considered as an artifact because one can simply consider (7.22) together with (7.23). Such a first-order system is studied in [119], where the new variables are called "stresses" and the corresponding first-order system is called the "stress–velocity–pressure" Stokes system, despite the fact that the new variables are not the components of the viscous stress tensor. This is not to be confused with the formulation of Section 7.1.2 for which the true viscous stress tensor is used.

Because it is not clear how (7.22) can be extended to the three-dimensional case and eventually to the Navier–Stokes equations, we do not consider this system when we formulate LSFEMs. □

7.2 Energy Balances in the Agmon–Douglis–Nirenberg Setting

Transformation of the primitive variable formulation (7.1) of the Stokes equations into a first-order system fulfills the first key to defining practical LSFEMs. The next step is to develop a collection of CLSPs for the three types of first-order systems defined in Section 7.1; these CLSPs serve as a basis for mathematically sound LSFEMs. We focus on defining data spaces $Y(\Omega)$ and $B(\partial\Omega)$ and solution spaces $X(\Omega)$ that verify the abstract energy balances (3.18) or (3.19) for the three types of first-order Stokes systems.[17]

[17] As usual, we adhere to the rule that data space components follow the order of the equations in the first order systems. Likewise, the order in which we list the variables of the first-order systems, e.g., $\{\mathbf{u},\omega,p\}$, determines the order of the component solution spaces in X.

Although there are myriad possible choices for these spaces, they all fall into two basic categories associated with the ADN setting and the vector-operator setting,[18] respectively. For reasons explained earlier, the vector-operator setting is used solely for the VVP first-order system with the normal velocity–tangential vorticity boundary condition defined in Section 7.1.1. This case is presented in Section 7.7.1. The ADN setting is used for all three first-order Stokes formulations with the understanding that formal application of the ADN theory to the VGVP system is left out from the discussion. The energy balances derived in the ADN setting are presented in Sections 7.2.1–7.2.3.

The Stokes operator with the velocity boundary condition (7.2), whether in the primitive variable form or in an equivalent first-order form, has positive nullity because the pressure p is determined only up to an additive constant. Using the terminology of Chapter 3, the zero mean pressure constraint (7.3) is the functional $\ell : X \mapsto \mathbb{R}$ that defines the algebraic complement space[19] X^C from (C.7). Because different Stokes systems have different sets of variables and (7.3) is always defined by using the pressure, throughout this chapter we use the notation

$$\ell(p) = \int_\Omega p \, d\Omega \tag{7.24}$$

instead of the generic $\ell(u)$.

7.2.1 The Velocity–Vorticity–Pressure System

The following theorem, which is a specialization of Theorem D.1, states the ADN energy balances for the two-dimensional VVP system (7.4) with the normal velocity–presssure (7.5), normal velocity–tangential vorticity (7.7), and the standard velocity (7.2) boundary conditions. The assumptions of Theorem D.1 for the VVP system are verified in Section D.2.4.

Theorem 7.4 *Assume that $\Omega \subset \mathbb{R}^2$ is a bounded domain of class C^{q+1}, where $q \geq 0$ is integer. Then, there exists a positive constant C_q such that, for any collection $\{\mathbf{u}, \omega, p\}$ of smooth functions,*

---

[18] Recall that the main distinction between these two settings is in how they treat the dependent variables in the PDE problem; see Remarks 3.6 and 5.6. In particular, the ADN setting views all variables as sets of independent scalar fields which leads to identification of, e.g., the solution space X with products of scalar Sobolev spaces $H^{q+t_j}(\Omega)$. In the vector-operator setting, the variables are identified with fields from primal (A.52) and/or dual (A.53) De Rham complexes.

[19] Recall that, according to Remark 5.8, another possible way to fix the pressure is to set $\ell(p) = \langle \delta(\mathbf{x}_0), p \rangle$; this corresponds to the common practice of fixing the value of the pressure at a point $\mathbf{x}_0$; see, e.g., [16,67]. The adverse effect on iterative solvers resulting from this choice is discussed in Section 7.6.4.

$$C_q\big(\|\mathbf{u}\|_{q+1}+\|\omega\|_{q+1}+\|p\|_{q+1}\big) \le \|\mathbf{u}\cdot\mathbf{n}\|_{q+1/2,\partial\Omega}+\|p\|_{q+1/2,\partial\Omega} \\ +\|\nabla^{\perp}\omega+\nabla p\|_q+\|\nabla\times\mathbf{u}-\omega\|_q+\|\nabla\cdot\mathbf{u}\|_q \tag{7.25}$$

for the normal velocity–pressure boundary condition and

$$C_q\big(\|\mathbf{u}\|_{q+1}+\|\omega\|_{q+1}+\|p\|_{q+1}\big) \le \|\mathbf{u}\cdot\mathbf{n}\|_{q+1/2,\partial\Omega}+\|\omega\|_{q+1/2,\partial\Omega} \\ +\|\nabla^{\perp}\omega+\nabla p\|_q+\|\nabla\times\mathbf{u}-\omega\|_q+\|\nabla\cdot\mathbf{u}\|_q+|\ell(p)| \tag{7.26}$$

for the normal velocity–tangential vorticity boundary condition. If Ω is of class C^{q+2}, where $q\ge 0$ is integer, there exists a positive constant C_q such that

$$C_q\big(\|\mathbf{u}\|_{q+2}+\|\omega\|_{q+1}+\|p\|_{q+1}\big) \le \|\mathbf{u}\|_{q+3/2,\partial\Omega} \\ +\|\nabla^{\perp}\omega+\nabla p\|_q+\|\nabla\times\mathbf{u}-\omega\|_{q+1}+\|\nabla\cdot\mathbf{u}\|_{q+1}+|\ell(p)| \tag{7.27}$$

for the velocity boundary condition. Moreover, Assumption D.2 *holds for the VVP system, i.e., the bounds* (7.25)–(7.27) *can be extended to all real values of q.*

Proof. The proof of (7.25)–(7.27) reduces to verifying the assumptions of the ADN theory for the VVP equations along with the boundary conditions (7.5), (7.7), and (7.2), respectively. This is done in Section D.2.4 so that all that remains is to specialize the solution and data spaces (D.5) from Theorem D.1 to the VVP boundary value problems. Because these spaces are determined by the ADN solution and equation weights, their identification in the present setting is simply a matter of selecting the proper ADN weights for each of the boundary condition, i.e., the weights for which the complementing condition holds.

Proposition D.16 asserts that (7.5) and (7.7) pass the complementing condition with the weights in (D.32). Thus, for the normal velocity–pressure boundary condition and the normal velocity–tangential vorticity boundary condition, the solution and data spaces (D.5) specialize to[20]

$$X_q=[H^{q+1}(\Omega)]^4 \qquad Y_q=[H^{q}(\Omega)]^4 \qquad B_q=[H^{q+1/2}(\partial\Omega)]^2 \tag{7.28}$$

and (7.25) and (7.26) follow from (D.6).

Propositions D.14 and D.15 assert that the VVP Stokes system with the velocity boundary condition (7.2) satisfies the complementing condition with the weights in (D.33) but not with the weights in (D.32). Consequently, the correct choice of weights for (7.2) is (D.33) and, for the velocity boundary condition, the solution and data spaces in (D.5) specialize to

$$X_q=[H^{q+2}(\Omega)]^2\times H^{q+1}(\Omega)\times H^{q+1}(\Omega) \\ Y_q=[H^{q}(\Omega)]^2\times H^{q+1}(\Omega)\times H^{q+1}(\Omega) \qquad B_q=[H^{q+3/2}(\partial\Omega)]^2\,. \tag{7.29}$$

The energy balance (7.27) follows again from (D.6) specialized to (7.29).

---

[20] Recall that, in two dimensions, the vorticity is viewed as a scalar-valued variable.

For a proof that (7.25)–(7.27) can be extended to all real values of q, see [53, Theorem 2]. □

In three dimensions, the VVP system (7.4) is appropriate for the ADN setting but its extended version[21] (7.10) is not. Accordingly, to obtain energy balances for the velocity boundary condition, we specialize Theorem D.1 to the original VVP formulation (7.4), whereas for the non-standard boundary conditions, we use the extended formulation (7.12) that does fit the ADN setting.

Theorem 7.5 *Assume that $\Omega \subset \mathbb{R}^3$ is a bounded domain of class C^{q+1}, where $q \geq 0$ is an integer. Then, there exists a positive constant C_q such that, for any collection $\{\mathbf{u}, \phi, \omega, p\}$ of smooth functions,*

$$\begin{aligned} C_q\big(\|\mathbf{u}\|_{q+1} + \|\phi\|_{q+1} + \|\omega\|_{q+1} + \|p\|_{q+1}\big) \leq \\ \|\mathbf{u}\cdot\mathbf{n}\|_{q+1/2,\partial\Omega} + \|p\|_{q+1/2,\partial\Omega} + \|\phi\|_{q+1/2,\partial\Omega} \\ + \|\nabla\times\omega + \nabla p\|_q + \|\nabla\cdot\omega\|_q + \|\nabla\times\mathbf{u} + \nabla\phi - \omega\|_q + \|\nabla\cdot\mathbf{u}\|_q \end{aligned} \tag{7.30}$$

for the normal velocity–pressure boundary condition and

$$\begin{aligned} C_q\big(\|\mathbf{u}\|_{q+1} + \|\phi\|_{q+1} + \|\omega\|_{q+1} + \|p\|_{q+1}\big) \leq \\ \|\mathbf{u}\cdot\mathbf{n}\|_{q+1/2,\partial\Omega} + \|\omega\times\mathbf{n}\|_{q+1/2,\partial\Omega} + \|\phi\|_{q+1/2,\partial\Omega} + |\ell(p)| \\ + \|\nabla\times\omega + \nabla p\|_q + \|\nabla\cdot\omega\|_q + \|\nabla\times\mathbf{u} + \nabla\phi - \omega\|_q + \|\nabla\cdot\mathbf{u}\|_q \end{aligned} \tag{7.31}$$

for the normal velocity–tangential vorticity boundary condition. If Ω is of class C^{q+2}, where $q \geq 0$ is an integer, there exists a positive constant C_q such that, for any collection $\{\mathbf{u}, \omega, p\}$ of smooth functions,

$$\begin{aligned} C_q\big(\|\mathbf{u}\|_{q+2} + \|\omega\|_{q+1} + \|p\|_{q+1}\big) \leq \|\mathbf{u}\|_{q+3/2,\partial\Omega} + |\ell(p)| \\ + \|\nabla\times\omega + \nabla p\|_q + \|\nabla\times\mathbf{u} - \omega\|_{q+1} + \|\nabla\cdot\mathbf{u}\|_{q+1} \end{aligned} \tag{7.32}$$

for the velocity boundary condition. Moreover, the bounds (7.30)–(7.32) *can be extended to all real q.*

Proof. As in the case of the two-dimensional VVP system, the proof comes down to verifying the assumptions of the ADN theory for the appropriate VVP boundary value problems. The main difference with Theorem 7.4 is that, depending on the boundary condition, these assumptions are checked using different formulations of the first-order VVP system.

Concretely, for the normal velocity–pressure (7.13) and the normal velocity–tangential vorticity (7.14) boundary conditions, we propose to check the ADN

[21] Recall that the redundant equation (7.9) is added to (7.4) in order to allow for a homogeneous elliptic principal part for the three-dimensional VVP Stokes operator. Theorem 7.5 shows that the two non-standard boundary conditions make it possible to take advantage of this improved ellipticity in the energy balances.

theory using the extended VVP system[22] (7.12). Proposition D.19 in Section D.2.4 shows that in this case the complementing condition holds with the weights in (D.45), i.e., (7.12) together with either of the non-standard boundary conditions is homogeneous elliptic. As a result, the solution and data spaces from (D.5) specialize to

$$X_q = [H^{q+1}(\Omega)]^8 \qquad Y_q = [H^q(\Omega)]^8 \qquad B_q = [H^{q+1/2}(\partial\Omega)]^4 \tag{7.33}$$

and (7.30)–(7.31) follow from (D.6).

On the other hand, Propositions D.17–D.18 tell us that the ellipticity of (7.12) with the velocity boundary condition cannot be improved.[23] Indeed, those propositions establish that the complementing condition holds for (7.12) and (7.2) with the weights in (D.46), but not with the weights in (D.45), i.e., this boundary value problem remains non-homogeneous elliptic. Because the same is true for the original VVP system, the use of the extended formulation is not justified for (7.2) and we verify the ADN theory using (7.4). According to Section D.2.4, the complementing condition holds for (7.4) and (7.2) with the weights in (D.42) so that the solution and data spaces from (D.5) specialize to

$$\begin{gathered} X_q = [H^{q+2}(\Omega)]^3 \times [H^{q+1}(\Omega)]^3 \times H^{q+1}(\Omega) \\ Y_q = [H^q(\Omega)]^3 \times [H^{q+1}(\Omega)]^3 \times H^{q+1}(\Omega) \qquad B_q = [H^{q+3/2}(\partial\Omega)]^3 . \end{gathered} \tag{7.34}$$

Therefore, (7.32) follows from (D.6). For extension of (7.30)–(7.32) to all real q see [53]. □

Remark 7.6 The role of the redundant equation (7.9) is to allow the existence of two principal parts for the three-dimensional VVP system. Being able to take advantage of the available homogeneous elliptic part depends on which boundary condition is imposed. In contrast, the original VVP formulation admits only one principal part and it is non-homogeneous elliptic. Using (D.43), which are the analogues of the weights in (D.45), results in a degenerate principal part: $\det\mathcal{L}^P(x,\xi) = 0$. As a result, regardless of the boundary condition, the ellipticity of (7.4) cannot be improved. □

7.2.2 The Velocity–Stress–Pressure System

We obtain the energy balances for the VSP Stokes system by again specializing Theorem D.1. The assumptions of that theorem for the VSP system are verified

[22] Formally, the use of the extended VVP system leads to the inclusion of the slack variable ϕ in the energy balances. However, it is easy to see that this variable is identically zero and can be omitted from the formulation of LSFEMs.

[23] Naturally, the velocity boundary condition for this system is extended by the Dirichlet condition (7.11) for the slack variable so that the boundary operator corresponds to (D.48).

in Section D.2.5. The unique principal part of the VSP first-order system is nonhomogeneous elliptic in $\mathbb{R}^2$ and $\mathbb{R}^3$. As a result, the structure of its energy balance does not change with the boundary conditions or the space dimension. Thus, as already stated in Section 7.1.2, attention is restricted to the velocity boundary condition.

Theorem 7.7 *Let q be a nonnegative integer. Assume that $\Omega \subset \mathbb{R}^d$, where $d = 2,3$ is the space dimension, is a bounded domain of class C^{q+2}. Then, there exists a positive constant C_q such that, for any collection $\{\mathbf{u}, \underline{\mathbf{T}}, p\}$ of smooth functions,*

$$\begin{aligned} C_q\big(\|\mathbf{u}\|_{q+2} + \|\underline{\mathbf{T}}\|_{q+1} + \|p\|_{q+1}\big) &\leq \|\mathbf{u}\|_{q+3/2,\partial\Omega} + |\ell(p)| \\ &+ \|\sqrt{2\nu}\,\nabla\cdot\underline{\mathbf{T}} - \nabla p\|_q + \|\nabla\cdot\mathbf{u}\|_{q+1} + \|\underline{\mathbf{T}} - \sqrt{2\nu}\,\varepsilon(\mathbf{u})\|_{q+1}. \end{aligned} \tag{7.35}$$

The bound (7.35) *can be extended to all real q.*

Proof. The size of the VSP system and the number of independent conditions in (7.2) in d dimensions equals $M = 4d-2$ and $m = d$, respectively. From Proposition D.22, we know that the velocity boundary condition and the VSP operator verify the complementing condition with the weights in (D.56). In three dimensions, the proper weights are defined in (D.57). Assume that X_q, Y_q, and B_q in (D.5) are defined using these weights. Then, (7.35) follows from (D.6). For a proof of the extension of (7.35) to all real q, we refer to [55]. □

7.2.3 The Velocity Gradient–Velocity–Pressure System

In this section, we state the energy balances for the VGVP system (7.16) and its extended version (7.20) without recourse to the ADN theory. The reasons for this, and why we continue to refer to the ensuing setting as the "ADN setting," were explained in Section 7.1.3. Also, for the reasons discussed in Remark 7.2, we only consider the energy balances for the VGVP system for the standard velocity boundary condition (7.2).

For the original VGVP system (7.16), we have the following a priori estimate.

Theorem 7.8 *Assume that $\Omega \subset \mathbb{R}^d$, where $d = 2,3$ is the space dimension, is a bounded domain of class C^1. Then, there exists a positive constant C such that for every*[24] $\{\mathbf{u}, \underline{\mathbf{U}}, p\} \in [H^1(\Omega)]^d \times [L^2(\Omega)]^{d^2} \times L^2(\Omega)$

$$\begin{aligned} C\big(\|\mathbf{u}\|_1 + \|\underline{\mathbf{U}}\|_0 + \|p\|_0\big) &\leq \|\mathbf{u}\|_{1/2,\partial\Omega} + |\ell(p)| \\ &+ \|-\nu\nabla\cdot\underline{\mathbf{U}} + \nabla p\|_{-1} + \|\underline{\mathbf{U}} - (\nabla\mathbf{u})^{\mathsf{T}}\|_0 + \|\nabla\cdot\mathbf{u}\|_0. \end{aligned} \tag{7.36}$$

Proof. See [98, Theorem 3.1]. □

[24] When we write $\underline{\mathbf{U}} \in [L^2(\Omega)]^{d^2}$, we mean that each of the d^2 elements of the tensor $\underline{\mathbf{U}}$ belongs to $L^2(\Omega)$, e.g., in two dimensions, we have that $(\underline{\mathbf{U}})_{1,1}$, $(\underline{\mathbf{U}})_{1,2}$, $(\underline{\mathbf{U}})_{2,1}$, and $(\underline{\mathbf{U}})_{2,2}$ belong to $L^2(\Omega)$.

The solution and data spaces implied by the a priori bound in this theorem are

$$\begin{aligned} X_{-1} &= [H^1(\Omega)]^d \times [L^2(\Omega)]^{d^2} \times L^2(\Omega) \\ Y_{-1} = H^{-1}(\Omega)^d \times [L^2(\Omega)]^{d^2} \times L^2(\Omega) &\qquad B_{-1} = H^{1/2}(\partial\Omega)^d , \end{aligned} \tag{7.37}$$

i.e., the original VGVP system (7.16) is non-homogeneous elliptic. The use of the index -1 in (7.37) is justified by the fact that the exact same spaces[25] would have resulted by setting $q = -1$ in a family of energy balances obtained through a formal application of the ADN theory to (7.16).

For the extended VGVP system, we have the following a priori estimate.

Theorem 7.9 *Assume that $\Omega \subset \mathbb{R}^d$, where $d = 2,3$ is the space dimension, is a bounded domain of class C^1. Then, there exists a positive constant C such that for every $\{\mathbf{u}, \underline{\mathbf{U}}, p\} \in [H^1(\Omega)]^d \times [H^1(\Omega)]^{d^2} \times H^1(\Omega)$*

$$\begin{aligned} C\big(\|\mathbf{u}\|_1 + \|\underline{\mathbf{U}}\|_1 + \|p\|_1\big) \le \|\mathbf{u}\|_{1/2,\partial\Omega} &+ \|\mathbf{n} \times \underline{\mathbf{U}}\|_{1/2,\partial\Omega} + |\ell(p)| \\ + \| -\nu\nabla\cdot\underline{\mathbf{U}} + \nabla p\|_0 &+ \|\underline{\mathbf{U}} - (\nabla\mathbf{u})^T\|_0 \\ &+ \|\nabla\times\underline{\mathbf{U}}\|_0 + \|\nabla\cdot\mathbf{u}\|_0 + \|\nabla(tr\underline{\mathbf{U}})\|_0 . \end{aligned} \tag{7.38}$$

Proof. See [98, Theorem 3.2]. □

Theorem 7.9 implies an energy balance for the extended VGVP system (7.20) in the following solution and data spaces:

$$\begin{aligned} X_0 &= [H^1(\Omega)]^d \times [H^1(\Omega)]^{d^2} \times H^1(\Omega) \\ Y_0 = [L^2(\Omega)]^d \times [L^2(\Omega)]^{d^2} &\times [L^2(\Omega)]^{(7d-12)} \times L^2(\Omega) \times [L^2(\Omega)]^d \\ B_0 &= [H^{1/2}(\partial\Omega)]^d \times [H^{1/2}(\partial\Omega)]^{(7d-12)} . \end{aligned} \tag{7.39}$$

Therefore, among the first-order Stokes formulations considered so far, (7.20) is the only one that *achieves homogeneous ellipticity with the velocity boundary condition.* This and the fact that (7.39) can be related[26] to spaces indexed by $q = 0$ in an ADN energy balance, justifies the use of the subscript 0 in the space designations above.

---

[25] For the VGVP system (7.16), the number of equations equals the number of the scalar unknowns in both two and three dimensions, i.e., that system does not have to be "padded" by slack variables to make it fit the ADN setting. Therefore, an ADN energy balance for (7.16) does not include additional solution space components corresponding to slack variables.

[26] Because the extended VGVP system has more equations than unknowns, it has to be padded by slack variables to make it fit the ADN setting. Therefore, solution and data spaces derived in the ADN setting would include extra components for the slack variables and their boundary conditions. When properly formulated, the new system is completely equivalent to the old one and so it follows that all slack variables have to be identically zero. Consequently, formulation of LSFEMs can proceed without these variables, i.e., using the energy balance implied by the spaces in (7.39).

Remark 7.10 Because in this section we did not use the ADN theory, the results in Theorems 7.8 and 7.9 are stated for concrete data and solution spaces, as opposed to families of spaces parameterized by a regularity index q appearing in Sections 7.2.1 and 7.2.2. The spaces implied by Theorems 7.8 and 7.9 correspond to $q = -1$ and $q = 0$, respectively, in an ADN energy balance. The abstract least-squares theory for ADN systems in Chapter 4 tells us that these are the only choices worth considering in the formulation of LSFEMs for first-order systems so that our ability to leverage this theory for the formulation and the analysis of least-squares methods for (7.16) and (7.20) has not been diminished by not formally using the ADN setting. □

7.3 Continuous Least-Squares Principles in the Agmon–Douglis–Nirenberg Setting

The next step in the development of LSFEMs is the formulation of well-posed CLSPs based on the results in Section 7.2. As usual, CLSPs are stated by giving the definition of a least-squares functional and a solution space over which this functional is to be minimized. Following the practice from Chapters 5 and 6, boundary conditions are imposed on the solution spaces so that boundary residuals do not have to be included in the least-squares functionals.

We use the ADN setting[27] to formulate CLSPs for the VVP, VSP, and VGVP Stokes systems with the velocity boundary condition and for the extended VVP system (7.12) with the two non-standard sets of boundary conditions. For the VVP and VSP systems, this yield energy balances parameterized by the regularity index q. The abstract least-squares theory for ADN systems in Chapter 4 tells us that the best settings for least-squares methods result from the choices $q = 0$ or $q = -1$, with the choice depending on whether the system is homogeneous or non-homogeneous[28] elliptic. Because any other choice of q does not improve the setting for LSFEMs, CLSPs are stated[29] only for these two cases.

7.3.1 The Velocity–Vorticity–Pressure System

We start with CLSPs derived from the ADN energy balances in Section 7.2.1. Consider first the two non-standard sets of boundary conditions. Combined with a suitable VVP system, (7.5)–(7.8) give rise to homogeneous elliptic problems. Because the form of the appropriate VVP system and the boundary conditions depends on the space dimension, we state the CLSPs in $\mathbb{R}^2$ and $\mathbb{R}^3$ separately. To this end, we

[27] CLSPs in the vector-operator setting are considered in Section 7.7.2 and only for the three-dimensional VVP system with the normal velocity–tangential vorticity boundary condition (7.8).

[28] For non-homogeneous elliptic systems, $q = -1$ ensures the validity of Assumption 3.20.

[29] Note that the energy balances for the VGVP systems in Theorems 7.8 and 7.9 already correspond to these indices.

need the spaces

$$\mathbf{H}^1_n(\Omega) = \{\mathbf{u} \in [H^1(\Omega)]^d \mid \mathbf{n}\cdot\mathbf{u} = 0 \text{ on } \partial\Omega\} \quad d = 2,3,$$

and

$$\mathbf{H}^1_t(\Omega) = \{\mathbf{u} \in [H^1(\Omega)]^3 \mid \mathbf{n}\times\mathbf{u} = 0 \text{ on } \partial\Omega\},$$

introduced in Section 6.3.2.

Non-standard boundary conditions

In two dimensions, the appropriate VVP system for (7.5) and (7.7) is given by the original formulation (7.4). Assuming that the unknowns are ordered as $\{\mathbf{u},\omega,p\}$ and setting $q=0$ in the energy balance (7.25) from Theorem 7.4 for the normal velocity–pressure boundary condition (7.5), gives the following well-posed CLSP:

$$\begin{cases} J_0(\mathbf{u},\omega,p;\mathbf{f}) = \|\nabla^\perp\omega + \nabla p - \mathbf{f}\|_0^2 + \|\nabla\times\mathbf{u} - \omega\|_0^2 + \|\nabla\cdot\mathbf{u}\|_0^2 \\ X_0 = \mathbf{H}^1_n(\Omega)\times H^1(\Omega)\times H^1_0(\Omega). \end{cases} \tag{7.40}$$

For the normal velocity–tangential vorticity boundary condition (7.7), we use (7.26) from the same theorem and, with the same q, obtain another well-posed CLSP:

$$\begin{cases} J_0(\mathbf{u},\omega,p;\mathbf{f}) = \\ \qquad \|\nabla^\perp\omega + \nabla p - \mathbf{f}\|_0^2 + \|\nabla\times\mathbf{u} - \omega\|_0^2 + \|\nabla\cdot\mathbf{u}\|_0^2 + |\ell(p)|^2 \\ X_0 = \mathbf{H}^1_n(\Omega)\times H^1_0(\Omega)\times H^1(\Omega). \end{cases} \tag{7.41}$$

In three dimensions, the appropriate ADN setting for the non-standard boundary conditions is given by the extended VVP system (7.12) along with the extended versions (7.13) and (7.14) of these boundary conditions. Assuming that the variables are ordered as $\{\mathbf{u},\phi,\omega,p\}$ and setting $q=0$ in the energy balance (7.30) from Theorem 7.5 gives the three-dimensional counterpart of (7.40) for the normal velocity–pressure boundary condition:

$$\begin{cases} J_0(\mathbf{u},\phi,\omega,p;\mathbf{f}) = \\ \qquad \|\nabla\times\omega + \nabla p - \mathbf{f}\|_0^2 + \|\nabla\cdot\omega\|_0^2 + \|\nabla\times\mathbf{u} + \nabla\phi - \omega\|_0^2 + \|\nabla\cdot\mathbf{u}\|_0^2 \\ X_0 = \mathbf{H}^1_n(\Omega)\times H^1_0(\Omega)\times \mathbf{H}^1_n(\Omega)\times H^1_0(\Omega). \end{cases} \tag{7.42}$$

The analogue of (7.41) in $\mathbb{R}^3$ for the normal velocity–tangential vorticity boundary condition is obtained in the same fashion from (7.31):

$$\begin{cases} J_0(\mathbf{u},\phi,\omega,p;\mathbf{f}) = |\ell(p)|^2 + \\ \qquad \|\nabla\times\omega+\nabla p-\mathbf{f}\|_0^2 + \|\nabla\cdot\omega\|_0^2 + \|\nabla\times\mathbf{u}+\nabla\phi-\omega\|_0^2 + \|\nabla\cdot\mathbf{u}\|_0^2 \\ X_0 = \mathbf{H}_n^1(\Omega)\times H_0^1(\Omega)\times\mathbf{H}_t^1(\Omega)\times H^1(\Omega)\,. \end{cases} \tag{7.43}$$

Remark 7.11 It is not difficult to show that the slack variable is identically zero; see, e.g., [34, 51]. As a result, when using (7.42) and (7.43) to formulate LSFEMs, we can drop this variable from the least-squares functional. On the other hand, the redundant equation (7.9) is necessary for the VVP system to have a homogeneous elliptic principal part and its residual has to be included in (7.42) and (7.43); see Remark 7.6. □

Velocity boundary conditions

For the velocity boundary condition (7.2) it is not necessary to differentiate between two and three space dimensions because, in either case, the ellipticity of the corresponding boundary value problem cannot be improved beyond non-homogeneous elliptic. As a result, we can use the same set of unknowns $\{\mathbf{u},\omega,p\}$ in $\mathbb{R}^2$ and $\mathbb{R}^3$. Setting $q=-1$ in (7.27) and (7.32) gives the following CLSP:

$$\begin{cases} J_{-1}(\mathbf{u},\omega,p;\mathbf{f}) = \\ \qquad \|\nabla\times\omega+\nabla p-\mathbf{f}\|_{-1}^2 + \|\nabla\times\mathbf{u}-\omega\|_0^2 + \|\nabla\cdot\mathbf{u}\|_0^2 + |\ell(p)|^2 \\ X_{-1} = [H_0^1(\Omega)]^d\times[L^2(\Omega)]^{d(d-1)/2}\times L^2(\Omega) \end{cases} \tag{7.44}$$

which is valid in both $\mathbb{R}^2$ (with $\nabla\times\omega = \nabla^{\perp}\omega$) and $\mathbb{R}^3$.

Remark 7.12 The CLSPs (7.40) and (7.42) correspond to the abstract principle (3.25) for problems with zero nullity. On the other hand, the CLSPs (7.41), (7.43), and (7.44) correspond to the abstract principle (3.26) for problems with positive nullity. In these CLSPs, the one-dimensional null-space is removed by adding the term $|\ell(p)|^2$ to the least-squares functional. An alternative approach would be to replace the minimization space by its algebraic complement with respect to $\ell(p)$; see (C.7) for a definition. Because $\ell(p)$ acts only on the pressure, this change affects only the pressure component of the minimization space. For example, changing X_{-1} in (7.44) to its algebraic complement gives the following well-posed CLSP:

$$\begin{cases} J_{-1}(\mathbf{u},\omega,p;\mathbf{f}) = \|\nabla\times\omega+\nabla p-\mathbf{f}\|_{-1}^2 + \|\nabla\times\mathbf{u}-\omega\|_0^2 + \|\nabla\cdot\mathbf{u}\|_0^2 \\ X_{-1}^C = [H_0^1(\Omega)]^d\times[L^2(\Omega)]^{d(d-1)/2}\times L_0^2(\Omega)\,. \end{cases} \tag{7.45}$$

The same idea can be applied to (7.41) and (7.43), resulting in least-squares functionals without the term $|\ell(p)|^2$. □

7.3.2 The Velocity–Stress–Pressure System

The VSP system is non-homogeneous elliptic in $\mathbb{R}^2$ and $\mathbb{R}^3$. The symmetric viscous stress variable $\underline{\mathbf{T}}$ has $d(d+1)/2$ independent components in $\mathbb{R}^d$. Assuming that the variables are ordered as $\{\mathbf{u},\underline{\mathbf{T}},p\}$ and setting $q=-1$ in (7.35), gives the following CLSP:[30]

$$\begin{cases} J_{-1}(\mathbf{u},\underline{\mathbf{T}},p;\mathbf{f}) = \\ \qquad \|\sqrt{2\nu}\nabla\cdot\underline{\mathbf{T}}-\nabla p-\mathbf{f}\|_{-1}^2+\|\nabla\cdot\mathbf{u}\|_0^2+\|\underline{\mathbf{T}}-\sqrt{2\nu}\,\varepsilon(\mathbf{u})\|_0^2+|\ell(p)|^2 \\ X_{-1}=[H_0^1(\Omega)]^d\times[L^2(\Omega)]^{d(d+1)/2}\times L^2(\Omega). \end{cases} \tag{7.46}$$

This CSLP is well-posed for the VSP system with the velocity boundary condition (7.2) in two and three dimensions and corresponds to the abstract CLSP (3.26) for problems with positive nullity. In the light of Remark 7.12, we can replace (7.46) by

$$\begin{cases} J_{-1}(\mathbf{u},\underline{\mathbf{T}},p;\mathbf{f}) = \\ \qquad \|\sqrt{2\nu}\nabla\cdot\underline{\mathbf{T}}-\nabla p-\mathbf{f}\|_{-1}^2+\|\nabla\cdot\mathbf{u}\|_0^2+\|\underline{\mathbf{T}}-\sqrt{2\nu}\,\varepsilon(\mathbf{u})\|_0^2 \\ X_{-1}^C=[H_0^1(\Omega)]^d\times[L^2(\Omega)]^{d(d+1)/2}\times L_0^2(\Omega) \end{cases} \tag{7.47}$$

for which the corresponding minimization is performed over the complement space with respect to the functional $|\ell(p)|$. The CLSP (7.47) is a specialization of the abstract principle (3.25) for problems with zero nullity.

7.3.3 The Velocity Gradient–Velocity–Pressure System

The two first-order VGVP systems use the same set of variables which we assume are ordered as $\{\mathbf{u},\underline{\mathbf{U}},p\}$. The non-symmetric velocity gradient tensor $\underline{\mathbf{U}}$ has d^2 independent components in $\mathbb{R}^d$.

In both two and three dimensions, the VGVP system (7.16) is not homogeneous elliptic. Using the energy balance (7.36) from Theorem 7.8 gives the following CLSP for this system for the velocity boundary condition (7.2):

[30] When we write that the symmetric tensor $\underline{\mathbf{T}}\in[L^2(\Omega)]^{d(d+1)/2}$, we mean that each of the independent $d(d+1)/2$ components of $\underline{\mathbf{T}}$ belongs to $L^2(\Omega)$, e.g., in two dimensions, we have that $(\underline{\mathbf{T}})_{1,1}$, $(\underline{\mathbf{T}})_{1,2}=(\underline{\mathbf{T}})_{2,1}$, and $(\underline{\mathbf{T}})_{2,2}$ belong to $L^2(\Omega)$.

$$\begin{cases} J_{-1}(\mathbf{u},\underline{\mathbf{U}},p;\mathbf{f}) = \\ \qquad \|-\nu\nabla\cdot\underline{\mathbf{U}}+\nabla p-\mathbf{f}\|_{-1}^2+\|\underline{\mathbf{U}}-(\nabla\mathbf{u})^{\mathsf{T}}\|_0^2+\|\nabla\cdot\mathbf{u}\|_0^2+|\ell(p)|^2 \\ X_{-1} = [H_0^1(\Omega)]^d\times[L^2(\Omega)]^{d^2}\times L^2(\Omega)\,. \end{cases} \tag{7.48}$$

This principle is well-posed in two and three dimensions.

The extended VGVP system (7.20) is homogeneous elliptic in both two and three dimensions and requires an additional boundary condition on $\underline{\mathbf{U}}$. Let

$$\underline{\mathbf{H}}_{\times}^1(\Omega) = \{\underline{\mathbf{U}}\in[H^1(\Omega)]^{d^2}\;|\;\mathbf{n}\times\underline{\mathbf{U}}=0 \text{ on } \partial\Omega\}\,.$$

Using the energy balance (7.38) from Theorem 7.9, we have the following CLSP for (7.20) with the velocity boundary condition:

$$\begin{cases} J_0(\mathbf{u},\underline{\mathbf{U}},p;) = \\ \qquad \|-\nu\nabla\cdot\underline{\mathbf{U}}+\nabla p-\mathbf{f}\|_0^2+\|\underline{\mathbf{U}}-(\nabla\mathbf{u})^{\mathsf{T}}\|_0^2 \\ \qquad\qquad +\|\nabla\times\underline{\mathbf{U}}\|_0^2+\|\nabla\cdot\mathbf{u}\|_0^2+\|\nabla(tr\underline{\mathbf{U}})\|_0^2+|\ell(p)|^2 \\ X_0 = [H_0^1(\Omega)]^d\times\underline{\mathbf{H}}_{\times}^1(\Omega)\times H^1(\Omega)\,. \end{cases} \tag{7.49}$$

The functionals in (7.48) and (7.49) correspond to the abstract CLSP (3.27) for problems with positive nullity. By changing the minimization spaces from X_{-1} and X_0 to their complements X_{-1}^C and X_0^C with respect to $\ell(p)$, respectively, we can drop the term $|\ell(p)|^2$ from the definition of the least-squares functionals. We have already seen plenty of examples of such functionals so that we omit their statement for VGVP systems.

7.4 Discrete Least-Squares Principles in the Agmon–Douglis–Nirenberg Setting

The next step after a set of well-posed CLSPs is secured is to transform them into practical DLSPs. Our inventory[31] of CLSPs for the Stokes equations derived in the ADN setting[32] includes the principles (7.40)–(7.44) for the VVP system, (7.46) for the VSP system, and (7.48) and (7.49) for the VGVP systems.[33] Some of these CLSPs correspond to homogeneous elliptic systems and some to non-homogeneous

[31] For the purposes of cataloguing the CLSPs, we treat principles that include $|\ell(p)|$ in the least-squares functional and their versions that use the algebraic complement with respect to $\ell(p)$ as the same CLSP.

[32] DLSPs in the vector-operator setting are discussed in Section 7.7.3.

[33] Although formally the energy balances for the VGVP systems were not derived using the ADN theory, the corresponding CLSPs fit the ADN setting of Chapter 4; see Section 7.2.3.

elliptic systems. In the former case, we follow the transformation process from Section 4.4. If the underlying system is non-homogeneous elliptic, transformation to a practical DLSP proceeds according to Section 4.5.

Approximation of the minimization space X from each CLSP also follows the recipes formulated in Sections 4.4 and 4.5. In principle, this allows us to approximate each scalar solution component by an independently chosen scalar nodal space G^r with $r \geq 1$. In practice, to optimize accuracy and performance of the LSFEMs, this possibility is never fully utilized and instead the degrees of the nodal spaces for the different variables are closely related.

7.4.1 The Velocity–Vorticity–Pressure System

Compliant DLSPs for non-standard boundary conditions

Homogeneous elliptic systems are the most straightforward setting for LSFEMs so we begin with transformations of (7.40)–(7.43) into practical compliant DLSPs. Obtaining these DLSPs is a simple matter of choosing a finite element subspace of X_0 and then restricting minimization of J_0 to that subspace. Without much difficulty, we see that the compliant discretizations of (7.40) and (7.41) correspond to

$$\begin{cases} J_0(\mathbf{u}^h, \omega^h, p^h; \mathbf{f}) = \|\nabla^\perp \omega^h + \nabla p^h - \mathbf{f}\|_0^2 + \|\nabla \times \mathbf{u}^h - \omega^h\|_0^2 + \|\nabla \cdot \mathbf{u}^h\|_0^2 \\ X_{r,0}^h = [G^r]^2 \cap \mathbf{H}_n^1(\Omega) \times G^r \times G^r \cap H_0^1(\Omega) \end{cases} \tag{7.50}$$

for the normal velocity–pressure boundary condition and

$$\begin{cases} J_0(\mathbf{u}^h, \omega^h, p^h; \mathbf{f}) = \\ \qquad \|\nabla^\perp \omega^h + \nabla p^h - \mathbf{f}\|_0^2 + \|\nabla \times \mathbf{u}^h - \omega^h\|_0^2 + \|\nabla \cdot \mathbf{u}^h\|_0^2 + |\ell(p^h)\|^2 \\ X_{r,0}^h = [G^r]^2 \cap \mathbf{H}_n^1(\Omega) \times G^r \cap H_0^1(\Omega) \times G^r \end{cases} \tag{7.51}$$

for the normal velocity–tangential vorticity boundary condition. Their three-dimensional counterparts, i.e., the compliant discretizations of (7.42) and (7.43), are given by

$$\begin{cases} J_0(\mathbf{u}^h, \omega^h, p^h; \mathbf{f}) = \\ \qquad \|\nabla \times \omega^h + \nabla p^h - \mathbf{f}\|_0^2 + \|\nabla \cdot \omega^h\|_0^2 + \|\nabla \times \mathbf{u}^h - \omega^h\|_0^2 + \|\nabla \cdot \mathbf{u}^h\|_0^2 \\ X_{r,0}^h = [G^r]^3 \cap \mathbf{H}_n^1(\Omega) \times [G^r]^3 \cap \mathbf{H}_n^1(\Omega) \times G^r \cap H_0^1(\Omega) \end{cases} \tag{7.52}$$

for the normal velocity–pressure boundary condition and

$$\begin{cases} J_0(\mathbf{u}^h,\omega^h,p^h;\mathbf{f}) = |\ell(p^h)\|^2 + \\ \qquad \|\nabla\times\omega^h+\nabla p^h-\mathbf{f}\|_0^2+\|\nabla\cdot\omega^h\|_0^2+\|\nabla\times\mathbf{u}^h-\omega^h\|_0^2+\|\nabla\cdot\mathbf{u}^h\|_0^2 \\ X^h_{r,0} = [G^r]^3\cap\mathbf{H}^1_n(\Omega)\times[G^r]^3\cap\mathbf{H}^1_t(\Omega)\times G^r \end{cases} \tag{7.53}$$

for the normal velocity–tangential vorticity boundary condition. In (7.52) and (7.53), we have taken advantage of Remark 7.11 to remove the slack variable from the formulation of the LSFEMs.

Non-compliant DLSPs for the velocity boundary condition

The last CLSP from the ADN inventory is (7.44) that corresponds to a non-homogenous elliptic system. As a result, it does not have a practical *compliant* DLSP. To get a practical LSFEM from (7.44), we follow the steps in Section 4.5. The first option is to use the transition diagram in (4.28), resulting in the DLSP

$$\begin{cases} J^h_{-1}(\mathbf{u}^h,\omega^h,p^h;\mathbf{f}) = \\ \qquad h^2\|\nabla\times\omega^h+\nabla p^h-\mathbf{f}\|_0^2+\|\nabla\times\mathbf{u}^h-\omega^h\|_0^2+\|\nabla\cdot\mathbf{u}^h\|_0^2+|\ell(p^h)|^2 \\ X^h_{r+,0} = [G^{r+1}]^d\cap[H^1_0(\Omega)]^d\times[G^r]^{d(d-1)/2}\times G^r\,. \end{cases} \tag{7.54}$$

Theorem 4.7 can be used to show that the functional in (7.54) satisfies a mesh-dependent discrete energy balance, i.e., this DLSP is *quasi-norm-equivalent*.

The second option is to use the transition diagram in (4.39), resulting in the DLSP

$$\begin{cases} J_{-h}(\mathbf{u}^h,\omega^h,p^h;\mathbf{f}) = \\ \qquad \|\nabla\times\omega^h+\nabla p^h-\mathbf{f}\|_{-h}^2+\|\nabla\times\mathbf{u}^h-\omega^h\|_0^2+\|\nabla\cdot\mathbf{u}^h\|_0^2+|\ell(p^h)|^2 \\ X^h_{\widehat{r},0} = [G^{r+1}]^d\cap[H^1_0(\Omega)]^d\times[G^{\widehat{r}}]^{d(d-1)/2}\times G^{\widehat{r}}\,. \end{cases} \tag{7.55}$$

Theorem 4.9 can be used to show that (7.55) satisfies a mesh-independent discrete energy balance, i.e., this DLSP is *norm-equivalent*.

The approximation spaces in (7.54) and (7.55) specialize the spaces defined in (4.35) and (4.44), respectively. We have more to say about these spaces in Remark 7.15.

7.4.2 The Velocity–Stress–Pressure System

The first-order VSP system is not homogeneous elliptic regardless of the space dimension and the boundary operator. As a result, it does not admit a practical compliant DLSP. Therefore, we proceed with the transition options from Section 4.5.

Using the diagram in (4.28) results in the DLSP

$$\begin{cases} J_{-1}^h(\mathbf{u}^h, \underline{\mathbf{T}}^h, p^h; \mathbf{f}) = |\ell(p^h)|^2 + \\ \qquad h^2 \|\sqrt{2\nu}\,\nabla\cdot\underline{\mathbf{T}}^h - \nabla p^h - \mathbf{f}\|_0^2 + \|\nabla\cdot\mathbf{u}^h\|_0^2 + \|\underline{\mathbf{T}}^h - \sqrt{2\nu}\,\varepsilon(\mathbf{u}^h)\|_0^2 \\ X_{r+,0}^h = [G^{r+1}]^d \cap [H_0^1(\Omega)]^d \times [G^r]^{d(d+1)/2} \times G^r . \end{cases} \tag{7.56}$$

As is the case with (7.54), one can show that the energy balance of the least-squares functional in (7.56) is mesh dependent and so this DLSP is quasi-norm-equivalent.

The alternative option is to use (4.39). Then, we have a discrete negative norm functional and the norm-equivalent DLSP

$$\begin{cases} J_{-h}(\mathbf{u}^h, \underline{\mathbf{T}}^h, p^h; \mathbf{f}) = |\ell(p^h)|^2 + \\ \qquad \|\sqrt{2\nu}\,\nabla\cdot\underline{\mathbf{T}}^h - \nabla p^h - \mathbf{f}\|_{-h}^2 + \|\nabla\cdot\mathbf{u}^h\|_0^2 + \|\underline{\mathbf{T}}^h - \sqrt{2\nu}\,\varepsilon(\mathbf{u}^h)\|_0^2 \\ X_{\widehat{r},0}^h = [G^{r+1}]^d \cap [H_0^1(\Omega)]^d \times [G^{\widehat{r}}]^{d(d+1)/2} \times G^{\widehat{r}} . \end{cases} \tag{7.57}$$

7.4.3 The Velocity Gradient–Velocity–Pressure System

The parent VGVP first-order system (7.16) is non-homogeneous elliptic and as such does not have a practical compliant DLSP. Therefore, it does not offer substantial benefits over the VSP and VVP systems which have fewer variables. Nevertheless, in Chapter 8, we need a norm-equivalent DLSP for this system in order to discuss some issues arising in the use of VGVP formulations for problems with rough solutions. We define this DLSP using (7.48) as a parent CLSP and then apply the transition diagram (4.39). The result is the well-posed DLSP

$$\begin{cases} J_{-h}(\mathbf{u}^h, \underline{\mathbf{U}}^h, p^h; \mathbf{f}) = |\ell(p^h)|^2 + \\ \qquad \|-\nu\nabla\cdot\underline{\mathbf{U}}^h + \nabla p^h - \mathbf{f}\|_{-h}^2 + \|\underline{\mathbf{U}}^h - (\nabla\mathbf{u}^h)^\top\|_0^2 + \|\nabla\cdot\mathbf{u}^h\|_0^2 \\ X_{\widehat{r},0}^h = [G^{r+1}]^d \cap [H_0^1(\Omega)]^d \times [G^{\widehat{r}}]^{d^2} \times G^{\widehat{r}} . \end{cases} \tag{7.58}$$

The extended VGVP system (7.20) is homogeneous elliptic so that it has a practical compliant DLSP. Obtaining this DLSP is simply a matter of choosing a finite element subspace of X_0 and then restricting minimization of J_0 from the parent principle (7.49) to that space:

$$\begin{cases} J_0(\mathbf{u}^h, \underline{\mathbf{U}}^h, p^h; \mathbf{f}) = \\ \qquad \| -\nu\nabla\cdot\underline{\mathbf{U}}^h + \nabla p^h - \mathbf{f}\|_0^2 + \|\underline{\mathbf{U}}^h - (\nabla\mathbf{u}^h)^\top\|_0^2 \\ \qquad\qquad + \|\nabla\times\underline{\mathbf{U}}^h\|_0^2 + \|\nabla\cdot\mathbf{u}^h\|_0^2 + \|\nabla(tr\underline{\mathbf{U}}^h)\|_0^2 + |\ell(p^h)|^2 \\ \\ X_{r,0}^h = [G^r]^d \cap [H_0^1(\Omega)]^d \times [G^r]^{d^2} \cap \underline{\mathbf{H}}_\times^1(\Omega) \times G^r \,. \end{cases} \tag{7.59}$$

7.5 Error Estimates in the Agmon–Douglis–Nirenberg Setting

The final step in the least-squares framework from Chapter 4 is the error analysis. For compliant DLSPs that correspond to homogeneous elliptic systems, the error estimates follow from the abstract result in Theorem 4.2. For quasi-norm-equivalent and norm-equivalent DLSPs corresponding to non-homogeneous elliptic systems, the error estimates follow from Theorem 4.8 and Theorem 4.10, respectively.

7.5.1 The Velocity–Vorticity–Pressure System

Error estimates for non-standard boundary conditions

For VVP systems, we have four compliant DLSPs for the non-standard boundary conditions: the two-dimensional formulations (7.50) and (7.51) and their three-dimensional versions (7.52) and (7.53). The following theorem is a specialization of Theorem 4.2 to these LSFEMs.

Theorem 7.13 *Let $r \geq 1$ and $\{\mathbf{u}^h, \omega^h, p^h\} \in X_{r,0}^h$ denote a minimizer of* (7.50). *Assume that the exact solution $\{\mathbf{u}, \omega, p\}$ of the two-dimensional VVP system* (7.4) *with the normal velocity–pressure boundary conditions* (7.5) *belongs to the space $X_q = [H^{q+1}(\Omega)]^2 \cap \mathbf{H}_n^1(\Omega) \times H^{q+1}(\Omega) \times H^{q+1}(\Omega) \cap H_0^1(\Omega)$ for some $q \geq 0$. Then, there exists a positive constant C such that*

$$\|\mathbf{u}-\mathbf{u}^h\|_1 + \|\omega-\omega^h\|_1 + \|p-p^h\|_1 \leq Ch^{\widetilde{r}}\left(\|\mathbf{u}\|_{\widetilde{r}+1} + \|\omega\|_{\widetilde{r}+1} + \|p\|_{\widetilde{r}+1}\right), \tag{7.60}$$

where $\widetilde{r} = \min\{r,q\}$. If, in addition, Ω is such that the first-order VVP system has full elliptic regularity, we also have the optimal $L^2(\Omega)$ estimate

$$\|\mathbf{u}-\mathbf{u}^h\|_0 + \|\omega-\omega^h\|_0 + \|p-p^h\|_0 \leq Ch^{\widetilde{r}+1}\left(\|\mathbf{u}\|_{\widetilde{r}+1} + \|\omega\|_{\widetilde{r}+1} + \|p\|_{\widetilde{r}+1}\right). \tag{7.61}$$

The error bounds (7.60) and (7.61) *hold for the minimizer of* (7.51) *provided the solution of the two-dimensional VVP system with the normal velocity–tangential vorticity boundary condition* (7.7) *belongs to the space* $X_q = [H^{q+1}(\Omega)]^2 \cap \mathbf{H}_n^1(\Omega) \times H^{q+1}(\Omega) \cap H_0^1(\Omega) \times H^{q+1}(\Omega) \cap L_0^2(\Omega)$ *for some* $q \geq 0$.

Furthermore, (7.60) and (7.61) *also hold for* (7.52) *and* (7.53) *provided that the exact solutions* $\{\mathbf{u}, \omega, p\}$ *of* (7.12) *with the normal velocity–pressure and the normal velocity–tangential vorticity boundary conditions belong to*

$$X_q = [H^{q+1}(\Omega)]^3 \cap \mathbf{H}_n^1(\Omega) \times [H^{q+1}(\Omega)]^3 \cap \mathbf{H}_n^1(\Omega) \times H^{q+1}(\Omega) \cap H_0^1(\Omega)$$

and

$$X_q = [H^{q+1}(\Omega)]^3 \cap \mathbf{H}_n^1(\Omega) \times [H^{q+1}(\Omega)]^3 \cap \mathbf{H}_t^1(\Omega) \times H^{q+1}(\Omega) \cap L_0^2(\Omega),$$

respectively. □

Error estimates for the velocity boundary condition

For VVP systems, we also have the quasi-norm-equivalent and norm-equivalent DLSPs (7.54) and (7.55), respectively, for the velocity boundary condition. The error estimates resulting from the specialization of Theorems 4.8 and 4.10, respectively, to these LSFEMs have similar structure and can be combined together in the following theorem.

Theorem 7.14 *Let* $\{\mathbf{u}^h, \omega^h, p^h\} \in X^h_{\widetilde{r},0}$, $r \geq 0$, *denote a solution of the norm-equivalent DLSP* (7.55). *Assume that the exact solution* $\{\mathbf{u}, \omega, p\}$ *of the VVP system* (7.4) *with the velocity boundary condition* (7.2) *belongs to the space* $X_q = [H^{q+2}(\Omega)]^d \cap [H_0^1(\Omega)]^d \times [H^{q+1}(\Omega)]^{d(d-1)/2} \times H^{q+1}(\Omega) \cap L_0^2(\Omega)$ *for some* $q \geq 0$. *Then, there exists a positive constant C such that*

$$\|\mathbf{u}-\mathbf{u}^h\|_1 + \|\omega-\omega^h\|_0 + \|p-p^h\|_0 \leq Ch^{\widetilde{r}+1}\left(\|\mathbf{u}\|_{\widetilde{r}+2} + \|\omega\|_{\widetilde{r}+1} + \|p\|_{\widetilde{r}+1}\right) \quad (7.62)$$

and

$$\|\omega-\omega^h\|_1 + \|p-p^h\|_1 \leq Ch^{\widetilde{r}}\left(\|\mathbf{u}\|_{\widetilde{r}+2} + \|\omega\|_{\widetilde{r}+1} + \|p\|_{\widetilde{r}+1}\right), \quad (7.63)$$

where $\widetilde{r} = \min\{r, q\}$.

Let $\{\mathbf{u}^h, \omega^h, p^h\} \in X^h_{r+,0}$, $r \geq 1$, *denote the solution of the quasi-norm-equivalent DLSP* (7.54). *The error estimates* (7.62) *and* (7.63) *hold for this solution.* □

Remark 7.15 The error estimates in Theorem 7.14 show that it pays to approximate the variables according to their ADN regularity indices t_j. For example, optimal convergence rates for all variables can be achieved when velocity components are approximated by finite element spaces of one degree higher than for the vorticity and the pressure. This stems from the fact that, for non-homogeneous elliptic systems, the variables are assigned unequal ADN indices.

We also draw attention to the difference in the minimal admissible degrees in the spaces $X^h_{\widetilde{r},0}$ and $X^h_{r+,0}$ used in the norm-equivalent DLSP (7.55) and the quasi-norm-equivalent DLSP (7.54), respectively. The former space allows approximation of the velocity field by the *lowest-order* nodal space G^1 whereas, in $X^h_{r+,0}$, the velocity has to be approximated by at least G^2 elements. This difference is caused by the use of duality arguments to prove the error estimates in the quasi-norm-equivalent case; see also Remark 4.11. The computational studies provided in Section 7.6.2 show that the minimal admissible degree requirement, as well as the presence of the weights in the quasi-norm-equivalent DLSP (7.54), are necessary to achieve the theoretical convergence rates in (7.62) and (7.63) and that their violation degrades the accuracy of the LSFEMs. □

Remark 7.16 By changing the finite element space for the pressure to $G^r \cap L^2_0(\Omega)$, the term $|\ell(p)|^2$ can be eliminated from the least-squares functionals in (7.54) and (7.55). The resulting DLSPs correspond to the application of the transition diagrams in (4.28) and (4.39) to the parent CLSP (7.45).

As has already been mentioned, in practice, an "honest" implementation of $G^r \cap L^2_0(\Omega)$ is seldom used. Instead, a widespread approach is to "pin" the pressure down at some point $\mathbf{x}_0 \in \Omega$. Effectively, this replaces the true complement space with respect to the zero mean constraint by another one with respect to the functional $\ell_{\mathbf{x}_0}(p) = \langle \delta(\mathbf{x}_0), p \rangle$. This functional causes an increase in the condition numbers [67] and should not be used with iterative solvers. We refer to Section 7.6.4 for computational examples. □

7.5.2 The Velocity–Stress–Pressure System

For the VSP system, we have the quasi-norm-equivalent and norm-equivalent DLSPs (7.54) and (7.55) , respectively. The following theorem specializes the abstract error estimates from Theorems 4.8 and 4.10, respectively, to those DLSPs.

Theorem 7.17 *Let* $\{\mathbf{u}^h, \underline{\mathbf{T}}^h, p^h\} \in X^h_{\widetilde{r},0}$, $r \geq 0$, *be a solution of the norm-equivalent DLSP* (7.57). *Assume that the exact solution* $\{\mathbf{u}, \underline{\mathbf{T}}, p\}$ *of the VSP system* (7.15) *with the velocity boundary condition* (7.2) *belongs to the space* $X_q = [H^{q+2}(\Omega)]^d \cap [H^1_0(\Omega)]^d \times [H^{q+1}(\Omega)]^{d(d+1)/2} \times H^{q+1}(\Omega)$, *where* $q \geq 0$. *Then, there exists a positive constant C such that*

$$\|\mathbf{u}-\mathbf{u}^h\|_1 + \|\underline{\mathbf{T}}-\underline{\mathbf{T}}^h\|_0 + \|p-p^h\|_0 \leq Ch^{\widetilde{r}+1}\left(\|\mathbf{u}\|_{\widetilde{r}+2} + \|\underline{\mathbf{T}}\|_{\widetilde{r}+1} + \|p\|_{\widetilde{r}+1}\right) \quad (7.64)$$

and

$$\|\underline{\mathbf{T}}-\underline{\mathbf{T}}^h\|_1 + \|p-p^h\|_1 \leq Ch^{\widetilde{r}}\left(\|\mathbf{u}\|_{\widetilde{r}+2} + \|\underline{\mathbf{T}}\|_{\widetilde{r}+1} + \|p\|_{\widetilde{r}+1}\right), \quad (7.65)$$

where $\widetilde{r} = \min\{r,q\}$. *The error estimates* (7.64) and (7.65) *hold for the solution* $\{\mathbf{u}^h, \underline{\mathbf{T}}^h, p^h\} \in X^h_{r+,0}$, $r \geq 1$, *of the quasi-norm-equivalent DLSP* (7.54). □

The observations made in Remark 7.15 regarding the minimal–degree requirement of the approximation for the velocity in norm-equivalent and quasi norm-equivalent

DLSPs for the VVP system also hold for (7.57) and (7.56). The comments made in Remark 7.16 about the zero mean pressure constraint also extend to the present case.

7.5.3 The Velocity Gradient–Velocity–Pressure System

For VGVP systems, we only present error estimates for the compliant[34] DLSP (7.59). The following theorem specializes the abstract a priori error estimates from Theorem 4.2 to that DLSP.

Theorem 7.18 *Let* $\{\mathbf{u}^h, \underline{\mathbf{U}}^h, p^h\} \in X^h_{r,0}$, $r \geq 1$, *denote a solution of* (7.59). *Assume that the exact solution* $\{\mathbf{u}, \underline{\mathbf{U}}, p\}$ *of the extended VGVP system* (7.20) *with the velocity boundary condition* (7.2) *belongs to the space* $X_q = [H^{q+1}(\Omega)]^d \cap [H^1_0(\Omega)]^d \times [H^{q+1}(\Omega)]^{d^2} \cap \underline{\mathbf{H}}^1_\times(\Omega) \times H^{q+1}(\Omega)$ *for some* $q \geq 0$. *Then, there exists a positive constant C such that*

$$\|\mathbf{u}-\mathbf{u}^h\|_1 + \|\underline{\mathbf{U}}-\underline{\mathbf{U}}^h\|_1 + \|p-p^h\|_1 \leq Ch^{\widetilde{r}}\left(\|\mathbf{u}\|_{\widetilde{r}+1} + \|\underline{\mathbf{U}}\|_{\widetilde{r}+1} + \|p\|_{\widetilde{r}+1}\right), \quad (7.66)$$

where $\widetilde{r} = \min\{r, q\}$. *The optimal* $L^2(\Omega)$ *estimate*

$$\|\mathbf{u}-\mathbf{u}^h\|_0 + \|\underline{\mathbf{U}}-\underline{\mathbf{U}}^h\|_0 + \|p-p^h\|_0 \leq Ch^{\widetilde{r}+1}\left(\|\mathbf{u}\|_{\widetilde{r}+1} + \|\underline{\mathbf{U}}\|_{\widetilde{r}+1} + \|p\|_{\widetilde{r}+1}\right) \quad (7.67)$$

holds, provided the extended VGVP system (7.20) *has full elliptic regularity.* □

7.6 Practicality Issues in the Agmon–Douglis–Nirenberg Setting

We close the presentation of LSFEMs for the Stokes equations in the ADN setting[35] with a discussion of practicality issues raised by their formulation and analysis. In particular, in this section, we provide answers to some questions that arise in the use and implementation of the LSFEMs developed in the ADN setting. After a brief discussion of solution methods for the discrete systems, we focus on the following list of issues that are not addressed by the theoretical results presented in Sections 7.4 and 7.5.

- Are the weights in (7.54) and (7.56) important for the corresponding LSFEMS?
- Can we safely use velocity spaces that do not meet the theoretical minimal-degree requirement stated in Remark 7.15?

[34] The error estimates for (7.58) are very similar to those in Theorems 7.14 and 7.17 and are not stated here.

[35] A discussion of practicality issues that arise in the vector-operator setting is given in Section 7.7.8.

- Provided the minimal-degree requirement for velocity approximates is met, can we use equal-order approximation spaces?
- Can LSFEMs, developed in the ADN setting, provide good mass conservation?
- Is it better to keep the term $|\ell(p)|^2$ in the least-squares functional or to "impose" the zero mean constraint by pinning the pressure at some point in $\overline{\Omega}$?

Many of these questions are relevant to LSFEMs in general and, therefore, improve our broader understanding of the practical issues pertaining to LSFEMs. For example, the question about the zero mean constraint is obviously relevant to LSFEMs in both the ADN and the vector-operator settings.

7.6.1 Solution of the Discrete Equations

The DLSPs described in Section 7.4 give rise to discrete variational equations of the form:

$$\textit{seek } u^h \in X^h \textit{ such that} \qquad Q(u^h, v^h) = F(v^h) \qquad \forall v^h \in X^h, \tag{7.68}$$

where X^h is one of the spaces $X^h_{r,0}$, $X^h_{r+,0}$, or $X^h_{\hat{r},0}$. From (7.68), one obtains, in the usual manner, a linear algebraic system

$$\mathsf{Q}\vec{u} = \vec{f}, \tag{7.69}$$

where Q is a *symmetric and positive definite matrix*. This linear system can be solved by direct methods such as the Cholesky factorization, or by robust iterative methods such as preconditioned conjugate gradient methods. The choice between direct and iterative solvers depends on the size of the problem and the type of underlying least-squares principle.

If the discrete problem (7.68) corresponds to one of the compliant DLSPs (7.50)–(7.53) or (7.59) or one of the quasi-norm-equivalent DLSPs (7.54) or (7.56), then the bilinear form $Q(\cdot,\cdot)$ involves only standard $L^2(\Omega)$ inner products and Q is a sparse matrix that can be assembled in a standard manner.

However, if (7.68) corresponds to one of the norm-equivalent DLSPs (7.55), (7.57), or (7.58), assembly of Q requires computation of minus one inner products and the matrix itself may not be sparse anymore. Furthermore, the operator $\mathcal{K}^h$ that enters the definition of the minus one inner product (B.100) may only be given in the form of a "black-box" algorithm. As a result, for such methods, we may not have access to a fully assembled discretization matrix so that iterative solution methods are the only option avaliable to solve (7.69). We discuss the implementation of discrete negative norm methods in more detail in Section 8.4.1. Here, we briefly state how the system (7.69) can be preconditioned for compliant and norm-equivalent DLSPs.

The case of compliant DLSPs is covered by Theorem 4.2 which asserts that Q is spectrally equivalent to the matrix K in (4.24). As a result, the matrices engendered by (7.50)–(7.53) and (7.59) are spectrally equivalent to

$$\mathsf{K}^{\mathrm{VVP}} = \begin{pmatrix} \mathsf{D}_{\mathbf{u}} & 0 & 0 \\ 0 & \mathsf{D}_{\omega} & 0 \\ 0 & 0 & \mathsf{D}_p \end{pmatrix} \quad \text{and} \quad \mathsf{K}^{\mathrm{VGVP}} = \begin{pmatrix} \mathsf{D}_{\mathbf{u}} & 0 & 0 \\ 0 & \mathsf{D}_{\underline{\mathbf{U}}} & 0 \\ 0 & 0 & \mathsf{D}_p \end{pmatrix}, \tag{7.70}$$

respectively. The blocks in (7.70) are defined as in Theorem 4.2, using the basis functions for the indicated variable. The linear systems associated with these methods can be preconditioned using the matrix $\widetilde{\mathsf{K}}$ from (4.25). The number of Poisson preconditioner blocks T in this matrix depends on the total number of the solution components. For the VVP-based systems, there are $d+(d(d-1)/2)+1$ blocks and for the VGVP-based systems, there are $d+d^2+1$ blocks.

Norm-equivalent DLSPs are covered by Theorem 4.10. This theorem asserts that Q is spectrally equivalent to the matrix[36] K_{-h} in (4.47). Consequently, the matrices produced by (7.55), (7.57), and (7.58) are spectrally equivalent to

$$\mathsf{K}^{\mathrm{VVP}}_{-h} = \begin{pmatrix} \mathsf{D}_{\mathbf{u}} & 0 & 0 \\ 0 & \mathsf{G}_{\omega} & 0 \\ 0 & 0 & \mathsf{G}_p \end{pmatrix} \qquad \mathsf{K}^{\mathrm{VSP}}_{-h} = \begin{pmatrix} \mathsf{D}_{\mathbf{u}} & 0 & 0 \\ 0 & \mathsf{G}_{\underline{\mathbf{T}}} & 0 \\ 0 & 0 & \mathsf{G}_p \end{pmatrix}$$
$$\mathsf{K}^{\mathrm{VGVP}}_{-h} = \begin{pmatrix} \mathsf{D}_{\mathbf{u}} & 0 & 0 \\ 0 & \mathsf{G}_{\underline{\mathbf{U}}} & 0 \\ 0 & 0 & \mathsf{G}_p \end{pmatrix}, \tag{7.71}$$

respectively. The blocks in these matrices are defined as in Theorem 4.10, using the basis functions for the indicated solution components. According to that theorem, the linear systems for the norm-equivalent LSFEMs can be preconditioned by the matrix $\widetilde{\mathsf{K}}_{-h}$ defined in (4.51). In the settings of (7.55), (7.57), and (7.58), the number of Poisson preconditioner blocks in this matrix equals the number of velocity components, i.e., the space dimension d. The number of lumped mass blocks depends on the particular first-order system used. For the VVP system it is $(d(d-1)/2)+1$, for the VSP system it is $(d(d+1)/2)+1$, and for the VGVP system there are d^2+1 such blocks.

7.6.2 Issues Related to Non-Homogeneous Elliptic Systems

For non-homogeneous elliptic systems, practical DLSPs have to be developed by using the transition diagrams in Section 4.5. Thus, for such systems we have the choice of

- a quasi-norm-equivalent *straightforward* $L^2(\Omega)$ functional obtained using (4.27)
- a quasi-norm-equivalent *weighted* $L^2(\Omega)$ functional obtained using (4.28)

[36] Recall that the blocks G and K in this matrix correspond to variables whose ADN indices are $t_j = 1$ and $t_j = 2$, respectively.

- a norm-equivalent *discrete negative norm* functional obtained using (4.39).

How one approximates the norm-generating operator $\mathcal{S}_Y = (-\Delta)^{-1/2}$ for $H^{-1}(\Omega)$ is the main difference among the corresponding least-squares functionals. In the first case, this operator is approximated by the identity $\mathcal{I}$, in the second by the scaled identity $h\mathcal{I}$, and in the third case by the operator $h\mathcal{I} + (\mathcal{K}^h)^{1/2}$.

In Chapter 4, we derived optimal error estimates for the second and third choices; for the first choice, such estimates were not available. Naturally, one asks if the lack of theory for the straightforward $L^2(\Omega)$ functional is merely a theoretical shortcoming or if it also has practical significance for LSFEMs based on that functional. An examination of this issue is especially important because, for the Stokes equations, straightforward $L^2(\Omega)$ functionals provide the most widely used setting for LS-FEMs [138, 235, 236, 261, 297, 298, 326, 329], even when the underlying first-order system is non-homogeneous elliptic. Specifically, we pose the following questions.

- The VVP system can change type from non-homogeneous to homogeneous elliptic, depending on the boundary condition. LSFEMs based on the weighted functional have provably optimal error estimates in the non-homogeneous elliptic case. LSFEMs based on the straightforward functionals have provably optimal error estimates in the homogeneous elliptic case but not in the non-homogeneous elliptic case. Is it therefore safe to view the inclusion of weights in (7.54) as merely being a mathematical "oddity" required to prove error estimates in the non-homogeneous elliptic case so that, in practice, it is safe to omit the weights? Phrasing this another way, we can ask: does the absence of formal error estimates automatically imply that straightforward $L^2(\Omega)$ functionals for non-homogeneous elliptic systems are inferior to their weighted and discrete negative norm counterparts?
- Computationally, the straightforward $L^2(\Omega)$ functional is more attractive than the two functionals for which there is a formal approximation theory: it is much easier to implement than the discrete negative norm functional and gives better conditioned matrices than the weighted $L^2(\Omega)$ functional. In addition, the validity of the error estimates for the weighted functional requires a minimal-degree for the approximating velocity space; the straightforward $L^2(\Omega)$ functional does not have this requirement in situations where provable error estimates are available, i.e, the homogeneous elliptic case. Is it therefore safe to violate the minimal-degree requirement and use piecewise linear finite element spaces for the velocity approximation, even for weighted functionals?
- In addition, the error estimates for the weighted and discrete negative norm functionals use different-order finite element spaces for the different variables; the straightforward $L^2(\Omega)$ functional does not have these requirement in situations where provable error estimates are available, i.e, the homogeneous elliptic case. Is it therefore safe, so long as the minimal-degree requirement is met, to use equal-order interpolation for all variables, even in the non-homogeneous elliptic case?

As we now see, the first two questions have negative answers and, although the third question has a positive answer, there may result in an underutilization of the approximation "power" of some of the component finite element spaces.

Loss of accuracy for straightforward $L^2(\Omega)$ functionals

Our first task is to demonstrate that the approximation theory from Chapter 4 is "sharp" and that for non-homogeneous elliptic systems the straightforward $L^2(\Omega)$ functional defined by (4.27) suffers from reduced convergence rates. The VVP and VSP systems with the velocity boundary condition are non-homogeneous elliptic and can be used in this study. We propose to compare the performance of (7.54) and (7.56) with and without the weights, using the "manufactured solutions"

$$\begin{cases} u_1 = e^x \cos y + \sin y \\ u_2 = -e^x \sin y + (1 - x^3) \\ \omega = \nabla \times \mathbf{u} \\ p = \sin y \cos x + xy^2 - \frac{1}{6} - \sin 1(1 - \cos 1) \end{cases} \tag{7.72}$$

for the VVP system and

$$\begin{cases} u_1 = u_2 = \sin(\pi x)\sin(\pi y) \\ T_{11} = T_{12} = T_{22} = \sin(\pi x)\exp(\pi y) \\ p = \cos(\pi x)\exp(\pi y) \end{cases} \tag{7.73}$$

for the VSP system.

In the computational examples, Ω is the unit square in $\mathbb{R}^2$ and the LSFEM solutions are computed on a sequence of uniform partitions $\mathcal{T}_h$ of Ω into triangles, consisting of up to 3,200 affine elements. The finite element spaces $X^h_{r+,0}$ in (7.54) and (7.56) are defined with $r = 1$, i.e., $\mathbf{u}$ is approximated by continuous piecewise quadratic finite elements and the rest of the unknowns are approximated by continuous piecewise linear elements.

Estimates of the convergence rates for the two examples are summarized in Table 7.1. The WLS column lists the rates when the weights are included in (7.54) and (7.56). The LS column lists the rates obtained with the weights removed from the functionals in these DLSPs. The column BA lists the theoretical best approximation rates obtainable out of the space $X^h_{r+,0}$. From the data in Table 7.1, it is clear that (7.54) and (7.56) perform better when the weights are included in the least-squares functionals. Without the weights, convergence rates for the velocity approximations in $L^2(\Omega)$ and $H^1(\Omega)$ norms drop by almost two and one orders, respectively. The $L^2(\Omega)$ convergence rates of the other variables are also reduced. This computational evidence suggests that for non-homogeneous elliptic systems a *straightforward* $L^2(\Omega)$ least-squares functional cannot produce optimally convergent solutions so that *the weights are critical for the accuracy of quasi-norm-equivalent DLSPs.*

	$L^2(\Omega)$ convergence rates			$H^1(\Omega)$ convergence rates		
VVP variable	WLS	LS	BA	WLS	LS	BA
u	3.51	1.23	3.00	2.21	1.25	2.00
v	3.50	1.31	3.00	2.06	1.25	2.00
ω	2.30	1.29	2.00	1.00	1.00	1.00
p	1.90	0.95	2.00	1.09	1.00	1.00
VSP variable	WLS	LS	BA	WLS	LS	BA
u	3.59	1.11	3.00	2.85	1.00	2.00
v	3.13	1.28	3.00	2.77	1.17	2.00
T_{11}	2.42	1.25	2.00	0.99	0.94	1.00
T_{12}	2.48	1.14	2.00	1.01	0.99	1.00
T_{22}	2.34	1.26	2.00	1.05	0.76	1.00
p	2.40	0.94	2.00	1.10	0.92	1.00

Table 7.1 Estimated $L^2(\Omega)$ and $H^1(\Omega)$ convergence rates for VVP and VSP-based quasi-norm-equivalent DLSPs (7.54) and (7.56) with (WLS) and without (LS) weights and the exact solutions (7.72) and (7.73), respectively. The benchmark rates are given in the column (BA) that gives the rates for the best approximation out of the corresponding spaces.

Loss of accuracy when minimal-degree requirements are not met

Having established the practical importance of the weights, we now consider the necessity of the condition in (7.54) and (7.56) that the velocity must be approximated by at least quadratic elements. This *minimal-degree* requirement is used for proving the optimal error estimates of Theorems 7.14 and 7.17. However, this leaves open the question of whether this condition is necessary in practical computations. This question is examined by minimizing the VSP least-squares functional from (7.56) over the "incorrect" equal-order approximating space

$$X^h_{1,0} = [G^1]^2 \cap [H^1_0(\Omega)]^2 \times [G^1]^3 \times G^1$$

that employs piecewise linear elements for all solution components.

Table 7.2 provides the numerical convergence rates for the problem with exact solution (7.73) obtained with this space as well as with the "correct" space

$$X^h_{1+,0} = [G^2]^2 \cap [H^1_0(\Omega)]^2 \times [G^1]^3 \times G^1 \,.$$

Because we do not have formal error estimates for $X^h_{1,0}$, the best we can do is to compare numerical results with the theoretical best approximation rates obtainable out of that space. Note that the rates for the velocity approximations using the two spaces are not compared because $X^h_{1,0}$ and $X^h_{1+,0}$ employ finite element spaces of different degrees for this variable. The data in Table 7.2 clearly show that convergence rates for the stress and the pressure variables are lower than those for the best approximation. Moreover, these rates are worse than the rates for the same variables computed using the "correct" space $X^h_{1+,0}$. From these data, it follows that

> *using the "wrong" minimization space $X^h_{1,0}$ that does not meet the minimal-degree requirement does have negative ramifications on the accuracy of quasi-norm-equivalent DLSPs.*

	$L^2(\Omega)$ convergence rates			$H^1(\Omega)$ convergence rates		
VSP variable	$X^h_{1,0}$	$X^h_{1+,0}$	BA	$X^h_{1,0}$	$X^h_{1+,0}$	BA
u	1.91	–	2.00	1.00	–	1.00
v	2.01	–	2.00	1.02	–	1.00
T_{11}	1.42	2.42	2.00	0.77	0.99	1.00
T_{12}	1.50	2.48	2.00	0.97	1.01	1.00
T_{22}	1.60	2.34	2.00	0.84	1.05	1.00
p	0.42	2.40	2.00	0.58	1.10	1.00

Table 7.2 Estimated $L^2(\Omega)$ and $H^1(\Omega)$ convergence rates for VSP-based quasi-norm-equivalent DLSP (7.56) for the problem with exact solution (7.73) when minimization is carried over the "correct" space $X^h_{1+,0}$ and the "incorrect" space $X^h_{1,0}$, compared to the convergence rates for the best approximation out of the "incorrect" space $X^h_{1,0}$.

Equal-order interpolation for weighted functionals

A related question of practical interest is whether there are any negative consequences for the quasi-norm-equivalent DLSP when an equal-order space *meets* the minimal-degree requirement. We do not expect that things are as bad as in the previous case, so the real question is whether using higher-order spaces in places where the theory does not call for them is wasteful. To test this scenario, we use the VVP system with the velocity boundary condition and the exact solution from Example D.20 with $n = 1$. Then, we minimize the functional from (7.54) over the equal-order space

$$X^h_{2,0} = [G^2]^2 \cap [H^1_0(\Omega)]^2 \times G^2 \times G^2$$

that meets the minimal-degree requirement but in which pressure and vorticity are approximated by elements of one degree higher than required by the theory. Our task is facilitated by the fact that, for other boundary conditions, the VVP system does admits a compliant DLSP for which this space happens to be the "correct" minimization space with provably optimal error estimates. We use this compliant DLSP as a reference when assessing convergence rates.

Table 7.3 compares the convergence rates of (7.54) obtained using $X^h_{2,0}$ with the convergence rates for the best approximation out of this space and for the compliant DLSP (7.50) for the VVP system with the normal velocity–pressure boundary condition (7.5). The rates for the reference compliant DLSP (7.50) match the rates for the best approximation; this is consistent with the assertions of Theorem 7.13. The rates for the quasi-norm-equivalent DLSP (7.54) are actually better than the rates for (7.50); however, there are no theoretical assurances that this always is the case. Nevertheless, Theorem 7.14 is valid for $X^h_{2,0}$ and at the least, the estimate (7.62) holds with suboptimal (with respect to the best approximation) order of convergence for ω and p. Thus,

> *except for a possible "underutilization" of the available accuracy in the vorticity and the pressure spaces, the use of equal-order interpolation spaces that meet the minimal-degree requirement is acceptable for weighted functionals.*

	$L^2(\Omega)$ norm			$H^1(\Omega)$ norm		
VVP variable	v-bc	BA	nv/p-bc	v-bc	BA	nv/p-bc
u	3.64	3.00	3.11	2.15	2.00	2.04
v	3.31	3.00	3.10	2.10	2.00	2.02
ω	3.57	3.00	3.00	2.35	2.00	1.93
p	3.11	3.00	2.98	2.37	2.00	1.97

Table 7.3 Estimated $L^2(\Omega)$ and $H^1(\Omega)$ convergence rates for the problem with exact solution (D.20) for the VVP-based quasi-norm-equivalent DLSP (7.54) for the velocity boundary condition when minimization is carried over the equal-order quadratic space $X^h_{2,0}$ (columns v-bc) compared to the convergence rates for the best approximation out of $X^h_{2,0}$ (columns BA) and for the VVP-based compliant DLSP (7.50) for the normal velocity–pressure boundary condition (columns nv/p-bc).

7.6.3 Mass Conservation

Local conservation and strong compatibility are two principal advantages of the vector-operator setting for LSFEMs; see Section 7.7.5. In the ADN setting, these properties are lost because the discrete Hodge decomposition results that are key to many of the proofs in Section 7.7 do not hold for the minimization spaces in Section 7.4. Thus, in general, for the ADN setting, $\nabla \cdot \mathbf{u}^h \neq 0$ either pointwise or averaged over each finite element. For quasi-norm-equivalent and norm-equivalent DLSPs, the size of the latter residual is guaranteed to be within the approximation order of the finite element spaces, provided all assumptions of the theory are met; otherwise there is no theoretical assurance that $\nabla \cdot \mathbf{u}^h$ is small.

In Section 7.6.2, we saw that violation of the minimal-degree condition for the velocity did not seem to seriously affect the convergence rates for that variable (see

Table 7.2), which may be the reason why this choice continues to be widely used in LSFEMs for the Stokes (and Navier–Stokes) equations. Unfortunately, violation of this condition has significant impact on mass conservation. In [118], it is noted that using the equal-order, piecewise linear space $X^h_{1,0}$ in a straightforward $L^2(\Omega)$ quasi-norm-eqivalent method is accompanied by a severe loss of mass; this also happens in a weighted $L^2(\Omega)$ quasi-norm-equivalent formulation. The solution to this behavior proposed in [118] is to enforce element-wise mass conservation by using Lagrange multipliers. This, of course, leads to an indefinite linear system and negates one of the principal advantages of LSFEMs. A brief discussion of this approach is given in Section 12.5.

Of course, using piecewise linear elements for the velocity disregards the assumptions of the approximation theory and so poor mass conservation with such spaces should not be counted against LSFEMs. In this section, we show that if one chooses to follow the guidelines of the approximation theory for the ADN setting, then LSFEMs deliver excellent mass conservation that can be further enhanced by using an additional weight for the continuity equation term in the least-squares functional.

To demonstrate mass conservation properties of LSFEMs when one employs finite element spaces that meet the minimal-degree requirement, we minimize (7.54) over the space $X^h_{2,0}$ from the last section, and use the same flow example as in [118]. In this example, the computational domain Ω is the rectangle $[-1,3]\times[-1,1]$ with a circular cutout of radius r centered at the origin. Figure 7.6.3 shows Ω for $r=0.3$ and $r=0.6$ and its finite element partition $\mathcal{T}_h$ into triangles. The boundary condition is set as follows:

$$\mathbf{u} = \begin{cases} (0,0)^\mathsf{T} & \text{on the surface of the cylinder} \\ (1,0)^\mathsf{T} & \text{on the boundary of the rectangle.} \end{cases} \tag{7.74}$$

The quasi-norm-equivalent functional in (7.54) is modified by multiplying the residual of the continuity equation by an additional weight μ; see [134]. The resulting least-squares functional

$$\begin{aligned} J^h_{-1}(\mathbf{u}^h,\omega^h,p^h;\mathbf{f}) &= |\ell(p^h)|^2 + \\ h^2\|\nu\nabla^\perp\omega^h+\nabla p^h-\mathbf{f}\|_0^2 &+ \mu\|\nabla\cdot\mathbf{u}^h\|_0^2+\|\nabla\times\mathbf{u}^h-\omega^h\|_0^2 \end{aligned} \tag{7.75}$$

allows one to adjust the weight of the continuity equation term separately from the other terms.

To assess how well mass is conserved, we measure the total mass flow and the average velocity across the narrowest openings between the cylinder and the wall. The amount of mass that must pass through the openings is the same in both cases so that the average velocity is higher for the larger cylinder. Because the flow is symmetric, it suffices to measure the mass flow and the average velocity through only one of the openings. The boundary condition (7.74) implies that one unit of mass flows through the opening. Table 7.4 compares the exact and the computed approximate values for these quantities. With the weight μ set to unity, the mass

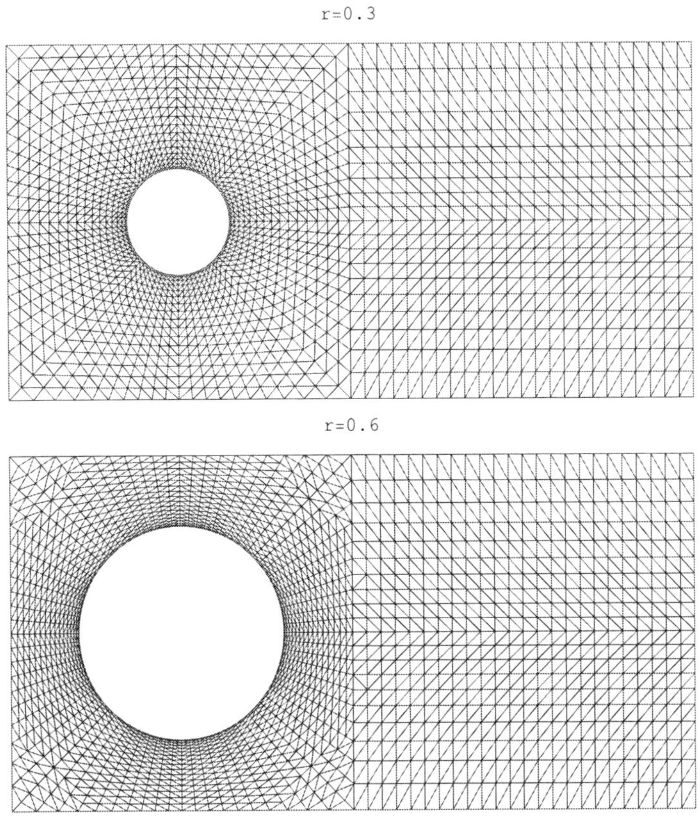

Fig. 7.1 Computational domains for testing mass conservation in LSFEMs and their partition into 3250 triangular elements.

r	μ	Exact mass flow	Computed mass flow	$\mathbf{u}_{av}$	$\mathbf{u}^h_{av}$
0.6	1	1.0000	0.9390	2.5000	2.3476
0.6	10	1.0000	0.9896	2.5000	2.4739
0.6	20	1.0000	0.9957	2.5000	2.4892
0.3	1	1.0000	0.99043	1.4285	1.4149
0.3	10	1.0000	0.99883	1.4285	1.4269
0.3	20	1.0000	0.99934	1.4285	1.4276

Table 7.4 Mass flow rates and average velocities using (7.75) and the equal-order space $X^h_{2,0}$.

loss for $r = 0.6$ is approximately 6%; for $r = 0.3$ it is only 1%. Using moderate values for μ reduces the loss to 0.5% and 0.1%, respectively. The average velocity of the finite element velocity field also improves and is in an excellent agreement with the exact values.

The data in Table 7.4 show that mass conservation does not suffer when the guidelines of the approximation theory with respect to the choice of the finite element spaces are followed. If further improvement[37] is needed, it can be achieved by increasing the weight assigned to the continuity equation in the least-squares functional, keeping in mind that improvements in how well the continuity equation $\nabla \cdot \mathbf{u} = 0$ is satisfied may be detrimental to how well the other equations in the VVP system are satisfied.

7.6.4 The Zero Mean Pressure Constraint

We handled the one-dimensional null-space of the Stokes operator using the approaches from Chapter 3. In the abstract setting, one option is to modify the original operator to an operator (C.10) that includes terms characterizing the null-space. This allowed us to formulate the least-squares principle (3.12) that uses standard spaces. Specialization of the abstract least-squares functional in (3.11) to the Stokes systems gave rise to least-squares functionals that included the term $|\ell(p)|^2$ and least-squares principles that use standard spaces for the pressure component of the solution.

A second option to remove the null-space is to change the domain of the operator to one of the abstract spaces in (C.4), (C.5), or (C.7). In the context of the Stokes system, this approach results in LSFEMs for which the pressure approximation is constrained to have zero mean. Unfortunately, the global nature of (7.3) means that any attempt to enforce it on the pressure space affects all variables and leads to a cumbersome finite element method; see [191]. As a result, practical implementations "cheat" by substituting enforcement of the true zero mean constraint by using procedures collectively known as *setting the pressure datum*. These procedures essentially amount to removing one degree of freedom from the pressure space.

Setting the pressure datum can be accomplished in many different ways, ranging from specifying a pressure value at a grid node to more complicated approaches in which a boundary traction is specified at a single node in lieu of the velocity condition; see [16, 24, 191] and the references cited therein for more details. Not surprisingly, in practice, the simplest approach of fixing the pressure value at a node also happens to be the most widely used in practice. Using the terminology of Chapter 3, this approach is equivalent to choosing $\ell(p) = \langle \delta(\mathbf{x}_0), p \rangle$ for some point $\mathbf{x}_0 \in \overline{\Omega}$, instead of choosing $\ell(p)$ to be the "correct" zero mean functional (7.24). Unfortunately, from a computational viewpoint, the former choice is not as good because it increases the condition number of the discretization matrix by a factor of $\ln h$ in two dimensions and by a factor of h^{-1} in three dimensions; see [67, Theorem 5.2]. As a result, the performance of iterative solvers may be degraded with this approach.

[37] Of course, if LSFEMs with exact mass conservation are desired, then one has to switch to the vector-operator setting and the least-squares formulations in Section 7.7. However, one should keep in mind that theoretical results in that setting are only available for some non-standard boundary conditions and, in particular, not for the standard velocity boundary condition (7.2).

Our first goal is to demonstrate that this possibility is by no means remote and that the reduced solver performance can be observed in practice. Our second objective is to convince the reader that using the first approach for which $|\ell(p)|^2$ with $\ell(p) = \int_\Omega p d\Omega$ is added to the least-squares functional, is better for iterative methods. We start by showing how to implement the term $|\ell(p)|^2$ with very little additional computational effort.

We again use (7.54) and the equal-order quadratic space $X_{2,0}^h$ defined in Section 7.6.2. Let us first consider implementation of the least-squares method. The necessary condition corresponding to (7.54) is: *seek* $\{\mathbf{u}^h, \omega^h, p^h\} \in X_{2,0}^h$ *such that*

$$Q(\{\mathbf{u}^h, \omega^h, p^h\}; \{\mathbf{v}^h, \xi^h, q^h\}) = F(\{\mathbf{v}^h, \xi^h, q^h\}) \qquad \forall \{\mathbf{v}^h, \xi^h, q^h\} \in X_{2,0}^h,$$

where

$$\begin{aligned} Q(\{\mathbf{u}^h, \omega^h, p^h\}; \{\mathbf{v}^h, \xi^h, q^h\}) &= h^2\left(\nabla \times \omega^h + \nabla p^h, \nabla \times \xi^h + \nabla q^h\right) \\ &+ \left(\nabla \times \mathbf{u}^h - \omega^h, \nabla \times \mathbf{v}^h - \xi^h\right) + \left(\nabla \cdot \mathbf{u}^h, \nabla \cdot \mathbf{v}^h\right) + \int_\Omega p^h \, d\Omega \int_\Omega q^h \, d\Omega \end{aligned} \tag{7.76}$$

and

$$F(\{\mathbf{v}^h, \xi^h, q^h\}) = h^2\left(\mathbf{f}, \nabla \times \xi^h + \nabla q^h\right).$$

Let $\{\mathbf{v}_i\}_{i=1}^N$, $\{\xi_i\}_{i=1}^M$, and $\{q_i\}_{i=1}^K$ denote the standard nodal bases (see Section B.2.1) for the velocity, the vorticity, and the pressure spaces in the approximating space $X_{2,0}^h$ and let $\vec{u} \in \mathbb{R}^{N+M+K}$ denote the coefficient vector corresponding to the finite element solution $\{\mathbf{u}^h, \omega^h, p^h\}$. The matrix generated by the bilinear form in (7.76) can be written as

$$\mathsf{A} = \widetilde{\mathsf{A}} + \vec{b}\vec{b}^\mathsf{T},$$

where $\widetilde{\mathsf{A}}$ is assembled from the $L^2(\Omega)$ inner product terms in (7.76) and

$$\vec{b} = \left(0, \ldots, 0, \int_\Omega q_1^h d\Omega, \ldots, \int_\Omega q_K^h d\Omega\right)^\mathsf{T} \in \mathbb{R}^{N+M+K},$$

where there are $N+M$ leading zeros, contains the integrals of the pressure basis functions. The term $\vec{b}\vec{b}^T$ is a rank-one matrix that contains a dense $K \times K$ block so that a direct method is not appropriate to solve the discrete least-squares problem. However, an iterative method only requires the computation of the matrix-vector product $\mathsf{A}\vec{u}$ which can be accomplished without explicitly forming $\vec{b}\vec{b}^\mathsf{T}$. Indeed,

$$\mathsf{A}\vec{u} = \widetilde{\mathsf{A}}\vec{u} + \vec{b}\vec{b}^\mathsf{T}\vec{u} = \widetilde{\mathsf{A}}\vec{u} + \vec{b}(\vec{b}^\mathsf{T}\vec{u}).$$

In other words, $\mathsf{A}\vec{u}$ is a rank-one update of the matrix-vector product $\widetilde{\mathsf{A}}\vec{u}$ that can be computed by the following procedure [67]:

1. form the vector $\vec{v} = \widetilde{\mathsf{A}}\vec{u}$
2. compute the scalar $\alpha = \vec{b}^\mathsf{T}\vec{u}$
3. update $\vec{v} \leftarrow \vec{v} + \alpha\vec{b}$.

The computation of $\widetilde{\mathbb{A}}\vec{u}$ is a standard component of any iterative solver, so the only additional work involved is the inner product in step 2 which involves $2(N+M+K)-1$ flops, and the update in step 3 which costs an additional $2(N+M+K)$ flops; these figures are small compared to the work necessary to compute the matrix-vector product $\widetilde{\mathbb{A}}\vec{u}$. The row vector $\vec{b}^{\mathsf{T}}$ can be precomputed and stored, thus resulting in an efficient computation of α and $\alpha\vec{b}$.

Let us now see how an iterative method performs when the term $|\ell(p)|^2$, with $\ell(p)$ chosen as in (7.24), is included in the least-squares functional compared to when p is "pinned" at a grid point $\mathbf{x}_0$. We use the same example of a flow past a cylinder as in Section 7.6.3 with the boundary condition specified in (7.74). The radius of the circular cutout is set to $r = 0.5$. The discrete problems are solved by the Jacobi preconditioned conjugate gradient method. Iterations are carried out until the relative residual error becomes smaller than 10^{-6}.

The LSFEM with the zero mean term $|\ell(p)|^2$ included in the functional took[38] 1000 iterations to converge. For the LSFEM with the pressure "pinned" at a point, the number of iterations depend on the location of the grid node $\mathbf{x}_0$ at which the pressure is pinned. The number of iterations for different choices of this node are shown in Figure 7.2, where, due to symmetry, only half the domain is shown.

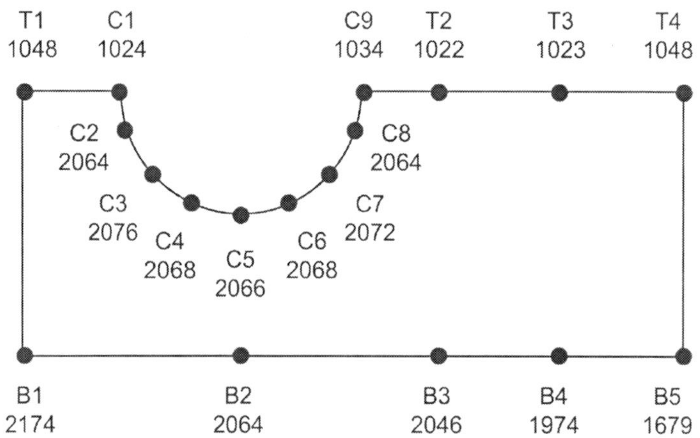

Fig. 7.2 Number of Jacobi preconditioned CG iterations for different choices of the pressure pinning point $\mathbf{x}_0$; the number of iterations when the term $|\ell(p)|^2$ is included in the functional is 1000.

[38] The fact that the number of iterations is large, even for the LSFEM with the zero mean term included in the functional, reflects the facts that first, the LSFEM corresponding to the mesh-weighted functional in (7.54) results in a matrix problem having a condition number of $O(h^{-4})$ and second, that Jacobi preconditioning is not a very efficient means of accelerating convergence of conjugate gradient methods.

The numbers of iterations given in Figure 7.2 show that the fastest convergence of the preconditioned conjugate gradient method occurs when $\mathbf{x}_0$ is placed along the line of symmetry $y = 0$; placing $\mathbf{x}_0$ at the corner points requires about twice as many conjugate gradient iterations, e.g., compare the number of iterations for the nodes labeled T1, C1, C9, T2, T3, and T4 with the number of iterations for the nodes labeled B1 and B5. In fact, placing $\mathbf{x}_0$ anywhere on the bottom $y = -1$ (or top $y = 1$) part of the boundary roughly doubles the number of iterations needed. Placing $\mathbf{x}_0$ on the boundary of the circular cutout has the same effect, except when the node also happens to lie on the line of symmetry. More importantly, comparing the pairs C1 and C2 and C8 and C9, we see that the number of conjugate gradient iterations can vary very significantly even between two choices for the pressure datum point that are close to each other. This suggests that, in general, it may be very difficult, if not impossible, to identify the nodes which optimize conjugate gradient performance. It may be somewhat easier to pinpoint some choices that should be avoided because, as we have found, region corners are consistently among the worst performers.

7.7 Least-Squares Finite Element Methods in the Vector-Operator Setting

We now turn to the LSFEMs for the VVP Stokes system in the vector-operator setting.[39] See Remarks 7.1 and 7.2 for an explanation why we do not consider VSP and VGVP systems in the vector-operator setting.

7.7.1 Energy Balances

Energy balances for VVP systems in the vector-operator setting are sought in terms of the spaces[40] $G(\Omega)$, $\mathbf{C}(\Omega)$, $\mathbf{D}(\Omega)$, and $S(\Omega)$ from the De Rham complex (A.52). Insofar as the variables in the VVP system are concerned, this means that we can regard p as an element of $G(\Omega)$ or $S(\Omega)$, whereas both $\mathbf{u}$ and ω can be regarded as elements of $\mathbf{C}(\Omega)$ and/or $\mathbf{D}(\Omega)$.

These associations *make the velocity boundary condition unnatural*[41] *for the vector-operator setting* because it imposes both the tangential and normal components

---

[39] Recall that the main distinction between these two ADN and vector-operator settings is in how they treat the dependent variables in the PDE problem; see Remarks 3.6 and 5.6.

[40] For clarity, we assume unit weights for all these spaces and drop the weights from the norm and the inner product designations. This is acceptable because, in this chapter, we work with the non-dimensional Stokes equations. In Chapters 5 and 6, the weights were needed in order to preserve the unit consistency implied by various constitutive laws, which in turn led to important properties of LSFEMs, such as the "splitting" of the discrete equations proved in Theorems 5.37 and 6.20.

[41] This does not mean that the vector-operator setting cannot be used with the velocity boundary condition, but rather that the advantages of doing so are unclear.

of the velocity field on the boundary. In contrast, a vector field in $\mathbf{C}(\Omega)$ or $\mathbf{D}(\Omega)$ can have its tangential or normal components imposed but not both at the same time. This is why the vector-operator setting is much more appropriate for the nonstandard normal velocity–pressure and the normal velocity–tangential vorticity boundary conditions, both of which can be imposed on the spaces in (A.52) without any difficulty. In what follows, the main focus is on the second nonstandard boundary condition because the combination of normal velocity and tangential vorticity proves to be particularly well-suited for the formulation of locally conservative LSFEMs.

To fully appreciate the role of the normal velocity–tangential vorticity boundary condition in the formation of locally conservative LSFEMs, we restrict attention to $\mathbb{R}^3$, where both $\mathbf{u}$ and ω are vector fields. The key results easily specialize to $\mathbb{R}^2$; however, the converse is not true, which is another reason to focus on the three-dimensional setting.

The first a priori bound concerns the original VVP Stokes system (7.4) with (7.8) and is given in the following theorem.

Theorem 7.19. *Assume that Ω is a simply connected bounded domain in $\mathbb{R}^3$ with a Lipschitz boundary. Then, there exists a positive constant C such that*[42]

$$C(\|\mathbf{u}\|_{DC}+\|\omega\|_C+\|p\|_G) \le \|\nabla\times\omega+\nabla p\|_0+\|\nabla\times\mathbf{u}-\omega\|_0+\|\nabla\cdot\mathbf{u}\|_0 \quad (7.77)$$

for all $\{\mathbf{u},\omega,p\}\in \mathbf{D}_0(\Omega)\cap\mathbf{C}(\Omega)\times\mathbf{C}_0(\Omega)\times G(\Omega)\cap L^2_0(\Omega)$.

Proof. Owing to the boundary condition $\omega\times\mathbf{n}=0$ on $\partial\Omega$,

$$\|\nabla\times\omega+\nabla p\|_0^2 = \|\nabla\times\omega\|_0^2+\|\nabla p\|_0^2 \quad (7.78)$$

and

$$\|\nabla\times\mathbf{u}-\omega\|_0^2 = \|\nabla\times\mathbf{u}\|_0^2+\|\omega\|_0^2-(\nabla\times\mathbf{u},\omega)-(\mathbf{u},\nabla\times\omega)\,. \quad (7.79)$$

Using this identity and the Cauchy inequality, it follows that

$$\begin{aligned}&\|\nabla\times\mathbf{u}-\omega\|_0^2+\|\nabla\cdot\mathbf{u}\|_0^2\\ &\qquad\ge \|\nabla\times\mathbf{u}\|_0^2+\|\nabla\cdot\mathbf{u}\|_0^2+\|\omega\|_0^2-\|\nabla\times\mathbf{u}\|_0\,\|\omega\|_0-\|\mathbf{u}\|_0\,\|\nabla\times\omega\|_0\,.\end{aligned}$$

Because $\mathbf{u}\in\mathbf{D}_0(\Omega)\cap\mathbf{C}(\Omega)$, the Poincaré-Friedrichs inequality (A.69) implies

$$\|\nabla\times\mathbf{u}\|_0^2+\|\nabla\cdot\mathbf{u}\|_0^2 \ge \frac{1}{C_P^2}\|\mathbf{u}\|_0^2\,. \quad (7.80)$$

In conjunction with the previous bound, we can therefore write

[42] The norm $\|\cdot\|_{DC}$ is defined in (A.61).

$$\|\nabla\times\mathbf{u}-\omega\|_0^2+\|\nabla\cdot\mathbf{u}\|_0^2 \geq \frac{1}{2}\left(\|\nabla\times\mathbf{u}\|_0^2+\|\nabla\cdot\mathbf{u}\|_0^2\right)$$
$$+\frac{1}{2C_P^2}\|\mathbf{u}\|_0^2+\|\omega\|_0^2-\|\nabla\times\mathbf{u}\|_0\|\omega\|_0-\|\mathbf{u}\|_0\|\nabla\times\omega\|_0 .$$

Using the ε-inequality for the last two terms gives

$$\|\nabla\times\mathbf{u}\|_0\|\omega\|_0 \leq \frac{\varepsilon_1}{2}\|\nabla\times\mathbf{u}\|_0^2+\frac{1}{2\varepsilon_1}\|\omega\|_0^2$$

and

$$\|\mathbf{u}\|_0\|\nabla\times\omega\|_0 \leq \frac{\varepsilon_2}{2}\|\mathbf{u}\|_0^2+\frac{1}{2\varepsilon_2}\|\nabla\times\omega\|_0^2,$$

respectively. Let $\beta>0$ denote a positive real parameter that is to be determined later. Combining the last three inequalities gives

$$\beta\|\nabla\times\omega+\nabla p\|_0^2+\|\nabla\times\mathbf{u}-\omega\|_0^2+\|\nabla\cdot\mathbf{u}\|_0^2$$
$$\geq \beta\left(\|\nabla\times\omega\|_0^2+\|\nabla p\|_0^2\right)+\frac{1}{2}\left(\|\nabla\times\mathbf{u}\|_0^2+\|\nabla\cdot\mathbf{u}\|_0^2\right)+\frac{1}{2C_P^2}\|\mathbf{u}\|_0^2+\|\omega\|_0^2$$
$$-\frac{\varepsilon_1}{2}\|\nabla\times\mathbf{u}\|_0^2-\frac{1}{2\varepsilon_1}\|\omega\|_0^2-\frac{\varepsilon_2}{2}\|\mathbf{u}\|_0^2-\frac{1}{2\varepsilon_2}\|\nabla\times\omega\|_0^2$$
$$\geq \left(\beta-\frac{1}{2\varepsilon_2}\right)\|\nabla\times\omega\|_0^2+\left(1-\frac{1}{2\varepsilon_1}\right)\|\omega\|_0^2$$
$$+\frac{1}{2}(1-\varepsilon_1)\|\nabla\times\mathbf{u}\|_0^2+\frac{1}{2}\left(\frac{1}{C_P^2}-\varepsilon_2\right)\|\mathbf{u}\|_0^2+\beta\|\nabla p\|_0^2+\frac{1}{2}\|\nabla\cdot\mathbf{u}\|_0^2 .$$

Choosing $\varepsilon_1=2/3$, $\varepsilon_2=1/(2C_P^2)$, and $\beta=1/\varepsilon_2=2C_P^2$ gives

$$1-\varepsilon_1=\frac{1}{3}\qquad 1-\frac{1}{2\varepsilon_1}=\frac{1}{4}\qquad \frac{1}{C_P^2}-\varepsilon_2=\frac{1}{2C_P^2}\qquad \beta-\frac{1}{2\varepsilon_2}=C_P^2 .$$

Combining all of the above and using the Poincaré-Friedrichs inequality (A.63) to bound $\|\nabla p\|_0$ from below gives

$$2C_P^2\|\nabla\times\omega+\nabla p\|_0^2+\|\nabla\times\mathbf{u}-\omega\|_0^2+\|\nabla\cdot\mathbf{u}\|_0^2$$
$$\geq \frac{1}{6}\|\nabla\times\mathbf{u}\|_0^2+\frac{1}{2}\|\nabla\cdot\mathbf{u}\|_0^2+\frac{1}{4C_P^2}\|\mathbf{u}\|_0^2$$
$$+C_P^2\|\nabla\times\omega\|_0^2+\frac{1}{4}\|\omega\|_0^2+\frac{2C_P^2}{(1+C_P)^2}\|p\|_G^2$$
$$\geq \min\left\{C_{\mathbf{u}},C_\omega,2C_P^2(1+C_P)^{-2}\right\}\left(\|\mathbf{u}\|_{DC}^2+\|\omega\|_C^2+\|p\|_G^2\right),$$

where

$$C_\omega=\min\left\{C_P^2,\frac{1}{4}\right\}\qquad\text{and}\qquad C_{\mathbf{u}}=\min\left\{\frac{1}{4C_P^2},\frac{1}{6},\frac{1}{2}\right\}.$$

The theorem follows by noting that

$$\|\nabla\times\omega+\nabla p\|_0^2+\|\nabla\times\mathbf{u}-\omega\|_0^2+\|\nabla\cdot\mathbf{u}\|_0^2$$
$$\geq\frac{1}{\max\{1,2C_P^2\}}\left(2C_P^2\|\nabla\times\omega+\nabla p\|_0^2+\|\nabla\times\mathbf{u}-\omega\|_0^2+\|\nabla\cdot\mathbf{u}\|_0^2\right). \qquad \square$$

Remark 7.20 The key junctures in the proof of this theorem are the identities (7.78) and (7.79) and the Poincaré-Friedrichs inequality (7.80). This observation is used later, in Section 7.7, for the analysis of LSFEMs based on (7.77). □

For the extended VVP system (7.10) with the normal velocity–tangential vorticity boundary condition (7.8), we have the following result.

Corollary 7.21 *Assume that Ω is as in Theorem* 7.19. *Then, there exists a positive constant C such that*

$$C\big(\|\mathbf{u}\|_{DC}+\|\omega\|_{DC}+\|p\|_G\big)$$
$$\leq\|\nabla\times\omega+\nabla p\|_0+\|\nabla\cdot\omega\|_0+\|\nabla\times\mathbf{u}-\omega\|_0+\|\nabla\cdot\mathbf{u}\|_0 \tag{7.81}$$

for all $\{\mathbf{u},\omega,p\}\in\mathbf{D}_0(\Omega)\cap\mathbf{C}(\Omega)\times\mathbf{D}(\Omega)\cap\mathbf{C}_0(\Omega)\times G(\Omega)\cap L_0^2(\Omega)$.

Proof. The results follows from (7.77) and the definition (A.61) of $\|\cdot\|_{DC}$. □

This result can be further strengthened under some additional assumptions on the domain.

Corollary 7.22 *Assume that Ω satisfies the hypotheses of Theorem* A.8. *Then, there exists a positive constant C such that*

$$C\big(\|\mathbf{u}\|_1+\|\omega\|_1+\|p\|_1\big)$$
$$\leq\|\nabla\times\omega+\nabla p\|_0+\|\nabla\cdot\omega\|_0+\|\nabla\times\mathbf{u}-\omega\|_0+\|\nabla\cdot\mathbf{u}\|_0 \tag{7.82}$$

for all $\{\mathbf{u},\omega,p\}\in\mathbf{D}_0(\Omega)\cap\mathbf{C}(\Omega)\times\mathbf{D}(\Omega)\cap\mathbf{C}_0(\Omega)\times G(\Omega)\cap L_0^2(\Omega)$.

Proof. The result follows from (7.81) and the equivalence bound (A.62) that holds when Ω is as prescribed by Theorem A.8. □

Except for the absence of a slack variable, the last result is identical to the ADN bound (7.31) with $q=0$.

Remark 7.23 The results of Theorem 7.19 and Corollaries 7.21 and 7.22 easily extend to the three-dimensional normal velocity–pressure boundary condition (7.6). We skip the details because this boundary condition does not lead to better LSFEMs than (7.8). □

7.7.2 Continuous Least-Squares Principles

The two energy balances from Section 7.7.1 that are of most interest[43] to us are stated in Theorem 7.19 and Corollary 7.21. Using the first one, i.e., (7.77), gives rise to the well-posed CLSP

$$\begin{cases} J(\mathbf{u},\omega,p;\mathbf{f}) = \|\nabla\times\omega+\nabla p-\mathbf{f}\|_0^2+\|\nabla\times\mathbf{u}-\omega\|_0^2+\|\nabla\cdot\mathbf{u}\|_0^2+|\ell(p)|^2 \\ X=\mathbf{D}_0(\Omega)\cap\mathbf{C}(\Omega)\times\mathbf{C}_0(\Omega)\times G(\Omega)\,. \end{cases} \tag{7.83}$$

The second balance, i.e., (7.81), gives an "extended" version of (7.83):

$$\begin{cases} J(\mathbf{u},\omega,p;\mathbf{f}) = \\ \qquad \|\nabla\times\omega+\nabla p-\mathbf{f}\|_0^2+\|\nabla\cdot\omega\|_0+\|\nabla\times\mathbf{u}-\omega\|_0^2+\|\nabla\cdot\mathbf{u}\|_0^2+|\ell(p)|^2 \\ X=\mathbf{D}_0(\Omega)\cap\mathbf{C}(\Omega)\times\mathbf{D}(\Omega)\cap\mathbf{C}_0(\Omega)\times G(\Omega)\,. \end{cases} \tag{7.84}$$

Of course, the comments made in Remark 7.12 remain in full force for (7.83) and (7.84). The versions of these CLSPs without the $|\ell(p)|^2$ term are

$$\begin{cases} J(\mathbf{u},\omega,p;\mathbf{f}) = \|\nabla\times\omega+\nabla p-\mathbf{f}\|_0^2+\|\nabla\times\mathbf{u}-\omega\|_0^2+\|\nabla\cdot\mathbf{u}\|_0^2 \\ X^C=\mathbf{D}_0(\Omega)\cap\mathbf{C}(\Omega)\times\mathbf{C}_0(\Omega)\times G(\Omega)\cap L^2_0(\Omega) \end{cases} \tag{7.85}$$

and

$$\begin{cases} J(\mathbf{u},\omega,p;\mathbf{f}) = \\ \qquad \|\nabla\times\omega+\nabla p-\mathbf{f}\|_0^2+\|\nabla\cdot\omega\|_0+\|\nabla\times\mathbf{u}-\omega\|_0^2+\|\nabla\cdot\mathbf{u}\|_0^2 \\ X^C=\mathbf{D}_0(\Omega)\cap\mathbf{C}(\Omega)\times\mathbf{D}(\Omega)\cap\mathbf{C}_0(\Omega)\times G(\Omega)\cap L^2_0(\Omega)\,, \end{cases} \tag{7.86}$$

respectively.

7.7.3 Discrete Least-Squares Principles

In Sections 7.7.1 and 7.7.2, we considered the vector-operator setting for the three-dimensional VVP system with the normal velocity–tangential vorticity boundary condition (7.8). We have a choice between the CLSPs (7.83) and (7.84) that include $|\ell(p)|$ in the functional and the CLSPs (7.85) and (7.86) that do not. All of these CLSPs only involve $L^2(\Omega)$ norms of the first-order equations so that, formally,

---

[43] The third energy balance in that section, i.e., (7.82), is equivalent to (7.31) with $q=0$ and the corresponding CLSP is essentially the same as (7.43).

they all admit practical compliant discretizations. However, what is always included in the corresponding minimization space is one or both of the intersection spaces $\mathbf{D}_0(\Omega)\cap\mathbf{C}(\Omega)$ and $\mathbf{D}(\Omega)\cap\mathbf{C}_0(\Omega)$. The problems caused by these spaces were first encountered in Section 5.6.2 when we attempted to discretize them conformally.

One of the lessons learned in Section 5.6.2 is that intersection spaces should be avoided, if possible, in variational settings for finite element methods. There, this lesson is not difficult to follow because we had an alternative CLSP that did not include $\mathbf{D}_0(\Omega)\cap\mathbf{C}(\Omega)$ or $\mathbf{D}(\Omega)\cap\mathbf{C}_0(\Omega)$. The situation with (7.83) and (7.84) is, however, closer to the setting in Section 6.3 where intersection spaces appeared naturally in the energy balance of the div–curl equations. To deal with their approximation in (7.83)–(7.84), we adopt the same approach[44] as in the formulation of the non-conforming DLSPs (6.50) and (6.51), i.e., we approximate conformally *only one of the component spaces* in $\mathbf{D}_0(\Omega)\cap\mathbf{C}(\Omega)$ or $\mathbf{D}(\Omega)\cap\mathbf{C}_0(\Omega)$.

As in Sections 6.3.4 and 6.3.5, the key approximation and structural properties of compatible finite element spaces impose some restrictions on the domain Ω and the finite element partition $\mathcal{T}_h$. In addition to the usual assumption that Ω is contractible and $\mathcal{T}_h$ is an affine simplicial mesh, in this section we assume that Ω is convex with a polyhedral connected boundary $\partial\Omega$. The first set of restrictions ensures that the De Rham complex (A.52) and its finite element approximation (B.45) are exact and that the approximation results from Theorems B.7 and B.9 hold.[45] The second set is necessary to secure some additional regularity required in the proofs.

For pedagogical reasons, we begin with the two CLSPs for which minimization is carried over the complement space with respect to $\ell(p)$. To discretize (7.85) and (7.86), we have to choose approximating spaces for

$$X^C = \mathbf{D}_0(\Omega)\cap\mathbf{C}(\Omega)\times\mathbf{C}_0(\Omega)\times G(\Omega)\cap L_0^2(\Omega)$$

and

$$X^C = \mathbf{D}_0(\Omega)\cap\mathbf{C}(\Omega)\times\mathbf{D}(\Omega)\cap\mathbf{C}_0(\Omega)\times G(\Omega)\cap L_0^2(\Omega),$$

respectively. From Chapters 5–6, we already know that only one of the component spaces in $\mathbf{D}_0(\Omega)\cap\mathbf{C}(\Omega)$ and $\mathbf{D}(\Omega)\cap\mathbf{C}_0(\Omega)$ should be approximated conformally. Insofar as the choice of this component is concerned, we follow the approach in Section 6.3.3 and pick the one that is constrained by the boundary condition, i.e.,

$$\mathbf{D}_0(\Omega)\cap\mathbf{C}(\Omega)\longrightarrow\mathbf{D}_0^h(\Omega)\qquad\mathbf{D}(\Omega)\cap\mathbf{C}_0(\Omega)\longrightarrow\mathbf{C}_0^h(\Omega). \tag{7.87}$$

Approximation of the pressure space $G(\Omega)\cap L_0^2(\Omega)$, which appears in both solution spaces above, merits some more scrutiny in the vector-operator setting. First, observe that $G(\Omega)\cap L_0^2(\Omega)$ is an intersection space on its own right, except that its conformal approximation is not as problematic as that for $\mathbf{D}_0(\Omega)\cap\mathbf{C}(\Omega)$ or $\mathbf{D}(\Omega)\cap\mathbf{C}_0(\Omega)$. Nevertheless, we could, in principle, consider approximating con-

[44] An alternative solution is the reformulation strategy in Section 6.2 in which the variable belonging to an intersection space is "split" into $\mathbf{C}(\Omega)$ and $\mathbf{D}(\Omega)$ components. To save space, we do not pursue this idea in the context of the VVP Stokes equations.

[45] We refer to Remarks 5.32, B.8, and B.10 for more details.

formally only one of its components. If this component is $G(\Omega)$, then the term $|\ell(p)|$ has to be added to the least-squares functional which effectively brings us to discretization of (7.83) or (7.84). A far less obvious choice is to approximate the $L^2_0(\Omega)$ component of $G(\Omega)\cap L^2_0(\Omega)$. Such an approximation would be hard to justify in the ADN setting because it entails trading the analytic gradient for a discrete weak[46] gradient. This is not so in the vector-operator setting for which this choice enables some additional conservation properties[47] in the LSFEMs. In summary, as far as the pressure is concerned, we have the following three approximation possibilities:

- conformal: $\quad G(\Omega)\cap L^2_0(\Omega) \longrightarrow \widetilde{G}^h = G^h \cap L^2_0(\Omega)$
- $L^2_0(\Omega)$-conformal: $\quad G(\Omega)\cap L^2_0(\Omega) \longrightarrow S^h_0$
- $G(\Omega)$-conformal: $\quad G(\Omega)\cap L^2_0(\Omega) \longrightarrow G^h$

For the original VVP Stokes system with the normal velocity–tangential vorticity boundary condition (7.8), the first choice gives the following DLSP:

$$\begin{cases} J^h(\mathbf{u}^h,\omega^h,p^h;\mathbf{f}) = \\ \qquad \|\nabla\times\omega^h+\nabla p^h-\mathbf{f}\|_0^2+\|\nabla_h^*\times\mathbf{u}^h-\omega^h\|_0^2+\|\nabla\cdot\mathbf{u}^h\|_0^2 \\ X^{h,C} = \mathbf{D}^h_0\times\mathbf{C}^h_0\times\widetilde{G}^h . \end{cases} \tag{7.88}$$

Clearly, this DLSP is a non-conforming discretization of (7.85). The second choice for the pressure approximation results in

$$\begin{cases} J^h(\mathbf{u}^h,\omega^h,p^h;\mathbf{f}) = \\ \qquad \|\nabla\times\omega^h+\nabla_h^* p^h-\mathbf{f}\|_0^2+\|\nabla_h^*\times\mathbf{u}^h-\omega^h\|_0^2+\|\nabla\cdot\mathbf{u}^h\|_0^2 \\ X^{h,C} = \mathbf{D}^h_0\times\mathbf{C}^h_0\times S^h_0 \end{cases} \tag{7.89}$$

which is another non-conforming discretization of (7.85). The last choice gives

$$\begin{cases} J^h(\mathbf{u}^h,\omega^h,p^h;\mathbf{f}) = \\ \qquad \|\nabla\times\omega^h+\nabla p^h-\mathbf{f}\|_0^2+\|\nabla_h^*\times\mathbf{u}^h-\omega^h\|_0^2+\|\nabla\cdot\mathbf{u}^h\|_0^2+|\ell(p^h)|^2 \\ X^h = \mathbf{D}^h_0\times\mathbf{C}^h_0\times G^h \end{cases} \tag{7.90}$$

which is a non-conforming discretization of (7.83).

For the extended VVP Stokes system (7.10) with the normal velocity–tangential vorticity boundary condition, we have the DLSP

---

[46] Recall that in order to prevent this from happening in LSFEMs based on the ADN setting the minimization space X^h_r, defined in (4.44), defaults to a product of the lowest order C^0 nodal spaces G^1 when $r=0$; see Remark 4.11.

[47] Approximation of the pressure by S^h_0 allows one to take advantage of the discrete Hodge decomposition in Theorem B.23. The details are presented in Section 7.7.5.

$$\begin{cases} J^h(\mathbf{u}^h, \omega^h, p^h; \mathbf{f}) = \\ \quad \|\nabla \times \omega^h + \nabla p^h - \mathbf{f}\|_0^2 + \|\nabla_h^* \cdot \omega^h\|_0 + \|\nabla_h^* \times \mathbf{u}^h - \omega^h\|_0^2 + \|\nabla \cdot \mathbf{u}^h\|_0^2 \\ X^{h,C} = \mathbf{D}_0^h \times \mathbf{C}_0^h \times \widetilde{G}^h \end{cases} \tag{7.91}$$

which is a non-conforming discretization of (7.86), the DLSP

$$\begin{cases} J^h(\mathbf{u}^h, \omega^h, p^h; \mathbf{f}) = \\ \quad \|\nabla \times \omega^h + \nabla_h^* p^h - \mathbf{f}\|_0^2 + \|\nabla_h^* \cdot \omega^h\|_0 + \|\nabla_h^* \times \mathbf{u}^h - \omega^h\|_0^2 + \|\nabla \cdot \mathbf{u}^h\|_0^2 \\ X^{h,C} = \mathbf{D}_0^h \times \mathbf{C}_0^h \times S_0^h, \end{cases} \tag{7.92}$$

which is another non-conforming discretization of (7.86), and the DLSP

$$\begin{cases} J^h(\mathbf{u}^h, \omega^h, p^h; \mathbf{f}) = |\ell(p^h)|^2 + \\ \quad \|\nabla \times \omega^h + \nabla p^h - \mathbf{f}\|_0^2 + \|\nabla_h^* \cdot \omega^h\|_0 + \|\nabla^* \times \mathbf{u}^h - \omega^h\|_0^2 + \|\nabla \cdot \mathbf{u}^h\|_0^2 \\ X^h = \mathbf{D}_0^h \times \mathbf{C}_0^h \times G^h, \end{cases} \tag{7.93}$$

which is a non-conforming discretization of (7.84).

The discrete weak gradient, curl, and divergence operators used in the above DLSPs were defined in (B.49), (B.50), and (B.51), respectively. Their appearance in the least-squares functionals is mandated by the non-conforming finite element approximation of the respective solution space components. Of course, the observation made in Section 6.3.3 about the importance of "matching" domains and ranges of these discrete operators remains in full force in the present context as well.

Specifically, to enable the additional conservation properties mentioned earlier in this section, the approximating spaces $\mathbf{D}_0^h$, $\mathbf{C}_0^h$, and S_0^h used to define (7.89) and (7.92) must belong to the same finite element De Rham complex. Then, the discrete Hodge decomposition in Theorem B.23 asserts that $\mathbf{D}_0^h = \nabla \times (\mathbf{C}_0^h) \oplus \nabla_h^*(S_0^h)$ which is pivotal for acquiring the additional conservation properties.

7.7.4 Stability of Discrete Least-Squares Principles

The DLSPs defined in the last section are non-conforming, which makes the setting for their analysis very similar to the one encountered in Section 6.3.5. Thus, our first task is to establish stability of (7.88)–(7.92) which no longer follows automatically from the stability[48] of the corresponding parent CLSPs.

[48] Recall that the parent CLSPs of these methods are stable with respect to solution spaces involving the norm $\|\cdot\|_{DC}$ which is not defined for finite element spaces that are only curl- or div-conforming.

We present the details for formulations without the $|\ell(p)|$ term in the least-squares functional, i.e., (7.88), (7.89), (7.91), and (7.92). Extending the results to LSFEMs that include $|\ell(p)|$ is straightforward. The stability of (7.88) and (7.91) follows from discrete analogues of (7.77) and (7.81) in Corollary 7.21 which are established in the next theorem.

Theorem 7.24 *Assume that* $\mathbf{C}_0^h$ *and* $\mathbf{D}_0^h$ *belong to the same finite element De Rham complex in* (B.45) *and* $\|\cdot\|_{D^*C}$ *and* $\|\cdot\|_{DC^*}$ *are the norms for these spaces defined in* (6.58) *and* (6.59)*, respectively. Then, there exists a positive constant C, independent of h, such that*

$$C\big(\|\mathbf{u}^h\|_{DC^*}+\|\omega^h\|_C+\|p^h\|_G\big)\\ \leq \|\nabla\times\omega^h+\nabla p^h\|_0+\|\nabla_h^*\times\mathbf{u}^h-\omega^h\|_0+\|\nabla\cdot\mathbf{u}^h\|_0 \tag{7.94}$$

and

$$C\big(\|\mathbf{u}^h\|_{DC^*}+\|\omega^h\|_{D^*C}+\|p^h\|_G\big)\\ \leq \|\nabla\times\omega^h+\nabla p^h\|_0+\|\nabla_h^*\cdot\omega^h\|_0+\|\nabla_h^*\times\mathbf{u}^h-\omega^h\|_0+\|\nabla\cdot\mathbf{u}^h\|_0 \tag{7.95}$$

for all $\{\mathbf{u}^h,\omega^h,p^h\}\in X^h\equiv\mathbf{D}_0^h\times\mathbf{C}_0^h\times\widetilde{G}^h$.

Proof. The bounds (7.94) and (7.95) are not implied by (7.77) and (7.81) because the finite element spaces in the present theorem are not conformal approximations of the spaces in Theorem 7.19 and Corollary 7.21. Fortunately, compatibility of the finite element spaces turns out to be powerful enough to offset the lack of conformity and to allow us to reuse the arguments of Theorem 7.19 to prove (7.94).

The pivotal factor which makes this possible is that the key junctures in the proof of Theorem 7.19, identified in Remark 7.20, can be reproduced by compatible finite element spaces. First, because pressure and vorticity are approximated by the curl- and grad-conforming spaces $\mathbf{C}_0^h$ and $\widetilde{G}^h$, respectively, we have that

$$\|\nabla\times\omega^h+\nabla p^h\|_0^2=\|\nabla\times\omega^h\|_0^2+\|\nabla p^h\|_0^2, \tag{7.96}$$

which provides the first key juncture (7.78). A discrete version of the second key juncture, i.e., (7.79), follows from the assumption that $\mathbf{C}_0^h$ and $\mathbf{D}_0^h$ are in the same exact sequence. This ensures that the range of $\nabla_h^*\times$ is in $\mathbf{C}_0^h$ and that

$$\big(\nabla_h^*\times\mathbf{v}^h,\mathbf{w}^h\big)=\big(\mathbf{v}^h,\nabla\times\mathbf{w}^h\big)\qquad\forall\mathbf{v}^h\in\mathbf{D}_0^h\quad\text{and}\quad\forall\mathbf{w}^h\in\mathbf{C}_0^h.$$

Using this identity, we easily see that

$$\|\nabla_h^*\times\mathbf{u}^h-\omega^h\|_0^2=\|\nabla_h^*\times\mathbf{u}^h\|_0^2+\|\omega^h\|_0^2-\big(\nabla_h^*\times\mathbf{u}^h,\omega^h\big)-\big(\mathbf{u}^h,\nabla\times\omega^h\big). \tag{7.97}$$

Finally, the discrete Poincaré-Friedrichs inequality (B.84) for div-conforming spaces gives us a discrete analogue of the third key juncture (7.80):

$$\|\nabla_h^* \times \mathbf{u}^h\|_0^2 + \|\nabla \cdot \mathbf{u}\|_0^2 \geq \frac{1}{C_P^2}\|\mathbf{u}^h\|_0^2. \tag{7.98}$$

The proof of (7.94) now follows without much difficulty by repeating the remaining steps from the proof of Theorem 7.19, taking into consideration the definition of $\|\cdot\|_{DC^*}$. The second bound follows from (7.94) and the definition of $\|\cdot\|_{D^*C}$. □

The discrete energy balances (7.94) and (7.95) assert the well-posedness of (7.88) and (7.91), respectively. To verify that the other two DLSPs, i.e., (7.89) and (7.92), are also well-posed, we need a second set of discrete analogues for (7.77) and (7.81). These are established in the following theorem.

Theorem 7.25 *Assume that* $\mathbf{C}_0^h$, $\mathbf{D}_0^h$, *and* S_0^h *belong to the same finite element De Rham complex in* (B.45) *and* $\|\cdot\|_{D^*C}$ *and* $\|\cdot\|_{DC^*}$ *are the norms for these spaces defined in* (6.58) *and* (6.59), *respectively. Then, there exists a positive constant C, independent of h, such that*

$$\begin{aligned} &C\big(\|\mathbf{u}^h\|_{DC^*} + \|\omega^h\|_C + \|\nabla_h^* p^h\|_0\big) \\ &\qquad \leq \|\nabla\times\omega^h + \nabla_h^* p^h\|_0 + \|\nabla_h^*\times\mathbf{u}^h - \omega^h\|_0 + \|\nabla\cdot\mathbf{u}^h\|_0 \end{aligned} \tag{7.99}$$

and

$$\begin{aligned} &C\big(\|\mathbf{u}^h\|_{DC^*} + \|\omega^h\|_{D^*C} + \|\nabla_h^* p^h\|_0\big) \\ &\leq \|\nabla\times\omega^h + \nabla_h^* p^h\|_0 + \|\nabla_h^*\cdot\omega^h\|_0 + \|\nabla_h^*\times\mathbf{u}^h - \omega^h\|_0 + \|\nabla\cdot\mathbf{u}^h\|_0 \end{aligned} \tag{7.100}$$

for all $\{\mathbf{u}^h, \omega^h, p^h\} \in X^h \equiv \mathbf{D}_0^h \times \mathbf{C}_0^h \times S_0^h$.

Proof. Our plan is to show that discrete versions of the key junctures in the proof of Theorem 7.19 continue to hold when the pressure is approximated by the discontinuous space S_0^h. Then, (7.99) and (7.100) follow as in the last theorem. Clearly, (7.97) and (7.98) remain valid because the finite element spaces for the velocity and the vorticity have not changed. Let us show that a discrete version of the first key juncture is also available. In the present setting, this follows from the assumption that $\mathbf{C}_0^h$, $\mathbf{D}_0^h$, and S_0^h belong to the same finite element exact sequence. On the one hand, this ensures that the range of the discrete gradient ∇_h^*, defined in (B.49), is in $\mathbf{D}_0^h$ and that

$$\big(\mathbf{w}^h, \nabla_h^* q^h\big) = -\big(\nabla\cdot\mathbf{w}^h, q^h\big) \quad \forall q^h \in S_0^h \qquad \text{and} \quad \forall \mathbf{w}^h \in \mathbf{D}_0^h.$$

On the other hand, because $\mathbf{C}_0^h$ and $\mathbf{D}_0^h$ are also in the same exact sequence, the range of $\nabla\times$ is in $\mathbf{D}_0^h$. As a result,

$$\big(\nabla\times\omega^h, \nabla_h^* p^h\big) = \big(\nabla\cdot\nabla\times\omega^h, p^h\big) = 0$$

which provides the following discrete version of (7.78):

$$\|\nabla\times\omega^h + \nabla_h^* p^h\|_0^2 = \|\nabla\times\omega^h\|_0^2 + \|\nabla_h^* p^h\|_0^2.$$

The proof now follows as in Theorem 7.24. □

7.7.5 Conservation of Mass and Strong Compatibility

With the stability of the LSFEMs cleared up, we proceed with their conservation properties. First we study a discrete VVP Stokes system.

Theorem 7.26 *Assume that $\mathbf{C}_0^h$, $\mathbf{D}_0^h$, and S_0^h belong to the same finite element De Rham complex in* (B.45) *and let π_D denote the $L^2(\Omega)$-projection defined in* (B.16). *Then, the discrete problem*

$$\begin{cases} \nabla\times\omega^h+\nabla_h^* p^h=\pi_D\mathbf{f} \\ \nabla_h^*\times\mathbf{u}^h-\omega^h=0 \\ \nabla\cdot\mathbf{u}^h=0 \end{cases} \tag{7.101}$$

has a unique solution $\{\mathbf{u}^h,\omega^h,p^h\}\in X^h\equiv\mathbf{D}_0^h\times\mathbf{C}_0^h\times S_0^h$.

Proof. The uniqueness of solutions of (7.101) follows from (7.99). To show that there is a non-trivial solution, we proceed as follows. From the discrete Hodge decomposition for fields belonging to $\mathbf{D}_0^h$ (see Theorem B.23) and the fact that the component spaces of X^h are in the same exact sequence, it follows that there exist $\omega^h\in\mathbf{C}_0^h$ and $p^h\in S_0^h$ such that

$$\nabla_h^*\cdot\omega^h=0 \qquad\text{and}\qquad \pi_D\mathbf{f}=\nabla\times\omega^h+\nabla_h^* p^h,$$

i.e., ω^h and p^h satisfy the first equation in (7.101). Using again the exactness of the finite element spaces and the discrete Hodge decomposition for fields belonging to $\mathbf{C}_0^h$ (see Theorem B.22), we have that

$$\omega^h=\nabla r^h+\nabla_h^*\times\mathbf{u}^h$$

for some $r^h\in G_0^h$ and $\mathbf{u}^h\in\mathbf{D}_0^h$. Furthermore, because $\nabla_h^*\cdot\omega^h=0$,

$$\nabla^*\cdot\nabla r^h=\nabla_h^*\cdot\omega=0\,.$$

From the definition (B.51) of $\nabla_h^*\cdot$, the last identity is equivalent to the weak equation

$$(\nabla r^h,\nabla q^h)=0 \qquad \forall q^h\in G_0^h\,.$$

It follows that $r^h\equiv 0$ and

$$\nabla_h^*\times\mathbf{u}^h-\omega^h=0\,,$$

i.e., $\{\mathbf{u}^h,\omega^h,p^h\}$ satisfy the first two equations in the discrete VVP system (7.101). To prove the theorem it remains to show that we can choose $\mathbf{u}^h$ to be divergence free. Using again (B.79), we can write the velocity as

$$\mathbf{u}^h = \mathbf{u}_N^h + \mathbf{u}_{N\perp}^h\,,$$

where $\nabla\cdot\mathbf{u}_N^h = 0$ and $\nabla_h^*\times\mathbf{u}_{N\perp}^h = 0$. Thanks to the last identity,

$$\omega^h = \nabla_h^*\times\mathbf{u}^h = \nabla_h^*\times\mathbf{u}_N^h\,.$$

Thus, all equations in (7.101) hold for $\{\mathbf{u}^h\equiv\mathbf{u}_N^h,\omega^h,p^h\}\in X^h$. □

Corollary 7.27 *Any solution* $\{\mathbf{u}^h,\omega^h,p^h\}$ *of the discrete VVP system* (7.101) *also solves its extended version*

$$\left\{\begin{aligned} \nabla\times\omega^h+\nabla_h^*p^h &= \pi_D\mathbf{f}\\ \nabla_h^*\cdot\omega^h &= 0\\ \nabla_h^*\times\mathbf{u}^h-\omega^h &= 0\\ \nabla\cdot\mathbf{u}^h &= 0\,. \end{aligned}\right. \tag{7.102}$$

□

Corollary 7.27 reveals that the second equation in (7.102), i.e., $\nabla_h^*\cdot\omega^h = 0$, is "redundant" because it is implied by the rest of the equations. Therefore, the relationship between (7.101) and (7.102) mimics the one between their continuous counterparts (7.4) and (7.10), respectively.

In the terminology of [314], discrete problems such as (7.101) and (7.102) are referred to as *mimetic*[49] discretizations of the corresponding VVP Stokes formulations. The following theorem establishes the connection between the discrete mimetic systems and the LSFEMs for which the approximating space for the pressure is S_0^h.

Theorem 7.28 *Assume that* $\mathbf{C}_0^h$, $\mathbf{D}_0^h$, *and* S_0^h *belong to the same finite element De Rham complex in* (B.45). *Then, the solution of the mimetic VVP system* (7.101) *coincides with the solution of the DLSP* (7.89).

Proof. Let $\{\mathbf{u}^h,\omega^h,p^h\}\in X^h = \mathbf{D}_0^h\times\mathbf{C}_0^h\times S_0^h$ denote the solution of (7.101). The DLSP (7.89) has a unique minimizer and to prove the theorem it suffices to show that $\{\mathbf{u}^h,\omega^h,p^h\}$ satisfies the first-order optimality condition of the least-squares formulation. To this end, we write the corresponding variational problem as

$$\left\{\begin{aligned} \left(\nabla_h^*\times\mathbf{u}^h-\omega^h,\nabla_h^*\times\mathbf{v}^h\right)+\left(\nabla\cdot\mathbf{u}^h,\nabla\cdot\mathbf{v}^h\right) &= 0 && \forall\mathbf{v}^h\in\mathbf{D}_0^h\\ \left(\nabla\times\omega^h+\nabla_h^*p^h,\nabla\times\xi^h\right)-\left(\nabla_h^*\times\mathbf{u}^h-\omega^h,\xi^h\right) &= \left(\mathbf{f},\nabla\times\xi^h\right) && \forall\xi^h\in\mathbf{C}_0^h\\ \left(\nabla\times\omega^h+\nabla_h^*p^h,\nabla_h^*q^h\right) &= \left(\mathbf{f},\nabla_h^*q^h\right) && \forall q^h\in S_0^h\,. \end{aligned}\right.$$

[49] Using the terminology of [286,288], these problems are *direct* discretizations of the VVP Stokes system in the sense that every field and differential operator has been replaced by a discrete version without going through a variational statement. See [287] for an example of such discretizations for incompressible flows using covolume methods.

The second and third equations in (7.101) imply that

$$\begin{cases} \left(\nabla_h^* \times \mathbf{u}^h - \omega^h, \xi^h\right) = 0 \\ \left(\nabla_h^* \times \mathbf{u}^h - \omega^h, \nabla_h^* \times \mathbf{v}^h\right) = 0 \\ \left(\nabla \cdot \mathbf{u}^h, \nabla \cdot \mathbf{v}^h\right) = 0 \end{cases}$$

for all $\xi^h \in \mathbf{C}_0^h$ and all $\mathbf{v}^h \in \mathbf{D}_0^h$. Thus, we are left with showing that

$$\begin{cases} \left(\nabla \times \omega^h + \nabla_h^* p^h, \nabla \times \xi^h\right) = \left(\mathbf{f}, \nabla \times \xi^h\right) & \forall \xi^h \in \mathbf{C}_0^h \\ \left(\nabla \times \omega^h + \nabla_h^* p^h, \nabla_h^* q^h\right) = \left(\mathbf{f}, \nabla_h^* q^h\right) & \forall q^h \in S_0^h \end{cases} \tag{7.103}$$

holds for the solution of (7.101). By assumption, $\mathbf{C}_0^h$ and $\mathbf{D}_0^h$ are in the same exact sequence so that $\nabla \times \xi^h \in \mathbf{D}_0^h$ for any $\xi^h \in \mathbf{C}_0^h$. This, in conjunction with the first equation in (7.101) and definition (B.16) of the $L^2(\Omega)$-projection π_D, implies that

$$\left(\mathbf{f}, \nabla \times \xi^h\right) = \left(\pi_D \mathbf{f}, \nabla \times \xi^h\right) = \left(\nabla \times \omega^h + \nabla_h^* p^h, \nabla \times \xi^h\right),$$

i.e., the first equation in (7.103) holds. Likewise, $\mathbf{D}_0^h$ and S_0^h are in the same sequence so that $\nabla_h^* q^h \in \mathbf{D}_0^h$ for any $q^h \in S_0^h$. Therefore,

$$\left(\mathbf{f}, \nabla_h^* q^h\right) = \left(\pi_D \mathbf{f}, \nabla_h^* q^h\right) = \left(\nabla \times \omega^h + \nabla_h^* p^h, \nabla_h^* q^h\right),$$

i.e., the second equation in (7.103) also holds. □

This theorem reveals an attractive property of (7.89) that cannot be achieved in the ADN setting. Namely, the velocity approximation is divergence free in every element, i.e., the LSFEM is locally conservative. Furthermore, by showing that (7.89) and the mimetic discrete system (7.101) have identical solutions, this theorem justifies calling the LSFEM (7.89) *strongly compatible* or *mimetic*. The following corollary establishes the analogous relationship between the DLSP corresponding to the extended VVP Stokes system and the mimetic discretization (7.102) of that system.

Corollary 7.29 *Assume that* $\mathbf{C}_0^h$, $\mathbf{D}_0^h$ *and* S_0^h *belong to the same finite element De Rham complex in* (B.45). *Then, the solution of the extended mimetic VVP system* (7.102) *coincides with the solution of the extended DLSP* (7.92).

Finally, we have the following equivalence result.

Corollary 7.30 *Assume that* $\mathbf{C}_0^h$, $\mathbf{D}_0^h$, *and* S_0^h *belong to the same finite element De Rham complex in* (B.45). *Then, the solutions of the DLSPs* (7.89) *and* (7.92) *are identical.*

Proof. From Theorem 7.28 and Corollary 7.29 we know that minimizers of (7.89) and (7.92) solve the mimetic VVP system (7.101) and its extended version (7.102), respectively. But, according to Corollary 7.27, solutions of the two mimetic systems coincide. □

Corollary 7.30 tells us that, in the vector-operator setting, the redundancy of $\nabla \cdot \omega = 0$ extends to the least-squares formulations and that the term $\|\nabla_h^* \cdot \omega^h\|$ can be safely removed[50] from the least-squares functional in (7.92). In other words, with an appropriate choice of the approximating spaces, (7.89) and (7.92) are completely equivalent. We add more to this observation later in this section.

Let us now examine the two DLSPs for which the pressure variable is approximated conformingly, i.e., by a finite element subspace $\widetilde{G}^h \subset G(\Omega) \cap L_0^2(\Omega)$. A quick inspection of (7.88) and (7.91) reveals that they cannot be strongly compatible in the sense of Theorem 7.28 or Corollary 7.29 because the ranges of ∇ and $\nabla\times$ belong to different[51] discrete spaces, even if their domains are in the same finite element De Rham complex.

Fortunately, it turns out that loss of strong compatibility in (7.88) and (7.91) is limited to that equation, and that most of the attractive properties of the previous setting, including divergence free velocity approximation, continue to hold. In fact, as the following theorem shows, all four LSFEMs compute *identical* velocity and vorticity approximations.

Theorem 7.31 *Let* $X_S^h = \mathbf{D}_0^h \times \mathbf{C}_0^h \times S_0^h$ *be the minimization space in* (7.89) *and* (7.92) *and let* $X_G^h = \mathbf{D}_0^h \times \mathbf{C}_0^h \times \widetilde{G}^h$ *be the minimization space in* (7.88) *and* (7.91). *The DLSPs* (7.89), (7.92), (7.88), *and* (7.91) *yield identical velocity and vorticity approximations; those approximations satisfy the weak problem*

$$\begin{cases} (\nabla\times\omega^h, \nabla\times\xi^h) + (\nabla_h^*\cdot\omega^h, \nabla_h^*\cdot\xi) = (\mathbf{f}, \nabla\times\xi^h) & \forall \xi^h \in \mathbf{C}_0^h \\ (\nabla_h^*\times\mathbf{u}^h, \nabla_h^*\times\mathbf{v}^h) + (\nabla\cdot\mathbf{u}^h, \nabla\cdot\mathbf{v}^h) = (\omega^h, \nabla_h^*\times\mathbf{v}^h) & \forall \mathbf{v}^h \in \mathbf{D}_0^h, \end{cases} \tag{7.104}$$

provided $\mathbf{C}_0^h$, $\mathbf{D}_0^h$, *and* S_0^h *are in the same finite element De Rham complex* (B.45).

Proof. Problem (7.104) is a coupled system of two non-conforming DLSPs for the div–curl equations. Using Theorem 6.11 and the material from Section 6.3.5, it is not difficult to prove that (7.104) is well-posed and has a unique solution $\{\mathbf{u}^h, \omega^h\}$. We show that this solution coincides with the velocity and the vorticity approximations computed by the four LSFEMs.

Owing to the fact that tangential components of ω^h vanish on $\partial\Omega$, we have

$$(\nabla\times\xi^h, \nabla q^h) = (\nabla\times\xi^h, \nabla_h^* s^h) = 0 \qquad \forall \xi^h \in \mathbf{C}_0^h, \quad \forall q^h \in \widetilde{G}^h, \quad \forall s^h \in S_0^h.$$

As a result, the optimality systems for (7.89), (7.92), (7.88), and (7.91) split into a separate equation for the pressure and a coupled velocity-vorticity problem. For (7.89) and (7.92), the pressure equation

$$(\nabla_h^* p^h, \nabla_h^* q^h) = (\mathbf{f}, \nabla_h^* q^h) \qquad \forall q^h \in S_0^h \tag{7.105}$$

---

[50] Note that this is in stark contrast with the ADN setting where the corresponding term cannot be dropped without demoting the VVP system to non-homogeneous elliptic; see Remark 7.11.

[51] The key to proving that the least-squares solution satisfies the mimetic momentum equation in (7.101) is the fact that the range of $\nabla\times$ acting on $\mathbf{C}_0^h$ and the range of ∇_h^* acting on S_0^h are both contained in $\mathbf{D}_0^h$.

is a non-conforming Galerkin approximation of a pure Neumann problem for the Poisson equation. For (7.88) and (7.91), the pressure equation

$$\left(\nabla p^h, \nabla q^h\right) = \left(\mathbf{f}, \nabla q^h\right) \qquad \forall q^h \in \widetilde{G}^h \tag{7.106}$$

is a conforming Galerkin discretization of the same Neumann problem. In either case, the determination of the pressure is completely autonomous and does not affect the velocity and vorticity approximations.

After the pressure has been uncoupled, the optimality systems for (7.88) and (7.89) reduce to

$$\begin{cases} \left(\nabla \times \omega^h, \nabla \times \xi^h\right) - \left(\nabla_h^* \times \mathbf{u}^h - \omega^h, \xi^h\right) = \left(\mathbf{f}, \nabla \times \xi^h\right) & \forall \xi^h \in \mathbf{C}_0^h \\ \left(\nabla_h^* \times \mathbf{u}^h, \nabla_h^* \times \mathbf{v}^h\right) + \left(\nabla \cdot \mathbf{u}^h, \nabla \cdot \mathbf{v}^h\right) = \left(\omega^h, \nabla_h^* \times \mathbf{v}^h\right) & \forall \mathbf{v}^h \in \mathbf{D}_0^h \end{cases} \tag{7.107}$$

whereas the optimality systems for (7.91) and (7.92) have the form

$$\begin{cases} \left(\nabla \times \omega^h, \nabla \times \xi^h\right) + \left(\nabla_h^* \cdot \omega^h, \nabla_h^* \cdot \xi^h\right) & \\ \qquad -\left(\nabla_h^* \times \mathbf{u}^h - \omega^h, \xi^h\right) = \left(\mathbf{f}, \nabla \times \xi^h\right) & \forall \xi^h \in \mathbf{C}_0^h \\ \left(\nabla_h^* \times \mathbf{u}^h, \nabla_h^* \times \mathbf{v}^h\right) + \left(\nabla \cdot \mathbf{u}^h, \nabla \cdot \mathbf{v}^h\right) = \left(\omega^h, \nabla_h^* \times \mathbf{v}^h\right) & \forall \mathbf{v}^h \in \mathbf{D}_0^h. \end{cases} \tag{7.108}$$

To complete the proof, it suffices to show that the solution $\{\mathbf{u}^h, \omega^h\}$ of (7.104) satisfies (7.107) and (7.108). We first use the orthogonal decomposition

$$\xi^h = \xi_N^h + \xi_{N\perp}^h$$

of $\xi^h \in \mathbf{C}_0^h$ that is defined in (B.77). Because $\nabla \times \xi_N^h = 0$ and $\nabla_h^* \cdot \xi_{N\perp}^h = 0$, the first equation in (7.104) splits in two:

$$\begin{cases} \left(\nabla \times \omega^h, \nabla \times \xi_{N\perp}^h\right) = \left(\mathbf{f}, \nabla \times \xi_{N\perp}^h\right) \\ \left(\nabla_h^* \cdot \omega^h, \nabla_h^* \cdot \xi_N^h\right) = 0. \end{cases} \tag{7.109}$$

Furthermore, from the identity

$$0 = \left(\nabla_h^* \cdot \omega^h, \nabla_h^* \cdot \xi_N^h\right) = \left(\nabla_h^* \cdot \omega^h, \nabla_h^* \cdot \xi^h\right) \qquad \forall \xi^h \in \mathbf{C}_0^h,$$

it easily follows that $\nabla_h^* \cdot \omega^h = 0$, i.e., $\omega^h \in N_0^h(\nabla \times)^\perp$. This takes care of the term $\left(\nabla_h^* \cdot \omega^h, \nabla_h^* \cdot \xi^h\right)$ in (7.108), thereby reducing it to the same weak formulation as (7.107). As a result, the theorem follows if we can show that

$$\left(\nabla_h^* \times \mathbf{u}^h - \omega^h, \xi^h\right) = 0 \qquad \forall \xi^h \in \mathbf{C}_0^h.$$

Using again the orthogonal decomposition of ξ^h,

$$\left(\nabla_h^* \times \mathbf{u}^h - \omega^h, \xi^h\right) = \left(\nabla_h^* \times \mathbf{u}^h - \omega^h, \xi_N^h\right) + \left(\nabla_h^* \times \mathbf{u}^h - \omega^h, \xi_{N\perp}^h\right). \tag{7.110}$$

The first term in the right-hand side vanishes because $\xi_N^h \in N_0^h(\nabla\times)$ and $\omega^h \in N_0^h(\nabla\times)^\perp$ so that[52]

$$\left(\nabla_h^* \times \mathbf{u}^h, \xi_N^h\right) = \left(\mathbf{u}^h, \nabla \times \xi_N^h\right) = 0 \qquad \text{and} \qquad \left(\omega^h, \xi_N^h\right) = 0\,.$$

To prove that the second term in the right-hand side of (7.110) is also zero we use the orthogonal decomposition

$$\mathbf{v}^h = \mathbf{v}_N^h + \mathbf{v}_{N\perp}^h$$

of $\mathbf{v}^h \in \mathbf{D}_0^h$ that is defined in (B.79). Because $\nabla \cdot \mathbf{v}_N^h = 0$, the second equation in (7.104) reduces to

$$\left(\nabla_h^* \times \mathbf{u}^h - \omega^h, \nabla_h^* \times \mathbf{v}_N^h\right) = 0 \qquad \forall \mathbf{v}_N^h \in N_0^h(\nabla\cdot)\,. \tag{7.111}$$

On the other hand, from (B.82), we know that fields in $N_0^h(\nabla\times)^\perp$ can be characterized as weak discrete curls of the fields in $N_0^h(\nabla\cdot)$, i.e., $\xi_{N\perp}^h = \nabla_h^* \times \mathbf{v}_N^h$. Combining these results shows that

$$\left(\nabla_h^* \times \mathbf{u}^h - \omega^h, \xi_{N\perp}^h\right) = \left(\nabla_h^* \times \mathbf{u}^h - \omega^h, \nabla_h^* \times \mathbf{v}_N^h\right) = 0\,.$$

This completes the proof. □

The first important conclusion of this theorem is that, insofar as velocity and vorticity approximations are concerned, it does not matter which one of the four least-squares formulations (7.89), (7.92), (7.88), or (7.91) is used in the computation. A second, equally important, conclusion is that all four LSFEMs provide *locally conservative* approximations of the velocity field. A third conclusion is that, as long as $\mathbf{C}_0^h$ and $\mathbf{D}_0^h$ are in the same exact sequence, the equation $\nabla_h^* \cdot \omega^h = 0$ is "redundant" and unimportant for the analysis. Therefore, to obtain error estimates, we can work with the simpler least-squares principles (7.88) and (7.89). However, the "redundant" term can be important for the performance of iterative solvers and should be included in the implementation of the methods. We return to this topic in Section 7.7.8 which discusses practical issues concerning the above LSFEMs.

The results in this section reveal the different roles played by the pressure approximating space on the one hand and the velocity and vorticity approximating spaces on the other hand. The impact of the former is limited to the momentum equation which can hold in either the "strong" form of the mimetic system (7.101) or in an alternative "weak" variational form. On the other hand, the divergence-free property of the least-squares velocity approximation strongly depends on $\mathbf{C}_0^h$ and $\mathbf{D}_0^h$ belonging to the same finite element De Rham complex. This is the setting that is assumed for the error analysis in the next section.

52 Note that in order to "integrate by parts" the discrete weak curl using definition (B.50), the spaces $\mathbf{D}_0^h$ and $\mathbf{C}_0^h$ must be in the same finite element De Rham complex. This is one of the hypotheses of the theorem.

7.7.6 Error Estimates

This section provides a priori error estimates for (7.88)–(7.92). For convenience, two key results needed in the analysis are quoted below. The first one, given in [181], asserts the regularity of a dual problem related to the VVP Stokes system.

Theorem 7.32 *Assume that Ω is a bounded convex polyhedron. Then, for every $\mathbf{g} \in [L^2(\Omega)]^3$, the problem: seek $\{\mathbf{w}, r\} \in [H^1(\Omega)]^3 \times H^1_0(\Omega)$ such that*

$$\begin{cases} \nabla \times \nabla \times \mathbf{w} + \nabla r = \mathbf{g} & \text{in } \Omega \\ \nabla \cdot \mathbf{w} = 0 & \text{in } \Omega \\ \mathbf{n} \times \mathbf{w} = 0 & \text{on } \partial\Omega \end{cases} \tag{7.112}$$

has the equivalent variational formulation: seek $\{\mathbf{w}, r\} \in \mathbf{C}_0(\Omega) \times G_0(\Omega)$ such that

$$\begin{cases} \big(\nabla \times \mathbf{w}, \nabla \times \mathbf{v}\big) + \big(\nabla r, \mathbf{v}\big) = \big(\mathbf{g}, \mathbf{v}\big) & \forall \mathbf{v} \in \mathbf{C}_0(\Omega) \\ \nabla \cdot \mathbf{w} = 0 & \text{in } \Omega\,. \end{cases} \tag{7.113}$$

The problem (7.113) *has a unique solution. Moreover, that solution satisfies the following bounds:*

$$\|\nabla \times \mathbf{w}\|_1 \le C\|\mathbf{g}\|_0 \qquad \|\nabla \times \nabla \times \mathbf{w}\|_0 \le \|\mathbf{g}\|_0 \qquad \|\nabla r\|_0 \le \|\mathbf{g}\|_0\,. \tag{7.114}$$

The second result, found in [283], provides useful information about the solenoidal component of $\mathbf{C}_0^h$ vector fields.

Theorem 7.33 *Assume that Ω is a bounded convex polyhedron. Then, there exists a real $s > 2$ such that every $\mathbf{v}^h \in \mathbf{C}_0^h$ can be written as*

$$\mathbf{v}^h = \mathbf{w} + \nabla p\,, \tag{7.115}$$

where $p \in H^1_0(\Omega)$ solves the weak equation

$$\big(\nabla p, \nabla q\big) = \big(\mathbf{v}^h, \nabla q\big) \qquad \forall q \in G_0(\Omega) \tag{7.116}$$

and $\mathbf{w} \in [W^{1,s}(\Omega)]^3 \cap \mathbf{C}_0$ satisfies

$$\nabla \times \mathbf{w} = \nabla \times \mathbf{v}^h \qquad \text{and} \qquad \nabla \cdot \mathbf{w} = 0\,. \tag{7.117}$$

Furthermore, if Π_C is the projection operator defined in Section B.2.1 *and G_0^h is in the same exact sequence as $\mathbf{C}_0^h$, there exists $p^h \in G_0^h$ such that*

$$\mathbf{v}^h = \Pi_C \mathbf{w} + \nabla p^h\,. \tag{7.118}$$

Finally, the error bound

$$\|\mathbf{w} - \Pi_C \mathbf{w}\|_0 \le C(s) h^{1+\frac{3}{s}-\frac{3}{2}} \|\nabla \times \mathbf{v}^h\|_0 \tag{7.119}$$

holds, provided $\mathbf{C}_0^h$ *is a curl-conforming space of the second kind.*[53] □

The error estimates for the LSFEMs based on the four DLSPs (7.89), (7.92), (7.88), and (7.91) are summarized below.

Theorem 7.34 *Assume that* $\mathbf{C}_0^h$ *and* $\mathbf{D}_0^h$ *belong to the same finite element De Rham complex in* (B.45). *Let* $\{\mathbf{u}^h, \omega^h, p_S^h\} \in X_S^h \equiv \mathbf{D}_0^h \times \mathbf{C}_0^h \times S_0^h$ *denote a minimizer of* (7.89) *and* (7.92) *and* $\{\mathbf{u}^h, \omega^h, p_G^h\} \in X_G^h \equiv \mathbf{D}_0^h \times \mathbf{C}_0^h \times \widetilde{G}^h$ *denote a minimizer of* (7.88) *and* (7.91). *Finally, assume that* $\{\mathbf{u}, \omega, p\} \in \mathbf{D}_0(\Omega) \times \mathbf{C}_0(\Omega) \times G(\Omega) \cap L_0^2(\Omega)$ *is a solution of the VVP Stokes system system* (7.4) *with the boundary condition* (7.8). *Then,*

$$\|\nabla \times (\omega - \omega^h)\|_0 \le \inf_{\xi^h \in \mathbf{C}_0^h} \|\nabla \times (\omega - \xi^h)\|_0 \tag{7.120}$$

$$\|\nabla \times \mathbf{u} - \nabla_h^* \times \mathbf{u}^h\|_0 \le 2\|\nabla \times \mathbf{u} - \pi_C(\nabla \times \mathbf{u})\|_0 + \|\omega - \omega^h\|_0 \tag{7.121}$$

$$\|\mathbf{u} - \mathbf{u}^h\|_0 \le \|\mathbf{u} - \pi_D(\mathbf{u})\|_0 + \|\pi_D(\mathbf{u}) - \Pi_D(\mathbf{u})\|_0 + \|\omega - \omega^h\|_0 . \tag{7.122}$$

If $\mathbf{C}_0^h$ *is a curl-conforming space of the first kind,*

$$\|\omega - \omega^h\|_0 \le C(1+h)\|\nabla \times (\omega - \omega^h)\|_0 + \inf_{\xi^h \in \mathbf{C}_0^h} \left\{ \|\omega - \xi^h\|_0 + \|\nabla \times (\omega - \xi^h)\|_0 \right\} \tag{7.123}$$

and if $\mathbf{C}_0^h$ *is of the second kind,*

$$\|\omega - \omega^h\|_0 \le C(h + h^{1-\alpha(s)})\|\nabla \times (\omega - \omega^h)\|_0 + \inf_{\xi^h \in \mathbf{C}_0^h} \left\{ \|\omega - \xi^h\|_0 + h^{1-\alpha(s)} \|\nabla \times (\omega - \xi^h)\|_0 \right\}, \tag{7.124}$$

where s is the real number from Theorem 7.33 *and* $\alpha(s) = 3/2 - 3/s$.

The pressure approximation in the strongly compatible formulations (7.89) *and* (7.92) *satisfies the error bound*

$$\|\nabla p - \nabla_h^* p_S^h\|_0 \le 2\|\nabla p - \pi_D(\nabla p)\|_0 + \|\nabla \times (\omega - \omega^h)\|_0 , \tag{7.125}$$

provided S_0^h *is in the same De Rham complex as* $\mathbf{D}_0^h$. *The pressure approximation in* (7.88) *and* (7.91) *satisfies the error bounds*

$$\|\nabla p - \nabla p_G^h\|_0 \le \inf_{q^h \in \widetilde{G}^h} \|\nabla p - \nabla q^h\|_0 \quad \text{and} \quad \|p - p_G^h\|_0 \le h\|\nabla p - \nabla p_G^h\|_0 , \tag{7.126}$$

[53] See (B.25) for the definition of these spaces.

independently of the choice of $\mathbf{C}_0^h$ *and* $\mathbf{D}_0^h$.

Proof. From the proof of Theorem 7.31, it follows that velocity and vorticity approximations, computed by the four LSFEMs, satisfy the following equations:

$$\begin{cases} \left(\nabla\times\omega^h,\nabla\times\xi^h\right)=\left(\mathbf{f},\nabla\times\xi^h\right) & \forall\xi^h\in\mathbf{C}_0^h \\ \left(\nabla_h^*\times\mathbf{u}^h,\xi^h\right)=\left(\omega^h,\xi^h\right) & \forall\xi^h\in\mathbf{C}_0^h \\ \left(\mathbf{u}^h,\nabla\times\xi^h\right)=\left(\omega^h,\xi^h\right) & \forall\xi^h\in\mathbf{C}_0^h. \end{cases} \tag{7.127}$$

To prove the individual error estimates from the statement of the theorem we proceed as follows.

Estimate of $\|\nabla\times(\omega-\omega^h)\|_0$

The error bound (7.120) follows by a standard elliptic argument. Using that $\mathbf{n}\times\xi^h=0$ on $\partial\Omega$ and setting $\mathbf{f}=\nabla\times\omega+\nabla p$ in the first equation of (7.127) gives

$$\left(\nabla\times\omega^h,\nabla\times\xi^h\right)=\left(\nabla\times\omega+\nabla p,\nabla\times\xi^h\right)=\left(\nabla\times\omega,\nabla\times\xi^h\right)$$

and the standard error orthogonality relation

$$\left(\nabla\times(\omega-\omega^h),\nabla\times\xi^h\right)=0 \qquad \forall\xi^h\in\mathbf{C}_0^h. \tag{7.128}$$

Using (7.128) and the Cauchy inequality,

$$\begin{aligned}\|\nabla\times(\omega-\omega^h)\|_0^2&=\left(\nabla\times(\omega-\omega^h),\nabla\times(\omega-\xi^h)\right)\\ &\le\|\nabla\times(\omega-\omega^h)\|_0\|\nabla\times(\omega-\xi^h)\|_0 \qquad \forall\xi^h\in\mathbf{C}_0^h\end{aligned}$$

from which (7.120) easily follows.

Estimate of $\|\omega-\omega^h\|_0$

As usual, this a priori estimate requires a duality argument. The one used here is adapted from [181] and relies on Theorems 7.32 and 7.33. In the estimation, we make use of the $H^1(\Omega)$-seminorm projection that is defined by

$$\left(\nabla r-\nabla r^h,s^h\right)=0 \quad \forall s^h\in G_0^h \tag{7.129}$$

(see [183, p. 101]) and that has the following property:

$$\|\nabla r-\nabla r^h\|_0\le C\|\nabla r\|_0. \tag{7.130}$$

We start by establishing two identities that hold for any $\mathbf{w}\in\mathbf{C}_0$ and any $r\in G_0(\Omega)$. The first one,

$$\left(\nabla\times\mathbf{w},\nabla\times(\omega-\omega^h)\right)=\left(\nabla\times(\mathbf{w}-\xi^h),\nabla\times(\omega-\omega^h)\right)\qquad\forall\xi^h\in\mathbf{C}_0^h,\tag{7.131}$$

follows from the orthogonality relation (7.128). The second identity,

$$\left(\nabla r,\omega-\omega^h\right)=\left(\nabla r-\nabla q^h,\omega-\omega^h\right)\qquad\forall q^h\in G_0^h,\tag{7.132}$$

holds because[54] $\nabla_h^*\cdot\omega^h=0$, $\nabla\cdot\omega=0$, and q^h vanishes on $\partial\Omega$.

The proof of (7.123) and (7.124) proceeds as follows. Given an arbitrary $\mathbf{g}\in[L^2(\Omega)]^3$, let $\{\mathbf{w},r\}\in\mathbf{C}_0(\Omega)\times G_0(\Omega)$ denote the corresponding solution of the dual problem (7.113). Setting $\mathbf{v}=\omega-\omega^h$ in (7.113) and using (7.131) and (7.132) yields

$$\begin{aligned}\left(\mathbf{g},\omega-\omega^h\right)&=\left(\nabla\times\mathbf{w},\nabla\times(\omega-\omega^h)\right)+\left(\nabla r,\omega-\omega^h\right)\\&=\left(\nabla\times(\mathbf{w}-\xi^h),\nabla\times(\omega-\omega^h)\right)+\left(\nabla r-\nabla q^h,\omega-\omega^h\right).\end{aligned}\tag{7.133}$$

To bound the first term in the right-hand side of (7.133), we use the Cauchy inequality, the approximation result (B.28), and the regularity result (7.114):

$$\begin{aligned}\left(\nabla\times(\mathbf{w}-\xi^h),\nabla\times(\omega-\omega^h)\right)&\le\|\nabla\times(\mathbf{w}-\xi^h)\|_0\,\|\nabla\times(\omega-\omega^h)\|_0\\&\le Ch\|\nabla\times\mathbf{w}\|_1\|\nabla\times(\omega-\omega^h)\|_0\le Ch\|\nabla\times(\omega-\omega^h)\|_0\,\|\mathbf{g}\|_0\,.\end{aligned}$$

For the second term on the right-hand side of (7.133), we set $q^h=r^h$, where $r^h\in G_0^h$ is the $H^1(\Omega)$-seminorm projection of r onto G_0^h. After adding and subtracting an arbitrary $\xi^h\in\mathbf{C}_0^h$,

$$\left(\nabla r-\nabla r^h,\omega-\omega^h\right)=\left(\nabla r-\nabla r^h,\omega-\xi^h\right)+\left(\nabla r-\nabla r^h,\xi^h-\omega^h\right).\tag{7.134}$$

For the first term on the right-hand side of (7.134), we use the Cauchy inequality, (7.130), and (7.114):

$$\begin{aligned}\left(\nabla r-\nabla r^h,\omega-\xi^h\right)&\le\|\nabla r-\nabla r^h\|\,\|\omega-\xi^h\|_0\\&\le C\|\nabla r\|_0\,\|\omega-\xi^h\|_0\le C\|\mathbf{g}\|_0\,\|\omega-\xi^h\|_0\,.\end{aligned}$$

To bound the second term on the right-hand side of (7.134), we apply the decomposition (7.118) from Theorem 7.115 to the field $\xi^h-\omega^h\in\mathbf{C}_0^h$. This allows us to write

$$\xi^h-\omega^h=\Pi_C\mathbf{v}+\nabla p^h\,,$$

where $p^h\in G_0^h$, $\nabla\times\mathbf{v}=\nabla\times(\xi^h-\omega^h)$, $\nabla\cdot\mathbf{v}=0$, and $\mathbf{n}\times\mathbf{v}=0$ on $\partial\Omega$. Using this decomposition, the projection property (7.129) of r^h, the fact that $\nabla\cdot\mathbf{v}=0$, the projection bound (7.130), and the regularity (7.114), we get

---

[54] Recall that $\nabla_h^*\cdot\omega^h=0$ if and only if $\left(\omega^h,\nabla q^h\right)=0$ for all $q^h\in G_0^h$.

$$
\begin{aligned}
(\nabla r - \nabla r^h, \xi^h - \omega^h) &= (\nabla r - \nabla r^h, \Pi_C \mathbf{v} + \nabla p^h) = (\nabla r - \nabla r^h, \Pi_C \mathbf{v}) \\
&= (\nabla r - \nabla r^h, \Pi_C \mathbf{v} - \mathbf{v}) \le C \|\nabla r\|_0 \|\Pi_C \mathbf{v} - \mathbf{v}\|_0 \le C \|\mathbf{g}\|_0 \|\Pi_C \mathbf{v} - \mathbf{v}\|_0 .
\end{aligned}
$$

The upper bound for $\|\Pi_C \mathbf{v} - \mathbf{v}\|_0$ depends on which kind of the curl-conforming space $\mathbf{C}_0^h$ we use. If $\mathbf{C}_0^h$ is of the first kind, we use that Π_C is a bounded projection, the Poincaré-Friedrichs inequality (A.64), and $\nabla \times \mathbf{v} = \nabla \times (\xi^h - \omega^h)$ to obtain

$$
\|\Pi_C \mathbf{v} - \mathbf{v}\|_0 \le C \|\nabla \times \mathbf{v}\|_0 = C \|\nabla \times (\xi^h - \omega^h)\|_0 .
$$

For curl-conforming spaces of the second kind, this bound can be improved by taking advantage of (7.119). In this case,

$$
\|\Pi_C \mathbf{v} - \mathbf{v}\|_0 \le C h^{1-\alpha(s)} \|\nabla \times (\xi^h - \omega^h)\|_0 ,
$$

where $\alpha(s)$ is the function defined in the statement of the theorem. After collecting all relevant results, it follows that

$$
\|\omega - \omega^h\|_0 \le C \left(\|\omega - \xi^h\|_0 + h \|\nabla \times (\omega - \omega^h)\|_0 + \|\nabla \times (\xi^h - \omega^h)\|_0 \right)
$$

for curl-conforming spaces of the first kind and

$$
\|\omega - \omega^h\|_0 \le C \left(\|\omega - \xi^h\|_0 + h \|\nabla \times (\omega - \omega^h)\|_0 + h^{1-\alpha(s)} \|\nabla \times (\xi^h - \omega^h)\|_0 \right)
$$

for curl-conforming spaces of the second kind. The estimates (7.123) and (7.124) follow without much difficulty from these results and the triangle inequality.

Estimate of $\|\nabla \times \mathbf{u} - \nabla_h^* \times \mathbf{u}^h\|_0$

The solution of the VVP Stokes system satisfies

$$
(\nabla \times \mathbf{u}, \xi^h) = (\omega, \xi^h) \qquad \forall \xi^h \in \mathbf{C}_0^h .
$$

This, and the second equation in (7.127), yield the "orthogonality" equation

$$
(\nabla_h^* \times \mathbf{u}^h - \nabla \times \mathbf{u}, \xi^h) = (\omega^h - \omega, \xi^h) \qquad \forall \xi^h \in \mathbf{C}_0^h .
$$

The rest of the proof is a simple variation of the standard elliptic argument. Using this "orthogonality" and the extension of $(\nabla_h^* \times)$ from Theorem B.19 gives

$$
\begin{aligned}
\|\nabla_h^* \times (\mathbf{u}^h - \mathbf{u})\|_0^2 &= (\nabla_h^* \times (\mathbf{u}^h - \mathbf{u}), \nabla_h^* \times \mathbf{u}^h - \nabla \times \mathbf{u} + \nabla \times \mathbf{u} - \nabla_h^* \times \mathbf{u}) \\
&= (\omega^h - \omega, \nabla_h^* \times (\mathbf{u}^h - \mathbf{u})) + (\nabla_h^* \times (\mathbf{u}^h - \mathbf{u}), \nabla \times \mathbf{u} - \nabla_h^* \times \mathbf{u})
\end{aligned}
$$

and an upper bound for the discrete error:

$$\|\nabla_h^* \times (\mathbf{u}^h - \mathbf{u})\|_0 \le \|\nabla \times \mathbf{u} - \nabla_h^* \times \mathbf{u}\|_0 + \|\omega^h - \omega\|_0 .$$

The error bound (7.121) follows from this estimate, the triangle inequality, and the identity[55] $\nabla_h^* \times \mathbf{u} = \pi_C(\nabla \times \mathbf{u})$ proved in Theorem B.19.

Estimate of $\|\mathbf{u} - \mathbf{u}^h\|_0$

By assumption, $\mathbf{C}_0^h$ and $\mathbf{D}_0^h$ belong to the same exact sequence. Therefore, $\mathbf{D}_0^h$ contains the range of the curl on $\mathbf{C}_0^h$, i.e, $\nabla \times \xi^h \in \mathbf{D}_0^h$ for all $\xi^h \in \mathbf{D}_0^h$. This allows us to write

$$\left(\omega, \xi^h\right) = \left(\nabla \times \mathbf{u}, \xi^h\right) = \left(\mathbf{u}, \nabla \times \xi^h\right) = \left(\pi_D(\mathbf{u}), \nabla \times \xi^h\right) \qquad \forall \xi^h \in \mathbf{C}_0^h .$$

Together with the last equation in (7.127), this yields an alternative “orthogonality” relation for the velocity approximation:

$$\left(\pi_D(\mathbf{u}) - \mathbf{u}^h, \nabla \times \xi^h\right) = \left(\omega - \omega^h, \xi^h\right) \qquad \forall \xi^h \in \mathbf{C}_0^h . \tag{7.135}$$

We use (7.135) to estimate the discrete error[56]

$$\|\pi_D(\mathbf{u}) - \mathbf{u}^h\|_0 = \sup_{\mathbf{w}^h \in \mathbf{D}_0^h} \frac{\left(\pi_D(\mathbf{u}) - \mathbf{u}^h, \mathbf{w}^h\right)}{\|\mathbf{w}^h\|_0} .$$

Let $\mathbf{w}^h$ be an arbitrary but fixed function in $\mathbf{D}_0^h$. From Theorem B.23 we know that

$$\mathbf{w}^h = \nabla \times \xi^h + \nabla_h^* s^h ,$$

where $s^h \in S_0^h$, $\xi^h \in \mathbf{C}_0^h$ and $\nabla_h^* \cdot \xi^h = 0$. Therefore,

$$\left(\pi_D(\mathbf{u}) - \mathbf{u}^h, \mathbf{w}^h\right) = \left(\pi_D(\mathbf{u}) - \mathbf{u}^h, \nabla \times \xi^h\right) + \left(\pi_D(\mathbf{u}) - \mathbf{u}^h, \nabla_h^* s^h\right) .$$

To estimate the first term we use Theorem B.22 to write $\xi^h = \xi_N^h + \xi_{N\perp}^h$, where

$$\xi_N^h = \nabla q^h, \qquad q^h \in G_0^h, \qquad \text{and} \qquad \|\xi_{N\perp}^h\|_0 \le \|\nabla \times \xi^h\|_0 .$$

But because $\nabla_h^* \cdot \xi^h = 0$, it follows that $\xi^h \in N_0^h(\nabla \times)^\perp$, i.e., $\xi_N^h = 0$ and $\xi^h = \xi_{N\perp}^h$. Using this and the orthogonality relation (7.135)

$$\left(\pi_D(\mathbf{u}) - \mathbf{u}^h, \nabla \times \xi_{N\perp}^h\right) = \left(\omega - \omega^h, \xi_{N\perp}^h\right)$$
$$\le \|\omega - \omega^h\|_0 \|\xi_{N\perp}^h\|_0 \le \|\omega - \omega^h\|_0 \|\nabla \times \xi^h\|_0 \le \|\omega - \omega^h\|_0 \|\mathbf{w}^h\|_0 .$$

---

[55] π_C is the $L^2(\Omega)$ projection onto $\mathbf{C}_0^h$ defined in (B.16).

[56] π_D is the $L^2(\Omega)$ projection onto $\mathbf{D}_0^h$ defined in (B.16).

To estimate the second term we add and subtract[57] $\Pi_D(\mathbf{u})$:

$$\left(\pi_D(\mathbf{u}) - \mathbf{u}^h, \nabla_h^* s^h\right) = \left(\pi_D(\mathbf{u}) - \Pi_D(\mathbf{u}), \nabla_h^* s^h\right) + \left(\Pi_D(\mathbf{u}) - \mathbf{u}^h, \nabla_h^* s^h\right).$$

From the definition of the weak discrete gradient (B.49), the fact that Π_D has the commuting diagram property (B.1), $\nabla \cdot \mathbf{u} = 0$, and $\nabla \cdot \mathbf{u}^h = 0$ it follows that

$$\left(\Pi_D(\mathbf{u}) - \mathbf{u}^h, \nabla_h^* s^h\right) = \left(\nabla \cdot \Pi_D(\mathbf{u}) - \nabla \cdot \mathbf{u}^h, s^h\right) = 0.$$

This gives a bound for the second term:

$$\begin{aligned}\left(\pi_D(\mathbf{u}) - \mathbf{u}^h, \nabla_h^* s^h\right) &= \left(\pi_D(\mathbf{u}) - \Pi_D(\mathbf{u}), \nabla_h^* s^h\right) \\ &\le \|\pi_D(\mathbf{u}) - \Pi_D(\mathbf{u})\|_0 \|\nabla_h^* s^h\|_0 \le \|\pi_D(\mathbf{u}) - \Pi_D(\mathbf{u})\|_0 \|\mathbf{w}^h\|_0,\end{aligned}$$

and a bound for the discrete error:

$$\|\pi_D(\mathbf{u}) - \mathbf{u}^h\|_0 \le \|\omega - \omega^h\|_0 + \|\pi_D(\mathbf{u}) - \Pi_D(\mathbf{u})\|_0.$$

The a priori estimate (7.122) follows from this result and the triangle inequality.

Estimate of $\|\nabla p - \nabla_h^* p_S^h\|_0$

In Theorem 7.31, we showed that p_S^h solves the non-conforming pressure equation (7.105). Because $(\nabla \times \omega^h, \nabla_h^* q^h) = 0$ for all $q^h \in S_0^h$, setting $\mathbf{f} = \nabla \times \omega + \nabla p$ in that equation gives

$$\begin{aligned}\left(\nabla_h^* p^h, \nabla_h^* q^h\right) &= \left(\nabla \times \omega + \nabla p, \nabla_h^* q^h\right) \\ &= \left(\nabla \times (\omega - \omega^h) + \nabla \times \omega^h + \nabla p, \nabla_h^* q^h\right) \\ &= \left(\nabla \times (\omega - \omega^h), \nabla_h^* q^h\right) + \left(\nabla p, \nabla_h^* q^h\right)\end{aligned}$$

and the "orthogonality" relation

$$\left(\nabla_h^* p^h - \nabla p, \nabla_h^* q^h\right) = \left(\nabla \times (\omega - \omega^h), \nabla_h^* q^h\right) \qquad \forall q^h \in S_0^h.$$

Using the latter and the extension of ∇_h^* defined in Theorem B.19 yields

$$\begin{aligned}\|\nabla_h^* p^h - \nabla_h^* p\|_0^2 &= \left(\nabla_h^* p^h - \nabla_h^* p, \nabla_h^* p^h - \nabla p + \nabla p - \nabla_h^* p\right) \\ &= \left(\nabla \times (\omega - \omega^h), \nabla_h^* p^h - \nabla_h^* p\right) + \left(\nabla_h^* p^h - \nabla_h^* p, \nabla p - \nabla_h^* p\right)\end{aligned}$$

and an upper bound for the discrete error:

$$\|\nabla_h^* p^h - \nabla_h^* p\|_0 \le \|\nabla p - \nabla_h^* p\|_0 + \|\nabla \times (\omega - \omega^h)\|_0.$$

---

[57] Π_D is the bounded projection operator onto $\mathbf{D}_0^h$; see (B.15), whose existence is postulated in Section B.2.

The a priori estimate (7.125) follows from this bound, the triangle inequality, and the identity[58] $\nabla_h^* p = \pi_D(\nabla p)$ from Theorem B.19.

Estimates of $\|\nabla p - \nabla p_G^h\|_0$ and $\|p - p_G^h\|_0$

Theorem 7.31 asserts that p_G^h solves the conforming pressure equation (7.106). This equation is a standard Galerkin formulation of a pure Neumann problem. Consequently, the first error bound in (7.126) follows by a standard elliptic argument. Likewise, the $L^2(\Omega)$-norm error follows by a standard duality argument (see e.g., [82, Theorem 5.4.8, p. 131]) and the assumption that Ω is a convex polyhedral domain.[59] □

For sufficiently smooth solutions of the VVP Stokes equations, the error bounds in Theorem 7.34 can be used to estimate the rates of convergence in the LSFEM approximation. Besides the regularity of the exact solution, these rates depend on the approximation orders of the finite element spaces for the velocity, vorticity, and the pressure. Considering that the errors of the velocity (and the pressure for the strongly compatible LSFEMs) depend on those for the vorticity, it is reasonable to fix $\mathbf{C}_0^h$ first. The two basic choices[60] for this space are given by curl-conforming elements of the first and the second kinds. Once $\mathbf{C}_0^h$ is selected, we have to pick an approximating space $\mathbf{D}_0^h$ for the velocity from the same finite element De Rham complex (B.45) that contains $\mathbf{C}_0^h$. The final step is the selection of an approximating space for the pressure. For the strongly compatible LSFEMs (7.89) and (7.92), the non-conforming pressure space S_0^h has to be in the same finite element complex (B.45) as $\mathbf{C}_0^h$ and $\mathbf{D}_0^h$, whereas for (7.88) and (7.91), the conforming pressure space $\widetilde{G}^h = G^h \cap L_0^2(\Omega)$ can be selected completely independently from the other spaces. It is not difficult to see that to each choice of $\mathbf{C}_0^h$ correspond exactly two types of least-squares minimization spaces.

If $\mathbf{C}_0^h$ is of the first kind, i.e., $\mathbf{C}_0^h = \mathbf{C}_0^{r-}$, it belongs to one of the last two sequences in (B.45) and we can set, in (7.88) and (7.91),

$$X_G^h = \mathbf{D}_0^{r-1} \times \mathbf{C}_0^{r-} \times \widetilde{G}^s \qquad \text{or} \qquad X_G^h = \mathbf{D}_0^{r-} \times \mathbf{C}_0^{r-} \times \widetilde{G}^s \tag{7.136}$$

where $\widetilde{G}^s = G^s \cap L_0^2(\Omega)$, and we can set, in (7.89) and (7.92),

$$X_S^h = \mathbf{D}_0^{r-1} \times \mathbf{C}_0^{r-} \times S_0^{r-2} \qquad \text{or} \qquad X_S^h = \mathbf{D}_0^{r-} \times \mathbf{C}_0^{r-} \times S_0^{r-1}\,. \tag{7.137}$$

If $\mathbf{C}_0^h$ is of the second kind, i.e., $\mathbf{C}_0^h = \mathbf{C}_0^r$, it belongs to one of the first two sequences in (B.45). In this case, the minimization space for (7.88) and (7.91) can be set to

---

[58] π_D is the $L^2(\Omega)$ projection onto $\mathbf{D}_0^h$ defined in (B.16).

[59] This ensures the full elliptic regularity of (7.106); see [183, Theorem 1.8, p. 12].

[60] We do not consider finite element spaces other than those defined in Section B.2.1.

$$X_G^h = \mathbf{D}_0^{r-1} \times \mathbf{C}_0^r \times \widetilde{G}^s \qquad \text{or} \qquad X_G^h = \mathbf{D}_0^{r-} \times \mathbf{C}_0^r \times \widetilde{G}^s \tag{7.138}$$

and the minimization space for (7.89) and (7.92) can be set to

$$X_S^h = \mathbf{D}_0^{r-1} \times \mathbf{C}_0^r \times S_0^{r-2} \qquad \text{or} \qquad X_S^h = \mathbf{D}_0^{r-} \times \mathbf{C}_0^r \times S_0^{r-1}\,. \tag{7.139}$$

Theorem 7.35 *Assume that $r \geq 2$. If X_G^h and X_S^h are defined by* (7.136) *and* (7.137), *respectively, and the exact solution of the VVP Stokes system* (7.4) *with the boundary condition* (7.8) *is such that $\mathbf{u} \in \mathbf{D}_0(\Omega) \cap [H^{r+1}(\Omega)]^3$ and $\omega \in \mathbf{C}_0(\Omega) \cap [H^{r+1}(\Omega)]^3$, then the error bounds*

$$\begin{aligned} \|\nabla \times (\omega - \omega^h)\|_0 &\leq Ch^r \|\nabla \times \omega\|_r \\ \|\omega - \omega^h\|_0 &\leq Ch^r \left(\|\omega\|_r + \|\nabla \times \omega\|_r\right) \\ \|\nabla \times \mathbf{u} - \nabla_h^* \times \mathbf{u}^h\|_0 &\leq Ch^r \left(\|\nabla \times \mathbf{u}\|_r + \|\omega\|_r + \|\nabla \times \omega\|_r\right) \\ \|\mathbf{u} - \mathbf{u}^h\|_0 &\leq Ch^r \left(\|\mathbf{u}\|_r + \|\omega\|_r + \|\nabla \times \omega\|_r\right) \end{aligned} \tag{7.140}$$

hold for the minimizers of (7.88)–(7.92).

If X_G^h and X_S^h are defined by (7.138) *and* (7.139), *respectively, and the exact solution is such that $\mathbf{u} \in \mathbf{D}_0(\Omega) \cap [H^{r+2}(\Omega)]^3$ and $\omega \in \mathbf{C}_0(\Omega) \cap [H^{r+1}(\Omega)]^3$, then the error bounds*

$$\begin{aligned} \|\nabla \times (\omega - \omega^h)\|_0 &\leq Ch^r \|\nabla \times \omega\|_r \\ \|\omega - \omega^h\|_0 &\leq Ch^{r+1} \left(\|\omega\|_{r+1} + h^{-\alpha(s)} \|\nabla \times \omega\|_r\right) \\ \|\nabla \times \mathbf{u} - \nabla_h^* \times \mathbf{u}^h\|_0 &\leq Ch^{r+1} \left(\|\nabla \times \mathbf{u}\|_{r+1} + \|\omega\|_{r+1} + h^{-\alpha(s)} \|\nabla \times \omega\|_r\right) \\ \|\mathbf{u} - \mathbf{u}^h\|_0 &\leq Ch^r \left(\|\mathbf{u}\|_r + h\|\omega\|_{r+1} + h^{1-\alpha(s)} \|\nabla \times \omega\|_r\right), \end{aligned} \tag{7.141}$$

where s is the real number from Theorem 7.33 *and $\alpha(s)$ is the function defined in Theorem* 7.34, *hold for the minimizers of* (7.88)–(7.92).

If $p \in L_0^2(\Omega) \cap H^{r+1}(\Omega)$, $r \geq 2$, then the pressure approximation in the strongly compatible DLSPs (7.89) *and* (7.92) *satisfies the error bound*

$$\|\nabla p - \nabla_h^* p\|_0 \leq Ch^r \left(\|\nabla p\|_r + \|\nabla \times \omega\|_r\right) \tag{7.142}$$

with either one of the two possible choices for X_S^h in (7.137) *and* (7.139). *The pressure approximation in* (7.88) *and* (7.91) *satisfies the error bound*

$$\| p - p^h\|_0 + h\|\nabla p - \nabla p^h\|_0 \leq Ch^{s+1} \|p\|_{s+1}, \tag{7.143}$$

provided $p \in L_0^2(\Omega) \cap H^{s+1}(\Omega)$ for some integer $s \geq 1$.

If $r = 1$, the estimates (7.140)–(7.143) *continue to hold with the second definition of X_G^h and X_S^h in* (7.136)–(7.139).

Proof. The results follow from the a priori estimates in Theorem 7.34, the rates of convergence in Theorems B.6–B.9, and the properties of the $L^2(\Omega)$-projection operators in Theorem B.5. □

An important conclusion that can be drawn from this theorem is that accuracy of the velocity approximation (and the pressure approximation for the strongly compatible LSFEMs) is locked in by the choice of the approximating space for the vorticity and does not change between the two possible realizations of X_S^h corresponding to that choice. We have more to say about this fact and its possible impact on the implementation of LSFEMs in Section 7.7.8.

It is instructive to compare the pressure errors for the strongly compatible LSFEMs with those for (7.88) and (7.91). In the former case, the error of $\nabla_h^* p^h$ depends on the velocity space $\mathbf{D}_0^h$ because the range of ∇_h^* is contained in $\mathbf{D}_0^h$, i.e., the accuracy of $\nabla_h^* p^h$ is governed by the approximation order of[61] $\mathbf{D}_0^h$ rather than that of S_0^h. In contrast, the pressure in (7.88) is approximated autonomously by a conforming subspace of $G(\Omega) \cap L_0^2(\Omega)$. As a result, the accuracy of its approximation depends solely on the approximation order of $\widetilde{G}^h$ and the regularity of the Neumann problem (7.106).

7.7.7 *Connection Between Discrete Least-Squares Principles and Mixed-Galerkin Methods*

In this section, we establish connections between (7.88) and (7.89) and two mixed-Galerkin formulations for the Stokes equations. That such connections exist should come as no surprise because, in Section 5.8, we already demonstrated links between LSFEMs in the vector-operator setting and standard finite element methods for div–grad systems. The relationship in this section is less direct in the sense that the mixed methods tied to (7.88) and (7.89) correspond to a first-order vector potential–vorticity–pressure formulation of the Stokes system rather than the VVP equations on which those LSFEMs are based.

We begin with a theorem that connects (7.88) and a mixed formulation proposed in [181].

Theorem 7.36 *Let* $X_M^h = \mathbf{C}_0^h \cap N_0^h(\nabla\times)^\perp \times \mathbf{C}_0^h \times \widetilde{G}^h$ *and let* $\{\eta^h, \omega^h, p^h\} \in X_M^h$ *solve the mixed-Galerkin formulation*

$$\begin{cases} (\nabla\times\omega^h, \nabla\times\xi^h) = (\mathbf{f}, \nabla\times\xi^h) & \forall \xi^h \in \mathbf{C}_0^h \\ (\nabla\times\eta^h, \nabla\times\xi^h) = (\omega^h, \xi^h) & \forall \xi^h \in \mathbf{C}_0^h \\ (\nabla p^h, \nabla q^h) = (\mathbf{f}, \nabla q^h) & \forall q^h \in \widetilde{G}^h . \end{cases} \quad (7.144)$$

[61] The same holds true for the approximation of $\nabla\times\mathbf{u}$ by $\nabla_h^*\times\mathbf{u}^h$. The range of $\nabla_h^*\times$ belongs in $\mathbf{C}_0^h$ which means that the accuracy of $\nabla_h^*\times\mathbf{u}^h$ is governed by the approximation order of $\mathbf{C}_0^h$ rather than that of $\mathbf{D}_0^h$.

Let $\mathbf{D}_0^h$ *denote a div-compatible finite element space that contains the range of* $\nabla\times$*, acting on* $\mathbf{C}_0^h$ *and define* $X^h = \mathbf{D}_0^h \times \mathbf{C}_0^h \times \widetilde{G}^h$*. Then,* $\{\mathbf{u}^h = \nabla\times\eta^h, \omega^h, p^h\} \in X^h$ *is a solution of the DLSP* (7.88).

Conversely, if $\mathbf{C}_0^h$ *and* $\mathbf{D}_0^h$ *in the definition of* (7.88) *are in the same finite element De Rham complex and* $\{\mathbf{u}^h, \omega^h, p^h\} \in X^h$ *is a solution of the DLSP* (7.88)*, then there exists a vector potential* $\eta^h \in \mathbf{C}_0^h$ *such that* $\mathbf{u}^h = \nabla\times\eta^h$*,* $\nabla_h^*\cdot\eta^h = 0$ *and* $\{\eta^h, \omega^h, p^h\} \in X_M^h$ *is solution of* (7.144).

Proof. To prove the first part of the theorem, let $\mathbf{u}^h = \nabla\times\eta^h$. Then, clearly

$$\left(\nabla\cdot\mathbf{u}^h, \nabla\cdot\mathbf{v}^h\right) = 0 \qquad \forall\mathbf{v}^h \in \mathbf{D}_0^h.$$

In terms of $\mathbf{u}^h$, the second equation of (7.144) assumes the form

$$\left(\mathbf{u}^h, \nabla\times\xi^h\right) = \left(\omega^h, \xi^h\right) \qquad \forall\xi^h \in \mathbf{C}_0^h$$

that, in light of definition (B.50), is the same as

$$\left(\nabla_h^*\times\mathbf{u}^h, \xi^h\right) = \left(\omega^h, \xi^h\right) \qquad \forall\xi^h \in \mathbf{C}_0^h.$$

Because $\nabla_h^*\times\mathbf{v}^h \in \mathbf{C}_0^h$ for any $\mathbf{v}^h \in \mathbf{D}_0^h$, it is also true that

$$\left(\nabla_h^*\times\mathbf{u}^h, \nabla_h^*\times\mathbf{v}^h\right) = \left(\omega^h, \nabla_h^*\times\mathbf{v}^h\right) \qquad \forall\mathbf{v}^h \in \mathbf{D}_0^h.$$

Combining the last three identities with the equations in the mixed formulation (7.144) and using that $(\nabla\times\mathbf{w}^h, \nabla q^h) = 0$ for all $\mathbf{w}^h \in \mathbf{C}_0^h$ and $q^h \in \widetilde{G}^h$ shows that $\{\mathbf{u}^h, \omega^h, p^h\} \in X^h$ solves the variational equation

$$\begin{aligned}\left(\nabla\times\omega^h + \nabla p^h, \nabla\times\xi^h + \nabla q^h\right) + \left(\nabla_h^*\times\mathbf{u}^h - \omega^h, \nabla_h^*\times\mathbf{v}^h - \xi^h\right) \\ +\left(\nabla\cdot\mathbf{u}^h, \nabla\cdot\mathbf{v}^h\right) = \left(\mathbf{f}, \nabla\times\xi^h + \nabla q^h\right) \qquad \forall\{\mathbf{v}^h, \xi^h, q^h\} \in X^h.\end{aligned}$$

This problem is exactly the first-order necessary condition for the DLSP (7.88).

To prove the second part of the theorem, assume that $\{\mathbf{u}^h, \omega^h, p^h\} \in X^h$ is a minimizer of (7.88). We have already shown in the proof of Theorem 7.31 that the first and the last equation in (7.144) hold for the LSFEM solution. Therefore, all that is left is to prove is that $\mathbf{u}^h$ has a discrete potential which satisfies the second equation of (7.144). The existence of $\eta^h \in \mathbf{C}_0^h$ such that $\mathbf{u}^h = \nabla\times\eta^h$ and $\nabla_h^*\cdot\eta^h = 0$ follows from Theorem B.23 and the fact that $\mathbf{u}^h$ is divergence free. It is easy to see that the second statement, i.e., that η^h satisfies the second equation (7.144), is also true. Indeed, from Theorem 7.31 we know that

$$\left(\nabla_h^*\times\mathbf{u}^h, \xi^h\right) = \left(\omega^h, \xi^h\right) \qquad \forall\xi^h \in \mathbf{C}_0^h$$

which, using the definition of $\nabla_h^*\times$, we can write as

$$\left(\mathbf{u}^h, \nabla\times\xi^h\right) = \left(\omega^h, \xi^h\right) \qquad \forall\xi^h \in \mathbf{C}_0^h,$$

which, after the substitution $\mathbf{u}^h = \nabla \times \eta^h$, gives

$$(\nabla \times \eta^h, \nabla \times \xi^h) = (\omega^h, \xi^h) \qquad \forall \xi^h \in \mathbf{C}_0^h.$$

This is precisely the second equation in (7.144). □

The LSFEM (7.89) is connected to a similar mixed-Galerkin formulation.

Theorem 7.37 *Assume that $\mathbf{C}_0^h$ and S_0^h are in the same finite element De Rham complex. Let $X_M^h = \mathbf{C}_0^h \cap N_0^h(\nabla\times)^\perp \times \mathbf{C}_0^h \times S_0^h$ and let $\{\eta^h, \omega^h, p^h\} \in X_M^h$ solve the mixed-Galerkin formulation*

$$\begin{cases} (\nabla \times \omega^h, \nabla \times \xi^h) = (\mathbf{f}, \nabla \times \xi^h) & \forall \xi^h \in \mathbf{C}_0^h \\ (\nabla \times \eta^h, \nabla \times \xi^h) = (\omega^h, \xi^h) & \forall \xi^h \in \mathbf{C}_0^h \\ (\nabla_h^* p^h, \nabla_h^* q^h) = (\mathbf{f}, \nabla_h^* q^h) & \forall q^h \in S_0^h. \end{cases} \tag{7.145}$$

Let $\mathbf{D}_0^h$ be a div-compatible finite element space that contains the range of $\nabla\times$, acting on $\mathbf{C}_0^h$ and define $X^h = \mathbf{D}_0^h \times \mathbf{C}_0^h \times S_0^h$. Then, $\{\mathbf{u}^h = \nabla \times \eta^h, \omega^h, p^h\} \in X^h$ is a solution of the DLSP (7.89).

Conversely, if $\mathbf{C}_0^h$, $\mathbf{D}_0^h$ and S_0^h in the definition of (7.89) *are in the same finite element De Rham complex and $\{\mathbf{u}^h, \omega^h, p^h\} \in X^h$ is a solution of the DLSP (7.89), then there exists a vector potential $\eta^h \in \mathbf{C}_0^h$ such that $\mathbf{u}^h = \nabla \times \eta^h$; $\nabla_h^* \cdot \eta^h = 0$ and $\{\eta^h, \omega^h, p^h\} \in X_M^h$ solves* (7.145).

Proof. The proof is analogous to that for Theorem 7.36. □

The two mixed[62] formulations are identical except for the equations that govern the pressure approximation. Computationally, the main disadvantage of (7.144) and (7.145) is in the structure of the approximating space for the vector potential, which is sought in $N_0^h(\nabla\times)^\perp$. The characterization of this subspace of $\mathbf{C}_0^h$ is not as straightforward as that of its orthogonal complement $N_0^h(\nabla\times)$ which renders it less suitable for computation. In particular, a basis for $N_0^h(\nabla\times)^\perp$ cannot be obtained in the same straightforward manner as that for $N_0^h(\nabla\times)$. In contrast, bases for all approximating spaces in the LSFEMs can be defined without much difficulty.

7.7.8 Practicality Issues in the Vector Operator Setting

The minimization spaces in (7.88)–(7.92) are composed of finite element spaces belonging to a discrete De Rham complex and are governed by the choice of the approximating space for the vorticity. These LSFEMs also involve discrete weak forms of the gradient, curl, and divergence operators. Therefore, the first two practical questions that arise with regard to (7.88)–(7.92) are

[62] For similar vector potential-vorticity formulations of the Stokes equations with the velocity boundary condition, we refer to [183, pp. 260-277].

- how to choose the minimization spaces?

and after these spaces are defined,

- how to form and solve the corresponding linear systems?

In light of the fact that the minimization spaces of (7.89), (7.92) and (7.88), (7.91) differ only by the pressure component, a logical follow up question is

- which one of the two types of methods is better?

Let us examine the first question, i.e., the definition of the least-squares minimization space. In both types of LSFEMs, construction of X^h starts with the vorticity component. The choice between a curl-compatible space of the first or of the second kind depends largely on whether or not the potential for improvement in some of the convergence rates, afforded by spaces of the second kind, is deemed valuable. Considering that one cannot expect to gain a full order of accuracy, and that spaces of the second kind are more expensive[63] to work with, the advantage of using them is not immediately obvious.

Nonetheless, assuming that $\mathbf{C}_0^h$ is fixed, the next step is to choose one of the two possible realizations of X^h corresponding to that space. Theorem 7.35 asserts that, for $r \geq 2$, either one of them provides the same rates of convergence; this removes accuracy as a factor in their selection. Also, for the same value of r, the two realizations of X^h have roughly the same number of degrees of freedom so that their costs are comparable. Thus, the only remaining decision factor is whether or not the implementation of LSFEMs is supposed to support the lowest possible order spaces from the De Rham complex. In this case, the second realization of X^h is preferable because it allows us to drop the value of r to one. For example, the lowest order version of the first space in (7.137) corresponds to $r = 2$ and is given by

$$X_S^h = \mathbf{D}_0^1 \times \mathbf{C}_0^{2-} \times S_0^0 .$$

The lowest order version of the second space is

$$X_S^h = \mathbf{D}_0^{1-} \times \mathbf{C}_0^{1-} \times S_0^0$$

and corresponds to $r = 1$.

After the minimization space is defined, one has to form and solve the linear systems corresponding to (7.88)–(7.92). Thanks to Theorem 7.31, most of the questions that arise in this context can be answered by the material already provided in Sections 6.5.1 and 6.5.2. Indeed, after the pressure computation is uncoupled, the vorticity and velocity approximations in all four formulations can be determined by solving the system (7.104). This system can be further split into a div–curl formulation for the vorticity

$$(\nabla \times \omega^h, \nabla \times \xi^h) + (\nabla_h^* \cdot \omega^h, \nabla_h^* \cdot \xi^h) = (\mathbf{f}, \nabla \times \xi^h) \qquad \forall \xi^h \in \mathbf{C}_0^h \tag{7.146}$$

---

[63] What we mean here is that $\mathbf{C}^r-$ is optimized with respect to the curl operator, i.e., it provides the same order of accuracy for that operator as $\mathbf{C}^r$, but requires fewer degrees of freedom; see (B.28).

that can be solved independently and a div–curl system for the velocity

$$\left(\nabla_h^* \times \mathbf{u}^h, \nabla_h^* \times \mathbf{v}^h\right) + \left(\nabla \cdot \mathbf{u}^h, \nabla \cdot \mathbf{v}^h\right) = \left(\omega^h, \nabla_h^* \times \mathbf{v}^h\right) \qquad \forall \mathbf{v}^h \in \mathbf{D}_0^h \qquad (7.147)$$

that can be solved once ω^h is known.

The problems (7.146) and (7.147) correspond to the optimality systems of the non-conforming DLSPs (6.50) and (6.51), respectively. In other words, each one is a well-posed formulation with a unique solution in the indicated spaces. As a result, minimizers of (7.88)–(7.92) can be computed by the following algorithm:

1. solve (7.146) to find the vorticity approximation $\omega^h \in \mathbf{C}_0^h$
2. using ω^h from step 1 as data, solve (7.147) to find $\mathbf{u}^h \in \mathbf{D}_0^h$
3. solve the pressure equation corresponding to the strongly compatible LSFEMs (7.89) and (7.92), or to the other pair (7.88) and (7.91) of LSFEMs.

In Section 6.5.2, we explain the implementation of the first two steps. The last step is completely independent from the first two and amounts to solving a pure Neumann problem for the pressure. In (7.88) and (7.91), this problem is approximated by a conforming Galerkin method (7.106) that can be solved by conventional multilevel algorithms. In the strongly compatible LSFEMs (7.89) and (7.92), the pressure equation is approximated by the non-conforming Galerkin formulation (7.105). Although this problem is not as straightforward to solve as (7.106), it can be handled using the techniques explained in Section 6.5.2.

At this point, it is clear that deciding which one of the two types of LSFEMs, i.e., the strongly compatible or the simply compatible, is better depends primarily on whether or not one is interested in having a solution that satisfies the "mimetic" form of the momentum equation in (7.101). If this is the case, then (7.89) and (7.92) are the "better" formulations. However, if the advantage of satisfying the mimetic momentum equation is outweighed by the added expense of solving (7.105), then (7.88) and (7.91) become more attractive.

7.8 A Summary of Conclusions and Recommendations

The LSFEMs in this chapter followed the key recipes for practicality and optimality. Their theoretical and practical properties varied significantly on the functional setting used to formulate them, and on some additional factors specific to that setting. For example, in the vector-operator setting, we had to develop non-conforming LSFEMs because the intersection space $\mathbf{C} \cap \mathbf{D}$ occurred naturally in the energy balances. In the ADN setting we had to distinguish between homogeneous elliptic and non-homogenous elliptic systems in the formulation and the analysis of LSFEMs.

Although computational experience with LSFEMs in the vector-operator setting is very limited, this type of methods appears to be well positioned to compete with mixed-Galerkin formulations in applications where local mass conservation is deemed of utmost importance. LSFEMs in the ADN setting have been used extensively and their computational properties are much better documented.

Our recommendations, based on the discussion in this chapter, regarding the use of the LSFEMs for the Stokes problem from each setting can be summarized as follows. In the vector-operator setting, the use of a strongly compatible least-squares formulation is justified if and only if one wants a solution that satisfies the "mimetic" momentum equation in (7.101). If this is not the case, (7.88) and (7.91) are better alternatives because they produce the same velocity and vorticity approximations as the strongly compatible formulations, but have a simpler pressure equation.

Standard nodal spaces are one of the main appeals of LSFEMs in the ADN setting. However, this setting is also more treacherous. In Sections 7.4 and 7.5, we saw that theoretical properties of LSFEMs in the ADN setting depended strongly on the classification of the first-order systems as homogeneous or non-homogeneous elliptic. In Section 7.6, we confirmed that these theoretical differences also hold in practice and that the key assumptions of the theory cannot be ignored in practical calculations without some detriments. In particular, we have the following observations.

- Using straightforward (unweighted) $L^2(\Omega)$ quasi-norm-equivalent DLSPs is not safe when the first-order Stokes system is non-homogeneous elliptic.
- The minimal approximation condition for weighted $L^2(\Omega)$ quasi-norm-equivalent DLSPs is sharp and cannot be ignored; using anything less than G^2 elements for the velocity approximations results in non-optimally accurate solutions for such methods.
- Violation of the minimal approximation condition also degrades the mass conservation in such DLSPs.
- On the other hand, the need to use different interpolation orders for different variables in weighted $L^2(\Omega)$ quasi-norm-equivalent DLSPs is not critical to the accuracy, as long as the minimal approximation condition is satisfied. Consequently, equal-order implementations of such methods are an acceptable alternative whenever one can benefit from more uniform data structures.
- Some implementations of the zero-mean pressure constraint can adversely affect performance of iterative solvers. The method described in Section 7.6.4 should be used in lieu of "pinning" the pressure at a node.

We conclude with a table that compares and contrasts the key properties of the LSFEMs presented in this chapter.

Method→ Property↓	SC (7.89) (7.92)	C (7.88) (7.91)	CMPL (7.50), (7.51) (7.52), (7.53) (7.59)	NE (7.55) (7.57) (7.58)	QNE (7.54) (7.56)
$\nabla \times \omega^h + \nabla_h^* p^h = \mathbf{f}$	√	–	–	–	–
$\nabla_h^* \times \mathbf{u}^h - \omega^h = 0$	√	√	–	–	–
$\nabla \times \mathbf{u}^h = 0$	√	√	–	–	–
Energy balance holds provided	$\partial\Omega$ is Lipschitz	$\partial\Omega$ is Lipschitz	Ω is convex polygon or C^1	Ω is convex polygon or C^1	Ω is C^2
Provably optimal	√	√	√	√	√
- on non-affine $\mathcal{T}_h$	–	–	√	√	√
- with velocity BC	–	–	only (7.59)	√	√
Condition number	$O(h^{-2})$	$O(h^{-2})$	$O(h^{-2})$	$O(h^{-2})$	$O(h^{-4})$
Preconditioner	√	√	√	√	–
Coding effort	not as simple	not as simple	simple	not as simple	simple

Table 7.5 Summary properties of LSFEMs for the Stokes equations in the vector-operator setting (strongly compatible (SC) and compatible (C) formulations) and in the ADN setting (compliant (CMPL), norm-equivalent (NE), and quasi-norm-equivalent (QNE) formulations).

Part IV
Least-Squares Finite Element Methods for Other Settings

Chapter 8
The Navier–Stokes Equations

The Navier–Stokes equations model flows of incompressible viscous fluids. Such flows arise in virtually all fields of science and engineering, either on their own or as part of more complex multi-physics phenomena. This makes the Navier–Stokes equations one of the most important modeling tools in science and engineering, and their numerical solutions one of the most important tasks in computational science.

Our interest in considering the Navier–Stokes equations is motivated not only by their intrinsic importance, but also to provide a setting for discussing how least-squares finite element methods (LSFEMs) can be applied to *nonlinear* problems. As such, this chapter provides useful insights on how LSFEMs can be used for other nonlinear partial differential equations (PDEs).

The Navier–Stokes equations in primitive variable form are given by[1,2]

$$\begin{cases} -\nu\Delta\mathbf{u}+(\mathbf{u}\cdot\nabla)\mathbf{u}+\nabla p=\mathbf{f} & \text{in } \Omega \\ \nabla\cdot\mathbf{u}=0 & \text{in } \Omega \end{cases} \tag{8.1}$$

[1] The first equation in (8.1) is a conservation of momentum statement so that it is often referred to as the *momentum equation*; some authors reserve the name Navier–Stokes for that equation alone and not the system (8.1). The second equation is a conservation of mass statement and is usually referred to as the *continuity equation.* The Navier–Stokes equations differ from the Stokes equations (7.1) through the inclusion of the *convection* term $(\mathbf{u}\cdot\nabla)\mathbf{u}$ in the momentum equation. In fact, the Stokes equations can be viewed as a simplification of the Navier–Stokes for settings, e.g., lubrication problems, for which convection effects are negligible compared to the diffusion effects modeled by the first term in the momentum equation. On the other hand, there are settings for which the reverse holds, i.e., convection dominates diffusion. Neglecting the diffusion term in the momentum equation leads to the Euler equations that model flows of inviscid fluids; see Chapter 10.

[2] Written out, the convection term $(\mathbf{u}\cdot\nabla)\mathbf{u}$ is a vector whose jth, $j=1,\dots,d$, component is given by

$$\sum_{i=1}^{d} u_i\frac{\partial u_j}{\partial x_i},$$

where d denotes the space dimension and u_i, $i=1,\dots,d$, denotes the ith component of $\mathbf{u}$.

P.B. Bochev, M.D. Gunzburger, *Least-Squares Finite Element Methods,*
Applied Mathematical Sciences 166, DOI: 10.1007/b13382_8,

along with the velocity boundary condition (7.2) and the zero mean pressure constraint (7.3). In Section 1.4.6, we noted that the standard mixed form (1.68) of (8.1) is an example of a variational problem that does not represent an optimality system corresponding to an optimization problem.

Most of the arguments in Chapter 7 used to justify the application of LSFEMs for the Stokes equations also apply to (8.1). The structure of the mixed-Galerkin Navier–Stokes problem (1.68) is very much like the structure of the mixed-Galerkin Stokes problem (1.59) and, not surprisingly, finite element methods for the former are subject to the same inf-sup compatibility conditions [183, 191] as those for the latter. Actually, the need for a better variational setting for finite element methods is even more acute in the case of (8.1) because their discretization leads to a nonlinear system of algebraic equations that must be solved in an iterative manner. For example, using the mixed-Galerkin method in conjunction with Newton linearization means that one has to solve multiple discrete saddle-point equations; in contrast, an LSFEM approach leads to a sequence of symmetric and positive definite algebraic equations.

The complications added to the mixed-Galerkin setting by the nonlinearity of (8.1) and the obvious advantages of the least-squares setting on the other hand, have prompted extensive practical [28, 51, 52, 117, 140, 141, 228, 234, 236, 237, 239, 261, 297, 298, 326–329] and theoretical [34–36, 38, 39, 44, 53–55, 68] studies of LSFEMs for the Navier–Stokes equations and related applications such as, e.g., fluid–structure interaction problems [202, 245, 301].

Having made a significant investment in a least-squares theory for the Stokes equations, we are now in an excellent position to extend the methods discussed in Chapter 7 to the Navier–Stokes setting and provide rigorous analyses of their properties. The formal extension of the continuous least-squares principles (CLSPs) and methods of Chapter 7 to (8.1) is simply a matter of including an appropriate form of the nonlinear term in the prototype Stokes least-squares functionals. The analysis of the resulting LSFEMs is, however, a completely different matter. Fortunately, to this end, it turns out that, for several important cases, we can apply the same abstract nonlinear approximation theory [91, 183] that has been successfully used for mixed-Galerkin finite element methods for the Navier–Stokes equations.

We implement this agenda only for the velocity–vorticity–pressure (VVP) and extended velocity gradient–velocity–pressure (VGVP) first-order reformulations of (8.1), and only using the Agmon–Douglis–Nirenberg (ADN) setting for the energy balances. The reasons for this are as follows. First, for the linear Stokes equations, these reformulations exemplified the two principal cases in the ADN setting, namely non-homogeneous elliptic and homogeneous elliptic first-order systems, respectively. Any other first-order system reformulation of (8.1) falls into one of these two categories and can be treated using similar approaches to those we are about to discuss. A second reason is that the extension of the abstract approximation theory discussed in [91, 183] to least-squares formulations for (8.1) is best understood in the context of the ADN setting where it has been used in the analysis of LSFEMs for the VVP [34, 35, 39] and VGVP [44, 68] versions of the Navier–Stokes equations.

8.1 First-Order System Formulations of the Navier–Stokes Equations

The specialization of the first-order VVP system (7.4) and its extended version (7.12) to the Navier–Stokes equations is a matter of choosing a form for the nonlinear term in (8.1). The most straightforward approach is to keep this term in its original form. However, for the analysis of the resulting LSFEMs, it is more convenient to replace this term by an expression that involves the vorticity variable. This can be accomplished through the use of the vector identity

$$(\mathbf{u}\cdot\nabla)\mathbf{u} = \frac{1}{2}\nabla|\mathbf{u}|^2 - \mathbf{u}\times\nabla\times\mathbf{u} = \frac{1}{2}\nabla|\mathbf{u}|^2 + \omega\times\mathbf{u}$$

and setting $s = p + 1/2|\mathbf{u}|^2$ to be the total pressure head. Using $\mathbf{u}$, ω, and s as dependent variables, we can transform (8.1) into the the first-order system[3]

$$\begin{cases} \nu\nabla\times\omega + \omega\times\mathbf{u} + \nabla s = \mathbf{f} & \text{in } \Omega \\ \nabla\times\mathbf{u} - \omega = 0 & \text{in } \Omega \\ \nabla\cdot\mathbf{u} = 0 & \text{in } \Omega\,. \end{cases} \tag{8.2}$$

We continue to refer to (8.2) as the "VVP" system and s as the "pressure." Naturally, in three-dimensions there is also the extended VVP system

$$\begin{cases} \nu\nabla\times\omega + \omega\times\mathbf{u} + \nabla s = \mathbf{f} & \text{in } \Omega \\ \nabla\cdot\omega = 0 & \text{in } \Omega \\ \nabla\times\mathbf{u} - \omega + \nabla\phi = 0 & \text{in } \Omega \\ \nabla\cdot\mathbf{u} = 0 & \text{in } \Omega \end{cases} \tag{8.3}$$

involving a redundant equation and a slack variable. Along with the systems (8.2) and (8.3), we have the zero mean pressure constraint (7.3) and suitable boundary conditions. As in Section 7.2, the velocity boundary condition (7.2) in two and three-dimensions or the normal velocity–pressure condition (7.5) in two dimensions are appropriate for the VVP system (8.2). The three-dimensional version (7.13) of the latter boundary condition[4] should be considered with the extended VVP system (8.3).

The extension of VGVP first-order formulations to the Navier–Stokes equations is even simpler. One of the variables in these systems is already the velocity gradient $\underline{\mathbf{U}} = (\nabla\mathbf{u})^\top$. As a result, the nonlinear term in (8.1) can be expressed as

$$(\mathbf{u}\cdot\nabla)\mathbf{u} = (\nabla\mathbf{u})^\top\cdot\mathbf{u} = \underline{\mathbf{U}}\cdot\mathbf{u}\,.$$

---

[3] Recall that in two-dimensions $\nabla\times\omega = \nabla^\perp\omega$ and $\omega\times\mathbf{u} = \omega\mathbf{u}^\perp = \omega(-u_2,u_1)^\top$.

[4] Because formulation and analysis of LSFEMs for the Navier-Stokes equations with the second non-standard boundary condition is very similar, the set (7.7) and (7.14) is not considered. Another reason to drop the normal velocity-tangential vorticity boundary condition from consideration is that its potential to define locally conservative LSFEMs cannot be realized in the ADN setting.

Then, using $\underline{\mathbf{U}}$, $\mathbf{u}$, and p as dependent variables, the Navier–Stokes equations in primitive variable form can be transformed into the first-order VGVP system

$$\begin{cases} -\nu\nabla\cdot\underline{\mathbf{U}}+\underline{\mathbf{U}}\cdot\mathbf{u}+\nabla p=\mathbf{f} & \text{in } \Omega \\ \underline{\mathbf{U}}-(\nabla\mathbf{u})^{\mathsf{T}}=0 & \text{in } \Omega \\ \nabla\cdot\mathbf{u}=0 & \text{in } \Omega\,. \end{cases} \tag{8.4}$$

The companion extended VGVP system is

$$\begin{cases} -\nu\nabla\cdot\underline{\mathbf{U}}+\underline{\mathbf{U}}\cdot\mathbf{u}+\nabla p=\mathbf{f} & \text{in } \Omega \\ \underline{\mathbf{U}}-(\nabla\mathbf{u})^{\mathsf{T}}=0 & \text{in } \Omega \\ \nabla\times\underline{\mathbf{U}}=0 & \text{in } \Omega \\ \nabla\cdot\mathbf{u}=0 & \text{in } \Omega \\ \nabla(tr\underline{\mathbf{U}})=0 & \text{in } \Omega\,. \end{cases} \tag{8.5}$$

As in Section 7.1.3, we add the velocity boundary condition (7.2), the redundant boundary condition (7.19), and the zero mean pressure constraint (7.3) to both (8.4) and (8.5).

8.2 Least-Squares Principles for the Navier–Stokes Equations

The abstract least-squares theory from Chapter 3 cannot be directly applied to the nonlinear Navier–Stokes equations because the notions of energy balance, norm-equivalent least-squares functionals, and least-squares principles were derived in the context of linear operator equations. The keystone of this theory, i.e., the connection between norm-equivalent least-squares functionals and equivalent inner products on Hilbert spaces, ceases to exist when some of the residuals contain nonlinear terms.

Our approach to least-squares principles for the first-order Navier–Stokes systems (8.2)–(8.5) is to first define CLSPs by adopting the data and solution spaces from the prototype CLSPs for the corresponding linear Stokes versions of these systems. Figuratively speaking, we define the CLSPs for the Navier–Stokes equations by simply "inserting" the nonlinear term into any well-defined *prototype* CLSP for the associated linear first-order Stokes system. We then define the associated[5] DLSPs by repeating the same transformation step that led to *prototype* DLSPs for the Stokes equations.

This approach to the development of LSFEMs for the Navier–Stokes equations is prompted by the abstract approximation theory of [91] that asserts that for nonlinear

[5] We continue to use the terms "compliant," "norm-equivalent," and "quasi-norm-equivalent" when referring to DLSPs for the Navier–Stokes equations, even though the least-squares functionals for these equations are not quadratic and do not define norms. The use of this terminology is justified by the fact that linearized versions of Navier–Stokes DLSPs inherit their type, i.e., compliant, norm-equivalent, or quasi-norm-equivalent, from the prototype Stokes DLSP.

equations that are compact perturbations of linear operator equations, the functional setting for the linear part governs the setting for the whole problem. As a result, using this theory, it is possible to study existence, uniqueness, and error estimation issues for LSFEMs for the Navier–Stokes equations by building upon many of the results established for the linear Stokes equations.

8.2.1 Continuous Least-Squares Principles

According to the template just outlined, we use (7.40) to define the appropriate CLSP for the first-order Navier–Stokes system (8.2) in two dimensions with the normal velocity–pressure boundary condition (7.25):

$$\begin{cases} J_0(\mathbf{u},\omega,s;\mathbf{f}) = \|\nu\nabla^{\perp}\omega+\omega\times\mathbf{u}+\nabla s-\mathbf{f}\|_0^2+\|\nabla\times\mathbf{u}-\omega\|_0^2+\|\nabla\cdot\mathbf{u}\|_0^2 \\ X_0 = \mathbf{H}_n^1\times H^1(\Omega)\times H_0^1(\Omega)\,. \end{cases} \tag{8.6}$$

In three dimensions, the relevant CLSP is obtained by specializing (7.42) to the extended Navier–Stokes system (8.3):[6]

$$\begin{cases} J_0(\mathbf{u},\omega,s;\mathbf{f}) = \\ \qquad \|\nu\nabla\times\omega+\omega\times\mathbf{u}+\nabla s-\mathbf{f}\|_0^2+\|\nabla\cdot\omega\|_0^2+\|\nabla\times\mathbf{u}-\omega\|_0^2+\|\nabla\cdot\mathbf{u}\|_0^2 \\ X_0 = \mathbf{H}_n^1\times\mathbf{H}_n^1\times H_0^1(\Omega)\,. \end{cases} \tag{8.7}$$

For (8.2) with the velocity boundary condition, we have the extension of (7.44):

$$\begin{cases} J_{-1}(\mathbf{u},\omega,s;\mathbf{f}) = \\ \qquad \|\nu\nabla\times\omega+\omega\times\mathbf{u}+\nabla s-\mathbf{f}\|_{-1}^2+\|\nabla\times\mathbf{u}-\omega\|_0^2+\|\nabla\cdot\mathbf{u}\|_0^2+|\ell(s)|^2 \\ X_{-1} = [H_0^1(\Omega)]^d\times[L^2(\Omega)]^{d(d-1)/2}\times L^2(\Omega)\,. \end{cases} \tag{8.8}$$

For (8.4) and (8.5), we proceed in the same fashion and "insert" the nonlinear term from these systems into (7.48) and (7.49), respectively. This results in the CLSP

$$\begin{cases} J_{-1}(\mathbf{u},\underline{\mathbf{U}},p;\mathbf{f}) = \\ \qquad \|-\nu\nabla\cdot\underline{\mathbf{U}}+\underline{\mathbf{U}}\cdot\mathbf{u}+\nabla p-\mathbf{f}\|_{-1}^2+\|\underline{\mathbf{U}}-(\nabla\mathbf{u})^{\mathsf{T}}\|_0^2+\|\nabla\cdot\mathbf{u}\|_0^2+|\ell(p)|^2 \\ X_{-1} = [H_0^1(\Omega)]^d\times[L^2(\Omega)]^{d^2}\times L^2(\Omega) \end{cases} \tag{8.9}$$

---

[6] In this chapter we completely drop the slack variable ϕ from the least-squares formulations because it is only needed for the formal application of the ADN theory to establish the norm-equivalence of the prototype CLSP (7.42) for the linear Stokes equations.

in which the momentum equation residual is measured in the H^{-1} norm and a CLSP in which this residual is measured in the $L^2(\Omega)$ norm:

$$\begin{cases} J_0(\mathbf{u},\underline{\mathbf{U}},p;\mathbf{f}) = \|-\nu\nabla\cdot\underline{\mathbf{U}}+\underline{\mathbf{U}}\cdot\mathbf{u}+\nabla p-\mathbf{f}\|_0^2+\|\underline{\mathbf{U}}-(\nabla\mathbf{u})^{\mathsf{T}}\|_0^2 \\ \qquad\qquad +\|\nabla\times\underline{\mathbf{U}}\|_0^2+\|\nabla\cdot\mathbf{u}\|_0^2+\|\nabla(tr\underline{\mathbf{U}})\|_0^2+|\ell(p)|^2 \\ X_0=[H_0^1(\Omega)]^d\times\underline{\mathbf{H}}^1_\times(\Omega)\times H^1(\Omega)\,. \end{cases} \tag{8.10}$$

Of course, the comments made in Remark 7.12 about the zero mean pressure constraint remain in full force for (8.8)–(8.10).

8.2.2 Discrete Least-Squares Principles

We restrict attention to the LSFEMs that are analyzed in Section 8.3. First, we have two compliant DLSPs for the VVP Navier–Stokes system with the normal velocity–pressure boundary condition in two and three dimensions. The two-dimensional compliant DLSP is given by

$$\begin{cases} J_0(\mathbf{u}^h,\omega^h,s^h;\mathbf{f}) = \\ \qquad \|\nu\nabla^\perp\omega^h+\omega^h\times\mathbf{u}^h+\nabla s^h-\mathbf{f}\|_0^2+\|\nabla\times\mathbf{u}^h-\omega^h\|_0^2+\|\nabla\cdot\mathbf{u}^h\|_0^2 \\ X^h_{r,0}=[G^r]^2\cap\mathbf{H}^1_n\times G^r\times G^r\cap H_0^1(\Omega)\,, \end{cases} \tag{8.11}$$

whereas, in three dimensions, the compliant DLSP is given by[7]

$$\begin{cases} J_0(\mathbf{u}^h,\omega^h,s^h;\mathbf{f}) = \|\nu\nabla\times\omega^h+\omega^h\times\mathbf{u}^h+\nabla s^h-\mathbf{f}\|_0^2 \\ \qquad\qquad +\|\nabla\cdot\omega^h\|_0^2+\|\nabla\times\mathbf{u}^h-\omega^h\|_0^2+\|\nabla\cdot\mathbf{u}^h\|_0^2 \\ X^h_{r,0}=[G^r]^3\cap\mathbf{H}^1_n\times[G^r]^3\cap\mathbf{H}^1_n\times G^r\cap H_0^1(\Omega)\,. \end{cases} \tag{8.12}$$

For the VVP Navier–Stokes system with the velocity boundary condition we have the norm-equivalent DLSP

$$\begin{cases} J_{-h}(\mathbf{u}^h,\omega^h,s^h;\mathbf{f}) = |\ell(s^h)|^2 \\ \quad +\|\nu\nabla\times\omega^h+\omega^h\times\mathbf{u}^h+\nabla s^h-\mathbf{f}\|_{-h}^2+\|\nabla\times\mathbf{u}^h-\omega^h\|_0^2+\|\nabla\cdot\mathbf{u}^h\|_0^2 \\ X^h_{\widehat{r},0}=[G^{r+1}]^d\cap[H_0^1(\Omega)]^d\times[G^{\widehat{r}}]^{2d-3}\times G^{\widehat{r}} \end{cases} \tag{8.13}$$

---

[7] Recall that, according to Remark 7.11, the slack variable can be dropped from the CLSP and so it does not have to be included in the associated DLSP.

which is an extension of the discrete negative norm DLSP (7.55) for the Stokes equations.

For the VGVP Navier–Stokes first-order system (8.4) we have the CLSP in (8.9). The associated DLSP

$$\begin{cases} J_{-h}(\mathbf{u}^h, \underline{\mathbf{U}}^h, p^h; \mathbf{f}) = |\ell(p^h)|^2 + \\ \qquad \| - \nu \nabla \cdot \underline{\mathbf{U}}^h + \underline{\mathbf{U}}^h \cdot \mathbf{u}^h + \nabla p^h - \mathbf{f}\|_{-h}^2 + \|\underline{\mathbf{U}}^h - (\nabla \mathbf{u}^h)^\mathsf{T}\|_0^2 + \|\nabla \cdot \mathbf{u}^h\|_0^2 \\ X_{\hat{r},0}^h = [G^{r+1}]^d \cap [H_0^1(\Omega)]^d \times [G^{\hat{r}}]^{d^2} \times G^{\hat{r}} \end{cases} \tag{8.14}$$

is based on the norm-equivalent DLSP (7.58) for the Stokes equations. Lastly, we extend (7.59) to the following compliant DLSP:

$$\begin{cases} J(\mathbf{u}^h, \underline{\mathbf{U}}^h, p^h; \mathbf{f}) = \\ \qquad \| - \nu \nabla \cdot \underline{\mathbf{U}}^h + \underline{\mathbf{U}}^h \cdot \mathbf{u}^h + \nabla p^h - \mathbf{f}\|_0^2 + \|\underline{\mathbf{U}}^h - (\nabla \mathbf{u}^h)^\mathsf{T}\|_0^2 \\ \qquad\qquad + \|\nabla \times \underline{\mathbf{U}}^h\|_0^2 + \|\nabla \cdot \mathbf{u}^h\|_0^2 + \|\nabla(tr\underline{\mathbf{U}}^h)\|_0^2 + |\ell(p^h)|^2 \\ X_{r,0}^h = [G^r]^d \cap [H_0^1(\Omega)]^d \times [G^r]^{d^2} \cap \underline{\mathbf{H}}_\times^1(\Omega) \times G^r . \end{cases} \tag{8.15}$$

We could have easily extended any of the quasi-norm-equivalent weighted $L^2(\Omega)$ functionals for the Stokes equations to the Navier–Stokes case. However, we restrict attention to compliant and norm-equivalent DLSPs because they offer the greatest potential in the context of the nonlinear Navier–Stokes equations. The reason for this is that the higher condition number of the linear systems in weighted $L^2(\Omega)$ LSFEMs makes them less competitive when such systems have to be solved multiple times in the course of, e.g., the Newton linearization.

The norm-equivalent DLSP (8.14) is needed in Section 8.4. There, we also state and use a straightforward $L^2(\Omega)$ norm version of (8.14). There is no need to state this method here because it cannot be treated by the approximation theories discussed in the next section.

8.3 Analysis of Least-Squares Finite Element Methods

The analysis of LSFEMs for the Navier–Stokes equations is presented in Sections 8.3.2–8.3.5. For convenience and completeness, the key results needed for that analysis are summarized in Section 8.3.1.

8.3.1 Quotation of Background Results

We provide a brief summary of approximation results for abstract nonlinear equations. This material is adapted to our needs from [91, 183]. Then, we review several inequalities and embedding results that are needed to estimate the nonlinear terms in the Navier–Stokes equations.

Approximation of abstract nonlinear equations

Let X and Y be Banach spaces and let $\Lambda \subset \mathbb{R}$ be a closed interval. We consider abstract nonlinear problems with the following canonical structure: given $\lambda \in \Lambda$ find $u \in X$ such that

$$\mathcal{F}(\lambda, u) \equiv u + \mathcal{T} \circ \mathcal{G}(\lambda, u) = 0, \tag{8.16}$$

where[8] $\mathcal{T} \in L(Y,X)$ is independent of λ and $\mathcal{G}$ is a C^2 map $\Lambda \times X \mapsto Y$. The set $\{\{\lambda, u(\lambda)\} \mid \lambda \in \Lambda;\ u(\lambda) \in X\}$ is called a *branch of solutions* of (8.16) if $\mathcal{F}(\lambda, u(\lambda)) = 0$ for $\lambda \in \Lambda$ and the map $\lambda \to u(\lambda)$ is continuous from Λ into X. If, in addition, the Frèchet derivative $D_u\mathcal{F}(\lambda, u(\lambda))$ of $\mathcal{F}$ with respect to u is an isomorphism of X for all $\lambda \in \Lambda$, then the branch $\{\{\lambda, u(\lambda)\} \mid \lambda \in \Lambda; u(\lambda) \in X\}$ is called *nonsingular* or *regular.*

In view of the canonical structure of (8.16), approximations of this problem can be obtained in the following manner. We choose a finite-dimensional subspace $X^h \subset X$ and define the discrete nonlinear problem by replacing $\mathcal{T}$ with a linear operator $\mathcal{T}^h \in L(Y, X^h)$ that approximates $\mathcal{T}$ and is independent of λ:

$$\mathcal{F}^h(\lambda, u^h) \equiv u^h + \mathcal{T}^h \circ \mathcal{G}(\lambda, u^h) = 0. \tag{8.17}$$

The approximation in (8.17) is effected solely through the discrete operator $\mathcal{T}^h$ and leaves the nonlinear term intact. We refer to such approximations of (8.16) as being *compliant.* This terminology is consistent with the decision to let Navier–Stokes DLSPs inherit their qualifiers from the prototype Stokes DLSPs. In particular, we see that the compliant DLSPs for the Navier–Stokes equations such as (8.11), (8.12), or (8.15) correspond to compliant discretizations of the abstract problem (8.16), whereas *non-compliant* DLSPs such as (8.13) or (8.14) give rise to *non-compliant* discretizations of (8.16) whose canonical form is similar to (8.17) but which also involves an approximation of the nonlinear term. Non-compliant approximations are induced by the use of the discrete negative norm $\|\cdot\|_{-h}$ to measure the residual of the momentum equation in (8.13) and (8.14). As a result, such DLSPs cannot be obtained solely through approximation of the linear parts of their respective parent CLSPs (8.8) and (8.9).

Error estimates for compliant approximations u^h of solutions u of (8.16) are derived under the following hypotheses. First, we assume that there exists a Banach

---

[8] The symbol $L(X,Y)$ is introduced in Appendix C and stands for the linear space of all bounded linear operators $X \mapsto Y$.

space V continuously embedded in Y such that

$$D_u\mathcal{G}(\lambda,u) \in L(X,V) \qquad \forall \lambda \in \Lambda \quad \text{and} \quad \forall u \in X\,. \tag{8.18}$$

We also assume that

$$\lim_{h\to 0} \|(\mathcal{T}-\mathcal{T}^h)g\|_X = 0 \qquad \forall g \in Y \tag{8.19}$$

and that

$$\lim_{h\to 0} \|\mathcal{T}-\mathcal{T}^h\|_{L(V,X)} = 0\,. \tag{8.20}$$

Under these assumptions, we have the following result for compliant discretizations of (8.16).

Theorem 8.1 [183, Theorem 3.3, p. 307] *Let X and Y be Banach spaces and let Λ be a compact subset of $\mathbb{R}$. Assume that $\mathcal{G}$ is a C^2 mapping from $\Lambda \times X$ into Y and that all second Frèchet derivatives of $\mathcal{G}$ are bounded on all bounded subsets of $\Lambda \times X$. Assume that* (8.18)–(8.20) *hold and that $\{\{\lambda,u(\lambda)\} \mid \lambda\in\Lambda;\, u(\lambda)\in X\}$ is a branch of regular solutions of* (8.16)*. Then, there exists a neighborhood $\mathcal{O}$ of the origin in X, and for h sufficiently small, a unique C^2 function $\lambda \to u^h \in X^h$ such that $\{\{\lambda,u^h(\lambda)\} \mid \lambda\in\Lambda;\, u^h(\lambda)\in X^h\}$ is a branch of regular solutions of* (8.17) *and $u(\lambda)-u^h(\lambda)\in\mathcal{O}$ for all $\lambda\in\Lambda$. Moreover, there exists a constant $C>0$ whose value is independent of h and λ such that*

$$\|u(\lambda)-u^h(\lambda)\|_X \le C\|(\mathcal{T}-\mathcal{T}^h)\circ\mathcal{G}(\lambda,u(\lambda))\|_X \qquad \forall \lambda\in\Lambda\,. \qquad \Box \tag{8.21}$$

For non-compliant discretizations of the abstract problem (8.16), we use another approach that deals directly with the discrete equations. The drawback of this approach is that it does not say much about the existence of regular branches of discrete solutions.

Let X^h denote a finite-dimensional subspace. We consider an abstract *discrete* nonlinear equation of the form

$$\textit{seek} \quad u^h\in X^h \quad \textit{such that} \quad \mathcal{F}^h(u^h) = 0\,. \tag{8.22}$$

The following theorem establishes conditions under which (8.22) has a solution and provides an estimate of the "discrete" error.

Theorem 8.2 [183, Theorem 3.6, p. 312] *Assume that there exists a function $\widetilde{u}^h\in X^h$ such that $D_u\mathcal{F}^h(\widetilde{u}^h)$ is an isomorphism of X^h onto itself. Let*

$$\varepsilon = \|\mathcal{F}^h(\widetilde{u}^h)\|_{X^h}\,, \tag{8.23}$$

$$\gamma = \|D_u\mathcal{F}^h(\widetilde{u}^h)^{-1}\|_{L(X^h,X^h)}\,, \tag{8.24}$$

and

$$\delta(\alpha) = \sup_{v^h\in B(\widetilde{u}^h,\alpha)} \|D_u\mathcal{F}^h(\widetilde{u}^h) - D_u\mathcal{F}^h(v^h)\|_{L(X^h,X^h)}\,, \tag{8.25}$$

where

$$B(\widetilde{u}^h,\alpha) = \{v^h \in X^h \mid \|\widetilde{u}^h - v^h\|_{X^h} \le \alpha\}\,.$$

Furthermore, assume that

$$2\gamma\delta(2\gamma\varepsilon) < 1\,. \tag{8.26}$$

Then, the problem $\mathcal{F}^h(u^h) = 0$ *has a solution* $u^h \in X^h$ *that belongs to the ball* $B(\widetilde{u}^h, 2\gamma\varepsilon)$*,* $D_u\mathcal{F}^h(u^h)$ *is an isomorphism of* X^h *onto itself, and* $\|D_u\mathcal{F}^h(u^h)^{-1}\|_{X^h} \le 2\gamma$*. Furthermore,* u^h *is the only solution in the ball* $B(\widetilde{u}^h,\alpha)$ *whose radius* α *satisfies* $\gamma\delta(\alpha) < 1$ *and we have the error estimate*

$$\|u^h - v^h\|_{X^h} \le \frac{\gamma}{1-\gamma\delta(\alpha)}\|\mathcal{F}^h(v^h)\|_{X^h} \qquad \forall\, v^h \in B(\widetilde{u}^h,\alpha)\,. \qquad \square \tag{8.27}$$

The proof of this theorem is based on a fixed-point argument; see [183].

Inequalities related to the Navier–Stokes equations

Recall (see, e.g., [334, p. 12]) the inequalities

$$\left|\int_\Omega uvwz\,d\Omega\right| \le C_1\|u\|_1\,\|v\|_1\,\|w\|_1\,\|z\|_1 \tag{8.28}$$

and

$$\left|\int_\Omega u(\partial v/\partial x_i)\,w\,d\Omega\right| \le C_2\|u\|_1\,\|v\|_1\,\|w\|_1 \tag{8.29}$$

that hold for all functions u, v, w, and z in $H^1(\Omega)$. Also recall that (see [183, Corollary 1.1])

$$\|uv\|_{3/2} \le C\|u\|_0\|v\|_1 \qquad \forall\, u \in L^2(\Omega)\,, \quad \forall\, v \in H^1(\Omega) \tag{8.30}$$

$$\|uv\|_2 \le C\|u\|_0\|v\|_2 \qquad \forall\, u \in L^2(\Omega)\,, \quad \forall\, v \in H^2(\Omega) \tag{8.31}$$

$$\|uv\|_2 \le C\|u\|_1\|v\|_1 \qquad \forall\, u,v \in H^1(\Omega) \tag{8.32}$$

$$\|uvw\|_{3/2} \le C\|u\|_1\|v\|_1\|w\|_1 \qquad \forall\, u,v,w \in H^1(\Omega)\,. \tag{8.33}$$

From (8.30)–(8.32), it follows that $\{u,v\} \mapsto uv$ is a continuous bilinear mapping $L^2(\Omega)\times H^1(\Omega) \mapsto L^{3/2}(\Omega)$, $L^2(\Omega)\times H^2(\Omega) \mapsto L^2(\Omega)$, and $H^1(\Omega)\times H^1(\Omega) \mapsto L^2(\Omega)$. The last inequality implies that $\{u,v,w\} \mapsto uvw$ is a continuous trilinear mapping $H^1(\Omega)\times H^1(\Omega)\times H^1(\Omega) \mapsto L^{3/2}(\Omega)$.

8.3.2 Compliant Discrete Least-Squares Principles for the Velocity–Vorticity–Pressure System

In this section, we apply Theorem 8.1 to compliant DLSPs for the VVP Navier–Stokes system with the normal velocity–pressure boundary condition.[9] Details are presented for the three-dimensional DLSP (8.12) with the slack variable ignored in the equations and the boundary conditions. Specialization to the two-dimensional case of (8.11) is fairly straightforward and requires the use of appropriate definitions for the curl operator and the vector product in $\mathbb{R}^2$.

Recall that functional settings for the Navier–Stokes CLSPs are inherited form their prototype Stokes CLSPs. We make the usual assumption that homogeneous boundary conditions are imposed on the solution spaces. Thus, the solution space of the nonlinear parent CLSP (8.7) of (8.12) is given by X_q in (7.33) restricted by the boundary condition (7.13). In view of Remark 7.11, the setting for (8.7) can be further simplified by dropping the slack variable. Without this variable and with the boundary condition (7.13) imposed on X_q, the relevant solution spaces for the VVP Navier–Stokes system (8.3) assume the form

$$X_q = [H^{q+1}(\Omega)]^3 \cap \mathbf{H}^1_n(\Omega) \times [H^{q+1}(\Omega)]^3 \cap \mathbf{H}^1_n(\Omega) \times H^{q+1}(\Omega) \cap H^1_0(\Omega), \quad (8.34)$$

where $q \geq 0$ is an integer regularity index. The data space Y_q from (7.33) is unchanged, whereas the boundary data space B_q is not needed because the boundary conditions are imposed on X_q. Some additional spaces required to cast (8.7) and (8.12) into the canonical forms of (8.16) and its discretization (8.17), respectively, are introduced in below when needed.

For the application of the abstract approximation theory it is convenient to reformulate (8.7) and (8.12) in the following way. Let $\mathbf{f} \in [L^2(\Omega)]^3$ denote the right-hand side of the momentum equation in (8.3) and let $\{\mathbf{u}_0, \omega_0, s_0\} \in X_0$ be the unique solution of the Stokes problem (7.12) with $\nu = 1$ and the normal velocity–pressure boundary condition (7.13). Then, (8.3) can be replaced by the system

$$\begin{cases} \nabla \times \omega + \lambda(\omega + \omega_0) \times (\mathbf{u} + \mathbf{u}_0) + \nabla s = 0 & \text{in } \Omega \\ \nabla \cdot \omega = 0 & \text{in } \Omega \\ \nabla \times \mathbf{u} - \omega = 0 & \text{in } \Omega \\ \nabla \cdot \mathbf{u} = 0 & \text{in } \Omega \end{cases} \quad (8.35)$$

in which $\mathbf{f}$ has been eliminated and[10] $\lambda = \nu^{-1}$. Evidently, if $\{\mathbf{u}_1, \omega_1, s_1\}$ solves the problem (8.35) with (7.13), then $\{\mathbf{u}_1 + \mathbf{u}_0, \omega_1 + \omega_0, s_1 + s_0\}$ solves the original VVP system (8.3) and (7.13).

To the system (8.35), there correspond reformulated versions of the CLSP (8.7)

[9] No theory is available for straightforward $L^2(\Omega)$ least-squares formulations for the Navier-Stokes equations with the velocity boundary condition.

[10] A rescaling of s is also applied, i.e., $\nu^{-1}s$ is replaced by s. Note also that, for nondimensional equations, λ is the Reynolds number.

$$\begin{cases} J_0(\mathbf{u},\omega,s;\mathbf{u}_0,\omega_0,\lambda) = \|\nabla\times\omega+\lambda(\omega+\omega_0)\times(\mathbf{u}+\mathbf{u}_0)+\nabla s\|_0^2 \\ \qquad +\|\nabla\cdot\omega\|_0^2+\|\nabla\times\mathbf{u}-\omega\|_0^2+\|\nabla\cdot\mathbf{u}\|_0^2 \\ X_0 = \mathbf{H}_n^1(\Omega)\times\mathbf{H}_n^1(\Omega)\times H_0^1(\Omega) \end{cases} \tag{8.36}$$

and the compliant DLSP (8.12):

$$\begin{cases} J_0(\mathbf{u}^h,\omega^h,s^h;\mathbf{u}_0,\omega_0,\lambda) = \|\nabla\times\omega^h+\lambda(\omega^h+\omega_0)\times(\mathbf{u}^h+\mathbf{u}_0)+\nabla s^h\|_0^2 \\ \qquad +\|\nabla\cdot\omega^h\|_0^2+\|\nabla\times\mathbf{u}^h-\omega^h\|_0^2+\|\nabla\cdot\mathbf{u}^h\|_0^2 \\ X_{r,0}^h = [G^r]^3\cap\mathbf{H}_n^1(\Omega)\times[G^r]^3\cap\mathbf{H}_n^1(\Omega)\times G^r\cap H_0^1(\Omega)\,. \end{cases} \tag{8.37}$$

Let $u=\{\mathbf{u},\omega,s\}$ and $v=\{\mathbf{v},\xi,q\}$. Minimizers of (8.36) are subject to the following first-order optimality condition:

$$\lim_{\varepsilon\to 0}\frac{d}{d\varepsilon}J_0(u+\varepsilon v;\mathbf{u}_0,\omega_0,\lambda)=0 \qquad \forall v\in X_0\,. \tag{8.38}$$

It is easy to see that the Euler–Lagrange equation (8.38) is equivalent to the following variational problem: *seek* $u=\{\mathbf{u},\omega,s\}\in X_0$ *such that*

$$Q(u,v;\mathbf{u}_0,\omega_0,\lambda)=0 \qquad \forall v\in X_0\,, \tag{8.39}$$

where

$$\begin{aligned} &Q(u,v;\mathbf{u}_0,\omega_0,\lambda) = \\ &\quad \lambda\big(\nabla\times\omega+\nabla s+\lambda(\omega+\omega_0)\times(\mathbf{u}+\mathbf{u}_0),\xi\times(\mathbf{u}+\mathbf{u}_0)+(\omega+\omega_0)\times\mathbf{v}\big)_0 \\ &\quad +\lambda\big((\omega+\omega_0)\times(\mathbf{u}+\mathbf{u}_0),\nabla\times\xi+\nabla q\big)_0 \\ &\quad +\big(\nabla\times\omega+\nabla s,\nabla\times\xi+\nabla q\big)_0+\big(\nabla\cdot\omega,\nabla\cdot\xi\big)_0 \\ &\quad +\big(\nabla\times\mathbf{u}-\omega,\nabla\times\mathbf{v}-\xi\big)_0+\big(\nabla\cdot\mathbf{u},\nabla\cdot\mathbf{v}\big)_0\,. \end{aligned} \tag{8.40}$$

We split $Q(u,v;\mathbf{u}_0,\omega_0,\lambda)$ into a bilinear part

$$\begin{aligned} Q^{\mathcal{T}}(u,v) = {} & \big(\nabla\times\omega+\nabla s,\nabla\times\xi+\nabla q\big)_0+\big(\nabla\cdot\omega,\nabla\cdot\xi\big)_0 \\ & +\big(\nabla\times\mathbf{u}-\omega,\nabla\times\mathbf{v}-\xi\big)_0+\big(\nabla\cdot\mathbf{u},\nabla\cdot\mathbf{v}\big)_0 \end{aligned}$$

that is identical with bilinear form associated with the prototype Stokes CLSP (7.42), and a non-linear part

$$\begin{aligned} &Q^{\mathcal{G}}(u,v;\mathbf{u}_0,\omega_0,\lambda) = \\ &\quad \lambda\big(\nabla\times\omega+\nabla s+\lambda(\omega+\omega_0)\times(\mathbf{u}+\mathbf{u}_0),\xi\times(\mathbf{u}+\mathbf{u}_0)+(\omega+\omega_0)\times\mathbf{v}\big)_0 \\ &\quad +\lambda\big((\omega+\omega_0)\times(\mathbf{u}+\mathbf{u}_0),\nabla\times\xi+\nabla q\big)_0 \end{aligned}$$

that contains contributions emanating from the nonlinear convection term in the Navier–Stokes equations. In terms of these forms (8.39) can be stated as: *seek* $u \in X_0$ *such that*

$$Q^{\mathcal{T}}(u,v) + Q^{\mathcal{G}}(u,v;\mathbf{u}_0,\omega_0,\lambda) = 0 \qquad \forall v \in X_0\,. \tag{8.41}$$

The discrete problem (8.37) is a compliant discretization of (8.36). Therefore, the minimizer of (8.37) is subject to the same necessary condition as the minimizer of (8.36), but restricted to the finite element space $X^h_{r,0}$. Let $u^h = \{\mathbf{u}^h, \omega^h, s^h\}$ and $v^h = \{\mathbf{v}^h, \xi^h, q^h\}$. Using the notation introduced above, the discrete problem assumes the form: *seek* $u^h \in X^h_{r,0}$ *such that*

$$Q^{\mathcal{T}}(u^h,v^h) + Q^{\mathcal{G}}(u^h,v^h;\mathbf{u}_0,\omega_0,\lambda) = 0 \qquad \forall v^h \in X^h_{r,0}\,. \tag{8.42}$$

Canonical form of the LSFEM

In this section, we establish that (8.41) and (8.42) have the canonical forms of (8.16) and (8.17), respectively. The additional spaces required to show this are the dual X_0^* of the solution space X_0 and the space

$$V = [L^{3/2}(\Omega)]^3 \times [L^{3/2}(\Omega)]^3 \times L^{3/2}(\Omega)\,. \tag{8.43}$$

Lemma 8.3 *The embedding* $V \subset X_0^*$ *is compact.*

Proof. By virtue of the Sobolev embedding theorem, in $\mathbb{R}^3$ each component of X_0 embeds compactly into $L^q(\Omega)$ for any $0 < q \le 6$. Using the fact that the adjoint of a compact operator is also compact, it follows that each component of V is compactly embedded into the respective component of the dual space X_0^*.[11] □

Referring to the notation of Section 8.3.1, we make the associations

$$X = X_0\,, \qquad X^h = X^h_{r,0}\,, \qquad Y = X_0^*, \qquad \text{and} \qquad V = V\,.$$

We define $\mathcal{T}$ to be the solution operator[12] of the prototype CLSP (7.42) for the Stokes equations:

$$\begin{cases} \mathcal{T} : X_0^* \mapsto X_0 \quad \text{with} \quad w = \mathcal{T}\cdot g \quad \text{for} \quad g \in X_0^* \quad \text{if and only if} \\ \qquad\qquad Q^{\mathcal{T}}(w,v) = (g,v)_0 \qquad \forall v \in X_0\,, \end{cases} \tag{8.44}$$

where

[11] The proof of this lemma was suggested to the authors by Michael Renardy.

[12] The *data* space Y_0 of (7.42) should not be confused with the *domain* X_0^* of the least-squares solution operator. The former serves to define the proper norms to measure the residual energy of the equations, whereas the latter defines the set of admissible arguments for $\mathcal{T}$. By writing the prototype Stokes least-squares problem as $Q^{\mathcal{T}}(w,v) = (\mathcal{L}^* f, v)_0$, where $\mathcal{L}$ is the VVP Stokes operator, we see that $\mathcal{L}^* f \in X_0^*$ for $f \in Y_0$ because $\mathcal{L}^*$ is a first-order operator.

$$(g,v)_0 = (\mathbf{g}_1,\mathbf{v})_0 + (\mathbf{g}_2,\xi)_0 + (g_3,q)_0 .$$

Likewise, we define $\mathcal{T}^h$ to be the solution operator for the compliant prototype DLSP (7.52) for the Stokes equations:

$$\begin{cases} \mathcal{T}^h : X_0^* \mapsto X^h \quad \text{with} \quad w^h = \mathcal{T}^h \cdot g \quad \text{for} \quad g \in X_0^* \quad \text{if and only if} \\ \qquad Q^{\mathcal{T}}(w^h, v^h) = (g, v^h)_0 \qquad \forall v^h \in X^h . \end{cases} \tag{8.45}$$

Owing to the scaling of momentum equation (8.35) by $\lambda = \nu^{-1}$, the operators $\mathcal{T}$ and $\mathcal{T}^h$ are independent of λ, as required by the abstract theory of Section 8.3.1. Finally, the nonlinear operator $\mathcal{G}$ is defined as follows:

$$\begin{cases} \mathcal{G} : \Lambda \times X_0 \to X_0^* \quad \text{with} \quad g = \mathcal{G}(\lambda, w) \quad \text{for} \quad w \in X_0 \quad \text{if and only if} \\ \qquad Q^{\mathcal{G}}(w, v; \mathbf{u}_0, \omega_0, \lambda) = (g, v)_0 \qquad \forall v \in X_0 . \end{cases} \tag{8.46}$$

The following Lemma establishes that, with these identifications, (8.41) and (8.42) have the desired canonical forms.

Lemma 8.4 *Problems* (8.41) *and* (8.42) *have the form of* (8.16) *and* (8.17), *respectively, with* $\mathcal{T}$, $\mathcal{T}^h$, *and* $\mathcal{G}$ *defined as in* (8.44), (8.45), *and* (8.46), *respectively.*

Proof. Assume that $u \in X_0$ solves (8.16). Then $-u = \mathcal{T} \cdot g$ if and only if

$$-(g,v)_0 = Q^{\mathcal{T}}(u,v) \qquad \forall v \in X_0$$

and $g = \mathcal{G}(\lambda, u)$ if and only if

$$(g,v)_0 = Q^{\mathcal{G}}(u, v; \mathbf{u}_0, \omega_0, \lambda) \qquad \forall v \in X_0 .$$

Thus, $u + \mathcal{T} \circ \mathcal{G}(\lambda, u) = 0$ if and only if (8.41) holds, i.e., $Q(u, v; \mathbf{u}_0, \omega_0, \lambda) = 0$ for all $v \in X_0$. The proof connecting (8.42) to (8.17) is identical. □

Verification of the assumptions of the abstract theory

The next step after showing that CLSPs and DLSPs have the requisite canonical forms is to verify the assumptions of Theorem 8.1. First, we need to show that $\mathcal{T}^h$ approximates $\mathcal{T}$ as required by (8.19) and (8.20) and, after that, we need to verify the assumptions on the nonlinear operator (8.46).

Lemma 8.5 *Let* $\mathcal{T}$ *and* $\mathcal{T}^h$ *denote the operators defined in* (8.44) *and* (8.45), *respectively. Then,* $\mathcal{T} \in L(X_0^*, X_0)$ *and* $\mathcal{T}^h \in L(X_0^*, X^h)$.

Proof. Recall that $\mathcal{T}$ is the solution operator of the prototype CLSP (7.42) for the Stokes equations. From Theorem 7.4, it follows that $\mathcal{T}$ is a continuous linear operator $X_0^* \mapsto X_0$. Likewise, $\mathcal{T}^h$ is the solution operator of the prototype compliant DLSP (7.52). Theorem 7.13 asserts that $\mathcal{T}^h$ is a continuous linear operator $X_0^* \mapsto X^h$. □

Lemma 8.6 *Assume that $X^h_{r,0}$, $r \geq 1$, is the space defined in* (8.37). *For any $g \in X_0^*$,*

$$\lim_{h \to 0} \|(\mathcal{T} - \mathcal{T}^h) g\|_{X_0} = 0. \tag{8.47}$$

If g is such that $u = \mathcal{T} \cdot g$ is in X_q with $q \geq 1$, then

$$\|(\mathcal{T} - \mathcal{T}^h) \cdot g\|_{X_0} \leq C h^{\widetilde{r}} \left(\|\mathbf{u}\|_{\widetilde{r}+1} + \|\omega\|_{\widetilde{r}+1} + \|s\|_{\widetilde{r}+1} \right), \tag{8.48}$$

where $\widetilde{r} = \min\{r, q\}$.

Proof. The error bound in (8.48) simply restates (7.60) from Theorem 7.13. However, this result does not imply (8.47) because in general $\mathcal{T}$ maps $g \in X_0^*$ to a function $u \in X_0$ for which $q = 0$ and $\widetilde{r} = 0$ in (8.48).

We prove (8.47) directly. Given $g \in X_0^*$, $u^h = \mathcal{T}^h \cdot g$, and $u = \mathcal{T} \cdot g$, this assertion is equivalent to

$$\lim_{h \to 0} \left(\|\mathbf{u} - \mathbf{u}^h\|_1 + \|\omega - \omega^h\|_1 + \|s - s^h\|_1 \right) = 0.$$

Because $\mathcal{T}^h$ is the solution operator of a compliant DLSP, from (3.63) in Theorem 3.28 it follows that

$$\begin{aligned} &\|\mathbf{u} - \mathbf{u}^h\|_1 + \|\omega - \omega^h\|_1 + \|s - s^h\|_1 \\ &\qquad \leq \inf_{v^h \in X^h_{r,0}} C \left(\|\mathbf{u} - \mathbf{v}^h\|_1 + \|\omega - \xi^h\|_1 + \|s - q^h\|_1 \right). \end{aligned} \tag{8.49}$$

Using that $D(\overline{\Omega})$ is dense in $H^1(\Omega)$, it follows that, for any fixed $\varepsilon > 0$, there exists $\omega_\varepsilon \in D(\overline{\Omega})$ such that $\|\omega - \omega_\varepsilon\|_1 < \varepsilon/2$. On the other hand, (B.20) implies the existence, for any $h > 0$, of a discrete function $\xi^h \in [G^r]^3$ such that

$$\|\omega_\varepsilon - \xi^h\|_1 \leq C h^r \|\omega_\varepsilon\|_{r+1}$$

with C independent of h. This guarantees the existence of $h_\varepsilon > 0$ and ξ^h_ε such that $\|\omega - \xi^h_\varepsilon\|_1 < \varepsilon/2$. Thus, we have established that, for any $\varepsilon > 0$, there exist $h_\varepsilon > 0$ and ξ^h_ε such that $\|\omega - \xi^h_\varepsilon\|_1 < \varepsilon$, i.e., $\lim_{h \to 0} \|\omega - \xi^h\|_1 = 0$. Identical arguments show that the rest of the terms also converge to zero so that (8.47) holds. □

There is one more assumption left to verify regarding $\mathcal{T}^h$ and $\mathcal{T}$.

Lemma 8.7 *The operator $\mathcal{T}^h$ converges to $\mathcal{T}$ in $L(V, X_0)$, i.e.,*

$$\lim_{h \to 0} \|\mathcal{T} - \mathcal{T}^h\|_{L(V, X_0)} = 0. \tag{8.50}$$

Proof. From (8.47) and the uniform boundedness theorem, it follows that $\mathcal{S}^h = \mathcal{T} - \mathcal{T}^h$ is uniformly bounded, i.e., there exists a constant $C_\mathcal{S} > 0$, independent of h, such that

$$\|\mathcal{S}^h\|_{L(X_0^*,X_0)} \leq C_{\mathcal{S}}\,. \tag{8.51}$$

Suppose that (8.50) does not hold. Then, there must be an $\varepsilon > 0$ such that for every $h = 1/n$, $n = 1,2,3,\ldots$, one can find $g_n \in V$ with $\|g_n\|_V = 1$ and $\|\mathcal{S}^h g_n\|_{X_0} \geq \varepsilon$. Because $\{g_n\}$ is bounded in V and the latter is compactly embedded in X_0^*, it follows that $g_n \to \tilde{g}$, where $\tilde{g} \in X_0^*$. Using (8.47) and (8.51), it follows that

$$\|\mathcal{S}^h g_n\|_{X_0} = \|\mathcal{S}^h(g_n - \tilde{g}) + \mathcal{S}^h\tilde{g}\|_{X_0} \leq C_{\mathcal{S}}\|g_n - \tilde{g}\|_{X_0^*} + \|\mathcal{S}^h\tilde{g}\|_{X_0} \to 0$$

as $n \to \infty$, which is a contradiction. □

Next, we verify the assumptions concerning the nonlinear operator $\mathcal{G}$.

Lemma 8.8 *Let $\mathcal{G}$ denote the operator defined by* (8.46) *and let Λ denote a compact subset of $\mathbb{R}^+$ and $\lambda = 1/\nu \in \Lambda$. Then,*

1. *$\mathcal{G}$ is a C^2 mapping from $\Lambda \times X_0$ into X_0^**
2. *$D_u\mathcal{G}(\lambda,\widehat{u}) \in L(X_0,V)$ for all $\widehat{u} \in X_0$*
3. *all second derivatives of $\mathcal{G}$ are bounded on bounded subsets of $\Lambda \times X_0$.*

Proof. Definition (8.46) implies that $g = \mathcal{G}(\lambda,u)$ if and only if

$$\begin{aligned}(g,v)_0 &= (\mathbf{g}_1,\mathbf{v})_0 + (\mathbf{g}_2,\xi)_0 + (g_3,q)_0 \\ &= \lambda\big(\nabla\times\omega + \nabla s + \lambda(\omega+\omega_0)\times(\mathbf{u}+\mathbf{u}_0), \xi\times(\mathbf{u}+\mathbf{u}_0) + (\omega+\omega_0)\times\mathbf{v}\big)_0 \\ &\quad + \lambda\big((\omega+\omega_0)\times(\mathbf{u}+\mathbf{u}_0), \nabla\times\xi + \nabla q\big)_0\end{aligned}$$

for all $v = \{\mathbf{v},\xi,q\} \in X_0$. After grouping terms according to their test function, we find that

$$\begin{aligned}\mathbf{g}_1 &= \lambda\big(\nabla\times\omega + \nabla s + \lambda(\omega+\omega_0)\times(\mathbf{u}+\mathbf{u}_0)\big)\times(\omega+\omega_0) \\ \mathbf{g}_2 &= -\lambda\big(\nabla\times\omega + \nabla s + \lambda(\omega+\omega_0)\times(\mathbf{u}+\mathbf{u}_0)\big)\times(\mathbf{u}+\mathbf{u}_0) \\ &\quad + \lambda\nabla\times\big((\omega+\omega_0)\times(\mathbf{u}+\mathbf{u}_0)\big) \\ g_3 &= -\lambda\nabla\cdot\big((\omega+\omega_0)\times(\mathbf{u}+\mathbf{u}_0)\big)\,.\end{aligned}$$

All right-hand side terms above have the form $u_i v_j w_k$ or $u_i(\partial v_j/\partial x_k)$, where u_i, v_j, and w_k are scalar components of $u_0 = \{\mathbf{u}_0,\omega_0,s_0\}$ and $u = \{\mathbf{u},\omega,s\}$. Both u_0 and u are in X_0, which is a product of $H^1(\Omega)$ spaces so that $u_i,v_j,w_k \in H^1(\Omega)$. Using (8.30) and (8.33), it follows that $\mathbf{g}_1 \in [L^{3/2}(\Omega)]^3$, $\mathbf{g}_2 \in [L^{3/2}(\Omega)]^3$, and $g_3 \in L^{3/2}(\Omega)$, i.e., the range of $\mathcal{G}(\lambda,u)$ is actually in the space V that, according to Lemma 8.3, is compactly embedded in X_0^*. The first statement of the lemma now follows by observing that $\mathcal{G}(\lambda,u)$ is a polynomial map in λ, the components of u, and their partial derivatives so that it is in fact a C^∞ map $\Lambda \times X_0 \mapsto X_0^*$.

Now, let $D_u\mathcal{G}(\lambda,\widehat{u})[u]$ denote the Frechèt derivative of $\mathcal{G}$ at $\widehat{u} = \{\widehat{\mathbf{u}},\widehat{\omega},\widehat{s}\}$ acting on $u = \{\mathbf{u},\omega,s\}$, where $\widehat{u},u \in X_0$. To prove the second statement, we need to show that $D_u\mathcal{G}(\lambda,\widehat{u})[u] \in V$ and

$$\|D_u\mathcal{G}(\lambda,\widehat{u})[u]\|_V \le C\|u\|_{X_0}\,. \tag{8.52}$$

Using (8.46), it is not difficult to show that $g = D_u\mathcal{G}(\lambda,\widehat{u})[u]$ if and only if

$$\begin{aligned}
(g,v)_0 &= (\mathbf{g}_1,\mathbf{v})_0 + (\mathbf{g}_2,\xi)_0 + (g_3,q)_0 \\
&= \lambda\big(\nabla\times\widehat{\omega} + \nabla\widehat{s} + \lambda(\widehat{\omega}+\omega_0)\times(\widehat{\mathbf{u}}+\mathbf{u}_0), \xi\times\mathbf{u} + \omega\times\mathbf{v}\big)_0 \\
&\quad +\lambda\big(\nabla\times\omega + \nabla s + \lambda(\omega\times(\widehat{\mathbf{u}}+\mathbf{u}_0) + (\widehat{\omega}+\omega_0)\times\mathbf{u}), \\
&\qquad\qquad\qquad \xi\times(\widehat{\mathbf{u}}+\mathbf{u}_0) + (\widehat{\omega}+\omega_0)\times\mathbf{v}\big)_0 \\
&\quad +\lambda\big(\omega\times(\widehat{\mathbf{u}}+\mathbf{u}_0) + (\widehat{\omega}+\omega_0)\times\mathbf{u}, \nabla\times\xi + \nabla q\big)_0
\end{aligned} \tag{8.53}$$

for all $v = \{\mathbf{v},\xi,q\} \in X_0$. Collecting terms by their test function shows that

$$\begin{aligned}
\mathbf{g}_1 &= \lambda\big(\nabla\times\omega + \nabla s + \lambda(\omega\times(\widehat{\mathbf{u}}+\mathbf{u}_0) + (\widehat{\omega}+\omega_0)\times\mathbf{u})\big)\times\big(\widehat{\omega}+\omega_0\big) \\
&\quad +\lambda\big(\nabla\times\widehat{\omega} + \nabla\widehat{s} + \lambda(\widehat{\omega}+\omega_0)\times(\widehat{\mathbf{u}}+\mathbf{u}_0)\big)\times\omega
\end{aligned}$$

$$\begin{aligned}
\mathbf{g}_2 &= -\lambda\big(\nabla\times\omega + \nabla s + \lambda(\omega\times(\widehat{\mathbf{u}}+\mathbf{u}_0) + (\widehat{\omega}+\omega_0)\times\mathbf{u})\big)\times\big(\widehat{\mathbf{u}}+\mathbf{u}_0\big) \\
&\quad -\lambda\big(\nabla\times\widehat{\omega} + \nabla\widehat{s} + \lambda(\widehat{\omega}+\omega_0)\times(\widehat{\mathbf{u}}+\mathbf{u}_0)\big)\times\mathbf{u} \\
&\quad +\lambda\nabla\times\big(\omega\times(\widehat{\mathbf{u}}+\mathbf{u}_0) + (\widehat{\omega}+\omega_0)\times\mathbf{u}\big)
\end{aligned}$$

$$g_3 = -\lambda\nabla\cdot\big(\omega\times(\widehat{\mathbf{u}}+\mathbf{u}_0) + (\widehat{\omega}+\omega_0)\times\mathbf{u}\big)\,.$$

As in the previous case, all right-hand side terms above have the form $u_iv_jw_k$ or $u_i(\partial v_j/\partial x_k)$, where u_i, v_j, and w_k are scalar components of $\widehat{u}$, u_0, and u. Therefore, $u_i, v_j, w_k \in H^1(\Omega)$ and using again (8.30) and (8.33) it follows that $\mathbf{g}_1 \in [L^{3/2}(\Omega)]^3$, $\mathbf{g}_2 \in [L^{3/2}(\Omega)]^3$, and $g_3 \in L^{3/2}(\Omega)$, i.e., $g \in V$. The same inequalities easily imply that (8.52) also holds.

Next, let $D_u^2\mathcal{G}(\lambda,\widehat{u})[u',u'']$ denote the Hessian of $\mathcal{G}$ at $\widehat{u}$ acting on the pair $u' = \{\mathbf{u}',\omega',s'\}$ and $u'' = \{\mathbf{u}'',\omega'',s''\}$, where $\widehat{u},u',u'' \in X_0$. To prove the last statement of the lemma it suffices to show that

$$\|D_u^2\mathcal{G}(\lambda,\widehat{u})[u',u'']\|_Y \le C\|u'\|_{X_0}\|u''\|_{X_0}\,.$$

From (8.53), it easily seen that $g = D_u^2\mathcal{G}(\lambda,\widehat{u})[u',u'']$ if and only if

$$\begin{aligned}
(\mathbf{g}_1,\mathbf{v})_0 &= \lambda\big(\nabla\times\omega'' + \nabla s'' + \lambda(\omega''\times(\widehat{\mathbf{u}}+\mathbf{u}_0) + (\widehat{\omega}+\omega_0)\times\mathbf{u}''), \omega'\times\mathbf{v}\big)_0 \\
&\quad +\lambda\big(\nabla\times\omega' + \nabla s' + \lambda(\omega'\times(\widehat{\mathbf{u}}+\mathbf{u}_0) + (\widehat{\omega}+\omega_0)\times\mathbf{u}'), \omega''\times\mathbf{v}\big)_0 \\
&\quad +\lambda^2\big(\omega''\times\mathbf{u}' + \omega'\times\mathbf{u}'', (\widehat{\omega}+\omega_0)\times\mathbf{v}\big)_0
\end{aligned}$$

$$\begin{aligned}
(\mathbf{g}_2,\xi)_0 &= \lambda\big(\nabla\times\omega'' + \nabla s'' + \lambda(\omega''\times(\widehat{\mathbf{u}}+\mathbf{u}_0) + (\widehat{\omega}+\omega_0)\times\mathbf{u}''), \xi\times\mathbf{u}'\big)_0 \\
&\quad +\lambda\big(\nabla\times\omega' + \nabla s' + \lambda(\omega'\times(\widehat{\mathbf{u}}+\mathbf{u}_0) + (\widehat{\omega}+\omega_0)\times\mathbf{u}'), \xi\times\mathbf{u}''\big)_0 \\
&\quad +\lambda\big(\omega''\times\mathbf{u}' + \omega'\times\mathbf{u}'', \nabla\times\xi + \lambda\xi\times(\widehat{\mathbf{u}}+\mathbf{u}_0)\big)_0
\end{aligned}$$

$$(g_3,q)_0 = \lambda\big(\omega''\times\mathbf{u}' + \omega'\times\mathbf{u}'', \nabla q\big)_0$$

for all $v = \{\mathbf{v}, \xi, q\} \in X_0$. In this case, the right-hand side terms have the form $u_i v_j w_k z_l$ or $u_i(\partial v_j/\partial x_l)w_k$, where u_i, v_j, w_k, and z_l are scalar components of $\widehat{u}$, u_0, u', u'', and v. All these functions belong to X_0 so that $u_i, v_j, w_k, z_l \in H^1(\Omega)$. Using (8.28) and (8.29) to estimate the right-hand terms yields

$$|(g,v)_0| \le C\big(\lambda, \|u_0\|_{X_0}, \|\widehat{u}\|_{X_0}\big)\, \|u'\|_{X_0}\, \|u''\|_{X_0}\, \|v\|_{X_0}\,,$$

where $C(\lambda, \|u_0\|_{X_0}, \|\widehat{u}\|_{X_0})$ is a polynomial function of λ, $\|u_0\|_{X_0}$, and $\|\widehat{u}\|_{X_0}$. Because $Y = X_0^*$ is the dual of X_0,

$$\|D_u^2\mathcal{G}(\lambda,\widehat{u})[u',u'']\|_Y = \sup_{v\in X_0} \frac{(g,v)_0}{\|v\|_{X_0}} \le C\big(\lambda, \|u_0\|_{X_0}, \|\widehat{u}\|_{X_0}\big)\, \|u'\|_{X_0}\, \|u''\|_{X_0}\,.$$

It follows that $D_u^2\mathcal{G}(\lambda,\widehat{u})$ is bounded on all bounded subsets of $\Lambda \times X_0$. □

We now have all the prerequisites to apply Theorem 8.1 and derive error estimates for the compliant DLSP (8.42).

Theorem 8.9 *Assume that Λ is a closed interval of $\mathbb{R}^+$ and, if $u = \{\mathbf{u}, \omega, s\}$, that $\{\{\lambda, u(\lambda)\} \mid \lambda \in \Lambda\,;\, u(\lambda) \in X_0\}$ is a branch of regular solutions of* (8.41). *Assume that the finite element space $X^h_{r,0}$, $r \ge 1$, is defined as in* (8.37). *Then, there is a neighborhood $\mathcal{O}$ of the origin in X_0 and, for h sufficiently small, a unique branch $\{\{\lambda, u^h(\lambda)\} \mid \lambda \in \Lambda\,;\, u^h(\lambda) \in X^h_{r,0}\}$, with $u^h = \{\mathbf{u}^h, \omega^h, s^h\}$, of regular solutions of the discrete problem* (8.42) *such that $u(\lambda) - u^h(\lambda) \in \mathcal{O}$ for all $\lambda \in \Lambda$. Moreover,*

$$\|\mathbf{u}(\lambda) - \mathbf{u}^h(\lambda)\|_1 + \|\omega(\lambda) - \omega^h(\lambda)\|_1 + \|s(\lambda) - s^h(\lambda)\|_1 \to 0 \tag{8.54}$$

as $h \to 0$, uniformly in λ. If, in addition, the solution $u(\lambda)$ of (8.41) *is in X_q for some $q \ge 1$, then there is a constant C, independent of h, such that*

$$\begin{aligned}\|\mathbf{u}(\lambda) - \mathbf{u}^h(\lambda)\|_1 &+ \|\omega(\lambda) - \omega^h(\lambda)\|_1 + \|s(\lambda) - s^h(\lambda)\|_1 \\ &\le Ch^{\widetilde{r}}\big(\|\mathbf{u}(\lambda)\|_{\widetilde{r}+1} + \|\omega(\lambda)\|_{\widetilde{r}+1} + \|s(\lambda)\|_{\widetilde{r}+1}\big)\,,\end{aligned} \tag{8.55}$$

where $\widetilde{r} = \min\{r, q\}$.

Proof. Lemma 8.8 verifies (8.18) whereas (8.19) and (8.20) have been verified in Lemmas 8.5–8.7. It follows that all the assumptions of Theorem 8.1 hold for the continuous and discrete least-squares problems (8.39) and (8.42). Thus, (8.54) follows from (8.21) and (8.47).

To prove (8.55), assume that $u(\lambda) \in X_q$ for some $q \ge 1$, and let $g = \mathcal{G}(\lambda, u(\lambda))$. Then, $\mathcal{T} \cdot g = -u(\lambda) \in X_q$ and (8.55) follows from (8.21) and (8.48). □

8.3.3 Norm-Equivalent Discrete Least-Squares Principles for the Velocity–Vorticity–Pressure System

This section specializes Theorem 8.2 to the non-compliant DLSP (8.13) in three dimensions. Existence, uniqueness, and error estimates for solutions of this DLSP are obtained assuming a fixed value of $\lambda = \nu^{-1}$. For this reason, dependence on this parameter is suppressed in the notation for operators and solutions. Also, because Theorem 8.2 is applied directly to the discrete nonlinear problem, we do not need the Euler–Lagrange equation for the parent Navier–Stokes CLSP (8.8).

As in Section 8.3.2, functional and approximation settings of LSFEMs for the Navier–Stokes equations are inherited from their prototype Stokes CLSPs and DLSPs. For added simplicity in the analysis, both the boundary condition and the zero mean pressure constraint are imposed on the solution spaces. With these assumptions, the spaces X_q and Y_q from (7.34) specialize[13] to

$$X_q = [H^{q+2}(\Omega)]^3 \cap [H_0^1(\Omega)]^3 \times [H^{q+1}(\Omega)]^3 \times H^{q+1}(\Omega) \cap L_0^2(\Omega) \tag{8.56}$$

and

$$Y_q = [H^q(\Omega)]^3 \times [H^{q+1}(\Omega)]^3 \times H^{q+1}(\Omega) \cap L_0^2(\Omega), \tag{8.57}$$

respectively, where $q \geq -1$ is an integer regularity index. The relevant prototype Stokes CLSP is (7.45) and the prototype Stokes DLSP is obtained by changing the pressure space in (7.55) to $G^r \cap L_0^2(\Omega)$; see Remark 7.16. The finite element approximation space inherited from this DLSP is a specialization of the approximation space (4.44) for non-homogeneous elliptic problems (see Section 4.5.2):

$$X^h = X_{\widehat{r},0}^h = [G^{r+1}]^3 \cap [H_0^1(\Omega)]^3 \times [G^{\widehat{r}}]^3 \times G^{\widehat{r}} \cap L_0^2(\Omega). \tag{8.58}$$

Recall that $\widehat{r} = \max\{1, r\}$ and $r \geq 0$; see Section 4.5.2. Finally, as in Section 8.3.2, the residual of the momentum equation in (8.13) is scaled by $\lambda = \nu^{-1}$, i.e., the least-squares functional assumes the form

$$\begin{aligned} &J_{-h}(\mathbf{u}^h, \omega^h, s^h; \mathbf{f}) = \\ &\quad \|\nabla \times \omega^h + \lambda \omega^h \times \mathbf{u}^h + \nabla s^h - \mathbf{f}\|_{-h}^2 + \|\nabla \times \mathbf{u}^h - \omega^h\|_0^2 + \|\nabla \cdot \mathbf{u}^h\|_0^2 . \end{aligned} \tag{8.59}$$

For simplicity, we continue to use the same symbols s^h and $\mathbf{f}$ to denote the rescaled total head λs^h and source term $\lambda \mathbf{f}$.

Let $u^h = \{\mathbf{u}^h, \omega^h, s^h\}$ and $v^h = \{\mathbf{v}^h, \xi^h, q^h\}$. The necessary condition for the discrete minimizer of (8.13) is obtained in the usual manner by setting the first variation of (8.59) to zero:

$$\lim_{\varepsilon \to 0} \frac{d}{d\varepsilon} J_{-h}(u^h + \varepsilon v^h; \mathbf{f}) = 0 \qquad \forall v^h \in X_{\widehat{r},0}^h . \tag{8.60}$$

[13] The boundary data space B_q is not needed because boundary condition is imposed in (8.56).

It is a simple matter to check that the Euler–Lagrange equation (8.60) is the following variational problem:

$$\textit{seek } u^h \in X^h_{\widehat{r},0} \quad \textit{such that} \quad Q_{-h}(u^h, v^h; \mathbf{f}) = 0 \qquad \forall v^h \in X^h_{\widehat{r},0}\,, \tag{8.61}$$

where

$$\begin{aligned} Q_{-h}(u^h, v^h; \mathbf{f}) &= \left(\nabla \times \mathbf{u}^h - \omega^h, \nabla \times \mathbf{v}^h - \xi^h\right) + \left(\nabla \cdot \mathbf{u}^h, \nabla \cdot \mathbf{v}^h\right) \\ &+ \left(\nabla \times \omega^h + \lambda \omega^h \times \mathbf{u}^h + \nabla s^h - \mathbf{f}, \nabla \times \xi^h + \nabla q^h + \lambda(\omega^h \times \mathbf{v}^h + \xi^h \times \mathbf{u}^h)\right)_{-h}. \end{aligned}$$

Problem (8.61) is a nonlinear system of algebraic equations that can be written as

$$\textit{seek } u^h \in X^h \quad \textit{such that} \quad \mathcal{F}_{-h}(u^h) = 0\,, \tag{8.62}$$

i.e., it has the form of (8.22). The rest of this section deals with showing that, under certain conditions, (8.61) has a unique solution which is "close" to the solution of the Navier–Stokes equations (8.2).

Application of Theorem 8.2 to the nonlinear least-squares problem (8.62) requires some additional assumptions. Let $\widetilde{u} \equiv \{\widetilde{\mathbf{u}}, \widetilde{\omega}, \widetilde{s}\}$ denote the solution of the velocity–vorticity–pressure form (8.2), (7.2), and (7.3) of the Navier–Stokes equations that is being approximated by (8.62). In what follows, we assume that $\widetilde{u}$ is regular solution and that $\widetilde{u} \in X_q$, $q \geq 0$. Recall that $\widetilde{u}$ is regular solution of (8.2) and (7.2); see [183, p. 299], if and only if the linearized VVP system

$$\begin{cases} \nabla \times \omega + \lambda(\omega \times \widetilde{\mathbf{u}} + \widetilde{\omega} \times \mathbf{u}) + \nabla s = \mathbf{f} & \text{in } \Omega \\ \nabla \times \mathbf{u} - \omega = 0 & \text{in } \Omega \\ \nabla \cdot \mathbf{u} = 0 & \text{in } \Omega \end{cases}$$

along with the boundary condition (7.2) and the zero mean pressure constraint (7.3) has a unique solution $u \equiv \{\mathbf{u}, \omega, s\} \in X_0$ for every $\mathbf{f} \in [H^{-1}(\Omega)]^3$ and $\mathbf{f} \mapsto u$ is continuous from $[H^{-1}(\Omega)]^3$ into X_{-1}. As a result, if $\widetilde{u}$ is regular solution of the VVP Navier–Stokes formulation, there exists a positive constant C such that, for any $u \in X_{-1}$,

$$\begin{aligned} &C\left(\|\mathbf{u}\|_1 + \|\omega\|_0 + \|s\|_0\right) \leq \\ &\quad \|\nabla \times \omega + \lambda(\omega \times \widetilde{\mathbf{u}} + \widetilde{\omega} \times \mathbf{u}) + \nabla s\|_{-1} + \|\nabla \times \mathbf{u} - \omega\|_0 + \|\nabla \cdot \mathbf{u}\|_0\,. \end{aligned} \tag{8.63}$$

Verification of the assumptions of the abstract theory

We now verify that, under the additional hypotheses on the solution of the VVP Navier–Stokes system (8.2), (7.2), and (7.3), all assumptions of Theorem 8.2 hold for the discrete least-squares problem (8.61). The asymptotic error estimates for the least-squares solution then follow from (8.27).

We begin by splitting $Q_{-h}(\cdot,\cdot;\mathbf{f})$ into a bilinear (Stokes) form

$$\begin{aligned} Q^{\mathcal{T}}_{-h}(u^h,v^h) &= (\nabla\times\omega^h+\nabla s^h, \nabla\times\xi^h+\nabla q^h)_{-h} \\ &\quad +(\nabla\times\mathbf{u}^h-\omega^h,\nabla\times\mathbf{v}^h-\xi^h)_0+(\nabla\cdot\mathbf{u}^h,\nabla\cdot\mathbf{v}^h)_0 \end{aligned} \tag{8.64}$$

that is independent of λ and $\mathbf{f}$, and a nonlinear form

$$\begin{aligned} Q^{\mathcal{G}}_{-h}(u^h,v^h;\mathbf{f}) &= \lambda\left(\nabla\times\omega^h+\lambda\omega^h\times\mathbf{u}^h+\nabla s^h-\mathbf{f},\, \omega^h\times\mathbf{v}^h+\xi^h\times\mathbf{u}^h\right)_{-h} \\ &\quad +\left(\lambda\omega^h\times\mathbf{u}^h-\mathbf{f},\nabla\times\xi^h+\nabla q^h\right)_{-h}, \end{aligned} \tag{8.65}$$

that includes[14] λ and $\mathbf{f}$. Clearly,

$$Q_{-h}(u^h,v^h;\mathbf{f}) = Q^{\mathcal{T}}_{-h}(u^h,v^h)+Q^{\mathcal{G}}_{-h}(u^h,v^h;\mathbf{f}).$$

Using this splitting, the nonlinear problem (8.62) can be cast in a form that is similar to the abstract form (8.17) for compliant DLSPs. To this end, let

$$Y = Y_{-1} = [H^{-1}(\Omega)]^3\times[L^2(\Omega)]^3\times L^2_0(\Omega)$$

and consider the linear operator defined[15] by

$$\begin{cases} \mathcal{T}_{-h}: Y\mapsto X^h \quad \text{with} \quad u^h=\mathcal{T}_{-h}\cdot g \quad \text{for} \quad g\in Y \quad \text{if and only if} \\ \qquad\qquad Q^{\mathcal{T}}_{-h}(u^h,v^h) = (g,v^h)_0 \qquad \forall v^h\in X^h. \end{cases}$$

We also have the nonlinear operator defined by

$$\begin{cases} \mathcal{G}_{-h}: X^h\mapsto Y \quad \text{with} \quad g=\mathcal{G}_{-h}(u^h) \quad \text{for} \quad u^h\in X^h \quad \text{if and only if} \\ \qquad\qquad Q^{\mathcal{G}}_{-h}(u^h,v^h;\mathbf{f}) = (g,v^h)_0 \qquad \forall v^h\in X^h, \end{cases}$$

and its Fréchet derivative at $\widehat{u}^h=\{\widehat{\mathbf{u}}^h,\widehat{\omega}^h,\widehat{s}^h\}$ acting on $u^h=\{\mathbf{u}^h,\omega^h,s^h\}$:

$$\begin{cases} D_u\mathcal{G}_{-h}(\widehat{u}^h): X^h\mapsto Y \quad \text{with} \quad g=D_u\mathcal{G}_{-h}(\widehat{u}^h)[u^h] \quad \text{for} \quad u^h\in X^h \\ \quad \text{if and only if} \qquad D_uQ^{\mathcal{G}}_{-h}(u^h,v^h;\mathbf{f},\widehat{u}^h) = (g,v^h)_0 \qquad \forall v^h\in X^h, \end{cases}$$

where $D_uQ^{\mathcal{G}}_{-h}(u^h,v^h;\mathbf{f},\widehat{u}^h)$ is the bilinear form

---

[14] Recall that the explicit dependence on λ is suppressed from the notation because, in this section, λ is assumed fixed.

[15] Note that the approximation space $X^h=X^h_{\widehat{r},0}$ defined in (8.58) is a proper subspace of the solution space X_{-1} from (8.56) and that the latter coincides with the dual of Y_{-1}, i.e., $Y^*_{-1}=X_{-1}$. Therefore, the right-hand side $(g,v^h)_0$ in the definition of $\mathcal{T}_{-h}$ is meaningful.

$$
\begin{aligned}
&D_u Q^{\mathcal{G}}_{-h}(u^h, v^h; \mathbf{f}, \widehat{u}^h) = \\
&\quad \lambda \left(\widehat{\omega}^h \times \mathbf{u}^h + \omega^h \times \widehat{\mathbf{u}}^h, \nabla \times \xi^h + \nabla q^h + \lambda(\widehat{\omega}^h \times \mathbf{v}^h + \xi^h \times \widehat{\mathbf{u}}^h)\right)_{-h} \\
&\quad + \lambda \left(\nabla \times \widehat{\omega}^h + \nabla \widehat{s}^h + \lambda \widehat{\omega}^h \times \widehat{\mathbf{u}}^h - \mathbf{f}, \omega^h \times \mathbf{v}^h + \xi^h \times \mathbf{u}^h\right)_{-h} \\
&\quad + \lambda \left(\nabla \times \omega^h + \nabla s^h, \widehat{\omega}^h \times \mathbf{v}^h + \xi^h \times \widehat{\mathbf{u}}^h\right)_{-h}.
\end{aligned}
\tag{8.66}
$$

With these definitions, the nonlinear equation (8.62) and the Fréchet derivative $D_u\mathcal{F}_{-h}$ assume the forms[16]

$$
\mathcal{F}_{-h}(u^h) \equiv u^h + \mathcal{T}_{-h} \circ \mathcal{G}_{-h}(u^h) = 0 \tag{8.67}
$$

and

$$
D_u\mathcal{F}_{-h}(\widehat{u}^h)[u^h] \equiv \left(\mathcal{I} + \mathcal{T}_{-h} \circ D_u\mathcal{G}_{-h}(\widehat{u})\right)[u^h], \tag{8.68}
$$

respectively. The weak variational form of (8.68) given by

$$
D_u Q_{-h}(u^h, v^h; \mathbf{f}, \widehat{u}^h) = Q^{\mathcal{T}}_{-h}(u^h, v^h) + D_u Q^{\mathcal{G}}_{-h}(u^h, v^h; \mathbf{f}, \widehat{u}^h). \tag{8.69}
$$

The bilinear form (8.64) is the same as one would obtain from the prototype norm-equivalent Stokes DLSP (7.55). Therefore, Theorem 4.9 implies that $Q^{\mathcal{T}}_{-h}(\cdot,\cdot)$ is continuous and coercive on $X^h \times X^h$.

To apply Theorem 8.2, it is critical to ensure the existence of $\widetilde{u}^h \in X^h$ such that $D_u\mathcal{F}_{-h}(\widetilde{u}^h)$ is invertible. From (8.69), it is clear that the existence of $(D_u\mathcal{F}_{-h}(\widetilde{u}^h))^{-1}$ is equivalent to the coercivity of $D_u Q_{-h}(\cdot,\cdot; \mathbf{f}, \widetilde{u}^h)$ on $X^h \times X^h$.

By assumption, $\widetilde{u}$ is a regular solution of the VVP Navier–Stokes system. As a result, it is reasonable to expect that $D_u Q_{-h}(\cdot,\cdot; \mathbf{f}, \widetilde{u}^h)$ is coercive on $X^h \times X^h$ provided $\widetilde{u}^h$ is a good approximation of $\widetilde{u}$. From the definition (8.58) of X^h and the fact that (B.20) holds for its component spaces, it follows that there exists $\widetilde{u}^h = \{\widetilde{\mathbf{u}}^h, \widetilde{\omega}^h, \widetilde{s}^h\}$ such that

$$
\|\widetilde{\mathbf{u}} - \widetilde{\mathbf{u}}^h\|_0 + h\|\widetilde{\mathbf{u}} - \widetilde{\mathbf{u}}^h\|_1 \le h^{\widetilde{r}+2} C \|\widetilde{\mathbf{u}}\|_{\widetilde{r}+2} \tag{8.70}
$$

$$
\|\widetilde{\omega} - \widetilde{\omega}^h\|_0 + h\|\widetilde{\omega} - \widetilde{\omega}^h\|_1 \le h^{\widetilde{r}+1} C \|\widetilde{\omega}\|_{\widetilde{r}+1} \tag{8.71}
$$

$$
\|\widetilde{s} - \widetilde{s}^h\|_0 + h\|\widetilde{s} - \widetilde{s}^h\|_1 \le h^{\widetilde{r}+1} C \|\widetilde{s}\|_{\widetilde{r}+1}, \tag{8.72}
$$

where $\widetilde{r} = \min\{r, q\}$. Owing to the assumption that $\widetilde{u} \in X_q$ for $q \ge 0$ and because $r \ge 0$, we have that $\widetilde{r} \ge 0$. Note that (8.70)–(8.72) also imply

$$
\|\widetilde{\mathbf{u}}^h\|_1 \le C\|\widetilde{\mathbf{u}}\|_{\widetilde{r}+2}, \quad \|\widetilde{\omega}^h\|_0 \le C\|\widetilde{\omega}\|_{\widetilde{r}+1}, \quad \text{and} \quad \|\widetilde{s}^h\|_0 \le C\|\widetilde{s}\|_{\widetilde{r}+1}. \tag{8.73}
$$

---

[16] Formally, (8.67) has the same canonical structure as the problems considered in Theorem 8.1. However, the nonlinear operator $\mathcal{G}_{-h}$ in (8.67) is not the same nonlinear operator that one would obtain from the parent CLSP (8.8) because, in (8.67), $\mathcal{G}_{-h}$ is defined by using a discrete negative norm.

We now proceed to show that $D_u Q_{-h}(\cdot,\cdot;\mathbf{f},\widetilde{u}^h)$ is indeed coercive, provided h is sufficiently small. The first step is the following technical result.

Lemma 8.10 *There exists a positive constant C such that, for $\mathbf{u} \in [H^1(\Omega)]^3$, $\omega \in [H^1(\Omega)]^3$, and $s \in H^1(\Omega)$,*

$$\|\nabla \times \omega\|_{-h} \leq C\big(h\|\nabla \times \omega\|_0 + \|\omega\|_0\big) \tag{8.74}$$

$$\|\nabla s\|_{-h} \leq C\big(h\|\nabla s\|_0 + \|s\|_0\big) \tag{8.75}$$

$$\|\omega \times \mathbf{u}\|_{-h} \leq C\big(h\|\omega\|_1 + \|\omega\|_0\big)\|\mathbf{u}\|_1 . \tag{8.76}$$

For $u^h = \{\mathbf{u}^h, \omega^h, s^h\} \in X^h$, the inequalities (8.74)–(8.76) *specialize to*

$$\|\nabla \times \omega^h\|_{-h} \leq C\|\omega^h\|_0 \tag{8.77}$$

$$\|\nabla s^h\|_{-h} \leq C\|s^h\|_0 \tag{8.78}$$

$$\|\omega^h \times \mathbf{u}^h\|_{-h} \leq C\|\omega^h\|_0\|\mathbf{u}^h\|_1 . \tag{8.79}$$

Proof. Property (B.101) of the discrete negative norm implies that

$$\|\nabla \times \omega\|_{-h} \leq C\big(h\|\nabla \times \omega\|_0 + \|\nabla \times \omega\|_{-1}\big)$$
$$\|\nabla s\|_{-h} \leq C\big(h\|\nabla s\|_0 + \|\nabla s\|_{-1}\big)$$
$$\|\omega \times \mathbf{u}\|_{-h} \leq C\big(h\|\omega \times \mathbf{u}\|_0 + \|\omega \times \mathbf{u}\|_{-1}\big) .$$

The bounds (8.74)–(8.76) now follow from the inequalities

$$\|\nabla \times \omega\|_{-1} \leq \|\omega\|_0, \qquad \|\nabla s\|_{-1} \leq \|s\|_0$$

$$\|\omega \times \mathbf{u}\|_0 \leq C\|\omega\|_1\|\mathbf{u}\|_1, \qquad \text{and} \qquad \|\omega \times \mathbf{u}\|_{-1} \leq C\|\omega\|_0\|\mathbf{u}\|_1 .$$

The remaining bounds (8.77)–(8.79) follow from (8.74)–(8.76) and the inverse inequality (B.86). □

The next step is to estimate the "truncation" error of the discrete nonlinear equation.

Lemma 8.11 *Assume that $\widetilde{u}^h$ satisfies* (8.70)–(8.72). *Then, $Q_{-h}(\widetilde{u}^h,\cdot;\mathbf{f})$ is a continuous linear functional $X^h \mapsto \mathbb{R}$ such that*

$$Q_{-h}(\widetilde{u}^h, v^h; \mathbf{f}, \lambda) \leq C(\widetilde{u}) h^{\widetilde{r}+1}\|v^h\|_{X_{-1}} \qquad \forall v^h \in X^h . \tag{8.80}$$

Proof. From the definition of Q_{-h}, it is clear that $Q_{-h}(\widetilde{u}^h, v^h; \mathbf{f})$ is linear with respect to v^h. Using the Cauchy inequality we obtain

$$\begin{aligned} Q(\widetilde{u}^h, v^h;\mathbf{f}) \leq\ & \|\nabla \times \widetilde{\omega}^h + \nabla\widetilde{s}^h + \lambda\widetilde{\omega}^h \times \widetilde{\mathbf{u}}^h - \mathbf{f}\|_{-h} \\ & \times \|\nabla \times \xi^h + \nabla q^h + \lambda(\widetilde{\omega}^h \times \mathbf{v}^h + \xi^h \times \widetilde{\mathbf{u}}^h)\|_{-h} \\ & + \|\nabla \times \widetilde{\mathbf{u}}^h - \widetilde{\omega}^h\|_0 \|\nabla \times \mathbf{v}^h - \xi^h\|_0 + \|\nabla \cdot \widetilde{\mathbf{u}}^h\|_0 \|\nabla \cdot \mathbf{v}^h\|_0 . \end{aligned} \tag{8.81}$$

Using that $\widetilde{u}$ solves (8.2) and $\widetilde{u} \in X_q$, $q \geq 0$, and (8.70)–(8.71) yields the bounds

$$\|\nabla \times \widetilde{\mathbf{u}}^h - \widetilde{\omega}^h\|_0 \, \|\nabla \times \mathbf{v}^h - \xi^h\|_0 \leq Ch^{\widetilde{r}+1}\big(\|\widetilde{\mathbf{u}}\|_{\widetilde{r}+2} + \|\widetilde{\omega}\|_{\widetilde{r}+1}\big)\|v^h\|_{X_{-1}} \tag{8.82}$$

$$\|\nabla \cdot \widetilde{\mathbf{u}}^h\|_0 \, \|\nabla \cdot \mathbf{v}^h\|_0 \leq Ch^{\widetilde{r}+1}\|\widetilde{\mathbf{u}}\|_{\widetilde{r}+2}\|v^h\|_{X_{-1}}\,. \tag{8.83}$$

For the first factor of the first term in the right-hand side of (8.81), the triangle inequality gives

$$\begin{aligned}&\|\nabla \times \widetilde{\omega}^h + \nabla \widetilde{s}^h + \lambda \widetilde{\omega}^h \times \widetilde{\mathbf{u}}^h - \mathbf{f}\|_{-h}\\ &\qquad\leq \|\nabla \times (\widetilde{\omega} - \widetilde{\omega}^h)\|_{-h} + \|\nabla(\widetilde{s} - \widetilde{s}^h)\|_{-h}\\ &\qquad\qquad + \|\widetilde{\omega}^h \times (\widetilde{\mathbf{u}} - \widetilde{\mathbf{u}}^h)\|_{-h} + \|(\widetilde{\omega} - \widetilde{\omega}^h) \times \widetilde{\mathbf{u}}\|_{-h}\,.\end{aligned}$$

From (8.74) and (8.71), it follows that

$$\|\nabla \times (\widetilde{\omega} - \widetilde{\omega}^h)\|_{-h} \leq C\big(h\|\widetilde{\omega} - \widetilde{\omega}^h\|_1 + \|\widetilde{\omega} - \widetilde{\omega}^h\|_0\big) \leq Ch^{\widetilde{r}+1}\|\widetilde{\omega}\|_{\widetilde{r}+1}\,.$$

Also, (8.75) and (8.72) imply

$$\|\nabla(\widetilde{s} - \widetilde{s}^h)\|_{-h} \leq C\big(h\|\widetilde{s} - \widetilde{s}^h\|_1 + \|\widetilde{s} - \widetilde{s}^h\|_0\big) \leq Ch^{\widetilde{r}+1}\|\widetilde{s}\|_{\widetilde{r}+1}$$

and (8.76), (8.71), (8.70), and (8.73) imply

$$\|\widetilde{\omega}^h \times (\widetilde{\mathbf{u}} - \widetilde{\mathbf{u}}^h)\|_{-h} + \|(\widetilde{\omega} - \widetilde{\omega}^h) \times \widetilde{\mathbf{u}}\|_{-h} \leq Ch^{\widetilde{r}+1}\|\widetilde{\omega}\|_{\widetilde{r}+1}\|\widetilde{\mathbf{u}}\|_{\widetilde{r}+2}\,.$$

To estimate the second factor of the first term in the right-hand side of (8.81), we use again the triangle inequality

$$\begin{aligned}&\|\nabla \times \xi^h + \nabla q^h + \lambda(\widetilde{\omega}^h \times \mathbf{v}^h + \xi^h \times \widetilde{\mathbf{u}}^h)\|_{-h}\\ &\qquad\leq \|\nabla \times \xi^h\|_{-h} + \|\nabla q^h\|_{-h} + \lambda\big(\|\widetilde{\omega}^h \times \mathbf{v}^h\|_{-h} + \|\xi^h \times \widetilde{\mathbf{u}}^h\|_{-h}\big)\end{aligned}$$

and (8.77), (8.78), (8.79), and (8.73) to obtain

$$\|\nabla \times \xi^h\|_{-h} \leq C\|\xi^h\|_0, \qquad \|\nabla q^h\|_{-h} \leq C\|q^h\|_0,$$

$$\|\widetilde{\omega}^h \times \mathbf{v}^h\|_{-h} \leq C\|\widetilde{\omega}\|_{\widetilde{r}+1}\|\mathbf{v}^h\|_1, \qquad \text{and} \qquad \|\xi^h \times \widetilde{\mathbf{u}}^h\|_{-h} \leq C\|\xi^h\|_0\|\widetilde{\mathbf{u}}\|_{\widetilde{r}+2}\,.$$

Collecting all results shows that

$$\begin{aligned}&\|\nabla \times \widetilde{\omega}^h + \nabla \widetilde{s}^h + \lambda \widetilde{\omega}^h \times \widetilde{\mathbf{u}}^h - \mathbf{f}\|_{-h}\\ &\qquad \times \|\nabla \times \xi^h + \nabla q^h + \lambda(\widetilde{\omega}^h \times \mathbf{v}^h + \xi^h \times \widetilde{\mathbf{u}}^h)\|_{-h} \leq C(\widetilde{u})h^{\widetilde{r}+1}\|v^h\|_{X_{-1}}\,.\end{aligned} \tag{8.84}$$

The lemma now follows from (8.82)–(8.84). □

We now have all the prerequisites to show that the weak variational form (8.69) of the Fréchet derivative is well-posed.

Lemma 8.12 *Assume that* $\widetilde{u}^h$ *satisfies* (8.70) and (8.72). *Then, there exists* $h_0 > 0$ *such that* $D_u Q_{-h}(\cdot,\cdot;\mathbf{f},\widetilde{u}^h)$ *is continuous and coercive on* $X^h \times X^h$ *for all* $h < h_0$.

Proof. To prove the lemma, we start by setting $v^h = u^h$ in (8.69):

$$\begin{aligned} D_u Q_{-h}(u^h,u^h;\mathbf{f},\widetilde{u}^h) &= Q^{\mathcal{T}}_{-h}(u^h,u^h) + D_u Q^{\mathcal{G}}_{-h}(u^h,u^h;\mathbf{f},\widetilde{u}^h) \\ &= \|\nabla\times\omega^h + \nabla s^h + \lambda(\widetilde{\omega}^h\times\mathbf{u}^h + \omega^h\times\widetilde{\mathbf{u}}^h)\|^2_{-h} + \|\nabla\times\mathbf{u}^h - \omega^h\|^2_0 \\ &\quad + \|\nabla\cdot\mathbf{u}^h\|^2_0 + 2\lambda\left(\nabla\times\widetilde{\omega}^h + \nabla\widetilde{s}^h + \lambda\widetilde{\omega}^h\times\widetilde{\mathbf{u}}^h - \mathbf{f}, \omega^h\times\mathbf{u}^h\right)_{-h}. \end{aligned} \tag{8.85}$$

Let $\widetilde{u}\in X_q$, $q\ge 0$, be the regular solution of the VVP Navier–Stokes equations whose existence is postulated in Section 8.3.3. Then, the first term on the right-hand side in (8.85) can be bounded from below as follows:

$$\begin{aligned} &\|\nabla\times\omega^h + \nabla s^h + \lambda(\widetilde{\omega}^h\times\mathbf{u}^h + \omega^h\times\widetilde{\mathbf{u}}^h)\|_{-h} \\ &\qquad \ge \|\nabla\times\omega^h + \nabla s^h + \lambda(\widetilde{\omega}\times\mathbf{u}^h + \omega^h\times\widetilde{\mathbf{u}})\|_{-h} \\ &\qquad\quad -\lambda\left(\|(\widetilde{\omega}-\widetilde{\omega}^h)\times\mathbf{u}^h\|_{-h} + \|\omega^h\times(\widetilde{\mathbf{u}}-\widetilde{\mathbf{u}}^h)\|_{-h}\right). \end{aligned}$$

Using that $\widetilde{u}\in X_q$ with $q\ge 0$ in conjunction with (8.76), (8.71) and (8.70) gives

$$\|(\widetilde{\omega}-\widetilde{\omega}^h)\times\mathbf{u}^h\|_{-h} \le C h^{\widetilde{r}+1}\|\mathbf{u}^h\|_1\|\widetilde{\omega}\|_{\widetilde{r}+1}$$

and

$$\|\omega^h\times(\widetilde{\mathbf{u}}-\widetilde{\mathbf{u}}^h)\|_{-h} \le C h^{\widetilde{r}+1}\|\omega^h\|_0\|\widetilde{\mathbf{u}}\|_{\widetilde{r}+2}.$$

Using these bounds and the ε-inequality it is not hard to see that

$$\begin{aligned} &\|\nabla\times\omega^h + \nabla s^h + \lambda(\widetilde{\omega}^h\times\mathbf{u}^h + \omega^h\times\widetilde{\mathbf{u}}^h)\|^2_{-h} \\ &\ge \frac{1}{2}\|\nabla\times\omega^h + \nabla s^h + \lambda(\widetilde{\omega}\times\mathbf{u}^h + \omega^h\times\widetilde{\mathbf{u}})\|^2_{-h} - C(\widetilde{u})h^{2(\widetilde{r}+1)}\|u^h\|^2_{X_{-1}}. \end{aligned} \tag{8.86}$$

Finally, using (8.84) and (8.79) for the last term on the right-hand side in (8.85)

$$\begin{aligned} &2\lambda\left(\nabla\times\widetilde{\omega}^h + \nabla\widetilde{s}^h + \lambda\widetilde{\omega}^h\times\widetilde{\mathbf{u}}^h - \mathbf{f}, \omega^h\times\mathbf{u}^h\right)_{-h} \\ &\qquad \le C(\widetilde{u})h^{\widetilde{r}+1}\|\omega^h\|_0\|\mathbf{u}^h\|_1 \le C(\widetilde{u})h^{\widetilde{r}+1}\|u^h\|^2_{X_{-1}}. \end{aligned} \tag{8.87}$$

Combining (8.86), (8.87), and (8.63), i.e., the assumption that $\widetilde{u}$ is regular solution of the VVP Navier–Stokes equations, with the lower bound from (B.101) yields

$$\begin{aligned} &D_u Q_{-h}(u^h,u^h;\mathbf{f},\widetilde{u}^h) \\ &\quad \ge \frac{1}{2}\|\nabla\times\omega^h + \nabla s^h + \lambda(\widetilde{\omega}\times\mathbf{u}^h + \omega^h\times\widetilde{\mathbf{u}})\|^2_{-h} \\ &\qquad\qquad + \|\nabla\times\mathbf{u}^h - \omega^h\|^2_0 + \|\nabla\cdot\mathbf{u}^h\|^2_0 - C(\widetilde{u})h^{\widetilde{r}+1}\|u^h\|^2_{X_{-1}} \\ &\quad \ge C_1\left(\|\mathbf{u}^h\|^2_1 + \|\omega^h\|^2_0 + \|s^h\|^2_0\right) - C(\widetilde{u})h^{\widetilde{r}+1}\|u^h\|^2_{X_{-1}} = \left(C_1 - C(\widetilde{u})h^{\widetilde{r}+1}\right)\|u^h\|^2_{X_{-1}}. \end{aligned}$$

It follows that there exists $h_0 > 0$ such that $D_u Q_{-h}(\cdot,\cdot;\mathbf{f},\widehat{u}^h)$ is coercive whenever $h < h_0$. The continuity of $D_u Q_{-h}$ follows by a straightforward application of the Cauchy and triangle inequalities. □

The next lemma establishes a Lipschitz bound for the Frechèt derivative $D_u\mathcal{F}_{-h}$.

Lemma 8.13 *There is a constant C, independent of h, such that for all $\widehat{u}_1^h, \widehat{u}_2^h \in X^h$,*

$$\begin{aligned}\|D_u\mathcal{F}_{-h}(\widehat{u}_1^h) - D_u\mathcal{F}_{-h}(\widehat{u}_2^h)\|_{L(X^h,X^h)} \\ \le C\|\widehat{u}_1^h - \widehat{u}_2^h\|_{X_{-1}}\big(1+\|\widehat{u}_1^h\|_{X_{-1}} + \|\widehat{u}_2^h\|_{X_{-1}}\big)\,.\end{aligned} \tag{8.88}$$

Proof. From (8.68) and the definitions of $\mathcal{T}_{-h}$ and $D_u\mathcal{G}_{-h}$, it follows that

$$u_i^h = D_u\mathcal{F}_{-h}(\widehat{u}_i^h)[u^h] \qquad \text{for } u^h \in X^h \text{ and } i = 1,2$$

if and only if u_i^h, $i = 1,2$, solves the problem: *seek $u_i^h \in X^h$ such that*

$$Q^{\mathcal{T}}_{-h}(u_i^h, v^h) = D_u Q_{-h}(u^h, v^h;\mathbf{f},\widehat{u}_i^h) \qquad \forall v^h \in X^h\,. \tag{8.89}$$

Therefore,

$$Q^{\mathcal{T}}_{-h}(u_1^h - u_2^h, v^h) = D_u Q_{-h}(u^h, v^h;\mathbf{f},\widehat{u}_1^h) - D_u Q_{-h}(u^h, v^h;\mathbf{f},\widehat{u}_2^h)\,. \tag{8.90}$$

The right-hand side in this identity expands to

$$\begin{aligned}&D_u Q_{-h}(u^h, v^h;\mathbf{f},\widehat{u}_1^h) - D_u Q_{-h}(u^h, v^h;\mathbf{f},\widehat{u}_2^h)\\
&\quad = \lambda\big(\nabla\times(\widehat{\omega}_1^h - \widehat{\omega}_2^h) + \nabla(\widehat{s}_1^h - \widehat{s}_2^h) + \lambda(\widehat{\omega}_1^h\times\widehat{\mathbf{u}}_1^h - \widehat{\omega}_2^h\times\widehat{\mathbf{u}}_2^h),\\
&\qquad\qquad \omega^h\times\mathbf{v}^h + \xi^h\times\mathbf{u}^h\big)_{-h}\\
&\quad +\lambda\big(\nabla\times\omega^h + \nabla s^h, (\widehat{\omega}_1^h - \widehat{\omega}_2^h)\times\mathbf{v}^h + \xi^h\times(\widehat{\mathbf{u}}_1^h - \widehat{\mathbf{u}}_2^h)\big)_{-h}\\
&\quad +\lambda\big((\widehat{\omega}_1^h - \widehat{\omega}_2^h)\times\mathbf{u}^h + \omega^h\times(\widehat{\mathbf{u}}_1^h - \widehat{\mathbf{u}}_2^h), \nabla\times\xi^h + \nabla s^h\big)_{-h}\\
&\quad +\lambda^2\big((\widehat{\omega}_1^h\times\mathbf{u}^h + \omega^h\times\widehat{\mathbf{u}}_1^h, \widehat{\omega}_1^h\times\mathbf{v}^h + \xi^h\times\widehat{\mathbf{u}}_1^h)_{-h}\\
&\qquad\qquad -(\widehat{\omega}_2^h\times\mathbf{u}^h + \omega^h\times\widehat{\mathbf{u}}_2^h, \widehat{\omega}_2^h\times\mathbf{v}^h + \xi^h\times\widehat{\mathbf{u}}_2^h)_{-h}\big)\,.\end{aligned} \tag{8.91}$$

For the first term in (8.91) we use (8.77)–(8.79):

$$\begin{aligned}&\lambda\big(\nabla\times(\widehat{\omega}_1^h - \widehat{\omega}_2^h) + \nabla(\widehat{s}_1^h - \widehat{s}_2^h) + \lambda(\widehat{\omega}_1^h\times\widehat{\mathbf{u}}_1^h - \widehat{\omega}_2^h\times\widehat{\mathbf{u}}_2^h), \omega^h\times\mathbf{v}^h + \xi^h\times\mathbf{u}^h\big)_{-h}\\
&\quad\le\lambda\Big[\|\nabla\times(\widehat{\omega}_1^h - \widehat{\omega}_2^h)\|_{-h} + \|\nabla(\widehat{s}_1^h - \widehat{s}_2^h)\|_{-h} + \lambda\|\widehat{\omega}_1^h\times\widehat{\mathbf{u}}_1^h - \widehat{\omega}_2^h\times\widehat{\mathbf{u}}_2^h\|_{-h}\Big]\\
&\qquad\times\Big[\|\omega^h\times\mathbf{v}^h\|_{-h} + \|\xi^h\times\mathbf{u}^h\|_{-h}\Big]\\
&\quad\le\lambda\Big[\|\widehat{\omega}_1^h - \widehat{\omega}_2^h\|_0 + \|\widehat{s}_1^h - \widehat{s}_2^h\|_0 + \lambda\sqrt{(\|\widehat{\omega}_1^h - \widehat{\omega}_2^h\|_0^2 + \|\widehat{\mathbf{u}}_1^h - \widehat{\mathbf{u}}_2^h\|_1^2)(\|\widehat{\omega}_1^h\|_0^2 + \|\widehat{\mathbf{u}}_2^h\|_1^2)}\Big]\\
&\qquad\times\Big[(\|\omega^h\|_0^2 + \|\mathbf{u}^h\|_1^2)(\|\xi^h\|_0^2 + \|\mathbf{v}^h\|_1^2)\Big]^{1/2}\\
&\quad\le C\|\widehat{u}_1^h - \widehat{u}_2^h\|_{X_{-1}}\|u^h\|_{X_{-1}}\|v^h\|_{X_{-1}}\Big[2 + \lambda(\|\widehat{u}_1^h\|_{X_{-1}} + \|\widehat{u}_2^h\|_{X_{-1}})\Big]\,.\end{aligned}$$

The second and third terms in (8.91) are estimated in a similar manner:

$$\begin{aligned}
&\lambda\big(\nabla\times\omega^h+\nabla s^h,(\widehat{\omega}_1^h-\widehat{\omega}_2^h)\times\mathbf{v}^h+\xi^h\times(\widehat{\mathbf{u}}_1^h-\widehat{\mathbf{u}}_2^h)\big)_{-h}\\
&\quad\le\lambda\Big[\|\nabla\times\omega^h\|_{-h}+\|\nabla s^h\|_{-h}\Big]\Big[\|(\widehat{\omega}_1^h-\widehat{\omega}_2^h)\times\mathbf{v}^h\|_{-h}+\|\xi^h\times(\widehat{\mathbf{u}}_1^h-\widehat{\mathbf{u}}_2^h)\|_{-h}\Big]\\
&\quad\le C\|\widehat{u}_1^h-\widehat{u}_2^h\|_{X_{-1}}\|u^h\|_{X_{-1}}\|v^h\|_{X_{-1}};
\end{aligned}$$

$$\begin{aligned}
&\lambda\big((\widehat{\omega}_1^h-\widehat{\omega}_2^h)\times\mathbf{u}^h+\omega^h\times(\widehat{\mathbf{u}}_1^h-\widehat{\mathbf{u}}_2^h),\nabla\times\xi^h+\nabla s^h\big)_{-h}\\
&\quad\le\lambda\Big[\|(\widehat{\omega}_1^h-\widehat{\omega}_2^h)\times\mathbf{u}^h\|_{-h}+\|\omega^h\times(\widehat{\mathbf{u}}_1^h-\widehat{\mathbf{u}}_2^h)\|_{-h}\Big]\Big[\|\nabla\times\xi^h\|_{-h}+\|\nabla s^h\|_{-h}\Big]\\
&\quad\le C\|\widehat{u}_1^h-\widehat{u}_2^h\|_{X_{-1}}\|u^h\|_{X_{-1}}\|v^h\|_{X_{-1}}.
\end{aligned}$$

To estimate the fourth term in (8.91), we add and subtract

$$\big(\widehat{\omega}_2^h\times\mathbf{u}^h+\omega^h\times\widehat{\mathbf{u}}_2^h,\widehat{\omega}_1^h\times\mathbf{v}^h+\xi^h\times\widehat{\mathbf{u}}_1^h\big)_{-h},$$

and then repeatedly apply the Cauchy and triangle inequalities along with (8.79):

$$\begin{aligned}
&\lambda^2\Big[(\widehat{\omega}_1^h\times\mathbf{u}^h+\omega^h\times\widehat{\mathbf{u}}_1^h,\widehat{\omega}_1^h\times\mathbf{v}^h+\xi^h\times\widehat{\mathbf{u}}_1^h)_{-h}\\
&\qquad\qquad-(\widehat{\omega}_2^h\times\mathbf{u}^h+\omega^h\times\widehat{\mathbf{u}}_2^h,\widehat{\omega}_2^h\times\mathbf{v}^h+\xi^h\times\widehat{\mathbf{u}}_2^h)_{-h}\Big]\\
&\quad=\lambda^2\Big[\big((\widehat{\omega}_1^h-\widehat{\omega}_2^h)\times\mathbf{u}^h+\omega^h\times(\widehat{\mathbf{u}}_1^h-\widehat{\mathbf{u}}_2^h),\widehat{\omega}_1^h\times\mathbf{v}^h+\xi^h\times\widehat{\mathbf{u}}_1^h\big)_{-h}\\
&\qquad\qquad+\big(\widehat{\omega}_2^h\times\mathbf{u}^h+\omega^h\times\widehat{\mathbf{u}}_2^h,(\widehat{\omega}_1^h-\widehat{\omega}_2^h)\times\mathbf{v}^h+\xi^h\times(\widehat{\mathbf{u}}_1^h-\widehat{\mathbf{u}}_2^h)\big)_{-h}\Big]\\
&\quad\le C\Big[\|\widehat{\omega}_1^h-\widehat{\omega}_2^h\|_0\|\mathbf{u}^h\|_1+\|\widehat{\mathbf{u}}_1^h-\widehat{\mathbf{u}}_2^h\|_1\|\omega^h\|_0\Big]\Big[\|\widehat{\omega}_1^h\|_0\|\mathbf{v}^h\|_1+\|\xi^h\|_0\|\widehat{\mathbf{u}}_1^h\|_1\Big]\\
&\qquad+C\Big[\|\widehat{\omega}_1^h-\widehat{\omega}_2^h\|_0\|\mathbf{v}^h\|_1+\|\widehat{\mathbf{u}}_1^h-\widehat{\mathbf{u}}_2^h\|_1\|\xi^h\|_0\Big]\Big[\|\widehat{\omega}_2^h\|_0\|\mathbf{u}^h\|_1+\|\omega^h\|_0\|\widehat{\mathbf{u}}_2^h\|_1\Big]\\
&\quad\le C\|\widehat{u}_1^h-\widehat{u}_2^h\|_{X_{-1}}\|u^h\|_{X_{-1}}\|v^h\|_{X_{-1}}\big(\|\widehat{u}_1^h\|_{X_{-1}}+\|\widehat{u}_2^h\|_{X_{-1}}\big).
\end{aligned}$$

From the above results, it is easily seen that

$$\begin{aligned}
&\big|D_uQ_{-h}(u^h,v^h;\mathbf{f},\widehat{u}_1^h)-D_uQ_{-h}(u^h,v^h;\mathbf{f},\widehat{u}_2^h)\big|\\
&\qquad\le C\|\widehat{u}_1^h-\widehat{u}_2^h\|_{X_{-1}}\|u^h\|_{X_{-1}}\|v^h\|_{X_{-1}}\big(1+\|\widehat{u}_1^h\|_{X_{-1}}+\|\widehat{u}_2^h\|_{X_{-1}}\big).
\end{aligned}\tag{8.92}$$

Using (8.90), the coercivity of $Q^{\mathcal{T}}_{-h}(\cdot,\cdot)$, and (8.92), we obtain

$$\begin{aligned}
\|(D_u\mathcal{F}_{-h}(\widehat{u}_1^h)-D_u\mathcal{F}_{-h}(\widehat{u}_2^h))[u^h]\|_{X_{-1}}^2&=\|u_1^h-u_2^h\|_{X_{-1}}^2\le CQ^{\mathcal{T}}_{-h}\big(u_1^h-u_2^h,u_1^h-u_2^h\big)\\
&\le\Big[C\|\widehat{u}_1^h-\widehat{u}_2^h\|_{X_{-1}}\big(1+\|\widehat{u}_1^h\|_{X_{-1}}+\|\widehat{u}_2^h\|_{X_{-1}}\big)\Big]\Big[\|u^h\|_{X_{-1}}\|u_1^h-u_2^h\|_{X_{-1}}\Big].
\end{aligned}$$

Therefore,

$$\|\big(D_u\mathcal{F}_{-h}(\widehat{u}_1^h)-D_u\mathcal{F}_{-h}(\widehat{u}_2^h)\big)[u^h]\|_{X_{-1}}\le C\|\widehat{u}_1^h-\widehat{u}_2^h\|_{X_{-1}}\big(1+\|\widehat{u}_1^h\|_{X_{-1}}+\|\widehat{u}_2^h\|_{X_{-1}}\big)\|u^h\|_{X_{-1}}.$$

The Lipschitz bound (8.88) easily follows from this inequality. □

We now proceed to verify the assumptions of Theorem 8.2 for the nonlinear least-squares variational equation (8.61).

Lemma 8.14 *Assume that* $\widetilde{u}^h = \{\widetilde{\mathbf{u}}^h, \widetilde{\omega}^h, \widetilde{s}^h\}$ *satisfies* (8.70)–(8.72). *Then, there exists a positive constant C such that*

$$\|\mathcal{F}_{-h}(\widetilde{u}^h)\|_{X_{-1}} \leq Ch^{\widetilde{r}+1}. \tag{8.93}$$

Proof. From (8.67) and the definitions of $\mathcal{T}_{-h}$ and $\mathcal{G}_{-h}$, it follows that

$$u^h = \mathcal{F}_{-h}(\widetilde{u}^h) \equiv \widetilde{u}^h + \mathcal{T}_{-h} \circ \mathcal{G}_{-h}(\widetilde{u}^h)$$

if and only if u^h solves the variational problem: *seek* $u^h \in X^h$ *such that*

$$Q^{\mathcal{T}}_{-h}(u^h, v^h) = Q_{-h}(\widetilde{u}^h, v^h; \mathbf{f}) \qquad \forall v^h \in X^h. \tag{8.94}$$

The bilinear form $Q^{\mathcal{T}}_{-h}(\cdot,\cdot)$ in (8.94) is the same as the one associated with the norm-equivalent Stokes DLSP (7.55). Theorem 4.9 asserts that this form is strongly coercive, wheras from Lemma 8.11, we know that $Q_{-h}(\widetilde{u}^h, \cdot; \mathbf{f})$ is a continuous linear functional. From Lemma 1.4 (the Lax–Milgram Lemma) it follows that (8.94) has a unique solution u^h. Furthermore, using (4.42) from Theorem 4.9 and (8.80), we obtain

$$C\|u^h\|^2_{X_{-1}} \leq Q^{\mathcal{T}}_{-h}(u^h, u^h) = Q_{-h}(\widetilde{u}^h, u^h; \mathbf{f}) \leq Ch^{\widetilde{r}+1}\|u^h\|_{X_{-1}}.$$

The lemma follows by noting that $\|\mathcal{F}_{-h}(\widetilde{u}^h)\|_{X_{-1}} = \|u^h\|_{X_{-1}}$. □

The following lemma verifies the final assumptions of Theorem 8.2.

Lemma 8.15 *Assume that* $\widetilde{u}^h$ *satisfies* (8.70)–(8.72). *Then,*

1. *there exists* $h_0 > 0$ *such that, for all* $h < h_0$, $D_u\mathcal{F}_{-h}(\widetilde{u}^h)$ *is an isomorphism of* X^h *onto itself*
2. $\|D_u\mathcal{F}_{-h}(\widetilde{u}^h)^{-1}\|_{L(X^h,X^h)}$ *is bounded from above independently of h.*

Proof. To prove the first assertion, we need show that

$$D_u\mathcal{F}_{-h}(\widetilde{u}^h)[u^h] = w^h$$

has a unique solution $u^h \in X^h$ for every $w^h \in X^h$. Using (8.68) and the definitions of $\mathcal{T}_{-h}$ and $D_u\mathcal{G}_{-h}$, it follows that the operator form of this equation

$$\big(\mathcal{I} + \mathcal{T}_{-h} \circ D_u\mathcal{G}_{-h}(\widetilde{u}^h)\big) \cdot u^h = w^h$$

is equivalent to the following variational problem: *seek* $u^h \in X^h$ *such that*

$$D_uQ_{-h}(u^h, v^h; \mathbf{f}, \widetilde{u}^h) = Q^{\mathcal{T}}_{-h}(w^h, v^h) \qquad \forall v^h \in X^h. \tag{8.95}$$

Let h_0 be the positive real number from Lemma 8.12. Then, for w^h fixed, $Q^{\mathcal{T}}_{-h}(w^h, \cdot)$ defines a continuous linear functional on X^h and, for $h < h_0$, $D_uQ_{-h}(\cdot,\cdot; \mathbf{f}, \widetilde{u}^h)$ is continuous and coercive. As a result, existence and uniqueness of solutions to (8.95), i.e., that $D_u\mathcal{F}_{-h}(\widetilde{u}^h)$ is an isomorphism, follows from Lemma 1.4.

To prove the second assertion, without loss of generality we may assume that $h < h_0$. Recall that

$$\gamma \equiv \|D_u\mathcal{F}_{-h}(\widetilde{u}^h)^{-1}\|_{L(X^h,X^h)} = \sup_{0\neq w^h\in X^h} \frac{\|D_u\mathcal{F}_{-h}(\widetilde{u}^h)^{-1}[w^h]\|_{X_{-1}}}{\|w^h\|_{X_{-1}}}.$$

Given an arbitrary $w^h \in X^h$, let u^h denote the solution of (8.95). Using the coercivity of $D_uQ_{-h}(\cdot,\cdot;\mathbf{f},\widetilde{u}^h)$ and the continuity of $Q^{\mathcal{T}}_{-h}(\cdot,\cdot)$, we obtain

$$C_1\|u^h\|^2_{X_{-1}} \leq D_uQ_{-h}(u^h,u^h;\mathbf{f},\widetilde{u}^h) = Q^{\mathcal{T}}_{-h}(w^h,u^h) \leq C_2\|w^h\|_{X_{-1}}\|u^h\|_{X_{-1}}\,.$$

Because $u^h = D_u\mathcal{F}_{-h}(\widetilde{u}^h)^{-1}[w^h]$, this inequality implies that

$$\|D_u\mathcal{F}_{-h}(\widetilde{u}^h)^{-1}[w^h]\|_{X_{-1}} \leq C\|w^h\|_{X_{-1}}\,.$$

Thus, $\gamma \leq C$, independently of h. □

The main result of this section is as follows.

Theorem 8.16 *Assume that $\widetilde{u} = \{\widetilde{\mathbf{u}},\widetilde{\omega},\widetilde{s}\}$ is a regular solution of the Navier–Stokes equations* (8.2), (7.2), *and* (7.3) *and assume that $\widetilde{u} \in X_q$ for $q \geq 0$. Then, for h sufficiently small, there exist $\alpha > 0$ and $C > 0$ such that the LSFEM* (8.61) *has a unique solution $u^h = \{\mathbf{u}^h,\omega^h,s^h\} \in X^h$ in the ball $B(\widetilde{u},\alpha)$ and*

$$\|\widetilde{\mathbf{u}}-\mathbf{u}^h\|_1 + \|\widetilde{\omega}-\omega^h\|_0 + \|\widetilde{s}-s^h\|_0 \leq Ch^{\widetilde{r}+1}\,, \tag{8.96}$$

where $\widetilde{r} = \min\{r,q\}$. □

Proof. We apply Theorem 8.2 with $\widetilde{u}^h$ chosen to be a discrete function satisfying (8.70)–(8.72). The choice of $\widetilde{u}^h$ and (8.88) in Lemma 8.13 imply that

$$\begin{aligned}\delta(\alpha) &= \sup_{v^h\in B(\widetilde{u}^h,\alpha)} \|D_u\mathcal{F}_{-h}(\widetilde{u}^h) - D_u\mathcal{F}_{-h}(v^h)\|_{L(X^h,X^h)}\\ &\leq \sup_{v^h\in B(\widetilde{u}^h,\alpha)} \|\widetilde{u}^h - v^h\|\left(1+\|\widetilde{u}^h\|_{X_0}+\|v^h\|_{X_0}\right) \leq C\alpha\,,\end{aligned}$$

i.e., $\delta(\alpha)$ is a decreasing function of α. From (8.93) and Lemma 8.15, we know that there exists $h_1 > 0$ such that, for $h < h_1$,

$$2\gamma\delta(2\gamma\varepsilon) < 1\,.$$

Furthermore, Lemma 8.15 guarantees the existence of $h_0 > 0$ such that, for $h < h_0$, $D_u\mathcal{F}_{-h}(\widetilde{u}^h)$ is an isomorphism of X^h onto itself. As a result, all the hypotheses of Theorem 8.2 are satisfied for (8.61) and the least-squares variational equation has a unique solution in $B(\widetilde{u}^h,\alpha)$, provided $h < \min\{h_0,h_1\}$. The error estimate (8.96) follows from (8.93) by choosing $v^h = \widetilde{u}^h$ in (8.27). □

8.3.4 Compliant Discrete Least-Squares Principles for the Velocity Gradient–Velocity–Pressure System

In this section, we specialize Theorem 8.1 to the compliant DLSP (8.15) for the extended VGVP Navier–Stokes equations (8.5). Because the prototype Stokes system (7.20) of this problem is homogeneous elliptic, the key steps are quite similar to those encountered in Section 8.3.2.[17] In particular, as in Section 8.3.2, it is more convenient to use the reformulated system

$$\left\{\begin{aligned} -(\nabla\cdot\underline{\mathbf{U}})+\lambda(\underline{\mathbf{U}}+\underline{\mathbf{U}}_0)\cdot(\mathbf{u}+\mathbf{u}_0)+\nabla p &= 0 && \text{in } \Omega\\ \underline{\mathbf{U}}-(\nabla\mathbf{u})^{\mathsf{T}} &= 0 && \text{in } \Omega\\ \nabla\times\underline{\mathbf{U}} &= 0 && \text{in } \Omega\\ \nabla\cdot\mathbf{u} &= 0 && \text{in } \Omega\\ \nabla(tr\underline{\mathbf{U}}) &= 0 && \text{in } \Omega\,, \end{aligned}\right. \tag{8.97}$$

where $\underline{\mathbf{U}}_0=(\nabla\mathbf{u}_0)^{\mathsf{T}}$ and $\{\mathbf{u}_0,p_0\}$ solves the scaled Stokes problem

$$\left\{\begin{aligned} -\Delta\mathbf{u}+\nabla p &= \lambda\mathbf{f} && \text{in } \Omega\\ \nabla\cdot\mathbf{u} &= 0 && \text{in } \Omega \end{aligned}\right.$$

with the velocity boundary condition (7.2) and the zero mean pressure constraint (7.3). The functional setting for LSFEMs for (8.97) is inherited from the prototype Stokes CLSP (7.49) with the usual assumption that solution spaces are constrained by the homogeneous boundary conditions (7.2), (7.19), and the pressure constraint (7.3). This allows us to drop the boundary data space B_0 from (7.39); the rest of the spaces in (7.39) specialize to

$$X_0=[H_0^1(\Omega)]^d\times\underline{\mathbf{H}}^1_\times(\Omega)\times H^1(\Omega)\cap L_0^2(\Omega) \tag{8.98}$$

and

$$Y_0=[L^2(\Omega)]^d\times[L^2(\Omega)]^{d^2}\times[L^2(\Omega)]^{(7d-12)}\times L^2(\Omega)\times[L^2(\Omega)]^d,$$

respectively.

For the reformulated system (8.97), the CLSP (8.10) becomes

$$\left\{\begin{aligned} &J_0(\mathbf{u},\underline{\mathbf{U}},p;\mathbf{u}_0,\underline{\mathbf{U}}_0,\lambda)=\|-\nabla\cdot\underline{\mathbf{U}}+\lambda(\underline{\mathbf{U}}+\underline{\mathbf{U}}_0)\cdot(\mathbf{u}+\mathbf{u}_0)+\nabla p\|_0^2\\ &\qquad+\|\underline{\mathbf{U}}-(\nabla\mathbf{u})^{\mathsf{T}}\|_0^2+\|\nabla\times\underline{\mathbf{U}}\|_0^2+\|\nabla\cdot\mathbf{u}\|_0^2+\|\nabla(tr\underline{\mathbf{U}})\|_0^2\\ &X_0=[H_0^1(\Omega)]^d\times\underline{\mathbf{H}}^1_\times(\Omega)\times H^1(\Omega)\cap L_0^2(\Omega) \end{aligned}\right. \tag{8.99}$$

[17] Recall that the prototype Stokes system of the VVP Navier–Stokes equations (8.3) with the normal velocity–pressure boundary condition is also homogeneous elliptic.

whereas the compliant DLSP (8.15) assumes the form

$$
\begin{cases}
J_0(\mathbf{u}^h, \underline{\mathbf{U}}^h, p^h; \mathbf{u}_0, \underline{\mathbf{U}}_0, \lambda) = \\
\qquad \| - \nabla \cdot \underline{\mathbf{U}}^h + \lambda (\underline{\mathbf{U}}^h + \underline{\mathbf{U}}_0) \cdot (\mathbf{u}^h + \mathbf{u}_0) + \nabla p^h \|_0^2 + \| \underline{\mathbf{U}}^h - (\nabla \mathbf{u}^h)^{\mathsf{T}} \|_0^2 \\
\qquad\qquad + \| \nabla \times \underline{\mathbf{U}}^h \|_0^2 + \| \nabla \cdot \mathbf{u}^h \|_0^2 + \| \nabla (tr \underline{\mathbf{U}}^h) \|_0^2 \\
\\
X_{r,0}^h = [G^r]^d \cap [H_0^1(\Omega)]^d \times [G^r]^{d^2} \cap \underline{\mathbf{H}}_\times^1(\Omega) \times G^r \cap L_0^2(\Omega) .
\end{cases}
\tag{8.100}
$$

Let $u = \{\mathbf{u}, \underline{\mathbf{U}}, p\}$ and $v = \{\mathbf{v}, \underline{\mathbf{V}}, q\}$. Minimizers of (8.99) are subject to the usual first-order optimality condition

$$
\lim_{\varepsilon \to 0} \frac{d}{d\varepsilon} J_0(u + \varepsilon v; \mathbf{u}_0, \underline{\mathbf{U}}_0, \lambda) = 0 \qquad \forall v \in X_0 \tag{8.101}
$$

that assumes the familiar form: *seek* $u \in X_0$ *such that*

$$
Q(u, v; \mathbf{u}_0, \underline{\mathbf{U}}_0, \lambda) = 0 \qquad \forall v \in X_0 . \tag{8.102}
$$

The nonlinear variational form in (8.102) is given by

$$
\begin{aligned}
&Q(u, v; \mathbf{u}_0, \underline{\mathbf{U}}_0, \lambda) \\
&\quad = \big(- \nabla \cdot \underline{\mathbf{U}} + \lambda (\underline{\mathbf{U}} + \underline{\mathbf{U}}_0) \cdot (\mathbf{u} + \mathbf{u}_0) + \nabla p, \\
&\qquad\qquad - \nabla \cdot \underline{\mathbf{V}} + \lambda \left((\underline{\mathbf{U}} + \underline{\mathbf{U}}_0) \cdot \mathbf{v} + \underline{\mathbf{V}} \cdot (\mathbf{u} + \mathbf{u}_0) \right) + \nabla q \big)_0 \\
&\quad + \big(\underline{\mathbf{U}} - \nabla \mathbf{u}^{\mathsf{T}}, \underline{\mathbf{V}} - \nabla \mathbf{v}^{\mathsf{T}} \big)_0 + \big(\nabla \cdot \mathbf{u}, \nabla \cdot \mathbf{v} \big)_0 \\
&\quad + \big(\nabla (tr \underline{\mathbf{U}}), \nabla (tr \underline{\mathbf{V}}) \big)_0 + \big(\nabla \times \underline{\mathbf{U}}, \nabla \times \underline{\mathbf{V}} \big)_0 .
\end{aligned}
$$

As in Section 8.3.2, we split this form into a bilinear part

$$
\begin{aligned}
Q^{\mathcal{T}}(u, v) = &\big(- \nabla \cdot \underline{\mathbf{U}} + \nabla p, - \nabla \cdot \underline{\mathbf{V}} + \nabla q \big)_0 + \big(\underline{\mathbf{U}} - \nabla \mathbf{u}^{\mathsf{T}}, \underline{\mathbf{V}} - \nabla \mathbf{v}^{\mathsf{T}} \big)_0 \\
&+ \big(\nabla \times \underline{\mathbf{U}}, \nabla \times \underline{\mathbf{V}} \big)_0 + \big(\nabla \cdot \mathbf{u}, \nabla \cdot \mathbf{v} \big)_0 + \big(\nabla (tr \underline{\mathbf{U}}), \nabla (tr \underline{\mathbf{V}}) \big)_0
\end{aligned}
\tag{8.103}
$$

that is identical with the bilinear form associated with the prototype Stokes CLSP (7.49), and a nonlinear part

$$
\begin{aligned}
&Q^{\mathcal{G}}(u, v; \mathbf{u}_0, \underline{\mathbf{U}}_0, \lambda) = \\
&\quad \lambda \big(- \nabla \cdot \underline{\mathbf{U}} + \nabla p + \lambda (\underline{\mathbf{U}} + \underline{\mathbf{U}}_0) \cdot (\mathbf{u} + \mathbf{u}_0), (\underline{\mathbf{U}} + \underline{\mathbf{U}}_0) \cdot \mathbf{v} + \underline{\mathbf{V}} \cdot (\mathbf{u} + \mathbf{u}_0) \big)_0 \\
&\quad + \lambda \big((\underline{\mathbf{U}} + \underline{\mathbf{U}}_0) \cdot (\mathbf{u} + \mathbf{u}_0), - \nabla \cdot \underline{\mathbf{V}} + \nabla q \big)_0 .
\end{aligned}
\tag{8.104}
$$

Thus, we can write (8.102) in the following split form: *seek* $u \in X_0$ *such that*

$$
Q^{\mathcal{T}}(u, v) + Q^{\mathcal{G}}(u, v; \mathbf{u}_0, \underline{\mathbf{U}}_0, \lambda) = 0 \qquad \forall v \in X_0 . \tag{8.105}
$$

The discrete principle (8.100) is a compliant discretization of the principle (8.99). Its minimizer satisfies the same first-order optimality condition, but restricted to the finite element space $X^h_{r,0}$. Let $u^h = \{\mathbf{u}^h, \underline{\mathbf{U}}^h, p^h\}$ and $v^h = \{\mathbf{v}^h, \underline{\mathbf{V}}^h, q^h\}$. The compliant LSFEM is given by: *seek* $u^h \in X^h$ *such that*

$$\mathcal{Q}^{\mathcal{T}}(u^h, v^h) + \mathcal{Q}^{\mathcal{G}}(u^h, v^h; \mathbf{u}_0, \underline{\mathbf{U}}_0, \lambda) = 0 \qquad \forall v^h \in X^h_{r,0}. \tag{8.106}$$

Canonical form of the LSFEM

In this section, we show that (8.105) and (8.106) have the canonical forms of (8.16) and (8.17), respectively. Owing to the fact that the prototype VGVP Stokes system is homogeneous elliptic, the functional setting for this task is very similar to the one encountered in Section 8.3.2. Specifically, the abstract spaces X, Y, and V are identified with X_0, its dual X_0^*, and the space

$$V = [L^{3/2}(\Omega)]^d \times [L^{3/2}(\Omega)]^{d^2} \times L^{3/2}(\Omega),$$

respectively. Of course, the approximating space X^h is identified with the equal order finite element space $X^h_{r,0}$. We continue to use the compact notation $u = \{\mathbf{u}, \underline{\mathbf{U}}, p\}$, $u^h = \{\mathbf{u}^h, \underline{\mathbf{U}}^h, p^h\}$, $v = \{\mathbf{v}, \underline{\mathbf{V}}, q\}$, and $v^h = \{\mathbf{v}^h, \underline{\mathbf{V}}^h, q^h\}$ and assume that λ belongs to a compact subset Λ of $\mathbb{R}^+$.

As in Section 8.3.2, the linear operator $\mathcal{T}$ is defined to be the solution operator of the prototype Stokes CLSP (7.49):

$$\begin{cases} \mathcal{T}: X_0^* \mapsto X_0 \quad \text{with} \quad w = \mathcal{T}\cdot g \quad \text{for} \quad g \in X_0^* \quad \text{if and only if} \\ \qquad \mathcal{Q}^{\mathcal{T}}(w, v) = (g, v)_0 \qquad \forall v \in X_0, \end{cases} \tag{8.107}$$

where now $\mathcal{Q}^{\mathcal{T}}(\cdot,\cdot)$ is the form defined in (8.103) and

$$(g, v)_0 = (\mathbf{g}_1, \mathbf{v})_0 + (\underline{\mathbf{g}}_2, \underline{\mathbf{V}})_0 + (g_3, q)_0.$$

Likewise, $\mathcal{T}^h$ is defined to be the compliant least-squares approximation of $\mathcal{T}$:

$$\begin{cases} \mathcal{T}^h: X_0^* \mapsto X^h \quad \text{with} \quad w^h = \mathcal{T}^h\cdot g \quad \text{for} \quad g \in X_0^* \quad \text{if and only if} \\ \qquad \mathcal{Q}^{\mathcal{T}}(w^h, v^h) = (g, v^h)_0 \qquad \forall v^h \in X^h. \end{cases} \tag{8.108}$$

Clearly, both $\mathcal{T}$ and $\mathcal{T}^h$ are independent of λ, as required by the abstract theory in Section 8.3.1. The definition of the nonlinear operator $\mathcal{G}$ is also similar to that in Section 8.3.2:

$$\begin{cases} \mathcal{G}: \Lambda \times X_0 \to X_0^* \quad \text{with} \quad g = \mathcal{G}(\lambda, w) \quad \text{for} \quad w \in X_0 \quad \text{if and only if} \\ \qquad \mathcal{Q}^{\mathcal{G}}(w, v; \mathbf{u}_0, \underline{\mathbf{U}}_0, \lambda) = (g, v)_0 \quad \forall v \in X_0 \end{cases} \tag{8.109}$$

except that $Q^{\mathcal{G}}(\cdot,\cdot;\mathbf{u}_0,\underline{\mathbf{U}}_0,\lambda)$ is the form defined in (8.104). With these definitions we have the following analogue of Lemma 8.4.

Lemma 8.17 *Problems* (8.105) *and* (8.106) *have the form of* (8.16) *and* (8.17), *respectively, with $\mathcal{T}$, $\mathcal{T}^h$, and $\mathcal{G}$ defined as in* (8.107), (8.108), *and* (8.109), *respectively.*

Proof. The proof is identical to the proof of Lemma 8.4.

Approximation result

Error estimates for the LSFEM (8.106) can now be derived by using the abstract theory of Section 8.3.1. We write (8.105) and (8.106) as $\mathcal{F}(\lambda,u)=0$ and $\mathcal{F}^h(\lambda,u^h)=0$, respectively. As usual, $D_u\mathcal{G}(\lambda,u)$ and $D_u\mathcal{F}(\lambda,u)$ respectively denote the Fréchet derivative of $\mathcal{G}$ and $\mathcal{F}$ with respect to u. Recall that $\{\{\lambda,u(\lambda)\}\mid\lambda\in\Lambda\,;\,u(\lambda)\in X\}$ is a regular branch of solutions of $\mathcal{F}(\lambda,u)=0$ if $u=u(\lambda)$ solves this equation for every $\lambda\in\Lambda$, $\lambda\mapsto u(\lambda)$ is continuous map $\Lambda\mapsto X$, and $D_u\mathcal{F}(\lambda,u)$ is an isomorphism of X.

The following result is a specialization of Theorem 8.1 to the compliant LSFEM (8.106) for the VGVP Navier–Stokes equations.

Theorem 8.18 *Assume that Λ is a closed interval of $\mathbb{R}^+$ and, if $u=\{\mathbf{u},\underline{\mathbf{U}},p\}$, that $\{\{\lambda,u(\lambda)\}\mid\lambda\in\Lambda\,;\,u(\lambda)\in X\}$ is a branch of regular solutions of* (8.102). *Assume that the finite element space $X^h_{r,0}$, $r\geq 1$ is defined as in* (8.100). *Then, there exists a neighborhood $\mathcal{O}$ of the origin in X_0 and, for h sufficiently small, a unique branch $\{\{\lambda,u^h(\lambda)\}\mid\lambda\in\Lambda\,;\,u^h(\lambda)\in X^h_{r,0}\}$, with $u^h=\{\mathbf{u}^h,\underline{\mathbf{U}}^h,p^h\}$, of regular solutions of the discrete problem* (8.106) *such that $u(\lambda)-u^h(\lambda)\in\mathcal{O}$ for all $\lambda\in\Lambda$. Moreover,*

$$\|\mathbf{u}(\lambda)-\mathbf{u}^h(\lambda)\|_1+\|\underline{\mathbf{U}}(\lambda)-\underline{\mathbf{U}}^h(\lambda)\|_1+\|p(\lambda)-p^h(\lambda)\|_1\to 0 \tag{8.110}$$

as $h\to 0$, uniformly in λ. If, in addition, the solution $u(\lambda)$ of (8.105) *belongs to the space*

$$X_q=[H^{q+1}(\Omega)]^d\cap[H^1_0(\Omega)]^d\times[H^{q+1}(\Omega)]^{d^2}\cap\underline{\mathbf{H}}^1_\times(\Omega)\times H^{q+1}(\Omega)\cap L^2_0(\Omega)$$

for some $q\geq 1$, then, there exists a constant C, independent of h, such that

$$\begin{aligned}\|\mathbf{u}(\lambda)-\mathbf{u}^h(\lambda)\|_1&+\|\underline{\mathbf{U}}(\lambda)-\underline{\mathbf{U}}^h(\lambda)\|_1+\|p(\lambda)-p^h(\lambda)\|_1\\&\leq Ch^{\widetilde{r}}\left(\|\mathbf{u}(\lambda)\|_{\widetilde{r}+1}+\|\underline{\mathbf{U}}(\lambda)\|_{\widetilde{r}+1}+\|p(\lambda)\|_{\widetilde{r}+1}\right).\end{aligned}\tag{8.111}$$

where $\widetilde{r}=\min\{r,q\}$.

Proof. Because the prototype Stokes systems of (8.99) and (8.36) are homogeneous elliptic, the nonlinear terms in these functionals are products of $H^1(\Omega)$ fields, and the discrete problems (8.100) and (8.37) are compliant, the functional setting for the application of the abstract approximation theory to (8.99) is very similar to the one encountered in Section 8.3.2. As a result, the techniques in Section 8.3.2 can be easily extended to LSFEMs for the modifed VGVP Navier–Stokes equations. In particular, the proofs of Lemmas 8.5–8.7 can be adapted to show that $\mathcal{T} \in L(Y,X)$, $\mathcal{T}^h \in L(Y,X^h)$,

$$\lim_{h\to 0} \|(\mathcal{T} - \mathcal{T}^h)g\|_X = 0 \qquad \forall g \in Y$$

and

$$\lim_{h\to 0} \|\mathcal{T} - \mathcal{T}^h\|_{L(V,X)} = 0\,,$$

where $\mathcal{T}$ and $\mathcal{T}^h$ are the linear operators defined in (8.107) and (8.108), respectively, and X, Y, and V are the spaces defined in Section 8.3.4.

It is equally easy, albeit somewhat tedious, to extend the proof of Lemma 8.8 to show that the nonlinear operator $\mathcal{G}$, defined in (8.109), is a C^2 map $\Lambda \times X \mapsto Y$ such that all second derivatives of $\mathcal{G}$ are bounded on bounded subsets of $\Lambda \times X$ and $D_u\mathcal{G}(\lambda, u) \in L(X,V)$.

In summary, the CLSP (8.99) and its compliant discretization (8.100) satisfy all assumptions of the abstract theory in Section 8.3.1. As a result, all statements of the present theorem, except for the error estimate (8.111), follow directly from Theorem 8.1. The proof of (8.111), assuming sufficient solution regularity, can be obtained in the same way as in Lemma 8.6. For complete details, see [44, 68]. □

8.3.5 A Norm-Equivalent Discrete Least-Squares Principle for the Velocity Gradient–Velocity–Pressure System

In this section, we apply Theorem 8.2 to the norm-equivalent DLSP (8.14) whose parent CLSP is the negative norm principle (8.9). This CLSP inherits its functional setting from the prototype Stokes CLSP (7.48). As usual, we assume that the velocity boundary condition and the zero mean pressure constraint are imposed on the solution space in (7.37), i.e., in this section we work with

$$X_{-1} = [H^1(\Omega)]^d \times [L^2(\Omega)]^{d^2} \times L^2_0(\Omega)\,.$$

The spaces

$$X_q = [H^{q+2}(\Omega)]^d \cap [H^1_0(\Omega)]^d \times [H^{q+1}(\Omega)]^{d^2} \times H^{q+1}(\Omega) \cap L^2_0(\Omega)$$

are used to state regularity assumptions for the exact solutions of (8.9) needed in the error analysis. Clearly, the least-squares solution space, i.e., the space where we seek the minimizer of (8.9), corresponds to $q = -1$.

The approximating space in (8.14) is inherited from the prototype Stokes DLSP (7.58) with the pressure component modified to account for the zero mean constraint:

$$X^h = X^h_{\widehat{r},0} = [G^{r+1}]^d \cap [H^1_0(\Omega)]^d \times [G^{\widehat{r}}]^{d^2} \times G^{\widehat{r}} \cap L^2_0(\Omega)\,.$$

Recall that $\widehat{r} = \max\{1, r\}$ and $r \geq 0$; see Section 4.5.2.

Following Section 8.3.3, the residual of the momentum equation in (8.14) is scaled by $\lambda = \nu^{-1}$:

$$J_{-h}(\mathbf{u}^h, \underline{\mathbf{U}}^h, p^h; \mathbf{f}) = \\ \| - \nabla \cdot \underline{\mathbf{U}}^h + \lambda \underline{\mathbf{U}}^h \cdot \mathbf{u}^h + \nabla p^h - \mathbf{f} \|_{-h}^2 + \| \underline{\mathbf{U}}^h - (\nabla \mathbf{u}^h)^\mathsf{T} \|_0^2 + \| \nabla \cdot \mathbf{u}^h \|_0^2 . \tag{8.112}$$

Because the analysis in this section applies to the case of a fixed λ, this parameter is suppressed from the notation.

Let $u^h = \{\mathbf{u}^h, \underline{\mathbf{U}}^h, p^h\}$ and $v = \{\mathbf{v}^h, \underline{\mathbf{V}}^h, q^h\}$. The first-order optimality condition for the discrete minimizers of (8.14) is obtained by setting the first variation of (8.112) to zero:

$$\lim_{\varepsilon \to 0} \frac{d}{d\varepsilon} J_{-h}(u^h + \varepsilon v^h; \mathbf{f}) = 0 \qquad \forall v^h \in X^h . \tag{8.113}$$

As usual, the Euler–Lagrange equation (8.113) is equivalent to the following variational problem: *seek $u^h \in X^h$ such that*

$$Q_{-h}(u^h, v^h; \mathbf{f}) = 0 \qquad \forall v^h \in X^h , \tag{8.114}$$

where

$$Q_{-h}(u^h, v^h; \mathbf{f}) = \\ \big(- \nabla \cdot \underline{\mathbf{U}}^h + \nabla p^h + \lambda \underline{\mathbf{U}}^h \cdot \mathbf{u}^h - \mathbf{f}, - \nabla \cdot \underline{\mathbf{V}}^h + \nabla q^h + \lambda (\underline{\mathbf{U}}^h \cdot \mathbf{v}^h + \underline{\mathbf{V}}^h \cdot \mathbf{u}^h) \big)_{-h} \\ + \big(\underline{\mathbf{U}}^h - (\nabla \mathbf{u}^h)^\mathsf{T}, \underline{\mathbf{V}}^h - (\nabla \mathbf{v}^h)^\mathsf{T} \big)_0 + \big(\nabla \cdot \mathbf{u}^h, \nabla \cdot \mathbf{v}^h \big)_0 .$$

As in Section 8.3.3, some additional hypotheses are necessary regarding the solution $\{\widetilde{\mathbf{u}}, \widetilde{\underline{\mathbf{U}}}, \widetilde{p}\}$ of the VGVP system (8.4) that is being approximated by the LSFEM. First, we assume that $\{\widetilde{\mathbf{u}}, \widetilde{\underline{\mathbf{U}}}, \widetilde{p}\} \in X_q$ for $q \geq 0$ and second, we assume that $\{\widetilde{\mathbf{u}}, \widetilde{\underline{\mathbf{U}}}, \widetilde{p}\}$ is a regular solution of (8.4). The latter implies that the linearized problem

$$\begin{cases} -\nabla \cdot \underline{\mathbf{U}} + \lambda (\widetilde{\underline{\mathbf{U}}} \cdot \mathbf{u} + \underline{\mathbf{U}} \cdot \widetilde{\mathbf{u}}) + \nabla p = \mathbf{f} \\ \underline{\mathbf{U}} - (\nabla \mathbf{u})^\mathsf{T} = 0 \\ \nabla \cdot \mathbf{u} = 0 \end{cases}$$

with the boundary condition (7.2) has a unique (weak) solution $\{\mathbf{u}, \underline{\mathbf{U}}, p\} \in X_{-1}$ for every $\mathbf{f} \in [H^{-1}(\Omega)]^d$ and that $\mathbf{f} \mapsto \{\mathbf{u}, \underline{\mathbf{U}}, p\}$ is continuous from $[H^{-1}(\Omega)]^d$ into X_{-1}. As a result, for any $u \in X_{-1}$,

$$C\big(\|\mathbf{u}\|_1 + \|\underline{\mathbf{U}}\|_0 + \|p\|_0 \big) \\ \leq \| - \nabla \cdot \underline{\mathbf{U}} + \lambda (\widetilde{\underline{\mathbf{U}}} \cdot \mathbf{u} + \underline{\mathbf{U}} \cdot \widetilde{\mathbf{u}}) + \nabla p \|_{-1} + \| \underline{\mathbf{U}} - (\nabla \mathbf{u})^\mathsf{T} \|_0 + \| \nabla \cdot \mathbf{u} \|_0 . \tag{8.115}$$

It is obvious that, so far, the functional setting for the LSFEMs in this section is quite similar to the setting encountered in Section 8.3.3. There are two reasons for this similarity. First, the prototype Stokes first-order systems for the VVP formulation (8.8) and the VGVP formulation (8.9) are both non-homogeneous elliptic. This explains why solution spaces of the two CLSPs have the same structure. Second, both (8.13) and (8.14) are norm-equivalent approximations of their respective parent CLSPs, which explains the similarity in the approximation spaces.

Owing to this similarity and the similar structure of the discrete nonlinear terms in (8.13) and (8.14), the techniques in Section 8.3.3 can be easily extended to the norm-equivalent DLSP (8.14) for the first-order VGVP Navier–Stokes system (8.4). To save space, the proof of the main result that specializes Theorem 8.2 to (8.14) is omitted; for complete details, see [68].

Theorem 8.19 *Assume that* $\widetilde{u} = \{\widetilde{\mathbf{u}}, \widetilde{\underline{\mathbf{U}}}, \widetilde{p}\}$ *is a regular solution of the Navier–Stokes equations* (8.4), (7.2), *and* (7.3) *and assume that* $\widetilde{u} \in X_q$ *for* $q \geq 0$. *Then, for* h *sufficiently small, there exist* $\alpha > 0$ *and* $C > 0$ *such that the LSFEM* (8.114) *has a unique solution* $u^h \equiv \{\mathbf{u}^h, \underline{\mathbf{U}}^h, p^h\} \in X^h$ *in the ball* $B(\widetilde{u}, \alpha)$ *and*

$$\|\widetilde{\mathbf{u}} - \mathbf{u}^h\|_1 + \|\widetilde{\underline{\mathbf{U}}} - \underline{\mathbf{U}}^h\|_0 + \|\widetilde{p} - p^h\|_0 \leq Ch^{\widetilde{r}+1}, \tag{8.116}$$

where $\widetilde{r} = \min\{r, q\}$. □

8.4 Practicality Issues

Much of the discussion in Section 7.6 about practicality issues for LSFEMs for the Stokes equations obtained in the ADN setting also applies to LSFEMs for the Navier–Stokes equations. Because the analysis of the latter relies on the well-posedness of the prototype least-squares principles for the Stokes equations, all comments pertaining to this issue from Section 7.6 remain in full force for the Navier–Stokes equations. The same is true about issues related to mass conservation and the manner in which the zero mean pressure constraint is enforced.

In this section, we examine additional practicality issues for LSFEMs for the Navier–Stokes equations, two of which are also relevant to the Stokes case. Specifically, the issues discussed in this section are as follows.

- DLSPs for the Navier–Stokes equations give rise to nonlinear systems of algebraic equations that can be solved using, e.g., Newton's method. However, for high values of the Reynolds number, this may not be straightforward because it is well known that the attraction ball for Newton's method decreases in size as the Reynolds number increases. What are efficient strategies to deal with this issue and the solution of the nonlinear equations in general?
- Norm equivalent DLSPs for the Navier–Stokes (and the Stokes) equations require the computation of discrete minus one inner products. As a result, formal assembly of the discrete equations leads to dense algebraic problems. How can

one implement negative norm methods in practice and how does the choice of the discrete negative norm affect the accuracy of the approximate solutions and the iterative solution process?

- The extended VGVP system (8.5) (and likewise for the extended Stokes VGVP system (7.20)) includes equations that are obtained by differentiation of other equations. By minimizing the $L^2(\Omega)$ residuals of these equations, the CLSP (8.10) makes an implicit regularity assumption about the original primitive variables. How does the compliant DLSP (8.15) perform for problems whose solutions do not posses this regularity?

Before tackling this list of issues, let us make some general comments about LSFEMs for (8.1). Most such methods are based on the VVP formulations (8.2) or (8.3) of the Navier–Stokes equations; see [35,39,51–54,228,235–237,261,298,326]. First-order VGVP systems are studied in [36,43,44,68] and a stress-based first-order reformulation is discussed in [236]. Particular methods also differ in their choice of discretization spaces and the form of the nonlinear term. For example, the methods described in [228, 235, 237, 261] use, as a prototype least-squares principle,[18] the straightforward $L^2(\Omega)$ quasi-norm-equivalent Stokes functional and piecewise linear elements. We already commented on these choices in Sections 7.6.2 and 7.6.3, where it is demonstrated that linear elements and straightforward $L^2(\Omega)$ functionals are not appropriate for non-homogeneous elliptic operators. This conclusion also applies to the corresponding LSFEMs for the Navier–Stokes equations.

Several authors have used least-squares principles in conjunction with the p-version of the finite element method or with spectral element methods; see [28,239, 298,300,346]. Numerical results reported in these papers suggest that higher-order finite element spaces can ameliorate some of the issues plaguing the straightforward $L^2(\Omega)$ least-squares functionals, including the poor mass conservation attained with linear elements.

The solution methods for the nonlinear discrete equations documented in the literature almost exclusively employ Newton linearization in conjunction with a preconditioned conjugate gradient method along with some form of continuation with respect to the Reynolds number. Incomplete Choleski factorization preconditioners were used in [261]. Many other authors rely on Jacobi or block-Jacobi preconditioners in order to take advantage of assembly-free implementations of the conjugate gradient method. Large-scale computations and parallelization issues have been considered in [140,236,261,326–329], among others. Numerical comparisons between VVP, VSP, and VGVP first-order formulations are given in [141].

[18] The LSFEMs in [228, 235, 237] also use the standard form $(\mathbf{u} \cdot \nabla)\mathbf{u}$ of the nonlinear term. Discussion of the relative advantages and disadvantages of different forms of the nonlinear term can be found in [230].

8.4.1 Solution of the Nonlinear Equations

Newton's method

All LSFEMs discussed in this section have the form:[19] *seek* $u^h \in X^h$ *such that*

$$Q(u^h, v^h; \mathbf{f}, \lambda) = 0 \qquad \forall v^h \in X^h, \tag{8.117}$$

where $Q(\cdot,\cdot;\mathbf{f},\lambda)$ is a functional on $X^h \times X^h$ that is *nonlinear* in its first argument and linear in the second argument. In (8.117), $\mathbf{f}$ is the given body force vector and, after appropriate nondimensionalizations, the scalar constant $\lambda = \nu^{-1}$ is the Reynolds number.[20] The problem (8.117) is equivalent to a nonlinear system of algebraic equations of generic form[21]

$$\mathsf{F}(\vec{u}^h; \mathbf{f}, \lambda) = 0 \tag{8.118}$$

for the coefficient vector $\vec{u}^h$ corresponding to the unknown LSFEM solution $u^h \in X^h$.

The nonlinear system (8.118) must be solved in an iterative manner. There are many methods that one might use for such a purpose; here we only consider Newton's method. For a given initial guess u_0^h, the sequence of Newton iterates

$$\vec{u}_k^h = \vec{u}_{k-1}^h + \delta\vec{u}_k^h, \qquad k = 1,2,\ldots, \tag{8.119}$$

is generated recursively by solving, for $k = 1,2,\ldots$, the linear system

$$D_{\vec{u}}\mathsf{F}(\vec{u}_{k-1}^h; \mathbf{f}, \lambda) \cdot \delta\vec{u}_k^h = -\mathsf{F}(\vec{u}_{k-1}^h; \mathbf{f}, \lambda), \tag{8.120}$$

where $D_{\vec{u}}\mathsf{F}(\vec{u}_{k-1}^h; \mathbf{f}, \lambda)$ denotes the Jacobian matrix of $\mathsf{F}(\cdot; \mathbf{f}, \lambda)$ evaluated at $\vec{u}_{k-1}^h$. The explicit form of the system of algebraic equations (8.120) is rather formidable.

---

[19] For VVP systems, $u^h = \{\mathbf{u}^h, \omega^h, s^h\}$; for VGVP systems, $u^h = \{\mathbf{u}^h, \underline{\mathbf{U}}^h, p^h\}$.

[20] For example, consider the DLSP (8.13) for the VVP formulation of the Navier–Stokes equations. After rescaling with respect to ν and assuming that the zero mean pressure constraint is enforced on the pressure approximation space, we have, for this DLSP, $X^h = [G^{r+1}]^d \cap [H_0^1(\Omega)]^d \times [G^r]^{2d-3} \times G^r \cap L_0^2(\Omega)$ and

$$J_{-h}(\mathbf{u}^h, \omega^h, s^h; \mathbf{f}, \lambda) = \|\nabla\times\omega^h + \lambda\omega^h \times \mathbf{u}^h + \nabla s^h - \mathbf{f}\|_{-h}^2 + \|\nabla\times\mathbf{u}^h - \omega^h\|_0^2 + \|\nabla\cdot\mathbf{u}^h\|_0^2.$$

Then, with $u^h = \{\mathbf{u}^h, \omega^h, s^h\}$ and $v^h = \{\mathbf{v}^h, \xi^h, q^h\}$, (8.117) is given by

$$\begin{aligned} Q(u^h, v^h; \mathbf{f}, \lambda) = {} & \left(\nabla\times\omega^h + \lambda\omega^h\times\mathbf{u}^h + \nabla s^h - \mathbf{f}, \nabla\times\xi^h + \lambda(\omega^h\times\mathbf{v}^h + \xi^h\times\mathbf{u}^h) + \nabla q^h\right)_{-h} \\ & + \left(\nabla\times\mathbf{u}^h - \omega^h, \nabla\times\mathbf{v}^h - \xi^h\right)_0 + \left(\nabla\cdot\mathbf{u}^h, \nabla\cdot\mathbf{v}^h\right)_0. \end{aligned}$$

[21] A finite element function is denoted by, e.g., u^h, and its corresponding vector of coefficients by $\vec{u}^h$. For example, for a LSFEM for a VVP system, if $u^h = \{\mathbf{u}^h, \omega^h, p^h\} \in X^h$ denotes a finite element function, then $\vec{u}^h = \{\vec{\mathbf{u}}^h, \vec{\omega}^h, \vec{p}^h\}$ denotes the block vector whose blocks are the coefficients of the individual functions in u^h.

However, it also has some very good features.[22] The coefficient matrix in (8.120) is *symmetric,* and, in a neighborhood of a solution of (8.118), it is necessarily *positive definite,*[23] independent of the value of the Reynolds number. These valuable properties of LSFEMs can be used to

> *define algorithms for the numerical solution of the Navier–Stokes equations that encounter only* symmetric *and* positive definite *algebraic systems in the solution process.*

Solution of these systems can be accomplished by efficient iterative solvers, such as preconditioned conjugate gradient methods for which the solution of (8.118) can be implemented without assembling the matrix in (8.120). Along with the guaranteed local and quadratic convergence of Newton's method, this makes LSFEMs very attractive for large-scale computations, a feature that has been exploited in [228, 236, 261].

Solution of the linearized equations

We defined LSFEMs for the Navier–Stokes equations by extending a well-posed *prototype* least-squares principle for the Stokes equations. Besides the fact that this enabled us to prove the error estimates of this chapter, it turns out that this choice is also essential for the efficient solution of the linearized equations (8.120).

The matrix of the linear system in (8.120) is generated by a bilinear form $D_u Q(\cdot,\cdot;\mathbf{f},\lambda,u^h_{k-1})$, i.e., (8.120) is the algebraic form of the following discrete weak problem: *seek* $\delta u^h_k \in X^h$ *such that*[24]

$$D_u Q(\delta u^h_k, v^h;\mathbf{f},\lambda,u^h_{k-1}) = -Q(u^h_{k-1},v^h;\mathbf{f},\lambda) \qquad \forall v^h \in X^h. \tag{8.121}$$

---

[22] These features follow from the obvious fact that the Jacobian matrix in (8.120) is also characterized as being the Hessian matrix for the corresponding discrete least-squares functional.

[23] Here is a sketch of the argument that can be used to prove these facts for the compliant DLSP (8.37) for the VVP Navier–Stokes equations. Assume that $\{(\lambda,\widetilde{u}(\lambda)) \mid \lambda \in \Lambda\,;\, \widetilde{u} \in X\}$ is regular branch of solutions of (8.35) and (7.13) that is being approximated by solving (8.39). Then, according to Theorem 8.9, the discrete problem has a regular branch of discrete solutions $\{(\lambda,\widetilde{u}^h(\lambda) \mid \lambda \in \Lambda\,;\, \widetilde{u}^h \in X^h\}$ that is unique in a neighborhood $\mathcal{O}$ of the origin in X_0. It then follows that the Jacobian matrix $D_u\mathsf{F}(\cdot;\mathbf{f},\lambda)$ is nonsingular in a nontrivial neighborhood of $\widetilde{u}^h(\lambda)$. As a result, the attraction ball of Newton's method is nontrivial and the matrix $D_u\mathsf{F}(u^h_{k-1};\mathbf{f},\lambda)$ is guaranteed to be nonsingular, provided u^h_{k-1} is close enough to $\widetilde{u}^h(\lambda)$. Moreover, because $D_u\mathsf{F}(\cdot;\mathbf{f},\lambda)$ is the Hessian matrix of the least-squares functional in (8.37), it follows that $D_u\mathsf{F}(u^h_{k-1};\mathbf{f},\lambda)$ is necessarily symmetric and, in a neighborhood of $\widetilde{u}^h(\lambda)$, positive definite.

[24] For the norm-equivalent DLSP (8.13), we have that

$$\begin{aligned}
D_u Q_{-h}(u^h,v^h;\mathbf{f},u^h_{k-1}) &= \left(\nabla\times\omega^h_{k-1}+\lambda\omega^h_{k-1}\times\mathbf{u}^h_{k-1}+\nabla s^h_{k-1}-\mathbf{f},\lambda(\omega^h\times\mathbf{v}^h+\xi^h\times\mathbf{u}^h)\right)_{-h}\\
&+\left(\nabla\times\omega^h+\lambda(\omega^h_{k-1}\times\mathbf{u}^h+\omega^h\times\mathbf{u}^h_{k-1})+\nabla s^h,\nabla\times\xi^h+\lambda(\omega^h_{k-1}\times\mathbf{v}^h+\xi^h\times\mathbf{u}^h_{k-1})+\nabla q^h\right)_{-h}\\
&+(\nabla\times\mathbf{u}^h-\omega^h,\nabla\times\mathbf{v}^h-\xi^h)_0+(\nabla\cdot\mathbf{u}^h,\nabla\cdot\mathbf{v}^h)_0.
\end{aligned}$$

From the approximation theory in this chapter it follows that if u^h_{k-1} is sufficiently close to a regular solution of the Navier–Stokes equations, then $D_u Q(\cdot,\cdot;\mathbf{f},\lambda,u^h_{k-1})$ has the same spectral equivalence properties as the bilinear form $Q^{\mathcal{T}}(\cdot,\cdot)$ from the prototype least-squares principle for the Stokes equations. As a result, for such functions u^h_{k-1} we have that[25]

$$C^{-1}\|v^h\|_X^2 \le D_u Q(v^h,v^h;\mathbf{f},\lambda,u^h_{k-1}) \le C\|v^h\|_X^2 \qquad \forall v^h \in X^h. \tag{8.122}$$

This means that the matrices in (8.120) are spectrally equivalent to exactly the same matrix as the matrix Q generated by the Stokes form $Q^{\mathcal{T}}(\cdot,\cdot)$.

In particular, for LSFEMs whose prototype Stokes principle is one of the compliant DLSPs (7.50), (7.52), or (7.59), the matrix in (8.120) is spectrally equivalent to one of the matrices defined in (7.70) so that this system can be preconditioned by the same preconditioners one would use for the discretized Stokes equations.[26]

For LSFEMs whose parent Stokes principle is one of the norm-equivalent DLSPs (7.55), (7.57), or (7.58), the matrix in (8.120) is spectrally equivalent to one of the matrices in (7.71). Again, this means that to precondition the linearized system (8.120), we can use exactly the same preconditioners as for the corresponding systems originating from the parent Stokes principles. In this case, the appropriate preconditioners are generated from the matrix $\widetilde{\mathsf{K}}_{-h}$ defined in (4.51).

Continuation methods

The need to incorporate continuation or some like strategy into a solution method for the discrete nonlinear problem corresponding to a LSFEM for the Navier–Stokes equations arises for the following reasons.[27] First, as the Reynolds number increases, the size of the attraction ball for Newton's method decreases. Second, the positive definiteness of the Hessian matrix is guaranteed only in a neighborhood of the minimizer. As a result, for an arbitrary initial guess, we may have that Newton's method does not converge and/or that the coefficient matrix in (8.120) is not positive definite.[28] In order to guarantee that the initial guess is within the attraction ball of

---

[25] For instance, for the VVP formulation (8.61), the argument goes as follows. If u^h_{k-1} is sufficiently close to $\widetilde{u}$, the form $D_u Q(\cdot,\cdot;\mathbf{f},u^h_{k-1})$ is coercive (see Lemma 8.12) and

$$C_1\|v^h\|_{X_{-1}}^2 \le D_u Q(v^h,v^h;\mathbf{f},u^h_{k-1}) \le C_2\|v^h\|_{X_{-1}}^2 \qquad \forall v^h \in X^h.$$

[26] Recall that these preconditioners are given by the block diagonal matrix $\widetilde{\mathsf{K}}$ from (4.25) and that the number of Poission preconditioner blocks T in that matrix depends on the total number of the solution components for (7.50), (7.52), and (7.59).

[27] Of course, the need to invoke a continuation or some other like method is necessary for other discretization methods for the Navier–Stokes equations as well.

[28] The symmetry property of the Hessian matrix always holds.

Newton's method and that the coefficient matrix in (8.120) is positive definite, one can use continuation or homotopy methods, among others.

A simple continuation method for (8.118) can be defined as follows; see e.g. [246, 305]. Let $\lambda = \nu^{-1}$ denote the target Reynolds number. Consider a sequence of increasing Reynolds numbers $\{\lambda_m\}_{m=1}^{M}$ with $\lambda_M = \lambda$. Let $\vec{u}^{m,h}$ denote solution of system (8.118) for λ_m. For any m, this solution is obtained by solving the sequence of linear systems (8.120) for $k = 1, 2, \ldots$ and a given initial approximation $\vec{u}_0^{m,h}$. To start the continuation procedure, we can choose λ_1 to be sufficiently small, e.g., $\lambda_1 = 1$, so that the iteration (8.120) converges if $\vec{u}_0^{1,h}$ is defined to be the solution of the prototype Stokes problem used to define the LSFEM for Navier–Stokes equations. The remaining initial guesses $\vec{u}_0^{m,h}$, $m = 2, 3, \ldots$, can be determined by "continuing along the tangent," i.e., by solving the linear system

$$D_{\vec{u}}\mathsf{F}(\vec{u}^{m-1,h};\mathbf{f},\lambda_{m-1}) \cdot \left(\vec{u}_0^{m,h} - \vec{u}^{m-1,h}\right) = -(\lambda_m - \lambda_{m-1}) D_\lambda \mathsf{F}(\vec{u}^{m-1,h};\mathbf{f},\lambda_{m-1})$$

or by even the simpler "continuation along a constant" method

$$\vec{u}_0^{m,h} = \vec{u}^{m-1,h}.$$

The combined Newton-continuation method is now completely defined by (8.120) and the choice of the continuation strategy. Because Newton's method is guaranteed to be locally convergent and because the neighborhood of a minimizer where the Hessian matrix is positive definite is also nontrivial, one can guarantee, by choosing $\lambda_m - \lambda_{m-1}$ sufficiently small, that the Newton-continuation method only encounters symmetric and positive definite matrices.

8.4.2 Implementation of Norm-Equivalent Methods

We have considered two norm-equivalent DLSPs for the Navier–Stokes equations: (8.13) was based on the VVP formulation (8.2) and (8.14) on the VGVP formulation (8.4). Compared to quasi-norm-equivalent DLSPs and compliant DLSPs, implementation of discrete negative norm methods is not so straightforward.

There are two main issues to consider here. First, how does one choose the operator $\mathcal{K}^h$ and the parameter α in the discrete negative norm definition (B.99). Second, how does one evaluate the discrete minus one inner product (B.100). In Section B.4.2, we indicate that any good preconditioner for the Poisson equation can be used for $\mathcal{K}^h$. However, such preconditioners usually come in "black-box" form rather than in a closed matrix form,[29] so that standard assembly procedures are not applicable.

We discuss the implementation of discrete negative norm methods using (8.14) as a prototype problem. Almost everything that is discussed in the context of this

[29] Even if we could somehow find the matrix equivalent of the "black-box," it may very likely be a dense matrix because $\mathcal{K}^h$ approximates the Galerkin solution operator for the Poisson equation.

DLSP readily applies to discrete negative norm LSFEMs based on the VVP system; see [39] for details. We choose (8.14) because, in Section 8.4.4, this method is used to illustrate some other issues that are specific to the VGVP formulation.

The nonlinear problem is solved using Newton's method as formulated in (8.119) and (8.120). However, owing to the use of a discrete minus one inner product, the assembly of $D_{\vec{u}}\mathsf{F}(\vec{u}^h_{k-1};\mathbf{f},\lambda)$ is not practical, either because $\mathcal{K}^h$ is not known explicitly or because it is a dense matrix. This means that iterative methods are the only option to solve (8.120). For reasons explained in Section 8.4.1, the matrix $D_{\vec{u}}\mathsf{F}(\vec{u}^h_{k-1};\mathbf{f},\lambda)$ is guaranteed to be symmetric and positive definite in some neighborhood of the exact solution. Thus, (8.120) can be solved by a preconditioned conjugate gradient method. Furthermore, that matrix is also guaranteed to be spectrally equivalent to the same matrix as the matrix from the prototype Stokes principle. Therefore, (8.120) can be preconditioned by the same preconditioners as one would use for the prototype least-squares principle (7.58). Recall that the matrix Q of the prototype principle is spectrally equivalent to the matrix $\mathsf{K}^{\text{VGVP}}_{-h}$ from (7.71). Therefore, the matrix

$$\widetilde{\mathsf{K}}_{-h} = \begin{pmatrix} \mathsf{T} & 0 & 0 \\ 0 & h^d \mathsf{I}_{\underline{\mathbf{U}}} & 0 \\ 0 & 0 & h^d \mathsf{I}_p \end{pmatrix}, \tag{8.123}$$

where T is formally identified with the "inverse" of $\mathcal{K}^h$, is spectrally equivalent to $D_{\vec{u}}\mathsf{F}(\vec{u}^h_{k-1};\mathbf{f},\lambda)$ and can be used to precondition (8.120).

Let us sketch an implementation of the conjugate gradient method, preconditioned by (8.123), for (8.120). The key steps at each iteration are the computation of the product of $D_{\vec{u}}\mathsf{F}(\vec{u}^h_{k-1};\mathbf{f},\lambda)$ with the conjugate direction vector and the application of the preconditioner.

The first step can be implemented as follows. Let $\{v^h_i\}_{i=1}^N$ with $v^h_i = \{\mathbf{v}^h_i, \underline{\mathbf{V}}^h_i, q^h_i\}$ denote a standard basis for $X^h = X^h_{\widehat{r},0}$. Then, the i,j element of the Jacobian matrix $D_{\vec{u}}\mathsf{F}(\vec{u}^h_{k-1};\mathbf{f},\lambda)$ is given by

$$[D_{\vec{u}}\mathsf{F}(\vec{u}^h_{k-1};\mathbf{f},\lambda)]_{ij} = D_u Q_{-h}(v^h_j, v^h_i;\mathbf{f}, u^h_{k-1}), \tag{8.124}$$

where $D_u Q_{-h}(\cdot,\cdot;\mathbf{f},u^h_{k-1})$ is the linearization of the form in (8.114) evaluated at u^h_{k-1}. Therefore, the elements of the matrix-vector product $D_{\vec{u}}\mathsf{F}(\vec{u}^h_{k-1};\mathbf{f},\lambda)\cdot\vec{u}^h$ can be computed according to the formula

$$[D_{\vec{u}}\mathsf{F}(\vec{u}^h_{k-1};\mathbf{f},\lambda)\cdot\vec{u}^h]_i = D_u Q_{-h}(u^h, v^h_i;\mathbf{f}, u^h_{k-1}),$$

where, as always, u^h is the finite element function corresponding to the coefficient vector $\vec{u}^h$.

For the norm-equivalent DLSP (8.14) we are considering here, $Q_{-h}(\cdot,\cdot;\mathbf{f})$ is the form in (8.114) so that, if $u^h = \{\mathbf{u}^h, \underline{\mathbf{U}}^h, p^h\} \in X^h$, $v^h_i = \{\mathbf{v}^h_i, \underline{\mathbf{V}}^h_i, q^h_i\} \in X^h$, and $u^h_{k-1} = \{\mathbf{u}^h_{k-1}, \underline{\mathbf{U}}^h_{k-1}, p^h_{k-1}\} \in X^h$, we have that

$$
\begin{aligned}
&D_u Q_{-h}(u^h, v^h; \mathbf{f}, u^h_{k-1}) = \\
&\quad \lambda\big(-\nabla\cdot\underline{\mathbf{U}}^h_{k-1} + \nabla p^h_{k-1} + \lambda\underline{\mathbf{U}}^h_{k-1}\cdot\mathbf{u}^h_{k-1} - \mathbf{f},\ \underline{\mathbf{U}}^h\cdot\mathbf{v}^h_i + \underline{\mathbf{V}}^h_i\cdot\mathbf{u}^h\big)_{-h} \\
&\quad +\big(-\nabla\cdot\underline{\mathbf{U}}^h + \nabla p^h + \lambda(\underline{\mathbf{U}}^h_{k-1}\cdot\mathbf{u}^h + \underline{\mathbf{U}}^h\cdot\mathbf{u}^h_{k-1}), \\
&\qquad\qquad -\nabla\cdot\underline{\mathbf{V}}^h_i + \nabla q^h_i + \lambda(\underline{\mathbf{U}}^h_{k-1}\cdot\mathbf{v}^h + \underline{\mathbf{V}}^h_i\cdot\mathbf{u}^h_{k-1})\big)_{-h} \\
&\quad + \big(\underline{\mathbf{U}}^h - (\nabla\mathbf{u}^h)^\mathsf{T}, \underline{\mathbf{V}}^h_i - (\nabla\mathbf{v}^h_i)^\mathsf{T}\big)_0 + \big(\nabla\cdot\mathbf{u}^h, \nabla\cdot\mathbf{v}^h_i\big)_0 .
\end{aligned}
\tag{8.125}
$$

The bilinear form (8.125) has two terms containing the discrete minus one inner product $(\cdot,\cdot)_{-h}$. According to (B.100), their computation requires the application of $\mathcal{K}^h$ to a suitably defined vector. In the implementation of the LSFEM, $\mathcal{K}^h$ is treated as a black-box algorithm. This black-box can represent several multigrid V-cycles, an algebraic multigrid algorithm, or, more generally, any approximation scheme for the Poisson equation. With these assumptions, the evaluation of, e.g., the first term in (8.125), can be implemented as follows. First, using (B.100), we have that

$$
\begin{aligned}
&\big(-\nabla\cdot\underline{\mathbf{U}}^h_{k-1} + \nabla p^h_{k-1} + \lambda\underline{\mathbf{U}}^h_{k-1}\cdot\mathbf{u}^h_{k-1} - \mathbf{f},\ \lambda(\underline{\mathbf{U}}^h\cdot\mathbf{v}^h_i + \underline{\mathbf{V}}^h_i\cdot\mathbf{u}^h)\big)_{-h} \\
&\quad = \big((\alpha h^2\mathcal{I} + \mathcal{K}^h)\mathbf{g}^h,\ \lambda(\underline{\mathbf{U}}^h\cdot\mathbf{v}^h_i + \underline{\mathbf{V}}^h_i\cdot\mathbf{u}^h)\big)_0 \\
&\quad = \alpha h^2\big(\mathbf{g}^h,\ \lambda(\underline{\mathbf{U}}^h\cdot\mathbf{v}^h_i + \underline{\mathbf{V}}^h_i\cdot\mathbf{u}^h)\big)_0 + \big(\mathcal{K}^h\mathbf{g}^h,\ \lambda(\underline{\mathbf{U}}^h\cdot\mathbf{v}^h_i + \underline{\mathbf{V}}^h_i\cdot\mathbf{u}^h)\big)_0,
\end{aligned}
$$

where

$$
\mathbf{g}^h = -\nabla\cdot\underline{\mathbf{U}}^h_{k-1} + \nabla p^h_{k-1} + \lambda\underline{\mathbf{U}}^h_{k-1}\cdot\mathbf{u}^h_{k-1} - \mathbf{f}.
$$

The term multiplied by αh^2 is a weighted $L^2(\Omega)$-inner product that can be computed in a standard manner. To compute the second term we need to apply $\mathcal{K}^h$ to $\mathbf{g}^h$.

Because $\mathcal{K}^h$ is an approximation of the Galerkin solution operator $\mathcal{S}^h$ for the Poisson equation, we can generate the components of $\vec{\mathbf{g}}^h$, the vector corresponding to the function projection of $\mathbf{g}^h$ onto the velocity approximating space, in the same way as one would generate the source term for the Galerkin method. If $\{\mathbf{v}^h_i\} \in [G^r \cap H^1_0(\Omega)]^d$ is a basis set for the velocity approximation space, we then have that the vector $\vec{\mathbf{g}}^h$ has the ith component $(\mathbf{g}^h, \mathbf{v}^h_i)_0$. This vector is the datum for the black-box evaluator[30] of $\mathcal{K}^h$. Upon completion, this black-box returns the coefficients of the finite element function $\widetilde{\mathbf{g}}^h = \mathcal{K}^h\mathbf{g}^h$. Thus,

$$
\big(\mathcal{K}^h\mathbf{g}^h, \lambda(\underline{\mathbf{U}}^h\cdot\mathbf{v}^h_i + \underline{\mathbf{V}}^h_i\cdot\mathbf{u}^h)\big)_0 = \big(\widetilde{\mathbf{g}}^h, \lambda(\underline{\mathbf{U}}^h\cdot\mathbf{v}^h_i + \underline{\mathbf{V}}^h_i\cdot\mathbf{u}^h)\big)_0.
$$

The right-hand side term of this expression is an $L^2(\Omega)$ inner product that can be computed in a standard manner. The remaining minus one discrete inner product terms are computed in the same way.

The action of the preconditioner in (8.123) can be implemented using the same approach. Given a vector $\vec{v}^h$ corresponding to a finite element function $v^h = (\mathbf{v}^h, \underline{\mathbf{V}}^h, q^h)$, application of $\widetilde{\mathsf{K}}_{-h}$ to $\vec{v}^h$ is straightforward, except for the first block

[30] If, for whatever reason, algebraic or other multigrid algorithms cannot be used in the implementation, one could design the black-box evaluator by solving the equation (B.97) that defines $\mathcal{S}^h$ (see Section B.4.2) approximately using few conjugate gradient iterations.

where we need to compute the action of $\mathcal{K}^h$ on the velocity vector $\vec{\mathbf{v}}^h$. We can generate the data vector for the black-box in the same way as before, i.e., by treating $\mathbf{v}^h$ as a source term for (B.97). Thus, we first form the data vectors with ith component $(\mathbf{v}^h, \mathbf{v}_i^h)_0$ and then apply the black-box to this vector.

8.4.3 The Utility of Discrete Negative Norm Least-Squares Finite Element Methods

Implementation of discrete negative norm methods is rather unorthodox and more complex than that of weighted $L^2(\Omega)$ norm least-squares. Thus, it is important to know if the additional effort can be offset by advantages over weighted methods in other spheres; it is also important to know just how robust negative norm methods are in practice.

Negative norm functionals have two theoretical advantages over their weighted counterparts. First, because they are norm-equivalent, the resulting algebraic equations have natural preconditioners given by (4.51). In contrast, weighted functionals have mesh-dependent energy balances (see Theorem 4.7), which means that weighted methods do not have such natural preconditioners.

A second advantage has to do with the minimal approximation condition required of DLSPs for non-homogeneous elliptic systems. The DLSP (8.14) is a discretization of (8.9) whose prototype Stokes CLSP (7.48) is associated with a non-homogeneous elliptic system. According to Remark 4.11, a quasi-norm-equivalent DLSP for such systems is subject to a minimal approximation condition for the velocity that states that this variable has to be approximated by at least quadratic (or biquadratic or triquadratic) finite elements. Numerical experiments in Section 7.6.2 showed that this condition is real, i.e., it is not just a theoretical expedient, and its violation reduces the accuracy of the LSFEM. According to the same remark, for norm-equivalent DLSPs, the minimal approximation condition for the velocity drops to linear (or bilinear or trilinear) elements. Thus, in theory, negative norm functionals can rehabilitate lower-order elements for non-homogeneous elliptic systems. Because of the popularity of linear and bilinear elements, if this property also holds in practice, it would increase the appeal of LSFEMs based on norm-equivalent DLSPs for such problems.

To see whether these advantages also hold in practice, we implement (8.14) in two-dimensions, using *bilinear* finite elements. If a bilinear implementation of (8.14) achieves the theoretical rates of convergence, this would mean that discrete negative norms do help to reduce the minimal approximation condition not only in theory, but also in practice. We also focus on the performance of the conjugate gradient methods preconditioned by (8.123).

By varying the spectral equivalence of $\mathcal{K}^h$ and the value of α, we can simultaneously examine the performance of iterative solvers and the robustness of negative norm methods, i.e., find the answer to the second question. To do this, $\mathcal{K}^h$ is evaluated by solving (B.97) by conjugate gradient methods with several choices for the convergence tolerance. A large value of the tolerance gives an operator $\mathcal{K}^h$ that is "less" spectrally equivalent to the Galerkin solution operator $\mathcal{S}^h$. A tighter tolerance gives an $\mathcal{K}^h$ that has better spectral equivalence. Then, we fix a reasonably good $\mathcal{K}^h$ and start changing the value of α.

To estimate convergence rates, we take Ω to be the unit square and use the following manufactured solution:

$$\begin{cases} \mathbf{u} = \big(\exp(x)\cos(y) + \sin(y), -\exp(x)\sin(x) + (1-x^3)\big)^T \\ \underline{\mathbf{U}} = \nabla \mathbf{u}^T \\ p = \sin(y)\cos(x) + xy^2 - \dfrac{1}{6} - \sin(1)(1-\cos(1)) . \end{cases} \tag{8.126}$$

Because the exact solution is a fictitious flow that does not depend on Re, all computations are carried with $Re = 1$. Convergence rates are estimated by solving the discrete equations on uniform grids with 17×17 and 33×33 grid lines, respectively.

Recall that the use of equal-order bilinear interpolation for all variables is acceptable for norm-equivalent functionals; see Remark 4.11. Therefore, the conclusions of Theorem 8.19 remain valid when $r = 0$ and $X^h_{\hat{r},0}$ is replaced by an equal-order space defined according to (4.44). In this case, the error estimate (8.116) specializes to

$$\|\mathbf{u} - \mathbf{u}^h\|_1 + \|\underline{\mathbf{U}} - \underline{\mathbf{U}}^h\|_0 + \|p - p^h\|_0 \leq Ch . \tag{8.127}$$

Let us see how changing $\mathcal{K}^h$ and α affects the validity of (8.127). In theory, this estimate should hold as long as $\mathcal{K}^h$ remains spectrally equivalent to $\mathcal{S}^h$ and α is positive (otherwise (B.99) reduces to the semi-norm (B.95)).

To assess the importance of (B.98) in the definition of the negative norm, we fix $\alpha = 0.2$ (this choice becomes clear shortly) and compute numerical rates of convergence for varying operators $\mathcal{K}^h$. As explained, we vary the degree of spectral equivalence of $\mathcal{K}^h$ by changing the tolerance with which we solve (B.97). The results of these experiments are summarized in Table 8.1. To find out how (8.127) is affected by the choice of α, the tolerance of the conjugate gradient method used for the evaluation of $\mathcal{K}^h$ is set to 10^{-5} and computations are carried with $\alpha = 0$, $\alpha = 0.2$, and $\alpha = 1$. The results are presented in Table 8.2. The rates asserted by the theoretical estimate (8.127) are shown in Tables 8.1 and 8.2 in boldface. As usual, the column marked by "BA" shows the best approximation error out of the finite element space.

From the rates reported in Table 8.1, we can conclude that (B.98) is quite important to the accuracy of the LSFEM. The first two columns in that table show a complete loss of convergence when $\mathcal{K}^h$ is calculated with a loose tolerance. When the tolerance is tightened, i.e., the adherence of $\mathcal{K}^h$ to (B.98) is improved, we see that LSFEM solutions converge as predicted by (8.127). Among other things, the match between theoretical and actual convergence rates affirms that *bilinear* (or *linear*) *elements do work* in a norm-equivalent DLSP.

The data in Table 8.2 suggests that the choice of α is less important to the accuracy of the method than the spectral equivalence condition (B.98). When $\alpha = 0$,

Δ^{-1} tolerance →	10^{-1}	10^{-2}	10^{-3}	10^{-4}	10^{-5}	BA
Variable↓	$L^2(\Omega)$-error rate					
u	-0.614	0.370	1.861	1.869	1.869	–
$\underline{\mathbf{U}}$	**-0.717**	**-0.052**	**1.535**	**1.586**	**1.587**	**1.000**
p	**-1.261**	**-0.066**	**1.559**	**1.580**	**1.584**	**1.000**
Variable↓	H^1-error rate					
u	**-0.632**	**0.791**	**1.014**	**1.016**	**1.016**	**1.000**
$\underline{\mathbf{U}}$	-1.110	-0.349	0.652	0.665	0.666	–
p	-1.439	-0.622	0.677	0.707	0.708	–

Table 8.1 Convergence rates of the norm-equivalent DLSP (8.14) for $\alpha = 0.2$ and varying tolerances in the conjugate gradient approximation of $\mathcal{K}^h$ viewed as an approximation to the Δ^{-1}, i.e., in the evaluation of the discrete minus one norm.

	$L^2(\Omega)$-error rate				H^1-error rate			
α	0	0.2	1	BA	0	0.2	1	BA
u	1.850	1.869	1.873	-	**1.018**	**1.016**	**1.015**	**1.000**
$\underline{\mathbf{U}}$	**1.559**	**1.587**	**1.605**	**1.000**	0.585	0.666	0.701	-
p	**1.219**	**1.584**	**1.584**	**1.000**	0.134	0.708	0.686	-

Table 8.2 Convergence rates of the norm-equivalent DLSP (8.14) with a fixed tolerance for $\mathcal{K}^h$ and varying α.

the only significant drop occurs in the asymptotic rate for the H^1-norm error of the pressure. However, this error is not included in (8.127) and, as a result, the error in the norm of X_{-1}, that is, $\|\mathbf{u}-\mathbf{u}^h\|_1 + \|\underline{\mathbf{U}}-\underline{\mathbf{U}}^h\|_0 + \|p-p^h\|_0$, remains of order $O(h)$.

The next issue of interest is how changes in $\mathcal{K}^h$ and α influence the convergence of the conjugate gradient method used to solve the linearized equations. The value of α defines the balance between the weighted $L^2(\Omega)$ norm and the discrete negative semi-norm terms in (B.99). By taking larger α, we can make the linear system appear more like a system generated by a weighted $L^2(\Omega)$, quasi-norm-equivalent DLSP. From Theorem 4.8, we know that for such DLSPs the condition number is bounded by $O(h^{-4})$. Thus, for larger α, we expect to see the need for more conjugate gradient iterations in order to converge to the same tolerance.

The situation with $\mathcal{K}^h$ is a little bit more delicate because we use the same operator to define both the negative norm and the preconditioner (8.123). Because now we are only interested in the effect of $\mathcal{K}^h$ as a preconditioner, in what follows we assume that the $\mathcal{K}^h$ in (B.99) is fixed and use $\mathcal{K}^h_{CG}$ to denote the operator employed in (8.123).

To determine the importance of α for the preconditioned conjugate gradient method used to solve the discretized linearized equations, we fix the tolerances for $\mathcal{K}^h$ and $\mathcal{K}^h_{CG}$ equal to 10^{-5}. Then, we compare the number of preconditioned conjugate gradient iterations needed to achieve the same relative error tolerance in the solution of the linear system for values of α between 0 and 1. Figure 8.1 shows the numerical results for 9×9, 17×17, and 33×33 uniform grids. From these plots, we conclude that unless $\alpha > 0$, performance of the preconditioned conjugate gradient method degrades significantly. Indeed, for all three grids, $\alpha = 0$ results in the highest number of iterations (101, 133, and 174, respectively), whereas taking even a small positive α, e.g., $\alpha = 0.05$, helps to reduce the number of iterations by a factor greater than two. Even more important, when $\alpha = 0$, we see that the number of iterations grows as the number of grid lines increases, i.e., convergence of the preconditioned conjugate gradient method is not independent of h. This behavior can be explained by noting that setting $\alpha = 0$ in (8.14) gives a least-squares functional equivalent to using the semi-norm (B.95) for which the lower bound in (B.101) does not hold.

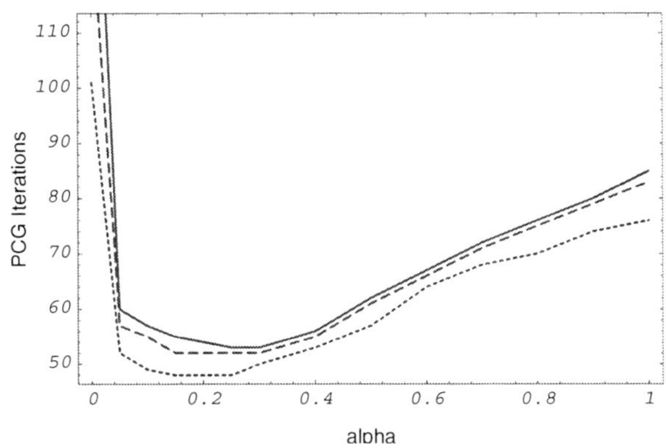

Fig. 8.1 Dependence on α of number of iterations needed in the preconditioned conjugate gradient method used to solve the discrete linearized Newton system. The solid, dashed, and dotted lines correspond to 33×33, 17×17, and 9×9 uniform grids, respectively.

From Figure 8.1, it also appears that for $0.1 < \alpha < 0.6$ the convergence of the preconditioned conjugate gradient method is independent of the grid size h, with the fastest convergence occurring in the range $0.2 \le \alpha \le 0.3$. The optimal value of α seems to depend mildly on the grid size; for the 9×9 grid it is $\alpha \approx 0.2$ whereas for the 33×33 grid it is $\alpha \approx 0.3$. The ability of (8.123) to provide convergence behavior that is independent of h for these values of α is also illustrated by the data in Table 8.3. This table compares the number of conjugate gradient iterations without preconditioning, using a Jacobi preconditioner, and using (8.123).

Lastly, to determine how the level of closeness $\mathcal{K}^h_{CG}$ has to the Galerkin–Poisson solution operator $\mathcal{S}^h$ affects the preconditioner, we fix $\alpha = 0.2$ and vary the tolerance in the discrete Poisson solver used to compute $\mathcal{K}^h_{CG}$. The results are summarized in Table 8.4. The data in that table suggest that the performance of the preconditioner (8.123) is not very sensitive with respect to the quality of $\mathcal{K}^h_{CG}$ and that the overall

	Grid size			
Preconditioner	5×5	9×9	17×17	33×33
none	122	220	419	759
Jacobi	37	52	102	185
(8.123)	44	48	52	53

Table 8.3 Number of conjugate gradient iterations with different preconditioners for the solution of the discrete linearized Newton system; $\alpha = 0.2$ and the tolerance for determining $\mathcal{K}^h$ is 10^{-5}.

	Δ^{-1} tolerance				
Grid	5×10^{-1}	1×10^{-1}	1×10^{-3}	1×10^{-4}	1×10^{-5}
9×9	137	49	48	48	48
17×17	258	56	52	52	52

Table 8.4 Number of preconditioned conjugate gradient iterations for $\alpha = 0.2$ and for different choices for the tolerance of the conjugate gradient iteration used to determine $\mathcal{K}^h_{CG}$.

performance of the conjugate gradient method for the solution of the discrete linearized Newton system depends more critically on α.

Driven cavity flow

A less quantitative test for the bilinear finite element implementation of (8.14) is to solve the two-dimensional lid-driven cavity problem. Even though the exact solution of this problem is not known in closed form, it is often used as a standard benchmark for Navier–Stokes solvers. As a result, there are many numerical studies available for comparison, including the very through numerical study of [180].

For the two-dimensional driven cavity flow problem, Ω is the unit square, the body force $\mathbf{f} = (0,0)^\top$, and $\mathbf{u} = (1,0)^\top$ on the top wall of Ω and zero on all other parts of the boundary. We computed the driven cavity flow for $Re = 100$ using 17×17, 33×33, and 45×45 uniform grids. Figure 8.2 shows the velocity field and vorticity contours for the 33×33 grid. The vorticity is computed from $\underline{\mathbf{U}}^h$ using the formula $\omega^h = \underline{\mathbf{U}}^h_{21} - \underline{\mathbf{U}}^h_{12}$. In the "eyeball" norm, these plots appear to be in good agreement with other studies, including [180].

A somewhat more quantitative measure for the quality of the LSFEM solution for the driven cavity problem is presented in Figure 8.3. The velocity profiles through the geometrical center of the cavity are compared to the the benchmark results of [180] obtained using a finite difference scheme on 129×129 uniform grid. The first velocity component u is plotted along the vertical line $x = 0.5$. The second velocity component v is plotted along the horizontal line $y = 0.5$. From the plots in Figure 8.3, we see that u is very close to the benchmark data, even for the fairly coarse 17×17 grid. The accuracy of the second velocity component on this grid is not as

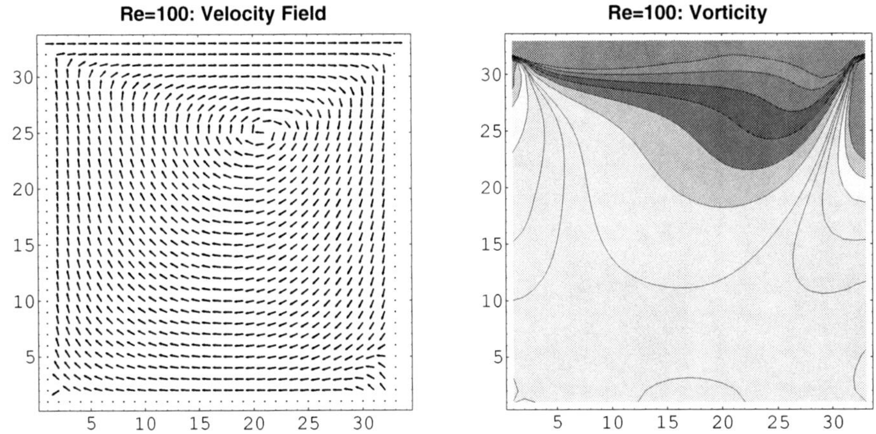

Fig. 8.2 Solution of the driven cavity flow problem for $Re = 100$ by the negative norm LSFEM (8.14) implemented with bilinear elements. The grid size is 33×33. Left: the velocity field; right: vorticity contors.

good. However, after increasing the grid size to 33×33 and then to 45×45, we see that v improves significantly and becomes very close to the benchmark results.

Earlier we observed that setting $\alpha = 0$ in the definition of the minus one inner product (B.99) affected the pressure variable most. This also turns out to be true for the driven cavity example. Figure 8.4 shows the onset of spurious oscillations in the pressure when $\alpha = 0$. However, even a small positive α is sufficient to restore the stability.

8.4.4 Advantages and Disadvantages of Extended Systems

The system (8.4) is associated with a non-homogeneous elliptic operator; we saw that for such problems, straightforward $L^2(\Omega)$ LSFEMs do not work well. This is caused by the fact that in non-homogeneous elliptic systems, there are two distinct groups of variables that have different differentiability properties. Straightforward $L^2(\Omega)$ functionals, on the other hands, implicitly treat all variables as if they have the same differentiability, i.e., as if the system were homogeneous elliptic.

The extended system (8.5) is designed to fix this problem by changing the type of the prototype Stokes linear operator from non-homogeneous to homogeneous elliptic. The idea is to uncouple dependencies between the variables until all their components can be viewed as scalar $H^1(\Omega)$ functions. Adding the "redundant" equations $\nabla \times \underline{\mathbf{U}} = 0$ and $\nabla(tr\underline{\mathbf{U}}) = 0$ and the boundary condition $\underline{\mathbf{U}} \times \mathbf{n} = 0$ on $\partial\Omega$ did the trick and enabled us to prove optimal error estimates in Theorem 8.18.

However, changing the type of the underlying elliptic system raises some questions about the equivalence of (8.5) and (8.1), the Navier–Stokes equations in the

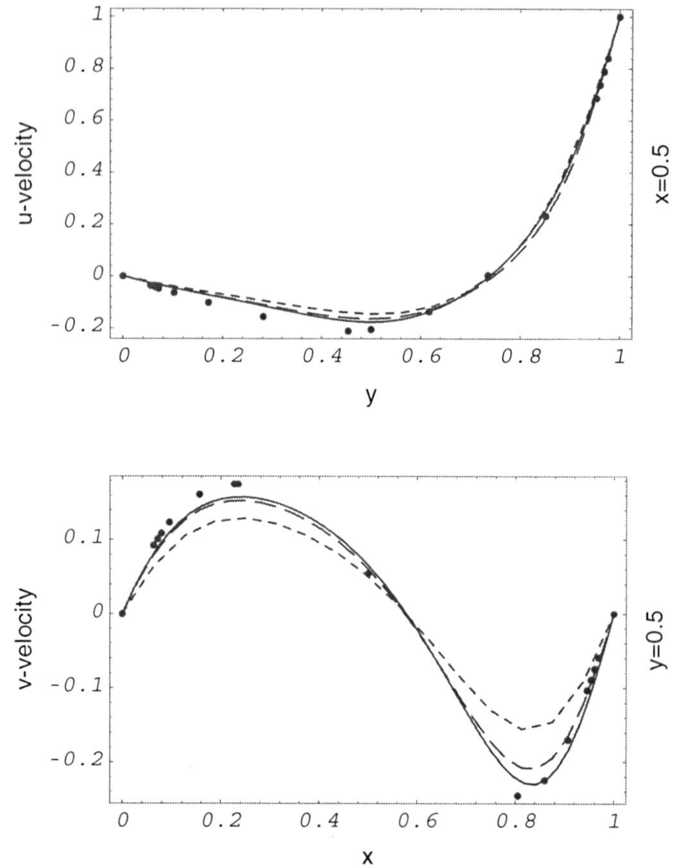

Fig. 8.3 Velocity profiles in the driven cavity problem for the negative norm LSFEM (8.14) for 17×17 (short dashes), 33×33 (long dashes), and 45×45 (solid line) bilinear elements vs. benchmark results of [180] (dots)

primitive variable form. Suppose first that the latter has a weak solution $\{\mathbf{u}, p\} \in [H^2(\Omega)]^d \times H^1(\Omega)$. Then, $\nabla \mathbf{u} \in [H^1(\Omega)]^{d^2}$ and the triple $\{\mathbf{u}, \underline{\mathbf{U}} = (\nabla \mathbf{u})^\mathsf{T}, p\}$ solves (8.5) in the $L^2(\Omega)$ sense. Assume now that $\{\mathbf{u}, p\}$ is strictly in $[H^1(\Omega)]^d \times L^2_0(\Omega)$. Then, $\nabla \mathbf{u}$ is strictly in $[L^2(\Omega)]^{d^2}$ and the triple $\{\mathbf{u}, \underline{\mathbf{U}} = (\nabla \mathbf{u})^\mathsf{T}, p\}$ does not satisfy (8.5) in the $L^2(\Omega)$ sense. The equations that are violated are precisely the "redundant" equations added to make the system homogeneous elliptic. Note that $\{\mathbf{u}, \underline{\mathbf{U}} = (\nabla \mathbf{u})^\mathsf{T}, p\}$ still satisfies the *original* VGVP system (8.5). In this system $\mathbf{u}$ and $\underline{\mathbf{U}}$ belonged to two different groups of variables which reflected the fact that originally $\underline{\mathbf{U}}$ was introduced as a *derivative* of the velocity.

This thought experiment seems to suggest that LSFEMs based on the extended VGVP system (8.5) should do fine as long as the original Navier–Stokes equations (8.1) have sufficiently regular solutions. However, it also looks as if such LSFEMs

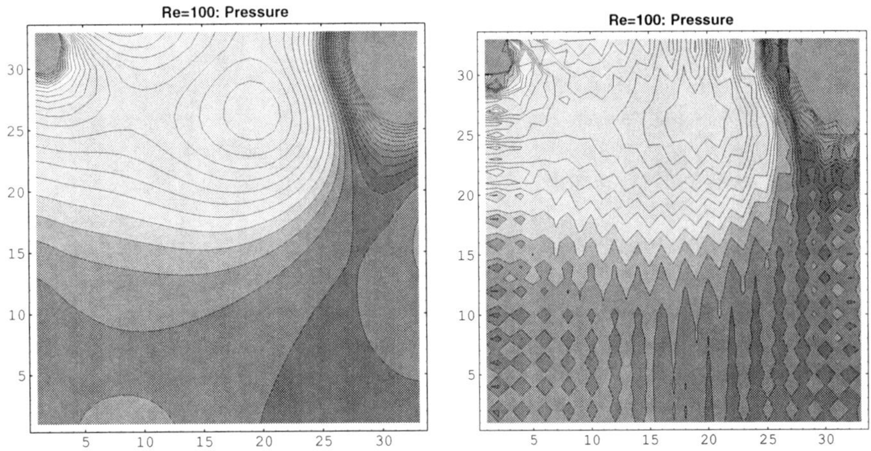

Fig. 8.4 Dependence of the least-squares solution on the choice of α in the minus one inner product (B.99) for the approximation of the solution of the driven cavity problem for $Re = 100$ obtained by the negative norm LSFEM (8.14) using a 33×33 grid. Left: pressure contours with $\alpha = 0.01$; right: pressure contours for $\alpha = 0$.

may run into trouble with less regular solutions of (8.1). On the other hand, a DLSP based on the original VGVP system (8.4) should work equally well in both cases. Our plan is to test these conjectures in practice by using the smooth manufactured solution (8.126) and a less regular solution of the Navier–Stokes equations (8.1). The driven cavity problem is a good choice for the latter because its associated velocity field is not in $[H^2(\Omega)]^d$.

We use two representatives of LSFEMs based on the original VGVP system (8.4). The first one is the norm-equivalent DLSP (8.14) implemented as described[31] in Section 8.4.3. This case falls under Theorem 8.19 and we have the error estimates (8.127). The second representative is the straightforward $L^2(\Omega)$ quasi-norm-equivalent DLSP

$$\begin{cases} J(\mathbf{u}^h, \underline{\mathbf{U}}^h, p^h; \mathbf{f}) = \\ \qquad \| -\nu\nabla\cdot\underline{\mathbf{U}}^h + \underline{\mathbf{U}}^h\cdot\mathbf{u}^h + \nabla p^h - \mathbf{f}\|_0^2 + \|\underline{\mathbf{U}}^h - (\nabla\mathbf{u}^h)^{\mathsf{T}}\|_0^2 + \|\nabla\cdot\mathbf{u}^h\|_0^2 \\ X^h_{r,0} = [G^r]^d \cap [H^1_0(\Omega)]^d \times [G^r]^{d^2} \times G^r \cap L^2_0(\Omega) \end{cases} \tag{8.128}$$

implemented using bilinear elements, i.e., $r = 1$ in (8.128). This method is included to provide further evidence about the poor performance of straightforward $L^2(\Omega)$ LSFEMs when the attendant system is not homogeneous elliptic.

Our third contender is the compliant DLSP (8.15) implemented using bilinear elements. The asymptotic error estimate of this DLSP is given by (8.111). For bilinear elements, we need to set $r = 1$ in (8.111) and the error bound specializes to

---

[31] The minus one inner product is computed with $\alpha = 0.2$ and the tolerance for $\mathcal{K}^h$ set to 10^{-5}.

$$\|\mathbf{u}-\mathbf{u}^h\|_1+\|\underline{\mathbf{U}}-\underline{\mathbf{U}}^h\|_1+\|p-p^h\|_1 \le Ch. \tag{8.129}$$

Smooth manufactured solution

Convergence rates for the three LSFEMs are estimated using a pair of uniform grids with 17×17 and 33×33 grid lines, respectively. The results are shown in Table 8.5. We continue to use boldface symbols to denote the data that correspond to the error norms included in (8.127) and (8.129). From Table 8.5, we see that as far as the theoretical error estimates are concerned, the norm-equivalent DLSP (8.14) and the compliant DLSP (8.15) perform in accordance with the theory. When compared against each other, the compliant DLSP has a slight edge over the norm-equivalent DLSP. In contrast, the straightforward $L^2(\Omega)$ DLSP consistently underperforms in all error measures. This mirrors the behavior observed in Section 7.6.2 for straightforward $L^2(\Omega)$ LSFEMs for the non-homogeneous elliptic VVP and VSP systems. Thus, if we were to rank our contenders based on their performance on smooth solutions, the compliant DLSP (8.15) is the clear winner, the norm-equivalent DLSP is close behind, and the straightforward $L^2(\Omega)$ principle is a distant third. The compliant DLSP is also the easiest to implement and so it ranks high in this regard as well.

	$L^2(\Omega)$ error rates			H^1 error rates		
Method	(8.14)	(8.15)	(8.128)	(8.14)	(8.15)	(8.128)
$\mathbf{u}$	1.869	1.921	0.944	**1.016**	**1.002**	0.918
$\underline{\mathbf{U}}$	**1.587**	1.776	0.881	0.666	**1.091**	0.594
p	**1.584**	1.741	0.706	0.708	**1.530**	0.531

Table 8.5 Comparison of LSFEMs based on (8.4) and its extended version (8.5) for the smooth manufactured solution (8.126). The original system is represented by the norm-equivalent DLSP (8.14), and the straightforward $L^2(\Omega)$ DLSP (8.128). The extended system is represented by the compliant DLSP (8.15).

Driven cavity flow

It turns out that the ranking of compliant and norm-equivalent DLSPs changes dramatically for less regular solutions.[32]

Recall that for the driven cavity problem, the velocity does not belong to $[H^2(\Omega)]^d$ so that the extended system (8.5) cannot hold in the $L^2(\Omega)$ sense. A compliant LSFEM based on this system then produces an "interpretation" of the rough solution by projecting the data onto the equal order finite element space $X^h_{r,0}$ using

[32] We do not further consider the straightforward $L^2(\Omega)$ method because it is clear that it should not be used at all.

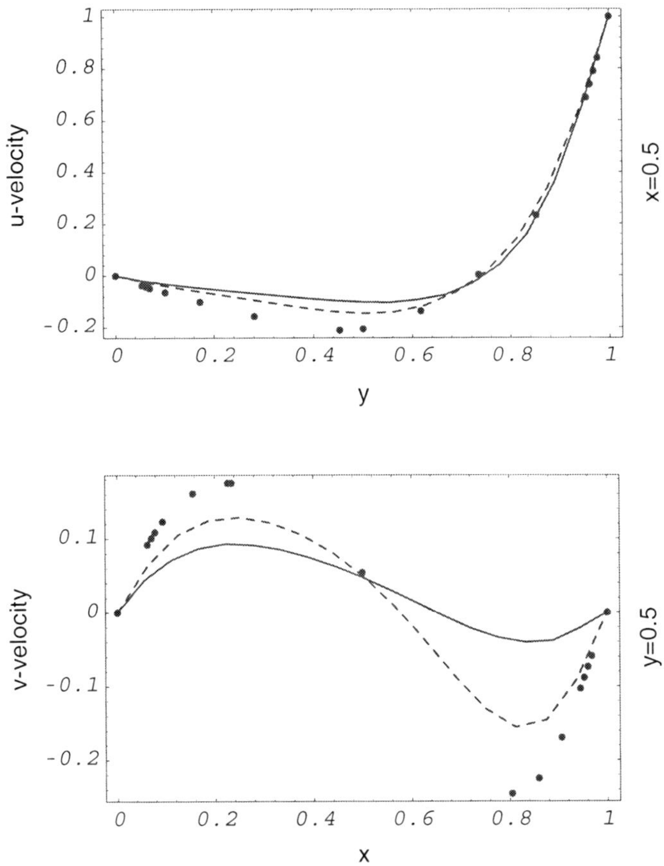

Fig. 8.5 Comparison of the norm-equivalent DLSP (8.14) and the compliant DLSP (8.15) for the driven cavity problem with $Re = 100$. The plots show the velocity component profiles (top: first component; bottom: second component) through the center of the cavity computed using a uniform 17×17 grid. Dashed lines correspond to (8.14), solid lines correspond to (8.15), and dots are the benchmark results from [180].

an inner product that is equivalent to the $H^1(\Omega)$ inner product. As a result, such a method "smooths out" the roughness of the original solution. We can clearly see this "smoothing" effect in Figure 8.5. Although (8.15) manages to compute a reasonably good approximation of the first velocity component, it significantly underestimates the second component.

An even better confirmation of this undesirable smoothing can be obtained by comparing velocity fields and vorticity contours computed through (8.15) and (8.14), respectively. The qualitative differences between the two calculations, shown in Figure 8.6, are clearly visible, even in the "eye-ball" norm. Especially striking are the differences between the vorticity fields computed by the two methods. The result for the norm-equivalent DLSP is consistent with the benchmarks in [180]; in

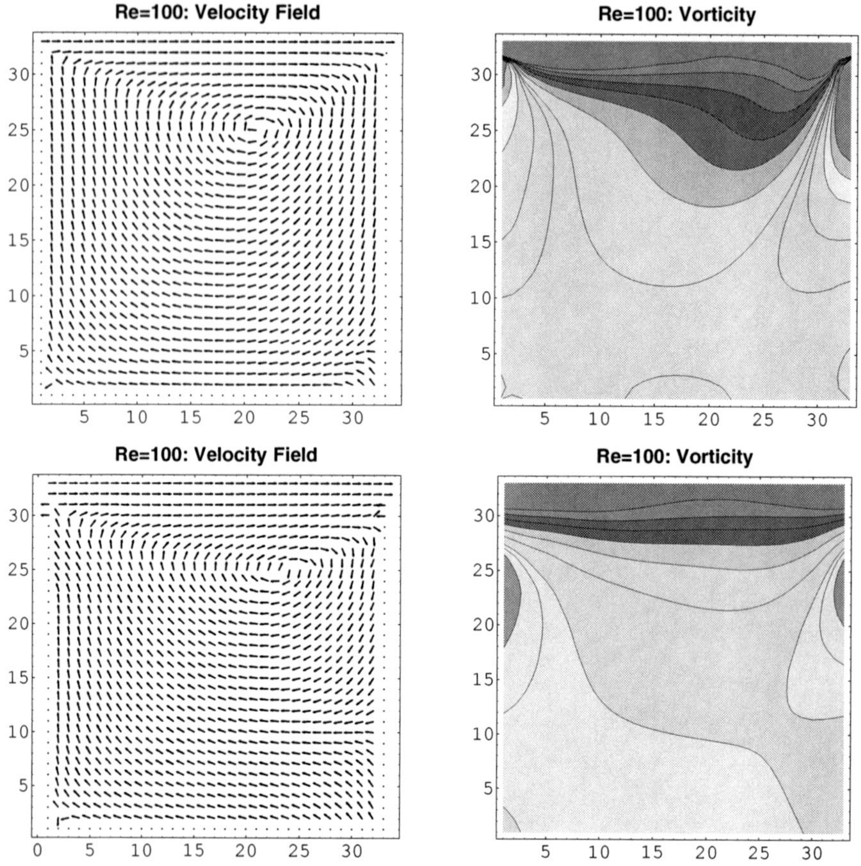

Fig. 8.6 Comparison of the norm-equivalent DLSP (8.14) and the compliant DLSP (8.15) for the driven cavity flow problem with $Re = 100$ using a uniform 33×33 grid. The top plots show the velocity field and vorticity contours for (8.14); the bottom plots shows the same data for (8.15).

contrast, the vorticity field computed by the compliant method is very different. There are substantial differences in the flow patterns as well. Most notably, the flow computed with the compliant method has a much less pronounced central vortex; this is in contrast to the norm-equivalent method which reproduces a flow pattern in good agreement with [180].

8.5 A Summary of Conclusions and Recommendations

In this chapter, we developed LSFEMs for the Navier–Stokes equations by including an appropriate form of the nonlinear term in a *prototype* least-squares principle for the linear Stokes equations. The analysis presented in Section 8.3 asserted that the properties of the thus obtained LSFEMs for the *nonlinear* Navier–Stokes equations closely mirror the theoretical properties of the prototype least-squares principles for the *linear* Stokes equations. Numerical examples from Section 8.4 confirmed again that for LSFEMs whose prototype Stokes principles are associated with non-homogeneous elliptic systems, the use of straightforward $L^2(\Omega)$ least-squares functionals is not recommended.

However, we also saw that the "curse" of non-homogenous elliptic systems cannot be always avoided by simply adding more redundant equations. This worked well if solutions of the original Navier–Stokes equations were smooth enough, but failed for problems with rough solutions such as the driven cavity problem.

In summary,

- compliant DLSPs based on the extended system (8.5) are safe to use only in cases when the solution to be approximated is known to be sufficiently regular
- in general cases for which solution regularity is not known a priori, the safest bet appears to be a norm-equivalent DLSP; in this case, the use of the VVP system is preferable because it has less unknowns than the VGVP first-order system
- the use of equal-order linear, bilinear, or trilinear elements is mathematically and practically justified for norm-equivalent DLSPs; this is not true for straightforward $L^2(\Omega)$ functionals associated with non-homogeneous elliptic systems.

The key properties of the LSFEMs presented in this chapter are compared and contrasted in Table 8.6.

Method→ Property↓	VVP compliant (8.11), (8.12)	VVP norm-equivalent (8.13)	VGVP compliant (8.15)	VGVP norm-equivalent (8.14)
Provably optimal	√	√	√	√
- with velocity BC	–	√	√	√
Robust for $\mathbf{u} \notin [H^2(\Omega)]^d$	√	√	–	√
Preconditioner	√	√	√	√
Variable count in $\mathbb{R}^2/\mathbb{R}^3$	4/7	4/7	7/13	7/13
Coding effort	simple	not as simple	simple	not as simple

Table 8.6 Summary properties of LSFEMs for the Navier-Stokes equations.

Chapter 9
Parabolic Partial Differential Equations

Evolution problems arise in all areas of science and engineering applications. Many evolution processes are dissipative in nature and can be modeled by *parabolic* partial differential equations (PDEs). Parabolic PDEs possess mathematical properties that have had a profound impact on the design of numerical methods for their approximate solution. Most notably, the cylindrical nature of the space–time domain has naturally led to discretizations by finite element methods in space and finite difference methods in time. As a rule, such *separated* discretizations lead to *marching* schemes, i.e., methods for which the approximate solution is obtained one time level at a time, requiring the solution of an elliptic PDE at each time level.

Because least-squares finite element methods (LSFEMs) are based on residual minimization, there is no conceptual difficulty in applying the least-squares paradigm to parabolic PDEs. Some examples of LSFEMs for time-dependent problems that use space–time elements can be found in, e.g., [28, 29, 146, 296, 299]. Although attractive analytically, space–time elements tend to be more complicated to implement, may lead to a substantial increase in the number of unknowns, and, if one is not careful, to fully coupled space–time approximations for which one cannot march one time step at a time.[1]

Space–time elements can be avoided by first using a standard finite difference scheme to discretize in time so that one is left with a sequence of compactly perturbed elliptic equations that determine the semi-discrete in time solution at each time level. These elliptic problems can be discretized, e.g., by a Galerkin method,[2] or, what is of interest to us, by a LSFEM. In the latter case, we refer to the overall, fully-discrete method as a *finite difference least-squares finite element method* (FD-LSFEM) This is the class of LSFEMs that is the main focus of this chapter. In Section 9.1, we consider FD-LSFEMs for a generalized heat equation. We also

[1] This is not to say that space–time LSFEMs are not of interest; in fact, using space–time adaptive methods, they hold (a yet largely unrealized) promise of providing a very efficient means for approximating solutions of parabolic PDEs.

[2] A detailed discussion of Galerkin finite element methods for heat equation-type problems can be found in [336]. Mixed-Galerkin finite element methods for the time dependent Stokes equation are discussed in, e.g., [182, 191, 333].

P.B. Bochev, M.D. Gunzburger, *Least-Squares Finite Element Methods*,
Applied Mathematical Sciences 166, DOI: 10.1007/b13382_9,

briefly consider, in Section 9.1.4, space-time least-squares principles for this setting. In Section 9.2, we consider FD-LSFEMs for the time-dependent Stokes equations. FD-LSFEMs for the advection–diffusion–reaction equations are considered in Section 12.7.

9.1 The Generalized Heat Equation

The problems considered in this section are of the form of the (generalized) heat equation:[3]

$$\begin{cases} c\dfrac{\partial\phi}{\partial t} - \nabla\cdot(\Theta_1\nabla\phi) = f & \text{for } (\mathbf{x},t)\in\Omega\times(0,T] \\ \phi = g_D & \text{for } (\mathbf{x},t)\in\Gamma\times(0,T] \\ (\Theta_1\nabla\phi)\cdot\mathbf{n} = g_N & \text{for } (\mathbf{x},t)\in\Gamma^*\times(0,T] \\ \phi = \phi_0(\mathbf{x}) & \text{for } t=0 \text{ and } \mathbf{x}\in\Omega\,, \end{cases} \tag{9.1}$$

where Γ and Γ^* are disjoint parts of the boundary $\partial\Omega$ of the domain Ω, $[0,T]$ is a given time interval, and $\mathbf{n}$ denotes the unit outward normal on the boundary $\partial\Omega$. The symmetric, positive definite matrix Θ_1 satisfying

$$\frac{1}{\alpha}\sum_{i=1}^{d}\xi_i^2 \le \sum_{i,j=1}^{d}\xi_i\xi_j\Theta_{1\,ij} \le \alpha\sum_{i=1}^{d}\xi_i^2\,, \tag{9.2}$$

the scalar $c>0$ satisfying

$$\frac{1}{\alpha}\le c\le\alpha\,, \tag{9.3}$$

where $\alpha>0$, as well as the boundary data g_D and g_N, the source data f, and the initial data ϕ_0 are all assumed to be given. The source and boundary data are generally functions of space and time. Often, however, c and Θ_1 are media properties and are therefore independent of time; we assume that this is true throughout this chapter.

As what has by now become a familiar first step towards defining LSFEMs, we recast (9.1) into first-order form. Of course, as we learned for the case of elliptic PDEs, there are many ways to effect such a transformation. Here, we use the specific first-order formulation

[3] See (5.1) for the elliptic counterpart of (9.1).

$$
\begin{cases}
c\dfrac{\partial \phi}{\partial t} + \nabla \cdot \mathbf{u} = f & \text{for } (\mathbf{x},t) \in \Omega \times (0,T] \\
\mathbf{u} + \Theta_1 \nabla \phi = 0 & \text{for } (\mathbf{x},t) \in \Omega \times (0,T] \\
\phi = g_D & \text{for } (\mathbf{x},t) \in \Gamma \times (0,T] \\
\mathbf{u} \cdot \mathbf{n} = -g_N & \text{for } (\mathbf{x},t) \in \Gamma^* \times (0,T] \\
\phi = \phi_0(\mathbf{x}) & \text{for } t = 0 \text{ and } \mathbf{x} \in \Omega\,.
\end{cases}
\tag{9.4}
$$

In Sections 9.1.1 and 9.1.2, we consider LSFEMs for which one uses an implicit finite difference scheme to discretize in time, thus obtaining a semi-discretization of the parabolic partial differential equation. Then, at each time step, one is faced with an elliptic partial differential equation whose approximate solution is obtained using a LSFEM.

It is of interest to us that (9.4) is also a first-order formulation for the following second-order problem for the vector field $\mathbf{u}$:

$$
\begin{cases}
{\Theta_1}^{-1}\dfrac{\partial \mathbf{u}}{\partial t} - \nabla\big(c^{-1}\nabla \cdot \mathbf{u}\big) = -\nabla\big(c^{-1} f\big) & \text{for } (\mathbf{x},t) \in \Omega \times (0,T] \\
\mathbf{u} \cdot \mathbf{n} = -g_N & \text{for } (\mathbf{x},t) \in \Gamma^* \times (0,T] \\
\mathbf{n} \times \mathbf{u} \times \mathbf{n} = \mathbf{n} \times \Theta_1 \nabla g_D \times \mathbf{n} & \text{for } (\mathbf{x},t) \in \Gamma \times (0,T] \\
{\Theta_1}^{-1}\mathbf{u} = -\nabla \phi_0(\mathbf{x}) & \text{for } t = 0 \text{ and } \mathbf{x} \in \Omega\,.
\end{cases}
\tag{9.5}
$$

9.1.1 Backward-Euler Least-Squares Finite Element Methods

Semi-discretization in time

One can take advantage of the cylindrical nature of the space–time domain $\Omega \times (0,T)$ by first discretizing (9.4) in time. To this end, we introduce a partition $\{[t_{k-1},t_k]\}_{k=1}^{K}$ of the time interval $[0,T]$ into K subintervals, where $t_0 = 0$, $t_K = T$, and $t_{k-1} < t_k$. We set $\Delta t_k = t_k - t_{k-1}$ for $k = 1,\ldots,K$. Then, if the *backward-Euler method* is used to effect the discretization in time, we have the sequence of semi-discrete in time systems: for $k = 1,\ldots,K$,

$$
\begin{cases}
\dfrac{c}{\Delta t_k}\phi^{(k)} + \nabla \cdot \mathbf{u}^{(k)} = f(\mathbf{x},t_k) + \dfrac{c}{\Delta t_k}\phi^{(k-1)} & \text{for } \mathbf{x} \in \Omega \\
\mathbf{u}^{(k)} + \Theta_1 \nabla \phi^{(k)} = \alpha\Big(\mathbf{u}^{(k-1)} + \Theta_1 \nabla \phi^{(k-1)}\Big) & \text{for } \mathbf{x} \in \Omega \\
\phi^{(k)} = g_D(\mathbf{x},t_k) & \text{for } \mathbf{x} \in \Gamma \\
\mathbf{u}^{(k)} \cdot \mathbf{n} = -g_N(\mathbf{x},t_k) & \text{for } \mathbf{x} \in \Gamma^*,
\end{cases}
\tag{9.6}
$$

where α is either 0 or 1,[4] $\phi^{(0)} = \phi_0(\mathbf{x})$ for $\mathbf{x} \in \Omega$, and, if $\alpha = 1$, $\mathbf{u}^{(0)} = -\Theta_1 \nabla \phi_0(\mathbf{x})$ for $\mathbf{x} \in \Omega$. In (9.6), $\phi^{(k)}(\mathbf{x})$ and $\mathbf{u}^{(k)}(\mathbf{x})$ denote the approximations to $\phi(\mathbf{x},t_k)$ and $\mathbf{u}(\mathbf{x},t_k)$, respectively.

Each member of the sequence of systems (9.6) is a perturbed elliptic problem.[5] Thus, we could now apply a least-square minimization principle to each and every member of that sequence. For simplicity, we assume that $g_D = g_N = 0$. Then, we define the sequence of continuous least-squares principles (CLSPs): for $k = 1,\dots,K$,

$$\left\{\begin{aligned} & J_k^{\mathsf{be}}(\phi^{(k)},\mathbf{u}^{(k)}; f(\mathbf{x},t_k),\phi^{(k-1)},\mathbf{u}^{(k-1)}) \\ & \qquad = \left\| {\Theta_1}^{-1/2}\left(\mathbf{u}^{(k)} + \Theta_1 \nabla \phi^{(k)} - \alpha\left(\mathbf{u}^{(k-1)} + \Theta_1 \nabla \phi^{(k-1)}\right)\right)\right\|_0^2 \\ & \qquad + \left\| \left(\frac{\Delta t_k}{c}\right)^{1/2}\left(\frac{c}{\Delta t_k}\phi^{(k)} + \nabla\cdot\mathbf{u}^{(k)} - f(\mathbf{x},t_k) - \frac{c}{\Delta t_k}\phi^{(k-1)}\right)\right\|_0^2 \\ & X = G_\Gamma(\Omega) \times \mathbf{D}_{\Gamma^*}(\Omega), \end{aligned}\right. \tag{9.7}$$

where $\phi^{(0)} = \phi_0$ and, if $\alpha = 1$, $\mathbf{u}^{(0)} = \Theta_1 \nabla \phi_0$.[6] Each CLSP in (9.7) corresponds to an energy balance for the first-order system (9.6) derived in the vector operator setting; see Remark 5.12 for more information on this topic. Because (9.6) is a particular case of the div–grad system (5.17), well-posedness of (9.7), for each $k = 1,\dots,K$, follows from Theorem 5.14 in Section 5.4.2. We use the vector operator setting to define CLSPs for (9.6) because it possesses some advantages that turn out to be particularly useful for the analysis of FD-LSFEMs. Specifically, the splitting property of the least-squares formulation and the existing connections between least-squares minimizers and solutions of Galerkin and mixed Galerkin methods, established in Theorem 5.37, greatly simplify error estimates for fully-discrete approximations based on (9.7). The other possible setting for the energy balances, the Agmon–Douglis–Nirenberg setting in Section 5.4.1, does not offer these advantages and is not used in this chapter.

It is important to note that (9.7) defines a marching method. For example, for the case $\alpha = 0$, starting with $\phi^{(0)} = \phi_0$, one sets $k = 1$ in (9.7) which then becomes a well-posed problem for determining $\phi^{(1)}$ and $\mathbf{u}^{(1)}$. Obviously, by incrementing the value of k, one sequentially determines the remaining pairs $\phi^{(k)}$ and $\mathbf{u}^{(k)}$, $k =$

---

[4] Thus, (9.6) defines two backward-Euler semi-discretization methods for (9.4). The reason for introducing two such methods is made apparent when we arrive at (9.11) and (9.12). However, we can note now that if $\alpha = 0$, there is no need to define an initial condition for $\mathbf{u}^{(0)}$.

[5] For example, it is easy to see that if $\alpha = 0$, (9.6) corresponds to the sequence of elliptic partial differential equations

$$-\nabla\cdot\Theta_1\nabla\phi^{(k)} + \frac{c}{\Delta t_k}\phi^{(k)} = f(\mathbf{x},t_k) + \frac{c}{\Delta t_k}\phi^{(k-1)}$$

with $\phi^{(0)}(\mathbf{x}) = \phi_0(\mathbf{x})$ as well as $\phi^{(k)} = g_D(\mathbf{x},t_k)$ on Γ and $\Theta_1\nabla\phi^{(k)}\cdot\mathbf{n} = g_N(\mathbf{x},t_k)$ on Γ^*.

[6] The weights ${\Theta_1}^{-1/2}$ and $(\Delta t_k/c)^{1/2}$ appearing in the two terms in the functional in (9.7) are not only useful for analysis purposes, but also ensure that the two terms have the same units.

$2,\ldots,K$. The same observation holds when $\alpha = 1$, but now one also has to supply the initial condition $\mathbf{u}^{(0)} = -\Theta_1 \nabla \phi_0$.

The necessary condition corresponding to the sequence of minimization problems (9.7) is given by: for $k = 1,\ldots,K$, find $\phi^{(k)} \in G_\Gamma(\Omega)$ and $\mathbf{u}^{(k)} \in \mathbf{D}_{\Gamma^*}(\Omega)$ that satisfy

$$Q_k^{\mathsf{be}}(\phi^{(k)}, \mathbf{u}^{(k)}; \psi, \mathbf{v}) = F_k^{\mathsf{be}}(\psi, \mathbf{v}; \phi^{(k-1)}, \mathbf{u}^{(k-1)}) \qquad \forall \psi \in G_\Gamma(\Omega),\ \mathbf{v} \in \mathbf{D}_{\Gamma^*}(\Omega), \tag{9.8}$$

where $\phi^{(0)} = \phi_0$ and $\mathbf{u}^{(0)} = -\Theta_1 \nabla \phi_0$. For all $\phi, \psi \in G_\Gamma(\Omega)$ and $\mathbf{u}, \mathbf{v} \in \mathbf{D}_{\Gamma^*}(\Omega)$, we have that

$$Q_k^{\mathsf{be}}(\phi, \mathbf{u}; \psi, \mathbf{v}) = \int_\Omega \left(\mathbf{u} + \Theta_1 \nabla \phi\right) \cdot {\Theta_1}^{-1}(\mathbf{v} + \Theta_1 \nabla \psi)\, d\Omega + \int_\Omega \frac{\Delta t_k}{c}\left(\frac{c}{\Delta t_k}\phi + \nabla \cdot \mathbf{u}\right)\left(\frac{c}{\Delta t_k}\psi + \nabla \cdot \mathbf{v}\right) d\Omega$$

and

$$F_k^{\mathsf{be}}(\psi, \mathbf{v}; \phi^{(k-1)}, \mathbf{u}^{(k-1)}) = \int_\Omega \frac{\Delta t_k}{c}\left(f(\mathbf{x}, t_k) + \frac{c}{\Delta t_k}\phi^{(k-1)}\right)\left(\frac{c}{\Delta t_k}\psi + \nabla \cdot \mathbf{v}\right) d\Omega + \alpha \int_\Omega \left(\mathbf{u}^{(k-1)} + \Theta_1 \nabla \phi^{(k-1)}\right) \cdot {\Theta_1}^{-1}\left(\mathbf{v} + \Theta_1 \nabla \psi\right) d\Omega\,.$$

Note that $Q_k^{\mathsf{be}}(\cdot,\cdot,\cdot,\cdot)$ does not depend on k whenever we have a constant time step size, i.e., if $\Delta t_k = \Delta t$ for all $k = 1,\ldots,K$.

Expanding the definition of $Q_k^{\mathsf{be}}(\cdot,\cdot;\cdot,\cdot)$, one obtains

$$\begin{aligned} Q_k^{\mathsf{be}}(\phi^{(k)}, \mathbf{u}^{(k)}; \psi, \mathbf{v}) = &\int_\Omega \left(\nabla \phi^{(k)} \cdot \Theta_1 \nabla \psi + \frac{c}{\Delta t_k}\phi^{(k)}\psi\right) d\Omega \\ &+ \int_\Omega \left(\frac{\Delta t_k}{c}\nabla \cdot \mathbf{u}^{(k)} \nabla \cdot \mathbf{v} + \mathbf{u}^{(k)} \cdot {\Theta_1}^{-1}\mathbf{v}\right) d\Omega \\ &+ \int_\Omega \left(\mathbf{u}^{(k)} \cdot \nabla \psi + \psi \nabla \cdot \mathbf{u}^{(k)}\right) d\Omega \\ &+ \int_\Omega \left(\phi^{(k)} \nabla \mathbf{v} + \mathbf{v} \cdot \nabla \phi^{(k)}\right) d\Omega \end{aligned} \tag{9.9}$$

so that, because $\psi \in G_\Gamma(\Omega)$ and $\mathbf{v} \in \mathbf{D}_{\Gamma^*}(\Omega)$, we have that[7,8]

---

7 We see here the usefulness of the weights introduced in the functional in (9.7) which allow for the cancellation of the mixed terms, i.e., the terms involving both a scalar function and a vector function, appearing in the last two lines of (9.9). The same cancellation property is used in the proof of Theorem 5.14 and is the key reason to prefer the vector operator setting for the semi-discrete in time system (9.6).

8 Note that (9.10) can be obtained even if $\phi^{(k)}$ and $\mathbf{u}^{(k)}$ satisfy inhomogeneous boundary conditions.

$$
\begin{aligned}
Q_k^{\mathrm{be}}(\phi^{(k)},\mathbf{u}^{(k)};\psi,\mathbf{v}) = \int_\Omega \left(\nabla\phi^{(k)}\cdot\Theta_1\nabla\psi + \frac{c}{\Delta t_k}\phi^{(k)}\psi\right) d\Omega \\
+ \int_\Omega \left(\frac{\Delta t_k}{c}\nabla\cdot\mathbf{u}^{(k)}\nabla\cdot\mathbf{v} + \mathbf{u}^{(k)}\cdot{\Theta_1}^{-1}\mathbf{v}\right) d\Omega\,.
\end{aligned}
\tag{9.10}
$$

We then see that (9.8), or equivalently (9.7), separates into the two problems: for $k = 1,\ldots,K$, find $\phi^{(k)} \in G_\Gamma(\Omega)$ that satisfies

$$
\begin{aligned}
&\int_\Omega \left(\nabla\phi^{(k)}\cdot\Theta_1\nabla\psi + \frac{c}{\Delta t_k}\phi^{(k)}\psi\right) d\Omega \\
&\quad = \int_\Omega \left(f(\mathbf{x},t_k) + \frac{c}{\Delta t_k}\phi^{(k-1)}\right)\psi\, d\Omega \\
&\qquad +\alpha\int_\Omega \left(\mathbf{u}^{(k-1)} + \Theta_1\nabla\phi^{(k-1)}\right)\cdot\nabla\psi\, d\Omega \qquad \forall\,\psi\in G_\Gamma(\Omega)\,,
\end{aligned}
\tag{9.11}
$$

where $\phi^{(0)} = \phi_0$, and $\mathbf{u}^{(k)} \in \mathbf{D}_{\Gamma^*}(\Omega)$ that satisfies

$$
\begin{aligned}
&\int_\Omega \left(c^{-1}\nabla\cdot\mathbf{u}^{(k)}\nabla\cdot\mathbf{v} + \frac{1}{\Delta t_k}\mathbf{u}^{(k)}\cdot{\Theta_1}^{-1}\mathbf{v}\right) d\Omega \\
&\quad = \int_\Omega c^{-1}f(\mathbf{x},t_k)\nabla\cdot\mathbf{v}\, d\Omega + \alpha\int_\Omega \frac{1}{\Delta t_k}\mathbf{u}^{(k-1)}\cdot{\Theta_1}^{-1}\mathbf{v}\, d\Omega \\
&\qquad +(1-\alpha)\int_\Omega \frac{1}{\Delta t_k}\phi^{(k-1)}\nabla\cdot\mathbf{v}\, d\Omega \qquad \forall\,\mathbf{v}\in\mathbf{D}_{\Gamma^*}(\Omega)\,,
\end{aligned}
\tag{9.12}
$$

where, for $\alpha = 1$, $\mathbf{u}^{(0)} = -\Theta_1\nabla\phi_0$.

The pair of equations (9.11) and (9.12) are weakly coupled in the following sense. First, consider the case $\alpha = 0$. Then, (9.11) can be solved for $\phi^{(k)}$ for $k = 1,\ldots,K$ without regard to $\mathbf{u}^{(k)}$ and, subsequently, (9.12) can be solved for $\mathbf{u}^{(k)}$ for $k = 1,\ldots,K$. On the other hand, if $\alpha = 1$, then (9.12) can be solved for $\mathbf{u}^{(k)}$ for $k = 1,\ldots,K$ without regard to $\phi^{(k)}$ and, subsequently, (9.11) can be solved for $\phi^{(k)}$ for $k = 1,\ldots,K$.[9]

We have the choices of determining $\phi^{(k)}$ and $\mathbf{u}^{(k)}$ from (9.11) and (9.12) with $\alpha =$ 0 or 1. In fact, we have the third choice[10] of using (9.11) with $\alpha = 0$ to determine $\phi^{(k)}$ and (9.12) with $\alpha = 1$ to determine $\mathbf{u}^{(k)}$. In this case, we can solve for both $\phi^{(k)}$ and $\mathbf{u}^{(k)}$ for $k = 1,\ldots,K$ totally independently of each other.

Remark 9.1 For the problems considered in this section, it is obviously advantageous to choose the value of α that effects the greatest uncoupling. However, it is still of great interest to consider values of α that do not result in uncoupling of the problems for ϕ and $\mathbf{u}$. The reason for this is that in many situations, e.g.,

[9] This reversal of roles when one changes the value of α from 0 to 1 is the reason we introduced the two variants of the backward-Euler semi-discretization of (9.4).

[10] The obvious fourth choice of using (9.11) with $\alpha = 1$ to determine $\phi^{(k)}$ and (9.12) with $\alpha = 0$ to determine $\mathbf{u}^{(k)}$ is not competitive.

problems with nonlinear reaction terms or problems with advective flux terms (see Section 12.7), uncoupling of the equations may not be possible for any value of α so that we want some indication of what happens to, e.g., error estimates, in such situations. For instance, analysis of FD-LSFEMs for transient advection-reaction-diffusion problems in [348] indicates some differences in the error estimates with $\alpha = 0$ and $\alpha = 1$; see Section 12.7 for further details. □

Least-squares finite element spatial discretization

For each k, the problem (9.7) is exactly of the form discussed in Section 5.17. Thus, LSFEMs can be defined by minimizing[11] $J_k^{\text{be}}(\cdot,\cdot;f(t_k),\phi^{(k-1)},\mathbf{u}^{(k-1)})$ over exactly the same choices of conforming finite element spaces as were used in Section 5.6.1. Specifically, we have the space $X^h = G_\Gamma^h \times \mathbf{V}_{\Gamma^*}^h$, where we couple the standard nodal finite element for the approximation of the scalar-valued function ϕ: for $r \in \{1,2,\ldots\}$,

$$G_\Gamma^h = G^r(\Omega) \cap G_\Gamma(\Omega) \tag{9.13}$$

with three choices for the approximation of the vector-valued function $\mathbf{u}$: for $s \in \{1,2,\ldots\}$,

$$\mathbf{V}_{\Gamma^*}^h = \begin{cases} [G^s(\Omega)]^d \cap \mathbf{D}_{\Gamma^*}(\Omega) & \text{for which we set } s_1 = s+1 \\ \mathbf{D}^{s}\text{-}(\Omega) \cap \mathbf{D}_{\Gamma^*}(\Omega) & \text{for which we set } s_1 = s \\ \mathbf{D}^s(\Omega) \cap \mathbf{D}_{\Gamma^*}(\Omega) & \text{for which we set } s_1 = s+1 . \end{cases} \tag{9.14}$$

Note that the spaces G_Γ^h and $\mathbf{V}_{\Gamma^*}^h$ are constrained to satisfy appropriate boundary conditions. The first choice in (9.14) corresponds to the nodal DLSP (5.63) in Remark 5.23. The second and third choices correspond to the compatible DLSP formulation (5.62). Recall that $\mathbf{D}^s$-(Ω) and $\mathbf{D}^s(\Omega)$ stand for div-conforming spaces of the first and the second kind, respectively; see Section B.2.1.

The approximation properties for the component finite element spaces are as follows: there exists $\phi^h \in G_\Gamma^h = G^r(\Omega) \cap G_\Gamma(\Omega)$ such that, for any[12] $\phi \in H^{r+1}(\Omega) \cap G_\Gamma(\Omega)$,

$$\|\phi - \phi^h\|_0 + h\|\nabla(\phi - \phi^h)\|_0 \le C h^{r+1} \|\phi\|_{r+1} \tag{9.15}$$

---

[11] From now on, whenever there is no chance of confusion, we suppress noting dependences on the spatial variable $\mathbf{x}$, e.g., we use the abbreviated notation $\phi(t_k)$ for $\phi(\mathbf{x},t_k)$.

[12] For the sake of simplicity, we assume that we have full regularity with respect to the polynomial degree, e.g., we use polynomials of degree r and the solution $\phi \in H^{r+1}(\Omega)$. If the regularity is given by, say, $\phi \in H^{q+1}(\Omega)$, we would instead have the approximation property

$$\|\phi - \phi^h\|_0 + h\|\nabla(\phi - \phi^h)\|_0 \le C h^{\min\{r,q\}+1} \|\phi\|_{\min\{r,q\}+1}$$

(see Theorem B.6) and, for any $\mathbf{u} \in [H^{s+1}(\Omega)]^d \cap \mathbf{D}_{\Gamma^*}(\Omega)$, there exists $\mathbf{u}^h \in \mathbf{V}^h_{\Gamma^*}$ such that

$$\|\mathbf{u}-\mathbf{u}^h\|_0 \le Ch^{s_1}\|\mathbf{u}\|_{s_1} \qquad \text{and} \qquad \|\nabla\cdot(\mathbf{u}-\mathbf{u}^h)\|_0 \le Ch^s\|\nabla\cdot\mathbf{u}\|_s \tag{9.16}$$

(see Theorem B.9).

We define approximate solutions of (9.7) or, equivalently, of (9.8) or (9.11)–(9.12) as follows. Let $\phi^{(h,0)} \in G^h_\Gamma$ and, if needed, $\mathbf{u}^{(h,0)} \in \mathbf{V}^h_{\Gamma^*}$ denote approximations of the initial data ϕ_0 and $-\Theta_1\nabla\phi_0$, respectively. Then, for $k=1,\dots,K$, we determine $\phi^{(h,k)}(\mathbf{x}) \in G^h_\Gamma$ and $\mathbf{u}^{(h,k)}(\mathbf{x}) \in \mathbf{V}^h_{\Gamma^*}$ from the DLSP

$$\begin{cases} J^h = J^{\mathsf{be}}_k(\phi^{(h,k)},\mathbf{u}^{(h,k)};f(t_k),\phi^{(h,k-1)},\mathbf{u}^{(h,k-1)}) \\ X^h = G^h_\Gamma \times \mathbf{V}^h_{\Gamma^*}\,. \end{cases} \tag{9.17}$$

Equivalently, $\phi^{(h,k)} \in G^h_\Gamma$ and $\mathbf{u}^{(h,k)} \in \mathbf{V}^h_{\Gamma^*}$ solve (9.11) and (9.12) restricted to those subspaces. As a result, we arrive at the following *fully-discrete backward-Euler FD-LSFEM.* Let $\phi^{(h,0)} \in G^h_\Gamma$ and $\mathbf{u}^{(h,0)} \in \mathbf{V}^h_{\Gamma^*}$ denote approximations of the initial data ϕ_0 and $-\Theta_1\nabla\phi_0$, respectively. Then, for $k=1,\dots,K$, determine $\phi^{(h,k)} \in G^h_\Gamma$ from

$$\begin{aligned} &\int_\Omega\left(\nabla\phi^{(h,k)}\cdot\Theta_1\nabla\psi^h + \frac{c}{\Delta t_k}\phi^{(h,k)}\psi^h\right)d\Omega \\ &\quad = \int_\Omega\left(f(t_k)+\frac{c}{\Delta t_k}\phi^{(h,k-1)}\right)\psi^h\,d\Omega \\ &\qquad\quad +\alpha\int_\Omega\left(\mathbf{u}^{(h,k-1)}+\Theta_1\nabla\phi^{(h,k-1)}\right)\cdot\nabla\psi^h\,d\Omega \qquad \forall\,\psi^h\in G^h_\Gamma \end{aligned} \tag{9.18}$$

and $\mathbf{u}^{(h,k)} \in \mathbf{V}^h_{\Gamma^*}$ from

$$\begin{aligned} &\int_\Omega\left(c^{-1}\nabla\cdot\mathbf{u}^{(h,k)}\nabla\cdot\mathbf{v}^h + \frac{1}{\Delta t_k}\mathbf{u}^{(h,k)}\cdot{\Theta_1}^{-1}\mathbf{v}^h\right)d\Omega \\ &\quad = \int_\Omega c^{-1}f(t_k)\nabla\cdot\mathbf{v}^h\,d\Omega + \alpha\int_\Omega\frac{1}{\Delta t_k}\mathbf{u}^{(h,k-1)}\cdot{\Theta_1}^{-1}\mathbf{v}^h d\Omega \\ &\qquad\quad +(1-\alpha)\int_\Omega\frac{1}{\Delta t_k}\phi^{(h,k-1)}\nabla\cdot\mathbf{v}^h\,d\Omega \qquad \forall\,\mathbf{v}\in\mathbf{V}^h_{\Gamma^*}\,. \end{aligned} \tag{9.19}$$

Setting $\alpha=0$, one recognizes (9.18) as being exactly the same sequence of equations that is obtained by a backward-Euler finite element discretization of the second-order form (9.1) of the heat equation, i.e.,

> *for the first-order system* (9.4), *the backward-Euler LSFEM* (9.17) *with* $\alpha=0$ *recovers exactly the same approximate solution for the scalar-valued variable* ϕ *as does the backward-Euler–Galerkin finite element method for the parent second-order equation* (9.1).

Analogously, setting $\alpha = 1$, one recognizes (9.19) as being exactly the same sequence of equations that is obtained by a backward-Euler finite element discretization of the alternate second-order form (9.5) corresponding to the (9.4), i.e.,

> *for the first-order system* (9.4), *the backward-Euler LSFEM* (9.17) *with* $\alpha = 1$ *recovers exactly the same approximate solution for the vector-valued variable* $\mathbf{u}$ *as does the backward-Euler–Galerkin finite element method for the parent second-order equation* (9.5).

This "splitting" property of the FD-LSFEM formulation is direct analogue of the splitting property in Theorem 5.37, and represents an important advantage of the vector operator setting.

As before, we have the choice of determining $\phi^{(h,k)}$ and $\mathbf{u}^{(h,k)}$ from (9.18) and (9.19) with $\alpha = 0$ or 1 and the third choice of using (9.18) with $\alpha = 0$ to determine $\phi^{(h,k)}$ and (9.19) with $\alpha = 1$ to determine $\mathbf{u}^{(h,k)}$. In the last case, we can solve for both $\phi^{(h,k)}$ and $\mathbf{u}^{(h,k)}$ for $k = 1,\ldots,K$ totally independently of each other. As a result, for this case, one could use different spatial grids and different time steps in each of (9.18) and (9.19). This remains true if we use (9.18) and (9.19) with $\alpha = 0$ or 1 in both equations; however, in these cases, one has to evaluate, e.g., by interpolation or projection, $\phi^{(h,k)}(\mathbf{x})$ on the grid used for $\mathbf{u}^{(h,k)}$ (for $\alpha = 0$) or $\mathbf{u}^{(h,k)}$ on the grid used for $\phi^{(h,k)}(\mathbf{x})$ (for $\alpha = 1$.) For the sake of simplicity, we assume, in all cases, that the same grid is used to determine $\phi^{(h,k)}$ and $\mathbf{u}^{(h,k)}$.

Remark 9.2 It also of interest to examine the relation between (9.12) with $\alpha = 1$ and the backward-Euler mixed-Galerkin method for (9.4) given by

$$\begin{aligned}\int_\Omega \mathbf{u}^{(h,k)} \cdot \Theta_1{}^{-1}\mathbf{v}^h d\Omega - \int_\Omega \phi^{(h,k)} \nabla\cdot\mathbf{v}^h d\Omega &= \int_\Omega \mathbf{u}^{(h,k-1)} \cdot \Theta_1{}^{-1}\mathbf{v}^h d\Omega \\ -\int_\Omega \phi^{(h,k-1)}\nabla\cdot\mathbf{v}^h d\Omega \qquad &\forall \mathbf{v}^h \in \mathbf{V}^h_{\Gamma^*}(\Omega)\subset \mathbf{D}_{\Gamma^*}(\Omega)\end{aligned} \tag{9.20}$$

and

$$\begin{aligned}\int_\Omega \frac{\phi^{(h,k)}-\phi^{(h,k-1)}}{\Delta t_k}\psi^h d\Omega + \int_\Omega c^{-1}\psi^h\nabla\cdot\mathbf{u}^{(h,k)} d\Omega &= \\ \int_\Omega c^{-1} f(t_k)\psi^h d\Omega \qquad &\forall \psi^h \in S^h(\Omega)\subset S(\Omega)\,.\end{aligned} \tag{9.21}$$

In case of a compatible FD-LSFEM, e.g., if $\mathbf{V}^h_{\Gamma^*}$ is defined using the second or the third space in (9.14), there exists a matching scalar space $S^h(\Omega)\subset S(\Omega)$ such that $\{\mathbf{V}^h_{\Gamma^*}(\Omega), S^h(\Omega)\}$ is a stable pair for mixed-Galerkin methods for the Kelvin principle. Furthermore, let us assume that, as in Theorem 5.37, the mesh partition $\mathcal{T}_h$ is affine and that the pair $\{\mathbf{V}^h_{\Gamma^*}(\Omega), S^h(\Omega)\}$ belongs to one of the finite element De Rham complexes in (B.45). Then, the divergence is a bounded surjection $\mathbf{V}^h_{\Gamma^*}(\Omega)\mapsto S^h(\Omega)$ so that we may choose, in (9.21), $\psi^h = \nabla\cdot\mathbf{v}^h$. Combining this result with (9.20), one easily sees that (9.20) and (9.21) are exactly the same as (9.19) with $\alpha = 1$. Thus, *for the first-order system* (9.4), *the backward-Euler FD-LSFEM with* $\alpha = 1$ *based on finite element spaces that are compatible with the Kelvin principle recovers exactly the same approximate solution for the vector variable as does the backward-Euler mixed-Galerkin finite element method* (9.20) *and* (9.21).

In the case of a non-compatible, nodal FD-LSFEM, e.g., if $\mathbf{V}^h_{\Gamma*}$ is defined using the first space in (9.14), then the identification of (9.19) with the Kelvin principle discretization (9.20) and (9.21) is no longer possible because that type of finite element space is not a stable space for approximating the Kelvin principle. This is one of the drawbacks associated with using standard nodal elements for the vector variable. Remark 5.24 lists other drawbacks which, of course, remain in full force in the present context. The adverse impact of nodal finite element approximations of the vector field on the accuracy of LSFEMs is illustrated in Section 5.9.1. □

Error estimates for fully-discrete approximations

We first consider error estimates for the approximations of ϕ determined from (9.18) with $\alpha = 0$. Because (9.18) with $\alpha = 0$ is exactly the same as the backward-Euler Galerkin finite element method for (9.1), we can use well-known results about such a discretization of (9.1) to ascertain the same results about approximations of the scalar variable ϕ obtained from the backward-Euler LSFEM with $\alpha = 0$. For example, the following error estimates for sufficiently smooth solutions of the heat equation (9.1) can be found in [336].

Theorem 9.3 *Let* $G^h_\Gamma = G^r(\Omega) \cap G_\Gamma(\Omega)$ *denote a Lagrange finite element space consisting of continuous piecewise polynomial functions of degree less than or equal to* r *so that* (9.15) *holds. Let* $\Delta t = \max_{k=1,\dots,K} \Delta t_k$. *Let* $\phi(\mathbf{x},t)$ *denote the solution of* (9.1). *Let* $\phi^{(h,0)}(\mathbf{x}) \in G^h_\Gamma$ *denote an approximation of the initial data* $\phi_0(\mathbf{x})$ *in* (9.1) *and assume that* $\|\phi_0 - \phi^{(h,0)}\|_0 \le Ch^{r+1}\|\phi_0\|_{r+1}$. *For* $k = 1,\dots,K$, *let* $\phi(t_k) = \phi(\mathbf{x},t_k)$ *and let* $\phi^{(h,k)}(\mathbf{x}) \in G^h_\Gamma$ *be the scalar-valued variable that solves* (9.17) *or, equivalently,* (9.18), *with* $\alpha = 0$. *Then, we have that*

$$\|\phi(t_k) - \phi^{(h,k)}\|_0 \le C\big(h^{r+1} + \Delta t\big) \tag{9.22}$$

and

$$\|\nabla\phi(t_k) - \nabla\phi^{(h,k)}\|_0 \le C\big(h^{r} + \Delta t\big) \tag{9.23}$$

for $k = 0,\dots,K$, *where* C *is independent of* h *and* Δt *but depends continuously on* $\|\phi_0\|_{r+1}$, $\|\phi_t\|_{L^2(0,T;H^{r+1}(\Omega))}$, *and* $\|\phi_{tt}\|_{L^2(0,T;L^2(\Omega))}$. □

Thus, we obtain optimal error estimates for ϕ and $\nabla\phi$ out of the spaces $G^r(\Omega)$ and $\nabla G^r(\Omega) = \{\mathbf{w}^h \in [L^2(\Omega)]^d \mid \mathbf{w}^h = \nabla\psi^h \ \forall\, \psi^h \in G^r(\Omega)\}$, respectively.

At this point, we have several choices for defining an approximation to $\mathbf{u}(\mathbf{x},t)$. We could, of course, use $-\Theta_1\nabla\phi^{(h,k)}$ as an approximation to $\mathbf{u}(\mathbf{x},t_k)$ for $k = 1,\dots,K$. Because Θ_1 is not necessarily constant, in general, $-\Theta_1\nabla\phi^{(h,k)}$ is not a finite element function.[13] However, in any case, we can use (9.23) to obtain the same error estimate (with a different value of C, of course) for $\|\mathbf{u}(t_k) - (-\Theta_1 \nabla \phi^{(h,k)})\|_0$. Note that, even if Θ_1 is a scalar constant, $\Theta_1 \nabla \phi^{(h,k)} \notin \mathbf{D}_{\Gamma^*}(\Omega)$ for $\phi^{(h,k)}$ if G^h_Γ consists of finite element functions that are merely continuous over Ω. Thus, $-\Theta_1 \nabla \phi^{(h,k)}$ may be viewed as a non-conforming approximation to $\mathbf{u}$.[14]

Alternate (conforming) approximations to $\mathbf{u}$ can be obtained from (9.19) with either $\alpha = 0$ or 1. The following two results are adapted from [309].

Theorem 9.4 *Let the hypotheses of Theorem* 9.3 *hold. Assume that* $\mathbf{V}^h_{\Gamma^*}$ *is a div-conforming finite element space, i.e., that* $\mathbf{V}^h_{\Gamma^*}$ *is defined using the second or the third space in* (9.14). *Let* $\mathbf{u}(\mathbf{x},t) = -\Theta_1 \nabla \phi$. *For* $k = 1,\ldots,K$, *let* $\mathbf{u}(t_k) = \mathbf{u}(\mathbf{x},t_k)$ *and let* $\mathbf{u}^{(h,k)}(\mathbf{x}) \in \mathbf{V}^h_{\Gamma^*}$ *be the vector-valued variable that solves* (9.17) *or, equivalently,* (9.19), *with* $\alpha = 0$. *Then, for* $k = 0,\ldots,K$, *we have that*

$$
\begin{aligned}
&\|\mathbf{u}(t_k) - \mathbf{u}^{(h,k)}\|_0 + \Delta t_k^{1/2} \|\nabla \cdot (\mathbf{u}(t_k) - \mathbf{u}^{(h,k)})\|_0 \\
&\qquad \le C\big(h^{s_1} + \Delta t^{1/2} h^s + \Delta t + \min\{h^r, \Delta t^{-1/2} h^{r+1}\}\big)\,,
\end{aligned}
\tag{9.24}
$$

where C is independent of h and Δt but depends continuously on $\|\phi_0\|_{r+1}$, $\|\phi_t\|_{L^2(0,T;H^{r+1}(\Omega))}$, $\|\phi_{tt}\|_{L^2(0,T;L^2(\Omega))}$, *and* $\|\mathbf{u}\|_{L^\infty(0,T;[H^{s+1}(\Omega)]^d)}$.

Proof. From (9.4), we have that

$$
\int_\Omega \frac{\Delta t_k}{c}\left(c\frac{\partial \phi}{\partial t} + \nabla \cdot \mathbf{u} - f\right) \nabla \cdot \mathbf{v}\, d\Omega = 0 \qquad \forall \mathbf{v} \in \mathbf{D}_{\Gamma^*}(\Omega)
$$

and

$$
\int_\Omega (\mathbf{u} + \Theta_1 \nabla \phi) \cdot \Theta_1{}^{-1} \mathbf{v}\, d\Omega = 0 \qquad \forall \mathbf{v} \in \mathbf{D}_{\Gamma^*}(\Omega)
$$

so that

$$
\begin{aligned}
&\int_\Omega \left(\frac{\Delta t_k}{c} \nabla \cdot \mathbf{u}(t_k) \nabla \cdot \mathbf{v} + \mathbf{u}(t_k) \cdot \Theta_1{}^{-1} \mathbf{v}\right) d\Omega \\
&\quad = \int_\Omega \left(\frac{\Delta t_k}{c} f(t_k) + \phi(t_{k-1}) + \Delta t_k \rho(t_k)\right) \nabla \cdot \mathbf{v}\, d\Omega \qquad \forall \mathbf{v} \in \mathbf{D}_{\Gamma^*}(\Omega)\,,
\end{aligned}
\tag{9.25}
$$

where

$$
\rho(\mathbf{x},t_k) = \frac{\phi(\mathbf{x},t_k) - \phi(\mathbf{x},t_{k-1})}{\Delta t_k} - \frac{\partial \phi}{\partial t}(\mathbf{x},t_k)
$$

and where we have used the fact that

$$
\int_\Omega (\phi(t_k) \nabla \cdot \mathbf{v} + \nabla \phi(t_k) \cdot \mathbf{v})\, d\Omega = 0
$$

---

[13] In fact, $-\Theta_1 \nabla \phi^{(h,k)} \in \nabla G^r(\Omega)$ only if Θ_1 is a scalar constant.

[14] In the terminology of Chapter 5, $-\Theta_1 \nabla \phi \in \mathbf{D}_{\Gamma^*}(\Omega)$ and $-\nabla \phi \in \mathbf{C}_\Gamma(\Omega)$ are the *flux* and the *intensity*, respectively; see Section 5.3.1. Properties of compatible finite elements (see Section B.3) imply that $\nabla \phi^{(h,k)} \in \mathbf{C}_\Gamma(\Omega)$, i.e., $\nabla \phi^{(h,k)}$ is a conforming approximation of the intensity.

for $\phi(t_k) \in G_\Gamma$ and $\mathbf{v} \in V_{\Gamma^*}$. Combining (9.19) with $\alpha = 0$ and (9.25), we obtain

$$\int_\Omega \left(\frac{\Delta t_k}{c} \nabla \cdot (\mathbf{u}(t_k) - \mathbf{u}^{(h,k)}) \nabla \cdot \mathbf{v}^h + (\mathbf{u}(t_k) - \mathbf{u}^{(h,k)}) \cdot \Theta_1{}^{-1} \mathbf{v}^h \right) d\Omega$$

$$= \int_\Omega (\phi(t_{k-1}) - \phi^{(h,k-1)}) \nabla \cdot \mathbf{v}^h \, d\Omega + \int_\Omega \Delta t_k \rho(t_k) \nabla \cdot \mathbf{v}^h \, d\Omega \qquad \forall \mathbf{v}^h \in \mathbf{V}^h_{\Gamma^*}.$$

Then, for any $\mathbf{w}^{(h,k)}(\mathbf{x}) \in \mathbf{V}^h_{\Gamma^*}$, we have that

$$\begin{aligned} &\int_\Omega \left(\frac{\Delta t_k}{c} \nabla \cdot (\mathbf{u}^{(h,k)} - \mathbf{w}^{(h,k)}) \nabla \cdot \mathbf{v}^h + (\mathbf{u}^{(h,k)} - \mathbf{w}^{(h,k)}) \cdot \Theta_1{}^{-1} \mathbf{v}^h \right) d\Omega \\ &= \int_\Omega \left(\frac{\Delta t_k}{c} \nabla \cdot (\mathbf{u}(t_k) - \mathbf{w}^{(h,k)}) \nabla \cdot \mathbf{v}^h + (\mathbf{u}(t_k) - \mathbf{w}^{(h,k)}) \cdot \Theta_1{}^{-1} \mathbf{v}^h \right) d\Omega \\ &- \int_\Omega (\phi(t_{k-1}) - \phi^{(h,k-1)}) \nabla \cdot \mathbf{v}^h \, d\Omega - \int_\Omega \Delta t_k \rho(t_k) \nabla \cdot \mathbf{v}^h \, d\Omega \quad \forall \mathbf{v}^h \in \mathbf{V}^h_{\Gamma^*}. \end{aligned} \tag{9.26}$$

Setting $\mathbf{v}^h = \mathbf{u}^{(h,k)} - \mathbf{w}^{(h,k)} \in \mathbf{V}^h_{\Gamma^*}$ and using standard inequalities, we then obtain, for any $\mathbf{w}^{(h,k)}(\mathbf{x}) \in \mathbf{V}^h_{\Gamma^*}$,

$$\begin{aligned} &\|\mathbf{u}(t_k) - \mathbf{u}^{(h,k)}\|_0 + \Delta t_k^{1/2} \|\nabla \cdot (\mathbf{u}(t_k) - \mathbf{u}^{(h,k)})\|_0 \\ &\quad \le C\Big(\|\mathbf{u}(t_k) - \mathbf{w}^{(h,k)}\|_0 + \Delta t_k^{1/2} \|\nabla \cdot (\mathbf{u}(t_k) - \mathbf{w}^{(h,k)})\|_0 \\ &\qquad + \Delta t_k^{-1/2} \|(\phi(t_{k-1}) - \phi^{(h,k-1)})\|_0 + \Delta t_k^{1/2} \|\rho(t_k)\|_0 \Big), \end{aligned} \tag{9.27}$$

where C is independent of h, Δt_k, ϕ, and $\mathbf{u}$. On the other hand, because

$$-\int_\Omega (\phi(t_{k-1}) - \phi^{(h,k-1)}) \nabla \cdot \mathbf{v}^h \, d\Omega = \int_\Omega \nabla (\phi(t_{k-1}) - \phi^{(h,k-1)}) \cdot \mathbf{v}^h \, d\Omega,$$

we can, in a similar manner, obtain from (9.26)

$$\begin{aligned} &\|\mathbf{u}(t_k) - \mathbf{u}^{(h,k)}\|_0 + \Delta t_k^{1/2} \|\nabla \cdot (\mathbf{u}(t_k) - \mathbf{u}^{(h,k)})\|_0 \\ &\quad \le C\Big(\|\mathbf{u}(t_k) - \mathbf{w}^{(h,k)}\|_0 + \Delta t_k^{1/2} \|\nabla \cdot (\mathbf{u}(t_k) - \mathbf{w}^{(h,k)})\|_0 \\ &\qquad + \|\nabla(\phi(t_{k-1}) - \phi^{(h,k-1)})\|_0 + \Delta t_k^{1/2} \|\rho(t_k)\|_0 \Big) \end{aligned} \tag{9.28}$$

for any $\mathbf{w}^{(h,k)}(\mathbf{x}) \in \mathbf{V}^h_{\Gamma^*}$, where again C is independent of h, Δt_k, ϕ, and $\mathbf{u}$. It is a standard result (see, e.g., [336]) that $\|\rho(t_k)\|_0 \le C(\Delta t_k)^{1/2}$, where C depends on $\|\phi_{tt}\|_{L^2(0,T;L^2(\Omega))}$. Then, using this result along with (9.16) and (9.22) in (9.27) and (9.28), we obtain (9.24). □

Theorem 9.5 *Assume that the hypotheses of Theorem* 9.3 *hold, and that* $\mathbf{V}^h_{\Gamma^*}$ *is a div-conforming finite element space so that* (9.16) *holds. Let* $\mathbf{u}(\mathbf{x},t) = -\Theta_1 \nabla \phi$. *For*

$k = 0,\dots,K$, let $\mathbf{u}(t_k) = \mathbf{u}(\mathbf{x},t_k)$ and let $\mathbf{u}^{(h,k)}(\mathbf{x}) \in \mathbf{V}^h_{\Gamma*}$ be the vector-valued variable that solves (9.17) *or, equivalently,* (9.19), *with $\alpha = 1$. Assume that $\mathbf{u}^{(h,0)}(\mathbf{x}) \in \mathbf{V}^h_{\Gamma*}$ satisfies*

$$\|\mathbf{u}_0 - \mathbf{u}^{(h,0)}\|_0 \le Ch^{s_1}\|\mathbf{u}_0\|_{s_1},$$

where $\mathbf{u}_0 = -\Theta_1{}^{-1}\nabla\phi_0$. Then, for $k = 0,\dots,K$, we have that

$$\|\mathbf{u}(t_k)-\mathbf{u}^{(h,k)}\|_0^2 + \Big(\sum_{m=1}^{k}\Delta t_m\|\nabla\cdot\big(\mathbf{u}(t_k)-\mathbf{u}^{(h,k)}\big)\|_0^2\Big)^{1/2} \le C\big(h^{s_1}+h^s+\Delta t\big), \quad (9.29)$$

where C is independent of h and Δt but is a continuous function of $\|\phi_{tt}\|^2_{L^2(0,T;L^2(\Omega))}$, $\|\mathbf{u}_0\|_{s_1}$, $\|\mathbf{u}\|_{L^\infty(0,T;[H^{s+1}(\Omega)]^d)}$, and $\|\mathbf{u}_t\|_{L^2(0,T;[H^{s_1}(\Omega)]^d)}$.

Proof. From (9.4) and (9.25), one easily obtains that

$$\begin{aligned}\int_\Omega\Big(\frac{\Delta t_k}{c}\nabla\cdot\mathbf{u}(t_k)\nabla\cdot\mathbf{v}+\mathbf{u}(t_k)\cdot\Theta_1{}^{-1}\mathbf{v}\Big)\,d\Omega \\ = \int_\Omega\Big(\frac{\Delta t_k}{c}f(t_k)+\phi(t_{k-1})+\Delta t_k\rho(t_k)\Big)\nabla\cdot\mathbf{v}\,d\Omega \\ + \int_\Omega\Big(\mathbf{u}(t_{k-1})\cdot\Theta_1{}^{-1}+\nabla\phi(t_{k-1})\Big)\cdot\mathbf{v}\,d\Omega \qquad \forall\mathbf{v}\in\mathbf{D}_{\Gamma^*}(\Omega)\end{aligned}$$

so that, using the fact that

$$\int_\Omega\Big(\phi(t_{k-1})\nabla\cdot\mathbf{v}+\nabla\phi(t_{k-1})\cdot\mathbf{v}\Big)\,d\Omega = 0$$

for $\phi(t_{k-1}) \in G_\Gamma(\Omega)$ and $\mathbf{v}\in\mathbf{D}_{\Gamma^*}(\Omega)$, we have that

$$\begin{aligned}\int_\Omega\Big(\frac{\Delta t_k}{c}\nabla\cdot\mathbf{u}(t_k)\nabla\cdot\mathbf{v}+\mathbf{u}(t_k)\cdot\Theta_1{}^{-1}\mathbf{v}\Big)\,d\Omega \\ = \int_\Omega\Big(\mathbf{u}(t_{k-1})\cdot\Theta_1{}^{-1}\mathbf{v}+\big(\frac{\Delta t_k}{c}f(t_k)+\Delta t_k\rho(t_k)\big)\nabla\cdot\mathbf{v}\Big)d\Omega \quad \forall\mathbf{v}\in\mathbf{D}_{\Gamma^*}(\Omega).\end{aligned}$$

Combining this result with (9.19) with $\alpha = 1$, one obtains

$$\begin{aligned}\int_\Omega\Big(\frac{\Delta t_k}{c}\nabla\cdot\big(\mathbf{u}(t_k)-\mathbf{u}^{(h,k)}\big)\nabla\cdot\mathbf{v}^h+\big(\mathbf{u}(t_k)-\mathbf{u}^{(h,k)}\big)\cdot\Theta_1{}^{-1}\mathbf{v}^h\Big)\,d\Omega \\ = \int_\Omega\Big(\big(\mathbf{u}(t_{k-1})-\mathbf{u}^{(h,k-1)}\big)\cdot\Theta_1{}^{-1}\mathbf{v}^h+\Delta t_k\rho(t_k)\nabla\cdot\mathbf{v}^h\Big)d\Omega \qquad \forall\mathbf{v}^h\in\mathbf{V}^h_{\Gamma*}\end{aligned}$$

or, for arbitrary $\mathbf{w}^{(h,k)}, \mathbf{w}^{(h,k-1)} \in \mathbf{V}^h_{\Gamma*}$,

$$\int_{\Omega}\left(\frac{\Delta t_k}{c}\nabla\cdot\left(\mathbf{u}^{(h,k)}-\mathbf{w}^{(h,k)}\right)\nabla\cdot\mathbf{v}^h+\left(\mathbf{u}^{(h,k)}-\mathbf{w}^{(h,k)}\right)\cdot{\Theta_1}^{-1}\mathbf{v}^h\right)d\Omega$$
$$=\int_{\Omega}\left(\left(\mathbf{u}^{(h,k-1)}-\mathbf{w}^{(h,k-1)}\right)\cdot{\Theta_1}^{-1}\mathbf{v}^h\right)d\Omega$$
$$+\int_{\Omega}\left(\frac{\Delta t_k}{c}\nabla\cdot\left(\mathbf{u}(t_k)-\mathbf{w}^{(h,k)}\right)\nabla\cdot\mathbf{v}^h-\Delta t_k\rho(t_k)\nabla\cdot\mathbf{v}^h\right)d\Omega$$
$$+\int_{\Omega}\left(\left(\mathbf{u}(t_k)-\mathbf{w}^{(h,k)}\right)-\left(\mathbf{u}(t_{k-1})-\mathbf{w}^{(h,k-1)}\right)\right)\cdot{\Theta_1}^{-1}\mathbf{v}^h\,d\Omega\,.$$

Setting $\mathbf{v}^h=\mathbf{u}^{(h,k)}-\mathbf{w}^{(h,k)}\in\mathbf{V}^h_{\Gamma*}$ and applying standard inequalities yields

$$\|{\Theta_1}^{-1/2}\left(\mathbf{u}^{(h,k)}-\mathbf{w}^{(h,k)}\right)\|_0^2+\Delta t_k\|c^{-1/2}\nabla\cdot\left(\mathbf{u}^{(h,k)}-\mathbf{w}^{(h,k)}\right)\|_0^2$$
$$\le\frac{1+\Delta t_k}{2}\|{\Theta_1}^{-1/2}\left(\mathbf{u}^{(h,k)}-\mathbf{w}^{(h,k)}\right)\|_0^2+\frac{\Delta t_k}{2}\|c^{-1/2}\nabla\cdot\left(\mathbf{u}^{(h,k)}-\mathbf{w}^{(h,k)}\right)\|_0^2$$
$$+\frac{1}{2}\|{\Theta_1}^{-1/2}\left(\mathbf{u}^{(h,k-1)}-\mathbf{w}^{(h,k-1)}\right)\|_0^2+C\Delta t_k\Big(\|\nabla\cdot\left(\mathbf{u}(t_k)-\mathbf{w}^{(h,k)}\right)\|_0^2$$
$$+\|\rho(t_k)\|_0^2+\frac{1}{\Delta t_k}\|\left(\mathbf{u}(t_k)-\mathbf{w}^{(h,k)}\right)-\left(\mathbf{u}(t_{k-1})-\mathbf{w}^{(h,k-1)}\right)\|_0^2\Big)$$

or

$$\|{\Theta_1}^{-1/2}\left(\mathbf{u}^{(h,k)}-\mathbf{w}^{(h,k)}\right)\|_0^2+\Delta t_k\|c^{-1/2}\nabla\cdot\left(\mathbf{u}^{(h,k)}-\mathbf{w}^{(h,k)}\right)\|_0^2$$
$$\le\|{\Theta_1}^{-1/2}\left(\mathbf{u}^{(h,k-1)}-\mathbf{w}^{(h,k-1)}\right)\|_0^2+\Delta t_k\|{\Theta_1}^{-1/2}\left(\mathbf{u}^{(h,k)}-\mathbf{w}^{(h,k)}\right)\|_0^2$$
$$+C\Delta t_k\Big(\|\nabla\cdot\left(\mathbf{u}(t_k)-\mathbf{w}^{(h,k)}\right)\|_0^2+\|\rho(t_k)\|_0^2$$
$$+\frac{1}{\Delta t_k}\|\left(\mathbf{u}(t_k)-\mathbf{w}^{(h,k)}\right)-\left(\mathbf{u}(t_{k-1})-\mathbf{w}^{(h,k-1)}\right)\|_0^2\Big)$$

so that by summing,[15]

$$\begin{aligned}&\|{\Theta_1}^{-1/2}\left(\mathbf{u}^{(h,k)}-\mathbf{w}^{(h,k)}\right)\|_0^2+\sum_{m=1}^{k}\Delta t_m\|c^{-1/2}\nabla\cdot\left(\mathbf{u}^{(h,m)}-\mathbf{w}^{(h,m)}\right)\|_0^2\\&\le\sum_{m=1}^{k}\Delta t_m\|{\Theta_1}^{-1/2}\left(\mathbf{u}^{(h,m)}-\mathbf{w}^{(h,m)}\right)\|_0^2+C\Big(\|\mathbf{u}^{(h,0)}-\mathbf{w}^{(h,0)}\|_0^2\\&\quad+\sum_{m=1}^{k}\Delta t_m\|\nabla\cdot\left(\mathbf{u}(t_m)-\mathbf{w}^{(h,m)}\right)\|_0^2+\sum_{m=1}^{k}\Delta t_m\|\rho(t_m)\|_0^2\\&\quad+\sum_{m=1}^{k}\|\left(\mathbf{u}(t_m)-\mathbf{w}^{(h,m)}\right)-\left(\mathbf{u}(t_{m-1})-\mathbf{w}^{(h,m-1)}\right)\|_0^2\Big).\end{aligned}\tag{9.30}$$

[15] In each of (9.30)–(9.34), the constant C is independent of h, Δt_k, and $\mathbf{u}$.

From the hypothesis of the theorem, we have that there exists $\mathbf{w}^{(h,0)} \in \mathbf{V}^h_{\Gamma*}$ such that

$$\|\mathbf{u}^{(h,0)} - \mathbf{w}^{(h,0)}\|_0 \le \|\mathbf{u}_0 - \mathbf{u}^{(h,0)}\|_0 + \|\mathbf{u}_0 - \mathbf{w}^{(h,0)}\|_0 \le Ch^{s_1}\|\mathbf{u}_0\|_{s_1}. \qquad (9.31)$$

From (9.16), we have that there exist $\mathbf{w}^{(h,m)} \in \mathbf{V}^h_{\Gamma*}$, $m = 1,\dots,K$, such that

$$\sum_{m=1}^{k} \Delta t_m \|\nabla\cdot\big(\mathbf{u}(t_m) - \mathbf{w}^{(h,m)}\big)\|_0^2 \le Ch^{2s}\|\mathbf{u}\|^2_{L^\infty(0,T;H^{s+1}(\Omega))}. \qquad (9.32)$$

It is shown in [336] that whenever (9.15) holds,

$$\sum_{m=1}^{k} \Delta t_m \|\rho(t_m)\|_0^2 \le C\Delta t^2\|\phi_{tt}\|^2_{L^2(0,T;L^2(\Omega))}. \qquad (9.33)$$

Using (9.16) (applied to $\mathbf{u}_t$), we have that there exist $\mathbf{w}^{(h,m)} \in \mathbf{V}^h_{\Gamma*}$, $m = 1,\dots,K$, such that

$$\begin{aligned}
\sum_{m=1}^{k} &\|\big(\mathbf{u}(t_m) - \mathbf{w}^{(h,m)}\big) - \big(\mathbf{u}(t_{m-1}) - \mathbf{w}^{(h,m-1)}\big)\|_0^2 \\
&= \Big\|\sum_{m=1}^{k}\int_{t_{m-1}}^{t_m} \big(\mathbf{u}(t) - \widetilde{\mathbf{w}}^{(h,m)}(t)\big)_{,t}\,dt\Big\|_0^2 \\
&\le Ch^{2s_1}\sum_{m=1}^{k}\int_{t_{m-1}}^{t_m} \|\mathbf{u}_t\|_{s_1}^2\,dt \le Ch^{2s_1}\|\mathbf{u}_t\|^2_{L^2(0,T;H^{s_1}(\Omega))},
\end{aligned} \qquad (9.34)$$

where $\widetilde{\mathbf{w}}^{(h,m)}(t)$ denotes the linear interpolant (in time) for the data $\{t_{m-1},\mathbf{w}^{(h,m-1)}\}$ and $\{t_m,\mathbf{w}^{(h,m)}\}$.

Combining (9.30)–(9.34), we obtain, for $k = 1,\dots,K$,

$$\begin{aligned}
\|\Theta_1{}^{-1/2}\big(\mathbf{u}^{(h,k)} - \mathbf{w}^{(h,k)}\big)\|_0^2 &+ \sum_{m=1}^{k}\Delta t_m\|c^{-1/2}\nabla\cdot\big(\mathbf{u}^{(h,m)} - \mathbf{w}^{(h,m)}\big)\|_0^2 \\
&\le \sum_{m=1}^{k}\Delta t_m\|\Theta_1{}^{-1/2}\big(\mathbf{u}^{(h,m)} - \mathbf{w}^{(h,m)}\big)\|_0^2 + C\big(h^{2s} + h^{2s_1} + \Delta t^2\big),
\end{aligned}$$

where C is independent of h and Δt_k but is a continuous function of $\|\phi_{tt}\|^2_{L^2(0,T;L^2(\Omega))}$, $\|\mathbf{u}_0\|_{s+1}$, $\|\mathbf{u}\|_{L^\infty(0,T;[H^{s+1}(\Omega)]^d)}$, and $\|\mathbf{u}_t\|_{L^2(0,T;[H^{s_1}(\Omega)]^d)}$. Then, an application of the discrete Gronwall inequality, the triangle inequality, and (9.16) results in (9.29). □

Of course, we could use (9.18) with $\alpha = 1$ instead of 0 to determine approximations of the scalar variable ϕ. Although there is no good reason to do so, for the sake of completeness, we state an error estimate for this choice; the proof can be found in [309].

Theorem 9.6 *Let the hypotheses of Theorems* 9.3 *and* 9.5 *hold, except that now* $\phi^{(h,k)}(\mathbf{x}) \in G^h_\Gamma$ *denotes the scalar-valued variable that solves* (9.17) *or, equivalently,* (9.18), *with* $\alpha = 1$. *Then, for* $k = 0,\dots,K$, *we have that*

$$\|\phi(t_k) - \phi^{(h,k)}\|_0 \le C\big(h^{r+1} + h^s + h^{s_1} + \Delta t\big), \tag{9.35}$$

where C *is independent of* h *and* Δt *but depends continuously on* $\|\phi_0\|_{r+1}$, $\|\phi_t\|_{L^2(0,T;H^{r+1}(\Omega))}$, $\|\phi_{tt}\|_{L^2(0,T;L^2(\Omega))}$, $\|\mathbf{u}\|_{L^\infty(0,T;[H^{s+1}(\Omega)]^d)}$, *and* $\|\mathbf{u}_t\|_{L^2(0,T;[H^{s_1}(\Omega)]^d)}$.

9.1.2 Second-Order Time Accurate Least-Squares Finite Element Methods

Crank–Nicolson least-squares finite element methods

A more accurate time discretization results from using the *Crank–Nicolson method* instead of a backward-Euler method. Now, there are three ways one can define a Crank–Nicolson scheme for (9.4). For $\alpha = -1$, 0, or 1, we have the sequence of semi-discrete in time systems: for $k = 1,\dots,K$,

$$\begin{cases} \dfrac{c}{\Delta t_k}\phi^{(k)} + \dfrac{1}{2}\nabla\cdot\mathbf{u}^{(k)} = \dfrac{1}{2}\Big(f(t_k) + f(t_{k-1})\Big) & \\ \qquad\qquad + \dfrac{c}{\Delta t_k}\phi^{(k-1)} - \dfrac{1}{2}\nabla\cdot\mathbf{u}^{(k-1)} & \text{for } \mathbf{x}\in\Omega \\ \mathbf{u}^{(k)} + \Theta_1\nabla\phi^{(k)} = \alpha\Big(\mathbf{u}^{(k-1)} + \Theta_1\nabla\phi^{(k-1)}\Big) & \text{for } \mathbf{x}\in\Omega \\ \phi^{(k)} = g_D(\mathbf{x},t_k) & \text{for } \mathbf{x}\in\Gamma \\ \mathbf{u}^{(k)}\cdot\mathbf{n} = g_N(\mathbf{x},t_k) & \text{for } \mathbf{x}\in\Gamma^*. \end{cases} \tag{9.36}$$

Note that we now not only have to set $\phi^{(0)} = \phi_0(\mathbf{x})$, but for all three values of α, we also have to set $\mathbf{u}^{(0)} = -{\Theta_1}^{-1}\nabla\phi_0$.

For (9.36), we define the sequence of least-squares functionals

$$\begin{aligned} &J^{\mathsf{cn}}_k(\phi^{(k)},\mathbf{u}^{(k)}; f(t_k), f(t_{k-1}), \phi^{(k-1)}, \mathbf{u}^{(k-1)}) \\ &\quad = \Bigg\| \left(\frac{\Delta t_k}{c}\right)^{1/2}\Bigg(\frac{c}{\Delta t_k}\phi^{(k)} + \frac{1}{2}\nabla\cdot\mathbf{u}^{(k)} \\ &\qquad\qquad - \frac{f(t_k)+f(t_{k-1})}{2} - \frac{c}{\Delta t_k}\phi^{(k-1)} + \frac{1}{2}\nabla\cdot\mathbf{u}^{(k-1)}\Bigg)\Bigg\|_0^2 \\ &\qquad + \Big\|{\Theta_1}^{-1/2}\Big(\mathbf{u}^{(k)} + \Theta_1\nabla\phi^{(k)} - \alpha\big(\mathbf{u}^{(k-1)} + \Theta_1\nabla\phi^{(k-1)}\big)\Big)\Big\|_0^2 \end{aligned} \tag{9.37}$$

for $k = 1,\dots,K$ and the associated sequence of CSLPs

$$\begin{cases} J_k^{\mathsf{cn}}(\phi^{(k)},\mathbf{u}^{(k)}; f(t_k), f(t_{k-1}), \phi^{(k-1)}, \mathbf{u}^{(k-1)}) \\ \\ X = G_\Gamma(\Omega)\times \mathbf{D}_{\Gamma^*}(\Omega)\,, \end{cases} \tag{9.38}$$

where $\phi^{(0)}(\mathbf{x}) = \phi_0(\mathbf{x})$ and $\mathbf{u}^{(0)}(\mathbf{x}) = -\Theta_1\nabla\phi_0(\mathbf{x})$.

The necessary conditions corresponding to the sequence of minimization problems (9.38) is given by: for $k = 1,\dots,K$,

$$\begin{aligned} Q_k^{\mathsf{cn}}(\phi^{(k)},\mathbf{u}^{(k)};\psi,\mathbf{v}) = F_k^{\mathsf{cn}}(\psi,\mathbf{v}; f(t_k), f(t_{k-1}), \phi^{(k-1)}, \mathbf{u}^{(k-1)}) \\ \forall\, \psi\in G_\Gamma(\Omega),\ \mathbf{v}\in \mathbf{D}_{\Gamma^*}(\Omega)\,, \end{aligned} \tag{9.39}$$

where, for each $k = 1,\dots,K$ and for all $\psi\in G_\Gamma(\Omega)$ and $\mathbf{v}\in\mathbf{D}_{\Gamma^*}(\Omega)$,

$$\begin{aligned} Q_k^{\mathsf{cn}}(\phi^{(k)},\mathbf{u}^{(k)};\psi,\mathbf{v}) &= \frac{1}{2}\int_\Omega \Big(\mathbf{u}^{(k)} + \Theta_1\nabla\phi^{(k)}\Big)\cdot{\Theta_1}^{-1}\Big(\mathbf{v}+\Theta_1\nabla\psi\Big)\,d\Omega \\ &\quad + \int_\Omega \frac{\Delta t_k}{c}\Big(\frac{c}{\Delta t_k}\phi^{(k)} + \frac{1}{2}\nabla\cdot\mathbf{u}^{(k)}\Big)\Big(\frac{c}{\Delta t_k}\psi + \frac{1}{2}\nabla\cdot\mathbf{v}\Big)\,d\Omega \end{aligned}$$

and

$$\begin{aligned} &F_k^{\mathsf{cn}}(\psi,\mathbf{v}; f(t_k), f(t_{k-1}), \phi^{(k-1)}, \mathbf{u}^{(k-1)}) \\ &= \frac{\alpha}{2}\int_\Omega \Big(\mathbf{u}^{(k-1)} + \Theta_1\nabla\phi^{(k-1)}\Big)\cdot{\Theta_1}^{-1}\Big(\mathbf{v}+\Theta_1\nabla\psi\Big)\,d\Omega \\ &\quad \int_\Omega \frac{\Delta t_k}{c}\Big(\frac{f(t_k)+f(t_{k-1})}{2} + \frac{c}{\Delta t_k}\phi^{(k-1)} - \frac{1}{2}\nabla\cdot\mathbf{u}^{(k-1)}\Big)\Big(\frac{c}{\Delta t_k}\psi + \frac{1}{2}\nabla\cdot\mathbf{v}\Big)\,d\Omega\,. \end{aligned}$$

Expanding the definition of $Q_k^{\mathsf{cn}}(\cdot,\cdot;\cdot,\cdot)$, one obtains

$$\begin{aligned} Q_k^{\mathsf{cn}}(\phi^{(k)},\mathbf{u}^{(k)};\psi,\mathbf{v}) &= \int_\Omega \Big(\frac{1}{2}\nabla\phi^{(k)}\cdot\Theta_1\nabla\psi + \frac{c}{\Delta t_k}\phi^{(k)}\psi\Big)\,d\Omega \\ &\quad +\frac{1}{2}\int_\Omega \Big(\frac{1}{2}\frac{\Delta t_k}{c}\nabla\cdot\mathbf{u}^{(k)}\nabla\cdot\mathbf{v} + \mathbf{u}^{(k)}\cdot{\Theta_1}^{-1}\mathbf{v}\Big)\,d\Omega \\ &\quad +\frac{1}{2}\int_\Omega \Big(\mathbf{u}^{(k)}\cdot\nabla\psi + \psi\nabla\cdot\mathbf{u}^{(k)}\Big)\,d\Omega \\ &\quad +\frac{1}{2}\int_\Omega \Big(\phi^{(k)}\nabla\mathbf{v} + \mathbf{v}\cdot\nabla\phi^{(k)}\Big)\,d\Omega \end{aligned}$$

so that, because $\psi\in G_\Gamma(\Omega)$ and $\mathbf{v}\in\mathbf{D}_{\Gamma^*}(\Omega)$,

$$
\begin{aligned}
Q_k^{\mathrm{cn}}(\phi^{(k)},\mathbf{u}^{(k)};\psi,\mathbf{v}) = &\int_\Omega \left(\frac{1}{2}\nabla\phi^{(k)}\cdot\Theta_1\nabla\psi + \frac{c}{\Delta t_k}\phi^{(k)}\psi\right)d\Omega \\
&+\frac{1}{2}\int_\Omega \left(\frac{1}{2}\frac{\Delta t_k}{c}\nabla\cdot\mathbf{u}^{(k)}\nabla\cdot\mathbf{v} + \mathbf{u}^{(k)}\cdot{\Theta_1}^{-1}\mathbf{v}\right)d\Omega\,.
\end{aligned}
\tag{9.40}
$$

We then see that (9.39) separates into the two weakly coupled problems: for $k = 1,\ldots,K$,

$$
\begin{aligned}
&\int_\Omega \left(\frac{1}{2}\nabla\phi^{(k)}\cdot\Theta_1\nabla\psi + \frac{c}{\Delta t_k}\phi^{(k)}\psi\right)d\Omega \\
&\quad = \frac{\alpha}{2}\int_\Omega \nabla\phi^{(k-1)}\cdot\Theta_1\nabla\psi\, d\Omega + \frac{\alpha+1}{2}\int_\Omega \mathbf{u}^{(k-1)}\cdot\nabla\psi\, d\Omega \\
&\qquad + \int_\Omega \left(\frac{f(t_k)+f(t_{k-1})}{2} + \frac{c}{\Delta t_k}\phi^{(k-1)}\right)\psi\, d\Omega \qquad \forall\,\psi\in G_\Gamma(\Omega)
\end{aligned}
\tag{9.41}
$$

and

$$
\begin{aligned}
&\int_\Omega \left(\frac{1}{2c}\nabla\cdot\mathbf{u}^{(k)}\nabla\cdot\mathbf{v} + \frac{1}{\Delta t_k}\mathbf{u}^{(k)}\cdot{\Theta_1}^{-1}\mathbf{v}\right)d\Omega \\
&\quad = \alpha\int_\Omega \frac{1}{\Delta t_k}\mathbf{u}^{(k-1)}\cdot{\Theta_1}^{-1}\mathbf{v}\, d\Omega + (\alpha-1)\int_\Omega \frac{1}{\Delta t_k}\nabla\phi^{(k-1)}\cdot\mathbf{v}\, d\Omega \\
&\qquad + \int_\Omega \frac{1}{c}\left(\frac{f(t_k)+f(t_{k-1})}{2} - \frac{1}{2}\nabla\cdot\mathbf{u}^{(k-1)}\right)\nabla\cdot\mathbf{v}\, d\Omega \qquad \forall\,\mathbf{v}\in\mathbf{D}_{\Gamma^*}(\Omega)\,.
\end{aligned}
\tag{9.42}
$$

The nature of the (weak) coupling between (9.41) and (9.42) is different for the three choices of α. For $\alpha = -1$, the terms involving $\mathbf{u}^{(k-1)}$ in the right-hand side of (9.41) cancel so that we can solve that sequence of equations for $\phi^{(k)}$, $k = 1,\ldots,K$, without regard to (9.42) or to $\mathbf{u}^{(k)}$. Subsequently, (9.42) may be used to determine $\mathbf{u}^{(k)}$ for $k = 1,\ldots,K$. On the other hand, for $\alpha = 1$, the terms involving $\phi^{(k-1)}$ in the right-hand side of (9.42) cancel so that we can solve that sequence of equations for $\mathbf{u}^{(k)}$, $k = 1,\ldots,K$, without regard to (9.41) or to $\phi^{(k)}$. Subsequently, (9.41) may be used to determine $\phi^{(k)}$ for $k = 1,\ldots,K$. For $\alpha = 0$, one cannot solve either (9.41) or (9.42) independently of the other for all $k = 1,\ldots,K$. However, at each time step, i.e., for each k, the two equations do uncouple.

Finite element discretizations can be defined in the usual manner by choosing conforming finite element space $G_\Gamma^h \subset G_\Gamma(\Omega)$ and $\mathbf{V}_{\Gamma^*}^h \subset \mathbf{D}_{\Gamma^*}(\Omega)$ and then minimizing the sequence of functionals $J_k^{\mathrm{cn}}(\cdot;\cdot)$ over these subspaces, i.e., by solving a sequence of compliant DLSPs corresponding to (9.38). Equivalently, the finite element approximations $\phi^{(h,k)}(\mathbf{x}) \in G_\Gamma^h$ and $\mathbf{u}^{(h,k)}(\mathbf{x}) \in \mathbf{V}_{\Gamma^*}^h$ to the exact solution $\phi^{(k)} \in G_\Gamma(\Omega)$ and $\mathbf{u}^{(k)} \in \mathbf{D}_{\Gamma^*}(\Omega)$ of (9.38), respectively, are determined from (9.41) and (9.42) restricted to the finite element spaces G_Γ^h and $\mathbf{V}_{\Gamma^*}^h$. The possible choices for G_Γ^h and $\mathbf{V}_{\Gamma^*}^h$ are the same as for the backward-Euler time discretization case; see (9.13) and (9.14), respectively.

Two interesting cases are given by (9.41) with $\alpha = -1$ and (9.42) with $\alpha = 1$ for which one obtains the fully-discrete systems

$$\int_\Omega \left(\frac{1}{2} \nabla \phi^{(h,k)} \cdot \Theta_1 \nabla \psi^h + \frac{c}{\Delta t_k} \phi^{(h,k)} \psi^h \right) d\Omega$$
$$= \int_\Omega \Big(\frac{f(t_k)+f(t_{k-1})}{2} \psi^h - \frac{1}{2} \nabla \phi^{(h,k-1)} \cdot \Theta_1 \nabla \psi^h \tag{9.43}$$
$$+ \frac{c}{\Delta t_k} \phi^{(h,k-1)} \psi^h \Big) d\Omega \qquad \forall \psi^h \in G_\Gamma^h \subset G_\Gamma(\Omega)$$

for $\phi^{(h,k)}$, $k = 1, \ldots, K$, and

$$\int_\Omega \left(\frac{1}{2c} \nabla \cdot \mathbf{u}^{(h,k)} \nabla \cdot \mathbf{v}^h + \frac{1}{\Delta t_k} \mathbf{u}^{(h,k)} \cdot \Theta_1{}^{-1} \mathbf{v}^h \right) d\Omega$$
$$= \int_\Omega \Big(\frac{1}{\Delta t_k} \mathbf{u}^{(h,k-1)} \cdot \Theta_1{}^{-1} \mathbf{v}^h + \frac{1}{c} \frac{f(t_k)+f(t_{k-1})}{2} \tag{9.44}$$
$$- \frac{1}{2c} \nabla \cdot \mathbf{u}^{(h,k-1)} \nabla \cdot \mathbf{v}^h \Big) d\Omega \qquad \forall \mathbf{v}^h \in \mathbf{V}_{\Gamma^*}^h \subset \mathbf{D}_{\Gamma^*}(\Omega)$$

for $\mathbf{u}^{(h,k)}$, $k = 1, \ldots, K$, respectively. After choosing the approximate initial data $\phi^{(h,0)}$, (9.43) may be used to determine $\phi^{(h,k)}$ for $k = 1, \ldots, K$. Likewise, after choosing the approximate initial data $\mathbf{u}^{(h,0)}$, (9.44) may be used to determine $\mathbf{u}^{(h,k)}$ for $k = 1, \ldots, K$.

One recognizes (9.43) as being exactly the same sequence of equations that are obtained by a Crank–Nicolson Galerkin finite element discretization of the second-order form (9.1) corresponding to (9.4), i.e., *the Crank–Nicolson FD-LSFEM with* $\alpha = -1$ *for the first-order system* (9.4) *recovers exactly the same approximate solution for the scalar-valued variable as does the Crank–Nicolson Galerkin finite element method for the parent second-order equation* (9.1). As a result, we can use well-known results about such a discretization of (9.1) to ascertain the same results about approximations of ϕ obtained from the FD-LSFEM. For example, the following optimal error estimate for sufficiently smooth solutions of (9.1) can be found in [336].

Theorem 9.7 *Assume that the hypotheses of Theorem* 9.3 *hold except that now, for each* $k = 1, \ldots, K$, $\phi^{(h,k)}(\mathbf{x}) \in G_\Gamma^h$ *solves* (9.41). *Then, for* $k = 1, \ldots, K$,

$$\|\phi(t_k) - \phi^{(h,k)}\|_0 \le C\big(h^{r+1} + \Delta t^2\big), \tag{9.45}$$

where C is independent of h and Δt *but depends continuously on* $\|\phi_0\|_{r+1}$, $\|\phi_t\|_{L^2(0,T;H^{r+1}(\Omega))}$, $\|\phi_{ttt}\|_{L^2(0,T;L^2(\Omega))}$, *and* $\|\Delta \phi_{tt}\|_{L^2(0,T;L^2(\Omega))}$. □

If one is instead interested in determining approximations to $\mathbf{u}$, one can obtain them directly from (9.42) without any need to determine approximations to ϕ. One also recognizes (9.44) as being exactly the same sequence of equations that are obtained by a Crank–Nicolson finite element discretization of the alternate second-order form (9.5) corresponding to (9.4), i.e., *the Crank–Nicolson FD-LSFEM with* $\alpha = 1$ *for the first-order system* (9.4) *recovers exactly the same vector-valued variable as does the Crank–Nicolson Galerkin finite element method for the parent second-order equation* (9.5). The proof of the following result for this case can be found in [309].

Theorem 9.8 *Assume that the hypotheses of Theorem* 9.4 *hold except that now, for each* $k = 1,\ldots,K$, $\mathbf{u}^{(h,k)}(\mathbf{x}) \in \mathbf{V}^h_{\Gamma^*}$ *solves* (9.42). *Then, for* $k = 1,\ldots,K$,

$$\|\mathbf{u}(t_k) - \mathbf{u}^{(h,k)}\|_0 \le C\big(h^{s_1} + h^s + \Delta t^2\big), \tag{9.46}$$

where C *is independent of* h *and* Δt *but depends continuously on* $\|\mathbf{u}_0\|_{s+1}$, $\|\mathbf{u}\|_{L^\infty(0,T;H^{s+1}(\Omega))}$, $\|\mathbf{u}_t\|_{L^\infty(0,T;H^{s_1}(\Omega))}$, $\|\phi_{ttt}\|_{L^2(0,T;L^2(\Omega))}$, *and* $\|\mathbf{u}_{tt}\|_{L^2(0,T;H^1(\Omega))}$. □

Two-Step Backward-Differentiation Least-Squares Finite Element Methods

Better time accuracy can also be obtained without the need for defining an initial condition for $\nabla\phi^{(0)}$ if one uses a backward-differentiation method. For example, assuming that the time step is fixed,[16] i.e., $\Delta t_k = \Delta t$ for all K, the same order of time-accuracy as that for the Crank–Nicolson method can be obtained using a *two-step backward-differentiation method.* We once again set $\phi^{(0)} = \phi_0(\mathbf{x})$ and then we use one step of the backward-Euler method (9.6) to determine $\phi^{(1)}$ and $\mathbf{u}^{(1)}$.[17] Then, for $\alpha = 0$ or 1, we have the sequence of systems

$$\begin{cases} \dfrac{3c}{2\Delta t}\phi^{(k)} + \nabla\cdot\mathbf{u}^{(k)} = f(t_k) + \dfrac{2c}{\Delta t}\phi^{(k-1)} - \dfrac{c}{2\Delta t}\phi^{(k-2)} & \text{for } \mathbf{x}\in\Omega \\ \dfrac{3}{2}\big(\mathbf{u}^{(k)} + \Theta_1\nabla\phi^{(k)}\big) = 2\alpha\big(\mathbf{u}^{(k-1)} + \Theta_1\nabla\phi^{(k-1)}\big) & \\ \qquad\qquad -\dfrac{1}{2}\alpha\big(\mathbf{u}^{(k-2)} + \Theta_1\nabla\phi^{(k-2)}\big) & \text{for } \mathbf{x}\in\Omega \\ \phi^{(k)} = g_D(\mathbf{x},t_k) & \text{for } \mathbf{x}\in\Gamma \\ \mathbf{u}^{(k)}\cdot\mathbf{n} = g_N(\mathbf{x},t_k) & \text{for } \mathbf{x}\in\Gamma^* \end{cases} \tag{9.47}$$

for $k = 2,\ldots,K$. We then define the sequence of least-square functionals

[16] The method can be easily generalized to the non-uniform time step case.

[17] The accuracy of the overall method is not compromised by the use of the Euler method for one time step because the local order of accuracy, i.e., over a single time step, of the Euler method is the same as the global order of accuracy, i.e., over K time steps, of the two-step method (9.47).

$$
\begin{aligned}
&J_k^{\text{bd}}(\phi^{(k)},\mathbf{u}^{(k)};f(t_k),\phi^{(k-1)},\phi^{(k-2)},\mathbf{u}^{(k-1)})\\
&\quad=\Big\|\Theta_1{}^{-1/2}\Big(\mathbf{u}^{(k)}+\Theta_1\nabla\phi^{(k)}\Big)-\frac{4}{3}\alpha\big(\mathbf{u}^{(k-1)}+\Theta_1\nabla\phi^{(k-1)}\big)\\
&\qquad\qquad+\frac{1}{3}\alpha\big(\mathbf{u}^{(k-2)}+\Theta_1\nabla\phi^{(k-2)}\big)\Big\|_0^2\\
&\quad+\frac{2}{3}\Big\|\Big(\frac{\Delta t}{c}\Big)^{1/2}\Big(\frac{3c}{2\Delta t}\phi^{(k)}+\nabla\cdot\mathbf{u}^{(k)}-f(t_k)-\frac{2c}{\Delta t}\phi^{(k-1)}+\frac{c}{2\Delta t}\phi^{(k-2)}\Big)\Big\|_0^2
\end{aligned}
\tag{9.48}
$$

for $k=2,\dots,K$ and the correponding sequence of CLSPs

$$
\begin{cases}
J_k^{\text{bd}}(\phi^{(k)},\mathbf{u}^{(k)};f(t_k),\phi^{(k-1)},\phi^{(k-2)},\mathbf{u}^{(k-1)},\mathbf{u}^{(k-2)})\\[2ex]
X=G_\Gamma(\Omega)\times\mathbf{D}_{\Gamma^*}(\Omega)\,.
\end{cases}
\tag{9.49}
$$

Note that now we need to specify both $\phi^{(0)}(\mathbf{x})=\phi(\mathbf{x},0)=\phi_0(\mathbf{x})$ and $\phi^{(1)}(\mathbf{x})\approx\phi(\mathbf{x},\Delta t)$; as is mentioned above, the latter is determined through one step of the backward-Euler method, i.e., from (9.7), or equivalently (9.11) and (9.12), with $k=1$. If $\alpha=1$, we also need to specify $\mathbf{u}^{(0)}=-\Theta_1\nabla\phi^{(0)}$ and $\mathbf{u}^{(1)}$. The latter can also be determined through one step of the backward-Euler method.

The necessary conditions corresponding to the sequence of minimization problems (9.49) are given by: for $k=2,\dots,K$,

$$
\begin{aligned}
Q_k^{\text{bd}}(\phi^{(k)},\mathbf{u}^{(k)};\psi,\mathbf{v})=F_k^{\text{bd}}(\psi,\mathbf{v};f(t_k),\phi^{(k-1)},\phi^{(k-2)},\mathbf{u}^{(k-1)},\mathbf{u}^{(k-1)})\\
\forall\,\psi\in G_\Gamma(\Omega),\ \mathbf{v}\in\mathbf{D}_{\Gamma^*}(\Omega)\,,
\end{aligned}
\tag{9.50}
$$

where, for each $k=2,\dots,K$ and for all $\psi\in G_\Gamma(\Omega)$ and $\mathbf{v}\in\mathbf{D}_{\Gamma^*}(\Omega)$,

$$
\begin{aligned}
Q_k^{\text{bd}}(\phi^{(k)},\mathbf{u}^{(k)};\psi,\mathbf{v})&=\int_\Omega\Big(\mathbf{u}^{(k)}+\Theta_1\nabla\phi^{(k)}\Big)\cdot\Theta_1{}^{-1}\Big(\mathbf{v}+\Theta_1\nabla\psi\Big)\,d\Omega\\
&\quad+\frac{2}{3}\int_\Omega\frac{\Delta t}{c}\Big(\frac{3c}{2\Delta t}\phi^{(k)}+\nabla\cdot\mathbf{u}^{(k)}\Big)\Big(\frac{3c}{2\Delta t}\psi+\nabla\cdot\mathbf{v}\Big)\,d\Omega
\end{aligned}
$$

and

$$
\begin{aligned}
&F_k^{\text{bd}}(\psi,\mathbf{v};f(t_k),\phi^{(k-1)},\phi^{(k-2)},\mathbf{u}^{(k-1)},\mathbf{u}^{(k-1)})\\
&\quad=\frac{4}{3}\alpha\int_\Omega\Big(\mathbf{u}^{(k-1)}+\Theta_1\nabla\phi^{(k-1)}\Big)\cdot\Theta_1{}^{-1}\Big(\mathbf{v}+\Theta_1\nabla\psi\Big)\,d\Omega\\
&\qquad-\frac{1}{3}\alpha\int_\Omega\Big(\mathbf{u}^{(k-2)}+\Theta_1\nabla\phi^{(k-2)}\Big)\cdot\Theta_1{}^{-1}\Big(\mathbf{v}+\Theta_1\nabla\psi\Big)\,d\Omega\\
&\qquad+\frac{2}{3}\int_\Omega\frac{\Delta t}{c}\Big(f(\mathbf{x},t_k)+\frac{2c}{\Delta t}\phi^{(k-1)}-\frac{c}{2\Delta t}\phi^{(k-2)}\Big)\Big(\frac{3c}{2\Delta t}\psi+\nabla\cdot\mathbf{v}\Big)\,d\Omega\,.
\end{aligned}
$$

Expanding the definition of $Q_k^{\text{bd}}(\cdot,\cdot;\cdot,\cdot)$, one obtains

$$Q_k^{\mathsf{bd}}(\phi^{(k)},\mathbf{u}^{(k)};\psi,\mathbf{v}) = \int_\Omega \left(\nabla\phi^{(k)}\cdot\Theta_1\nabla\psi + \frac{3c}{2\Delta t}\phi^{(k)}\psi\right)d\Omega + \int_\Omega \left(\frac{2\Delta t}{3c}\nabla\cdot\mathbf{u}^{(k)}\nabla\cdot\mathbf{v} + \mathbf{u}^{(k)}\cdot{\Theta_1}^{-1}\mathbf{v}\right)d\Omega + \int_\Omega \left(\mathbf{u}^{(k)}\cdot\nabla\psi + \psi\nabla\cdot\mathbf{u}^{(k)}\right)d\Omega + \int_\Omega \left(\phi^{(k)}\nabla\cdot\mathbf{v} + \mathbf{v}\cdot\nabla\phi^{(k)}\right)d\Omega$$

or, because $\psi \in G_\Gamma(\Omega)$ and $\mathbf{v} \in \mathbf{D}_{\Gamma^*}(\Omega)$,

$$Q_k^{\mathsf{bd}}(\phi^{(k)},\mathbf{u}^{(k)};\psi,\mathbf{v}) = \int_\Omega \left(\nabla\phi^{(k)}\cdot\Theta_1\nabla\psi + \frac{3c}{2\Delta t}\phi^{(k)}\psi\right)d\Omega + \int_\Omega \left(\frac{2\Delta t}{3c}\nabla\cdot\mathbf{u}^{(k)}\nabla\cdot\mathbf{v} + \mathbf{u}^{(k)}\cdot{\Theta_1}^{-1}\mathbf{v}\right)d\Omega\,. \tag{9.51}$$

We then see that (9.50) separates into the two weakly coupled problems: for $k = 2,\ldots,K$,

$$\begin{aligned}\int_\Omega &\left(\nabla\phi^{(k)}\cdot\Theta_1\nabla\psi + \frac{3c}{2\Delta t}\phi^{(k)}\psi\right)d\Omega \\ &= \int_\Omega \left(f(t_k) + \frac{2c}{\Delta t}\phi^{(k-1)} - \frac{c}{2\Delta t}\phi^{(k-2)}\right)\psi\, d\Omega \\ &\quad + \frac{4}{3}\alpha\int_\Omega \left(\mathbf{u}^{(k-1)} + \Theta_1\nabla\phi^{(k-1)}\right)\cdot\nabla\psi\, d\Omega \\ &\quad - \frac{1}{3}\alpha\int_\Omega \left(\mathbf{u}^{(k-2)} + \Theta_1\nabla\phi^{(k-2)}\right)\cdot\nabla\psi\, d\Omega \qquad \forall\,\psi\in G_\Gamma(\Omega)\end{aligned} \tag{9.52}$$

and

$$\begin{aligned}\int_\Omega &\left(c^{-1}\nabla\cdot\mathbf{u}^{(k)}\nabla\cdot\mathbf{v} + \frac{3}{2\Delta t}\mathbf{u}^{(k)}\cdot{\Theta_1}^{-1}\mathbf{v}\right)d\Omega \\ &= \int_\Omega c^{-1}f(t_k)\nabla\cdot\mathbf{v}\, d\Omega + 2\alpha\int_\Omega \frac{1}{\Delta t}\mathbf{u}^{(k-1)}\cdot{\Theta_1}^{-1}\mathbf{v}\, d\Omega \\ &\quad + 2(\alpha-1)\int_\Omega \frac{1}{\Delta t}\nabla\phi^{(k-1)}\cdot\mathbf{v}\, d\Omega - \frac{1}{2}\alpha\int_\Omega \frac{1}{\Delta t}\mathbf{u}^{(k-2)}\cdot{\Theta_1}^{-1}\mathbf{v}\, d\Omega \\ &\quad - \frac{1}{2}(\alpha-1)\int_\Omega \frac{1}{\Delta t}\nabla\phi^{(k-2)}\cdot\mathbf{v}\, d\Omega \qquad \forall\,\mathbf{v}\in\mathbf{D}_{\Gamma^*}(\Omega)\,.\end{aligned} \tag{9.53}$$

In particular, with $\alpha = 0$, (9.52) reduces to

$$
\begin{aligned}
&\int_\Omega \left(\nabla \phi^{(k)} \cdot \Theta_1 \nabla \psi + \frac{3c}{2\Delta t} \phi^{(k)} \psi \right) d\Omega \\
&\quad = \int_\Omega \left(f(t_k) + \frac{2c}{\Delta t} \phi^{(k-1)} - \frac{c}{2\Delta t} \phi^{(k-2)} \right) \psi \, d\Omega \qquad \forall \psi \in G_\Gamma(\Omega)
\end{aligned}
\tag{9.54}
$$

and, with $\alpha = 1$, (9.53) reduces to

$$
\begin{aligned}
&\int_\Omega \left(\frac{3}{2\Delta t} \mathbf{u}^{(k)} \cdot \Theta_1{}^{-1} \mathbf{v} + c^{-1} \nabla \cdot \mathbf{u}^{(k)} \nabla \cdot \mathbf{v} \right) d\Omega \\
&\quad = \int_\Omega \left(\frac{2}{\Delta t} \mathbf{u}^{(k-1)} - \frac{1}{2\Delta t} \mathbf{u}^{(k-2)} \right) \cdot \Theta_1{}^{-1} \mathbf{v} \, d\Omega \\
&\qquad + \int_\Omega c^{-1} f(t_k) \nabla \cdot \mathbf{v} \, d\Omega \qquad \forall \mathbf{v} \in \mathbf{D}_{\Gamma^*}(\Omega) .
\end{aligned}
\tag{9.55}
$$

We can now proceed as for the Crank–Nicolson FD-LSFEM, after noting that (9.54) and (9.55) are nothing more than the two-step backward-discretization method for (9.1) and (9.5), respectively. Again, following [336], we have that the error estimate (9.45) holds for the two-step backward-differentiation method as well, but with C now depending continuously on $\|\phi_0\|_{r+1}$, $\|\phi_t\|_{L^2(0,T;H^{r+1}(\Omega))}$, $\|\phi_{tt}\|_{L^\infty(0,T;L^2(\Omega))}$, and $\|\phi_{ttt}\|_{L^2(0,T;L^2(\Omega))}$. In addition, the proofs of Theorems 9.4 and 9.5 for the backward-Euler case can be amended to show that the error estimates (9.24), (9.29), and (9.35) all hold, with Δt replaced by Δt^2 and with an appropriate change in the dependences of constants C on ϕ and $\mathbf{u}$.

9.1.3 Comparison of Finite-Difference Least-Squares Finite Element Methods

In Tables 9.1 and 9.2, we summarize the results presented in Sections 9.1.1 and 9.1.2 for FD-LSFEMs, i.e., formulations in which one first uses a standard finite-difference method to effect temporal discretization and then uses a LSFEM to effect spatial discretization.

The first pattern one notices from Tables 9.1 and 9.2 is that whenever an approximation to a variable can be determined independently of the other, its error estimate only depends on the approximation properties of its own approximating space. Thus, whenever ϕ can be determined independently of $\mathbf{u}$, its error estimates only depend on the approximation properties of G^h_Γ (see 9.15) and, conversely, whenever $\mathbf{u}$ can be determined independently of ϕ, its error estimates only depend on the approximation properties of $\mathbf{V}^h_{\Gamma^*}$ (see 9.16). Whenever an approximation to a variable cannot be determined independently of the other, its error estimates depend on the approximation properties of both G^h_Γ and $\mathbf{V}^h_{\Gamma^*}$. These observations clearly also apply to grid sizes and time steps, should different ones be used for each variable. Thus, whenever an approximation to a variable can be determined independently, the error only depends on the grid and time steps used for its approximation but if

Method	Variable	Independence?	$\mathbf{V}^h_{\Gamma*} = [G^s(\Omega)]^d \cap \mathbf{D}_{\Gamma*}(\Omega)$ or $\mathbf{D}^s(\Omega) \cap \mathbf{D}_{\Gamma*}(\Omega)$
BE $\alpha = 0$	ϕ	yes	$h^{r+1} + \Delta t$
	$\mathbf{u}$	no	$h^{s+1} + \Delta t^{1/2} h^s + \Delta t + \min\{h^r, \Delta t^{-1/2} h^{r+1}\}$
BE $\alpha = 1$	ϕ	no	$h^{r+1} + \Delta t + h^s$
	$\mathbf{u}$	yes	$h^s + \Delta t$
CN $\alpha = -1$	ϕ	yes	$h^{r+1} + \Delta t^2$
BD $\alpha = 0$	$\mathbf{u}$	no	$h^{s+1} + \Delta t^{1/2} h^s + \Delta t^2 + \min\{h^r, \Delta t^{-1/2} h^{r+1}\}$
CN $\alpha = 1$	ϕ	no	$h^{r+1} + \Delta t^2 + h^s$
BD $\alpha = 1$	$\mathbf{u}$	yes	$h^s + \Delta t^2$

Table 9.1 Comparison of $L^2(\Omega)$ error estimates for FD-LSFEMs for parabolic equations; BE=backward-Euler, CN=Crank–Nicolson, BD=two-step backward-differentiation methods; the value of α selects the variant of the method; the "Variable" column indicates to what variable the rest of the row applies; the "Independence?" column answers the question: can approximations to the variable be determined independently of the other variable?; the last column gives the convergence rates with respect to the spatial grid size parameter h and the maximal time step Δt when the approximating spaces for $\mathbf{u}$ and ϕ are given by $\mathbf{V}^h_{\Gamma*} = \mathbf{D}^s(\Omega) \cap \mathbf{D}_{\Gamma*}(\Omega)$ and $G^h_\Gamma = G^r(\Omega) \cap G_\Gamma(\Omega)$, respectively; here, $r \in \{1,2,\dots\}$ and $s \in \{1,2,\dots\}$; full regularity of the exact solutions is assumed.

Method	Variable	Independence?	$\mathbf{V}^h_{\Gamma*} = \mathbf{D}^s\text{-}(\Omega)$
BE $\alpha = 0$	ϕ	yes	$h^{r+1} + \Delta t$
	$\mathbf{u}$	no	$h^s + \Delta t^{1/2} h^s + \Delta t + \min\{h^r, \Delta t^{-1/2} h^{r+1}\}$
BE $\alpha = 1$	ϕ	no	$h^{r+1} + \Delta t + h^s$
	$\mathbf{u}$	yes	$h^s + \Delta t$
CN $\alpha = -1$	ϕ	yes	$h^{r+1} + \Delta t^2$
BD $\alpha = 0$	$\mathbf{u}$	no	$h^s + \Delta t^{1/2} h^s + \Delta t^2 + \min\{h^r, \Delta t^{-1/2} h^{r+1}\}$
CN $\alpha = 1$	ϕ	no	$h^{r+1} + \Delta t^2 + h^s$
BD $\alpha = 1$	$\mathbf{u}$	yes	$h^s + \Delta t^2$

Table 9.2 Same as for Table 9.1 except that now the approximating space for $\mathbf{u}$ is $\mathbf{V}^h_{\Gamma*} = \mathbf{D}^s\text{-}(\Omega) \cap \mathbf{D}_{\Gamma*}(\Omega)$.

it cannot be determined independently, the error depends on the grid and time steps used to approximate both variables.

Whenever approximations to the scalar-valued variable ϕ can be determined independently of that for the vector-valued variable $\mathbf{u}$, the entries of Tables 9.1 and 9.2 show that optimal error estimates can be obtained for the approximation of the former. However, when approximations to the $\mathbf{u}$ can be determined independently, optimal error estimates are obtained for div-conforming spaces of the first kind $\mathbf{D}^{s}\text{-}(\Omega)$ but not for div-conforming spaces $\mathbf{D}^{s}(\Omega)$ of the second kind. For example, for the backward-Euler LSFEM with $\alpha = 1$, we have that for the lowest-order versions of all three methods that the $L^2(\Omega)$ norm of the error is of $O(h)$; this is only optimal for $\mathbf{D}^{1}\text{-}(\Omega)$ but not for $\mathbf{D}^{1}(\Omega)$ because $\mathbf{D}^{1}\text{-}(\Omega)$ contains only the piecewise constant complete polynomial space whereas the other space contains the piecewise linear complete polynomial space.

Note that if $s = r$, the $L^2(\Omega)$ error estimates for approximations to ϕ and $\mathbf{u}$ indicate different convergence rates, even when continuous piecewise polynomial spaces are used for both variables.

Remark 9.9 The entries in the table for the rows corresponding to choices of α for which the determination of the approximate solution for ϕ or $\mathbf{u}$ do not uncouple[18] give an indication of what one can expect, with respect to error estimates, for problems for which uncoupling in not possible; see Remark (9.1). □

9.1.4 Space–Time Least-Squares Principles

Global space–time least-squares principles

Instead of first discretizing in time, one could directly apply a residual minimization principle to (9.4). This approach leads to a space–time least-squares functional. For example, assuming the boundary and initial conditions are enforced strongly on the approximate solution, one could define the functional

$$\int_0^T \left(\left\| c^{-1/2}\big(c\frac{\partial \phi}{\partial t} + \nabla\cdot\mathbf{u} - f\big)\right\|_0^2 + \|\Theta_1{}^{-1/2}(\mathbf{u} + \Theta_1 \nabla\phi)\|_0^2 \right) dt \tag{9.56}$$

that measures the residuals of the first two equations in (9.4). A space–time least-squares principle can then be defined by minimizing (9.56) over an appropriate space–time solution space for $\{\phi, \mathbf{u}\}$. Discretization can be effected by minimizing (9.56) over a finite dimensional subspace of that solution space. This leads to the use of space–time finite element methods and, most probably, fully coupled space–time discrete systems for which approximate solutions are not determined through a marching in time procedure.

We do not pursue this approach further in this book, other than to say that although globally coupled space–time least-squares finite elements method may, at first glance, seem to be hopelessly impractical, they do enable the use of spatially

[18] These are the rows with an entry "no" in the "Independence?" column.

nonuniform time steps. Incorporating this feature into an adaptive strategy based on space–time residual error estimation may, in the end, result in methods that are indeed practical.

Local space–time least-squares principles

Somewhere between the discretize in time before applying a least-squares principle approach of Sections 9.1.1 and 9.1.2 and the global space–time least squares principle that precedes any discretization, one can define a local, space–time least squares principle for (9.4), i.e., for $k = 1,\dots,K$, we define the sequence of local space–time least-square functionals

$$J_k(\phi,\mathbf{u};f) = \int_{t_{k-1}}^{t_k} \left(\left\| \left(\frac{\Delta t_k}{c}\right)^{1/2} \left(c\frac{\partial\phi}{\partial t} + \nabla\cdot\mathbf{u} - f\right) \right\|_0^2 + \|\Theta_1{}^{-1/2}(\mathbf{u}+\Theta_1\nabla\phi)\|_0^2 \right) dt \tag{9.57}$$

and the sequence of least-squares principles

$$\begin{cases} J^{(k)}(\phi^{(k)},\mathbf{u}^{(k)};f) \\ X = H^1((0,T),G_\Gamma(\Omega)) \times L^2((0,T),\mathbf{D}_{\Gamma^*}(\Omega)) \end{cases} \tag{9.58}$$

with $\phi(\cdot,0) = \phi_0$. Note that (9.58) defines a time-marching procedure; starting with the initial condition $\phi(\cdot,0) = \phi_0$, $\phi(\cdot,t)$ and $\mathbf{u}(\cdot,t)$ can be determined sequentially in the intervals $(t_{k-1},t_k]$, $k = 1,\dots,K$.

As in Sections 9.1.1 and 9.1.2, we can exploit the cylindrical nature of the space–time domain $\Omega\times(0,T)$ to discretize separately in time and space. Temporal discretization is effected by choosing subspaces

$$S^K((0,T),G_\Gamma(\Omega)) \subset H^1((0,T),G_\Gamma(\Omega))$$

and

$$\mathbf{V}^K((0,T),\mathbf{D}_{\Gamma^*}(\Omega)) \subset L^2((0,T),\mathbf{D}_{\Gamma^*}(\Omega))$$

that are defined with respect to the partition $\{(t_{k-1},t_k)\}_{k=1}^K$ of $(0,T)$, and then minimizing the sequence of functionals (9.57) with respect to these subspaces. Equivalently, if, for $k = 1,\dots,K$, we let

$$S_k^K = S^K((0,T),G_\Gamma(\Omega))|_{(t_{k-1},t_k]}$$

and

$$\mathbf{V}_k^K = \mathbf{V}^K((0,T),\mathbf{D}_{\Gamma^*}(\Omega))_{(t_{k-1},t_k)},$$

we solve the sequence of CLSPs

$$\begin{cases} J_k(\phi^{(k)}, \mathbf{u}^{(k)}; f) \\ X = S_k^K \times \mathbf{V}_k^K \end{cases} \tag{9.59}$$

with $\phi(\cdot,0) = \phi_0$.

In [271, 272], continuous piecewise linear functions in time, with respect to the partition $\{(t_{k-1},t_k)\}_{k=1}^K$, were chosen for $S^K((0,T),G_\Gamma(\Omega))$ and discontinuous piecewise linear functions, with respect to the same partition, were chosen for $\mathbf{V}^K((0,T),\mathbf{D}_{\Gamma^*}(\Omega))$. For each $k = 1,\dots,K$, it is assumed that an approximation $\phi^{(k-1)} \approx \phi(t_{k-1})$ is known. Obviously this is true for $k = 1$ for which we have that $\phi^{(0)} = \phi_0$. The degrees of freedom in the corresponding interval $(t_{k-1},t_k]$ are chosen to be the "nodal" values of the piecewise linear functions used to approximate ϕ and $\mathbf{u}$, i.e., the values of the piecewise linear approximations evaluated at the end points of the time interval. We denote those degrees of freedom by

$$\phi^{(k)} \approx \phi(\cdot,t_k), \qquad \mathbf{u}_-^{(k)} \approx \mathbf{u}(\cdot,t_{k-1}^+) \qquad \mathbf{u}_+^{(k)} \approx \mathbf{u}(\cdot,t_k^-),$$

where $\phi^{(k)} \in G_\Gamma(\Omega)$, $\mathbf{u}_-^{(k)} \in \mathbf{D}_{\Gamma^*}(\Omega)$, and $\mathbf{u}_+^{(k)} \in \mathbf{D}_{\Gamma^*}(\Omega)$. In terms of these degrees of freedom, the functional in (9.57) can be written as

$$\begin{aligned} &J_k(\phi^{(k)}, \mathbf{u}_-^{(k)}, \mathbf{u}_+^{(k)}; f, \phi^{(k-1)}) = \\ &\int_{t_{k-1}}^{t_k} \Bigg(\bigg\| \left(\frac{\Delta t_k}{c}\right)^{1/2} \bigg(c\frac{\phi^{(k)} - \phi^{(k-1)}}{\Delta t_k} + \frac{t - t_{k-1}}{\Delta t_k} \nabla \cdot \mathbf{u}_+^{(k)} \\ &\qquad + \frac{t_k - t}{\Delta t_k} \nabla \cdot \mathbf{u}_-^{(k)} - f \bigg) \bigg\|_0^2 \\ &+ \bigg\| \Theta_1^{-1/2} \bigg(\frac{t - t_{k-1}}{\Delta t_k} \left(\mathbf{u}_+^{(k)} + \Theta_1 \nabla \phi^{(k)}\right) \\ &\qquad + \frac{t_k - t}{\Delta t_k} \left(\mathbf{u}_-^{(k)} + \Theta_1 \nabla \phi^{(k)}\right) \bigg) \bigg\|_0^2 \Bigg) dt \end{aligned} \tag{9.60}$$

for $k = 1,\dots,K$. If we assume that $f(\cdot,t)$ can be approximated by its piecewise linear interpolant[19] with respect to the partition $\{(t_{k-1},t_k)\}_{k=1}^K$, i.e.,

---

[19] If the interpolant is not well defined, we can instead use a piecewise linear approximation to f that is defined with respect to well-defined quantities. For example, if the moments $\int_{t_{k-1}}^{t_k} f\,dt$ and $\int_{t_{k-1}}^{t_k} t f\,dt$ are known, then one can determine a linear approximation $f \approx at + b$ in the interval (t_{k-1},t_k) by solving

$$\begin{cases} \dfrac{t_k^2 - t_{k-1}^2}{2} a + (t_k - t_{k-1}) b = \displaystyle\int_{t_{k-1}}^{t_k} f\,dt \\ \dfrac{t_k^3 - t_{k-1}^3}{3} a + \dfrac{t_k^2 - t_{k-1}^2}{2} b = \displaystyle\int_{t_{k-1}}^{t_k} t f\,dt \end{cases}$$

for a and b.

$$f(\mathbf{x},t) \approx \frac{t_k - t}{\Delta t_k} f(\mathbf{x},t_{k-1}) + \frac{t - t_{k-1}}{\Delta t_k} f(\mathbf{x},t_k) \quad \text{for } t \in [t_{k-1},t_k],\ k = 1,\dots,K,$$

then (9.60) is approximated by

$$\begin{aligned}
&J_k\big(\phi^{(k)},\mathbf{u}_-^{(k)},\mathbf{u}_+^{(k)};f,\phi^{(k-1)}\big)\\
&\approx \widetilde{J}_k\big(\phi^{(k)},\mathbf{u}_-^{(k)},\mathbf{u}_+^{(k)};f(t_k),f(t_{k-1}),\phi^{(k-1)}\big)\\
&= \int_{t_{k-1}}^{t_k} \Bigg(\bigg\| \Big(\frac{\Delta t_k}{c}\Big)^{1/2} \Big(c\frac{\phi^{(k)}-\phi^{(k-1)}}{\Delta t_k} + \frac{t-t_{k-1}}{\Delta t_k}\big(\nabla\cdot\mathbf{u}_+^{(k)} - f(t_k)\big)\\
&\qquad\qquad + \frac{t_k - t}{\Delta t_k}\big(\nabla\cdot\mathbf{u}_-^{(k)} - f(t_{k-1})\big)\Big)\bigg\|_0^2\\
&\qquad + \bigg\| \Theta_1^{-1/2}\Big(\frac{t-t_{k-1}}{\Delta t_k}\big(\mathbf{u}_+^{(k)} + \Theta_1\nabla\phi^{(k)}\big)\\
&\qquad\qquad + \frac{t_k - t}{\Delta t_k}\big(\mathbf{u}_-^{(k)} + \Theta_1\nabla\phi^{(k)}\big)\Big)\bigg\|_0^2\Bigg)\,dt\,.
\end{aligned}$$

The integrand of the approximate functionals $\widetilde{J}_k\big(\cdot,\cdot,\cdot;f(t_k),f(t_{k-1}),\phi^{(k-1)}\big)$ is a quadratic function of t and thus may be integrated exactly, e.g., by applying Simpon's rule. The result is

$$\begin{aligned}
&\widetilde{J}_k\big(\phi^{(k)},\mathbf{u}_-^{(k)},\mathbf{u}_+^{(k)};f(t_k),f(t_{k-1}),\phi^{(k-1)}\big) = \frac{\Delta t_k}{6}\Big\|\Theta_1^{-1}\big(\mathbf{u}_+^{(k)} + \Theta_1\nabla\phi^{(k)}\big)\Big\|_0^2\\
&\quad + \frac{\Delta t_k}{3}\Big\|\Theta_1^{-1}\big(\mathbf{u}_+^{(k)} + \Theta_1\nabla\phi^{(k)}\big) + \Theta_1^{-1}\big(\mathbf{u}_-^{(k)} + \Theta_1\nabla\phi^{(k-1)}\big)\Big\|_0^2\\
&\quad + \frac{\Delta t_k}{6}\Big\|\Theta_1^{-1}\big(\mathbf{u}_-^{(k)} + \Theta_1\nabla\phi^{(k-1)}\big)\Big\|_0^2\\
&\quad + \frac{1}{6}\Big\|c^{-1/2}\Big(c\big(\phi^{(k)} - \phi^{(k-1)}\big) + \Delta t_k\big(\nabla\cdot\mathbf{u}_-^{(k)} - f(t_{k-1})\big)\Big)\Big\|_0^2\\
&\quad + \frac{2}{3}\Big\|c^{-1/2}\Big(c\big(\phi^{(k)} - \phi^{(k-1)}\big)\\
&\qquad + \frac{\Delta t_k}{2}\big(\nabla\cdot\mathbf{u}_+^{(k)} - f(t_k) + \nabla\cdot\mathbf{u}_-^{(k)} - f(t_{k-1})\big)\Big)\Big\|_0^2\\
&\quad + \frac{1}{6}\Big\|c^{-1/2}\Big(c\big(\phi^{(k)} - \phi^{(k-1)}\big) + \Delta t_k\big(\nabla\cdot\mathbf{u}_+^{(k)} - f(t_k)\big)\Big)\Big\|_0^2.
\end{aligned} \tag{9.61}$$

One determines the semi-discrete in time approximations to ϕ and $\mathbf{u}$ in the interval (t_{k-1},t_k) by minimizing $\widetilde{J}_k\big(\cdot,\cdot,\cdot;f(t_k),f(t_{k-1}),\phi^{(k-1)}\big)$ over $G_\Gamma(\Omega)\times\mathbf{D}_{\Gamma^*}(\Omega)\times\mathbf{D}_{\Gamma^*}(\Omega)$. Note again that this defines a time marching method because the minimization problem for the interval (t_{k-1},t_k) is completely defined once the solution of the minimization problem for the interval (t_{k-2},t_{k-1}) has been obtained and, to start with, the minimization problem for the interval (t_0,t_1) is well defined once the initial condition is specified. For brevity, we do not write down the necessary condition for each minimization problem.

Full discretization is effected by minimizing, for each $k = 1,2,\ldots$, the functional $\widetilde{J}_k\big(\cdot,\cdot,\cdot; f(t_k), f(t_{k-1}), \phi^{(k-1)}\big)$ over (conforming) finite element subspaces of $G_\Gamma(\Omega)\times \mathbf{D}_{\Gamma^*}(\Omega)\times \mathbf{D}_{\Gamma^*}(\Omega)$. See [271, 272] for details and for an analysis of this approach.

9.1.5 Practical Issues

The material in this section complements the examples discussed in Section 5.9.1. There we examine how approximation of the flux by standard nodal elements in (5.63) backfires when the exact solution of the div-grad system (5.18) is not smooth enough. Here we show that the perils associated with the use of C^0 approximating spaces persist for FD-LSFEMs for the generalized heat-equation. For brevity, only the backward Euler FD-LSFEM based on (9.7) with $\alpha = 0$, is considered.

To this end, we use as a test example the "five-spot problem" from [222]. In this problem, we solve (9.4) for $\mathbf{x}\in\Omega = [0,1]^2$, $t\in(0,1]$, $\Gamma^* = \partial\Omega$, $c = 1$, $\Theta_1 = \mathbb{I}$, initial value $\phi_0 = 0$, homogeneous Neumann boundary condition, i.e., $g_N = 0$, and source term

$$f(\mathbf{x}) = \begin{cases} \dfrac{1}{4} & \text{if } \mathbf{x} = (0,0) \\ -\dfrac{1}{4} & \text{if } \mathbf{x} = (1,1) \\ 0 & \text{in all other cases}. \end{cases} \tag{9.62}$$

Following [222], in the actual implementation the source term is replaced by an equivalent distribution of the normal velocity. The five-spot problem models a simple injection/production well configuration for a single phase flow; see [222] for further details.

The backward Euler FD-LSFEM for this problem is implemented using a uniform partition of the time interval $[0,1]$ into $K = 100$ subintervals and a uniform partition $\mathcal{T}_h$ of Ω into 625 square elements. The scalar variable ϕ is approximated by the lowest-order bilinear nodal space $G^1(\Omega)$ whereas for the flux $\mathbf{u}$ we consider both $\mathbf{V}^h_{\Gamma^*} = [G^1(\Omega)]^d \cap \mathbf{D}_{\Gamma^*}(\Omega)$ and $\mathbf{V}^h_{\Gamma^*} = \mathbf{D}^1\text{-}(\Omega)\cap\mathbf{D}_{\Gamma^*}(\Omega)$, respectively. Thus, we compare FD-LSFEM solutions corresponding to spatial discretization by a nodal LSFEM with that for a compatible (mimetic) LSFEM.

We conduct two series of experiments. In the first one, we solve the *unperturbed* five-spot problem with source term given by (9.62). In the second series, f is perturbed by the function $n\phi_{n,n}$, where

$$\phi_{n,n} = \cos(n\pi x)\cos(n\pi y),$$

i.e., we set $f_n = f + n\phi_{n,n}$. Note that $\phi_{n,n}$ is the eigenfunction of the Laplacian $-\Delta$ corresponding to the eigenvalue $\lambda_{n,n} = 2n^2\pi$. Thus, a right-hand side $n\phi_{n,n}$

corresponds to a solution $\phi = 1/(2n\pi^2)\phi_{n,n}$ of the differential equation. For our experiments, we choose $n = 25$. As a result, the net variation added to the solution of the five-spot problem by perturbing the right hand side to $f + n\phi_{n,n}$ is less than

$$\frac{1}{2n\pi^2}\left|\phi_{n,n}\right| \approx 0.002\,.$$

Based on this estimate, the difference between the solutions of the perturbed and unperturbed problems should be barely visible to the naked eye. To test this conjecture, we examine visually snapshots of the flux vector field and its Cartesian components at the final time step for the nodal and the mimetic version of the FD-LSFEM. Figure 9.1 shows the results for the unperturbed five-spot problem. There we see that the two implementations produce qualitatively similar, at least with respect to the eye-ball norm, flux approximations. Although there are some differences between the Cartesian components in the two FD-LSFEM solutions (in particular, note the "small" wiggles for the nodal implementation), the overall vector fields are quite close.

Next, we examine mimetic and nodal FD-LSFEM solutions for the perturbed problem. Insofar as the mimetic solution is concerned, Figure 9.2 shows that, as expected, there is no perceptible difference between the plots for that solution in this figure and those in Figure 9.1. In fact, any differences between the results for the perturbed and unperturbed problems are entirely consistent with the size of the perturbation. However, the situation for the nodal FD-LSFEM solution is completely different. From Figure 9.2, we see that by the time the final time $T = 1$ is reached, the small oscillations added to the source have been significantly amplified to a point where the approximate solution has become grossly inaccurate. The snapshot of the nodal vector field at $T = 1$ shows huge differences not only with the mimetic field but also with the nodal field shown in Figure 9.1.

9.2 The Time-Dependent Stokes Equations

The primitive variable formulation of the time-dependent Stokes system with the standard velocity boundary condition is given by[20]

$$\begin{cases} \dfrac{\partial \mathbf{u}}{\partial t} - \nu\Delta\mathbf{u} + \nabla p = \mathbf{f} & \text{for } (\mathbf{x},t) \in \Omega \times (0,T] \\ \nabla\cdot\mathbf{u} = 0 & \text{for } (\mathbf{x},t) \in \Omega \times (0,T] \\ \mathbf{u} = 0 & \text{for } (\mathbf{x},t) \in \partial\Omega \times (0,T] \\ \displaystyle\int_\Omega p\,d\Omega = 0 & \text{for } t \in (0,T] \\ \mathbf{u}(\mathbf{x},0) = \mathbf{u}_0(\mathbf{x}) & \text{for } \mathbf{x} \in \Omega\,, \end{cases} \tag{9.63}$$

[20] The corresponding steady-state Stokes system is given by (7.1)–(7.3).

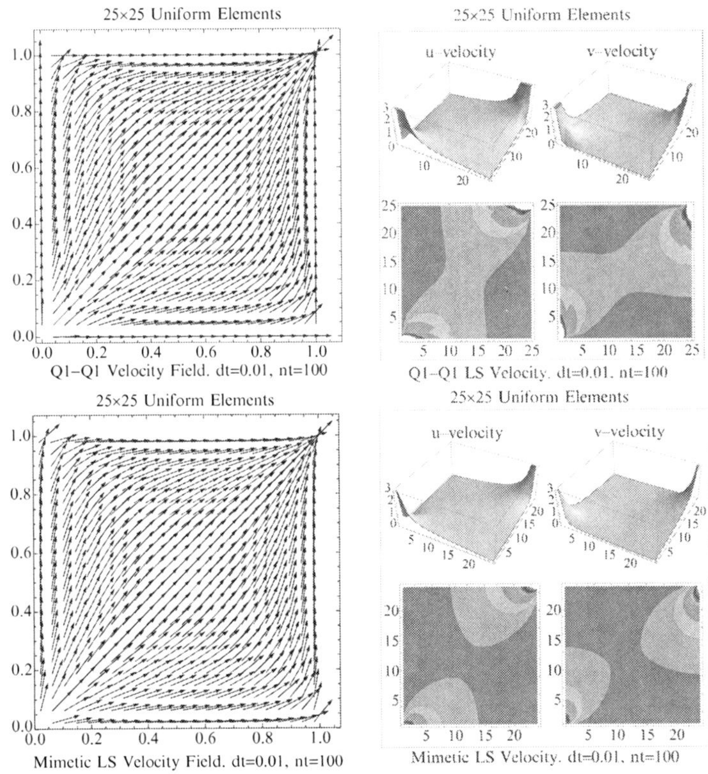

Fig. 9.1 Snapshots of flux approximations at the final time $T = 1$ computed by two different implementations of the backward Euler FD-LSFEM for the *unperturbed* five-spot problem. The top row shows the flux field and its x and y components for the nodal implementation, where $\mathbf{V}^h_{\Gamma*} = [G^1(\Omega)]^d \cap \mathbf{D}_{\Gamma*}(\Omega)$. The bottom row shows the same data for the compatible (mimetic) implementation for which $\mathbf{V}^h_{\Gamma*} = \mathbf{D}^{1}\text{-}(\Omega) \cap \mathbf{D}_{\Gamma*}(\Omega)$.

where $\mathbf{u}_0(\cdot)$ denotes the initial velocity field.[21] Mixed-Galerkin finite element methods for the approximate solution of (9.63) are discussed in, e.g., [182, 191, 333].

As is by now the common first step in the development of LSFEMs, we recast (9.63) into an equivalent first-order system. To this end, one can use time-dependent versions of any of the first-order system reformulations of the steady-state Stokes equations given in Section 7.1. Here, we only consider the velocity–vorticity–pressure (VVP) formulation[22,23]

[21] See Chapter 7 for an explanation of the necessity for imposing a zero mean pressure constraint and for a description of the physical meaning of the variables and other data appearing in (9.63).

[22] See (7.4) for the VVP formulation of the steady-state Stokes equations.

[23] FD-LSFEMs for the time-dependent velocity–stress-pressure and velocity gradient–velocity–pressure systems can be formulated in an entirely analogous manner. Here, we choose to focus on the time-dependent VVP system because it is, by far, the one most used in engineering practice.

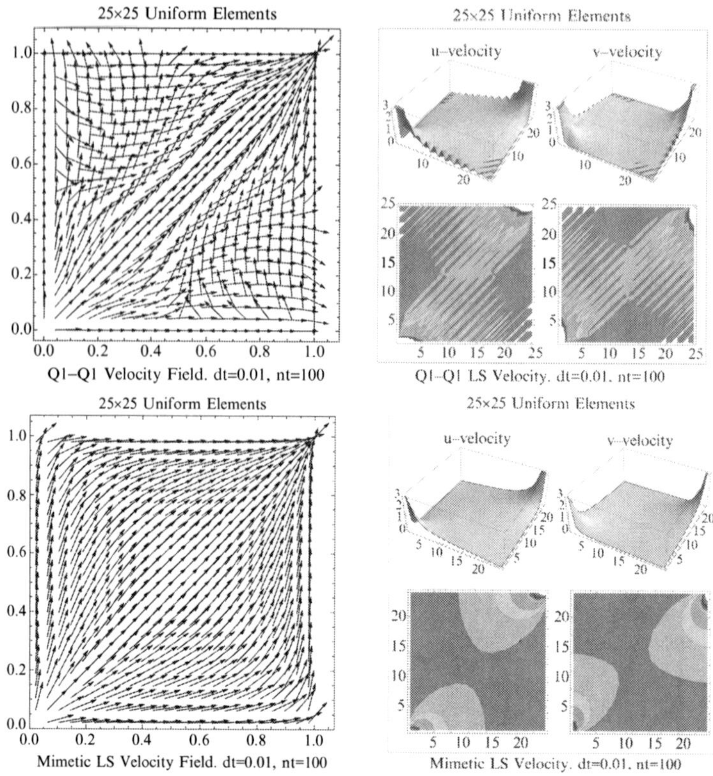

Fig. 9.2 Snapshots of flux approximations at the final time $T = 1$ computed by two different implementations of the backward Euler FD-LSFEM for the *perturbed* five-spot problem. The top row shows the flux field and its x and y components for the nodal implementation, where $\mathbf{V}^h_{\Gamma*} = [G^1(\Omega)]^d \cap \mathbf{D}_{\Gamma^*}(\Omega)$. The bottom row shows the same data for the compatible (mimetic) implementation for which $\mathbf{V}^h_{\Gamma*} = \mathbf{D}^1\text{-}(\Omega) \cap \mathbf{D}_{\Gamma^*}(\Omega)$.

$$
\left\{
\begin{array}{rl}
\dfrac{\partial \mathbf{u}}{\partial t} + \nu \nabla \times \omega + \nabla p = \mathbf{f} & \text{for } (\mathbf{x},t) \in \Omega \times (0,T] \\
\nabla \times \mathbf{u} - \omega = 0 & \text{for } (\mathbf{x},t) \in \Omega \times (0,T] \\
\nabla \cdot \mathbf{u} = 0 & \text{for } (\mathbf{x},t) \in \Omega \times (0,T] \\
\mathbf{u} = 0 & \text{for } (\mathbf{x},t) \in \partial\Omega \times (0,T] \\
\displaystyle\int_\Omega p \, d\Omega = 0 & \text{for } t \in (0,T] \\
\mathbf{u}(\mathbf{x},0) = \mathbf{u}_0(\mathbf{x}) & \text{for } \mathbf{x} \in \Omega\,.
\end{array}
\right.
\tag{9.64}
$$

FD-LSFEMs for (9.64) can be defined in much the same manner as for the generalized heat equation; see Sections 9.1.1 and 9.1.2[24]: first we discretize (9.64) in time and then solve the resulting sequence of elliptic problem by least-squares minimization. Recall that well-posed CLSPs for the semi discrete in time VVP system can be defined using either the Agmon–Douglis–Nirenberg (ADN) or the vector operator settings for the energy balances. Section 7.7.1 explains why the velocity boundary condition does not fit naturally in the latter setting.[25] This is why in this section we restrict attention to CLSPs corresponding to ADN energy balances.

Semi-discretization in time and energy balances

For brevity, we only consider the backward-Euler method for time discretization; the second-order methods discussed in Section 9.1.2 can be applied to the Stokes problem in an entirely similar manner.

We again introduce a partition $\{[t_{k-1},t_k]\}_{k=1}^K$ of the time interval $[0,T]$ into K subintervals, where $t_0 = 0$, $t_K = T$, and $t_{k-1} < t_k$. We set $\Delta t_k = t_k - t_{k-1}$ for $k = 1,\dots,K$. Then, for the backward-Euler method for discretization in time, we have the sequence of semi-discrete in time systems: for $k = 1,\dots,K$,

$$
\begin{cases}
\dfrac{1}{\Delta t_k}\mathbf{u}^{(k)} + \nu\nabla\times\omega^{(k)} + \nabla p^{(k)} = \mathbf{f}(\mathbf{x},t_k) + \dfrac{1}{\Delta t_k}\mathbf{u}^{(k-1)} & \text{in } \Omega \\
\nabla\times\mathbf{u}^{(k)} - \omega^{(k)} = 0 & \text{in } \Omega \\
\nabla\cdot\mathbf{u}^{(k)} = 0 & \text{in } \Omega \\
\mathbf{u}^{(k)} = 0 & \text{on } \partial\Omega \\
\displaystyle\int_\Omega p^{(k)}\,d\Omega = 0\ ,
\end{cases}
\tag{9.65}
$$

where $\mathbf{u}^{(0)} = u_0(\mathbf{x})$.[26]

The ADN theory classifies, for each $k = 1,\dots,K$, (9.65) as a non-homogeneous elliptic system. For simplicity, we assume that the boundary condition on the

---

[24] Unlike the case for Sections 9.1.1 and 9.1.2, there is no advantage accruing from the introduction, through the parameter α, of multiple methods of each type. Thus, for the Stokes system, we only use methods analogous to the case $\alpha = 0$ in Sections 9.1.1 and 9.1.2.

[25] For brevity, we do not consider the two non-standard boundary conditions in Section 7.1.1 for which the vector operator setting is more natural.

[26] Note that, in (9.64), an initial condition is given only for the velocity $\mathbf{u}$; no initial conditions need be imposed on the pressure p or vorticity ω. The backward-Euler scheme (9.65) also only requires an initial condition for the velocity. In contrast, the Crank-Nicolson method for (9.64) for which the first equation in (9.65) is replaced by

$$
\begin{aligned}
&\frac{1}{\Delta t_k}\mathbf{u}^{(k)} + \frac{1}{2}\nu\nabla\times\omega^{(k)} + \frac{1}{2}\nabla p^{(k)} \\
&\qquad = \frac{1}{2}\mathbf{f}(\mathbf{x},t_k) + \frac{1}{2}\mathbf{f}(\mathbf{x},t_{k-1}) + \frac{1}{\Delta t_k}\mathbf{u}^{(k-1)} - \frac{1}{2}\nu\nabla\times\omega^{(k-1)} - \frac{1}{2}\nabla p^{(k-1)}
\end{aligned}
$$

velocity and the zero mean pressure condition are imposed on the spaces for $\mathbf{u}^{(k)}$ and $p^{(k)}$, respectively. Then, if $\Omega \subset \mathbb{R}^d$ is of class C^1, the ADN theory also provides the energy balances: for $k = 1, \ldots, K$,

$$C\big(\|\mathbf{u}^{(k)}\|_1 + \|\omega^{(k)}\|_0 + \|p^{(k)}\|_0\big) \leq$$

$$\Big\|\frac{1}{\Delta t_k}\mathbf{u}^{(k)} + \nu\nabla \times \omega^{(k)} + \nabla p^{(k)}\Big\|_{-1} + \|\nabla \times \mathbf{u}^{(k)} - \omega^{(k)}\|_0 + \|\nabla \times \mathbf{u}^{(k)}\|_0 \quad (9.66)$$

$$\forall \{\mathbf{u}^{(k)}, \omega^{(k)}, p^{(k)}\} \in X^C_{-1} = [H^1_0(\Omega)]^d \times [L^2(\Omega)]^{d(d-1)/2} \times L^2_0(\Omega)$$

for some positive constant C. To see this, one need only note that the term $\mathbf{u}^{(k)}/\Delta t_k$ does not participate in the principal part of the differential operator in (9.65). In fact, that principal part is identical to that for the steady-state Stokes problem, so that the energy balance for (9.65) is the same as for (7.4). Then, (9.66) follows in the same way as (7.32)[27] and (7.27)[28] followed, for $d = 3$ and $d = 2$, respectively, for the steady-state VVP system.

Continuous least-squares principles

Each member of the sequence of systems (9.65) is a perturbed steady-state Stokes problem.[29] Thus, we now apply a least-squares minimization principle to each and every member of that sequence. The sequence of continuous least-squares principles (CLSPs) corresponding to the energy balance (9.66) are given by:[30] for $k = 1, \ldots, K$,

---

also requires an initial condition for the pressure. This is not so easy to determine, so that if a higher-order accurate time discretization is needed, then the backward differentiation method is preferable. In that case the first equation in (9.65) is replaced by

$$\frac{3}{2\Delta t_k}\mathbf{u}^{(k)} + \nu\nabla \times \omega^{(k)} + \nabla p^{(k)} = \mathbf{f}(\mathbf{x}, t_k) + \frac{2}{\Delta t_k}\mathbf{u}^{(k-1)} - \frac{1}{2\Delta t_k}\mathbf{u}^{(k-2)}$$

so that no initial condition on the pressure is needed. Note however, that we have to perform a single step of the backward-Euler method to get the two-level backward differentiation method started.

[27] Actually, we compare to (7.32) specialized to $q = -1$ and, because we have assumed that the spaces satisfy the zero velocity boundary condition and zero mean pressure condition, with the terms $\|\mathbf{u}\|_{1/2,\partial\Omega}$ and $|\ell(p)|$ omitted.

[28] For $d = 2$, we do not explicitly note, in (9.66), the differences between the $\nabla\times$ and $\nabla^\perp$ curl operators.

[29] For example, it is easy to see that (9.65) corresponds to the sequence of perturbed steady-state Stokes problems

$$-\nu\Delta\mathbf{u}^{(k)} + \nabla p^{(k)} + \frac{1}{\Delta t_k}\mathbf{u}^{(k)} = f(\mathbf{x}, t_k) + \frac{1}{\Delta t_k}\mathbf{u}^{(k-1)}$$

with $\mathbf{u}^{(0)}(\mathbf{x}) = \mathbf{u}_0(\mathbf{x})$ as well as $\nabla \cdot \mathbf{u}^{(k)} = 0$ on Ω, $\mathbf{u}^{(k)} = 0$ on $\partial\Omega$, and $\int_\Omega p^{(k)}\, d\Omega = 0$.

[30] The least-squares principles (9.67) follow from (9.66) in exactly the same way as (7.45) follows from (7.32) for the steady-state VVP system.

$$\begin{cases} J^k_{-1}\left(\mathbf{u}^{(k)},\omega^{(k)},p^{(k)};f(\mathbf{x},t_k),\mathbf{u}^{(k-1)}\right) = \\ \quad = \dfrac{T}{\nu}\left\| \dfrac{1}{\Delta t_k}\mathbf{u}^{(k)} + \nu\nabla\times\omega^{(k)} + \nabla p^{(k)} - \mathbf{f}(\mathbf{x},t_k) - \dfrac{1}{\Delta t_k}\mathbf{u}^{(k-1)}\right\|^2_{-1} \\ \qquad\qquad +\|\nabla\times\mathbf{u}^{(k)} - \omega^{(k)}\|^2_0 + \|\nabla\cdot\mathbf{u}^{(k)}\|^2_0 \\ X^C_{-1} = [H^1_0(\Omega)]^d \times [L^2(\Omega)]^{d(d-1)/2} \times L^2_0(\Omega), \end{cases} \tag{9.67}$$

where $\mathbf{u}^{(0)} = \mathbf{u}_0$. On the right-hand side of (9.67), the weight in the first term is introduced so that all three terms have the same units.[31] The sequence of least-squares principles (9.67) defines a marching method. Starting with $\mathbf{u}^{(0)} = \mathbf{u}_0$, one sets $k=1$ in (9.67) which then becomes a well-posed problem for determining $\mathbf{u}^{(1)}$, $\omega^{(1)}$, and $p^{(1)}$. Obviously, by incrementing the value of k, one sequentially determines the remaining triplets $\mathbf{u}^{(k)}$, $\omega^{(k)}$, and $p^{(k)}$, $k=2,\dots,K$. Note that initial conditions $\omega^{(0)}$, and $p^{(0)}$ on the vorticity and pressure, respectively, are not needed to determine $\mathbf{u}^{(1)}$, $\omega^{(1)}$, and $p^{(1)}$.

The functionals $J^k_{-1}(\cdot,\cdot,\cdot;\cdot,\cdot)$ are norm-equivalent, i.e., for $k=1,\dots,K$, we have, using (9.66),

$$C\left(\|\mathbf{u}^{(k)}\|^2_1 + \|\omega^{(k)}\|^2_0 + \|p^{(k)}\|^2_0\right) \le J^k_{-1}(\mathbf{u}^{(k)},\omega^{(k)},p^{(k)};0,0)\,. \tag{9.68}$$

Discrete least-squares principles

Because of the appearance of the $H^{-1}(\Omega)$ norm, the CLSP (9.67) does not lead to a sequence of practical compliant discrete least-squares principles (DLSPs). Instead, we have the choice of following the paths taken for the steady-state VVP system that led from the CLSP (7.45) to the quasi-norm equivalent DLSP (7.54) or the norm-equivalent DLSP (7.55). For brevity, we only follow the second path so that we are led to the sequence of *practical norm-equivalent* DLSPs[32] for the approximation $\{\mathbf{u}^{(h,k)},\omega^{(h,k)},p^{(h,k)}\}$ to $\{\mathbf{u}(\mathbf{x},t_k),\omega(\mathbf{x},t_k),p(\mathbf{x},t_k)\}$: for $k=1,\dots,K$,

$$\begin{cases} J^k_{-h}\left(\mathbf{u}^{(h,k)},\omega^{(h,k)},p^{(h,k)};f(\mathbf{x},t_k),\mathbf{u}^{(h,k-1)}\right) \\ \quad = \dfrac{T}{\nu}\left\| \dfrac{1}{\Delta t_k}\mathbf{u}^{(h,k)} + \nu\nabla\times\omega^{(h,k)} + \nabla p^{(h,k)} - \mathbf{f}(\mathbf{x},t_k) - \dfrac{1}{T}\mathbf{u}^{(h,k-1)}\right\|^2_{-h} \\ \qquad\qquad +\|\nabla\times\mathbf{u}^{(h,k)} - \omega^{(h,k)}\|^2_0 + \|\nabla\cdot\mathbf{u}^{(h,k)}\|^2_0 \\ X^h_{\widehat{r},0} = [G^{r+1}]^d \cap [H^1_0(\Omega)]^d \times [G^{\widehat{r}}]^{d(d-1)/2} \times G^{\widehat{r}} \cap L^2_0(\Omega), \end{cases} \tag{9.69}$$

[31] There may be some advantage accruing from using the weight $\Delta t_k/\nu$ instead of T/ν in (9.67). However, this issue has not been studied.

[32] Because we are constraining the pressure space to satisfy the zero mean pressure condition, we can omit the $|\ell(p)|$ term that appears in (7.55).

where $\mathbf{u}^{(h,0)} \in [G^{r+1}]^d \cap [H_0^1(\Omega)]^d$ is an approximation to $\mathbf{u}_0$.

The necessary conditions corresponding to the sequence of least-squares principles (9.69) is given by: for $k = 1,\dots,K$, find $\{\mathbf{u}^{(h,k)}, \omega^{(h,k)}, p^{(h,k)}\} \in X^h_{\widehat{r},0}$ that satisfy

$$\begin{aligned} Q^k(\mathbf{u}^{(h,k)}, \omega^{(h,k)}, p^{(h,k)}; \mathbf{v}^h, \xi^h, q^h) = F^k(\mathbf{v}^h, \xi^h, q^h; f(\mathbf{x}, t_k), \mathbf{u}^{(h,k-1)}) \\ \forall \{\mathbf{v}^h, \xi^h, q^h\} \in X^h_{\widehat{r},0}, \end{aligned} \tag{9.70}$$

where $\mathbf{u}^{(h,0)} \in [G^{r+1}]^d \cap [H_0^1(\Omega)]^d$ is an approximation to $\mathbf{u}_0$. In (9.70), we have, for all $\{\mathbf{u}^h, \omega^h, p^h\} \in X^h_{\widehat{r},0}$ and $\{\mathbf{v}^h, \xi^h, q^h\} \in X^h_{\widehat{r},0}$,

$$\begin{aligned} Q^k(\mathbf{u}^h, \omega^h, p^h; \mathbf{v}^h, \xi^h, q^h) &= \big(\nabla\times\mathbf{u}^h - \omega^h, \nabla\times\mathbf{v}^h - \xi^h\big)_0 + \big(\nabla\cdot\mathbf{u}^h, \nabla\cdot\mathbf{v}^h\big)_0 \\ &\quad + \frac{T}{\nu}\left(\frac{1}{\Delta t_k}\mathbf{u}^h + \nu\nabla\times\omega^h + \nabla p^h\,,\ \frac{1}{\Delta t_k}\mathbf{v}^h + \nu\nabla\times\xi^h + \nabla q^h\right)_{-h} \end{aligned}$$

and

$$\begin{aligned} &F^k(\mathbf{v}^h, \xi^h, q^h; \mathbf{f}(\mathbf{x}, t_k), \mathbf{u}^{(h,k-1)}) \\ &\qquad = \frac{T}{\nu}\left(\mathbf{f}(\mathbf{x}, t_k) + \frac{1}{\Delta t_k}\mathbf{u}^{(h,k-1)}\,,\ \frac{1}{\Delta t_k}\mathbf{v}^h + \nu\nabla\times\xi^h + \nabla q^h\right)_{-h}. \end{aligned}$$

Note that $Q_k(\cdot,\cdot,\cdot;\cdot,\cdot)$ does not depend on k whenever we have a constant time step, i.e., if $\Delta t_k = \Delta t$ for all $k = 1,\dots,K$.

Remark 9.10 Perhaps curiously, FD-LSFEMs for the Stokes problems such as (9.70) have not been analyzed so that error estimates are not available. However, there is no reason to believe that if $\{\mathbf{u}^{(h,k)}, \omega^{(h,k)}, p^{(h,k)}\} \in X^h_{\widehat{r},0}$, $r \ge 0$, denotes a sequence of solutions of (9.70) and if the exact solution $\{\mathbf{u}, \omega, p\}$ of the time-dependent VVP system (9.64) is sufficiently smooth, then, there exists a positive constant C such that

$$\|\mathbf{u}(\cdot,t_k) - \mathbf{u}^{(h,k)}\|_1 + \|\omega(\cdot,t_k) - \omega^{(h,k)}\|_0 + \|p(\cdot,t_k) - p^{(h,k)}\|_0 \le Ch^{r+1}$$

and

$$\|\omega(\cdot,t_k) - \omega^{(h,k)}\|_1 + \|p(\cdot,t_k) - p^{(h,k)}\|_1 \le Ch^r,$$

where C depends on $\mathbf{u}$, ω, and p. These estimates are analogous to the estimates of Theorem 7.14 for the steady-state VVP system. □

Remark 9.11 Algorithmically, one can, without difficulty, extend the DLSP (9.69) to the Navier-Stokes equations. □

Chapter 10
Hyperbolic Partial Differential Equations

In contrast to the great success of variational methods, especially finite element methods, for elliptic and even parabolic problems, their application to hyperbolic partial differential equations (PDEs) has met with somewhat less spectacular success. It is fair to say that even today, the most advanced finite element methods for hyperbolic PDEs are not completely satisfactory when compared with specialized finite volume and finite difference schemes. For example, computing monotone finite element solutions that capture solution discontinuities over a narrow band of cells remains a challenging and, for the most part, unresolved task.

The status of least-squares finite element methods (LSFEMs) for hyperbolic problems to a large degree mirrors this situation. Although the idea of replacing a hyperbolic PDE by an attractive Rayleigh–Ritz-like formulation[1] is very appealing, its straightforward application, without proper accounting for the distinctions between elliptic and hyperbolic PDEs, may lead to less than satisfactory methods.

The formulation of good LSFEMs for hyperbolic PDEs is further complicated by the fact that, for that class of PDEs, Hilbert spaces are not the only nor necessarily the best choice for defining energy balances. As a result, the numerical solution of hyperbolic PDEs by residual minimization has followed two parallel pathways, one for which residuals are minimized in Banach space norms[2] and the other for which minimization is effected with respect to norms in Hilbert spaces.

The detailed analysis of minimization problems in Banach spaces is beyond the scope of this book. However, the great promise shown by finite element methods derived in this setting warrants their inclusion in our discussion. We do this by applying the formal pattern, established in Chapter 3, to minimization problems in both Hilbert and Banach spaces: to define a finite element method, we start with an "energy" principle in an infinite-dimensional space which is used to obtain a well-posed residual minimization problem which is then restricted to a suitable finite

[1] Recall that least-squares principles allow for the possibility of recasting *any* PDE into unconstrained minimization of a quadratic "energy" functional.

[2] Strictly speaking, because in general Banach space settings we are not employing squares of Hilbert space norms to measure the size of residuals, the appellation "least squares" for the corresponding residual minimization principles does not make sense.

P.B. Bochev, M.D. Gunzburger, *Least-Squares Finite Element Methods*,
Applied Mathematical Sciences 166, DOI: 10.1007/b13382_10,

dimensional subspace. Of course, one aspect of the abstract theory of Chapter 3 that cannot be automatically extended to Banach spaces is the formal error analysis of LSFEMs. Instead, here we rely directly on the few theoretical results available in the literature.

We also restrict the main focus of this chapter to conservation laws which are one of the most important examples of hyperbolic PDEs. The model equations are introduced in Section 10.1, followed, in Section 10.2, by derivation of energy balances in Hilbert and Banach spaces. In Section 10.4, conforming and non-conforming finite element methods derived from "true" least-squares functionals are discussed. Methods derived from principles that minimize the $L^1(\Omega)$ norm of the residual are considered in Section 10.5. In Section 10.6, a class of finite element methods that try to recover the benefits of $L^1(\Omega)$ principles but that do not give up the differentiability of bona fide least-squares principles is considered. The chapter concludes with a brief discourse on practical issues and a summary of conclusions and recommendations.

10.1 Model Conservation Law Problems

Conservation laws are PDEs that can be written in the following canonical abstract form:

$$\begin{cases} \dfrac{\partial u}{\partial t} + \nabla \cdot F(u) = f & \text{in } \Omega \times (0,T) \\ \mathcal{B}u = h & \text{on } \Gamma_- \times (0,T) \\ u = u_0 & \text{in } \Omega \text{ at } t = 0, \end{cases} \tag{10.1}$$

where F is a *flux* function, u is a *conserved* variable, and Γ_- is a subset of $\partial\Omega$ with a positive measure. The simplest example of a conservation law is the linear scalar advection equation

$$\frac{\partial \phi}{\partial t} + \nabla \cdot (\mathbf{b}\phi) = f \tag{10.2}$$

for which $u = \phi$ and $F(\phi) = \mathbf{b}\phi$ with $\mathbf{b}(\mathbf{x})$ denoting a given vector. This equation models the transport of a conserved scalar quantity, such as the concentration, by the given velocity field $\mathbf{b}$.

An example of a nonlinear conservation law is the inviscid Burger equation

$$\frac{\partial \phi}{\partial t} + \phi \frac{\partial \phi}{\partial x} = 0 \tag{10.3}$$

for which $u = \phi$ and $F(\phi) = \frac{1}{2}\phi^2$. The Burger equation is often used as a simple model of fluid dynamics systems. A more realistic model for such systems is the compressible Euler equations

$$\begin{cases} \dfrac{\partial \rho}{\partial t} + \nabla \cdot (\rho \mathbf{u}) = 0 \\ \dfrac{\partial \rho \mathbf{u}}{\partial t} + \nabla \cdot (\rho \mathbf{u} \otimes \mathbf{u} + \mathsf{I} p) = 0 \\ \dfrac{\partial E}{\partial t} + \nabla \cdot \big(\mathbf{u}(E+p)\big) = 0, \end{cases} \tag{10.4}$$

where $E = \rho(e + \mathbf{u}\cdot\mathbf{u}/2)$ is the total energy per unit volume (e is the internal energy per unit mass for the fluid), p is the pressure, $\mathbf{u} = (u_1, u_2, u_3)^\top$ the fluid velocity, and ρ the fluid density. The Euler equations have the canonical form (10.1) with $u = \{\rho, \rho\mathbf{u}, E\}$ and

$$F(u) = \begin{pmatrix} \rho\mathbf{u} \\ \rho\mathbf{u}\otimes\mathbf{u} + \mathsf{I}p \\ \mathbf{u}(E+p) \end{pmatrix}.$$

The equations in the system (10.4) express the fundamental laws of conservation of mass, momentum, and energy, respectively, for a compressible, inviscid fluid flow. It can be thought of as a simplification of the compressible Navier–Stokes equations (8.1) in which viscosity effects are neglected; see Chapter 8. The Euler equations are the primary flow model in computational gas dynamics and form the basis for more complicated models in applications areas such as combustion.

In some cases, it is convenient to write conservation laws in the *nonconservative*, first-order system form

$$\frac{\partial v}{\partial t} + \mathsf{A}_1 \frac{\partial v}{\partial x} + \mathsf{A}_2 \frac{\partial v}{\partial y} + \mathsf{A}_3 \frac{\partial v}{\partial z} + \vec{b} = 0, \tag{10.5}$$

where v denotes a vector of n primitive variables, A_i, $i = 1,2,3$, denote square $n \times n$ matrices, and $\vec{b}$ denotes an n-vector. In general, both the A_is and $\vec{b}$ depend on $\mathbf{x}$ and u. For example, in two dimensions, the nonconservative form of the Euler equations is given by (10.5) with $v = (\rho, \mathbf{u}, p)^\top$ and

$$\mathsf{A}_1 = \begin{pmatrix} u_1 & \rho & 0 & 0 \\ 0 & u_1 & 0 & \rho^{-1} \\ 0 & 0 & u_1 & 0 \\ 0 & \gamma\rho & 0 & u_1 \end{pmatrix} \quad \text{and} \quad \mathsf{A}_2 = \begin{pmatrix} u_2 & 0 & \rho & 0 \\ 0 & u_2 & 0 & 0 \\ 0 & 0 & u_2 & \rho^{-1} \\ 0 & 0 & \gamma\rho & u_2 \end{pmatrix},$$

where u_1 and u_2 denote the components of $\mathbf{u}$.

There have been many computational studies of LSFEMs for the model problems described above. However, only the linear scalar advection equation has received adequate theoretical attention. For this reason, the bulk of this chapter deals with least-squares methods for (10.2) in which, as in Chapter 9, the time discretization step precedes the spatial discretization step. This allows us to limit theoretical discussions to the following version of (10.2):

$$\begin{cases} \nabla \cdot (\mathbf{b}\phi) + c\phi = f & \text{in } \Omega \\ \phi = g & \text{on } \Gamma_- . \end{cases} \tag{10.6}$$

The second term in the differential equation can be interpreted as either coming from a finite difference discretization of the time derivative or as a (linear) reaction term. Thus, we refer to (10.6) as the *advection–reaction* equation. The reaction coefficient $c(\mathbf{x})$ is a bounded measurable function on Ω and

$$\Gamma_- = \{\mathbf{x} \in \partial\Omega \mid \mathbf{n}(\mathbf{x}) \cdot \mathbf{b}(\mathbf{x}) < 0\} \tag{10.7}$$

denotes the inflow part of $\partial\Omega$. The outflow boundary is given by $\Gamma_+ = \partial\Omega \setminus \Gamma_-$ and $\nabla_b \phi = \mathbf{b} \cdot \nabla\phi$ refers to the streamwise derivative, i.e., the derivative of ϕ in the direction of $\mathbf{b}$.

The following inner product and norm on Γ_- are needed:

$$(\phi, \psi)_{-b} = \int_{\Gamma_-} \phi\psi\, \mathbf{n} \cdot \mathbf{b}\, d\Gamma \qquad \|\phi\|^2_{-b} = \int_{\Gamma_-} \phi^2\, |\mathbf{n} \cdot \mathbf{b}|\, d\Gamma .$$

Finally, recall the definition of a non-characteristic surface.

Definition 10.1 The surface s defined by the implicit equation $s(\mathbf{x}) = 0$ is non-characteristic at a point $\widetilde{\mathbf{x}}$ for the first-order system (10.5) if

$$\det\left(\mathsf{A}_1 \frac{\partial s}{\partial x} + \mathsf{A}_2 \frac{\partial s}{\partial y} + \mathsf{A}_3 \frac{\partial s}{\partial z} \right) \neq 0 \qquad \text{at } \mathbf{x} = \widetilde{\mathbf{x}} .$$

A surface is non-characteristic if it is non-characteristic at every point $\mathbf{x}$. □

It is easy to see that a surface s is non-characteristic for (10.6) if $\mathbf{b} \cdot \nabla s \neq 0$. Geometrically, this means that, at any point of s, the tangent plane cannot contain the advective vector $\mathbf{b}$.

10.2 Energy Balances

The advection–reaction equation is a limit case, as $\varepsilon \to 0$, of the advection–diffusion–reaction problem (1.65). However, the absence of the diffusion term requires a completely different analytic setting for the reduced problem. Indeed, solutions of (1.65) may have internal and/or boundary layers, but they remain in $H^1(\Omega)$ as long as $\varepsilon > 0$. In contrast, solutions of (10.6) propagate any discontinuity in the boundary data along characteristics so that they are not, in general, $H^1(\Omega)$ regular.

In this section, we consider two possible settings for the energy balance of (10.6). In both cases, the solution space X is a graph space of the advection–reaction operator $\mathcal{L}\phi \equiv \nabla \cdot (\mathbf{b}\phi) + c\phi$, i.e., solutions of (10.6) are not required to have weak first derivatives, except along the streamlines. The main difference between the two settings is in the type of spaces used to construct X. In Section 10.2.1, we assume that $\mathcal{L}\phi$ is *square integrable*, which results in a Hilbertian graph space. In Section 10.2.2, this requirement is relaxed by assuming that $\mathcal{L}\phi$ is only *integrable*, leading to a Banach graph space. In both cases, energy balances are derived under the following assumption on the coefficients of (10.6):

Assumption 10.2 *The advective field* $\mathbf{b}(\mathbf{x})$ *is of class* $C^1(\overline{\Omega})$ *and the reaction coefficient* $c(\mathbf{x})$ *is a bounded measurable function on* Ω.

10.2.1 Energy Balances in Hilbert Spaces

The Hilbertian setting for the energy balances in this section is similar to the settings employed in [46, 155, 209, 243, 244, 324]. The advection–reaction operator is considered as a map $\mathcal{L} : X_H \mapsto L^2(\Omega)$, where

$$X_H = \{\phi \in L^2(\Omega) \mid \nabla \cdot (\mathbf{b}\phi) + c\phi \in L^2(\Omega)\} \tag{10.8}$$

is the graph space of $\mathcal{L}$. X_H is Hilbert space when equiped with the graph norm

$$\|\phi\|_H = \left(\|\phi\|_0^2 + \|\mathcal{L}\phi\|_0^2\right)^{1/2} = \left(\|\phi\|_0^2 + \|\nabla \cdot (\mathbf{b}\phi) + c\phi\|_0^2\right)^{1/2}. \tag{10.9}$$

The main goal of this section is to show that this graph norm is equivalent to the "energy" norm

$$|||\phi||| = \left(\|\nabla \cdot (\mathbf{b}\phi) + c\phi\|_0^2 + \|\phi\|_{-b}^2\right)^{1/2}. \tag{10.10}$$

This equivalence follows from a trace inequality and a Poincaré-type inequality for functions in X_H.

Lemma 10.3 (Trace inequality) *Assume that* $\mathbf{b}$ *and* c *are as in Assumption* 10.2, *that* $\partial\Omega$ *is of class* C^1, *and* Γ_- *is non-characteristic. Then, there exists a continuous trace operator* $\gamma : X_H \mapsto L^2(\Gamma_-)$. *In particular, there exists a positive constant* C_T *such that, for all* $\phi \in X_H$,

$$\|\phi\|_{0,\Gamma_-} \le C_T \|\phi\|_H . \tag{10.11}$$

Furthermore, for any $\phi \in X_H$, *we have the Green's formula*

$$\int_\Omega \phi \nabla \cdot (\mathbf{b}\phi)\, d\Omega = \frac{1}{2}\left(\int_\Omega \phi^2 \nabla \cdot \mathbf{b}\, dx + \int_{\partial\Omega} \phi^2 \mathbf{n}\cdot\mathbf{b}\, d\partial\Omega\right). \tag{10.12}$$

Proof.[3] For brevity, the main idea is outlined for $\mathbf{b} = (1,0)^\mathsf{T}$, $\mathcal{L}\phi = \phi_x$, and $\Omega = [0,1]^2$. In this case, $\Gamma_- = \{(x,y)\,|\,x = 0,\ 0 \le y \le 1\}$ and for any $0 \le x \le 1$ and $\phi \in D(\overline{\Omega})$,

$$\phi(0,y) = \phi(x,y) - \int_0^x \phi_x(\xi,y)\, d\xi .$$

[3] The authors thank E. Suli who communicated the proof of this lemma to them.

Squaring both sides and using Cauchy's inequality yields

$$\phi^2(0,y) \le 2\phi^2(x,y) + 2\int_0^x \phi_x^2(\xi,y)d\xi\,.$$

Integration over $\overline{\Omega}$ gives the upper bound

$$\|\phi(0,\cdot)\|^2_{0,\Gamma_-} \le 2\big(\|\phi\|_0^2 + \|\phi_x\|_0^2\big) = 2|\!|\!|\phi|\!|\!|_H^2\,,$$

i.e., (10.11) holds with $C_T = \sqrt{2}$. The existence of the trace operator and the validity of (10.11) and (10.12) for functions belonging to X_H follow from the density of $D(\overline{\Omega})$ in X_H. □

Lemma 10.4 (Poincaré-type inequality) *Assume that $\partial\Omega$ is of class C^1, Γ_- is non-characteristic, and that*[4]

$$c + \frac{1}{2}\nabla\cdot\mathbf{b} \ge \gamma_0 > 0\,. \tag{10.13}$$

Then, there exists a positive constant C_P such that, for all $\phi \in X_H$,

$$\|\phi\|_0 \le C_P|\!|\!|\phi|\!|\!|\,, \tag{10.14}$$

where $|\!|\!|\cdot|\!|\!|$ is the energy norm defined in (10.10).

Proof. Using Green's formula (10.12), we obtain

$$\int_\Omega \Big(\nabla\cdot(\mathbf{b}\phi) + c\phi\Big)\phi\, d\Omega = \int_\Omega \Big(\phi\nabla\cdot(\mathbf{b}\phi) + c\phi^2\Big)\, d\Omega$$
$$= \int_\Omega \Big(\frac{1}{2}\nabla\cdot\mathbf{b} + c\Big)\phi^2\, d\Omega + \frac{1}{2}\Big(\int_{\Gamma_-}\phi^2\,\mathbf{n}\cdot\mathbf{b}\, d\Gamma + \int_{\Gamma_+}\phi^2\,\mathbf{n}\cdot\mathbf{b}\, d\Gamma\Big)$$

or, equivalently,

$$\int_\Omega \Big(\frac{1}{2}\nabla\cdot\mathbf{b} + c\Big)\phi^2\, d\Omega + \frac{1}{2}\int_{\Gamma_+}\phi^2\,\mathbf{n}\cdot\mathbf{b}\, d\Omega$$
$$= \int_\Omega \Big(\nabla\cdot(\mathbf{b}\phi) + c\phi\Big)\phi\, d\Omega + \frac{1}{2}\int_{\Gamma_-}\phi^2\,|\mathbf{n}\cdot\mathbf{b}|\, d\Omega\,.$$

After dropping the (non-negative) term on Γ_+, using the hypotheses (10.13) about $\mathbf{b}$ and c, and Cauchy's inequality for the first term on the right-hand side, we see that

$$\gamma_0\|\phi\|_0^2 \le \|\nabla\cdot(\mathbf{b}\phi) + c\phi\|_0\|\phi\|_0 + \frac{1}{2}\|\phi\|^2_{-b}\,.$$

Using the ε-inequality with $\varepsilon = \sqrt{\gamma_0}$ for the first term on the right-hand side gives

[4] In many, if not most applications, $\nabla\cdot\mathbf{b} = 0$ in which case (10.13) reduces to the simple condition $c \ge \gamma_0 > 0$.

$$\|\phi\|_0^2 \le \frac{1}{\gamma_0^2}\|\nabla\cdot(\mathbf{b}\phi)+c\phi\|_0^2+\frac{2}{\gamma_0}\|\phi\|_{-b}^2 \le C_P^2 |||\phi|||^2,$$

where $C_P = \max\left\{1/\gamma_0, \sqrt{2/\gamma_0}\right\}$. □

The following theorem establishes the energy balance for (10.6) in the graph space X_H.

Theorem 10.5 *Assume that $\partial\Omega$ is of class C^1, Γ_- is non-characteristic, and that* (10.13) *holds. Then, there exist positive constants C_1 and C_2 such that, for all $\phi \in X_H$,*

$$C_1\|\phi\|_H \le |||\phi||| \le C_2\|\phi\|_H. \tag{10.15}$$

Proof. Using the Poincaré inequality (10.14) gives

$$\|\phi\|_H^2 = \|\phi\|_0^2 + \|\mathcal{L}\phi\|_0^2 \le C_P^2|||\phi|||^2 + \|\mathcal{L}\phi\|_0^2 \le (1+C_P^2)|||\phi|||^2.$$

Therefore, the lower bound in (10.15) holds with $C_1 = (1+C_P^2)^{-1/2}$. To prove the upper bound, we use the trace inequality (10.11) and the obvious upper bound

$$\|\phi\|_{-b} \le \nu\|\phi\|_{0,\Gamma_-},$$

where $\nu = \max_{x\in\Gamma_-}|\mathbf{n}\cdot\mathbf{b}|$, to obtain

$$\begin{aligned} |||\phi|||^2 &= \|\mathcal{L}\phi\|_0^2 + \|\phi\|_{-b}^2 \le \|\mathcal{L}\phi\|_0^2 + \nu^2\|\phi\|_{0,\Gamma_-}^2 \\ &\le \|\mathcal{L}\phi\|_0^2 + \nu^2 C_T^2\|\phi\|_H^2 \le \left(1+\nu^2C_T^2\right)\|\phi\|_H^2 \end{aligned}$$

so that the upper bound holds with $C_2 = (1+\nu^2C_T^2)^{1/2}$. □

Remark 10.6 Lemma 10.3 implies that functions belonging to X_H have well-defined traces in $L^2(\Gamma_-)$, provided the boundary is C^1 regular. This assumption can be dropped by using a somewhat more complicated boundary norm (see [319, 320]) defined by

$$\|\phi\|_{-\ell}^2 = \int_{\Gamma_-}\phi^2\,\ell(\mathbf{x}(s))|\mathbf{b}\cdot\mathbf{n}|/\|\mathbf{b}\|\,d\Gamma, \tag{10.16}$$

where $\ell(\mathbf{x})$ denotes the length of the streamline defined by the vector field $\mathbf{b}$ connecting Γ_- to Γ_+. This boundary norm (10.16) satisfies a trace inequality similar to (10.11) and, if used in (10.10) (in lieu of $\|\cdot\|_{-b}$), gives rise to an energy norm that satisfies a Poincaré inequality similar to (10.14). As a result, the energy balance in Theorem 10.5 also holds for this modified energy norm. □

10.2.2 Energy Balances in Banach Spaces

In this section, we state an energy balance for (10.6) assuming that $\mathcal{L}\phi$ is only integrable. In this setting, the advection–reaction operator is considered as a mapping

$\mathcal{L} : X_B \mapsto L^1(\Omega)$, where now the graph space is defined by

$$X_B = \{\phi \in L^1(\Omega) \mid \nabla\cdot(\mathbf{b}\phi) + c\phi \in L^1(\Omega)\}. \tag{10.17}$$

When X_B is equipped with the graph norm

$$\|\phi\|_B = \|\phi\|_{L^1(\Omega)} + \|\mathcal{L}\phi\|_{L^1(\Omega)}, \tag{10.18}$$

it becomes a Banach space. The Banach space setting is potentially attractive for problems that have discontinuous solutions. However, because residual minimization in $L^1(\Omega)$ leads to non-differentiable functionals, direct use of this setting for the numerical solution of hyperbolic PDEs has been very limited; see [188, 229, 256].

In light of the discussion in Section 2.2.2, we can view the resulting $L^1(\Omega)$ minimization problem as an example of a mathematically correct but impractical minimization setting for finite element methods. An important difference between impractical "true" LSFEMs and $L^1(\Omega)$ "least-squares" finite element methods is that the latter cannot be turned into a practical discrete minimization problem by following the transformation recipes formulated in Section 3.4.1. Instead, in Section 10.5, we discuss approaches that replace direct minimization in $L^1(\Omega)$ by a sequence of regularized $L^2(\Omega)$ minimization problems.

It can be shown that the advection–reaction operator is an isomorphism $X_B \mapsto L^1(\Omega)$; see [188] and the references cited therein. Using this fact, one can prove the following result.

Theorem 10.7 *There exists a positive constant C such that, for all $\phi \in X_B$,*

$$C\|\phi\|_B \le \|\mathcal{L}\phi\|_{L^1}, \tag{10.19}$$

where $\|\cdot\|_B$ is the graph norm defined in (10.18).

10.3 Continuous Least-Squares Principles

The energy balances developed in Sections 10.2.1 and 10.2.2 offer two quite different pathways for recasting the conservation law (10.6) into an optimization problem. Each path has its own advantages and disadvantages.

The most obvious advantage of the Hilbertian setting from Section 10.2.1 is its compatibility with the abstract least-squares theory of Section 3.2. As a result, we can use the energy balance (10.15) to define a well-posed continuous least-squares principle (CLSP) for the advection-reaction problem by following the exact same procedure as described in Section 3.2.2. The resulting CLSP

$$\begin{cases} J_H(\phi; f, g) = \|\nabla\cdot(\mathbf{b}\phi) + c\phi - f\|_0^2 + \|\phi - g\|_{-b}^2 \\ X = X_H \end{cases} \tag{10.20}$$

is guaranteed, by Theorem 3.10, to have a unique minimizer $\phi \in X_H$ and that minimizer satisfies the Euler–Lagrange equation: *seek* $\phi \in X_H$ *such that*

$$Q(\phi;\psi) = F(\phi) \qquad \forall\, \psi \in X_H\,, \tag{10.21}$$

where

$$Q(\phi;\psi) = \int_\Omega \big(\nabla\cdot(\mathbf{b}\phi) + c\phi\big)\big(\nabla\cdot(\mathbf{b}\psi) + c\psi\big)\,d\Omega + \int_{\Gamma_-} \phi\,\psi\,|\mathbf{n}\cdot\mathbf{b}|\,d\Gamma$$

and

$$F(\psi) = \int_\Omega f\big(\nabla\cdot(\mathbf{b}\psi) + c\psi\big)\,d\Omega + \int_{\Gamma_-} g\psi\,|\mathbf{n}\cdot\mathbf{b}|\,d\Gamma\,,$$

respectively.

Because the least-squares functional in (10.20) uses only standard $L^2(\Omega)$ norms, the weak problem (10.21) is practical. This is another important advantage of the Hilbertian setting for (10.6). As is shown in Section 10.7.2, the main drawback of Hilbert spaces is the strong diffusivity of the least-squares weak equation that has to be compensated by additional "adjustments" to (10.20).

According to Remark 10.6,

$$\begin{cases} J_H(\phi;f,g) = \|\nabla\cdot(\mathbf{b}\phi) + c\phi - f\|_0^2 + \|\phi - g\|_{-\ell}^2 \\[1ex] X = X_H, \end{cases} \tag{10.22}$$

where $\|\cdot\|_{-\ell}$ is the norm defined in (10.16), is another well-posed CLSP for the advection–reaction equation (10.6) whose norm equivalence does not require the C^1 assumption on $\partial\Omega$. However, because the characteristic length $\ell(\mathbf{x})$ required in the definition of $\|\cdot\|_{-\ell}$ may be difficult to compute for some problems, this CLSP is less practical than (10.20). Thus, in the following sections, we only consider LSFEMs based on (10.20).

The most obvious disadvantage of the Banach space setting of Section 10.2.2 is that $L^1(\Omega)$ minimization is much more difficult to implement compared to $L^2(\Omega)$ minimization. Also, "least-squares" principles in the Banach space setting do not fit in the abstract framework of Section 3.2. Nevertheless, based on the result of Theorem 10.7, we can consider the following continuous "least-squares" principle for (10.6):

$$\begin{cases} J_B(\phi;f) = \|\nabla\cdot(\mathbf{b}\phi) + c\phi - f\|_{L^1(\Omega)} \\[1ex] X = X_{B,g} = \{\phi \in X_B \mid \phi = g \text{ on } \Gamma_-\}\,. \end{cases} \tag{10.23}$$

Although the abstract results of Section 3.2.2 are not applicable to this minimization problem, it is equivalent to (10.6) in the sense that they have the same solution.

10.3.1 Extension to Time-Dependent Conservation Laws

Both (10.20) and (10.23) can be easily extended to time-dependent conservation laws by discretizing first in time and then minimizing the residual of the semi-discrete[5] equation in X_H or X_B, respectively. However, for practical reasons, most existing methods have only treated the Hilbertian setting.

One of the first attempts at solving hyperbolic PDEs by LSFEMs is found in [231] which treats (10.2) in one space dimension; in this case, $\mathbf{b} = b$, a scalar function. The semi-discrete equation is derived by the generalized θ-method and its solution is advanced in time by using a sequence of CLSPs, i.e., at each time step we solve the minimization problem

$$\begin{cases} J(\phi) = \|(\phi - \overline{\phi}) + \Delta t\big(\theta(b\phi)_x + (1-\theta)(b\overline{\phi})_x\big)\|_0^2 \\ X = X_H\,. \end{cases} \tag{10.24}$$

In (10.24), $\overline{\phi}$ denotes the solution at time t and ϕ the unknown solution at time $t+\Delta t$. In [231], this approach is also extended to one-dimensional nonlinear conservation laws by using the same discretization in time and the following (non-quadratic) functional:

$$J(\phi) = \|(\phi - \overline{\phi}) + \Delta t\,\big(\theta F(\phi)_x + (1-\theta)F(\overline{\phi})_x\big)\,\|_0^2\,. \tag{10.25}$$

It can be shown (see [231]) that the resulting LSFEMs are equivalent to a particular form of the Taylor–Galerkin approach [145, 267] and are unconditionally stable for linear problems.

A further example is the method of [233] for the two-dimensional compressible Euler equations. This method uses the nonconservative form (10.5) of the Euler equations and the backward-Euler method in time, i.e., given a solution $\overline{v}$ at time t, the solution v at time $t+\Delta t$ is determined by solving the CLSP

$$\begin{cases} J(v) = \left\|(v - \overline{v}) + \Delta t\left(\mathsf{A}_1\dfrac{\partial v}{\partial x} + \mathsf{A}_2\dfrac{\partial v}{\partial y}\right)\right\|_0^2 \\ X = X_H\,. \end{cases} \tag{10.26}$$

Obviously, extensions of (10.23) to time-dependent problems can be obtained from that CLSP by measuring residuals in the $L^1(\Omega)$ norm and changing the minimization space to X_B.

[5] For space-time least-squares formulations, see [28, 29, 285, 295, 320].

10.4 Least-Squares Finite Element Methods in a Hilbert Space Setting

In this section, we consider LSFEMs derived from the CLSP (10.20). The first group of methods is based on conforming discrete least-squares principles (DLSPs) that use proper finite element subspaces of the graph space X_H. The second group includes several non-conforming methods.

10.4.1 Conforming Methods

A compliant DLSP for the advection–reaction equation is given by

$$\begin{cases} J_H(\phi^h; f, g) = \|\nabla\cdot(\mathbf{b}\phi^h) + c\phi^h - f\|_0^2 + \|\phi^h - g\|_{-b}^2 \\ X^h = X_H^h\,, \end{cases} \tag{10.27}$$

where X_H^h is a finite element subspace of the graph space X_H. Because (10.6) is a first-order problem and the least-squares functional involves only standard $L^2(\Omega)$ norms, the principle (10.27) passes the practicality tests from Section 2.2.2.

The standard C^0 nodal finite element space $G^r(\Omega)$ is trivially a subspace of the graph space X_H, and so it can be used in (10.27). We refer to the resulting LSFEMs for (10.6) as being *straightforward*. The following theorem specializes the abstract results of Theorem 3.28 to straightforward LSFEMs for the advection–reaction equation.

Theorem 10.8 *Let $X_H^h = G^r(\Omega)$ for some integer $r \geq 1$. The compliant DLSP (10.27) has a unique minimizer ϕ^h. Moreover,*

$$|||\phi - \phi^h||| \leq \inf_{\psi^h\in X_H^h} |||\phi - \psi^h|||\,. \tag{10.28}$$

If $\phi \in X_H \cap H^{r+1}(\Omega)$, then

$$|||\phi - \phi^h||| \leq Ch^r|\phi|_{r+1}\,. \qquad \Box \tag{10.29}$$

It is clear that each one of the semi-discrete in time CLSPs discussed in Section 10.3.1 can also be discretized by standard C^0 finite element spaces. In fact, the DLSPs corresponding to the original formulations of these methods in [231, 233] are respectively given by

$$\begin{cases} J(\phi^h) = \left\|(\phi^h - \overline{\phi}^h) + \Delta t\big(\theta(b\phi^h)_x + (1-\theta)(b\overline{\phi}^h)_x\big)\right\|_0^2 \\ X^h = G^r(\Omega) \end{cases} \tag{10.30}$$

and

$$\begin{cases} J(v^h) = \left\| (v^h - \overline{v}^h) + \Delta t \left(\mathsf{A}_1 \dfrac{\partial v^h}{\partial x} + \mathsf{A}_2 \dfrac{\partial v^h}{\partial y} \right) \right\|_0^2 \\ X^h = [G^r(\Omega)]^4 \,. \end{cases} \tag{10.31}$$

10.4.2 Non-Conforming Methods

There are several reasons to extend the range of the available LSFEMs for conservation laws beyond the straightforward method defined by (10.27). Clearly, if the exact solution of (10.6) is sufficiently smooth, Theorem 10.8 indicates that standard C^0 finite element spaces are completely satisfactory. However, if the boundary data have a discontinuity, it is shown in [268] that residual minimization in the norms of X_H or X_B over C^0 finite element spaces lead to solutions that necessarily spread the discontinuity over several elements and are not monotone; see Section 10.7.2 for a further discussion. One potential remedy is to replace the standard finite element space by a discontinuous space and a non-conforming version of (10.27).

Nonlinear conservation laws can develop discontinuous solutions even if the data are smooth. It has been observed that, in such cases, the straightforward methods defined by (10.30) and (10.31) may break down. In [232], it was observed that for high-speed compressible flow problems, the presence of nonlinear instabilities as the developing shock steepens leads to a failure of the straightforward method based on (10.30). For the Euler equations, it was noted in [233] that the straightforward method defined by (10.31) is stable and gives non-oscillatory shock profiles, as long as the time step is sufficiently large to maintain the Courant number in the range 10 to 50. However, outside this range, the straightforward method was prone to instabilities.

The fix proposed in [232, 233] was to augment the functionals in (10.30) and (10.26) by an $H^1(\Omega)$-seminorm of the nonlinear residual. The resulting $H^1(\Omega)$ least-squares method serves as our second example of non-conforming DLSPs.

Discontinuous least-squares finite element methods

Our first example of a non-conforming LSFEM is the discontinuous method proposed in [319] in which minimization is carried over the finite element space[6] $S^r(\Omega) = \cup_\kappa P_r(\kappa)$. Let $\Sigma^h = \mathcal{C}_{d-1}(\mathcal{T}_h)$, where d is the space dimension, denote the set of all oriented[7] inter-element interfaces for a given finite element partition $\mathcal{T}_h$. For $\phi^h \in S^r(\Omega)$, we define the interface semi-norm

[6] See Section B.2.1 for definition of discontinuous finite element spaces and Section 12.10 for further examples of discontinuous LSFEMs. Here, $\cup_\kappa P_r(\kappa)$ refers to the union of finite element spaces P_r defined over the individual elements κ.

[7] The sets $\mathcal{C}_m(\mathcal{T}_h)$, $0 \le m \le d$ are defined in (B.4). Recall that for $d = 2,3$ the interface $C \in \mathcal{C}_{d-1}(\mathcal{T}_h)$ is oriented by choosing a unit normal vector $\mathbf{n}_c$ at some interior point $\mathbf{x} \in C$.

$$\|\phi^h\|^2_{\Sigma^h} = \sum_{C \in \mathcal{C}_{d-1}(\mathcal{T}_h)} \int_C [\phi^h]^2 \, \omega^h_c \, |\mathbf{b} \cdot \mathbf{n}_c| \, d\Gamma \,, \tag{10.32}$$

where ω^h_c is a mesh-dependent weight and $[\cdot]$ is the jump function; see (12.83) for a definition. Note that (10.32) vanishes if ϕ^h is a C^0 function.

The discontinuous least-squares method of [319] is defined by the following mesh-dependent DLSP:

$$\begin{cases} J^h_H(\phi^h; f, g) = \sum_{\kappa \in \mathcal{T}_h} \|\nabla \cdot (\mathbf{b}\phi^h) + c\phi^h - f\|^2_{0,\kappa} + \|\phi^h - g\|^2_{-b} + \|\phi^h\|^2_{\Sigma^h} \\ X^h = S^r(\Omega) \,. \end{cases} \tag{10.33}$$

The role of the interface term $\|\phi^h\|_{\Sigma^h}$ is to promote interelement continuity in a direction parallel to $\mathbf{b}$, while at the same time allowing for larger jumps in a direction perpendicular to $\mathbf{b}$. This mimics the continuity properties of solutions to (10.6) which may jump across the characteristics, but remain continuous along them.

$H^1(\Omega)$ least-squares finite element methods

Our second example of a non-conforming LSFEM is the $H^1(\Omega)$ method introduced in [231]. There, this approach was motivated by drawing parallels between least-squares and multi-objective optimization problems. In this context, the term added to the least-squares functional can be likened to the second term in (11.2) whose purpose is to limit the growth of the controls; see Chapter 11 for a general discussion of control and optimization problems.

For a one-dimensional nonlinear conservation law, the $H^1(\Omega)$ method amounts to replacing (10.25) by the following least-squares functional:

$$J_1(\phi) = \|R(\phi)\|^2_0 + |\alpha R(\phi)|^2_1 \,, \tag{10.34}$$

where $0 < \alpha \ll 1$ is a penalty parameter, the seminorm $|\cdot|_1$ is defined in (A.6), and

$$R(\phi) = (\phi - \overline{\phi}) + \Delta t \big(\theta F(\phi)_x + (1-\theta) F(\overline{\phi})_x\big)$$

is the residual of the semi-discrete equation. The penalty term involving the $H^1(\Omega)$ norm contributes additional artificial dissipation proportional to Δt times the linearized flux function at $\overline{\phi}$. Conforming discretization of this functional requires C^1 finite element spaces so that it is not practical[8] in more than one space dimension.

Thus, to extend the $H^1(\Omega)$ approach to the Euler equations, in [233] the least-squares functional in (10.26) was replaced by the non-conforming functional

$$\widehat{J}_1(v^h) = \|R(v^h)\|^2_{0,\Omega} + \sum_{\kappa \in \mathcal{T}_h} |\alpha R(v^h)|^2_{1,\kappa} \,,$$

[8] Impracticality of C^1 finite element spaces is discussed in Section 2.2.2.

where now $R(v^h)$ stands for the residual of the non-conservative form of the Euler equations. This functional is non-conforming in the sense that it can be viewed as a triangulation-dependent version of the functional

$$J_1(v) = \|R(v)\|_{0,\Omega}^2 + |\alpha R(v)|_{1,\Omega}^2 .$$

10.5 Residual Minimization Methods in a Banach Space Setting

Methods for the approximate solution of PDEs based on residual minimization in Banach spaces are extremely rare. Arguably, the first example of this approach is the $L^1(\Omega)$ method for one-dimensional conservation laws proposed in [255, 256]. Despite the initial promise shown by $L^1(\Omega)$ methods, their reliance on mathematical programming to compute the minimizers meant that they could not compete with more conventional approaches. As a result, $L^1(\Omega)$ methods remained dormant until, in [188], a regularization procedure was formulated that allowed to approximate the minimizer by gradient methods. Although it is not clear if $L^1(\Omega)$ methods will ever become a practical alternative to other methods, they are an intriguing example of residual minimization that has influenced the development of LSFEMs for conservation laws. Thus, in this section, we offer a brief survey of $L^1(\Omega)$ methods and then, in Section 10.6, examine a class of LSFEMs motivated by them.

10.5.1 An $L^1(\Omega)$ Minimization Method

We describe the $L^1(\Omega)$ approach of [255, 256] using the following simple one dimensional problem:

$$\begin{cases} \dfrac{\partial \phi}{\partial x} = 0 & \text{on } (0,1) \\ \phi(0) = g_0 \quad \text{and} & \phi(1) = g_1 . \end{cases} \tag{10.35}$$

For $g_0 \neq g_1$, the "physically" meaningful solution of (10.35) is a "compression shock" at $x = 1$ given by $\phi(x) = g_0$ for $0 \le x < 1$ and $\phi(1) = g_1$. This solution is the limit, as $\varepsilon \to 0$, of the solution of the singularly perturbed second-order elliptic equation $-\varepsilon\phi_{xx} + \phi_x = 0$ along with the boundary conditions in (10.35).

The first step in the $L^1(\Omega)$ approach is a conventional finite volume discretization of the following regularly perturbed version of (10.35):

$$\begin{cases} \dfrac{\partial \phi}{\partial x} + 2\varepsilon\phi = 0 & \text{on } (0,1) \\ \phi(0) = g_0 \quad \text{and} & \phi(1) = g_1 . \end{cases} \tag{10.36}$$

Assuming that $(0,1)$ has been partitioned into N (not necessarily uniform) subintervals (x_{i-1},x_i), $i=1,2,\ldots,N$, the discrete equations are given by

$$\begin{cases} \phi_i - \phi_{i-1} + \varepsilon h_i(\phi_i + \phi_{i-1}) = 0 & \text{for } i = 1,2,\ldots,N, \\ \phi_0 = g_0, \qquad \phi_N = g_1\,, \end{cases} \tag{10.37}$$

where $h_i = x_i - x_{i-1}$ is the length of the ith interval. The next step is to seek solution of this over-determined $N \times (N-1)$ linear system for the unknown values ϕ_i, $i = 1,\ldots,N-1$, by minimizing the ℓ_1 norm of the residual in (10.36):

$$J_{\ell_1}(\vec{\phi}) = \sum_{i=1}^{N} |\phi_i - \phi_{i-1} + \varepsilon h_i(\phi_i + \phi_{i-1})|\,. \tag{10.38}$$

It is easy to see that without the perturbation term, any monotone grid function $\{\psi_i\}_{i=0}^N$ that satisfies $\psi_0 = g_0$ and $\psi_N = g_1$ is a minimizer of this functional.

The last step in the $L^1(\Omega)$ approach is to recast (10.38) as a linear programming problem and solve it by a discrete optimization algorithm. We can write (10.38) in the form

$$\mathsf{A}\vec{\phi} = \vec{g}\,,$$

where

$$\vec{\phi} = (\phi_1,\ldots,\phi_{N-1})^\mathsf{T}\,, \qquad \vec{g} = \left((1-\varepsilon h_1)g_0, 0, \ldots, 0, -(1+\varepsilon h_N)g_1\right)^\mathsf{T}\,,$$

and A is an $N \times (N-1)$ matrix. Let $\vec{r} = \vec{g} - \mathsf{A}\vec{\phi}$ denote the residual of this linear system. Finding the minimizer of the ℓ_1 functional (10.38) is equivalent to finding the solution $\vec{\phi}$ of the following linear programing problem (see [256]):

$$\text{minimize } \|\vec{r}\|_{\ell_1} \qquad \text{subject to} \qquad \vec{z}^\mathsf{T}\vec{r} = \vec{z}^\mathsf{T}\vec{g}\,, \tag{10.39}$$

where $\vec{z}$ is a vector that spans the null space of A and

$$\|\vec{r}\|_{\ell^1} = \sum_{i=1}^{N} |r_i| \tag{10.40}$$

is the ℓ_1 vector norm on $\mathbb{R}^N$. The ℓ_1 algorithm consists of finding the null-space vector[9] $\vec{z}$ and solving (10.39). It turns out that this solution is given by

$$\phi_i = g_0 \prod_{k=1}^{i} \frac{1-\varepsilon h_k}{1+\varepsilon h_k}\,, \qquad i = 1,2,\ldots,N-1\,,$$

so that the discontinuity is confined to the last cell (x_{N-1},x_N). In [256], this method was also applied to a time-independent version of the inviscid Burger equation.

[9] For the system (10.37), this vector can be found explicitly; this is not the case for a system obtained with the usual perturbation choice $-\varepsilon\phi_{xx}$; see [256].

10.5.2 Regularized $L^1(\Omega)$ Minimization Method

In the approach of [256], the $L^1(\Omega)$ minimization problem is solved by linear programming methods that do not require differentiability. To enable application of more conventional (and potentially more efficient) gradient-based algorithms, a regularization of the $L^1(\Omega)$ functional that renders it Fréchet differentiable was proposed in [188]. The starting point is the discrete $L^1(\Omega)$ principle

$$\begin{cases} J_B(\phi^h; f) = \|\nabla\cdot(\mathbf{b}\phi^h) + c\phi^h - f\|_{L^1(\Omega)} \\ X^h_{B,g} = G^1 \cap X_{B,g} \end{cases} \tag{10.41}$$

obtained by restriction of (10.23) to a standard C^0 finite element space. Because the inclusion $X^h_{B,g} \subset X_{B,g}$ clearly holds, problem (10.41) is a compliant discrete $L^1(\Omega)$ principle. However, this "DLSP" cannot be treated by the abstract framework in Section 3.2. The following theorem is proved in [188].

Theorem 10.9 *The following holds true for the compliant $L^1(\Omega)$ principle (10.41).*

1. *The problem* (10.41) *has no local minimizers and at least one global minimizer.*
2. *All minimizers have the stability property*

$$\|\phi^h\|_B \le C\|f\|_{L^1(\Omega)} \,.$$

3. *All minimizers satisfy the a priori error bound*

$$\|\phi - \phi^h\|_B \le \inf_{\psi^h \in X^h_{B,g}} \|\phi - \psi^h\|_B \,. \qquad \square$$

The last assertion of Theorem 10.9 is akin to the best approximation property (3.63) for compliant DLSPs, i.e., a minimizer of (10.41) also minimizes the graph norm distance between the exact solution and the finite element space $X^h_{B,g}$.

However, just as in Section 10.5.1, the $L^1(\Omega)$ principle (10.41) is ill-suited for numerical computation. It can have multiple minimizers[10] and the $L^1(\Omega)$ functional is not differentiable. In Section 10.5.1, the first issue was dealt with by using the non-singularly perturbed equation (10.36) and the second issue was circumvented by using linear programming.

The approach in [188] allows the simultaneous resolution of both of these issues by using the discrete functional

$$\begin{cases} J_{B,\varepsilon}(\phi^h_\varepsilon; f) = \displaystyle\int_\Omega \|\nabla\cdot(\mathbf{b}\phi^h_\varepsilon) + c\phi^h_\varepsilon - f\|_{\ell^1,\varepsilon}\, d\Omega \\ X^h_{B,g} = G^1 \cap X_{B,g}\,, \end{cases} \tag{10.42}$$

[10] Recall that the parent principle of (10.41), i.e., the $L^1(\Omega)$ minimization problem (10.23), does have a unique minimizer. The fact that its conforming discretization can have multiple minimizers is another key distinction between the Banach and Hilbertian settings for conservation laws.

where, for $\varepsilon > 0$ and $\vec{x} \in \mathbb{R}^n$,

$$\|\vec{x}\|_{\ell^1,\varepsilon} = \sum_{i=1}^{n} \frac{|\vec{x}_i|^2}{|\vec{x}_i| + \varepsilon}$$

is a regularization of the ℓ_1 norm defined in (10.40). The regularized norm is Fréchet differentiable so that (10.42) can be solved by gradient methods. The solution of the original problem is obtained as a limit of the sequence $\{\phi_\varepsilon^h\}$ as $\varepsilon \to 0$. The following results are proved in [188].

Theorem 10.10 *The solutions of the regularized problem* (10.42) *possess the stability property*

$$\|\phi - \phi_\varepsilon^h\|_B \leq C\Big(\varepsilon c(\|f\|_{L^1}) + \min_{\psi^h \in X_{B,g}^h} \|\phi - \psi^h\|_B\Big),$$

where $c(\cdot)$ is a continuous function. Moreover, up to a subsequence, every limit ϕ_0^h of $\{\phi_\varepsilon^h\}$ is a solution of the non-regularized discrete $L^1(\Omega)$ principle (10.41). □

For further discussions of theoretical and computational properties of $L^1(\Omega)$-based finite element methods, including a definition of an iterative algorithm that determines minimizers of (10.41) by solving a sequence of regularized problems, see [188, 189].

10.6 Least-Squares Finite Element Methods Based on Adaptively Weighted $L^2(\Omega)$ Norms

In this section, we present a class of methods that attempt to reproduce the benefits of residual minimization in Banach spaces without giving up the differentiability of the functionals found in bona fide least-squares principles. A common feature of these algorithms is their use of adaptively weighted $L^2(\Omega)$ norms to reduce the smearing of discontinuities and over/undershoots.

10.6.1 An Iteratively Re-Weighted Least-Squares Finite Element Method

Our first example is a LSFEM proposed in [229]. There it was observed that although a standard $L^2(\Omega)$ LSFEM may significantly smear discontinuities, it can be used as a reliable error indicator to detect the discontinuity, i.e., to determine which elements are near the discontinuity. Then, contributions from these elements to the least-squares functional can be reduced by suitably defined weights. By unweighting the "shocked" elements, the least-squares principle essentially disregards the equations associated with these elements. The unweighting process can be repeated in an iterative manner until some desired convergence tolerance is satisfied. The net result is a finite element solution that approximates well smooth portions of the exact solution without having the typical over/undershoots near the discontinuities, i.e., it resembles a minimizer of an $L^1(\Omega)$ functional.

The iteratively re-weighted least-squares method of [229] consists of solving, for $k = 1,2,\ldots$, the sequence of mesh-dependent DLSPs

$$\begin{cases} J_{\omega_k}(\phi_k^h; f) = \|\sqrt{\omega_k}(\nabla\cdot(\mathbf{b}\phi_k^h) + c\phi_k^h - f)\|_0^2 + \|\sqrt{\omega_k}(\phi - g)\|_{0,\Gamma_-}^2 \\ X_k^h = G^1 \end{cases} \tag{10.43}$$

until $\|\phi_k^h - \phi_{k-1}^h\|_0 \leq tol$. For $k = 1$, the weight ω_1 is set to 1, i.e., (10.43) is a straightforward $L^2(\Omega)$ least-squares principle. For subsequent steps,

$$\omega_k = \frac{1}{|R(\phi_{k-1}^h)|^6} \qquad \text{for } k = 2,3,\ldots,$$

where $R(\phi_{k-1}^h) = \nabla\cdot(\mathbf{b}\phi_{k-1}^h) + c\phi_{k-1}^h - f$ is the residual of the least-squares solution determined from the previous step. The result is an iterative solution procedure that resembles the solution algorithm of [188] for (10.42). However, the re-weighted method uses bona fide least-squares principles rather than a regularized ℓ_1 norm.

It is possible to implement the iteratively re-weighted procedure using other error monitor functions; see [229]. There are no theoretical results regarding the convergence properties and accuracy of this method. Nevertheless, computational results given in [229] show essentially monotone solutions and shock resolution within a narrow layer of elements.

10.6.2 A Feedback Least-Squares Finite Element Method

The feedback least-squares method of [120] for the model advection–reaction problem (10.6) is a relative of the iteratively re-weighted method of Section 10.6.1. The differences are that the feedback method uses a statistical approach, adopted from [107], to locate the discontinuity and a different definition of the weight function.

Assume that g has a jump discontinuity at $\mathbf{x} \in \Gamma_-$ that is propagated by the solution ϕ along the characteristic $\chi \subset \Omega$. Similarly to the iteratively re-weighted method from Section 10.6.1, the feedback LSFEM relies on the residual of the finite element solution ϕ^h to locate χ. However, in the feedback method, the unweighting of the least-squares functional is confined to a *discontinuity set* M_χ containing the elements that are near χ. To define this set, each element $\kappa \in \mathcal{T}_h$ is ranked using the following quantities.

1. The mean residual for the element κ:

$$\ll R \gg_\kappa = \int_\kappa |R_\kappa(\phi^h)|/\mu(\kappa)\, d\Omega \qquad \forall \kappa \in \mathcal{T}_h,$$

where $\mu(\kappa)$ is the element measure and $R_\kappa(\phi) = \big(\nabla \cdot (\mathbf{b}\phi^h) + c\phi^h - f\big)|_\kappa$ is the element residual.

2. The mean residual for the finite element partition $\mathcal{T}_h$:

$$\ll R \gg_h = \frac{1}{N_e} \sum_{\kappa \in \mathcal{T}_h} \ll R \gg_\kappa,$$

where N_e is the number of elements in $\mathcal{T}_h$.

3. The mean deviation for $\mathcal{T}_h$:

$$\ll \delta \gg_h = \Big(\frac{1}{N_e} \sum_{\kappa \in \mathcal{T}_h} \big(\ll R \gg_h - \ll R \gg_\kappa \big)^2 \Big)^{1/2}.$$

An initial set M_0 is defined by including all elements κ whose mean element residual exceeds the mean residual for $\mathcal{T}_h$, plus a term proportional to the mean deviation:

$$M_0 = \{\kappa \in \mathcal{T}_h \mid \ll R \gg_\kappa \, \geq \, \ll R \gg_h + \varepsilon \ll \delta \gg_h\}. \tag{10.44}$$

Here, ε is a positive parameter that may be used to adjust the sensitivity of this detection criterion. For triangular partitions $\mathcal{T}_h$, the set M_χ is constructed from M_0 using the following recursive process (see Figure 10.1).

1. Tag all elements in M_0 by 1 and all elements in $\mathcal{T}_h \setminus M_0$ by 0.
2. Update element tags according to the following rules: initialize $m_\Omega = 3$;
   - if κ has tag 0 and m_Ω adjacent elements have tags 1, set the tag of κ to 1;
   - if κ has tag 1 and all adjacent elements have tag 0, set the tag of κ to 0;
   - set $m_\Omega = 2$ and repeat until no tags change.
3. M_χ is the set of all elements whose tag equals 1.

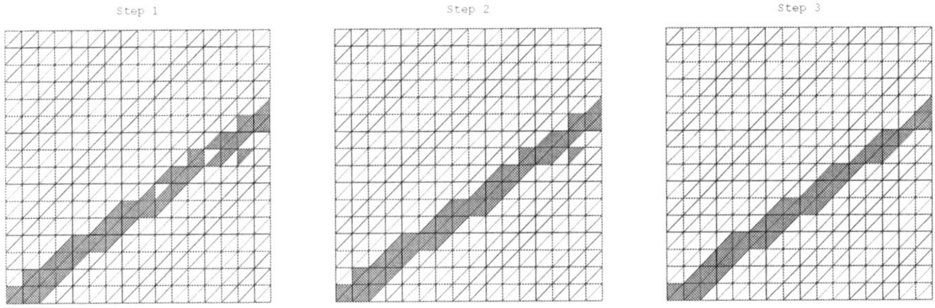

Fig. 10.1 The shaded triangles in the leftmost plot are in the set M_0. The center plot shows an intermediate set obtained at step 2. The rightmost plot shows the set M_χ.

This procedure is used to define the feedback LSFEM as follows.

1. Set $\omega_\kappa = 1$ for all $\kappa \in \mathcal{T}_h$ and compute a minimizer ϕ_0^h of (10.43).
2. Use ϕ_0^h to construct the set M_χ.
3. Set

$$\omega_\kappa = \begin{cases} 1 & \text{for} \quad \kappa \notin M_\chi \\ \dfrac{1}{1+\alpha\|\nabla \phi_0^h\|_\kappa} & \text{for} \quad \kappa \in M_\chi, \end{cases}$$

where α is an amplifying constant (see [229]) and solve (10.43).

If necessary, step 3 or steps 2–3 can be repeated until the relative error between two consecutive solutions becomes smaller than some prescribed tolerance.

Usually, the construction of M_χ from M_0 takes just few steps which are cheaper than solving the LSFEM problem itself. As a result, the feedback method tends to be more efficient than the iteratively re-weighted approach for which the first couple of iterations are essentially used to determine the same information as provided by the discontinuity set M_χ.

10.7 Practicality Issues

We now give a brief discourse on the practical aspects of using LSFEMs to determine approximations of solutions of hyperbolic PDEs.

10.7.1 Approximation of Smooth Solutions

Theorem 10.5 asserts that for sufficiently smooth solutions, compliant LSFEMs for (10.6) are optimal in the energy norm (10.10). Recall that for elliptic PDEs, energy norm estimates can be supplemented by optimal error bounds in $L^2(\Omega)$. Because least-squares principles for (10.6) embed this hyperbolic equation into an elliptic problem, it would be natural to expect that optimal $L^2(\Omega)$ estimates also hold for (10.20). As far as computations are concerned, this indeed turns out to be the case, as can be seen from the data in Table 10.1.

	$L^2(\Omega)$ error	$H^1(\Omega)$ error	$\|\nabla_b\phi - \nabla_b\phi^h\|_0$
SUPG	3.0000	2.0024	1.9970
LSFEM	3.0055	2.0087	2.0377

Table 10.1 Comparison of convergence rates for streamline-upwind Petrov–Galerkin (SUPG) and the compliant LSFEM (10.27) for (10.6) using the manufactured solution $\phi = \exp(x)\sin(x)\sin(y)$, finite element partition of $\Omega = [0,1]^2$ into triangles, and $X^h = G^2$ (piecewise quadratic elements on triangles).

Unfortunately, proving optimal $L^2(\Omega)$ estimates by a standard duality argument is impossible because the least-squares dual problem: *seek* $\psi \in X_H$ *such that*

$$Q(\phi;\psi) = (f,\phi) \qquad \forall \phi \in X_H\,, \tag{10.45}$$

where $Q(\cdot;\cdot)$ is the bilinear form in (10.21), is not sufficiently regular. To see this, assume for simplicity that $\mathbf{b}$ is solenoidal, $c = 1$, $g = 0$, and X_H is constrained by the boundary condition on Γ_- and define

$$\Gamma_{++} = \{\mathbf{x} \in \partial\Omega \mid \mathbf{n}(\mathbf{x}) \cdot \mathbf{b}(\mathbf{x}) > 0\}\,.$$

Formal integration by parts in (10.45) shows that the strong formulation of this dual problem is given by the boundary value problem

$$\begin{cases} -\nabla\cdot\big(\mathbf{b}(\nabla_b\phi+\phi)\big)+\big(\nabla_b\phi+\phi\big) = f & \text{in } \Omega \\ \phi = 0 & \text{on } \Gamma_- \\ \nabla_b\phi+\phi = 0 & \text{on } \Gamma_{++}\,, \end{cases} \tag{10.46}$$

where $\nabla_b\phi$ is the streamwise derivative defined in Section 10.1. The differential operator in (10.46) only controls the second derivatives along the streamlines, i.e., it is degenerate elliptic. Thus, (10.46) does not possess the full elliptic regularity necessary to prove optimal $L^2(\Omega)$ error estimates. Nevertheless, in some cases, using a perturbed version of (10.46), it is possible to demonstrate *improved* $L^2(\Omega)$ estimates for the straightforward LSFEM; see [46].

Theorem 10.11 *Assume that* Ω *is a rectangular domain,* $\mathbf{b} = (1,0)^\top$, (10.6) *has a solution* $\phi \in H^{r+1}(\Omega)$*, and* ϕ^h *is minimizer of the compliant DLSP* (10.27). Then,

$$\|\phi - \phi^h\|_0 \le Ch^{k+1/3}\|\phi\|_{k+1}\,. \tag{10.47}$$

The conclusion of this theorem can be extended to more general domains and advective vectors. However, at present, it is not clear if and how the exponent $k+1/3$ can be improved to $k+1$.

10.7.2 Approximation of Discontinuous Solutions

For discontinuous solutions of (10.6), the analysis of LSFEMs becomes complicated. Not much is known about the error behavior except when the flow is grid-aligned. In this case, error estimates were derived in [320].

A different aspect of LSFEMs was investigated in [268] where the ability of residual minimization principles to provide sharp resolution and correct placement of solution discontinuities was examined. The main result showed that a LSFEM based on residual minimization in *fixed norms* cannot simultaneously resolve the discontinuity within a single mesh interval and compute a monotone solution. The

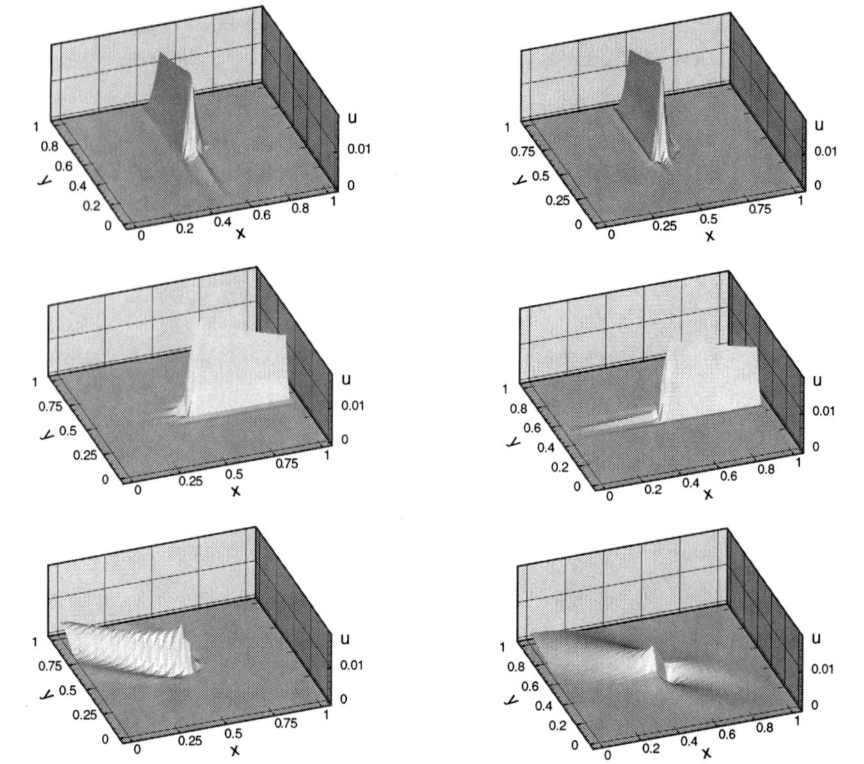

Fig. 10.2 Discontinuity smearing by straightforward LSFEMs on uniform triangular grids. The left plots show a reference SUPG solution; $\mathbf{b} = (0,1)^{\mathrm{T}}$ for the top row; $\mathbf{b} = (1,0)^{\mathrm{T}}$ for the middle row; and $\mathbf{b} = (-\sqrt{2}/2, \sqrt{2}/2)^{\mathrm{T}}$ for the bottom row. The vertical axis scale is the same in all plots.

conclusion drawn in [268] was that this may be possible for least-squares principles defined along the lines described in Section 10.6, i.e., by using adaptively weighted $L^2(\Omega)$ norms.

Computational studies and analyses of LSFEMs on uniform grids [45, 120] also reveal that discontinuity smearing and the amount of dissipation in straightforward LSFEMs increases as the angle between the grid lines and the flow direction approaches $\pi/4$. Figure 10.2 shows some typical results for a cone advected in three different directions. Plots in the top and the middle row show that, for advection along the grid lines, the straightforward LSFEM and SUPG method produce comparable solutions. However, when $\mathbf{b}$ is aligned with the bisector of the right angle in each triangle, the LSFEM diffuses the cone at a higher rate than does SUPG. These results further reinforce the conclusion of [268] that straightforward LSFEMs defined in "static" Hilbert spaces, without further modifications, are not appropriate for discontinuous solutions.

Fortunately, the LSFEMs of Section 10.6 can deal successfully with these problems and provide non-oscillatory, essentially monotone approximations of discon-

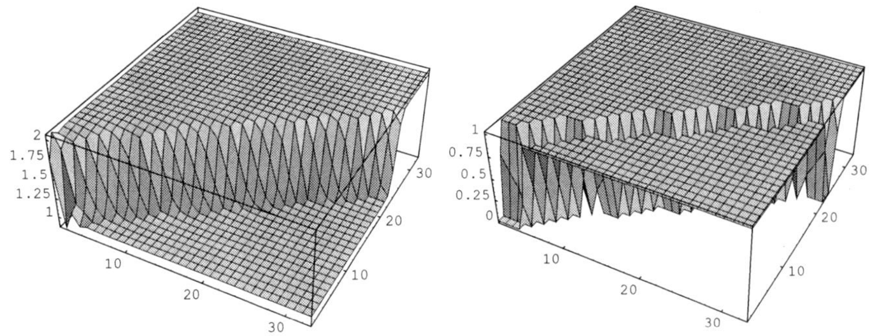

Fig. 10.3 Feedback LSFEM solution (left plot) and adaptive weight function (right plot) for constant advection (10.48). The mesh $\mathcal{T}_h$ is defined using 17×17 uniform grid lines and has $2 \times 16 \times 16 = 512$ triangle elements with a total of $33 \times 33 = 1089$ nodes.

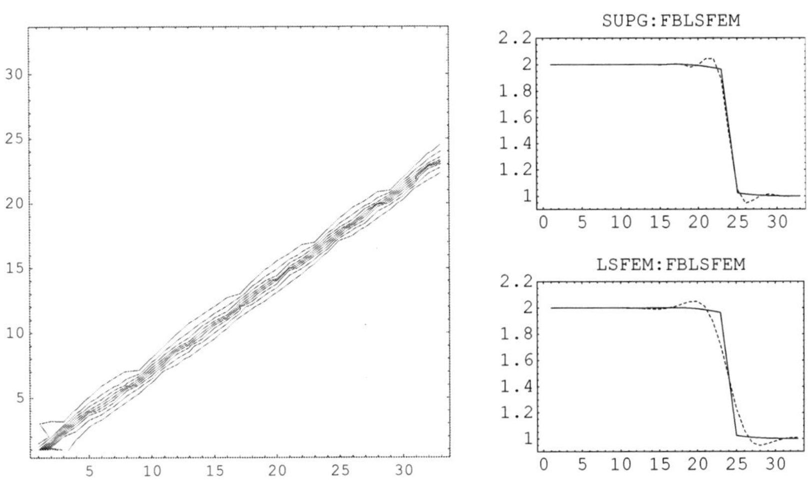

Fig. 10.4 Feedback LSFEM solution for constant advection (10.48). The mesh is defined as in Figure 10.3. The left plot shows the feedback LSFEM solution contours. The right plots compare horizontal profiles at $y = 0.5$ of the feedback LSFEM solution (solid line) with the SUPG and straightforward LSFEM solutions (dotted lines).

tinuous solutions that sharply resolve the discontinuities. Two examples of such solutions by the feedback LSFEM are shown in Figures 10.3–10.6 which also provide comparisons to approximations determined by the SUPG method and straightforward LSFEMs. In these examples Ω is the unit square, $\mathcal{T}_h$ is a uniform partition of Ω into triangles, and $X^h = G^2$ (piecewise quadratic finite elements). The grid is defined by partitioning Ω into squares using $n \times n$ uniformly spaced grid lines in the x and y directions after which each square is divided into two triangles formed by the sides of the square and the diagonal from the bottom left to the top right vertex. Figures 10.3–10.4 shows results for a constant advection case where

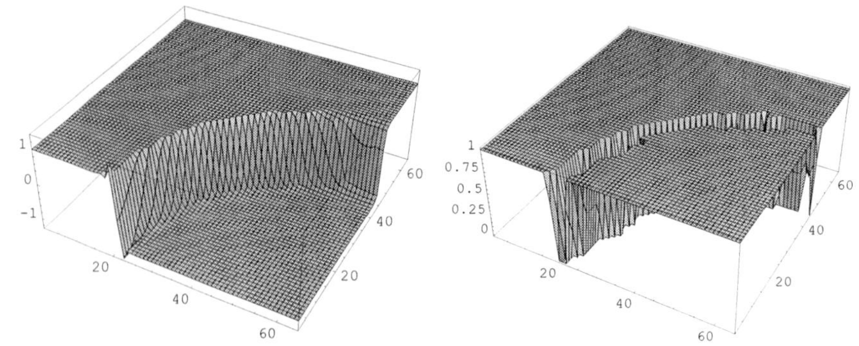

Fig. 10.5 Feedback LSFEM solution (left plot) and adaptive weight function (right plot) for circular advection (10.49). The mesh $\mathcal{T}_h$ is defined using 33×33 uniform grid lines and has $2 \times 32 \times 32 = 2048$ triangle elements with a total of $65 \times 65 = 4225$ nodes.

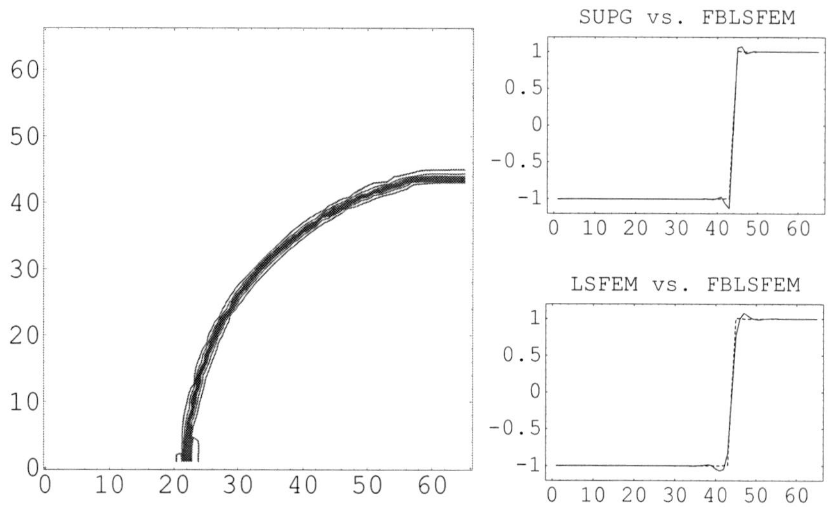

Fig. 10.6 Feedback LSFEM solution for circular advection (10.49) on uniform grids. The mesh is defined as in Figure 10.5 The left plot shows the feedback LSFEM solution contours. The right plots compare outflow profiles of the feedback LSFEM solution (dashed line) with the SUPG and straightforward LSFEM solutions.

$$\mathbf{b} = (1, \tan 35^\circ)^\mathsf{T} \quad \text{and} \quad g(x,y) = \begin{cases} 2 \text{ on } \Gamma_L \\ 1 \text{ on } \Gamma_B . \end{cases} \tag{10.48}$$

and Figures 10.5–10.6 correspond to circular advection:

$$\mathbf{b} = (-y, x)^\mathsf{T} \quad \text{and} \quad g(x,y) = \begin{cases} -1 \text{ on } \Gamma_B \text{ and } x < 43/64 \\ \ \ 1 \text{ on } \Gamma_B \text{ and } x \geq 43/64 \\ \ \ 1 \text{ on } \Gamma_R . \end{cases} \tag{10.49}$$

In (10.48) and (10.49),

$$\Gamma_L = \{(x,y)\,|\,x=0;\; 0 \le y \le 1\}, \qquad \Gamma_R = \{(x,y)\,|\,x=1;\; 0 \le y \le 1\},$$

and

$$\Gamma_B = \{(x,y)\,|\,y=0;\; 0 \le x \le 1\}.$$

Figures 10.3–10.6 show that the feedback LSFEM is able to resolve the discontinuity over a fairly narrow band of elements. To fully appreciate the improvement in the feedback LSFEM, one should carefully examine the plot in Figure 10.4 that compares horizontal profiles of this solution with the straightforward LSFEM and SUPG solutions. These profiles also show that the feedback LSFEM solution is essentially monotone and does not exhibit the over/undershoots that are prominent features of SUPG solutions.

Finally, we note that solutions obtained using the iteratively re-weighted LSFEM in Section 10.6.1 are comparable in quality with those for the feedback LSFEM.

10.8 A Summary of Conclusions and Recommendations

The main message of this chapter is that solving hyperbolic PDEs by LSFEMs is an achievable, but by no means simple, task. Of course, this is also true for other methods such as SUPG, finite difference, and finite volume methods, and merely reflects the fact that hyperbolic PDEs are more difficult to solve than elliptic PDEs.

The theoretical and computational results presented in this chapter strongly suggest that, for hyperbolic PDEs, least-squares principles defined with respect to "static" Hilbert spaces are not the most appropriate choice without further modifications. This applies with equal force to both straightforward and non-conforming LSFEMs because such methods, even combined with grid refinement and higher-order elements [320, 325], are not able to produce monotone solutions. Indeed, a comparison between the compliant (10.27) and discontinuous (10.33) LSFEMs in [320] reveals that both methods tend to smear discontinuities at about the same rate. In other words, merely switching to discontinuous elements is not enough to offset the natural dissipation present in least-squares formulations based on "static" Hilbert spaces. The same study indicates that discontinuity smear can be reduced by using higher-order elements but that such elements do not eliminate the over/undershoots in the straightforward least-squares solution.

The most promising approaches appear to be either explicitly or implicitly related to minimization problems in Banach spaces. Computationally, the $L^1(\Omega)$ method of [188] recovers what is essentially a viscosity solution of the conservation law. The adaptively weighted $L^2(\Omega)$ norm LSFEMs of Section 10.6 also perform very well by using "dynamic" Hilbert spaces, and are easier to implement and use than the $L^1(\Omega)$ method. However, both $L^1(\Omega)$ and "dynamic" LSFEMs are not yet at a stage where they can truly compete with more established approaches that have a much longer history of use in practice.

Table 10.2 compares and contrasts properties of select methods from this chapter.

Method→ Property↓	Compliant (10.27)	Re-weighted LSFEM Section 10.6.1	Feedback LSFEM Section 10.6.2	Regularized $L^1(\Omega)$ (10.42)
Provably optimal	$\surd$	–	–	$\surd$
Monotone	–	$\surd$	$\surd$	$\surd$
Solution cost	low	high	medium	medium
Coding effort	simple	simple	simple	not as simple

Table 10.2 Summary properties of select LSFEMs for the advection–reaction problem (10.6).

Chapter 11
Control and Optimization Problems

Optimization and control problems for systems governed by partial differential equations (PDEs) arise in many applications. Experimental studies of such problems go back at least 100 years [312] and computational approaches have been applied since the advent of the computer age. Most of the efforts in the latter direction have employed elementary optimization strategies but, more recently, there has been considerable practical and theoretical interest in the application of sophisticated local and global optimization strategies, e.g., Lagrange multiplier methods, sensitivity or adjoint-based gradient methods, quasi-Newton methods, evolutionary algorithms, and so on.

The optimal control or optimization problems of interest here consist of

- *state variables* – variables that describe the system being modeled
- *control variables* or *design parameters* – variables at our disposal that can be used to affect the state variables
- a *state system* – PDEs relating the state and control variables
- an *objective* or *cost functional* of the state and control variables whose minimization is the goal.

Then, optimal control or optimization problems require

> *finding state and control variables that minimize the given objective functional, subject to the state system being satisfied.*

We restrict attention to linear elliptic state systems and to convex quadratic functionals.

The classical Lagrange multiplier rule is a standard approach for solving finite-dimensional constrained optimization problems. It is not surprising then that several popular approaches to solving optimization and control problems constrained by PDEs are also based on solving optimality systems deduced from the application of the Lagrange multiplier rule. In the linear constraints/quadratic functional context we consider here, the optimality system, viewed as a coupled system, is a symmetric and weakly coercive linear system in the state, Lagrange multiplier, and control variables. In the context of finite element methods, optimality systems are

P.B. Bochev, M.D. Gunzburger, *Least-Squares Finite Element Methods*, Applied Mathematical Sciences 166, DOI: 10.1007/b13382_11,

usually discretized using mixed-Galerkin methods, resulting in typical saddle-point type matrix problems that are symmetric and indefinite; we review this approach in Section 11.2.1; for additional details, see, e.g., [87, 183, 191].

Another means for determining approximate solutions of optimality systems is to apply a least-squares finite element method (LSFEM); see, e.g., [61, 195]. This approach, which is discussed in Section 11.2.2, not only offers the by now well-catalogued advantages of LSFEMs over mixed-Galerkin finite element methods, e.g., symmetry and positive definiteness, but in addition has advantages specific to optimal control and optimization settings. In particular, discretized optimality systems, in practice, offer formidable computational challenges, not the least because they involve at least double the number of degrees of freedom compared to what one encounters for forward simulation problems. For this reason, many strategies have been suggested for uncoupling the different component equations in optimality systems. In this respect, the internal structure of the discretized optimality systems obtained through least-squares principles offer important advantages over those obtained through mixed-Galerkin principles.

In the Lagrange multiplier-based approach, one usually constrains the cost functional by a Galerkin weak formulation of the constraint equations. Instead, one can constrain by a least-squares minimization form of these equations. This leads to different optimality systems that have advantages over using the Galerkin form of the constraints. This approach was considered in [62] and is discussed here in Section 11.4.

Penalty methods are another popular approach for finite-dimensional optimization problems, mostly because, for the setting considered here, they result in positive definite discrete formulations and require fewer degrees of freedom than that required by Lagrange multiplier-based approaches. Penalty methods have not, however, generated much interest for infinite-dimensional PDE-constrained optimization and control problems, probably because, in practical implementations, they suffer from conditioning problems. Using methodologies in modern LSFEMs such as those discussed in the earlier chapters, the penalty approach can be rehabilitated to yield practical and efficient algorithms for optimal control problems. These algorithms, hereafter referred to as least-squares/penalty methods, enforce the PDE constraints by using well-posed least-squares functionals as penalty terms that are added to the original cost functional. This type of penalty method offers certain efficiency-related advantages compared to methods based on the solution of the Lagrange multiplier-based optimality system either by mixed-Galerkin methods or LSFEMs. Least-squares/penalty methods have been used, e.g., for optimal shape design problems [25, 37, 42, 196]. In [56], several of the ideas and concepts discussed in Section 11.3 were presented in the concrete context of the Stokes equations; a more abstract discussion was provided in [60].

We use the framework and results established in Chapter 1 about constrained optimization problems and mixed-Galerkin formulations and the results established in Chapter 3 about LSFEMs to study optimal control problems.

11.1 Quadratic Optimization and Control Problems in Hilbert Spaces with Linear Constraints

In this section, we apply the results reviewed in Sections 1.2.4 and 1.3.2 concerning mixed variational and constrained optimization problems to the study of a class of optimization and control problem. We begin with four Hilbert spaces[1] Θ, Φ, $\widehat{\Phi}$, and $\widetilde{\Phi}$ along with their dual spaces denoted by $(\cdot)^*$. We assume that $\Phi \subseteq \widehat{\Phi} \subseteq \widetilde{\Phi}$ with continuous embeddings and that $\widetilde{\Phi}$ acts as the pivot space for both the pair $\{\Phi^*, \Phi\}$ and the pair $\{\widehat{\Phi}^*, \widehat{\Phi}\}$ so that we not only have that $\Phi \subseteq \widehat{\Phi} \subseteq \widetilde{\Phi} \subseteq \widehat{\Phi}^* \subseteq \Phi^*$, but also

$$\langle \psi, \phi \rangle_{\Phi^*,\Phi} = \langle \psi, \phi \rangle_{\widehat{\Phi}^*,\widehat{\Phi}} = (\psi, \phi)_{\widetilde{\Phi}} \qquad \forall \psi \in \widehat{\Phi}^* \subseteq \Phi^*,\ \phi \in \Phi \subseteq \widehat{\Phi}, \tag{11.1}$$

where $(\cdot,\cdot)_{\widetilde{\Phi}}$ denotes the inner product on $\widetilde{\Phi}$ and $\langle \cdot,\cdot \rangle$ denotes the duality pairing.

Next, we define the *quadratic functional*

$$K(\phi, \theta) = a_1(\phi - \widehat{\phi}, \phi - \widehat{\phi}) + a_2(\theta, \theta) \qquad \forall \phi \in \Phi,\ \theta \in \Theta, \tag{11.2}$$

where $a_1(\cdot,\cdot)$ and $a_2(\cdot,\cdot)$ are symmetric bilinear forms on $\widehat{\Phi} \times \widehat{\Phi}$ and $\Theta \times \Theta$, respectively, and $\widehat{\phi} \in \widehat{\Phi}$ is a given function.[2] We make the following assumptions about the bilinear forms $a_1(\cdot,\cdot)$ and $a_2(\cdot,\cdot)$:

$$\begin{cases} a_1(\phi, \mu) \le \alpha_{a_1} \|\phi\|_{\widehat{\Phi}} \|\mu\|_{\widehat{\Phi}} & \forall \phi, \mu \in \widehat{\Phi} \\ a_2(\theta, \nu) \le \alpha_{a_2} \|\theta\|_{\Theta} \|\nu\|_{\Theta} & \forall \theta, \nu \in \Theta \\ a_1(\phi, \phi) \ge 0 & \forall \phi \in \widehat{\Phi} \\ a_2(\theta, \theta) \ge \beta_{a_2} \|\theta\|^2_{\Theta} & \forall \theta \in \Theta, \end{cases} \tag{11.3}$$

where α_{a_1}, α_{a_2}, and β_{a_2} are all positive constants. The second term in the functional (11.2) can be interpreted as a penalty term[3] which limits the size of the control θ.

Given another Hilbert space Λ, the additional bilinear forms $b_1(\cdot,\cdot)$ on $\Phi \times \Lambda$ and $b_2(\cdot,\cdot)$ on $\Theta \times \Lambda$, and the linear functional $G(\cdot)$ on Λ, we define the *linear*

---

[1] Due to a shortage of symbols and to avoid confusion with the repeated use of the same symbols for different contexts, in this chapter we depart, at least partially, from the use of upper-case italic letters to denote Hilbert spaces.

[2] In the language of control theory, Φ is called the *state space,* ϕ the *state variable,* Θ the *control space,* and θ the *control variable.* In many applications, the control space is finite dimensional in which case θ is often referred to as the vector of *design variables* or *design parameters.* Often Θ is chosen to be a bounded set in a Hilbert space but, for our purposes, we consider the less general situation of Θ itself being a Hilbert space.

[3] The usage of the terminology "penalty term" in conjunction with the second term in (11.2) should not be confused with the usage of that terminology below, e.g., in Section 11.3.

constraint equation[4]

$$b_1(\phi,\psi)+b_2(\theta,\psi)=G(\psi) \qquad \forall\,\psi\in\Lambda\,. \tag{11.4}$$

We make the following assumptions about the bilinear forms $b_1(\cdot,\cdot)$ and $b_2(\cdot,\cdot)$:

$$\begin{cases} b_1(\phi,\psi)\le\alpha_{b_1}\|\phi\|_\Phi\|\psi\|_\Lambda & \forall\,\phi\in\Phi,\ \psi\in\Lambda \\ b_2(\theta,\psi)\le\alpha_{b_2}\|\theta\|_\Theta\|\psi\|_\Lambda & \forall\,\theta\in\Theta,\ \psi\in\Lambda \\ \sup\limits_{\psi\in\Lambda,\psi\neq0}\dfrac{b_1(\phi,\psi)}{\|\psi\|_\Lambda}\ge\beta_{b_1}\|\phi\|_\Phi & \forall\,\phi\in\Phi \\ \sup\limits_{\phi\in\Phi,\phi\neq0}\dfrac{b_1(\phi,\psi)}{\|\phi\|_\Phi}>0 & \forall\,\psi\in\Lambda\,, \end{cases} \tag{11.5}$$

where α_{b_1}, α_{b_2}, and β_{b_1} are all positive constants.

We consider the *optimal control problem*:

$$\min_{\{\phi,\theta\}\in\Phi\times\Theta} K(\phi,\theta) \quad \textit{subject to} \quad b_1(\phi,\psi)+b_2(\theta,\psi)=G(\psi) \quad \forall\,\psi\in\Lambda\,. \tag{11.6}$$

It is easy to verify that the problem (11.6) falls into the framework of Section 1.2.4 concerning constrained optimization problems. To this end, we let $V\equiv\Phi\times\Theta$, $S\equiv\Lambda$, $\|\{\phi,\theta\}\|_V=(\|\phi\|_\Phi^2+\|\theta\|_\Theta^2)^{1/2}$ for all $\{\phi,\theta\}\in V$, and

$$\begin{cases} a(\{\phi,\theta\},\{\mu,\nu\})=a_1(\phi,\mu)+a_2(\theta,\nu) & \forall\,\phi,\mu\in\Phi,\ \theta,\nu\in\Theta \\ b(\{\phi,\theta\},\{\psi\})=b_1(\phi,\psi)+b_2(\theta,\psi) & \forall\,\phi\in\Phi,\ \theta\in\Theta,\ \psi\in\Lambda \\ D(\{\mu,\nu\})=a_1(\mu,\widehat{\phi}) & \forall\,\mu\in\Phi,\ \nu\in\Theta\,. \end{cases} \tag{11.7}$$

Then, with the obvious identifications $v=\{\phi,\theta\}$, $r=\{\mu,\nu\}$, and $s=\{\psi\}$, the functional in (1.31) is equivalent[5] to (11.2) as are the constraint equations in (1.31) and (11.4). The constrained optimization problem (1.31) and the optimal control problem (11.6) are also equivalent.

11.1.1 Existence of Optimal States and Controls

We begin with the following preliminary result.

---

[4] One should view (11.4) as a Galerkin weak formulation of a given PDE constraint, i.e., of the operator constraint equation (11.16). In fact, one usually formulates the PDE constraint in the operator form (11.16) and then derives a (Galerkin) weak formulation of the form (11.4).

[5] Actually, the functional in (1.31) differs from (11.2) by the constant $a_1(\widehat{\phi},\widehat{\phi})$ which, of course, has no bearing on the solution of the problem (11.6).

Lemma 11.1 *Let the assumptions* (11.3) *and* (11.5) *hold. Then, the spaces* $V \equiv \Phi \times \Theta$ *and* $S \equiv \Lambda$ *and the bilinear forms* $a(\cdot,\cdot)$ *and* $b(\cdot,\cdot)$ *defined in* (11.7) *satisfy the assumptions* (1.22), (1.23), (1.25), (1.29), *and* (1.30).

Proof. Using (11.3) and the continuous embedding $\Phi \subset \widehat{\Phi}$, we have that

$$\begin{aligned} a(\{\phi,\theta\},\{\mu,\nu\}) &= a_1(\phi,\mu) + a_2(\theta,\nu) \le \alpha_{a_1}\|\phi\|_{\widehat{\Phi}}\|\mu\|_{\widehat{\Phi}} + \alpha_{a_2}\|\theta\|_\Theta\|\nu\|_\Theta \\ &\le \alpha_{a_1}\|\phi\|_\Phi\|\mu\|_\Phi + \alpha_{a_2}\|\theta\|_\Theta\|\nu\|_\Theta \\ &\le \max\{\alpha_{a_1},\alpha_{a_2}\}\big(\|\phi\|_\Phi^2 + \|\theta\|_\Theta^2\big)^{1/2}\big(\|\mu\|_\Phi^2 + \|\nu\|_\Theta^2\big)^{1/2} \end{aligned}$$

for all $\phi,\mu \in \Phi$ and $\theta,\nu \in \Theta$ so that (1.22) holds with $\alpha_a = \max\{\alpha_{a_1},\alpha_{a_2}\}$. Similarly, we have, using (11.5), that (1.23) holds with $\alpha_b = \max\{\alpha_{b_1},\alpha_{b_2}\}$.

The symmetry of $a(\cdot,\cdot)$ follows from that of $a_1(\cdot,\cdot)$ and $a_2(\cdot,\cdot)$. Next, from (11.3) we have that

$$a(\{\phi,\theta\},\{\phi,\theta\}) = a_1(\phi,\phi) + a_2(\theta,\theta) \ge a_2(\theta,\theta) \ge \beta_{a_2}\|\theta\|_\Theta^2 \quad \forall \phi \in \Phi,\ \theta \in \Theta$$

so that (1.29) holds.

We next define the subspace $Z \subset \Phi \times \Theta$ by

$$Z = \Big\{\{\phi,\theta\} \in \Phi\times\Theta \mid b_1(\phi,\psi) + b_2(\theta,\psi) = 0 \quad \forall \psi \in \Lambda\Big\}. \tag{11.8}$$

The assumptions (11.5) imply that, given any $\theta \in \Theta$, the problem

$$b_1(\phi,\psi) = -b_2(\theta,\psi) \qquad \forall \psi \in \Lambda \tag{11.9}$$

has a unique solution ϕ_θ and, moreover, that solution satisfies

$$\|\phi_\theta\|_\Phi \le \frac{\alpha_{b_2}}{\beta_{b_1}}\|\theta\|_\Theta\,; \tag{11.10}$$

see Theorem 1.3. Thus, Z can be completely characterized by $\{\phi_\theta,\theta\} \in \Phi\times\Theta$, where, for arbitrary $\theta \in \Theta$, ϕ_θ is the solution of (11.9). Then, (11.10) and (11.3) imply that

$$\begin{aligned} a(\{\phi_\theta,\theta\},\{\phi_\theta,\theta\}) &= a_1(\phi_\theta,\phi_\theta) + a_2(\theta,\theta) \ge a_2(\theta,\theta) \ge \beta_{a_2}\|\theta\|_\Theta^2 \\ &\ge \frac{\beta_{a_2}}{2}\|\theta\|_\Theta^2 + \frac{\beta_{b_1}^2\beta_{a_2}}{2\alpha_{b_2}^2}\|\phi_\theta\|_\Phi^2 \ge \frac{\beta_{a_2}}{2}\min\left\{1,\frac{\beta_{b_1}^2}{\alpha_{b_2}^2}\right\}\|\{\phi_\theta,\theta\}\|_V^2 \end{aligned}$$

for all $\{\phi_\theta,\theta\} \in Z$. As a result, (1.30) holds with $\beta_a = \frac{\beta_{a_2}}{2}\min\{1,\frac{\beta_{b_1}^2}{\alpha_{b_2}^2}\}$.

To verify (1.25), note that

$$\sup_{\phi\in\Phi,\phi\neq 0}\frac{b_1(\phi,\psi)}{\|\phi\|_\Phi} \ge \beta_{b_1}\|\psi\|_\Lambda \qquad \forall \psi \in \Lambda\,. \tag{11.11}$$

Indeed, the assumptions in (11.5) imply (see Theorem 1.3) that for any $\psi \in \Lambda$, the problem

$$b_1(\phi,\mu) = (\psi,\mu)_\Lambda \qquad \forall \mu \in \Lambda \tag{11.12}$$

has a unique solution ϕ_ψ and, moreover, that solution satisfies

$$\|\phi_\psi\|_\Phi \leq \frac{1}{\beta_{b_1}} \|\psi\|_\Lambda \,. \tag{11.13}$$

Using (11.12) and (11.13), it is easy to see that

$$\frac{b_1(\phi_\psi,\psi)}{\|\phi_\psi\|_\Phi} = \frac{\|\psi\|_\Lambda^2}{\|\phi_\psi\|_\Phi} \geq \beta_{b_1} \|\psi\|_\Lambda \qquad \forall \psi \in \Lambda$$

which immediately implies (11.11). Finally, using (11.11),

$$\sup_{\{\phi,\theta\}\in\Phi\times\Theta,\,\{\phi,\theta\}\neq\{0,0\}} \frac{b(\{\phi,\theta\},\{\psi\})}{\left(\|\phi\|_\Phi^2 + \|\theta\|_\Theta^2\right)^{1/2}} \geq \sup_{\phi\in\Phi,\,\phi\neq 0} \frac{b_1(\phi,\psi)}{\|\phi\|_\Phi} \geq \beta_{b_1} \|\psi\|_\Lambda$$

for all $\psi \in \Lambda$ so that (1.25) holds with $\beta_b = \beta_{b_1}$. □

We immediately have the following result.

Theorem 11.2 *Let the assumptions* (11.3) *and* (11.5) *hold. Then, the optimal control problem* (11.6) *has a unique solution* $\{\phi,\theta\} \in \Phi\times\Theta$.

Proof. The result immediately follows from Theorem 1.5, Proposition 1.7, and Lemma 11.1. □

It is instructive to rewrite the functional (11.2), the constraint (11.4), and the optimal control problem (11.6) in operator notation. To this end, we note that the bilinear forms serve to define operators

$$\begin{array}{lll} \mathcal{A}_1 : \widehat{\Phi} \to \widehat{\Phi}^* & \mathcal{A}_2 : \Theta \to \Theta^* & \mathcal{B}_1 : \Phi \to \Lambda^* \\ \mathcal{B}_1^* : \Lambda \to \Phi^* & \mathcal{B}_2 : \Theta \to \Lambda^* & \mathcal{B}_2^* : \Lambda \to \Theta^* \end{array}$$

through the following relations:

$$\begin{cases} a_1(\phi,\mu) = \langle \mathcal{A}_1\phi,\mu\rangle_{\widehat{\Phi}^*,\widehat{\Phi}} & \forall \phi,\mu \in \widehat{\Phi} \\ a_2(\theta,\nu) = \langle \mathcal{A}_2\theta,\nu\rangle_{\Theta^*,\Theta} & \forall \theta,\nu \in \Theta \\ b_1(\phi,\psi) = \langle \mathcal{B}_1\phi,\psi\rangle_{\Lambda^*,\Lambda} = \langle \mathcal{B}_1^*\psi,\phi\rangle_{\Phi^*,\Phi} & \forall \phi \in \Phi,\ \psi \in \Lambda \\ b_2(\psi,\theta) = \langle \mathcal{B}_2\theta,\psi\rangle_{\Lambda^*,\Lambda} = \langle \mathcal{B}_2^*\psi,\theta\rangle_{\Theta^*,\Theta} & \forall \theta \in \Theta,\ \psi \in \Lambda\,. \end{cases} \tag{11.14}$$

We also have the function $g \in \Lambda^*$ defined through $G(\psi) = \langle g,\psi\rangle_{\Lambda^*,\Lambda}$ for all $\psi \in \Lambda$. Then, the functional (11.2) and the constraint (11.4) respectively take the forms

$$K(\phi,\theta) = \frac{1}{2}\big\langle \mathcal{A}_1(\phi-\widehat{\phi}),(\phi-\widehat{\phi})\big\rangle_{\widehat{\Phi}^*,\widehat{\Phi}} + \frac{1}{2}\langle \mathcal{A}_2\theta,\theta\rangle_{\Theta^*,\Theta} \tag{11.15}$$

for all $\phi \in \Phi$ and $\theta \in \Theta$ and

$$\mathcal{B}_1\phi + \mathcal{B}_2\theta = g \qquad \text{in } \Lambda^* \tag{11.16}$$

and the optimal control problem (11.6) takes the form

$$\min_{\{\phi,\theta\}\in\Phi\times\Theta} K(\phi,\theta) \qquad \text{subject to (11.16)}. \tag{11.17}$$

Assumptions (11.3) and (11.5) imply that $\mathcal{A}_1$, $\mathcal{A}_2$, $\mathcal{B}_1$, $\mathcal{B}_2$, $\mathcal{B}_1^*$, and $\mathcal{B}_2^*$ are bounded with

$$\|\mathcal{A}_1\|_{\widehat{\Phi}\to\widehat{\Phi}^*} \le \alpha_{a_1} \qquad \|\mathcal{A}_2\|_{\Theta\to\Theta^*} \le \alpha_{a_2} \qquad \|\mathcal{B}_1\|_{\Phi\to\Phi^*} \le \alpha_{b_1}$$

$$\|\mathcal{B}_1^*\|_{\Phi\to\Phi^*} \le \alpha_{b_1} \qquad \|\mathcal{B}_2\|_{\Phi\to\Theta^*} \le \alpha_{b_2} \qquad \|\mathcal{B}_2^*\|_{\Theta\to\Phi^*} \le \alpha_{b_2}$$

and that the operator $\mathcal{B}_1$ is invertible with $\|\mathcal{B}_1^{-1}\|_{\Lambda^*\to\Phi} \le 1/\beta_{b_1}$. Note also that the subspace $Z \subset V = \Phi\times\Theta$ can be defined by

$$Z = \Big\{\{\phi,\theta\} \in \Phi\times\Theta \mid \phi = -\mathcal{B}_1^{-1}\mathcal{B}_2\theta \quad \forall\theta\in\Theta\Big\}.$$

11.1.2 Least-Squares Formulation of the Constraint Equation

For later use, we consider least-squares principles for the constraint equation (11.16). In this section, we take the view that the control $\theta \in \Theta$ is a given function as is, of course, the data function $g \in \Lambda^*$, and we use the constraint equation to find the corresponding state $\phi \in \Phi$.

The constraint equation is given in variational form in (11.4) and in equivalent operator form in (11.16). In this section, we show how it can also be characterized through a least-squares principle. In order to keep the exposition simple, we only apply the compliant discrete principles studied in Section 3.5.1; these are the most straightforward means for defining a LSFEM. For the sake of clarity, some of the material of Chapter 3 is repeated here, but specialized to the context of the constraint equation (11.4) of the optimization problem (11.6).

Given $\theta \in \Theta$ and $g \in \Lambda^*$, consider the least-squares principle

$$\begin{cases} J_c(\phi;\theta,g) = \|\mathcal{B}_1\phi + \mathcal{B}_2\theta - g\|^2_{\Lambda^*} \qquad \forall\phi\in\Phi,\ \theta\in\Theta,\ g\in\Lambda^* \\ X = \Phi. \end{cases} \tag{11.18}$$

Clearly, this problem is equivalent to (11.4) and (11.16), i.e., solutions of (11.18) are solutions of (11.4) or (11.16) and conversely. The Euler–Lagrange equation corresponding to the least-squares principle (11.18) is given, in variational form, by

$$\widetilde{b}_1(\phi,\mu) = \widetilde{G}_1(\mu) - \widetilde{b}_2(\theta,\mu) \qquad \forall\mu\in\Phi, \tag{11.19}$$

where

$$\widetilde{b}_1(\phi,\mu) = \big(\mathcal{B}_1\mu, \mathcal{B}_1\phi\big)_{\Lambda^*} \qquad \forall \phi,\mu \in \Phi, \tag{11.20}$$

$$\widetilde{b}_2(\theta,\mu) = \big(\mathcal{B}_1\mu, \mathcal{B}_2\theta\big)_{\Lambda^*} \qquad \forall \theta \in \Theta, \mu \in \Phi, \tag{11.21}$$

and

$$\widetilde{G}_1(\mu) = \big(\mathcal{B}_1\mu, g\big)_{\Lambda^*} \qquad \forall \mu \in \Phi. \tag{11.22}$$

The properties of these bilinear forms and linear functional are given in the following result.

Proposition 11.3 *Assume that* (11.5) *holds. Then, the bilinear form* $\widetilde{b}_1(\cdot,\cdot)$ *is symmetric and*

$$\begin{cases} \widetilde{b}_1(\phi,\mu) \le \alpha_{b_1}^2 \|\phi\|_\Phi \|\mu\|_\Phi & \forall \phi,\mu \in \Phi \\ \widetilde{b}_2(\theta,\mu) \le \alpha_{b_1}\alpha_{b_2} \|\mu\|_\Phi \|\theta\|_\Theta & \forall \theta \in \Theta,\ \mu \in \Phi \\ \widetilde{b}_1(\phi,\phi) \ge \beta_{b_1}^2 \|\phi\|_\Phi^2 & \forall \phi \in \Phi. \end{cases} \tag{11.23}$$

Moreover, $\widetilde{G}_1(\mu) \le \alpha_{b_1}\|g\|_{\Lambda^*}\|\mu\|_\Phi$ *for all* $\mu \in \Phi$ *and the problem* (11.19)*, or equivalently* (11.18)*, has a unique solution.*

Proof. The symmetry of the bilinear form $\widetilde{b}_1(\cdot,\cdot)$ follows immediately from its definition. Because $b_1(\phi,\psi) \le \alpha_{b_1}\|\phi\|_\Phi\|\psi\|_\Lambda$ for all $\phi \in \Phi$ and $\psi \in \Lambda$, we have that

$$\langle \mathcal{B}_1\phi, \psi\rangle_{\Lambda^*,\Lambda} = b_1(\phi,\psi) \le \alpha_{b_1}\|\phi\|_\Phi\|\psi\|_\Lambda \qquad \forall \phi \in \Phi,\ \psi \in \Lambda$$

from which it easily follows that $\|\mathcal{B}_1\phi\|_{\Lambda^*} \le \alpha_{b_1}\|\phi\|_\Phi$ for all $\phi \in \Phi$. We then have that

$$\widetilde{b}_1(\phi,\mu) = \big(\mathcal{B}_1\mu, \mathcal{B}_1\phi\big)_{\Lambda^*} \le \|\mathcal{B}_1\phi\|_{\Lambda^*}\|\mathcal{B}_1\mu\|_{\Lambda^*} \le \alpha_{b_1}^2\|\phi\|_\Phi\|\mu\|_\Phi$$

and

$$\widetilde{G}_1(\mu) = \big(\mathcal{B}_1\mu, g\big)_{\Lambda^*} \le \|\mathcal{B}_1\mu\|_{\Lambda^*}\|g\|_{\Lambda^*} \le \alpha_{b_1}\|g\|_{\Lambda^*}\|\mu\|_\Phi .$$

In a similar way, one shows that $\|\mathcal{B}_2\theta\|_{\Lambda^*} \le \alpha_{b_2}\|\theta\|_\Theta$ for all $\theta \in \Theta$ and

$$\widetilde{b}_2(\theta,\mu) = \big(\mathcal{B}_1\mu, \mathcal{B}_2\theta\big)_{\Lambda^*} \le \alpha_{b_1}\alpha_{b_2}\|\theta\|_\Theta\|\mu\|_\Phi .$$

Next, because $\sup_{\psi\in\Lambda,\psi\neq 0}\big(b_1(\phi,\psi)/\|\psi\|_\Lambda\big) \ge \beta_{b_1}\|\phi\|_\Phi$ for all $\phi \in \Phi$, we have that

$$\|\mathcal{B}_1\phi\|_{\Lambda^*} = \sup_{\psi\in\Lambda,\psi\neq 0} \frac{\langle \mathcal{B}_1\phi,\psi\rangle_{\Lambda^*,\Lambda}}{\|\psi\|_\Lambda} = \sup_{\psi\in\Lambda,\psi\neq 0} \frac{b_1(\phi,\psi)}{\|\psi\|_\Lambda} \ge \beta_{b_1}\|\phi\|_\Phi \qquad \forall \phi \in \Phi.$$

We then have that

$$\widetilde{b}_1(\phi,\phi) = \big(\mathcal{B}_1\phi, \mathcal{B}_1\phi\big)_{\Lambda^*} = \|\mathcal{B}_1\phi\|_{\Lambda^*}^2 \ge \beta_{b_1}^2\|\phi\|_\Phi^2 .$$

The unique solvability of (11.19) then follows from Corollary 1.4 (the Lax-Milgram lemma). □

As an immediate consequence of Proposition 11.3, we have that the least-squares functional in (11.18) is norm equivalent in the following sense.

Corollary 11.4 *Assume that the conditions on the bilinear form $b_1(\cdot,\cdot)$ in* (11.5) *hold. Then, for all $\phi \in \Phi$*

$$\beta_{b_1}^2 \|\phi\|_\Phi^2 \le J_c(\phi;0,0) = \widetilde{b}_1(\phi,\phi) = \big(\mathcal{B}_1\phi,\mathcal{B}_1\phi\big)_{\Lambda^*} \le \alpha_{b_1}^2 \|\phi\|_\Phi^2 . \qquad \square \tag{11.24}$$

Note that the bilinear form $\widetilde{b}_1(\cdot,\cdot)$ is symmetric and strongly coercive even when the operator $\mathcal{B}_1$ in (11.16) is indefinite and/or non-symmetric.

Discretization of (11.19) is accomplished in the standard manner. One chooses a conforming subspace $\Phi^h \subset \Phi$ and then, given $\theta \in \Theta$ and $g \in \Lambda^*$, one solves the problem

$$\widetilde{b}_1(\phi^h,\mu^h) = \widetilde{G}_1(\mu^h) - \widetilde{b}_2(\theta,\mu^h) \qquad \forall \mu^h \in \Phi^h . \tag{11.25}$$

Then, (11.24), the Lax–Milgram lemma (Corollary 1.4), and Céa's lemma (Theorem 1.9) immediately imply the following results.

Proposition 11.5 *Assume that* (11.5) *holds. Then, the problem* (11.25) *has a unique solution $\phi^h \in \Phi^h$ and, if ϕ denotes the solution of the problem* (11.19), *there exists a constant $C > 0$ whose value is independent of h, ϕ, and ϕ^h such that*

$$\|\phi - \phi^h\|_\Phi \le C \inf_{\widetilde{\phi}^h \in \Phi^h} \|\phi - \widetilde{\phi}^h\|_\Phi . \qquad \square$$

If $\{\phi_j^h\}_{j=1}^J$ denotes a basis for Φ^h, then the discretized problem (11.25) is equivalent to the matrix problem

$$\widetilde{\mathsf{B}}_1 \vec{\phi} = \vec{g}_0 , \tag{11.26}$$

where $\vec{\phi}$ is the vector of coefficients for ϕ^h,

$$(\widetilde{\mathsf{B}}_1)_{ij} = \widetilde{b}_1(\phi_i^h,\phi_j^h) = \big(\mathcal{B}_1\phi_i^h,\mathcal{B}_1\phi_j^h\big)_{\Lambda^*} ,$$

and

$$(\vec{g}_0)_i = \widetilde{G}_1(\phi_i^h) - \widetilde{b}_2(\theta,\phi_i^h) = \big(\mathcal{B}_1\phi_i^h, g - \mathcal{B}_2\theta\big)_{\Lambda^*} .$$

The following result follows easily from Proposition 11.3.

Corollary 11.6 *Assume that the conditions on the bilinear form $b_1(\cdot,\cdot)$ in* (11.5) *hold. Then, the matrix $\widetilde{\mathsf{B}}_1$ is symmetric and uniformly (with respect to h) positive definite.* □

11.2 Solution via Lagrange Multipliers of the Optimal Control Problem

For all $\{\mu,\nu\} \in V = \Phi \times \Theta$ and $\psi \in S = \Lambda$, we introduce the Lagrangian functional[6]

$$
\begin{aligned}
L(\{\mu,\nu\},\{\psi\}) &= K(\{\mu,\nu\}) + 2\big(b(\{\mu,\nu\},\{\psi\}) - G(\psi)\big) \\
&= a_1(\mu-\widehat{\phi},\mu-\widehat{\phi}) + a_2(\nu,\nu) + 2\big(b_1(\mu,\psi) + b_2(\nu,\psi) - G(\psi)\big)\,.
\end{aligned}
$$

Then, (11.6) is equivalent to the unconstrained optimization problem of finding saddle points $\{\{\phi,\theta\},\{\lambda\}\}$ in $V \times S$ of the Lagrangian functional. These saddle points may be found by solving the first-order necessary conditions

$$
\begin{cases}
a_1(\phi,\mu) \qquad\qquad\qquad + b_1(\mu,\lambda) = a_1(\widehat{\phi},\mu) & \forall \mu \in \Phi \\
\qquad\qquad a_2(\theta,\nu) \ + b_2(\nu,\lambda) = 0 & \forall \nu \in \Theta \\
b_1(\phi,\psi) + b_2(\theta,\psi) \qquad\qquad = G(\psi) = \langle g,\psi\rangle_{\Lambda^*,\Lambda} & \forall \psi \in \Lambda\,.
\end{cases}
\tag{11.27}
$$

The third equation in the *optimality system* (11.27) is simply the constraint equation. The first equation is commonly referred to as the *adjoint* or *co-state equation* and the Lagrange multiplier λ is referred to as the *adjoint* or *co-state* variable. The second equation in (11.27) is referred to as the *optimality condition* because it is merely a statement that the gradient of the functional $K(\cdot,\cdot)$ defined in (11.2) vanishes at the optimum.

Using results listed in Section 1.2.3 of Chapter 1, the following result is immediate.

Theorem 11.7 *Let the assumptions* (11.3) *and* (11.5) *hold. Then, the optimality system* (11.27) *has a unique solution* $\{\phi,\theta,\lambda\} \in \Phi \times \Theta \times \Lambda$. *Moreover, that solution satisfies*

$$
\|\phi\|_\Phi + \|\theta\|_\Theta + \|\lambda\|_\Lambda \le C\big(\|g\|_{\Lambda^*} + \|\widehat{\phi}\|_{\widehat{\Phi}}\big)
$$

and $\{\phi,\theta\} \in \Phi \times \Theta$ *is the unique solution of the optimal control problem* (11.6).

Proof. With the associations $V = \Phi \times \Theta$, $S = \Lambda$, $v = \{\phi,\theta\}$, and $p = \{\lambda\}$, the results immediately follow from Theorem 1.5, Proposition 1.7, and Lemma 11.1. □

Using the operators introduced in (11.14), the optimality system (11.27) takes the form

$$
\begin{cases}
\mathcal{A}_1\phi \qquad\qquad + \mathcal{B}_1^*\lambda = \mathcal{A}_1\widehat{\phi} & \text{in } \Phi^* \\
\qquad\quad \mathcal{A}_2\theta + \mathcal{B}_2^*\lambda = 0 & \text{in } \Theta^* \\
\mathcal{B}_1\phi + \mathcal{B}_2\theta \qquad\qquad = g & \text{in } \Lambda^*\,.
\end{cases}
\tag{11.28}
$$

[6] The factor 2 in the constraint term is introduced so that the resulting optimality does not contain such a factor.

11.2.1 Galerkin Finite Element Methods for the Optimality System

We choose (conforming) finite-dimensional subspaces $\Phi^h \subset \Phi$, $\Theta^h \subset \Theta$, and $\Lambda^h \subset \Lambda$ and then restrict (11.27) to the subspaces, i.e., we seek $\{\phi^h, \theta^h, \lambda^h\} \in \Phi^h \times \Theta^h \times \Lambda^h$ that satisfies

$$\begin{cases} a_1(\phi^h, \mu^h) \qquad\qquad\qquad\quad + b_1(\mu^h, \lambda^h) = a_1(\widehat{\phi}, \mu^h) & \forall\, \mu^h \in \Phi^h \\ \qquad\qquad\quad a_2(\theta^h, \nu^h) \;\; + b_2(\nu^h, \lambda^h) = 0 & \forall\, \nu^h \in \Theta^h \\ b_1(\phi^h, \psi^h) + b_2(\theta^h, \psi^h) \qquad\qquad\qquad = \langle g, \psi^h \rangle_{\Lambda^*,\Lambda} & \forall\, \psi^h \in \Lambda^h. \end{cases} \tag{11.29}$$

This is also the optimality system for the minimization of (11.2) over $\Phi^h \times \Theta^h$ subject to the constraint $b_1(\phi^h, \psi^h) + b_2(\theta^h, \psi^h) = G(\psi^h) = \langle g, \psi^h \rangle_{\Lambda^*,\Lambda}$ for all $\psi^h \in \Lambda^h$.

From the discussion in Sections 1.3.1 and 1.3.2, we know that the continuity conditions in (11.3) and (11.5) imply that those conditions also hold with respect to the conforming subspaces $\Phi^h \subset \Phi$, $\Theta^h \subset \Theta$, and $\Lambda^h \subset \Lambda$. Likewise, the strong coercivity assumption on the bilinear form $a_2(\cdot,\cdot)$, i.e., the third assumption in (11.3), implies that that condition holds with respect to the conforming subspace $\Theta^h \subset \Theta$. However, the weak coercivity assumptions on the bilinear from $b_1(\cdot,\cdot)$, i.e., the third and fourth inequalities in (11.5), do not imply that those inequalities hold with respect to the conforming subspaces $\Phi^h \subset \Phi$ and $\Lambda^h \subset \Lambda$; see Figure 1.1. Thus, we make the following additional assumptions about Φ^h, Λ^h, and $b_1(\cdot,\cdot)$:

$$\begin{cases} \sup\limits_{\psi^h \in \Lambda^h, \psi^h \neq 0} \dfrac{b_1(\phi^h, \psi^h)}{\|\psi^h\|_\Lambda} \geq \beta^h_{b_1} \|\phi^h\|_\Phi & \forall\, \phi^h \in \Phi^h \\ \sup\limits_{\phi^h \in \Phi^h, \phi^h \neq 0} \dfrac{b_1(\phi^h, \psi^h)}{\|\phi^h\|_V} > 0 & \forall\, \psi^h \in \Lambda^h, \end{cases} \tag{11.30}$$

where $\beta^h_{b_1}$ is a positive constant whose value is independent of h. Also, note that, in general, for the subspace $Z^h \subset \Phi^h \times \Theta^h$, defined by

$$Z^h = \Big\{ \{\phi^h, \theta^h\} \in \Phi^h \times \Theta^h \;\Big|\; b_1(\phi^h, \psi^h) + b_2(\theta^h, \psi^h) = 0 \quad \forall\, \psi^h \in \Lambda^h \Big\}, \tag{11.31}$$

$Z^h \not\subset Z$ even though $\Phi^h \subset \Phi$, $\Theta^h \subset \Theta$, and $\Lambda^h \subset \Lambda$; see Figure 1.2.

Analogous to Lemma 11.1, we have the following result.

Lemma 11.8 *Let the assumptions* (11.3), (11.5), *and* (11.30) *hold. Then, the spaces* $V^h = \Phi^h \times \Theta^h$ *and* $S^h = \Phi^h$ *and the bilinear forms* $a(\cdot,\cdot)$ *and* $b(\cdot,\cdot)$ *defined in* (11.7) *satisfy the assumptions* (1.42) *and* (1.44).

Proof. The proof proceeds exactly as that for Lemma 11.1; the constants in (1.42) and (1.44) are given by $\beta^h_b = \beta^h_{b_1}$ and

$$\beta_a^h = \frac{\beta_{a_2}}{2} \min\left\{1, \frac{(\beta_{b_1}^h)^2}{\alpha_{b_2}^2}\right\},$$

respectively. □

We then easily obtain the following results.

Theorem 11.9 *Let the assumptions* (11.3), (11.5), *and* (11.30) *hold. Then, the discrete optimality system* (11.29) *has a unique solution* $\{\phi^h, \theta^h, \lambda^h\} \in \Phi^h \times \Theta^h \times \Lambda^h$ *and, moreover, that solution satisfies*

$$\|\phi^h\|_\Phi + \|\theta^h\|_\Theta + \|\lambda^h\|_\Lambda \le C\big(\|g\|_{\Lambda^*} + \|\widehat{\phi}\|_{\widehat{\Phi}}\big).$$

Furthermore, let $\{\phi, \theta, \lambda\} \in \Phi \times \Theta \times \Lambda$ *denote the unique solution of the optimality system* (11.27), *or, equivalently, of the optimal control problem* (11.6). *Then,*

$$\begin{aligned} &\|\phi - \phi^h\|_\Phi + \|\theta - \theta^h\|_\Theta + \|\lambda - \lambda^h\|_\Lambda \\ &\le C\Big(\inf_{\mu^h \in \Phi^h} \|\phi - \mu^h\|_\Phi + \inf_{\xi^h \in \Theta^h} \|\theta - \xi^h\|_\Theta + \inf_{\psi^h \in \Lambda^h} \|\lambda - \psi^h\|_\Lambda\Big). \end{aligned} \tag{11.32}$$

Proof. The results immediately follow from Theorem 1.11 and Lemma 11.8. □

If $\{\mu_j^h\}_{j=1}^J$, $\{\nu_k^h\}_{k=1}^K$, and $\{\psi_\ell^h\}_{\ell=1}^L$ denote bases for Φ^h, Θ^h, and Λ^h, respectively, we then have

$$\phi^h = \sum_{j=1}^J \phi_j \mu_j^h, \quad \theta^h = \sum_{k=1}^K \theta_k \nu_k^h, \quad \text{and} \quad \lambda^h = \sum_{\ell=1}^L \lambda_\ell \psi_\ell^h$$

for some sets of coefficients $\{\phi_j\}_{j=1}^J$, $\{\theta_k\}_{k=1}^K$, and $\{\lambda_\ell\}_{\ell=1}^L$. Let $\vec{\phi} = (\phi_1, \ldots, \phi_J)^\mathsf{T}$, $\vec{\theta} = (\theta_1, \ldots, \theta_K)^\mathsf{T}$, $\vec{\lambda} = (\lambda_1, \ldots, \lambda_L)^\mathsf{T}$,

$$\begin{aligned}
&(\mathsf{A}_1)_{ij} = a_1(\mu_j^h, \mu_i^h) = \langle \mathcal{A}_1 \mu_j^h, \mu_i^h \rangle_{\Phi^*, \Phi} && \text{for } i,j = 1,\ldots,J \\
&(\mathsf{A}_2)_{ij} = a_2(\nu_j, \nu_i) = \langle \mathcal{A}_2 \nu_j, \nu_i \rangle_{\Theta^*, \Theta} && \text{for } i,j = 1,\ldots,K \\
&(\mathsf{B}_1)_{ij} = b_1(\mu_j, \psi_i) = \langle \mathcal{B}_1 \mu_j, \psi_i \rangle_{\Lambda^*, \Lambda} && \text{for } i = 1,\ldots,L,\ j = 1,\ldots,J \\
&(\mathsf{B}_2)_{ij} = b_2(\nu_j, \psi_i) = \langle \mathcal{B}_2 \nu_j, \psi_i \rangle_{\Lambda^*, \Lambda} && \text{for } i = 1,\ldots,L,\ j = 1,\ldots,K \\
&(\vec{d})_i = a_1(\widehat{\phi}, \mu_i^h) = \langle \mathcal{A}_1 \widehat{\phi}, \mu_i^h \rangle_{\Phi^*, \Phi} && \text{for } i = 1,\ldots,J \\
&(\vec{g})_i = G(\psi_i) = \langle g, \psi_i \rangle_{\Lambda^*, \Lambda} && \text{for } i = 1,\ldots,L.
\end{aligned}$$

Then, the discrete optimality system (11.29) is equivalent to the linear system

$$\begin{pmatrix} \mathsf{A}_1 & 0 & \mathsf{B}_1^T \\ 0 & \mathsf{A}_2 & \mathsf{B}_2^T \\ \mathsf{B}_1 & \mathsf{B}_2 & 0 \end{pmatrix} \begin{pmatrix} \vec{\phi} \\ \vec{\theta} \\ \vec{\lambda} \end{pmatrix} = \begin{pmatrix} \vec{f} \\ 0 \\ \vec{g} \end{pmatrix}. \tag{11.33}$$

Remark 11.10 There are two sets of inf–sup conditions associated with the problems (11.27) and (11.29). First, we have the "inner" conditions in (11.5) and (11.30) on the bilinear form $b_1(\cdot,\cdot)$ that involve only the state variable (but not the control variable) and that guarantee, for a given control function, the unique solvability of the state equation and the discrete state equation, respectively, i.e., of the third equations in (11.27) and (11.29). Second, we have the "outer" conditions in (1.25) and (1.42) for the bilinear form $b(\cdot,\cdot)$ defined in (11.7) and that involve both the state and control variables. These latter conditions help guarantee the unique solvability of the optimality system (11.27) and the discrete optimality system (11.29), respectively. Note that the outer conditions and the related saddle point nature of the optimality systems occur regardless of the nature of the inner problem, i.e., the state equations. For example, even if the state equations involve a strongly coercive bilinear form $b_1(\cdot,\cdot)$ so that the last two inequalities in (11.5) can be replaced by $b_1(\phi,\phi) \geq \beta_{b_1}\|\phi\|_\Phi^2$ for all $\phi \in \Phi$, i.e., the inner inf–sup conditions are replaced by a strong coercivity condition, we would still have the outer inf–sup condition (1.25). □

Remark 11.11 Because of the nature of the assumptions (11.3) and (11.5), the outer discrete inf–sup condition (1.42) on the bilinear form $b(\cdot,\cdot)$ with respect to the discrete spaces is satisfied merely by assuming that inner discrete inf–sup condition (11.30) holds. Thus, by merely guaranteeing that the discrete constraint equations within the discretized optimal control problem are uniquely solvable for any given discrete control, i.e., assuming that the "inner" inf–sup conditions hold, we have that the "outer" discrete inf–sup condition on the bilinear form $b(\cdot,\cdot)$ holds. The latter, of course, is crucial to the stability and convergence of finite element approximations determined from the discrete optimality system (11.29) or its matrix equivalent (11.33).

On the other hand, if (11.30) does not hold, then there exists a $\phi_0^h \in \Phi^h$ such that $\phi_0^h \neq 0$ and $b_1(\phi_0^h, \psi^h) = 0$ for all $\psi^h \in \Lambda^h$. Then, $b(\{\phi_0^h, 0\}, \{\psi^h\}) = b_1(\phi_0^h, \psi^h) = 0$ for all $\psi^h \in \Lambda^h$ so that

$$\sup_{\psi^h \in \Lambda^h,\, \psi^h \neq 0} \frac{b(\{\phi_0^h, 0\}, \{\psi^h\})}{\|\psi^h\|_\Lambda} = 0.$$

This implies that the discrete inf–sup condition (1.42) does not hold so that (11.29) or its matrix equivalent (11.33) may not be solvable, i.e., the coefficient matrix in (11.33) may not be invertible.

In fact, the assumptions in (11.30) imply that B_1 is uniformly invertible. This, and the facts (which follow from (11.3)) that the symmetric matrices A_1 and A_2 are positive semi-definite and positive definite, respectively, are enough to guarantee that the coefficient matrix in (11.33) is invertible. On the other hand, if (11.30) does not hold so that the matrix B_1 has a nontrivial null space, then, under the other assumptions that have been made, one cannot guarantee the invertibility of the coefficient matrix in (11.33). □

Remark 11.12 Solving the discrete optimality system (11.29), or equivalently, the linear system (11.33), is often a formidable task. If the constraint equations (11.4) are a system of PDEs, then the last (block) row of (11.33) represents a Galerkin finite element discretization of that system. The discrete adjoint equations, i.e., the first row in (11.33), are also a discretization of a system of PDEs. Moreover, the dimension of the discrete adjoint vector $\vec{\lambda}$ is essentially the same as that of discrete state vector $\vec{\phi}$. Thus, (11.33) is at least twice the size (we have yet to account for the discrete control variables in $\vec{\theta}$) of the algebraic system corresponding to the discretization of the PDE constraints. Thus, if the latter is difficult or costly to solve, it is even more difficult to deal with the discrete optimality system. For this reason, there have been many approaches suggested for uncoupling the three components of discrete optimality systems such as (11.29), or equivalently, (11.33); see, e.g., [193], for a discussion of several of these approaches. We note that these approaches rely on the invertibility of the matrices B_1 and A_2, properties that follow from (11.30) and (11.3), respectively. □

11.2.2 Least-Squares Finite Element Methods for the Optimality System

Even if the state equation (11.4) or, equivalently (11.16), involves a symmetric, positive definite operator $\mathcal{B}_1$, i.e., even if the bilinear form $b_1(\cdot,\cdot)$ is symmetric and strongly coercive, the discrete optimality system (11.29) or, equivalently (11.33), obtained through a Galerkin discretization is indefinite. For example, if $\mathcal{B}_1 = -\Delta$ with zero boundary conditions, then B_1 is a symmetric, positive definite matrix, but the coefficient matrix in (11.33) is indefinite.

It follows that, with respect to optimality systems arising in control and optimization problems, the answer to the first question from Section 2.2.4 is affirmative: the use of least-squares principles for these systems is justified by their natural association with mixed variational formulations and saddle-point problems. Thus, in this section we focus on least-squares principles for the optimality system (11.28). It is clear that the desirable properties of least-squares principles applied to (11.28) remain in place even if the state system bilinear form $b_1(\cdot,\cdot)$ is only weakly coercive, i.e., even if the operator $\mathcal{B}_1$ is merely invertible and not necessarily positive definite.

As is seen in Chapter 3, given a system of PDEs, there are many ways to define LSFEMs for determining approximate solutions. In order to keep the exposition relatively simple, here we only consider the compliant discrete least-squares principles (DLSPs) of Section 3.5.1 that are the most straightforward means for defining a LSFEM.

A least-squares functional can be defined by summing the squares of the norms of the residuals of the three equations in (11.28) resulting in the least-squares principle

$$\begin{cases} J(\phi,\theta,\lambda;\widehat{\phi},g) = \\ \qquad \|\mathcal{A}_1\phi+\mathcal{B}_1^*\lambda-\mathcal{A}_1\widehat{\phi}\|_{\Phi^*}^2+\|\mathcal{A}_2\theta+\mathcal{B}_2^*\lambda\|_{\Theta^*}^2+\|\mathcal{B}_1\phi+\mathcal{B}_2\theta-g\|_{\Lambda^*}^2 \\ X=\Phi\times\Theta\times\Lambda\,. \end{cases} \tag{11.34}$$

The unique solution of (11.28) is also a solution of this least-squares principle. The first-order necessary conditions corresponding to this principle are given by: *seek* $\{\phi,\theta,\lambda\}\in\Phi\times\Theta\times\Lambda$ *such that*

$$Q(\{\phi,\theta,\lambda\};\{\mu,\nu,\psi\}) = F\big(\{\mu,\nu,\psi\};\{\mathcal{A}_1\widehat{\phi},g\}\big) \tag{11.35}$$

for all $\{\mu,\nu,\psi\}\in\Phi\times\Theta\times\Lambda$, where

$$\begin{aligned} Q(\{\phi,\theta,\lambda\};\{\mu,\nu,\psi\}) &= (\mathcal{A}_1\mu+\mathcal{B}_1^*\psi,\mathcal{A}_1\phi+\mathcal{B}_1^*\lambda)_{\Phi^*} \\ &\quad +(\mathcal{A}_2\nu+\mathcal{B}_2^*\psi,\mathcal{A}_2\theta+\mathcal{B}_2^*\lambda)_{\Theta^*}+(\mathcal{B}_1\mu+\mathcal{B}_2\nu,\mathcal{B}_1\phi+\mathcal{B}_2\theta)_{\Lambda^*} \end{aligned} \tag{11.36}$$

for all $\{\phi,\theta,\lambda\}$, $\{\mu,\nu,\psi\}\in\Phi\times\Theta\times\Lambda$ and

$$F\big(\{\mu,\nu,\psi\};\{\mathcal{A}_1\widehat{\phi},g\}\big) = (\mathcal{A}_1\mu+\mathcal{B}_1^*\psi,\mathcal{A}_1\widehat{\phi})_{\Phi^*}+(\mathcal{B}_1\mu+\mathcal{B}_2\nu,g)_{\Lambda^*} \tag{11.37}$$

for all $\{\mu,\nu,\psi\}\in\Phi\times\Theta\times\Lambda$. We then have the following result.

Lemma 11.13 *Let the assumptions* (11.3) *and* (11.5) *hold. Then, the bilinear form* $Q(\cdot,\cdot)$ *is symmetric and continuous on* $\{\Phi\times\Theta\times\Lambda\}\times\{\Phi\times\Theta\times\Lambda\}$ *and the linear functional* $F(\cdot)$ *is continuous on* $\{\Phi\times\Theta\times\Lambda\}$. *Moreover, the bilinear form* $Q(\cdot,\cdot)$ *is strongly coercive on* $\{\Phi\times\Theta\times\Lambda\}$, *i.e.,*

$$Q\big(\{\phi,\theta,\lambda\},\{\phi,\theta,\lambda\}\big)\geq C\big(\|\phi\|_\Phi^2+\|\theta\|_\Theta^2+\|\lambda\|_\Lambda^2\big) \tag{11.38}$$

for all $\{\phi,\theta,\lambda\}\in\Phi\times\Theta\times\Lambda$.

Proof. The symmetry and continuity of the form $Q(\cdot,\cdot)$ and the continuity of the form $F(\cdot)$ are easily proved.

To prove (11.38), consider the following generalized form of the optimality system (11.28) written in operator form:

$$\begin{cases} \mathcal{A}_1\phi \qquad\qquad + \mathcal{B}_1^*\lambda = d & \text{in } \Phi^* \\ \qquad\quad \mathcal{A}_2\theta + \mathcal{B}_2^*\lambda = q & \text{in } \Theta^* \\ \mathcal{B}_1\phi + \mathcal{B}_2\theta \qquad\quad = g & \text{in } \Lambda^*, \end{cases} \tag{11.39}$$

where $\{d,q,g\}\in\Phi^*\times\Theta^*\times\Lambda^*$ is a general data triple and $\{\phi,\theta,\lambda\}\in\Phi\times\Theta\times\Lambda$ is the corresponding solution triple. In the same way that Theorem 11.7 is proved, we have that if (11.3) and (11.5) hold, then, for any $\{d,q,g\}\in\Phi^*\times\Theta^*\times\Lambda^*$, the generalized optimality system (11.39) has a unique solution $\{\phi,\theta,\lambda\} \in \Phi \times \Theta \times \Lambda$ and that solution satisfies

$$\|\phi\|_\Phi + \|\theta\|_\Theta + \|\lambda\|_\Lambda \leq C\big(\|d\|_{\Phi^*} + \|q\|_{\Theta^*} + \|g\|_{\Lambda^*}\big)\,. \tag{11.40}$$

From (11.36), we have that for all $\{\phi,\theta,\lambda\} \in \Phi \times \Theta \times \Lambda$

$$\begin{aligned} &Q\big(\{\phi,\theta,\lambda\},\{\phi,\theta,\lambda\}\big) \\ &\quad = \|\mathcal{A}_1\phi + \mathcal{B}_1^*\lambda\|^2_{\Phi^*} + \|\mathcal{A}_2\theta + \mathcal{B}_2^*\lambda\|^2_{\Theta^*} + \|\mathcal{B}_1\phi + \mathcal{B}_2\theta\|^2_{\Lambda^*}\,. \end{aligned} \tag{11.41}$$

Clearly, for any $\{\phi,\theta,\lambda\} \in \Phi \times \Theta \times \Lambda$, there exists $\{d,q,g\} \in \Phi^* \times \Theta^* \times \Lambda^*$ such that $\{\phi,\theta,\lambda\}$ is a solution of (11.39). This observation, (11.39), and (11.41) then yield that

$$\begin{aligned} &\|\phi\|^2_\Phi + \|\theta\|^2_\Theta + \|\lambda\|^2_\Lambda \leq C\big(\|d\|^2_{\Phi^*} + \|q\|^2_{\Theta^*} + \|g\|^2_{\Lambda^*}\big) \\ &\qquad = C\big(\|\mathcal{A}_1\phi + \mathcal{B}_1^*\lambda\|^2_{\Phi^*} + \|\mathcal{A}_2\theta + \mathcal{B}_2^*\lambda\|^2_{\Theta^*} + \|\mathcal{B}_1\phi + \mathcal{B}_2\theta\|^2_{\Lambda^*}\big) \end{aligned} \tag{11.42}$$

for all $\{\phi,\theta,\lambda\} \in \Phi \times \Theta \times \Lambda$. Combining (11.41) and (11.42) yields (11.38). □

Corollary 11.14 *The least-squares functional in* (11.34) *is norm equivalent, i.e., there exist constants* $\gamma_1 > 0$ *and* $\gamma_2 > 0$ *such that*

$$\gamma_1\big(\|\phi\|^2_\Phi + \|\theta\|^2_\Theta + \|\lambda\|^2_\Lambda\big) \leq J(\phi,\theta,\lambda;0,0) \leq \gamma_2\big(\|\phi\|^2_\Phi + \|\theta\|^2_\Theta + \|\lambda\|^2_\Lambda\big) \tag{11.43}$$

for all $\{\phi,\theta,\lambda\} \in \Phi \times \Theta \times \Lambda$.

Proof. The result follows from

$$\begin{aligned} J(\phi,\theta,\lambda;0,0) &= \|\mathcal{A}_1\phi + \mathcal{B}_1^*\lambda\|^2_{\Phi^*} + \|\mathcal{A}_2\theta + \mathcal{B}_2^*\lambda\|^2_{\Theta^*} + \|\mathcal{B}_1\phi + \mathcal{B}_2\theta\|^2_{\Lambda^*} \\ &= Q\big(\{\phi,\theta,\lambda\},\{\phi,\theta,\lambda\}\big) \end{aligned}$$

and the coercivity and continuity of the bilinear form $Q(\cdot,\cdot)$. □

The norm equivalence of the least-squares functional in (11.34) makes it easy to define and analyze LSFEMs for the approximation of solutions of the optimality system (11.27) or, equivalently, (11.28). Indeed, following Section 3.5.1, one merely chooses conforming finite element subspaces $\Phi^h \subset \Phi$, $\Theta^h \subset \Theta$, and $\Lambda^h \subset \Lambda$ and then restricts the least-squares principle (11.34) to the subspaces, i.e., one solves the compliant DLSP

$$\begin{cases} J_h(\phi^h,\theta^h,\lambda^h;\widehat{\phi},g) = J(\phi^h,\theta^h,\lambda^h;\widehat{\phi},g) \\ X^h = \Phi^h \times \Theta^h \times \Lambda^h\,. \end{cases} \tag{11.44}$$

Equivalently, one can require that $\{\phi^h,\theta^h,\lambda^h\} \in \Phi^h \times \Theta^h \times \Lambda^h$ satisfy

$$Q\big(\{\phi^h,\theta^h,\lambda^h\},\{\mu^h,\nu^h,\psi^h\}\big) = F\big(\{\mu^h,\nu^h,\psi^h\};\{\mathcal{A}_1\widehat{\phi},g\}\big) \tag{11.45}$$

for all $\{\mu^h, v^h, \psi^h\} \in \Phi^h \times \Theta^h \times \Lambda^h$. Due to the norm equivalence of the functional, the following result is a direct consequence of Theorem 3.28 in Section 3.5.1.

Theorem 11.15 *Let the assumptions* (11.3) *and* (11.5) *hold. Then, for any* $\widehat{\phi} \in \widehat{\Phi}$ *and* $g \in \Lambda^*$, *the problem* (11.44) *or, equivalently,* (11.45) *has a unique solution* $\{\phi^h, \theta^h, \lambda^h\} \in \Phi^h \times \Theta^h \times \Lambda^h$. *Moreover, we have the optimal error estimate: there exists a constant* $C > 0$ *whose value is independent of h, such that*

$$\begin{aligned}
&\|\phi - \phi^h\|_\Phi + \|\theta - \theta^h\|_\Theta + \|\lambda - \lambda^h\|_\Lambda \\
&\quad \le C\Big(\inf_{\widetilde{\phi}^h \in \Phi^h} \|\phi - \widetilde{\phi}^h\|_\Phi + \inf_{\widetilde{\theta}^h \in \Theta^h} \|\theta - \widetilde{\theta}^h\|_\Theta + \inf_{\widetilde{\lambda}^h \in \Lambda^h} \|\lambda - \widetilde{\lambda}^h\|_\Lambda \Big),
\end{aligned} \tag{11.46}$$

where $\{\phi, \theta, \lambda\} \in \Phi \times \Theta \times \Lambda$ *is the unique solution of the problem* (11.35) *or, equivalently, of the problems* (11.27) *or* (11.28). *Note also that* $\{\phi, \theta\} \in \Phi \times \Theta$ *is the unique solution of the problem* (11.6). □

The discrete problem (11.45) is equivalent to the linear algebraic system

$$\begin{pmatrix} \mathsf{K}_1 & \mathsf{H}_1^\mathsf{T} & \mathsf{H}_2^\mathsf{T} \\ \mathsf{H}_1 & \mathsf{K}_2 & \mathsf{H}_3^\mathsf{T} \\ \mathsf{H}_2 & \mathsf{H}_3 & \mathsf{K}_3 \end{pmatrix} \begin{pmatrix} \vec{\phi} \\ \vec{\theta} \\ \vec{\lambda} \end{pmatrix} = \begin{pmatrix} \vec{f} \\ \vec{q} \\ \vec{g} \end{pmatrix}. \tag{11.47}$$

Indeed, if one chooses bases $\{\mu_j^h\}_{j=1}^J$, $\{v_k^h\}_{k=1}^K$, and $\{\psi_\ell^h\}_{\ell=1}^L$ for Φ_h, Θ_h, and Λ_h, respectively, we then have $\phi^h = \sum_{j=1}^J \phi_j \mu_j^h$, $\theta^h = \sum_{k=1}^K \theta_k v_k^h$, and $\lambda^h = \sum_{\ell=1}^L \lambda_\ell \psi_\ell^h$ for some sets of coefficients $\{\phi_j\}_{j=1}^J$, $\{\theta_k\}_{k=1}^K$, and $\{\lambda_\ell\}_{\ell=1}^L$ that are determined by solving (11.47). In (11.47), we have that $\vec{\phi} = (\phi_1, \ldots, \phi_J)^\mathsf{T}$, $\vec{\theta} = (\theta_1, \ldots, \theta_K)^\mathsf{T}$, $\vec{\lambda} = (\lambda_1, \ldots, \lambda_L)^\mathsf{T}$,

$$\begin{aligned}
&\big(\mathsf{K}_1\big)_{ij} = (\mathcal{A}_1\mu_i, \mathcal{A}_1\mu_j)_{\Phi^*} + (\mathcal{B}_1\mu_i, \mathcal{B}_1\mu_j)_{\Lambda^*} && \text{for } i,j = 1,\ldots,J \\
&\big(\mathsf{K}_2\big)_{ik} = (\mathcal{A}_2 v_i, \mathcal{A}_1 v_k)_{\Theta^*} + (\mathcal{B}_2 v_i, \mathcal{B}_2 v_k)_{\Lambda^*} && \text{for } i,k = 1,\ldots,K \\
&\big(\mathsf{K}_3\big)_{i\ell} = (\mathcal{B}_1^*\psi_i, \mathcal{B}_1^*\psi_\ell)_{\Phi^*} + (\mathcal{B}_2^*\psi_i, \mathcal{B}_2^*\psi_\ell)_{\Theta^*} && \text{for } i,\ell = 1,\ldots,L \\
&\big(\mathsf{H}_1\big)_{ij} = (\mathcal{B}_2 v_i, \mathcal{B}_1\mu_j)_{\Lambda^*} && \text{for } i = 1,\ldots,K,\ j = 1,\ldots,J \\
&\big(\mathsf{H}_2\big)_{ij} = (\mathcal{B}_1^*\psi_i, \mathcal{A}_1\mu_j)_{\Phi^*} && \text{for } i = 1,\ldots,L,\ j = 1,\ldots,J \\
&\big(\mathsf{H}_3\big)_{ik} = (\mathcal{B}_2^*\psi_i, \mathcal{A}_2 v_k)_{\Theta^*} && \text{for } i = 1,\ldots,L,\ k = 1,\ldots,K \\
&\big(\vec{f}\big)_i = (\mathcal{A}_1\mu_i, \mathcal{A}_1\widehat{\phi})_{\Phi^*} + (\mathcal{B}_1\mu_i, g)_{\Lambda^*} && \text{for } i = 1,\ldots,J \\
&\big(\vec{q}\big)_i = (\mathcal{B}_2 v_i, g)_{\Lambda^*} && \text{for } i = 1,\ldots,K \\
&\big(\vec{g}\big)_i = (\mathcal{B}_1^*\psi_i, \mathcal{A}_1\widehat{\phi})_{\Phi^*} && \text{for } i = 1,\ldots,L.
\end{aligned}$$

Remark 11.16 It easily follows from Lemma 11.13 that the coefficient matrix of (11.47) is *symmetric and positive definite.* This should be compared to the linear system (11.33) that results from a Galerkin finite element discretization of the optimality system (11.27) for which the coefficient matrix is symmetric and *indefinite.* □

Remark 11.17 The stability of the discrete problem (11.45), the convergence and optimal accuracy of the approximate solution $\{\phi^h, \theta^h, \lambda^h\}$, and the symmetry and positive definiteness of the discrete system (11.47) obtained by the LSFEM follow from the assumptions (11.3) and (11.5) that guarantee the well-posedness of the infinite-dimensional optimization problem (11.6) and its corresponding optimality system (11.27). It is important to note that all of these desirable properties of the LSFEM do not require that the bilinear form $b_1(\cdot,\cdot)$ and the finite element spaces Φ^h and Λ^h satisfy the inner inf–sup conditions (11.30) that are necessary for the well posedness of the Galerkin finite element discretization (11.29) of the optimality system (11.27) □

Remark 11.18 The need to, in practice, uncouple the equations in (11.33) holds as well for the linear system (11.47). Uncoupling approaches for (11.33) rely on the invertibility of the matrices B_1 and A_2; the first of these is, in general, non-symmetric and indefinite, even when the necessary discrete inf–sup conditions in (11.30) are satisfied. For (11.47), uncoupling strategies rely on the invertibility of the matrices K_1, K_2, and K_3; all three of these matrices are symmetric and positive definite even when (11.30) is not satisfied. An example of a simple uncoupling strategy is to apply a block-Gauss Seidel method to (11.47), which would proceed as follows:

start with initial guesses $\vec{\phi}^{(0)}$ and $\vec{\theta}^{(0)}$ for the discretized state and control; then, for $k = 1,2,\ldots$, successively solve the linear systems

$$\begin{aligned} \mathsf{K}_3\vec{\lambda}^{(k+1)} &= \vec{g} - \mathsf{H}_2\vec{\phi}^{(k)} - \mathsf{H}_3\vec{\theta}^{(k)} \\ \mathsf{K}_1\vec{\phi}^{(k+1)} &= \vec{d} - \mathsf{H}_1^T\vec{\theta}^{(k)} - \mathsf{H}_2^T\vec{\lambda}^{(k+1)} \\ \mathsf{K}_2\vec{\theta}^{(k+1)} &= \vec{q} - \mathsf{H}_1\vec{\phi}^{(k+1)} - \mathsf{H}_3^T\vec{\lambda}^{(k+1)} \end{aligned} \tag{11.48}$$

until satisfactory convergence is achieved, e.g., until some norm of the difference between successive iterates is less than some prescribed tolerance.

Because the coefficient matrix in (11.47) is symmetric and positive definite, this iteration converges. Moreover, all three coefficient matrices K_3, K_1, and K_2 of the linear systems in (11.48) are themselves symmetric and positive definite so that very efficient solution methodologies, including parallel ones, can be applied for their solution. We also note that, in order to obtain faster convergence rates, better uncoupling iterative methods, e.g., over-relaxation schemes or a conjugate gradient method, can be applied instead of the Gauss–Seidel iteration of (11.48). □

11.3 Methods Based on Direct Penalization by the Least-Squares Functional

A straightforward way to use least-squares principles in the optimization setting of problem (11.6) is to enforce the constraint equations (11.4), or equivalently (11.16), by penalizing the functional (11.2), or its equivalent form (11.15), by the least-squares functional in (11.18); see [56, 196] for examples of the use of this approach in concrete settings. Thus, instead of solving the constrained problem (11.6) or its equivalent form (11.17), we solve the unconstrained problem

$$\min_{\{\phi,\theta\}\in\Phi\times\Theta} K_\varepsilon(\phi,\theta), \tag{11.49}$$

where, for given $\widehat{\phi}\in\widehat{\Phi}$ and $g\in\Lambda^*$,

$$K_\varepsilon(\phi,\theta) = K(\phi,\theta) + \frac{1}{\varepsilon}J_c(\phi;\theta,g) \qquad \forall\phi\in\Phi,\ \theta\in\Theta \tag{11.50}$$

so that, for all $\phi\in\Phi$, $\theta\in\Theta$,

$$K_\varepsilon(\phi,\theta) = a_1(\phi-\widehat{\phi},\phi-\widehat{\phi}) + a_2(\theta,\theta) + \frac{1}{\varepsilon}\|\mathcal{B}_1\phi+\mathcal{B}_2\theta-g\|^2_{\Lambda^*}.$$

The Euler–Lagrange equations corresponding to the unconstrained minimization problem (11.49) are given by

$$\begin{cases} \langle\mathcal{A}_1\phi_\varepsilon,\mu\rangle_{\widehat{\Phi}^*,\widehat{\Phi}} + \frac{1}{\varepsilon}\big(\mathcal{B}_1\mu,(\mathcal{B}_1\phi_\varepsilon+\mathcal{B}_2\theta_\varepsilon)\big)_{\Lambda^*} \\ \qquad = \langle\mathcal{A}_1\widehat{\phi},\mu\rangle_{\widehat{\Phi}^*,\widehat{\Phi}} + \frac{1}{\varepsilon}\big(\mathcal{B}_1\mu,g\big)_{\Lambda^*} & \forall\mu\in\Phi \\ \langle\mathcal{A}_2\theta_\varepsilon,\nu\rangle_{\Theta^*,\Theta} + \frac{1}{\varepsilon}\big(\mathcal{B}_2\nu,(\mathcal{B}_1\phi_\varepsilon+\mathcal{B}_2\theta_\varepsilon)\big)_{\Lambda^*} \\ \qquad = \frac{1}{\varepsilon}\big(\mathcal{B}_2\nu,g\big)_{\Lambda^*} & \forall\nu\in\Theta. \end{cases} \tag{11.51}$$

For $\phi_\varepsilon\in\Phi$ and $\theta_\varepsilon\in\Theta$, the Riesz representation theorem guarantees that there exists a unique $\lambda_\varepsilon\in\Lambda$ such that

$$\varepsilon\big\langle\upsilon,\lambda_\varepsilon\big\rangle_{\Lambda^*,\Lambda} = \big(\upsilon,\mathcal{B}_1\phi_\varepsilon+\mathcal{B}_2\theta_\varepsilon-g\big)_{\Lambda^*} \qquad \forall\upsilon\in\Lambda^*. \tag{11.52}$$

Let $\mathcal{R}_\Lambda:\Lambda\to\Lambda^*$ denote the *Riesz operator*. Then, if $\upsilon=\mathcal{R}_\Lambda\psi$ and $\chi=\mathcal{R}_\Lambda\lambda$ for $\{\psi,\lambda\}\in\Lambda$ and $\{\upsilon,\chi\}\in\Lambda^*$, we have that $\|\psi\|_\Lambda=\|\upsilon\|_{\Lambda^*}$, $\|\lambda\|_\Lambda=\|\chi\|_{\Lambda^*}$, and

$$(\psi,\lambda)_\Lambda = \langle\mathcal{R}_\Lambda\psi,\lambda\rangle_{\Lambda^*,\Lambda} = \langle\upsilon,\lambda\rangle_{\Lambda^*,\Lambda} = \langle\upsilon,\mathcal{R}_\Lambda^{-1}\chi\rangle_{\Lambda^*,\Lambda} = (\upsilon,\chi)_{\Lambda^*}.$$

Then, one easily sees that (11.51) can be expressed in the equivalent form

$$\begin{cases} a_1(\phi_\varepsilon,\mu) & + b_1(\mu,\lambda_\varepsilon) & = a_1(\widehat{\phi},\mu) & \forall \mu \in \Phi \\ & a_2(\theta_\varepsilon,\nu) + b_2(\nu,\lambda_\varepsilon) & = 0 & \forall \nu \in \Theta \\ b_1(\phi_\varepsilon,\psi) + b_2(\theta_\varepsilon,\psi) - \varepsilon\big\langle \mathcal{R}_\Lambda \psi,\lambda_\varepsilon\big\rangle_{\Lambda^*,\Lambda} & & = G(\psi) & \forall \psi \in \Lambda\,. \end{cases} \tag{11.53}$$

The system (11.53) is a *regular perturbation* of the system (11.28) that is the Euler–Lagrange equations for the minimization problem (11.6) or its equivalent form (11.17), where $\lambda \in \Lambda$ is the Lagrange multiplier introduced to enforce the constraint equation in (11.17).

The system (11.53) can be expressed in equivalent operator form as

$$\begin{cases} \mathcal{A}_1\phi_\varepsilon & + \mathcal{B}_1^*\lambda_\varepsilon & = \mathcal{A}_1\widehat{\phi} & \text{in } \Phi^* \\ & \mathcal{A}_2\theta_\varepsilon + \mathcal{B}_2^*\lambda_\varepsilon & = 0 & \text{in } \Theta^* \\ \mathcal{B}_1\phi_\varepsilon + \mathcal{B}_2\theta_\varepsilon & - \varepsilon\mathcal{R}_\Lambda\lambda_\varepsilon & = g & \text{in } \Lambda^*. \end{cases} \tag{11.54}$$

Concerning the penalized control problem (11.6), we have the following results.

Theorem 11.19 *Let the assumptions* (11.3) *and* (11.5) *hold. Then, for each* $0 < \varepsilon \leq 1$, (11.53) *or, equivalently,* (11.51) *and* (11.52), *or, equivalently, the penalized optimal control problem* (11.49), *has a unique solution* $\{\phi_\varepsilon,\theta_\varepsilon,\lambda_\varepsilon\} \in \Phi\times\Theta\times\Lambda$. *Let* $\{\phi,\theta,\lambda\} \in \Phi\times\Theta\times\Lambda$ *denote the unique solution of the optimality system* (11.28) *or, equivalently, of the optimal control problem* (11.6). *Then, for some constant* $C > 0$ *whose value is independent of* ε,

$$\|\phi-\phi_\varepsilon\|_\Phi + \|\theta-\theta_\varepsilon\|_\Theta + \|\lambda-\lambda_\varepsilon\|_\Lambda \leq C\varepsilon\big(\|g\|_{\Lambda^*} + \|\widehat{\phi}\|_{\widehat{\Phi}}\big)\,. \tag{11.55}$$

Proof. Let the bilinear forms $a(\cdot,\cdot)$ and $b(\cdot,\cdot)$ be defined as in (11.7). Then, (11.28) and (11.53) can be respectively written as

$$\begin{cases} a(\{\phi,\theta\},\{\mu,\nu\}) + b(\{\mu,\nu\},\{\lambda\}) = a_1(\widehat{\phi},\mu) & \forall\{\mu,\nu\} \in \Phi\times\Theta \\ b(\{\phi,\theta\},\{\psi\}) = G(\psi) & \forall\psi\in\Lambda \end{cases} \tag{11.56}$$

and, because $\langle\mathcal{R}_\Lambda\psi,\lambda_\varepsilon\rangle_{\Lambda^*,\Lambda} = (\lambda_\varepsilon,\psi)_\Lambda$ for all $\psi\in\Lambda$,

$$\begin{cases} a(\{\phi_\varepsilon,\theta_\varepsilon\},\{\mu,\nu\}) + b(\{\mu,\nu\},\{\lambda_\varepsilon\}) = a_1(\widehat{\phi},\mu) & \forall\{\mu,\nu\} \in \Phi\times\Theta \\ b(\{\phi_\varepsilon,\theta_\varepsilon\},\{\psi\}) \quad - \varepsilon(\lambda_\varepsilon,\psi)_\Lambda = G(\psi) & \forall\psi\in\Lambda\,. \end{cases} \tag{11.57}$$

In Lemma 11.1, it is shown that if (11.3) and (11.5) hold, then the bilinear forms $a(\cdot,\cdot)$ and $b(\cdot,\cdot)$ satisfy (1.22), (1.29), and (1.30) and (1.23) and (1.25), respectively. Then, the results of the theorem follow from well-known results about the systems (11.56) and (11.57); see, e.g., [183, Theorem 4.3, p.65] and [55, 76, 83, 87, 191]. □

Now, let us return to the system (11.51) that can be written in more compact form as

$$Q_\varepsilon(\{\phi_\varepsilon,\theta_\varepsilon\},\{\mu,\nu\}) = F_\varepsilon(\{\mu,\nu\}) \qquad \forall\{\mu,\nu\}\in\Phi\times\Theta, \tag{11.58}$$

where, for all $\{\phi,\theta\},\{\mu,\nu\}\in\Phi\times\Theta$,

$$Q_\varepsilon(\{\phi,\theta\},\{\mu,\nu\}) = a_1(\phi,\mu) + a_2(\theta,\mu) + \frac{1}{\varepsilon}\big(\mathcal{B}_1\mu + \mathcal{B}_2\nu, \mathcal{B}_1\phi + \mathcal{B}_2\theta\big)_{\Lambda^*} \quad (11.59)$$

and

$$F_\varepsilon(\{\mu,\nu\}) = a_1(\widehat{\phi},\mu) + \frac{1}{\varepsilon}\big(\mathcal{B}_1\mu + \mathcal{B}_2\nu, g\big)_{\Lambda^*}. \quad (11.60)$$

Concerning the bilinear form $Q_\varepsilon(\cdot,\cdot)$ and the linear functional $F_\varepsilon(\cdot)$, we have the following results.

Lemma 11.20 *Let the bilinear form $Q_\varepsilon(\cdot,\cdot)$ and the linear functional $F_\varepsilon(\cdot)$ be defined by* (11.59) *and* (11.60)*, respectively. Let the assumptions* (11.3) *and* (11.5) *hold and let $0 < \varepsilon \le 1$. Then, there exist positive constants α_{q_1}, α_{q_2}, α_{f_1}, α_{f_2}, and β_q whose values do not depend on ε such that, for all $\{\phi,\theta\}$, $\{\mu,\nu\} \in \Phi \times \Theta$,*

$$Q_\varepsilon(\{\phi,\theta\},\{\mu,\nu\}) \le \Big(\alpha_{q_1} + \frac{\alpha_{q_2}}{\varepsilon}\Big)\|\{\phi,\theta\}\|_{\Phi\times\Theta}\|\{\mu,\nu\}\|_{\Phi\times\Theta}, \quad (11.61)$$

$$Q_\varepsilon(\{\phi,\theta\},\{\phi,\theta\}) \ge \beta_q\|\{\phi,\theta\}\|^2_{\Phi\times\Theta}, \quad (11.62)$$

and

$$F_\varepsilon(\{\mu,\nu\}) \le \Big(\alpha_{f_1}\|\widehat{\phi}\|_{\widehat{\Phi}} + \frac{\alpha_{f_2}}{\varepsilon}\|g\|_{\Lambda^*}\Big)\|\{\mu,\nu\}\|_{\Phi\times\Theta}. \quad (11.63)$$

Proof. We have that

$$\begin{aligned}
Q_\varepsilon(\{\phi,\theta\},\{\mu,\nu\}) &\le \alpha_{a_1}\|\phi\|_\Phi\|\mu\|_\Phi + \alpha_{a_2}\|\theta\|_\Theta\|\nu\|_\Theta \\
&\quad + \frac{1}{\varepsilon}\Big(\alpha_{b_1}\|\phi\|_\Phi + \alpha_{b_2}\|\theta\|_\Theta\Big)\Big(\alpha_{b_1}\|\mu\|_\Phi + \alpha_{b_2}\|\nu\|_\Theta\Big) \\
&\le \Big(\max\{\alpha_{a_1},\alpha_{a_2}\} + \frac{2}{\varepsilon}\max\{\alpha^2_{b_1},\alpha^2_{b_2}\}\Big)\|\{\phi,\theta\}\|_{\Phi\times\Theta}\|\{\mu,\nu\}\|_{\Phi\times\Theta}
\end{aligned}$$

so that (11.61) holds with $\alpha_{q_1} = \max\{\alpha_{a_1},\alpha_{a_2}\}$ and $\alpha_{q_2} = 2\max\{\alpha^2_{b_1},\alpha^2_{b_2}\}$.

The proof of (11.63) proceeds in a similar manner; one obtains that $\alpha_{f_1} = \alpha_{a_1}$ and $\alpha_{f_2} = 2^{1/2}\max\{\alpha_{b_1},\alpha_{b_2}\}$.

Next, suppose $\{\phi,\theta\} \in Z$ so that by (11.9) and (11.10), $\mathcal{B}_1\phi + \mathcal{B}_2\theta = 0$ and $\|\phi\|_\Lambda \le \frac{\alpha_{b_2}}{\beta_{b_1}}\|\theta\|_\Theta$. Then,

$$\begin{aligned}
Q_\varepsilon(\{\phi,\theta\},\{\phi,\theta\}) &= a_1(\phi,\phi) + a_2(\theta,\theta) \ge a_2(\theta,\theta) \ge \beta_{a_2}\|\theta\|^2_\Theta \\
&\ge \frac{\beta_{a_2}}{2}\left(\|\theta\|^2_\Theta + \frac{\beta^2_{b_1}}{\alpha^2_{b_2}}\|\phi\|^2_\Phi\right) \\
&\ge \frac{\beta_{a_2}}{2}\min\left\{1,\frac{\beta^2_{b_1}}{\alpha^2_{b_2}}\right\}\|\{\phi,\theta\}\|^2 \qquad \forall\{\phi,\theta\}\in Z.
\end{aligned} \quad (11.64)$$

Now, let the bilinear form $b(\cdot,\cdot)$ be defined as in (11.7). Then, by Lemma 11.1, we have that (1.25) holds with $\beta_b = \beta_{b_1}$ in which case it is well known (see, e.g., [183]) that

$$\sup_{\{\psi\}\in\Lambda,\{\psi\}\neq\{0\}} \frac{b(\{\phi,\theta\},\{\psi\})}{\|\{\psi\}\|_\Lambda} \geq \beta_{b_1}\|\{\phi,\theta\}\|_{\Phi\times\Theta} \qquad \forall\{\phi,\theta\}\in Z^\perp.$$

Then, because for all $\{\phi,\theta\}\in\Phi\times\Theta$,

$$b(\{\phi,\theta\},\{\psi\}) = b_1(\phi,\psi)+b_2(\theta,\psi) = \langle\mathcal{B}_1\phi+\mathcal{B}_2\theta,\psi\rangle_{\Lambda^*,\Lambda},$$

we have that

$$\|\mathcal{B}_1\phi+\mathcal{B}_2\theta\|_{\Lambda^*} \geq \beta_{b_1}\|\{\phi,\theta\}\|_{\Phi\times\Theta} \qquad \forall\{\phi,\theta\}\in Z^\perp$$

so that, for $0<\varepsilon\leq 1$,

$$\begin{aligned} Q_\varepsilon(\{\phi,\theta\},\{\phi,\theta\}) &= a_1(\phi,\phi)+a_2(\theta,\theta)+\frac{1}{\varepsilon}\|\mathcal{B}_1\phi+\mathcal{B}_2\theta\|^2_{\Lambda^*} \\ &\geq \beta^2_{b_1}\|\{\phi,\theta\}\|^2 \qquad \forall\{\phi,\theta\}\in Z^\perp. \end{aligned} \tag{11.65}$$

Because $Z\subset\Phi\times\Theta$ is a closed subspace, combining (11.64) and (11.65) yields (11.62) with $\beta_q=\min\left\{\frac{\beta_{a_2}}{2}\min\left\{1,\frac{\beta^2_{b_1}}{\alpha^2_{b_2}}\right\},\beta^2_{b_1}\right\}$. □

We see that (11.51) and (11.53) are completely equivalent. One may then proceed to discretize either of these systems.[7] It is important to note that the two resulting discrete systems are *not equivalent* and can, in fact, have significantly different properties.

11.3.1 Discretization of the Perturbed Optimality System

We consider obtaining a discretization of (11.51) by first discretizing (11.53), or equivalently (11.54), and then eliminating the Lagrange multiplier. Discretization can be effected by choosing conforming finite element spaces $\Phi^h\subset\Phi$, $\Theta^h\subset\Theta$, and $\Lambda^h\subset\Lambda$ and then restricting (11.53) to the subspaces to obtain

$$\left\{\begin{array}{llll} a_1(\phi^h_\varepsilon,\mu^h) & +\,b_1(\mu^h,\lambda^h_\varepsilon) & = a_1(\widehat{\phi},\mu^h) & \forall\mu^h\in\Phi^h \\ & a_2(\theta^h_\varepsilon,\nu^h)+b_2(\nu^h,\lambda^h_\varepsilon) & = 0 & \forall\nu^h\in\Theta^h \\ b_1(\phi^h_\varepsilon,\psi^h)+b_2(\theta^h_\varepsilon,\psi^h) & -\,\varepsilon(\mathcal{R}_\Lambda\lambda^h_\varepsilon,\psi^h)_\Lambda & = G(\psi^h) & \forall\psi^h\in\Lambda^h. \end{array}\right. \tag{11.66}$$

[7] The results of Lemma 11.20 provide an alternate means for proving, for any $0<\varepsilon\leq 1$, that the system (11.51) has a unique solution. Indeed, those results assert that the symmetric bilinear form $Q_\varepsilon(\cdot,\cdot)$ is continuous and strongly coercive and that the linear functional $\mathcal{F}_\varepsilon(\cdot)$ is continuous so that the existence and uniqueness of the solution of (11.58), or equivalently, of (11.51) follows by the Lax–Milgram lemma. However, due to the ε^{-1} coefficient in the right-hand side of (11.61), the results of Lemma 11.20 cannot be used to derive the estimate (11.55) for the solution of (11.51); this is done indirectly by using the equivalence of (11.51) and (11.52) with (11.53).

In the usual way, the discrete system (11.66) is equivalent to a matrix problem. In addition to the matrices and vectors appearing in (11.33), we define the $L \times L$ matrix $(\mathsf{R}_\Lambda)_{ij} = (\mathcal{R}_\Lambda \psi_i^h, \psi_j^h)_\Lambda$, where $\{\psi_i^h\}_{i=1}^L$ denotes a basis for $\Lambda^h \subset \Lambda$. Then, the discrete regularized problem (11.66) is equivalent to the linear system

$$\begin{pmatrix} \mathsf{A}_1 & 0 & \mathsf{B}_1^T \\ 0 & \mathsf{A}_2 & \mathsf{B}_2^T \\ \mathsf{B}_1 & \mathsf{B}_2 & -\varepsilon \mathsf{R}_\Lambda \end{pmatrix} \begin{pmatrix} \vec{\phi}_\varepsilon \\ \vec{\theta}_\varepsilon \\ \vec{\lambda}_\varepsilon \end{pmatrix} = \begin{pmatrix} \vec{f} \\ \vec{0} \\ \vec{g} \end{pmatrix}, \tag{11.67}$$

where $\vec{\phi}_\varepsilon$, $\vec{\theta}_\varepsilon$, and $\vec{\lambda}_\varepsilon$ are vectors of coefficients of the representation of ϕ_ε^h, θ_ε^h, and λ_ε^h in terms of the bases for Φ^h, Θ^h, and Λ^h, respectively.

It is now easy to see how one can eliminate $\vec{\lambda}_\varepsilon$ from (11.67), or equivalently, λ_ε^h from (11.66). Indeed, one easily deduces from (11.67) that

$$\begin{cases} \left(\mathsf{A}_1 + \dfrac{1}{\varepsilon}\mathsf{B}_1^T \mathsf{R}_\Lambda^{-1} \mathsf{B}_1\right) \phi_\varepsilon + \dfrac{1}{\varepsilon}\mathsf{B}_1^T \mathsf{R}_\Lambda^{-1} \mathsf{B}_2 \theta_\varepsilon = \vec{f} + \dfrac{1}{\varepsilon}\mathsf{B}_1^T \mathsf{R}_\Lambda^{-1} \vec{g} \\ \left(\mathsf{A}_2 + \dfrac{1}{\varepsilon}\mathsf{B}_2^T \mathsf{R}_\Lambda^{-1} \mathsf{B}_2\right) \theta_\varepsilon + \dfrac{1}{\varepsilon}\mathsf{B}_2^T \mathsf{R}_\Lambda^{-1} \mathsf{B}_1 \phi_\varepsilon = \dfrac{1}{\varepsilon}\mathsf{B}_2^T \mathsf{R}_\Lambda^{-1} \vec{g}. \end{cases} \tag{11.68}$$

Note that (11.68) only involves the approximations $\phi_\varepsilon^h \in \Phi^h$ and $\theta_\varepsilon^h \in \Theta^h$ of the state variable $\phi \in \Phi$ and the control variable $\theta \in \Theta$, respectively, and does not involve the approximation $\lambda_\varepsilon^h \in \Psi^h$ of the adjoint variable $\lambda \in \Psi$. Once (11.68) is used to determine $\vec{\phi}_\varepsilon$ and $\vec{\theta}_\varepsilon$, $\vec{\lambda}_\varepsilon$ may be determined from the last equation in (11.67).

Now, consider what is required to guarantee that the coefficient matrix of the linear system (11.67) or, equivalently, of (11.68) is stably invertible as either or both the grid size h and the penalty parameter ε tend to zero. It is not difficult to show, based on the assumptions (11.3) and (11.5) that have been made about the bilinear forms appearing in (11.66), that a necessary and sufficient condition for the stable invertibility of (11.67) or (11.68) is that the matrix B_1 be stably invertible. As noted in Section 11.2, the conditions imposed in (11.5) on the bilinear form $b_1(\cdot,\cdot)$ are not suffcient to guarantee the stable invertibility of B_1; to do so, we have to assume that that form and the subspaces Φ^h and Λ^h satisfy the discrete stability conditions in (11.30). These are exactly the conditions required for the stable invertibility of the analogous discretization (11.29), or equivalently (11.33), of the unpenalized optimality system (11.28). In other words, despite the fact that

> (11.53) *is equivalent to enforcing the constraint* (11.16) *by penalizing the functional* (11.15) *by the* well-posed *least-squares functional* in (11.18)

and despite the fact that

> *given a control θ, stable approximations of the state ϕ may be obtained by minimizing the least-squares functional* in (11.18) without *having to assume that the discrete spaces Φ^h and Λ^h satisfy* (11.30),

the stable solution of (11.66), or equivalently (11.67) or (11.68), requires that (11.30) is satisfied. Thus, one of the main advantages of using LSFEMs, i.e., being able to circumvent (11.30), is lost.[8]

The following error estimate is easily derived using well-known techniques.

Theorem 11.21 *Let* (11.3), (11.5), *and* (11.30) *hold. Then,* (11.66), *or equivalently* (11.68), *has a unique solution* $\{\phi_\varepsilon^h, \theta_\varepsilon^h, \lambda_\varepsilon^h\} \in \Phi^h \times \Theta^h \times \Lambda^h$. *Moreover, if* $\{\phi, \theta, \lambda\} \in \Phi \times \Theta \times \Lambda$ *denotes the unique solution of the optimization problem* (11.6) *or equivalently, of* (11.51), *or equivalently, of* (11.53), *then there exist a constant* $C > 0$ *whose value is independent of* ε *and* h *such that*

$$\begin{aligned} \|\phi - \phi_\varepsilon^h\|_\Phi + \|\theta - \theta_\varepsilon^h\|_\Theta + \|\lambda - \lambda_\varepsilon^h\|_\Lambda &\le C\varepsilon\left(\|g\|_{\Lambda^*} + \|\widehat{\phi}\|_{\widehat{\Phi}}\right) \\ +C\Big(\inf_{\widetilde{\phi}^h \in \Phi^h} \|\phi_\varepsilon - \widetilde{\phi}^h\|_\Phi + \inf_{\widetilde{\theta}^h \in \Theta^h} \|\theta_\varepsilon - \widetilde{\theta}^h\|_\Theta &+ \inf_{\widetilde{\lambda}^h \in \Lambda^h} \|\lambda_\varepsilon - \widetilde{\lambda}^h\|_\Lambda\Big). \end{aligned} \tag{11.69}$$

Proof. Standard finite element analyses [76,83,87,183] yield, for the pair of systems (11.53) and (11.66), that $\|\phi_\varepsilon - \phi_\varepsilon^h\|_\Phi + \|\theta_\varepsilon - \theta_\varepsilon^h\|_\Theta + \|\lambda_\varepsilon - \lambda_\varepsilon^h\|_\Lambda$ is bounded by the second term on the right-hand side of (11.69). Then, (11.55) and the triangle inequality yields (11.69). □

Remark 11.22 Our discussion serves to reinforce the important observation about penalty methods that is made in Section 2.1.1, namely that they are not stabilization methods, i.e., penalty methods do not circumvent the discrete inf–sup conditions (11.30).[9] Penalty methods are properly viewed as being *methods for facilitating the solution of* (11.29) *or* (11.33). Because here we are primarily interested in retaining the advantage that LSFEMs provide for circumventing conditions such as (11.30), we do not consider discretizations of (11.53) as the best way to incorporate least-squares notions into the optimization problems we are considering. □

Remark 11.23 It is usually the case that the approximation-theoretic terms on the right-hand side of (11.69) satisfy inequalities of the type

$$\left\{ \begin{array}{ll} \inf_{\widetilde{\phi}^h \in \Phi^h} \|\phi_\varepsilon - \widetilde{\phi}^h\|_\Phi \le Ch^r & \inf_{\widetilde{\theta}^h \in \Theta^h} \|\theta_\varepsilon - \widetilde{\theta}^h\|_\Theta \le Ch^s \\ \multicolumn{2}{c}{\inf_{\widetilde{\lambda}^h \in \Lambda^h} \|\lambda_\varepsilon - \widetilde{\lambda}^h\|_\Lambda \le Ch^t\,,} \end{array} \right. \tag{11.70}$$

where $r,s,t > 0$ depend on the degree of the polynomials used for the spaces Φ^h, Θ^h, and Λ^h and the regularity of the solution $\{\phi_\varepsilon, \theta_\varepsilon, \lambda_\varepsilon\}$ of (11.53), or equivalently, of (11.51). Then, (11.69) implies that

[8] Although discretizations of (11.28) and (11.53) both require the imposition of (11.30) on the finite element spaces Φ^h and Λ^h, the latter system still has some advantages. Foremost among these is that one can reduce the number of variables by eliminating $\vec{\lambda}_\varepsilon$ from (11.67) to obtain (11.68). Furthermore, as long as (11.30) is satisfied, the system (11.68) is symmetric and positive definite whereas (11.33) is symmetric but indefinite.

[9] The fact that discretizations of (11.28) and (11.53) both require the imposition of (11.30) should not be surprising, given that the latter is a regular perturbation of the former.

$$\|\phi - \phi_\varepsilon^h\|_\Phi + \|\theta - \theta_\varepsilon^h\|_\Theta + \|\lambda - \lambda_\varepsilon^h\|_\Lambda \leq C\left(\varepsilon + h^r + h^s + h^t\right). \qquad \Box \tag{11.71}$$

11.3.2 Discretization of the Eliminated System

Instead of discretizing (11.53) and then eliminating the approximation of the Lagrange multiplier to obtain (11.68), one can directly discretize the eliminated system (11.51) or, equivalently, minimize the functional $K_\varepsilon(\cdot,\cdot)$ over $\{\phi^h, \theta^h\} \in \Phi^h \times \Theta^h$. Choosing approximating subspaces $\Phi^h \subset \Phi$ and $\Theta^h \subset \Theta$, the discrete problem is then given by

$$\begin{cases} a_1(\phi_\varepsilon^h, \mu^h) + \dfrac{1}{\varepsilon}\big(\mathcal{B}_1\mu^h, (\mathcal{B}_1\phi_\varepsilon^h + \mathcal{B}_2\theta_\varepsilon^h)\big)_{\Lambda^*} \\ \qquad = a_1(\widehat{\phi}, \mu^h) + \dfrac{1}{\varepsilon}\big(\mathcal{B}_1\mu^h, g\big)_{\Lambda^*} & \forall \mu^h \in \Phi^h \\ a_2(\theta_\varepsilon^h, \nu^h) + \dfrac{1}{\varepsilon}\big(\mathcal{B}_2\nu^h, (\mathcal{B}_1\phi_\varepsilon^h + \mathcal{B}_2\theta_\varepsilon^h)\big)_{\Lambda^*} \\ \qquad = \dfrac{1}{\varepsilon}\big(\mathcal{B}_2\nu^h, g\big)_{\Lambda^*} & \forall \nu^h \in \Theta^h. \end{cases} \tag{11.72}$$

This system can be written in the more compact form

$$\mathcal{Q}_\varepsilon(\{\phi_\varepsilon^h, \theta_\varepsilon^h\}, \{\mu^h, \nu^h\}) = \mathcal{F}_\varepsilon(\{\mu^h, \nu^h\}) \qquad \forall \{\mu^h, \nu^h\} \in \Phi^h \times \Theta^h, \tag{11.73}$$

where the bilinear form $\mathcal{Q}_\varepsilon(\cdot,\cdot)$ and linear functional $\mathcal{F}_\varepsilon(\cdot)$ are defined in (11.59) and (11.60), respectively.

In the usual way, the discrete system (11.72) is equivalent to a matrix problem. In addition to the matrices A_1 and A_2 and the vector $\vec{f}$ defined in Section 11.2.1 and the matrix $\widetilde{\mathsf{B}}_1$ defined in Section 11.1.2, we define the matrices

$$\begin{cases} (\widetilde{\mathsf{B}}_2)_{jk} = \widetilde{b}_2(\theta_k, \phi_j) = \big(\mathcal{B}_2\theta_k, \mathcal{B}_1\phi_j\big)_{\Lambda^*} & \text{for } k = 1,\dots,K,\ j = 1,\dots,J \\ (\widetilde{\mathsf{C}})_{k\ell} = \big(\mathcal{B}_2\theta_k, \mathcal{B}_2\theta_\ell\big)_{\Lambda^*} & \text{for } k,\ell = 1,\dots,K \end{cases}$$

and the vectors

$$\begin{cases} (\vec{g}_1)_j = \big(\mathcal{B}_1\phi_j, g\big)_{\Lambda^*} & \text{for } j = 1,\dots,J \\ (\vec{g}_2)_k = \big(\mathcal{B}_2\theta_k, g\big)_{\Lambda^*} & \text{for } k = 1,\dots,K, \end{cases}$$

where $\{\phi_j^h\}_{j=1}^J$ and $\{\theta_k\}_{k=1}^K$ denote the chosen basis sets for Φ^h and Θ^h, respectively. Then, (11.72) is equivalent to the matrix problem

$$\begin{pmatrix} \mathsf{A}_1 + \dfrac{1}{\varepsilon}\widetilde{\mathsf{B}}_1 & \dfrac{1}{\varepsilon}\widetilde{\mathsf{B}}_2 \\ \dfrac{1}{\varepsilon}\widetilde{\mathsf{B}}_2^T & \mathsf{A}_2 + \dfrac{1}{\varepsilon}\widetilde{\mathsf{C}} \end{pmatrix} \begin{pmatrix} \vec{\phi}_\varepsilon \\ \vec{\theta}_\varepsilon \end{pmatrix} = \begin{pmatrix} \vec{f} + \dfrac{1}{\varepsilon}\vec{g}_1 \\ \dfrac{1}{\varepsilon}\vec{g}_2 \end{pmatrix}, \tag{11.74}$$

where $\vec{\phi}_\varepsilon$ and $\vec{\theta}_\varepsilon$ are the vectors of coefficients for ϕ_ε^h and θ_ε^h, respectively.

It is clear that (11.68) and (11.74) are different, i.e., the discretize-then-eliminate approach yields a different, non-equivalent discrete system from that obtained using the eliminate-then-discretize approach, despite the fact that their respective parent continuous systems (11.53) and (11.51) are equivalent.

The following error estimate is derived using well-known techniques.

Theorem 11.24 *Let* (11.3) *and* (11.5) *hold. Then, for* $0 < \varepsilon \le 1$, (11.72), *or equivalently,* (11.73) *has a unique solution* $\{\phi_\varepsilon^h, \theta_\varepsilon^h\} \in \Phi^h \times \Theta^h$. *Moreover, if* $\{\phi, \theta\} \in \Phi \times \Theta$ *denotes the unique solution of the optimization problem* (11.6) *or equivalently, of* (11.51), *or equivalently, of* (11.53), *then there exist a constant* $C > 0$ *whose value is independent of* ε *and h such that*

$$\begin{aligned}\|\phi - \phi_\varepsilon^h\|_\Phi + \|\theta - \theta_\varepsilon^h\|_\Theta &\le C\varepsilon\left(\|g\|_{\Lambda^*} + \|\widehat{\phi}\|_{\widehat{\Phi}}\right)\\ &+C\left(1+\frac{1}{\varepsilon}\right)\left(\inf_{\widetilde{\phi}^h\in\Phi^h}\|\phi_\varepsilon - \widetilde{\phi}^h\|_\Phi + \inf_{\widetilde{\theta}^h\in\Theta^h}\|\theta_\varepsilon - \widetilde{\theta}^h\|_\Theta\right).\end{aligned} \tag{11.75}$$

Proof. Because of Lemma 11.20, the existence and uniqueness of the solution of (11.73) follows from the Lax–Milgram lemma. Moreover, standard finite element analyses for the problem (11.58) and its discretization (11.73) yield that

$$\|\phi_\varepsilon - \phi_\varepsilon^h\|_\Phi + \|\theta_\varepsilon - \theta_\varepsilon^h\|_\Theta \le C\left(1+\frac{1}{\varepsilon}\right)\left(\inf_{\widetilde{\phi}^h\in\Phi^h}\|\phi_\varepsilon - \widetilde{\phi}^h\|_\Phi + \inf_{\widetilde{\theta}^h\in\Theta^h}\|\theta_\varepsilon - \widetilde{\theta}^h\|_\Theta\right).$$

Then, (11.55) and the triangle inequality yields (11.75). □

Remark 11.25 Note that (11.74) is determined without the need for choosing a subspace $\Lambda^h \subset \Lambda$ for the approximation of the Lagrange multiplier. As a result, unlike the case for (11.68), for a fixed value of ε, the stable invertibility of the system (11.74) does not require the state approximation space Φ^h to satisfy (11.30). In fact, because of (11.61) and (11.62), for a fixed value of ε, the coefficient matrix in (11.74) is uniformly (with respect to h) positive definite for any choices for Φ^h and Θ^h. Correspondingly, the results of Theorem 11.24 also do not require that the discrete inf–sup conditions in (11.30) holds. □

Remark 11.26 The approximation-theoretic terms on the right-hand side of (11.75) satisfy inequalities of the type (11.70). Then, (11.75) implies that

$$\|\phi - \phi_\varepsilon^h\|_\Phi + \|\theta - \theta_\varepsilon^h\|_\Theta \le C\left(\varepsilon + \frac{h^r + h^s}{\varepsilon}\right), \tag{11.76}$$

where the value of $C > 0$ is independent of h and ε. The estimate (11.76) shows that nothing ontoward happens as $h \to 0$ for fixed ε. In fact, as $h \to 0$, the error in ϕ_ε^h and θ_ε^h is of order ε which is the best one can hope for a fixed value of ε.

However, (11.76) suggests that something bad may[10] happen as $\varepsilon \to 0$. In fact, this effect is well known as *locking* and indeed does happen for at least some choices of Φ^h; see, e.g., [76] and Remark 2.2 for a discussion of locking phenomena. Thus, to be safe, (11.76) suggests that as $\varepsilon \to 0$, h should be chosen to depend on ε in such a way that the right-hand side tends to zero as ε and h tend to zero. For example, if $s \geq r$, as is often the case, then to equilibrate the two terms in the right-hand side of (11.76), we choose $h = \varepsilon^{2/r}$ so that

$$\|\phi - \phi_\varepsilon^h\|_\Phi + \|\theta - \theta_\varepsilon^h\|_\Theta \leq C\varepsilon = Ch^{r/2}.$$

In this case, convergence is guaranteed for any choice for Φ^h and Θ^h, but the rate of convergence (with respect to h) may be suboptimal. This should be compared to the results for the discretization of the perturbed optimality system (see (11.71)) for which optimal rates of convergence with respect to h are obtained and locking does not occur. Of course, the estimate (11.71) requires that the finite element spaces satisfy the discrete stability conditions in (11.30), whereas the estimate (11.76) holds without the need to impose those stability conditions. □

11.4 Methods Based on Constraining by the Least-Squares Functional

Another means of incorporating least-squares notions into a solution method for the constrained optimization problem (11.6) is to solve, instead of (11.6) or its equivalent form (11.17), the bilevel minimization problem

$$\min_{\{\phi,\theta\}\in\Phi\times\Theta} K(\phi,\theta) \qquad \text{subject to} \qquad \min_{\{\phi,\theta\}\in\Phi\times\Theta} J_c(\phi;\theta,g), \tag{11.77}$$

where $J_c(\cdot;\cdot,\cdot)$ is the least-squares functional defined in (11.18). Thus, instead of imposing the constraint equation (11.16) through the Galerkin weak formulation (11.4), we now impose that constraint through the least-squares principle (11.18). From (11.19), one sees that (11.77) is equivalent to the problem

$$\begin{cases} \min_{\{\phi,\theta\}\in\Phi\times\Theta} K(\phi,\theta) \qquad \text{subject to} \\ \widetilde{b}_1(\phi,\mu) + \widetilde{b}_2(\theta,\mu) = \widetilde{G}_1(\mu) \qquad \forall \mu \in \Phi, \end{cases} \tag{11.78}$$

where $\widetilde{b}_1(\cdot,\cdot)$, $\widetilde{b}_2(\cdot,\cdot)$, and $\widetilde{G}_1(\cdot)$ are defined in (11.20)–(11.22). The Euler–Lagrange equations corresponding to the minimization problem (11.78) are given, in operator form, by

[10] Because (11.76) only provides an upper bound for the error, it does not with certainty predict what happens as $\varepsilon \to 0$.

$$\begin{cases} \mathcal{A}_1\phi & & + \mathcal{B}_1^*\mathcal{R}_\Lambda^{-1}\mathcal{B}_1\mu = \mathcal{A}_1\widehat{\phi} & \text{in } \Phi^* \\ & \mathcal{A}_2\theta & + \mathcal{B}_2^*\mathcal{R}_\Lambda^{-1}\mathcal{B}_1\mu = 0 & \text{in } \Theta^* \\ \mathcal{B}_1^*\mathcal{R}_\Lambda^{-1}\mathcal{B}_1\phi + \mathcal{B}_1^*\mathcal{R}_\Lambda^{-1}\mathcal{B}_2\theta & & = \mathcal{B}_1^*\mathcal{R}_\Lambda^{-1}g & \text{in } \Phi^*, \end{cases} \tag{11.79}$$

where $\mu \in \Phi$ is the Lagrange multiplier introduced to enforce the constraint in (11.78).

The constrained optimization problem (11.78) should be contrasted with the constrained optimization problem (11.17). Both problems involve the same functional $K(\cdot,\cdot)$, but are constrained differently. As a result, the former leads to the optimality system (11.27) or, equivalently, (11.28), whereas the latter leads to the optimality system (11.79). Although both optimality systems are of saddle-point type, their internal structures are significantly different. For example, the operator $\mathcal{B}_1$ that plays a central role in (11.28) may be non-symmetric and indefinite; on the other hand, the operator $\widetilde{\mathcal{B}}_1 = \mathcal{B}_1^*\mathcal{R}_\Lambda^{-1}\mathcal{B}_1$ that plays the analogous role in (11.79) is always symmetric and positive definite whenever (11.5) holds. Note also that in (11.79), the Lagrange multiplier μ naturally belongs to the same space Φ as does the state variable ϕ whereas in (11.28), the Lagrange multiplier λ, in general, belongs to the different space Λ. These differences between constraining the objective functional $K(\cdot,\cdot)$ by a least-squares functional instead of a Galerkin form of the constraint equations lead to the important advantages for the former approach.

Penalization can be used to facilitate the solution of the system (11.79) in just the same way as (11.53) is related to (11.28). Consider the penalized functional

$$\widetilde{K}_\varepsilon(\phi,\theta) = K(\phi,\theta) + \frac{1}{2\varepsilon}\|\mathcal{B}_1^*\mathcal{R}_\Lambda^{-1}\mathcal{B}_1\phi + \mathcal{B}_1^*\mathcal{R}_\Lambda^{-1}\mathcal{B}_2\theta - \mathcal{B}_1^*\mathcal{R}_\Lambda^{-1}g\|_{\Phi^*}^2$$

and the unconstrained optimization problem

$$\min_{\{\phi,\theta\}\in\Phi\times\Theta} \widetilde{K}_\varepsilon(\phi,\theta). \tag{11.80}$$

In the same way that (11.54) resulted from the optimization problem (11.49), the problem (11.80) leads to the following regular perturbation of the optimality system (11.79):

$$\begin{cases} \mathcal{A}_1\phi_\varepsilon & & + \mathcal{B}_1^*\mathcal{R}_\Lambda^{-1}\mathcal{B}_1\mu_\varepsilon = \mathcal{A}_1\widehat{\phi} & \text{in } \Phi^* \\ & \mathcal{A}_2\theta_\varepsilon & + \mathcal{B}_2^*\mathcal{R}_\Lambda^{-1}\mathcal{B}_1\mu_\varepsilon = 0 & \text{in } \Theta^* \\ \mathcal{B}_1^*\mathcal{R}_\Lambda^{-1}\mathcal{B}_1\phi_\varepsilon + \mathcal{B}_1^*\mathcal{R}_\Lambda^{-1}\mathcal{B}_2\theta_\varepsilon & - & \varepsilon\mathcal{R}_\Phi\mu_\varepsilon = \mathcal{B}_1^*\mathcal{R}_\Lambda^{-1}g & \text{in } \Phi^*, \end{cases} \tag{11.81}$$

where $\mu_\varepsilon \in \Phi$ and $\mathcal{R}_\Phi : \Phi \to \Phi^*$ is again a Riesz operator.

Just as is the case for the problem (11.17), we have four ways to define discretized problems from which approximations to the solutions of (11.78) can be obtained. Three of the approaches are discussed in Section 11.4.1 to Section 11.4.3. A fourth approach is to use a LSFEM to discretize (11.79), just as (11.47) resulted from applying a LSFEM to (11.28). We do not further consider this approach because, having already used a least-square functional to constrain the objective functional, we do not need to further use a least-squares approach for discretizing the resulting optimality system. Indeed, the desirable properties of LSFEMs are recovered by one of the other approaches discussed below.

The three remaining discretization approaches for the optimization problem (11.78) are completely analogous to three of the discretization approaches previously discussed for the equivalent optimization problem (11.17). However, the resulting discrete systems corresponding to the optimization problem (11.78) have some very different and advantageous properties compared to the analogous discrete systems for the optimization problem (11.17).

11.4.1 Discretization of the Optimality System

A Galerkin finite element method can be used to effect a discretization of the optimality system (11.79) and leads to an indefinite discrete problem, just as is the case for (11.28) which led to the indefinite discrete problem (11.67). After choosing conforming subspaces[11] $\Phi^h \subset \Phi$ and $\Theta^h \subset \Theta$, discretization of the weak form of the unperturbed optimality system (11.79) yields the discrete system

$$\begin{pmatrix} \mathsf{A}_1 & 0 & \widetilde{\mathsf{B}}_1 \\ 0 & \mathsf{A}_2 & \widetilde{\mathsf{B}}_2^T \\ \widetilde{\mathsf{B}}_1 & \widetilde{\mathsf{B}}_2 & 0 \end{pmatrix} \begin{pmatrix} \vec{\phi} \\ \vec{\theta} \\ \vec{\mu} \end{pmatrix} = \begin{pmatrix} \vec{f} \\ \vec{0} \\ \vec{g}_1 \end{pmatrix}, \tag{11.82}$$

where the matrices A_1, A_2, $\widetilde{\mathsf{B}}_1$, and $\widetilde{\mathsf{B}}_2$ and the vectors $\vec{f}$ and $\vec{g}_1$ are as in (11.74).

The system (11.82) is necessarily indefinite, regardless of the properties of the matrix $\widetilde{\mathsf{B}}_1$. However, we have that the matrices $\widetilde{\mathsf{B}}_1$ and A_2 are symmetric and positive definite merely if (11.3) and (11.5) hold. Thus, the coefficient matrix in (11.82) is uniformly invertible with respect to h without any conditions on the discrete subspaces $\Phi^h \subset \Phi$ and $\Theta^h \subset \Theta$. This should be contrasted with the situation for (11.67) whose uniform invertibility required that the discrete spaces satisfy the discrete compatibility conditions in (11.30).

11.4.2 Discretize-Then-Eliminate Approach for the Perturbed Optimality System

The system (11.81) that results from a regular perturbation of the optimality system (11.79) can be discretized directly; subsequently, the discrete Lagrange multiplier

[11] Note that, in contrast to (11.67), there is no need to choose a third subspace $\Lambda^h \subset \Lambda$.

can be eliminated to obtain a discrete system that is analogous to the discrete system (11.68) that results from the regular perturbation (11.54) of (11.28).

Choosing conforming subspaces $\Phi^h \subset \Phi$ and $\Theta^h \subset \Theta$, discretization of the weak formulation corresponding to (11.81) results in the matrix problem[12]

$$\begin{pmatrix} \mathsf{A}_1 & 0 & \widetilde{\mathsf{B}}_1 \\ 0 & \mathsf{A}_2 & \widetilde{\mathsf{B}}_2^T \\ \widetilde{\mathsf{B}}_1 & \widetilde{\mathsf{B}}_2 & -\varepsilon \mathsf{R}_\Phi \end{pmatrix} \begin{pmatrix} \vec{\phi}_\varepsilon \\ \vec{\theta}_\varepsilon \\ \vec{\mu}_\varepsilon \end{pmatrix} = \begin{pmatrix} \vec{f} \\ \vec{0} \\ \vec{g}_1 \end{pmatrix}, \tag{11.83}$$

where the matrix R_Φ corresponds to the bilinear form $(\phi,\mu)_\Phi = \langle \mathcal{R}_\Phi \mu, \phi\rangle_{\Phi^*,\Phi}$ for $\phi,\mu \in \Phi$. The system (11.83) is symmetric and indefinite, but it is uniformly (with respect to h) invertible without regard to (11.30). Indeed, we again have that the matrices $\widetilde{\mathsf{B}}_1$ and A_2 are symmetric and positive definite whenever (11.3) and (11.5) hold. This should be contrasted with the situation for (11.67) whose uniform invertibility required that the discrete spaces satisfy the discrete compatibility conditions contained in (11.30).

The vector of coefficients $\vec{\mu}_\varepsilon$ may be eliminated from (11.83) to yield

$$\begin{cases} \left(\mathsf{A}_1 + \dfrac{1}{\varepsilon}\widetilde{\mathsf{B}}_1 \mathsf{R}_\Phi{}^{-1}\widetilde{\mathsf{B}}_1\right)\vec{\phi}_\varepsilon + \dfrac{1}{\varepsilon}\widetilde{\mathsf{B}}_1 \mathsf{R}_\Phi{}^{-1}\widetilde{\mathsf{B}}_2\vec{\theta}_\varepsilon = \vec{f} + \dfrac{1}{\varepsilon}\widetilde{\mathsf{B}}_1 \mathsf{R}_\Phi{}^{-1}\vec{g}_1 \\ \left(\mathsf{A}_2 + \dfrac{1}{\varepsilon}\widetilde{\mathsf{B}}_2^T \mathsf{R}_\Phi{}^{-1}\widetilde{\mathsf{B}}_2\right)\vec{\theta}_\varepsilon + \dfrac{1}{\varepsilon}\widetilde{\mathsf{B}}_2^T \mathsf{R}_\Phi{}^{-1}\widetilde{\mathsf{B}}_1\vec{\phi}_\varepsilon = \dfrac{1}{\varepsilon}\widetilde{\mathsf{B}}_2^T \mathsf{R}_\Phi{}^{-1}\vec{g}_1 \,. \end{cases} \tag{11.84}$$

The following results can be obtained using standard techniques applied to the systems (11.82) and (11.83).

Theorem 11.27 *Let* (11.3) *and* (11.5) *hold. Then,* (11.83) *has a unique solution* $\{\vec{\phi}_\varepsilon, \vec{\theta}_\varepsilon, \vec{\mu}_\varepsilon\}$. *Let* $\{\phi,\theta,\mu\} \in \Phi\times\Theta\times\Phi$ *denote the unique solution of the optimization problem* (11.77) *so that* $\{\phi,\theta\}$ *also uniquely solves the optimization problem* (11.6). *Let* $\{\phi_\varepsilon^h,\theta_\varepsilon^h,\mu_\varepsilon^h\} \in \Phi^h\times\Theta^h\times\Phi^h$ *denote the finite element functions corresponding to the coefficient vectors* $\vec{\phi}_\varepsilon$, $\vec{\theta}_\varepsilon$, *and* $\vec{\mu}_\varepsilon$. *Then, there exists a constant* $C>0$ *whose value is independent of* ε *and* h *such that*

$$\begin{aligned} &\|\phi-\phi_\varepsilon^h\|_\Phi + \|\theta-\theta_\varepsilon^h\|_\Theta + \|\mu-\mu_\varepsilon^h\|_\Phi \le C\varepsilon\left(\|g\|_{\Lambda^*} + \|\widehat{\phi}\|_{\widehat{\Phi}}\right) \\ &+C\Big(\inf_{\widetilde{\phi}^h\in\Phi^h}\|\phi_\varepsilon-\widetilde{\phi}^h\|_\Phi + \inf_{\widetilde{\theta}^h\in\Theta^h}\|\theta_\varepsilon-\widetilde{\theta}^h\|_\Theta + \inf_{\widetilde{\mu}^h\in\Phi^h}\|\mu_\varepsilon-\mu^h\|_\Phi\Big). \qquad \square \end{aligned} \tag{11.85}$$

Using (11.70), we have from (11.85) that

$$\|\phi-\phi_\varepsilon^h\|_\Phi + \|\theta-\theta_\varepsilon^h\|_\Theta + \|\mu-\mu_\varepsilon^h\|_\Phi \le C(\varepsilon+h^r+h^s) \tag{11.86}$$

so that if $r\ge s$ and one chooses $\varepsilon = h^r$, one obtains the optimal error estimate

[12] We recognize that the discrete system (11.83) is a regular perturbation of the system (11.82).

$$\|\phi - \phi_\varepsilon^h\|_\Phi + \|\theta - \theta_\varepsilon^h\|_\Theta + \|\mu - \mu_\varepsilon^h\|_\Phi \le C\varepsilon = Ch^r . \tag{11.87}$$

Note that unlike for Theorem 11.21, the results of Theorem 11.27 do not require that (11.30) be satisfied. Also, unlike for Theorem 11.24, we get better convergence rates and locking cannot occur.

11.4.3 Eliminate-Then-Discretize Approach for the Perturbed Optimality System

The Lagrange multiplier $\mu \in \Phi$ can be eliminated from the perturbed optimality system (11.81) after which a finite element discretization step can be applied. This process is analogous to how the discrete system (11.74) is deduced from (11.81) and in the present setting results in the following linear system of algebraic equations:

$$\begin{pmatrix} \mathsf{A}_1 + \frac{1}{\varepsilon}\widetilde{\mathsf{K}}_1 & \frac{1}{\varepsilon}\widetilde{\mathsf{K}}_2 \\ \frac{1}{\varepsilon}\widetilde{\mathsf{K}}_2^T & \mathsf{A}_2 + \frac{1}{\varepsilon}\widetilde{\mathsf{C}} \end{pmatrix} \begin{pmatrix} \vec{\phi}_\varepsilon \\ \vec{\theta}_\varepsilon \end{pmatrix} = \begin{pmatrix} \vec{f} + \frac{1}{\varepsilon}\widetilde{\vec{g}}_1 \\ \frac{1}{\varepsilon}\widetilde{\vec{g}}_2 \end{pmatrix} . \tag{11.88}$$

Details about the definitions of the matrices and vectors appearing in (11.88) can be found in [60]; we do not provide them here because, as discussed in Section 11.5, this particular approach is not competitive. However, we point out that (11.84) and (11.88) are not the same, even though they represent discretizations of equivalent formulations of the optimization problem (11.6). However, the coefficient matrices of both discrete systems are symmetric and uniformly (with respect to h) positive definite without regard to (11.30). This is different from the situation for the equivalent systems (11.51) and (11.54) that respectively led to the discrete systems (11.74) and (11.67) whose uniform invertibility required the condition (11.30).

The following results can be obtained using standard techniques; again, see [60] for details.

Theorem 11.28 *Let* (11.3) *and* (11.5) *hold. Then, for* $0 < \varepsilon \le 1$, (11.88) *has a unique solution* $\vec{\phi}_\varepsilon$ *and* $\vec{\theta}_\varepsilon$. *Let* $\{\phi, \theta\} \in \Phi \times \Theta$ *denote the unique solution of the optimization problem* (11.6) *and let* $\{\phi_\varepsilon^h, \theta_\varepsilon^h\} \in \Phi^h \times \Theta^h$ *denote the finite element functions corresponding to the coefficient vectors* $\vec{\phi}_\varepsilon$ *and* $\vec{\theta}_\varepsilon$. *Then, there exists a constant* $C > 0$ *whose value is independent of* ε *and* h *such that*

$$\begin{aligned} \|\phi - \phi_\varepsilon^h\|_\Phi + \|\theta - \theta_\varepsilon^h\|_\Theta &\le C\varepsilon \left(\|g\|_{\Lambda^*} + \|\widehat{\phi}\|_{\widehat{\Phi}} \right) \\ &+ C\Big(1 + \frac{1}{\varepsilon}\Big)\Big(\inf_{\widetilde{\phi}^h \in \Phi^h} \|\phi_\varepsilon - \widetilde{\phi}^h\|_\Phi + \inf_{\widetilde{\theta}^h \in \Theta^h} \|\theta_\varepsilon - \widetilde{\theta}^h\|_\Theta \Big) . \end{aligned} \tag{11.89}$$

□

11.5 Relative Merits of the Different Approaches

In the preceding sections, we have discussed several ways to incorporate least-squares ideas into optimal control problems. We now provide a summary list of the various possibilities. In addition to the various least-squares-related methods, we include the standard approach of applying a Galerkin finite element method to the optimality system obtained after applying the Lagrange multiplier rule to the optimization problem; see Section 11.2.1.

0. Lagrange multiplier rule applied to the optimization problem followed by a mixed-Galerkin finite element discretization of the resulting optimality system; see Section 11.2.1.

1. Lagrange multiplier rule applied to the optimization problem followed by a least-squares formulation of the resulting optimality system followed by a finite element discretization; see Section 11.2.2.

2. Lagrange multiplier rule applied to the optimization problem followed by a penalty perturbation of the resulting optimality system followed by a finite element discretization followed by the elimination of the discrete Lagrange multiplier; see Section 11.3.1.

3. Penalization of the cost functional by a least-squares functional followed by optimization followed by a finite element discretization of the resulting optimality equations; see Section 11.3.2.

4. Constraining the cost functional by a least-squares formulation of the state equations to obtain a modified optimization problem followed by the Lagrange multiplier rule to obtain an optimality system followed by a finite element discretization; see Section 11.4.1.

5. Constraining the cost functional by a least-squares formulation of the state equations to obtain a modified optimization problem followed by the Lagrange multiplier rule followed by a penalty perturbation of the resulting optimality system followed by a finite element discretization followed by the elimination of the discrete Lagrange multiplier; see Section 11.4.2.

6. Constraining the cost functional by a least-squares formulation of the state equations to obtain a modified optimization problem followed by penalization of the cost functional followed by optimization followed by a finite element discretization of the resulting optimality equations; see Section 11.4.3.

In Table 11.1, we compare the seven methods just listed with respect to several desirable properties. The properties are posed in the form of the following questions:

discrete inf–sup not required – are the finite element spaces required to satisfy (11.30) in order that the resulting discrete systems be stably invertible as $h \to 0$?

locking impossible – is it possible to guarantee that the discrete systems are stably invertible as $\varepsilon \to 0$ with fixed h?

optimal error estimate – are optimal estimates for the error in the approximate solutions obtainable, possibly after choosing ε to depend on h?

symmetric matrix system – are the discrete systems symmetric?

reduced number of unknowns – is it possible to eliminate unknowns to obtain a smaller discrete system?

positive definite matrix system – do the discrete systems, possibly after the elimination of unknowns, have a positive definite coefficient matrix?

	Method						
Property↓	0	1	2	3	4	5	6
Discrete inf–sup not required	–	√	–	√	√	√	√
Locking impossible	√	√	√	–	√	√	–
Optimal error estimate	√	√	√	–	√	√	–
Symmetric matrix system	√	√	√	√	√	√	√
Reduced number of unknowns	–	–	√	√	–	√	√
Positive definite matrix system	–	√	√	√	–	√	√

Table 11.1 Properties of different approaches that use least-squares notions for the approximate solution of the optimization problem (11.6).

From Table 11.1, we see that only approach 5 has all its boxes checked, so that as far as the six properties used for comparison purposes in that table, that approach seems preferable. Method 1 is also a good candidate, suffering only from an increased number of unknowns due to the need to introduce Lagrange multipliers. On the other hand, unlike Method 5, Method 1 does not involve a penalty parameter which may be advantageous when using iterative solution methods.

11.6 Example: Optimization Problems for the Stokes Equations

We illustrate the application of least-squares notions to optimization and control problems constrained by PDEs by considering two optimization problems for which the constraint equations are the Stokes system. We focus on Methods 1 and 5 from among the methods listed in Section 11.5 (see Sections 11.2.2 and 11.4.2, respectively) because those are the most promising.

For illustrative purposes, we only consider the velocity–vorticity–pressure (VVP) first-order formulation of the Stokes problem given by (7.4) along with the velocity boundary condition (7.2) and the zero mean pressure constraint (7.3). For simplicity, we also only consider distributed controls, i.e., the control acts through the body force term in (7.4). Furthermore, we have to define functionals whose minimization is the objective of control. Specifically, in the control and optimization setting we study here, we let Ω denote an open bounded domain in $\mathbb{R}^d$, $d = 2$ or 3, with boundary $\partial\Omega$.[13] As in Chapter 7, the symbols $\mathbf{u}$, ω, and p stand for the velocity, vorticity[14], and pressure fields, respectively, and θ denotes a distributed control function. We then consider the VVP Stokes system (7.4) with the distributed control function included in the momentum equation

$$\begin{cases} \nabla \times \omega + \nabla p + \theta = \mathbf{g} & \text{in } \Omega \\ \nabla \cdot \mathbf{u} = 0 & \text{in } \Omega \\ \nabla \times \mathbf{u} - \omega = 0 & \text{in } \Omega \\ \mathbf{u} = 0 & \text{on } \partial\Omega, \end{cases} \quad \text{and} \quad \int_\Omega p \, d\Omega = 0, \tag{11.90}$$

and the following objective functionals

$$\text{Case I:} \quad K_1(\omega, \theta) = \frac{1}{2}\int_\Omega |\omega|^2 \, d\Omega + \frac{\delta}{2}\int_\Omega |\theta|^2 \, d\Omega \tag{11.91}$$

$$\text{Case II:} \quad K_2(\mathbf{u}, \theta; \widehat{\mathbf{u}}) = \frac{1}{2}\int_\Omega |\mathbf{u} - \widehat{\mathbf{u}}|^2 \, d\Omega + \frac{\delta}{2}\int_\Omega |\theta|^2 \, d\Omega \,, \tag{11.92}$$

where $\mathbf{g}$ and $\widehat{\mathbf{u}}$ are given functions. Note that, because $\mathbf{u} = 0$ on $\partial\Omega$ and $\nabla \cdot \mathbf{u} = 0$ in Ω, the functional (11.91) can be written as

$$K_1(\mathbf{u}, \theta) = \frac{1}{2}\int_\Omega |\nabla \mathbf{u}|^2 \, d\Omega + \frac{\delta}{2}\int_\Omega |\theta|^2 \, d\Omega \,. \tag{11.93}$$

The optimization problems that we use as examples are to find $\{\mathbf{u}, \omega, p, \theta\}$ that minimizes either of the functionals in (11.91) or (11.92), subject to the Stokes system in the velocity–vorticity–pressure form (11.90) being satisfied. In the first case, i.e., for the functional (11.91), the problem we study is to find a distributed control function θ that minimizes, in the $[L^2(\Omega)]^{d(d-1)/2}$ sense, the vorticity, or, by (11.93), the gradient of the velocity vector over the flow domain Ω; it can be shown, e.g., see [179], that this is equivalent to minimizing the drag of the flow. In the second case, i.e., for the functional (11.92), the problem we study is to find a distributed control function θ such that the flow velocity $\mathbf{u}$ matches as well as possible, in the $[L^2(\Omega)]^d$ sense, a given velocity field $\widehat{\mathbf{u}}$. As mentioned in conjunction with the functional (11.2), the second terms in the functionals (11.91) and (11.92) are penalty terms that are used to limit the size of the control θ.

---

[13] We assume that Ω is a convex polygon in $\mathbb{R}^2$ or a convex polyhedron in $\mathbb{R}^3$. This assumption substantially simplifies some of the proofs given below.

[14] Recall that vorticity field has $d(d-1)/2$ components in $\mathbb{R}^d$, $d = 2, 3$, i.e., it is a scalar function in $\mathbb{R}^2$ and a vector field in $\mathbb{R}^3$; see Section 7.1.1. Because in this section we treat the two space dimension cases simultaneously, for brevity we use the same notation $\nabla\times$ to denote the curl operator acting on ω.

Insofar as least-squares principles for the state equations (11.90) are concerned, from Section 7.3 we know that the Agmon-Douglis-Nirenberg (ADN) setting for the energy balances is more appropriate for VVP systems augmented with the velocity boundary condition. As a result, the least-squares notions used in this section to solve optimization problems constrained by (11.90) correspond to that setting.[15]

11.6.1 The Optimization Problems and Galerkin Finite Element Methods

To cast the optimization problems studied in this section in the abstract framework of Section 11.1, we make the identifications

$$\Phi = \Lambda = [H_0^1(\Omega)]^d \times [L^2(\Omega)]^{d(d-1)/2} \times L_0^2(\Omega) \quad \text{and} \quad \Theta = [L^2(\Omega)]^d$$

so that

$$\Phi^* = \Lambda^* = [H^{-1}(\Omega)]^d \times [L^2(\Omega)]^{d(d-1)/2} \times L_0^2(\Omega) \quad \text{and} \quad \Theta^* = [L^2(\Omega)]^d .$$

Let

$$\widehat{\Phi} = \widetilde{\Phi} = [L^2(\Omega)]^d \times [L^2(\Omega)]^{d(d-1)/2} \times L_0^2(\Omega) .$$

Then, $\Phi \subset \widehat{\Phi} = \widetilde{\Phi} = \widehat{\Phi}^* \subset \Phi^*$. For $\phi = \{\mathbf{u}, \omega, p\} \in \Phi$, we define the norm

$$\|\phi\|_\Phi = \left(\|\mathbf{u}\|_1^2 + \|\omega\|_0^2 + \|p\|_0^2\right)^{1/2}$$

and likewise for the other product spaces. Furthermore, we make the associations of

trial functions:	$\phi = \{\mathbf{u}, \omega, p\} \in \Phi$	$\theta = \{\theta\} \in \Theta$	$\lambda = \{\mathbf{v}, \xi, q\} \in \Lambda$
test functions:	$\mu = \{\widetilde{\mathbf{u}}, \widetilde{\omega}, \widetilde{p}\} \in \Phi$	$\nu = \{\widetilde{\theta}\} \in \Theta$	$\psi = \{\widetilde{\mathbf{v}}, \widetilde{\xi}, \widetilde{r}\} \in \Lambda$
data:	$g = \{\mathbf{g}, 0, 0\} \in \Lambda^*$	$\widehat{\phi} = \{\widehat{\mathbf{u}}, 0, 0\} \in \widehat{\Phi}$.	

We next define the bilinear forms corresponding to the objective functional:

$$a_1(\phi, \mu) = \begin{cases} (\omega, \widetilde{\omega})_0 & \text{for Case I} \\ (\mathbf{u}, \widetilde{\mathbf{u}})_0 & \text{for Case II} \end{cases} \qquad \forall \phi = \{\mathbf{u}, \omega, p\} \in \widehat{\Phi}, \quad \mu = \{\widetilde{\mathbf{u}}, \widetilde{\omega}, \widetilde{p}\} \in \widehat{\Phi}$$

$$a_2(\theta, \nu) = \delta(\theta, \widetilde{\theta})_0 \qquad \forall \theta = \{\theta\} \in \Theta, \quad \nu = \{\widetilde{\theta}\} \in \Theta ,$$

and the bilinear forms corresponding to the state equations:

---

[15] Recall that the ADN setting views all variables as sets of independent scalar fields which leads to identification of, e.g., the least-squares minimization space X with products of scalar Sobolev spaces $H^{q+t_j}(\Omega)$; see Remarks 3.6 and 5.6 for additional discussion of the distinctions between the two settings for the energy balances.

$$b_1(\phi,\psi) = \big\langle \nu\nabla\times\omega + \nabla p, \widetilde{\mathbf{v}}\big\rangle_{[H^{-1}(\Omega)]^d,[H^1_0(\Omega)]^d} + \big(\nabla\times\mathbf{u} - \omega, \widetilde{\xi}\big)_0 - \big(\nabla\cdot\mathbf{u}, \widetilde{r}\big)_0$$
$$\forall \phi = \{\mathbf{u},\omega,p\}\in\Phi, \quad \psi = \{\widetilde{\mathbf{v}},\widetilde{\xi},\widetilde{r}\}\in\Lambda$$

$$b_2(\theta,\psi) = \big(\theta\,,\widetilde{\mathbf{v}}\big)_0 \qquad \forall \theta = \{\theta\}\in\Theta, \quad \psi = \{\widetilde{\mathbf{v}},\widetilde{\xi},\widetilde{r}\}\in\Lambda\,.$$

For $\mathbf{g}\in[H^{-1}(\Omega)]^d$, we also define the linear functional

$$G(\psi) = \big\langle \mathbf{g}\,,\widetilde{\mathbf{v}}\big\rangle_{[H^{-1}(\Omega)]^d,[H^1_0(\Omega)]^d} \qquad \forall \psi = \{\widetilde{\mathbf{v}},\widetilde{\xi},\widetilde{r}\}\in\Lambda\,.$$

The operators associated with the bilinear forms are then

$$\mathcal{A}_1 = \begin{cases} \begin{pmatrix} 0 & 0 & 0 \\ 0 & \mathcal{I} & 0 \\ 0 & 0 & 0 \end{pmatrix} & \text{for Case I} \\ \begin{pmatrix} \mathcal{I} & 0 & 0 \\ 0 & 0 & 0 \\ 0 & 0 & 0 \end{pmatrix} & \text{for Case II} \end{cases} \qquad \mathcal{A}_2 = \delta\mathcal{I} \tag{11.94}$$

$$\mathcal{B}_1 = \begin{pmatrix} 0 & \nabla\times & \nabla \\ \nabla\times & -\mathcal{I} & 0 \\ -\nabla\cdot & 0 & 0 \end{pmatrix} \qquad \mathcal{B}_2 = \begin{pmatrix} \mathcal{I} \\ 0 \\ 0 \end{pmatrix} \qquad g = \begin{pmatrix} \mathbf{g} \\ 0 \\ 0 \end{pmatrix}.$$

It is now easily seen that the functionals $K_1(\cdot,\cdot)$ and $K_2(\cdot,\cdot;\cdot)$ defined in (11.91) and (11.92), respectively, can be written in the form (11.2). Likewise, the Stokes system (11.90) can be written in the form (11.4). Thus, the two optimization problems for the Stokes system can both be written in the form (11.6), with $K(\cdot,\cdot)$ being either $K_1(\cdot,\cdot)$ or $K_2(\cdot,\cdot)$ as appropriate. We can also show that, for these problems, the assumptions (11.3) and (11.5) are satisfied.

Proposition 11.29 *Let the spaces Φ, $\widehat{\Phi}$, Θ, and Λ and the bilinear forms $a_1(\cdot,\cdot)$, $a_2(\cdot,\cdot)$, $b_1(\cdot,\cdot)$, and $b_2(\cdot,\cdot)$ be defined as above in this section. Then, the assumptions* (11.3) *and* (11.5) *are satisfied.*

Proof. The four inequalities in (11.3) and the first two inequalities in (11.5) are easily verified with $\alpha_{a_1}=1$, $\alpha_{a_2}=\delta$, $\beta_{a_2}=\delta$, $\alpha_{b_1}=3$, and $\alpha_{b_1}=1$. The third inequality in (11.5) is verified if, for any $\phi=\{\mathbf{u},\omega,p\}\in\Phi$, one can find a $\widetilde{\psi}=\{\widetilde{\mathbf{v}},\widetilde{\xi},\widetilde{r}\}\in\Lambda$ such that

$$b_1(\phi,\widetilde{\psi}) = \langle\nabla\times\omega+\nabla p,\widetilde{\mathbf{v}}\rangle_{[H^{-1}(\Omega)]^d,[H^1_0(\Omega)]^d} + (\nabla\times\mathbf{u}-\omega,\widetilde{\xi})_0 - (\nabla\cdot\mathbf{u},\widetilde{r})_0$$
$$\geq \beta_{b_1}\|\phi\|_\Phi\|\widetilde{\psi}\|_\Lambda = \beta_{b_1}\big(\|\mathbf{u}\|_1^2+\|\omega\|_0^2+\|p\|_0^2\big)^{1/2}\big(\|\widetilde{\mathbf{v}}\|_1^2+\|\widetilde{\xi}\|_0^2+\|\widetilde{r}\|_0^2\big)^{1/2}$$

for some constant $\beta_{b_1}>0$. To this end, for any $\phi=\{\mathbf{u},\omega,p\}\in\Phi$, let $\widetilde{\psi}=\{\widetilde{\mathbf{v}},\widetilde{\xi},\widetilde{r}\}\in\Lambda$ satisfy the system

$$\nabla \times \widetilde{\mathbf{v}} = \omega \qquad \nabla \cdot \widetilde{\mathbf{v}} = -p; \qquad \widetilde{\xi} = \nabla \times \mathbf{u} \qquad \widetilde{r} = -\nabla \cdot \mathbf{u} \tag{11.95}$$

in Ω. Recall that we have assumed that Ω is a convex polygon in $\mathbb{R}^2$ or a convex polyhedron in $\mathbb{R}^3$ so that, by Theorem A.8, there exist constants $\gamma_1 > 0$ and $\gamma_2 > 0$ such that

$$\gamma_1 \|\mathbf{v}\|_1^2 \leq \|\nabla \times \mathbf{v}\|_0^2 + \|\nabla \cdot \mathbf{v}\|_0^2 \leq \gamma_2 \|\mathbf{v}\|_1^2 \qquad \forall \mathbf{v} \in [H_0^1(\Omega)]^d . \tag{11.96}$$

Clearly, from (11.96) and the last two equations in (11.95) we have that

$$\|\widetilde{\xi}\|_0^2 + \|\widetilde{r}\|_0^2 = \|\nabla \times \mathbf{u}\|_0^2 + \|\nabla \cdot \mathbf{u}\|_0^2 \leq \gamma_2 \|\mathbf{u}\|_1^2 .$$

Also, because $\widetilde{\mathbf{v}} \in [H_0^1(\Omega)]^d$, we have from (11.96) and the first two equations in (11.95) that

$$\gamma_1 \|\widetilde{\mathbf{v}}\|_1^2 \leq \|\nabla \times \widetilde{\mathbf{v}}\|_0^2 + \|\nabla \cdot \widetilde{\mathbf{v}}\|_0^2 = \|\omega\|_0^2 + \|p\|_0^2 .$$

Combining the last two results yields that

$$\|\widetilde{\mathbf{v}}\|_1^2 + \|\widetilde{\xi}\|_0^2 + \|\widetilde{r}\|_0^2 \leq \max\left\{\gamma_1^{-1}, \gamma_2\right\}\left(\|\mathbf{u}\|_1^2 + \|\omega\|_0^2 + \|p\|_0^2\right) . \tag{11.97}$$

We also have that (see, e.g., [183]), for all $\widetilde{\mathbf{v}} \in [H_0^1(\Omega)]^d$, $\omega \in [L^2(\Omega)]^{d(d-1)/2}$, and $p \in L_0^2(\Omega)$,

$$\langle \nabla \times \omega + \nabla p, \widetilde{\mathbf{v}} \rangle_{[H^{-1}(\Omega)]^d, [H_0^1(\Omega)]^d} = (\omega, \nabla \times \widetilde{\mathbf{v}})_0 - (p, \nabla \cdot \widetilde{\mathbf{v}})_0$$

so that

$$b_1(\phi, \widetilde{\psi}) = (\omega, \nabla \times \widetilde{\mathbf{v}})_0 - (p, \nabla \cdot \widetilde{\mathbf{v}})_0 + (\nabla \times \mathbf{u} - \omega, \widetilde{\xi})_0 - (\nabla \cdot \mathbf{u}, \widetilde{r})_0 .$$

Then, with $\phi = \{\mathbf{u}, \omega, p\} \in \Phi$ and $\widetilde{\psi} = \{\widetilde{\mathbf{v}}, \widetilde{\xi}, \widetilde{r}\} \in \Lambda$, we have that

$$\begin{aligned}
b_1(\phi, \widetilde{\psi}) &= \|\omega\|_0^2 + \|p\|_0^2 + \|\nabla \times \mathbf{u}\|_0^2 + \|\nabla \cdot \mathbf{u}\|_0^2 - (\omega, \nabla \times \mathbf{u}) \\
&\geq \|\omega\|_0^2 + \|p\|_0^2 + \|\nabla \times \mathbf{u}\|_0^2 + \|\nabla \cdot \mathbf{u}\|_0^2 - \|\omega\|_0 \|\nabla \times \mathbf{u}\|_0 \\
&\geq \frac{1}{2}\|\omega\|_0^2 + \|p\|_0^2 + \frac{1}{2}\|\nabla \times \mathbf{u}\|_0^2 + \|\nabla \cdot \mathbf{u}\|_0^2 \\
&\geq \frac{1}{2}\min\{1, \gamma_1\}\left(\|\mathbf{u}\|_1^2 + \|\omega\|_0^2 + \|p\|_0^2\right) \\
&\geq \beta_{b_1}\left(\|\mathbf{u}\|_1^2 + \|\omega\|_0^2 + \|p\|_0^2\right)^{1/2}\left(\|\widetilde{\mathbf{v}}\|_1^2 + \|\widetilde{\xi}\|_0^2 + \|\widetilde{r}\|_0^2\right)^{1/2} \\
&= \beta_{b_1}\|\phi\|_\Phi \|\widetilde{\psi}\|_\Lambda ,
\end{aligned}$$

where $\beta_{b_1} = \min\{1, \gamma_1\} / \left(2(\max\left\{\frac{1}{\gamma_1}, \gamma_2\right\})^{1/2}\right)$ and where we have used (11.96) and (11.97). Thus, the third inequality in (11.5) is verified. Note that β_{b_1} depends only on the comparability constants in (11.96). □

Remark 11.30 We have verified the assumptions (11.3) and (11.5) for the two optimization problems that consist of finding $\{\mathbf{u}, \omega, p, \theta\}$ that minimize either the functional (11.91) or (11.92), subject to the Stokes system in the form (11.90) being satisfied. As a result, all the results of Sections 11.1, 11.2.2, and 11.4 apply to those problems as do the results of Sections 11.2 and 11.3 that require (11.3) and (11.5) but not (11.30). In particular, we have that each of the two optimization problems has a unique solution and that the Lagrange multiplier rule may be used to characterize their solutions as solutions of the optimality system (11.28). □

Remark 11.31 A Galerkin finite element method for determining approximate solutions of the optimality system (11.28) could be used. Such an approach, unlike least-squares finite element discretizations, does not involve the "squaring" of operators which interferes with the practicality requirements set forth in Section 2.2.2 so that there is no need to transform the standard primitive variable Stokes system

$$\begin{cases} -\Delta \mathbf{u} + \nabla p + \theta = \mathbf{g} & \text{in } \Omega \\ \nabla \cdot \mathbf{u} = 0 & \text{in } \Omega \\ \mathbf{u} = 0 & \text{on } \partial\Omega \end{cases} \qquad \text{and} \qquad \int_\Omega p \, d\Omega = 0 \tag{11.98}$$

into an equivalent first-order form as in (11.90); one would then also use the form (11.93) for the functional K_1 instead of the form (11.91). We then would have $\phi = \{\mathbf{u}, p\}$, $\theta = \{\theta\}$, and so on, and use, instead of the operators defined in (11.94), the operators

$$\begin{gathered} \mathcal{A}_1 = \begin{pmatrix} \nabla\times & 0 \\ 0 & 0 \end{pmatrix} \quad \text{for Case I} \qquad \mathcal{A}_1 = \begin{pmatrix} \mathcal{I} & 0 \\ 0 & 0 \end{pmatrix} \quad \text{for Case II} \\ \mathcal{A}_2 = \delta\mathcal{I} \qquad B_1 = \begin{pmatrix} -\Delta & \nabla \\ -\nabla\cdot & 0 \end{pmatrix} \quad \text{and} \quad B_2 = \begin{pmatrix} \mathcal{I} \\ 0 \end{pmatrix}. \end{gathered} \tag{11.99}$$

The assumptions (11.3) and (11.5) can also be verified for the bilinear forms associated with these operators.

As noted in Section 11.2.1, a Galerkin discretization of the optimality system (11.28) using either of the forms (11.94) or (11.99) for the operators requires that the assumptions in (11.30) hold. If one uses (11.99), the finite element spaces for the velocity and pressure approximations have to satisfy the inf–sup condition (1.42) which, for the Stokes system (11.98) is given by

$$\inf_{q^h \in S^h, q^h \neq 0} \sup_{\mathbf{v}^h \in \mathbf{V}^h, \mathbf{v}^h \neq \mathbf{0}} \frac{\int_\Omega q^h \nabla \cdot \mathbf{v}^h \, d\Omega}{\|q^h\|_0 \|\mathbf{v}^h\|_1} \geq \beta \tag{11.100}$$

for some constant $\beta > 0$. Theorem 1.11 asserts that this condition guarantees the unique solvability of the discrete Stokes system and restricts the choice of finite element spaces used for the velocity and pressure approximations; see Section 1.4.3

and [83, 87, 191] for details. In particular, one cannot use piecewise polynomial spaces of the same order and defined with respect to the same grid for the velocity and pressure approximations. If one instead uses (11.94), an even more onerous inf–sup condition is required of the finite element spaces for the velocity, vorticity, and pressure.

Note that (11.100) is a *third* level of discrete inf–sup conditions that we have encountered in our deliberations: (11.100) is necessary and sufficient to guarantee that the inf–sup condition (11.30) holds; the latter is necessary and sufficient to guarantee that the inf–sup condition (1.42) holds. □

11.6.2 Least-Squares Finite Element Methods for the Constraint Equations

Using the spaces, variables, bilinear forms, operators, and so on defined above, it is easy to see that the least-squares functional in (11.18) is given by, for the example problems we are considering,[16]

$$K_c(\{\mathbf{u},\omega,p\};\theta,\mathbf{g}) = \|\nabla\times\omega+\nabla p+\theta-\mathbf{g}\|_{-1}^2+\|\nabla\times\mathbf{u}-\omega\|_0^2+\|\nabla\cdot\mathbf{u}\|_0^2. \tag{11.101}$$

The Stokes system (11.90) can then be reformulated as an optimization problem; indeed, solutions of (11.90) are characterized as minimizers of the least-squares functional (11.101). The Euler–Lagrange equation corresponding to this minimization problem now takes the form: for given $\theta\in[L^2(\Omega)]^d$ and $\mathbf{g}\in[H^{-1}(\Omega)]^d$, find $\{\mathbf{u},\omega,p\}\in[H_0^1(\Omega)]^d\times[L^2(\Omega)]^{d(d-1)/2}\times L_0^2(\Omega)$ such that

$$\begin{aligned}\widetilde{b}_1\big(\{\mathbf{u},\omega,p\},\{\widetilde{\mathbf{u}},\widetilde{\omega},\widetilde{p}\}\big) &= \widetilde{G}_1(\{\widetilde{\mathbf{u}},\widetilde{\omega},\widetilde{p}\})-\widetilde{b}_2\big(\{\theta,\{\widetilde{\mathbf{u}},\widetilde{\omega},\widetilde{p}\}\big)\\ &\forall\{\widetilde{\mathbf{u}},\widetilde{\omega},\widetilde{p}\}\in[H_0^1(\Omega)]^d\times[L^2(\Omega)]^{d(d-1)/2}\times L_0^2(\Omega),\end{aligned}\tag{11.102}$$

where the bilinear forms $\widetilde{b}_1(\cdot,\cdot)$ and $\widetilde{b}_2(\cdot,\cdot)$ and $\widetilde{G}_1(\cdot)$ are defined in (11.20), (11.21), and (11.22), respectively. We then have the following results.

Proposition 11.32 *Let the spaces Φ, $\widehat{\Phi}$, Θ, and Λ and the bilinear forms $a_1(\cdot,\cdot)$, $a_2(\cdot,\cdot)$, $b_1(\cdot,\cdot)$, and $b_2(\cdot,\cdot)$ be defined as above. Let $K_c(\cdot,\cdot,\cdot;\cdot,\cdot)$ be defined by* (11.101). *Then, we have the norm equivalence result*

$$\gamma_1\big(\|\mathbf{u}\|_1^2+\|\omega\|_0^2+\|p\|_0^2\big)\le K_c(\mathbf{u},\omega,p;0,0)\le\gamma_2\big(\|\mathbf{u}\|_1^2+\|\omega\|_0^2+\|p\|_0^2\big) \tag{11.103}$$

[16] As discussed earlier in Chapter 7 the least-squares functional (11.101) does not satisfy all keys to practicality postulated in Section 2.2.2. The use of this functional allows us to keep the exposition simple and focus on issues relevant to the application of least-squares notions in optimization and control problems. Of course, to implement the methods in this section one would rely on the practical functionals discussed in Chapter 7, such as the quasi-norm-equivalent functional in (7.54) or its norm-equivalent cousin in (7.55).

for some constants $\gamma_1, \gamma_2 > 0$. *Moreover, the bilinear form* $\widetilde{b}_1(\{\cdot,\cdot,\cdot\},\{\cdot,\cdot,\cdot\})$ *is symmetric, continuous, and coercive and the problem* (11.102) *has a unique solution; that solution is the unique minimizer of the least-squares functional* (11.101).

Proof. The results follow from Lemma 11.4 and Propositions 11.3 and 11.29. □

To define LSFEM approximations of the constraint equations, we first choose conforming finite element subspaces $\mathbf{V}^h \subset [H^1_0(\Omega)]^d$, $\mathbf{W}^h \subset [L^2(\Omega)]^{d(d-1)/2}$, and $S^h \subset L^2_0(\Omega)$. We then minimize the functional in (11.101) over the subspaces, or equivalently, solve the problem: for given $\theta \in [L^2(\Omega)]^d$ and $\mathbf{g} \in [H^{-1}(\Omega)]^d$, find $\{\mathbf{u}^h, \omega^h, p^h\} \in \mathbf{V}^h \times \mathbf{W}^h \times S^h$ such that

$$\begin{aligned}
&\widetilde{b}_1\big(\{\mathbf{u}^h,\omega^h,p^h\},\{\widetilde{\mathbf{u}}^h,\widetilde{\omega}^h,\widetilde{p}^h\}\big)\\
&\qquad = \widetilde{G}_1(\{\widetilde{\mathbf{u}}^h,\widetilde{\omega}^h,\widetilde{p}^h\}) - \widetilde{b}_2\big(\{\theta,\{\widetilde{\mathbf{u}}^h,\widetilde{\omega}^h,\widetilde{p}^h\}\big)
\end{aligned} \tag{11.104}$$

for all $\{\widetilde{\mathbf{u}}^h,\widetilde{\omega}^h,\widetilde{p}^h\} \in \mathbf{V}^h \times \mathbf{W}^h \times S^h$. We then have the following results.

Proposition 11.33 *Let* $\mathbf{V}^h \subset [H^1_0(\Omega)]^d$, $\mathbf{W}^h \subset [L^2(\Omega)]^{d(d-1)/2}$, *and* $S^h \subset L^2_0(\Omega)$. *Then, the discrete problem* (11.104) *has a unique solution* $\{\mathbf{u}^h, \omega^h, p^h\} \in \mathbf{V}^h \times \mathbf{W}^h \times S^h$. *Let* $\{\mathbf{u}, \omega, p\} \in [H^1_0(\Omega)]^d \times [L^2(\Omega)]^{d(d-1)/2} \times L^2_0(\Omega)$ *denote the unique solution of* (11.102). *Then,*

$$\begin{aligned}
&\|\mathbf{u}-\mathbf{u}^h\|_1 + \|\omega-\omega^h\|_0 + \|p-p^h\|_0\\
&\quad \le C\Big(\inf_{\widetilde{\mathbf{u}}^h\in\mathbf{V}^h}\|\mathbf{u}-\widetilde{\mathbf{u}}^h\|_1 + \inf_{\widetilde{\omega}^h\in\mathbf{W}^h}\|\omega-\widetilde{\omega}^h\|_0 + \inf_{\widetilde{p}^h\in S^h}\|p-\widetilde{p}^h\|_0\Big).
\end{aligned} \tag{11.105}$$

Proof. The results follow in a straightforward manner from Proposition 11.32. □

11.6.3 Least-Squares Finite Element Methods for the Optimality Systems

We now consider Method 1 from the list in Section 11.5 and which is discussed in Section 11.2.2. Using the associations of spaces and variables defined in Section 11.6.1 and the operators defined in (11.94), it is easy to see that the least-squares funtional in (11.34) is given by, for the example problems we are considering,

$$\begin{aligned}
K\big(\{\mathbf{u},\omega,p\},\theta,\{\mathbf{v},\xi,q\};\widehat{\mathbf{u}},\mathbf{g}\big) &= \|\nabla\times\xi + \nabla q + \delta_2(\mathbf{u}-\widehat{\mathbf{u}})\|^2_{-1}\\
&+\|\nabla\times\mathbf{v} - \xi + \delta_1\omega\|^2_0 + \|\nabla\cdot\mathbf{v}\|^2_0 + \|\delta\theta + \mathbf{v}\|^2_0\\
&+\|\nabla\times\omega + \nabla p + \theta - \mathbf{g}\|^2_{-1} + \|\nabla\times\mathbf{u} - \omega\|^2_0 + \|\nabla\cdot\mathbf{u}\|^2_0,
\end{aligned} \tag{11.106}$$

where

$$\delta_1 = \begin{cases} 1 & \text{for Case I}\\ 0 & \text{for Case II}\end{cases} \qquad \text{and} \qquad \delta_2 = \begin{cases} 0 & \text{for Case I}\\ 1 & \text{for Case II}.\end{cases}$$

As in (11.35), we have that the unique minimizer of the least-squares functional (11.106) can be characterized as being the solution of the problem: find $\{\mathbf{u},\omega,p\} \in [H_0^1(\Omega)]^d \times [L^2(\Omega)]^{d(d-1)/2} \times L_0^2(\Omega)$, $\theta \in [L^2(\Omega)]^d$, and $\{\mathbf{v},\xi,q\} \in [H_0^1(\Omega)]^d \times [L^2(\Omega)]^{d(d-1)/2} \times L_0^2(\Omega)$ such that

$$
\begin{aligned}
Q\big(\{\mathbf{u},\omega,p\},\theta,\{\mathbf{v},\xi,q\};\{\widetilde{\mathbf{u}},\widetilde{\omega},\widetilde{p}\},\widetilde{\theta},\{\widetilde{\mathbf{v}},\widetilde{\xi},\widetilde{q}\}\big)& \\
= F\big(\{\widetilde{\mathbf{u}},\widetilde{\omega},\widetilde{p}\},\widetilde{\theta},\{\widetilde{\mathbf{v}},\widetilde{\xi},\widetilde{q}\};\widehat{\mathbf{u}},\mathbf{g}\big)& \\
\forall\{\widetilde{\mathbf{u}},\widetilde{\omega},\widetilde{p}\} \in [H_0^1(\Omega)]^d \times [L^2(\Omega)]^{d(d-1)/2} \times L_0^2(\Omega),\ \widetilde{\theta} \in [L^2(\Omega)]^d,& \\
\forall\{\widetilde{\mathbf{v}},\widetilde{\xi},\widetilde{q}\} \in [H_0^1(\Omega)]^d \times [L^2(\Omega)]^{d(d-1)/2} \times L_0^2(\Omega),&
\end{aligned}
\tag{11.107}
$$

where we have the bilinear form

$$
\begin{aligned}
&Q\big(\{\mathbf{u},\omega,p\},\theta,\{\mathbf{v},\xi,q\};\{\widetilde{\mathbf{u}},\widetilde{\omega},\widetilde{p}\},\widetilde{\theta},\{\widetilde{\mathbf{v}},\widetilde{\xi},\widetilde{q}\}\big) \\
&\quad = \Big(\nabla\times\xi+\nabla q+\delta_2\mathbf{u}\,,\,\nabla\times\widetilde{\xi}+\nabla\widetilde{q}+\delta_2\widetilde{\mathbf{u}}\Big)_{-1} \\
&\qquad +\Big(\nabla\times\mathbf{v}-\xi+\delta_1\omega\,,\,\nabla\times\widetilde{\mathbf{v}}-\widetilde{\xi}+\delta_1\widetilde{\omega}\Big)_0 \\
&\qquad +\Big(\nabla\cdot\mathbf{v}\,,\,\nabla\cdot\widetilde{\mathbf{v}}\Big)_0+\Big(\delta\theta+\mathbf{v}\,,\,\delta\widetilde{\theta}+\widetilde{\mathbf{v}}\Big)_0 \\
&\qquad +\Big(\nabla\times\omega+\nabla p+\theta\,,\,\nabla\times\widetilde{\omega}+\nabla\widetilde{p}+\widetilde{\theta}\Big)_{-1} \\
&\qquad +\Big(\nabla\times\mathbf{u}-\omega\,,\,\nabla\times\widetilde{\mathbf{u}}-\widetilde{\omega}\Big)_0+\Big(\nabla\cdot\mathbf{u}\,,\,\nabla\cdot\widetilde{\mathbf{u}}\Big)_0
\end{aligned}
\tag{11.108}
$$

and the linear functional

$$
\begin{aligned}
&F\big(\{\widetilde{\mathbf{u}},\widetilde{\omega},\widetilde{p}\},\widetilde{\theta},\{\widetilde{\mathbf{v}},\widetilde{\xi},\widetilde{q}\};\widehat{\mathbf{u}},\mathbf{g}\big) \\
&\quad = \Big(\delta_2\widehat{\mathbf{u}}\,,\,\nabla\times\widetilde{\xi}+\nabla\widetilde{q}+\delta_2\widetilde{\mathbf{u}}\Big)_{-1}+\Big(\mathbf{g}\,,\,\nabla\times\widetilde{\omega}+\nabla\widetilde{p}+\widetilde{\theta}\Big)_{-1},
\end{aligned}
\tag{11.109}
$$

where $(\cdot,\cdot)_{-1}$ denotes the inner product in $[H^{-1}(\Omega)]^d$.

To define least-squares finite element approximations of the optimization problems, we first choose conforming finite element subspaces $\mathbf{V}^h \subset [H_0^1(\Omega)]^d$, $\mathbf{W}^h \subset [L^2(\Omega)]^{d(d-1)/2}$, $S^h \subset L_0^2(\Omega)$, and $\mathbf{T}^h \subset [L^2(\Omega)]^d$. We then minimize the functional in (11.106) over the subspaces, or equivalently, solve the problem: find $\{\mathbf{u}^h,\omega^h,p^h\} \in \mathbf{V}^h \times \mathbf{W}^h \times S^h$, $\theta^h \in \mathbf{T}^h$, and $\{\mathbf{v}^h,\xi^h,q^h\} \in \mathbf{V}^h \times \mathbf{W}^h \times S^h$ such that

$$
\begin{aligned}
&Q(\{\mathbf{u}^h,\omega^h,p^h\},\theta^h,\{\mathbf{v}^h,\xi^h,q^h\};\{\widetilde{\mathbf{u}}^h,\widetilde{\omega}^h,\widetilde{p}^h\},\widetilde{\theta}^h,\{\widetilde{\mathbf{v}}^h,\widetilde{\xi}^h,\widetilde{q}^h\})\\
&\qquad\qquad = F\big(\{\widetilde{\mathbf{u}}^h,\widetilde{\omega}^h,\widetilde{p}^h\},\widetilde{\theta}^h,\{\widetilde{\mathbf{v}}^h,\widetilde{\xi}^h,\widetilde{q}^h\};\widehat{\mathbf{u}},\mathbf{g}\big)\\
&\quad \forall\{\widetilde{\mathbf{u}}^h,\widetilde{\omega}^h,\widetilde{p}\}^h\in\mathbf{V}^h\times\mathbf{W}^h\times S^h,\ \widetilde{\theta}^h\in\mathbf{T}^h,\\
&\quad \forall\{\widetilde{\mathbf{v}}^h,\widetilde{\xi}^h,\widetilde{q}^h\}\in\mathbf{V}^h\times\mathbf{W}^h\times S^h.
\end{aligned}
\tag{11.110}
$$

Proposition (11.29) and the results of Section 11.2.2 allow us to prove the following results.

Theorem 11.34 *Let $\Phi=\Lambda=[H_0^1(\Omega)]^d\times[L^2(\Omega)]^{d(d-1)/2}\times L_0^2(\Omega)$ and let $\Theta=[L^2(\Omega)]^d$. Then,*

(i) the bilinear form $Q(\cdot;\cdot)$ defined in (11.108) *is symmetric, continuous, and coercive on $\{\Phi\times\Theta\times\Lambda\}\times\{\Phi\times\Theta\times\Lambda\}$.*

Let $\widehat{\mathbf{u}}\in[L^2(\Omega)]^d$ and $\mathbf{g}\in[H^{-1}(\Omega)]^d$ be given. Then,

(ii) the linear functional $F(\cdot)$ defined in (11.109) *is continuous on $\{\Phi\times\Theta\times\Lambda\}$*
(iii) the problem (11.107) *has a unique solution $\big\{\{\mathbf{u},\omega,p\},\,\theta,\,\{\mathbf{v},\xi,q\}\big\}\in\Phi\times\Theta\times\Lambda$.*

Let $\mathbf{V}^h\subset[H_0^1(\Omega)]^d$, $\mathbf{W}^h\subset[L^2(\Omega)]^{d(d-1)/2}$, $S^h\subset L_0^2(\Omega)$, and $\mathbf{T}^h\subset[L^2(\Omega)]^d$ and let $\Phi^h=\Lambda^h=\mathbf{V}^h\times\mathbf{W}^h\times S^h$ and $\Theta^h=\mathbf{T}^h$. Then,

(iv) the discrete problem (11.110) *has a unique solution*

$$
\big\{\{\mathbf{u}^h,\omega^h,p^h\},\,\theta^h,\,\{\mathbf{v}^h,\xi^h,q^h\}\big\}\in\Phi^h\times\Theta^h\times\Lambda^h
$$

(v) we have the error estimate:

$$
\begin{aligned}
&\|\mathbf{u}-\mathbf{u}^h\|_1+\|\omega-\omega^h\|_0+\|p-p^h\|_0+\|\theta-\theta^h\|_0\\
&\qquad\qquad +\|\mathbf{v}-\mathbf{v}^h\|_1+\|\xi-\xi^h\|_0+\|q-q^h\|_0\\
&\quad\le C\Big(\inf_{\widetilde{\mathbf{u}}^h\in\mathbf{V}^h}\|\mathbf{u}-\widetilde{\mathbf{u}}^h\|_1+\inf_{\widetilde{\omega}^h\in\mathbf{W}^h}\|\omega-\widetilde{\omega}^h\|_0\\
&\qquad\qquad +\inf_{\widetilde{p}^h\in S^h}\|p-\widetilde{p}^h\|_0+\inf_{\widetilde{\theta}^h\in\mathbf{T}^h}\|\theta-\widetilde{\theta}^h\|_0\\
&\qquad\qquad +\inf_{\widetilde{\mathbf{v}}^h\in\mathbf{V}^h}\|\mathbf{v}-\widetilde{\mathbf{v}}^h\|_1+\inf_{\widetilde{\xi}^h\in\mathbf{W}^h}\|\xi-\widetilde{\xi}^h\|_0\\
&\qquad\qquad +\inf_{\widetilde{q}^h\in S^h}\|q-\widetilde{q}^h\|_0\Big).
\end{aligned}
\tag{11.111}
$$

Proof. The results follow in a straightforward manner from Proposition 11.29 along with Lemma 11.13 and Theorem 11.15. □

Remark 11.35 Following the remarks at the end of Section 11.2.2, the discrete problem (11.110) is equivalent to a linear algebraic system having a symmetric, positive definite coefficient matrix. In the case of a Galerkin discretization of the optimality system, the coefficient matrix is indefinite. Also, the results in Theorem 11.34 about the solution of the discrete problem (11.110) follow merely from the conformity of the finite element subspaces, i.e., merely from the inclusions $\mathbf{V}^h \subset [H_0^1(\Omega)]^d$, $\mathbf{W}^h \subset [L^2(\Omega)]^{d(d-1)/2}$, $S^h \subset L_0^2(\Omega)$, and $\mathbf{T}^h \subset [L^2(\Omega)]^d$. In particular, unlike the case of Galerkin finite element discretizations of the optimality system, they do not require that the finite element spaces satisfy additional conditions such as (11.100). In particular, in (11.110), one can choose the same degree piecewise polynomial space G^r defined with respect to the same grid for all variables. □

Remark 11.36 A Galerkin finite element discretization of the optimality system can use a formulation in terms of the operators defined in (11.99) whereas the least-squares based discretization employs a formulation in terms of the operators defined in (11.94). Thus, the latter approach involves more unknowns, i.e., 10 scalar fields in two dimensions and 17 scalar fields in three dimensions, compared to the former that involves 8 and 11 scalar fields, respectively. This apparent disadvantage of the least-squares approach should be balanced against the advantages already discussed, including the ability to use equal order interpolation for all variables, the positive definiteness of the discrete systems, and the possiblity for using more efficient uncoupling strategies. □

Remark 11.37 Suppose one chooses the continuous, piecewise polynomial finite element spaces G^r of degree r for the approximation of all variables; this is permissible for the LSFEMs we are considering because these spaces are not required to satisfy (11.100). Suppose also that the solution of the optimality system satisfies $\mathbf{u}, \mathbf{v} \in [H^{r+1}(\Omega)]^d \cap [H_0^1(\Omega)]^d$, $\omega, \xi \in [H^r(\Omega)]^{d(d-1)/2}$, $p, q \in H^r(\Omega) \cap L_0^2(\Omega)$, and $\theta \in [H^r(\Omega)]^d$. Then, the error estimate (11.111) implies that

$$\begin{aligned} &\|\mathbf{u}-\mathbf{u}^h\|_1 + \|\omega-\omega^h\|_0 + \|p-p^h\|_0 + \|\theta-\theta^h\|_0 \\ &\qquad + \|\mathbf{v}-\mathbf{v}^h\|_1 + \|\xi-\xi^h\|_0 + \|q-q^h\|_0 = O(h^r), \end{aligned} \tag{11.112}$$

where h is a measure of the grid size. □

11.6.4 Constraining by the Least-Squares Functional for the Constraint Equations

We now consider specialization of Method 5 of Section 11.4.2 (see Section 11.5) to the Stokes optimization problems. In Sections 11.1.2–11.4, additional bilinear forms, linear functionals, operators, and functions are introduced. In the context of the optimization problems considered in this section, the bilinear forms (11.20) and (11.21) are respectively given by

$$\widetilde{b}_1(\{\mathbf{u},\omega,p\},\{\widetilde{\mathbf{u}},\widetilde{\omega},\widetilde{p}\})$$
$$= (\nabla\times\omega+\nabla p,\nabla\times\widetilde{\omega}+\nabla\widetilde{p})_{-1}+(\nabla\times\mathbf{u}-\omega,\nabla\times\widetilde{\mathbf{u}}-\widetilde{\omega})_0+(\nabla\cdot\mathbf{u},\nabla\cdot\widetilde{\mathbf{u}})_0$$
$$= (\nabla\times\omega+\nabla p,(-\Delta)^{-1}(\nabla\times\widetilde{\omega}+\nabla\widetilde{p}))_0$$
$$+(\nabla\times\mathbf{u}-\omega,\nabla\times\widetilde{\mathbf{u}}-\widetilde{\omega})_0+(\nabla\cdot\mathbf{u},\nabla\cdot\widetilde{\mathbf{u}})_0$$

and

$$\widetilde{b}_2(\{\theta,\{\widetilde{\mathbf{u}},\widetilde{\omega},\widetilde{p}\}) = (\theta,\nabla\times\widetilde{\omega}+\nabla\widetilde{p})_{-1} = (\theta,(-\Delta)^{-1}(\nabla\times\widetilde{\omega}+\nabla\widetilde{p}))_0.$$

We also have that

$$(\{\mathbf{u},\omega,p\},\{\widetilde{\mathbf{u}},\widetilde{\omega},\widetilde{p}\})_\Lambda = (\mathbf{u},\widetilde{\mathbf{u}})_{-1}+(\omega,\widetilde{\omega})_0+(p,\widetilde{p})_0$$
$$= (\mathbf{u},(-\Delta)^{-1}\widetilde{\mathbf{u}})_0+(\omega,\widetilde{\omega})_0+(p,\widetilde{p})_0$$

and that the linear functional (11.22) is given by

$$\widetilde{G}_1(\{\widetilde{\mathbf{u}},\widetilde{\omega},\widetilde{p}\}) = (\mathbf{g},\nabla\times\widetilde{\omega}+\nabla\widetilde{p})_{-1}$$
$$= (\mathbf{g},(-\Delta)^{-1}(\nabla\times\widetilde{\omega}+\nabla\widetilde{p}))_0$$
$$= (\nabla\times(-\Delta)^{-1}\mathbf{g},\widetilde{\omega})_0-(\nabla\cdot(-\Delta)^{-1}\mathbf{g},\widetilde{p})_0.$$

Using these definitions, we can specialize Method 5 of Section 11.4.2 (see Section 11.5) to the Stokes optimization problems.

Recall Proposition 11.29. We then easily have that Proposition 11.3, Corollaries 11.4 and 11.6, and Theorem 11.27 all hold for the Stokes optimization problems without the need for the discrete inf–sup condition (11.100) to hold.

Discrete systems for the Stokes control problem

Let $\{\mathbf{u}_j\}_{j=1}^{J_1}$, $\{\omega_j\}_{j=1}^{J_2}$, $\{p_j\}_{j=1}^{J_3}$, and $\{\theta_k\}_{k=1}^{K}$ respectively denote basis sets for the finite element spaces $\mathbf{V}^h\subset[H_0^1(\Omega)]^d$, $\mathbf{W}^h\subset[L^2(\Omega)]^{d(d-1)/2}$, $S^h\subset L_0^2(\Omega)$, and $\Theta^h\subset[L^2(\Omega)]^d$, where Θ^h is the finite element space introduced to approximate the control θ. Then, the matrices and vectors appearing in the discrete systems (11.83) and (11.84) are easily identified to be as follows:[17]

[17] Note that in the present case, we have the Riesz operator

$$\mathcal{R}_\Phi = \begin{pmatrix} -\Delta & 0 & 0 \\ 0 & \mathcal{I} & 0 \\ 0 & 0 & \mathcal{I} \end{pmatrix}.$$

$$
\mathsf{A}_1 = \begin{pmatrix} \mathsf{A}_{1,11} & 0 & 0 \\ 0 & \mathsf{A}_{1,22} & 0 \\ 0 & 0 & 0 \end{pmatrix} \qquad \widetilde{\mathsf{B}}_1 = \begin{pmatrix} \widetilde{\mathsf{B}}_{1,11} & \widetilde{\mathsf{B}}_{1,12} & 0 \\ \widetilde{\mathsf{B}}^T_{1,12} & \widetilde{\mathsf{B}}_{1,22} & \widetilde{\mathsf{B}}_{1,23} \\ 0 & \widetilde{\mathsf{B}}^T_{1,23} & \widetilde{\mathsf{B}}_{1,33} \end{pmatrix} \qquad \widetilde{\mathsf{B}}_2 = \begin{pmatrix} 0 \\ \widetilde{\mathsf{B}}_{2,21} \\ \widetilde{\mathsf{B}}_{2,31} \end{pmatrix}
$$

$$
\mathsf{R}_\Phi = \begin{pmatrix} \mathsf{R}_{,11} & 0 & 0 \\ 0 & \mathsf{I} & 0 \\ 0 & 0 & \mathsf{I} \end{pmatrix} \qquad \vec{f} = \begin{pmatrix} \mathbf{f}_{,1} \\ 0 \\ 0 \end{pmatrix} \qquad \vec{g}_1 = \begin{pmatrix} 0 \\ \mathbf{g}_{1,2} \\ \mathbf{g}_{1,3} \end{pmatrix},
$$

where

$$
\left(\mathsf{A}_{1,11}\right)_{ij} = \begin{cases} 0 & \text{(Case I)} \\ \left(\mathbf{u}_j, \mathbf{u}_i\right)_0 & \text{(Case II)} \end{cases} \qquad \text{for } i,j = 1,\dots,J_1
$$

$$
\left(\mathsf{A}_{1,22}\right)_{ij} = \begin{cases} \left(\omega_j, \omega_i\right)_0 & \text{(Case I)} \\ 0 & \text{(Case II)} \end{cases} \qquad \text{for } i,j = 1,\dots,J_2
$$

$$
\left(\mathsf{A}_2\right)_{k\ell} = \delta\left(\theta_\ell, \theta_k\right)_0 \qquad \text{for } k,\ell = 1,\dots,K,
$$

$$
\left(\widetilde{\mathsf{B}}_{1,11}\right)_{ij} = \left(\nabla\times\mathbf{u}_j, \nabla\times\mathbf{u}_i\right)_0 + \left(\nabla\cdot\mathbf{u}_j, \nabla\cdot\mathbf{u}_i\right)_0 \quad \text{for } i,j = 1,\dots,J_1
$$

$$
\left(\widetilde{\mathsf{B}}_{1,12}\right)_{ij} = -\left(\omega_j, \nabla\times\mathbf{u}_i\right)_0 \qquad \text{for } i = 1,\dots,J_1,\ j = 1,\dots,J_2
$$

$$
\left(\widetilde{\mathsf{B}}_{1,22}\right)_{ij} = \left(\nabla\times\omega_j, \nabla\times\omega_i\right)_{-1} + \left(\omega_j, \omega_i\right)_0 \qquad \text{for } i,j = 1,\dots,J_2
$$

$$
\left(\widetilde{\mathsf{B}}_{1,23}\right)_{ij} = \left(\nabla p_j, \nabla\times\omega_i\right)_0 \qquad \text{for } i = 1,\dots,J_2,\ j = 1,\dots,J_3
$$

$$
\left(\widetilde{\mathsf{B}}_{1,33}\right)_{ij} = \left(\nabla p_j, \nabla p_i\right)_{-1} \qquad \text{for } i,j = 1,\dots,J_3
$$

$$
\left(\widetilde{\mathsf{B}}_{2,21}\right)_{jk} = \left(\theta_k, \nabla\times\omega_j\right)_0 \qquad \text{for } j = 1,\dots,J_2,\ k = 1,\dots,K
$$

$$
\left(\widetilde{\mathsf{B}}_{2,31}\right)_{jk} = \left(\theta_k, \nabla p_j\right)_0 \qquad \text{for } j = 1,\dots,J_3,\ k = 1,\dots,K
$$

$$
\left(\mathsf{R}_{,11}\right)_{ij} = \left(\nabla\mathbf{u}_j, \nabla\mathbf{u}_i\right)_0 \qquad \text{for } i,j = 1,\dots,J_1
$$

$$
\left(\mathbf{f}_{,1}\right)_i = \begin{cases} 0 & \text{(Case I)} \\ \left(\widehat{\mathbf{u}}, \mathbf{u}_i\right)_0 & \text{(Case II)} \end{cases} \qquad \text{for } i = 1,\dots,J_1
$$

$$
\left(\mathbf{g}_{1,2}\right)_i = \left(\mathbf{g}, \nabla\times\omega_i\right)_{-1} \qquad \text{for } i = 1,\dots,J_2
$$

$$
\left(\mathbf{g}_{1,3}\right)_i = \left(\mathbf{g}, \nabla p_i\right)_{-1} \qquad \text{for } i = 1,\dots,J_3\,.
$$

In this way, the matrix system (11.84) for Method 5 is completely defined.

For Method 5, Theorem 11.27 implies that, if one uses the continuous finite element spaces G^r of degree r for all variables and if one chooses $\varepsilon = h^r$, then

$$\|\mathbf{u}-\mathbf{u}_\varepsilon^h\|_1 + \|\omega - \omega_\varepsilon^h\|_0 + \|p - p_\varepsilon^h\|_0 + \|\theta - \theta_\varepsilon^h\|_0 = O(h^r),$$

provided that the solutions of the optimization problems we are considering in this section are sufficiently smooth. One could use finite element spaces of one degree lower[18] for the approximations of the vorticity ω, the pressure p, and the control θ than that used for the velocity $\mathbf{u}$ and still obtain the same error estimate. However, one of the strengths of using LSFEMs in the context of optimization and control problems is that one can use any conforming finite element spaces and, in particular, one can use the same degree finite element spaces for all variables; we see that Method 5 inherits this strength.

Note also that the LSFEM discussed in this section requires fewer unknowns than that required for the LSFEM for the optimality systems considered in Section 11.6.3. Of course, this results from the fact that, unlike the case for the method of Section 11.6.3, there is no need to introduce Lagrange multiplier variables. On the other hand, the method considered in this section does involve a penalty parameter (see Section 11.4.2) whereas that of Section 11.6.3 does not.

[18] An example of such an approximating space is the space $X_{r,0}^h$ used in the norm-equivalent DLSP (7.55). See Remark 7.15 for further discussion of finite element spaces for the VVP Stokes system.

Chapter 12
Variations on Least-Squares Finite Element Methods

Least-squares finite element methods (LSFEMs) for partial differential equations (PDEs) and related ideas have arisen in many other contexts.[1] In this chapter, we examine several additional examples of methods that use least-squares notions. Some sections deal with additional least-squares finite element methods, some of which do not readily fit into the framework of Chapter 3. Others deal with additional applications of least-squares finite element methods. The last two sections discuss least-squares methods based on other approximation paradigms. All of these topics deserve mention; however, to keep the book from becoming prohibitively long, the discussion of these additional topics is brief. Our goal here is to acquaint the reader with the wide scope of the least-squares finite element universe; details about the topics discussed may be found in the cited references.

12.1 Weak Enforcement of Boundary Conditions

Recall that for finite element methods that are based on the minimization of energy functionals or on residual orthogonalization principles, boundary conditions are divided into two categories. First, one has *essential* boundary conditions that have to be imposed on candidate solutions of the variational principle or equations, i.e., the test and trial spaces are constrained to satisfy such boundary conditions. Second, one has boundary conditions that are *natural* to the variational principle, i.e., the boundary conditions are satisfied (weakly) without having to constrain the test and trial spaces.[2]

---

[1] We have already encountered one such notion in Section 2.1.1, namely the Galerkin least-squares stabilized method.

[2] For example, consider the Dirichlet principle (1.50) for the Poisson problem (1.52) or, equivalently, the corresponding variational equation (1.51); note that the standard Galerkin principle applied to (1.52) also results in (1.51). For this problem, the Dirichlet boundary condition $\phi = 0$ on Γ is essential; this is why candidate minimizers in (1.50) or, equivalently, test and trial functions in (1.51), were constrained to satisfy this boundary condition. On the other hand, the Neumann

P.B. Bochev, M.D. Gunzburger, *Least-Squares Finite Element Methods*, Applied Mathematical Sciences 166, DOI: 10.1007/b13382_12,

LSFEMs offer a choice of how boundary conditions are treated, a choice that is not available for other finite element methods. On the one hand,

> by constraining the space of candidate minimizers of the least-squares functional and after possibly reformulation into a first-order system, *any boundary condition can be treated as an essential boundary condition*;

on the other hand, because least-squares principles are artificial "energy" principles,

> one can include boundary condition residual norms in the least-squares functionals so that *any boundary condition can be made natural for the corresponding least-squares principle.*

The ability of least-squares principles to treat any boundary condition as essential or natural is further evidence of the great flexibility offered by LSFEMs and is viewed as one of the important advantages of LSFEMs compared to other finite element methods.[3]

In this book, for the sake of brevity and clarity, we have almost exclusively taken the first approach to handling boundary conditions, i.e., we have constrained the spaces of candidate minimizers of least-squares functionals; here, we briefly consider the second approach.

The abstract problem (3.16) included the data space B, the boundary condition $\mathcal{B}u = g$ in B, the norm-equivalent least-squares functional (3.24) that included the boundary residual term $\|\mathcal{B}u - g\|_B^2$, and the straightforward continuous least-squares principle (CLSP) (3.25). Unfortunately, this recipe for dealing with boundary conditions through the inclusion of boundary residual terms in the least-squares functionals invariably leads to impractical LSFEMs. The cause of this situation is that norm-equivalence is achieved for boundary data spaces B that involve *fractional-order* Sobolev spaces. The remedies available are very similar to those that we have discussed for treating least-squares principles that involve negative norms in the least-squares functional. Basically, one has three choices, all of which involve replacing the impractical $\|\mathcal{B}u - g\|_B^2$ term in the functional by a more practical term:

- replace $\|\mathcal{B}u - g\|_B$ by the $L^2(\partial\Omega)$ norm, i.e., by $\|\mathcal{B}u - g\|_{L^2(\partial\Omega)}$
- replace $\|\mathcal{B}u - g\|_B$ by a mesh-weighted $L^2(\partial\Omega)$ norm $h^\alpha\|\mathcal{B}u - g\|_{L^2(\partial\Omega)}$ for an appropriate value of α
- replace $\|\mathcal{B}u - g\|_B$ by an equivalent norm whose definition requires a spectrally equivalent approximation to the solution of a PDE.

The first choice usually leads to a loss of accuracy, the second can lead to ill-conditioned discrete systems, and the third requires additional work.

---

boundary condition $\partial\phi/\partial n = 0$ on Γ^* is natural because it is "automatically" satisfied without the need for further constraining candidate minimizers or test and trial functions.

[3] This is especially true when one recalls that for Galerkin finite element methods, essential boundary conditions are often difficult to implement, e.g., for curved boundaries or for inhomogeneous data. In LSFEMs, one can arrange for such boundary conditions to be natural for the least-square principle.

To illustrate these observations, consider the div–grad first-order system formulation of the Poisson equation given by[4]

$$\begin{cases} \nabla \cdot \mathbf{u} = f & \text{in } \Omega \\ \mathbf{u} + \nabla \phi = 0 & \text{in } \Omega \\ \phi = g & \text{on } \partial\Omega \end{cases} \tag{12.1}$$

for which we have the norm-equivalent least-squares functional[5]

$$J_{1/2}(\phi,\mathbf{u};f) = \|\mathbf{u}+\nabla\phi\|^2_{[L^2(\Omega)]^d} + \|\nabla\cdot\mathbf{u}-f\|^2_{L^2(\Omega)} + \|\phi-g\|^2_{H^{1/2}(\partial\Omega)}. \tag{12.2}$$

The corresponding CLSP is impractical because it results in a discrete system whose construction requires the evaluation of $H^{1/2}(\partial\Omega)$ inner products.

A practical least-squares functional is defined by[6]

$$J_0(\phi,\mathbf{u};f) = \|\mathbf{u}+\nabla\phi\|^2_{[L^2(\Omega)]^d} + \|\nabla\cdot\mathbf{u}-f\|^2_{L^2(\Omega)} + \|\phi-g\|^2_{L^2(\partial\Omega)}, \tag{12.3}$$

i.e., by replacing the $H^{1/2}(\partial\Omega)$ term in (12.2) by a $L^2(\partial\Omega)$ term. The functional (12.3) corresponds to a transition diagram (3.54) in which the norm-generating operator $\mathcal{S}_{1/2,\partial\Omega}$ of the $H^{1/2}(\partial\Omega)$ norm is replaced by the identity operator $\mathcal{I}$. This is the same type of straightforward approximation of the norm generating operator that led us to the quasi-norm-equivalent LSFEM (4.27) in Section 4.5.1. As is the case with that LSFEM, methods based on $J_0(\cdot,\cdot;\cdot)$ suffer from a loss of accuracy.

Motivated by inverse inequalities that hold for finite element functions, a practical discrete least-squares functional is defined by

$$\begin{aligned} &J^h_{1/2}(\phi^h,\mathbf{u}^h;f) = \\ &\qquad \|\mathbf{u}^h+\nabla\phi^h\|^2_{[L^2(\Omega)]^d} + \|\nabla\cdot\mathbf{u}^h-f\|^2_{L^2(\Omega)} + h^{-1}\|\phi^h-g\|^2_{L^2(\partial\Omega)}, \end{aligned} \tag{12.4}$$

i.e., by weighting the $L^2(\partial\Omega)$ term in (12.3). This choice corresponds to a transition diagram (3.54) in which $\mathcal{S}_{1/2,\partial\Omega}$ is approximated by a scaled identity operator $h^{-1/2}\mathcal{I}$. This parallels the formulation, in Section 4.5.1, of the weighted LSFEM (4.28). Just like in that case, the resulting LSFEMs are optimally accurate. However, unlike (4.28), LSFEMs based on (12.4) do not suffer from conditioning problems;[7] thus LSFEMs based on this functional are truly practical.

---

[4] The boundary condition $\phi = g$ is an essential boundary condition for Rayleigh–Ritz and Galerkin formulations of the second-order Poisson problem.

[5] In order to avoid confusion, we are purposely very precise in designating the norms use.

[6] The boundary condition $\phi = g$ is an essential boundary condition for Rayleigh–Ritz and Galerkin formulations of the second-order Poisson problem; using (12.3), it becomes a natural boundary condition in the least-squares setting.

[7] In (4.28), the impractical $H^{-1}(\Omega)$ norm term in the functional is replaced by a term involving the *stronger* $L^2(\Omega)$ norm, whereas in (12.4), the impractical $H^{1/2}(\partial\Omega)$ term in the functional (12.3) has been replaced by a term involving the *weaker* $L^2(\partial\Omega)$ norm. This difference accounts for the

Both (12.3) and (12.4) are quasi-norm-equivalent functionals because they are based on approximations of $\mathcal{S}_{1/2,\partial\Omega}$ that are not spectrally equivalent to that operator. As a result, even though the associated linear systems do not suffer from higher condition numbers, finding efficient preconditioners for them may be hindered by the lack of norm-equivalence in the parent least-squares functional.

A solution to this problem, similar to the approach that led us to the norm-equivalent DLSP (4.39) in Section 4.5.2, is to replace $\mathcal{S}_{1/2,\partial\Omega}$ by a spectrally equivalent approximation. We describe one possible approach, suggested in [318], which relies on multi-level decompositions

$$G^{h,0} \subset G^{h,1} \subset \cdots \subset G^{h,l} \subset G(\Omega) \quad \text{and} \quad \mathbf{D}^{h,0} \subset \mathbf{D}^{h,1} \subset \cdots \subset \mathbf{D}^{h,l} \subset \mathbf{D}(\Omega)$$

of compatible approximating spaces for the dependent variables, defined on a sequence of uniformly regular partitions $\mathcal{T}_{h,l}$ of Ω into finite elements. Let $\Pi_{\partial\Omega,i}$ denote the $L^2(\partial\Omega)$ projection onto the trace space of $G^{h,i}$ defined by

$$(\Pi_{\partial\Omega,i}\phi, \psi^{h,i})_{0,\partial\Omega} = (\phi, \psi^{h,i})_{0,\partial\Omega} \qquad \forall\, \psi^{h,i} \in G^{h,i}.$$

Then, one can show that the operator

$$\mathcal{S}_{h/2,\partial\Omega} = \left(h_0^{-2}\Pi_{\partial\Omega,0} + \sum_{i=1}^{l} h_i^{-2}(\Pi_{\partial\Omega,i} - \Pi_{\partial\Omega,i-1}) \right)^{1/4} \tag{12.5}$$

is spectrally equivalent to the norm-generating operator $\mathcal{S}_{1/2,\partial\Omega}$ on $G^{h,l}$; see [80, 318]. As a result, the associated discrete trace norm

$$\|\phi^{h,l}\|_{h/2,\partial\Omega} = \|(\mathcal{S}_{h/2,\partial\Omega})\phi^{h,l}\|_{0,\partial\Omega} = \big((\mathcal{S}_{h/2,\partial\Omega})\phi^{h,l}, (\mathcal{S}_{h/2,\partial\Omega})\phi^{h,l}\big)^{1/2}_{0,\partial\Omega}$$

is norm-equivalent approximation of the $H^{1/2}(\partial\Omega)$ norm on $G^{h,l}$. It follows that the least-squares functional defined by

$$\begin{aligned} J_{h/2}(\phi^h, \mathbf{u}^h; f) &= \\ &\|\mathbf{u}^h + \nabla\phi^h\|^2_{[L^2(\Omega)]^d} + \|\nabla\cdot\mathbf{u}^h - f\|^2_{L^2(\Omega)} + \|\phi^h - g\|^2_{h/2,\partial\Omega}, \end{aligned} \tag{12.6}$$

is both practical and norm-equivalent.

The minimization of any of the functionals in (12.2)–(12.6) is effected with respect to *unconstrained* continuous and/or discrete spaces. For example, for the functional (12.4), one can choose discrete spaces such that $X^h \subset G(\Omega) \times \mathbf{D}(\Omega)$. In this manner, the boundary condition $\phi = g$ which is essential to Rayleigh–Ritz and Galerkin principles, becomes a natural boundary condition for least-squares principles.

---

better behavior, e.g., with respect to conditioning, of LSFEMs based on (12.4) as compared to the situation for (4.28) which is based on weighted $L^2(\Omega)$ replacements of $H^{-1}(\Omega)$ norms.

Next, instead of (12.1), we consider the problem[8]

$$\begin{cases} \nabla \cdot \mathbf{u} = f & \text{in } \Omega \\ \mathbf{u} + \nabla \phi = 0 & \text{in } \Omega \\ \mathbf{u} \cdot \mathbf{n} = g & \text{on } \partial\Omega \end{cases} \tag{12.7}$$

for which we have the norm-equivalent least-squares functional

$$J_{-1/2}(\phi, \mathbf{u}; f) = \|\mathbf{u} + \nabla\phi\|^2_{[L^2(\Omega)]^d} + \|\nabla \cdot \mathbf{u} - f\|^2_{L^2(\Omega)} + \|\mathbf{u} \cdot \mathbf{n} - g\|^2_{H^{-1/2}(\partial\Omega)} \tag{12.8}$$

that is impractical, the straightforward functional

$$J_0(\phi, \mathbf{u}; f) = \|\mathbf{u} + \nabla\phi\|^2_{[L^2(\Omega)]^d} + \|\nabla \cdot \mathbf{u} - f\|^2_{L^2(\Omega)} + \|\mathbf{u} \cdot \mathbf{n} - g\|^2_{L^2(\partial\Omega)}$$

that, at best, leads to a loss of accuracy, and the mesh-weighted functional

$$\begin{aligned} &J^h_{-1/2}(\phi^h, \mathbf{u}^h; f) = \\ &\qquad \|\mathbf{u}^h + \nabla\phi^h\|^2_{[L^2(\Omega)]^d} + \|\nabla \cdot \mathbf{u}^h - f\|^2_{L^2(\Omega)} + h\|\mathbf{u} \cdot \mathbf{n}^h - g\|^2_{L^2(\partial\Omega)} \end{aligned} \tag{12.9}$$

that results in linear systems with higher condition numbers.[9]

Again, (12.8) correspond to a transition diagram (3.54) in which the norm-generating operator $\mathcal{S}_{-1/2,\partial\Omega}$ of the $H^{-1/2}(\partial\Omega)$ norm is replaced by the identity operator $\mathcal{I}$, whereas in (12.9) this operator is approximated by the scaled identity $h^{1/2}\mathcal{I}$. Because these operators are not spectrally-equivalent to $\mathcal{S}_{-1/2,\partial\Omega}$, the resulting least-squares functionals are only quasi-norm-equivalent.

A norm-equivalent alternative to (12.8)–(12.9) can be defined by using the fact that $H^{-1/2}(\partial\Omega)$ is dual to $H^{1/2}(\partial\Omega)$. As a result, a suitably defined negative adjoint of the multilevel operator (12.5) provides a spectrally equivalent approximation of $\mathcal{S}_{-1/2,\partial\Omega}$. In the present context the relevant negative adjoint is given by

$$\big((\mathcal{S}_{-h/2,\partial\Omega})\mathbf{n} \cdot \mathbf{u}^{h,l}, \psi^{h,l}\big)_{0,\partial\Omega} = \big(\mathbf{n} \cdot \mathbf{u}^{h,l}, (\mathcal{S}_{h/2,\partial\Omega})^{-1}\psi^{h,l}\big)_{0,\partial\Omega}$$

for all $\mathbf{u}^{h,l} \in \mathbf{D}^{h,l}$ and $\psi^{h,l} \in G^{h,l}$. It can be shown that the associated norm

$$\|\mathbf{n} \cdot \mathbf{u}^{h,l}\|_{-h/2,\partial\Omega} = \|(\mathcal{S}_{-h/2,\partial\Omega})\mathbf{n} \cdot \mathbf{u}^{h,l}\|_{0,\partial\Omega}$$

is equivalent to the $H^{-1/2}(\partial\Omega)$ norm on $\mathbf{D}^{h,l}$; see [318]. As a result,

---

[8] The boundary condition $\mathbf{u} \cdot \mathbf{n} = \partial\phi/\partial n = g$ is a natural boundary condition for Rayleigh–Ritz and Galerkin formulations of the second-order Poisson problem.

[9] Note that, in (12.9), we have replaced the $H^{-1/2}(\partial\Omega)$ norm term in (12.8) by the term involving the *stronger* $L^2(\partial\Omega)$ norm so that this situation is entirely analogous to the replacement of $H^{-1}(\Omega)$ norms by stronger $L^2(\Omega)$ norms in (4.28).

$$
\begin{aligned}
J_{-h/2}(\phi^h, \mathbf{u}^h; f) = \\
\|\mathbf{u}^h + \nabla\phi^h\|^2_{[L^2(\Omega)]^d} + \|\nabla\cdot\mathbf{u}^h - f\|^2_{L^2(\Omega)} + \|\mathbf{u}\cdot\mathbf{n}^h - g\|^2_{-h/2,\partial\Omega}
\end{aligned}
\tag{12.10}
$$

is a norm-equivalent least-squares functional. This approach has been analyzed in [318] where it is shown that it yields optimally accurate solutions without increasing the conditioning of the discrete system. In particular, much of the discussion of Section B.4.2 about the efficient implementation of this approach in the context of replacing $H^{-1}(\Omega)$ norms with spectrally equivalent discrete norms applies with simple changes, to the current context.

The minimization of any of the functionals in (12.8)–(12.10) is effected with respect to *unconstrained* continuous and/or discrete spaces. For example, for the functional (12.4), one can again choose discrete spaces such that $X^h \subset G(\Omega) \times \mathbf{D}(\Omega)$. In this manner, the boundary condition $\mathbf{u}\cdot\mathbf{n} = g$ which is natural to Rayleigh–Ritz and Galerkin principles, is also a natural boundary condition for least-squares principles. However, one can omit the boundary norm terms in these functionals and use constrained discrete spaces, e.g., if $g = 0$, $X^h \subset G(\Omega) \times \mathbf{D}_0(\Omega)$, so that this "natural" boundary condition can be treated as an "essential" one within the LSFEM framework.

A comprehensive account of the treatment of boundary conditions within LS-FEMs is given in [342] for Petrovski elliptic systems in two dimensions with Lopatinski boundary conditions. Although the discussion there is for a specialized case, much of what is said applies to more general situations.

12.2 LL* Finite Element Methods

According to the general theory of Chapter 4, homogeneous elliptic Agmon–Douglis–Nirenberg (ADN) systems offer one of the most attractive settings for LS-FEMs. For such systems, all equation residuals can be measured by standard $L^2(\Omega)$ norms, and the resulting least-squares functionals are norm-equivalent in the space X_0 defined in (4.13). This means that the least-squares variational formulation is equivalent to a loosely coupled system of Laplace-like operators and that the algebraic equations can be preconditioned by standard preconditioners for the Poisson equation; see Theorem 4.2.

If, on the other hand, the ADN theory classifies a first-order system as inhomogeneous elliptic, then the ADN setting for the energy balances leads to LSFEMs with less attractive computational properties. One alternative is to augment the first-order operator by redundant equations until it becomes homogeneous elliptic. As we saw from the examples in Section 5.9.3, this approach fails when the solution is not smooth enough.

Another alternative, discussed in Chapters 5–7, is to abandon the ADN setting and develop energy balances in terms of Hilbert spaces such as $\mathbf{D}(\Omega)$ or $\mathbf{C}(\Omega)$ using the vector-operator setting. This led to practical, optimally accurate, and locally conservative LSFEMs but required the compatible finite element spaces described in Section B.2.

LL* least-squares principles (see [100, 258, 274]) offer a third alternative that allows us, in the case of inhomogeneous elliptic systems, to continue to use standard C^0 spaces and still achieve good results. Essentially, the LL* approach can be viewed as a least-squares principle formulated in terms of a problem-dependent negative norm. The resulting variational formulation is not necessarily practical. However, under some assumptions, the LL* least-squares solution u can be determined indirectly by solving a practical variational formulation for a "potential" y in the domain of the adjoint operator $\mathcal{L}^*$ and then setting $u = \mathcal{L}^* y$. This results in a practical procedure which forms the basis of LL* LSFEMs.

In the rest of this section, we provide a brief introduction to LL* LSFEMs. In addition, in Section 12.12, we consider LL* LSFEMs in connection with problems whose solutions display singular behavior.

Abstract LL* formulation

We describe the basic idea of LL* LSFEMs in the abstract setting of (3.16) with homogeneous boundary data. The operator $\mathcal{L}$ has domain $D(\mathcal{L})$ and range $R(\mathcal{L})$ and is assumed to satisfy hypothesis (3.7). Furthermore, we assume that its adjoint $\mathcal{L}^* : D(\mathcal{L}^*) \mapsto R(\mathcal{L}^*)$ is such that the exact solution u of (3.16) belongs to $R(\mathcal{L}^*)$ and

$$(\mathcal{L}u, v) = (u, \mathcal{L}^* v) \tag{12.11}$$

for all sufficiently smooth functions. The final assumption is that $\mathcal{L}^*$ has a bounded inverse. Then,

$$\|v\|_* = \sup_{w \in D(\mathcal{L}^*)} \frac{(v,w)}{\|\mathcal{L}^* w\|_0} \tag{12.12}$$

defines a *problem-dependent negative norm.*[10] Using (12.12), we define the problem-dependent least-squares functional

$$J_*(v,f) = \|\mathcal{L}v - f\|_*^2 \tag{12.13}$$

and the LL* least-squares principle:

$$\min_{v \in R(\mathcal{L}^*)} J_*(v,f)\,. \tag{12.14}$$

Note that, in (12.14), minimization is over the range of $\mathcal{L}^*$. Formally, the LL* principle (12.14) is not practical because it requires computation of a problem-dependent negative inner product. However, because of the following result, in actuality the minimizer of (12.14) can be computed in a very practical manner.

---

[10] Similarly to the proof of Theorem A.1, one can show that $\|v\|_* = (v,s)$, where $s \in D(\mathcal{L}^*)$ solves the weak equation $(\mathcal{L}^* s, \mathcal{L}^* w) = (v,w)$ for all $w \in D(\mathcal{L}^*)$.

Proposition 12.1 *Assume that u is a solution of* (3.16). *Then,*

$$\min_{v \in R(\mathcal{L}^*)} \|\mathcal{L}v - f\|_*^2 = \min_{w \in D(\mathcal{L}^*)} \|\mathcal{L}^* w - u\|_0^2 . \tag{12.15}$$

Proof. Using $\mathcal{L}u = f$ and the negative norm definition (12.12), we have that

$$\|\mathcal{L}v - f\|_* = \sup_{w \in D(\mathcal{L}^*)} \frac{(\mathcal{L}(v-u), w)}{\|\mathcal{L}^* w\|_0} = \sup_{w \in D(\mathcal{L}^*)} \frac{((v-u), \mathcal{L}^* w)}{\|\mathcal{L}^* w\|_0} = \|v-u\|_0 .$$

The proposition follows by noting that because $v \in R(\mathcal{L}^*)$, then $v = \mathcal{L}^* w$ for some $w \in D(\mathcal{L}^*)$. □

This result shows that minimizing the residual of (3.16) in the problem-dependent negative norm (12.12) is the same as minimizing the error $(v-u)$ in the $L^2(\Omega)$ norm; see [100]. Thus, instead of minimizing the negative norm functional (12.13) over the range of $\mathcal{L}^*$, we can minimize the $L^2(\Omega)$ functional

$$J_{\mathcal{L}^*}(w,u) = \|\mathcal{L}^* w - u\|_0^2 \tag{12.16}$$

over w belonging to the domain of $\mathcal{L}^*$. The following theorem shows that this can be done by solving a Galerkin-like weak equation.

Theorem 12.2 *Assume that u is a solution of* (3.16). *The minimizer of* $\|\mathcal{L}^* w - u\|_0^2$ *out of* $D(\mathcal{L}^*)$ *solves the problem: seek* $y \in D(\mathcal{L}^*)$ *such that*

$$(\mathcal{L}^* y, \mathcal{L}^* w) = (f, w) \qquad \forall w \in D(\mathcal{L}^*) . \tag{12.17}$$

Proof. The Euler-Lagrange equation for (12.16) is: seek $y \in D(\mathcal{L}^*)$ such that

$$(\mathcal{L}^* y, \mathcal{L}^* w) = (u, \mathcal{L}^* w) \qquad \forall w \in D(\mathcal{L}^*) .$$

The theorem follows by using (12.11) and the fact that $\mathcal{L}u = f$. □

Corollary 12.3 *The minimizer of the LL* least-squares principle* (12.14) *is given by* $u = \mathcal{L}^* y$, *where* $y \in D(\mathcal{L}^*)$ *is the solution of the weak equation* (12.17). □

It is instructive to compare the strong equation (3.32) that arises from a CLSP with the strong form of (12.17). In the former case, the differential operator whose weak Galerkin form coincides with the least-squares variational problem is given by $\mathcal{L}^*\mathcal{L}$; see Section 3.2.2. In contrast, the strong form of (12.17) is the differential equation

$$\mathcal{L}\mathcal{L}^* y = f . \tag{12.18}$$

Therefore, an alternative interpretation of LL* least-squares principles is that they are Galerkin methods applied to a reformulation of (3.16) in terms of "potential" or "dual" variables, i.e., y in (12.18).

Discrete LL* least-squares principles

Formally, a compliant LL* discrete least-squares principle (DLSP) is given by $\{J_*, X^h\}$, where X^h is a finite element subspace of $R(\mathcal{L}^*)$. As already noted, this DLSP is not practical because it requires computation of a problem-dependent negative inner product.

Corollary 12.3 suggests an alternative approach for computing an approximate minimizer of (12.14). We first solve (12.17) by a finite element method and then apply $\mathcal{L}^*$ to its solution to generate an approximate solution of (12.14). If $\mathcal{L}$ is a first-order operator, then so is $\mathcal{L}^*$ and the weak problem (12.17) can be approximated by standard finite element spaces.

The main advantage of this approach is that we approximate the "potential" y rather than the original solution u. We expect y to be more regular than u and so, even if $\mathcal{L}^*$ is not homogeneous elliptic, it can be augmented by redundant equations without compromising its ability to recover the desired solution. In contrast, if the primal operator $\mathcal{L}$ is augmented by additional equations to make it homogeneous elliptic, and u happens to be less regular, the least-squares formulation may not be able to recover the original solution.

In summary, to approximate the minimizer of (12.14), we proceed as follows. First, if $\mathcal{L}^*$ is not homogeneous elliptic, it is replaced by an augmented version, still denoted by $\mathcal{L}^*$, that is homogeneous elliptic. Second, we choose a finite element subspace W^h of $D(\mathcal{L}^*)$ and solve the discrete weak problem: *seek* $y^h \in W^h$ *such that*

$$\left(\mathcal{L}^* y^h, \mathcal{L}^* w^h\right) = \left(f, w^h\right) \qquad \forall w^h \in W^h .$$

The last step is to set $u^h = \mathcal{L}^* y^h$.

For examples of LL* methods for scalar elliptic equations and Maxwell's equations, see [258, 274].

12.3 Mimetic Reformulation of Least-Squares Finite Element Methods

We now describe an approach, developed in [71], that can be used to mitigate the deterioration in the accuracy of LSFEMs that employ the lowest-order Raviart–Thomas element on non-affine quadrilateral grids. We saw a firsthand example of this issue in Section 5.9.2, where implementation of a compatible LSFEM, using the space $\mathbf{D}^1$-, for the div–grad system (5.17) experienced a severe drop in the convergence rate for the flux variable approximation on such grids. The loss of convergence is even more severe for the div–grad system (5.18) without the reaction term for which both variables were affected.

The main goal of this section is to show that a simple modification of the compatible LSFEM (5.62) can completely eliminate these problems and make the compatible method safe to use on general shape regular quadrilateral grids, including those that are not affine. The reformulation of (5.62) is motivated by an observation that mimetic finite difference methods [314] for (5.17) do not seem to experience any difficulties on such grids. A mimetic discretization of this first-order system uses the *natural mimetic* divergence DIV and *adjoint* gradient $\overline{\text{GRAD}}$ operators; see [223, 224]. Of particular interest to us is DIV which is constructed using the coordinate-invariant definition [14, p. 188]

$$\nabla\cdot\mathbf{u}(\mathbf{x}) = \lim_{V\ni x;\,\mu(V)\to 0} \frac{\int_{\partial V}\mathbf{u}\cdot\mathbf{n}\,dS}{\mu(V)} \tag{12.19}$$

of the divergence operator.[11] The resulting discrete operator, termed "natural divergence," maps face-based values (the fluxes of **u**) onto cell-based constants, i.e., DIV operates on the same set of degrees-of-freedom as used to define the lowest-order Raviart–Thomas space $\mathbf{D}^{1}$-. This observation is key to the proposed *mimetic* reformulation of (5.62) because it allows us to extend the action of DIV in a natural way to vector fields in $\mathbf{D}^{1}$-. The reformulated compatible LSFEM is then obtained by replacing every occurrence of the analytic divergence operator $\nabla\cdot$ in (5.62) by the natural divergence DIV.

Extension of the mimetic divergence to the lowest-order Raviart–Thomas space

Let $\mathcal{T}_h$ denote a finite element partition of a bounded region $\Omega\subset\mathbb{R}^2$ into shape regular but not necessarily affine quadrilaterals and let $\mathcal{C}_1(\mathcal{T}_h)$ denote the set of all oriented faces[12] in $\mathcal{T}_h$; see (B.4) for details. In what follows, we restrict attention to the lowest-order Raviart–Thomas space $\mathbf{D}^{1}$-(Ω) defined by (B.31) with $r=1$. Recall that the unisolvent set of degrees of freedom for this space is given by the average flux across element faces; see (B.32).

The definition of the natural divergence DIV is based on the coordinate-independent characterization of $\nabla\cdot\mathbf{u}$ given in (12.19), applied to each element $\kappa\in\mathcal{T}_h$. Let $\mathcal{C}_1(\mathcal{T}_h)^*$ and $\mathcal{T}_h^*$ denote the duals of $\mathcal{C}_1(\mathcal{T}_h)$ and $\mathcal{T}_h$, respectively, i.e., collections of real numbers $\{F_i\}$ and $\{K_\kappa\}$, respectively, associated with the oriented faces and elements in the mesh. The natural divergence [224] is a mapping $\text{DIV}:\mathcal{C}_1(\mathcal{T}_h)^*\mapsto\mathcal{T}_h^*$ defined by

$$\text{DIV}(\mathbf{u}^h)|_\kappa = \frac{1}{\mu(\kappa)}\sum_{C_i\in C_1(\kappa)}\sigma_{C_i}F_i \qquad \forall\,\kappa\in\mathcal{T}_h\,, \tag{12.20}$$

where $\mathbf{u}^h\in\mathcal{C}_1(\mathcal{T}_h)^*$ and

[11] In this formula, V is a bounded region and $\mu(V)$ denotes its measure.

[12] Recall that, in two-dimensions, the "faces" of an element are one-dimensional entities.

$$\sigma_{C_i} = \begin{cases} 1 & \text{if } \mathbf{n}_{C_i} = \mathbf{n}_\kappa \\ -1 & \text{if } \mathbf{n}_{C_i} = -\mathbf{n}_\kappa . \end{cases}$$

Note that, from (B.32), it follows that the domain $\mathcal{C}_1(\mathcal{T}_h)^*$ of the natural divergence and $\mathbf{D}^{1\text{-}}(\Omega)$ have the same dimension, i.e., they are isomorphic. Similarly, it is easy to see that the range space $\mathcal{T}_h^*$ of DIV is isomorphic with the piecewise constant space $S^0(\Omega)$ defined by (B.42) with $r=0$. Finally, it is clear that the F_is coincide with the degrees of freedom that define the fields in $\mathbf{D}^{1\text{-}}(\Omega)$:

$$\mathbf{u}^h = \sum_{C_i \in \mathcal{C}_1(\mathcal{T}_h)} F_i \mathbf{u}_{C_i} \quad \forall \mathbf{u}^h \in \mathbf{D}^{1\text{-}}(\Omega),$$

where $\{\mathbf{u}_{C_i}\}$ is the basis of $\mathbf{D}^{1\text{-}}(\Omega)$ dual to the unisolvency set in (B.32). It follows that the action of DIV can be extended to the lowest-order Raviart–Thomas space $\mathbf{D}^{1\text{-}}(\Omega)$ by simply adopting (12.20) to compute the discrete divergence of $\mathbf{u}^h \in \mathbf{D}^{1\text{-}}(\Omega)$. This defines an operator DIV : $\mathbf{D}^{1\text{-}}(\Omega) \mapsto S^0(\Omega)$ that is used to reformulate the compatible LSFEM. It is easy to see that for the dual basis of $\mathbf{D}^{1\text{-}}(\Omega)$,

$$\mathrm{DIV}(\mathbf{u}_{C_i}|_\kappa) = \frac{\sigma_{C_i}}{\mu(\kappa)} \qquad \forall C_i \in \mathcal{C}_1(\kappa). \tag{12.21}$$

One can show that the natural divergence has a pointwise commuting diagram property and is a surjective mapping $\mathbf{D}^{1\text{-}}(\Omega) \mapsto S^0(\Omega)$ with a continuous lifting from $S^0(\Omega)$ into $\mathbf{D}^{1\text{-}}(\Omega)$; see [71]. This is to be contrasted with the analytic divergence that is not a surjection $\mathbf{D}^{1\text{-}}(\Omega) \mapsto S^0(\Omega)$ and satisfies a weaker form of the commuting diagram property; see [87, p. 138] and the discussion in Section B.2.3. Thus, in a sense, DIV has better properties than $\nabla\cdot$ on non-affine quadrilateral grids. We next see that these properties allow us to define more robust LSFEMs for such grids.

The reformulated compatible least-squares method

We consider the lowest-order compatible LSFEM, i.e., $X^h = G_0^1(\Omega) \times \mathbf{D}^{1\text{-}}(\Omega)$ in (5.62). The mimetic reformulation of (5.62) is then obtained by replacing the analytic divergence with DIV. The resulting *reformulated* DLSP is given by[13]

$$\begin{cases} J_{\mathrm{DIV}}(\phi^h, \mathbf{u}^h; f) \\ \qquad = \|\mathrm{DIV}(\mathbf{u}^h) + \gamma\Theta_0\phi^h - \Theta_0 f\|^2_{0,\Theta_0^{-1}} + \|\nabla\phi^h + \Theta_1^{-1}\mathbf{u}^h\|^2_{0,\Theta_1} \\ X^h = G_0^1(\Omega) \times \mathbf{D}^{1\text{-}}(\Omega) . \end{cases} \tag{12.22}$$

Minimizers of (12.22) are subject to the usual first-order optimality condition that, in the present case, takes the form: *seek* $\{\phi^h, \mathbf{u}^h\} \in G_0^1(\Omega) \times \mathbf{D}^{1\text{-}}(\Omega)$ *such that*

[13] Recall that γ is a non-dimensional real parameter that equals 0 or 1; see Chapter 5.

$$\begin{aligned}\int_\Omega (\nabla\phi^h + \Theta_1^{-1}\mathbf{u}^h)\Theta_1(\nabla\psi^h + \Theta_1^{-1}\mathbf{v}^h)\,d\Omega \\ + \int_\Omega (\mathrm{DIV}(\mathbf{u}^h) + \gamma\Theta_0\phi^h)\Theta_0^{-1}(\mathrm{DIV}(\mathbf{v}^h) + \gamma\Theta_0\psi^h)\,d\Omega \\ = \int_\Omega f\Theta_0^{-1}(\mathrm{DIV}(\mathbf{v}^h) + \gamma\Theta_0\psi^h)\,d\Omega\end{aligned} \tag{12.23}$$

for all $\{\psi^h, \mathbf{v}^h\} \in G_0^1(\Omega)\times \mathbf{D}^1\text{-}(\Omega)$.

Unlike (5.62), the reformulated LSFEM is not a compliant method. One important consequence of this fact is that the splitting in Theorem 5.37 does not extend to the mimetic reformulation (12.22). This would have required a discrete Green's identity

$$\int_\Omega \phi^h \mathrm{DIV}(\mathbf{u}^h)\,d\Omega + \int_\Omega \mathbf{u}^h\cdot\nabla\phi^h\,d\Omega = 0 \qquad \forall\{\phi^h,\mathbf{u}^h\}\in G_0^1(\Omega)\times\mathbf{D}^1\text{-}(\Omega)$$

that, in general, does not hold. Therefore, the flux computed by (12.22) is not locally conservative in the sense defined in Remark 5.38. However, this can be fixed by using the flux-correction procedure described in Section 5.9.1.

Stability and error analyses of the reformulated LSFEM are summarized in the following theorem. For details, see [71].

Theorem 12.4 *Assume that $\mathcal{T}_h$ is a shape-regular but not necessarily affine finite element partition of Ω. Then, there is a positive constant C such that*

$$\begin{aligned}C\big(\|\mathrm{DIV}(\mathbf{u}^h)\|_0 + \|\mathbf{u}^h\|_0 + \|\phi^h\|_1\big) \\ \le \|\mathrm{DIV}(\mathbf{u}^h) + \gamma\Theta_0\phi^h\|_{0,\Theta_0^{-1}} + \|\nabla\phi^h + \Theta_1^{-1}\mathbf{u}^h\|_{0,\Theta_1}\end{aligned} \tag{12.24}$$

for all $\{\phi^h,\mathbf{u}^h\}\in G_0^1(\Omega)\times\mathbf{D}^1\text{-}(\Omega)$. *If the exact solution of* (5.17) *is such that* $\phi \in H_0^1(\Omega)\cap H^2(\Omega)$ *and* $\mathbf{u}\in\mathbf{D}(\Omega)\cap[H^2(\Omega)]^2$, *then the solution obtained through the reformulated least-squares finite element method* (12.22) *satisfies the error bound*

$$\|\nabla\cdot\mathbf{u} - \mathrm{DIV}(\mathbf{u}^h)\|_0 + \|\mathbf{u}-\mathbf{u}^h\|_0 + \|\phi-\phi^h\|_1 \le Ch\big(\|\phi\|_2 + \|\mathbf{u}\|_2\big)\,. \qquad \Box \tag{12.25}$$

Practicality issues

To demonstrate the effectiveness of the mimetic reformulation approach, we compare and contrast the compatible LSFEM (5.62) with (12.22), using the same setup as in Section 5.9.2. Numerical results are summarized in Tables 12.1 and 12.2. From the data in these tables, it is clear that the mimetic reformulation restores the optimal order of convergence for all variables regardless of whether or not the reaction terms is included in the div–grad system.

To verify that the reduced order of convergence of the flux in the compatible LSFEM is real, we provide, in addition to the convergence rates with respect to the analytic divergence, the estimated convergence rates of the flux with respect to

Error	Method	65 × 65	129 × 129	Order
$\|\phi - \phi^h\|_0$	LS	0.216246E-06	0.541219E-07	1.9984
	RLS	0.219502E-06	0.549429E-07	1.9982
$\|\nabla(\phi - \phi^h)\|_0$	LS	0.7994094E-02	0.4005637E-02	0.9969
	RLS	0.7994033E-02	0.4005629E-02	0.9969
$\|\mathbf{u} - \mathbf{u}^h\|_0$	LS	0.3620386E-01	0.2254463E-01	0.6834
	RLS	0.3247333E-01	0.1620301E-01	1.0030
$\|\nabla\cdot\mathbf{u} - \nabla\cdot\mathbf{u}^h\|_0$	LS	0.1089084E+01	0.1073177E+01	0.0212
$\|\nabla\cdot\mathbf{u} - \mathrm{DIV}(\mathbf{u}^h)\|_0$	LS	0.2849267E+00	0.2101464E+00	0.4392
	RLS	0.2197499E+00	0.1098694E+00	1.0001

Table 12.1 Error data and estimated orders of convergence for the compatible LSFEM (LS) and its mimetic reformulation (RLS): problem (5.17) with reaction term ($\gamma = 1$).

Error	Method	65 × 65	129 × 129	Order
$\|\phi - \phi^h\|_0$	LS	0.1497621E-02	0.1436716E-02	0.0599
	RLS	0.9470421E-06	0.2369894E-06	1.9986
$\|\nabla(\phi - \phi^h)\|_0$	LS	0.1358298E-01	0.1138904E-01	0.2542
	RLS	0.8010383E-02	0.4007615E-02	0.9991
$\|\mathbf{u} - \mathbf{u}^h\|_0$	LS	0.3630391E-01	0.2269303E-01	0.6779
	RLS	0.3247372E-01	0.1620305E-01	1.0030
$\|\nabla\cdot\mathbf{u} - \nabla\cdot\mathbf{u}^h\|_0$	LS	0.1089083E+01	0.1073176E+01	0.0212
$\|\nabla\cdot\mathbf{u} - \mathrm{DIV}(\mathbf{u}^h)\|_0$	LS	0.2854252E+00	0.2108029E+00	0.4372
	RLS	0.2197499E+00	0.1098694E+00	1.0001

Table 12.2 Error data and estimated orders of convergence for the compatible LSFEM (LS) and its mimetic reformulation (RLS): problem (5.18) without reaction term ($\gamma = 0$).

DIV. Note that application of the mimetic divergence to the finite element solution of (5.62) can be interpreted as a post-processing step. Interestingly, the data in Tables 12.1 and 12.2 show that this step does somewhat improve convergence of the flux variable. However, it is important to remember that post-processing of the flux does not help in any way to alleviate the loss of convergence in the potential variable when $\gamma = 0$; see Table 12.2.

Implementation of the reformulated least-squares method

An existing finite element program for the compatible LSFEM (5.62) can be trivially converted to its mimetic reformulation by changing just a few lines of code. For the lowest-order Raviart–Thomas space, the mapping F_K that takes the reference quadrilateral (2-cube) $\hat{\kappa} = [-1,1]^2$ into $\kappa \in \mathcal{T}_h$ is usually a bilinear function. In this case, there holds the identity (see [183, p. 105])

$$\mu(\kappa) = \det(DF_\kappa(0,0))\mu(\hat{\kappa}).$$

From this identity, (12.20), and the fact that $\mu(\hat{\kappa}) = 4$, it follows that

$$\mathrm{DIV}(\mathbf{u}_{C_i}|_\kappa) = \frac{\sigma_{C_i}}{\mu(\kappa)} = \frac{\sigma_{C_i}}{4\det(DF(0,0))}.$$

As a result, conversion of a finite element code for the compatible LSFEM to one that implements the mimetic reformulation (12.22) amounts to replacing multiple calls to the function that computes $\nabla \cdot \mathbf{u}_{C_i}(\mathbf{x})$ at quadrature points, along with the computation of $\det(DF)$ at those points, by a single call to compute $\det(DF(0,0))$ combined with a few Boolean operations related to the orientation choice σ_{C_i}.

12.4 Collocation Least-Squares Finite Element Methods

In the abstract least-squares framework of Section 3.2, the least-squares miminization step precedes the discretization step. The main reason for doing so is to ensure that the resulting LSFEMs are always connected to a well-posed CLSP. In this section, we review LSFEMs for which the discretization step is taken prior to the least-squares miminization step. Such methods are commonly known as *least-squares collocation, point least-squares, point matching,* or *overdetermined collocation* methods; see [153, 277].

Point matching collocation methods

The main idea is as follows. Consider the linear boundary value problem (3.16) and a set of sufficiently smooth functions $\{\phi_i\}_{i=1}^N$ defined on $\overline{\Omega}$. The approximate solution u^h is sought as a linear combination of these functions:

$$u(\mathbf{x}) \approx u^h(\vec{a},\mathbf{x}) = \sum_{i=1}^N a_i\phi_i(\mathbf{x}).$$

To determine the unknown coefficient vector $\vec{a} = (a_1, a_2, \ldots, a_N)^\mathsf{T}$, a least-squares collocation method proceeds as follows. Let

$$R_i\big(\mathcal{L}(\vec{a},\mathbf{x})\big), \quad i = 1,\ldots,M, \qquad \text{and} \qquad R_i\big(\mathcal{B}(\vec{a},\mathbf{x})\big), \quad l = 1,\ldots,L,$$

denote residuals of the equation and the boundary condition in (3.16), respectively. One then chooses a finite set of points $\{\mathbf{x}_s\}_{s=1}^{S_1}$ in Ω and another set of points $\{\mathbf{x}_s\}_{s=S_1+1}^{S}$ on $\partial\Omega$. Then, a least-squares functional is defined by summing the weighted squares of the residuals evaluated at the points $\mathbf{x}_s$:

$$J(\vec{a}) = \sum_{i=1}^{M}\sum_{s=1}^{S_1} \alpha_{is}\Big(R_i\big(\mathcal{L}(\vec{a},\mathbf{x}_s)\big)\Big)^2 + \sum_{i=1}^{L}\sum_{s=S_1+1}^{S} \beta_{is}\Big(R_i\big(\mathcal{B}(\vec{a},\mathbf{x}_s)\big)\Big)^2 \tag{12.26}$$

and an $S\times N$ linear system

$$\mathsf{A}\vec{a} = \vec{b} \tag{12.27}$$

for the unknown coefficients is obtained by setting to zero the derivatives of (12.26) with respect to a_is. Typically, the number S of collocation points is much larger than the number N of "basis" functions so that (12.27) is an overdetermined linear system. The final step is to solve (12.27) by algebraic least-squares, i.e., the collocation least-squares solution of (3.16) is determined by solving the normal equations

$$\mathsf{A}^T\mathsf{A}\vec{a} = \mathsf{A}^T\vec{b}.$$

Methods formulated along these lines have been used for the numerical solution of the Navier–Stokes equations [277] and hyperbolic problems, including the shallow water equations [263,264,353,354]. For numerous other applications of collocation least-squares methods, see [153].

Evidently, when the number of collocation points S equals the number of degrees of freedom N in $u^h(\vec{a},\mathbf{x})$, these methods reduce to a standard collocation procedure. Similarly, if $u^h(\vec{a},\mathbf{x})$ is defined using a finite element space and the collocation points and weights correspond to a quadrature rule, then collocation least-squares formulation is equivalent to a bona fide LSFEM formulation in which integration has been replaced by quadrature; see [230].

Collocation LSFEMs offer some specific advantages. For example, because only a finite set of points $\mathbf{x}_s$ in the domain Ω need be specified, collocation LSFEMs are attractive for problems posed on irregularly shaped domains; see [263]. On the other hand, because the normal equations tend to be ill-conditioned, such methods require additional techniques such as scaling or orthonormalization, in order to obtain a reliable solution; see [153]. Alternately, and perhaps more stably, (12.27) can be solved using the singular value decomposition of A.

Subdomain collocation methods

Collocation LSFEMs, just as standard collocation methods, use point-by-point matching criteria to define the discrete problem. Instead of using a set of points, we can also "collocate" over a set of subdomains of Ω.

In such *subdomain* collocation methods, the discrete problems are obtained by averaging the differential equations over each subdomain. Here, for an illustration of this approach, we consider the subdomain collocation LSFEM of [116] in two space dimensions. This method uses standard $C^0(\Omega)$ finite element basis functions and so the first step is to replace (3.16) by an equivalent first-order system, if the problem is not already in this form. Thus, in what follows, $\mathcal{L}$ is a square $M\times M$

matrix of first-order differential expressions; see Section D.1 for reduction of PDEs to first-order form.

Assume that $\mathcal{T}_h$ is a regular partition of Ω into triangles κ_k and let $X^h = G^1$. The finite elements κ_k also serve as collocation subdomains. We let K denote the number of triangles in $\mathcal{T}_h$. For simplicity, we assume that the finite element functions in X^h satisfy the essential boundary conditions in (3.16). Then, a set of discrete equations is formed by averaging[14] the boundary value problem (3.16) over each of the triangles $\kappa_k \in \mathcal{T}_h$:

$$\int_{\kappa_k} \mathcal{L}u^h \, d\Omega = \int_{\kappa_k} f \, d\Omega \quad \text{for } k = 1, \ldots, K. \tag{12.28}$$

Note that (12.28) not only involves a separate average over each triangle, but if (3.16) represents a system of equations, each equation in that system is averaged separately as well. After choosing a basis for X^h, it is not difficult to see that (12.28) is equivalent to a linear algebraic system

$$\mathsf{A}\vec{a} = \vec{f} \tag{12.29}$$

having MK equations in approximately MN unknowns, where M denotes the number of equations and unknown variables in the system (12.28) and N denotes the number vertices in $\mathcal{T}_h$. Thus, in (12.29), there are roughly twice as many equations as unknowns. The subdomain-Galerkin/least-squares method of [116] consists of forming the matrix A and subsequently solving (12.29) by using the normal equations or some other means for solving overdetermined systems. If the data f are sufficiently smooth and the first-order system is homogeneous elliptic, it can be shown (see [116]) that the resulting method is optimal in the sense that

$$\|u - u^h\|_1 \le C_1 h \|f\|_1 \qquad \text{and} \qquad \|u - u^h\|_0 \le C_0 h^2 \|f\|_1 .$$

The discretization step in (12.28) can also be interpreted as an application of a nonstandard Galerkin method to the system (3.16) in which the test space consists of piecewise constant test functions with respect to $\mathcal{T}_h$. Alternatively, one can think of the equations in (12.28) as being obtained by a finite volume discretization procedure. Similar subdomain collocation LSFEMs have also been developed for the numerical solution of Maxwell's equations; see [109].

12.5 Restricted Least-Squares Finite Element Methods

LSFEMs for incompressible flow problems, defined using the ADN setting for the energy balances, satisfy the continuity equation only approximately. Such methods

---

[14] If higher-order and presumably more accurate finite element spaces are used, then one would not just average the residual over each triangle, but also set higher moments of the residual to zero over each triangle.

were discussed in Section 7.4 of Chapter 7 and in Chapter 8. From the examples in Sections 7.6.2 and 7.6.3, we know that the minimal-order condition is a prerequisite for these LSFEM to have optimal accuracy and good conservation properties. This means, e.g., that for the quasi-norm equivalent DLSP (7.54) that is based on the velocity–vorticity–pressure (VVP) systems (7.4) or (7.12), the velocity field must be approximated by finite element spaces G^{r+1} of at least second degree; see Remark 7.15. For such spaces, Theorem 7.14 implies that

$$\|\nabla \cdot \mathbf{u}^h\|_0 \leq Ch^{r+1},$$

where $r \geq 1$ is the integer in the definition of $X^h_{r+,0}$ in (7.54). Moreover, in Section 7.6, we saw that mass conservation for these methods can be further enhanced by using moderately sized multiplicative weights in the continuity equation residual term in the least-squares functional.

In contrast, equal-order finite elements of first degree not only violate the minimal order condition that is necessary for Theorem 7.14 to hold, but also exhibit poor mass conservation, as demonstrated in [118]. This observation motivated the development, in that paper, of the *restricted* LSFEM as a way to improve mass conservation for low-order velocity finite element spaces.

The main idea of the restricted LSFEM is to retain the first-degree finite element space G^1 for all variables in the least-squares functional but to impose the continuity equation in (7.4) or (7.12) as an explicit constraint on each finite element. To satisfy the constraint, the restricted LSFEM uses discrete Lagrange multipliers. As a result, this approach does not fit into the framework of Chapter 3 because the resulting weak equation is of saddle-point type.

To describe the restricted LSFEM for (7.4) or (7.12), let $\mathcal{L}$ denote the first-order VVP Stokes operator and $u^h = \{\mathbf{u}^h, \omega^h, p^h\}$. We consider a finite element partition $\mathcal{T}_h$ of Ω into K elements κ_k and set $X^h = [G^1]^d \cap [H^1_0(\Omega)]^d \times [G^1]^{2d-3} \times G^1$. With $J_0(u;f) = \|\mathcal{L}u - f\|_0^2$, the restricted least-squares principle is then given by the following constrained minimization problem:

$$\left\{ \begin{array}{l} \displaystyle \min_{u^h \in X^h} J_0(u^h;f) \\ \text{subject to} \quad \displaystyle\int_{\kappa_k} \nabla \cdot \mathbf{u}^h \, d\Omega = 0 \qquad \text{for } k = 1,\ldots,K. \end{array} \right. \tag{12.30}$$

To solve (12.30), we introduce a piecewise constant Lagrange multiplier space $P^h = S^0$ and seek the stationary point $\{u^h, \lambda^h\}$ of the Lagrangian

$$L(u^h, \lambda^h) = J_0(u^h;f) - \int_\Omega \lambda^h \nabla \cdot \mathbf{u}^h \, d\Omega. \tag{12.31}$$

The first-order optimality condition for (12.31) is: *seek* $\{u^h, \lambda^h\} \in X^h \times P^h$ *such that*

$$\int_\Omega \mathcal{L}u^h \cdot \mathcal{L}v^h \, d\Omega - \int_\Omega \lambda^h \nabla \cdot \mathbf{v}^h \, d\Omega - \int_\Omega \mu^h \nabla \cdot \mathbf{u}^h \, d\Omega$$
$$= \int_\Omega \mathbf{f} \cdot \mathcal{L}v^h \, d\Omega \qquad \forall \{v^h, \mu^h\} \in X^h \times P^h.$$

The main advantage of the restricted LSFEM is that one obtains exact element-wise mass conservation for piecewise linear velocity approximations. The main drawbacks are the saddle-point structure of the resulting linear system of equations and the larger number of unknowns.

12.6 Optimization-Based Least-Squares Finite Element Methods

The methods discussed in this section should not be confused with the LSFEMs for control and optimization problems discussed in Chapter 11. There, we used least-squares principles to *solve optimization and control problems* constrained by PDEs whereas here we use optimal control or optimization methods to *define a LSFEM*. Specifically, we consider LSFEMs for PDEs derived by transforming the original boundary value problem into an optimal control or optimization problem whose objective functional is a least-squares type functional.

To describe the main idea of these methods, consider the following nonlinear problem:

$$-\Delta\phi - G(\phi) = 0 \qquad \text{in } \Omega \tag{12.32}$$

along with the boundary condition $\phi = 0$ on $\partial\Omega$. Recall that an H^{-1} least-squares functional for (12.32) is given by

$$J_{-1}(\phi) = \|\Delta\phi + G(\phi)\|_{-1}^2 . \tag{12.33}$$

From Section 5.2 we know that the minimization of (12.33) over $H_0^1(\Omega)$ results in a CLSP that (when linearized) fits the abstract theory of Chapter 3.

In contrast, the optimization LSFEM method for (12.32) is based on the minimization of the negative norm functional

$$J_{-1}(\phi,\xi) = \|\Delta(\phi - \xi)\|_{-1}^2 \tag{12.34}$$

subject to the constraints

$$-\Delta\xi = G(\phi) \quad \text{in } \Omega \qquad \text{and} \qquad \xi = 0 \quad \text{on } \partial\Omega . \tag{12.35}$$

In the context of optimization and control problems governed by PDEs (see Chapter 11), one can identify ϕ with the control variable, ξ with the state variable, (12.35) with the state equation, and (12.34) with the objective functional.

Formally, (12.34) does not appear to be a practical formulation because it involves the minus one norm. The key difference that does make (12.34) practical is that the Laplacian operator in that functional also acts on the "state" variable ξ.

Then, we can use the identity (A.14) that states that

$$\|\Delta\phi\|_{-1} = \|\nabla\phi\|_0 \qquad \forall \phi \in H_0^1(\Omega).$$

As a result, (12.34) is equivalent to the practical objective functional

$$J_0(\phi,\xi) = \|\nabla(\phi-\xi)\|_0^2. \tag{12.36}$$

To summarize, the optimization-based least-squares principle for (12.32) can be stated as follows:

minimize $J_0(\phi,\xi)$*, given by* (12.36)*, over* $\phi \in H_0^1(\Omega)$*, subject to the state equation* (12.35).

To solve this optimization problem, one can use an abstract version of the conjugate gradient method; see [92]. At each iteration, this method would require the solution of two Dirichlet problems (12.35) for the computation of the descent direction.

The optimization-based LSFEM approach has been extended to nonlinear flow problems, including compressible flows (see [92, 93, 185]) and the Navier–Stokes equations (see [92, 184].) We show how this approach can be applied to the Navier–Stokes equations (1.67). We define the objective functional similarly to (12.34) as

$$J_{-1}(\mathbf{u},\xi) = \nu\|\Delta(\xi-\mathbf{u})\|_{-1}^2. \tag{12.37}$$

The role of the control is now played by the velocity field $\mathbf{u}$ and ξ is a state variable governed by the following Stokes equation:

$$\begin{cases} -\nu\Delta\xi + \nabla q = -\mathbf{u}\cdot\nabla\mathbf{u} & \text{in } \Omega \\ \nabla\cdot\xi = 0 & \text{in } \Omega \\ \xi = 0 & \text{on } \partial\Omega. \end{cases} \tag{12.38}$$

Finally, we take the space

$$Z = \left\{ \mathbf{v} \in [H_0^1(\Omega)]^d \mid \int_\Omega s\nabla\cdot\mathbf{v}\,d\Omega = 0 \quad \forall s \in L_0^2(\Omega) \right\}$$

defined in (1.60), as the minimization space. The optimization-based least-squares principle for (1.67) is then given by:

minimize $J_{-1}(\mathbf{u},\xi)$ *given by* (12.37) *over* $\mathbf{u} \in Z$*, subject to the state equation* (12.38).

Again using the identity (A.14), it follows that

$$J_{-1}(\mathbf{u},\xi) = \nu\|\nabla(\xi-\mathbf{u})\|_0^2$$

which is a practical functional. The optimal control problem can be solved as before, by an abstract conjugate gradient algorithm. In the present case, the computation of the descent direction at each iteration involves the solution of several Stokes problems; see [92, 184].

12.7 Least-Squares Finite Element Methods for Advection–Diffusion–Reaction Problems

This section surveys least-squares finite element methods for the scalar advection–diffusion–reaction equation (1.65), first encountered in Section 1.4.5, and a time-dependent version of that problem. In what follows we work with (1.65) written in the conservative form[15]

$$\begin{cases} -\nabla\cdot(\varepsilon\nabla\phi - \mathbf{b}\phi) + c\phi = f & \text{in } \Omega \\ \phi = g & \text{on } \partial\Omega\,. \end{cases} \tag{12.39}$$

The term $\varepsilon\nabla\phi$ represents the diffusive flux and $\mathbf{b}\phi$ is the advective flux. For simplicity we assume that $\varepsilon > 0$ is constant and that a homogeneous boundary condition is imposed, i.e., $g = 0$. Regarding $c(\mathbf{x})$ and $\mathbf{b}(\mathbf{x})$, we retain the hypotheses made in Assumption 10.2 as well as assumption (10.13), repeated here for convenience: there exists a positive constant γ_0 such that

$$c(\mathbf{x}) + \frac{1}{2}\nabla\cdot\mathbf{b}(\mathbf{x}) \geq \gamma_0 \qquad \forall \mathbf{x}\in\Omega\,.$$

This assumption (see [244]) guarantees that (12.39) has a unique weak solution $\phi \in H_0^1(\Omega)$ and that that solution satisfies the stability bound

$$\sqrt{\varepsilon}\|\nabla\phi\|_0 + \|\phi\|_0 \leq \|f\|_0\,.$$

The time-dependent version of the advection–diffusion–reaction problem is given by

$$\begin{cases} \dfrac{\partial\phi}{\partial t} - \nabla\cdot(\varepsilon\nabla\phi - \mathbf{b}\phi) + c\phi = f & \text{for } (\mathbf{x},t)\in\Omega\times(0,T] \\ \phi = g & \text{for } (\mathbf{x},t)\in\partial\Omega\times(0,T] \\ \phi = \phi_0(\mathbf{x}) & \text{for } t=0 \text{ and } \mathbf{x}\in\Omega\,, \end{cases} \tag{12.40}$$

where ϕ_0 is a given initial data function.

The use of least-squares principles for (12.39) and (12.40) is easily justified by noting that, even in the case when one is interested only in approximating the primal variable ϕ, application of Galerkin methods results in non-symmetric linear systems. In principle, the LSFEMs formulated in Chapters 5 and 9 can be extended without much difficulty to the steady state problem (12.39) and its time-dependent version

[15] For clarity we retain the notations ε and c commonly used in the literature for the diffusivity and the reaction coefficients, respectively, instead of the notation used in Chapters 5 and 9.

(12.40), respectively, by simply including the advective term in the least-squares functional. The resulting LSFEMs are perfectly appropriate for settings in which the diffusivity coefficient ε is sufficiently large, at least compared to the mesh size h used to effect the discretization.

Based on the examples presented in Section 10.7.2, one can also expect that these LSFEMs perform reasonably well when the problem becomes advection-dominated, or when ε is small compared to the mesh size h. Indeed, Figures 10.2–10.6 show that solutions of straightforward $L^2(\Omega)$-norm LSFEMs remain largely free of spurious oscillations in the limit case $\varepsilon = 0$ when (12.39) reduces to the hyperbolic advection-reaction problem (10.6). In particular, from these figures we see that the oscillations remain confined to a narrow strip near the solution discontinuity and are no worse than the localized oscillations in the streamline–upwinding–Petrov Galerkin (SUPG) formulation.

In light of these observations the goals of this section are twofold. On the one hand, we demonstrate that important properties of the LSFEMs formulated earlier such as, e.g., the optimal $L^2(\Omega)$ error estimates for compatible LSFEMs in Theorem 5.35, continue to hold in the presence of the advective term. On the other hand, the alternative LSFEMs presented in this section illustrate some of the possible ways in which the performance of conventional LSFEMs can be further improved for advection-dominated problems. The necessarily brief exposition of these topics is based primarily on material from [95, 160, 210, 257, 349]. Details can be found in these papers and the references cited therein.

Energy balances for the steady-state problem

Formulation of LSFEMs for the advection–diffusion–reaction problem (12.39) follows the familiar pattern from Chapter 5 and begins with transformation of (12.39) into an equivalent first-order system:

$$\begin{cases} \nabla\cdot\mathbf{u} + c\phi = f & \text{in } \Omega \\ \varepsilon\nabla\phi - \mathbf{b}\phi + \mathbf{u} = 0 & \text{in } \Omega \\ \phi = 0 & \text{on } \partial\Omega\,. \end{cases} \tag{12.41}$$

This system is similar to the div–grad formulation (5.17) in Chapter 5 except that in (12.41) the new variable $\mathbf{u}$ represents the negative *total flux*.[16]

The next step is to derive energy balances for (12.41) that serve as a basis for obtaining well-posed continuous least-squares principles (CLSPs). As in Chapter 5 we are led to two types of energy balances that correspond to the ADN and the vector-operator settings, respectively. Assuming, for simplicity, that solution spaces are constrained by the Dirichlet boundary condition in (12.41), an ADN energy

[16] An alternative first-order reformulation of (12.39) can be obtained by keeping $\mathbf{u}$ to be the negative diffusive flux, i.e., $\mathbf{u} = -\varepsilon\nabla\phi$. In this case, the first equation in (12.41) is replaced by $\nabla\cdot\mathbf{u} + \nabla\cdot(\mathbf{b}\phi) + c\phi = f$. Because properties of LSFEMs based on this reformulation are comparable to those obtained from (12.41), this setting is not discussed.

balance for (12.41) is given by

$$\varepsilon C\big(\|\phi\|_1 + \|\mathbf{u}\|_0\big) \le \|\nabla\cdot\mathbf{u} + c\phi\|_{-1} + \|\varepsilon\nabla\phi - \mathbf{b}\phi + \mathbf{u}\|_0\,, \tag{12.42}$$

whereas the vector-operator setting gives rise to the energy balance

$$\varepsilon C\big(\|\phi\|_G + \|\mathbf{u}\|_D\big) \le \|\nabla\cdot\mathbf{u} + c\phi\|_0 + \|\varepsilon\nabla\phi - \mathbf{b}\phi + \mathbf{u}\|_0\,; \tag{12.43}$$

see [95,160,210,257,349]. Clearly, (12.42) and (12.43) are analogues of (5.32) and (5.41), respectively, except that here we have exposed the dependence of the lower comparability constant on the diffusivity parameter ε.

Least-squares finite element methods in the Agmon–Douglis–Nirenberg setting

LSFEMs for (12.41), based on the ADN energy balance (12.42), fall into the same two categories as the corresponding methods from Section 5.6.1. We have the choice of a *weighted* $L^2(\Omega)$-norm quasi-norm-equivalent DLSP which extends (5.60) to the advection–diffusion–reaction problem, or a *discrete negative norm* DLSP which extends method (5.61). Properties of the resulting LSFEMs for (12.41) do not differ substantially from those of (5.60) and (5.61) to warrant further consideration.

Instead, we proceed with a more interesting *streamline-upwind* reformulation of (12.41), proposed in [257], which improves the original energy balance (12.42) of the advection–diffusion–reaction problem by a streamline diffusion term $\|\mathbf{b}\cdot\nabla\phi\|_0$ and reduction of the factor ε in (12.42) to $\sqrt{\varepsilon}$. The resulting strengthened energy balance leads to LSFEMs with improved stability for advection-dominated problems. The key idea is to modify the differential equation in (12.39) to

$$-\nabla\cdot(\varepsilon\nabla\phi - \mathbf{b}\phi) + c\phi + \tau\nabla\cdot\big(\mathbf{b}(\mathcal{L}\phi - f)\big) = f\,, \tag{12.44}$$

where $\mathcal{L}$ is the differential operator in (12.39) and τ is stabilization parameter. This equation can be interpreted as strong form of a consistently stabilized method for (12.39) (see Section 2.1.2) in the sense that a weak Galerkin formulation of (12.44) is similar to the classical SUPG method of [215]. The next step is to replace (12.44) by the "first-order" system[17]

$$\begin{cases} \nabla\cdot\mathbf{u} + c\phi = f_\tau & \text{in } \Omega \\ \varepsilon\nabla\phi - \mathbf{b}\phi - \tau\mathbf{b}\mathcal{L}\phi + \mathbf{u} = 0 & \text{in } \Omega\,, \end{cases} \tag{12.45}$$

where[18] $f_\tau = f + \tau\nabla\cdot(\mathbf{b}f)$.

[17] We use quotation marks because (12.45) includes the second order operator $\mathcal{L}$ and so it is not a true first-order system.

[18] Alternatively, one can view (12.45) as a modification of (12.41) in which the total flux variable is augmented by the term $\tau\mathbf{b}\mathcal{L}\phi$ and simultaneously correcting the right-hand side in the first equation to maintain consistency.

Below (see (12.53)) we present LSFEMs that use (12.45) directly and deal with the second-order term in $\mathcal{L}$ much in the same way as described in Remark 2.3. The approach pursued in [257] differs in that it discards the diffusive flux from $\mathcal{L}$ to obtain a true first-order version of (12.45):

$$\begin{cases} \nabla\cdot\mathbf{u}+c\phi = f_\tau & \text{in } \Omega \\ \varepsilon\nabla\phi-\mathbf{b}\phi-\tau\mathbf{b}\big(\nabla\cdot(\mathbf{b}\phi)+c\phi\big)+\mathbf{u}=0 & \text{in } \Omega\,. \end{cases} \tag{12.46}$$

Because the diffusive flux is deleted from $\mathcal{L}$, the first-order system (12.46) is not an equivalent reformulation of (12.39). However, one can show that solutions of (12.46) are close to solutions of (12.39) and that its energy balance is better than that for (12.42) which holds for the original, equivalent first-order system (12.41). To this end we need the parameter-dependent norm

$$\|\phi\|_{1,\tau}^2 = \varepsilon\|\nabla\phi\|_0^2+\|\phi\|_0^2+\tau\|\mathbf{b}\cdot\nabla\phi\|_0^2,$$

defined for all $\phi\in H_0^1(\Omega)$, and its dual norm

$$\|f\|_{-1,\tau} = \sup_{\phi\in H_0^1(\Omega)}\frac{(f,\phi)}{\|\phi\|_{1,\tau}},$$

defined for all $f\in H^{-1}(\Omega)$. It is shown in [257] that, for sufficiently small τ, the first-order system (12.46) admits the following energy balance

$$\begin{aligned} &C\big(\|\phi\|_{1,\tau}+\|\nabla\cdot\mathbf{u}\|_{-1,\tau}\big) \\ &\quad\le\|\nabla\cdot\mathbf{u}+c\phi\|_{-1,\tau}+\frac{1}{\sqrt{\varepsilon}}\|\varepsilon\nabla\phi-\mathbf{b}\phi-\tau\mathbf{b}\big(\nabla\cdot(\mathbf{b}\phi)+c\phi\big)+\mathbf{u}\|_0\,. \end{aligned} \tag{12.47}$$

Furthermore, if $\{\phi_\tau,\mathbf{u}_\tau\}$ is solution of (12.46) and ϕ is solution of (12.39),

$$\|\phi_\tau-\phi\|_{1,\tau}\le C\varepsilon\sqrt{\tau}\|\Delta\phi\|_0\,.$$

When $\varepsilon<h$ the choice $\tau\approx h$ makes the consistency error comparable to the discretization error; see [257].

The energy balance (12.47) gives rise to a well-posed *negative-norm* CLSP

$$\begin{cases} J_{-1,\tau}(\phi,\mathbf{u};f)=\|\nabla\cdot\mathbf{u}+c\phi-f_\tau\|_{-1,\tau}^2 \\ \qquad\qquad +\dfrac{1}{\varepsilon}\|\varepsilon\nabla\phi-\mathbf{b}\phi-\tau\mathbf{b}\big(\nabla\cdot(\mathbf{b}\phi)+c\phi\big)+\mathbf{u}\|_0^2 \\ X_{-1}=H_0^1(\Omega)\times[L^2(\Omega)]^d, \end{cases} \tag{12.48}$$

which is an analogue of (5.50). Of course, just like that method, (12.48) is not practical because it requires computation of a negative norm. Transformation of (12.48) into a practical norm-equivalent DLSP can be accomplished along the same lines as it is done for (5.50). First, using the same techniques as in the proof of Theorem

A.1, one can show that

$$\|f\|_{-1,\tau}^2 = \left(\mathcal{L}_\tau^{-1} f\,,\,f\right)_0,$$

where $\mathcal{L}_\tau^{-1} : H^{-1}(\Omega) \mapsto H_0^1(\Omega)$ is the solution operator of the weak problem: *seek* $\phi \in H_0^1(\Omega)$ *such that*

$$\varepsilon\big(\nabla\phi,\nabla\psi\big) + \big(\phi,\psi\big) + \tau\big(\mathbf{b}\cdot\nabla\phi,\mathbf{b}\cdot\nabla\psi\big) = \big(f,\psi\big) \quad \forall\,\psi \in H_0^1(\Omega)\,. \tag{12.49}$$

Then, one proceeds as in Section B.4.2 and constructs a discrete negative norm using definition (B.99):

$$\|f\|_{-h,\tau} = \left((\alpha h^2\mathcal{I} + \mathcal{K}_\tau^h)f\,,\,f\right)^{1/2}$$

The only difference is that, in the present setting, $\mathcal{K}_\tau^h$ is a spectrally equivalent approximation of a finite element Galerkin solution operator for (12.49).

Replacement of the negative norm in (12.48) by its discrete approximation $\|\cdot\|_{-h,\tau}$ completes transformation of that CLSP into a practical norm-equivalent DLSP which is a counterpart of (5.61). In [257], it is shown that solutions of the corresponding LSFEM satisfy the error estimates

$$\|\phi - \phi^h\|_{1,\tau} \sim O\big(h\varepsilon^{-\frac{1}{2}} + \tau h\varepsilon^{-\frac{1}{2}} + h\big)$$

if $\phi \in H^2(\Omega)$ and

$$\|\phi - \phi^h\|_{1,\tau} \sim O\big(h^2\varepsilon^{-\frac{1}{2}} + \tau h\varepsilon^{-\frac{1}{2}} + h^2\big)$$

if $\phi \in H^3(\Omega)$. As a result, choosing $h \approx \varepsilon$ and $\tau \approx h$ recovers the $O(h^{\frac{3}{2}})$ estimate in the classical SUPG method; see [244]. Choosing $h \approx \sqrt{\varepsilon}$ results in a more practical setting for which the error in both ϕ and $\mathbf{u}$ is of order $O(h+\tau)$; see [257].

Least-squares finite element methods in the vector-operator setting

The energy balance (12.43) resulting from the vector-operator setting, gives rise to a compliant DLSP

$$\begin{cases} J(\phi^h,\mathbf{u}^h;f) = \|\nabla\cdot\mathbf{u}^h + c\phi^h - f\|_0^2 + \dfrac{1}{\varepsilon}\|\varepsilon\nabla\phi^h - \mathbf{b}\phi^h + \mathbf{u}^h\|_0^2 \\ X^h = G_0^h(\Omega)\times\mathbf{D}^h(\Omega) \end{cases} \tag{12.50}$$

that extends (5.62) to the advection–diffusion–reaction first-order system (12.41). The presence of the advective flux prevents this DLSP from having the splitting property established in Theorem 5.37. As a result, the scalar and vector fields computed by (12.50) cannot be related to solutions of Galerkin and mixed Galerkin formulations, respectively, and $\mathbf{u}^h$ is not locally conservative[19] despite the presence of a reaction term. However, one important property of (5.62), namely optimal $L^2(\Omega)$-norm error estimates for *both* variables, is inherited by (12.50).

Theorem 12.5. *Let* $\{\phi^h, \mathbf{u}^h\} \in X^h$ *denote a solution of* (12.50) *and assume that* $\mathbf{D}^h(\Omega)$ *is such that*

$$\left(\nabla\cdot(\mathbf{u} - \Pi_D \mathbf{u}), \nabla\cdot\mathbf{v}^h\right) = 0 \qquad \forall \mathbf{v}^h \in \mathbf{D}^h(\Omega). \tag{12.51}$$

Then, there is a positive constant C_ε*, independent of h, such that*

$$\|\phi - \phi^h\|_0 + \|\mathbf{u} - \mathbf{u}^h\|_0 \le C_\varepsilon \left(\inf_{\psi^h \in G_0^h} \|\phi - \psi^h\|_0 + \inf_{\mathbf{v}^h \in \mathbf{D}^h} \|\mathbf{u} - \mathbf{v}^h\|_0 \right). \tag{12.52}$$

□

For a proof of this theorem, see [349]. The error bound (12.52) extends the optimal $L^2(\Omega)$-norm estimate (5.88) from Theorem 5.35 to advection–diffusion–reaction problems. However, the proofs of these two theorems are not equivalent.

Assumption (12.51) restricts the scope of (12.52) to div-conforming spaces for which $\nabla\cdot(\mathbf{D}^h)$ is a piecewise polynomial space; such a restriction is not required[20] in the proof of Theorem 5.35 which is based on the existence of a discrete Hodge decomposition for $\mathbf{D}^h$. On the other hand, it is not known whether the proof of Theorem 5.35 can be extended to advection–diffusion–reaction problems.

It is easy to see that the constant C_ε in (12.52) behaves like $1/\varepsilon$ and that the error bound may deteriorate as (12.39) becomes more advection-dominated, or if $\varepsilon < h$. We review three possible alternatives to (12.50) which offer improved stability in these regimes. The first one is to use the equivalent "first-order" system reformulation (12.45) of the modified advection–diffusion–reaction problem (12.44). The resulting *streamline-diffusion* DLSP

$$\begin{cases} J_\tau(\phi^h, \mathbf{u}^h; f) = \|\nabla\cdot\mathbf{u}^h + c\phi^h - f_\tau\|_0^2 \\[1ex] \quad + \displaystyle\sum_{\kappa\in\mathcal{T}_h} \frac{1}{\varepsilon}\|\varepsilon\nabla\phi^h - \mathbf{b}\phi^h - \tau\mathbf{b}\big(-\varepsilon\Delta\phi^h + \nabla\cdot(\mathbf{b}\phi^h) + c\phi^h\big) + \mathbf{u}^h\|_{0,\kappa}^2 \\[1ex] X^h = G_0^h(\Omega)\times\mathbf{D}^h(\Omega) \end{cases} \tag{12.53}$$

was proposed in [210] and uses "broken" $L^2(\Omega)$-norm for the second residual in (12.44) in order to avoid the need for C^1 finite element spaces.

Another version of the streamline-diffusion LSFEM can be obtained by using the non-equivalent first-order reformulation (12.46) of the modified equation (12.44):

---

[19] Of course, the local conservation can be restored using the post-processing algorithm formulated in Section 5.8.2.

[20] The assumption that $\mathcal{T}_h$ is an affine finite element partition in Theorem 5.35 is required only in order to assert the rates of convergence in (5.88).

$$\begin{cases} J_\tau(\phi^h, \mathbf{u}^h; f) = \|\nabla\cdot\mathbf{u}^h + c\phi^h - f_\tau\|_0^2 \\ \qquad\qquad + \dfrac{1}{\varepsilon}\|\varepsilon\nabla\phi^h - \mathbf{b}\phi^h - \tau\mathbf{b}\big(\nabla\cdot(\mathbf{b}\phi^h) + c\phi^h\big) + \mathbf{u}^h\|_0^2 \\ X^h = G_0^h(\Omega)\times\mathbf{D}^h(\Omega), \end{cases} \tag{12.54}$$

This DLSP can be viewed as the vector-operator setting analogue of a discrete negative norm method based on (12.48).

Finally, one can "stabilize" the compliant DLSP (12.50) by adding the residual of the second-order equation (12.39):

$$\begin{cases} J_\tau(\phi^h, \mathbf{u}^h; f) = \|\nabla\cdot\mathbf{u}^h + c\phi^h - f\|_0^2 + \dfrac{1}{\varepsilon}\|\varepsilon\nabla\phi^h - \mathbf{b}\phi^h + \mathbf{u}^h\|_0^2 \\ \qquad\qquad + \displaystyle\sum_{\kappa\in\mathcal{T}_h}\tau\| - \varepsilon\Delta\phi^h + \nabla\cdot(\mathbf{b}\phi^h) + c\phi^h - f\|_{0,\kappa}^2 \\ X^h = G_0^h(\Omega)\times\mathbf{D}^h(\Omega). \end{cases} \tag{12.55}$$

As in (12.53), the broken $L^2(\Omega)$-norm preempts the need for C^1 finite element spaces for ϕ. This method was considered in [210]. For further details about (12.53) and the stabilized LSFEM (12.55) we refer to that paper.

Least-squares finite element methods for time-dependent problems

We now briefly consider finite difference/least-squares finite element methods (FD-LSFEMs) for the time-dependent advection–diffusion–reaction problem (12.40). As is already noted at the beginning of this section, the FD-LSFEMs for the generalized heat equation (9.1) in Chapter 9 can be extended without much difficulty to (12.40) by simply inserting the advective flux term into the least-squares functional.

Let us illustrate this process using as a prototype the backward Euler method (9.7). Referring to the notation in Chapter 9, we set $c = 1$ and $\Theta_1 = \varepsilon$ in (9.7) and construct a partition $\{[t_{k-1}, t_k]\}_{k=1}^K$ of the time interval $[0, T]$ into K subintervals, where $t_0 = 0$, $t_K = T$, and $t_{k-1} < t_k$. Then, we define the following sequence of least-squares principles: for $k = 1, \ldots, K$,

$$\begin{cases} J_k^{\mathsf{be}}(\phi^{(k)}, \mathbf{u}^{(k)}; f(\mathbf{x}, t_k), \phi^{(k-1)}, \mathbf{u}^{(k-1)}) = \\ \dfrac{1}{\varepsilon}\left\|\varepsilon\nabla\phi^{(k)} - \mathbf{b}\phi^{(k)} + \mathbf{u}^{(k)} - \alpha\big(\varepsilon\nabla\phi^{(k-1)} - \mathbf{b}\phi^{(k-1)} + \mathbf{u}^{(k-1)}\big)\right\|_0^2 \\ \qquad + \Delta t_k\left\|\left(c + \dfrac{1}{\Delta t_k}\right)\phi^{(k)} + \nabla\cdot\mathbf{u}^{(k)} - f(\mathbf{x}, t_k) - \dfrac{1}{\Delta t_k}\phi^{(k-1)}\right\|_0^2 \\ X = G_0(\Omega)\times\mathbf{D}(\Omega), \end{cases} \tag{12.56}$$

where $\phi^{(0)} = \phi_0$ and, if $\alpha = 1$, $\mathbf{u}^{(0)} = -\varepsilon\nabla\phi_0 + \mathbf{b}\phi_0$. Just like its prototype, (12.56) defines a marching method for the semi-discrete in time approximation $\{\phi^{(k)}, \mathbf{u}^{(k)}\}$ of the solution to the following equivalent first-order reformulation of (12.40):

$$\begin{cases} \dfrac{\partial\phi}{\partial t} + \nabla\cdot\mathbf{u} + c\phi = f & \text{for } (\mathbf{x},t) \in \Omega\times(0,T] \\ \varepsilon\nabla\phi - \mathbf{b}\phi + \mathbf{u} = 0 & \text{for } (\mathbf{x},t) \in \Omega\times(0,T] \\ \phi = 0 & \text{for } (\mathbf{x},t) \in \partial\Omega\times(0,T] \\ \phi = \phi_0(\mathbf{x}) & \text{for } t = 0 \text{ and } \mathbf{x}\in\Omega\,, \end{cases} \tag{12.57}$$

i.e., we can obtain (12.56) by using first the backward-Euler method to discretize (12.57) in time, followed by application of least-squares minimization to the resulting semi-discrete in time perturbed elliptic problem. A fully discrete FD-LSFEM formulation for the approximate solutions of (12.57) is then obtained by replacing (12.56) with a compliant DLSP, i.e., by restricting minimization in (12.56) to a conforming finite element subspace $X_h \subset X$.

This example shows that extension of FD-LSFEMs from Chapter 9 to the advection–diffusion–reaction problem (12.40) is indeed simple and straightforward. Unfortunately, this statement does not apply to the analysis of the resulting FD-LSFEMs. The culprit is the advective flux term which inhibits the splitting property of the least-squares formulation and prevents us from reusing the error analysis in Chapter 9.

Fortunately, it turns out that although the simpler and more straightforward techniques from Chapter 9 cannot be used with the time-dependent advection–diffusion–reaction problem, approximation properties of FD-LSFEMs for (12.57) remain similar to those of their prototypes from Chapter 9. This is shown in the following theorem which is adapted from [348].

Theorem 12.6. *Assume that the spatial discretization in* (12.56) *is effected using the compatible finite element space* $X_h = G_0^r(\Omega)\times\mathbf{D}^{(s)}(\Omega)$, *where* $r \ge 1$ *and* $\mathbf{D}^{(s)}$ *is one of the spaces* $\mathbf{D}^s(\Omega)$ *or* $\mathbf{D}^{s}\text{-}(\Omega)$, $s \ge 1$. *Let the initial data* $\phi^{(h,0)}$ *be specified as in Theorem* 9.3 *and, if* $\alpha = 1$, $\mathbf{u}^{(h,0)} = -\Pi_D(\varepsilon\nabla\phi_0 - \mathbf{b}\phi_0)$.

If $\alpha = 0$, *the fully discrete FD-LSFEM solution satisfies the error bound*

$$\begin{aligned} &\|\phi(t_k) - \phi^{(h,k)}\|_0 + \sqrt{\Delta t_k}\left(\|\mathbf{u}(t_k) - \mathbf{u}^{(h,k)}\|_0 + \|\nabla(\phi(t_k) - \phi^{(h,k)})\|_0\right) \\ &\quad \le C\big(h^r + h^s + \Delta t\big) \end{aligned} \tag{12.58}$$

for $k = 0,\dots,K$ *and, if* $\alpha = 1$, *the error bound is given by*

$$\|\phi(t_k) - \phi^{(h,k)}\|_0 + \|\mathbf{u}(t_k) - \mathbf{u}^{(h,k)}\|_0 \le C\big(h^{r+1} + h^{(s)} + \Delta t\big) \tag{12.59}$$

for $k = 0,\dots,K$, *where* $(s) = s+1$ *if* $\mathbf{D}^{(s)} = \mathbf{D}^s(\Omega)$ *and* $(s) = s$ *for* $\mathbf{D}^{(s)} = \mathbf{D}^{s}\text{-}$. *The constant* C *in* (12.58) *and* (12.59) *is independent of* h *and* Δt *but depends continuously on the appropriate solution norms and the diffusivity parameter* ε. □

Even though the dependence of the constant C in (12.58)–(12.59) on ε is not stated explicitly, it is clear that as ε becomes small, or at least small compared to the grid size h, the quality of FD-LSFEM approximations based on the equivalent first-order system (12.57) may deteriorate. Thus it might be of interest to consider extensions of the streamline diffusion and stabilized LSFEMs from Section 12.7 to the time-dependent problem (12.40).

Formally, extension of these methods proceeds along the same lines as for the FD-LSFEM described above in this section. For example, to extend the streamline diffusion LSFEM (12.53) to (12.40), we start with a time-dependent version

$$\begin{cases} \dfrac{\partial \phi}{\partial t} + \nabla\cdot\mathbf{u} + c\phi = f & \text{for } (\mathbf{x},t) \in \Omega\times(0,T] \\ \varepsilon\nabla\phi - \mathbf{b}\phi - \tau\mathbf{b}\mathcal{L}\phi + \mathbf{u} = 0 & \text{for } (\mathbf{x},t) \in \Omega\times(0,T] \\ \phi = 0 & \text{for } (\mathbf{x},t) \in \partial\Omega\times(0,T] \\ \phi = \phi_0(\mathbf{x}) & \text{for } t=0 \text{ and } \mathbf{x}\in\Omega \end{cases} \tag{12.60}$$

of the equivalent "first-order" system (12.45), apply backward Euler discretization in time to obtain a sequence of semi-discrete in time problems

$$\begin{cases} \left(c + \dfrac{1}{\Delta t_k}\right)\phi^{(k)} + \nabla\cdot\mathbf{u}^{(k)} = f(\mathbf{x},t_k) + \dfrac{1}{\Delta t_k}\phi^{(k-1)} & \text{for } \mathbf{x}\in\Omega \\ \varepsilon\nabla\phi^{(k)} - \mathbf{b}\phi^{(k)} - \dfrac{\tau}{\Delta t_k}\mathbf{b}\mathcal{L}\phi^{(k)} + \mathbf{u}^{(k)} = -\dfrac{\tau}{\Delta t_k}\mathbf{b}\mathcal{L}\phi^{(k-1)} & \text{for } \mathbf{x}\in\Omega \\ \phi^{(k)} = \phi_0(\mathbf{x},t_k) & \text{for } \mathbf{x}\in\partial\Omega\,, \end{cases} \tag{12.61}$$

and then define the sequence of streamline-diffusion least-squares principles: for $k = 1,\dots,K$,

$$\begin{cases} J^{\mathsf{be}}_{k,\tau}(\phi^{(k)},\mathbf{u}^{(k)}; f(\mathbf{x},t_k),\phi^{(k-1)},\mathbf{u}^{(k-1)}) = \\ \quad \Delta t_k \left\| \left(c + \dfrac{1}{\Delta t_k}\right)\phi^{(k)} + \nabla\cdot\mathbf{u}^{(k)} - f(\mathbf{x},t_k) - \dfrac{1}{\Delta t_k}\phi^{(k-1)} \right\|_0^2 \\ \quad + \displaystyle\sum_{\kappa\in\mathcal{T}_h}\left(\frac{\Delta t_k}{\varepsilon}\right)\left\| \varepsilon\nabla\phi^{(k)} - \mathbf{b}\phi^{(k)} - \frac{\tau}{\Delta t_k}\mathbf{b}\mathcal{L}(\phi^{(k)} - \phi^{(k-1)}) + \mathbf{u}^{(k)} \right\|_0^2 \\ X = G_0(\Omega)\times\mathbf{D}(\Omega)\,, \end{cases} \tag{12.62}$$

where $\phi^{(0)} = \phi_0$. Note that in (12.61) the stabilizing term $\tau\mathbf{b}\mathcal{L}\phi$ is also differenced in time because $\mathcal{L}$ is the second-order operator corresponding to (12.40), i.e., it contains the time derivative of ϕ.

Overall, we see that extension of (12.53) to the time-dependent advection–diffusion–reaction setting is, indeed, fairly straightforward. Unfortunately, analysis of the resulting FD-LSFEMs remains an open question that, so far, has not been addressed in the literature. One obvious complication for the analysis is that the stabilizing term $\tau \mathbf{b} \mathcal{L} \phi$ has to be differenced in time which makes the structure of the streamline diffusion FD-LSFEM more complicated than that of the standard FD-LSFEM based on (12.62). Dropping the time derivative of ϕ from $\mathcal{L}$ could certainly simplify things but at the same time would also lead to a loss of consistency.

12.8 Least-Squares Finite Element Methods for Higher-Order Problems

The LSFEMs developed in Chapters 5–7 dealt with boundary value problems for second-order elliptic differential operators. To satisfy the practicality requirements stated in Section 2.2.2 these problems were replaced by equivalent first-order reformulations of the governing equations and the boundary conditions. The same approach can be easily extended to higher-order boundary value problems. For instance, the reduction algorithm described in Section 4.1 applies to elliptic systems of arbitrary order and can be used to that end.

The goal of this section is to briefly describe how practical and optimally accurate LSFEMs can be defined for higher-order boundary value problems. The leading prototype for such problems is the fourth-order biharmonic equation

$$\Delta^2 \phi = f \qquad \text{in } \Omega \tag{12.63}$$

along with the homogeneous Dirichlet boundary conditions

$$\phi = 0 \qquad \text{and} \qquad \frac{\partial \phi}{\partial n} = 0 \qquad \text{on } \partial\Omega\,. \tag{12.64}$$

The fourth-order biharmonic problem (12.63)–(12.64) can be recast as a first-order system by essentially repeating twice the steps needed to transform Δ into a div–grad first-order operator; see Section 5.3.1. Not surprisingly, the outcome is also a div–grad first-order system but with twice the number of equations and the scalar and vector variables. Following [167] we introduce the pair of scalar variables

$$\phi_0 = \phi\,, \qquad \phi_1 = \Delta \phi_0\,,$$

and the pair of vector variables

$$\mathbf{v}_0 = \nabla \phi_0\,, \qquad \mathbf{v}_1 = \nabla \phi_1\,.$$

Then, a first-order problem corresponding to (12.63)–(12.64) is given by the "double" div–grad system

$$\begin{cases} \nabla \cdot \mathbf{v}_1 &= f \\ \nabla \cdot \mathbf{v}_0 - \phi_1 &= 0 \\ \mathbf{v}_0 - \nabla \phi_0 &= 0 \\ \mathbf{v}_1 - \nabla \phi_1 &= 0 \end{cases} \quad \text{in } \Omega \tag{12.65}$$

along with the boundary conditions

$$\phi_0 = 0 \qquad \text{and} \qquad \mathbf{v}_0 \cdot \mathbf{n} = 0 \qquad \text{on } \partial\Omega\,. \tag{12.66}$$

In the vector-operator setting the appropriate solution space for (12.65)–(12.66) is given by

$$X = \big\{\{\phi_0, \phi_1, \mathbf{v}_0, \mathbf{v}_1\} \in G_0 \times G \times \mathbf{D}_0(\Omega) \times \mathbf{D}(\Omega)\big\}\,.$$

The proof of Theorem 5.15 can be extended to the "double" div–grad system to show that it satisfies the following energy balance:

$$\begin{aligned} &C\big(\|\phi_0\|_G + \|\phi_1\|_G + \|\mathbf{v}_0\|_D + \|\mathbf{v}_1\|_D\big) \\ &\qquad \le \|\nabla \cdot \mathbf{v}_1\|_0 + \|\nabla \cdot \mathbf{v}_0 - \phi_1\|_0 + \|\mathbf{v}_0 - \nabla\phi_0\|_0 + \|\mathbf{v}_1 - \nabla\phi_1\|_0\,. \end{aligned}$$

The energy balances in the ADN setting are similar to those obtained in Theorem 5.9. Once the energy balances for (12.65)–(12.66) are identified, formulation of CLSPs and their conversion to practical DLSPs can proceed as in Sections 5.5–5.6.

Of course, the first-order operator in (12.65) has an extended version that is analogous to the extended div–grad system (5.25) from Section 5.3.2. Least-squares principles for this extended version follow in the footsteps of the formulations discussed in Section 5.6.2 and share all their advantages and disadvantages.

The transformation scheme applied to the biharmonic problem (12.63)–(12.64) can be generalized to scalar elliptic problems of order $2m$; see [167]. Starting with $\phi_0 = \phi$ and $\mathbf{v}_0 = \nabla\phi_0$, one introduces the new variables $V = \{\mathbf{v}_0, \dots, \mathbf{v}_{m-1}\}$ and $\Phi = \{\phi_0, \dots, \phi_{m-1}\}$ recursively according to the formulas

$$\phi_i = \nabla \cdot \mathbf{v}_{i-1} \qquad \text{and} \qquad \mathbf{v}_i = \nabla\phi_i\,, \quad i = 1, \dots, m-1\,.$$

The resulting first order system has the structure

$$\begin{pmatrix} \mathsf{I} & \mathsf{G} \\ \mathsf{D} & \mathsf{C} \end{pmatrix} \begin{pmatrix} V \\ \Phi \end{pmatrix} = F \tag{12.67}$$

where in (12.67) G is matrix of gradients, D is matrix of divergences, and C is a bounded (with respect to $L^2(\Omega)$) multiplication operator.

12.9 Least-Squares Finite Element Methods for Div–Grad–Curl Systems

In Chapters 5 and 6, we considered least-squares finite element methods for div–grad, div–curl, and curl–curl systems. The div–grad–curl system

$$\begin{cases} \nabla\times\mathbf{u}+\nabla\phi=\mathbf{g} \\ \nabla\cdot\mathbf{u}=f \end{cases} \quad \text{in } \Omega \tag{12.68}$$

is another basic first-order elliptic problem that arises in practice.[21,22]

In three dimensions, the div–curl–grad system (12.68) is well posed with either the boundary conditions[23]

$$\mathbf{u}\times\mathbf{n}=0 \qquad \text{on } \partial\Omega \tag{12.69}$$

or

$$\mathbf{u}\cdot\mathbf{n}=0 \qquad \text{and} \qquad \phi=0 \qquad \text{on } \partial\Omega\,. \tag{12.70}$$

We view the two-dimensional case as a reduction of the three-dimensional case in situations for which all data and unknowns are only functions of x_1 and x_2 and Ω is an infinite cylinder, i.e.,

$$\Omega=\Big\{\{x_1,x_2\}\in\omega \quad \text{and} \quad x_3\in(-\infty,\infty)\Big\}$$

for some $\omega\in\mathbb{R}^2$. No assumptions are made about how many components of $\mathbf{u}$ and $\mathbf{g}$ are non-zero. It is not difficult to see that (12.68)–(12.69) reduce to the two uncoupled systems

$$\begin{pmatrix} \phi_{,x_1}+u_{3,x_2} \\ \phi_{,x_2}-u_{3,x_1} \end{pmatrix}=\mathbf{k}\times\mathbf{g}\times\mathbf{k} \quad \text{in } \omega \qquad \text{and} \qquad \begin{cases} \phi=0 \\ \text{or} \\ u_3=0 \end{cases} \quad \text{on } \partial\omega\,, \tag{12.71}$$

where $\mathbf{k}$ is the Cartesian unit vector defined in Section A.1, and

$$\begin{cases} u_{2,x_1}-u_{1,x_2}=\mathbf{g}\cdot\mathbf{k} \\ u_{1,x_1}+u_{2,x_2}=f \end{cases} \quad \text{in } \omega \qquad \text{and} \qquad \begin{cases} u_1n_1+u_2n_2=0 \\ \text{or} \\ u_1n_2-u_2n_1=0 \end{cases} \quad \text{on } \partial\omega \tag{12.72}$$

---

[21] For example, (12.68) is representative of a potential formulation of static electromagnetics, with $\mathbf{u}$ and ϕ denoting the magnetic and electric potentials, respectively.

[22] The system (12.68) also arises in the treatment of div–curl systems in three dimensions for which $\nabla\cdot\mathbf{g}=0$ and ϕ arises as a slack variable; see Section D.2.3. In fact, if $\nabla\cdot\mathbf{g}=0$ and either of the boundary conditions (12.70) or (12.69) hold, it is a simple matter to show that $\nabla\phi=0$ so that (12.68) reduces to a div–curl system.

[23] For simplicity, we only consider homogeneous boundary conditions.

for $\{\phi, u_3\}$ and $\{u_1, u_2\}$, respectively.[24,25]

The system (12.68) with either (12.69) or (12.70) in three dimensions and the systems (12.71) and (12.72) in two dimensions are all homogeneous elliptic. Therefore, it is a straightforward matter to define compliant LSFEMs that satisfy all the keys to practicality. For the sake of brevity, we only consider the three-dimensional case and the system (12.68) with either (12.69) or (12.70). The ADN energy balances for this problem are given in (6.31) and (6.32) that reduce to, in the case of homogeneous boundary conditions and for $q = 0$,[26]

$$\|\mathbf{u}\|_1 + \|\phi\|_1 \le C\big(\|\nabla\times\mathbf{u} + \nabla\phi\|_0 + \|\nabla\cdot\mathbf{u}\|_0 + \|\mathbf{u}\times\mathbf{n}\|_{1/2,\partial\Omega}\big)$$

that applies to (12.68) and (12.69) and

$$\|\mathbf{u}\|_1 + \|\phi\|_1 \le C\big(\|\nabla\times\mathbf{u} + \nabla\phi\|_0 + \|\nabla\cdot\mathbf{u}\|_0 + \|\phi\|_{1/2,\partial\Omega} + \|\mathbf{u}\cdot\mathbf{n}\|_{1/2,\partial\Omega}\big),$$

that applies to (12.68) and (12.70). These energy balances immediately lead to the CLSPs. For example, assuming that the solution space is constrained to satisfy the boundary conditions, for (12.68) and (12.70) we have the CLSP

$$\begin{cases} J(\mathbf{u},\phi;f,\mathbf{g}) = \|\nabla\times\mathbf{u} + \nabla\phi - \mathbf{g}\|_0^2 + \|\nabla\cdot\mathbf{u} - f\|_0^2 \\ X = \big\{\mathbf{v}\in[H^1(\Omega)]^3 \mid \mathbf{v}\cdot\mathbf{n} = 0 \text{ on } \partial\Omega\big\}\times H_0^1(\Omega). \end{cases} \tag{12.73}$$

At this point we can define DLSPs in the usual manner.

The CLSP (12.73) corresponds to an energy balance derived in the ADN setting. As a result, the minimization space in this formulation is a direct product of scalar Sobolev spaces $H^1(\Omega)$ constrained by the appropriate boundary conditions. Least-squares methods for div-grad-curl systems can also be developed using the vector-operator setting for the energy balances. Such formulations are better suited for problem configurations that do not satisfy the assumptions stated in Theorem

[24] The differential operators in (12.71) and (12.72) are the same; in fact, they are simply the Cauchy–Reimann operator. However, the boundary conditions are fundamentally different. In (12.71), one or the other variable is constrained on the boundary whereas in (12.72), linear combinations of the variables are constrained.

[25] Two special cases arise in practice. First, $\mathbf{g} = g_3\mathbf{k}$. In this case, (12.71) only has the trivial solutions $\phi = 0$ and u_3 =constant for the boundary condition $\phi = 0$ on $\partial\omega$ or ϕ =constant and $u_3 = 0$ for the boundary condition $u_3 = 0$ on $\partial\omega$ so that we need only consider the system (12.72) which becomes a standard two-dimensional div–curl system for $\{u_1, u_2\}$. The second special case is $f = 0$ and $\mathbf{g}\cdot\mathbf{k} = 0$ for which (12.72) only has the trivial solution $u_1 = u_2 = 0$ so that we need only consider the two-dimensional problem (12.71). In the general case we have to consider both systems (12.71) and (12.72) but these are uncoupled.

[26] In Section 6.3, the variable ϕ is a slack variable introduced to turn the 4×3 div–curl system into a square 4×4 div–grad–curl system that can be handled by the ADN theory. Because of the compatibility requirements on the data of div–curl systems, the slack variable can be shown to be trivial, i.e., either zero or a constant. In the general setting of div–grad–curl systems for which the data does not have to satisfy the compatibility requirements of div–curl systems, ϕ is not trivial. This again shows that div–grad–curl systems can be thought of as parent systems of div–curl systems.

A.8, i.e., when $\mathbf{D}(\Omega)\cap\mathbf{C}_0(\Omega)$ and $\mathbf{D}_0(\Omega)\cap\mathbf{C}(\Omega)$ are not continuously embedded in $[G(\Omega)]^d=[H^1(\Omega)]^d$. If this is the case, then the norm on $\mathbf{D}(\Omega)\cap\mathbf{C}(\Omega)$ is not equivalent to a product of $H^1(\Omega)$ norms. Recall that in the absence of such an equivalence, nodal vector finite element spaces loose their approximability property; see Theorem B.13, the discussion in Section B.2.2, and the numerical examples in Section 5.9.3.

The techniques from the proofs of Theorem 7.19 and Corollary 7.21 in Section 7.7.1 can be easily adapted to show that

$$C\big(\|\mathbf{u}\|_{DC}+\|\phi\|_G\big)\le\|\nabla\times\mathbf{u}+\nabla\phi\|_0+\|\nabla\cdot\mathbf{u}\|_0 \tag{12.74}$$

for all $\{\mathbf{u},\phi\}\in\mathbf{D}(\Omega)\cap\mathbf{C}_0(\Omega)\times G(\Omega)\cap L^2_0(\Omega)$ and all $\{\mathbf{u},\phi\}\in\mathbf{D}_0(\Omega)\cap\mathbf{C}(\Omega)\times G_0(\Omega)$. It follows that

$$\begin{cases} J(\mathbf{u},\phi;f,\mathbf{g})=\|\nabla\times\mathbf{u}+\nabla\phi-\mathbf{g}\|_0^2+\|\nabla\cdot\mathbf{u}-f\|_0^2\\[2ex] X=\mathbf{D}(\Omega)\cap\mathbf{C}_0(\Omega)\times G(\Omega)\cap L^2_0(\Omega), \end{cases} \tag{12.75}$$

is a well-posed CLSP for (12.68) with the boundary condition (12.69), whereas

$$\begin{cases} J(\mathbf{u},\phi;f,\mathbf{g})=\|\nabla\times\mathbf{u}+\nabla\phi-\mathbf{g}\|_0^2+\|\nabla\cdot\mathbf{u}-f\|_0^2\\[2ex] X=\mathbf{D}_0(\Omega)\cap\mathbf{C}(\Omega)\times G_0(\Omega), \end{cases} \tag{12.76}$$

is a well-posed CLSP for the div–grad–curl system with (12.70).

Formulation and analysis of LSFEMs based on (12.75) and (12.76) can now proceed much along the same lines as in Section 7.7.3. In particular, with an appropriate choice of compatible approximating spaces we obtain strongly compatible LSFEMs whose solutions satisfy discrete div–grad–curl systems similar to the discrete VVP systems (7.101) and (7.102) encountered in Section 7.7.5.

12.10 Domain Decomposition Least-Squares Finite Element Methods

The methods discussed in this section can be viewed as least-squares counterparts of traditional domain decomposition [340], substructuring [159], discontinuous Galerkin [10], and domain bridging [30] methods. We start with an example of a LSFEM for transmission problems, and then extended it to a discontinuous LSFEM and a LSFEM for domain bridging (also known as mesh-tying).[27]

These types of methods require some specific notations. For simplicity, we state the notation in the case of an open bounded domain Ω with Lipschitz continuous boundary $\partial\Omega$, composed of two subdomains; $\overline{\Omega}_1\cup\overline{\Omega}_2=\overline{\Omega}$ and $\Omega_1\cap\Omega_2=\emptyset$. The

[27] Transmission problems can also be treated by the methods discussed in Section 12.11.

interface between the two domains, $\Sigma = \overline{\Omega}_1 \cap \overline{\Omega}_2$, is a connected nonempty set, and, for $i = 1,2$, $\Gamma_i = \partial\Omega \cap \overline{\Omega}_i$; see Figure 12.1. Note that if $\overline{\Omega}_i \subset \Omega$, then $\Gamma_i = \emptyset$.

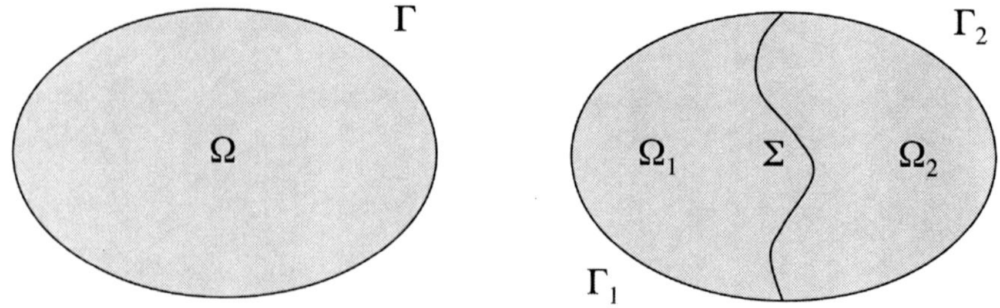

Fig. 12.1 Typical configuration for domain decomposition LSFEMs: the domain Ω is composed of two subdomains, shown on the right.

Given a partition of Ω into subdomains Ω_i, $i = 1,2$, we define the space

$$G_{12} = \{\varphi = \{\varphi_1, \varphi_2\} \mid \varphi_i \in G(\Omega_i),\ i = 1,2\},$$

its subspace $G_{12,0}$ consisting of pairs $\{\varphi_1, \varphi_2\} \in G_{\Gamma_1}(\Omega_1) \times G_{\Gamma_2}(\Omega_2)$ that vanish on Γ_i, and the space

$$\mathbf{D}_{12} = \{\mathbf{v} = \{\mathbf{v}_1, \mathbf{v}_2\} \mid \mathbf{v}_i \in \mathbf{D}(\Omega_i),\ i = 1,2\}.$$

The spaces G_{12} and $\mathbf{D}_{12}$, equipped with the inner products

$$((\varphi,\psi))_1 = \sum_{i=1}^{2} (\varphi_i, \psi_i)_{1,\Omega_i} \qquad \text{and} \qquad ((\mathbf{u},\mathbf{v}))_D = \sum_{i=1}^{2} (\mathbf{u}_i, \mathbf{v}_i)_{D,\Omega_i},$$

respectively, are Hilbert spaces with corresponding induced norms $|||\cdot|||_1$ and $|||\cdot|||_D$.

Let $\mathcal{T}_h^i$ denote a finite element partition of Ω_i, $i = 1,2$. The discrete subdomains induce approximations Γ_1^h, Γ_2^h, Σ_1^h, and Σ_2^h of Γ_1, Γ_2, and the interface Σ, respectively. Discretization of Ω is given by $\Omega^h = \Omega_1^h \cup \Omega_2^h$. Note that Σ_1^h and Σ_2^h are not necessarily spatially coincident. When $\Sigma_1^h = \Sigma_2^h = \Sigma^h$, the interface is called *matching*; if $\Sigma_1^h \neq \Sigma_2^h$, the interface is *non-matching*. Matching interfaces arise in problems where Σ can be approximated exactly by the inter-element boundaries or is defined in terms of these boundaries, e.g., in domain-decomposition methods. Non-matching interfaces arise in situations where the problem definition naturally leads to curved interfaces that cannot be represented exactly[28] or when, for effi-

[28] Approximation of a curved interface Σ by non-matching interfaces Σ_1^h and Σ_2^h is fundamentally different from approximation of a curved boundary $\partial\Omega$ by, e.g., polygons or lines. Both cases are variational crimes in the sense of [323, p. 193]; the latter leads to a perturbation of the original problem that can be estimated by the Strang lemma [323, Lemma 4.1]. In contrast, non-matching

ciency, mesh generation on a complex shape is replaced by independent meshing of its subdomains [30, 47, 143, 254, 294].

Extension of these definitions to more than two subdomains is straightforward.

Methods for transmission problems and discontinuous methods

For clarity, the weight notation Θ_i, introduced in Chapter 5, is not used in this section to avoid potential confusion between the weight index and the subdomain index. Instead we adopt the symbols A and α to denote a symmetric, positive definite matrix-valued function and a positive scalar-valued function, respectively. Both A and α are assumed to be continuous in Ω except across a piecewise smooth surface Σ that divides Ω into two simply connected subdomains Ω_1 and Ω_2; see Figure 12.1. We consider the following *transmission* problem:

$$
\begin{aligned}
-\nabla \cdot \mathsf{A}_i \nabla \varphi_i + \alpha_i \varphi_i &= f_i \qquad \text{in } \Omega_i, \ \ i = 1,2 && (12.77)\\
\varphi_i &= 0 \qquad \text{on } \Gamma_i, \ \ i = 1,2 && (12.78)\\
\varphi_1 - \varphi_2 &= 0 \qquad \text{on } \Sigma && (12.79)\\
\mathsf{A}_1 \nabla \varphi_1 \cdot \mathbf{n}_1 + \mathsf{A}_2 \nabla \varphi_2 \cdot \mathbf{n}_2 &= 0 \qquad \text{on } \Sigma\,, && (12.80)
\end{aligned}
$$

where, for $i = 1,2$, $\mathbf{n}_i$ is the normal to Σ that points away from Ω_i, f_i is a given source term, $\alpha_i = \alpha|_{\Omega_i}$, and $\mathsf{A}_i = \mathsf{A}|_{\Omega_i}$. As usual, least-squares principles are applied to an equivalent first-order formulation of (12.77):

$$
\begin{cases}
\nabla \cdot \mathbf{u}_i + \alpha_i \varphi_i = f_i \\
\mathbf{u}_i + \mathsf{A}_i \nabla \varphi_i = 0
\end{cases}
\qquad \text{in } \Omega_i, \ \ i = 1,2 \qquad (12.81)
$$

along with the boundary condition (12.78), the interface condition (12.79), and the interface condition (12.80) rewritten in terms of $\mathbf{u}$:

$$
\mathbf{u}_1 \cdot \mathbf{n}_1 + \mathbf{u}_2 \cdot \mathbf{n}_2 = 0 \qquad \text{on } \Sigma\,. \qquad (12.82)
$$

A LSFEM for (12.81) is formulated under the assumption that Ω_1^h and Ω_2^h have a matching interface Σ^h. In this case, the jump terms

$$
[\psi] = \psi_1 - \psi_2 \qquad \text{and} \qquad [\mathbf{v}] = \mathbf{v}_1 \cdot \mathbf{n}_1 + \mathbf{v}_2 \cdot \mathbf{n}_2 \qquad (12.83)
$$

are well defined on Σ^h. Note that $\Sigma_1^h = \Sigma_2^h = \Sigma^h$ does not imply that the grids on Σ_1^h and Σ_2^h match. The spaces $G_{12,0}$ and $\mathbf{D}_{12}$ can be approximated by the conforming finite element subspaces

$$
G_{12,0}^h = \{\psi^h = \{\psi_1^h, \psi_2^h\} \mid \psi_i^h \in G^h(\Omega_i^h) \text{ and } \psi_i|_{\Gamma_i} = 0, \ i = 1,2\}
$$

---

interfaces correspond to a computational domain that has gaps and overlaps where the original PDE ceases to be well-defined.

and

$$\mathbf{D}_{12}^h = \{\mathbf{v}^h = \{\mathbf{v}_1^h, \mathbf{v}_2^h\} \mid \mathbf{v}_i^h \in \mathbf{D}^h(\Omega_i^h),\ i = 1,2\}$$

defined with respect to the finite element partitions $\mathcal{T}_h^i$ of the sub-domains Ω_i, $i = 1,2$. As a result, the solution of the transmission problem (12.78), (12.79), (12.81), and (12.82) or, equivalently, of (12.77)–(12.80), can be approximated by the following DLSP, originally proposed in [102]:

$$\begin{cases} J^h(\psi^h, \mathbf{v}^h; f) = \sum_{i=1}^{2} \left(\|\nabla \cdot \mathbf{v}_i^h + \alpha_i \psi_i - f_i\|_{0,\Omega_i^h}^2 + \|\mathbf{v}_i^h + \mathsf{A}_i \nabla \psi_i^h\|_{0,\Omega_i^h}^2 \right) \\ \qquad\qquad + \dfrac{1}{h^{1+\varepsilon_0}} \displaystyle\int_{\Sigma^h} [\psi^h]^2 ds + \dfrac{1}{h^{\varepsilon_1}} \displaystyle\int_{\Sigma^h} [\mathbf{v}^h]^2 ds \\ X^h = G_{12,0}^h \times \mathbf{D}_{12}^h, \end{cases} \tag{12.84}$$

where the values of the real parameters ε_0 and ε_1 can be adjusted to improve convergence. A similar method, with a DLSP based on the second-order problem (12.77), was considered in [18]. That method, however, does not meet the practicality requirements stated in Section 2.2.2.

The optimality condition for (12.84) is: seek $\{\varphi^h, \mathbf{u}^h\} \in G_{12,0}^h \times \mathbf{D}_{12}^h$ such that:

$$Q^h(\varphi^h, \mathbf{u}^h; \psi^h, \mathbf{v}^h) = F(\psi^h, \mathbf{v}^h) \qquad \forall \{\psi^h, \mathbf{v}^h\} \in G_{12,0}^h \times \mathbf{D}_{12}^h, \tag{12.85}$$

where

$$Q^h(\varphi^h, \mathbf{u}^h; \psi^h, \mathbf{v}^h) = \frac{1}{h^{1+\varepsilon_0}} \int_{\Sigma^h} [\varphi^h][\psi^h]\, ds + \frac{1}{h^{\varepsilon_1}} \int_{\Sigma^h} [\mathbf{u}^h][\mathbf{v}^h]\, ds$$
$$+ \sum_{i=1}^{2} \left(\nabla \cdot \mathbf{u}_i^h + \alpha_i \varphi_i, \nabla \cdot \mathbf{v}_i^h + \alpha_i \psi_i\right)_{0,\Omega_i^h} + \left(\mathbf{u}_i^h + \mathsf{A}_i \nabla \varphi_i^h, \mathbf{v}_i + \mathsf{A}_i \nabla \psi_i\right)_{0,\Omega_i^h}$$

and

$$F(\psi^h, \mathbf{v}^h) = \sum_{i=1}^{2} \left(f_i, \nabla \cdot \mathbf{v}_i^h + \alpha_i \psi_i^h\right)_{0,\Omega_i^h}.$$

It can be shown that the weak formulation (12.85) has a unique solution that converges optimally to a solution of (12.78), (12.79), (12.81), and (12.82) or, equivalently, of (12.77)–(12.80); see [102]. To implement this method, the jump terms in (12.84) are computed using field extensions from each sub-domain, and so the grids on Σ_1^h and Σ_2^h are not required to match. This is in contrast to mortar methods, e.g., see [90, 169], for which a separate grid on Σ^h may be required for a discrete Lagrange multiplier.

It is straightforward to extend (12.84) to a *discontinuous* LSFEM. Let $\mathcal{T}_h$ denote a partition of Ω into finite elements $\{\kappa_k\}$ and let

$$\widetilde{\Sigma}^h = \mathcal{C}_{d-1}(\mathcal{T}_h) \setminus \mathcal{C}_{d-1}(\partial\Omega),$$

where d is the space dimension and $\mathcal{C}_m(\cdot)$ is the set defined in (B.4), denote the set of all inter-element interfaces. For example, in two dimensions, $d = 2$ and

$$\widetilde{\Sigma}^h = \mathcal{C}_1(\mathcal{T}_h) \setminus \mathcal{C}_1(\partial\Omega)$$

is the set of all interior element edges. Clearly, we can then define an approximate interface $\Sigma^h \approx \Sigma$ such that $\Sigma^h \subset \widetilde{\Sigma}^h$ and the resulting configuration corresponds to a matching interface Σ^h between the approximate subdomains. As a result, the jump terms (12.83) are well-defined on the interface Σ^h. Then, if the function spaces $G(\kappa_k) \times \mathbf{D}(\kappa_k)$ defined over each finite element are approximated by finite element subspaces $G^h(\kappa_k) \times \mathbf{D}^h(\kappa_k)$, we define the following DLSP:

$$\begin{cases} J^h(\psi^h, \mathbf{v}^h; f) = \displaystyle\sum_{\kappa_k \in \mathcal{T}_h} \Big(\|\nabla \cdot \mathbf{v}_k^h + \alpha_k \psi_k - f_k\|_{0,\kappa_k}^2 + \|\mathbf{v}_k^h + \mathrm{A}_k \nabla \psi_k^h\|_{0,\kappa_k}^2 \Big) \\ \qquad\qquad\qquad + \dfrac{1}{h^{1+\varepsilon_0}} \displaystyle\int_{\Sigma^h} [\psi^h]^2 ds + \dfrac{1}{h^{\varepsilon_1}} \int_{\Sigma^h} [\mathbf{v}^h]^2 ds \\ X^h = \Big\{ \{\psi^h, \mathbf{v}^h\} \in \displaystyle\prod_{\kappa_k \in \mathcal{T}_h} G^h(\kappa_k) \times \mathbf{D}^h(\kappa_k) \mid \psi^h = 0 \text{ on } \partial\Omega \Big\}. \end{cases} \tag{12.86}$$

For obvious reasons, the method defined by this least-squares principle can be referred to as a *discontinuous* LSFEM. We note that conventional discontinuous Galerkin methods employ similar jump terms to stabilize their weak equations.

Mesh-tying methods

We now consider the extension of (12.84) to non-matching interfaces. This configuration arises in, e.g., the meshing of complex engineering structures and contact problems. Methods for non-matching interfaces are often called *mesh-tying* or *domain-bridging* methods [30, 143, 294] because these methods use independently meshed subdomains that must be "tied" together. Broadly speaking, mesh-tying is the opposite of substructuring and domain decomposition for which the subdomains are formed by unions of finite elements from a given partition $\mathcal{T}_h$ of the whole computational domain.

A minimal requirement for any mesh-tying or domain-bridging method is a consistency condition called the *patch test*. A method passes a patch test of order r if it can recover any solution of the transmission problem (12.77)–(12.80) that is a polynomial of degree r. Surprisingly, this turns out to be very difficult to accomplish if one uses Lagrange multiplier methods,[29] even if the patch test is limited to $r = 1$; see [254] for a further discussion of such methods.

[29] Existing approaches start by selecting one of the non-matching interfaces as a master and the other as a slave surface. One class of methods (see [169–171]) defines Lagrange multipliers on the slave surface and uses a projection operator from the master surface. Another class (see [143, 144, 254]) builds additional mesh structures between the slave and master interfaces using tools that range from mesh imprinting to local L^2 projections. A third type of methods, developed in [294],

Interestingly, by using least-squares principles, one can define mesh-tying methods that pass a patch test of *arbitrary* order. The only precondition imposed in defining such methods is to ensure that there are no void regions between the subdomains. Assuming that Ω_i^h represent independently meshed contiguous parts of the same body, this can be accomplished by perturbing the discrete interfaces until there are no void regions. The perturbation required to overlap the meshes is similar in magnitude to the perturbations used in either r-adaptivity or Lagrangian algorithms. LSFEMs defined along these lines resemble the Arelquin method [30] and the overlapping domain bridging method [190]. However, by measuring the residual energy and not the physical energy, a least-squares functional is allowed to measure its energy *redundantly* in subdomain intersections without causing any physical inconsistencies.[30] This fact not only greatly simplifies the algorithm, but is also the reason why mesh-tying LSFEMs can pass a patch test for any $r \geq 1$.

In what follows, we assume that Ω_1^h and Ω_2^h are such that the void region $\Omega_V = \Omega/(\overline{\Omega}_1^h \cup \overline{\Omega}_2^h)$ is empty and the overlap region $\Omega_O = \Omega_1^h \cap \Omega_2^h$ is not. Let Σ^h denote the set of spatially coincident interface segments. We define the mesh-tying "region" as

$$\Omega_T^h = \Omega_O \cup \Sigma^h = \overline{\Omega}_1^h \cap \overline{\Omega}_2^h, \tag{12.87}$$

i.e., Ω_T^h is the union of the overlap region and any spatially coincident segments of the discrete interface. Note that the standard jump terms (12.83) are defined on Σ^h and can be "reused" in a mesh-tying least-squares principle. On the rest of Ω_T^h, i.e., on Ω_O, they are replaced by the "generalized" jump terms

$$\|\psi_1^h - \psi_2^h\|_{1,\Omega_O}^2 \qquad \text{and} \qquad \|\mathbf{v}_1^h - \mathbf{v}_2^h\|_{D,\Omega_O}^2, \tag{12.88}$$

respectively. Thus, we obtain the following DLSP:

$$\begin{cases} J^h(\psi^h, \mathbf{v}^h; f) = \sum_{i=1}^{2} \left(\|\nabla \cdot \mathbf{v}_i^h + \psi_i^h - f_i\|_{0,\Omega_i^h}^2 + \|\mathbf{v}_i^h + \mathsf{A}_i \nabla \psi_i^h\|_{0,\Omega_i^h}^2 \right) \\ \quad + \dfrac{1}{h^{1+\varepsilon_0}} \displaystyle\int_{\Sigma^h} [\psi^h]^2 \, ds + \dfrac{1}{h^{\varepsilon_1}} \displaystyle\int_{\Sigma^h} [\mathbf{v}^h]^2 \, ds + \omega_\varphi \|\psi_1^h - \psi_2^h\|_{1,\Omega_O}^2 + \omega_\mathbf{v} \|\mathbf{v}_1^h - \mathbf{v}_2^h\|_{D,\Omega_O}^2 \\ X^h = G_{12,0}^h \times \mathbf{D}_{12}^h . \end{cases}$$

This mesh-tying LSFEM was proposed in [47] and analyzed in [48]. In particular, the analysis of [48] shows that if Σ^h is empty, i.e., Ω_T^h consists only of the overlap region Ω_O, then the weights ω_φ and $\omega_\mathbf{v}$ are mesh independent and can be chosen equal to 3. Among other things, this implies that for the mesh-tying LSFEM, there is no particular advantage in trying to match the interfaces which can be a com-

---

avoids mesh refinement but requires an *interface balancing procedure* to cancel out the signed areas of the gaps and overlaps.

[30] In contrast, methods that minimize the physical energy, subject to appropriate constraints on the interfaces, must invoke special efforts to avoid counting the physical energy twice in the overlap regions; see [30].

putationally difficult task. Instead, it pays more dividends to have larger[31] overlap regions Ω_O which can be much easier to arrange. For further details and numerical examples, see [47,48].

12.11 Least-Squares Finite Element Methods for Multi-Physics Problems

A wide range of important applications in science and engineering are described in terms of multi-physics or multi-model problems. One class of problems of this type is characterized by requiring the use of different mathematical models in different nonoverlapping subdomains.[32] The multiple models are tied together through interface conditions that hold along the interfaces between the different subdomains.

Abstractly, we have the following setting, in the case of a two-model multi-physics system. The domain Ω is composed of two nonoverlapping subdomains Ω_i, $i=1,2$, that abut at an interface $\Sigma = \overline{\Omega}_1 \cap \overline{\Omega}_2$; see the sketch in Figure 12.1. We then have

$$\mathcal{L}_1 u_1 = f_1 \quad \text{in } \Omega_1 \qquad \text{and} \qquad \mathcal{L}_2 u_2 = f_2 \quad \text{in } \Omega_2\,, \tag{12.89}$$

where, for $i=1,2$, we have the solution and data spaces X_i and Y_i, respectively, the operators $\mathcal{L}_i : X_i \to Y_i$, and the data $f_i \in Y_i$. In addition to boundary conditions imposed along $\Gamma_i = \partial\Omega \cap \overline{\Omega}_i$, $i=1,2$, which we assume are satisfied by functions belonging to X_i, we also have the coupling, interface conditions

$$\mathcal{T}_1(u_1,u_2) = 0 \qquad \text{and} \qquad \mathcal{T}_2(u_1,u_2) = 0 \qquad \text{on } \Sigma\,, \tag{12.90}$$

where, for $i=1,2$, the operators $\mathcal{T}_i$ are well-defined acting on traces of function belonging to X_1 and X_2.

There are two basic computational approaches to solving problems of the type (12.89) and (12.90): one is non-iterative and monolithic and the other is iterative and is based on operator splitting. The latter would solve the two equations in (12.89) separately by using specialized numerical methods for each type of subproblem. The coupling information contained in the interface conditions in (12.90) is transferred between the subproblems after each solution iteration. As with classical operator-split methods, these schemes tend to be only conditionally stable. Monolithic approaches offer an alternative approach where the two equations in (12.89) along

---

[31] Recall that a least-squares functional measures residual rather than physical energy, and so there is no need to subtract the energy from Ω_O.

[32] Another class of multi-physics problems involves *concurrent* models for which coupled, multiple model equations hold simultaneously at all points in the domain. An example of such problems are thermo-mechanical systems in which PDEs governing the mechanical behavior of a system or process are coupled to PDEs governing the thermal behavior. The coupled system of PDEs that define a concurrent multi-physics model can be treated as a single (larger) system to which one can apply any type of discretization method, including a LSFEM. For this reason, we do not discuss this case any further.

with the interface conditions in (12.90) are treated as components of a single coupled system of PDEs.

One easily notices that the problem (12.89) and (12.90) looks a lot like the transmission problem considered in Section 12.10. Indeed, transmission problems are special "multimodel" problems for which the underlying mathematical model, i.e., the PDE, is the same in both domains but certain data, e.g., coefficients in the PDE, differ in the two domains. Thus, the methods described in Section 12.10 for transmission problem can be extended[33] to true multi-physics problems having the structure given by (12.89) and (12.90). For example, emulating (12.84), one could define a DLSP

$$\begin{cases} J^h(u_1,u_2;f_1,f_2) = \sum_{i=1}^{2} \left(\|\mathcal{L}_1 u_1^h - f_1\|_{Y_i,h}^2 + \|\mathcal{T}_i(u_1,u_2)\|_{R(\mathcal{T}_i),h}^2 \right) \\ X^h = X_1^h \times X_2^h \end{cases}$$

for appropriate discrete spaces X_i^h and discrete norms $\|\cdot\|_{Y_i,h}$ and $\|\cdot\|_{R(\mathcal{T}_i),h}$, $i = 1,2$, where $R(\mathcal{T}_i)$ denotes the range space of the operator $\mathcal{T}_i(\cdot,\cdot)$. We already discussed this approach in Section 12.10; therefore, here, we examine two other approaches.

Optimization-based methods

Multi-physics and multi-model problems of the type (12.89) and (12.90) can be treated by any of the many domain decomposition approaches available in the literature. Relevant to LSFEMs is an optimization-based approach described as follows.

Suppose problem (12.89) is supplemented by the "boundary" conditions along the iterface Σ

$$\mathcal{B}_1(u_1) = g_1 \qquad \text{and} \qquad \mathcal{B}_2(u_2) = g_2 \qquad \text{on } \Sigma\,, \tag{12.91}$$

where g_1 and g_2 are given data. Next, suppose that (12.89) and (12.91) determine solutions $\widetilde{u}_{g_1}$ and $\widetilde{u}_{g_2}$. Clearly, these solutions satisfy (12.89), but for *arbitrary* choices of g_1 and g_2, there is no reason to expect that they satisfy the interface conditions in (12.90). On the other hand, we have tacitly assumed that the problems (12.89) and (12.90) do have a solution, so that there exist *particular* g_1 and g_2 so that (12.90) is satisfied. Indeed, if a solution $\{u_1,u_2\}$ of (12.89) and (12.90) were known, the relevant g_1 and g_2 are simply given by $g_1 = \mathcal{B}_1(u_1)$ and $g_2 = \mathcal{B}_2(u_2)$. Of course, the solution $\{u_1,u_2\}$ of (12.89) and (12.90) is not known so that we propose to find it by solving the following optimization problem:

$$\begin{cases} \min_{g_1,g_2} K\big(\{u_1,u_2\},\{g_1,g_2\};f_1,f_2\big) \\ \text{subject to (12.89) and (12.91) being satisfied}\,, \end{cases} \tag{12.92}$$

[33] Conversely, the approaches considered in this section can be applied to the transmission problem discussed in Section 12.10.

where

$$K(\{u_1,u_2\},\{g_1,g_2\};f_1,f_2) = \|\mathcal{T}_1(u_1,u_2)\|^2_{R(\mathcal{T}_1)} + \|\mathcal{T}_2(u_1,u_2)\|^2_{R(\mathcal{T}_2)}\,. \quad (12.93)$$

In this problem, $\{u_1,u_2\}$ are "state" variables, $\{g_1,g_2\}$ are "control" variables, and (12.89) and (12.91) are constraint equations that candidate states and controls are required to satisfy.[34] Details about this approach to solving domain decomposition and multi-physics problems may be found in [148, 194, 197–199].

It is not difficult to see that the minimization problem (12.92) is exactly of the form discussed in Chapter 11 and may therefore be treated by any of the methods, including the various LSFEMs, discussed in that chapter.

Other monolithic and operator splitting methods

In this section, we use the concrete setting of fluid–structure interaction problems to briefly discuss other approaches for solving multi-physics problems. Fluid–structure interaction problems are perhaps the most studied of multi-physics problems. They arise in applications that include blood flows in arteries, the design of artificial hearts, and flow-induced structural vibrations in aircraft and cars, just to name a few. As the name suggests, a general fluid–structure interaction problem consists of at least two sets of coupled governing equations, i.e., one component of (12.89) describes the structural response and the other the fluid flow. The fluid and the structure usually share a common physical interface through which interactions takes place by means of momentum and energy exchanges.

Several features make fluid–structure problems extremely challenging to solve numerically. First, they involve individual mathematical models that are difficult to solve by themselves but that now must be coupled. Second, the numerical treatment of interfaces and interface conditions between a fluid and a moving body is far from trivial. Lastly, in most applications, the shapes of the fluid and solid regions are not known in advance, i.e., they are unknowns in the simulation. Both monolithic and operator splitting approaches have been used for fluid–structure interaction problems. For a review of traditional methods for fluid–structure interaction problems, see [291]. In the remainder of this section, we consider monolithic and operator

---

[34] In many, if not most situations, it is possible to choose g_1 and g_2 so that one of the interface conditions in (12.90) is satisfied. Say it is $\mathcal{T}_1(u_1,u_2) = 0$ that is satisfied for all candidate "controls" g_1 and g_2; then, the functional in (12.93) simplifies to

$$\begin{cases} \min\limits_{g_1,g_2} K\big(\{u_1,u_2\},\{g_1,g_2\};f_1,f_2\big) \\ \text{subject to (12.89) and (12.91) being satisfied}\,, \end{cases}$$

where

$$K(\{u_1,u_2\},\{g_1,g_2\}\};f_1,f_2) = \|\mathcal{T}_2(u_1,u_2)\|^2_{R(\mathcal{T}_2)}\,.$$

For example, suppose the first interface condition in (12.90) is given by $u_1 - u_2 = 0$ on Σ. Then, if we choose $g_1 = g_2 = g$ and $u_1 = g$ and $u_2 = g$ for (12.91), we have that the first interface condition is satisfied for any g. Note that in this case, the number of "control" functions reduces to one.

splitting approaches for such problems based on least-squares principles; our presentation follows largely the developments outlined in [202, 245].

The model problem in two dimensions consists of a linear elastic solid and a Newtonian fluid occupying the regions Ω_E and Ω_F, respectively, that are separated by an interface Σ. The governing equation for the linear elastic solid in terms of the displacement $\mathbf{u}$ is given by

$$-\mu\Delta_\xi \mathbf{u} - (\lambda+\mu)\nabla_\xi\nabla_\xi\cdot\mathbf{u} = 0 \qquad \text{in } \Omega_E\,, \tag{12.94}$$

where λ and μ are the Lamé constants. Equation (12.94) is defined on the undeformed (reference) configuration using Lagrangian coordinates $\xi = (\xi,\eta)$. The fluid flow is governed by the Navier–Stokes equations (1.67) in the Eulerian frame $\mathbf{x} = (x,y)$, repeated here for convenience:

$$\begin{cases} -\nu\Delta_x\mathbf{v} + \mathbf{v}\cdot\nabla_x\mathbf{v} + \nabla_x p = \mathbf{f} \\ \nabla_x\cdot\mathbf{v} = 0 \end{cases} \qquad \text{in } \Omega_F\,. \tag{12.95}$$

By Σ_ξ and Σ_x we denote the interface in the reference and the deformed configurations, respectively. Below we restrict attention to coupling of the fluid and the solid via the following traction matching condition (see [202]):

$$\mathbf{n}\cdot\sigma_E(\mathbf{u}) = \mathbf{n}\cdot\sigma_F(\mathbf{v}) \qquad \text{on } \Sigma_x\,.$$

Further examples of coupling terms can be found in [245].

The solid and fluid equations (12.94) and (12.95) are defined with respect to different coordinate systems. To couple them, elliptical grid generation equations (EGG)

$$\Delta_x\xi = 0 \qquad \text{in } \Omega_x\,, \tag{12.96}$$

are used to change the variables in the fluid equations to a mapped reference domain $\Omega_{F,\xi}$; see [202]. After this, all equations are replaced by their first-order equivalent forms. The coupled fluid–structure model then consists of the first-order equations for the solid:

$$\begin{cases} \underline{\mathrm{U}} - \nabla\mathbf{u} = 0 \\ -\nu(\nabla\cdot\underline{\mathrm{U}})^{\mathsf{T}} - \lambda\nabla\mathrm{tr}(\underline{\mathrm{U}}) = 0 \\ \nabla\times\underline{\mathrm{U}} = 0 \end{cases} \qquad \text{in } \Omega_E\,, \tag{12.97}$$

the first-order EGG system in reference coordinates:

$$\begin{cases} J - \nabla\mathbf{x} = 0 \\ (J^{-\mathsf{T}}J^{-1}\nabla)\cdot J = 0 \\ \nabla\times J = 0 \end{cases} \qquad \text{in } \Omega_\xi\,, \tag{12.98}$$

and the first-order velocity gradient–velocity–pressure form (see Section 8.1) of the Navier–Stokes equations transformed to Ω_ξ, i.e., also in reference coordinates:

$$\left\{\begin{aligned} \underline{\mathbf{V}} - \nabla \mathbf{v} &= 0 \\ \nu(J^{-\mathrm{T}}J^{-1}\nabla)\cdot\underline{\mathbf{V}} - (J^{-1}\nabla p)^{\mathrm{T}} - (\mathbf{v}\cdot J^{-1}\underline{\mathbf{V}}) &= 0 \\ \omega\nabla\mathrm{tr}(J^{-1}\underline{\mathbf{V}}) &= 0 \qquad \text{in } \Omega_\xi\,. \\ (J^{-1}\nabla)\cdot\mathbf{v} &= 0 \\ \nabla\times\underline{\mathbf{V}} &= 0 \end{aligned}\right. \tag{12.99}$$

The weight ω plays the same role as in (7.75) and can be used to improve mass conservation in the least-squares formulation of (12.99).

The components of the first-order system (12.97)–(12.99) are coupled by the stress matching condition which, when written in terms of the new variables, has the form

$$\begin{aligned} &J^{-1}\mathbf{n}\cdot\left(\underline{\mathbf{U}}+\underline{\mathbf{U}}^{\mathrm{T}}+\frac{\lambda}{\mu}\mathrm{tr}(\underline{\mathbf{U}})\delta_{\mathrm{ij}}\right) \\ &\qquad = J^{-1}\mathbf{n}\cdot\left(J^{-1}\underline{\mathbf{V}}+(J^{-1}\underline{\mathbf{V}})^{\mathrm{T}}-p\delta_{ij}\right) \qquad \text{on } \Sigma_\xi\,. \end{aligned} \tag{12.100}$$

A monolithic LSFEM for the fluid–structure problem can be defined by summing up residuals of (12.97)–(12.100). For the sake of brevity, we omit the definition of this functional, as it repeats the patterns already seen in previous chapters.

The coupled problem (12.97)–(12.100) can also be solved by an "operator-splitting" LSFEM in which individual least-squares functionals for (12.97)–(12.99) are defined and minimized separately, but following a predetermined order. Specifically, we begin by solving a least-squares problem for the fluid equations (12.99), followed by the minimization of a combined least-squares functional for the solid equation (12.97) and the matching condition (12.100), and finally, solving a least-squares problem for the EGG equation (12.98). This process is repeated until satisfactory convergence is achieved; see [202] for details.

12.12 Least-Squares Finite Element Methods for Problems with Singular Solutions

Solutions of elliptic boundary value problems may develop singularities if the boundary and/or data are not smooth. Problem configurations that lead to singular solutions are not at all uncommon in practical applications. Examples include domains with cracks, domains with corners such as an L-shaped membrane, and transmission problems with discontinuous material parameters.[35]

---

[35] The LSFEMs for transmission problems considered in Section 12.10 use independently defined finite element spaces with respect to subregions determined by the material discontinuity. As a result, the finite element solution is not constrained to be continuous along the interface. The methods considered in this section use a single, conforming finite element space defined over the whole domain.

Standard piecewise polynomial finite element spaces tend to converge poorly in the presence of singularities irrespective of the approximation scheme, be it Galerkin or least-squares, used. The cause for this poor convergence is the inability of polynomials to capture well the behavior of the leading terms of the exact solution near the singular points. One logical possibility would be to consider finite element spaces enriched by "singular" functions that match the relevant leading terms.

The idea to supplement finite element spaces by singular functions is not new; it was proposed in [161, 323] to improve the accuracy of Galerkin methods for the Poisson equation for L-shaped domains, domains with cracks, and transmission problems. Assume that ϕ is a solution of the Poisson equation (1.52) that has one or more singularities. The approach suggested in [323, p. 263] is based on the fact that one can find a set of "singular functions" $\{\psi_k\}_{k=1}^K$ such that, for some c_ks, $\phi - \sum c_k \psi_k$ is smooth. Because these functions are needed only near the singular points, they are defined to act only near the points at which the singular behavior occurs, i.e., they smoothly reduce to zero within a specified distance of those points. Then, an enriched finite element space[36] is defined by $\widetilde{X}^h = X^h \cup \{\psi_k\}_{k=1}^K$, where X^h is a standard $C^0(\Omega)$ finite element subspace of $H^1(\Omega)$. The enriched space is also a subspace of $H^1(\Omega)$, and so it can be used in the Galerkin method in lieu of the standard piecewise polynomial one. The use of enriched spaces can be easily adapted to least-squares principles.

There are, however, some issues with the direct use of singular functions in the approximation space. One is the computation of various integrals in the discrete problem that involve ψ_k. This should be done either analytically or through the use of very accurate quadrature rules; see [323, p. 264]. A second, more serious issue is the conditioning of the resulting algebraic system, especially if the discrete problem is solved by iterative methods.

Weighted least-squares functionals

It turns out that by using suitably weighted least-squares functionals, the use of singular functions can be completely avoided. This idea was pioneered in [129, 166] and later used in [259] for problems with more general boundary conditions. To describe the main idea, we consider the first-order Poisson system (2.35) and assume that Ω is a planar region with a corner at P. Recall that the standard least-squares functional for this system is given by (2.36). The associated DLSP leads to approximations achieving the full accuracy possible with the finite element spaces used if and only if the exact solution is sufficiently smooth. To define a least-squares functional that works for solutions with singularity at P, we introduce the weighted $L^2(\Omega)$ norm

---

[36] A further extension of this idea is the partition of unity method [22] for which the singular functions are added by using the fact that the standard Lagrangian nodal basis forms a partition of unity. The result is a conforming finite element space. The Treffetz elements discussed in Section 12.13 can be viewed as an extreme example of enrichment where the (non-conforming) approximation space contains only enrichment functions.

$$\|\psi\|_{0,\beta}^2 = \int_\Omega r^{2\beta} |\psi|^2 d\Omega \,,$$

where $\beta \geq 0$ and r is the distance to the corner at P. Then, we replace (2.36) by the following weighted least-squares functional:

$$J(\phi, \mathbf{v}; f) = \|\nabla \cdot \mathbf{v} - f\|_{0,\beta}^2 + \|\nabla\phi + \mathbf{v}\|_0^2 \,. \tag{12.101}$$

The choice of β is dictated by the strength of the singularity. If θ_P is the interior angle at the corner P, then

$$\phi \approx O(r^{\pi/\theta_P}) \qquad \mathbf{v} \approx O(r^{\pi/\theta_P - 1}) \qquad \nabla \cdot \mathbf{v} \approx O(r^{\pi/\theta_P - 2}) \,.$$

Assume that the corner at P is reentrant so that $\theta_P > \pi$. Then, the first term in (12.101) is finite only if

$$\beta > 1 - \pi/\theta_P \,.$$

The analysis given in [129] suggests one should use the larger value $\beta = 1 - \pi/2\theta_P$ in order to obtain an optimal rate of convergence with respect to the *unweighted* $L^2(\Omega)$ norm. It should be pointed out that optimal convergence rates do require a specially constructed mesh that follows the refinement rules described in [21].

The weighted LSFEM in [259] uses similar ideas in the context of a two-stage solution process for the extended first-order Poisson system (5.25). For problems with singularities, the first stage involves the minimization of the weighted functional

$$J_\omega(\mathbf{v}; f) = \|\omega(\nabla \cdot \mathbf{v} - f)\|_0^2 + \|\omega\nabla \times \mathbf{v}\|_0^2 \,,$$

where $\omega = r^\beta$ for some $\beta > 0$. The values of β are selected to make $[H^1(\Omega)]^2$ dense in $\mathbf{C}(\Omega) \cap \mathbf{D}(\Omega)$ with respect to a weighted $L^2(\Omega)$ norm induced by J_ω. The minimization of J_ω is with respect to the space $[H_0^1(\Omega)]^2$, constrained by the homogeneous tangential and normal boundary conditions in (5.25). The minimizer $\mathbf{u}$ of J_ω is then used in the second stage to compute ϕ by minimizing the functional

$$J_2(\phi; \mathbf{u}) = \|\mathbf{u} + \nabla\phi\|_0^2 \,.$$

For more details about this method and computational examples, see [259, 260].

LL* methods

The LL* version of LSFEMs discussed in Section 12.2 can be used to advantage for problems with singular solutions; see [258]. As is done in that paper, we consider the eddy current problem (see (6.14) and (6.15)) written in terms of the electric field $\mathbf{E}$ and magnetic field $\mathbf{H}$. After time discretization by the backward-Euler method, we have, at each time step, the system

$$\mathcal{L}u = \begin{pmatrix} -\sigma & \nabla\times \\ 0 & \nabla\cdot\widetilde{\mu} \\ \nabla\times & \widetilde{\mu} \\ \nabla\cdot\sigma & 0 \end{pmatrix} \begin{pmatrix} \mathbf{E} \\ \mathbf{H} \end{pmatrix} = \begin{pmatrix} 0 \\ 0 \\ \widetilde{\mu}\overline{\mathbf{H}} \\ 0 \end{pmatrix} = F\,, \tag{12.102}$$

where

$$u = \{\mathbf{E},\mathbf{H}\} \in \big\{\mathbf{C}(\Omega)\cap\mathbf{D}_0(\Omega,\sigma)\big\}\times\big\{\mathbf{C}_0(\Omega)\cap\mathbf{D}(\Omega,\widetilde{\mu})\big\}\,, \tag{12.103}$$

$\widetilde{\mu} = \mu/\Delta t$, and $\overline{\mathbf{H}}$ denotes the magnetic field determined at the previous time step. We assume that the system (12.102) is supplemented with the boundary conditions $\sigma\mathbf{E}\cdot\mathbf{n} = 0$ and $\mathbf{H}\times\mathbf{n} = 0$ on $\partial\Omega$.

The corresponding adjoint formulation is given by

$$\mathcal{L}^*y = \begin{pmatrix} -\sigma & 0 & \nabla\times & -\sigma\nabla \\ \nabla\times & -\widetilde{\mu}\nabla & \widetilde{\mu}I & 0 \end{pmatrix} \begin{pmatrix} \mathbf{E}^* \\ p \\ \mathbf{H}^* \\ q \end{pmatrix} = \begin{pmatrix} \mathbf{E} \\ \mathbf{H} \end{pmatrix} = u\,, \tag{12.104}$$

where[37]

$$y = \{\mathbf{E}^*, p, \mathbf{H}^*, q\} \in \mathbf{C}(\Omega)\times G_0(\Omega)\times\mathbf{C}_0(\Omega)\times G(\Omega)\,. \tag{12.105}$$

One can add additional equations to the system (12.104) that serve to make the solution unique.[38] In [258], for problems with singular solutions, the additional equations are chosen so that one obtains the system

$$\mathcal{L}^*y = \begin{pmatrix} -\sigma & 0 & \nabla\times & -\sigma\nabla \\ 0 & 0 & r^\beta\nabla\cdot & 0 \\ \nabla\times & -\widetilde{\mu}\nabla & \widetilde{\mu}I & 0 \\ r^\beta\nabla\cdot & 0 & 0 & 0 \end{pmatrix} \begin{pmatrix} \mathbf{E}^* \\ p \\ \mathbf{H}^* \\ q \end{pmatrix} = \begin{pmatrix} \mathbf{E} \\ r^\alpha s_1 \\ \mathbf{H} \\ r^\alpha s_2 \end{pmatrix} = u\,, \tag{12.106}$$

where

$$y = \{\mathbf{E}^*, p, \mathbf{H}^*, q\} \in \big\{\mathbf{C}(\Omega)\cap\mathbf{D}_0(\Omega)\big\}\times G_0(\Omega)\times\big\{\mathbf{C}_0(\Omega)\cap\mathbf{D}(\Omega)\big\}\times G(\Omega)$$

and where β and r are defined as in the discussion above for weighted least-squares functionals.[39] Note that the weight functions are introduced only in the gauge equa-

[37] Note that in (12.104), the constitutive parameters σ and $\widetilde{\mu}$ are always undifferentiated, i.e., they appear outside the differentiation operators, whereas in (12.102) they are acted on by those operators. Also note that the spaces used in (12.105) are all unweighted whereas some of those used in (12.103) are weighted. These features possessed by (12.104) and (12.105) offer advantages over (12.102) and (12.103) for interface or transmission problems and for problems with singular solutions.

[38] In the language of electromagnetics, adding equations for this purpose is referred to as "choosing a gauge."

[39] The operator $\mathcal{L}^*$ given in (12.106) is adjoint to the operator $\mathcal{L}$ of the system

tions. Also note that, once y is determined, one can find the variables of interest from $u = \mathcal{L}^* y$ so that

$$\begin{cases} \mathbf{E} = -\sigma \mathbf{E}^* + \nabla \times \mathbf{H}^* - \sigma \nabla q \\ \mathbf{H} = -\widetilde{\mu} \mathbf{H}^* + \nabla \times \mathbf{E}^* - \widetilde{\mu} \nabla p. \end{cases} \quad (12.107)$$

We may now proceed as in Section 12.2 to define a least-squares functional and determine approximations $y^h = \{\mathbf{E}^*, p^h, \mathbf{H}^*, q^h\}$ of $y = \{\mathbf{E}^*, p, \mathbf{H}^*, q\}$, after which approximations of the variables of interest can be recovered by applying (12.107). See [258] for error estimates and other details about the LL* finite element approach to problems with singular solutions.

12.13 Treffetz Least-Squares Finite Element Methods

The LSFEMs considered thus far are based on the approximation of function spaces by finite element spaces that are characterized by the use of "general purpose" polynomial functions; see Appendix B. The use of such simple approximating spaces is what makes finite element methods so powerful. However, there are problems where polynomials, and especially low-order polynomials, are not accurate enough without also invoking excessive mesh refinement. One typical example is the Helmholtz equation (1.63). For large values of k, this equation has highly oscillatory solutions that are difficult to resolve by linear or even higher-order polynomials. A practical rule is that the mesh has to provide at least six grid points per wavelength, but even this may not be enough in some cases. For large problems where Ω is many wavelengths in diameter, this can quickly lead to intractable systems.

One alternative for Helmholtz and other like problems is to replace the "general-purpose" finite element space by a specialized[40] "finite element" space that is defined using exact solutions of the PDE on each element. By using such an approximating space, the governing differential equations are automatically satisfied within

---

$$\mathcal{L}u = \begin{pmatrix} -\sigma & 0 & \nabla \times & -\nabla r^\beta \\ 0 & 0 & \nabla \cdot \widetilde{\mu} & 0 \\ \nabla \times & -\nabla r^\beta & \widetilde{\mu} & 0 \\ \nabla \cdot \sigma & 0 & 0 & 0 \end{pmatrix} \begin{pmatrix} \mathbf{E} \\ s_1 \\ \mathbf{H} \\ s_2 \end{pmatrix} = \begin{pmatrix} \mathbf{0} \\ 0 \\ \widetilde{\mu}\overline{\mathbf{H}} \\ 0 \end{pmatrix} = F$$

that is found from (12.102) by introducing the slack variables s_1 and s_2 and additional terms in the equations that render the resulting system into one that has a solution for any right-hand side. Note that (12.102) has a solution only if the first and third components of a general right-hand side F are divergence free. If this is indeed the case, the slack variables in (12.106) vanish and one recovers (12.102). Note also the difference in how the weight function r^β and the constitutive parameters appear in this system as compared to (12.106), i.e., the weight function and constitutive parameters are differentiated in the former and are not differentiated in the latter.

[40] The use of singular functions [323, p. 262–270] in finite element methods is another example of the applications of this approach; in this case, one uses specialized basis functions to help resolve singularities in the solution. The LSFEMs considered in Section 12.12 illustrate the use of this approach in connection with least-squares principles.

the finite elements. The key issue in the practical use of this approach is how to glue together the local element spaces into a global approximation that at least weakly satisfies the continuity requirements of the underlying solution space for the PDE.

Numerical methods that are based on this idea are usually called *Treffetz element methods* or *T–element methods*; see [201] for a general discussion of this approach and [242] for examples of T–element methods and their applications. T–element methods differ in the manner in which the inter-element continuity and the boundary conditions are enforced. In this section, we review an approach that uses least-squares principles for that purpose. The origins of this approach can be traced to [241]; our discussion is based on the Treffetz LSFEM for the Helmholtz equation considered in [278, 321].

To describe the least-squares T–element method, we use the following model scattering problem:

$$\begin{cases} -\Delta\phi - k^2\phi = 0 & \text{in } \Omega \\ \phi = 0 & \text{on } \partial\Omega \end{cases} \quad \text{and} \quad \frac{\partial\phi}{\partial\mathbf{n}} - ik\phi = 0 \quad \text{on } \partial\Omega_\infty\,, \tag{12.108}$$

with a constant wavenumber $k > 0$. In (12.108), Ω denotes a bounded domain in $\mathbb{R}^2$, $\partial\Omega$ is the near boundary and $\partial\Omega_\infty$ is the absorbing boundary. The boundary condition on $\partial\Omega_\infty$ is a simple absorbing condition; other conditions, e.g., the Sommerfeld radiation condition [321] at infinity, are also possible.

To define a T–element space for (12.108), Ω is divided into N subdomains Ω_ℓ. Because inter-domain continuity is imposed via the variational formulation, the subdomains can be quite general and are not required to follow the rules for finite element partitions listed in Section B.1. However, standard finite element partitions are also acceptable as T–elements.

On each subdomain Ω_ℓ, we consider a basis consisting of M smooth functions $\{\phi_{\ell j}\}_{j=1}^M$ such that each $\phi_{\ell j}$ solves the Helmholtz equation (12.108) in Ω_ℓ:

$$-\Delta\phi_{\ell j} - k^2\phi_{\ell j} = 0 \quad \text{on } \Omega_\ell, \qquad j = 1, \ldots, M\,.$$

Outside of Ω_ℓ, the basis functions $\{\phi_{\ell j}\}_{i=1}^M$ vanish. These basis functions can be defined by truncating a complete set of local solutions to the Helmholtz operator such as plane waves; see [278, 321]. Note that, in general, the set of all basis functions is discontinuous across the subdomain interfaces so that the resulting T–element space is not a conforming subspace of $H^1(\Omega)$.

The approximate solution ϕ_T is defined as a linear combination of the subdomain basis functions:

$$\phi_T(\mathbf{x}) = \sum_{\ell=1}^{N}\sum_{j=1}^{M} c_{\ell i}\phi_{\ell j}(\mathbf{x})\,. \tag{12.109}$$

Because the T–element space is not a conforming subspace of $H^1(\Omega)$, this solution cannot be used in a conventional weak Galerkin or least-squares formulation of (12.108). Instead, one defines T–element methods in a way that resembles discontinuous Galerkin methods by using jump terms to impose the desired continuity between the subdomains and penalty terms to impose the boundary conditions.[41]

Treffetz LSFEMs can also be viewed as an example, using T–element spaces, of the discontinuous LSFEM (12.86) discussed in Section 12.10. Let Σ denote union of all the interfaces between all subdomains Ω_ℓ. The use of T–elements means that, in (12.86), all subdomain residual terms are zero and the least-squares functional only involves the interface jump terms and boundary residual terms on the Dirichlet boundary $\partial\Omega$ and the absorbing boundary $\partial\Omega_\infty$. As a result, we have the following Treffetz DLSP:

$$\begin{cases} J_T(\psi_T) = \omega_1 \int_\Sigma [\psi_T]^2 ds + \omega_2 \int_\Sigma |[\mathbf{v}_T]|^2 ds \\ \qquad\qquad + \omega_3 \int_{\partial\Omega} |\phi_T|^2 ds + \omega_4 \int_{\partial\Omega_\infty} \left| \frac{\partial \phi_T}{\partial \mathbf{n}} - \mathrm{i}k\phi_T \right|^2 ds \\ X_T = \bigcup_{\ell=1}^{N} \{\phi_{\ell j}\}_{j=1}^{M}, \end{cases} \tag{12.110}$$

where $[\cdot]$ denotes the full jump across the interface, i.e., $[\mathbf{v}] = \mathbf{v}_1 - \mathbf{v}_2$ on Σ. Note that this imposes a stronger inter-element continuity condition than the normal jump (12.82) residual that is used in (12.86). The Treffetz LSFEM (12.110) is very similar to the one proposed in [321], except that the latter uses the Sommerfeld radiation condition instead of the simple absorbing condition in (12.108).

For further details about Treffetz LSFEMs, see [241, 278, 321] and the refences cited therein.

12.14 A Posteriori Error Estimation and Adaptive Mesh Refinement

One of the most attractive features of LSFEMs is the transparent availability of effective a posteriori residual error indicators that can used as a foundation for effective adaptive mesh refinement strategies. The availability of such error estimators is easy to demonstrate.

Consider the abstract problem

$$\begin{cases} \mathcal{L}u = f & \text{in } \Omega \\ \mathcal{B}u = 0 & \text{on } \partial\Omega \end{cases} \tag{12.111}$$

the CLSP $\{J(u;f), X_0\}$, where

[41] The main difference between discontinuous Galerkin and Treffetz methods is that, in the latter case, T–elements are used so that the governing equations are automatically satisfied in the interiors of the subdomains Ω_ℓ and, as a result, weak formulations consist of only the jump and penalty terms.

$$J(u;f) = \|\mathcal{L}u - f\|^2_{Y,\Omega}\,.$$

For simplicity, we assume that the solutions space X_0 is constrained to satisfy the boundary condition $\mathcal{B}u = 0$. We also assume, again for simplicity, that the problem (12.111) is uniquely solvable for any $f \in Y$, i.e., that the operator has zero nullity and deficiency.[42] As a result, we have that if $u \in X_0$ denotes a solution of (12.111), then $J(u;f) = 0$. We define the "energy" norm $|||u||| = (J(u;0))^{1/2} = \|\mathcal{L}u\|^2_{Y,\Omega}$. We suppose that a finite element subspace $X_0^h \subset X_0$ is constructed using a triangulation $\mathcal{T}_h$ of Ω. Let $\{\kappa_m\}_{m=1}^M$ denote the set of finite elements in $\mathcal{T}_h$ and, for each $m = 1,\dots,M$, define the local "energy norm" $|||\cdot|||^2_{\kappa_m} = \|\mathcal{L}(\cdot)\|_{Y,\kappa_m}$. Note that $|||\cdot|||^2 = \sum_{m=1}^M |||\cdot|||^2_{\kappa_m}$. If u denotes the exact solution and u^h an approximation to that solution, we let $e^h = u - u^h$ denote the error.

We consider compliant DLSPs $\{J(u^h;f), X_0^h\}$ so that $|||\cdot||| = \|\cdot\|_{Y,\Omega}$ and $|||\cdot|||_{\kappa_m} = \|\cdot\|_{Y,\kappa_m}$. We then define the global a posteriori residual error indicator

$$r = \|\mathcal{L}u^h - f\|_{Y,\Omega}$$

and, for $m = 1,\dots,M$, the local a posteriori residual error indicators

$$r_m = \|\mathcal{L}u^h - f\|_{Y,\kappa_m}\,.$$

Note that both r and the r_ms are computable once an approximate solution u^h is determined. We then define the global and local effectivity indices θ and θ_m, respectively, corresponding to the error indicators by

$$\theta = \frac{r}{|||e^h|||} \qquad \text{and} \qquad \theta_m = \frac{r_m}{|||e^h|||_{\kappa_m}} \qquad \text{for } m = 1,\dots,M\,.$$

Effectivity indices tell how well an error indicator performs as an estimator of the true error, e.g., in the case of θ_m, it tells how well the residual error indicator r_m performs as an estimator of the true error e^h measured in the local "energy" norm $|||\cdot|||_{\kappa_m}$. In general, the goal[43] in designing a posteriori error indicators is to be able to show that the corresponding effectivity indices are bounded from above and below by positive constants whose values are independent of the mesh parameter h. It is easy to show that the error indicators r and r_m defined above are, in fact, *exact*, i.e., we have $\theta = 1$ and $\theta_m = 1$ for all j. To show that $\theta = 1$, note that

$$r^2 = J(u^h;f) = \|\mathcal{L}u^h - f\|^2_{Y,\Omega} = \|\mathcal{L}(u-u^h)\|^2_{Y,\Omega} = J(u-u^h;0) = |||u-u^h|||^2$$

so that

$$\theta = \frac{\|\mathcal{L}u^h - f\|_{Y,\Omega}}{|||u-u^h|||} = 1\,.$$

[42] More general cases, e.g., problems with non-zero nullities or deficiencies and functionals that impose inhomogeneous boundary conditions weakly can be treated in a similar manner.

[43] If this goal is met, efficient and effective mesh refinement strategies can be designed based on the error indicator.

Thus, the global residual norm $\|\mathcal{L}u^h - f\|_{Y,\Omega}$ is an exact indicator for the error $u - u^h$ measured in the natural global "energy" norm $|||\cdot|||$. Similarly, we have that

$$r_m^2 = \|\mathcal{L}u^h - f\|_{Y,\kappa_m}^2 = \|\mathcal{L}(u-u^h)\|_{Y,\kappa_m}^2 = |||u-u^h|||_{\kappa_m}^2 \qquad \text{for } m = 1,\ldots,M$$

so that

$$\theta_m = \frac{\|\mathcal{L}u^h - f\|_{Y,\kappa_m}}{|||u-u^h|||_{\kappa_m}} = 1 \qquad \text{for } m = 1,\ldots,M\,.$$

Thus,

> *the local residual norm* $\|\mathcal{L}u^h - f\|_{Y,\kappa_m}$ *is an* exact *indicator for the error measured in the local "energy" norm* $|||\cdot|||_{\kappa_m}$.

One may prefer to measure the error in the local solution norm $\|\cdot\|_{X,\kappa_m}$. Because

$$C_1\|\cdot\|_{X,\kappa_m} \le |||\cdot|||_{\kappa_m} \le C_2\|\cdot\|_{X,\kappa_m}\,,$$

and $r_m = |||e^h|||_{\kappa_m}$, we see that if the local effectivity indices are now defined by

$$\theta_m = \frac{r_m}{\|e^h\|_{X,\kappa_m}} \qquad \text{for } m = 1,\ldots,M\,, \tag{12.112}$$

then

$$C_1 \le \theta_m \le C_2\,.$$

The least-squares residual error indicator with respect to the solution norm of the error is no longer exact, but the corresponding effectivity index can be bounded from above and below by known constants.

Determining the effectiveness of residual error indicators for norm-equivalent and quasi-norm equivalent DLSPs is not so straightforward and has not been extensively studied in the literature, although some efforts in this direction can be found in [15].

Adaptive mesh refinement strategy

An adaptive mesh refinement strategy based on the residual error estimators r_m, $m = 1,\ldots,M$, can be designed in the usual manner.

Given an initial triangulation $\mathcal{T}_h^0$ *of* Ω *and the corresponding finite element space* $X_0^{h,0}$, *for* $n = 0,1,\ldots$ *until convergence,*

1. *determine the approximate solution* $u^{h,n}$ *of the DLSP* $\{J(u_h,f), X_0^{h,n}\}$
2. *for* $m = 1,\ldots,M$, *compute the local residual error indicators* $r_m = \|\mathcal{L}u^{h,n} - f\|_{Y,\kappa_m}$
3. *define a new triangulation* $\mathcal{T}_h^{n+1}$ *of* Ω *and the corresponding finite element space* $X_0^{h,n+1}$ *by refining elements for which* r_m *is "large" and coarsening element patches for which* r_m *is "small."*

Step 3 can be effected by any of the many means that have been developed in adaptive mesh generation based on error indicators. For the specific setting of LSFEMs, see [15, 104, 266]. In [104], residual error estimators based on least-squares functionals are discussed in the context of div–grad systems and a specific recipe for effecting Step 3 is provided and tested. Similar treatments in the context of the velocity–vorticity–pressure formulation of the Stokes system are provided in [266]. Abstract linear and nonlinear problems are considered in [15] as are the velocity–vorticity–pressure and velocity gradient–velocity–pressure formulations of the Stokes equations.

12.15 Least-Squares Wavelet Methods

Wavelet methods are cousins to finite element methods; they share many similarities yet also exhibit significant differences. Here, because of the similarities, we provide a brief discussion of the use of wavelet bases for the discretization of least-squares principles. We follow the more detailed presentation of [132].

Fundamental to wavelet methods are appropriate *wavelet bases* for function spaces $H \subseteq L^2(\Omega)$ defined with respect to a bounded domain $\Omega \subset \mathbb{R}^d$ and possessing certain properties; see, e.g., [130, 251] and the references cited therein for such constructions.

Wavelets are indexed by a parameter $\lambda = \{j, k, \mathbf{e}\}$ that encodes information such as the *resolution level* or *scale* $|\lambda| = j$, the *spatial location* k, and possibly the type $\mathbf{e}$ of wavelet in the multivariate case. For univariate domains $\Omega = (a, b)$, the indices are simply of the form $\lambda = \{j, k\}$. In view of the finite domain Ω, there is a coarsest level j_0. We denote the infinite set of all possible indices by $\mathbb{I}$. A wavelet basis is a collection of functions

$$\Psi = \{\psi_\lambda : \lambda \in \mathbb{I}\} \subset H$$

having the following properties.

(R) Ψ constitutes a *Riesz basis* for H: every $v \in H$ has a unique expansion

$$v = \sum_{\lambda \in \mathbb{I}} v_\lambda \, \psi_\lambda \tag{12.113}$$

in terms of Ψ and its expansion coefficients satisfy a *norm-equivalence* relation, i.e., there exist constants $0 < c \leq C < \infty$ such that

$$c\|\vec{v}\|_{\ell^2(\mathbb{I})} \leq \Big\| \sum_{\lambda \in \mathbb{I}} v_\lambda \, \psi_\lambda \Big\|_0 \leq C\|\vec{v}\|_{\ell^2(\mathbb{I})} \qquad \forall \vec{v} \in \ell^2(\mathbb{I}) \tag{12.114}$$

holds, where $\vec{v} = \{v_\lambda\}_{\lambda \in \mathbb{I}}$ is the vector representation of the coefficients in (12.113). In other words, wavelet expansions induce isomorphisms between the space of functions H and the sequence space $\ell^2(\mathbb{I})$.

(L) The functions ψ_λ are local: for each $\lambda \in \mathbb{I}$ one has

$$\operatorname{diam}(\operatorname{supp}\psi_\lambda) \sim 2^{-|\lambda|}.$$

(CP) There exists an integer $\widetilde{m}$ such that

$$\langle v, \psi_\lambda \rangle \le C 2^{-|\lambda|(d/2+\widetilde{m})} |v|_{W^{\widetilde{m}}_\infty(\operatorname{supp}\psi_\lambda)},$$

for some positive constant $C < \infty$, where $\langle \cdot, \cdot \rangle$ denotes the duality pairing between H and its topological dual H'.

The *locality property* (L) is, of course, a key to the efficiency of wavelet methods for the approximation of solutions of PDEs, just as it is for finite element methods. The *cancellation property* (CP) implies that integrating a function against a wavelet is like taking an $\widetilde{m}$th order difference which annihilates the smooth part of v. The underlying moment conditions entail a direct approximation estimate for a finite-dimensional subspace $V_{|\lambda|}$ that is spanned by all wavelet functions up to a highest refinement level $|\lambda|$. The *Riesz property*[44] (R) has several important theoretical and practical consequences. For example, in addition to the norm-equivalence relation (12.114), we have similar norm equivalences between Sobolev norms (with both positive and negative indices) of a function v and $\ell^2(\mathbb{I})$ norms of simple scalings of its wavelet coefficients $\{v_\lambda\}$.[45]

Properties (R), (L), and (CP) allow one to prove strong theoretical statements such as (asymptotically) optimal condition number estimates for linear elliptic operators or convergence results for adaptive methods for linear and nonlinear variational problems [125, 126, 131, 251]. At the same time, they still allow one to work computationally with piecewise polynomials. Concrete constructions of biorthogonal wavelet bases on bounded Euclidean tensor product domains based on B-splines can be found in [251]. Constructions for multivariate box domains can be handled by tensor products. For more general domains or manifolds, see the constructions in [133, 252] and the references cited therein.

Once a wavelet basis has been chosen, it can be used in much the same manner as a finite element basis to define and analyze a discretization of a least-squares principle. However, one can also take advantage of certain additional consequences of the Riesz basis property possessed by wavelet bases. For one thing, it allows for the straightforward construction of infinite matrix problems that are equivalent to the first-order necessary condition corresponding to the (undiscretized) least-squares principle. This includes the construction of these matrices for problems involving negative and factional inner products. Discretizations are then easily defined by truncations of the infinite-dimensional matrix problems. In addition, the fact that equivalent norms can be defined in terms of $\ell^2(\mathbb{I})$ norms of scaled wavelet coefficients

[44] Hierarchical finite element bases [351, 352, 355] are also multiresolution bases. However, they do not possess the Riesz basis property.

[45] This is, of course, entirely analogous to the situation for another Riesz basis expansion of square integrable functions, namely Fourier series. For example, if $v = \sum_{\xi=-\infty}^{\infty} c_\xi e^{i\xi x}$, we then have that $\|v\|_0^2$ is equivalent to $\sum_{\xi=-\infty}^{\infty} c_\xi^2$ whereas $\|v\|_1^2$ is equivalent to $\sum_{\xi=-\infty}^{\infty} \xi^2 c_\xi^2$ and $\|v\|_{-1}^2$ is equivalent to $\sum_{\xi=-\infty}^{\infty} \xi^{-2} c_\xi^2$.

results in straightforward preconditioning strategies for the matrix problems corresponding to discretized least-squares principles. See [132] for details.

12.16 Meshless Least-Squares Methods

Another cousin of finite element methods are *meshless* approaches to the discretization of PDEs that eschew the use of grids in the discretization process; instead, those processes are based on points so that meshfree methods are related to particle and other such methods. Several meshless method have been proposed, yielding a variety of approximation spaces from which to choose.[46] Not surprisingly, meshless approaches to residual mininimization approximation of solutions for PDEs have also been considered; see, e.g., [269, 292, 293]. Here, for the sake of concreteness, we consider the partition of unity method discussed in [269].

We begin by describing how meshfree basis functions are constructed. Let[47]

$$\Omega_i = \{\mathbf{x} \in \mathbb{R}^2 \mid |\mathbf{x} - \mathbf{z}_i| \le r_i\} \qquad \text{for } i = 1, \ldots, N,$$

where $\{\mathbf{z}_i\}_{i=1}^N$ is a given set of points in a domain $\Omega \subset \mathbb{R}^2$ and the associated radii $\{r_i\}_{i=1}^N$ are chosen so that the set $\{\Omega_i\}_{i=1}^N$ covers Ω, i.e., $\cup_{i=1}^N \Omega_i \supset \Omega$; see [149] for an effective, truly meshless means for determining point sets and support radii.[48] Following [151], let

$$\phi_i(\mathbf{x}) = \Phi\left(\frac{|\mathbf{x} - \mathbf{z}_i|}{r_i}\right) \qquad \text{for } i = 1, \ldots, N,$$

where

$$\Phi(s) = \begin{cases} 1 - 6s^2 + 8s^3 - 3s^4 & \text{for } 0 \le s \le 1 \\ 0 & \text{for } s \ge 1\,. \end{cases}$$

Thus, $\phi_i(\mathbf{x}) = 0$ outside of Ω_i. Then, if[49]

$$\psi_i(\mathbf{x}) = \frac{\phi_i(\mathbf{x})}{\sum_{j=1}^N \phi_j(\mathbf{x})} \qquad \text{for } i = 1, \ldots, N,$$

---

[46] These include the diffuse element, element free Galerkin, finite point, HP cloud, meshfree local Petrov-Galerkin, smooth particle hydrodynamics, moving least-squares, material-point, partition of unity, and reproducing kernel particle methods; see, e.g., [178,262], for classifications and reviews.

[47] Subdomains other than circles, e.g., rectangles, can be used instead. Also, meshfree methods in three dimensions can be developed in an entirely analogous manner.

[48] The supporting radii have to satisfy certain conditions; see [149, 269] for details.

[49] Note that the summation appearing in the denominator need only range over indices j such that $\Omega_i \cap \Omega_j \neq \emptyset$.

the set $\{\psi_i(\mathbf{x})\}_{i=1}^N$ defines a partition of unity of Ω with respect to the covering set $\{\Omega_i\}_{i=1}^N$, i.e., we have that $\sum_i \psi_i(\mathbf{x}) = 1$ for all $\mathbf{x} \in \Omega$. This immediately defines a global approximating space $V^0 = \mathrm{span}\{\psi_i(\mathbf{x})\}_{i=1}^N$. Due to the smoothness of the quartic spline window function $\Phi(\cdot)$, we have that $V^0 \subset H^1(\Omega)$.

The partition of unity provides a means for transferring local approximation properties to the entire domain. For example, global approximation can be enhanced by considering, for $i = 1,\ldots,N$, the local spaces

$$V_i^q = \mathrm{span}\Big\{ \psi_i \ , \ \Big(\frac{x_1 - z_{1,i}}{r_i}\Big)\psi_i \ , \ \Big(\frac{x_2 - z_{2,i}}{r_i}\Big)\psi_i \ , \ \cdots \ , \\ \Big(\frac{x_1 - z_{1,i}}{r_i}\Big)^q \psi_i \ , \ \Big(\frac{x_1 - z_{1,i}}{r_i}\Big)^{q-1}\Big(\frac{x_2 - z_{2,i}}{r_i}\Big)\psi_i \ , \ \cdots, \ \Big(\frac{x_2 - z_{2,i}}{r_i}\Big)^q \psi_i \Big\},$$

where x_1 and x_2 denote the components of $\mathbf{x}$ and $z_{1,i}$ and $z_{2,i}$ denote the components of $\mathbf{z}_i$. Through direct summation, these local spaces give rise to the global partition of unity approximating space

$$V^q = V_1^q \oplus V_2^q \oplus \cdots \oplus V_N^q \, .$$

For vector-valued functions, we define, for $i = 1,\ldots,N$, the local approximating spaces $[V_i^q]^2$. For example, for $q = 1$ and for $i = 1,\ldots,N$, we have

$$[V_i^1]^2 = \mathrm{span}\Big\{ \psi_i^{(1)} \ , \ \psi_i^{(2)} \ , \ \Big(\frac{x_1 - z_{1,i}}{r_i}\Big)\psi_i^{(1)} \ , \ \Big(\frac{x_2 - z_{2,i}}{r_i}\Big)\psi_i^{(1)} \ , \\ \Big(\frac{x_1 - z_{1,i}}{r_i}\Big)\psi_i^{(2)} \ , \ \Big(\frac{x_2 - z_{2,i}}{r_i}\Big)\psi_i^{(2)} \Big\},$$

where

$$\psi_i^{(1)}(\mathbf{x}) = \begin{pmatrix} \psi_i(\mathbf{x}) \\ 0 \end{pmatrix} \qquad \text{and} \qquad \psi_i^{(2)}(\mathbf{z}) = \begin{pmatrix} 0 \\ \psi_i(\mathbf{x}) \end{pmatrix}.$$

Again, direct summation yields the global partition of unity approximating space

$$[V^1]^2 = [V_1^1]^2 \oplus \cdots \oplus [V_N^1]^2 \, .$$

An approximation theory for the spaces V^q is established in [22]. The straightforward generalization of that theory to the spaces $[V^q]^2$ is discussed in [269].

In [269], meshless least-squares methods for div–curl systems in two dimensions with homogeneous normal boundary conditions are considered. If the discrete solution space is constrained to satisfy the boundary condition, optimal error estimates are derived. For example, provided the boundary is smooth enough, it is shown that

$$\|\mathbf{u} - \mathbf{u}^r\|_1 = O(r) \, ,$$

where $r = \max_i\{r_i\}$ and $\mathbf{u} \in [H^2(\Omega)]^2$ and $\mathbf{u}^r \in \{\mathbf{v} \in [V^1]^2 \mid \mathbf{v}\cdot\mathbf{n} = 0 \text{ on } \partial\Omega\}$ denote the exact and approximate solutions, respectively.

Note that, unlike standard nodal finite element bases, the partition of unity basis functions do not possess the Kronecker delta property so that the unknown associated with a node on the boundary is not the function value at that node and interior nodes contribute to the discrete solution on the boundary. This makes the imposition of essential boundary conditions a somewhat difficult matter for meshless methods based on, e.g., Galerkin formulations. On the other hand, this makes least-squares approaches especially well suited for meshless methods because one can impose essential boundary conditions weakly through the least-squares functional. For example, for div–curl systems in two dimensions with a homogenous normal boundary condition, one can use the least-squares functional[50]

$$J_{-r/2}(\mathbf{u}) = \|\nabla\cdot\mathbf{u}\|_0^2 + \|\nabla\times\mathbf{u} - f\|_0^2 + r\|\mathbf{n}\cdot\mathbf{u}\|_{0,\partial\Omega}^2\,. \tag{12.115}$$

In [269], several implementation issues connected with meshless least-squares methods are discussed. One important issue is quadrature rules for the approximate assembly of the discrete system. Here, two difficulties arise. First, due to the high degree of the basis functions,[51] higher-order rules have to be employed compared to what one uses for nodal finite element bases. This adds considerably to the cost of the assembly process. Second, if circles are used as support regions, the need for quadrature rules over lenticular regions, i.e., over the intersection of two circles, arise. This difficulty makes the use of rectangular support regions relatively more attractive. See [292] for further discussions of quadrature rule effects in meshless methods.

---

[50] The weight r in (12.115) plays the same role as the mesh dependent weight h does in the weak treatment of boundary conditions within least-squares finite element methods; see Section 12.1 and in particular, (12.9). Note that the last term in (12.115) replaces the $\|\mathbf{n}\cdot\mathbf{u}\|_{-1/2,\partial\Omega}^2$ term that arises from energy balances in the ADN setting.

[51] Recall that the construction of the basis functions starts with the fourth degree polynomial $\Phi(\cdot)$.

Part V
Supplementary Material

Appendix A
Analysis Tools

This appendix provides a brief summary of definitions and facts about basic function spaces arising in the study of many partial differential equations (PDEs). Our main focus is on standard Sobolev spaces as well as some Hilbert spaces related to the gradient, curl, and divergence operators because these spaces are the most relevant ones to the subject matter of the book, namely the numerical solution of PDEs by least-squares finite element methods (LSFEMs). For more comprehensive treatments of the various function spaces arising in the theory of PDEs, see, e.g., [1, 17, 183, 187, 250, 265, 304, 331, 342].

A.1 General Notations and Symbols

Throughout the book, we make use of many standard symbols and notational conventions. The majority are listed in this section, with the most notable exceptions being notations and symbols specific to finite element methods that are introduced in Appendix B.

Usually, we consider PDE problems posed on a simply connected open bounded region $\Omega \subset \mathbb{R}^d$, $d = 2$ or 3. Unless stated otherwise, it is assumed that the boundary $\partial\Omega$ of Ω is Lipschitz continuous. Given an oriented surface Σ in $\mathbb{R}^d$ the symbol $\mathbf{n}(\mathbf{x})$ stands for the unit outer normal vector at a point $\mathbf{x} \in \Sigma$. The positive unit tangent vector at a point $\mathbf{x}$ on an oriented curve λ in $\mathbb{R}^d$ is denoted by $\mathbf{t}(\mathbf{x})$. When there is no possible ambiguity, we simply write $\mathbf{n}$ and $\mathbf{t}$.

The majority of the boundary value problems discussed in the book have at most two different types of boundary conditions imposed on $\partial\Omega$. In such cases, we assume that $\partial\Omega$ consists of two disjoint pieces Γ and Γ^*, i.e.,

$$\overline{\Gamma \cup \Gamma^*} = \overline{\partial\Omega} \qquad \text{and} \qquad \Gamma \cap \Gamma^* = \emptyset\,.$$

P.B. Bochev, M.D. Gunzburger, *Least-Squares Finite Element Methods*,
Applied Mathematical Sciences 166, DOI: 10.1007/b13382_13,

In some instances, we refer to Γ and Γ^* as the *Dirichlet* and *Neumann* parts of the boundary $\partial\Omega$, respectively. For boundary value problems that have only one type of boundary condition imposed, one of Γ and Γ^* is assumed to be empty.

Function spaces are denoted using uppercase italic Roman letters such as U, W, X, and Y. Functions belonging to such spaces are denoted by lowercase italic Roman and Greek symbols, e.g., ϕ, ξ, u, and v, and functionals and bilinear forms again use uppercase italic Roman letters, e.g., $F(u)$ and $Q(\phi,\psi)$. Operators acting between abstract spaces are denoted by calligraphic letters, e.g., $\mathcal{A}$, $\mathcal{B}$. Dual spaces and adjoint operators are marked by an asterisk, e.g., V^* is the dual of V and $\mathcal{A}^*$ is the adjoint of $\mathcal{A}$.

The notations $(\cdot,\cdot)_V$ and $\|\cdot\|_V$ are reserved for inner products and norms on the Hilbert space V; in some cases, when it is clear what space the inner product and the norm refer to, the space designation is omitted. For the duality pairing between V and its dual V^*, we use the symbol $\langle\cdot,\cdot\rangle_{V^*,V}$ or simply $\langle\cdot,\cdot\rangle$ when there is no chance for ambiguity.

For specific application problems, it is sometimes more convenient to adopt different notations for scalar, vector, and tensor-valued functions. In such cases, we reserve the use of lowercase italic Roman and Greek symbols for scalar functions, lowercase boldface Roman letters such as $\mathbf{u}$ for vector fields, and underlined boldface uppercase Roman letters such as $\underline{\mathbf{T}}$ for tensor-valued functions. One exception is the zero element in these space which we always write as 0. When such distinctions are necessary, the notation for the function spaces consisting of such functions follows similar conventions with, e.g., X, Y, Z denoting scalar spaces, $\mathbf{X}, \mathbf{Y}, \mathbf{Z}$ denoting vector spaces, and $\underline{\mathbf{X}}$, $\underline{\mathbf{Y}}$, $\underline{\mathbf{Z}}$ tensor spaces. However, we retain the same notations for functionals, bilinear forms, and operators as in the abstract case.

Matrices and (column) vectors in Euclidean spaces are denoted by symbols such as A and $\vec{u}$, respectively. The Euclidean norm of a vector $\vec{u}$ is $|\vec{u}|$ and $\|\mathsf{A}\|$ denotes a (subordinate) matrix norm.

When a vector is used to provide indices, we refer to it as a multi-index and drop the arrow notation, i.e., we have $\mu = (\mu_1,\ldots,\mu_k)$. The order of μ is defined by

$$|\mu| = \sum_{i=1}^{k} |\mu_i| \,.$$

Integer multi-indices provide convenient notation for partial derivatives; see [304, p. 38]:

$$D^\alpha u = \frac{\partial^{|\alpha|} u}{\partial^{\alpha_1} x_1 \cdots \partial^{\alpha_n} x_n} \,. \tag{A.1}$$

At times it is convenient to use ∂_{x_i} instead of $\partial/\partial x_i$.

In some cases, we choose to follow traditional and well-established notations such as $\mathbf{E}$, $\mathbf{J}$, $\mathbf{B}$, and $\mathbf{H}$ for electromagnetic variables or ε and σ for the strain and stress tensors in linear elasticity.

We use many of the standard notations from vector calculus such as the inner product and vector product operations $\mathbf{u}\cdot\mathbf{v}$ and $\mathbf{u}\times\mathbf{v}$, respectively, the Cartesian triple $\{\mathbf{i},\mathbf{j},\mathbf{k}\}$ of unit vectors, the partial derivative notation $\mathbf{u}_{i,j} = \partial\mathbf{u}_i/\partial x_j$, the nabla or gradient operator $\nabla = (\partial_{x_1}, \partial_{x_2}, \partial_{x_3})^\top$, and the divergence and curl operators $\nabla\cdot$ and $\nabla\times$, respectively. Two-dimensional vector calculus operations are inherited from the three-dimensional case in an obvious manner with the exception of the curl operator. Consider the three-dimensional vector fields $\mathbf{u} = u_3\mathbf{k}$ and $\mathbf{v} = v_1\mathbf{i} + v_2\mathbf{j}$. Then,

$$\nabla \times \mathbf{u} = u_{3,2}\mathbf{i} - u_{3,1}\mathbf{j} \qquad \text{and} \qquad \nabla \times \mathbf{v} = (v_{2,1} - v_{1,2})\mathbf{k}.$$

In $\mathbb{R}^2$, these identities give rise to two "versions" of the curl operator. The first one maps a scalar-valued function u into the two-dimensional vector field

$$\nabla \times u = u_{,2}\mathbf{i} - u_{,1}\mathbf{j}. \tag{A.2}$$

Note that $\nabla \times u = \nabla u \times \mathbf{k}$ which is why some authors [96,98] refer to this version of the curl operator in two dimensions as the *rotated gradient* and denote it by $\nabla^\perp$. The second curl operator in $\mathbb{R}^2$ maps a two-dimensional vector field into a scalar-valued function

$$\nabla \times \mathbf{u} := u_{2,1} - u_{1,2}. \tag{A.3}$$

Restriction of the vector product to two dimensions is obtained in the same way. Let $\mathbf{u} = u_1\mathbf{i} + u_2\mathbf{j}$, $\mathbf{v} = v_1\mathbf{i} + v_2\mathbf{j}$, and $\mathbf{p} = p\mathbf{k}$. Then, we have that

$$\mathbf{u} \times \mathbf{v} = (u_1v_2 - u_2v_1)\mathbf{k} \qquad \text{and} \qquad \mathbf{u} \times \mathbf{p} = p(u_2\mathbf{i} - u_1\mathbf{j}).$$

These identities produce two versions of the cross product in $\mathbb{R}^2$. The first one has two vector-valued arguments and produces a scalar-valued function:

$$\mathbf{u} \times \mathbf{v} = u_1v_2 - u_2v_1 \tag{A.4}$$

and the second acts on vector-valued and a scalar-valued function pairs and produces a vector field:

$$\mathbf{u} \times p = (pu_2)\mathbf{i} - (pu_1)\mathbf{j}. \tag{A.5}$$

A.2 Function Spaces

Modern theories for PDEs rely heavily on the notions of weak or generalized solutions and the Sobolev spaces to which these solutions belong. In addition to providing a very brief survey of definitions and results connected with Sobolev spaces, this section also presents some facts about Hilbert spaces related to the basic differential operators gradient, curl, and divergence. These operators and their domains form a mathematical structure called a De Rham complex which plays an important role in the design of compatible (or mimetic) discretization methods [7, 8, 13, 75, 137, 139, 204, 206, 314, 317].

A.2.1 The Sobolev Spaces $H^s(\Omega)$

For any real $s \geq 0$, we use the standard notation and definition (see [1]) for the Sobolev spaces $H^s(\Omega)$ with inner products and norms denoted by $(\cdot,\cdot)_{s,\Omega}$ and $\|\cdot\|_{s,\Omega}$, respectively. The space $H^0(\Omega)$ is the Hilbert space $L^2(\Omega)$ of all square integrable functions. If s is a positive integer, the spaces $H^s(\Omega)$ consist of all square integrable functions whose derivatives up to order s are also square integrable. For s a negative integer, the spaces $H^s(\Omega)$ are identified with the duals of $H^{-s}(\Omega)$. Similar definitions apply to Sobolev spaces for functions defined on the boundary $\partial\Omega$ of Ω. Those definitions can be extended to nonempty subsets γ of $\partial\Omega$; see, e.g., [1, 265, 304] for details. The inner product and norm on $H^s(\gamma)$ are denoted by $(\cdot,\cdot)_{s,\gamma}$ and $\|\cdot\|_{s,\gamma}$, respectively. Whenever there is no chance for ambiguity, the measures Ω and γ are omitted from inner product and norm designations.

Several subspaces of $H^1(\Omega)$ and $L^2(\Omega)$ play special roles in boundary value problems for PDEs. The set of all L^2 functions having zero mean with respect to Ω is a subspace of $L^2(\Omega)$ denoted by $L_0^2(\Omega)$, i.e,

$$L_0^2(\Omega) = \left\{ u \in L^2(\Omega) \mid \int_\Omega u\, d\Omega = 0 \right\}.$$

Let γ denote a nonempty subset of the boundary $\partial\Omega$. By $H_\gamma^1(\Omega)$ we denote the subspace of $H^1(\Omega)$ that consists of all functions $u \in H^1(\Omega)$ that vanish on γ. If $\gamma = \partial\Omega$, then $H_{\partial\Omega}^1(\Omega)$ is denoted by $H_0^1(\Omega)$. Thus, we have that

$$H_\gamma^1(\Omega) = \{ u \in H^1(\Omega) \mid u = 0 \text{ on } \gamma \}$$

and

$$H_0^1(\Omega) = \{ u \in H^1(\Omega) \mid u = 0 \text{ on } \partial\Omega \}.$$

For any function u belonging to $H_\gamma^1(\Omega)$ or $H_0^1(\Omega)$, the semi-norm

$$|u|_{1,\Omega} = \left(\sum_{i=1}^d \|u_{,x_i}\|_{0,\Omega}^2 \right)^{1/2} \tag{A.6}$$

defines a norm that is equivalent to the standard norm $\|u\|_{1,\Omega}$.

Another space related to $H^1(\Omega)$ is the quotient space

$$H^1(\Omega)/\mathbb{R} = \{ \widehat{u} \subset H^1(\Omega) \mid u, v \in \widehat{u} \text{ if and only if } u - v = \text{constant} \}. \tag{A.7}$$

$H^1(\Omega)/\mathbb{R}$ is a Hilbert space when equipped with the quotient norm

$$\|\widehat{u}\|_{H^1(\Omega)/\mathbb{R}} = \inf_{u \in \widehat{u}} \|u\|_1 \tag{A.8}$$

and, for $u \in \widehat{u}$, the mapping $\widehat{u} \mapsto |u|_1$ is norm equivalent to (A.8) [183, p. 13].

Following standard practice, we make one exception when using negative indices for the dual Sobolev spaces. Specifically, $H^{-1}(\Omega)$ denotes the dual space of either

$H_0^1(\Omega)$ or $H_\gamma^1(\Omega)$, where it is clear from the context which one of these choices applies. A norm for $f \in H^{-1}(\Omega)$ is defined by

$$\|f\|_{-1} = \sup_{u \in H_\gamma^1(\Omega)} \frac{\langle f, u \rangle_{H^{-1}(\Omega), H_\gamma^1(\Omega)}}{\|u\|_1}, \tag{A.9}$$

where $\langle f, u \rangle_{H^{-1}(\Omega), H_\gamma^1(\Omega)} = \int_\Omega f u \, d\Omega$. We denote by $(H^1(\Omega))^*$ the dual of $H^1(\Omega)$.

Under some assumptions on Ω, functions in $H^s(\Omega)$, $s > 1/2$, have well-defined traces in $H^{s-1/2}(\partial\Omega)$; see [183, Theorem 1.5, p. 8]. The norm on $H^{s-1/2}(\partial\Omega)$ can be defined as follows:

$$\|\phi\|_{s-1/2,\partial\Omega} = \inf_{\psi \in H^s(\Omega), \psi = \phi \text{ on } \partial\Omega} \|\psi\|_{s,\Omega}. \tag{A.10}$$

Of most interest to us are the trace spaces $H^{1/2}(\partial\Omega)$ and $H^{3/2}(\partial\Omega)$. The dual space of $H^{1/2}(\partial\Omega)$, denoted by $H^{-1/2}(\partial\Omega)$, is equipped with a dual norm similar to (A.9):

$$\|\phi\|_{-1/2,\partial\Omega} = \sup_{\psi \in H^{1/2}(\partial\Omega)} \frac{\langle \phi, \psi \rangle_{H^{-1/2}(\partial\Omega), H^{1/2}(\partial\Omega)}}{\|\psi\|_{1/2,\partial\Omega}}, \tag{A.11}$$

where $\langle \phi, \psi \rangle_{H^{-1/2}(\partial\Omega), H^{1/2}(\partial\Omega)} = \int_{\partial\Omega} \phi \psi \, d\Gamma$. Trace spaces can also be defined on portions of the boundary, e.g., we have $H^{1/2}(\Gamma)$ and $H^{-1/2}(\Gamma^*)$ with the associated norms

$$\|\phi\|_{1/2,\Gamma} = \inf_{\psi \in H^1(\Omega), \psi = \phi \text{ on } \Gamma} \|\psi\|_1 \tag{A.12}$$

and

$$\|\phi\|_{-1/2,\Gamma^*} = \sup_{\psi \in H^{1/2}(\Gamma^*)} \frac{\langle \phi, \psi \rangle_{H^{-1/2}(\Gamma^*), H^{1/2}(\Gamma^*)}}{\|\psi\|_{1/2,\Gamma^*}}, \tag{A.13}$$

respectively.

Notations for vector and tensor Sobolev spaces follow the conventions established in Section A.1. For example, the tensor product space

$$H^{s_1}(\Omega) \times \cdots \times H^{s_d}(\Omega)$$

is denoted by the symbol $\mathbf{H}^\alpha(\Omega)$, where $\alpha = (\alpha_1, \ldots, \alpha_d)$ is a multi-index, and the inner product and norm on this space are denoted by $(\cdot,\cdot)_{\mathbf{H}^\alpha}$ and $\|\cdot\|_{\mathbf{H}^\alpha}$, respectively. If $\alpha_1 = \alpha_2 = \cdots = \alpha_d = a$, we simply write $[H^a(\Omega)]^d$ and use the abbreviated notation $(\cdot,\cdot)_a$ and $\|\cdot\|_a$. For example,

$$[H^1(\Omega)]^d = \prod_{i=1}^d H^1(\Omega), \qquad [L^2(\Omega)]^d = \prod_{i=1}^d L^2(\Omega),$$

and, if $\mathbf{u}, \mathbf{v} \in [L^2(\Omega)]^d$ with $\mathbf{u} = (u_1, \ldots, u_d)^T$ and $\mathbf{v} = (v_1, \ldots, v_d)^T$,

$$(\mathbf{u},\mathbf{v})_0 = \sum_{i=1}^{d} (u_i, v_i)_0 \qquad \text{and} \qquad \|\mathbf{u}\|_0^2 = (\mathbf{u},\mathbf{u})_0 = \sum_{i=1}^{d} \|u_i\|_0^2 .$$

On several occasions we use the Banach space $L^p(\Omega)$ of functions whose pth power is integrable, with norm $\|\cdot\|_{0,p}$. Note that $\|\cdot\|_{0,2}$ is the same as $\|\cdot\|_0$.

Characterization of the minus one norm and inner product

The following theorem provides an equivalent representation of the negative norm (A.9) that provides a means for its efficient computational evaluation.

Theorem A.1 *For any $\psi \in H^{-1}(\Omega)$,*

$$\|\psi\|_{-1}^2 = \|(-\Delta)^{-1/2}\psi\|_0^2 , \tag{A.14}$$

where $(-\Delta)^{-1}$ is the solution operator of the Poisson equation with homogeneous Dirichlet boundary conditions, i.e., of the problem (5.7).

Proof. We provide a proof for the case of $\gamma = \partial\Omega$; the proof for the case $\gamma \subset \partial\Omega$ follows similar lines.

Note that, by definition, $u = (-\Delta)^{-1}\psi$ if and only if $-\Delta u = \psi$ in Ω and $u = 0$ on $\partial\Omega$. Therefore, u solves the weak formulation of the Poisson problem, i.e., we have that

$$(\nabla u, \nabla v) = -(\Delta u, v) = (\psi, v) \qquad \forall v \in H_0^1(\Omega).$$

In particular,

$$\|\nabla u\|_0^2 = (\nabla u, \nabla u) = -(\Delta u, u) = (\psi, u).$$

From (A.9) and the Poincaré–Friedrichs inequality (A.63), we have that

$$\sup_{v \in H_0^1(\Omega)} \frac{(\psi, v)^2}{\|v\|_1^2} \geq \frac{(\psi, u)^2}{\|u\|_1^2} = \frac{\|\nabla u\|_0^4}{\|u\|_1^2} \geq C\|\nabla u\|_0^2 = (\psi, u) = (\psi, (-\Delta)^{-1}\psi).$$

To prove the upper bound, let $v \in H_0^1(\Omega)$. Then,

$$\frac{(\psi, v)^2}{\|v\|_1^2} = \frac{(\nabla u, \nabla v)^2}{\|v\|_1^2} \leq \|\nabla u\|_0^2 = (\nabla u, \nabla u) = (\psi, u) = (\psi, (-\Delta)^{-1}\psi).$$

Therefore,

$$\|\psi\|_{-1}^2 = (\psi, (-\Delta)^{-1}\psi) = ((-\Delta)^{-1/2}\psi, (-\Delta)^{-1/2}\psi) = \|(-\Delta)^{-1/2}\psi\|_0^2 . \qquad \square$$

We also have an equivalent representation of the negative inner product that provides a means for its efficient computational evaluation.

Corollary A.2 *For any $\psi, \eta \in H^{-1}(\Omega)$,*

$$(\psi,\eta)_{-1} = ((-\Delta)^{-1}\psi,\eta)_0 = ((-\Delta)^{-1/2}\psi,(-\Delta)^{-1/2}\eta)\,. \qquad \Box \qquad \text{(A.15)}$$

Weighted Sobolev spaces

Many PDE problems involve coefficient functions or parameters that describe material properties or constitutive relations. It is often convenient to view these functions as weights that enter into the definition of the function spaces related to the problem being considered.[1]

We use Θ, possibly with subscripts, e.g., Θ_k, as the standard notation for a weight function which may be scalar or matrix-valued. We make exceptions to this rule when discussing concrete problems where a commonly employed notation has already been established. For example, in Maxwell's equations, we adhere to the ubiquitous use of μ and σ for the magnetic permeability and the electric conductivity.

Remark A.3 Because a weight Θ describes a physical property, it is natural to think of it as having the same units as the physical quantity with which it is associated with. For example, if Θ is the permeability of a porous medium, it has units of m^2 (square meters), if Θ is the magnetic permeability μ, it has units of H/m (henrys per meter), if Θ is the conductivity σ it is measured in S/m (siemens per meter), and so on. As a result, we can think of weighted function spaces as models for physical quantities for which the unit is inseparable from the quantity itself, e.g., a number that describes area is useless without giving the unit of area. $\Box$

To avoid tedious technical details, we consider weight functions Θ that are piecewise $C^0(\overline{\Omega})$ so that they are necessarily bounded. If Θ is a scalar weight function, we also assume that it is bounded away from zero, i.e., for some θ, we have

$$0 < \theta \le \Theta(\mathbf{x}) \quad \forall \mathbf{x} \in \overline{\Omega}\,. \qquad \text{(A.16)}$$

In the definition of weighted vector Sobolev spaces, Θ is a $d \times d$ *matrix* of scalar weight functions.[2] We only consider symmetric and uniformly positive definite weight functions, i.e., there must be a positive constant θ, such that

$$0 < \theta \vec{\xi}^T \vec{\xi} \le \vec{\xi}^T \Theta(\mathbf{x}) \vec{\xi} \qquad \forall \mathbf{x} \in \Omega, \quad \forall \vec{\xi} \in \mathbb{R}^d\,. \qquad \text{(A.17)}$$

---

[1] Another reason for using weighted spaces that is of particular importance to us is to achieve dimensional and/or unit consistency in least-squares functionals. Often, these functionals are defined by adding equation residuals measured in standard unweighted norms, without paying any particular attention to the consistency of their units. This is often used to criticize least-squares methods because of a lack of physical consistency between the terms in the least-squares functionals. By using weighted spaces to measure equation residuals, we can avoid this pitfall and define dimensionally and unit consistent least-squares functionals.

[2] Here, we intentionally avoid the matrix notation from Section A.1 to emphasize that the primary attribute of Θ is being a weight function. However, in some concrete examples, we do use the notation of Section A.1 to denote matrix-valued coefficient functions.

The definition of the weighted Sobolev spaces $H^s(\Omega,\Theta)$ follows the same rules as for their standard counterparts. Specifically, the weighted space $L^2(\Omega,\Theta)$ is defined to be the closure of $C^0(\Omega)$ with respect to the norm $\|\cdot\|_{0,\Omega,\Theta}$ induced by the *weighted* L^2 inner product and norm

$$(u,v)_{0,\Omega,\Theta} = \int_\Omega u\Theta v\, d\Omega \qquad \text{and} \qquad \|u\|^2_{0,\Omega,\Theta} = (u,u)_{0,\Omega,\Theta}\,, \tag{A.18}$$

respectively, and $H^s(\Omega,\Theta)$ is defined as

$$H^s(\Omega,\Theta) = \left\{u \in L^2(\Omega,\Theta_0) \mid D^\alpha u \in L^2(\Omega,\Theta_\alpha)\ \forall\, |\alpha| \le s\right\}, \tag{A.19}$$

where the Θ_αs are weights corresponding to the multi-index $\alpha = (\alpha_1,\ldots,\alpha_n)$. Thus, we have that $H^s(\Omega,\Theta)$ is the space consisting of all $L^2(\Omega,\Theta)$ functions whose derivatives up to order s also belong to $L^2(\Omega,\Theta)$. For vector valued functions, we have the weighted L^2 inner product and norm

$$(\mathbf{u},\mathbf{v})_{0,\Omega,\Theta} = \int_\Omega \mathbf{u}\cdot\Theta\cdot\mathbf{v}\, d\Omega \qquad \text{and} \qquad \|\mathbf{u}\|^2_{0,\Omega,\Theta} = (\mathbf{u},\mathbf{u})_{0,\Omega,\Theta}\,, \tag{A.20}$$

respectively.

To simplify notations for weighted Sobolev spaces, we omit the domain designation from norms and inner products.

A.2.2 *Spaces Related to the Gradient, Curl, and Divergence Operators*

The gradient, curl, and divergence operators are the three basic first-order differential operators of vector calculus. Combinations of these operators also define key second-order differential operators such as the scalar and vector Laplace operators, the curl–curl operator, and the grad–div operator. Conversely, when second-order PDEs are written as first-order systems, they lead to the first-order problems in Chapters 5–7 that are expressed in terms of the gradient, curl, and divergence operators.

For uniformity, we use the nabla notation ∇, $\nabla\times$, and $\nabla\cdot$ to denote the gradient, curl, and divergence operators. In this section, we review several sets of function spaces relevant to these operators. Because vector calculus "works best" in three dimensions, we state most of the facts and definitions for three-dimensional vector fields and briefly indicate their specialization to the two-dimensional case. To make the presentation easier to follow, we consider only topologically "simple" domains Ω, such as simply connected or contractible domains[3] in $\mathbb{R}^d$, $d = 2,3$.

[3] A domain is *simply connected* if it is path-connected and for every two paths with the same endpoints there is a continuous transformation of one to the other. A domain is *contractible* if it is homotopy equivalent to a point, i.e., it can be continuously deformed to a point. Every contractible

The first set of spaces consists of the domains of the gradient, curl, and divergence operators, considered as operators between spaces of square integrable functions, and also the range of the divergence:[4]

$$G(\Omega) = \{u \in L^2(\Omega) \mid \nabla u \in L^2(\Omega)\} \tag{A.21}$$

$$\mathbf{C}(\Omega) = \{\mathbf{u} \in \mathbf{L}^2(\Omega) \mid \nabla \times \mathbf{u} \in \mathbf{L}^2(\Omega)\} \tag{A.22}$$

$$\mathbf{D}(\Omega) = \{\mathbf{u} \in \mathbf{L}^2(\Omega) \mid \nabla \cdot \mathbf{u} \in L^2(\Omega)\} \tag{A.23}$$

$$S(\Omega) = \{u \in L^2(\Omega) \mid u = \nabla \cdot \mathbf{u};\ \mathbf{u} \in \mathbf{D}(\Omega)\}. \tag{A.24}$$

The second set consists of subspaces of (A.21)–(A.23) constrained by homogeneous boundary conditions on $\gamma = \Gamma$ or $\gamma = \Gamma^*$ and the associated range of the divergence:

$$G_\gamma(\Omega) = \{u \in G(\Omega) \mid u = 0 \quad \text{on } \gamma\} \tag{A.25}$$

$$\mathbf{C}_\gamma(\Omega) = \{\mathbf{u} \in \mathbf{C}(\Omega) \mid \mathbf{u} \times \mathbf{n} = 0 \quad \text{on } \gamma\} \tag{A.26}$$

$$\mathbf{D}_\gamma(\Omega) = \{\mathbf{u} \in \mathbf{D}(\Omega) \mid \mathbf{u} \cdot \mathbf{n} = 0 \quad \text{on } \gamma\} \tag{A.27}$$

$$S_\gamma(\Omega) = \{u \in L^2(\Omega) \mid u = \nabla \cdot \mathbf{u};\ \mathbf{u} \in \mathbf{D}_\gamma(\Omega)\}. \tag{A.28}$$

When $\Gamma = \partial\Omega$ or $\Gamma^* = \partial\Omega$, we write $G_0(\Omega)$, $\mathbf{C}_0(\Omega)$, $\mathbf{D}_0(\Omega)$, and $S_0(\Omega)$ instead of $G_\gamma(\Omega)$, $\mathbf{C}_\gamma(\Omega)$, $\mathbf{D}_\gamma(\Omega)$, and $S_\gamma(\Omega)$. In this case, $S_0(\Omega) = L_0^2(\Omega)$ is the subspace of $L^2(\Omega)$ consisting of functions having zero mean; in all other cases, $S_\gamma(\Omega) = L^2(\Omega)$.

Weighted analogues of (A.21)–(A.24) and (A.25)–(A.28) are obtained by using weighted L^2 spaces in place of standard L^2 spaces. In fact, we have need for two weighted versions of (A.21)–(A.24). In the first, the domains of ∇, $\nabla\times$, and $\nabla\cdot$ are $L^2(\Omega, \Theta_0)$, $\mathbf{L}^2(\Omega, \Theta_1)$, and $\mathbf{L}^2(\Omega, \Theta_2)$, respectively, and the range of $\nabla\cdot$ is $L^2(\Omega, \Theta_3)$:

$$G(\Omega, \Theta_0) = \{u \in L^2(\Omega, \Theta_0) \mid \nabla u \in \mathbf{L}^2(\Omega, \Theta_1)\} \tag{A.29}$$

$$\mathbf{C}(\Omega, \Theta_1) = \{\mathbf{u} \in \mathbf{L}^2(\Omega, \Theta_1) \mid \nabla \times \mathbf{u} \in \mathbf{L}^2(\Omega, \Theta_2)\} \tag{A.30}$$

$$\mathbf{D}(\Omega, \Theta_2) = \{\mathbf{u} \in \mathbf{L}^2(\Omega, \Theta_2) \mid \nabla \cdot \mathbf{u} \in L^2(\Omega, \Theta_3)\} \tag{A.31}$$

$$S(\Omega, \Theta_3) = \{u \in L^2(\Omega, \Theta_3) \mid u = \nabla \cdot \mathbf{u};\ \mathbf{u} \in \mathbf{D}(\Omega, \Theta_2)\}. \tag{A.32}$$

The constrained versions $G_\gamma(\Omega, \Theta_0)$, $\mathbf{C}_\gamma(\Omega, \Theta_1)$, $\mathbf{D}_\gamma(\Omega, \Theta_2)$, and $S_\gamma(\Omega, \Theta_3)$ of these spaces are defined in an obvious manner. The second set of weighted spaces results from taking $L^2(\Omega, \Theta_3^{-1})$, $\mathbf{L}^2(\Omega, \Theta_2^{-1})$, and $\mathbf{L}^2(\Omega, \Theta_1^{-1})$ for the domains of ∇, $\nabla\times$, and $\nabla\cdot$, and $L^2(\Omega, \Theta_0^{-1})$ for the range of $\nabla\cdot$:

---

domain is simply connected, but the converse is not true. A counterexample is a ball without its center which is simply connected but not contractible.

[4] In common practice, the spaces $G(\Omega)$, $\mathbf{C}(\Omega)$, and $\mathbf{D}(\Omega)$ are denoted by $H^1(\Omega)$, $H(\Omega, \text{curl})$, and $H(\Omega, \text{div})$, respectively. The notation of (A.21)–(A.23) is more consistent across the different operators. See also Remark A.5. In addition, the Ladyzhenskaya theorem [253] shows that $S(\Omega) = L^2(\Omega)$.

$$G(\Omega,\Theta_3^{-1}) = \{u \in L^2(\Omega,\Theta_3^{-1}) \mid \nabla u \in \mathbf{L}^2(\Omega,\Theta_2^{-1})\} \tag{A.33}$$
$$\mathbf{C}(\Omega,\Theta_2^{-1}) = \{\mathbf{u} \in \mathbf{L}^2(\Omega,\Theta_2^{-1}) \mid \nabla \times \mathbf{u} \in \mathbf{L}^2(\Omega,\Theta_1^{-1})\} \tag{A.34}$$
$$\mathbf{D}(\Omega,\Theta_1^{-1}) = \{\mathbf{u} \in \mathbf{L}^2(\Omega,\Theta_1^{-1}) \mid \nabla \cdot \mathbf{u} \in L^2(\Omega,\Theta_0^{-1})\} \tag{A.35}$$
$$S(\Omega,\Theta_0^{-1}) = \{u \in L^2(\Omega,\Theta_0^{-1}) \mid u = \nabla \cdot \mathbf{u};\ \mathbf{u} \in \mathbf{D}(\Omega,\Theta_1^{-1})\}. \tag{A.36}$$

The constrained versions $G_\gamma(\Omega,\Theta_3^{-1})$, $\mathbf{C}_\gamma(\Omega,\Theta_2^{-1})$, $\mathbf{D}_\gamma(\Omega,\Theta_1^{-1})$, and $S_\gamma(\Omega,\Theta_0^{-1})$ of these spaces are again defined in an obvious manner.

The weighted spaces $L^2(\Omega,\Theta_i)$ and $L^2(\Omega,\Theta_i^{-1})$ are related in the following way. It is easy to verify that if $u \in L^2(\Omega,\Theta_i)$, then $\Theta_i u \in L^2(\Omega,\Theta_i^{-1})$ and vice versa. As a result, Θ_i and Θ_i^{-1} respectively define mappings

$$\Theta_i : L^2(\Omega,\Theta_i) \mapsto L^2(\Omega,\Theta_i^{-1}) \quad \text{and} \quad \Theta_i^{-1} : L^2(\Omega,\Theta_i^{-1}) \mapsto L^2(\Omega,\Theta_i). \tag{A.37}$$

Remark A.4 In view of Remark A.3, the mappings (A.37) can be interpreted as conversions of units between different physical quantities. To illustrate this point, consider the material constitutive law

$$\mathbf{B} = \mu \mathbf{H}$$

that relates the magnetic flux density $\mathbf{B}$ and the magnetic field strength $\mathbf{H}$. Recall that the units of $\mathbf{B}$, $\mathbf{H}$, and μ are Wb/m^2 (weber per meter squared), A/m (ampere per meter), and H/m (henry per meter), respectively, where $H = Wb/A$. Taking the product of the units of μ and $\mathbf{H}$, we see that

$$\frac{A}{m}\frac{H}{m} = \frac{A}{m}\frac{Wb}{Am} = \frac{Wb}{m^2}$$

i.e., the units on both sides of the equation $\mathbf{B} = \mu\mathbf{H}$ are the same so that relation is dimensionally consistent.

On the other hand, because the energy density of the magnetic field is given by

$$\frac{1}{2}\frac{\mathbf{B}^2}{\mu} = \frac{1}{2}\mu\mathbf{H}^2,$$

it follows that $\mathbf{B} \in L^2(\Omega,\mu^{-1})$ and $\mathbf{H} \in L^2(\Omega,\mu)$. Then,

$$\mu : L^2(\Omega,\mu) \mapsto L^2(\Omega,\mu^{-1})$$

is the operator that converts units of magnetic field strength into units of magnetic flux density. □

The spaces (A.21)–(A.23), (A.25)–(A.27), (A.29)–(A.31), and (A.33)–(A.35) are Hilbert spaces when equipped with appropriate graph norms. For example, for (A.21)–(A.23), we have the norms

$$\|u\|_G^2 = \|u\|_0^2 + \|\nabla u\|_0^2 \tag{A.38}$$

$$\|\mathbf{u}\|_C^2 = \|\mathbf{u}\|_0^2 + \|\nabla \times \mathbf{u}\|_0^2 \tag{A.39}$$

$$\|\mathbf{u}\|_D^2 = \|\mathbf{u}\|_0^2 + \|\nabla \cdot \mathbf{u}\|_0^2 \tag{A.40}$$

whereas for (A.29)–(A.31) we have

$$\|u\|_{G(\Theta_0)}^2 = \|u\|_{0,\Theta_0}^2 + \|\nabla u\|_{0,\Theta_1}^2 \tag{A.41}$$

$$\|\mathbf{u}\|_{C(\Theta_1)}^2 = \|\mathbf{u}\|_{0,\Theta_1}^2 + \|\nabla \times \mathbf{u}\|_{0,\Theta_2}^2 \tag{A.42}$$

$$\|\mathbf{u}\|_{D(\Theta_2)}^2 = \|\mathbf{u}\|_{0,\Theta_2}^2 + \|\nabla \cdot \mathbf{u}\|_{0,\Theta_3}^2 \,. \tag{A.43}$$

The norms on (A.24), (A.28), (A.32), and (A.36) are simply appropriately weighted L^2 norms.

When there is no possibility for confusion, we omit the weight designations from (A.29)–(A.36) and simply write $G(\Omega)$, $\mathbf{C}(\Omega)$, $\mathbf{D}(\Omega)$, and $S(\Omega)$; we do likewise for norms and inner products on weighted spaces.

Weak differential operators

Variational methods deal with weak solutions of PDEs and sometimes require "weaker" or generalized versions of the grad, curl, and div operators as well as of their pairings that produce second-order elliptic operators. For example, the gradient is not defined for functions belonging to $S(\Omega)$, and the curl and divergence cannot be applied to $\mathbf{v} \in \mathbf{D}(\Omega)$ and $\mathbf{u} \in \mathbf{C}(\Omega)$, respectively. Here, we define weak versions of the gradient, curl, and divergence operators that act as their surrogates for functions belonging to $S(\Omega)$, $\mathbf{D}(\Omega)$, and $\mathbf{C}(\Omega)$, respectively. The definition of weak operators for the two sets (A.29)–(A.32) and (A.33)–(A.36) of weighted spaces is identical so that we present the details only for the first set.

Let $X \in \{G(\Omega), \mathbf{C}(\Omega), \mathbf{D}(\Omega), S(\Omega)\}$ and $\beta \in \{\emptyset, \gamma\}$. In what follows,

$$X_\beta = \begin{cases} X & \text{if } \beta = \emptyset \\ X_\gamma & \text{if } \beta = \gamma \end{cases} \quad \text{and} \quad \Gamma_\beta = \begin{cases} \partial\Omega & \text{if } \beta = \emptyset \\ \partial\Omega \setminus \gamma & \text{if } \beta = \gamma. \end{cases}$$

The weak gradient $\nabla^* : S_\beta(\Omega) \mapsto \mathbf{D}_\beta(\Omega)$ is defined by[5]

$$\left(\nabla^* u, \widehat{\mathbf{v}}\right)_{\Theta_2} = \left(u, -\nabla \cdot \widehat{\mathbf{v}}\right)_{\Theta_3} + \int_{\Gamma_\beta} (\Theta_3 u)\widehat{\mathbf{v}} \cdot \mathbf{n}\, d\Gamma \qquad \forall \widehat{\mathbf{v}} \in \mathbf{D}_\beta(\Omega)\,. \tag{A.44}$$

The weak curl $\nabla^* \times : \mathbf{D}_\beta(\Omega) \mapsto \mathbf{C}_\beta(\Omega)$ is defined by

$$\left(\nabla^* \times \mathbf{v}, \widehat{\mathbf{u}}\right)_{\Theta_1} = \left(\mathbf{v}, \nabla \times \widehat{\mathbf{u}}\right)_{\Theta_2} - \int_{\Gamma_\beta} (\mathbf{n} \times \Theta_2 \mathbf{v}) \cdot \widehat{\mathbf{u}}\, d\Gamma \qquad \forall \widehat{\mathbf{u}} \in \mathbf{C}_\beta(\Omega)\,. \tag{A.45}$$

[5] Following the usual nomenclature, we refer to $\widehat{\mathbf{v}}$, $\widehat{\mathbf{u}}$, and $\widehat{q}$ in (A.44)–(A.46) as *test* functions.

The weak divergence $\nabla^* \cdot : \mathbf{C}_\beta(\Omega) \mapsto G_\beta(\Omega)$ is defined by

$$\left(\nabla^* \cdot \mathbf{u}, \widehat{q}\right)_{\Theta_0} = \left(\mathbf{u}, -\nabla \widehat{q}\right)_{\Theta_1} + \int_{\Gamma_\beta} (\mathbf{n} \cdot \Theta_1 \mathbf{u}) \widehat{q} \, d\Gamma \qquad \forall \widehat{q} \in G_\beta(\Omega). \tag{A.46}$$

Note that the arguments of ∇^*, $\nabla^* \times$, and $\nabla^* \cdot$ cannot be constrained by the same boundary conditions on Γ_β as their standard counterparts in (A.25)–(A.27) because the traces of u, $\mathbf{v} \times \mathbf{n}$, and $\mathbf{u} \cdot \mathbf{n}$ on $\partial\Omega$ are not defined for general $u \in S(\Omega)$, $\mathbf{v} \in \mathbf{D}(\Omega)$, and $\mathbf{u} \in \mathbf{C}(\Omega)$, respectively. Instead, ∇^*, $\nabla^* \times$, and $\nabla^* \cdot$ impose these conditions in a weak or generalized variational sense through the boundary terms included in their definitions. For example, to impose weakly homogeneous boundary conditions, one simply deletes the boundary terms from (A.44)–(A.46). Unless stated otherwise, we always assume homogeneous conditions on Γ_β.

To clarify further the meaning of the weak operators, note that if $u \in S_\beta(\Omega)$ is sufficiently smooth,

$$\begin{aligned} \left(\nabla^* u, \widehat{\mathbf{v}}\right)_{\Theta_2} &= \left(u, -\nabla \cdot \widehat{\mathbf{v}}\right)_{\Theta_3} = \int_\Omega \widehat{\mathbf{v}} \cdot \nabla(\Theta_3 u) \, d\Omega \\ &= \int_\Omega (\widehat{\mathbf{v}} \Theta_2) \Theta_2^{-1} \nabla(\Theta_3 u) \, d\Omega = \left(\nabla(\Theta_3 u), \Theta_2 \widehat{\mathbf{v}}\right)_{\Theta_2^{-1}}. \end{aligned} \tag{A.47}$$

Similarly, for all sufficiently smooth $\mathbf{v} \in \mathbf{D}_\beta(\Omega)$,

$$\left(\nabla^* \times \mathbf{v}, \widehat{\mathbf{u}}\right)_{\Theta_1} = \int_\Omega \nabla \times (\Theta_2 \mathbf{v}) \cdot \widehat{\mathbf{u}} \, d\Omega = \left(\nabla \times (\Theta_2 \mathbf{v}), \Theta_1 \widehat{\mathbf{u}}\right)_{\Theta_1^{-1}} \tag{A.48}$$

and, for all sufficiently smooth $\mathbf{u} \in \mathbf{C}_\beta(\Omega)$,

$$\left(\nabla^* \cdot \mathbf{u}, \widehat{q}\right)_{\Theta_0} = \int_\Omega \nabla \cdot (\Theta_1 \mathbf{u}) \widehat{q} \, d\Omega = \left(\nabla \cdot (\Theta_1 \mathbf{u}), \Theta_0 \widehat{q}\right)_{\Theta_0^{-1}}. \tag{A.49}$$

Thus, for smooth functions, the action of ∇^*, $\nabla^* \times$, and $\nabla^* \cdot$ coincides with the action of the standard gradient, curl, and divergence operators on functions whose weighted traces are specified on Γ_β:

$$\begin{cases} \nabla^* u = \Theta_2^{-1} \nabla(\Theta_3 u) & \text{in } \Omega \quad \text{and} \quad \Theta_3 u \text{ is given on } \Gamma_\beta \\ \nabla^* \times \mathbf{v} = \Theta_1^{-1} \nabla \times (\Theta_2 \mathbf{v}) & \text{in } \Omega \quad \text{and} \quad \mathbf{n} \times (\Theta_2 \mathbf{v}) \text{ is given on } \Gamma_\beta \\ \nabla^* \cdot \mathbf{u} = \Theta_0^{-1} \nabla \cdot (\Theta_1 \mathbf{u}) & \text{in } \Omega \quad \text{and} \quad \mathbf{n} \cdot (\Theta_1 \mathbf{u}) \text{ is given on } \Gamma_\beta. \end{cases} \tag{A.50}$$

These identities imply that, for smooth functions, the action of ∇^*, $\nabla^* \times$, and $\nabla^* \cdot$ can be constructed by combining the action of the standard operators ∇, $\nabla \times$, and $\nabla \cdot$ on the inversely weighted spaces (A.33)–(A.36) with the action of the mappings (A.37) defined by Θ_i:

$$
\begin{array}{ccc}
D(\Theta_2) \stackrel{\nabla^*}{\longleftarrow} S(\Theta_3) & C(\Theta_1) \stackrel{\nabla^*\times}{\longleftarrow} D(\Theta_2) & G(\Theta_0) \stackrel{\nabla^*\cdot}{\longleftarrow} C(\Theta_1) \\
\Theta_2^{-1}\uparrow \qquad\qquad \downarrow\Theta_3 & \Theta_1^{-1}\uparrow \qquad\qquad \downarrow\Theta_2 & \Theta_0^{-1}\uparrow \qquad\qquad \downarrow\Theta_1 \\
C(\Theta_2^{-1}) \stackrel{\nabla}{\longleftarrow} G(\Theta_3^{-1}) & D(\Theta_1^{-1}) \stackrel{\nabla\times}{\longleftarrow} C(\Theta_2^{-1}) & S(\Theta_0^{-1}) \stackrel{\nabla\cdot}{\longleftarrow} D(\Theta_1^{-1}).
\end{array}
$$

This, along with (A.50), justifies the use of ∇^*, $\nabla^*\times$, and $\nabla^*\cdot$ as surrogates for the standard operators when their arguments are not sufficiently smooth.

De Rham differential complex

From now on, unless stated otherwise, we assume that Ω is a contractible domain in $\mathbb{R}^3$. In this case, the four spaces defined in (A.29)–(A.32) have the following properties: the null-space of the gradient is the one-dimensional space $\mathbb{R}$ of all constant functions; the null-space of the curl is the range of the gradient; the null-space of the divergence is the range of the curl; and the range of the divergence is the space[6] $S(\Omega,\Theta_3)$. Succinctly,

$$N(\nabla) = \mathbb{R} \qquad N(\nabla\times) = R(\nabla) \qquad N(\nabla\cdot) = R(\nabla\times) \qquad S(\Omega) = R(\nabla\cdot). \quad \text{(A.51)}$$

A collection of spaces and operators linked by such relationships is called an *exact sequence* [7, 8, 13, 69, 75, 139] and is represented by the diagram:

$$\mathbb{R} \hookrightarrow G(\Omega,\Theta_0) \stackrel{\nabla}{\longrightarrow} \mathbf{C}(\Omega,\Theta_1) \stackrel{\nabla\times}{\longrightarrow} \mathbf{D}(\Omega,\Theta_2) \stackrel{\nabla\cdot}{\longrightarrow} S(\Omega,\Theta_3) \longrightarrow 0. \qquad \text{(A.52)}$$

The sequence (A.52) is a *differential De Rham complex.* Another version of the De Rham complex can be obtained using the inversely weighted spaces (A.29)–(A.32):

$$\mathbb{R} \hookrightarrow G(\Omega,\Theta_3^{-1}) \stackrel{\nabla}{\longrightarrow} \mathbf{C}(\Omega,\Theta_2^{-1}) \stackrel{\nabla\times}{\longrightarrow} \mathbf{D}(\Omega,\Theta_1^{-1}) \stackrel{\nabla\cdot}{\longrightarrow} S(\Omega,\Theta_0^{-1}) \longrightarrow 0. \quad \text{(A.53)}$$

We refer to (A.52) and (A.53) as the *primal* and *dual* complexes, respectively. Clearly, any construction or fact valid for one of the complexes also holds for the other. To save time and space, our subsequent discussion focuses only on the primal complex.

The spaces $G_\beta(\Omega)$, $\mathbf{C}_\beta(\Omega)$, $\mathbf{D}_\beta(\Omega)$, and $S_\beta(\Omega)$ with $\beta \in \{\emptyset, \gamma\}$ form *exact sequences* relative to β:

$$\mathbb{R} \hookrightarrow G_\beta(\Omega) \stackrel{\nabla}{\longrightarrow} \mathbf{C}_\beta(\Omega) \stackrel{\nabla\times}{\longrightarrow} \mathbf{D}_\beta(\Omega) \stackrel{\nabla\cdot}{\longrightarrow} S_\beta(\Omega) \longrightarrow 0. \qquad \text{(A.54)}$$

---

[6] The last relationship follows from the aforementioned Ladyzhenskaya theorem [253] which asserts that, for every $u \in S(\Omega)$, there exists $\mathbf{u} \in \mathbf{D}(\Omega)$ such that $u = \nabla\cdot\mathbf{u}$ and $\|\mathbf{u}\|_D \le C\|u\|_S$, i.e., the divergence is bounded surjection $\mathbf{D}(\Omega) \mapsto S(\Omega)$. The other two relationships follow from the Poincaré lemma [343, p. 23] which asserts that, for contractible domains and $\mathbf{u} \in \mathbf{C}(\Omega)$ and $\mathbf{v} \in \mathbf{D}(\Omega)$, $\nabla\times\mathbf{u} = 0$ and $\nabla\cdot\mathbf{v} = 0$ if and only if $\mathbf{u} = \nabla u$ and $\mathbf{v} = \nabla\times\mathbf{w}$ for some $u \in G(\Omega)$ and $\mathbf{w} \in \mathbf{C}(\Omega)$, respectively.

When $\gamma = \partial\Omega$, the De Rham complex (A.54) is given by

$$\mathbb{R} \hookrightarrow G_0(\Omega) \xrightarrow{\nabla} C_0(\Omega) \xrightarrow{\nabla\times} D_0(\Omega) \xrightarrow{\nabla\cdot} S_0(\Omega) \longrightarrow 0, \tag{A.55}$$

where $S_0(\Omega) = L_0^2(\Omega, \Theta_3)$ is the subspace of functions belonging to $L^2(\Omega, \Theta_3)$ that have zero mean.

Remark A.5 $G(\Omega)$ and $G_\gamma(\Omega)$ coincide with the Sobolev spaces $H^1(\Omega)$ and $H^1_\gamma(\Omega)$, respectively. The use of the alternative notation underscores the role of these spaces as part of the exact sequences in (A.52)–(A.55). □

The weak operators (A.44)–(A.46) together with the spaces $G_\beta(\Omega)$, $\mathbf{C}_\beta(\Omega)$, $\mathbf{D}_\beta(\Omega)$, and $S_\beta(\Omega)$ form another exact sequence where operators act from right to left:

$$0 \longleftarrow G_\beta(\Omega) \xleftarrow{\nabla^*\cdot} \mathbf{C}_\beta(\Omega) \xleftarrow{\nabla^*\times} \mathbf{D}_\beta(\Omega) \xleftarrow{\nabla^*} S_\beta(\Omega) \hookleftarrow \mathbb{R}. \tag{A.56}$$

The exactness of (A.56) follows from the definitions of the weak operators and the properties of the standard operators. For brevity, we demonstrate this for the sequence consisting of the weak operators and the spaces $G_0(\Omega)$, $\mathbf{C}_0(\Omega)$, $\mathbf{D}_0(\Omega)$, and $S_0(\Omega)$.

To show that $N(\nabla^*) = \{0\}$, assume that $\nabla^* u = 0$. Because the standard divergence is a surjective operator $\mathbf{D}_0(\Omega) \mapsto S_0(\Omega)$, there exists $\widehat{\mathbf{v}}_u \in \mathbf{D}_0(\Omega)$ such that $\nabla \cdot \widehat{\mathbf{v}}_u = u$. Together with the definition (A.44), this implies that[7]

$$0 = \left(\nabla^* u, \widehat{\mathbf{v}}_u\right)_{\Theta_2} = \left(u, -\nabla \cdot \widehat{\mathbf{v}}_u\right)_{\Theta_3} = \left(u, u\right)_{\Theta_3} = \|u\|^2_{\Theta_3}$$

so that $u \equiv 0$. To show that $\nabla^* \times \nabla^* = 0$, we apply successively the definitions (A.45) and (A.44):

$$\left(\nabla^* \times \nabla^* u, \widehat{\mathbf{u}}\right)_{\Theta_1} = \left(\nabla^* u, \nabla \times \widehat{\mathbf{u}}\right)_{\Theta_2} = \left(u, \nabla \cdot \nabla \times \widehat{\mathbf{u}}\right)_{\Theta_3} = 0.$$

The proof of $\nabla^* \cdot \nabla^* \times = 0$ is similar. To prove that $\nabla^*\cdot$ is a surjective mapping $\mathbf{C}_0(\Omega) \mapsto G_0(\Omega)$, choose an arbitrary $q \in G_0(\Omega)$. The statement $\nabla^* \cdot \mathbf{u} = q$ is equivalent to

$$\left(\mathbf{u}, \nabla \widehat{q}\right)_{\Theta_1} = -\left(q, \widehat{q}\right)_{\Theta_0} \qquad \forall \widehat{q} \in G_0(\Omega). \tag{A.57}$$

To find $\mathbf{u} \in \mathbf{C}_0(\Omega)$ that solves (A.57), we restrict the search to the range of ∇ in $\mathbf{C}_0(\Omega)$, i.e., we set $\mathbf{u} = \nabla p$, where $p \in G_0(\Omega)$. This transforms (A.57) into the following variational problem: *seek* $p \in G_0(\Omega)$ *such that*

$$\left(\nabla p, \nabla \widehat{q}\right)_{\Theta_1} = -\left(q, \widehat{q}\right)_{\Theta_0} \qquad \forall \widehat{q} \in G_0(\Omega).$$

This equation is the weak form of the Poisson problem that has a unique solution $p \in G_0(\Omega)$.

---

[7] Recall that, in the present case, $\Gamma = \partial\Omega$ and so $\Gamma_\beta = \emptyset$. As a result, the definition of weak operators does not include boundary terms.

The weak operators provide a useful characterization for the orthogonal complements of $N(\nabla\times)$ and $N(\nabla\cdot)$ in $\mathbf{C}(\Omega)$ and $\mathbf{D}(\Omega)$, respectively. From (A.51), we know that $N(\nabla\times) = R(\nabla)$ and $N(\nabla\cdot) = R(\nabla\times)$ so that

$$N(\nabla\times)^{\perp} = \{\mathbf{u} \in \mathbf{C}(\Omega) \mid (\mathbf{u}, \nabla\widehat{q})_{\Theta_1} = 0 \quad \forall \widehat{q} \in G(\Omega)\}$$

and

$$N(\nabla\cdot)^{\perp} = \{\mathbf{v} \in \mathbf{D}(\Omega) \mid (\mathbf{v}, \nabla\times\widehat{\mathbf{u}})_{\Theta_2} = 0 \quad \forall \widehat{\mathbf{u}} \in \mathbf{C}(\Omega)\},$$

respectively. In light of the definitions (A.46) and (A.45), these are the same as

$$N(\nabla\times)^{\perp} = \{\mathbf{u} \in \mathbf{C}(\Omega) \mid \nabla^{*}\cdot\mathbf{u} = 0\} = N(\nabla^{*}\cdot) \tag{A.58}$$

and

$$N(\nabla\cdot)^{\perp} = \{\mathbf{v} \in \mathbf{D}(\Omega) \mid \nabla^{*}\times\mathbf{v} = 0\} = N(\nabla^{*}\times), \tag{A.59}$$

respectively. The orthogonal complements $N_0^{\perp}(\nabla\times)$ and $N_0^{\perp}(\nabla\cdot)$, i.e., the nullspaces of curl and divergence in $\mathbf{C}_0(\Omega)$ and $\mathbf{D}_0(\Omega)$, have the same characterizations.

Remark A.6 The orthogonal complement of $N(\nabla)$ is

$$N(\nabla)^{\perp} = \{u \in G(\Omega) \mid (u,1)_{\Theta_0} = 0\} = G(\Omega)\cap L_0^2(\Omega). \tag{A.60}$$

We also have the trivial identity $N_0(\nabla)^{\perp} = G_0(\Omega)$. □

A.3 Properties of Function Spaces

In this section, some basic facts and results about the spaces introduced in Section A.2 are presented; these facts are needed in proofs of energy balances for least-squares principles. In particular, we survey several embedding theorems, Poincaré–Friedrichs inequalities, orthogonal decomposition results, and trace theorems.

A.3.1 Embeddings of $\mathbf{C}(\Omega)\cap\mathbf{D}(\Omega)$

The space $X = \mathbf{C}(\Omega)\cap\mathbf{D}(\Omega)$ arises naturally as a solution space for problems governed by PDEs that involve the curl and divergence of vector fields. Two important practical examples are Maxwell's equations and equations involving the Cauchy-Riemann operator. When equipped with the norm

$$\|\mathbf{u}\|_{DC}^2 = \|\mathbf{u}\|_0^2 + \|\nabla\cdot\mathbf{u}\|_0^2 + \|\nabla\times\mathbf{u}\|_0^2, \tag{A.61}$$

X is a Hilbert space. Whether this space is embedded in the Sobolev space $[G(\Omega)]^d = [H^1(\Omega)]^d$ depends on the shape and smoothness of Ω and the boundary conditions used as constraints in the definition of X. Without any additional assumptions, we have the following result.

Theorem A.7 ([183, Corollary 2.10, p. 36]) *Let Ω be an arbitrary open subset of $\mathbb{R}^d$, $d = 2,3$, and $\mathbf{u} \in \mathbf{D}(\Omega) \cap \mathbf{C}(\Omega)$. Then $\mathbf{u} \in [H^1(\Omega)_{loc}]^d$.* □

When X is constrained by homogeneous normal or tangential boundary conditions, the embedding in $[G(\Omega)]^d = [H^1(\Omega)]^d$ is contingent on the shape of Ω and/or the smoothness of its boundary. The following theorem combines results from [183, Proposition 3.1, p. 44; Theorem 3.7, p. 52; Theorem 3.8, p. 54; Theorem 3.9, p. 55].

Theorem A.8 *Assume that Ω is a connected, bounded, open subset of $\mathbb{R}^d$, $d = 2,3$. The spaces $\mathbf{D}(\Omega) \cap \mathbf{C}_0(\Omega)$ and $\mathbf{D}_0(\Omega) \cap \mathbf{C}(\Omega)$ are continuously embedded in $[G(\Omega)]^d = [H^1(\Omega)]^d$ provided one of the following is true:*

- *$\partial\Omega$ is of class $C^{1,1}$*
- *$d = 2$ and $\partial\Omega$ is piecewise smooth with no reentrant corners*
- *$d = 3$ and Ω is a convex polyhedron.* □

If the assumptions of Theorem A.8 are satisfied, there exists a positive constant C such that

$$\frac{1}{C}\|\mathbf{u}\|_1 \le \|\nabla\times\mathbf{u}\|_0 + \|\nabla\cdot\mathbf{u}\|_0 \le C\|u\|_1 \tag{A.62}$$

for all u in $\mathbf{D}(\Omega) \cap \mathbf{C}_0(\Omega)$ or $\mathbf{D}_0(\Omega) \cap \mathbf{C}(\Omega)$, i.e., the norm defined in (A.61) is equivalent to the standard Sobolev norm on $[H^1(\Omega)]^d$. As the following result shows, this is not true when the conditions of Theorem A.8 are violated.

Theorem A.9 ([127]) *Assume that Ω is a polyhedron in $\mathbb{R}^3$. If Ω is not convex, $\mathbf{C}_0(\Omega) \cap [H^1(\Omega)]^3$ and $\mathbf{D}_0(\Omega) \cap [H^1(\Omega)]^3$ are closed, infinite-codimensional subspaces of $\mathbf{C}_0(\Omega) \cap \mathbf{D}(\Omega)$ and $\mathbf{C}(\Omega) \cap \mathbf{D}_0(\Omega)$, respectively.* □

The results of Theorems A.8 and A.9 have important consequences for finite element approximations of weak formulations of PDEs posed in $\mathbf{C}(\Omega) \cap \mathbf{D}(\Omega)$; see Section B.2.2.

A.3.2 Poincaré–Friedrichs Inequalities

Poincaré–Friedrichs inequalities assert that $\|\nabla u\|_0$, $\|\nabla\times\mathbf{u}\|_0$, and $\|\nabla\cdot\mathbf{u}\|_0$ define equivalent norms on the orthogonal complements of the null-spaces of these operators in $G(\Omega)$, $\mathbf{C}(\Omega)$, and $\mathbf{D}(\Omega)$, respectively; see, e.g., [13] for details.

Theorem A.10 *Let $\Omega \subset \mathbb{R}^d$, $d = 2,3$, be a contractible domain with a Lipschitz-continuous boundary. Then, there exists a positive constant C_P such that*

$$\|u\|_0 \le C_P\|\nabla u\|_0 \quad \textit{for} \quad u \in N(\nabla)^\perp \quad \textit{or} \quad u \in N_0(\nabla)^\perp \tag{A.63}$$

$$\|\mathbf{u}\|_0 \le C_P\|\nabla\times\mathbf{u}\|_0 \quad \textit{for} \quad \mathbf{u} \in N(\nabla\times)^\perp \quad \textit{or} \quad \mathbf{u} \in N_0(\nabla\times)^\perp \tag{A.64}$$

$$\|\mathbf{u}\|_0 \le C_P\|\nabla\cdot\mathbf{u}\|_0 \quad \textit{for} \quad \mathbf{u} \in N(\nabla\cdot)^\perp \quad \textit{or} \quad \mathbf{u} \in N_0(\nabla\cdot)^\perp. \;\square \tag{A.65}$$

Poincaré–Friedrichs inequalities also hold for weak operators.

Theorem A.11 *Assume the hypotheses of Theorem* A.10. *Then, there exists a positive constant* C_P *such that*

$$\|u\|_0 \le C_P\|\nabla^* u\|_0 \quad \textit{for} \quad u \in N(\nabla^*)^\perp \quad \textit{or} \quad u \in N_0(\nabla^*)^\perp \tag{A.66}$$

$$\|\mathbf{u}\|_0 \le C_P\|\nabla^* \times \mathbf{u}\|_0 \quad \textit{for} \quad \mathbf{u} \in N(\nabla^* \times)^\perp \quad \textit{or} \quad \mathbf{u} \in N_0(\nabla^* \times)^\perp \tag{A.67}$$

$$\|\mathbf{u}\|_0 \le C_P\|\nabla^* \cdot \mathbf{u}\|_0 \quad \textit{for} \quad \mathbf{u} \in N(\nabla^* \cdot)^\perp \quad \textit{or} \quad \mathbf{u} \in N_0(\nabla^* \cdot)^\perp . \tag{A.68}$$

Proof. For brevity, we provide proofs only for the spaces $G_0(\Omega)$, $\mathbf{C}_0(\Omega)$, $\mathbf{D}_0(\Omega)$, and $S_0(\Omega)$. To show (A.66), note that $N_0(\nabla^*)^\perp$ is simply the whole space $S_0(\Omega)$. The exactness of (A.55) implies that, for $0 \ne u \in S_0(\Omega)$, there exists $\mathbf{v}_u \in \mathbf{D}_0(\Omega)$ such that $u = \nabla \cdot \mathbf{v}_u$ and $\mathbf{v}_u \in N_0(\nabla \cdot)^\perp$. As a result, (A.65) holds for $\mathbf{v}_u$ and

$$\begin{aligned}\|\nabla^* u\|_0 &= \sup_{\mathbf{v} \in \mathbf{L}^2(\Omega)} \frac{(\nabla^* u, \mathbf{v})}{\|\mathbf{v}\|_0} \ge \frac{(\nabla^* u, \mathbf{v}_u)}{\|\mathbf{v}_u\|_0} \\ &= \frac{\|\nabla \cdot \mathbf{v}_u\|_0^2}{\|\mathbf{v}_u\|_0} \ge C_P^{-1}\|\nabla \cdot \mathbf{v}_u\|_0 = C_P^{-1}\|u\|_0 .\end{aligned}$$

To prove (A.67), let $0 \ne \mathbf{u} \in N_0(\nabla^* \times)^\perp$. From (A.59) it follows that $N_0(\nabla^* \times)^\perp = N_0(\nabla \cdot)$. Therefore, the exactness of (A.55) implies the existence of $\mathbf{v}_\mathbf{u} \in \mathbf{C}_0(\Omega)$ such that $\mathbf{u} = \nabla \times \mathbf{v}_\mathbf{u}$, whereas $\mathbf{u} \ne 0$ implies that $\mathbf{v}_\mathbf{u} \in N_0(\nabla \times)^\perp$. As a result, (A.64) holds for $\mathbf{v}_\mathbf{u}$ and

$$\begin{aligned}\|\nabla^* \times \mathbf{u}\|_0 &= \sup_{\mathbf{v} \in \mathbf{L}^2(\Omega)} \frac{(\nabla^* \times \mathbf{u}, \mathbf{v})}{\|\mathbf{v}\|_0} \ge \frac{(\nabla^* \times \mathbf{u}, \mathbf{v}_\mathbf{u})}{\|\mathbf{v}_\mathbf{u}\|_0} \\ &= \frac{\|\nabla \times \mathbf{v}_\mathbf{u}\|_0^2}{\|\mathbf{v}\|_0} \ge C_P^{-1}\|\nabla \times \mathbf{v}_\mathbf{u}\|_0 = C_P^{-1}\|\mathbf{u}\|_0 .\end{aligned}$$

The proof of (A.68) is similar and uses the fact that $N_0(\nabla^* \cdot)^\perp = N_0(\nabla \times)$, the exactness of (A.55), and the inequality (A.63). □

A Poincaré–Friedrichs inequality also exists for functions in $\mathbf{C}(\Omega) \cap \mathbf{D}(\Omega)$ whose normal or tangential components vanish on $\partial\Omega$.

Theorem A.12 *Let* $\Omega \subset \mathbb{R}^3$ *be a simply connected domain with a Lipschitz-continuous boundary. Then, there exists a positive constant* C_P *such that*

$$\|\mathbf{u}\|_0 \le C_P\big(\|\nabla \times \mathbf{u}\|_0 + \|\nabla \cdot \mathbf{u}\|_0\big) \tag{A.69}$$

for all $\mathbf{u}$ *belonging to* $\mathbf{D}(\Omega) \cap \mathbf{C}_0(\Omega)$ *or* $\mathbf{D}_0(\Omega) \cap \mathbf{C}(\Omega)$. □

Finally, recall a Poincaré–Friedrichs-like inequality for functions in $G(\Omega)$; see [281].

Theorem A.13 *Assume that $\partial\Omega = \Gamma \cup \Gamma^*$ and that Γ has positive measure. Then, there exists a positive constant C_P such that, for any $u \in G(\Omega)$,*

$$\|u\|_1 \leq C_P(\|\nabla u\|_0 + \|u\|_{1/2,\Gamma}). \qquad \square \tag{A.70}$$

A.3.3 Hodge Decompositions

The classical orthogonal decomposition of $L^2(\Omega)$ vector fields into irrotational and solenoidal parts is an example of a Hodge decomposition. The discussion in this section is derived from a generalized notion of Hodge decompositions that has its roots in the exterior calculus (see, e.g., [13, 168, 343]), specialized to our needs. In particular, we restrict attention to orthogonal decompositions of functions in $\mathbf{C}_0(\Omega)$ and $\mathbf{D}_0(\Omega)$ when Ω is a contractible domain so that it does not support non-trivial harmonic functions. For other settings, see, e.g., [13]. In this form, the Hodge decomposition states that $\mathbf{C}_0(\Omega)$ and $\mathbf{D}_0(\Omega)$ are direct sums of the null-space of the associated differential operator and its orthogonal complement.[8]

Theorem A.14 (Decomposition of $\mathbf{C}_0(\Omega)$) *Assume that $\Omega \in \mathbb{R}^3$ is a contractible domain with a Lipschitz-continuous boundary. Then, every $\mathbf{u} \in \mathbf{C}_0(\Omega)$ has the orthogonal decomposition*

$$\mathbf{u} = \mathbf{u}_N + \mathbf{u}_{N^\perp} \qquad \text{with} \qquad \mathbf{u}_N \in N_0(\nabla\times) \quad \text{and} \quad \mathbf{u}_{N^\perp} \in N_0(\nabla\times)^\perp. \tag{A.71}$$

Furthermore, there exist $p \in G_0(\Omega)$ such that $\mathbf{u}_N = \nabla p$ and a positive constant C_H such that

$$\|p\|_0 \leq C_H\|\mathbf{u}\|_0 \qquad \text{and} \qquad \|\mathbf{u}_{N^\perp}\|_0 \leq C_H\|\nabla\times\mathbf{u}\|_0. \qquad \square \tag{A.72}$$

Theorem A.15 (Decomposition of $\mathbf{D}_0(\Omega)$) *Assume that $\Omega \subset \mathbb{R}^3$ is a contractible domain with a Lipschitz-continuous boundary. Then, every $\mathbf{u} \in \mathbf{D}_0(\Omega)$ has the orthogonal decomposition*

$$\mathbf{u} = \mathbf{u}_N + \mathbf{u}_{N^\perp} \qquad \text{with} \qquad \mathbf{u}_N \in N_0(\nabla\cdot) \quad \text{and} \quad \mathbf{u}_{N^\perp} \in N_0(\nabla\cdot)^\perp. \tag{A.73}$$

Furthermore, there exist $\mathbf{v} \in \mathbf{C}_0(\Omega)$ such that $\mathbf{u}_N = \nabla\times\mathbf{v}$ and a positive constant C_H such that

$$\|\mathbf{v}\|_0 \leq C_H\|\mathbf{u}\|_0 \qquad \text{and} \qquad \|\mathbf{u}_{N^\perp}\|_0 \leq C_H\|\nabla\cdot\mathbf{u}\|_0. \qquad \square \tag{A.74}$$

---

[8] This statement also extends to $G(\Omega)$ and $G_0(\Omega)$, where the straightforward characterization of $N(\nabla)$ and $N_0(\nabla)$ makes it particularly simple. For instance, the Hodge decomposition of $G_0(\Omega)$ is trivial because $N_0(\nabla) = \{0\}$ and $N_0^\perp(\nabla) = G_0(\Omega)$; see Remark A.6. From the same remark, it follows that

$$G(\Omega) = \mathbb{R} \oplus G(\Omega) \cap L_0^2(\Omega),$$

i.e., every function in $G(\Omega)$ can be represented as a sum of a constant and a zero mean function.

A.3.4 Trace Theorems

We finish this appendix with two trace theorems, i.e., theorems that relate norms of functions on the interior of Ω with norms of their restrictions to the boundary $\partial\Omega$. We have the following well-known result that states that, for $s > 1/2$, functions belonging to $H^s(\Omega)$ have well-defined traces on $\partial\Omega$.

Theorem A.16 ([183, Theorem 1.5, p. 8]) *Assume that $s > 1/2$. Then, there exists a positive constant C such that*

$$\|u\|_{s-1/2,\partial\Omega} \le C\|u\|_s \qquad \forall u \in H^s(\Omega), \tag{A.75}$$

i.e., the mapping $\phi \mapsto \phi|_{\partial\Omega}$ is a continuous linear map $H^s(\Omega) \mapsto H^{s-1/2}(\partial\Omega)$. □

We also have a trace theorem for functions in $\mathbf{D}(\Omega)$.

Theorem A.17 ([183, Theorem 2.5, p. 27]) *There exists a positive constant C such that*

$$\|\mathbf{v}\cdot\mathbf{n}\|_{-1/2,\partial\Omega} \le \|\mathbf{v}\|_D \qquad \forall \mathbf{u} \in \mathbf{D}(\Omega), \tag{A.76}$$

i.e., the mapping $\mathbf{v} \mapsto \mathbf{v}\cdot\mathbf{n}|_{\partial\Omega}$ is a continuous linear map $\mathbf{D}(\Omega) \mapsto H^{-1/2}(\partial\Omega)$. □

When $\partial\Omega = \Gamma \cup \Gamma^*$, both trace theorems remain valid for the modified norms (A.12) and (A.13).

Appendix B
Compatible Finite Element Spaces

This appendix is intended to serve as a quick reference guide to the key finite element spaces used in this book and their basic properties. Its content is mostly devoted to finite element approximation of the De Rham complex (A.52), i.e., finite element spaces G^h, $\mathbf{C}^h$, $\mathbf{D}^h$, and S^h that, in addition to being proper subspaces of $G(\Omega)$, $\mathbf{C}(\Omega)$, $\mathbf{D}(\Omega)$, and $S(\Omega)$, respectively, also form an exact sequence.

The chief reason to consider simultaneous approximation of $G(\Omega)$, $\mathbf{C}(\Omega)$, $\mathbf{D}(\Omega)$, and $S(\Omega)$, instead of dealing with them one by one, is the recent appreciation of the fact that the stability and/or accuracy of many finite element methods,[1] including some least-squares finite element methods (LSFEMs), are contingent upon the existence of a *discrete* De Rham complex $\{G^h(\Omega), \mathbf{C}^h(\Omega), \mathbf{D}^h(\Omega), S^h(\Omega)\}$ and bounded projection operators Π such that for $\beta \in \{\emptyset, \gamma\}$ the diagram

$$\begin{array}{ccccccc} G_\beta(\Omega) & \xrightarrow{\nabla} & \mathbf{C}_\beta(\Omega) & \xrightarrow{\nabla\times} & \mathbf{D}_\beta(\Omega) & \xrightarrow{\nabla\cdot} & S_\beta(\Omega) \\ \Pi_G \downarrow & & \Pi_C \downarrow & & \Pi_D \downarrow & & \Pi_S \downarrow \\ G^h_\beta(\Omega) & \xrightarrow{\nabla} & \mathbf{C}^h_\beta(\Omega) & \xrightarrow{\nabla\times} & \mathbf{D}^h_\beta(\Omega) & \xrightarrow{\nabla\cdot} & S^h_\beta(\Omega) \end{array} \tag{B.1}$$

commutes; see, e.g., [7,69,73,137,204,206]. In what follows, we refer to finite element spaces that form a discrete De Rham complex as *compatible* approximations of $G(\Omega)$, $\mathbf{C}(\Omega)$, $\mathbf{D}(\Omega)$, and $S(\Omega)$, or simply as *compatible finite element spaces.*

In the finite element literature, the term "standard" or "nodal" finite elements is often used to describe approximations of the Sobolev space $G(\Omega) = H^1(\Omega)$. Because $H^1(\Omega)$ coincides with the first space in the De Rham complex, its approximations are provided by $G^h(\Omega)$ which obviates the need for a separate discussion of standard spaces.

[1] The same realization emerged at about the same time and independently in finite difference methods [139,224,314] and finite volume methods [276,287,289] which further underscores the relevance of this viewpoint in numerical methods for partial differential equations. More information about the use of *homological* ideas in discretizations is found in, e.g., [8,13,40,75,87,121,137,204].

P.B. Bochev, M.D. Gunzburger, *Least-Squares Finite Element Methods*,
Applied Mathematical Sciences 166, DOI: 10.1007/b13382_14,

B.1 Formal Definition and Properties of Finite Element Spaces

A *finite element* in $\mathbb{R}^d$, $d = 1,2,3$, (see [123, p.78]) is a triple $\{\kappa, P, \Lambda\}$, where

1. κ is a closed subset of $\mathbb{R}^d$ with nonempty interior and a Lipschitz-continuous boundary
2. $P(\kappa)$ is an n-dimensional space of real scalar or vector-valued functions defined over κ
3. $\Lambda(\kappa)$ is a unisolvent set of n linear functionals $l_i : P(\kappa) \mapsto \mathbb{R}$, i.e., l_i are linearly independent and, for every set $\{a_1,\dots,a_n\}$ of n real numbers, there exists a unique $u \in P(\kappa)$ such that $l_i(u) = a_i$.

The set Λ constitutes the *degrees of freedom* of the element $\{\kappa, P, \Lambda\}$. The functions $\{u_i\}_{i=1}^n$ such that $l_i(u_j) = \delta_{ij}$ form a basis for $P(\kappa)$ that is dual to $\{l_i\}_{i=1}^n$. Let $X(\kappa)$ be a space of smooth functions so that $l_i(v)$ are defined for all $v \in X(\kappa)$ and $P(\kappa) \subset X(\kappa)$. The local *canonical* projection operator $X(\kappa) \mapsto P(\kappa)$ is defined by the formula

$$\Pi_\kappa(v) = \sum_{i=1}^{n} l_i(v) u_i . \tag{B.2}$$

A key feature of finite element methods is that every finite element $\{\kappa, P, \Lambda\}$ is completely defined by its *reference* element $\{\widehat{\kappa}, \widehat{P}, \widehat{\Lambda}\}$ and a diffeomorphism[2] $F_\kappa : \widehat{\kappa} \mapsto \kappa$. As a result, the shape of κ must be such that one can easily construct the diffeomorphism F_κ. For this reason, reference regions $\widehat{\kappa}$ are drawn from a relatively small set of standard shapes for which there exist polynomial diffeomorphisms F_κ. Examples include 3-simplices (tetrahedrons), 3-cubes (hexahedrons), and prisms in three dimensions and 2-simplices (triangles) and 2-cubes (quadrilaterals) in two dimensions.[3]

If F_κ and $\widehat{P}(\widehat{\kappa})$ are defined using polynomials of the same degree, the element $\{\kappa, P, \Lambda\}$ is called *isoparametric*; see [123, Remark 4.3.1, p. 226]. A set $\{\{\kappa, P, \Lambda\}\}$ of finite elements is referred to as an *affine family* if $F_\kappa(\widehat{\mathbf{x}}) = \mathsf{A}_\kappa \widehat{\mathbf{x}} + \vec{b}_\kappa$ for every element in the set; here, the A_κs and $\vec{b}_\kappa$s are $d \times d$ matrices and d vectors, respectively. In all other cases, $\{\{\kappa, P, \Lambda\}\}$ is called a *non-affine family*.

Given a reference element $\{\widehat{\kappa}, \widehat{P}, \widehat{\Lambda}\}$ with a basis $\{\widehat{u}_i\}$, the basis $\{u_i\}$ of $\{\kappa, P, \Lambda\}$ is defined by pullback:[4]

$$u_i = \Phi^*(\widehat{u}_i), \qquad i = 1,\dots,n . \tag{B.3}$$

---

[2] Recall that F_κ is a diffeomorphism if it is a bijective map such that F_κ and F_κ^{-1} are differentiable.

[3] In principle, $\widehat{P}(\widehat{\kappa})$ can be any finite dimensional space such as the first n Fourier modes, a finite set of wavelet functions, and so on. In practice, the prevalent choice for finite element methods has been to use polynomial functions. This choice combines simplicity with good approximation properties and leads to the easy determination of basis functions. For this reason, we restrict attention to polynomial finite element spaces.

[4] The actual form of Φ^* is discussed in Section B.2 and depends on which one of the four function spaces $G(\Omega)$, $\mathbf{C}(\Omega)$, $\mathbf{D}(\Omega)$, or $S(\Omega)$ is being approximated.

The space $\widehat{P}(\widehat{\kappa})$ is affine-invariant if it is preserved under the pullback of an affine transformation of $\mathbb{R}^d$, i.e., $\text{span}\{\widehat{u}_i\} = \text{span}\{u_i\}$.[5]

The symbol $\mathcal{C}_m(\kappa)$, $0 \le m \le d$, refers to the set of all m-dimensional subcells of an element κ. Thus, in three dimensions, $\mathcal{C}_0(\kappa)$ is the set of all vertices in κ, $\mathcal{C}_1(\kappa)$ is the set of all edges, $\mathcal{C}_2(\kappa)$ is the set of all faces, and $\mathcal{C}_3(\kappa)$ is the element itself. The set of all m-dimensional subcells in $\mathcal{T}_h$ is the union of the element sets:

$$\mathcal{C}_m(\mathcal{T}_h) = \bigcup_{\kappa\in\mathcal{T}_h} \mathcal{C}_m(\kappa). \tag{B.4}$$

In particular, $\mathcal{C}_d(\mathcal{T}_h) = \mathcal{T}_h$.

A conforming[6] finite element partition $\mathcal{T}_h$ of Ω is built by breaking Ω into subdomains κ that are images of the standard shapes and satisfy the following set of rules (see [123, p. 38]):

1. every $\kappa \in \mathcal{T}_h$ is closed, the interior $\overset{\circ}{\kappa}$ is non-empty, and the boundary $\partial\kappa$ is Lipschitz continuous
2. $\overline{\Omega} = \cup_{\kappa\in\mathcal{T}_h} \kappa$
3. the interiors of two distinct elements have empty intersection: $\overset{\circ}{\kappa}_1 \cap \overset{\circ}{\kappa}_2 = \emptyset$
4. for any element $\kappa \in \mathcal{T}_h$ any subcell in $\mathcal{C}_{d-1}(\kappa)$ is either a subset of $\partial\Omega$ or belongs to $\mathcal{C}_{d-1}(\kappa')$ for some other element $\kappa' \in \mathcal{T}_h$.

Given a Banach space $X(\Omega)$ and a partition $\mathcal{T}_h$ of Ω into finite elements $\{\{\kappa, P, \Lambda\}\}$, its proper subspace

$$X^h(\Omega) = \{u^h \in X(\Omega) \mid u^h|_\kappa \in P(\kappa) \quad \forall\, \kappa \in \mathcal{T}_h\} \tag{B.5}$$

is called a *conforming finite element* approximation of $X(\Omega)$, defined with respect to the partition $\mathcal{T}_h$. To obtain a conforming finite element space, the local element spaces must be glued together in a way that provides a sufficient level of interelement continuity for the inclusion $X^h(\Omega) \subset X(\Omega)$ to hold. This is achieved by selecting an appropriate set of local degrees of freedom $\widehat{\Lambda}(\widehat{\kappa})$ and then requiring that all degrees of freedom that live on shared subcells between two elements are single valued. The set of *global* degrees of freedom that define the global space $X^h(\Omega)$ is $\Lambda(\mathcal{T}_h) = \{l_i\}_{i=1}^N$ and $\{u_i\}_{i=1}^N$ is the associated dual basis. The global canonical projection operator $X(\Omega) \mapsto X^h(\Omega)$ is defined in a manner similar to (B.2):

$$\Pi_X(v) = \sum_{i=1}^N l_i(v) u_i\,. \tag{B.6}$$

---

[5] If $\{\kappa, P, \Lambda\}$ is an affine family, $P(\kappa)$ is also a polynomial space, and if $\widehat{P}(\widehat{\kappa})$ is affine-invariant, these two spaces coincide. However, for non-affine families, $P(\kappa)$ is not necessarily a polynomial space. An important example is finite elements defined on general quadrilateral and hexahedral cells for which F_κ is not affine.

[6] The use of the adjective "conforming" to describe geometrical properties of the finite element partition $\mathcal{T}_h$ should not be confused with the use of the same adjective to describe properties of finite element approximating spaces; see (B.5).

In general, Π_X may not be defined for all $u \in X(\Omega)$ so that it is not necessarily a bounded operator $X(\Omega) \mapsto X^h(\Omega)$. In what follows, we restrict attention to settings for which one can guarantee the existence of alternative projection operators $\widetilde{\Pi}_X : X(\Omega) \mapsto X^h(\Omega)$ that are bounded, i.e., for which

$$\|\widetilde{\Pi}_X(v)\|_X \leq C\|v\|_X \qquad \forall\, v \in X(\Omega). \tag{B.7}$$

An important characteristic of conforming finite element spaces is their ability to approximate functions in X with respect to the norm in that space. The following definition is adapted from [156, Definition 2.14, p. 89].

Definition B.1 The conforming finite element space $X^h(\Omega)$ is said to have the *approximability property* if

$$\lim_{h\to 0}\left(\inf_{u^h\in X^h(\Omega)} \|u-u^h\|_X\right) = 0 \qquad \forall u \in X(\Omega). \qquad \square \tag{B.8}$$

An important factor for the approximation properties of finite element spaces is the size and the shape of the elements κ that form the partition $\mathcal{T}_h$. Because $P(\kappa)$ is defined by a transformation of the reference space $\widehat{P}(\widehat{\kappa})$, its quality depends on the "quality" of the mapping F_κ which, in turn, depends on the geometry of κ. The size and the shape of a simplicial region κ can be described by two parameters:

$$h_\kappa = \inf_{\omega\supseteq\kappa} \operatorname{diam}\omega \qquad \text{and} \qquad \rho_\kappa = \sup_{\omega\subseteq\kappa} \operatorname{diam}\omega, \tag{B.9}$$

where ω denotes an open ball.[7] The following definition summarizes [156, Definition 1.107, p.60, Definition 1.140, p76].

Definition B.2 A parameterized family of triangulations $\{\mathcal{T}_h\}_{h>0}$ is called *regular* if there exists a positive constant σ, independent of h and κ, such that[8]

$$\sigma_\kappa = \frac{h_\kappa}{\rho_\kappa} \leq \sigma \qquad \forall\, \kappa \in \mathcal{T}_h, \quad \forall\, h > 0.$$

We say that $\{\mathcal{T}_h\}_{h>0}$ is *uniformly regular* if there exists a positive constant τ such that

$$\tau h \leq h_\kappa \leq \sigma\rho_\kappa \qquad \forall\, \kappa \in \mathcal{T}_h, \quad \forall\, h > 0. \qquad \square$$

---

[7] For non-simplicial regions, ρ_κ can be calculated by approximating κ by simplices $\widetilde{\kappa}$ and setting

$$\rho_\kappa = C \min_{\widetilde{\kappa}\subset\kappa} \rho_{\widetilde{\kappa}},$$

where $\rho_{\widetilde{\kappa}}$ is defined by (B.9) and C is a fixed constant; see [183, p. 105].

[8] Regular finite element partitions are also called *shape regular* because they consist of similarly "shaped" elements that cannot become "too flat". However, the elements in a regular partition can differ widely in their sizes. In contrast, uniformly regular meshes consist of elements that are roughly of the same size and shape. For this reason, such grids are sometimes called *quasi-uniform*.

B.2 Finite Element Approximation of the De Rham Complex

From now on, we assume that $\mathcal{T}_h$ is a regular partition of an open bounded domain $\Omega \subset \mathbb{R}^3$ so that F_κ and F_κ^{-1} are defined for all elements $\kappa \in \mathcal{T}_h$, the Jacobian DF_κ is invertible at every point $\widehat{\mathbf{x}} \in \widehat{\kappa}$, and $J_\kappa = \det(DF_\kappa)$ is non-zero for all $\widehat{\mathbf{x}} \in \widehat{\kappa}$.

For $\kappa \in \mathcal{T}_h$ with a reference element $\widehat{\kappa}$, we define the pullback as follows:

$$
\begin{aligned}
&\Phi_G^* : G(\widehat{\kappa}) \mapsto G(\kappa) && \Phi_G^*(\widehat{u}) = \widehat{u} \circ F_\kappa^{-1} \\
&\Phi_C^* : \mathbf{C}(\widehat{\kappa}) \mapsto \mathbf{C}(\kappa) && \Phi_C^*(\widehat{\mathbf{u}}) = \left((DF_\kappa)^{-\mathsf{T}} \cdot \widehat{\mathbf{u}}\right) \circ F_\kappa^{-1} \\
&\Phi_D^* : \mathbf{D}(\widehat{\kappa}) \mapsto \mathbf{D}(\kappa) && \Phi_D^*(\widehat{\mathbf{u}}) = \left(J_\kappa^{-1} DF_\kappa \cdot \widehat{\mathbf{u}}\right) \circ F_\kappa^{-1} \\
&\Phi_S^* : S(\widehat{\kappa}) \mapsto S(\kappa) && \Phi_S^*(\widehat{u}) = \left(J_\kappa^{-1} \widehat{u}\right) \circ F_\kappa^{-1}.
\end{aligned}
\tag{B.10}
$$

Definition (B.10) provides rules for "changing variables" in $G(\Omega)$, $\mathbf{C}(\Omega)$, $\mathbf{D}(\Omega)$, and $S(\Omega)$ that commute with the gradient, curl, and divergence operators:

$$
\begin{aligned}
\nabla \Phi_G^*(\widehat{u}) &= \Phi_C^*(\widehat{\nabla} \widehat{u}) = \left((DF_\kappa)^{-\mathsf{T}} \cdot \widehat{\nabla} \widehat{u}\right) \circ F_\kappa^{-1} \\
\nabla \times \Phi_C^*(\widehat{\mathbf{u}}) &= \Phi_D^*(\widehat{\nabla} \times \widehat{\mathbf{u}}) = \left(J_\kappa^{-1} DF_\kappa \cdot \widehat{\nabla} \times \widehat{\mathbf{u}}\right) \circ F_\kappa^{-1} \\
\nabla \cdot \Phi_D^*(\widehat{\mathbf{u}}) &= \Phi_S^*(\widehat{\nabla} \cdot \widehat{\mathbf{u}}) = \left(J_\kappa^{-1} \widehat{\nabla} \cdot \widehat{\mathbf{u}}\right) \circ F_\kappa^{-1}.
\end{aligned}
\tag{B.11}
$$

In (B.11), $\widehat{\nabla}$ is the nabla operator with respect to the reference variables $\widehat{\mathbf{x}} \in \widehat{\kappa}$.

To construct finite element approximations of the global spaces $G(\Omega)$, $\mathbf{C}(\Omega)$, $\mathbf{D}(\Omega)$, and $S(\Omega)$, we first specify *reference* spaces $G^h(\widehat{\kappa})$, $\mathbf{C}^h(\widehat{\kappa})$, $\mathbf{D}^h(\widehat{\kappa})$, and $S^h(\widehat{\kappa})$ that approximate $G(\widehat{\kappa})$, $\mathbf{C}(\widehat{\kappa})$, $\mathbf{D}(\widehat{\kappa})$, and $S(\widehat{\kappa})$, respectively. The local *element* spaces are defined by pullback of their parent reference spaces:

$$
\begin{aligned}
G^h(\kappa) &= \{u^h \in G(\kappa) \mid u^h = \Phi_G^*(\widehat{u}^h), \quad \widehat{u}^h \in G^h(\widehat{\kappa})\} \\
\mathbf{C}^h(\kappa) &= \{\mathbf{u}^h \in \mathbf{C}(\kappa) \mid \mathbf{u}^h = \Phi_C^*(\widehat{\mathbf{u}}^h), \quad \widehat{\mathbf{u}}^h \in \mathbf{C}^h(\widehat{\kappa})\} \\
\mathbf{D}^h(\kappa) &= \{\mathbf{u}^h \in \mathbf{D}(\kappa) \mid \mathbf{u}^h = \Phi_D^*(\widehat{\mathbf{u}}^h), \quad \widehat{\mathbf{u}}^h \in \mathbf{D}^h(\widehat{\kappa})\} \\
S^h(\kappa) &= \{u^h \in S(\kappa) \mid u^h = \Phi_S^*(\widehat{u}^h), \quad \widehat{u}^h \in S^h(\widehat{\kappa})\}.
\end{aligned}
\tag{B.12}
$$

Finally, the local element spaces are assembled into global spaces that are conforming approximations of $G(\Omega)$, $\mathbf{C}(\Omega)$, $\mathbf{D}(\Omega)$, and $S(\Omega)$:

$$
\begin{aligned}
G^h(\Omega) &= \{u^h \in G(\Omega) \mid u^h|_\kappa \in G^h(\kappa) \quad \forall \kappa \in \mathcal{T}_h\} \\
\mathbf{C}^h(\Omega) &= \{\mathbf{u}^h \in \mathbf{C}(\Omega) \mid \mathbf{u}^h|_\kappa \in \mathbf{C}^h(\kappa) \quad \forall \kappa \in \mathcal{T}_h\} \\
\mathbf{D}^h(\Omega) &= \{\mathbf{u}^h \in \mathbf{D}(\Omega) \mid \mathbf{u}^h|_\kappa \in \mathbf{D}^h(\kappa) \quad \forall \kappa \in \mathcal{T}_h\} \\
S^h(\Omega) &= \{u^h \in S(\Omega) \mid u^h|_\kappa \in S^h(\kappa) \quad \forall \kappa \in \mathcal{T}_h\}.
\end{aligned}
\tag{B.13}
$$

Several conditions must be met in order to ensure that this process gives rise to a discrete De Rham complex that is a good approximation of (A.52). First, the degrees of freedom on the reference element must be selected in a way that makes it possible to obtain the desired interelement continuity by making shared degrees of freedom single valued. Second, the global finite element spaces in (B.13) must form an exact sequence so that the parent reference-element spaces must be chosen in a way that guarantees that[9]

$$R(\nabla|_{G^h}) \subset \mathbf{C}^h(\Omega) \qquad R(\nabla\times|_{\mathbf{C}^h}) \subset \mathbf{D}^h(\Omega) \qquad R(\nabla\cdot|_{\mathbf{D}^h}) = S^h(\Omega). \quad \text{(B.14)}$$

The last condition that needs to be satisfied is the existence of bounded projection operators

$$\begin{aligned}
&\Pi_G : G(\Omega) \mapsto G^h(\Omega) && \|\Pi_G u\|_G \le C\|u\|_G && \forall u \in G(\Omega)\\
&\Pi_C : \mathbf{C}(\Omega) \mapsto \mathbf{C}^h(\Omega) && \|\Pi_C \mathbf{u}\|_C \le C\|\mathbf{u}\|_C && \forall \mathbf{u} \in \mathbf{C}(\Omega)\\
&\Pi_D : \mathbf{D}(\Omega) \mapsto \mathbf{D}^h(\Omega) && \|\Pi_D \mathbf{u}\|_D \le C\|\mathbf{u}\|_D && \forall \mathbf{u} \in \mathbf{D}(\Omega)\\
&\Pi_S : S(\Omega) \mapsto S^h(\Omega) && \|\Pi_S u\|_S \le C\|u\|_S && \forall u \in S(\Omega)
\end{aligned} \quad \text{(B.15)}$$

such that the diagram (B.1) commutes. The construction of bounded projections that support a commuting diagram property is a non-trivial and rather technical task. Further details about various approaches can be found in [13,121,122,136,137,317].

The global spaces $G^h(\Omega)$, $\mathbf{C}^h(\Omega)$, $\mathbf{D}^h(\Omega)$, and $S^h(\Omega)$ are piecewise smooth with respect to the finite element partition $\mathcal{T}_h$. As a result, $S^h(\Omega)$ is always a proper subspace of $S(\Omega) = L^2(\Omega)$. Insofar as the interelement continuity conditions for $G^h(\Omega)$, $\mathbf{C}^h(\Omega)$, and $\mathbf{D}^h(\Omega)$ are concerned, we have the following result (see [13, Lemma 5.1], [123, Theorem 2.1.1, p. 39], or [317, section 1.1.4]).

Theorem B.3 (Conformity requirements) *Let $\mathcal{T}_h$ be a conforming finite element partition of a bounded open region Ω in $\mathbb{R}^d$, $d = 2,3$, and let $G^h(\widehat{\kappa})$, $\mathbf{C}^h(\widehat{\kappa})$, and $\mathbf{D}^h(\widehat{\kappa})$ denote polynomial reference spaces.*

1. *A necessary and sufficient condition for the inclusion $G^h(\Omega) \subset G(\Omega)$ to hold is that all $u^h \in G^h(\Omega)$ must be single-valued on all $(d-1)$-dimensional subcells $C \in \mathcal{C}_{d-1}(\mathcal{T}_h)$ of the mesh $\mathcal{T}_h$.*
2. *A necessary and sufficient condition for the inclusion $\mathbf{C}^h(\Omega) \subset \mathbf{C}(\Omega)$ to hold is that all $\mathbf{u}^h \in \mathbf{C}^h(\Omega)$ must have a single-valued tangential component $\mathbf{n}\times\mathbf{u}^h$ on all $(d-1)$-dimensional subcells $C \in \mathcal{C}_{d-1}(\mathcal{T}_h)$ of the mesh $\mathcal{T}_h$.*
3. *A necessary and sufficient condition for the inclusion $\mathbf{D}^h(\Omega) \subset \mathbf{D}(\Omega)$ to hold is that all $\mathbf{u}^h \in \mathbf{D}^h(\Omega)$ must have a single-valued normal component $\mathbf{n}\cdot\mathbf{u}^h$ on all $(d-1)$-dimensional subcells $C \in \mathcal{C}_{d-1}(\mathcal{T}_h)$ of the mesh $\mathcal{T}_h$.*

According to this theorem, conformity with respect to $G(\Omega)$ can be achieved by gluing the element spaces $G^h(\kappa)$ so that $G^h(\Omega)$ is of class C^0. Likewise, conformity

---

[9] In some cases, the last condition only holds in the weaker sense that $\Pi_S(\nabla\cdot\mathbf{u}) = \Pi_s(\nabla\cdot\Pi_D\mathbf{u})$.

of $\mathbf{C}^h(\Omega)$ and $\mathbf{D}^h(\Omega)$ requires one to assemble the element spaces in a way that respectively ensures tangential and normal continuity across all $(d-1)$-dimensional subcells of the resulting piecewise smooth vector fields.

Remark B.4 Theorem B.3 actually holds for a wider range of spaces that consist of piecewise smooth, with respect to the partition $\mathcal{T}_h$, functions. □

$L^2(\Omega)$ projections

In addition to the bounded projection operators (B.15), on occasion we need $L^2(\Omega)$ projections onto $G^h(\Omega)$, $\mathbf{C}^h(\Omega)$, $\mathbf{D}^h(\Omega)$, and $S^h(\Omega)$. These operators are defined in the usual manner:

$$\begin{aligned}
&\pi_G : G(\Omega) \mapsto G^h(\Omega) && \left(\pi_G u, v^h\right)_{\Theta_0} = \left(u, v^h\right)_{\Theta_0} && \forall v^h \in G^h(\Omega)\\
&\pi_C : \mathbf{C}(\Omega) \mapsto \mathbf{C}^h(\Omega) && \left(\pi_C \mathbf{u}, \mathbf{v}^h\right)_{\Theta_1} = \left(\mathbf{u}, \mathbf{v}^h\right)_{\Theta_1} && \forall \mathbf{v}^h \in \mathbf{C}^h(\Omega)\\
&\pi_D : \mathbf{D}(\Omega) \mapsto \mathbf{D}^h(\Omega) && \left(\pi_D \mathbf{u}, \mathbf{v}^h\right)_{\Theta_2} = \left(\mathbf{u}, \mathbf{v}^h\right)_{\Theta_2} && \forall \mathbf{v}^h \in \mathbf{D}^h(\Omega)\\
&\pi_S : S(\Omega) \mapsto S^h(\Omega) && \left(\pi_S u, v^h\right)_{\Theta_3} = \left(u, v^h\right)_{\Theta_3} && \forall v^h \in S^h(\Omega)\,.
\end{aligned} \tag{B.16}$$

The $L^2(\Omega)$ projections return the finite element function that minimizes the $L^2(\Omega)$ distance between the operator's argument and the respective finite element space. This statement is formalized in the following theorem.

Theorem B.5 *Assume that $X(\Omega)$ is one of the spaces $G(\Omega)$, $\mathbf{C}(\Omega)$, $\mathbf{D}(\Omega)$, $S(\Omega)$, that $X^h(\Omega)$ is a conforming approximation of $X(\Omega)$, and that π_X is the corresponding $L^2(\Omega)$ projection defined in* (B.16). *Then,*

$$\|u - \pi_X u\|_0 \le \inf_{v^h \in X^h(\Omega)} \|u - v^h\|_0 \,. \quad \square \tag{B.17}$$

The proof of this theorem is straightforward.

B.2.1 Examples of Compatible Finite Element Spaces

Section B.2.3 discusses arrangement of three-dimensional simplicial finite element spaces into exact sequences, whereas in this section we catalogue some of the most widely used compatible spaces. For completeness we present compatible finite elements on both d-simplices and d-cubes, $d = 2,3$. Because $\{\kappa, P, \Lambda\}$ is completely defined by its reference element $\{\widehat{\kappa}, \widehat{P}, \widehat{\Lambda}\}$, each finite element space is described in terms of the associated reference element. For brevity, degrees of freedom are stated only for the lowest-order spaces. Complete details, including definitions of alternative basis functions, can be found in, e.g., [13, 84–87, 123, 137, 183, 188, 204, 317].

In two dimensions, the reference simplex is the triangle with vertices $(0,0)$, $(1,0)$, and $(0,1)$. In three dimensions, it is the tetrahedron with vertices $(0,0,0)$, $(1,0,0)$, $(0,1,0)$, and $(0,0,1)$. The reference d-cube is $[-1,1]^d$.

To define the reference finite element spaces, we need some additional notation. The symbol P_r stands for the set of all polynomials in $\mathbb{R}^d$ of degree less than or equal to r. The dimension of P_r equals $(d+r)!/(d!r!)$. The set of all homogeneous[10] polynomials of degree r is denoted by H_r. For elements defined on d-cubes, we need the spaces $P_{r_1,\dots,r_d}$ of polynomials whose degree in the ith coordinate direction does not exceed r_i. In the case for which all r_i are equal, we write $Q_r = P_{r,\dots,r}$. For clarity, the order of polynomials defined with respect to reference, local, and global finite element spaces is always indicated by a single superscript which replaces the generic mesh parameter h in the finite element space designation. Whenever necessary we continue to use subscripts to indicate boundary conditions imposed on the spaces.

The nodal element space $G^h(\Omega)$

Finite element approximations of $G(\Omega)$ are among the earliest examples of finite element spaces. They include, among others, the spaces used by Courant [128] at the very inception of the finite element methods.

On a d-simplex, the reference space consists of complete polynomial functions of degree at least one:

$$G^r(\widehat{\kappa}) = P_r, \qquad r \geq 1. \tag{B.18}$$

The reference space on a d-cube is a tensor product polynomial space of degree one or higher:

$$G^r(\widehat{\kappa}) = P_{r,\dots,r} = Q_r, \qquad r \geq 1. \tag{B.19}$$

For the lowest-order elements $G^1(\widehat{\kappa})$, the unisolvent set $\widehat{\Lambda}^1(\widehat{\kappa})$ contains the following functionals:

$$l_i(\widehat{u}) = \int_{C_i} \widehat{u}\delta(C_i)\, d\Omega \qquad \text{with} \qquad C_i \in \mathcal{C}_0(\widehat{\kappa}),$$

where $\delta(\cdot)$ denotes the Dirac delta function and $\mathcal{C}_0(\widehat{\kappa})$ are the element nodes. Thus, the degrees of freedom in $G^1(\widehat{\kappa})$ are the values of $\widehat{u}$ at the element nodes. This is why $G^r(\widehat{\kappa})$ is often called a "nodal" finite element, even though for higher-order spaces they can certainly be endowed with other types of unisolvent sets.

The dual of $\widehat{\Lambda}^1(\widehat{\kappa})$ is comprised of polynomials whose value equals 1 at one of the nodes and zero at all other nodes. Therefore, the canonical projection of a smooth function into $G^1(\widehat{\kappa})$ is the Lagrange interpolant at the nodes of the element. For this reason, these types of approximations of $G(\Omega)$ are also often called "Lagrangian elements."

[10] Recall that a multivariate polynomial $p(\mathbf{x})$ is called *homogeneous* if all its terms have the same degree, e.g., $p(\mathbf{x}) = x^2 + xy + yz + z^2$ is element of H_2.

The definition of a global finite element space in the lowest-order case $r = 1$ is particularly simple because global degrees of freedom are restricted to the nodes $\mathcal{C}_0(\mathcal{T}_h)$ of the mesh. In this case, local elements are glued together by matching their nodal values on all shared nodes. According to Theorem B.3, this is sufficient to ensure that $G^1(\Omega) \subset G(\Omega)$.

The order of the approximation by nodal finite element spaces is summarized in the following theorem.

Theorem B.6 *Let $\mathcal{T}_h$ be a regular partition of Ω into finite elements and let $G^r(\Omega)$ denote a nodal space defined for some $r \geq 1$. Then, there exists a positive constant C, independent of h, such that, for every $u \in H^{r+1}(\Omega)$,*

$$\left.\begin{array}{l} \|u - \Pi_G u\|_0 \\ \\ \|u - \pi_G u\|_0 \end{array}\right\} \leq Ch^{r+1}\|u\|_{r+1} \tag{B.20}$$

$$\|\nabla(u - \Pi_G u)\|_0 \leq Ch^r \|\nabla u\|_r . \qquad \square$$

The edge element space $\mathbf{C}^h(\Omega)$

Finite element subspaces of $\mathbf{C}(\Omega)$ were discovered much later than nodal elements. The first curl-conforming elements were developed in [282,284]. In the engineering literature, an alternative construction on 3-cubes that bypasses the reference element was given in [344].

On a d-simplex, the original curl-conforming element, introduced in [282], contains the following polynomial vector fields:

$$\mathbf{C}^r\text{-}(\widehat{\kappa}) = [P_{r-1}]^d \oplus \mathbf{H}'_r , \tag{B.21}$$

where

$$\mathbf{H}'_r = \{\mathbf{p} \in [H_r]^d \mid \mathbf{p}(\widehat{\mathbf{x}}) \cdot \widehat{\mathbf{x}} = 0\} \quad \text{with} \qquad r \geq 1 .$$

A simple argument shows that, in the lowest-order case $r = 1$,

$$\mathbf{H}'_1 = \widehat{\mathbf{x}} \times [P_0]^{2d-3} = \{\mathbf{p} \in [H_1]^d \mid \mathbf{p} = \widehat{\mathbf{x}} \times \vec{b}; \quad \vec{b} \in \mathbb{R}^{2d-3}\},$$

where the vector product in $d = 2$ is defined in (A.5). As a result, an equivalent characterization of the lowest-order space in (B.21) is given by

$$\mathbf{C}^1\text{-}(\widehat{\kappa}) = \{\vec{a} + \widehat{\mathbf{x}} \times \vec{b} \mid \vec{a} \in \mathbb{R}^d, \ \vec{b} \in \mathbb{R}^{2d-3}\}. \tag{B.22}$$

For a d-cube, the analogue of (B.21) is

$$\mathbf{C}^r\text{-}(\widehat{\kappa}) = \begin{cases} P_{r-1,r} \times P_{r,r-1} & \text{for } d = 2 \\ \\ P_{r-1,r,r} \times P_{r,r-1,r} \times P_{r,r,r-1} & \text{for } d = 3 \end{cases} \qquad \text{with} \quad r \geq 1 . \tag{B.23}$$

Note that the spaces defined in (B.21) and (B.23) contain $[P_{r-1}]^d$ and $[Q_{r-1}]^d$, respectively, but are proper subspaces of $[P_r]^d$ and $[Q_r]^d$. Following [13], we signified this fact by using r_- in the space designation.

The finite element spaces defined in (B.21) and (B.23) are usually called *Nédélec elements of the first kind*. In the lowest-order case, (B.21) and (B.23) share the same unisolvent set $\widehat{\Lambda}^{1}\text{-}(\widehat{\kappa})$:

$$l_i(\widehat{\mathbf{u}}) = \int_{C_i} \widehat{\mathbf{u}} \cdot \mathbf{t}\, d\ell \qquad \text{with} \qquad C_i \in \mathcal{C}_1(\widehat{\kappa}), \tag{B.24}$$

i.e., the degrees of freedom in $\mathbf{C}^1\text{-}(\widehat{\kappa})$ are the edge circulations of the vector field $\widehat{\mathbf{u}}$. The dual of $\widehat{\Lambda}^{1}\text{-}(\widehat{\kappa})$ consists of the elements in $\mathbf{C}^1\text{-}(\widehat{\kappa})$ whose circulation equals one along one of the edges and is zero along all other edges. For this reason, curl-conforming spaces are collectively known as "edge elements" even though for higher-order spaces their degrees of freedom live on a larger set of subcells.

In [284], a second family of curl-conforming elements was proposed that is usually called *Nédélec elements of the second kind*. The reference element for d-simplices is

$$\mathbf{C}^r(\widehat{\kappa}) = [P_r]^d \qquad \text{with} \qquad r \geq 1 \tag{B.25}$$

whereas for d-cubes it uses tensor product polynomial spaces

$$\mathbf{C}^r(\widehat{\kappa}) = [Q_r]^d \qquad \text{with} \qquad r \geq 1\,. \tag{B.26}$$

Note that the spaces $\mathbf{C}^r\text{-}(\widehat{\kappa})$ are properly contained in (B.25) or (B.26), as appropriate.[11]

To contrast the two types of Nédélec elements, let us compare the lowest order spaces $\mathbf{C}^1\text{-}(\widehat{\kappa})$ and $\mathbf{C}^1(\widehat{\kappa})$ on a 3-simplex (tetrahedron). The set $\widehat{\Lambda}^1(\widehat{\kappa})$ for $\mathbf{C}^1(\widehat{\kappa})$ consists of the following functionals (see [284]):

$$l_i(\widehat{\mathbf{u}}) = \int_{C_i} (\widehat{\mathbf{u}} \cdot \mathbf{t}) p(\widehat{\mathbf{x}})\, d\ell \qquad \text{with} \qquad C_i \in \mathcal{C}_1(\widehat{\kappa}) \quad \text{and} \quad p(\widehat{\mathbf{x}}) \in P_1(C_i)\,, \tag{B.27}$$

i.e., it contains zeroth- and first-order moments of the tangential component $\widehat{\mathbf{u}} \cdot \mathbf{t}$ along the edges of the element. Therefore, $\widehat{\Lambda}^1(\widehat{\kappa})$ specifies two degrees of freedom per edge and has dimension 12. In contrast, (B.24) contains only zeroth-order moments of $\widehat{\mathbf{u}} \cdot \mathbf{t}$, specifies one degree-of-freedom per edge, and has dimension 6. These dimensions correspond to the dimensions of $\mathbf{C}^1(\widehat{\kappa})$ and $\mathbf{C}^1\text{-}(\widehat{\kappa})$, respectively, and it

---

[11] One should not confuse (B.25) and (B.26) with the spaces $[G^r(\widehat{\kappa})]^d$ formed by direct products of *scalar spaces* $G^r(\widehat{\kappa})$, even though formally, they consist of the same polynomials. Although the reference spaces in (B.25) and (B.26) are certainly direct products of precisely the same polynomial spaces as in (B.18) and (B.19), the former are endowed with completely different unisolvent sets $\widehat{\Lambda}(\widehat{\kappa})$ that are designed to *provide conformity with respect to* $\mathbf{C}(\Omega)$; see Theorem B.3.

is not hard to show that $\widehat{\Lambda}^1(\widehat{\kappa})$ and $\widehat{\Lambda}^{1-}(\widehat{\kappa})$, respectively, are unisolvent sets for these spaces.[12]

In simple terms, on each edge, the space $\mathbf{C}^{1-}(\widehat{\kappa})$ approximates the tangential component $\widehat{\mathbf{u}}\cdot\mathbf{t}$ of $\widehat{\mathbf{u}}$ by a piecewise constant, whereas $\mathbf{C}^1(\widehat{\kappa})$ approximates $\widehat{\mathbf{u}}\cdot\mathbf{t}$ by a linear polynomial. Because of these differences, the approximation properties of the two types of Nédélec spaces are different.

Theorem B.7 *Let $\mathcal{T}_h$ be a uniformly regular partition of Ω into affine finite elements and let $\mathbf{C}^{(r)}(\Omega)$ denote one of the spaces $\mathbf{C}^r(\Omega)$ or $\mathbf{C}^{r-}(\Omega)$ with $r \geq 1$. Then, there exists a positive constant C, independent of h, such that, for every vector field $\mathbf{u} \in \mathbf{C}(\Omega)\cap[H^{r+1}(\Omega)]^d$,*

$$\left.\begin{array}{l}\|\mathbf{u}-\Pi_C\mathbf{u}\|_0\\ \\ \|\mathbf{u}-\pi_C\mathbf{u}\|_0\end{array}\right\} \leq C\left\{\begin{array}{ll}h^r\|\mathbf{u}\|_r & \text{if } \mathbf{C}^{(r)}(\Omega)=\mathbf{C}^{r-}(\Omega)\\ \\ h^{r+1}\|\mathbf{u}\|_{r+1} & \text{if } \mathbf{C}^{(r)}(\Omega)=\mathbf{C}^{r}(\Omega)\end{array}\right. \tag{B.28}$$

$$\|\nabla\times(\mathbf{u}-\Pi_C\mathbf{u})\|_0 \leq Ch^r\|\nabla\times\mathbf{u}\|_r\,. \qquad \square$$

For a proof and other details, see [282, 284]. The error estimate in Theorem B.7 reveals that for Nédélec elements of the first kind, the error in the $L^2(\Omega)$ norm of the vector field and its curl have the same order that matches the order of the curl approximation for Nédélec elements of the second kind. Considering that $\mathbf{C}^r(\Omega)$ has more degrees of freedom, this means that $\mathbf{C}^{r-}(\Omega)$ is "optimized" for the curl operator. For instance, in the lowest-order case on 3-simplices (tetrahedrons), both $\mathbf{C}^{1-}(\Omega)$ and $\mathbf{C}^1(\Omega)$ provide first-order accurate curl approximations. However, $\mathbf{C}^1(\Omega)$ has 2 degrees of freedom per edge whereas $\mathbf{C}^{1-}(\Omega)$ has only one.

Remark B.8 The estimates in Theorem B.7 do not automatically extend to non-affine families of finite elements. For example, on general unstructured d-cubes, $d \geq 2$, one can expect reduced convergence in the curl approximation; see [9] and the references cited therein. □

The face element space $\mathbf{D}^h(\Omega)$

Finite element approximations of $\mathbf{D}(\Omega)$ preceded curl-conforming elements by just a few years. The first div-conforming finite element was developed for 2-simplices (triangles) in [302]. In [282, 284], generalizations and extensions to three dimensions and other element shapes were provided. At about the same time, several other families of div-conforming finite elements were developed in [84–86].

For brevity, we present only two families of div-conforming spaces that closely resemble the two families of curl-conforming spaces. Our discussion follows [87, 282, 284]. The first family uses reference spaces that are proper subspaces of $[P_r]^d$

---

12 The set (B.27) is also unisolvent for the lowest-order space on cubes. Let $d = 3$. For $r = 1$, the dimension of the space in (B.26) equals 24. A 3-cube has 12 edges and (B.27) specifies 2 moments per edge for a total of 24 independent degrees of freedom.

and $[Q_r]^d$ and is similar to the Nédélec elements of the first kind. Because the first instance of this family was given in [302], these elements are usually called *Raviart–Thomas spaces*, even though [302] dealt only with elements on 2-simplices. On a d-simplex, the Raviart–Thomas reference space is given by

$$\mathbf{D}^{r}\text{-}(\widehat{\kappa}) = [P_{r-1}]^d + \widehat{\mathbf{x}} H_{r-1} \qquad \text{with} \qquad r \geq 1\,. \tag{B.29}$$

An equivalent representation for the lowest-order element is (compare with (B.22))

$$\mathbf{D}^{1}\text{-}(\widehat{\kappa}) = \{\vec{a} + b\widehat{\mathbf{x}} \mid \vec{a} \in \mathbb{R}^d,\ b \in \mathbb{R}\}\,. \tag{B.30}$$

The Raviart–Thomas element on a d-cube is given by (compare with (B.23))

$$\mathbf{D}^{r}\text{-}(\widehat{\kappa}) = \begin{cases} P_{r,r-1} \times P_{r-1,r} & \text{for } d = 2 \\ P_{r,r-1,r-1} \times P_{r-1,r,r-1} \times P_{r-1,r-1,r} & \text{for } d = 3 \end{cases} \qquad \text{with} \quad r \geq 1\,. \tag{B.31}$$

Similar to the first family of curl-conforming elements, the lowest-order Raviart–Thomas elements share the same unisolvent set $\widehat{\Lambda}^1\text{-}(\widehat{\kappa})$

$$l_i(\widehat{\mathbf{u}}) = \int_{C_i} \widehat{\mathbf{u}} \cdot \mathbf{n}\, dS \qquad \text{with} \qquad C_i \in \mathcal{C}_{d-1}(\widehat{\kappa}) \tag{B.32}$$

on d-simplices and d-cubes. The degrees of freedom prescribed by (B.32) for $\mathbf{D}^1\text{-}(\widehat{\kappa})$ are the fluxes of the vector field $\widehat{\mathbf{u}}$ across the $(d-1)$-dimensional subcells of the reference element $\widehat{\kappa}$. Thus, the dual of $\widehat{\Lambda}^1\text{-}(\widehat{\kappa})$ consists of vector fields in $\mathbf{D}^1\text{-}(\widehat{\kappa})$ that have unit flux across one of the subcells in $\mathcal{C}_{d-1}(\widehat{\kappa})$ and zero flux across all other subcells in that set. Because in three-dimensions $\mathcal{C}_{d-1}(\widehat{\kappa})$ is the set of all faces[13] of $\widehat{\kappa}$, div-conforming elements are also called "face elements" although, for higher-order spaces, their degrees of freedom can live on a larger set of subcells.

The second family of div-conforming elements was proposed in [284] and uses exactly the same reference spaces as the Nédélec elements of the second kind (B.25) and (B.26). Accordingly, we refer to the elements below as Nédélec div-conforming elements of the second kind or, when the context is clear, simply as Nédélec elements of the second kind. On a d-simplex, the reference space is given by

$$\mathbf{D}^{r}(\widehat{\kappa}) = [P_r]^d \qquad \text{with} \qquad r \geq 1 \tag{B.33}$$

and on a d-cube we have

$$\mathbf{D}^{r}(\widehat{\kappa}) = [Q_r]^d \qquad \text{with} \qquad r \geq 1\,. \tag{B.34}$$

[13] In two-dimensions $\mathcal{C}_{d-1}(\widehat{\kappa}) = \mathcal{C}_1(\widehat{\kappa})$ is the set of all edges in $\widehat{\kappa}$, i.e., the same set that is used to define the degrees of freedom for curl-conforming elements in (B.24). The difference between (B.24) and (B.32) is that the former uses the tangential component of $\widehat{\mathbf{u}}$ along $C \in \mathcal{C}_1(\widehat{\kappa})$, whereas the latter uses the normal component of that field.

Of course, now these reference spaces must be glued together to form an approximation of $\mathbf{D}(\Omega)$ which, according to Theorem B.3, calls for a different set of degrees of freedom. Again, we restrict attention to the lowest-order spaces in (B.33) and (B.34). For a d-simplex, the set $\widehat{\Lambda}^1(\widehat{\kappa})$ is

$$l_i(\widehat{\mathbf{u}}) = \int_{C_i} (\widehat{\mathbf{u}} \cdot \mathbf{n}) p(\widehat{\mathbf{x}})\, dS \qquad \text{with} \qquad C_i \in \mathcal{C}_{d-1}(\widehat{\kappa}) \quad \text{and} \quad p(\widehat{\mathbf{x}}) \in P_1(C_i) \tag{B.35}$$

whereas for the space $\mathbf{D}^1(\widehat{\kappa})$ on cubes, the unisolvent set is

$$l_i(\widehat{\mathbf{u}}) = \int_{C_i} (\widehat{\mathbf{u}} \cdot \mathbf{n}) q(\widehat{\mathbf{x}})\, dS \quad \text{with} \quad C_i \in \mathcal{C}_{d-1}(\widehat{\kappa}) \quad \text{and} \quad q(\widehat{\mathbf{x}}) \in Q_1(C_i)\,. \tag{B.36}$$

In two-dimensions the integrals in (B.35)–(B.36) define identical moments of the normal component $\widehat{\mathbf{u}} \cdot \mathbf{n}$ because $\mathcal{C}_{d-1}(\widehat{\kappa}) = \mathcal{C}_1(\widehat{\kappa})$ and $P_1(C_i) = Q_1(C_i)$ on one-dimensional subcells of $\widehat{\kappa}$. However, in three-dimensions $\mathcal{C}_{d-1}(\widehat{\kappa}) = \mathcal{C}_2(\widehat{\kappa})$ and so, $P_1(C_i)$ and $Q_1(C_i)$ are different spaces. As a result, unlike $\mathbf{C}^1(\widehat{\kappa})$, the elements $\mathbf{D}^1(\widehat{\kappa})$ do not share the same unisolvent set on 3-simplices and 3-cubes.[14]

The approximation properties of the two types of div-conforming spaces are given in the following theorem.[15]

Theorem B.9 *Let $\mathcal{T}_h$ be a uniformly regular partition of Ω into affine finite elements and let $\mathbf{D}^{(r)}(\Omega)$ denote one of the spaces $\mathbf{D}^r(\Omega)$ or $\mathbf{D}^{r}\text{-}(\Omega)$, $r \geq 1$. Then, there exists a positive constant C, independent of h, and such that, for every vector field $\mathbf{u} \in \mathbf{D}(\Omega) \cap [H^{r+1}(\Omega)]^d$,*

$$\left.\begin{aligned} &\|\mathbf{u} - \Pi_D \mathbf{u}\|_0 \\ &\|\mathbf{u} - \pi_D \mathbf{u}\|_0 \end{aligned}\right\} \leq C \begin{cases} h^r \|\mathbf{u}\|_r & \text{if } \mathbf{D}^{(r)}(\Omega) = \mathbf{D}^{r}\text{-}(\Omega) \\ h^{r+1} \|\mathbf{u}\|_{r+1} & \text{if } \mathbf{D}^{(r)}(\Omega) = \mathbf{D}^r(\Omega) \end{cases} \tag{B.37}$$

$$\|\nabla \cdot (\mathbf{u} - \Pi_D \mathbf{u})\|_0 \leq C h^r \|\nabla \cdot \mathbf{u}\|_r\,. \qquad \square$$

[14] The Brezzi–Douglas–Marini (BDM) [84, 86] div-conforming element on d-simplices has the same reference space as that in (B.33), but uses a different unisolvent set. In the lowest-order case, this set coincides with the unisolvent set (B.35) of the Nédélec elements of the second kind, i.e., the lowest-order versions of these elements are the same. However, on d-cubes, the BDM elements differ significantly from the Nédélec elements of the second kind (B.34). For example, on 2-cubes, the reference BDM space is given by (see [87, p.120])

$$BDM^r = (P_r)^2 + a\nabla \times (\widehat{x}_1{}^{r+1}\widehat{x}_2) + b\nabla \times (\widehat{x}_1\widehat{x}_2{}^{r+1})\,, \qquad a,b \in \mathbb{R}\,.$$

[15] The differences between Raviart–Thomas and Nédélec div-conforming elements are very similar to the differences between the two families of curl-conforming elements. For instance, in the lowest-order case in $\mathbb{R}^3$, a simplicial Raviart–Thomas element approximates the normal component $\widehat{\mathbf{u}} \cdot \mathbf{n}$ of $\widehat{\mathbf{u}}$ by a constant on each face whereas Nédélec's div-conforming element approximates $\widehat{\mathbf{u}} \cdot \mathbf{n}$ by a linear or bilinear polynomial. Not surprisingly, the error estimates for the two types of div-conforming elements also resemble the estimates in Theorem B.7.

For details, see [87,284]. From Theorem B.9, it is clear that the Raviart–Thomas elements are "optimized" with respect to the divergence operator; in this respect they resemble the Nédélec curl-conforming elements of the first kind.

Remark B.10 Just as in the curl-conforming case, Theorem B.9 does not extend to non-affine elements. A careful study of the two-dimensional case in [9] reveals that, on non-affine 2-cubes (quadrilaterals), the Raviart–Thomas element $\mathbf{D}^{r}\text{-}(\Omega)$ retains its $L^2(\Omega)$ accuracy, but the order of the divergence approximation is reduced:

$$\begin{aligned} \|\mathbf{u} - \Pi_D\mathbf{u}\|_0 &\leq Ch^r\|\mathbf{u}\|_r \\ \|\nabla\cdot(\mathbf{u} - \Pi_D\mathbf{u})\|_0 &\leq Ch^{r-1}\|\nabla\cdot\mathbf{u}\|_r\,. \end{aligned} \tag{B.38}$$

Furthermore, it is shown in [9] that for other families of div-conforming elements in $\mathbb{R}^2$, such as the Brezzi–Douglas–Marini [84, 86] and Brezzi–Douglas–Fortin–Marini [85] elements, *both* the $L^2(\Omega)$ error and the divergence error have reduced orders of convergence. The situation in three-dimensions is even worse because the loss of accuracy extends to the $L^2(\Omega)$ error as well [279]. These observations are particularly troubling for methods that employ low-order elements because convergence could be completely lost. □

The volume element space $S^h(\Omega)$

Any piecewise smooth function space is trivially a subspace of $L^2(\Omega)$. As a result, finite element approximations of $S(\Omega)$ are not subject to any interelement continuity conditions, i.e., the global space $S^h(\Omega)$ can be assembled from the local element spaces by simply taking their union. Because degrees of freedom are no longer required to be single-valued on shared interfaces, we have much greater flexibility in the choice of the reference polynomial space and its degrees of freedom. In particular, we can set

$$S^r(\widehat{\kappa}) = P_r \qquad \text{with} \qquad r \geq 0 \tag{B.39}$$

regardless of the shape of the reference cell. In the lowest-order case, the set $\widehat{\Lambda}^0(\widehat{\kappa})$ consists of a single functional that is the cell average of the scalar function $\widehat{u}$:

$$l(\widehat{u}) = \int_{\widehat{\kappa}} \widehat{u}\, d\Omega\,. \tag{B.40}$$

For this reason, approximations of $S(\Omega)$ are sometimes called *volume* elements. Another terminology often used is *discontinuous* elements. We use both terms because there is no risk for confusion with the rest of the elements considered so far.

The $L^2(\Omega)$ projection operator $\Pi_S = \pi_S : S(\Omega) \mapsto S^r(\Omega)$ is a natural choice of projection for volume elements. The operator π_S is bounded and has the following approximation property; see [156, Proposition 1.135, p.73].

Theorem B.11 *There exists a positive constant C, independent of h, such that for all* $0 \leq s \leq r+1$ *and* $u \in H^s(\Omega)$,

$$\|u - \Pi_S u\|_0 \le C h^s |u|_s \, . \qquad \square \tag{B.41}$$

Remark B.12 Given a reference volume element, the local element space is defined by pullback; see (B.10). When the element is non-affine, this results in non-polynomial spaces. Thus, for non-affine d-cubes one often forgoes the reference space and defines the local spaces directly:

$$S^r(\kappa) = P_r \qquad \text{with} \qquad r \ge 0 \, . \tag{B.42}$$

This has the advantage of producing a piecewise polynomial global space $S^r(\Omega)$. $\square$

B.2.2 Approximation of $\mathbf{C}(\Omega) \cap \mathbf{D}(\Omega)$

Finite element subspaces of $\mathbf{C}_0(\Omega) \cap \mathbf{D}(\Omega)$ or $\mathbf{C}(\Omega) \cap \mathbf{D}_0(\Omega)$ are needed for the approximate solution of variational equations stated in terms of the bilinear form

$$Q_{DC}(\mathbf{u}, \mathbf{v}) = \int_\Omega (\nabla \cdot \mathbf{u})(\nabla \cdot \mathbf{v}) + (\nabla \times \mathbf{u}) \cdot (\nabla \times \mathbf{v}) \, d\Omega \, . \tag{B.43}$$

If the assumptions of Theorem A.8 hold, then both the spaces $\mathbf{C}_0(\Omega) \cap \mathbf{D}(\Omega)$ and $\mathbf{C}(\Omega) \cap \mathbf{D}_0(\Omega)$ are continuously embedded in the Sobolev space $[H^1(\Omega)]^d$ so that they can be approximated by the *nodal vector* finite element space $\mathbf{G}^h(\Omega) = [G^h(\Omega)]^d$.

However, for settings in which Theorem A.8 is not applicable, the result in Theorem A.9 has some far-reaching consequences for the approximation of $\mathbf{C}_0(\Omega) \cap \mathbf{D}(\Omega)$ or $\mathbf{C}(\Omega) \cap \mathbf{D}_0(\Omega)$ and the finite element solution of problems involving the bilinear form (B.43). In particular, it implies loss of approximability for nodal vector finite element spaces; see e.g., [156, Corollary 3.20, p. 97].

Theorem B.13 *Let* $\{\mathbf{G}^h(\Omega)\}_{h>0}$ *be a family of finite element spaces conforming in* $[H^1(\Omega)]^3$ *and let* $\mathbf{G}^h_n(\Omega) = \{\mathbf{u}^h \in \mathbf{G}^h(\Omega) \mid \mathbf{u}^h \times \mathbf{n}|_{\partial\Omega} = 0\}$. *Then, under the hypotheses of Theorem* A.9, $\{\mathbf{G}^h_n(\Omega)\}_{h>0}$ *cannot have the approximability property* (B.8) *in* $\mathbf{C}_0(\Omega) \cap \mathbf{D}(\Omega)$. $\square$

This result implies that *nodal finite element spaces cannot be used to approximate* $\mathbf{C}_0(\Omega) \cap \mathbf{D}(\Omega)$. In particular, unless the exact solution of a problem stated in terms of (B.43) is $H^1(\Omega)$-regular, discretization of this weak equation by $\mathbf{G}^h_n(\Omega)$ fails to produce convergent approximations. Analogous observations holds for the approximation of $\mathbf{C}(\Omega) \cap \mathbf{D}_0(\Omega)$. The situation is further complicated by the following result; see e.g., [13, p. 25].

Proposition B.14 *Assume that* $X^h(\Omega)$ *is a piecewise smooth subspace of* $\mathbf{C}_0(\Omega) \cap \mathbf{D}(\Omega)$, *defined with respect to a conforming finite element partition* $\mathcal{T}_h$ *of* Ω. *Then,* $X^h(\Omega)$ *is a subspace of* $[H^1(\Omega)]^d$.

Proof. A piecewise smooth vector field in $\mathbf{C}(\Omega)$ is necessarily tangentially continuous on element faces; see Remark B.4 and Theorem B.3. Likewise, the normal component of a piecewise smooth vector field in $\mathbf{D}(\Omega)$ is necessarily continuous across element faces. It follows that a piecewise smooth vector field that is simultaneously in $\mathbf{C}(\Omega)$ and $\mathbf{D}(\Omega)$ is necessarily continuous on element faces, i.e., it belongs to $[H^1(\Omega)]^d$. □

Thus, Theorem B.13 tells us that, in general, we *cannot approximate* problems posed in $\mathbf{C}_0(\Omega)\cap\mathbf{D}(\Omega)$ by nodal elements, whereas, on the other hand, Proposition B.14 tells us that nodal elements are *the only conforming* finite element subspaces of $\mathbf{C}_0(\Omega)\cap\mathbf{D}(\Omega)$. This means that for insufficiently regular solutions, the weak form (B.43) cannot be used directly in a finite element method. Again, analogous observations holds for the approximation of $\mathbf{C}(\Omega)\cap\mathbf{D}_0(\Omega)$.

Remark B.15 For smooth vector fields the differential operator

$$-\Delta_{DC}\mathbf{u} = -\nabla\nabla\cdot\mathbf{u}+\nabla\times\nabla\times\mathbf{u} \tag{B.44}$$

coincides with the vector Laplacian

$$-\vec{\Delta}\mathbf{u} = (-\Delta u_1, -\Delta u_2, -\Delta u_3)^\mathsf{T}\,.$$

This fact is often misconstrued as implying the identity of the weak forms of these differential operators, i.e., that

$$Q_{DC}(\mathbf{u},\mathbf{v}) = \int_\Omega (\nabla\cdot\mathbf{u})(\nabla\cdot\mathbf{v}) + (\nabla\times\mathbf{u})\cdot(\nabla\times\mathbf{v})\,d\Omega = \int_\Omega \nabla\mathbf{u} : \nabla\mathbf{v}\,d\Omega\,.$$

However, even for smooth vector fields in $\mathbf{C}(\Omega)\cap\mathbf{D}_0(\Omega)$ or $\mathbf{C}_0(\Omega)\cap\mathbf{D}(\Omega)$ this is true only in some very special cases, e.g., when Ω is a rectangle [165], and fails to hold for general domains. A counterexample for $\mathbf{C}(\Omega)\cap\mathbf{D}_0(\Omega)$ is as follows (see [115]). Take Ω to be the unit ball in $\mathbb{R}^3$ with center at the origin and $\mathbf{u} = (y-z, z-x, x-y)^\mathsf{T}$. Then $\mathbf{u}\cdot\mathbf{n} = 0$ on $\partial\Omega$,

$$\nabla\times\mathbf{u} = -2(1,1,1)^\mathsf{T} \qquad \text{and} \qquad \nabla\cdot\mathbf{u} = 0\,.$$

As a result, for this smooth vector field

$$Q_{DC}(\mathbf{u},\mathbf{u}) = \frac{4}{3}\pi\times 12 \qquad \text{but} \qquad \int_\Omega \nabla\mathbf{u} : \nabla\mathbf{u}\,d\Omega = \frac{4}{3}\pi\times 6\,.$$

A counterexample for $\mathbf{C}_0(\Omega)\cap\mathbf{D}(\Omega)$ can be devised in a similar fashion. Take the same domain Ω and $\mathbf{u} = (x,y,z)^\mathsf{T}$. Then $\mathbf{u}\times\mathbf{n} = 0$ on $\partial\Omega$,

$$\nabla\times\mathbf{u} = (0,0,0)^\mathsf{T} \qquad \text{and} \qquad \nabla\cdot\mathbf{u} = 3\,.$$

In this case

$$Q_{DC}(\mathbf{u},\mathbf{u}) = \frac{4}{3}\pi \times 9 \qquad \text{but} \qquad \int_\Omega \nabla\mathbf{u} : \nabla\mathbf{u}\, d\Omega = \frac{4}{3}\pi \times 3 .$$

Furthermore, the inner products corresponding to $Q_{DC}(\cdot,\cdot)$ and the weak form of the vector Laplacian are equivalent only for the special settings defined in Theorem A.8. □

Remark B.16 The dilemma posed by Theorem B.13 and Proposition B.14 is the reason why the bilinear form (B.43) should not be used in finite element methods for $-\Delta_{DC}$, even though formally it provides an attractive Rayleigh–Ritz setting. Mixed-Galerkin methods for (B.44) are one way to avoid approximation of $\mathbf{C}_0(\Omega) \cap \mathbf{D}(\Omega)$ and $\mathbf{C}(\Omega) \cap \mathbf{D}_0(\Omega)$; see [13, p. 78]. These methods give up conformity with respect to one of the component spaces and treat $\mathbf{u}$ as an element of the other space. □

B.2.3 Exact Sequences of Finite Element Spaces

To obtain a finite element approximation of the De Rham complex in $\mathbb{R}^3$, we need to select G^h, $\mathbf{C}^h$, $\mathbf{D}^h$, and S^h so that the diagram in (B.1) commutes.[16] Because we are dealing with piecewise polynomial spaces, it is clear that the inclusions in (B.14), necessary for exactness, do not hold unless we carefully match the degrees and the definitions of the finite element spaces.

It turns out that this process is most straightforward for affine finite element spaces on simplicial grids[17] $\mathcal{T}_h$, partly because the global spaces are piecewise polynomials with well-understood properties. In [13, section 3.5], it is shown that for any fixed integer $r \geq 1$, there are at most four different exact sequences of affine finite element spaces:

$$\begin{array}{l}
\mathbb{R} \hookrightarrow G^r(\Omega) \xrightarrow{\nabla} \mathbf{C}^{r-1}(\Omega) \xrightarrow{\nabla\times} \mathbf{D}^{r-2}(\Omega) \xrightarrow{\nabla\cdot} S^{r-3}(\Omega) \longrightarrow 0 \\
\mathbb{R} \hookrightarrow G^r(\Omega) \xrightarrow{\nabla} \mathbf{C}^{r-1}(\Omega) \xrightarrow{\nabla\times} \mathbf{D}^{(r-1)-}(\Omega) \xrightarrow{\nabla\cdot} S^{r-2}(\Omega) \longrightarrow 0 \\
\mathbb{R} \hookrightarrow G^r(\Omega) \xrightarrow{\nabla} \mathbf{C}^{r-}(\Omega) \xrightarrow{\nabla\times} \mathbf{D}^{r-1}(\Omega) \xrightarrow{\nabla\cdot} S^{r-2}(\Omega) \longrightarrow 0 \\
\mathbb{R} \hookrightarrow G^r(\Omega) \xrightarrow{\nabla} \mathbf{C}^{r-}(\Omega) \xrightarrow{\nabla\times} \mathbf{D}^{r-}(\Omega) \xrightarrow{\nabla\cdot} S^{r-1}(\Omega) \longrightarrow 0 .
\end{array} \tag{B.45}$$

The finite element sequences in (B.45), together with suitably defined bounded projection operators, satisfy the commuting diagram in (B.1). If $r \geq 3$, all four sequences in (B.45) are distinct; for $1 \leq r < 3$, some of them are empty. In particular, for $r = 1$, (B.45) reduces to the single exact sequence

$$\mathbb{R} \hookrightarrow G^1(\Omega) \xrightarrow{\nabla} \mathbf{C}^{1-}(\Omega) \xrightarrow{\nabla\times} \mathbf{D}^{1-}(\Omega) \xrightarrow{\nabla\cdot} S^0(\Omega) \longrightarrow 0 \tag{B.46}$$

---

[16] In two-dimensions, the De Rham complex splits into two "truncated" exact sequences corresponding to the two versions (A.2)–(A.3) of the curl operator; see [69, 135, 137] for further details.

[17] For more information about exact sequences on d-cubes, see [3, 69, 137].

consisting of the lowest-order nodal element, the lowest order Nédélec element of the first-kind, the lowest-order Raviart–Thomas element, and the piecewise constant element.[18]

In many ways, simplicial elements provide the best setting for a discrete De Rham complex. Although the individual spaces in the complex can be defined for most standard element shapes, putting them together into an exact sequence is not as straightforward as for simplicial grids, even when the elements are affine. Switching to non-affine, non-simplicial elements further complicates matters.

To explain the difficulties with non-simplicial elements, it suffices to examine them on a mesh $\mathcal{T}_h$ consisting of a single 3-cube κ. The analogue of the third sequence in (B.45) is

$$\mathbb{R} \hookrightarrow G^r(\kappa) \xrightarrow{\nabla} \mathbf{C}^{r\text{-}}(\kappa) \xrightarrow{\nabla\times} \mathbf{D}^{r-1}(\kappa) \xrightarrow{\nabla\cdot} S^{r-2}(\kappa) \longrightarrow 0\,.$$

Assume first that κ is the reference 3-cube $[-1,1]^3$. Then, $\mathbf{D}^{r-1}(\kappa) = [Q_{r-1}]^3$, $S^{r-2}(\kappa) = Q_{r-2}$, and it is easy to see[19] that the divergence operator maps $\mathbf{D}^{r-1}(\kappa)$ onto $S^{r-2}(\kappa)$. But this combination of spaces makes it impossible to have the commuting diagram property

$$\Pi_S(\nabla\cdot\mathbf{u}) = \nabla\cdot(\Pi_D\mathbf{u}) \tag{B.47}$$

because the left-hand side belongs to Q_{r-2} whereas the divergence of a vector field belonging to $[Q_{r-1}]^3$ on the right-hand side is not necessarily in that space. Instead of the "strong" commuting diagram (B.47), one instead has the "weak" property

$$\int_\kappa p^h\,\nabla\cdot\mathbf{u}\,d\Omega = \int_\kappa p^h\,\nabla\cdot\Pi_D\mathbf{u}\,d\Omega \qquad \forall p^h\in Q_{r-2}; \tag{B.48}$$

see [284, Remark 2].

Furthermore, for non-affine elements, in addition to the strong commuting diagram property, one may also have to give up the surjective property of the divergence operator. To explain why this may be necessary, consider an analogue of the Whitney complex (B.46) on a single non-affine 3-cube κ. The last two spaces in this complex are the lowest-order Raviart–Thomas space

$$\mathbf{D}^{1\text{-}}(\kappa) = \{\mathbf{u}^h\in\mathbf{D}(\kappa)\mid \mathbf{u}^h = (J_\kappa^{-1}DF_\kappa\cdot\widehat{\mathbf{u}}^h)\circ F_\kappa^{-1},\ \widehat{\mathbf{u}}^h\in P_{1,0,0}\times P_{0,1,0}\times P_{0,0,1}\}$$

---

[18] A remarkable fact about (B.46) is that this sequence was conceived in [345] as a way to approximate the De Rham complex, many years before its importance for finite element methods was recognized. The spaces in (B.46) are called *Whitney forms* or *Whitney elements*; here we use the latter term.

Although Whitney elements were rediscovered one by one in the finite element literature, the fact that they can be linked together to form an exact sequence was not appreciated until the late 1980s when, in [72–74], connections were established between the stability of mixed methods and the exact sequence property. Eventually, these papers prompted the use of homological ideas in finite element methods; see, e.g., [7, 8, 13, 58, 69, 121, 122, 137, 204, 206, 276, 332], and the references cited therein.

[19] It suffices to show this for a monomial $p^h = x_1^{i_1}x_2^{i_2}x_3^{i_3}$ where $0\le i_j\le r-2$. Define the vector field $\mathbf{u}^h = (p^h/3)\big(x_1/(i_1+1), x_2/(i_2+1), x_3/(i_3+1)\big)^{\mathrm{T}}$. Then $\mathbf{u}^h\in[Q_{r-1}]^3$ and $\nabla\cdot\mathbf{u}^h = p^h$.

(see (B.31)) and the pullback of the constant space P_0

$$S^0(\kappa) = \{u^h \in S(\kappa) \mid u^h = (J_\kappa^{-1}\widehat{u}^h) \circ F_\kappa^{-1};\ \widehat{u}^h \in P_0\}$$

(see (B.10) and (B.12)). It is not difficult to show that $\nabla\cdot : \mathbf{D}^1\text{-}(\kappa) \mapsto S^0(\kappa)$ is surjective; see [69]. The problem is that, as defined, $S^0(\kappa)$ cannot reproduce constant functions because J_κ is not a constant. As a result, projection onto $S^0(\kappa)$ is not even first-order accurate. Coupled with the fact that div-conforming spaces already experience difficulties for non-affine cubes (see Remark B.10), this prompts the use of the direct definition (see Remark B.12) $S^0(\kappa) = P_0$.

B.3 Properties of Compatible Finite Element Spaces

Throughout this section, we assume that Ω is a contractible region in $\mathbb{R}^3$, $\mathcal{T}_h$ is uniformly regular partition of Ω into affine simplicial elements, and $G^h(\Omega)$, $\mathbf{C}^h(\Omega)$, $\mathbf{D}^h(\Omega)$, and $S^h(\Omega)$ form a discrete De Rham complex, i.e.,

- $\{G^h(\Omega), \mathbf{C}^h(\Omega), \mathbf{D}^h(\Omega), S^h(\Omega)\}$ is an exact sequence
- there exist bounded projection operators Π_G, Π_C, Π_D, and Π_S such that (B.1) commutes.

Many of the results in this section can be extended to more general mesh partitions and other discretization settings. We refer to [8, 13, 75, 87, 135–137, 139, 204, 223, 224, 226, 286, 289, 314, 317] for further details and related results in finite difference and finite volume methods.

To simplify notation, the polynomial order is deleted from space designations. For $\beta \in \{\emptyset, \gamma\}$, where $\gamma \subset \partial\Omega$, the symbols $N_\beta^h(\nabla) \subset G_\beta^h(\Omega)$, $N_\beta^h(\nabla\times) \subset \mathbf{C}_\beta^h(\Omega)$, and $N_\beta^h(\nabla\cdot) \subset \mathbf{D}_\beta^h(\Omega)$ denote the null-spaces of the gradient, curl, and divergence operators, respectively; $R_\beta^h(\nabla) \subset \mathbf{C}_\beta^h(\Omega)$, $R_\beta^h(\nabla\times) \subset \mathbf{D}_\beta^h(\Omega)$, and $R_\beta^h(\nabla\cdot) = S_\beta^h(\Omega)$ are their respective ranges.

B.3.1 Discrete Operators

The standard gradient, curl, and divergence operators provide one set of vector calculus operators on a finite element De Rham complex. These operators are well-defined by virtue of the fact that the discrete complex is a conforming approximation of the continuum one. This section introduces a second set of vector calculus operators that approximates the weak operators (A.44)–(A.46) and, together with the first set, gives rise to discrete versions of all basic second-order elliptic operators.

Remark B.17 It is important to keep in mind that the valuable properties of these operators can only be realized when they are defined using a discrete De Rham complex, i.e., when the domain and the range of each operator belong to one of the sequences in (B.45). □

Discrete weak operators

Compatible discretizations[20] of ∇^*, $\nabla^*\times$, and $\nabla^*\cdot$ are defined by restricting the definitions in (A.44)–(A.46) to the compatible finite element spaces in the discrete complex $\{G^h(\Omega),\, \mathbf{C}^h(\Omega), \mathbf{D}^h(\Omega), S^h(\Omega)\}$. Thus, we have the discrete weak gradient operator $\nabla_h^* : S_\beta^h(\Omega) \mapsto \mathbf{D}_\beta^h(\Omega)$ defined by

$$\left(\nabla_h^* u^h, \mathbf{v}^h\right)_{\Theta_2} = \left(u^h, -\nabla\cdot\mathbf{v}^h\right)_{\Theta_3} + \int_{\Gamma_\beta} (\Theta_3 u^h)\mathbf{v}^h \cdot \mathbf{n}\, d\Gamma \qquad \forall \mathbf{v}^h \in \mathbf{D}_\beta^h(\Omega), \quad \text{(B.49)}$$

the discrete weak curl operator $\nabla_h^*\times : \mathbf{D}_\beta^h(\Omega) \mapsto \mathbf{C}_\beta^h(\Omega)$ defined by

$$\left(\nabla_h^* \times \mathbf{v}^h, \mathbf{u}^h\right)_{\Theta_1} = \left(\mathbf{v}^h, \nabla\times\mathbf{u}^h\right)_{\Theta_2} - \int_{\Gamma_\beta} (\mathbf{n}\times\Theta_2\mathbf{v}^h)\mathbf{u}^h\, d\Gamma \quad \forall \mathbf{u}^h \in \mathbf{C}_\beta^h(\Omega), \quad \text{(B.50)}$$

and the discrete weak divergence operator $\nabla_h^*\cdot : \mathbf{C}_\beta^h(\Omega) \mapsto G_\beta^h(\Omega)$ defined by

$$\left(\nabla_h^* \cdot \mathbf{u}^h, q^h\right)_{\Theta_0} = \left(\mathbf{u}^h, -\nabla q^h\right)_{\Theta_1} + \int_{\Gamma_\beta} (\mathbf{n}\cdot\Theta_1\mathbf{u}^h) q\, d\Gamma \qquad \forall q^h \in G_\beta^h(\Omega). \quad \text{(B.51)}$$

The null-spaces of these operators are denoted by $N^h(\nabla_h^*) \subset S^h(\Omega)$, $N^h(\nabla_h^*\times) \subset \mathbf{D}^h(\Omega)$, and $N^h(\nabla_h^*\cdot) \subset \mathbf{C}^h(\Omega)$, respectively. Using (B.51) and (B.50), we have the following analogues of (A.58) and (A.59):

$$N^h(\nabla\times)^\perp = N^h(\nabla_h^*\cdot) \qquad \text{and} \qquad N^h(\nabla\cdot)^\perp = N^h(\nabla_h^*\times). \quad \text{(B.52)}$$

The discrete weak operators defined in (B.49)–(B.51) give rise to another exact sequence that is an analogue of (A.56):

$$0 \longleftarrow G_\beta^h(\Omega) \stackrel{\nabla_h^*\cdot}{\longleftarrow} \mathbf{C}_\beta^h(\Omega) \stackrel{\nabla_h^*\times}{\longleftarrow} \mathbf{D}_\beta^h(\Omega) \stackrel{\nabla_h^*}{\longleftarrow} S_\beta^h(\Omega) \hookleftarrow \mathbb{R}. \quad \text{(B.53)}$$

Proof that (B.53) is exact follows along the same lines as for (A.56) in Section A.2.2. For brevity, we only show that the weak discrete divergence is bounded surjection $\mathbf{C}_0^h(\Omega) \mapsto G_0^h(\Omega)$.

Proposition B.18 *For every $p^h \in G_0^h(\Omega)$, there exists $\mathbf{u}^h \in \mathbf{C}_0^h(\Omega)$ such that*

$$p^h = \nabla_h^* \cdot \mathbf{u}^h \qquad \textit{and} \qquad \left(\|\mathbf{u}^h\|_0^2 + \|\nabla_h^*\cdot\mathbf{u}^h\|_0^2\right)^{1/2} \le C\|p^h\|_0. \quad \text{(B.54)}$$

[20] In mimetic finite difference methods [314] and support-operator methods [310, 311], similar constructions are used to define the "adjoint" gradient, curl, and divergence operators.

Proof. We show that, given $p^h \in G_0^h(\Omega)$, there exists $\mathbf{u}_p^h = -\nabla \widetilde{p}^h$ such that $\nabla_h^* \cdot \mathbf{u}_p^h = p^h$. From (B.51), it follows that $\mathbf{u}_p^h$ solves the equation

$$(\mathbf{u}_p^h, -\nabla q^h) = (p^h, q^h) \qquad \forall q^h \in G_0^h(\Omega)$$

or, equivalently, $\widetilde{p}^h$ solves the equation

$$(\nabla \widetilde{p}^h, \nabla q^h) = (p^h, q^h) \qquad \forall q^h \in G_0^h(\Omega).$$

Clearly, this equation has a unique solution which proves that $\nabla_h^* \cdot$ is surjection. To prove the second part of the proposition, observe that $\mathbf{u}_p^h \in N_0^h(\nabla\times)$ so that (B.78) implies that

$$\|\mathbf{u}_p^h\|_0 \le C \|\nabla_h^* \cdot \mathbf{u}_p^h\|_0 .$$

This completes the proof. □

It is easy to see that the domains of ∇_h^*, $\nabla_h^* \times$, and $\nabla_h^* \cdot$ in (B.49)–(B.51) can actually be extended to $S(\Omega)$, $\mathbf{D}(\Omega)$, and $\mathbf{C}(\Omega)$, respectively. When the operands are sufficiently regular, it turns out that the action of the weak discrete operators can be described in terms of the $L^2(\Omega)$ projections (B.16). We state this result for the setting that is used in the book.

Theorem B.19 *Assume that ∇_h^*, $\nabla_h^* \times$, and $\nabla_h^* \cdot$ are constructed using a finite element complex $\{G_0^h(\Omega), \mathbf{C}_0^h(\Omega), \mathbf{D}_0^h(\Omega), S_0^h(\Omega)\}$ defined on a uniformly regular partition $\mathcal{T}_h$ of Ω into affine simplices. Then,*

$$\begin{aligned}
\nabla_h^* u &= \pi_D(\nabla u) && \forall u \in S(\Omega, \Theta_3) \cap G(\Omega, \Theta_3^{-1}) \\
\nabla_h^* \times \mathbf{u} &= \pi_C(\nabla \times \mathbf{u}) && \forall \mathbf{u} \in \mathbf{D}(\Omega, \Theta_2) \cap \mathbf{C}(\Omega, \Theta_2^{-1}) \\
\nabla_h^* \cdot \mathbf{u} &= \pi_G(\nabla \cdot \mathbf{u}) && \forall \mathbf{u} \in \mathbf{C}(\Omega, \Theta_1) \cap \mathbf{D}(\Omega, \Theta_1^{-1}) .
\end{aligned} \tag{B.55}$$

Proof. To prove the first statement, let $u \in S(\Omega, \Theta_3) \cap G(\Omega, \Theta_3^{-1})$. Then, (B.49) can be restated as

$$\left(\nabla_h^* u, \mathbf{v}^h\right)_{\Theta_2} = \left(u, -\nabla \cdot \mathbf{v}^h\right)_{\Theta_3} = \left(\nabla u, \mathbf{v}^h\right)_{\Theta_2} \qquad \forall \mathbf{v}^h \in \mathbf{D}_0^h(\Omega).$$

The right hand side corresponds to the definition of π_D in (B.16). The rest of the results of the theorem follow along the same lines. □

Finally, using the approximation properties of $L^2(\Omega)$ projections, we have the following result.

Corollary B.20 *Assume the hypotheses of Theorem* B.19 *hold. Then,*

$$\|\nabla u - \nabla_h^* u\|_0 \le C \begin{cases} h^r \|\nabla u\|_r & \text{if } \mathbf{D}^{(r)}(\Omega) = \mathbf{D}^{r}\text{-}(\Omega) \\ h^{r+1} \|\nabla u\|_{r+1} & \text{if } \mathbf{D}^{(r)}(\Omega) = \mathbf{D}^{r}(\Omega), \end{cases} \tag{B.56}$$

$$\|\nabla\times\mathbf{u}-\nabla_h^*\times\mathbf{u}\|_0\le C\begin{cases} h^r\|\nabla\times\mathbf{u}\|_r & \text{if } \mathbf{C}^{(r)}(\Omega)=\mathbf{C}^{r}\text{-}(\Omega)\\ h^{r+1}\|\nabla\times\mathbf{u}\|_{r+1} & \text{if } \mathbf{C}^{(r)}(\Omega)=\mathbf{C}^{r}(\Omega)\,,\end{cases}\tag{B.57}$$

and

$$\|\nabla\cdot\mathbf{u}-\nabla_h^*\cdot\mathbf{u}\|_0\le Ch^{r+1}\|\nabla\cdot\mathbf{u}\|_{r+1}\tag{B.58}$$

for all sufficiently smooth u and $\mathbf{u}$. □

Discrete second-order operators

The ranges of the standard gradient, curl, and divergence operators, restricted to a finite element De Rham complex, are respectively contained in the domains of the weak divergence, curl, and gradient operators defined in (B.49)–(B.51). Therefore, these operators can be paired together to produce discrete versions of the basic second-order elliptic operators. Because there are two versions of each first-order operator, depending on the order in which they are combined, we obtain two versions of each second-order operator.

In particular, we have a discrete Laplace operator acting on nodal finite element functions:

$$\Delta_G^h=\nabla_h^*\cdot\nabla\ :G_0^h\mapsto G_0^h\tag{B.59}$$

and a second discrete Laplace operator acting on volume elements:

$$\Delta_S^h=\nabla\cdot\nabla_h^*\ :S_0^h\mapsto S_0^h.\tag{B.60}$$

From (B.51), it follows that

$$\left(-\nabla_h^*\cdot\nabla u^h,q^h\right)_{\Theta_0}=\left(\nabla u^h,\nabla q^h\right)_{\Theta_1}\qquad\forall q^h\in G_0^h(\Omega)\tag{B.61}$$

so that Δ_G^h is the same operator as in the standard Galerkin method for the Poisson equation (1.52). The discrete Laplace operator in (B.60) is, in a sense, dual to Δ_G^h because it acts on discontinuous scalar fields. This operator can be related to mixed methods for the Poisson equation based on the Kelvin principle (1.54).

We also have two versions of the curl–curl operator, one for curl-conforming fields:

$$\nabla_h^*\times\nabla\times:\mathbf{C}_0^h(\Omega)\mapsto\mathbf{C}_0^h(\Omega)\tag{B.62}$$

and the other for div-conforming fields:

$$\nabla\times\nabla_h^*\times\ :\mathbf{D}_0^h(\Omega)\mapsto\mathbf{D}_0^h(\Omega)\tag{B.63}$$

and two versions of the grad–div operator, one for div-conforming fields:

$$\nabla_h^*\nabla\cdot\ :\mathbf{D}_0^h(\Omega)\mapsto\mathbf{D}_0^h(\Omega)\tag{B.64}$$

and the other for curl-conforming fields:

$$\nabla \nabla_h^* \cdot \; : \mathbf{C}_0^h(\Omega) \mapsto \mathbf{C}_0^h(\Omega) . \tag{B.65}$$

The operators in (B.62)–(B.65) can be combined to produce two different versions of the vector Laplace operator, one for curl-conforming functions:

$$\vec{\Delta}_C^h = \nabla_h^* \times \nabla \times - \nabla \nabla_h^* \cdot \; : \mathbf{C}_0^h(\Omega) \mapsto \mathbf{C}_0^h(\Omega) \tag{B.66}$$

and the other for div-conforming functions:

$$\vec{\Delta}_D^h = \nabla \times \nabla_h^* \times - \nabla_h^* \nabla \cdot \; : \mathbf{D}_0^h(\Omega) \mapsto \mathbf{D}_0^h(\Omega) . \tag{B.67}$$

Matrix forms of discrete operators

For any grid, standard and weak discrete operators acting on the lowest-order finite element De Rham complex admit particularly simple and elegant expressions in terms of the *mesh incidence matrices* and the mass matrices of the compatible finite element spaces [66].

The incidence matrix D_k between the k- and $k+1$-dimensional subcells of the mesh $\mathcal{T}_h$ is a rectangular matrix containing only the numbers -1, 0, and 1. The number of rows and columns in D_k equals $\dim C_{k+1}(\mathcal{T}_h)$ and $\dim C_k(\mathcal{T}_h)$, respectively. The actual entries in D_k depend on the orientations of the k- and $k+1$-subcells in the mesh. The action of the standard differential operators on compatible finite element functions of the lowest orders is given by (see [66]) the action of the incidence matrix on their coefficients, i.e.,

$$(\nabla) = \mathsf{D}_0 \qquad (\nabla \times) = \mathsf{D}_1 \qquad (\nabla \cdot) = \mathsf{D}_2 . \tag{B.68}$$

Furthermore, let M_G, M_C, M_D, and M_S denote the matrices giving the $L^2(\Omega)$ inner products on G^h, $\mathbf{C}^h$, $\mathbf{D}^h$, and S^h, respectively, defined in the usual manner. Then, the action of the weak discrete operators on finite element functions is equivalent to the action of the following matrices on their coefficient vectors (see [66]):

$$(\nabla_h^* \cdot) = \mathsf{M}_G^{-1} \mathsf{D}_0^\mathsf{T} \mathsf{M}_C \qquad (\nabla_h^* \times) = \mathsf{M}_C^{-1} \mathsf{D}_1^\mathsf{T} \mathsf{M}_D \qquad (\nabla_h^*) = \mathsf{M}_D^{-1} \mathsf{D}_2^\mathsf{T} \mathsf{M}_S . \tag{B.69}$$

Finally, using the identities in (B.68)–(B.69), we can express discrete second-order operators in terms of the same basic matrices, i.e.,

$$(\Delta_G^h) = \mathsf{M}_G^{-1} \mathsf{D}_0^\mathsf{T} \mathsf{M}_C \mathsf{D}_0 \tag{B.70}$$

$$(\vec{\Delta}_C^h) = \mathsf{M}_C^{-1} \mathsf{D}_1^\mathsf{T} \mathsf{M}_D \mathsf{D}_1 + \mathsf{D}_0 \mathsf{M}_G^{-1} \mathsf{D}_0^\mathsf{T} \mathsf{M}_C \tag{B.71}$$

$$(\vec{\Delta}_D^h) = \mathsf{D}_1 \mathsf{M}_C^{-1} \mathsf{D}_1^\mathsf{T} \mathsf{M}_D + \mathsf{M}_D^{-1} \mathsf{D}_2^\mathsf{T} \mathsf{M}_S \mathsf{D}_2 \tag{B.72}$$

$$(\Delta_S^h) = \mathsf{D}_2 \mathsf{M}_D^{-1} \mathsf{D}_2^\mathsf{T} \mathsf{M}_S . \tag{B.73}$$

A remarkable fact about (B.68)–(B.73) is that they hold for any finite element complex in which the degrees of freedom in $G^h(\Omega)$, $\mathbf{C}^h(\Omega)$, $\mathbf{D}^h(\Omega)$, and $S^h(\Omega)$ are associated only with the nodes, edges, faces, and cells, respectively, of the mesh; see [66]. One example is the Whitney complex (B.46) but (B.68)–(B.73) also holds for non-affine, non-simplicial grids as well as other *compatible discretizations* of these operators obtained by finite volume or mimetic finite difference methods; see, e.g., [223,224,276,287,332] and the recent collection [8].[21,22]

B.3.2 Discrete Poincaré–Friedrichs Inequalities

The discrete Poincaré–Friedrichs inequalities considered in this section mimic (A.63)–(A.65) and assert that $\|\nabla u^h\|_0$, $\|\nabla\times\mathbf{u}^h\|_0$, and $\|\nabla\cdot\mathbf{u}^h\|_0$ are equivalent norms on the orthogonal complements of the null-spaces of the gradient, curl, and divergence operators in $G^h(\Omega)$, $\mathbf{C}^h(\Omega)$, and $\mathbf{D}^h(\Omega)$, respectively. The following theorem is a simplified version of [13, Theorem 5.11, p. 74].

Theorem B.21 *Let $\Omega\subset\mathbb{R}^d$, $d=2,3$, be a contractible domain with a Lipschitz-continuous boundary. Then, there exists a positive constant C_P such that*

$$\|u^h\|_0\le C_P\|\nabla u^h\|_0 \quad \textit{for}\quad u^h\in N^h(\nabla)^\perp \quad \textit{or}\quad u^h\in N^h_0(\nabla)^\perp \tag{B.74}$$

$$\|\mathbf{u}^h\|_0\le C_P\|\nabla\times\mathbf{u}^h\|_0 \quad \textit{for}\quad \mathbf{u}^h\in N^h(\nabla\times)^\perp \quad \textit{or}\quad \mathbf{u}^h\in N^h_0(\nabla\times)^\perp \tag{B.75}$$

---

[21] Up to a multiplicative matrix factor, the matrix representations (B.70)–(B.73) of second-order operators are equivalent to finite element stiffness matrices obtained in the usual manner from weak forms. Consider, for example, the stiffness matrix for the Galerkin method for the Poisson equation (1.52):

$$\mathsf{K}_{ij}=\big(\nabla u_i^h,\nabla u_j^h\big)_{\Theta_1},$$

where $\{u_i^h\}$ is a basis for $G_0^1(\Omega)$. Let $\vec{u}_i$ be the coefficient vector of u_i^h. Recall that for an exact sequence of finite element spaces, the range of the gradient is a proper subspace of $\mathbf{C}_0^1\text{-}(\Omega)$. It follows that ∇u_i^h is the element of $\mathbf{C}_0^1\text{-}(\Omega)$ whose coefficient vector is $\mathsf{D}_0\vec{u}_i$ and

$$\mathsf{K}=\mathsf{D}_0^T\mathsf{M}_C\mathsf{D}_0=\mathsf{M}_G\Delta_G^h.$$

The appearance of the multiplicative matrix factor M_G in the above formula is owed to the fact that definition of the stiffness matrix includes the action of the inner product on finite element functions whereas definition of Δ_G^h does not include that action, i.e.,

$$\big(\Delta_G^h u_i^h,u_j^h\big)_{\Theta_0}=\big(\nabla u_i^h,\nabla u_j^h\big)_{\Theta_1}.$$

[22] The matrix representations in (B.69)–(B.73) can be viewed as factorizations of discrete differential operators into topological components D that depend only on mesh connectivity and metric components M that depend on the choice of inner products and account for the "shape" of the elements. By a judicious choice of basis functions and/or quadrature points, it may be possible to obtain diagonal mass matrices (see, e.g., [270]). In this case, weak operators are simply scaled versions of D_k^T, i.e., up to a scale factor, ∇_h^*, $\nabla_h^*\times$ and $\nabla^h\cdot$ are algebraic adjoints, i.e., transposes, of D_k, $k=0,1,2$. Likewise, the discrete second-order operators are scaled versions of $\mathsf{D}_k^T\mathsf{D}_k$.

$$\|\mathbf{u}^h\|_0 \le C_P \|\nabla\cdot\mathbf{u}^h\|_0 \quad \textit{for} \quad \mathbf{u}^h \in N^h(\nabla\cdot)^\perp \quad \textit{or} \quad \mathbf{u}^h \in N_0^h(\nabla\cdot)^\perp. \square \tag{B.76}$$

The first Poincaré–Friedrichs inequality (B.74) follows trivially from its analytic counterpart (A.63) because $N^h(\nabla) = N(\nabla)$ so that

$$N^h(\nabla)^\perp \subset N(\nabla)^\perp .$$

Proofs of (B.75) and (B.76) are more complicated because $N^h(\nabla\times) \subset N(\nabla\times)$ and $N^h(\nabla\cdot) \subset N(\nabla\cdot)$. As a result, $N^h(\nabla\times)^\perp$ and $N^h(\nabla\cdot)^\perp$ are not properly contained in $N(\nabla\times)^\perp$ and $N(\nabla\cdot)^\perp$, respectively. The existence of bounded projection operators is essential for these inequalities; see [13].

B.3.3 Discrete Hodge Decompositions

In this section, we focus on orthogonal decompositions of compatible finite element spaces with homogeneous boundary conditions; we assume that $\Gamma = \partial\Omega$ and $\Gamma^* = \emptyset$. Decompositions for more general boundary configurations can be handled in a similar manner. Because $G_0^h(\Omega)$, $\mathbf{C}_0^h(\Omega)$, $\mathbf{D}_0^h(\Omega)$, and $S_0^h(\Omega)$ are finite-dimensional, it is clear that

$$\begin{aligned}
G_0^h(\Omega) &= N_0^h(\nabla) \oplus N_0^h(\nabla)^\perp \\
\mathbf{C}_0^h(\Omega) &= N_0^h(\nabla\times) \oplus N_0^h(\nabla\times)^\perp \\
\mathbf{D}_0^h(\Omega) &= N_0^h(\nabla\cdot) \oplus N_0^h(\nabla\cdot)^\perp \\
S_0^h(\Omega) &= R_0^h(\nabla\cdot) \oplus R_0^h(\nabla\cdot)^\perp .
\end{aligned}$$

The first decomposition is trivial because $N_0^h(\nabla) = \{0\}$ and $N_0^h(\nabla)^\perp = G_0^h(\Omega)$. The last decomposition is also trivial because, for an exact sequence, $R_0^h(\nabla\cdot) = S_0^h(\Omega)$. Thus we focus on the decompositions of curl-conforming and div-conforming spaces.

Theorem B.22 (Hodge decomposition of $\mathbf{C}_0^h$) *Every $\mathbf{u}^h \in \mathbf{C}_0^h(\Omega)$ can be written as*

$$\mathbf{u}^h = \mathbf{u}_N^h + \mathbf{u}_{N\perp}^h \qquad \textit{with} \qquad \mathbf{u}_N^h \in N_0^h(\nabla\times) \quad \textit{and} \quad \mathbf{u}_{N\perp}^h \in N_0^h(\nabla\times)^\perp, \tag{B.77}$$

where $\nabla_h^* \cdot \mathbf{u}_{N\perp}^h = 0$ and $\mathbf{u}_N^h = \nabla p^h$ for some $p^h \in G_0^h(\Omega)$. Furthermore, there exists a positive constant C, independent of h, such that

$$\begin{gathered}
\|\mathbf{u}_{N\perp}^h\|_0 \le C\|\nabla\times\mathbf{u}^h\|_0 , \\
\|\mathbf{u}_N^h\|_0 \le C\|\nabla_h^*\cdot\mathbf{u}^h\|_0 , \qquad \textit{and} \qquad \|\mathbf{u}_N^h\|_0 \le C\|\mathbf{u}^h\|_0 .
\end{gathered} \tag{B.78}$$

Proof. Let $p^h \in G_0^h(\Omega)$ solve the equation

$$\left(\nabla p^h, \nabla q^h\right) = \left(\mathbf{u}^h, -\nabla q^h\right) \qquad \forall q^h \in G_0^h(\Omega).$$

Set $\mathbf{u}_N^h = \nabla p^h$ and $\mathbf{u}_{N\perp}^h = \mathbf{u}^h + \mathbf{u}_N^h$. Then $\mathbf{u}_{N\perp}^h \in N_0^h(\nabla\times)^\perp$ and the first inequality in (B.78) follows from the discrete Poincaré–Friedrichs inequality (B.75) applied to $\mathbf{u}_{N\perp}^h$ and the obvious identity $\|\nabla\times\mathbf{u}_{N\perp}^h\|_0 = \|\nabla\times\mathbf{u}^h\|_0$. The second inequality follows from the definitions of $\mathbf{u}_N^h$ and the discrete weak divergence (B.51) and the discrete Poincaré–Friedrichs inequality (B.74):

$$\begin{aligned}\|\mathbf{u}_N^h\|_0^2 &= \left(\nabla p^h, \nabla p^h\right) = \left(\mathbf{u}^h, -\nabla p^h\right) = \left(\nabla_h^* \cdot \mathbf{u}^h, p^h\right) \\ &\le \|\nabla_h^* \cdot \mathbf{u}^h\|_0 \|p^h\|_0 \le C_P \|\nabla_h^* \cdot \mathbf{u}^h\|_0 \|\nabla p^h\|_0 = C_P \|\nabla_h^* \cdot \mathbf{u}^h\|_0 \|\mathbf{u}_N^h\|_0 .\end{aligned}$$

The third inequality in (B.78) is a simple consequence of the definition of $\mathbf{u}_N^h$:

$$\|\mathbf{u}_N^h\|_0^2 = \left(\nabla p^h, \nabla p^h\right) = \left(\mathbf{u}^h, -\nabla p^h\right) \le \|\mathbf{u}^h\|_0 \|\nabla p^h\|_0 = \|\mathbf{u}^h\|_0 \|\mathbf{u}_N^h\|_0 . \qquad \square$$

An analogous result holds for div-compatible spaces.

Theorem B.23 (Hodge decomposition of $\mathbf{D}_0^h$) *Every $\mathbf{u}^h \in \mathbf{D}_0^h(\Omega)$ can be written as*

$$\mathbf{u}^h = \mathbf{u}_N^h + \mathbf{u}_{N\perp}^h \qquad \textit{with} \qquad \mathbf{u}_N^h \in N_0^h(\nabla\cdot) \qquad \textit{and} \qquad \mathbf{u}_{N\perp}^h \in N_0^h(\nabla\cdot)^\perp, \tag{B.79}$$

where $\nabla_h^* \times \mathbf{u}_{N\perp}^h = 0$, $\mathbf{u}_N^h = \nabla\times\mathbf{w}^h$, and $\mathbf{w}^h \in \mathbf{C}_0^h(\Omega)$ is such that $\nabla_h^* \cdot \mathbf{w} = 0$. Furthermore, there exists a positive constant C, independent of h, such that

$$\begin{gathered}\|\mathbf{u}_{N\perp}^h\|_0 \le C\|\nabla\cdot\mathbf{u}^h\|_0, \\ \|\mathbf{u}_N^h\|_0 \le C\|\nabla_h^* \times \mathbf{u}^h\|_0, \qquad \textit{and} \qquad \|\mathbf{u}_N^h\|_0 \le C\|\mathbf{u}^h\|_0 .\end{gathered} \tag{B.80}$$

Proof. Let $\{\mathbf{w}^h, p^h\} \in \mathbf{C}_0^h(\Omega) \times G_0^h(\Omega)$ solve the weak equation

$$\begin{cases} \left(\nabla\times\mathbf{w}^h, \nabla\times\mathbf{v}^h\right) + \left(\nabla p^h, \mathbf{v}^h\right) = \left(\mathbf{u}^h, \nabla\times\mathbf{v}^h\right) & \forall \mathbf{v}^h \in \mathbf{C}_0^h(\Omega) \\ \left(\nabla q^h, \mathbf{w}^h\right) = 0 & \forall q^h \in G_0^h(\Omega). \end{cases} \tag{B.81}$$

This equation has the structure of the abstract mixed variational problem (1.14) in Section 1.1. Its well-posedness, therefore, is subject to the assumptions in Theorem 1.11, including the discrete inf–sup conditions (1.41) and (1.42). In the present case, $b(q^h, \mathbf{v}^h) = (\nabla q^h, \mathbf{v}^h)$ and the space Z^h from (1.40) specializes to

$$Z^h = \{\mathbf{v}^h \in \mathbf{C}_0^h(\Omega) \mid (\nabla q^h, \mathbf{v}^h) = 0 \quad \forall q^h \in G_0^h(\Omega)\} = N_0^h(\nabla\times)^\perp.$$

As a result, the coercivity of $a(\mathbf{w}^h, \mathbf{v}^h) = (\nabla\times\mathbf{w}^h, \nabla\times\mathbf{v}^h)$ on Z^h follows from the discrete Poincaré–Friedrichs inequality (B.75):

$$a(\mathbf{w}^h, \mathbf{w}^h) = \|\nabla\times\mathbf{w}^h\|_0^2 \ge \frac{1}{C_P^2}\|\mathbf{w}^h\|_0^2 \qquad \forall \mathbf{w}^h \in Z^h.$$

Because $\nabla q^h \in \mathbf{C}_0^h(\Omega)$, we easily obtain the inf–sup condition for $b(q^h, \mathbf{v}^h)$:

$$\sup_{\mathbf{v}^h \in \mathbf{C}_0^h(\Omega)} \frac{(\nabla q^h, \mathbf{v}^h)}{\|\mathbf{v}^h\|_C} \geq \frac{(\nabla q^h, \nabla q^h)}{\|\nabla q^h\|_C} = \frac{\|\nabla q^h\|_0^2}{\|\nabla q^h\|_0} = \|\nabla q^h\|_0 .$$

It is easy to see that $a(\cdot\,,\cdot)$ and $b(\cdot\,,\cdot)$ are also continuous. As a result, existence and uniqueness of the solution to (B.81) follows from Theorem 1.11.

Let $\mathbf{u}_N^h = \nabla \times \mathbf{w}^h$ and $\mathbf{u}_{N\perp}^h = \mathbf{u}^h - \mathbf{u}_N^h$. From the second equation in (B.81), it follows that $\mathbf{w}^h \in Z^h$ and, choosing $\mathbf{v}^h = \nabla q^h$ in the first equation, gives $p^h = 0$. This result and the first equation in (B.81) imply that

$$\left(\mathbf{u}_{N\perp}^h, \nabla \times \mathbf{v}^h\right) = \left(\mathbf{u}^h - \nabla \times \mathbf{w}^h, \nabla \times \mathbf{v}^h\right) = 0$$

so that $\mathbf{u}_{N\perp}^h$ is indeed in the orthogonal complement of $N_0^h(\nabla\cdot)$. As a result, the first inequality in (B.80) follows from the discrete Poincaré–Friedrichs inequality (B.76) applied to $\mathbf{u}_{N\perp}^h$ and $\|\nabla\cdot\mathbf{u}_{N\perp}^h\|_0 = \|\nabla\cdot\mathbf{u}^h\|_0$. The second and third inequalities follow along the same lines as in the proof of Theorem B.22. From (B.81) and (B.50), we have that

$$\begin{aligned}\|\mathbf{u}_N^h\|_0^2 &= \left(\nabla \times \mathbf{w}^h, \nabla \times \mathbf{w}^h\right) = \left(\mathbf{u}^h, \nabla \times \mathbf{w}^h\right) = \left(\nabla_h^* \times \mathbf{u}^h, \mathbf{w}^h\right) \\ &\leq \|\nabla_h^* \times \mathbf{u}^h\|_0 \|\mathbf{w}^h\|_0 \leq C_P \|\nabla_h^* \times \mathbf{u}^h\|_0 \|\nabla \times \mathbf{w}^h\|_0 ,\end{aligned}$$

where we used the relation $\mathbf{w}^h \in Z^h = N_0^h(\nabla\times)^\perp$. Then, the Poincaré–Friedrichs inequality (B.75) holds for $\mathbf{w}^h$. The proof of the last inequality in (B.80) repeats the same steps.[23] □

Remark B.24 From (B.52), it follows that

$$\mathbf{C}_0^h(\Omega) = N_0^h(\nabla\times) \oplus N_0^h(\nabla_h^*\cdot) \qquad \text{and} \qquad \mathbf{D}_0^h(\Omega) = N_0^h(\nabla\cdot) \oplus N_0^h(\nabla_h^*\times)$$

and, because the discrete weak operators form another exact sequence,

$$N_0^h(\nabla_h^*\cdot) = R_0^h(\nabla_h^*\times) \qquad \text{and} \qquad N_0^h(\nabla_h^*\times) = R_0^h(\nabla_h^*) .$$

---

[23] The "grid decomposition property" (GDP), formulated in [162–164], is the earliest example of a discrete Hodge decomposition for div-conforming spaces. A finite element vector field $\mathbf{u}^h \in \mathbf{D}^h(\Omega)$ satisfies the GDP if there exist $\mathbf{w}^h, \mathbf{z}^h \in \mathbf{D}^h(\Omega)$ such that

$$\mathbf{u}^h = \mathbf{z}^h + \mathbf{w}^h , \qquad \nabla\cdot\mathbf{z}^h = 0 , \qquad (\mathbf{w}^h, \mathbf{z}^h) = 0 , \qquad \|\mathbf{w}^h\|_0 \leq C\|\nabla\cdot\mathbf{u}^h\|_{-1} .$$

In [163, 164], it was shown that GDP is necessary and sufficient for the stability of mixed finite element discretizations of the Kelvin principle (1.54).

With the exception of the required upper bound on $\mathbf{w}^h$, the GDP is identical with the discrete Hodge decomposition of Theorem B.23. Originally, the GDP was established only for vector fields on a criss-cross grid; see Figure 5.1 and [163]. Later, it was shown in [59] that the GDP holds for a wide range of div-conforming elements.

Consequently, the Hodge decompositions (B.77) and (B.79) can be respectively written in the following equivalent forms:

$$\forall \mathbf{u}^h \in \mathbf{C}_0^h(\Omega): \begin{cases} \mathbf{u}^h = \nabla p^h + \nabla_h^* \times \mathbf{w}^h, \\ p^h \in G_0^h(\Omega), \qquad \mathbf{w}^h \in \mathbf{D}_0^h(\Omega) \end{cases}$$

$$\forall \mathbf{u}^h \in \mathbf{D}_0^h(\Omega): \begin{cases} \mathbf{u}^h = \nabla_h^* p^h + \nabla \times \mathbf{w}^h, \\ p^h \in S_0^h(\Omega), \qquad \mathbf{w}^h \in \mathbf{C}_0^h(\Omega), \qquad \nabla_h^* \cdot \mathbf{w}^h = 0. \end{cases} \tag{B.82}$$

We conclude this section with a finite element version of the Poincaré–Friedrichs inequality (A.69) for vector fields in $\mathbf{C}(\Omega) \cap \mathbf{D}(\Omega)$.

Theorem B.25 *There exists a positive constant C_P, independent of h, such that*

$$\|\mathbf{u}^h\|_0 \le C_P\big(\|\nabla \times \mathbf{u}^h\|_0 + \|\nabla_h^* \cdot \mathbf{u}^h\|_0\big) \qquad \forall \mathbf{u}^h \in \mathbf{C}_0^h(\Omega) \tag{B.83}$$

$$\|\mathbf{u}^h\|_0 \le C_P\big(\|\nabla_h^* \times \mathbf{u}^h\|_0 + \|\nabla \cdot \mathbf{u}^h\|_0\big) \qquad \forall \mathbf{u}^h \in \mathbf{D}_0^h(\Omega). \tag{B.84}$$

Proof. Let $\mathbf{u}^h \in \mathbf{C}_0^h(\Omega)$. To prove (B.83), we use the discrete Hodge decomposition (B.77) for curl-conforming vector fields and the upper bounds in (B.78):

$$\|\mathbf{u}^h\|_0 = \|\mathbf{u}_N^h + \mathbf{u}_{N\perp}^h\|_0 = \|\mathbf{u}_N^h\|_0 + \|\mathbf{u}_{N\perp}^h\|_0 \le C\big(\|\nabla_h^* \cdot \mathbf{u}^h\|_0 + \|\nabla \times \mathbf{u}^h\|_0\big).$$

The proof of (B.84) follows from the Hodge decomposition in Theorem B.23. □.

B.3.4 Inverse Inequalities

Inverse inequalities relate different norms of finite element functions. For brevity, we state results for nodal finite element spaces $G^r(\Omega)$.

Theorem B.26 *Assume that Ω is a bounded open region in $\mathbb{R}^d$, $d = 2,3$, and T_h is uniformly regular partition of Ω into finite elements. Then, there exists a constant $C_I > 0$, independent of h, such that for all $u \in G^r(\Omega)$,*

$$\|u^h\|_1 \le C_I h^{-1} \|u^h\|_0, \tag{B.85}$$

$$\|u^h\|_0 \le C_I h^{-1} \|u^h\|_{-1}, \tag{B.86}$$

and

$$\|u^h\|_{1/2,\partial\Omega} \le C h^{-1/2} \|u^h\|_{0,\partial\Omega}. \qquad \square \tag{B.87}$$

One can also relate the $L^2(\Omega)$ norm $\|u^h\|_0$ of $u \in G^r(\Omega)$ with the Euclidean norm $|\vec{u}|$ of its coefficients; this result is useful, e.g., in the estimation of the condition numbers of finite element discretization matrices.

Theorem B.27 *Assume that Ω is a bounded open region in $\mathbb{R}^d$, $d = 2,3$, and $\mathcal{T}_h$ is regular partition of Ω into finite elements. Let $\{u_1^h, \ldots, u_N^h\}$ denote a Lagrangian basis for G^r and $\vec{u} = (\vec{u}_1, \ldots, \vec{u}_N)^\top$ denote the coefficient vector of $u^h \in G^r$, i.e.,*

$$u^h = \sum_{i=1}^{N} \vec{u}_i u_i^h .$$

Then, there exists a constant $C > 0$, independent of h, such that

$$C^{-1} h^d |\vec{u}|^2 \le \|u^h\|_0^2 \le C h^d |\vec{u}|^2 . \qquad \square \tag{B.88}$$

B.4 Norm Approximations

We conclude this appendix with several examples of discrete norms that find application in LSFEMs and other finite element methods. In LSFEMs, these norms are primarily needed to effect the transition from a given continuous least-squares principle to a practical discrete least-squares principle. The abstract transition diagram (3.54) shows that the definition of approximate discrete norms is tantamount to the approximation of their associated norm-generating operators $\mathcal{S}_{(*)}$; see Section 3.4.1 for examples. The type of the resulting norm, i.e., quasi-equivalent[24] or norm-equivalent, is largely dependent on the choice made in the approximation of these operators.

B.4.1 Quasi-Norm-Equivalent Approximations

The most straightforward approximation of a norm-generating operator $\mathcal{S}_{(*)}$ is based on scaling arguments. The basic idea is to replace the powers of $\mathcal{S}_{(*)}$ (see Section 3.4.1) by an identity operator scaled by a related power of the mesh parameter h. For Sobolev space norms generated by the powers of $-\Delta$, this corresponds to the substitution rule

$$(-\Delta)^s \mapsto h^{-2s}\mathcal{I} . \tag{B.89}$$

This rule is applied on an element-by-element basis leading to the *weighted* $L^2(\Omega)$ norms

$$\|\phi\|_{2s,h} = \Big(\sum_{\kappa \in \mathcal{T}_h} h_\kappa^{-4s} \|\phi\|_{0,\kappa}^2 \Big)^{1/2} . \tag{B.90}$$

If $\mathcal{T}_h$ is uniformly regular, a representative value of h can be factored out. For instance, using the expressions for $\|\cdot\|_{-1}$ and $\|\cdot\|_1$ from Example 3.23, we can define the following discrete approximations for these norms:

[24] Recall that quasi-equivalent norms have mesh-dependent equivalence constants.

$$\|\phi\|_{-1,h} = h\|\phi\|_0 \qquad \text{and} \qquad \|\phi\|_{1,h} = h^{-1}\|\phi\|_0 \,. \tag{B.91}$$

Equivalence properties of the norms in (B.91) are stated below.

Proposition B.28 *Assume that $\mathcal{T}_h$ is uniformly regular. Then, there exists a positive constant C, independent of h, such that*

$$\frac{1}{C}\|\phi^h\|_1 \le \|\phi^h\|_{1,h} \le Ch^{-1}\|\phi^h\|_1 \tag{B.92}$$

and

$$\frac{h}{C}\|\phi^h\|_{-1} \le \|\phi^h\|_{-1,h} \le C\|\phi^h\|_{-1} \,. \tag{B.93}$$

Proof. The results follow from the inverse inequalities (B.85) and (B.86). □.

Proposition B.28 characterizes the discrete norms in (B.91) as quasi-norm-equivalent approximations of $\|\cdot\|_{-1}$ and $\|\cdot\|_1$, respectively.

Scaling arguments can also be used to replace (impractical) trace norms by weighted $L^2(\partial\Omega)$ norms. Two frequently used cases are

$$\|\phi\|_{-1/2,h} = h^{1/2}\|\phi\|_{0,\partial\Omega} \qquad \text{and} \qquad \|\phi\|_{1/2,h} = h^{-1/2}\|\phi\|_{0,\partial\Omega} \,. \tag{B.94}$$

The quasi-equivalence of these discrete norms follows from the trace inverse inequality (B.87).

B.4.2 Norm-Equivalent Approximations

We now consider approximations of the minus one norm (A.9) that have better equivalence properties than the simple weighted analogue in (B.91). This norm arises in LSFEMs for non-homogeneous elliptic systems (see Section 4.5) and its norm-equivalent approximation can lead to discrete least-squares principles with better computational properties.

To avoid dependence on h in the equivalence bounds (B.93), we need an approximation of the generating operator $(-\Delta)^{-1/2}$ that captures its spectral behavior better than the scaled identity in (B.89). The key idea, proposed in [77], is to approximate the norm-generating operator according to the rule

$$(-\Delta)^{-1/2} \mapsto h\mathcal{I} + (\mathcal{K}^h)^{1/2},$$

where $\mathcal{K}^h$ is a spectrally equivalent preconditioner of the Laplace operator; see, e.g., [347] for further details about such preconditioners. To justify this choice, consider the seminorm

$$|\psi|_{-h} = \sup_{v^h \in G_0^h(\Omega)} \frac{(\psi, v^h)}{\|v^h\|_1} \qquad \text{for} \quad \psi \in H^{-1}(\Omega) \,. \tag{B.95}$$

Because $G_0^h(\Omega) \subset H_0^1(\Omega)$, (A.9) implies that

$$|\psi|_{-h} \leq \|\psi\|_{-1} \qquad \forall\, \psi \in H^{-1}(\Omega). \tag{B.96}$$

Let $\mathcal{S}^h : H^{-1}(\Omega) \mapsto G_0^h(\Omega)$ denote the Galerkin solution operator for (5.7), i.e., $\mathcal{S}^h\psi = u^h$ if and only if

$$\left(\nabla u^h, \nabla v^h\right) = \left(\psi, v^h\right) \qquad \forall\, v^h \in G_0^h(\Omega). \tag{B.97}$$

The same argument as in the proof of Theorem A.1 shows that

$$|\psi|^2_{-h} = \left(\mathcal{S}^h\psi, \psi\right),$$

i.e., we could set $\mathcal{K}^h = \mathcal{S}^h$. Of course, this choice would require the solution of a discrete Poisson equation every time one needs to compute the negative norm so that it is hardly practical. The key observation is that $\mathcal{K}^h$ has to capture the spectral properties of $\mathcal{S}^h$ rather than its accuracy as a solution operator for (5.7). As a result, $\mathcal{K}^h$ can be any operator that is spectrally equivalent to $\mathcal{S}^h$, i.e., such that

$$C^{-1}(\mathcal{S}^h v^h, v^h) \leq (\mathcal{K}^h v^h, v^h) \leq C(\mathcal{S}^h v^h, v^h) \qquad \forall\, v^h \in G_0^h(\Omega), \tag{B.98}$$

including any good preconditioner for the Poisson equation. Assuming that (B.98) holds for $\mathcal{K}^h$, we define the discrete negative norm for $\psi \in L^2(\Omega)$ according to

$$\|\psi\|_{-h} = \left((\alpha h^2\mathcal{I} + \mathcal{K}^h)\psi, \psi\right)^{1/2} = \left(\|\psi\|^2_{-1,h} + |\psi|^2_{-h}\right)^{1/2}, \tag{B.99}$$

where α is a positive real parameter. The associated inner product is given by

$$\left(\psi, \phi\right)_{-h} = \left((\alpha h^2\mathcal{I} + \mathcal{K}^h)\psi, \phi\right)^{1/2}. \tag{B.100}$$

The role of α is to balance the operators that comprise the discrete negative norm. In some cases, α can be tuned to optimize the performance of iterative solvers.

The following lemma shows that $\|\cdot\|_{-h}$ is equivalent to the restriction of $\|\cdot\|_{-1}$ to $G_0^h(\Omega)$.

Lemma B.1. *Assume that Ω is a bounded open region in $\mathbb{R}^d$, $d = 2,3$, and $\mathcal{T}_h$ is uniformly regular partition of Ω into finite elements. Then, there exists a constant $C > 0$, independent of h, such that, for all $u \in L^2(\Omega)$,*

$$C^{-1}\|u\|_{-1} \leq \|u\|_{-h} \leq C\left(h\|u\|_0 + \|u\|_{-1}\right) \tag{B.101}$$

and for all $u^h \in G_0^h(\Omega)$

$$C^{-1}\|u^h\|_{-1} \leq \|u^h\|_{-h} \leq C\|u^h\|_{-1}. \tag{B.102}$$

Proof. The upper bound follows from (B.99), (B.98), and (B.96):

$$\|u\|^2_{-h} = \left((\alpha h^2\mathcal{I} + \mathcal{K}^h)u, u\right)$$

$$= \alpha h^2\|u\|_0^2 + (\mathcal{K}^h u,u) \le C\Big(\alpha h^2\|u\|_0^2 + (\mathcal{S}^h u,u)\Big)$$

$$= C\big(h^2\|u\|_0^2 + |u|_{-h}^2\big) \le C\big(h^2\|u\|_0^2 + \|u\|_{-1}^2\big)\,.$$

To prove the lower bound, let $v \in H_0^1(\Omega)$ be arbitrary. From (B.20),

$$\|v - \Pi_G v\|_0 \le Ch\|v\|_1\,.$$

Adding and subtracting $\Pi_G v$ yields

$$\big(u,v\big) \le \|u\|_0\,\|v-\Pi_G v\|_0 + \big(u,\Pi_G v\big) \le Ch\|u\|_0\,\|v\|_1 + \big(u,\Pi_G v\big)\,.$$

Because $\|\Pi_G v\|_1 \le C\|v\|_1$ (see (B.15)), we have that

$$\frac{(u,v)}{\|v\|_1} \le Ch\|u\|_0 + \frac{(u,\Pi_G v)}{\|v\|_1} \le C\left(h\|u\|_0 + \frac{(u,\Pi_G v)}{\|\Pi_G v\|_1}\right)\,.$$

The result follows from the fact that $\Pi_G v \in G_0^h(\Omega)$, (B.95), and (B.98). For a finite element function $u^h \in G_0^h(\Omega)$, the norm-equivalence (B.102) follows by using the inverse inequality (B.86) on the right-hand side of (B.101). □

Similar ideas can be used to define norm-equivalent trace norms (see [318]) that can be used in lieu of the simpler weighted approximations in (B.94). We refer to [318] and the references cited therein for further details; see also Section 12.1.

Appendix C
Linear Operator Equations in Hilbert Spaces

This appendix contains a brief review of certain results about abstract operator equations in Hilbert spaces that are central to the development of the mathematical theory for least-squares finite element methods in Chapter 3.

Given two Hilbert spaces X and Y, we denote by $L(X,Y)$ the set of all bounded linear operators $\mathcal{Q} : X \mapsto Y$ and by $D(\mathcal{Q})$, $N(\mathcal{Q})$, and $R(\mathcal{Q})$ the domain, null space, and range of $\mathcal{Q}$, respectively. The notation $\{\mathcal{Q}, S\}$ is used to indicate a particular domain choice S for $\mathcal{Q}$. In what follows, we are concerned with Fredholm (or Noetherian) operators; see [304, p. 279].

Definition C.1 Let X and Y be Hilbert spaces.[1] The operator $\mathcal{Q} \in L(X,Y)$ is Fredholm if its range is closed, $\dim N(\mathcal{Q}) < \infty$, and $\operatorname{codim} R(\mathcal{Q}) < \infty$. The dimension of $N(\mathcal{Q})$, denoted by $\operatorname{null} \mathcal{Q}$, is called the nullity of $\mathcal{Q}$, the co-dimension of $R(\mathcal{Q})$, denoted by $\operatorname{def} \mathcal{Q}$, is called the deficiency of $\mathcal{Q}$, and

$$\operatorname{ind} \mathcal{Q} = \operatorname{null} \mathcal{Q} - \operatorname{def} \mathcal{Q}$$

is the Fredholm index of $\mathcal{Q}$. □

Given two Hilbert spaces X and Y, a Fredholm operator $\mathcal{Q} \in L(X,Y)$, and a function $f \in Y$, consider the operator equation

$$\textit{find } u \in X \textit{ such that } \quad \mathcal{Q}u = f\,. \tag{C.1}$$

Depending on the deficiency and the nullity of $\mathcal{Q}$, the problem (C.1) may or may not be solvable for arbitrary $f \in Y$ or may have multiple solutions. Our goal here is to find a solution space S and a data space H such that the estimate[2]

$$\alpha_1 \|u\|_S \le \|\mathcal{Q}u\|_H \le \alpha_2 \|u\|_S \qquad \forall u \in S \tag{C.2}$$

[1] The definitions that follow also hold on Banach spaces, but we restrict our attention to Hilbert spaces because the least-squares framework we develop operates in that setting.

[2] In (C.2), $\|u\|_S$ and $\|\mathcal{Q}u\|_H = \|f\|_H$ may be viewed as being "energies" of the solution and the data, respectively, and (C.2) may be viewed as an *"energy" balance* relation.

P.B. Bochev, M.D. Gunzburger, *Least-Squares Finite Element Methods*, Applied Mathematical Sciences 166, DOI: 10.1007/b13382_15,

holds for some positive constants α_1 and α_2 having values independent of u.

The upper bound in (C.2) holds trivially for any $\mathcal{Q} \in L(X,Y)$ with $S = X$ and $H = Y$. We refer to this as the *standard* choice for the data and solution spaces. The lower bound in (C.2) is equivalent to $\mathcal{Q}$ having a bounded inverse. Obviously, if $\{\mathcal{Q},X\}$ has positive nullity, this bound cannot hold with the standard choice of spaces. In this case, we have to replace $\{\mathcal{Q},X\}$ by another operator that results in a well-posed problem. On the other hand, if $\{\mathcal{Q},X\}$ has zero nullity but positive deficiency, the lower bound in (C.2) holds for the standard choice of spaces but (C.1) has a solution only if $f \in R(\mathcal{Q})$. In this case, we have to redefine the notion of a "solution" of (C.1). In both cases, we are led to consider auxiliary operator equations in lieu of (C.1).

C.1 Auxiliary Operator Equations

Given the operator equation (C.1) which may not be solvable or may have multiple solutions, we focus on defining a "nearby" problem

$$\widehat{\mathcal{Q}}\widehat{u} = \widehat{f} \tag{C.3}$$

that is uniquely solvable for $u \in S$, where $\widehat{\mathcal{Q}} : S \mapsto H$ and $\widehat{f} \in H$ denote an *auxiliary* operator and data, respectively.

Injective mappings by domain redefinition

Assume that $\mathcal{Q} : X \mapsto Y$ has positive nullity. Then, there exist $K = \operatorname{null}\mathcal{Q} < \infty$ linearly independent functions $u_1,\ldots,u_K$ belonging to X such that

$$\mathcal{Q}u_k = 0 \qquad \text{for } k = 1,\ldots,K$$

and $N(\mathcal{Q}) = \operatorname{span}\{u_1,\ldots,u_K\}$. There are several ways to redefine the domain X of $\mathcal{Q}$ so that $N(\mathcal{Q}) = \{0\}$. Instead of X, we can use the *factor* or *quotient* space

$$X^N = X/N = \{\widehat{u} \subset X \mid u,v \in \widehat{u} \Leftrightarrow u - v \in N(\mathcal{Q})\} \tag{C.4}$$

or the *orthogonal complement* space

$$X^\perp = \{u \in X \mid (u,u_k)_X = 0,\ k = 1,\ldots,K\}. \tag{C.5}$$

Another choice is the *algebraic complement* of $N(\mathcal{Q})$. Let $\ell_k : X \mapsto \mathbb{R}$, $k = 1,\ldots,K$, denote K bounded linear functionals such that the matrix

$$\mathsf{L} = \begin{pmatrix} \ell_1(u_1) & \dots & \ell_1(u_K) \\ \vdots & & \vdots \\ \ell_K(u_1) & \dots & \ell_K(u_K) \end{pmatrix} \tag{C.6}$$

is nonsingular. Then, the algebraic complement of $N(\mathcal{Q})$ is the subspace

$$X^C = \{u \in X \mid \ell_k(u) = 0, \ k = 1,\dots,K\}. \tag{C.7}$$

The algebraic complement subspace induces a similar decomposition as does the orthogonal complement subspace.

Lemma C.2 *For any $u \in X$, there exist unique $u_C \in X^C$ and $u_N \in N(\mathcal{Q})$ such that*

$$u = u_C + u_N.$$

Proof. Consider the linear system

$$\ell_j\big(u - \sum_{k=1}^{K} \beta_k u_k\big) = 0, \quad j,k = 1,\dots,K,$$

for the unknown coefficient vector $\vec{b} = (\beta_1,\dots,\beta_K)^\mathsf{T}$. In matrix form, we have that

$$\mathsf{L}\vec{b} = \vec{\ell}(u), \tag{C.8}$$

where $\vec{\ell}(u) = (\ell_1(u),\dots,\ell_K(u))^\mathsf{T}$. By assumption, L is invertible so that (C.8) has a unique solution $\vec{b} \in \mathbb{R}^K$. Then,

$$u_N = \sum_{k=1}^{K} \beta_k u_k \qquad \text{and} \qquad u_C = u - u_N$$

are the two functions needed to complete the proof. □

An obvious consequence is the following result.

Corollary C.3 *If $u \in N(\mathcal{Q})$ and $\vec{\ell}(u) = 0$, then $u \equiv 0$.* □

We can now prove that $\mathcal{Q}$ has trivial kernel on any of (C.4), (C.5), and (C.7).

Lemma C.4 *The operator*

$$\{\mathcal{Q},\breve{X}\}, \qquad \textit{where} \qquad \breve{X} \in \{X^N, X^\perp, X^C\}, \tag{C.9}$$

has zero nullity and the same deficiency as $\{\mathcal{Q},X\}$.

Proof. The range of $\{\mathcal{Q},\breve{X}\}$ is the same as the range of $\{\mathcal{Q},X\}$ so that the two operators have the same deficiency. To prove that $\{\mathcal{Q},\breve{X}\}$ has zero nullity, consider first the space $\breve{X} = X^N$ and note that $\widehat{0} = N(\mathcal{Q})$. Let $\widehat{u} \in X^N$ be such that $\mathcal{Q}\widehat{u} = 0$. If $\widehat{u} \neq \widehat{0}$, the class $\widehat{u}$ must contain a function $u \notin \widehat{0}$. This contradicts the fact that, for all $u \in \widehat{u}$, $Q\widehat{u} = 0$ if and only if $Qu = 0$. Consider next the space $\check{X} = X^C$ and assume that $u \in X^C$ is such that $Qu = 0$. Then, $u \in X^C \cap N(Q)$, i.e., both $\vec{\ell}(u) = 0$ and $u \in N(Q)$. From Corollary C.3, it follows that $u = 0$. The result for $\check{X} = X^\perp$ is obvious. □

Note that (C.9) changes the domain of Q but not the operator itself.

Injective mappings by operator redefinition

The spaces (C.4), (C.5), and (C.7) are nonstandard domains for Q. The elements of (C.4) are equivalence classes of functions and the elements of $X^\perp$ and X^C are subject to constraints. Approximating in such nonstandard spaces can be more difficult than approximating in X. Therefore, it is desirable to have procedures that eliminate the null-space by changing the operator Q instead of its domain X.

Because $N(Q)$ is isomorphic to $\mathbb{R}^K$, to eliminate the null-space it suffices to augment Q by an operator having domain X and range $\mathbb{R}^K$ whose restriction to $N(Q)$ is invertible. According to Corollary C.3, $\vec{\ell} : X \mapsto \mathbb{R}^K$ is one such mapping so that we are led to the operator $Q_\ell \in L(X, Y \times \mathbb{R}^K)$ defined by

$$Q_\ell : \quad X \ni u \mapsto \begin{pmatrix} Qu \\ \vec{\ell}(u) \end{pmatrix} \in Y \times \mathbb{R}^K \tag{C.10}$$

and for which we have the following result.

Lemma C.5 *The operator $\{Q_\ell, X\}$ is Fredholm with zero nullity and the same deficiency as $\{Q, X\}$.*

Proof. Let $u \in X$ be such that $Q_\ell u = 0$. Then, $Qu = 0$, $\vec{\ell}(u) = 0$, and, from Corollary C.3, it follows that $u = 0$.

To prove the second assertion, we must show that the range of $\{Q_\ell, X\}$ is $R(Q) \times \mathbb{R}^K$, i.e., that for any $f \in R(Q)$ and $\vec{c} \in \mathbb{R}^K$, there exists $u \in X$ such that $Q_\ell u = \{f, \vec{c}\}$. Because $f \in R(Q)$, there exists u_f such that $Qu_f = f$. On the other hand, the linear system

$$\mathsf{L}\vec{b} = \vec{c} - \vec{\ell}(u_f),$$

where L is the matrix defined in (C.6), has a unique solution $\vec{b} = (\beta_1, \ldots, \beta_N)^\mathsf{T}$ for any right-hand side. Then, the function

$$u_N = \sum_{i=1}^{K} \beta_i u_i$$

belongs to $N(Q)$ and has the property that $\vec{\ell}(u_N) = \vec{c} - \vec{\ell}(u_f)$. Therefore,

$$Q_\ell(u_f + u_N) = \{f, \vec{c}\}. \qquad \square$$

Both the operators defined in (C.9) and (C.10) have trivial kernels. The difference is that for (C.10), we changed the operator and retained the standard domain X whereas for (C.9), we retained the operator and changed the domain.

Surjective mappings by operator and domain redefinition

Assume now that $\mathcal{Q}$ has positive deficiency. Then, there exist $K^* = \mathrm{def}\mathcal{Q} < \infty$ linearly independent functions $v_1, \ldots, v_{K^*}$ belonging to Y such that, for any $u \in X$,

$$(\mathcal{Q}u, v_k)_Y = 0 \qquad \text{and} \qquad (v_j, v_k)_Y = \delta_{jk} \qquad \text{for } i,k = 1,\ldots,K^*.$$

Such an operator $\mathcal{Q}$ is not surjective and (C.1) is solvable if and only if f satisfies the K^* compatibility conditions

$$(f, v_k)_Y = 0 \qquad \text{for } k = 1,\ldots,K^*. \tag{C.11}$$

We seek a modification of $\{\mathcal{Q}, X\}$ that reduces its deficiency to zero. To define such a modification, note that any $f \in Y$ can be written as

$$f = \mathcal{Q}u + \sum_{i=1}^{K^*} \alpha_k v_k = \mathcal{Q}u + \vec{v} \cdot \vec{a}$$

for some $u \in X$, $\vec{v} = (v_1, \ldots, v_{K^*})^\mathsf{T}$, and $\vec{a} = (\alpha_1, \ldots, \alpha_{K^*})^\mathsf{T}$, where $\alpha_k = (f, v_k)_Y$. This prompts the extension of $\mathcal{Q}$ to the following operator:

$$\mathcal{Q}_v : \quad X \times \mathbb{R}^{K^*} \ni \{u, \vec{a}\} \mapsto \mathcal{Q}u + \vec{v} \cdot \vec{a} \in Y. \tag{C.12}$$

As a result, we have proved the following result.

Lemma C.6 $\{\mathcal{Q}_v, X \times \mathbb{R}^{K^*}\}$ *is a Fredholm operator having zero deficiency and the same nullity as* $\{\mathcal{Q}, X\}$. □

Note that in this case we have changed both the operator and the domain.

C.2 Energy Balances

We now derive energy balances of the type (C.2) for the auxiliary operators defined in Section C.1.

Operators having non-trivial kernels

The operators $\mathcal{Q} : \check{X} \mapsto Y$ and $\mathcal{Q}_\ell : X \mapsto Y \times \mathbb{R}^K$ defined in (C.9) and (C.10), respectively, have the same deficiency as $\mathcal{Q}$ but their kernels are trivial. Therefore, (C.9) and (C.10) are injective mappings with closed ranges which makes it possible to establish energy balances for both of them.

Theorem C.7 *Given Hilbert spaces X and Y and a Fredholm operator $\mathcal{Q}: X \mapsto Y$, let the space $\breve{X}$ and operator $\mathcal{Q}_\ell$ be defined as in* (C.9) *and* (C.10), *respectively. Then, there exist positive constants C_1 and C_2 such that*

$$C_1 \|\breve{u}\|_{\breve{X}} \leq \|\mathcal{Q}\breve{u}\|_Y \leq C_2 \|\breve{u}\|_{\breve{X}} \qquad \forall \breve{u} \in \breve{X} \tag{C.13}$$

and

$$C_1 \|u\|_X \leq \|\mathcal{Q}_\ell u\|_{Y \times \mathbb{R}^K} \leq C_2 \|u\|_X \qquad \forall u \in X\,, \tag{C.14}$$

where K denotes the nullity of $\mathcal{Q}$.

Proof. The arguments for both cases are similar so we only provide the proof of (C.14). The range of $\mathcal{Q}_\ell$ is $R(\mathcal{Q}) \times \mathbb{R}^K$. By assumption, $R(\mathcal{Q})$ is a closed subspace of Y. Therefore, $R(\mathcal{Q})$, endowed with the inner product of Y, is also a Hilbert space; see, e.g., [250, Theorem 1.4-7, p. 30]. As a result, $R(\mathcal{Q}) \times \mathbb{R}^K$ is a Hilbert space as well. Moreover, considered as a mapping $X \mapsto R(\mathcal{Q}) \times \mathbb{R}^K$, the operator $\mathcal{Q}_\ell$ is a bijection. Then, (C.14) follows from the bounded inverse theorem; see [250, Theorem 4.12-2]. □

Deficient operators

We extended the original operator $\mathcal{Q}$ to the operator $\mathcal{Q}_v$ of (C.12) having the same nullity and a trivial co-range. By itself, this operator is not very useful, but if the same modification is applied to $\{\mathcal{Q}, \breve{X}\}$ or $\{\mathcal{Q}_\ell, X\}$ defined in (C.9) and (C.10), respectively, they become *bijective* operators.

Lemma C.8 *Let the operator $\{\mathcal{Q}_v, \breve{X}\}$ be defined by*

$$\mathcal{Q}_v : \breve{X} \times \mathbb{R}^{K^*} \mapsto Y \qquad \textit{with} \qquad \mathcal{Q}_v\{\breve{u}, \vec{a}\} = \mathcal{Q}\breve{u} + \vec{v} \cdot \vec{a}\,. \tag{C.15}$$

Then, the operator equation

$$\mathcal{Q}_v\{\breve{u}, \vec{a}\} = f \tag{C.16}$$

has a unique solution $\{\breve{u}, \vec{a}\} \in \breve{X} \times \mathbb{R}^{K^*}$ for any $f \in Y$. Here, $\vec{a} = (f, \vec{v})_Y$. □

Lemma C.9 *Let the operator $\{\mathcal{Q}_{\ell,v}, X \times \mathbb{R}^{K^*}\}$ be defined by*

$$\mathcal{Q}_{\ell,v} : X \times \mathbb{R}^{K^*} \mapsto Y \times \mathbb{R}^K \qquad \textit{with} \qquad \mathcal{Q}_{\ell,v}\{u, \vec{a}\} = \begin{pmatrix} \mathcal{Q}u + \vec{v} \cdot \vec{a} \\ \vec{\ell}(u) \end{pmatrix}. \tag{C.17}$$

Then, the operator equation

$$\begin{cases} \mathcal{Q}u + \vec{v} \cdot \vec{a} = f \\ \vec{\ell}(u) = \vec{c} \end{cases} \tag{C.18}$$

has a unique solution $\{u,\vec{a}\} \in X \times \mathbb{R}^{K^*}$ *for any* $\{f,\vec{c}\} \in Y \times \mathbb{R}^{K}$. □

The main purpose of (C.15) and (C.17) is to provide the auxiliary problems (C.16) and (C.18), respectively. The relevant energy balance for those problems are as follows.

Theorem C.10 *Assume the hypotheses of Lemmas* C.8 *and* C.9 *hold. Then, there exist two positive constants*[3] C_1 *and* C_2 *such that*

$$C_1\big(\|\breve{u}\|_{\breve{X}} + |\vec{a}|\big) \le \|Q\breve{u} + \vec{v}\cdot\vec{a}\|_Y \le C_2\big(\|\breve{u}\|_{\breve{X}} + |\vec{a}|\big) \tag{C.19}$$

for all $\{u,\vec{a}\} \in \breve{X} \times \mathbb{R}^{K^*}$ *and*

$$C_1\big(\|u\|_X + |\vec{a}|\big) \le \|Qu + \vec{v}\cdot\vec{a}\|_Y + |\vec{\ell}(u)| \le C_2\big(\|u\|_X + |\vec{a}|\big) \tag{C.20}$$

for all $\{u,\vec{a}\} \in X \times \mathbb{R}^{K^*}$. □

[3] Recall that $|\vec{b}|$ denotes the Euclidean norm of a vector $\vec{b} \in \mathbb{R}^n$; see Section A.1.

Appendix D
The Agmon–Douglis–Nirenberg Theory and Verifying its Assumptions

In this appendix, we present basic notions of the Agmon–Douglis–Nirenberg (ADN) theory [2] for partial differential equations (PDEs) and then show how, in some concrete settings, the assumptions of that theory are verified.

D.1 The Agmon–Douglis–Nirenberg Theory

We consider the abstract boundary value problem (3.16) from Section 3.2: find a function $u = \{u_1, u_2, \ldots, u_M\} \in X(\Omega)$ such that

$$\begin{cases} \mathcal{L}(\mathbf{x}, D)u = f & \text{in } Y(\Omega) \\ \mathcal{B}(\mathbf{x}, D)u = g & \text{in } B(\partial\Omega) . \end{cases}$$

In the ADN theory, u is viewed as a collection of independent scalar functions. Accordingly, the operator $\mathcal{L}(\mathbf{x}, D)$ assumes the form of a square $M \times M$ matrix whose ijth element is a scalar differential expression:

$$\mathcal{L}_{ij}(\mathbf{x}, D) = \sum_{|\alpha_{ij}| \le r_{ij}} a_{\alpha_{ij}}(\mathbf{x}) D^{\alpha_{ij}} , \qquad i, j = 1, \ldots, M , \tag{D.1}$$

and $\mathcal{B}(\mathbf{x}, D)$ is a rectangular $L \times M$ matrix with elements

$$\mathcal{B}_{lj}(\mathbf{x}, D) = \sum_{|\beta_{lj}| \le q_{lj}} b_{\beta_{lj}}(\mathbf{x}) D^{\beta_{lj}} , \qquad l = 1, \ldots, L, \; j = 1, \ldots, M . \tag{D.2}$$

Here, α_{ij} and β_{lj} are multi-indices[1], r_{ij} and q_{lj} are nonnegative integers, and D^α is the partial derivative operator defined in (A.1).

To discuss the ADN theory, recall the notion of the *symbol* of a differential operator; see, e.g., [304, p. 38]. Let $\vec{\xi} \in \mathbb{R}^d$. The symbol $\mathcal{L}(\mathbf{x}, \vec{\xi})$ of $\mathcal{L}(\mathbf{x}, D)$ is the square

[1] See Section A.1 for definition of this notation.

P.B. Bochev, M.D. Gunzburger, *Least-Squares Finite Element Methods*, Applied Mathematical Sciences 166, DOI: 10.1007/b13382_16,

$M \times M$ matrix with elements

$$\mathcal{L}_{ij}(\mathbf{x}, \vec{\xi}) = \sum_{|\alpha_{ij}| \le r_{ij}} a_{\alpha_{ij}}(\mathbf{x}) \vec{\xi}^{\alpha_{ij}}, \qquad i, j = 1, \ldots, M,$$

where $\vec{\xi}^{\alpha} = \xi_1^{\alpha_1} \cdots \xi_d^{\alpha_d}$ for a multi-index $\alpha = (\alpha_1, \ldots, \alpha_d)$. Similarly, the symbol of $\mathcal{B}(\mathbf{x}, D)$ is the rectangular $L \times M$ matrix $\mathcal{B}(\mathbf{x}, \vec{\xi})$ with elements

$$\mathcal{B}_{lj}(\mathbf{x}, \vec{\xi}) = \sum_{|\beta_{lj}| \le q_{lj}} b_{\beta_{lj}}(\mathbf{x}) \vec{\xi}^{\beta_{lj}}, \qquad l = 1, \ldots, L, \; j = 1, \ldots, M.$$

Note that the symbol components are polynomials in $\vec{\xi}$. Therefore, the symbol is an algebraic operation.

We focus on differential operators that are elliptic in the sense of the following definition [2]:

Definition D.1. (ADN ellipticity) The system (3.16) is *ADN-elliptic* if there exist integer weights $\{s_i\}$ and $\{t_j\}$ for the equations and unknowns, respectively, such that[2]

1. $\deg \mathcal{L}_{ij}(\mathbf{x}, \vec{\xi}) \le s_i + t_j$
2. $\mathcal{L}_{ij}(\mathbf{x}, \vec{\xi}) \equiv 0$ whenever $s_i + t_j < 0$
3. $\det \mathcal{L}^P_{ij}(\mathbf{x}, \vec{\xi}) \neq 0$ for all real $\vec{\xi} \neq 0$, where the *principal part* $\mathcal{L}^P$ of $\mathcal{L}$ is defined as all terms $\mathcal{L}_{ij}$ for which $\deg \mathcal{L}_{ij}(\mathbf{x}, \vec{\xi}) = s_i + t_j$. □

Furthermore, we assume that $\mathcal{L}$ is *uniformly elliptic* operator of order $2m$, where $m \ge 1$ is integer. This means that there exists a positive constant C such that

$$C^{-1} |\vec{\xi}|^{2m} \le |\det \mathcal{L}^P(\mathbf{x}, \vec{\xi})| \le C |\vec{\xi}|^{2m}. \tag{D.3}$$

A necessary condition for a well-posed boundary value problem is for $\mathcal{B}$ to provide exactly m boundary conditions; therefore, in what follows we always take $L = m$.

The final condition on $\mathcal{L}$ which is satisfied for all elliptic systems in three or more space dimensions but must be assumed in two dimensions is the *supplementary condition* of [2].

Definition D.2. (Supplementary condition on $\mathcal{L}$) The operator $\mathcal{L}$ satisfies the *supplementary condition* if, for every pair of linearly independent vectors $\vec{\xi}$ and $\vec{\xi}'$, the polynomial[3] $\det \mathcal{L}^P(\mathbf{x}, \vec{\xi} + \tau \vec{\xi}')$ in the complex variable τ has exactly m roots with positive imaginary part. □

[2] It can be shown that for non-degenerate elliptic systems, one can always find s_i and t_j such that the principal part $\mathcal{L}^P$ does not vanish identically; see [341]. For such systems, the degree r of the determinant $\det \mathcal{L}(\mathbf{x}, \vec{\xi})$ equals the maximal degree r_{max} of the terms forming $\det \mathcal{L}(\mathbf{x}, \vec{\xi})$; in general, $r \le r_{max}$. Furthermore, in [341] it is shown that Definition D.1 is equivalent to ellipticity in the following sense: $r = r_{max}$ and $\vec{\xi} \equiv 0$ is the only real root of $(\det \mathcal{L}(\mathbf{x}, \vec{\xi}))' = 0$, where $(\det \mathcal{L}(\mathbf{x}, \vec{\xi}))'$ denotes the part of $\det \mathcal{L}(\mathbf{x}, \vec{\xi})$ of order r.

[3] Note that $\det \mathcal{L}^P(\mathbf{x}, \vec{\xi})$ is of even degree $2m$ with respect to $\vec{\xi}$.

An elliptic system that satisfies the supplementary condition is also called *regular elliptic*; see [307].

In addition to the sets $\{s_i\}$ and $\{t_j\}$, we introduce another set of integer weights $\{r_l\}, l = 1, \ldots, L$, for the boundary operator $\mathcal{B}$. Each r_l is attached to the lth boundary condition in (3.16) and must satisfy the inequality

$$\deg \mathcal{B}_{lj}(\mathbf{x}, \vec{\xi}) \le r_l + t_j$$

with the understanding that $\mathcal{B}_{lj} \equiv 0$ when $r_l + t_j < 0$. Finally, the principal part $\mathcal{B}^P$ of the boundary operator is defined as all terms $\mathcal{B}_{lj}$ such that $\deg \mathcal{B}_{lj}(\mathbf{x}, \vec{\xi}) = r_l + t_j$. The three sets of ADN weights can always be normalized in such a way that $s_i \le 0$, $r_l \le 0$, and $t_j \ge 0$. However, the weights may not be unique, even with such a normalization, i.e., there are examples of operators $\mathcal{L}$ for which one can define more than one principal part $\mathcal{L}^P$ that satisfies Definition D.1.

An important subset of ADN-elliptic systems is the class of Petrovski systems; see [307].

Definition D.3. A system is *elliptic in the sense of Petrovski* if it is elliptic in the sense of ADN and $s_1 = \cdots = s_M = 0$. If, in addition $t_1 = \cdots = t_M$, the system is called *homogeneous elliptic*. □

Given an ADN-elliptic operator $\mathcal{L}$, the problem (3.16) is well-posed if and only if the boundary operator $\mathcal{B}$ "complements" $\mathcal{L}$ in a proper way. In [2], it is shown that this is equivalent to an algebraic condition, called the *complementing condition*, on the principal parts $\mathcal{L}^P$ and $\mathcal{B}^P$.

To state this condition, let $\mathbf{n}$ denote the unit normal to $\partial\Omega$ at $\mathbf{x}$, $\tau_k^+(\mathbf{x}, \vec{\xi})$ denote the m roots of $\det \mathcal{L}^P(\mathbf{x}, \vec{\xi} + \tau\mathbf{n})$ having positive imaginary part, and

$$M^+(\mathbf{x}, \vec{\xi}, \tau) = \prod_{k=1}^{m} \left(\tau - \tau_k^+(\mathbf{x}, \vec{\xi})\right).$$

Finally, let $\mathcal{L}'$ denote the adjugate[4] matrix of $\mathcal{L}^P$.

Definition D.4. (Complementing condition) For any point $\mathbf{x} \in \partial\Omega$ and any real nonzero vector $\vec{\xi}$ tangent to $\partial\Omega$ at $\mathbf{x}$, regard $M^+(\mathbf{x}, \vec{\xi}, \tau)$ and the elements of the matrix $\mathcal{B}^P(\mathbf{x}, \vec{\xi} + \tau\mathbf{n})\mathcal{L}'(\mathbf{x}, \vec{\xi} + \tau\mathbf{n})$ given by

$$\sum_{j=1}^{M} \mathcal{B}_{lj}^P(\mathbf{x}, \vec{\xi} + \tau\mathbf{n})\mathcal{L}'_{jk}(\mathbf{x}, \vec{\xi} + \tau\mathbf{n}) \quad l = 1, \ldots, m \quad k = 1, \ldots, M$$

as polynomials in τ. The operators $\mathcal{L}$ and $\mathcal{B}$ satisfy the complementing condition if and only if the rows of the latter matrix are linearly independent modulo $M^+(\vec{\xi}, \tau)$, i.e.,

[4] Recall that the *adjugate* or *cofactor* matrix of A is the matrix $\mathsf{A}' = (\det \mathsf{A})\mathsf{A}^{-1}$; see [322, p.232].

$$\sum_{l=1}^{m} C_l \left(\sum_{j=1}^{M} \mathcal{B}^P_{lj}(\mathbf{x}, \vec{\xi} + \tau\mathbf{n}) \mathcal{L}'_{jk}(\mathbf{x}, \vec{\xi} + \tau\mathbf{n}) \right) \equiv 0 \qquad \bmod M^+ \tag{D.4}$$

for $k = 1, \ldots, M$, if and only if the constants C_l are all zero.[5] □

For brevity, in what follows we call the boundary value problem (3.16) *elliptic* if:

1. $\mathcal{L}$ is elliptic in the sense of ADN
2. $\mathcal{L}$ is regular elliptic
3. $\mathcal{L}$ is uniformly elliptic
4. $\mathcal{B}$ satisfies the complementing condition.

The main result of the ADN theory, relevant to least-squares methods, is a priori estimates that give rise to energy balances required to define well-posed least-squares principles for (3.16). These estimates are stated in terms of solution and data spaces that are direct products of the scalar spaces $H^k(\Omega)$ and $H^{k\pm1/2}(\partial\Omega)$, parameterized by a nonnegative integer *regularity index* q:

$$X_q = \prod_{j=1}^{M} H^{q+t_j}(\Omega) \qquad Y_q = \prod_{i=1}^{M} H^{q-s_i}(\Omega) \qquad B_q = \prod_{l=1}^{m} H^{q-r_l-1/2}(\partial\Omega). \tag{D.5}$$

Theorem D.1. *Let* $t' = \max t_j$, $q \geq r' = \max(0, \max r_l + 1)$, *and assume that* Ω *is a bounded domain of class* $C^{q+t'}$. *Furthermore, assume that the coefficients of* $\mathcal{L}_{ij}$ *are of class* $C^{q-s_i}(\overline{\Omega})$ *and that the coefficients of* $\mathcal{B}_{lj}$ *are of class* $C^{q-r_l}(\partial\Omega)$. *If* (3.16) *is elliptic and* $f \in Y_q$ *and* $g \in B_q$, *then*

1. *every solution* $u \in X_{r'}$ *in fact belongs to* X_q
2. *there is a positive constant* C, *independent of* u, f, *and* g, *such that, for every solution* $u \in X_q$,

$$\sum_{j=1}^{M} \|u_j\|_{q+t_j} \leq C\Big(\sum_{i=1}^{M} \|f_i\|_{q-s_i} + \sum_{l=1}^{m} \|g_l\|_{q-r_l-1/2,\partial\Omega} + \sum_{j=1}^{M} \|u_j\|_0 \Big). \tag{D.6}$$

Moreover, if the problem (3.16) *has a unique solution, then the* $L^2(\Omega)$*-norm on the right-hand side of* (D.6) *can be omitted.* □

[5] An alternative formulation of the complementing condition is as follows [304]. Assume that in a neighborhood of a point $\mathbf{x} \in \partial\Omega$ the boundary $\partial\Omega$ is flattened so that it lies on the plane $z = 0$. Then, on $z \geq 0$, consider a homogeneous, constant coefficient boundary value problem obtained by taking the principal parts of the differential and boundary operators of the original problem (3.16) and freezing their coefficients at $\mathbf{x}$:

$$\mathcal{L}^P(\mathbf{x}, D)u = 0 \quad \text{in } z \geq 0 \qquad \text{and} \qquad \mathcal{B}^P(\mathbf{x}, D)u = 0 \quad \text{on } z = 0.$$

The complementing condition holds for (3.16) if all solutions of these equations of the form $u = \exp(i\mathbf{x} \cdot \vec{\xi})w(z)$, where $\mathbf{x} = (x, y, 0)$, are identically zero, i.e., $w \equiv 0$. The ansatz for u reduces the above equations to a system of ordinary differential equations for the unknown function w and provides an alternative way to verifying that the complementing condition holds.

It can be shown that ADN-elliptic operators are of Fredholm type (see [307, 308, 342]) so that their range is closed and both the kernel and the co-range are finite dimensional. Therefore, Assumption 3.7 holds and $\{\mathcal{L}, \mathcal{B}\}$ can be extended to a bijective operator given by (3.17). Thus, without loss of generality, we can assume that (D.6) holds without the $L^2(\Omega)$ terms on the right-hand side.

To deal with conforming but non-compliant least-squares finite element methods (LSFEMs), we need one more, less straightforward, assumption about (3.16). To state this assumption, note that for all sufficiently smooth functions $u \in X_q$, the ADN a priori bound (D.6) implies the inequality

$$\|u\|_{X_q} \le C_q \left(\|\mathcal{L}u\|_{Y_q} + \|\mathcal{B}u\|_{B_q} \right) \tag{D.7}$$

with q a nonnegative integer. For the analysis of some LSFEMs, it is necessary to assume that this a priori bound can be extended to all real values of q.

Assumption D.2 *For every real q, there exists a constant C_q such that* (D.7) *holds for all smooth functions in Ω, i.e.,*

$$\|u\|_{X_q} \le C_q \left(\|\mathcal{L}u\|_{Y_q} + \|\mathcal{B}u\|_{B_q} \right) \qquad \forall q \in \mathbb{R} \quad \textit{and} \quad \forall u \in C^\infty(\overline{\Omega}). \tag{D.8}$$

Assumption D.2 is equivalent to the existence of a complete set of homeomorphisms for (3.16). It can be shown that (D.8) holds for homogeneous elliptic systems; see [307, 308]. However, it is not guaranteed to hold for general ADN systems. Thus, whenever dealing with non-homogeneous elliptic ADN problems, we have to verify Assumption D.2.

D.2 Verifying the Assumptions of the Agmon-Douglis-Nirenberg Theory

The application of the abstract least-squares theory of Chapter 4 to a given PDE problem requires one to verify the assumptions of the ADN elliptic theory presented in Section D.1. In this section, we show how to do this for several of the model problems considered in the book. Because the coefficients of the PDEs are irrelevant for this task (as long as they meet the smoothness requirements of the theory), for the sake of clarity, here we work with "coefficientless" forms of the PDEs.[6] We refer to these forms as *basic* PDEs. We begin, in Sections D.2.1–D.2.3, by checking the assumptions of the ADN theory for the basic div–grad, div–grad–curl, and div–curl problems. Then, in Sections D.2.4 and D.2.5, we show how to check these assumptions for two first-order versions of the Stokes equations. This material expands the discussions found in [34, 35, 53, 55].

---

[6] The algebraic conditions of the ADN theory are not difficult to verify, but the calculations can be very tedious. Allowing for non-constant coefficients in the PDE in this context introduces additional complexity that only distracts from the main ideas without contributing much in terms of insight. For this reason, all PDEs considered in this appendix have unit coefficients.

Some of the first-order systems considered in this section have principal parts whose ellipticity properties remain the same in two and three-dimensions. For such systems, verifying the ADN theory in $\mathbb{R}^2$ or $\mathbb{R}^3$ is very similar so that we consider only the shorter, two-dimensional case. However, in three dimensions, we also encounter systems that fail to be elliptic in the sense of ADN or have principal parts that change their ellipticity properties. In $\mathbb{R}^3$, such systems need to be "augmented" by additional "redundant" equations and/or "slack" variables. For such systems, we check the assumptions of the ADN theory separately in $\mathbb{R}^2$ and $\mathbb{R}^3$.

Recall that, in the ADN theory, differential and boundary operators are viewed as matrices of differential expressions acting on vector fields that are viewed as collections of independent scalar coordinate functions. For this reason, in this section, PDEs are often written using component notation instead of the basic vector calculus operators grad, curl, and div.

D.2.1 Div–Grad Systems

The div–grad system (5.18) is an example of a system that is ADN elliptic in two and three dimensions. In component form, this system with homogeneous Dirichlet or Neumann boundary conditions is given by, in two-dimensions,

$$\begin{cases} u_{1,x}+u_{2,y}=f \\ \phi_x+u_1=0 \\ \phi_y+u_2=0 \end{cases} \quad \text{in } \Omega \tag{D.9}$$

and

$$\begin{cases} \phi=0 & \text{Dirichlet} \\ \text{or} & \\ u_1n_1+u_2n_2=0 & \text{Neumann} \end{cases} \quad \text{on } \partial\Omega\,, \tag{D.10}$$

where $\mathbf{n}=(n_1,n_2)^{\mathsf{T}}$ denotes the unit outer normal to the boundary $\partial\Omega$. Assuming that the equations are ordered as in (D.9) and that the variables are ordered as $\{\phi,u_1,u_2\}$, the symbols of the differential and boundary operators in (D.9) and (D.10) are given by

$$\mathcal{L}(\mathbf{x},\vec{\xi})=\begin{pmatrix} 0 & \xi_1 & \xi_2 \\ \xi_1 & 1 & 0 \\ \xi_2 & 0 & 1 \end{pmatrix} \quad \text{and} \quad \mathcal{B}(\mathbf{x},\vec{\xi})=\begin{cases} (1,0,0) & \text{Dirichlet} \\ \text{or} & \\ (0,n_1,n_2) & \text{Neumann}\,, \end{cases}$$

respectively.

To show that the div–grad system is elliptic in the sense of ADN, we select the following weights for the equations, unknowns, and boundary conditions:

0	0	$u_{1,x}$	$u_{2,y}$
-1	ϕ_x	u_1	0
-1	ϕ_y	0	u_2
$s \uparrow\ t \rightarrow$	2	1	1

Dirichlet	-2	1	0	0
Neumann	-1	0	n_1	n_2
	$r \uparrow\ t \rightarrow$	2	1	1

Table D.1 Determination of the principal parts $\mathcal{L}^P$ (left) and $\mathcal{B}^P$ (right) for the div–grad system using the weights from (D.11). For all entries in the table on the left, we have that $\deg \mathcal{L}_{ij} = s_i + t_i$ so that $\mathcal{L}^P = \mathcal{L}$. For all entries in the first and second rows of the table on the right we have that $\deg \mathcal{B}_{ij} = r_i + t_i$ so that for both boundary conditions $\mathcal{B}^P = \mathcal{B}$.

$$s_1 = 0, s_2 = s_3 = -1 \qquad t_1 = 2, t_2 = t_3 = 1 \qquad r_1 = \begin{cases} -2 & \text{Dirichlet} \\ \text{or} & \\ -1 & \text{Neumann}. \end{cases} \tag{D.11}$$

Using Table D.1, one can easily determine that for the weights in (D.11), we have the principal parts[7] $\mathcal{L}^P(\mathbf{x}, \vec{\xi}) = \mathcal{L}(\mathbf{x}, \vec{\xi})$ and $\mathcal{B}^P(\mathbf{x}, \vec{\xi}) = \mathcal{B}(\mathbf{x}, \vec{\xi})$ so that[8]

$$\det \mathcal{L}^P(\mathbf{x}, \vec{\xi}) = \begin{vmatrix} 0 & \xi_1 & \xi_2 \\ \xi_1 & 1 & 0 \\ \xi_2 & 0 & 1 \end{vmatrix} = -|\vec{\xi}|^2 .$$

Thus, (D.3) is satisfied with $m = 1$ and $C = 1$, i.e., the two-dimensional div–grad system is uniformly elliptic of order 2. It can be easily verified that the three-dimensional div–grad system has the same order.

Remark D.3 The div–grad system illustrates an important distinction between the ADN definition of ellipticity and the "usual" definition where $\mathcal{L}^P$ is defined by taking only the differentiated terms. Such a principal part would correspond to $s_1 = s_2 = s_3 = 0$, $t_1 = t_2 = t_3 = 1$, and

$$\det \mathcal{L}^P(\mathbf{x}, \vec{\xi}) = \begin{vmatrix} 0 & \xi_1 & \xi_2 \\ \xi_1 & 0 & 0 \\ \xi_2 & 0 & 0 \end{vmatrix} \equiv 0 .$$

Thus, the div–grad system is not elliptic in the "usual" sense. □

In two dimensions, we also need to verify the supplementary condition.

Proposition D.4 *The div–grad system* (D.9) *satisfies the supplementary condition given in Definition* D.2.

---

[7] Recall that the principal part $\mathcal{L}^P$ of $\mathcal{L}$ is defined as all terms $\mathcal{L}_{ij}$ for which $\deg \mathcal{L}_{ij}(\mathbf{x}, \vec{\xi}) = s_i + t_j$ and that the principal part $\mathcal{B}^P$ of the boundary operator $\mathcal{B}$ is defined as all terms $\mathcal{B}_{lj}$ such that $\deg \mathcal{B}_{lj}(\mathbf{x}, \vec{\xi}) = r_l + t_j$.

[8] Throughout this chapter, for any two vectors $\vec{\xi}$ and $\vec{\xi}'$, $\vec{\xi} \cdot \vec{\xi}'$ denotes their Euclidean inner product; $|\vec{\xi}| = (\vec{\xi} \cdot \vec{\xi})^{1/2}$ denotes the Euclidean length of $\vec{\xi}$ only if its components are real.

Proof. The supplementary condition holds if for every pair of linearly independent real vectors $\vec{\xi}$ and $\vec{\xi}'$, the polynomial $\det \mathcal{L}^P(\mathbf{x}, \vec{\xi} + \tau\vec{\xi}')$ in the complex variable τ has exactly one root with a positive imaginary part. Let $\vec{\xi}$ and $\vec{\xi}'$ be two linearly independent vectors. Then,

$$\det \mathcal{L}^P(\mathbf{x}, \vec{\xi} + \tau\vec{\xi}') = -|\vec{\xi}|^2 - 2\tau\vec{\xi}\cdot\vec{\xi}' - \tau^2|\vec{\xi}'|^2 .$$

Therefore, the roots of $\det \mathcal{L}^P(\mathbf{x}, \vec{\xi} + \tau\vec{\xi}') = 0$ are given by

$$\tau_{1,2} = \frac{-\vec{\xi}\cdot\vec{\xi}' \pm \sqrt{(\vec{\xi}\cdot\vec{\xi}')^2 - |\vec{\xi}|^2|\vec{\xi}'|^2}}{|\vec{\xi}'|^2} .$$

Because $\vec{\xi}$ and $\vec{\xi}'$ are linearly independent, $\vec{\xi}\cdot\vec{\xi}' < |\vec{\xi}||\vec{\xi}'|$ so that

$$(\vec{\xi}\cdot\vec{\xi}')^2 - |\vec{\xi}|^2|\vec{\xi}'|^2 < 0 .$$

Therefore, $\det \mathcal{L}^P(\mathbf{x}, \vec{\xi} + \tau\vec{\xi}') = 0$ has a pair of complex conjugate roots, one of which must have a positive imaginary part. □

The last condition to verify is the complementing condition.

Proposition D.5 *The div–grad system* (D.9) *with either the Dirichlet or Neumann boundary condition in* (D.10) *satisfies the complementing condition given in Definition* D.4.

Proof. Let $\vec{\eta} = \vec{\xi} + \tau\mathbf{n}$, where $\mathbf{n}$ is the unit outer normal to $\partial\Omega$ at $\mathbf{x}$ and $\vec{\xi}$ is a unit vector tangent to $\partial\Omega$ at the same point. To verify the complementing condition, we need to determine the adjugate matrix[9] $\mathcal{L}'$ of the principal part matrix $\mathcal{L}^P$, the matrix $\mathcal{B}^P(\mathbf{x}, \vec{\eta})\mathcal{L}'(\mathbf{x}, \vec{\eta})$, and the polynomial $M^+(\mathbf{x}, \vec{\xi}, \tau)$ whose roots coincide with the m roots of $\det \mathcal{L}^P(\mathbf{x}, \vec{\eta}) = 0$ having positive imaginary parts. A short calculation shows that

$$\mathcal{L}'(\mathbf{x}, \vec{\eta}) = \begin{pmatrix} 1 & -\eta_1 & -\eta_2 \\ -\eta_1 & -\eta_2^2 & \eta_1\eta_2 \\ -\eta_2 & \eta_1\eta_2 & -\eta_1^2 \end{pmatrix}$$

so that

$$\mathcal{B}^P(\mathbf{x}, \vec{\eta})\mathcal{L}'(\mathbf{x}, \vec{\eta}) = \big(1, -\eta_1, -\eta_2\big) \tag{D.12}$$

for the Dirichlet boundary condition and

$$\mathcal{B}^P(\mathbf{x}, \vec{\eta})\mathcal{L}'(\mathbf{x}, \vec{\eta}) = \big(-\eta_1 n_1 - \eta_2 n_2,\ \eta_1\eta_2 n_2 - \eta_2^2 n_1,\ \eta_1\eta_2 n_1 - \eta_1^2 n_2\big) \tag{D.13}$$

for the Neumann boundary condition. Also, because the unit vectors $\vec{\xi}$ and $\mathbf{n}$ are orthogonal, it follows that

[9] Recall that the adjugate matrix is given by $\mathsf{A}' = (\det \mathsf{A})\mathsf{A}^{-1}$.

$$\det \mathcal{L}^P(\mathbf{x},\vec{\eta}) = -(\vec{\xi}+\tau\mathbf{n})\cdot(\vec{\xi}+\tau\mathbf{n}) = -(1+\tau^2) = -(\tau+\mathrm{i})(\tau-\mathrm{i})\,.$$

Therefore, $\tau^+ = \mathrm{i}$ is the only root of $\det \mathcal{L}^P(\mathbf{x},\vec{\eta}) = 0$ with a positive imaginary part and $M^+(\mathbf{x},\vec{\xi},\tau) = (\tau-\mathrm{i})$.

Let us first verify the complementing condition for the Dirichlet boundary operator. For this condition to hold, the row vector in (D.12) must be linearly independent modulo M^+, i.e., we need to show that

$$(1,-\eta_1,-\eta_2) \neq (0,0,0) \mod (M^+)\,.$$

Recall that $(1,-\eta_1,-\eta_2) = (0,0,0) \mod (M^+)$ if and only if there exist polynomials $p_i(\tau)$, $i = 1,2,3$, at least one of which is not identically zero, such that

$$\begin{aligned} 1 &= (\tau-\mathrm{i})p_1(\tau) \\ -(\xi_1+\tau n_1) &= (\tau-\mathrm{i})p_2(\tau) \\ -(\xi_2+\tau n_2) &= (\tau-\mathrm{i})p_3(\tau)\,. \end{aligned}$$

It is easy to see that the above identities cannot hold for any choice of the polynomials $p_i(\tau)$. In the first identity the left-hand side is a constant and the right-hand side is at least a linear polynomial. The left-hand sides in the last two identities are linear polynomials with real roots; the right-hand sides have at least one purely imaginary root.

To verify the complementing condition for the Neumann boundary operator, we need to show that the right-hand side of (D.13) is not equal to zero mod (M^+). This reduces to showing that the identities

$$\begin{aligned} -(\xi_1+\tau n_1)n_1 - (\xi_2+\tau n_2)n_2 &= (\tau-\mathrm{i})p_1(\tau) \\ (\xi_1+\tau n_1)(\xi_2+\tau n_2)n_2 - (\xi_2+\tau n_2)^2 n_1 &= (\tau-\mathrm{i})p_2(\tau) \\ (\xi_1+\tau n_1)(\xi_2+\tau n_2)n_1 - (\xi_1+\tau n_1)^2 n_2 &= (\tau-\mathrm{i})p_3(\tau) \end{aligned}$$

cannot hold for any choice of non-zero polynomials $p_i(\tau)$. Note that because $\vec{\xi}$ and $\mathbf{n}$ are mutually orthogonal unit vectors, $-(\xi_1+\tau n_1)n_1 - (\xi_2+\tau n_2)n_2 = -\tau$. Therefore, $p_1(\tau)$ can be at most of degree zero. After setting $p_1(\tau) = \text{constant}$, we see that the right-hand side is a linear polynomial with a complex root and that the left-hand side has the real root $\tau = 0$, i.e., the first identity can never hold. To show that the last two identities are also impossible, we proceed as follows. Without loss of generality, we may assume that, at $\mathbf{x}$, the coordinate axes are aligned with $\mathbf{n}$ and $\vec{\xi}$ so that $\mathbf{n} = (0,-1)^\mathsf{T}$ and $\vec{\xi} = (1,0)^\mathsf{T}$. Then, $(\xi_1+\tau n_1) = 1$, $(\xi_2+\tau n_2) = -\tau$, and the last two relations respectively simplify to

$$\tau = (\tau-\mathrm{i})p_2(\tau) \qquad \text{and} \qquad 1 = (\tau-\mathrm{i})p_3(\tau)\,.$$

It is clear that these relations cannot hold for any $p_2(\tau) \neq 0$ and $p_3(\tau) \neq 0$. □

D.2.2 Div–Grad–Curl Systems

The basic version of the div–grad–curl system in three dimensions (see Section 12.9) is given by

$$\begin{cases} \nabla\times\mathbf{u}+\nabla\phi=\mathbf{g} \text{ in } \Omega \\ \nabla\cdot\mathbf{u}=f \text{ in } \Omega \end{cases} \quad \text{and} \quad \begin{cases} \mathbf{n}\cdot\mathbf{u}=0 \text{ and } \phi=0 \text{ on } \partial\Omega \\ \text{or} \\ \mathbf{n}\times\mathbf{u}=\mathbf{0} \qquad \text{on } \partial\Omega. \end{cases} \tag{D.14}$$

As is discussed in Section 12.9, reduction to two dimensions leads to two possible basic systems that collectively are included in the system

$$\begin{cases} -v_{1,x_2}+v_{2,x_1}=g \\ v_{1,x_1}+v_{2,x_2}=f \end{cases} \quad \text{in } \Omega \tag{D.15}$$

and one of

$$\begin{cases} v_1=0 \\ v_2=0 \\ v_1n_1+v_2n_2=0 \\ -v_1n_2+v_2n_1=0 \end{cases} \quad \text{on } \partial\Omega. \tag{D.16}$$

Note that if the first boundary condition in (D.14) or the third boundary condition in (D.16) is applied, f must satisfy the compatibility condition

$$\int_\Omega f\,d\Omega=0. \tag{D.17}$$

Div–grad–curl system in two dimensions

In two dimensions, the component form of the div–grad–curl system is given by (D.15) and (D.16). Assuming that the equations and unknowns are ordered as in (D.15) and $\{v_1,v_2\}$, respectively, the symbols of the differential and boundary operators are given by

$$\mathcal{L}(\mathbf{x},\vec{\xi})=\begin{pmatrix} -\xi_2 & \xi_1 \\ \xi_1 & \xi_2 \end{pmatrix} \quad \text{and one of} \quad \mathcal{B}(\mathbf{x},\vec{\xi})=\begin{cases} (1,0) \\ (0,1) \\ (n_1,n_2) \\ (-n_2,n_1), \end{cases}$$

respectively. To show that the div–grad–curl system (D.15) is ADN elliptic, we select the following weights:

$$s_1=s_2=0 \qquad t_1=t_2=1 \qquad r_1=-1.$$

Then, $\mathcal{L}^P(\mathbf{x},\vec{\xi}) = \mathcal{L}(\mathbf{x},\vec{\xi})$, $\det\mathcal{L}^P(\mathbf{x},\vec{\xi}) = -|\vec{\xi}|^2$, and (D.3) holds with $C = 1$ and $m = 1$. Therefore, (D.15) is uniformly elliptic of order two. We next verify that the supplementary and complementing conditions are satisfied.

Proposition D.6 *The two-dimensional div–grad–curl system* (D.15) *satisfies the supplementary condition given in Definition* D.2.

Proof. The proof is the same as that for Proposition D.4. □

Proposition D.7 *The two-dimensional div–grad–curl system* (D.15) *with any of the boundary conditions in* (D.16) *satisfies the complementing condition given in Definition* D.4.

Proof. Let $\vec{\eta} = \vec{\xi} + \tau\mathbf{n}$, where $\mathbf{n}$ is the unit outer normal to $\partial\Omega$ at $\mathbf{x}$ and $\vec{\xi}$ is a unit vector tangent to $\partial\Omega$ at the same point. Because the unit vectors $\vec{\xi}$ and $\mathbf{n}$ are orthogonal, it follows that $(\vec{\eta}\cdot\vec{\eta}) = (1+\tau^2)$ and

$$\det\mathcal{L}^P(\mathbf{x},\vec{\eta}) = -(\vec{\eta}\cdot\vec{\eta}) = -(1+\tau^2)\,.$$

Therefore, $\tau^+ = \mathrm{i}$ is the only root of $\det\mathcal{L}^P(\mathbf{x},\vec{\eta}) = 0$ with a positive imaginary part and $M^+(\mathbf{x},\vec{\xi},\tau) = (\tau-\mathrm{i})$.

The adjugate of $\mathcal{L}^P(\mathbf{x},\vec{\eta})$ is given by

$$\mathcal{L}'(\mathbf{x},\vec{\eta}) = \begin{pmatrix} \eta_2 & -\eta_1 \\ -\eta_1 & -\eta_2 \end{pmatrix}$$

so that, for the four choices for the boundary condition given in (D.16), we have that

$$\mathcal{B}^P(\mathbf{x},\vec{\eta})\cdot\mathcal{L}'(\mathbf{x},\vec{\eta}) = \begin{cases} (\eta_2\,,\,-\eta_1) \\ (-\eta_1\,,\,-\eta_2) \\ (n_1\eta_2 - n_2\eta_1\,,\,-n_1\eta_1 - n_2\eta_2) \\ (-n_2\eta_2 - n_1\eta_1\,,\,n_2\eta_1 - n_1\eta_2)\,, \end{cases}$$

respectively.

For the sake of brevity, we supply a proof only for the third boundary condition. The proof for the fourth boundary condition is essentially the same because that condition can be obtained from the third boundary condition through a change in sign and a reversal of positions. The proofs for the first and second boundary conditions are entirely similar.

For the third boundary condition in (D.16), the complementing condition holds if

$$(n_1\eta_2 - n_2\eta_1, -n_1\eta_1 - n_2\eta_2) \neq 0 \mod (M^+)\,.$$

Therefore, we need to show that the identities

$$\begin{aligned} (n_1\eta_2 - n_2\eta_1) &= (\tau-\mathrm{i})p_1(\tau) \\ -(n_1\eta_1 + n_2\eta_2) &= (\tau-\mathrm{i})p_2(\tau) \end{aligned}$$

cannot hold. Because the unit vectors $\vec{\xi}$ and $\mathbf{n}$ are orthogonal, we have that

$$n_1\eta_1 + n_2\eta_2 = n_1(\xi_1 + \tau n_1) + n_2(\xi_2 + \tau n_2) = \tau$$

so that the second identity simplifies to $-\tau = (\tau - \mathrm{i})p_2(\tau)$. The left-hand side is a linear polynomial with root $\tau = 0$. The right-hand side has the imaginary root $\tau = \mathrm{i}$ and so equality is impossible. To show that the first identity also cannot hold, we proceed as in Proposition D.5 and set $\mathbf{n} = (0,-1)$ and $\vec{\xi} = (1,0)$. Then, the first identity turns into $1 = (\tau - \mathrm{i})p_1(\tau)$ which is also impossible. □

Div–grad–curl system in three dimensions

The component form of the div–grad–curl system (D.14) in three dimensions is

$$\begin{cases} u_{3,y} - u_{2,z} + \phi_x = f_1 \\ u_{1,z} - u_{3,x} + \phi_y = f_2 \\ u_{2,x} - u_{1,y} + \phi_z = f_3 \\ u_{1,x} + u_{2,y} + u_{3,z} = g \end{cases} \quad \text{in } \Omega. \tag{D.18}$$

Assuming the ordering of the equations and unknowns as in (D.18) and $\{u_1, u_2, u_3, \phi\}$, respectively, the symbol of the differential operator is given by

$$\mathcal{L}(\mathbf{x}, \vec{\xi}) = \begin{pmatrix} 0 & -\xi_3 & \xi_2 & \xi_1 \\ \xi_3 & 0 & -\xi_1 & \xi_2 \\ -\xi_2 & \xi_1 & 0 & \xi_3 \\ \xi_1 & \xi_2 & \xi_3 & 0 \end{pmatrix}. \tag{D.19}$$

We verify the ADN theory using the following weights:

$$s_1 = s_2 = s_3 = s_4 = 0 \qquad t_1 = t_2 = t_3 = t_4 = 1 \qquad r_1 = r_2 = -1. \tag{D.20}$$

In this case, $\mathcal{L}^P(\mathbf{x}, \vec{\xi}) = \mathcal{L}(\mathbf{x}, \vec{\xi})$, $\det \mathcal{L}^P(\mathbf{x}, \vec{\xi}) = -|\vec{\xi}|^4$, and (D.3) holds with $C = 1$ and $m = 2$. Therefore, in three dimensions, the div–grad–curl system is uniformly elliptic of order four and needs two independent boundary conditions. Either set of boundary conditions in (D.14) meets this requirement and satisfies the complementing condition.

Proposition D.1. *The three-dimensional div–grad–curl system* (D.18) *with either of the boundary conditions given in* (D.14) *satisfies the complementing condition given in Definition* D.4.

Proof. For the sake of brevity, we only provide a proof for the first set of boundary conditions in (D.14). The proof for the second set of boundary conditions follows along the same lines.[10]

As before, let $\vec{\eta} = \vec{\xi} + \tau\mathbf{n}$, where $\mathbf{n}$ is the unit outer normal to $\partial\Omega$ at $\mathbf{x}$ and $\vec{\xi}$ is a unit vector tangent to $\partial\Omega$ at the same point. Then,

$$\det \mathcal{L}^P(\mathbf{x}, \vec{\eta}) = -(\vec{\eta}\cdot\vec{\eta})^2 = -(1+\tau^2)^2 .$$

Therefore, $\tau_1^+ = \tau_2^+ = \mathrm{i}$ are the two roots of $\det \mathcal{L}^P(\mathbf{x}, \vec{\eta}) = 0$ with positive imaginary parts and $M^+(\mathbf{x}, \vec{\xi}, \tau) = (\tau - \mathrm{i})^2$.

The component form of the first set of boundary conditions in (D.14) is given by

$$u_1 n_1 + u_2 n_2 + u_3 n_3 = 0 \qquad \text{and} \qquad \phi = 0 \quad \text{on } \partial\Omega . \tag{D.21}$$

Therefore, the symbol of the boundary operator is

$$\mathcal{B}(\mathbf{x}, \vec{\xi}) = \begin{pmatrix} n_1 & n_2 & n_3 & 0 \\ 0 & 0 & 0 & 1 \end{pmatrix}, \tag{D.22}$$

and, for the weights defined in (D.20), $\mathcal{B}^P(\mathbf{x}, \vec{\xi}) = \mathcal{B}(\mathbf{x}, \vec{\xi})$. A simple calculation shows that the adjugate matrix of $\mathcal{L}^P(\mathbf{x}, \vec{\eta})$ is given by

$$\mathcal{L}'(\mathbf{x}, \vec{\eta}) = (\vec{\eta}\cdot\vec{\eta}) \begin{pmatrix} 0 & -\eta_3 & \eta_2 & -\eta_1 \\ \eta_3 & 0 & -\eta_1 & -\eta_2 \\ -\eta_2 & \eta_1 & 0 & -\eta_3 \\ -\eta_1 & -\eta_2 & -\eta_3 & 0 \end{pmatrix}$$

so that

$$\mathcal{B}^P(\mathbf{x}, \vec{\xi})\mathcal{L}'(\mathbf{x}, \vec{\eta}) =$$

$$(\vec{\eta}\cdot\vec{\eta}) \begin{pmatrix} n_2\eta_3 - n_3\eta_2 & n_3\eta_1 - n_1\eta_3 & n_1\eta_2 - n_2\eta_1 & -n_1\eta_1 - n_2\eta_2 - n_3\eta_3 \\ -\eta_1 & -\eta_2 & -\eta_3 & 0 \end{pmatrix}.$$

The complementing condition holds if the rows of this matrix are linearly independent $\mathrm{mod}(M^+)$. Thus, we need to show that the identities

$$\begin{aligned} (1+\tau^2)\left(C_1(n_2\eta_3 - n_3\eta_2) - C_2\eta_1\right) &= (\tau-\mathrm{i})^2 p_1(\tau) \\ (1+\tau^2)\left(C_1(n_3\eta_1 - n_1\eta_3) - C_2\eta_2\right) &= (\tau-\mathrm{i})^2 p_2(\tau) \\ (1+\tau^2)\left(C_1(n_1\eta_2 - n_2\eta_1) - C_2\eta_3\right) &= (\tau-\mathrm{i})^2 p_3(\tau) \\ (1+\tau^2)C_1(n_1\eta_1 + n_2\eta_2 + n_3\eta_3) &= (\tau-\mathrm{i})^2 p_4(\tau) \end{aligned}$$

[10] The proof for the second set of boundary conditions in (D.14) is greatly simplified if one chooses, without loss of generality, the coordinates at a point $\mathbf{x}$ so that the normal vector is aligned with one of the coordinate axes.

can hold if and only if $C_1 = C_2 = 0$ and $p_i(\tau) = 0$. Without loss of generality, we may assume that, at $\mathbf{x}$, the normal is aligned with the third coordinate so that $\mathbf{n} = (0,0,-1)^\mathsf{T}$ and $\eta_3 = -\tau$. Then, we can chose $\vec{\xi} = (\xi_1, \xi_2, 0)$ so that, after factoring out the common term $(\tau - \mathrm{i})$ from both sides of the identities, they simplify to

$$\begin{aligned}(\tau+\mathrm{i})\,(C_1\eta_2 - C_2\eta_1) &= (\tau-\mathrm{i})p_1(\tau)\\ (\tau+\mathrm{i})\,(-C_1\eta_1 - C_2\eta_2) &= (\tau-\mathrm{i})p_2(\tau)\\ C_2(\tau+\mathrm{i})\tau &= (\tau-\mathrm{i})p_3(\tau)\\ C_1(\tau+\mathrm{i})\tau &= (\tau-\mathrm{i})p_4(\tau).\end{aligned}$$

The right-hand sides in the last two identities are polynomials that have $\tau = \mathrm{i}$ as one of their roots. The left-hand sides are quadratic polynomials with roots $\tau = 0$ and $\tau = -\mathrm{i}$. Therefore, these identities cannot hold unless $C_1 = C_2 = 0$ and $p_i(\tau) = 0$. □

D.2.3 Div–Curl Systems

The basic version of either of the div–curl systems (6.8) and (6.9) is given by

$$\begin{cases} \nabla\times\mathbf{u} = \mathbf{g} & \text{in } \Omega \\ \nabla\cdot\mathbf{u} = f & \text{in } \Omega \end{cases} \quad \text{and} \quad \mathbf{n}\cdot\mathbf{u} = 0 \ \text{ or } \ \mathbf{n}\times\mathbf{u} = \mathbf{0} \quad \text{on } \partial\Omega. \tag{D.23}$$

The data $\mathbf{g}$ and f in (D.23) are required to satisfy the compatibility conditions

$$\nabla\cdot\mathbf{g} = 0 \quad \text{in } \Omega \tag{D.24}$$

and

$$\int_\Omega \mathbf{g}\, d\Omega = \mathbf{0} \quad \text{and} \quad \mathbf{g}\cdot\mathbf{n} = 0 \text{ on } \partial\Omega \qquad \text{if} \quad \mathbf{n}\times\mathbf{u} = \mathbf{0} \text{ on } \partial\Omega \tag{D.25}$$

or

$$\int_\Omega f\, d\Omega = 0 \qquad \text{if} \quad \mathbf{n}\cdot\mathbf{u} = 0 \text{ on } \partial\Omega. \tag{D.26}$$

In Section 12.9, it is shown that div–curl systems are special cases of div–grad–curl systems; we take advantage of this connection and the results of Section D.2.2 to verify that the problem (D.23) satisfies all the requirements of the ADN theory.

Div–curl system in two-dimensions

The component form of the div–curl system with homogeneous tangential or normal boundary conditions is given by

$$\begin{cases} -u_{1,y} + u_{2,x} = g \\ u_{1,x} + u_{2,y} = f \end{cases} \quad \text{in } \Omega \tag{D.27}$$

and

$$\begin{cases} u_1 n_1 + u_2 n_2 = 0 \quad \text{normal} \\ \quad \text{or} \\ -u_1 n_2 + u_2 n_1 = 0 \quad \text{tangential} \end{cases} \quad \text{on } \partial\Omega\,. \tag{D.28}$$

In two dimensions, the only compatibility conditions that apply are

$$\int_\Omega f\, d\Omega = 0 \qquad \text{if} \quad \mathbf{n}\cdot\mathbf{u} = 0 \text{ on } \partial\Omega$$

or

$$\int_\Omega g\, d\Omega = 0 \qquad \text{if} \quad \mathbf{n}\times\mathbf{u} = 0 \text{ on } \partial\Omega\,.$$

The system (D.27)–(D.28) is exactly the same as (D.15) with one of the third or fourth boundary conditions of (D.16). Thus, we can use the results of Section D.2.2 to conclude that the div–curl system (D.27) is uniformly elliptic of order two and that the following results hold.

Proposition D.8 *The two-dimensional div–curl system* (D.27) *satisfies the supplementary condition given in Definition* D.2.

Proposition D.9 *The two-dimensional div–curl system* (D.27) *with either the normal or tangential boundary condition in* (D.28) *satisfies the complementing condition given in Definition* D.4.

Div–curl systems in three-dimensions

In three dimensions, the div–curl system (D.23) has three unknowns and four equations. Because the ADN theory requires the number of equations to be the same as the number of unknowns, it is not applicable to the three-dimensional div–curl system in the form (D.23). Thus, in three dimensions, we introduce into (D.23) an additional "slack" variable ϕ in such a way that we obtain the div–grad–curl problem (D.14).

Proposition D.10 *Assume that the right-hand side functions f and $\mathbf{g}$ in* (D.14) *satisfy the compatibility conditions* (D.24)–(D.26). *Then,* $\nabla\phi = 0$.

Proof. Taking the divergence of the first equation shows that, because $\nabla\cdot\mathbf{g} = 0$, ϕ solves the Laplace problem $-\Delta\phi = 0$ in Ω. For the first set of boundary conditions in (D.14), we have that $\phi = 0$ on $\partial\Omega$ so that one easily concludes that $\phi = 0$ in Ω. For the second type of boundary condition in (D.14), we have that $\partial\phi/\partial n = 0$ on $\partial\Omega$. Therefore $\phi =$ constant and $\nabla\phi = 0$. □

Thus, we see that, in three dimensions, if the compatibility conditions (D.24)–(D.26) are satisfied, the div–grad–curl system (D.14) reduces to the div–curl system (D.23). Now, ϕ in (D.14) plays the role of a "slack" variable. As a result, we can use the results of Section D.2.2 to conclude that, in three dimensions, the div–curl system (D.23) is uniformly elliptic of order four and that the following result holds.

Proposition D.11 *The div–curl system* (D.23) *with either the normal or tangential boundary condition satisfies the complementing condition given in Definition* D.4.

Remark D.12 From (A.61) and the Poincaré inequality (A.69), one can show that, for $\mathbf{u} \in \mathbf{C}_0(\Omega) \cap \mathbf{D}(\Omega)$ or for $\mathbf{u} \in \mathbf{D}_0(\Omega) \cap \mathbf{C}(\Omega)$,

$$\|\mathbf{u}\|_{DC} \le C\big(\|\nabla \times \mathbf{u}\|_0 + \|\nabla \cdot \mathbf{u}\|_0\big) \tag{D.29}$$

without the need to introduce the "slack" variable ϕ. The result (D.29) which requires us to view $\mathbf{u}$ as a vector field with square integrable curl and divergence cannot be established using the ADN theory because it instead views $\mathbf{u}$ as a collection of three unrelated scalar coordinate functions. However, (D.29) does not contradict the ADN theory because it implies a "weaker" notion of ellipticity for the div–curl system. Indeed, (D.29) asserts that $\|\nabla \times \mathbf{u}\|_0 + \|\nabla \cdot \mathbf{u}\|_0$ can bound *certain but not all* combinations of the first derivatives of u_1, u_2, and u_3. In contrast, using Theorem D.1 with $q = 0$, we obtain that, if $\Omega \subset \mathbb{R}^3$ is a bounded domain of class C^1,

$$\sum_{i=1}^{3} \|u_i\|_1 + \|\phi\|_1 \le C\big(\|\nabla \times \mathbf{u} + \nabla\phi\|_0 + \|\nabla \cdot \mathbf{u}\|_0\big) \tag{D.30}$$

for any $\mathbf{u} \in [H^1(\Omega)]^3$ and any $\phi \in H^1(\Omega)$ satisfying either of the boundary conditions in (D.14).

From Theorem A.8 in Section A.3.1 we know that (D.30) holds only for special types of domains Ω. In contrast, (D.29) holds for a wider range of domains. □

D.2.4 The Velocity–Vorticity–Pressure Formulation of the Stokes System

In this section and the next, we respectively verify the assumptions of the ADN theory for the velocity–vorticity–pressure (VVP) and velocity–stress–pressure (VSP) systems from Sections 7.1.1 and 7.1.2, respectively. With regard to the ADN theory, these two systems have quite different properties. In particular, the VVP system admits two different principal parts, one of which corresponds to a homogeneous elliptic operator and the other to a non-homogeneous elliptic operator. In contrast, the VSP system has a single principal part that corresponds to a non-homogeneous elliptic operator. The possibility for a differential operator to have more than one principal part is another interesting aspect of the ADN notion of ellipticity.

The VVP system is an example of a problem that requires different treatment in two and three space dimensions. In two-dimensions, the VVP operator (7.4) has two elliptic principal parts. In three-dimensions, it has only one, and it is non-homogeneous elliptic. However, if the three-dimensional VVP system is augmented by the "redundant" equation (7.9) and a slack variable, the resulting extended VVP system (7.12) does admit two different principal parts, one of which is homogeneous elliptic. Because of the importance of homogeneous elliptic operators to LSFEMs, in three-dimensions we check the ADN theory using the extended VVP system.

The velocity–vorticity–pressure system in two dimensions

We first consider the VVP system (7.4) with the velocity boundary condition (7.2). Assuming the equations and unknowns are ordered as in (7.4) and $\{u_1, u_2, \omega, p\}$, respectively, the symbols of the differential and boundary operators are

$$\mathcal{L}(\mathbf{x},\vec{\xi}) = \begin{pmatrix} 0 & 0 & \xi_2 & \xi_1 \\ 0 & 0 & -\xi_1 & \xi_2 \\ -\xi_2 & \xi_1 & -1 & 0 \\ \xi_1 & \xi_2 & 0 & 0 \end{pmatrix} \quad \text{and} \quad \mathcal{B}(\mathbf{x},\vec{\xi}) = \begin{pmatrix} 1\,0\,0\,0 \\ 0\,1\,0\,0 \end{pmatrix}, \tag{D.31}$$

respectively. We verify the assumptions of the ADN theory for the two weight choices

$$t_1 = \cdots = t_4 = 1 \qquad s_1 = \cdots = s_4 = 0 \qquad r_1 = r_2 = -1 \tag{D.32}$$

and

$$t_1 = t_2 = 2;\ t_3 = t_4 = 1 \qquad s_1 = s_2 = 0;\ s_3 = s_4 = -1 \qquad r_1 = r_2 = -2. \tag{D.33}$$

Denote the principal parts corresponding to (D.32) and (D.33) by $\mathcal{L}_1^P$ and $\mathcal{L}_2^P$, respectively. Using the left chart in Table D.2, we see that, for the weights in (D.32),

$$\mathcal{L}_1^P(\mathbf{x},\vec{\xi}) = \begin{pmatrix} 0 & 0 & \xi_2 & \xi_1 \\ 0 & 0 & -\xi_1 & \xi_2 \\ -\xi_2 & \xi_1 & 0 & 0 \\ \xi_1 & \xi_2 & 0 & 0 \end{pmatrix} \quad \text{and} \quad \det \mathcal{L}_1^P(\mathbf{x},\vec{\xi}) = -|\vec{\xi}|^4 .$$

For (D.33), we use the right chart in Table D.2 to find that

$$\mathcal{L}_2^P(\mathbf{x},\vec{\xi}) = \mathcal{L}(\mathbf{x},\vec{\xi}) \quad \text{and} \quad \det \mathcal{L}_2^P(\mathbf{x},\vec{\xi}) = -|\vec{\xi}|^4 .$$

Therefore, both $\mathcal{L}_1^P$ and $\mathcal{L}_2^P$ verify Definition D.1 and the uniform ellipticity condition (D.3) with $m = 2$ and $C = 1$, i.e., the total order of the VVP system in two dimensions is four. This is the same as for the primitive variable formulation of the

0	0	0	ω_y	p_x
0	0	0	$-\omega_x$	p_y
0	$-u_{1y}$	u_{2x}	$\boxed{-\omega}$	0
0	u_{1x}	u_{2y}	0	0
$s\uparrow\ t\rightarrow$	1	1	1	1

0	0	0	ω_y	p_x
0	0	0	$-\omega_x$	p_y
-1	$-u_{1y}$	u_{2x}	$-\omega$	0
-1	u_{1x}	u_{2y}	0	0
$s\uparrow\ t\rightarrow$	2	2	1	1

Table D.2 Determination of the principal parts for the VVP system in two-dimensions for the two choices of weights for the equations and unknowns. For both tables, we have that $\deg \mathcal{L}_{ij} = s_i + t_i$ for all nontrivial entries except for the boxed term in the table on the left for which $\deg \mathcal{L}_{33} = 0 \neq s_3 + t_3 = 1$ so that that term is not present in the principal part.

Stokes system and means that the two-dimensional VVP system needs two independent boundary conditions. The velocity condition (7.2) meets this requirement.

Proposition D.13 *In two dimensions, both choices $\mathcal{L}_1^P$ and $\mathcal{L}_2^P$ for the principal part of the velocity–vorticity–pressure system* (7.4) *satisfy the supplementary condition given in Definition* D.2.

Proof. The supplementary condition holds if, for every pair of linearly independent real vectors $\vec{\xi}$ and $\vec{\xi}'$, the polynomial $\det \mathcal{L}_i^P(\mathbf{x}, \vec{\xi} + \tau\vec{\xi}')$ in the complex variable τ has exactly $m = 2$ roots with positive imaginary parts. It is easy to see that for linearly independent $\vec{\xi}$ and $\vec{\xi}'$,

$$\det \mathcal{L}_i^P(\mathbf{x}, \vec{\xi} + \tau\vec{\xi}') = -\big((\vec{\xi} + \tau\vec{\xi}')\cdot(\vec{\xi} + \tau\vec{\xi}')\big)^2 = -\big(|\vec{\xi}|^2 + 2\tau\vec{\xi}\cdot\vec{\xi}' + \tau^2|\vec{\xi}'|^2\big)^2.$$

Using the argument in the proof of Proposition D.4, it follows that $\det \mathcal{L}_i^P(\mathbf{x}, \vec{\xi} + \tau\vec{\xi}') = 0$ has two pairs of complex conjugate roots and thus, exactly two roots with positive imaginary parts. □

What makes the existence of two *valid* principal parts interesting for LSFEMs is that they correspond to two different types of elliptic systems that lead to continuous least-squares principles (CLSPs) with very different properties. If a given boundary operator satisfies the complementing condition with $\mathcal{L}_1^P$, then the VVP system can be classified as *homogeneous elliptic*. If the complementing condition is only satisfied with $\mathcal{L}_2^P$, then the VVP system has to be classified as *non-homogeneous elliptic*.

We proceed to check the complementing condition for the velocity boundary operator. The charts in Table D.3 show that $\mathcal{B}^P = \mathcal{B}$ for both choices (D.32) and (D.33) for the weights.

Proposition D.14 *In two dimensions, the velocity–vorticity–pressure system* (7.4) *with principal part $\mathcal{L}_2^P$ and the velocity boundary condition satisfies the complementing condition given in Definition* D.4.

Proof. Let $\vec{\eta} = \vec{\xi} + \tau\mathbf{n}$, where $\mathbf{n}$ is the unit outer normal to $\partial\Omega$ at $\mathbf{x}$ and $\vec{\xi}$ is unit vector tangent to $\partial\Omega$ at the same point. Then, the adjugate matrix of $\mathcal{L}_2^P(\mathbf{x}, \vec{\eta})$ is given by

−1	1	0	0	0
−1	0	1	0	0
$r\uparrow\ t\rightarrow$	1	1	1	1

−2	1	0	0	0
−2	0	1	0	0
$r\uparrow\ t\rightarrow$	2	2	1	1

Table D.3 Determination, in two dimensions, of the principal part $\mathcal{B}^P$ for the velocity boundary operator (7.2) for the two choices of weights for the equations and the unknowns. For both tables, we have that $\deg \mathcal{B}_{ij} = r_i + t_i = 0$ for all nontrivial entries so that $\mathcal{B}^P = \mathcal{B}$.

$$\mathcal{L}'(\mathbf{x},\vec{\eta}) = \begin{pmatrix} \eta_2^2 & -\eta_1\eta_2 & \eta_2(\vec{\eta}\cdot\vec{\eta}) & -\eta_1(\vec{\eta}\cdot\vec{\eta}) \\ -\eta_1\eta_2 & \eta_1^2 & -\eta_1(\vec{\eta}\cdot\vec{\eta}) & -\eta_2(\vec{\eta}\cdot\vec{\eta}) \\ -\eta_2(\vec{\eta}\cdot\vec{\eta}) & \eta_1(\vec{\eta}\cdot\vec{\eta}) & 0 & 0 \\ -\eta_1(\vec{\eta}\cdot\vec{\eta}) & -\eta_2(\vec{\eta}\cdot\vec{\eta}) & 0 & 0 \end{pmatrix}. \tag{D.34}$$

Again, because the unit vectors $\vec{\xi}$ and $\mathbf{n}$ are orthogonal, $(\vec{\eta}\cdot\vec{\eta}) = (1+\tau^2)$ and

$$\det \mathcal{L}_2^P(\mathbf{x},\vec{\eta}) = -(\vec{\eta}\cdot\vec{\eta})^2 = -(1+\tau^2)^2.$$

Therefore, $\tau_1^+ = \tau_2^+ = \mathrm{i}$ are the two roots of $\det \mathcal{L}_2^P(\mathbf{x},\vec{\eta})$ with positive imaginary parts, $M^+(\mathbf{x},\vec{\xi},\tau) = (\tau-\mathrm{i})^2$, and

$$\mathcal{B}^P(\mathbf{x},\vec{\eta})\cdot\mathcal{L}'(\mathbf{x},\vec{\eta}) = \begin{pmatrix} \eta_2^2 & -\eta_1\eta_2 & \eta_2(1+\tau^2) & -\eta_1(1+\tau^2) \\ -\eta_1\eta_2 & \eta_1^2 & -\eta_1(1+\tau^2) & -\eta_2(1+\tau^2) \end{pmatrix}.$$

The complementing condition holds if the rows of this matrix are linearly independent mod $(\tau-\mathrm{i})^2$. To prove this, we need to show that the identities

$$\begin{aligned} C_1(\xi_2+\tau n_2)^2 - C_2(\xi_1+\tau n_1)(\xi_2+\tau n_2) &= (\tau-\mathrm{i})^2 p_1(\tau) \\ -C_1(\xi_1+\tau n_1)(\xi_2+\tau n_2) + C_2(\xi_1+\tau n_1)^2 &= (\tau-\mathrm{i})^2 p_2(\tau) \\ (1+\tau^2)(C_1(\xi_2+\tau n_2) - C_2(\xi_1+\tau n_1)) &= (\tau-\mathrm{i})^2 p_3(\tau) \\ -(1+\tau^2)(C_1(\xi_1+\tau n_1) + C_2(\xi_2+\tau n_2)) &= (\tau-\mathrm{i})^2 p_4(\tau) \end{aligned} \tag{D.35}$$

cannot be satisfied unless $C_1 = C_2 = 0$ and $p_i(\tau)\equiv 0$. Let us show that at least one of (D.35) cannot hold for any other choice of C_1 and C_2. Given a point $\mathbf{x}\in\partial\Omega$ we may assume, without loss of generality, that at this point $\vec{\xi} = (1,0)^\mathsf{T}$ and $\mathbf{n} = (0,-1)^\mathsf{T}$. Then, the second identity in (D.35) simplifies to

$$C_1\tau + C_2 = (\tau-\mathrm{i})^2 p_2(\tau).$$

The left-hand side is a linear polynomial in τ and the right-hand side is at least a quadratic polynomial. Therefore, the second identity in (D.35) can only hold if $C_1 = C_2 = 0$ and $p_2 \equiv 0$. □

Proposition D.14 asserts that with the velocity boundary condition, the VVP system can be classified as *non-homogeneous* elliptic. This means that the boundary value problem (7.2)–(7.4) is well posed in the spaces (D.5) defined with the weights in (D.33).

The next proposition shows that the same boundary value problem is not well posed in the spaces (D.5) when they are defined according to the weights in (D.32). In other words, the VVP system with the velocity boundary condition cannot be classified as *homogeneous* elliptic.

Proposition D.15 *In two dimensions, the velocity–vorticity–pressure system* (7.4) *with principal part* $\mathcal{L}_1^P$ *and the velocity boundary condition does not satisfy the complementing condition given in Definition* D.4.

Proof. With $\vec{\eta}$ defined as in Proposition D.14, the adjugate of $\mathcal{L}_1^P(\mathbf{x},\vec{\eta})$ is

$$\mathcal{L}'(\mathbf{x},\vec{\eta}) = -(\vec{\eta}\cdot\vec{\eta})\begin{pmatrix} 0 & 0 & -\eta_2 & \eta_1 \\ 0 & 0 & \eta_1 & \eta_2 \\ \eta_2 & -\eta_1 & 0 & 0 \\ \eta_1 & \eta_2 & 0 & 0 \end{pmatrix} \tag{D.36}$$

so that

$$\mathcal{B}^P(\mathbf{x},\vec{\eta})\cdot\mathcal{L}'(\mathbf{x},\vec{\eta}) = -(1+\tau^2)\begin{pmatrix} 0 & 0 & -\eta_2 & \eta_1 \\ 0 & 0 & \eta_1 & \eta_2 \end{pmatrix}.$$

Because $\det\mathcal{L}_2^P(\mathbf{x},\vec{\eta}) = \det\mathcal{L}_1^P(\mathbf{x},\vec{\eta}) = -(\vec{\eta}\cdot\vec{\eta})^2$, the polynomial M^+ is the same as in Proposition D.14. As a result, to prove that the complementing condition fails for the velocity boundary condition with the principal part $\mathcal{L}_1^P$ we need to show that there are constants $C_1 \neq 0$, $C_2 \neq 0$, and at least one non-trivial polynomial $p_i(\tau)$, such that

$$\begin{aligned} (1+\tau^2)(-C_1(\xi_2+\tau n_2)+C_2(\xi_1+\tau n_1)) &= (\tau-\mathrm{i})^2 p_1(\tau) \\ (1+\tau^2)(C_1(\xi_1+\tau n_1)+C_2(\xi_2+\tau n_2)) &= (\tau-\mathrm{i})^2 p_2(\tau) \\ 0 &= (\tau-\mathrm{i})^2 p_3(\tau) \\ 0 &= (\tau-\mathrm{i})^2 p_4(\tau)\,. \end{aligned} \tag{D.37}$$

First of all, we see that the last two identities in (D.37) are trivially satisfied for any C_1 and C_2 by choosing $p_3(\tau) = p_4(\tau) = 0$. For the first two identities, the common term $(\tau-\mathrm{i})$ can be factored to obtain

$$\begin{aligned} (\tau+\mathrm{i})(-C_1(\xi_2+\tau n_2)+C_2(\xi_1+\tau n_1)) &= (\tau-\mathrm{i})p_1(\tau) \\ (\tau+\mathrm{i})(C_1(\xi_1+\tau n_1)+C_2(\xi_2+\tau n_2)) &= (\tau-\mathrm{i})p_2(\tau)\,. \end{aligned} \tag{D.38}$$

The left-hand sides in (D.38) are quadratic polynomials that have $\tau = -\mathrm{i}$ as one of their roots. To match this root on the right-hand sides, we choose $p_1(\tau) = A_1(\tau+\mathrm{i})$ and $p_2(\tau) = A_2(\tau+\mathrm{i})$ for some complex numbers A_1 and A_2. After factoring the common term $(\tau+\mathrm{i})$, (D.38) simplifies to

$$\begin{aligned} -C_1(\xi_2+\tau n_2)+C_2(\xi_1+\tau n_1) &= (\tau-\mathrm{i})A_1 \\ C_1(\xi_1+\tau n_1)+C_2(\xi_2+\tau n_2) &= (\tau-\mathrm{i})A_2\,. \end{aligned} \tag{D.39}$$

The complementing condition is violated if we succeed in finding a nontrivial combination of the constants C_1, C_2, A_1, and A_2 such that (D.39) holds. By comparing the like terms on both sides of (D.39), we see that C_1, C_2, A_1, and A_2 must solve the system

$$\begin{cases} -C_1\xi_2 + C_2\xi_1 = -\mathrm{i}A_1 \\ -C_1 n_2 + C_2 n_1 = A_1 \end{cases} \quad \text{and} \quad \begin{cases} C_1\xi_1 + C_2\xi_2 = -\mathrm{i}A_2 \\ C_1 n_1 + C_2 n_2 = A_2 \end{cases}. \tag{D.40}$$

Let $A_1 \neq 0$ be an arbitrary complex number. Using that $\vec{\xi}$ and $\mathbf{n}$ are orthogonal it is straightforward to check that (D.40) holds for

$$A_2 = -\mathrm{i}A_1, \qquad C_1 = A_1\xi_1 - \mathrm{i}A_1 n_1, \qquad \text{and} \qquad C_2 = A_1\xi_2 - \mathrm{i}A_1 n_2$$

so that the complementing condition fails. □

We now consider the VVP system with the two non-standard boundary conditions defined in Section 7.1.1.

Proposition D.16 *In two dimensions, the velocity–vorticity–pressure system* (7.4) *with principal part* $\mathcal{L}_1^P$ *and the normal velocity–pressure* (7.5) *or the normal velocity–tangential vorticity* (7.7) *boundary conditions satisfies the complementing condition given in Definition* D.4.

Proof. With $\vec{\eta}$ defined as in Proposition D.14, the adjugate of $\mathcal{L}_1^P(\mathbf{x}, \vec{\eta})$ is given by the matrix in (D.36) and the symbol of the normal velocity–pressure boundary condition (7.5) is given by

$$\mathcal{B}(\mathbf{x}, \vec{\xi}) = \begin{pmatrix} n_1 & n_2 & 0 & 0 \\ 0 & 0 & 0 & 1 \end{pmatrix}.$$

The principal part $\mathcal{L}_1^P$ is defined using the weights from (D.32) for which $\mathcal{B}^P = \mathcal{B}$. As a result,

$$\mathcal{B}^P(\mathbf{x}, \vec{\eta}) \cdot \mathcal{L}'(\mathbf{x}, \vec{\eta}) = -(1+\tau^2) \begin{pmatrix} 0 & 0 & -\eta_2 n_1 + \eta_1 n_2 & \eta_1 n_1 + \eta_2 n_2 \\ \eta_1 & \eta_2 & 0 & 0 \end{pmatrix}.$$

The complementing condition requires the rows of this matrix to be linearly independent mod M^+ where M^+ is the same polynomial as in Proposition D.14. Taking into account the orthogonality of $\vec{\xi}$ and $\mathbf{n}$ gives the identities

$$-\eta_2 n_1 + \eta_1 n_2 = 2\tau n_1 n_2 - 1 \quad \text{and} \quad \eta_1 n_1 + \eta_2 n_2 = \tau.$$

Thus, verification of the complementing condition reduces to showing that

$$\begin{aligned} C_2(\xi_1 + \tau n_1)(1+\tau^2) &= (\tau - \mathrm{i})^2 p_1(\tau) \\ C_2(\xi_2 + \tau n_2)(1+\tau^2) &= (\tau - \mathrm{i})^2 p_2(\tau) \\ C_1(2\tau n_1 n_2 - 1)(1+\tau^2) &= (\tau - \mathrm{i})^2 p_3(\tau) \\ C_1\tau(1+\tau^2) &= (\tau - \mathrm{i})^2 p_3(\tau) \end{aligned}$$

cannot hold unless $C_1 = C_2 = 0$ and $p_i(\tau) = 0$. After factoring out the common term $(\tau - \mathrm{i})$ from the first equation it assumes the form

$$C_2(\xi_1 + \tau n_1)(\tau + \mathrm{i}) = (\tau - \mathrm{i})p_1(\tau).$$

The left-hand side is a quadratic polynomial in τ with a real root $\tau_1 = -\xi_1/n_1$ and an imaginary root $\tau_2 = -\mathrm{i}$. The polynomial on the right-hand side has the root $\tau = \mathrm{i}$. Therefore, equality of the two sides is impossible unless $C_2 = 0$ and $p_1(\tau) = 0$. The same argument applied to one of the last two equations implies that $C_1 = 0$.

The symbol of the normal velocity–tangential vorticity boundary condition (7.7)

$$\mathcal{B}(\mathbf{x}, \vec{\xi}) = \begin{pmatrix} n_1 & n_2 & 0 & 0 \\ 0 & 0 & 1 & 0 \end{pmatrix}$$

is very similar to the symbol of the normal velocity–pressure boundary condition. As a result, checking the complementing condition for (7.7) and $\mathcal{L}_1^P$ relies on basically identical arguments, and is not presented here. □

Proposition D.16 asserts that the VVP system can be classified as *homogeneous* elliptic with the two non-standard boundary conditions. This means that the corresponding boundary value problems are well posed in the spaces (D.5) defined with the weights in (D.32).

The velocity–vorticity–pressure system in three dimensions

In three dimensions, the VVP system (7.4) has seven equations and seven unknowns. Assuming that the variables are ordered as $\{u_1, u_2, u_3, \omega_1, \omega_2, \omega_3, p\}$, the symbol of (7.4) is

$$\mathcal{L}(\mathbf{x}, \vec{\xi}) = \begin{pmatrix} 0 & 0 & 0 & 0 & -\xi_3 & \xi_2 & \xi_1 \\ 0 & 0 & 0 & \xi_3 & 0 & -\xi_1 & \xi_2 \\ 0 & 0 & 0 & -\xi_2 & \xi_1 & 0 & \xi_3 \\ 0 & -\xi_3 & \xi_2 & -1 & 0 & 0 & 0 \\ \xi_3 & 0 & -\xi_1 & 0 & -1 & 0 & 0 \\ -\xi_2 & \xi_1 & 0 & 0 & 0 & -1 & 0 \\ \xi_1 & \xi_2 & \xi_3 & 0 & 0 & 0 & 0 \end{pmatrix}. \tag{D.41}$$

If we choose

$$\begin{cases} t_1 = t_2 = t_3 = 2; \; t_4 = \cdots = t_7 = 1; \\ s_1 = s_2 = s_3 = 0; \; s_4 = \cdots = s_7 = -1, \end{cases} \tag{D.42}$$

then $\mathcal{L}^P(\mathbf{x}, \vec{\xi}) = \mathcal{L}(\mathbf{x}, \vec{\xi})$ and

$$\det \mathcal{L}^P(\mathbf{x}, \vec{\xi}) = |\vec{\xi}|^6.$$

Therefore, the VVP system (7.4) is uniformly elliptic of order 6; this is the same order as for the primitive variable formulation of the Stokes equations. Although we do not provide a proof, it is known that the velocity boundary operator satisfies the complementing condition with the above weights. However, if instead we choose

$$t_1 = \cdots = t_7 = 1 \quad \text{and} \quad s_1 = \cdots = s_7 = 0 \tag{D.43}$$

it is easy to see that $\det \mathcal{L}^P(\mathbf{x}, \vec{\xi}) = 0$. This means that, in three-dimensions, the VVP system (7.4) has only one principal part, and it is non-homogeneous elliptic.

To obtain a system that also admits a homogeneous elliptic principal part, we add the "redundant" equation (7.9), the "slack" variable ϕ, and the boundary condition (7.11). This results in the extended VVP system (7.12). For the remainder of this section, we focus on this problem. Assuming the equation order in (7.12) and the ordering $\{u_1, u_2, u_3, \phi, \omega_1, \omega_2, \omega_3, p\}$ of the unknowns, the symbol of (7.12) is

$$\mathcal{L}(\mathbf{x}, \vec{\xi}) = \begin{pmatrix} 0 & 0 & 0 & 0 & 0 & -\xi_3 & \xi_2 & \xi_1 \\ 0 & 0 & 0 & 0 & \xi_3 & 0 & -\xi_1 & \xi_2 \\ 0 & 0 & 0 & 0 & -\xi_2 & \xi_1 & 0 & \xi_3 \\ 0 & 0 & 0 & 0 & \xi_1 & \xi_2 & \xi_3 & 0 \\ 0 & -\xi_3 & \xi_2 & \xi_1 & -1 & 0 & 0 & 0 \\ \xi_3 & 0 & -\xi_1 & \xi_2 & 0 & -1 & 0 & 0 \\ -\xi_2 & \xi_1 & 0 & \xi_3 & 0 & 0 & -1 & 0 \\ \xi_1 & \xi_2 & \xi_3 & 0 & 0 & 0 & 0 & 0 \end{pmatrix}. \tag{D.44}$$

As in the two-dimensional case, we consider two sets of indices. The set

$$t_1 = \cdots = t_8 = 1 \qquad s_1 = \cdots = s_8 = 0 \tag{D.45}$$

is the analogue of (D.32) and corresponds to a homogeneous elliptic principal part. The analogue of (D.33) is the set

$$\begin{cases} t_1 = \cdots = t_4 = 2 & t_5 = \cdots = t_8 = 1 \\ s_1 = \cdots = s_4 = 0 & s_5 = \cdots = s_8 = -1 \end{cases} \tag{D.46}$$

that corresponds to a non-homogeneous elliptic principal part. We denote the principal parts for (D.45) and (D.46) by $\mathcal{L}_1^P$ and $\mathcal{L}_2^P$, respectively.[11] $\mathcal{L}_1^P$ and $\mathcal{L}_2^P$ satisfy the ADN-ellipticity definition and

$$\det \mathcal{L}_1^P(\mathbf{x}, \vec{\xi}) = \det \mathcal{L}_2^P(\mathbf{x}, \vec{\xi}) = |\vec{\xi}|^8, \tag{D.47}$$

i.e., the order of the extended VVP system (7.12) is eight and (D.3) holds for $\mathcal{L}_1^P$ and $\mathcal{L}_2^P$ with $m = 4$ and $C = 1$.

The increase in the total order is caused by the additional "redundant" equation (7.9) and "slack" variable ϕ. The extended system needs four independent boundary

[11] We have that $\mathcal{L}_2^P = \mathcal{L}$ and that $\mathcal{L}_1^P$ consists of $\mathcal{L}$ with the three -1 entries changed to zero.

conditions; in contrast, the original VVP system (7.4) and the Stokes system in the primitive variables need only three. We first consider the extended system with a combined velocity-slack variable boundary operator. The symbol of this operator is

$$\mathcal{B}(\mathbf{x},\vec{\xi}) = \begin{pmatrix} 1 & 0 & 0 & 0 & 0 & 0 & 0 & 0 \\ 0 & 1 & 0 & 0 & 0 & 0 & 0 & 0 \\ 0 & 0 & 1 & 0 & 0 & 0 & 0 & 0 \\ 0 & 0 & 0 & 1 & 0 & 0 & 0 & 0 \end{pmatrix}. \tag{D.48}$$

The boundary weights for $\mathcal{L}_1^P$ are $r_1 = r_2 = r_3 = r_4 = -1$. For $\mathcal{L}_2^P$ these weights are $r_1 = r_2 = r_3 = r_4 = -2$. Thus, $\mathcal{B}^P = \mathcal{B}$ in both cases.

In three dimensions, it is not necessary to verify the supplementary condition so that we proceed with the complementing condition.

Proposition D.17 *The three-dimensional extended VVP system* (7.12) *with principal part $\mathcal{L}_2^P$ and the velocity-slack variable boundary operator* (D.48) *satisfies the complementing condition given in Definition* D.4.

Proof. Let $\vec{\eta} = \vec{\xi} + \tau\mathbf{n}$, where $\mathbf{n}$ is the unit outer normal to $\partial\Omega$ at $\mathbf{x}$ and $\vec{\xi}$ is a unit vector tangent to $\partial\Omega$ at the same point. The adjugate matrix of $\mathcal{L}_2^P(\mathbf{x},\vec{\eta})$ is given by

$$\mathcal{L}'(\mathbf{x},\vec{\eta}) = (\vec{\eta}\cdot\vec{\eta})^3 \begin{pmatrix} -\widetilde{\eta}_{23} & \widetilde{\eta}_1\widetilde{\eta}_2 & \widetilde{\eta}_1\widetilde{\eta}_3 & 0 & 0 & \eta_3 & -\eta_2 & \eta_1 \\ \widetilde{\eta}_1\widetilde{\eta}_2 & -\widetilde{\eta}_{13} & \widetilde{\eta}_2\widetilde{\eta}_3 & 0 & -\eta_3 & 0 & \eta_1 & \eta_2 \\ \widetilde{\eta}_1\widetilde{\eta}_3 & \widetilde{\eta}_2\widetilde{\eta}_3 & -\widetilde{\eta}_{12} & 0 & \eta_2 & -\eta_1 & 0 & \eta_3 \\ 0 & 0 & 0 & 1 & \eta_1 & \eta_2 & \eta_3 & 0 \\ 0 & \eta_3 & -\eta_2 & \eta_1 & 0 & 0 & 0 & 0 \\ -\eta_3 & 0 & \eta_1 & \eta_2 & 0 & 0 & 0 & 0 \\ \eta_2 & -\eta_1 & 0 & \eta_3 & 0 & 0 & 0 & 0 \\ \eta_1 & \eta_2 & \eta_3 & 0 & 0 & 0 & 0 & 0 \end{pmatrix}, \tag{D.49}$$

where

$$\widetilde{\eta}_{kl} = (\eta_k^2 + \eta_l^2)/(\vec{\eta}\cdot\vec{\eta}) \quad \text{for } k \neq l \qquad \text{and} \qquad \widetilde{\eta}_i\widetilde{\eta}_j = \eta_i\eta_j/(\vec{\eta}\cdot\vec{\eta}).$$

Recall that $(\vec{\eta}\cdot\vec{\eta}) = (1+\tau^2)$. Then,

$$\det\mathcal{L}_2^P(\mathbf{x},\vec{\eta}) = (\vec{\eta}\cdot\vec{\eta})^4 = (1+\tau^2)^4,$$

$\tau_1^+ = \cdots = \tau_4^+ = \mathrm{i}$ are the four roots of $\det\mathcal{L}_2^P(\mathbf{x},\vec{\eta}) = 0$ with positive imaginary parts, and $M^+(\mathbf{x},\vec{\xi},\tau) = (\tau - \mathrm{i})^4$. To verify the complementing condition, we need to show that the rows of

$$\mathcal{B}^P(\mathbf{x},\vec{\eta})\cdot\mathcal{L}'(\mathbf{x},\vec{\eta}) = (\vec{\eta}\cdot\vec{\eta})^3 \begin{pmatrix} -\widetilde{\eta}_{23} & \widetilde{\eta}_1\widetilde{\eta}_2 & \widetilde{\eta}_1\widetilde{\eta}_3 & 0 & 0 & \eta_3 & -\eta_2 & \eta_1 \\ \widetilde{\eta}_1\widetilde{\eta}_2 & -\widetilde{\eta}_{13} & \widetilde{\eta}_2\widetilde{\eta}_3 & 0 & -\eta_3 & 0 & \eta_1 & \eta_2 \\ \widetilde{\eta}_1\widetilde{\eta}_3 & \widetilde{\eta}_2\widetilde{\eta}_3 & -\widetilde{\eta}_{12} & 0 & \eta_2 & -\eta_1 & 0 & \eta_3 \\ 0 & 0 & 0 & 1 & \eta_1 & \eta_2 & \eta_3 & 0 \end{pmatrix}$$

are linearly independent mod(M^+). To prove this, we focus on the relations connecting the first four coefficients in each row. After canceling the powers of $(\tau-\mathrm{i})$ that are common to these terms and the polynomial M^+ that appears in the right-hand side, the first four relations simplify to

$$\begin{aligned} (\tau+\mathrm{i})^2(-C_1\eta_{23}+C_2\eta_1\eta_2+C_3\eta_1\eta_3) &= (\tau-\mathrm{i})^2 p_1(\tau) \\ (\tau+\mathrm{i})^2(C_1\eta_1\eta_2-C_2\eta_{13}+C_3\eta_2\eta_3) &= (\tau-\mathrm{i})^2 p_2(\tau) \\ (\tau+\mathrm{i})^2(C_1\eta_1\eta_3+C_2\eta_2\eta_3-C_3\eta_{12}) &= (\tau-\mathrm{i})^2 p_3(\tau) \\ (\tau+\mathrm{i})^3 C_4 &= (\tau-\mathrm{i})p_4(\tau), \end{aligned} \tag{D.50}$$

where $\eta_{kl} = \eta_k^2+\eta_l^2$. Our goal is to show that (D.50) can only hold for $C_1 = C_2 = C_3 = C_4 = 0$ and $p_i(\tau) = 0$.

The left-hand side of the last equation in (D.50) is a cubic polynomial with a triple root $\tau = -\mathrm{i}$; the right-hand side has the root $\tau = \mathrm{i}$. Hence, this equation cannot hold unless $C_4 = 0$ and $p_4 = 0$. Without loss of generality, we may assume that $\mathbf{n} = (0,0,-1)^\mathsf{T}$. By setting $\vec{\xi} = (0,1,0)^\mathsf{T}$, we have that $\vec{\eta} = (0,1,-\tau)^\mathsf{T}$ and

$$\eta_{23} = \eta_2^2+\eta_3^2 = (1+\tau^2) = (\tau+\mathrm{i})(\tau-\mathrm{i}).$$

As a result, the first equation in (D.50) reduces to

$$-(\tau+\mathrm{i})^3(\tau-\mathrm{i})C_1 = (\tau-\mathrm{i})^2 p_1(\tau).$$

After canceling the common factor $(\tau-\mathrm{i})$ we obtain

$$-(\tau+\mathrm{i})^3 C_1 = (\tau-\mathrm{i})p_1(\tau).$$

Exactly as is the case for the last equation in (D.50), we conclude that $C_1 = 0$ and $p_1(\tau) = 0$. To prove that $C_2 = 0$, let $\mathbf{n}$ be as before and $\vec{\xi} = (1,0,0)^\mathsf{T}$ so that $\vec{\eta} = (1,0,-\tau)^\mathsf{T}$ and $\eta_{13} = (1+\tau^2)$. With this choice the second equation in (D.50) reduces to

$$-(\tau+\mathrm{i})^3(\tau-\mathrm{i})C_2 = (\tau-\mathrm{i})^2 p_2(\tau).$$

Canceling the common factor $(\tau-\mathrm{i})$ results in

$$-(\tau+\mathrm{i})^3 C_2 = (\tau-\mathrm{i})p_2(\tau).$$

Therefore, we must have $C_2 = 0$ and $p_2(\tau) = 0$. To show that $C_3 = 0$, note that $\eta_{12} = \xi_1^2+\xi_2^2$ and, because $C_1 = C_2 = 0$, the third equation in (D.50) reduces to

$$-(\tau+\mathrm{i})^2 C_3(\xi_1^2+\xi_2^2) = (\tau-\mathrm{i})^2 p_3(\tau).$$

This is only possible if $C_3 = 0$ and $p_3(\tau) = 0$. □

The following result is the analogue of Proposition D.15 in three-dimensions.

Proposition D.18 *The three-dimensional extended VVP system* (7.12) *with principal part* $\mathcal{L}_1^P$ *and the velocity-slack variable boundary operator* (D.48) *does not satisfy the complementing condition given in Definition* D.4.

Proof. The adjugate matrix of the principal part $\mathcal{L}_1^P$ is given by

$$\mathcal{L}'(\mathbf{x},\vec{\eta}) = (\vec{\eta}\cdot\vec{\eta})^3 \begin{pmatrix} 0 & 0 & 0 & 0 & 0 & \eta_3 & -\eta_2 & \eta_1 \\ 0 & 0 & 0 & 0 & -\eta_3 & 0 & \eta_1 & \eta_2 \\ 0 & 0 & 0 & 0 & \eta_2 & -\eta_1 & 0 & \eta_3 \\ 0 & 0 & 0 & 0 & \eta_1 & \eta_2 & \eta_3 & 0 \\ 0 & \eta_3 & -\eta_2 & \eta_1 & 0 & 0 & 0 & 0 \\ -\eta_3 & 0 & \eta_1 & \eta_2 & 0 & 0 & 0 & 0 \\ \eta_2 & -\eta_1 & 0 & \eta_3 & 0 & 0 & 0 & 0 \\ \eta_1 & \eta_2 & \eta_3 & 0 & 0 & 0 & 0 & 0 \end{pmatrix}. \quad \text{(D.51)}$$

As a result,

$$\mathcal{B}^P(\mathbf{x},\vec{\eta})\cdot\mathcal{L}'(\mathbf{x},\vec{\eta}) = (\vec{\eta}\cdot\vec{\eta})^3 \begin{pmatrix} 0 & 0 & 0 & 0 & 0 & \eta_3 & -\eta_2 & \eta_1 \\ 0 & 0 & 0 & 0 & -\eta_3 & 0 & \eta_1 & \eta_2 \\ 0 & 0 & 0 & 0 & \eta_2 & -\eta_1 & 0 & \eta_3 \\ 0 & 0 & 0 & 0 & \eta_1 & \eta_2 & \eta_3 & 0 \end{pmatrix}.$$

The structure of this matrix is similar to the structure of the matrix encountered in the two-dimensional case considered in Proposition D.15. As a result, the proof that the complementing condition fails can be constructed from the same arguments as in the proof of that proposition. For this reason we leave the details out. □

Proposition D.18 shows that the velocity boundary condition cannot be paired with the principal part $\mathcal{L}_1^P$. This means that the boundary value problem comprising of the extended VVP system with the velocity–slack variable boundary condition is not well-posed in the spaces (D.5) when they are defined according to the weights in (D.45). In order to benefit from the availability of a homogeneous elliptic principal part in the extended VVP system one has to switch to non-standard boundary conditions. This fact is established in the following proposition which is a three-dimensional analogue of Proposition D.16.

Proposition D.19 *The three-dimensional extended VVP system* (7.12) *with principal part* $\mathcal{L}_1^P$ *and the extended normal velocity–pressure* (7.13) *or the extended normal velocity–tangential vorticity* (7.14) *boundary conditions satisfies the complementing condition given in Definition* D.4.

Proof. The proof relies on what is essentially the same argument as in the proof of Proposition D.16. For brevity details are not presented here. □

According to this proposition, the boundary value problem obtained by pairing the extended VVP system (7.12) with either one of the non-standard boundary conditions in (7.13) and (7.14) is well-posed in the spaces (D.5) when they are defined according to the weights in (D.45).

Further comments about velocity–vorticity–pressure systems

Propositions D.15 and D.18 assert that the velocity boundary condition violates the complementing condition with the principal part $\mathcal{L}_1^P$. Therefore, the conclusions of Theorem D.1 are not valid if s_i and t_j are selected according to (D.32) or (D.45). This implies that the boundary value problem obtained by pairing the VVP Stokes system with the velocity boundary condition is non-homogeneous elliptic. For instance, in two-dimensions this means that the a priori estimate

$$C\big(\|\mathbf{u}\|_1 + \|\omega\|_1 + \|p\|_1\big) \le \|\nabla^\perp\omega + \nabla p\|_0 + \|\nabla\times\mathbf{u} - \omega\|_0 + \|\nabla\cdot\mathbf{u}\|_0 \quad \text{(D.52)}$$

cannot hold for $\mathbf{u}\in[H_0^1(\Omega)]^2$, $\omega\in H^1(\Omega)$, and $p\in H^1(\Omega)\cap L_0^2(\Omega)$. The following example from [35, 53] demonstrates this.

Example D.20 Let Ω be an arbitrary bounded domain in $\mathbb{R}^2$ with smooth boundary $\partial\Omega$. Set $\mathbf{u}_n = 0$, $\omega_n = -\sin(nx)\exp(-ny)$, $p_n = \cos(nx)\exp(-ny)$. If (D.52) were to be true, it would imply that, for any $n>0$, the term

$$\|\mathbf{u}_n\|_1 + \|\omega_n\|_1 + \|p_n\|_1 \sim O(n\exp(n))$$

is bounded from above by the term

$$\|\nabla^\perp\omega_n + \nabla p_n\|_0 + \|\nabla\times\mathbf{u}_n - \omega_n\|_0 + \|\nabla\cdot\mathbf{u}_n\|_0 \sim O(\exp(n))$$

which is a contradiction. A counterexample in three-dimensions is provided by $\mathbf{u}_n = 0$, $\omega_n = \big(0,0,-(z^2-1)\cos(nx)\exp(ny)\big)^\mathsf{T}$, $p_n = (z^2-1)\sin(nx)\exp(ny)$ and a bounded domain $\Omega\subset\mathbb{R}^3$ with smooth boundary $\partial\Omega$. □

On the other hand, Propositions D.16 and D.19 show that the normal velocity–pressure and normal velocity–tangential vorticity boundary conditions pass the complementing condition with the principal part $\mathcal{L}_1^P$. This means that the conclusions of Theorem D.1 are valid if s_i and t_j are selected according to (D.32) or (D.45), i.e., the boundary value problem obtained by pairing the VVP Stokes formulation with either one of the two non-standard boundary conditions is homogeneous elliptic.

According to Chapter 4, homogeneous elliptic systems offer the most straightforward case for developing LSFEMs in the ADN setting because they admit practical compliant DLSPs. In contrast, non-homogeneous elliptic systems do not admit practical compliant DLSPs. As a result, practical LSFEMs for the VVP system and the velocity (or the velocity–slack variable) boundary condition have to follow the approaches for non-homogeneous systems formulated in Section 4.5, whereas using the non-standard boundary conditions leads to straightforward practical LSFEMs.

Boundary conditions		$\mathbb{R}^3$	$\mathbb{R}^2$	Classification
BC1	Velocity	$\mathbf{u}$	$\mathbf{u}$	NHE
	Slack variable	ϕ	–	
BC1A	Velocity	$\mathbf{u}$	$\mathbf{u}$	NHE
	Normal vorticity	$\omega \cdot \mathbf{n}$	–	
BC2	Normal velocity	$\mathbf{u} \cdot \mathbf{n}$	$\mathbf{u} \cdot \mathbf{n}$	
	Normal vorticity	$\omega \cdot \mathbf{n}$	–	HE
	Pressure	p	p	
	Slack variable	ϕ	–	
BC2A	Normal velocity	$\mathbf{u} \cdot \mathbf{n}$	$\mathbf{u} \cdot \mathbf{n}$	
	Tangential vorticity	$\omega \times \mathbf{n}$	ω	HE
	Slack variable	ϕ	–	
BC2B	Normal velocity	$\mathbf{u} \cdot \mathbf{n}$	$\mathbf{u} \cdot \mathbf{n}$	not well-posed in $\mathbb{R}^3$
	Vorticity	ω	ω	HE in $\mathbb{R}^2$
BC3	Tangential velocity	$\mathbf{u} \times \mathbf{n}$	$\mathbf{u} \cdot \mathbf{t}$	NHE in $\mathbb{R}^3$
	Pressure	p	p	HE in $\mathbb{R}^2$
	Slack variable	ϕ	–	
BC3A	Tangential velocity	$\mathbf{u} \times \mathbf{n}$	$\mathbf{u} \cdot \mathbf{t}$	
	Normal vorticity	$\omega \cdot \mathbf{n}$	–	HE
	Pressure	p	p	
BC3B	Tangential velocity	$\mathbf{u} \times \mathbf{n}$	$\mathbf{u} \cdot \mathbf{t}$	HE
	Tangential vorticity	$\omega \times \mathbf{n}$	ω	

Table D.4 Classification of boundary conditions for the Stokes (and Navier–Stokes equations) for the VVP formulation. "HE" indicates that the system can be classified as homogeneous elliptic, i.e., the boundary operator passes the complementing condition with the homogeneous elliptic principal part $\mathcal{L}_1^P$. "NHE" indicates that the system has to be classified as non-homogeneous elliptic, i.e., the boundary operator passes the complementing condition with the non-homogeneous elliptic principal part $\mathcal{L}_2^P$ but not with $\mathcal{L}_1^P$.

One can show (see [34, 35]) that in addition to the two non-standard boundary conditions considered here there are other examples of boundary operators for the VVP Stokes system that pass the complementing condition with the homogeneous elliptic principal part $\mathcal{L}_1^P$. Table D.4 presents a summary of boundary condition sets for the VVP system in two and three-dimensions along with their classification as homogeneous or non-homogeneous elliptic.

Remark D.21 The complementing condition allows us to quickly check various boundary conditions and determine the classification of the VVP system as homogeneous or non-homogeneous elliptic. However, by itself, the complementing condition does not give much insight as to why different boundary operators lead to different classifications of the VVP system. An informal approach that sheds some light on the classification process is to inspect the boundary value problems obtained by pairing a principal part operator with a boundary condition.

Consider, for example, the two-dimensional principal part operator $\mathcal{L}_1^P$ with the velocity boundary condition (7.2) or the normal velocity–pressure condition (7.5). The corresponding boundary value problems are respectively given by

$$\left\{\begin{aligned} \nabla^{\perp}\omega + \nabla p &= 0 \\ \nabla \times \mathbf{u} &= 0 \\ \nabla \cdot \mathbf{u} &= 0 \\ \mathbf{u}_{|\partial\Omega} &= 0 \end{aligned}\right\} \quad \text{and} \quad \left\{\begin{aligned} \nabla^{\perp}\omega + \nabla p &= 0 \\ \nabla \times \mathbf{u} &= 0 \\ \nabla \cdot \mathbf{u} &= 0 \\ \mathbf{u}\cdot\mathbf{n}_{|\partial\Omega} = p_{|\partial\Omega} &= \mathbf{0} \end{aligned}\right\}. \tag{D.53}$$

The first boundary value problem uncouples into the following two independent problems:

$$\left\{\, \nu\nabla^{\perp}\omega + \nabla p = \mathbf{f} \,\right\} \quad \text{and} \quad \left\{\begin{aligned} \nabla \times \mathbf{u} &= 0 \\ \nabla \cdot \mathbf{u} &= 0 \\ \mathbf{u}_{|\partial\Omega} &= 0 \end{aligned}\right\}. \tag{D.54}$$

Neither one of these two systems is a well-posed boundary value problem. The first system in (D.54) is underdetermined and admits (with $\mathbf{f} = 0$) as solutions the functions ω_n and p_n defined in Example D.20. These functions have finite $L^2(\Omega)$ norms on any bounded domain Ω, but their $H^1(\Omega)$ semi-norm is unbounded as $n \to \infty$. The second problem in (D.54) is the basic div–curl system considered in Section D.2.3. The order of this system is two and a well-posed boundary value problem requires a single boundary condition; specification of $\mathbf{u}$ on $\partial\Omega$ provides two. Therefore, this problem is overdetermined.

In contrast, the second boundary value problem in (D.54) uncouples into the two well-posed problems

$$\left\{\begin{aligned} \nu\nabla^{\perp}\omega + \nabla p &= \mathbf{f} \\ p_{|\partial\Omega} &= 0 \end{aligned}\right\} \quad \text{and} \quad \left\{\begin{aligned} \nabla \times \mathbf{u} &= 0 \\ \nabla \cdot \mathbf{u} &= 0 \\ \mathbf{u}\cdot\mathbf{n}_{|\partial\Omega} &= 0 \end{aligned}\right\}.$$

A three-dimensional example is given as follows. Consider $\mathcal{L}_1^P$ with the velocity–normal vorticity boundary operator (labeled by (BC1A) in Table D.4.) This boundary value problem decouples into the two div–grad–curl boundary value problems

$$\left\{\begin{aligned} \nabla \times \omega + \nabla p &= \mathbf{f}_1 \\ \nabla \cdot \omega &= f_2 \\ \omega \cdot \mathbf{n}_{|\partial\Omega} &= 0 \end{aligned}\right\} \quad \text{and} \quad \left\{\begin{aligned} \nabla \times \mathbf{u} + \nabla \phi &= \mathbf{f}_3 \\ \nabla \cdot \mathbf{u} &= f_4 \\ \mathbf{u}_{|\partial\Omega} &= 0 \end{aligned}\right\}. \tag{D.55}$$

Neither one of these two systems is a well-posed boundary value problem in the sense of ADN. From Section D.2.3, we know that, in three-dimensions, the div–

curl system has order four and requires two independent boundary conditions. In the first system in (D.55), $\omega \cdot \mathbf{n}$ provides only one condition so that system is under-determined. In contrast, the second system is over-determined because $\mathbf{u}$ specifies three independent conditions. These examples explain why with the velocity boundary condition, the VVP system can never be classified as homogeneous elliptic. □

The informal arguments in Remark D.21 can be applied to all boundary conditions sets presented in Table D.4. It is not hard to see that the boundary operators that result in homogeneous elliptic systems are characterized by an equal "split" of the boundary conditions between the velocity variable (and the slack variable in $\mathbb{R}^3$) and the rest of the variables in the VVP system. For example, in three-dimensions the homogeneous elliptic principal part $\mathcal{L}_1^P$ splits into the two div–grad–curl sub-systems shown in (D.55) which require two independent boundary conditions each. All boundary operators from Table D.4 that are classified as homogeneous elliptic in $\mathbb{R}^3$ provide the requisite 2 independent conditions for each sub-system.

Because changing the space dimension from 2 to 3 changes the vorticity variable from a scalar function to a vector field, and the number of tangential components in a vector field from 1 to 2, classification of some boundary conditions, or even their well-posedness, can depend on the space dimension. For example, the boundary condition labeled by (BC2B) in Table D.4 results in a homogeneous elliptic VVP system in two-dimensions but becomes ill-posed in three-dimensions. This is caused by the fact that specification of ω on the boundary of a three-dimensional domain Ω provides 3 independent boundary conditions as opposed to just 1 in two-dimensions. Another example is provided by the boundary condition set labeled (BC3) in Table D.4. This set has different classifications in $\mathbb{R}^2$ and $\mathbb{R}^3$ because the number of independent conditions provided by the tangential velocity is different in two and three-dimensions.

D.2.5 The Velocity–Stress–Pressure Formulation of the Stokes System

In this section, we check the conditions of the ADN theory for the first-order VSP form (7.15) of the Stokes system. In both two and three-dimensions, this system has only one principal part and is *non-homogeneous elliptic*. The total order of the VSP operator coincides with the total order of the primitive variable formulation (7.1) of the Stokes equations. As a result, the VSP system is an example of a problem for which checking the ADN theory in two and three-dimensions differs only by the length of the calculations. To save time and space, we only consider the two-dimensional case with the velocity boundary condition.

Let $T_1 = T_{11}$, $T_2 = T_{12} = T_{21}$, and $T_3 = T_{22}$. Assuming that the equations are ordered as in (7.15) and that the ordering of the variables is $\{u_1, u_2, T_1, T_2, T_3, p\}$, the symbols of the differential and boundary operators are

$$\mathcal{L}(\mathbf{x},\vec{\xi}) = \begin{pmatrix} 0 & 0 & \xi_1 & \xi_2 & 0 & -\xi_1 \\ 0 & 0 & 0 & \xi_1 & \xi_2 & -\xi_2 \\ \xi_1 & \xi_2 & 0 & 0 & 0 & 0 \\ -\xi_1 & 0 & 1 & 0 & 0 & 0 \\ -\xi_2 & -\xi_1 & 0 & 2 & 0 & 0 \\ 0 & -\xi_2 & 0 & 0 & 1 & 0 \end{pmatrix}$$

and

$$\mathcal{B}(\mathbf{x},\vec{\xi}) = \begin{pmatrix} 1 & 0 & 0 & 0 & 0 & 0 \\ 0 & 1 & 0 & 0 & 0 & 0 \end{pmatrix},$$

respectively. We verify the assumptions of the ADN theory using the following indices:

$$\begin{cases} t_1 = t_2 = 2 & t_3 = \cdots = t_6 = 1 \\ s_1 = s_2 = 0 & s_3 = \cdots = s_6 = -1\,. \end{cases} \tag{D.56}$$

Then, $\mathcal{L}^P(\mathbf{x},\vec{\xi}) = \mathcal{L}(\mathbf{x},\vec{\xi})$ and $\det \mathcal{L}^P(\mathbf{x},\vec{\xi}) = -|\vec{\xi}|^4$. Therefore, (D.3) holds with $m = 2$ and $C = 1$ and the VSP system is uniform elliptic of order four, the same as for the primitive variable formulation of the Stokes equations. The arguments of Proposition D.13 can be used to show that $\mathcal{L}^P$ also satisfies the supplementary condition given in Definition D.2. A well-posed boundary value problem requires two independent boundary conditions; the velocity boundary condition (7.2) meets this requirement. For this operator, we set $r_1 = r_2 = -2$.

The choice (D.56) of the weights t_j for the unknowns means that the VSP system is classified as *non-homogeneous* elliptic. Let us show that this system *cannot* be classified as *homogeneous* elliptic. To this end, let

$$t_1 = \cdots = t_6 = 1 \qquad \text{and} \qquad s_1 = \cdots = s_6 = 0.$$

With this choice, the principal part is given by

$$\mathcal{L}^P(\mathbf{x},\vec{\xi}) = \begin{pmatrix} 0 & 0 & \xi_1 & \xi_2 & 0 & -\xi_1 \\ 0 & 0 & 0 & \xi_1 & \xi_2 & -\xi_2 \\ \xi_1 & \xi_2 & 0 & 0 & 0 & 0 \\ -\xi_1 & 0 & 0 & 0 & 0 & 0 \\ -\xi_2 & -\xi_1 & 0 & 0 & 0 & 0 \\ 0 & -\xi_2 & 0 & 0 & 0 & 0 \end{pmatrix}$$

It is easy to see that $\det \mathcal{L}^P(\mathbf{x},\vec{\xi}) = 0$ so that the third condition of the ADN ellipticity definition D.1 is violated.

We proceed with the verification of the complementing condition for the weight choices in (D.56).

Proposition D.22 *The VSP system* (7.15) *with the velocity boundary condition and the weights in* (D.56) *satisfies the complementing condition given in Definition* D.4.

Proof. Let $\mathbf{n}$ be the outer unit normal vector to $\partial\Omega$ at some point $\mathbf{x}$ and let $\vec{\xi}$ be a unit tangent vector to $\partial\Omega$ at the same point. Then

$$\det \mathcal{L}^P(\mathbf{x}, \vec{\xi} + \tau\mathbf{n}) = -(1+\tau^2)^2$$

and $M^+(\mathbf{x}, \vec{\xi}, \tau) = (\tau - i)^2$. Without loss of generality, we may assume that at $\mathbf{x}$ the coordinate axes are aligned with the directions of $\vec{\xi}$ and $\mathbf{n}$ so that $\vec{\xi} = (1,0)^\top$ and $\mathbf{n} = (0,-1)^\top$. A straightforward calculation shows that

$$\mathcal{B}^P(\mathbf{x}, \vec{\xi} + \tau\mathbf{n}) \cdot \mathcal{L}'(\mathbf{x}, \vec{\xi} + \tau\mathbf{n}) = \begin{pmatrix} -2\tau^2 & -2\tau & -(1+\tau^2) & 2\tau^2 & \tau-\tau^3 & -2\tau^2 \\ -2\tau & -2 & \tau(1+\tau^2) & 2\tau & 1-\tau^2 & -2\tau \end{pmatrix}.$$

As a result, verification of the complementing condition reduces to showing that the identities

$$\begin{aligned} C_1\tau^2 + C_2\tau &= (\tau - i)^2 p_1(\tau) \\ C_1\tau + C_2 &= (\tau - i)^2 p_2(\tau) \\ -C_1(1+\tau^2) + C_2\tau(1+\tau^2) &= (\tau - i)^2 p_3(\tau) \\ C_1\tau(1-\tau^2) + C_2(1-\tau^2) &= (\tau - i)^2 p_4(\tau) \end{aligned}$$

can only hold with $C_1 = C_2 = 0$ and $p_i(\tau) = 0$. We can easily arrive at the conclusion that $C_1 = C_2 = 0$ by examining the second equation above. The left-hand side in that equation is a linear polynomial in the complex variable τ whereas the right hand side is at least a quadratic polynomial in τ. As a result, this equation cannot hold unless $C_1 = C_2 = 0$ and $p_2(\tau) = 0$. □

In three dimensions, the variables are $\{u_1, u_2, u_3, T_1, T_2, T_3, T_4, T_5, T_6, p\}$, where $T_1 = T_{11}$, $T_2 = T_{12}$, $T_3 = T_{13}$, $T_4 = T_{22}$, $T_5 = T_{23}$, and $T_6 = T_{33}$. The complementing condition can be verified for the VSP system and the velocity boundary condition with the following weights:

$$\begin{cases} t_1 = t_2 = t_3 = 2 & t_4 = \cdots = t_{10} = 1 \\ s_1 = s_2 = s_3 = 0 & s_4 = \cdots = s_{10} = -1\,. \end{cases} \tag{D.57}$$

Because the VSP system is always classified as non-homogeneous elliptic, it does not admit practical compliant DLSPs. As a result, practical LSFEMs for this system have to be developed using the approaches for non-homogeneous systems described in Section 4.5.

References

1. R. Adams. *Sobolev Spaces*. Academic Press, New York, 1975.
2. S. Agmon, A. Douglis, and L. Nirenberg. Estimates near the boundary for solutions of elliptic partial differential equations satisfying general boundary conditions II. *Comm. Pure Appl. Math.*, 17:35–92, 1964.
3. M. Ainsworth and J. Coyle. Conditioning of hierarchic p-version Nédélec elements on meshes of curvilinear quadrilaterals and hexahedra. *SIAM J. Numer. Anal.*, 41(2):731–750, 2003.
4. R. C. Almeida and R. S. Silva. A stable Petrov-Galerkin method for convection-dominated problems. *Comput. Meth. Appl. Mech. Eng.*, 140:291–304, 1997.
5. P. Alotto and I. Perugia. A field-based finite element method for magnetostatics derived from an error minimization approach. *Int. J. Numer. Meth. Eng.*, 49(4):573–598, 2000.
6. E. Anderson, Z. Bai, C. Bischof, S. Blackford, J. Demmel, J. Dongarra, J. Du Croz, A. Greenbaum, S. Hammarling, A. McKenney, and D. Sorensen. *LAPACK Users' Guide*. SIAM, Philadelphia, third edition, 1999.
7. D. N. Arnold. Differential complexes and numerical stability. In Li Tatsien, editor, *Proceedings of the International Congress of Mathematicians, Beijing*, volume I, pages 137–157, 2002.
8. D. N. Arnold, P. Bochev, R. Lehoucq, R. Nicolaides, and M. Shashkov, editors. *Compatible Discretizations. Proceedings of IMA Hot Topics workshop on Compatible discretizations*. IMA Volume 142. Springer Verlag, 2006.
9. D. N. Arnold, D. Boffi, and R. S. Falk. Quadrilateral $H(div)$ finite elements. *SIAM J. Numer. Anal.*, 42:2429–2451, 2005.
10. D. N. Arnold, F. Brezzi, B. Cockburn, and L. D. Marini. Unified analysis of discontinuous Galerkin methods for elliptic problems. *SIAM J. Numer. Anal.*, 39:1749–1779, 2002.
11. D. N. Arnold, R. Falk, and R. Winther. Preconditioning in $H(div)$ and applications. *Math. Comp.*, 66:957–984, 1997.
12. D. N. Arnold, R. S. Falk, and R. Winther. Multigrid in $H(div)$ and $H(curl)$. *Numerische Mathematik*, 85:197–217, 2000.
13. D. N. Arnold, R. S. Falk, and R. Winther. Finite element exterior calculus, homological techniques, and applications. Technical Report 2094, Institute for Mathematics and Its Applications, 2006.
14. V. Arnold. *Mathematical Methods of Classical Mechanics*. Springer, New York, 1989.
15. O. Axelsson and I. Kaporin. A posteriori error estimates in L^2-norm for the least-squares finite element method applied to a first-order system of differential equations. Technical Report 9945, Department of Mathematics, University of Nijmegen, The Netherlands, 1999.
16. A.Yeckel and J. Derby. On setting a pressure datum when computing incompressible flows. *Int. J. Num. Meth. Fluids*, 29:19–34, 1999.

17. A. Aziz, editor. *The Mathematical Foundations of the Finite Element Method with Applications to Partial Differential Equations*. Academic Press, New York, 1972.
18. A. Aziz, R. Kellogg, and A. Stephens. Least-squares methods for elliptic systems. *Math. Comp.*, 44(169):53–70, 1985.
19. I. Babuška. Error bounds for finite element method. *Numerische Mathematik*, 16:322–333, 1971.
20. I. Babuška and A. Aziz. Survey lectures on the mathematical foundations of the finite element method. In A. Aziz, editor, *The Mathematical Foundations of the Finite Element Method with Applications to Partial Differential Equations*, pages 1–359. Academic Press, New York, 1972.
21. I. Babuška, R. B. Kellogg, and J. Pitkäranta. Direct and inverse error estimates for finite elements with mesh refinements. *Numerische Mathematik*, 33:447–471, 1979.
22. I. Babuška and J. M. Melenk. The partition of unity method. *Int. J. Numer. Meth. Eng.*, 40:727–758, 1997.
23. T. Barth, P. Bochev, M. Gunzburger, and J.N.Shadid. A taxonomy of consistently stabilized finite element methods for the Stokes problem. *SIAM J. Sci. Comp.*, 25(5):1585–1607, 2004.
24. E. Becker, G. Carey, and T. Oden. *Finite Elements, an Introduction, Vol. 1*. Prentice Hall, Upper Saddle River, NJ, 1981.
25. D. Bedivan and G. Fix. Least-squares methods for optimal shape design problems. *Comput. Math. Appl.*, 30:17–25, 1995.
26. M. A. Behr. *Stabilized Finite Element Methods for Incompressible Flows with Emphasis on Moving Boundaries and Interfaces*. PhD thesis, Department of Aerospace Engineering and Mechanics, University of Minnesota, 1992.
27. M. A. Behr, L. P. Franca, and T. E. Tezduyar. Stabilized finite element methods for the velocity-pressure-stress formulation of incompressible flows. *Comput. Meth. Appl. Mech. Eng.*, 104:31–48, 1993.
28. B. C. Bell and K. S. Surana. A space-time coupled p-version least-squares finite element formulation for unsteady fluid dynamics problems. *Int. J. Numer. Meth. Eng.*, 37:3545–3569, 1994.
29. B. C. Bell and K. S. Surana. A space-time coupled p-version least-squares finite element formulation for unsteady two-dimensional Navier-Stokes equations. *Int. J. Numer. Meth. Eng.*, 39:25932618, 1996.
30. H. Ben-Dia and G.Rateau. The Arlequin method as a flexible engineering design tool. *Int. J. Numer. Meth. Eng.*, 62:1442–1462, 2005.
31. M. Bercovier. Perturbation of mixed variational problems. Application to mixed finite element methods. *RAIRO Anal. Numer.*, 12:211–236, 1978.
32. C. Bernardi and N. Chorfi. Spectral discretization of the vorticity, velocity, and pressure formulation of the Stokes problem. *SIAM J. Numer. Anal.*, 44(2):826–850, 2006.
33. J. Blasco and R. Codina. Stabilized finite element method for the transient Navier-Stokes equations based on a pressure gradient projection. *Comput. Meth. Appl. Mech. Eng.*, 182:277–300, 2000.
34. P. Bochev. *Least-squares finite element methods for the Stokes and Navier-Stokes equations*. PhD thesis, Virginia Polytechnic Institute and State University, Blacksburg, Virginia, 1994.
35. P. Bochev. Analysis of least-squares finite element methods for the Navier-Stokes equations. *SIAM J. Numer. Anal.*, 34(5):1817–1844, 1997.
36. P. Bochev. Experiences with negative norm least-squares methods for the Navier-Stokes equations. *Electron. Trans. Numer. Anal.*, 6:44–62, 1997.
37. P. Bochev. Least-squares methods for optimal control. *Nonlinear Anal.*, 30(3):1875–1885, 1997.
38. P. Bochev. Least-squares methods for the Navier-Stokes equations: trends and developments. In H. Bock, G. Kansacht, R. Rannacher, F. Brezzi, R. Glowinski, Y. Kuznetsov, and J. Periaux, editors, *Proceedings of the 2nd European Conference on Numerical Methods and Applications ENUMATH 97, Heidelberg, September 29- October 3, 1997*, pages 172–179, River Edge, NJ, 1998. World Sci.

39. P. Bochev. Negative norm least-squares methods for the velocity-vorticity-pressure Navier-Stokes equations. *Numer. Meth. PDEs*, 15:237–256, 1999.
40. P. Bochev. A discourse on variational and geometric aspects of stability of discretizations. In H. Deconinck, editor, *33rd Computational Fluid Dynamics Lecture Series*, VKI LS 2003-05, Chaussee de Waterloo, 72, B-1640 Rhode Saint Genese, Belgium, 2003. Von Karman Institute for Fluid Dynamics. 90 pages.
41. P. Bochev. Least-squares finite element methods for first-order elliptic systems. *Int. J. Num. Anal. Model.*, 1(1):49–64, 2004.
42. P. Bochev and D. Bedivan. Least-squares methods for Navier-Stokes boundary control problems. *Int. J. Comput. Fluid Dyn.*, 9(1):43–58, 1997.
43. P. Bochev, Z. Cai, T. Manteuffel, and S. McCormick. First-order least squares for the Navier-Stokes equations. In *Proceedings of the Seventh Copper Mountain Conference on Multigrid Methods, Part 1*, number 3339 in NASA Conference Publication, pages 41–55, 1996.
44. P. Bochev, Z. Cai, T. Manteuffel, and S. McCormick. Analysis of velocity-flux least squares methods for the Navier-Stokes equations, Part-I. *SIAM J. Numer. Anal.*, 35(3):990–1009, 1998.
45. P. Bochev and J. Choi. A comparative study of least-squares, SUPG and Galerkin methods for convection problems. *Int. J. Comput. Fluid Dyn.*, 15(2):127–146, 2001.
46. P. Bochev and J. Choi. Improved least-squares error estimates for scalar hyperbolic problems. *Comput. Meth. Appl. Math.*, 1(2):115–124, 2001.
47. P. Bochev and D. Day. A least-squares method for consistent mesh tying. *Int. J. Num. Anal. Model.*, 4:342–352, 2007.
48. P. Bochev and D. Day. Analysis and computation of a least-squares method for consistent mesh tying. *J. Comp. Appl. Math*, 218:21–33, 2008.
49. P. Bochev, C. Dohrmann, and M. Gunzburger. Stabilization of low-order mixed finite elements for the Stokes equations. *SIAM J. Numer. Anal.*, 44(1):82–101, 2006.
50. P. Bochev, C. Garasi, J. Hu, A. Robinson, and R. Tuminaro. An improved algebraic multigrid method for solving Maxwell's equations. *SIAM J. Sci. Comp.*, 25(2):623–642, 2004.
51. P. Bochev and M. Gunzburger. Accuracy of least-squares methods for the Navier-Stokes equations. *Comput. & Fluids*, 22(4-5):549–563, 1993.
52. P. Bochev and M. Gunzburger. A least-squares finite element method for the Navier-Stokes equations. *Appl. Math. Lett.*, 6(2):27–30, 1993.
53. P. Bochev and M. Gunzburger. Analysis of least-squares finite element methods for the Stokes equations. *Math. Comp.*, 63:479–506, 1994.
54. P. Bochev and M. Gunzburger. Analysis of weighted least-squares finite element methods for the Navier-Stokes equations. In W.F. Ames, editor, *Proceedings of the 14th IMACS World Congress*, volume 2, pages 584–587, Atlanta, 1994. Georgia Tech.
55. P. Bochev and M. Gunzburger. Least-squares for the velocity-pressure-stress formulation of the Stokes equations. *Comput. Meth. Appl. Mech. Eng.*, 126(3-4):267–287, 1995.
56. P. Bochev and M. Gunzburger. Least-squares finite element methods for optimization and control problems for the Stokes equations. *Comp. Math. Appl.*, 48(7):1035–1057, 2004.
57. P. Bochev and M. Gunzburger. An absolutely stable Pressure-Poisson stabilized method for the Stokes equations. *SIAM J. Numer. Anal.*, 42(3):1189–1207, 2005.
58. P. Bochev and M. Gunzburger. Compatible discretizations of second-order elliptic equations. In *Boundary-Value Problems of Mathematical Physics and Related Problems of Function Theory*, volume 318 of *Zapiski Nauchnyh Seminarov POMI*, pages 75–99, 2005.
59. P. Bochev and M. Gunzburger. On least-squares finite element methods for the Poisson equation and their connection to the Dirichlet and Kelvin principles. *SIAM J. Numer. Anal.*, 43(1):340–362, 2005.
60. P. Bochev and M. Gunzburger. On least-squares variational principles for the discretization of optimization and control problems. *Meth. Appl. Anal.*, 12(4):395–426, 2005.
61. P. Bochev and M. Gunzburger. Least-squares finite element methods for optimality systems arising in optimization and control problems. *SIAM J. Numer. Anal.*, 43(6):517–543, 2006.

62. P. Bochev and M. Gunzburger. Least-squares/penalty finite element methods for optimization and control problems. In L. Biegler, O. Ghattas, M. Heinkenlschloss, D. Keyes, and B. van Bloemen Waanders, editors, *Real-Time PDE-Constrained Optimization*, volume 3 of *Computational Science and Engineering*, pages 73–92, Santa Fe, 2007. SIAM, Philadelphia.
63. P. Bochev and M. Gunzburger. A locally conservative least-squares method for Darcy flows. *Commun. Numer. Meth. Eng.*, 24:97–110, 2008.
64. P. Bochev, J. Hu, A. Robinson, and R. Tuminaro. Towards robust 3D Z-pinch simulations: discretization and fast solvers for magnetic diffusion in heterogeneous conductors. *Electron. Trans. Numer. Anal.*, 15:186–210, 2003.
65. P. Bochev, J. Hu, C. Siefert, and R. Tuminaro. An algebraic multigrid approach based on a compatible gauge reformulation of Maxwell's equations. *SIAM J. Sci. Comp.*, 31(1): 557–583, 2008. Published online 10/31/2008.
66. P. Bochev and M. Hyman. Principles of compatible discretizations. In D. N. Arnold, P. Bochev, R. Lehoucq, R. Nicolaides, and M. Shashkov, editors, *Compatible Discretizations, Proceedings of IMA Hot Topics Workshop on Compatible Discretizations*, volume IMA 142, pages 89–120. Springer Verlag, 2006.
67. P. Bochev and R. Lehoucq. On the finite element solution of the pure Neumann problem. *SIAM Rev.*, 47(1):50–66, 2005.
68. P. Bochev, T. Manteuffel, and S. McCormick. Analysis of velocity-flux least squares methods for the Navier-Stokes equations, Part-II. *SIAM J. Numer. Anal.*, 36(4):1125–1144, 1999.
69. P. Bochev and A. C. Robinson. Matching algorithms with physics: exact sequences of finite element spaces. In D. Estep and S. Tavener, editors, *Preservation of stability under discretization*, pages 145–165, Philadelphia, 2001. Colorado State University, SIAM.
70. P. Bochev, C. Siefert, R. Tuminaro, J. Xu, and Y. Zhu. Compatible gauge approaches for $H(div)$ equations. In M. Parks and S. Collis, editors, *Computer Science Research Institute Summer Proceedings*, pages 3–14, 2007.
71. P. B. Bochev and D. Ridzal. Rehabilitation of the lowest-order Raviart–Thomas element on quadrilateral grids. *SIAM J. Numer. Anal.*, 47(1):487–507, 2008.
72. A. Bossavit. Mixed finite elements and the complex of Whitney forms. In J.R. Whiteman, editor, *The Mathematics of Finite Elements and Applications*, volume VI, pages 37–44, London, 1988. Academic Press.
73. A. Bossavit. A rationale for "edge-elements" in 3-D fields computations. *IEEE Trans. Mag.*, 24:74–79, 1988.
74. A. Bossavit. Whitney forms: A class of finite elements for three-dimensional computations in electromagnetism. *IEE Proc.*, 135(8):493–500, 1988.
75. A. Bossavit. *Computational Electromagnetism.* Academic Press, San Diego, 1998.
76. D. Braess. *Finite Elements.* Cambridge University Press, Cambridge, 1997.
77. J. Bramble, R. Lazarov, and J. Pasciak. A least-squares approach based on a discrete minus one inner product for first order systems. Technical Report 94-32, Mathematical Science Institute, Cornell University, 1994.
78. J. Bramble, R. Lazarov, and J. Pasciak. Least-squares for second order elliptic problems. *Comput. Meth. Appl. Mech. Eng.*, 152:195–210, 1998.
79. J. Bramble and J. Pasciak. Least-squares methods for Stokes equations based on a discrete minus one inner product. *J. Comp. Appl. Math.*, 74:155–173, 1996.
80. J. Bramble, J. Pasciak, and J. Xu. Parallel multilevel preconditioners. *Math. Comp.*, 55:1–22, 1990.
81. J. H. Bramble and P. Lee. On variational formulations for the Stokes equations with nonstandard boundary conditions. *RAIRO Model. Math. Anal. Numer.*, 28:903–919, 1994.
82. S. Brenner and R. Scott. *The Mathematical Theory of Finite Element Methods*. Number 15 in Texts in Applied Mathematics. Springer Verlag, New York, 2002.
83. F. Brezzi. On existence, uniqueness and approximation of saddle-point problems arising from Lagrange multipliers. *Model. Math. Anal. Numer.*, 21, 1974.
84. F. Brezzi, J. Douglas, R. Durán, and M. Fortin. Mixed finite elements for second order elliptic problems in three variables. *Numerische Mathematik*, 51(2):237–250, 1987.

85. F. Brezzi, J. Douglas, M. Fortin, and D. Marini. Efficient rectangular mixed finite elements in two and three space variables. *M^2AN Math. Model. Numer. Anal.*, 21:581–604, 1987.
86. F. Brezzi, J. Douglas, and D. Marini. Two families of mixed finite elements for second order elliptic problems. *Numerische Mathematik*, 47:217–235, 1985.
87. F. Brezzi and M. Fortin. *Mixed and Hybrid Finite Element Methods*. Springer, Berlin, 1991.
88. F. Brezzi, L. P. Franca, T. J. R. Hughes, and A. Russo. $b = \int g$. *Comput. Meth. Appl. Mech. Eng.*, 145:329–339, 1997.
89. F. Brezzi, D. Marini, I. Perugia, D. Di Barba, and A. Savini. A novel field-based mixed formulation of magnetostatics. *IEEE Trans. Mag.*, 32:635–638, 1996.
90. F. Brezzi and L. D. Marini. The three-field formulation for elasticity problems. *GAMM Mitteilungen*, pages 124–153, 2005.
91. F. Brezzi, J. Rappaz, and P.-A. Raviart. Finite-dimensional approximation of nonlinear problems, Part I: Branches of nonsingular solutions. *Numerische Mathematik*, 36:1–25, 1980.
92. M. Bristeau, O. Pironneau, R. Glowinski, J. Periaux, P. Perrier, and G. Poirier. Application of optimal control and finite element methods to the calculation of transonic flows. In B. Hunt, editor, *Numerical Methods in Applied Fluid Dynamics*, pages 203–312. Academic Press, San Diego, 1980.
93. M. Bristeau, O. Pironneau, R. Glowinski, J. Periaux, P. Perrier, and G. Poirier. Finite element methods for transonic flow calculations. In W. G. Habashi, editor, *Recent Advances in Numerical Methods in Fluids*, volume 4, pages 703–731. Pineridge Press, 1985.
94. Z. Cai and J. Douglas. Stabilized finite element methods with fast iterative solution algorithms for the Stokes problem. *Comput. Meth. Appl. Mech. Eng.*, 166:115–129, 1998.
95. Z. Cai, R. Lazarov, T. Manteuffel, and S. McCormick. First order system least-squares for second order partial differential equations: Part I. *SIAM J. Numer. Anal.*, 31:1785–1799, 1994.
96. Z. Cai, T. Manteuffel, , and S. McCormick. First-order system least-squares for second order partial differential equations: Part II. *SIAM J. Numer. Anal.*, 34:425–454, 1997.
97. Z. Cai, T. Manteuffel, and S. McCormick. First-order system least-squares for velocity-vorticity-pressure form of the Stokes equations, with application to linear elasticity. *Electron. Trans. Numer. Anal.*, 3:150–159, 1995.
98. Z. Cai, T. Manteuffel, and S. McCormick. First-order system least-squares for the Stokes equations, with application to linear elasticity. *SIAM J. Numer. Anal.*, 34:1727–1741, 1997.
99. Z. Cai, T. Manteuffel, S. McCormick, and S. Parter. First-order system least-squares for planar elasticity - pure traction. *SIAM J. Numer. Anal.*, 35:320–335, 1998.
100. Z. Cai, T. Manteuffel, S. McCormick, and J. Ruge. First-order system $\mathcal{LL}^*$ (FOSLL)*: scalar elliptic partial differential equtions. *SIAM J. Numer. Anal.*, 39:1418–1445, 2001.
101. S. S. Cairns. *Introductory topology*. Ronald Press, New York, 1961.
102. Y. Cao and M. Gunzburger. Least-squares finite element approximations to solutions of interface problems. *SIAM J. Numer. Anal.*, 35(1):393–405, 1998.
103. G. Carey and A. Pehlivanov. Error estimates for least-squares mixed finite elements. *Math. Mod. Numer. Anal.*, 28:499–516, 1994.
104. G. Carey and A. Pehlivanov. Local error estimation and adaptive remeshing scheme for least-squares finite elements. *Comput. Meth. Appl. Mech. Eng.*, 150:125–131, 1997.
105. G. Carey, A. Pehlivanov, and P. Vassilevski. Least-squares mixed finite element methods for non-self-adjoint problems: I. Error estimates. *Numerische Mathematik*, pages 501–522, 1996.
106. E. G. Do Carmo and A. Galeao. A consistent upwind Petrov-Galerkin method for convection-dominated problems. *Comput. Meth. Appl. Mech. Eng.*, 68:83–95, 1988.
107. E. G. Do Carmo and A. Galeao. Feedback Petrov-Galerkin methods for convection-dominated problems. *Comput. Meth. Appl. Mech. Eng.*, 88:1–16, 1991.
108. J. Céa. Approsimation variationnelle des problemes aux limites. *Ann. Inst. Fourier*, 14: 345–444, 1964.
109. C. Chang. Finite element method for the solution of Maxwell's equations in multiple media. *Appl. Math. Comp.*, 25:89–99, 1988.

110. C. Chang. A least-squares finite element method for the Helmholtz equation. *Comput. Meth. Appl. Mech. Eng.*, 83:1–7, 1990.
111. C. Chang. A mixed finite element method for the Stokes problem: an acceleration-pressure formulation. *Appl. Math. Comp.*, 36:135–146, 1990.
112. C. Chang. Finite element approximation for grad-div type systems in the plane. *SIAM J. Numer. Anal.*, 29:452–461, 1992.
113. C. Chang. An error estimate of the least-squares finite element method for the Stokes problem in three dimensions. *Math. Comp.*, 63:41–50, 1994.
114. C. Chang. An error analysis of least-squares finite element method of velocity-pressure-vorticity formulation for Stokes problem: correction. Technical Report 95-53, Cleveland State University, Cleveland, 1995.
115. C. Chang and M. Gunzburger. A finite element method for first order elliptic systems in three dimensions. *Appl. Math. Comp.*, 23:171–184, 1987.
116. C. Chang and M. Gunzburger. A subdomain-Galerkin/least-squares method for first order elliptic systems in the plane. *SIAM J. Numer. Anal.*, 27:1197–1211, 1990.
117. C. Chang and B.-N Jiang. An error analysis of least-squares finite element method of velocity-pressure-vorticity formulation for the Stokes problem. *Comput. Meth. Appl. Mech. Eng.*, 84:247–255, 1990.
118. C. Chang and J. Nelson. Least squares finite element method for the Stokes problem with zero residual of mass conservation. *SIAM J. Numer. Anal.*, 34:480–489, 1997.
119. C. Chang, S. Y. Yang, and C. H. Hsu. A least-squares finite element method for incompressible flow in stress-velocity-pressure version. *Comput. Meth. Appl. Mech. Eng.*, 128:1–9, 1995.
120. J. Choi. *The least-squares method for hyperbolic problems*. PhD thesis, Department of Mathematics, The University of Texas at Arlington, 2000.
121. S. H. Christiansen. Stability of Hodge decompositions in finite element spaces of differential forms in arbitrary dimension. *Numerische Mathematik*, 107(1):87–106, 2007.
122. S. H. Christiansen and R. Winther. Smoothed projections in finite element exterior calculus. *Math. Comp.*, 77:813–829, 2007.
123. P. Ciarlet. *The Finite Element Method for Elliptic Problems*. SIAM Classics in Applied Mathematics. SIAM, Philadelphia, 2002.
124. R. Codina, M. Vázquez, and O. C. Zienkiewicz. A general algorithm for compressible and incompressible flows. Part III: the semi-implicit form. *Int. J. Numer. Meth. Fluids*, 27:13–32, 1998.
125. A. Cohen, W. Dahmen, and R. DeVore. Adaptive wavelet methods for elliptic operator equations – Convergence rates. *Math. Comp.*, 70:27–75, 2001.
126. A. Cohen, W. Dahmen, and R. DeVore. Adaptive wavelet scheme for nonlinear variational problems with convergence rates. *SIAM J. Numer. Anal.*, 41:1785–1823, 2003.
127. M. Costabel. A coercive bilinear form for the Maxwell's equations. *J. Math. Anal. Appl.*, 157:527–541, 1991.
128. R. Courant. Variational methods for the solution of problems of equilibrium and vibrations. *Bull. Amer. Math. Soc.*, 49:1–23, 1943.
129. C. L. Cox and G. J. Fix. On the accuracy of least-squares methods in the presence of corner singularities. *Comput. Math. Appl.*, 10(6):463–476, 1984.
130. W. Dahmen. Wavelet and multiscale methods for operator equations. In *Acta Numerica*, pages 55–228. Cambridge University Press, 1997.
131. W. Dahmen. Multiscale and wavelet methods for operator equations. In *C.I.M.E. Lecture Notes*. Springer, New York, 2003.
132. W. Dahmen, A. Kunoth, and R. Schnieder. Wavelet least-squares methods for boundary value problems. *SIAM J. Numer. Anal.*, 39:1985–2013, 2002.
133. W. Dahmen and R. Stevenson. Element by element construction of wavelets satisfying stability and moment conditions. *SIAM J. Numer. Anal.*, 37:319–352, 1999.
134. J. Deang and M. Gunzburger. Issues related to least-squares finite element methods for the Stokes equations. *SIAM J. Sci. Comp.*, 20:878–906, 1998.

135. L. Demkowicz and I. Babuška. Optimal p-interpolation error estimates for edge finite elements of variable order in 2D. *SIAM J. Numer. Anal.*, 41(4):1195–1208, 2003.
136. L. Demkowicz and A. Buffa. H^1, $H(curl)$ and $H(div)$-conforming projection-based interpolation in three dimensions. Quasi-optimal p-interpolation estimates. *Comput. Meth. Appl. Mech. Eng.*, 194:267–296, 2005.
137. L. Demkowicz, P. Monk, L. Vardapetyan, and W. Rachowicz. De Rham diagram for hp-finite element spaces. *Comput. & Math. Applic.*, 40:589–605, 2000.
138. B. Dennis and G. Dulikravich. Optimization of magneto-hydrodynamic control of diffuser flows using micro-genetic algorithms and least-squares finite element method. *Finite elements analysis and design*, 37:349–363, 2001.
139. A. Dezin. *Multidimensional Analysis and Discrete Models*. CRC Presss, Boca Raton, FL, 1995.
140. X. Ding and T. T. H. Tsang. Large eddy simulation of turbulent flows by a least-squares finite element method. *Int. J. Numer. Meth. Fluids*, 37:297–319, 2001.
141. X. Ding and T. T. H. Tsang. On first-order formulations of the least-squares finite element method for incompressible flows. *Int. J. Comput. Fluid Dynamics*, 17(3):183–197, 2003.
142. C. Dohrmann and P. Bochev. A stabilized finite element method for the Stokes problem based on polynomial pressure projections. *Int. J. Numer. Meth. Fluids*, 46:183–201, 2004.
143. C. R. Dohrmann, S. W. Key, and M. W. Heinstein. A method for connecting dissimilar finite element meshes in two dimensions. *Int. J. Numer. Meth. Eng.*, 48:655–678, 2000.
144. C. R. Dohrmann, S. W. Key, and M. W. Heinstein. Methods for connecting dissimilar three-dimensional finite element meshes. *Int. J. Numer. Meth. Eng.*, 47:1057–1080, 2000.
145. I. Donea. A Taylor-Galerkin method for convective transport problems. *Int. J. Numer. Meth. Eng.*, 20:101–119, 1984.
146. C. A. Dorao and H. A. Jakobsen. A parallel time-space least-squares spectral element solver for incompressible flow problems. *Appl. Math. Comp.*, 185:45–58, 2007.
147. J. Douglas and J. Wang. An absolutely stabilized finite element method for the Stokes problem. *Math. Comp.*, 52, 1989.
148. Q. Du and M. Gunzburger. A gradient method approach to optimization-based multidisciplinary simulations and nonoverlapping domain decomposition algorithms. *SIAM J. Numer. Anal.*, 37:1513–1541, 2000.
149. Q. Du, M. Gunzburger, and L. Ju. Meshfree, probabilistic determination of point sets and support regions for meshless computing. *Comput. Meth. Appl. Mech. Eng.*, 191:1349–1366, 2002.
150. H.-Y. Duan and G.-P. Liang. Normal and tangential continuous elements for least-squares mixed finite element methods. *Numer. Meth. PDEs*, 20(4):609–623, 2004.
151. C. Duarte and J. Oden. $h-p$ clouds – an $h-p$ meshless method. *Numer. Meth. PDEs*, 12:673–705, 1996.
152. F. Dubois. Vorticity-velocity-pressure formulation for the Stokes problem. *Math. Methods Appl. Sciences*, 25(13):1091–1119, 2002.
153. E. Eason. A review of least-squares methods for solving partial differential equations. *Int. J. Numer. Meth. Eng.*, 10:1021–1046, 1976.
154. H. C. Elman, D. J. Silvester, and A. J. Wathen. *Finite Elements and Fast Iterative Solvers with Applications in Incompressible Fluid Dynamics*. Numerical Mathematics and Scientific Computation. Oxford University Press, 2005.
155. K. Eriksson, D. Estep, P. Hansbo, and C. Johnson. *Computational Differential Equations*. Cambridge University Press, 1996.
156. A. Ern and J.-L. Guermond. *Theory and Practice of Finite Elements*. Number 159 in Applied Mathematical Sciences. Springer Verlag, New York, 2004.
157. R. D. Falgout and U. M. Yang. hypre: A library of high performance preconditioners. In *ICCS '02: Proceedings of the International Conference on Computational Science-Part III*, pages 632–641, London, 2002. Springer-Verlag.
158. R. Falk. An analysis of the penalty method and extrapolation for the stationary Stokes equations. In R. Vichnevetsky, editor, *Advances in Computer Methods for Partial Differential Equations*, pages 66–69, New Brunswick, NJ, 1975. AICA.

159. C. Farhat and F.-X. Roux. A method of finite element tearing and interconnecting and its parallel solution algorithm. *Int. J. Numer. Meth. Eng.*, 32:1205–1227, 1991.
160. J. M. Fiard, T. Manteuffel, and S. McCormick. First-order system least squares (FOSLS) for convection-diffusion problems: Numerical results. *SIAM J. Sci. Comput.*, 19(6):1958–1979, 1998.
161. G. J. Fix, S. Gulati, and G. I. Wakoff. On the use of singular functions with finite element approximations. *J. Comp. Phys.*, 13:209–228, 1973.
162. G. J. Fix and M. Gunzburger. On least-squares approximations to indefinite problems of the mixed type. *Inter. J. Numer. Meth. Eng.*, pages 453–469, 1978.
163. G. J. Fix, M. Gunzburger, and R. Nicolaides. On finite element methods of the least-squares type. *Comput. Math. Appl.*, 5:87–98, 1979.
164. G. J. Fix, M. Gunzburger, and R. Nicolaides. On mixed finite element methods for first-order elliptic systems. *Numerische Mathematik*, 37:29–48, 1981.
165. G. J. Fix and M. E. Rose. A comparative study of finite element and finite difference methods for Cauchy-Riemann type equations. *SIAM J. Numer. Anal.*, 22(2):250–260, 1985.
166. G. J. Fix and E. Stephan. Finite element methods of least-squares type for regions with corners. Technical Report 81–41, NASA Langley Research Center, Hampton, VA 23665, 1981.
167. G. J. Fix and E. Stephan. On the finite element least-squares approximation to higher order elliptic systems. *Arch. Rat. Mech. Anal.*, 91(2):137–151, 1986.
168. H. Flanders. *Differential Forms with Applications to the Physical Sciences*. Dover, New York, 1989.
169. B. Flemisch, J. M. Melenk, and B. I. Wohlmuth. Mortar methods with curved interfaces. *Appl. Numer. Math.*, 54(3-4):339–361, 2005.
170. B. Flemisch, M. A. Puso, and B. I. Wohlmuth. A new dual mortar method for curved interfaces: 2D elasticity. *Inter. J. Numer. Meth. Eng.*, 63(6):813–832, 2005.
171. B. Flemisch and B. I. Wohlmuth. Stable Lagrange multipliers for quadrilateral meshes of curved interfaces in 3D, IANS preprint 2005/005. Technical report, University of Stuttgart, 2005.
172. M. Fortin and A. Fortin. A generalized Usawa's algorithm for the solution of the Navier-Stokes equations. *Comm. Appl. Numer. Meth.*, 1:205–208, 1985.
173. M. Fortin and R. Glowinski. Augmented Lagrangian methods: Applications to numerical solution of boundary value problems. In J. L. Lions, G. Papanicolaou, R. T. Rockafellar, and H. Fujita, editors, *Studies in Mathematics and Its Applications*, volume 15. North-Holland, 1983.
174. L. P. Franca and C. Farhat. Bubble functions prompt unusual stabilized finite element methods. *Comput. Meth. Appl. Mech. Eng.*, 123:299–308, 1995.
175. L. P. Franca, S. Frey, and T. J. R. Hughes. Stabilized finite element methods: I. Application to the advective–diffusive model. *Comput. Meth. Appl. Mech. Eng.*, 95:253–276, 1992.
176. L. P. Franca, T. J. R. Hughes, and R. Stenberg. Stabilized finite element methods. In M. Gunzburger and R. Nicolaides, editors, *Incompressible Computational Fluid Dynamics*, pages 87–108. Cambridge, University Press, 1993.
177. L. P. Franca and R. Stenberg. Error analysis of some Galerkin least-squares methods for the elasticity equations. *SIAM J. Numer. Anal.*, 28:1680–1697, 1991.
178. T. Fries and H. Matthies. Classification and overview of meshfree methods. Technical report. *Numer. Meth. PDEs*, 12:673–705, 1996.
179. A. Fursikov, M. Gunzburger, and L. Hou. Boundary value problems and optimal boundary control for the Navier-Stokes system: The two-dimensional case. *SIAM J. Cont. Optim.*, 36:852–894, 1998.
180. U. Ghia, K. N. Ghia, and C. T. Shin. High-Re solutions for incompressible flow using the Navier-Stokes equations and a multigrid method. *J. Comp. Phys.*, 48:387–411, 1982.
181. V. Girault. Incompressible finite element methods for Navier-Stokes equations with nonstandard boundary conditions in $\mathbf{R}^3$. *Math. Comp.*, 51(183):55–74, 1988.
182. V. Girault and P. Raviart. *Finite Element Approximation of the Navier-Stokes Equations*. Springer, Berlin, 1979.

183. V. Girault and P. Raviart. *Finite Element Methods for Navier-Stokes Equations*. Springer, Berlin, 1986.
184. R. Glowinski, B. Mantel, J. Periaux, P. Perrier, and O.Pironneau. On an efficient new preconditioned conjugate gradient method. Application to the in-core solution of the Navier-Stokes equations via non-linear least-squares and finite element methods. In R. H. Gallagher, editor, *Finite Elements in Fluids*, volume 4, pages 365–401. Willey, New York, 1982.
185. R. Glowinski, J. Periaux, and O. Pironneau. Transonic flow simulation by the finite element method via optimal control. In R. H. Gallagher, editor, *Finite Elements in Fluids*, volume 3, pages 205–217. Wiley, New York, 1978.
186. P. Gresho and R. Sani. *Incompressible Flow and the Finite Element Method*. Wiley, Chichester, 1998.
187. P. Grisvard. *Elliptic Problems in Nonsmooth Domains*. Pitman, Boston, 1985.
188. J.-.L. Guermond. A finite element technique for solving first order PDE's in L^1. *SIAM J. Numer. Anal.*, 42(2):714–737, 2004.
189. J.-.L. Guermond and B. Popov. Linear advection with ill-posed boundary conditions via L^1-minimization. *SIAM J. Numer. Anal.*, 42(2):714–737, 2004.
190. P.-A. Guidault and T. Belytschko. On the L^2 and the H^1 couplings for an overlapping domain decomposition method using Lagrange multipliers. *Int. J. Numer. Meth. Eng.*, 70:322–350, 2007.
191. M. Gunzburger. *Finite Element Methods for Viscous Incompressible Flows*. Academic, Boston, 1989.
192. M. Gunzburger. Iterative penalty methods for the Stokes and Navier-Stokes equations. In *Finite Element Analysis in Fluids*, pages 1040–1045. University of Alabama, Huntsville, 1989.
193. M. Gunzburger. *Perspectives in Flow Control and Optimization*. SIAM, Philadelphia, 2003.
194. M. Gunzburger, M. Heinkenschloss, and H.-K. Lee. Solution of elliptic partial differential equations by an optimization-based domain decomposition method. *Appl. Math. Comp.*, 113:111–139, 2000.
195. M. Gunzburger and H. C. Lee. Analysis and approximation of optimal control problems for first-order elliptic systems in three dimensions. *Appl. Math. Comp.*, 100:49–70, 1999.
196. M. Gunzburger and H. C. Lee. A penalty/least-squares method for optimal control problems for first-order elliptic systems. *Appl. Math. Comp.*, 107:57–75, 2000.
197. M. Gunzburger and H.-K. Lee. An optimization-based domain decomposition method for the Navier-Stokes equations. *SIAM J. Numer. Anal.*, 37:1455–1480, 2000.
198. M. Gunzburger, H.-K. Lee, and J. Peterson. An optimization based domain decomposition method for partial differential equations. *Comp. Math. Appl.*, 37:77–93, 1999.
199. M. Gunzburger and J. Lee. A domain decomposition method for optimization problems for partial differential equations. *Comp. Math. Appl.*, 40:177–192, 2000.
200. M. A. Heroux, R. A. Bartlett, V. E. Howle, R. J. Hoekstra, J. J. Hu, T. G. Kolda, R. B. Lehoucq, K. R. Long, R. P. Pawlowski, E. T. Phipps, A. G. Salinger, H. K. Thornquist, R. S. Tuminaro, J. M. Willenbring, A. Williams, and K. S. Stanley. An overview of the Trilinos project. *ACM Trans. Math. Softw.*, 31(3):397–423, 2005.
201. I. Herrera. *Boundary Methods. An Algebraic Theory*. Pitman, London, 1984.
202. J. J. Heys, T. Manteuffel, S. McCormick, and J. Ruge. First-order system least-squares (FOSLS) for coupled fluid-elastic problems. *J. Comp. Phys.*, 195:560–575, 2004.
203. R. Hiptmair. Multigrid method for $H(div)$ in three dimensions. *Electron. Trans. Numer. Anal.*, 6:7–77, 1997.
204. R. Hiptmair. Canonical construction of finite elements. *Math. Comp.*, 68:1325–1346, 1999.
205. R. Hiptmair. Multigrid method for Maxwell's equations. *SIAM J. Numer. Anal.*, 36:204–225, 1999.
206. R. Hiptmair. Discrete Hodge operators. *Numerische Mathematik*, 90:265–289, 2001.
207. R. Hiptmair, G. Widmer, and J. Zou. Auxiliary space preconditioning in $H_0(curl;\Omega)$. *Numerische Mathematik*, 103(3):435–459, 2006.
208. R. Hiptmair and J. Xu. Nodal auxiliary space preconditioning in $H(curl)$ and $H(div)$ spaces. *SIAM J. Numer. Anal.*, 45(6):2483–2509, 2007.

209. P. Houston, J. A. Mackenzie, E. Suli, and G. Warnecke. A posteriori error analysis for numerical approximation of Friedrichs systems. *Numerische Mathematik*, 82:433–470, 1999.
210. P.-W. Hsieh and S.-Y. Yang. On efficient least-squares finite element methods for convection-dominated problems. *Comput. Meth. Appl. Mech. Eng.*, 2008. Submitted.
211. J. Hu, R. Tuminaro, P. Bochev, C. Garasi, and A. Robinson. Toward an h-independent algebraic multigrid method for Maxwells equations. *SIAM J. Sci. Comp.*, 27(5):1669–1688, 2006.
212. T. J. R. Huges and M. Mallet. A new finite element formulation for computational fluid dynamics: III. The generalized streamline operator for multidimensional advection–diffusion systems. *Comput. Meth. Appl. Mech. Eng.*, 58:305–328, 1986.
213. T. J. R. Hughes. Multiscale phenomena: Green's function, the Dirichlet-to-Neumann map, subgrid scale models, bubbles and the origins of stabilized methods. *Comput. Meth. Appl. Mech. Eng.*, 127:387–401, 1995.
214. T. J. R. Hughes and A. Brooks. Streamline upwind/ Petrov-Galerkin formulation for convection dominated flows with particular emphasis on the incompressible Navier-Stokes equations. *Comput. Meth. Appl. Mech. Eng.*, 32:199–259, 1982.
215. T. J. R. Hughes and A. Brooks. A theoretical framework for Petrov-Galerkin methods with discontinuous weighting functions: Application to the streamline-upwind procedure. In R. H. Gallagher et al, editor, *Finite Elements in Fluids*, volume 4, pages 47–65, New York, 1982. J. Wiley & Sons.
216. T. J. R. Hughes, G. R. Feijoo, L. Mazzei, and J. B. Quincy. The variational multiscale method: A paradigm for computational mechanics. *Comput. Meth. Appl. Mech. Eng.*, 166:3–24, 1998.
217. T. J. R. Hughes and L. P. Franca. A new finite element formulation for computational fluid dynamics: VII. The Stokes problem with various well-posed boundary conditions: symmetric formulations that converge for all velocity pressure spaces. *Comput. Meth. Appl. Mech. Eng.*, 65:85–96, 1987.
218. T. J. R. Hughes, L. P. Franca, and M. Balestra. A new finite element formulation for computational fluid dynamics: V. Circumventing the Babuška-Brezzi condition: A stable Petrov-Galerkin formulation of the Stokes problem accommodating equal-order interpolations. *Comput. Meth. Appl. Mech. Eng.*, 59:85–99, 1986.
219. T. J. R. Hughes, L. P. Franca, and G. Hulbert. A new finite element formulation for computational fluid dynamics: VIII. The Galerkin/least-squares method for advective–diffusive equations. *Comput. Meth. Appl. Mech. Eng.*, 73:173–189, 1989.
220. T. J. R. Hughes, W. Liu, and A. Brooks. Finite element analysis of incompressible viscous flows by the penalty function formulation. *J. Comp. Phys.*, 30:1–60, 1979.
221. T. J. R. Hughes, M. Mallet, and A. Mizukami. A new finite element formulation for computational fluid dynamics: II. Beyond SUPG. *Comput. Meth. Appl. Mech. Eng.*, 54:341–355, 1986.
222. T. J. R. Hughes, A. Masud, and J. Wan. A stabilized mixed Discontinuous Galerkin method for Darcy flow. *Comput. Meth. Appl. Mech. Eng.*, 195:3347–3381, 2006.
223. J. Hyman and M. Shashkov. Adjoint operators for the natural discretizations of the divergence, gradient and curl on logically rectangular grids. *Appl. Num. Math.*, 25:413–442, 1997.
224. J. Hyman and M. Shashkov. Natural discretizations for the divergence, gradient and curl on logically rectangular grids. *Comput. Math. Appl.*, 33:88–104, 1997.
225. J. Hyman and M. Shashkov. Mimetic discretizations for Maxwell's equations. *J. Comp. Phys.*, 151:881–909, 1999.
226. J. Hyman and M. Shashkov. The orthogonal decomposition theorems for mimetic finite difference schemes. *SIAM J. Numer. Anal.*, 36:788–818, 1999.
227. D. Jesperson. A least-squares decomposition method for solving elliptic equations. *Math. Comp.*, 31:873–880, 1977.
228. B.-N. Jiang. A least-squares finite element method for incompressible Navier-Stokes problems. *Int. J. Numer. Meth. Fluids*, 14:843–859, 1992.

229. B.-N. Jiang. Non-oscillatory and non-diffusive solution of convection problems by the iteratively reweighted least-squares finite element method. *J. Comp. Phys.*, 105(1):108–121, 1993.
230. B.-N. Jiang. *The Least-Squares Finite Element Method. Theory and Applications in Computational Fluid Dynamics and Electromagnetics.* Springer-Verlag, New York, 1998.
231. B.-N. Jiang and G. F. Carey. Least-squares finite elements for first-order hyperbolic systems. *Int. J. Num. Meth. Eng.*, 26:81–93, 1988.
232. B.-N. Jiang and G. F. Carey. A stable least-squares finite element method for nonlinear hyperbolic problems. *Int. J. Num. Meth. Fluids*, 8:933–942, 1988.
233. B.-N. Jiang and G. F. Carey. Least-squares finite element methods for compressible Euler equations. *Int. J. Num. Meth. Fluids*, 10:557–568, 1990.
234. B.-N. Jiang and C. Chang. Least-squares finite elements for the Stokes problem. *Comput. Meth. Appl. Mech. Eng.*, 78:297–311, 1990.
235. B.-N. Jiang, T. Lin, and L. Povinelli. A least-squares finite element method for 3D incompressible Navier-Stokes equations. Technical Report 93-0338, AIAA, 1993.
236. B.-N. Jiang, T. Lin, and L. Povinelli. Large-scale computation of incompressible viscous flow by least-squares finite element method. *Comput. Meth. Appl. Mech. Eng.*, 114:213–231, 1994.
237. B.-N. Jiang and L. Povinelli. Least-squares finite element method for fluid dynamics. *Comput. Meth. Appl. Mech. Eng.*, 81:13–37, 1990.
238. B.-N. Jiang and L. Povinelli. Optimal least-squares finite element methods for elliptic problems. *Comput. Meth. Appl. Mech. Eng.*, 102:199–212, 1993.
239. B.-N. Jiang and V. Sonnad. Least-squares solution of incompressible Navier-Stokes equations with the p-version of finite elements. Technical Report 91-14, NASA, Cleveland, 1991.
240. J. Jin. *The Finite Element Method in Electromagnetics.* John Wiley and Sons, Inc., 1993.
241. J. Jirousek and A. Wroblewski. Least-squares T-elements: Equivalent FE and BE forms of a substructure oriented boundary solution approach. *Comm. Numer. Meth. Eng.*, 10:21–32, 1994.
242. J. Jirousek and A. Wroblewski. T-elements: State of the art and future trends. *Arch. Comp. Meth. Eng.*, 3:323–434, 1996.
243. C. Johnson. *Numerical Solution of Partial Differential Equations by the Finite Element Method.* Cambridge University Press, 1992.
244. C. Johnson, U. Navert, and J. Pitkäranta. Finite element methods for linear hyperbolic problems. *Comput. Meth. Appl. Mech. Eng.*, 45:285–312, 1984.
245. O. Kayser-Herold. *Least-Squares Methods for the Solution of Fluid-Structure Interaction Problems.* PhD thesis, Department of mathematics and informatics, Technical University of Braunschweig, Germany, 2005.
246. H. Keller. *Numerical Methods in Bifurcation Problems.* Springer Verlag, Berlin, 1987.
247. S.-D. Kim, T. Manteuffel, and S. McCormick. First-order system least-squares FOSLS for spatial linear elasticity: pure traction. *SIAM J. Numer. Anal.*, 38:1454–1482, 2001.
248. T. V. Kolev and P. S. Vassilevski. Parallel H^1-based auxiliary space AMG solver for $H(curl)$ problems. Technical Report UCRL-TR-222763, Lawrence Livermore National Laboratory, Livermore, CA, 2006.
249. S. G. Krein and Yu. I. Petunin. Scales of Banach spaces. *Russian Math. Surv.*, 21(2):85–160, 1966.
250. E. Kreyszig. *Introductory Functional Analysis with Applications.* John Wiley & Sons, New York, 1989.
251. A. Kunoth. *Wavelet Methods – Elliptic Boundary Value Problems and Control Problems.* Vieweg+Teubner, 2001.
252. A. Kunoth and J. Sahner. Wavelets on manifolds: An optimized construction. *Math. Comp.*, 75:1319–1349, 2006.
253. O. Ladyzhenskaya. *The Mathematical Theory of Viscous Incompressible Flows.* Gordon and Breech, New York, 1969.

254. T. A. Laursen and M. W. Heinstein. Consistent mesh tying methods for topologically distinct discretized surfaces in non-linear solid mechanics. *Int. J. Numer. Meth. Eng.*, 57:1197–1242, 2003.
255. J. E. Lavery. Nonoscillatory solution of the steady-state inviscid Burgers equation by mathematical programming. *J. Comp. Phys.*, 79:436–448, 1988.
256. J. E. Lavery. Solution of steady-state one-dimensional conservation laws by mathematical programming. *SIAM J. Numer. Anal.*, 26(5):1081–1089, 1989.
257. R. Lazarov, L. Tobiska, and P. Vassilevski. Streamline-diffusion least-squares mixed finite element methods for convection-diffusion problems. *East-West J. Numer. Math*, 5(4): 249–264, 1997.
258. E. Lee and T. Manteuffel. $FOSLL^*$ method for the eddy current problem with three-dimensional edge singularities. *SIAM J. Numer. Anal.*, 45:787–809, 2007.
259. E. Lee, T. Manteuffel, and C. R. Westphal. Weighted-norm first-order system least squares (FOSLS) for problems with corner singularities. *SIAM J. Numer. Anal.*, 44(5):1974–1996, 2006.
260. E. Lee, T. Manteuffel, and C. R. Westphal. Weighted-norm first-order system least-squares (FOSLS) for Div/Curl systems with three dimensional edge singularities. *SIAM J. Numer. Anal.*, 46(3):1619–1639, 2008.
261. D. Lefebvre, J. Peraire, and K. Morgan. Least-squares finite element solution of compressible and incompressible flows. *Int. J. Num. Meth. Heat Fluid Flow*, 2:99–113, 1992.
262. S. Li and W. Liu. Meshfree and particle methods and their applications. *Appl. Mech. Rev.*, 55:1–34, 2002.
263. J. Liable and G. Pinder. Least-squares collocation solution of differential equations on irregularly shaped domains using orthogonal meshes. *Numer. Meth. PDE's*, 5:347–361, 1989.
264. J. Liable and G. Pinder. Solution of shallow water equations by least-squares collocation. *Water Resources Res.*, 29:445–456, 1993.
265. J. Lions and E. Magenes. *Nonhomogeneous Elliptic Boundary Value Problems and Applications*, volume I. Springer-Verlag, Berlin, 1972.
266. J.-L. Liu. Exact a posteriori error analysis of the least-squares finite element method. *Appl. Math. Comp.*, 116:297–305, 2000.
267. R. Lohner, K. Morgan, and O. C. Zienkiewicz. The solution of nonlinear hyperbolic equation systems by the finite element method. *Int. J. Numer. Meth. Fluids*, 4:1043–1063, 1984.
268. R. B. Lowrie and P. L. Roe. On the numerical solution of conservation laws by minimizing residuals. *J. Comp. Phys.*, 113:304–308, 08 1994.
269. H. MacMillan, M. Gunzburger, and J. Burkardt. Meshfree first-order system least-squares. *Numer. Math. Theor. Meth. Appl.*, 1:29–43, 2008.
270. Y. Maday and A. T. Patera. Spectral element methods for the incompressible Navier-Stokes equations. In A. K. Noor and J. T. Oden, editors, *State of the Art Surveys on Computational Mechanics*, pages 71–143, New York, 1987. ASME.
271. M. Majidi and G. Starke. Least-squares Galerkin methods for parabolic problems I: Semidiscretization in time. *SIAM J. Numer. Anal.*, 39(4):1302–1323, 2001.
272. M. Majidi and G. Starke. Least-squares Galerkin methods for parabolic problems II: The fully discrete case and adaptive algorithms. *SIAM J. Numer. Anal.*, 39(5):1648–1666, 2002.
273. D. S. Malkus and T. J. R. Hughes. Mixed finite element methods—reduced and selective integration techniques: a unification of concepts. *Comput. Meth. Appl. Mech. Eng.*, pages 63–81, 1990.
274. T. Manteuffel, S. McCormick, J. Ruge, and J. G. Schmidt. First-order system $\mathcal{LL}^*$ ($FOSLL^*$) for general scalar elliptic problems in the plane. *SIAM J. Numer. Anal.*, 43:2098–2120, 2005.
275. A. Masud and T. J. R. Hughes. A stabilized finite element method for Darcy flow. *Comput. Meth. Appl. Mech. Eng.*, 191, 2002.
276. C. Mattiussi. An analysis of finite volume, finite element and finite difference methods using some concepts from algebraic topology. *J. Comp. Phys.*, 133:289–309, 1997.

277. J. Milthorpe and G. Steven. On a least-squares approach to the integration of the Navier-Stokes equations. In R. H. Gallagher, editor, *Finite Elements in Fluids*, volume 3, pages 89–103, New York, 1978. Wiley.
278. P. Monk and D.-Q. Wang. A least-squares method for the Helmholtz equation. *Comput. Meth. Appl. Mech. Eng.*, 175:121–136, 1999.
279. R. L. Naff, T. F. Russell, and J. D. Wilson. Shape functions for velocity interpolation in general hexahedral cells. *Comput. Geosci.*, 6(3–4):285–314, 2002.
280. J. Nécas. Sur un méthode pur résourde les equations dérivées partielles du type elliptique vosine de la variationnelle. *Ann. Soc. Norm. Sup. Pisa, Ser. III*, 16:305–326, 1962.
281. J. Nécas. *Les Methodes Directes en Theorie des Equations Elliptiques*. Masson, Paris, 1967.
282. J. C. Nédélec. Mixed finite elements in $\mathbf{R}^3$. *Numerische Mathematik*, 35:315–341, 1980.
283. J. C. Nédélec. Elements finis mixtes incompressibles pour l'equation de Stokes dans $\mathbf{R}^3$. *Numerische Mathematik*, 39:97–112, 1986.
284. J. C. Nédélec. A new family of finite element methods in $\mathbf{R}^3$. *Numerische Mathematik*, 50:57–81, 1986.
285. H. Nguen and J. Reynen. A space-time least-squares finite element scheme for advection-diffusion equations. *Comput. Meth. Appl. Mech. Eng.*, 42:331–342, 1984.
286. R. Nicolaides. Direct discretization of planar div-curl problems. *SIAM J. Numer. Anal.*, 29:32–56, 1992.
287. R. Nicolaides. The covolume approach to computing incompressible flows. In M. Gunzburger and R. Nicolaides, editors, *Incompressible computational fluid dynamics, Trends and advances*, pages 295–334. Cambridge University Press, 1993.
288. R. Nicolaides and K. Trapp. Covolume discretizations of differential forms. In D. N. Arnold, P. Bochev, R. Lehoucq, R. Nicolaides, and M. Shashkov, editors, *Compatible Discretizations, Proceedings of IMA Hot Topics workshop on Compatible Discretizations*, volume IMA 142, pages 161–172. Springer Verlag, 2006.
289. R. Nicolaides and X. Wu. Covolume solutions of three-dimensional div-curl equations. *SIAM J. Numer. Anal.*, 34:2195–2203, 1997.
290. J. T. Oden. RIP-methods for Stokesian flow. In *Finite Elements in Fuids*, volume 4, pages 305–318. John Wiley, Chichester, 1982.
291. K. C. Park and C. A. Felippa. Recent development in coupled field analysis methods. In R. W. Lewis, P. Bettess, and E. Hinton, editors, *Numerical Methods in Coupled Systems*, pages 327–352. John Wiley & Sons, New York, London, Sydney, 1984.
292. S.-H. Park, K.-C. Kwon, and S.-K. Youn. A study on the convergence of least-squares meshfree method under inaccurate integration. *Int. J. Numer. Meth. Eng.*, 56:1397–1419, 2003.
293. S.-H. Park and S.-K. Youn. The least-squares meshfree method. *Int. J. Numer. Meth. Eng.*, 52:997–1012, 2001.
294. M. Parks, L. Romero, and P. Bochev. A novel Lagrange-multiplier based method for consistent mesh tying. *Comput. Meth. Appl. Mech. Eng.*, 196(35–36):3335–3347, 2007.
295. P. Perrochet and P. Azerad. Space-time integrated least-squares: solving a pure advection equation with a pure diffusion operator. *J. Comp. Phys.*, 117:183–193, 1995.
296. J. P. Pontaza. A least-squares finite element formulation for unsteady incompressible flows with improved velocity-pressure coupling. *J. Comp. Phys.*, 217:563–588, 2006.
297. J. P. Pontaza, X. Diao, J. N. Reddy, and K. S. Surana. Least-squares finite element models of two-dimensional compressible flows. *Finite elements in analysis and design*, 40:629–644, 2004.
298. J. P. Pontaza and J. N. Reddy. Spectral/hp least-squares finite element formulation for the Navier-Stokes equations. *J. Comp. Phys.*, 190:523–549, 2003.
299. J. P. Pontaza and J. N. Reddy. Space-time coupled spectral/hp least-squares finite element formulation for the incompressible Navier-Stokes equations. *J. Comp. Phys.*, 197:418–459, 2004.
300. M. M. J. Proot and M. Gerritsma. Least-squares spectral elements applied to the Stokes problem. *J. Comp. Phys.*, pages 454–477, 2002.

301. C. C. Rasmussen, R. A. Canfield, and J. N. Reddy. The least-squares finite element method applied to fluid-structure interaction problems. In *48th AIAA/ASME/ASCE/AHS/ASC Structures, Structural Dynamics, and Materials Conference*, number 2007-2407 in AIAA, 2007.
302. P. A. Raviart and J. M. Thomas. A mixed finite element method for second order elliptic problems. In Galligani and E. Magenes, editors, *Mathematical Aspects of the Finite Element Method, I*, number 606 in Lecture Notes in Math. Springer-Verlag, New York, 1977.
303. J. N. Reddy. *Introduction to the Finite Element Method*. McGraw-Hill College, 2004.
304. M. Renardy and R. Rogers. *Introduction to Partial Differential Equations*. Number 13 in Texts in Applied Mathematics. Springer-Verlag, Berlin, 1993.
305. W. Rheinboldt. Solution fields of nonlinear equations and continuation methods. *SIAM J. Numer. Anal.*, 17:221–246, 1980.
306. J. Rikabi, C. F. Bryant, and E. M. Freeman. An error-based approach to complementary formulations of static field solutions. *Int. J. Numer. Meth. Eng.*, 26(9):1963–1987, 1988.
307. Ya. A. Roitberg. A theorem about the complete set of isomorphisms for systems elliptic in the sense of Douglis and Nirenberg. *Ukrain. Mat. J.*, 25:396–405, 1973.
308. Ya. A. Roitberg and Z. Seftel. A theorem on homeomorphisms for elliptic systems and its applications. *Math. USSR Sbornik*, 7:439–465, 1969.
309. H. Rui, S.-D. Kim, and S. Kim. Split least-squares Galerkin finite element methods for linear and nonlinear parabolic problems. *J. Comp. Appl. Math.*, 223:938–952, 2008.
310. A. A. Samarskii, V. F. Tishkin, A. P. Favorskii, and M. Yu. Shashkov. Operational finite-difference schemes. *Diff. Equat.*, 17:854–862, 1981.
311. A. A. Samarskii, V. F. Tishkin, A. P. Favorskii, and M. Yu. Shashkov. Employment of the reference-operator method in the construction of finite difference analogs of tensor operations. *Diff. Equat.*, 18:881–885, 1982.
312. H. Schlichting and K. Gersten. *Boundary Layer Theory*. Springer, Berlin, 2000.
313. F. Shakib. *Finite element analysis of the compressible Euler and Navier-Stokes equations*. PhD thesis, Division of Applied Mathematics, Stanford University, 1989.
314. M. Shashkov. *Conservative Finite Difference Methods on General Grids*. CRC Press, Boca Raton, FL, 1996.
315. D. J. Silvester and N. Kechkar. Stabilized bilinear-constant velocity-pressure finite elements for the conjugate gradient solution of the Stokes problem. *Comput. Meth. Appl. Mech. Eng.*, 79:71–86, 1990.
316. D.J. Silvester. Optimal low order finite element methods for incompressible flow. *Comput. Meth. Appl. Mech. Eng.*, 111:357–368, 1994.
317. P. Solin, K. Segeth, and I. Dolezel. *Higher-Order Finite Element Methods*. CRC Press, Boca Raton, FL, 2004.
318. G. Starke. Multilevel boundary functionals for least-squares mixed finite element methods. *SIAM J. Numer. Anal.*, 36:1065–1077, 1999.
319. H. De Sterck, T. Manteuffel, S. McCormick, and L. Olson. Least-squares finite element methods and algebraic multigrid solvers for linear hyperbolic PDEs. *SIAM J. Sci. Comput.*, 26(1):31–54, 2005.
320. H. De Sterck, T. Manteuffel, S. McCormick, and L. Olson. Numerical conservation properties of $H(div)$-conforming least-squares finite element methods for the Burgers equation. *SIAM J. Sci. Comput.*, 26(5):1573–1597, 2005.
321. M. Stojek. Least-squares Trefftz-type elements for the Helmholtz equation. *Int. J. Numer. Meth. Eng.*, 41:831–849, 1998.
322. G. Strang. *Linear Algebra and Its Applications*. Harcourt Brace Jovanovich, New York, third edition, 1986.
323. G. Strang and G. Fix. *An Analysis of the Finite Element Method*. Prentice Hall, Englewood Cliffs, NJ, 1973.
324. E. Suli. A posteriori error analysis and adaptivity for finite element approximations of hyperbolic problems. Technical Report 97/21, Oxford University Computing Laboratory, Oxford, England, 1997.

325. F. Taghaddosi, W. Habashi, G. Guevremont, and D. Ait-Ali-Yahia. An adaptive least-squares method for the compressible Euler equations. *Int. J. Numer. Meth. Fluids*, 31:1121–1139, 1999.
326. L. Tang, T. Cheng, and T. T. H. Tsang. Transient solutions for three-dimensional lid-driven cavity flows by a least-squares finite element method. *Int. J. Numer. Meth. Fluids*, 21: 413–432, 1995.
327. L. Tang and T. T. H. Tsang. A least-squares finite element method for time dependent incompressible flows with thermal convection. *Int. J. Numer. Meth. Fluids*, 17:271–289, 1993.
328. L. Tang and T. T. H. Tsang. Transient solutions by a least-squares finite element method and Jacobi conjugate gradient technique. *Numer. Heat Trans. Part B*, 28:183–198, 1995.
329. L. Tang and T. T. H. Tsang. Temporal, spatial and thermal features of 3-D Rayleigh-Benard convection by least-squares finite element method. *Comput. Meth. Appl. Mech. Eng.*, 140:201–219, 1997.
330. T. Tarhasaari, L. Kettunen, and A. Bossavit. Some realizations of a discrete Hodge operator: a reinterpretation of finite element techniques. *IEEE Trans. Mag.*, 35(3):1494–1497, 1999.
331. M. Taylor. *Partial Differential Equations. Basic Theory*. Springer, New York, 1999.
332. F. L. Teixeira and W. C. Chew. Lattice electromagnetic theory from a topological viewpoint. *J. Math. Phys.*, 40:169–187, 1999.
333. R. Témam. *Navier-Stokes Equations*. North-Holland, Amsterdam, 1970.
334. R. Témam. *Nonlinear Functional Analysis and Navier-Stokes Equations*. SIAM, Philadelphia, 1983.
335. T. Tezduyar and Y. Osawa. Finite element stabilization parameters computed from element matrices and vectors. *Comput. Meth. Appl. Mech. Eng.*, 190:411–430, 2000.
336. V. Thomee. *Galerkin Finite Element Methods for Parabolic Problems*. Springer, 1997.
337. A. N. Tikhonov and V. A. Arsenin. *Solution of Ill-posed Problems*. John Wiley and Sons, New York, 1977.
338. E. Tonti. On the mathematical structure of a large class of physical theories. *Lincei, Rend. Sc. Fis. Mat. e Nat.*, 52(1):51–56, 1972.
339. E. Tonti. The algebraic-topological structure of physical theories. In *Proc. Conf. on Symmetry, Similarity and Group Theoretic Meth. in Mech*, pages 441–467, Calgari, Canada, 1974.
340. A. Toselli and O. Widlund. *Domain Decomposition Methods - Algorithms and Theory*. Springer Verlag, New York, 2005.
341. L. R. Volevich. A problem of linear programming arising in differential equations. *Uspekhi Mat. Nauk*, 18(3):155–162, 1963.
342. W. Wedland. *Elliptic Systems in the Plane*. Pitman, London, 1979.
343. S. Weintraub. *Differential Forms. A Complement to Vector Calculus*. Academic Press, San Diego, 1997.
344. J. S. Van Welij. Calculation of eddy currents in terms of H on hexahedra. *IEEE Trans. Magnetics*, 21:2239–2241, 1985.
345. H. Whitney. *Geometric Integration Theory*. Princeton University Press, Princeton, NJ, 1957.
346. D. Winterscheidt and K. S. Surana. *p*-version least-squares finite element formulations for two-dimensional, incompressible fluid flow. *Int. J. Numer. Meth. Fluids*, 18:43–69, 1994.
347. J. Xu. Iterative methods by space decomposition and subspace correction. *SIAM Review*, 34:581–613, 1992.
348. D.-P. Yang. Some least-squares Galerkin procedures for first-order time-dependent convection–diffusion system. *Comput. Meth. Appl. Mech. Eng.*, 180:81–95, 1999.
349. D.-P. Yang. Analysis of least-squares mixed finite element methods for nonlinear nonstationary convection-diffusion problems. *Math. Comp.*, 69(231):929–963, 2000.
350. I. Yotov. Mixed finite element methods for flow in porous media. Technical Report TR96-09, Dept. Comp. Appl. Math., Rice University, 1996.
351. H. Yserentant. On the multilevel splitting of finite element spaces. *Numerische Mathematik*, 49:379–412, 1986.
352. H. Yserentant. Hierarchical bases. In R. O'Malley, editor, *Proc. ICIAM 91*, pages 256–276. SIAM, Philadelphia, 1992.

353. D. Zeitoun, J. Liable, and G. Pinder. A weighted least-squares method for first order hyperbolic systems. *Int. J. Numer. Meth. Fluids*, 20(3):191–212, 1995.
354. D. Zeitoun and G. Pinder. A least-squares approach for solving remediation problems of contaminated aquifers. In *Comput. Meth. Water Resources*, volume 1 of *Computational Mechanics Publications*, pages 329–336. Elsevier Applied Science, New York, 1992.
355. O. Zienkiewicz, D. Kelly, J. Gago, and I. Babuška. Hierarchical finite element approaches, error estimates and adaptive refinement. In J. Whiteman, editor, *The Mathematics of Finite Elements and Applications*, volume IV, pages 313–346. Academic, New York, 1982.

Acronyms

We collect acronyms that are used in several places in the book. For convenience, the acronyms listed here are also redefined in each chapter in which they are used. In addition, several other acronyms appear in the book, but because they appear only locally, often within only a single section, we define them when they are used.

ADN	Agmon-Douglis–Nirenberg
CLSP	continuous least-squares principle
DLSP	discrete least-squares principle
FD-LSFEM	finite difference least-squares finite element method
FOSLS	first-order system least-squares
LSFEM	least-squares finite element method
PDE	partial differential equation
SUPG	streamwise-upwind-Petrov Galerin
VGVP	velocity gradient–velocity–pressure
VSP	velocity–stress–pressure
VVP	velocity–vorticy–pressure

Glossary

We gather definitions and notations that are used several times in the book. Of course, there are many other definitions and notations used, but, for the most part, those are more locally used so that it should be relatively easy to locate where they are introduced.

Notational conventions

dual spaces and adjoint operators	marked by an asterisk	V^*, $\mathcal{A}^*$
functionals and bilinear forms	uppercase italic Roman	$F(\cdot)$, $Q(\cdot,\cdot)$
function spaces	uppercase italic Roman	U, W, X, Y
functions	lowercase italic Roman and Greek	ϕ, ξ, u, v
matrices	upper case sans-serif	A, B
operators acting between abstract spaces	upper case calligraphy	$\mathcal{A}$, $\mathcal{B}$
tensor-valued functions	marked with underline	$\underline{\mathbf{U}}$, $\underline{\mathbf{V}}$
vectors in $\mathbb{R}^n$	marked with an arrow	$\vec{a}$, $\vec{\xi}$
vector-valued functions	lower case boldface	$\mathbf{u}$, $\mathbf{v}$, $\boldsymbol{\xi}$

Sobolev spaces

$L_0^2(\Omega)$	page 536
$H_\gamma^1(\Omega)$	page 536
$H_0^1(\Omega)$	page 536
$H^1(\Omega)/\mathbb{R}$	page 536
$H^{-1}(\Omega)$	page 536
$[H^a(\Omega)]^d$	page 537
$\mathbf{H}_n^1(\Omega)$	page 254
$\mathbf{H}_t^1(\Omega)$	page 254

Norms

$\lvert\cdot\rvert_{1,\Omega}$	page 536
$\lVert\cdot\rVert_{H^1(\Omega)/\mathbb{R}}$	page 536
$\lVert\cdot\rVert_{-1}$	page 537
$\lVert\cdot\rVert_{1/2}$	page 537
$\lVert\cdot\rVert_{-1/2}$	page 537
$\lVert\cdot\rVert_{DC}$	page 547

Vector operators

$\nabla^\perp$	page 535
∇^*	page 543
$\nabla^*\times$	page 543
$\nabla^*\cdot$	page 544

Spaces associated with operators	
$N(\cdot)$	null space
$R(\cdot)$	range space
$D(\cdot)$	domain space

Orthogonal complements spaces are denoted by, e.g., $N(\cdot)^{\perp}$.

Spaces related to vector operators

Notation used in book	Definition	Common notation
$G(\Omega)$	$\{u \in L^2(\Omega) \mid \nabla u \in L^2(\Omega)\}$	$H^1(\Omega)$
$\mathbf{C}(\Omega)$	$\{\mathbf{u} \in \mathbf{L}^2(\Omega) \mid \nabla \times \mathbf{u} \in \mathbf{L}^2(\Omega)\}$	$H(\Omega, \text{curl})$
$\mathbf{D}(\Omega)$	$\{\mathbf{u} \in \mathbf{L}^2(\Omega) \mid \nabla \cdot \mathbf{u} \in L^2(\Omega)\}$	$H(\Omega, \text{div})$
$S(\Omega)$	$\{u \in L^2(\Omega) \mid u = \nabla \cdot \mathbf{u};\ \mathbf{u} \in \mathbf{D}(\Omega)\}$	$L^2(\Omega)$

Corresponding *constrained* spaces, e.g., $G_\gamma(\Omega)$, constrained by boundary conditions, where $\gamma = \Gamma$ or Γ^*, are defined in (A.25)–(A.27).
Corresponding *constrained* spaces, e.g., $G_0(\Omega)$, constrained by boundary conditions on $\partial\Omega$ are defined on page 541.
Corresponding *weighted* spaces, e.g., $G(\Omega, \Theta_0)$, are defined in (A.29)–(A.35).
Norms corresponding to these spaces, e.g., $\|u\|_G^2$ and $\|u\|_{G(\Theta_0)}^2$, are defined in (A.38)–(A.43).

Geometric sets associated with a finite element κ	
$C_0(\kappa)$	set of vertices
$C_1(\kappa)$	set of edges
$C_2(\kappa)$	set of faces
$C_3(\kappa)$	κ itself

The sets of all vertices, edges, faces, and elements associated with a finite element partition $\mathcal{T}_h$ of a domain are denoted by $C_m(\mathcal{T}_h)$, $m = 0, 1, 2, 3$, respectively; see page 555.

Finite element spaces associated with a finite element κ

space	definition
$G^h(\kappa)$	$\{u^h \in G(\kappa) \mid u^h = \Phi_G^*(\widehat{u}^h), \quad \widehat{u}^h \in G^h(\widehat{\kappa})\}$
$\mathbf{C}^h(\kappa)$	$\{\mathbf{u}^h \in \mathbf{C}(\kappa) \mid \mathbf{u}^h = \Phi_C^*(\widehat{\mathbf{u}}^h), \quad \widehat{\mathbf{u}}^h \in \mathbf{C}^h(\widehat{\kappa})\}$
$\mathbf{D}^h(\kappa)$	$\{\mathbf{u}^h \in \mathbf{D}(\kappa) \mid \mathbf{u}^h = \Phi_D^*(\widehat{\mathbf{u}}^h), \quad \widehat{\mathbf{u}}^h \in \mathbf{D}^h(\widehat{\kappa})\}$
$S^h(\kappa)$	$\{u^h \in S(\kappa) \mid u^h = \Phi_S^*(\widehat{u}^h), \quad \widehat{u}^h \in S^h(\widehat{\kappa})\}$

The pullbacks $\Phi_G^*(\cdot)$, etc., are defined in (B.10).

Finite element spaces associated with the domain Ω

space	definition
$G^h(\Omega)$	$\{u^h \in G(\Omega) \mid u^h\|_\kappa \in G^h(\kappa) \quad \forall \kappa \in \mathcal{T}_h\}$
$\mathbf{C}^h(\Omega)$	$\{\mathbf{u}^h \in \mathbf{C}(\Omega) \mid \mathbf{u}^h\|_\kappa \in \mathbf{C}^h(\kappa) \quad \forall \kappa \in \mathcal{T}_h\}$
$\mathbf{D}^h(\Omega)$	$\{\mathbf{u}^h \in \mathbf{D}(\Omega) \mid \mathbf{u}^h\|_\kappa \in \mathbf{D}^h(\kappa) \quad \forall \kappa \in \mathcal{T}_h\}$
$S^h(\Omega)$	$\{u^h \in S(\Omega) \mid u^h\|_\kappa \in S^h(\kappa) \quad \forall \kappa \in \mathcal{T}_h\}$

Compatible finite element spaces on the reference element $\widehat{\kappa}$ and their degrees of freedom in the lowest-order case $r = 1$

name	symbol	definition	degrees of freedom	conformity
nodal elements				
Lagrangian	$G^r(\widehat{\kappa})$	(B.18) for simplices	nodal values	gradient–
		(B.19) for cubes		conforming
edge elements				
Nédélec, 1st kind	$\mathbf{C}^{r\text{-}}(\widehat{\kappa})$	(B.21) for simplices		
		(B.23) for cubes	circulations	curl–
Nédélec, 2nd kind	$\mathbf{C}^r(\widehat{\kappa})$	(B.25) for simplices	along edges	conforming
		(B.26) for cubes		
face elements				
Raviart–Thomas	$\mathbf{D}^{r\text{-}}(\widehat{\kappa})$	(B.29) for simplices		
		(B.31) for cubes	fluxes across	divergence–
Nédélec, 2nd kind	$\mathbf{D}^r(\widehat{\kappa})$	(B.33) for simplices	faces	conforming
		(B.34) for cubes		
volume elements				
discontinuous	$S^r(\widehat{\kappa})$	(B.39)	volume averages	–

The definition of these spaces are given in terms of the spaces
P_r = all polynomials in $\mathbb{R}^d$ of degree less than or equal to r
Q_r = all polynomials whose degree in each and every coordinate direction does not exceed r
$\mathbf{H}_r' = \{\mathbf{p} \in [H_r]^d \mid \mathbf{p}(\widehat{\mathbf{x}}) \cdot \widehat{\mathbf{x}} = 0\}$.

Compatible finite element spaces associated with a domain Ω are defined in the usual manner, i.e., $G^r(\Omega) = \{u^h \in G(\Omega) \mid u^h|_\kappa \in G^r(\kappa)\}$. Constrained versions of compatible finite element spaces associated with a domain Ω are also defined in the usual manner, i.e., $G_0^r(\Omega) = G^r(\Omega) \cap G_0(\Omega)$.

Index

Breinigsville, PA USA
03 June 2010
239100BV00006B/17/P